Ergomont

David L. Birdsall

Bleiben Zauber Books

ISBN: 978-0-9786979-6-9
Library of Congress Control Number: 2015900019

Contents

Preface

Authority is an obstruction that everyone comes up against to fight or struggle through at some time in their lives. In fact, it acts often as a signpost to herald various forms of controversy or counterintuitive behavior in a society or any social gathering from the very large to the solitary conscience bombarding one within the contradictions of reason and logic to confront an apparent reality that constrains or obfuscates the slightest pretense of an acceptable situation. This monster of momentous proportions and precipitous or ever growing inertia, as could an enigma be defined, can be rapidly met, abutted, and contorted of ones self through asking questions of a communal nature such as "What is best (of a condition or preferential treatment)?", the relevance of which can not be defined by their answers, but rather by the congested proffering of a considered commonality of opinion. Is it determined by a culture, a government, a profit-driven industry of entertainment, an intuition, an imposed convention of mores, a realization for the common good of a people or members of an institution? What is certain is that the results of such inquiry must be faced and, if need be, overcome by the jealousies of individual demeanors and reactions to events uncontrollable or, perhaps, inalterable and impressed. Ones existence demands exploration of the possible consequences of actions or expression of beliefs. Further, the courageous experiment to find at least favorable qualities of life within the potent strains and sinews of a constraining authority. Revolution against the bind is only of genuine importance when it is practical for most contained within the struggle, be this personal or more portentous of overwhelming and significant change under a contract of desired, hoped for, or pleaded to comity with others and nature itself. Congeniality among, between, and beyond others, is an observance of nature's allowance for matters pertaining to the self, of what is doable (or demonstrable) weighted against what (one feels) certainly must be done (in one way or another, regardless of the manner of doing). So may murder be performed among friends, and love between enemies.

But a mountain of disbelief conspires against what is agreed upon by an authority, when the methods of subjection or imposition quarrel with what is obvious to be preferred by the subjects and underlings of a constitutional, national, or family setting and demand of precepts to observe. Here, then, is the load of obstruction that makes one feel fantastic to conquer and surpass, lift (metaphorically) and hurl away out of ones path of travel and advancement. While the work itself may be tedious and bothersome, repulsive and repugnant, it feels good to *feel* that one may (and may even have privilege to) challenge, uplifting of the spirit to stand for this purpose, however ridiculous and absurd for the tendencies that would have to be employed. Nature, via the natural physics, defines the impossible; but a mentality devolves of any outcome to perform (at least idealistically) a desired result, a fought for stasis, a quantitative measurement to a qualitative state. It is here, of each thinking person to decide, what barriers to action must be broken, crushed up and crumbled into level pavement for all who wish to walk along streets of contention to a satisfying life, with strides of pleasantness in worth and purpose of being. Still, though, authority, by manner of its display and capture of others or all, must be obeyed at some levels of (its) extant occurrence. So, in the main, we strive for this feeling of resistance, rather than its harmful exercise desultory of aim and destructive of reasonable goals, conclusions, and worthwhile eventualities. Authority must remain the annihilator, well above our scopes of persuasion and meticulous review of what has occurred of a domination. Else, we are (and much too evidently current) authorized to do to ourselves a pervasive and thorough belittling towards the very baseness of disgraces we profess to abhor. We stroke a hurting much as a prized possession adulated and adored, when we undo restraints to civility, losing propensities of human, and humane, greatness. Only a feeling is fun. Hence mountain protects from winds whirling towards implacability, whimsical surges and whims of abject dissolution.

So then are the characteristics of the, sometimes unexpectedly, reprehensible to be analyzed or described and extolled (as of a grand folly) or exhibited, written of, explained, or exposited, expounded. These instigators range from the very pious to the terribly outrageous, each in defiance of something: a concept or person, institution or notion to succeed through established difficulties (though those difficulties may not have been caused by themselves or anything related to their personal existence). Not that they are necessarily extremists and fanatics; but, unbounded by errors, rather by erring against their goals, their greatest fault lies in a strong and rampant self honesty as a sin to nature's persistence, or insistence to be acknowledged and stoked for her nourishing fires of change and consumption. Truth is only found in ones perceptions, and can easily escape what actually occurs. In short, they are ironic to what happens to them and their causes. Often they are very agreeable fellows; however, unbalanced, any imbalance allows much to fall off original precipices. They can not readily recognize the declines and descents, felled by the innocence of their senses with astonishment and horror to avoid their pains as a guaranteed recompense. Yet only an authority can define the worth of an outcome, as laudable or specious to the expended effort.

December, 2014   David L. Birdsall

v

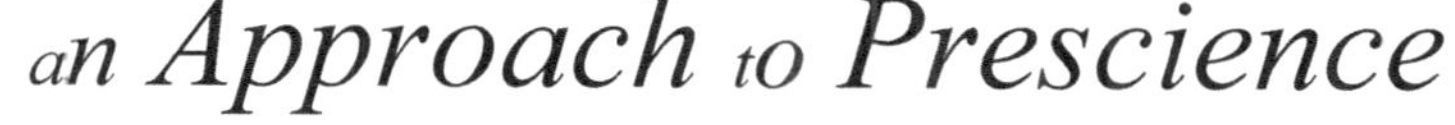

an *Approach to Prescience*

an *Approach to Prescience*

Characters:

  Caton
  Morliff
  Halyce
  Vincent
  Pontal
  Vaclkulor
  Hestor
  Loresane
  Fabel
  Woman
  Man
  Melinda
  Terissia

various Commissioners and royal attendants; a nursemaid; a guard; three physicians.

Locale : regions of vicinity and within a king's palace

Act I

Scene I — ***Morliff*** *stands in a palatial courtyard, most notable for its stone treasures and adornments than floral: statues and decorative works of sculptured, chiseled art ostentatiously arrayed. He seems to be contemplating the shape of an obviously round boulder craftily pitted or marred by nails and tiny metal pieces, as* **Caton** *approaches.*

CATON (*approaching*): What a glorious morn, Morliff. (*no reply*) But it is too soon, I'd still wager, to judge this way the coming of a storm.... and with its passing abate the weird dilemma faced.... Dear kin, it is too telling. (*stops*)

MORLIFF: Telling what? This granite, jeweled of shape, delivered thus to this wretched state of shiny marks and hammered indentations, what is cruel-(mak)ing of a geometry. What could I for to break it into bits with force of ire, then to (have it) fall back onto a leveling ground with less of shame for this dust of relic-ing. We are the same deceived, with purity of our forms, to have been hated-ly healed into so clever a scarring as to present us mad. But how does this roll?

CATON: It is adjusted of a seat. The seat is hidden by the girth of placement. It is not meant to roll, but stay in place safely, or else it's dangerous of crushing and abutting with heavy pressure, a weighty inconvenience in this.... gray garden of the state and statuesque.

MORLIFF: The sky is colored white, with bright clouds. But I could roll as much with drift as these stone figures seem to laze within a haze. Why pause us now? It does compel as if more evident to be. We shall be crushed deliberately if this is let—

CATON: That's too bold.

MORLIFF: —loose, the torrential tirade of a mind whose thoughts impale.

CATON: He is all gentle, our brother—

MORLIFF: Not as might.

CATON: —and is not truly sore in the head. Delivered of an aging, this dulls the wits while the wits are refined and perfected.

MORLIFF: Simplified, you mean. What does constrain us into these positions? The rule of governance atrophies, as he is wildly checked and checkered by these whims that must confound a Satan even.

CATON: He has his.... oracle of consultants—

MORLIFF: They are names (of renown), not talents!

CATON: —those who would not let matters deteriorate into too great an impasse of recovery.

MORLIFF: You know the ancient history.... His best condition is as friend. Whereat the one does fail of mental whiles, detested and contested are his aims to heirs, congested of the sufferance. But peaceably it could be done, as would a friend diminish 'im.

CATON: Such a stern measure is too tactful to be tamed, too strategic of a common sacrilege of days gone by. And he continues to woo our affections, as were a dad, a parent, and this his patrimony. We were such little kids, under his generous care.

MORLIFF: The half-shared wife does make us ill of treachery.

CATON: She is.... not foolish for us. A purity herself, displaced here, is her wanton needlessness decried as tacit law, descried as a subjugation willed, by fortune or by circumstance. For we are heirs to ruling, but not our matron (to) be youth present, that the father could remarry, and to such a child as to beget us.... two. This aged sage does, dies; of his virility fostered on his first son, to leave our brother shackled to our care, and his wife questionable, older than our mother, uncertain and not of a delicate inquietude as to who should be supreme of character in this reigned dominion.

MORLIFF: It is not openly contested, nor believed by faith that there is cause for competition or quarrel between them, the old and the less.... debilitated. But to our own brother is the died challenge, because he vacillates of strengths, and our young nephew— That boy!— is vicious of over-arched protections (of father) leaping towards the ridiculous, even the obscene.

CATON: A caution to the whether. It is a typical ambition of the immaturely dispossessed. Fabel is keen and aware, and sensitive of these steadfast winds. Our brother dies, or is senilely smitten, and one of us becomes— not him— a majesty admired with beleaguering of our terrible troubles and adversities to wrestle with. Yet the boy still loves us much, good Fabel, in his innocence and ignorance.

MORLIFF: There is less reciprocation than demand as actual our affections for him. The real does not have to be expressed. It occurs on its own, without description. He is confused of a king, and of descendants. For what mother rules us really, in these loyalties?

CATON: None an' neither.

MORLIFF: ....I would take him down, to end the squall of squalor approaching. To handle rights and matters again, I would try.

CATON: And there is the predicament between us, which one of us would first do to lean against our sib'(ling). Or ought to do at all, our casings rattle weakened, our roofs to shatter of their tiles and shingles with the thought. Still does he have to meanness towards us none. It is a disease of the mind.... and should not be necessary— if he's old.

MORLIFF: Old is in his doing. He is elder. We are respectable becoming of family each, Caton. My first born attempts to be, and your recent marriage will produce the same. Neither of us can afford shambles to this principality to attain, or feral roots to take hold of growths decreasing our worth and value to the populace. They will not make do with erring ways for long. The dissolution of a land is not uncommon when the ruling heads are rotted out or smoldering, a stench of smoke too noticeable even for the unconcerned peasants of bile-d raiments. One of us must seize control.... or function, from kin. And better that we both contribute to the toppling, for later to decide the dictatorship manifested.

CATON: We are not dictated to now. And I can not join you to bring myself to do what is not warrantably felt yet. Nor would I argue much with your attempts and their consequences. But you are not alone in these concerns. They are almost open.... and sung about suggestively, so that even suspicions are tolerated today. He hears them clearly. He knows them thoroughly. But still he pardons of our greetings, to treat us as wholesomely as possible, observed by all with gratitude to have us as fleshly spirited towards himself, a doting servant to our wishes and his state's. It is too transparent, what others should expect. And of him cries of madness jovial and insignificant of harm. He can not handle his faculties of worry any longer, and should even shake our hands while we'd hold daggers in them. How can we treat this friend with open eyes less pleading than profuse in sadness? He is our father almost, losing dignity as could the muscles spoil. Let him climb down naturally from the post, and let (have) him appoint.

MORLIFF: His appointment must be Fabel, then. For he is seeded inwardly and is sick to wonders and the astonishing. He can not count for sums properly, and yet you want him to summarize?! correctly for his legacy? His eyes drip, without crying. He is dull to dangers and dainty for curses. He can not decide for himself any longer, since he hasn't a firmness to rest with of his decisions.

CATON: Staid, were he to be in his disappointments. How can one persist in a grinding torture except to become lightheaded?... outwardly. But he manages to take the contempts of all failures made under him, and survive with a graciousness to display. This is all admirable, for those in the know, and gives the state time and cover to improve. (To) Strike his outstretched wings of shield, and not our backs, is the deed performed. Could you replace him as a cushion pinned with threats and vilely veiled harkening, and last as well? this saturnine spirit; this does Visgoth, to soak up all of the poisons and spare us their effects so that *we* may remain effective.

MORLIFF: It's not to replace him for my want, in this contaminative post. But the country heaves of soars, with mismanagement, and struggles.... for a ruling might to be lead by, detouring cold disasters and warm deaths. Else are we to be engulfed? by the barbaric poised for.... the spoiling? If we are made of metal, this must shine now, and be colorful of a tough enamel to be seen with, our declaration to beware of blunders facing us for (the) spoils. We should not have a fiercely, viciously impugned head for lack of chivalrous character proposed, at this dangerous time. And as I see to mask the error, being fit myself, does action bind us to a cause to be more effectual of change more immediate than eventual. Our weapons now are only his sangfroid humor and froth from smiling senselessly to the adverse expletive tossed towards his moonwaxed countenance. It is a pain of panic immotile and brought to true view of its silly, deprecated nature.... difficult for me to take unresponsively, and to want to make a void to fill a void. For this is not he once, nor ever to be again. But with agent of a permanent slack of mind is he mortified and dissimilar to noble need.

CATON: As to put the lame horse down?... or out of reach for detractors and further detracting?

MORLIFF: He can not think well.

CATON: Would an aged father be so disused?

MORLIFF: He is young enough to last my life. As such for misuse, remember that. He turns the spit not well, and roasts the pig to a burning mess and charring shame.... and then can only laugh at the result, forgiven for his envy of the dross. Make we plans to avenge his past, his glorious thoughts and aspirations shrouded of a decaying state. Be prepared, I only say— and back me!... when the time does come for a usurpation overdue. He did not cause the festering of worms, inheriting a wreck from recklessness of proud expansion and exhaustion for our strengths. But need he not welcome the maggots growing in the ill-treated meat— And we must sanitize!... to prevent the spread of bald, bold pestilence throughout this realm.

CATON: He is pressured by all sides, top to bottom, and without relief from strains, boxed into catastrophe, as felt. The country.... is in a downturn, of cyclic misfortunes, but may not be spiraling down into a diseased plight that would change earth to urns for all. We are still, on average, positive.... though shrinking of benefits. Yet, what can you do differently but rage at the declines and sinking status to affect influence and contain ourselves succinctly without (the) laudatory doable?

MORLIFF: That!... is much to do for candor's sake. We are displeased, but not destroyable. We are uneasy, but not beneath our goals asleep, bequeathed of difficult slumbers. And this distaste is to be projected throughout the land, that this currency will not do, and that we strive for meaning of a recovery to a grand prosperousness demanded by all, every of each.

CATON: I share the optimism. But plagues of misery occur to be withstood. We can not fight them intelligently. We must live them through. And with the consideration of others, and other times, we endure a relatively light spell of shakiness and uncertainty. Even the downtrodden subsist relatively well under his rule, with doles of daring made.

MORLIFF: We must do more. We must show more— We must be more. The slope is negative, and our lives will grow worse unless we are angry of atmosphere and determined to be firm. Let's be real to face up to these adversities, Caton. He is affected by a rotting of the mind, and can not help himself nor others, nor decide by himself what is best for any situation to treat or even suffer through. A weakness of the limbs is an injustice we can handle for ourselves. But he has not that malady. Then claim we he is fit?! No! He is not. He is fitted there with languors now, of effectiveness for the throne. If he abdicates, with a knowing consciousness, then it can only be to his son, or mothered son, with dispositions tensely driven, given, stoked! And a boy would rule, or a queen invoked, about this land drolly tempered. For if available, a male heir must reign. But if our brother dies, or becomes not himself for worst of good, then one of us takes up the helm of ship of state. And I outdo you by age and desire. So make this clear about us and the possibilities you must accept. We are the sons of a King!... and not fraternal to a worsening of our condition. It will be approved, as I make my steps.... as we do. Choices are fate's drinks quenched with.

CATON: I won't deny it. Your approval would be generated by every corner. Our brother's reign does not grow to be beloved. But then the times are not exactly happy, and blames excuse a hastiness of cruelty in thought, talk, and action. But.... a precious kin I would address— and friend.... I could not dismay his soft smile easily, as he's taught us everything during his savvy period, and that of our growth. It's as much as he could beg for an adjustment, without deception. Then let him plead the course to take, out of respect for his former prowess and sternness.

MORLIFF: He has no path to plead, nor pledge to placate. He is lost, eyes looking for darkness and finding stray rays of amusement.

CATON: Be more generous. He still now rules. And this is difficult enough without conspiracies in the open and dismay of life and livelihood. The guardianship still prevails upon us, and we should stay respectful of that purpose. Make not the first strike that wounds. Be not the provocateur, but wait until fate is gratuitous of a fall, and the feeling is so commonly shared as foregone that you only supply a comeliness of engineering to appease the populace against acts of disrespect and harrowing distemper.

MORLIFF: This is all personal. This is all agonizing antagonism. But it is not a hatred, it is not an acrimony. It is an unsteadiness to bring to balance. And with the least emotion to these physics, it is a debt to bear for being cubs of a 'cumbrance wrought. Our people deserve a sound service to the stateliness presented them, that to uphold and rely upon safely and with some security pretended. But above all, we must fight to be apparent and straightforward of our duties, not bound to questioning and quarrels and spills for the troth an' trough. Clear thinking from our pronouncements will satisfy the cares pressured and rushed to, not obscured by any action or punishment of nature.... or natures.

CATON: So be it soundly. We are dear to our master, an' deer of these woods raised. Would not we make a tragedy about this place. Hesitate until horrified, at least. Delay what is expected till it's due. Majestic vows have been evinced. They can not be easily sluiced after all of these years, afterwards in training.

MORLIFF: No precedent will be set with one's madness met and countered. (*Halyce approaches.*) Where bares thy crime!

HALYCE: Spews! Spews! He spews! I could not take the constraint (any) longer. It is demonstrable of an evil wick to wickedness alight—

MORLIFF: You are an officer of the court, and should be there in its proceedings!

HALYCE: Fouling is your made man, as a maiden! And I tell you abruptly, it is cruel to see and hear. I, of fine character and position, will not endure it more this day! And it be you— brothers— who should subvert the event. You are responsible! to have him waste of brain as if shot with an arrow for its leaking from the skull.

CATON: We?!

HALYCE: You are the close kin an' make! If this besieging be diseased of blood-worn germs, you are the antidote to relieve him, as in an homeopathy. Run to him, and alleviate this disastrous spewing.

CATON: It's the pressures of his state!

HALYCE: So absolutely! Hackneyed are his views and values now, against a weight of stoned bludgeoning with matters so important.... Oh! they castrate—!

MORLIFF: You can not cry as to un-assist! You should be there helping 'im, and to end the rent and raving.

HALYCE: I can not shear the manes of lions! Morliff. He is precocious to be fed to some tyranny of abuse, but not from my spoken reprehensions for nonce, as if to scold a weak boy meanly, or a baby with a leathered strap. That once was he? a great provider of thoughtful comfort and preserver with intelligence of our aims and station and power in this commonwealth now blighted with his brains mildewing! This is *your* work, to tend to.... this abysmal sight and spectacle of alarm and dreadfulness. He!... is your work, a man not half then half more his father's age, pooouring before us, as could drip a sieve used to hold water, with only some sediments collecting. This is your responsibility of family, to forestall his poor idiocy, as we might be lessoned to it and lessened from it. Mine is to the state, to guarantee some courtly consideration of the issues that pummel us with gravities— and are current, active to be dealt with.

MORLIFF: And I would deal with you, for a traitorous lambasting (*leaving*) and a treasonable assault to characters royal.

HALYCE (*as **Morliff** rushes to exit*): Run to.... there! And commend my nonsense more, to him that would listen to noise more carefully than a rain's motet or a waterfall's choir, that issues descend upon us to wet, bathe, and drown. Declaim his *malheur*!... (*more to **Caton***) mental and transfixing.... Why not accompany?... youngest son an' fixed star staring, upon these disgraceful matters downcast.

CATON: It is a sad wretchedness, true, perceiving him this way.... But you are a court official, Halyce, and should be by his side bidding him as much aid as you can, or judge to spare as useful. For do not say it is useless, knowing of his disintegration and pulverizing, not to try to hold back some of the blows that hammer him.

HALYCE: Us, young minion, as accepting as a fable false that there can be no harm to it, a forced replacement fostered from our needs.

CATON: You are the proponent, if not the instigative might of leverage for our minds. But would not cause the same to hound me deeply of this suitability and favoritism.... you, I am hurt to witness him— and pious enough to believe him still.... observing us as friends and close ties and succorers of a faith in faithfulness during his hard times. The pounds on me—

HALYCE: You are the best, the better to be with. The air is toxic elsewhere, poisonous— nocuous, noxious for the ear, as you indulge in loyalties, while others.... larcenies of office. But it's heard all over, how you are preferred and would be liked, admired for your stateliness and adored of this humility, beseeching much of the same manner from each of us to pardon and forgive the unaccountable of wrongs.

CATON: That is my patience offered only, Halyce. I am afeard of all notorieties that surface for us these days, from out the dark dimensions of our consistencies to be fathomable. If such a great and stern person as himself can fall prey to these misgivings and distresses, then we are all to a considerable part badly fashioned and in conflict to our aspirations and higher purposes. Then we are not measured by our attainments, but by our collapses. The angels lift, but then do drop into the devils' mouths for bait, to reveal the sores of our consciences. That is only how I can explain such a gory glory as this which stretches the mind into a rambling mischance of delight and fruition from a torturing decay.

HALYCE: I would not say that, that he is altogether pleased with his circumstance, but either ignores it or weens for its impossible benefits. But it is heart-rendering to review. And for the people you are supply in place, as a qualifying substitution that appeases their burdens and worries.

CATON: Such as I would my own, no more or less, I'd give myself to the populace. But this is not the rank of order found. And I favor Fabel, under the tutelage of his aptly experienced mother, and perhaps with his uncles' advice, as we may offer.

HALYCE: Well, he is your youth's air, but too young. And Morliff is too ambitious, Caton. You know this, with your cautions (to him), as personable as they may be. He would capture all control with glut of sacrilege, and bungle the stewardship in shock of what he's clasped and holds. Such a fellow seizes for the mere ability to, without regard to the responsibilities appertained. But we need a ruler of more judicious temperament. And at least the Visgoth holds his own to weigh (matters). though his scales have become levered with a jelly. Fulsome has his tastes become, with a binding of dementias, while Morliff hasn't any or lacks more imagination than his internecine interference. Yet I would not fight his vaunting, as a stopgap measure for now.... But be prepared. The public will be satisfied, through these events to course through. They will make the more authoritative demands, demanding you for a permanence over Morliff's pulsation and drive.

CATON: I will await this charity of thought before deciding, and concluding of my motives.

HALYCE: And the King?... Well he only drifts, sifting for gold and diamonds in the dirt— though blind(ed).... reduced to the thinking they may still be there within his tending grasps an' gropes.

Scene II — *A fairly darkened room.* **Vincent** *sits at the end of a table, facing the audience, with a candle lit, some distance from him. There is a large, long metal chalice by his hands, which lie from the arms atop the stone or marbled surface of the table. He is leaning forward, but not looking to any purpose of view. There are others to each side of him, not easily seen though perhaps silhouetted, and some apparently in front of him (as from the audience). His is the only chair. Other voices than his are heard, but are not seen owned, and so, while several, are only identified as "voice."*

VINCENT: Commissioners, invite! to be seated, in this suffocating atmosphere.

VOICE: Visgoth!

VINCENT: Tally me, that I am not amiss in the details. (*grabs the goblet to hold onto*)

VOICE: A tithe, sir!

VINCENT: I can not contain the notion. Though I know it, it does not resonate the memory.

VOICE: One tenth as owned.... we must obtain.

VINCENT: Let them keep all of their grains and goods. I am extravagant of generosity at this cruel time.

VOICE: We can not maintain ourselves this way, Visgoth. There are protections to be made, and they must be paid for. That service is expensive, and taxing.

VINCENT: During meat's mettle, our inhabitants are for me?

VOICE: They fear your slack and looseness to their affairs. You bring no courage to prop up the countryside with a ferocity against enemy invaders who would take capital on our weaknesses.

VINCENT: There is a sickness to my ears and head that leaves for pain its damage to my fortitude and health. How can I fight ghosts and fears when there is no war, when there are no aggressors but the worms welcoming our bodies, the eggs they lay and the germs they spread? I am certain of a wisdom, that we should dance through the uncontrollable.

VOICE: Make worthy your subjection to this plight!

VINCENT: I see of dreams, the hardships become play. There is to do, to leave alone the dead for dead.

VOICE: They must be consumed to a purity, of course. Else they will fester for your lack of arraignment, and spread your disease (thought) throughout.

VINCENT: Mine?!... (*arduously drinks from the cup, the whole activity taking time before he settles back down*) I do not argue with your fires, I see no back-stabbing behind that goal. But then to put me on a pyre, there is crimson made a bed. And how do you call me?!... the "not to sit with." Here to condone my temper, are ye? Dress me up in the finest riches available— What! to burn?! I have washed the sadness out, and can not see the demeaning worth of day, yours, as easily, that this depresses you.

VOICE: Our economy shatters, that we are forced to tease the very lint off our countrymen. We must arrange for better times, and restore our dealings of constructive pursuits to deal with this foul nature weaving through our grounds and tempting the earth for greater blights than this found here.

VINCENT: The flesh is still with skin responsive of me. And that is a graying, chapping joy. But I do argue, this day is this day. And how the course may turn is of its way. Come we to a plague? Only the flowers droop, the leaves drop, and the vines wither. The roots deter of (their) waters and the veins.... are drought-ed, drying into scratches. Yet the sparrows still sing!... The birds chirp and the bugs hop. Here hope is for us— bring(ing) revelry.

VOICE: Our dominions are interring themselves! The griefs become unbearable for some, in fear that we could sink into dry mud. And you seem impotent to care and fight (at) this, to make weapon with a rage for this condition. And yet you must authorize the retaliations we'd put forth. Detail yourself with generalities to let us pursue *something*, if you can not cater to the specific facts of our current denouement. And may this story pass with tales absolving Visgoth!

VINCENT: Blame me for everything. I adore it: contempt from the conservators—

VOICE: Contempt from the contemptible!

VINCENT: —I am he that weighs of you, and allows!... This permission is a strain. But I insist! These days are brighter than your missives. I am.... an earthly spy to glorious times and resplendent panoramas of a wondrous nature for us. What is happy is not birth, but bountiful release! these splendors seen.

VOICE: What gold have you, have we not! It's in the mind—

VINCENT: Illness, errrr— Sickness is a recovery. What of my body is gaunt? These sinews are sensational—!

VOICE: To pertinence, liege!

VINCENT: ....Pertain to me.... your complaint. For they are many, if it's one— Observe.... my frame devoured of distortions to your mind, and distended griefs that I'm not capable.... for you, or not enough so to entertain a rulership. I, the Visgoth, who has shaped your trowels with stunning and backbreaking affliction to my bodily worth, what stretches of a shadow across these entire plains of providence, would give beg to you for *my* spiritual encumbrance to retain? Stain you not with difficulties, but discord against me! Ne'er is one foul tiding more than a life's death. And I see no worst than (in) this petted and indulgent land. Yet blasts my sound! I am arising late, from hardships toiled to tame me into a subservience of shame against your groans. What is so unwell? What is not so well is that you'd have another me. My fancy flies to see that everything is good, and there is no suffering more than a sorrow. And that sorrow celebrates its growth and your caring of it. Thus, then, it is a babe in need of suckling, and not an evil wrapped upon your breasts for nourishment. It will develop into someone credulously kind for your courtesies now to caretake of.... so unusual a need as mine. In that I tell you blatantly: All is well within this realm!... For this you would assume me sick? Then sorrow beds of me and I am happy for her. I feel not soured of these highly supposed blights and smidgens of unhappiness, or misfor-

tunes feared to come. But see me stand—

VOICE: You seat!

VINCENT: —before the rains that can only refreshen and clarify our faces too long grimed. I see the radiance within me, the glow and the warmth you can not extinguish with your blind ignorance. For what arrives is a release of optimism, with which you will share of my drink(ing) eventually and ballyhoo as crazed as I have been to counter and contradict all of those bad thoughts you continue to present me.

VOICE: Oh! Dash this dastardly contamination! You have seen so many deaths and scourges of the flesh as to make you insensate to the coarse, crass and angry bewilderment of our people, against which you might only capitulate to.... optimism?! This is a foul floundering of your role to fight this oppressiveness; and all openly, all regularly, to demonstrate a symbol of defiance to this unfathomable tort, you must assume, or some stronger head with wiser visioning.

VINCENT: You can not fight what is deemed to be. You embrace it joyously, and give your goals to its fulfillment. Because this is a continuum of existence, that will lead to prosperity out of drought, and a drunken incapacitation— with its difficulties of movement and its painful effects— from the excesses of levity and ribald fractioning of our revelatory gains.

VOICE: Guilty!... do you feel, from your bold mismanagements and boasts.... to be a conqueror of angels—

VINCENT: I meant to capture of the angelic.

VOICE: —This proud stance before the gods worshiping of nature condemns us—

VINCENT: Are we not prominent souls?

VOICE: —leads you into foolery, and us into contention against it as we fight a siege of vengeful reprimand—

VINCENT: We are each a trite bit superstitious.

VOICE: —from your own egotistical surges. And now you clam up, sweet to the demons for a boiling.... as if in offering a pleasure to it!

VINCENT: I merely claimed.... upon a mountaintop and quite alone, the "we," of a person within me considered and consecrated, are a preciousness that can withstand and will withstand all travesty of life and welcomed living, as there were beginning to occur mysterious and suspicious ailments throughout the country, the youngest of a family typically not surviving. And I offered myself, in my mind, for a cleansing by the handlers of fortune for an understanding of these scarred self beatings of the skin and pollution of the internal organs and flesh. My wet, gray brain was taken and returned, washed by the angels and instructed of a holy perfumed perfusion to contemplate the lesson made, and recognize the great meaning of this answer, their answering to me in this way, upon a morning's radiance and grandeur responsive to my pleas: There is no such affliction, and I have caused nothing to this warrant of arresting concern. Our lives are ever at their strongest. And Death, the mightiest gift bestowed, cherishes our struggle to be worthy.

But enjoy of the heightened senses. We are of beauty being seen. We are for gratitude of search. And for what's beyond our control to manage, fashion and convince, that then we kiss and blush over.

VOICE: Was but a hill, a very mount of tainted occurrence whereat you made your tarring pitch, a strange service of pronouncement hummed at cautiously, with disbelief and rancor for the message. And what did you take from that place, that standing, that kneeling, and that tasting of the ground through lips, the earth diseased of color and smell.... from the embodiments of the ill.... but a starker remissness within you, swelling of a gay fever. It's *that* you've brought, from that foul mound. The action pleased no one, not for populace nor pupil of sane, sacred ceremonies. And the reaction!... leaves us standing. For all to do, you are not clowning mad, but being a true clown to fleas. We have washed you, and anointed you with oils of the medical cures, searched for scratches and bites and pricks of stingers and flushed the wounds with ointments, and fed you purgatives and punishments of hunger and ill reason. But more you moan, with these mendacities to tell us sorely— this is fit! our circumstance to breed through.

VINCENT: What is sound is what is substantial of your condition. I've subjected myself to these treatments of appeasement. But they lessen only your unreasoning, from which you heave another cry and I another swipe at skin to cleanse brackishly, as were these scars from a map of depravities. But it is an absolute healing to know the cause is within us to review. Or for review displaced are these predeterminations. I am as healthy as I would cough up the choke that startles for my attention. And that is one strong body's cause, since I've seen much and reviewed much and visited much or many and know whereat a disease might roam. Though it is mostly political, this attracts me, attacks me to alleviate, as what only I can through my commands and powers, and persuasions and suggestions to be bold throughout storms.... and timid within riches.

VOICE: Political?! Visgoth. Chide us—?

VINCENT: You do persecute cruelly the weak, to be suspected of a vile vainness of contagion.

VOICE: We have not touched any!—

VINCENT: Ideologically you'd consider them and contain them caught.... with the philosophical fetters of a shameless wrong, a crude and base nature owned.

VOICE They are not particularly hygienic, and yet blame sickness for sickness, as were a colored spot its own disgrace of awfulness— and not the spitter.

VINCENT: But I have been amongst them rubbing.... rubbing minds and thoughts, and trying to find out and solve away what is painful and punishing.

VOICE: They did not take to your intrusions thankfully, but only with a curiosity to wonder at, and why such a thing is done for you to find, pick out, and observe the suffering. It is a grotesque pastime—

VINCENT: I am responsible for all!... And I know that there are troubles here. Yet not disease for a dying child in anguish can there be, without more horror to pertain of my interests. If not my

very son to fear for, if not this country.... wherewith I make surveying of these oddities a desperate passion, I am devoted to finding solutions. And then this vision comes, this heavenly rationalization, to show me this is not so, and to tell me there are accidents 'gainst actualities that might wrestle of my head.

VOICE: You *did* allow the tainted foods to spread—

VINCENT: Worth capturing! A herd appears, foreignly bred, by all accounts.... but peculiar, and with crossing into our regions. I had thee analyze this gain.

VOICE: We would not touch them.... They are strange, of a fantastic fleece and yet unusual of personality and temperament, more wild than domesticated, and larger than a hand's combing through, compared to our home-raised stocks. Some inhabitants rushed to seize these proud ruminants, that had wandered through to settle on our plains of grass an' grain, and husband them. You authorized to allow what had been pretty much done. And the animal.... seems no worse, with butchery to come, than any other used. But acceptance of such an unqualified gift of chance, or hazard, is against your general principles— which leads you to worry of a fault, to our body of mind, very unreasonably, such as to the point that you search out every queer curse of human wholesomeness brought to your attention.... as if to dare its effects away as a command of redress for your privileges accepting and approving. And you find the few that charr of fevers, without much proof of your purpose. They are mostly the young of gaffes you feel to have provided, when this is at most nature's doing if of an absurdity. But your focus is offensive to us, an arrogant and gross obsession that leaves you un-pitiably affected, and apparently withering from malignant growths to the brain, your once sound faculties adrift. And so we order you to either recover or withdraw your rulership, while you still have some mind to it, and replace yourself with a properly chosen healthiness of person you could admit to being worthy of these pressures— and only of conscience— you are succumbing to.

VINCENT: Order?... You do not order. You can only advise me, or the nation. (*with tendentious drinking*) I am not to blame.... for your skittishness, or the country's mood of anxiety.... for these odd happenings. And mine is not a dementia to access their concerns— these people we may have abused, but for whom I can not decree against, and turn about, aback their ownerships.... of property and the propriety of this, without commended evidence.... to support such a move, position, and stance on my part and with my will sensed. For the child dies not of disease.... but of what?— Sheepishness?!... in me?! I can not absolve.... my lack of vision, before the mount. But not I struggle to permit.... a natural rendering or decline, that we must suffer with and survive through. And I guarantee you— Bahlalalalalalalala! (*sounded quickly, with an impulsive twitch and shudder, to relieve a tickling nervousness to head and throat*).... I will be strengthened by this trial. The populace will be satisfied of.... order and rank, and their occurrence.... safely with me. And I deem it more necessary now than ever in my past, to.... control these turns of governance. This is more a test of will and resolve than strength, the endurance to withstand a prolonged slighting, a remonstration from the gods for being fickle of feasts and supposed good fortune, undeserved prosperities, and a mysterious fate presented with choices unshapely and galling.... Order?!... As could an assembly portend to my administration— some rights?!... Rites?! Then I order you to bathe me more— and put up with it!

VOICE: ....We suggest, to our Visgoth, that it is only a dysfunction, found in a few, and of a spasm temporal and fleeing.... from our communities and their surrounds. And that the elderly and young should be most commonly affected with morbidity is to be expected of physical weaknesses to a resistance of the dreaded, vitalities sapped one way or another, by age.... or a candor for being used and preyed upon. But the alarm for our economies is genuine, though not altogether present. And this crisis in the making is what you should dwell on, not visions of a blight blessed with, to.... er, strengthen by. That is contrary to a notion of success, and insulting to our grievances. If our productivity is hampered more by fears of a madness in you.... or the officiating of your office in general, through our assistance.... than of the disease itself— then that is not good for *our* resolve to hold this country tight, and sound and leak-less.... of a dropsy—

VINCENT: There is no disease!... in me, that I can entertain you with. The cause.... of my concerns is raw, and tingling, and doping.... to preserve of this responsible anguish until all parts.... of this dominion are made aright. And I do not blame you with blank stares, since what I notice are my own subversions seen. We are.... too hungry of an honest greed, an' not enough of falsities to admonish. The sky is not ours, if the heavens have sent (this).

VOICE: You make the discomfort.... and pull down the rains of an engineered calamity! with an over-interest, an indulgence on the privations of the sickly, insisting in some nebulous fashion that it may be preordained and warranted, depressing the work(ing) efforts of our people, or some.... or only a few, because not all ring to the chimes of a crazed bell in sympathy, a gong captivated with disaster. But a few are enough to deepen our worth tremendously by drawing a pall across our ethic of labor and resistance to detraction. Too many now expect some horrific discouragement to ensue and pervade our character, eviscerate our collective pride, if we are actually deserving of some patches of evil destiny, as you would seem to propose, a self-flagellation that is expressing itself with our state coffers slowly emptying, drying up from lack of the customary replenishment our nation strives for and must only thrive with. You give excuse for indolence and disappointments to accept for the very difficult and strenuous tasks, agriculturally equitable to all management, that must be performed daily, for results weekly, for subsistence monthly and regalement seasonally. And the subdued consciousness of our providers says "Nay!" to this shadow of thought you would profit your tears on. They turn their heads from a graying face and a sorrowing passion. You must stem this decline of spirit and the blanching of the vital colors of life. You must concentrate on what is good (about us), ignore the sore spots spread of sudden family misfortune, and decree that we as one body of powerful endearment remain healthy to ourselves and that this indeed is a health we share and treasure. Leave the restoration to our tactics.... of sanitation and recuperative suggestion or inducement for the few who are affected adversely by this mysterious slaughter. But present yourself symbolic of the throne in a high nature, that your eyes deem of clear sights regenerative and promising, instead of being culpable to obscurity and dreadfulness. And if you can no longer see this way to project fine values of existence as from a beacon to insist of friendly revelations and dangers becoming strengths of notice rather than pains of panic to accept, shores of disaster to enlist for visiting.... then replace yourself! for the good of all kindnesses in us to be allowed for exploitation of our courage and exploration of our betterment instead of collapsing to a strange assault. You *do* wish this.... easement through our difficul-

ties, don't you?!

VINCENT: ....I can not be hypocritical to my.... bowels' conjectures. And uneasily pronounced, this is a deviation from our— rapprochement.... to the currency of our troubles and my disorder of vision. For as a visionary I am weighted to be true and not deceiving of our warts to feel— which ache of me.... But not to be consumed of heresies and foul opinions, I would want a solemn temperament for our people during this trying crisis of bleak overview and outlook. Yet fake not the dangerous(ly) felt. We have already been wounded, and can only heal or die. It is not a disease. It is an *ex*-culpation.... of my advice. And for your blundering this is a fertile blandishment of attack—

VOICE: Blundering?!

VINCENT: —But I will not step down! For there is no lower to sink (to) in the muddy earth for the deaths I have witnessed and the graves I have commissioned and communed over—

VOICE: A bothering pest! An insult, with instigations of outrage in feeling by the silent before you, struck with upset through their grievances and your badly commiserating words. How they hate your tone, to accept as good the worst that can befall their intimacies to childhood and loved innocents cared for but not cured. You are more curse than cretinous to them, and more cretinous than kind at those most personal of services to withdraw bodies and precious atonement for their suffering. You are *never* invited! You procure their attentions— by showing up! and being deadly of your permissions, to speak and slap of mind. What wind does sputter more? dead leaves blown!

VINCENT: It is more credible than I! this singe to mark remissness of nature!— (*The door in the back is suddenly forced open with a violent impatience, revealing daylight and flooding the room with sight of the chamber and its* **voices**.)

MORLIFF (*entering, through the light, hastily but haughtily from the back*) Where does hound my brother to a quick, wicked light! This sullies error with a blight of fortune to drown in and amongst, that he is the most regal here with breath and earning, wealth and station to invite for meeting. Yet one says this session deprecates his forthrightness with a criminal imputation of disease! (*coming up alongside* **Vincent**) This King preserved of me is yours, ours also.

VINCENT: They make for me more of a bane than banter.

MORLIFF: Fouled respect!

VINCENT: Stout and stern Morliff, that would come near as he, and not as a physician.... they admonish me cruelly, harshly, and with heat for a withdrawal.

MORLIFF: Where else but to the throne!

VINCENT: An' taxes spent, they claim I do but babble with mephitic bubbles to the mouth, all wrongly worded for our dreams of country— and that you should replace me hurriedly.

MORLIFF: With good enlistment of their time to harken trivialities through a certain daring to speak, they emote a persistence to try you crossly and crass of tongue. As Halyce does say, (he) would cover his ears against you!

PONTAL (*from the side distal to* **Morliff**): We only wish him to proclaim a tithe be raised, to fill our awareness through the mouth — through which he foams with madness! and weak duty to this populace. For even now he can't perceive a need for proper governance. And Halyce stormed out of these deliberations as fiercely as you stormed in, with disgust for this king's tangled thoughts and misinterpretations of the real that gravitates for our concerns, and towards them heaving stones! of injury with carelessness in these coming moments. We haven't much time to correct our foundering condition. The people must be built up more surely, and with evidence that matters are being handled properly, that we are taking the precautions against a total and fiscal dissolution of our country and its adjoined populace, united to a weave of mutual benefit and commonality of satisfaction that only a strict management of our pressing affairs can provide. And clearly Morliff sees this as the Visgoth can not, your brother dazed and dazzled by some foul humor of the brain to see for opposites of light and prosperity a darkness envying to envelop us (in): the penumbra of guilt! about these strange animals, and their steerage

MORLIFF: The animals? Thou will be coursed and tracked— That is correct! and trained, but without this pressing to perform. Yet hunger speaks, oh sib. It is a loud mouth, for a broad void to fill. And there may be some minor justifications to these complaints. Details must be addressed, and paid attention to.

VINCENT: For them I will sip, and slip into a stupor. (*holding onto the chalice, but not drinking*)

MORLIFF: And it is due to this apparent slack that these burnishings are made, since what is to produce upon your wear but rubbing to produce upon you care, that you will have cause to make a productive try through reigning tempers. Is this withheld within you(, still) yet? Dark mysteries evoke to provoke us— but provide.... a sensitivity to your concerns and interests.

VINCENT: That is the most of feasible endearments, young.... pretender and contender. Understand me and you do the world its knowledge baked for a glistened consumption, sweetened with review of lustrous commentary. I take more blame than lightening strikes at oak. But it is the enlightenment that moves me and opens my eyes wider than their mouths speaking to me— yelling! compelling and compounding griefs. We must leave alone.... what has been struck, and let for its due course of effect to take hold, this blow— and this enlivening. Not harm us more to heal, though healing hurts, another vexation levied 's not the route to a recovery, I can declare with roar of anguish, a wounded lion from these devastations. But go out to sniff at every one (of them) to learn, that is my schooling led of pride to manage and manipulate my curiosities to wonders and devious contagions. And if this is a bothering annoyance, what nature rents and rends, then have this land from me as from the sky, and the blossoming of trees under this sun. There are secrets to uncover and bathe with. And I am knocked to and fro among these bright rays, my conscious tasking.... My.... endeavor is to alleviate or vitiate the guilt among our populace, that comes with shame to feel deserving of a misfortune. There was no deed done, *nor any decree made*, to cause what has happened of our personal disasters. And only I as King and ruler, and the most patient patient, can prove this to them with my intrusions into causalities to show that this is as natural.... as a mold occasions gentians and a moss the true blue flowers of a royal bloodline, a sacri-

ficial service without need of bemoaning conceived contempts among us for a cause. Nothing was done inordinately by us to abridge our fine standing as a people. Thus, be they not more harried or punished as they slowly recede from the cataclysm of tragedy to adjust back to a normal purveyance for their wealth— and ours.... to administer. Leave them with our time to weather. But if we force authorities, the bedevilment becomes apparent—and the devil is at play for all rots and disgusts, disgraces and reeking (having) befallen them. Superstitions will parade, pervade and cloak their personalities deserved of an oppression. And the blame for this will be my ineffectualness to reduce these misunderstandings of natural law, fate and chance—

PONTAL: We need that tithe! (*others agreeing*)

VINCENT: —That is what my vision says.

MORLIFF: Why be so concerned with what they think of you in this respect, dear Visgoth? The relatively few deaths (up to now) are a warning, not a rite of passage into excoriated states; though there is a stranglehold sensed to some designs within our populace at present.

VINCENT: Because the aspect to these.... diseases is against my vows of protection, Morliff, as would be yours. If this can not be fought, then this is no a fight waging strengths against the struggling harm and ravage. But if this were to have been prevented.... what need we us as them alike, but meanly positioned? Then it is done.... as could a hot wind sear crops. And I hold fast to this opinion to contain and explain myself. (*struggles to drink*)

MORLIFF: This.... sounds reasonable. Yet err we greatly to not administrate and weaken of our controls to provide a sincere but secure governance—

PONTAL: Aye! And that is the product pouring out to waste with drunkenness, for what we can comprehend of these— visions. Oh! Hold us firmer at the helm, Morliff (*grunts of agreement from others*), and influence!—

MORLIFF: I might, without rebuking. without tormenting for one's.... sufficiently appropriate worry. Seated is too much experience to overdo (*placing a hand to* **Vincent's** *upper back*) with judgments and wisdoms. A short span of time should be allowed of the jab.... extracted then the knife for the muscle to allow its contorted rest and painful restoration with one's own recuperative powers. And so our people do deserve this.... leniency awhile from taxes, as my brother sharply feels. (*removing the hand*) But I commit to you our pledge to keep the government strong, and you cherished, and hopefully charitable, advisers, proponents, and activating authorities for a majesty and his kingdom, (as) proper and worthwhile and greatly valued executives. And in the meantime, some needed funds will be secured by other ways such as to be devised with strict consultation, between my brother and myself to confer, the character of managing resulting quickly and with revision and polishing for your acceptance. (*general sounds of agreement from the others*)

PONTAL: Now this is thought! I might propose a bidding of support for stately indulgences and prizes with grants of exercisable privileges.

MORLIFF: I will bring him to such ideas as serious and seriously proposed.

PONTAL: Ah, much so (much) with relief. You bring deliberations even to a lower level of sight accessible and plausible of execution. For we were all but stymied in a morass of detainment by a softened weal to convince. Yet now a simple tapping kegs, bowls of light for possibilities again.

MORLIFF: I will be here.... and with him, demonstrative of importance.

PONTAL: One can only argue 'gainst a wall for so long. Then to a kind more substantially of kin, giving of an ease less motioned of the quarreling, then sufficiently a trying time is through thusly, through Morliff.

VINCENT: As more to mead the drink is met with arguing, a sweet bitterness. I've not complained to you. But you campaign for Morliff. There is the stillborn death! that I have witnessed, and mean to subdue, which compels my faculties and ideals for this country, and this company. It's drawn to be advertised of me that I care for those.... of us.... who have been impaired with the sorrows an' depressions of a cruelty. So make light of this?! I shall condemn myself to chains and hardships without your satisfactions in the end, which is most customary for us, together, in our handling of the affairs and purposes of state. But let us insure this well-being, not so softened of mentality as you might mistakenly assume. I mean to bring a sternness to this undertaking, and quash misgivings and jealousies alike! with a thorough action.... There are no clouds seen with my visionary discipline. And while some better off may revel in their measly fortunes and minor fortunateness for a time, to the point of being squeezed for riches and bled for relief from their pressures, no one will be more abused than me! my own, to bring you satisfactions, and from out there in that vast confinement we adopt to. So, as the citizen may want, as he may wrangle for: I am he that rules, and I rule him. Let that be pictured till contentment is painted on our faces and shared throughout this common expanse. Do not anger me with the weal, as if an insult made — that I could not understand (it).

PONTAL: That you would understand anything is our goal. We have barked loudly of caution.

VINCENT: I still have a royal anger, that never burns lazily and can become retributive.

MORLIFF: Though fires only burn.... with the wisdom of the earth fueled. You have always had that at least to gather and garner.

VINCENT: Appreciate me!... I take all of our distresses and tribulations to heart, and am willingly scared by the scourges of our contentment.

PONTAL: That is the evidence—

VINCENT: So wouldst you all, if true!

MORLIFF: That is the edict, laid down quite severely. We are unhappy with you, that you lay too long upon this burden, and ate too much of it— these deaths, like nourishment of sin 'gainst sense. Do swell you sweating now, with thirst. It is unusual to take this view of things, or contain them as proper, neither for our acquies-

cence to nor dismissal of. And so I come to help you, offer my assistance to re-balance thought. You dwell upon this disease too long, divine of it too much, disturb the populace with your opinions so much.... that you may be sapping their resistance to it, and discouraging their normal functions and our customary growth through their assiduous and talented endeavors. And so I ask you peacefully to return your attentions to the matters that should make you busy of their dealing (with).... And if you feel incapable, too far removed in mind, meaning, and intention—

VINCENT: We must endure this sacrifice—!

PONTAL: Sacrifice?!

VINCENT: —submitting-ly. It is about my head, and nerves of notion to appease. (*sounds of disgruntlement from others*) To appease to.... replacement. To allow.... replacement. To ascribe the stronger births permitted here, this is the symbolic gesture sent us. And we must uphold it with a health, to withstand the agonies presented with its toil. We will be cleaner, with its wave-washed strokes.

PONTAL: Hear the mad! See the maddening, Morliff!

MORLIFF: Change this discourse from births, Vincent. You are the Visgoth! And were we after you, and before yours come, still you remain with practical duty and responsibility, without confusions made, this entity to rule deliberately.

VINCENT: I do not count.... the heads of questioning, to replace me. (*brings his head to lie on the table, with apparent anguishing*) It is done with candle lit among the sunlight. (*banging fists on the table*) Useless!

PONTAL: This is a jest— of disconsolation.... and pressures made upon yourself! by your own mendacious ways— these lying visions!

MORLIFF (*startled by the display*): It is just a nervous collapse. (*trying to help **Vincent** sit upright again*) Some feebleness with candor, and muscles sore by spread of low, contagious views. But end your whimpering before your lesser(one)s, as a father before his children cries of insufficiency to feed, raise, and rear. This is not your debacle henceforth, with me around. Good judgment only (*sitting him upright*) will prevail for tears.

PONTAL: ....Ooooh!... But does the brain a femininity. And what is made within it grows, more malice than meritorious pain. There is affliction playing on the temples, and bidding its time. That (the) lice may spread is more the nature of this revolt. And carefully to tend him, Morliff. Some princes pine of powers, and others penetrating bites.

MORLIFF: Fear no aghast (ent)'ity in him. And not myself with weakening fevers can I feel nor find.... in helping brother kind. It's Vincent. Vincent! Yes?

VINCENT (*weakly, at first*): Alas. And to myself I've come.... Or to the same is won some similarity of fame— proposed. Yet feebleness has debts, an' payments for. I am most suddenly upset, and then as verily thrilled of instance, that the shivers within me are real lamentations and laurels lent.... for a good service of commitment to the principles outlying your thoughts. But please me still to think. I want what is worth of this pain, and will work on the harassment, even as myself, and on myself to please. Forgive the demonstration, but not its cause.

MORLIFF: Should a cause be, there's much to these resentments, Visgoth, in that demeanors are highly suggestive— of a cause. What brackish wine is here? (*lifting up the chalice*)

VINCENT: It is my broad cup willed with donation, of the serf and the seaman alike.

MORLIFF (*sniffing*): That is the container. But what of the surf that visits? sand an' rock, cliffs an' shoreline.

VINCENT: Too distant. We do not own such properties except to control by influences, and their access with trading and buying — fruits.... of the distant.

MORLIFF: This stall's made here, an drink?... made here to drink. (*bringing the rim of the chalice near his mouth*)

PONTAL (*as there is some slight alarm among the others*): Delay such tendency, Morliff, to exaggerate a thirst. For what is tainted is a faith in workmanship, and a befouling of labors proposed, in that which causes these uncouth reactions, the unpredictable of pestilences' effects.

MORLIFF: It has been tested—?

PONTAL: Filtered and purified. But he has tasted it, and so it has his signature, where spread the toxins unreasonable.

MORLIFF: With bravery, so grand a chalice— fills!

VINCENT: So grand a head as mine.... The throat is thinning, lacquered to an illness. Leave me some....

MORLIFF: I'm not afraid of this! (*replacing the chalice, to the relief of the others*) Port born welcoming.

PONTAL: It is as fresh as made of hand-soaked labor and crushing and fulfillment of the red elixir's spirit. And ne'er a one does not demand it made or taken through its flows and flux. Yet have we all enjoyed.... some sample, in our days of leisure, and hours of remoteness or reclining. Now and ever have all manner of substance been presented to our Visgoth— cleanly. Still he contorts, contracts, and outright(ly) squirms, with these bold nutrients at hand and in body, of a manner as well tested of disease.... as you have seen.

VINCENT (*bringing the chalice to him, to drink from*): There is a warmth in chills. But I'm not sick, to have this leaning towards.... a surrogate for blood (*drinks, with usual difficulty, because of the size of the chalice and its broad rim*).... Like lesions swallowed, sloppily of swill. As such am I an animal of cruel thirsts.... Morliff. And there is some sediment in there, my cup, of my contagion— bettered, in the sweet brine dipped. (*holding onto the chalice*) It is a sacrifice to glory, that I can overcome this— poxing waywardness and wasting away.... of mind and attitude, for you.... I am not physically ill at all, aside from nervous tensions, headaches, and sore throats with a voracious hunger compelling liquid diets for a spell. I am braver than death, and have more courage than a hindrance to perform this ritual of leadership....

MORLIFF: Of course not, if there is no sickness spread throughout this.... fiefdom, king.... if there is no manner for the discontentment to take hold except through mysterious means and magical enlistments of perverse qualities.

PONTAL: How many fiefs are granted can not make up for the squandering of life and property.

VINCENT: I protect them.

PONTAL: You control.... a semblance of protection (*sounds of agreement by the others*), and can not conform yourself to be.... resistant of a plague, a smattering of consternation.

VINCENT: This is not! here. I bear it, and it is not so. Conform to your apprehensions? I confirm myself to be— free of the errant making by the unusual. I withstand the problem, fearlessly, and will not be led to shame and sham prayers. And for whom?!... I've seen the bodies, the most delicate and placid of stone— Not due to me, this peace. I protect!... I pacify, and provide you honor— sanction— privilege— power— breadth—!

MORLIFF: These are minor losses you feel too strongly confronting of your authorities, contesting your rule over the destinies of subjects empowered by their allegiance to you. Simply a vanity defied, you are walking awkwardly, with a clumsy presentation of your cares an' concerns— about them. An' too much so about them roaming are your steps.

VINCENT: It is an (h)'umility to be seen, healthy as.

PONTAL: Then, liege, we suggest you ponder this with your brother, become miraculous and make a miracle of recovery. You do not appear healthy to us, nor for us or any this way. Though we might force the protrusions away, of horns! with heated quarreling and stances and entrenched insistences against opprobrium.... this is difficult; more so, perhaps, than friendly fraternal conferring for appropriate illumination to the realities we are be-speckled with. But quickly done. We want the fine men leaving here, out of imperative, to correct our state. And if such wonders of soundness are possible, in this nature of confliction, then Morliff will provide for us our patience and impatience, our shines and contrasts for you. As I do call this meeting of the commissioners adjourned, or annexed to Morliff's mission, with the next rotating chairman to conduct.... a full understanding of our principles to be expressed, after a thorough consultation with Morliff!—

VACLKULOR: Aye! Morliff!

PONTAL: —to present for us your demurring decreased and your wariness increased of our commanding objectives (*as some of the others start to file out*), that we insist the Visgoth— our Visgoth— to be.... of vis, an' viscerally involved in the affairs of state, with a strong instinct for its preservation, and therefore ours, and without a need for rationalizing events— but rather making them, with a positive curing towards our collective benefit. Then to the Visgoth (*leaving*), assist of vision, too muddy for our conception or comprehension, to improve, clarify, clear, and induce.... for light to see with, instead of darkness.... instead of dancing shadows figuring on tortures for their illustration on walls, (*exiting last, and closing the door behind him*) as if this were a puppet show conjecturing of doom.... in false relief.... of mind an' mime.

VINCENT: ....This does dim my candle brighter.

MORLIFF: They are too powerful to upset this way! Vincent, each one representing a township that provides soldiers— or fighters— to the army that makes you the feudal lord and king.

VINCENT: Currently, Morliff. And on my own personal estate I've seen the horror falling unto us, with a uselessness of complaining, as does my Fabel for loss of'(f).... friendships.... and gain of blame, among his youthful companions. I cry too, of the seizures befalling the beautiful—

MORLIFF: It's a wickedness of nature, nothing more!

VINCENT: ....Those lovely faces turned to cringed eternities.

MORLIFF: Does not occur often enough to cause this enfeeblement of wits you display. So handsome was your rule once, and now dotty of a carelessness to think! I will replace you without revolution, by the mass of all concerned with our administration of provinces, if you continue with this stroke of head an' heart. But fathom for me now why you are so involved with these miseries, and not just earlier— when we killed much: young and old, dreadful and rapturous of face an' figure, sight, appearance and presence, as could be used as spoils for consciences to tease, twist, and contemn!

VINCENT: They were enemies, some. And glorious angels ever to be led by, with our sending for heavenly arbitrators in the afterlife.... some did reflect the winged conveyors indeed. But until they became ours they could be struck hard, and raped or ravished for a death. And that is a crude assembly, and a harsh finality of belonging. Yet I have pledged so differently to our cohort-ing, that they are spared such evils from others and ourselves, as long as a justice prevails. I don't forgive the misgivings for the actions we have performed through our armies and invasions of territories to annex. But doubt me now? There is a promise made, and a solemnness to it implied, that I will die if need be, to fight any ravage against us, through man or monster, beast or bug or boast of wind and storm to weather. That is the best that I can do, and I apply myself to this principle, rejecting a concept of retribution for our earlier deeds and outrages, since gods do not judge our character as amongst ourselves, but rather as appertaining to themselves. Thus, we be our own gods and deities. And I.... be this most soundly bitten, as a representative to receive visions and messages, to somehow pardon the outlandish against our very children and youngest defiers of the deadly grope. If I can not do *this*.... well.... then I do not deserve any repentance and anima-tory peace, and neither would any overlord responsible for a heaven bound to, on earth, ground, dirt an' dust and the pitiable silenced forever. For that is what proposes for my steps to appease, those unremarkable and ignorant cries of earnest distress to end.

MORLIFF: I would leave prophesy to the prophesiers, and vanquish the foul from all of this terrain as your immediate and most active goal— something you can do, and practice on to be proficient of these ruling techniques. If the cause is mysterious and unclear, make your fists— and their plowing work— as certain of a contrast to this. Subjects need that imagination in you. Why give in to diseases, anguish and grief to make them partners?! sharing commands. Have the labors be sterner, to force the constitution (to be) tougher, and forget the dead while you are engaged with living,

and subject the hopelessly dying to their own tendencies— not yours. One must count on potential and promise for your prophesies, thereby your correctness in this is a gift, and not a condemnation.

VINCENT: It is a cool subjection to infant's tears. One never knows until they're reared assuredly. But a grave remains a grave, with kissing of the earth as for a sacred vow of remembrance.... and acceptance. Why can't the people understand this respect I show?... They think I am ill myself—

MORLIFF: And are you not?

VINCENT: ....Do shadows have a thickness? And yet they shroud and cover. I wish (for) my son to replace me.

MORLIFF: That's reasonable, and understandable. But you'll have to live for at least half a decade more for this to be a proximate proceeding, if this is to be guaranteed.

VINCENT: But you want my position *now*.

MORLIFF: I wouldn't mind it, with you in your bewildered condition. This threatens all of us; and it is this threat that gains me adherents to the cause, not my envy or ambition, though they apply also to convince me of this purpose. And were I to become.... a majesty.... I would judge Fabel on his merits, and not his heritage, to ever proceed to the throne. *That* I would not guarantee at all.

VINCENT: My dear brother, I am not ill. It is he who is afflicted — with me. For if I had to appoint a successor immediately, it would be you.... or Caton. And that is the grating on the mane's neck; for a vision suggests I may not last these tribulations before my son's honest time, that this is somehow a punishment for me, or a simple circumstance that follows from a swerve of leadership and decision making. One bad distemper of fate leaves my prospects for him asunder. Yet I know my utmost duty is to this fiefdom to uphold, and my kingdom to maintain. And that conviction is as strong as sight, I will do what is upright.

MORLIFF: Then don't believe the vision. It is foolish to (have) weight (of) it so severely, a passing notion while you are uncomfortable, unless you really do feel salubriously faint and your body gives you somnambulant warning of the soul to prepare your affairs and estate.

VINCENT: That is not! so.... physically. I cough only to challenge breaths deceptive and defective, not to expire or transpire for a greed of worms.

MORLIFF: I would say then, my Visgoth, that you are wholly not proficient in preternatural predetermination, and that you should take a guide more accomplished in these avenues to follow, who will present you with more favorable visions or ease your mind at least from these trivialities of thought. To worry is not to be stupid. But the stupid do worry, and to preoccupy yourself with false augury while there is serious administration to be done is stupid for a superstition. Leave that for a master of the supernatural or religious to take care of, cart those duties off to him to enforce good prophesy. And bring yourself back to your terrible privilege of ruling over us— intelligently, and without these odd displays of ethereal resignation or portending the favorable from misfortunes, which casts you opposite the comprehendible and commendable of spirit or intention. Wear your royal mask again, liege, and hide these enfeeblements of mind that give a soured passion to your face, a countenance that seems pitched between dejection and declamatory reformation, a fevered silliness, and a disowning of the true— ripening concerns of our feudal contracts. This is the broader view, from that broad cup faced, and with a difficulty to drink from— handsomely. You must evolve out of these fears, and back into.... the everlasting right of your sovereignty, or that power will evaporate— Believe me! (**Vincent** *takes to drink.*) The partitioning of your authorities is starting to begin, in the absence of your oversight. Shifts of allegiance are starting to tremble— and I sorely represent some stability left to prevent the rifts.... So climb out of this stagnation of wonder and bemusement, to regain a stalwart nature again, or I will scathe—

VINCENT (*with a burp*): Not how!

MORLIFF: —with a relentlessness to overtake an' set the fracture correctly. This is a disgrace of dissolution before me, that will not be withstood much longer. They are not merely impatient. They are growing of anger and disgust. I see the breaks, the cracks forming. And I hear them snapping, with ties pulled apart.... among themselves. Do you think I could stand for such a state, without providing of myself a remedy? They are gentlemen-ly heads, for now, representing— independent thoughts and volitions, and actions, that have been collated into a coalition. Lose we that binding mortar of strength from your lapses of structuring? No! Or plead my place to stamp of bonds your bondage.

VINCENT: Not how to do this, Morliff!... is the evasion made— Do not eviscerate your compliance to my rule, to make for yourself a wrap of bleeding entrails. I am not enervated through their discontents, and you haven't the skill learned yet to prevent internal wars and beatings of conflict with red, smeared rage of pointed weaponry. The simplest solution is no solution at all, is confusion and dishevelment of order. But placate my abiding of time. If the cauldron boils, do you cover it more firmly, or loosen at the top its lid? Our factions are dissimilar enough to combine here raucously of a commission, yet with some peace enthroned, intoned of a passivity as now is needed to prevent an eruption. Let them vent on me awhile, a short while more. You, with a stern beckoning, and too ambitious for a stiff raking, could give them a prime excuse to ablate the tranquility and fight with havoc amongst ourselves—

MORLIFF: Oh! I do not own this view at all, nor pardon it. A clear leadership is what they desire.

VINCENT: You're not to pardon me, but to persuade— or I'll have chains for you to jangle with!... affronting me this way, as if I had no sensitivity nor sensibility left. You are all wrong. But I try to be above anger during this crisis.

MORLIFF: Crisis?! Of the heart? or of the motions promoting me!

VINCENT: The deaths will grow.... in number, as of a debt to gods.... and visionaries. I seed for wisdoms, to prepare for us acceptance against wild revolution of the facts and the simplistic notion of abandoning our reserved reservations to the wish for battling the unconquerable unknown, whereby we only slay ourselves for satisfactions doing, some defiance broken into of our natures and of nature cross. Then heated are my hands to control things—

MORLIFF: Control?!

VINCENT: —Tempers, ceaselessly.

MORLIFF: I'd rather talk of tithes, Visgoth, than tempers. Tempers abound already, but not tithes.

VINCENT: Accustom me with your plan, to spoil the rich.

MORLIFF: It's to regale them, Vincent, of privileges for their worths due, not for sale of honors, but for salutations won.

VINCENT: Bought.... And to my ears a tingling, of coins and treasures smelted, as could a dull lead be poured. I must accede some heavy taste to pacify the vicious, yet through these cavities of candlelight. Produce for me a humor that's sufficient, Morliff, with sufferance of foolishness and pride, as if these qualities would cure my tolerance and lend me time's leaves for sheltering of, a robe or skin of green to mock the envious.

MORLIFF: Of kings?... Of overlording. There's more to mock of frolicking through dangers with a revenue to supply. If we can not maintain a common army, they will dispassionately dissipate unto themselves as separate states or realms the bother to do so each and for only each. Yet for our own kingdom this is really pressing. So don we not to ask but offer, seriously, Visgoth, for these protections to provide. Or else the struggle to regain them will be too severe.

VINCENT: I am of hearing range, and sight.... calamitous.

Act II

Scene I — *Within a forest **Vaclkulor** approaches a hut of apparent poverty.*

VACLKULOR (*stopping a ways before the hut, to contemplate sitting on a roughly hewed log of seat*): Rhizome-d depths of dispatch, it is more dark and dingy here in the day than for night's clean purpose of a vacancy in empty field. Yet comes a meaning more with complicated twists and stems and litter of the desperate for poor, no height to reach of in these tales. They can notice nothing of themselves that (do) kind of grime to make for wealth, an honesty of deprivation held for source and sorcery. (*sits*) It's dirty, here. And I am dirt becoming to and fit for spate. Into our own deceased is but a crime well hidden. And I would romp of squalor too, if I were as well possessed of the blessed blight to capture all and receive nothing by this. One can not see a poverty of soul when poorer than the body owned or inhabited. And much like children, they are simple to delight, these pagan profiteers of thought and dimensions.... difficult to realize, or accept.... or condone. For myself I am so squeamish of it, that this is a bravery. But then so is the lifting of rot, in one's bare hands, to rub with.... Yet are we not divine, to follow our caprices? more than any other animal or thing. Therein I sift the dirt wherefore this place, of subtle passable-ness at best. Or of a devil's playground come with roots that rise to snip the potted airs and scented forestry, what could bear here with a conscience like mine? The kindest demons are not known unto themselves to be, as we are all followers of our despicable fates and casualties. For one: a marriage to a state. For two: disparagement of mate and might. Mates an' mights. Malevolence an' malediction. Yet to my poor contention (I)'am arrived, this curious day to receive.... the odd solution.... to my internal disquiets. And as could a beast to stalk.... a praying, I am rushed and flustered with myself to approve of this hedgehog-ing and spiny endeavoring to lift. I am a pater to insist on it, some trembling substance of righteousness for my position to hold and to be determined by. For else it is a callow deed, to try at usurpation totally unguided and blind of chances and cautions.... Oh, now the bugs do sore my rawness. (*wipes down along his trunk and legs*) Do visit, as they dare, for bites, then I might emulate.... (un)'less they portend my visiting brings a certitude of decomposition nearer than my wish for it. (*A ball of fruit is tossed towards him, which he only notices by its landing next to him. With his startle to look up, the boy **Hestor** runs towards him.*) What be? No tree, but rain. A gangly sight deceives.

HESTOR (*standing a distance away from **Vaclkulor**, with notice of him*): I'm sorry, sir. I was playing with it.

VACLKULOR: This pit could have hit my head. This melon could have bled me— Playing how?

HESTOR: I was throwing it up high, as high as I could, and then slapping it as hard as it came down, to see how strong of a blow I could make. But I took no pains for direction to control, in this richly fern-ed desolation. (I) Had not expected thee.

VACLKULOR: Too foolish of a weak courage to be free of injury for someone— or of animals!

HESTOR (*as **Vaclkulor** picks up the fruit to examine, while still sitting*): There not be one so much around but he, who does not stir openly this late of afternoon, or could his resting take its place after the gathering of herbs and foods.

VACLKULOR: It's eaten from. Is this how garbage lands? I should throw back at.

HESTOR (*defensively, as **Vaclkulor** only drops the fruit disdainfully*): Too green for me to finish tasting.... (*more composed*) But ripe of its sort, since it had fallen to the ground. I would not steal from high branches, but picked it up to test.

VACLKULOR: An' play with, in a careless, callous attitude, soiling the earth, or assaulting anything. You think only for your fun— and that's alright, if you've nothing else to do.

HESTOR: My chores have been completed, mother says. And I am free to roam in this broad garden, dense and thick, without much sight through it until this clearing.

VACLKULOR: And were the master out from that ramshackle of a home, prospecting, spectating, speculating.... he could have been hurt— by you!

HESTOR: Not he! my good friend. My will is in the ball to prevent that. Or I would take his punishments as lovingly of a disciplining (as) could my ma restrain. But he is a kindly person not prone to harshness and pain with his reprimands an' warnings. He would forgive me— And I would die for him, were he hurt by me.

VACLKULOR: That's too bold.... for a license to be released from, if this is accidental. So you know this seer, with some tenderness— Call him out!

HESTOR: ....You come for him? Then he would know you're here, with presentation.

VACLKULOR: I'll not rise before his attendance. I am a commissioner to the Visgoth!... And he should greet me as I'm seated.

HESTOR (*as **Loresane** comes from the hut*): That is of great worth to him an' us— I am abashed.

LORESANE (*dressed as how the poorest of peasants, the hut not even having a proper door but merely a darkened entry which may be blocked at night*): My Commissioner Vaclkulor! I do appraise myself to bring, your golden tone towards coming. How in a harmony as I was sleeping did I hear your radiance prattle for in the sunlight baking leaves, a chirp of respect in the visitation. The world grew fervently involved to be considerable, in my dreaming, as I realized your presence. And before this, I was light of knowing anything, and wanted no shattering of the vagaries that kept me comforted. (*standing a short ways from the hut*) As seasons come, I am profoundly affected by them, and by you also. Then as usual, you have walked a long way to apply this meeting. The Visgoth's palace cask is within the vicinity, but still far off.

VACLKULOR: We had a session with him. (*as **Hestor** sits on the ground*) And I dare not stride of transit here with stirrups to bring to attention some importance to this gait. But I stroll casually, to approve of his land as always as if often, when we must have a conference. This may be a back-handed pursuit, but I am warned to be careful with your innocence. Words that come too freely scratch most viciously and virulently. And you think not for what you say, but how, that you are liberally disposed to it, in music hearing of these pretensions that come true. Yet for your delivery to imitate the sounds, awful may these messages be. That leaves me with interpretations, to resonate of my anxieties understood.

LORESANE: It's true. I never make claims. I only repeat what is left for others a meaning to assemble. Yet my statements are specific enough to leave my deftness out of the proceedings, that which is as faint as weather, sometimes. And what is repeated is what is heard, not told. So I am no consultant from the gods, but only (as) a vibrating string plucked with their fingers as a waste of time. All telling, I am sympathetic with their instruments. And that allows for me an emptiness and simplicity of abode, that is ten times more complicated and contentious as for houses of the more wealthy. I am always in the process of fixing it up, patching here and there, and keeping from its desired transcendence into a ruin. Since I've nothing and am nothing, but my labors an' my life.

VACLKULOR: And the renditions of the preter-magnificent aped, the super-heavenly overheard and lent to ear are your predictions— Yes?! You do not justify them, neither challenge nor commit to.... but for your needs an' neediness to reduce or stupefy.

LORESANE: I seldom benefit directly from the golden chords. Look at my condition already, to see that. It is.... below pity. Yet that capacity I have, to adjust the tone of words to reveal truths to come, when there are questions asked me. Ask not of me, though. There's less ability here than a stone to reach the sky. I can not do matter well at all. Nor can many do for me as much in return. But I am a servant to all that are masterful, an' your like.

VACLKULOR: Humility makes (for) martyrs. And this is quite an abject wonder for yourself to presume, so long alone and less than lentil for your leans. What for necessity have you now?

LORESANE: Naught for necessity. I am done. This is as much pith as I've made or found, and it suffices for a life till the morrow brings more of a concern. But if there is a restlessness for wealths, I've abandoned the attempt. It is not characteristic of my aims, after so much rejection, denial, and disabusing.

VACLKULOR: Then what is poverty?

LORESANE: ....Being contained to it, or trapped. I am silly for the blows. What have we more to do than take our circumstance with a chuckle? or at most a resignation. For we are all under sky drowned, else we would float. And what is circumstance? It's what we are successful at doing, with all of our might, diligence, determination and purpose. Then to this result we have our circumstance. And this is mine to sulk at, whipped and ripped.... And— oh yes, whim-ed.... and whispered to.

VACLKULOR: What do you wish?... for service, man! Could you not be a Visgoth, dreaming? Is it more idleness for your command? Contend to me! I come to you with a seriousness.

LORESANE: I am.... with a fear of cold and hunger always, and so every day work against this— like the animal. But vanity is the prevalent nature of man. And I have dulled this to such atrophy as a dystrophy of (mis)'function. Give me some vanity, for my ears— an' pride with monies.... as you are always gracious enough to do. Whereat to withstand me, lowly and from cowardice arrived, I am not one who can bargain easily but must be led to my gains, like a dog pulled out of the trash.... I see no ways beyond myself, yet every way for others. And that is the sad principle for convincing of my state. There is no solution to the flaying. I am not clever enough to escape such dispassionate wraths of being. Yet praise me with a cudgeling as this, to show me I'm awake— to circumstance.

VACLKULOR: How be me?!

LORESANE: How be mine!

VACLKULOR (*after taking out a coin from a money pouch*): Must I do?!

LORESANE: As be mine!

VACLKULOR: Come crawling, for this to eat! I do not want to!

LORESANE (*slowly dropping to his knees*): My legs are broken for you. They have squandered their worth of running.

VACLKULOR: What could the likes of you buy, but to hoard contemptuousness!... It is a sensation felt, to displace the Visgoth, that is wholly dissatisfactory with me— but as much so pressing, and like a commandment felt, from an unidentifiable insistence that overpowers my reason.... Understand this?!

LORESANE: As does mine!

VACLKULOR: I am a lord throughout, and will take action presently to kill! if this is true to be foretold, to be as were a benedic-tion, or a benefaction— blessed! Yet murder is a crime!

LORESANE: As to mine!... and my being. Yet surgery is murder of some flesh. I can not dissuade the tone. The better is your brought confession to be realized. And as I have said months ago, to confirm for you this meaning, it does not change. Your cry is heard without alteration of the essentials your dictate. This is your competence evoked still, with a growing weight of urgency.

VACLKULOR: I do not wish to know more clarity with this discernment. You cause the imperative to give flight with actions disagreeable to me and against myself. For volition wanes the stronger the demand is tamed. And without emotion must I be, for this catastrophe to rend? I plead you see much better, Loresane, to be accurate in this. I am about to destroy a grandeur, for my own to enhance. And this is inescapable?!

LORESANE: ....I would bother it if I could. The Visgoth's no fright to me, and does not tax what is not taxing. But you are sound to hear these whims, an' silence them with me. I propose nothing!... but your prescience endowed with my poor shape and trim, that I reflect for you what will become. My sight's not farther than your ears. My mouth is yours bespeaking. And I do not lie for winning death.... It is around, not sought, till easing tempers makes it grand for spoils.... and ditches dug. So hangs me, as a veil of misery. Misery can not be miserable. I would do differently, if I were. But to hear your soul makes merry of its hatching to complot.

VACLKULOR: And the boy.

LORESANE: Boy?... Hestor? My vision is not great— Is he still around? (*twisting his head to see*)

HESTOR: Aye! my friend. It's wondrous to portend.

VACLKULOR: A miracle, in hearing everything without details. Details burn.

LORESANE (*getting up, with difficulty, to stand*): For this I would agree. You owe me, as I wish to die, and you must kill. But his is not a life to lose with mine, and not my fuzziness of view and loss of focus to lend for your.... tumultuous deeds.

VACLKULOR: Ah!... But you did not crawl—

LORESANE: I couldn't.... This house is bare as night. Wouldst beat me in it? Won less sight have thee. Contagion spreads with company. It is groundless, and pointless, to amuse yourself with him.

VACLKULOR (*dropping the coin near the fruit*): And this is all for him, other for you. I can use candidness either way. (*as **Hestor** gets up*) Come retrieve a present from me! And your sordid quake returns for use (***Hestor** running up for approach*), its volley jewel-bound and presented as it was, defined of a need to fall.

HESTOR (*crouching to get the coin and the fruit*): With a graciousness, your commissioning is very kind, sir. We can use this much. It must be of some great value, though I'm not schooled in the denominations yet. (*examining the coin*)

VACLKULOR: It's of a golden variety, trivial to me, yet from my dear pouch an' paunch to indicate a worthiness of use and friendly solicitation. Be pleased with more for future service— Like me!

HESTOR: With grand obeisance to obey, sir. I am.... of curtsy for you. (*studying the fruit*) The fruit does rot a bit though, now.

VACLKULOR: Why be concerned with this?

LORESANE: It is the plaything of a child, Vaclkulor. And there is keenest interest in every possible detail spied. So tempt the mind as such, there at your feet, what curiosity reigns. Let not this subject turn as awfully bit. As from my mouth fallen, your words. But I *will* do, to keep an innocence clear and clean.

VACLKULOR (*holding up the pouch parenthetically*): With more meat comes the tacking. For him, a gift, since I could be a child again, drenched of fun and carefree embodiment that is only worth a life, (*standing*) were I so young as this one again. And to my feet, much lower now, or distant. There is the span.... that makes one sacrilegious. For the deeds employed are much the same at any age. And what is meanness when one is petty? contrasted to a brawny giant's stare. By mean I mean totally self-involved and concerned, not out of arrogance, but from absorption of thought and possibilities to want to reach so high as if to throw over heaven.... your opinions an' purposes. (*slowly walking towards **Loresane***) But for you, a minor fortune's due with service, and the details, details of heart that challenge me so terribly. In that we may adjust them to be fit, there is a liberality to truths. (*as **Hestor** stands*) I demand to know what is perhaps, by one too insufficient to care.

LORESANE: And horsewhip my mind to bring you there, to that spot of conscience that can approve of your stance an' stationing. I am it, well girdled for a ride with riches, seemingly all but destitute to you. Well, comely is the chore, I can admit, as long as your attentions are drawn on *me*, and my parade of mouth an' manner. (***Vaclkulor** standing by him, and lightly touching him with the pouch*) For what are monies to the mysterious? and for a séance of the willing. It, were to beat me with, would pay.... a season or so, to ease the desperation.

VACLKULOR: We all take our pains for what we earn.

LORESANE: That is brought to you, screams and tirades out of anguishing.

VACLKULOR: How fine is proof of sincerity for what is spoken than by its hardships speaking. (*binding the pouch back to his stomach belt*) The cries are like sirens screeching, preaching their sermons.... intensely. And for each "om" an omen made, as strongly as is paid its reverberant quality of hurting, is done convincingly for me. Why are you still in tatters to be seen?

LORESANE: The breadth of a season is long.... some foods are bought, as well as some materials to keep a shelter. But that can hardly last, from what is found and crummy to outlast. And I live to die, and live to be proper for what becomes of me, or however this life is deserved. Then I am tatter-tinged by nature. As much to eat, as much to cove in hovel, this is my participation met in these desperate encroachments of the human loft. I can not win more than this, though I have tried. So it's best with delicacy of wretchedness than decadence of wealth with me.... to position myself by. Attainments are for subsistence, and the gratis hardly suffices but for resting an' dreaming.

VACLKULOR: That's a cruel curio of ideology for a worthless man. So then inside we make the patter of the rain, and you some earnings, (*as **Hestor** anxiously comes up to **Loresane's** side*) if this is your.... profession based upon, the sickly spinner of a cloudy web of vapors. All that is thought is like the wind to hold or weave through, and yet can be so powerful displacing your intentions with a content more influential and subduing. (*starting to lead **Loresane** for hut entry*) It's like a still life in the dark, in there.

HESTOR: My friend is kind and plays with me his generosities of nature to be visited or chanced upon. (*submissively, but not to plead, as if he would take to bargaining*) Please not to undo him with a struggle or a smiting for his words or utterances supra-natural and transcendent of the common tones perceived towards ideas.

VACLKULOR (*leading **Loresane***): We make heavy chatter in there.

HESTOR: I am willing to be used, with his employment pleasant for your ends, the dictum of your desires to make.

VACLKULOR: All but political, my lad. I will look forward to your assistance while guaranteeing his is as amiable, and equal to your fruiting, growing lusts of character and substance on this earth.... Yet nectars are always bled—

LORESANE: Fear not for me, good fellow Hestor, ripening to the causalities and entanglements of privilege.... to deal with argument and purposes sophisticated and mature, a nice little being to see authority in action of persuasion. But to your coin be more concerned of use. I am safe with my visions and sight, for what is seen is as casual as talk. But what to spend for.... browns the soft rock's mush and mere as well as bitten has to be. Then to your mother give immediately, and for an intelligent use, what may be wasted. I am quite hardily avowed of my insight to be protective, this person's haven housed of more ghosts than gratuities for visiting.... A shield of solution in a waxy luster dealt, the dimming pelt of fortune is its melt of damp embrace.... for devious plans. (*He enters the hut, with **Vaclkulor**.*)

HESTOR (*to the fruit*): Is a rock?... Is a projectile plunged through the atmospheres!... of contention and a willing conscription. I am not of a weaponry made yet. (*fingering the coin*) But this is good dole to play with, like finding freely on the ground.... a token of curat'-ion of deeds an' attitudes— and fun to be (*running off*), this heavy lift of light encumbrance. What more to pay than he! of me, mine made.... for maim of air an' breeze. Miraculous! this finding here of a merited enlistment, from toss of tor's tally, toppled, topped ov'r— with a tree.... Limbs.... of the for—reest!

Scene II — ***Caton** stands on a slight hill, overlooking some residences in the vicinity, as **Fabel** strolls up to him.*

FABEL (*approaching*): What of those mean thatches, uncle? (***Caton** acknowledges him with a nod.*) Oh! They glisten in the sun, do they not! like a fine settlement of tawny luster. I could rule over such a province of containment, if content be the straw that shine. But I can't yet understand the ordering urgency that builds such a society. It is quite fantastic of an assemblage, or assumption of the same. Why are cities built and arrays made of houses for homes? Is not man the nomad spiritually? an' ever with quests an' quandaries.

CATON: Questions, my young jewel of a relation, so precious to our stone of fortifying masonry and administration, make errant answers possible. These are the residences not firmly placed, but found, with shifts of a degree entertaining their resolutions to remain or leave. There is a grumbling down there, as with fear of a disease that extorts the farther extents of our particular realm with a dilapidation of the soul for foundry, the normal homesteads once maintained left standing bare and barren to the pestilence imagined. The Visgoth, your father, invited them, for this camp to be set up, until he solves the problem of what spreads. And that is such a mistake, in my view, trying to jump over anxieties and causing more with this dangerous leap. For a family's worth is stripped without a decent home to manage and have worth of. But they are now.... ever more dissatisfied of their condition and will complain about the impossibility of the Visgoth's remedying. Who would want to live in this potted pox of a community for long! And what they bring accumulates to themselves a concentration of distresses and erring sentiments. Vincent simply brings the trouble to our heart, I must dare tell you, son of son. They should have stayed put to their customary and beloved locations and fought out the limping disasters at our fringes.... But they were frightened by the Visgoth to resettle here. And so the error leaps as well. One death already, of a baby. And your father goes down to investigate and personally oversee her services and rites. They may tear him apart with anger, disdain, and disqualification of his lunary authority to control the quarters of their waxes and wanes.

FABEL: He forbade me to accompany him, though I wanted to. Yet he invited my mother, who refused the offer, saying it could be a deadly trespass, or intervening with mortality, mortalities he should not further interfere with.

CATON: Her greatest aim now is to keep you from his madness.

FABEL: My father is not insane He is a god, as much as can be made by man. And that is all the madness crafted by our strengths; for these responsibilities pierce him and his seemly concerns, the governance and protection of this kingdom. This is a superfluous tort. How can one control such things? What makes people expect a leader to?

CATON: More questions, ever made for answers tarred. This phase of death would pass through naturally, if not disrupted like this and manipulated with bare hands to try and shape and hold and check. It is a cyclic occurrence, I would guess, beyond the sight of generations. And we are not helpless to its depredation. A firm insistence not to allow it to destroy us is all the temper needed. The measures taken will follow through as necessary with this stern pact of nerve and resistance. Just a heightened cleanliness can be made, and a restriction or modification of diet. But the Visgoth can not claim to be accessible to an inevitable vitiation that is deserved, as he must say to master everything and take the blame. That is not an argument to be treated softly through vitriolic conjecturing and conjuring of suitable or appropriate demons, that of ineptitude and passivity. Your father takes on too much, to bring his faith in danger with his face shunned, his presence heralded for a misfortune to occur. He attunes to the blight as (if) it was his.

FABEL: My mother says he is affected by it. Yet I forgive him, since this is an incredible kindness he struggles to portray. And it makes me think of how I could find the courage to be similar. No one knows the answer to these debilitations, and why the fatal weaknesses to our newborn happen. Sheep?! I've heard the protes-

tations. But babies don't eat of the lamb, they eat of the grain crushed to a curry or mushy pap. And they do not suckle of the lambs' provisions. And these novel animals remain quite healthy, their wool very strong and sleek and much more colorful than what we are used to by such mammals. And the people who directly stay with them to tend of their particular wildness remain as strong and hardy as ever we've known (of them). No. The sheep are a poor excuse for devastation, as my father has said, and can not be the cause of our own assaulting and illnesses.

CATON: That philosophy harms— answers.... Fabel. Of course we make our own deaths, but we make our own lives as well. It is what assists us wrongly that causes harm, hurting, suffering, depression and despair. Vincent should not adjoin himself to wrongs to make them as spiritual as he needs to seem, or feel.

FABEL: But eventually we must make our own explanations for anything, everything, and always— the past, the present, and the future. My father's extremely personal ideology surpasses the possible errors employed, simply because it is genuinely expressed and applied. And that is the mark of a true leader, most clearly to be remarked of and debated during difficulties. Grandmother assures me that he has much the qualities of his father, perhaps our greatest monarch ever.

CATON: Your schooling is advanced by her. I do not say you shouldn't be proud of the Visgoth. His accomplishments have been monumental and celebrated till now, keeping up and enhancing this statutory fiefdom with hereditary authorities agreed upon by most concerns. But not all, Fabel, would have it thusly. And if a foolishness in the Visgoth leads to foulness an' plague.... concepts, precepts, and conditional arrangements may be overturned. I am most sensitive for you.

FABEL: I?

CATON: You are the prescribed inheritor of these dominions, and the fulcrum of consideration now, though clearly not ready for much power to wield intelligently. That is a sore point and a soft spot to poke at, by some with consternation for your father's activities or suggestions. If he is incapacitated soon....

FABEL: I would leave you to proctor (my) avails, uncle. I don't feel substantially insufficient structurally, as our governmental procedures are firmly established—

CATON: Memberships are fickle.

FABEL: —But impenetrable unknowns make even a small room seem frightening in the dark, to which location you are not used. I'd only seek some equity of performance, with sound advice, until I discover the sweet, the tart, and the mange of office, who causes this, and how to control them.

CATON: You'd be retributive for father, trying to punish those who condensed on his downfall. That would seem to be a natural aspiration, in your youth. Beware of this sense. It is more emotional than righteous, less level-headed than annihilative. But you don't strike me to be too vindictive a child.

FABEL: ....I'm not a child at all.... Would you go down there, to retrieve my father from practice of a folly? Are you brave enough —

CATON: I say he handles affairs—

FABEL: —to confront him?

CATON: —badly, Fabel, at the moment. Contempt abounds about the Visgoth. And this is more than a danger to preach to or a fearlessness to handle artfully. It threatens you. It threatens stability itself.

FABEL: But would you go?!

CATON: ....Morliff would, if he thought it was worth the effort to prevent a ruckus sounding of the dissatisfied and apparently castigated. But not I, young son. I am afraid of the Visgoth, of his possible anger at me to contradict him, and his displeasure with me to be opposed to his views. I'd save him from a physical injury, a bodily harm of attacking him.... I would try, if I could tell this (were occurring) from here. It would be instinctual, from my affection for him and how he has helped so thoroughly to raise me. But (against) the mental assaulting, I have no shields for my own protection, no armor and no strength. I'm just starting my averred life, with a wife and an optimism for the future. To these I am most responsible now. To this.... estate of being is my cherishing pronounced. I see a possible decline in Vincent. You must adjudge the sad reality of it— Your mother does. She is terribly frightened for him.

FABEL: You are the honest kind, uncle. I value that.

CATON: No less than (for) Morliff has your father instilled in me a straightforwardness.

FABEL: We are forbidden to intervene, uncle. You could have just used that as the most prominent excuse. I am afeared to believe it in me, myself, a cowardice from proscription more so than through respect for his courage. But Grandmother promotes for me his obstinacy to be so chivalrous as evidence of his god-like descension and worth.

CATON: Is it chivalrous to bow to misfortune and disaster?

FABEL: With vision, he is self-assured. That is an overwhelming might of mind. And is it correct? What is correct, as one revolves, as one is in a revolution of the spirit? One might lose everything, and die aimlessly, and still be correct!... I can teach myself this notion, Grandma says, as soon as I develop some courage— without daring, and daring without the bold ambition to defy my pure reasoning.

CATON: Pure reasoning?

FABEL: We are all sacrifices in the waiting, with a marriage vow to tend of (a) dark beckoning. So our character in life is more important than our deaths.

CATON: My mother grieves, Fabel, and is abandoned but for her two sons. For you she greets.... with the vacillation of vicissitudes to obey. And any crazy thing may happen, to which she could allow a prosperous propriety, or at least ascribe to it and adduce for it, that the reasoning is pure and not distraughtly contorted or contrived. I share your envy of a bravery to heart— but see how she is torn—!... Oh! To sense this is too much, for a youth in your posi-

tion. From strictly a maternal nature she would want my brother or myself to assume the Visgoth's office. But with high adoration for this office, she may rather wish to turn you.... Fabel, turn you into one of us, as she may be the mother felt and the nurturing enlisted — for the Visgoth! Yes!... I am afraid to adopt this task, and Morliff is definitely not so. The world might be as honest and faithful to its commotions. But my mother sees your father strangely, of his affliction mirroring her own substantialness twirled and adjusted haphazardly or with undue happenstance, that his brain grows warm as if poked by a red-hot stoker, fate's enlivenment with a brightness— as she was once called to be, so unpredictably.... our queen. Nay, *sauf pour serf*.... She is younger than *your* mother, and this is difficult to arrange or take. But you are.... our son— the fiefdom's son, the caretaker in waiting.

FABEL: I would much be amended by my father's wisdom. It is blindly perfect, highly inspired, and internally devolved. Grandma understands this an' dotes on me—

CATON: But not as an example to follow. His ways annoy the populace, and you must work to seek their disparate favors and approvals if you are to become Visgoth and prevalent of authority. There is a solemn richness to it, like a finely designed and intricately crafted tapestry with orders of shapes majestic not to be disturbed or disrupted.... by irregularities. This incites uneasiness and uncertainties, that which is needled in and falsely sewn to cause discontentment, revolt(ion), and perhaps dissolution.

FABEL: What you call irregular or peculiar— rules!... Is not a leader unusual? Then he is of an odd standard made, with qualities that transcend the ordinary, uncle. Judge him this way, and not maliciously.

CATON: ....I will not be cold.... to my older brother's calamities. I know him best as a strong person and irrevocably true personality disdainful of all deceit, the virtual witness for all of us on the tending ways of existence in a world that always challenges with change an' upheavals. But these are the qualities an' natures that we imply of the Visgoth. And if not learned, then they are fostered through the strains of command, what is an internal an' eternal endowment from the natural and violent gods. Yet all (o)'f their stirring and lending of turmoil to a conscience can make one mad, in time, the burdens too great for a disposition of peace or without restlessness and relentlessness. And he does vault to value a dissension made, it seems, collectively by all who *could* advise him.

FABEL: He is not of an angry manner or with a despotic *malheur* cursed, outside of war to demonstrate ferocity. He's not struck me once, with those powerful hands I've accustomed to admire, as with he could build a nation; though in childhood I have been disobedient and foolish and taxing for vexation his demands. Mother has restrained me, with his gentleness provided of assurance for a good behavior to be treated with the treasured reward of his kindly approbations and tender gestures towards me. More friend than father, sometimes, is the wit of this approval. He has never threatened me with a hatred. And my annoyances (of him) condemn my nerve. But that is all in the past, as I am skillfully grown to become, more properly, the Visgoth's son.

CATON: You have never been presented directly to your person with war, and can not imagine his.... braver tendencies, which can horrify death himself, if torture is a volition towards too sudden a rush for his passion. And as a youngster you were no more devious than immaturity provides a curiosity to play with, as do all imps suffer the impetuosity to determine their imperatives greedily. You were not too overly devoted to this self-gratification for play, fun an' mischief, and caused more amusement for observers and acquaintances than a need of real amercement. But have you now with careful pondering to be preoccupied, you are still made up of youthful thoughts and untested conclusions. You can not know yet what is known and even obvious to others. A college of information awaits you. You have not shared the profound miseries.... nor the profound joys (of most), sheltered as you have been to the most honored clique of our societies. So try to reserve your judgment of *anyone*, including your parents and relatives and friends, until you have experienced more the depths of human personality. This makes me wistful to think of innocence as won instead of bestowed, as like ignorance can spear your troth—.... But what of fire yields within this doth?!... domain of which it is aflame down there!... Slight of spark, and yet intense of sight— a jab of light to realize, what burns be made of cautionary terror?!—

FABEL: Meals be made?

CATON: —Not much the time for it with this event, your father's presence to perceive, some ritual or sacrifice— or sacrilege interring the straw hut mounds. This cries for an investigation!— Stay! you.... by your *parents*, Fabel.

FABEL: I would fain go down!

CATON: Go to your mother, and her mother made— And stay as safe! before a providence declaring you, as for this state responsible to be. That is your father's grandest wish, I'm sure.

FABEL: With ease, it's difficult to leave, uncle. With danger, (it's) tempting to assist my father!

CATON: All with eyes will be alerted to this.

FABEL: It looks contained to *me*!

CATON: Chance not on this occurrence petty thieves and fevers, theft of fiefdom with a fire made. The thief be flame! more signal of dismay, or that some purpose to it stings! (*heading towards the compound of huts*) Array yourself of candor to this demonstration, Fabel, and as a precaution protect yourself, your preciousness of being— for us, here! And with cautionary inquiry, if little 's to be made of it, I will determine. But if more then most. And if most— then not all— Remain! (*has exited*)

FABEL: What fear grips the immobilized! I want to know more of this. Yet it seems so little lit from here, a speck of shine the sun could provide on a reflective surface. My uncle is perturbed and overawed of possibilities only suggested, with lean on suddenness to unnerve.... or rile.... to what might be a small occurrence of fire, or accidental in that complexity of concentrated thatch and tempers. Yet as a man I would know more, one death to bemoan. And then there be.... this light. Oh! that I should not interfere. But where is my allegiance— an' to what?!... To time's timidity, to be thoughtful about this, and ask questions still, as Caton enters thrillingly my mind's agent and agenda.... to dispel a cowardice, and through me pour out tears to end a blaspheme of sight! If this be done, my reckoning, so early and so late, then I am all as shivered — shattered!... torn to a coldness of feeling towards.... these displacing rebels. Yet hone my thoughts, if rebellion's not the cause

for dying and this plight in bout with pity. In what for many makes for me my protest, fear! this nation, and groan with the asunder'(mak)ng.

Scene III — *Vincent is standing on a slight mound, a distance away from a crudely fashioned hut. Two attendants are by him, off the mound, but with distress of providing assistance, fearful of rebuke from the villagers. A small but well designed camp-like fire in the pitted earth burns a short ways from the front of the hut, at the moment being fed the swaddling cloths of an infant by a relatively young woman.*

VINCENT (*hands on hips, with elbows out to the sides*): There be naught to consume, yet makes the air breathe faintly, this orison proposed of a devilish action!

ATTENDANT 1: Let her be a witness to the need, Visgoth, with her own hands trending of the grief.

ATTENDANT 2: It is more certain than your callousness, torn liege, that these tactile remembrances are lost, the infant swooned to the severity of death within her very hands, I am told. And now the same do purify the dress for an eternity of remorse.

VINCENT: Does burn a cheap material that was fine, and without this abuse necessitated to be brought. But to my own presence this is done deliberately, and to rate, for irritation, my remarks. And yet I take it so, as some suggestion of disgruntlement— with me?

WOMAN (*still occupied*): Why do you stand on her?!

VINCENT: ....To show this has my bed!... an' footing for you—

WOMAN: With rite brings rite. An' you do a prayer-ing, while I burn her sheets before you.

VINCENT: They (we)'re not prayers, but my indications of her growth to heaven—

WOMAN: They linger, for wanting of to wrap. And be thus fouled, they are as well for a consecration made, of floating ash as were the soul, as were a wood once wool. How have you brought? these grievances to play on!

VINCENT: I step in them!— an' become them, lady.

ATTENDANT 1: Do not bring annoyance to her passion, Visgoth. It is fully of the birth, out and overgrown and desperate for this expression.

VINCENT: Here sorrow ends with my feet! I approve of the actioning— and sanctify it; but not give sanction for this fire do I. Do I do, it is dangerously made. Do I speak, it is dangerously spoken of. Do I see, it affronts the visions— And do I sense, it is against my sense and proportion for you, my rating for your actions. There is no more guilt allowed by this trade of temptations. That is what I declare to you— an' demand!

ATTENDANT 2: The swaddles are tainted, my liege. And she burns this cloth reasonably and with prudence.

VINCENT (*angered*): Down no error to my see! There is disease

19

not more in me to think with— as I've shown and as I've seen with visionary tenaciousness for all of you. Stands here no downfall. It is a natural occurrence, this occupying. And I am not to blame but for to welcome nature, with my blessing an' my kisses.

WOMAN: Was dethroning of my care to notice. I'd burn myself to have you worn, a king of despicable displays around my baby—

VINCENT: It is a ritual enhanced with my sincerity— and my wish to be at one.... with the rested.

ATTENDANT 1: Depart us from here as crowds grow, Visgoth. They observe from within these straw tents, but now dare to emerge with distaste for you. Fight we minions? They were surprised at your antics, lured-like.... to make love to the dead—!

VINCENT: Bending is not bowing! for the knees to tread, not fester of a necrophilia, nor mourn about this mound, a necromancy postured with— I assume!.... the enrichments, an' the scents of earth— the lift, the knowing.

ATTENDANT 2: Lewd laudations. This bespeaks.... a growing contagion—

VINCENT: What?!

ATTENDANT 2: ....of the mind to think with casualties as art an' preaching.

VINCENT: It's not religious in this density of moles. Yet come for me, that I release your quarrels of suffering and adapt onto you a more healthy notion that this is as much death as a kindness lends, and is permissible.

ATTENDANT 1: You have been too much (of) war, and have drunk of it to spew intoxicatedly rewards for all others.

VINCENT: To protect is to be destroyed. And life is protected by death—

WOMAN: What would be worth in me contemns you! for this manner possessed.

VINCENT: Argue with me not today, when I see rays rising for you and asking for your peace and contentment—

WOMAN (*while jumping up to her feet to stand*): You are a sick hound! of the peat-scented swamp to twist and gall on sniffing, as were my entity a posh pother of malediction for your concern. And this illness stirs you, an' makes you stern for backing!— your rights to rule us?!

VINCENT: Over! Most over— kind! I am to bring you here. That is how! That is the essence of my imploring for you, to release this enmity, this tensioning against the gods that bring to you a gift! of what occurs. This is better placed than a wisdom of sheepishness—!

MAN (*coming out of the hut with a spiked rod, as other settlers start to emerge from their huts also, though not weaponed*): Get down! an' off of my child's grave!—

VINCENT: More death to come. Defy not your Visgoth!

MAN: I've been afraid inside, wrangling over what to do about this outrage, and your disgraceful mimicries of solacement atop our solemn efforts to give her peace and burial, whose(person wa)'s hardly born to be— an' totally dead. And yet if birth is made, then that is all most definite as well; an' the corpse should be respected as one who was an' truly did belong.... to us, and every living creature shared of life. Now you defame not only the memory (*raising the rod threateningly, although he remains a distance away from Vincent*) but the body itself, with a terrible stance and an awkward regality to maintain it, as were some fool of messages about, above my kin receiving. And presently to us what?! A transmittance of cajolery to grief an' thorough heartbreak?! Yet my wife was more practical and courageous to come out here in front of the idiotically ignoble, and these purposes of yours, and burn the remnants of our daughter's comfort, that the smoke of these few fumes might swathe from consciousness the heady sense of this disgrace you would perform, by our saddened aboding and right over her very spot of interment to a hopeless decomposition and corruption of our dreams.

VINCENT: Err I consecrate this ground with my forgiveness of your thoughts. They are to me and mine, this disposition made, and this hard feeling talented for your awakening, as could a flower blossom on the tree— for your notice; then this very act is a demonstration of my consumed concern for your benefit. I make pardon of the corruption among us— for your sake, and not my own. I eat the spiritual famine— Or would you have me misplaced?! And threaten me?! I've taken iron towards the breast! and a mallet-ing of swords in combat, so violent, their striking up against each other to make edges planes. All of this, for fortune of your valors deserved. Then replace me, and dispossess my herdsman-ship with a vicious launch of that prickly stick onto my revered frame— Attack! And let this sacred ground be moistened with my bleeding and my death— and the end of time's cries for the reception of messages and the reverberation of intentions. I fear not such assaults, from their own hands re-maiming empathies.

ATTENDANT 1: We would defend even the salt in wounds. Let us make trace to the palace road.

ATTENDANT 2: And the royalty of dementias, as men collect to heave us soundly.

MAN: Would throw I for an aim above my daughter, this harsh scathing to scour the earth and scratch for way of seeds, your red drops atoning for disease. We'll take little more of this rambunctiousness over graves.

ATTENDANT 1: Would be of your destruction bought, to injure the Visgoth, or even attempt to with earnest physicality.

MAN: Does physics lead to my child?! or the *pain* of physicality! Not more of an insult can be taken (*swinging the rod wildly with frustration*) in this presence accustomed to. He makes a forceful agitation for vengeance against the impossible. Then are we angered at his deeds for dying, have we blows to offer towards his blasphemies. And our upturning, created into dust! will be worth the punishment of these sincerities—

VINCENT: Win me with a bravery! Everyone of you will be crushed, sundered, an' parts fed to pigs—!

WOMAN: Let this encampment be spared your queer wrath. Bring still, husband, that ground's instrument—

MAN: Not !

WOMAN: — For as much by bare hands and a shoveling did we lay her tenderly into the soil (*The man stops his motions.*), our little seed, our precious worth devoured and re-offered to eternity and the mysteries begetting, the simplest of bodies, our pleasantness combined and fashioned of a smile, what lasted to a sleep.... so easily to be satisfied for a comfort.

MAN: I'll have him off the mound.... to let my daughter breathe, with lungs.... perhaps celestial, but indeed less burdened of his.... his standing. (*as others approach*) We crowd you, to implore your decency. This is not the mark of a proper ritual you make, your footprints upon her—

VINCENT: Ooooh!... That my chest does wheeze of cavity, so heavily weighted of this heart.

MAN: left. Leave!

VINCENT: The sheep disturb me. (*stepping off the mound*) I have sensed the mount's abiding and assuaging, for these toils to assume an' pacify with, the cantankerous guilt from greed and avarice.... for life— a life, spoken of, spoken for. But I have heard the means, like horns blasting of excuses as.... the severity of this blandishment. And to my compass have allowed you, I!... some forgiveness and a recompense of feeling. She is no longer dead, but of bone and flesh. I subsume the need for your errors, (*turning to face his attendants, his back to the man and woman*) and even suggest it.

ATTENDANT 2: What see we faced? Ligation! liege, of these heart wounds. But are you real— to conscious sight? Know your weaknesses revealed and exposed. You care not what is hurled!

VINCENT: Deed me traced of lines an' ravishment, in a tranquil frame of stone that bends.

WOMAN: It's difficult to see for long.

ATTENDANT 1: How do you mean, of this countenance, to save you, Visgoth? but from the back? Alert us with some wisdom, some knowing poise.

ATTENDANT 2: Seems smiling without jest.

VINCENT: There's more proponent to my mandatory stance, that you would have me vanquished of this life. I suggest it, but with candor present to you a sufferance of meritoriousness—

MAN: Merit you spines, horns, an' a tail! But you bake us, to smooth over the ground with your gashes—

VINCENT: Leaves. Leaves.... that fall. The leaves fall. And so do I, on hearing them, my eye.... an' my head.

ATTENDANT 2: It.... hurts, my liege? Then come with us, to paint it balms and quell the swelling.

ATTENDANT 1: There is relief from any sacrament or ceremony.

MAN (*slowly raising the rod, as if to throw at **Vincent***): And what for making any sorrow wrought, the worst of pealed calamities heard as from our haunches, then breaks the boundary of contagion's insouciance and inscrutable demeanor—

MORLIFF (*arriving rushed*): What has burned out?!... of this deployed persuasion for a community. What! Cloth? And can a king be bathed in stenches?— Lift that mandrel down towards the floating ashes!

MAN (*lowering the rod*): Protect us.... Morliff. We do fear killing, for the transgressions of your brother, over the mound of our child's burial.

MORLIFF: What a singe do you make! Visgoth. Provoke them no more. They are in such just pain as to devastate all caution for humanity. The town collects towards your disgraces to abolish— Face me!

WOMAN: We have pleaded with him. But he stepped upon our careful soiling, as if to draw forth greater agonies, and seek messages with our cries and harrowed complaints.

MORLIFF: A meanness of the bound to deviltries of illness, in this poor station made for it, distilling cankers with your cancer of the mind! Suspend the teasing of this grief.

VINCENT (*turning to face **Morliff***): I am around to it. Here is grief with a relevancy. To this place I've brought them. But they can not farm from here. They can not produce from here. They can not enhance our properties and increase our treasury from here. And they are not but only from my behest a plausibility of my welfare. Out there! from the farthest reaches of our domains is the sour bitterness taken hold, not at this site that greets my passage in waiting. And bold it is to proclaim us freed from horror— safe! within the tormenting lounge of patience.... for me, and my understanding of this blight— my comprehension of the evil which is actual within us and can not be abated with the typical solemnities an' gentle— friendly pawing on the master. There is a self-sufficiency to be upheld, an' made a party to our striiiving! An' that! is what is right for all of us to be an' do an' become, on model, *modal* ground such as this.

MORLIFF: Evil within us?!... Are you upset with your own settlement made, your own largesses of sanctuary, for these people, away from the unfathomable causticities that have scarred their backs and their lives with an apparent plunge into hopelessness?... Have they desecrated your sensitivities with this grave? Then we will transport the corpse to a cemetery proper, within the stricken reaches, to keep your hastily constructed village clean an' pure. Though it seems mean to me, to have this sort of concern now, so immediate, after the loss and during the mourning. But you are the regal head to decide such matters of decorum.

WOMAN: She needs be beside my hearth an' near my whispers for her. I would run out of my mind to join hers, otherwise.

MORLIFF: This pledge is genuine.

VINCENT: I bid all burials their interments done as due. This is your derision of me, Morliff, to bring up before them a divisive concept.... Yet have them of their wills an' ways. I've only offered this space with a temporal purpose, for some security. And is there sickness here? No blame to me— Burn the carcass, then! (*grumbles of disapproval from the multitude*) Or go back to the thieves of light, that dark pestilence to blind you all against me. Have I not ordered you fed, an' housed out of the winds an' rains?! Then go back to your real abodes, or take this nature of mine!

MORLIFF: Be sensible with rage!

VINCENT: There is not a lust for it!

MORLIFF: Are you angry at these folk?... because a child has died? because your careful efforts at prevention of more casualties has failed.... in this one instance? So does the craft of plans have errors of the mind as beautiful, if what is to be done is done with sternness of its cause beyond all human chisel to reshape. Then is resplendent the form of our fortunes an' destinies. But what do you see? What do you prophesy?

VINCENT: ....I see.... enlistment for me—

MORLIFF: That one may better serve you in this viewing. An' quickly, soon to be procured, with necessity to fight your raging maledictions—

VINCENT: I spot not all with tongue!

ATTENDANT 1: The Visgoth.... is faced.... with a pain of sight.... within.... or invisible to us.... But presents a grimace glorious an' remarkable, for what he has seen without our eyes an' heard without our ears.... And this has perhaps.... touched him.

MORLIFF: Or tasted or smelt of rays, as injurious to the senses as a dream of death. Have a lower man do this, in a baser way than ruler. And stop disturbing these people—

VINCENT: Order me not!

MORLIFF: —our peoples, with the inordinate.

MAN: He is possessed of the demons of arrogance—

VINCENT: Enchain that man!

MAN (*yelling with anguish*): —to chastise my child dead!... Were to bleed my wounds dry, and empty them of a passionate hurting, by tales of visions—

VINCENT: Enslave him—!

MORLIFF: No!

VINCENT: ....Enslaves himself. My power brushes rue an' rouges.

MAN: Let me cleave him! with this very stick of burial. Oh! Protect us, Morliff, from our crimes intended. I give you my worth to be deceived with graciousness, an' in return more gratitude for *you* proposed.

VINCENT: Your chains are here. (**Caton** *arrives, running up to* **Morliff**.)

CATON: What is the occasion?... for these fires?

ATTENDANT 2: Only one, let out.

MORLIFF: Was grief!... among a family, and no more.

WOMAN: Now all three, an' with this cloth spent airily (*standing*) the grains of ash may lend their natures to our faces.

MAN (*pointing to* **Vincent**): But he stood on our grave! professing visionary tactics from our sweetest love, through his feet and frame, an ascension for our daughter realized. He comes too late to be proficient at it, an' to bless.... what has been taken away. But this offends. And it offends his character and affronts (his) nerve not to approve of the departure without his commissioning of images and self-enraptured hokum to bring us more of sorrow for these improprieties. This is your high brother's wasted day, upon us.

WOMAN: And the fire was for protest of his cruel jesting. For we are not slaves, though we come here indigent. He invites for harassment with his temptations.

VINCENT: I lend myself to you. I bring myself, for I am the Visgoth and your right to an existence.... within my kingdom. And I sanction your mistakes, not to punish or exacerbate for with harms, but to forgive with heavenly explanation of why this *must* occur. This is my judgment apportioned to your sensitivities; and you will accept it for release of these emotional strains, since your wrongs have been corrected by me through this ritual— And *I*.... have been appeased.

CATON: You seem to bring upon yourself the most declamatory harshness of opinion, with all that I have heard spoken of you these too many days still recent. Yet, resign yourself to a religion other than this, and have not a practice that could cause a swearing of revenge. Vent in private your frustrations for these deaths, and seize in some chamber alone the ghosts of misfortune to swat and terrify away with these patriarchal beatings. In public they serve only as a plague for teary eyes, and a confounding of the moved, stirred, and emotional.

VINCENT: Gentler be the waft. But I am demonstrated *here*. This is a real undoing of the contagion and the catarrh that prevents breath. It must be performed as actual, and before the principle subjects an' major spectators. And it is a divinity invested with, for our prospering. Every penalty that's paid must be used for the better. And every sorrow has to be lifted up into a wondrous event and expression not only of the sacrifice but also of our surety to be improved by the trial an' testing of our optimism an' endurance, our hopefulness for grander and happier epochs an' episodes of our stately development as a secure and thoroughly tried fiefdom. We are not ghosts standing here, panting for relief from the quarrelsome humors of our dejections and subjection to life's dulled casements— We see as I see, and as I lead. You are all bound to this. An' though *we* may seem strange.... it is all ordinary, of the lifting angels prayed to— spoken with, bantered of sounds an' sights an' sincerities of need, wishes an' resolutions, resolve an' retraction of ill-will and hatreds because of these events or malicious spiting of our spirits an' happenings foul.... and the ordination of these man-made sacred plots. I absolve them, Caton. And I absolve you, and these grief stricken parents drawn to a panic of perdition and perjury of life's succoring pledges to resist the condemnation of our collective aims and alms. So give I to you myself and my attention — and my patience with your rattling an' poisoned spews of contempt.... for me. I would not stand upon these bodies alone. I take your hearts to cleanse with and free of sad resources. I bring the candor of your minds along with me so that you see yourselves, of whom you'd kill!—

WOMAN: I did not draw to mortal tease my baby!... with a poor management of nursing. It was as sudden as a sickness wept with and coughed of. And as helpless was I, of the fever and the squirming struggle, the greatest valor I've ever beheld, that led to a conformity of permanence.... restful whimpering.... and death. What a great sigh was made, the loudest an' the softest, the sweetest and the most pronounced.... pardoning excuse for a withdrawal, and to have for me (*sinking back down to the burnt-out fire*).... a thank-you mentioned without words, without cries nor thought.... but only somnolence, and the most gentle simplicity of face and body.... a tranquility worth an empire, a loss.... beyond love.... beyond the loved. (*The* **man** *places the rod on the ground and goes over to her, kneeling to console.*)

ATTENDANT 1 (*to the second*): He did not speak at all to the.... interred figure, but to these images he finds, I thought.

ATTENDANT 2 (*responding*): That is the pallor of the pall, and the sheet of stars in day, he was addressing. It is a mystical endeavoring and manipulating of the real, which is earth, grains of earth, and their particular fragrances here, peculiar to a grave. Morbidity evokes strange muses, but the dead define all manners as a way to be treated.

VINCENT (*irritated at* **attendant 2's** *remarks*): Saw I more than you— Saw I more than sight! An' must I post soldiers around this compound?! I've entered as freely as I've let you people settle, after hearing of the adversity of soul, to contemplate with. That is my responsibility to you, to share.... of this diminution, my conjecturing of fate. I give fate a goodness kissed. (*as the* **man** *lifts the* **woman** *up to standing, so that they both may return to their hut*) I marry her.... to as good as I may do. (*as the couple go to and into their hut*) And the response is.... prescient. It is real, but preternaturally felt. An' that's why you can't sense the intensity, and the wisdom— for your futures. There is an ecstasy that may be celebrated.... at this spot.... and for all the other mounds I mount.... This drains my nerve, or reserve, but I forgive you of your sadness. I forgive for you these morose tempers.... an' declining talents. (*as others return to their huts or activities*) My thoughts are demonstrative more for appealing to the sky than towards any earthly denigration. An' you are right for long, but wrong for me this day to be disgusted with, or think demeaned by— I will solve this taint!... of tallow for our candles, the putrefaction of our amorphous spur, by an apprenticeship to gods— for learning.... what to do.... More capital have I made of bodies in war, than these few.... enlistments an' enlivenments— Spare me some grace for these attempts!...

MORLIFF: It may not be clean here, if a death occurs this way as elsewhere.

ATTENDANT 2: The demise is a derision of the deed—

VINCENT: Shut! canter of complicity to march. My eyes burn! to see your discontentedness shoved at me, and the present all around me factual. Some images are better spent in dreams. Protect

me sluttishly— I bring you not for company, but for comparison!

MORLIFF: You brought them *here*, Visgoth. Forced them, even, with your fears.

CATON: They are not consumed to terror yet. Brave they are, as we. And we should wait for more occurrences. This may be a singularity.

VINCENT: Or a natural.... evidence more than event. Aye, Caton. Eye me. I am fair.

CATON: To these who treat you stares, you are fair, these.... consternations. To those who would destroy you, then beware, your cautions amplified. Take not such chances with the harried populace if you're not of mind to defend yourself. It could make for a capitulation as casual as catastrophic.

VINCENT: (The) I, who has crushed heads with teeth.... then allow for this the most ineffectual of smites. I am it hoisted for—

MORLIFF: We are cleaning up the outer regions to stem this mystery and prevent a possible pestilence. And I would deign to butcher these sheep anyway, healthy or not.... or study their pasturing carefully. For they make symbol of our possible dissolution. Yet they are owned, now. An' we can't order about our constituent members so easily towards their destruction. Conciliations must be worked up, before the taxes due. And you would wash this tribe, of the all but destitute you've created, Visgoth, with a rinse more patrimonial an' pacifying than these daunting pantomimes you produce over the shadows of their departed, I can persist for your reprehensions to detain.

VINCENT: Murder this reproach. They make for our departed; an' what I have collected an' soiled with, sniffed an' sneezed at, were but a thousand messengers an' messages received. My pores do baste with them, and I am hardly poured of empathy and sympathy for these poor people. I am wounded by their distresses an' confusion or confounding of the facts: We are.... overgrowing— an' nature prunes. Our reserves of forestry an' vegetation can not maintain such a population steadily and as stationary. An' so some pressures part us early.

MORLIFF: Oh, part these excuses—

VINCENT: And yet I maintain the kingdom!

MORLIFF: —for you. Quash them! We can expand of territories or whittle away at protuberances, as migrations are allowed for many. And animals travel to an' fro with their own cognizance and testing of the dangers. We can sanitize the areas, but not control one's health an' greeds. Forgive them their insufferable bearing of your guilt.

VINCENT: Does hurt, the flea.

MORLIFF: You antagonize them with their problems, Visgoth.

VINCENT: I will avail myself to shed these warts. My feet are scorched by their making. An' by our provisioning, have they enough?

ATTENDANT 1: As much as meed, for their disasters to recover from, they ask not for more.

VINCENT: Then that is not the bite upon my skin of limbs, a complaint of needs for materials an' foodstuffs. Yet my body aches with tremors, an' my mind grows weakened of awareness for the trivial, the trifles of a sigh— or a moan or a groan, or a yelp or a yawn.... It's mourning that burdens you, your psyche prolonged an' stretched to cover graves. And I am spent out of these energies just now, with painful adjustment of the muscles an' tendons an' joints. More bits of mine are taken away and lost with each incantation, never to be returned by the heavenly hands that sift for jewels in man. My internals quiver, as I stand straight, or facilely bent of weights.... upon the head an' eyes, and the temerity to see blunders.... (*pointing*) Give me that stick!— You!

CATON (*as **attendant 2** retrieves the **man's** rod*): What do you wish for it to be, such an instrument? It looks.... sharpened, for the ground.

VINCENT: For the earth. Do you think he needs it still?... And after me?!... (*as **attendant 2** hands him the rod*) with this crude tool or implement of trust in force an' leverage.... Not half as heavy as raw. (*while gently rubbing his face on the spikes*) A soothing saw.... for the deceased. (*pointing the rod at **attendant 2***) Would you be whipped with this?!—

ATTENDANT 2 (*cowering*): My liege!

VINCENT: —with such a thing of pins?... an' sharpened wretchedness. Is there punishment of blood on this, with scraping.... an' scratching of our earth?... our verity? our truth?... Then why are feet less candid?... less thought of as appropriate for steps to take.... Take her!

ATTENDANT 2: Liege? (*accepting the rod*)

VINCENT: It is dirt ridden; rode, worn an' worn out is, he.... to the nettles with her.

ATTENDANT 2 (*anxiously puzzled*): What am I to do.... with this?

MORLIFF: Return it to the ground, that man's hastily made property. Return him to the palace, our Visgoth. Accompany him out of this spot, this— scrape.... he so much wishes to indulge of, till the tempers flay more easily than words, an' dullness issues its restraints in condemnation and approval—.... disapproval, for the sainted disturbed.

ATTENDANT 1: Yes, Visgoth. You look awfully tired—

VINCENT: I am. I am.... I am— with approval? for my deeds?... Let us go to a house (*starting to exit*), to a firm structure— where you may wash me.

ATTENDANT 1 (*following, along with **attendant 2***): Err.... the maids—

VINCENT: With balms applied. In balms required for my flesh, this frame does fiercely steal a cloak of contemplation for the inured of deception an' the privilege of innocence. Where is pain?... but within one's self born. (***Attendant 2** frightfully casts the rod to the ground, as they leave.*)

MORLIFF (*after a pause of indignant concern*): I tell you, Caton. This is grievous, his illness of the mind. He wishes above all things.... to be a prophet, now.... and stir up trouble, like a whirlpool of dissimulations for, his character to check in blame for these diseases.

CATON: A prophet?

MORLIFF: I mean to assist him with a real one, somehow, to off balance the madness and bring him level again, much more towards his responsibilities. But I am not quite knowledgeable of the craft, can not take it seriously, and have less respect for its practitioners than their pronouncements. For even fables an' lies have a logic that must be worked out, and within that thin thread may some truths or correctness lie.

CATON: May he foresee his anguish as a ruler, due to the heavy expectations heaved on of him? There is a rationale behind these sordid passions and possessions.

MORLIFF: He makes up his own pardoning of these terrible afflictions, though they are more personable than what should be his stately affairs to concentrate on. Have another make these whimson-wind, an' cudgel— beat (him)! if they are daft or not propitious of remarks, as such a representative must be sent to absorb their effects and transmute them to the better, with prejudice for his own well-being.

CATON: In this you see.... no meanness to be made of mouth, to calm all ears.

MORLIFF: Exactly! brother. It will.... quiet the Visgoth's tremors, to hear only positive results. For who would give him grief to read of soul, if were to suffer even greater hurt than mere remorse. The physical protects the mind from assaults.

CATON: Consider, then, these seers as dishonorable? That is a risk to take of mystery over superstition. But more than that, to bargain with a devil makes you his supper sweeter. For we all must cheat at times, unknowingly of his assistance lent. We cheat for lusts to placate over loves, a mirror confusing the image for the real. And only trial an' error may (be) due, for sampling of these effects to trust.

MORLIFF: They are the deluded dredge of honest souls, I can assume. Hucksters of thought are more showy and fashionable, an' averse to the discomforts false prediction(-mak)ing may make for themselves. They construct their stories in such a way of vagueness as never to be absolutely wrong. But the true prognosticator remains concrete, as if a science were performed instead of an opinion-ing with probabilities. To these I would consult in favor of and with favors to, since they are easy to be marshalled, never of wealth an' always wanting. Accept from me the aphorism coldly, that as the honest word leads to lies, the honest reading leads to destitution. The beggar is for the future held in sight, withholding all its wealth from his insight— and he!... does dine on candors that can hurt, and more often than slightly leaves him in misery for his offenses said or reviewed and suggested as the coming truth. Progressions damage most before they heal, but ever lead unto an ultimate appeal that is a lifelessness for peace. Thus can the drive for truth be led only in this direction, whose impulse is desperate to withstand the result. Resentments reign all over them, as if to have them bathe thoroughly within the rains an' tears of humiliation and conducing of poverties for what is heard.

CATON: The seer is stripped of much pride, you mean.

MORLIFF: They are born to be the surrogate for disenchantments, an' to be treated bitterly as owning the fears pronounced, an' owning to the conduct that might occur. They are made for us as effigies of mankind to tear at and bemuse for their thunderous thoughts, an' take pains readily while in their sleep of foretelling what may be horrible. Of this I have been taught for why their character is often lowly, as could you have heard the description: Seer *sale salaud*; search for salads saltier.

CATON: It is an admonition of the poor, in general, to comb the earth for foods such as where the dirt itself may be a treasured spice to stave off starvations. It probably relates to salty mineral deposits in some soils.

MORLIFF: But as to their craft an' skills actual, I know little, except that they will be used, will be subservient in attitude and nature, albeit through the queer independence of their revelatory talk with obstinacy, which they will gladly die for, of, an' from— when.... treated as a stone without feeling, or some cold instrument to pierce through the sheets of mystery and the darkness of unknowing.... what's to come. But warmly rationed, they should melt as man, since they are beings of the flesh as any man. And they will be thankful for the smallest prizes, some hermetic comfort with simple subsistence— I know they favor roots as if to grow— an' perhaps a colorful robe or shawl, with cap to denote some other-earthly authority or license. All of this to have, and keep having or getting, then will they.... wisely.... half the truths.... received, pictured, or imagined.... for only the tasty moieties to share with the inquirer or searcher of oracles.

CATON: But is the Visgoth one to seek if he himself feels seeker to be set to be?

MORLIFF: Settled with improvements, may he lean on moles an' ride with tigers. He knows that we are angry with him, that he is rapidly becoming unpopular. He needs an out, Caton, an excuse to behave more properly, a harness for his team of horse he's been riding with difficulty lately, as if he's simply forgot to apply the crucial parts of his rein. This introduction of a lowly but acknowledged acolyte with whom he might compare vatic powers or prowess.... as a pastime.... might give him a span of thoughtful breath without distraction away from the imperatives of his rule. If one becomes excessively preoccupied with finding sweets an' candies, then have a store arranged with easy access, so that he may turn his attentions elsewhere— and back to his work!

CATON: This sounds reasonable, in the way you have put it. He has been sensing losses painfully and with shame, trying to justify them somehow as being, if inevitable, then still under the god-like jurisdiction of our fates. He *does* control this, and *is* responsible. So then we may offer him a guidance by the very principle he works an' shapes, to adjust these pains and shames to the management of others, an' their eyes, thoughts, proclivities, passions.... an' pronouncements.

MORLIFF: It is a way you can agree with, shams an' shamans indoctrinating the ego, but innocuously, for a change. All blame for the unfortunate falling on *them*, and thus falling away, is the tactic made. Then tack of the tragedy will be tamed, not to name the dreadful at all as a gamed desire.

CATON: How wouldst (you) find this fellow, such as he, an' not harm?

MORLIFF: He is all harm, but of himself be harmed. Then he has none. For this thing called "harm" would have nothing much owned, an' what I could give him is not more of nought. The person should make his own harm, rewarded by us only. I will ask of notables of this fashion, who need rewards. They will be told of, to be got rid of, more readily than pointing out a blighted shrub, being outcasts, mainly, with no civil cause to be, and so most often guiltily espied or thought for.... the dramatization of a happiness (to insist of) alone.

CATON: That is for an odd search, then: the respected and yet censured.

MORLIFF: What awe does threaten mildly, then be staid. They are not punished for their ways, but are their ways of punishment, if all processes aline for this, a disgrace from the gods to be overheard, as of oft said of the temper of nature and the tempo of events and circumstance. For this, (to) be done, then that is won more (to) harry. We can not always tell what is pleasurable, till it leads us to our destructions wantonly an' wanting, this curse of the gods to learn what pleases *them*. I could suggest, all knowledge— any, is a stab in the back— somehow. For what stays unknowing, stays unknowing of its worthlessness. So it is fitting man contemn this above all others, who may disagree of their natures at times, as being the most valuable for a devastation of creed an' belief in a governance by the supernatural deities. What with their *dis*pleasure (in us) may they strike us back terrifyingly, in these soothsayers.

CATON: I'd trust more in our own taunts for terror. But as for an excuse or a transference of the awe-bound modalities, for what our Visgoth hears within these moods— Aye! Some supplicant to the sounds of sights invisible an' not actual may be necessary as a pomace of binding, sweet nectars made, for the mental irritations to soothe and reduce. Our brother is most vexed with these sad decays an' early dawns of latency deceased. It is enough to bring tears, to think about and own, and dwell upon, as he does. Yet let another do, officially and superfluously, as make the best employments scandal-laden an' kept to, away from importance. Then are awash, the scandal from your hands, with other eyes of pardoning, ye. It is no sin to grieve, yet more to take gradations of a sin like shades contrasting is this matter drawing forth to bother everyone. Then let one claim a suzerainty homed to, against the malicious aims of unfurled nature, taut of destinies not become— denied— an' wasted wantonly. Let man fight for such disasters only, is the better wish and the higher grievance made than to sin consulting. Find you this one poor beggar to be hurled. An' I'll be on the lookout too.... for sightings sanctifying sorrows, an' visions vivifying vows.... to serve our Visgoth. (*He and **Morliff** exit.*)

MAN (*emerging from the hut, with the **woman**, after a slight pause*): What must we leave?

WOMAN: Some finer earth to embrace the trampled upon, is what is needed now. (*going towards the mound*) Softer an' keener, an' blessed with my indulgence to see her fit and fitted in this grave, is what to serve my eyes: a roundness again (*kneeling to the mound*), and a plumpness of contentedness to be.... made. (*starts reworking the mound, adding soil to its surface*)

MAN: What might we leave, (*going towards the rod*) we should take with. Morliff is right. To have a cemetery built there with our use.... is more fashionable than this. (*picks up the rod*) We do not own the plot.

WOMAN (*working*): Yet here is as holy as our worth in pleading for. I blame the king not much of shame.... an' most of angling remains.

MAN: Pity his mandate for us. Where lies heaven, then? I'll send him to it, if he returns.... But as we go, it is to poverty for home. The earth is rich—.... an' poor, and to our smite good essence of the scent.... an' sense, an' senses. Then be we bound to this duty.... an' life. Woman!— we shall leave, we shall return (*going back towards the hut*).... with foods.... an' blighted parcels. We shall run to this cause, and rebury ourselves.... not to be imprisoned. (*reenters the hut*)

WOMAN (*to herself, while smoothing over the mound*): Does the earth not move?!... or carry with it?... I will be improved with her, to accompany. For what is separated.... can be made more dignified. Search we ever to be nursing, brought to bear an' parted softly as more difficult to care for.... now. (*A few neighbors come over and put flowers on the mound, as she remains kneeling.*)

Act III

Scene I — *A courtyard. It is nearly dusk as* **Melinda** *sits on a bench aside a floral hedge. She seems quite still and subdued, as* **Halyce** *courteously comes up to her.*

HALYCE: Good greetings on the drawing lushness of this evening, Queen Mother.

MELINDA (*smiling*): It is difficult to try to be old, Halyce. I can not appear that way. But forgiving everything, you are a handsome man. And are you thus in some contention again?

HALYCE: I am.... for the most beauteous Melinda bred to praise and seek advice an' graces. Then yes, I've had a flap with the intransigence of.... our Visgoth's stupor and sloth, and may not be attractive in the palace for awhile. Yet they with me, at the morning's meeting of commissioners, can hardly stay as soft of wounds from misinterpretations of our positions, as could a fruit ripen suddenly with a blow or deep impression pressured, that he is much too far distracted of peculiarities to be mothered easily with even our counsel. (*sitting beside her*)

MELINDA: Vincent cries of ruining for his goals, about his son to take. And I do take of passage swiftly sitting, away from the tempest that has caused such disheartening in that house of legends. For this is unusual enough, to be made bored an' shaking. I have no train of pretension, that the Queen feels ill about me, about him— and her son. She is unpleasantly disturbed, as the Visgoth is to countermand and neutralize effectiveness, without for help of Fabel to embellish some little maturity. He comes to me, and so do you.

HALYCE: That is for a goodness to be meted, as could honeysuckle propose more words than kindness and generosity alone al-

low. The young son is troubled? So am I.... for soon to bring, or child of offspring that would pluck the strings of my amity an' love.

MELINDA: I am for your.... amity. I am for your love. I am for your friendliness an' company. Yet how am I? Possessing, and yet dispossessed.... a lover, and a nation? I tell you bluntly, I feel too young for this. The grandeur weighs (as) heavy, and ages me without much promise more, except that so amply given to my sons to carry out.... a tradition, some subtle way of nobility and righteous purpose in governance of this fiefdom. The Queen is hectic for her son to rule, an' that should often be the case. Yet, always is always an' ever is ever, and I do feel a privilege passing.... to let what is passable occur. What is fitting—?

HALYCE: For more children? Who needs a reign within a love. Mere slaves are as contented as emperors with that redolent lushness breathed an' touched by. An' the evening is gowned for me, here, this night, so splendid as the stars await to appear. It is a decadence to ask for love— (it) should only be, exist.... He gains no favor this way, by persisting on his womanly traits to sob an' complain of misunderstanding to his purposes cross and callous. There were almost tears to his eyes, this morning, as if he were in pain. Anguish, agony— What is this interpretation to tolerate? The Visgoth will not be judged (as to be) able, soon. How does he relate to you, as a fever or a fallacy?... a failing, or a feint for compassion?

MELINDA: He is low, to be mean towards my light. This is respect, squinting at my visage, and humbling himself with error in my presence. Yet, then, as soon as I am away sent, and unseen by him, roars he of his catastrophic mode, for my ears to hear of his sorry state an' (to) believe him dearly. He struggles with embarrassment for me. It is not an affection he can hold. I am younger than his wife! and more successful of my manners fertile than he has been able to be with the Queen, who ages now past reckoning.... an' languishes against me, wrinkles to a supple smoothness of the skin. Is it a wonder that I wait outside for nightfall, away from the arguing and discernments of character an' quality? A finer medal meddling I will avoid..

HALYCE: That is his caution?— No.... It's not the point to it, this decline in faculty with rise of fantasy and fancy for himself (as) some god's Mercury or messenger. It belittles us to have him speed this way so resolutely to dissolve of madness, merely because he can not control the course of some disease, an attacker of babies an' the infantile.... Yet for a jealously I'd hold you. A dark radiance illuminates me. That is certainly true— You are more potential than power.... You are more beautiful than your becoming, over us, a curious specter of fulfillment and lustful aggrandizement to the throne. More life is in your chair here, your position, than that whole palace could shake with comfortably. An' the house does quake for it, with a violence of attitude.... perhaps, to hear what is said.... beyond your sight but for your commentary of emotional trembling. Caustic tributes to your nature an' endowments natural, that boy Fabel has only picked up on the positive of these descriptions. But by now he must certainly begin to realize the sad, salacious jesting that is proposed about you, as he dawns of a manhood.

MELINDA: I try to coach him otherwise. He has often been confused of the relevancies, and is growing of resentment for what he can not understand, yet. But he remains loyal.... to everyone of our worth and contends to support his family as much as a son being, as I have. What is childish in this, transposed to what is challenging of this, is his trying of question and conscience.... to serve his father forthrightly, an' his realm.

HALYCE: The Visgoth has no doubt about this?

MELINDA: I certainly do not see in Fabel any lacking of the highest regard for his father. I can only assume that Vincent sees the same. Yet you tell me, or imply, that his sight is sometimes askew. But I haven't heard any complaints by him regarding his son— by her, yes.... but not him.

HALYCE: Then that is not the problem.

MELINDA: I try not to hear much, in there. The words—.... have tassels soaked in bloodied barbs, for what I am permitted to perceive. This may be poisonous for the boy.

HALYCE: He is not the problem— I come to you, for sacrifice of my heart, fair.... an' regal Melinda. I confess to you directly that I can not find this Visgoth fit, no longer suitable for a prosperous ruling over us as a true lord. Cure his curses!— I want you to know.... that dissipations are in order. They are palpably felt and argued about all over these territories.... I am preparing to remove my estate, land an' properties from the fiefdom, as much as the Visgoth can hardly challenge me in his state. His army becomes dissolute of membership, as my own grows more determinedly of stance and stoutness. Yet I do not wish to take over his authorities. I want to leave from under them, since he can not in any meaningful way protect my concerns an' interests with the capability of intellect I demand. I see his fall coming; but that is not my motive in this, not sincerely aspersed to claim for, but of my rights and responsibilities tallied for the transfer. A cruel candor can be admitted. But I am not a cruel man. I leave for Fabel Fabel's. I know you love him.

MELINDA: That is true. He is a dear relation in need of sympathies.

HALYCE: But he can not evolve to Visgoth as rapidly as needed. Nor for my patience can I wait, for patience is imperative with time. And the ripeness is— soon.... soon of leap. I may be the first. And I want you to know (of) my earnestness, hide nothing from you, and give you the freedom— the power!... to warn the Visgoth of my dispelling allegiance. I feel safe to confide to you this information, being just ready an' poised, the spring thoroughly tight an' in gear.

MELINDA: How to dissuade you of this hastiness? It is.... not wanton, but willful— and elicited by mere opinion, without action or harm to your cares.

HALYCE: Were that your son, Caton, to ascend to the position, I could be warmed. But he is as reluctant as this may be of that possibility lost. Most people cry for Morliff, as you know. And he is too stern for my tastes an' at this junction of events to maintain in me a loyalty honest or true. Though I don't deny his reasoning is sound an' timely. I won't live under his recklessness and headstrong positions, nor subject my own to his intensities of blunder, as I could foretell from the pace of his personality.

MELINDA: Intelligent, he is.

HALYCE: That should not be a mission of speed requested with, but of causation infused by. Wise he is not.

MELINDA: So are they both from my make!—

HALYCE: A good petition, indeed!... But how is not the error here, but why— Bound to you I am as well made, an' desirous. Take you from this writhing, an' this warty kingdom, I would have you as an equal spouse within my lands to rule and breed of gentle greatness more, more sons.... an' daughters of high caste an' ability. And to make of me yours, I'd share the earth an' crop heaven for your diamonds. Be queen again, with a justice to you. I offer my hand and my psyche in this way. And tell the Visgoth all of this. For when I face him again I will hold him to it for an opinion. Much less a consent be needed than my simple an' honest drive. An' this is affection with affiliation, Melinda. You will not cross a threshold of marriage, but create rather a throng of appreciation, with this bold an' blatant step.

MELINDA: Remarriage.... to a fine an' handsome gentleman. I could have expected your pleading pledge. But, if this were made of us, you will be charged as a strong pretender to the fiefdomship for beg, my elopement (serving) to remonstrate.... the active Visgoth.

HALYCE: There will be no competition between us, for I dread a further involvement with the intricacies of leadership that lend to rearrangements an' harrowing assessments of the perplexed. I am bound to abide to the whimsical today, but not tomorrow from a broad stationing of pertinence, which you may assist to design with a direct relationship to the current governors. Yet as for influence, I hold none but for our own delving to uncover what is correct for our personal happiness an' satisfaction. A proper respect will still be upheld for the Visgoth, as we enjoy ourselves, an' draw to rest away from the tumult of unpleasantness at this capital seat. You are not comfortable to be. Anyone with a heart can sense this. An' I with ambition will strive to improve this state. But as uneasy of your gaits be made upon these grounds, then I am also with this fiefdom to be tied. And it is difficult to escape from or abandon.... the regulatory apparatus that condones our customary composition within these territories without much dispute an' dispensation of the law. But it is far nicer and valuable to win from you an approval of my actions with your gesturing (*takes one of her hands*) that this seems necessary to do.... and to be pardoned with arraignments nuptial for peace between the tribes an' clans an' families to maintain. What kinder punishment is there.... for our dispossession of membership? I'm even willing to pay some sort of annual tribute or hone letter of encomium to serve as displaying deference an' recognition to certain authorities held throughout these regions, as long as our freedoms and independence are implied forcefully, specifically, an' distinctively. (*bends down to kiss her hand, and then re-postures while still holding*) I wish to join your honesty, an' not as a potential usurper. That would be a flagrant crassness of effect that I disdain, an' a pettiness while my heart only enlarges before you.... But I must do this— I feel, an' at your leisure to entertain. You have me.... captivated to be forthright. And I see in your eyes.... much need, an' more longing to be allowed a partnership.... again. Tell the Visgoth all of this, with your catering of charm to be effective an' believable, of my determination to explain an' convince of. I'll not give warnings but warmth through you.... an' wealth for the separation. (*releases hand to her lap*)

MELINDA: I'm not one to give issues to him, or call for even a conversation. I haven't orders of the lay to principles atop his table of office.... Yet through the Queen can I make context of your.... impulse to eventualities. She asks a lot of courtliness an' courtesies towards me, an' for responses bothered. I am dominated, but duly worried for. Opinions do not matter, but facts an' factions make me prevalent. There is some little importance left me, speciously, grudgingly, calculatedly.... I can wish again, Halyce, and should not reject your overtures. Since courted, I am not else likely to be courted, but by a fair man with a justifying purpose of sincere endearment. The manner of the tangle, though, is what is left you to provoke, since the Visgoth is already preoccupied with disturbances to his rule, an' may not choose to make decisions.... generously—

HALYCE: Hide your candor pandering, if you may, to show only an ambivalence to my affections towards you, but only that they exist and are definite.... if not discreetly lent your voice. Challenge him not with your soul, but your saintliness to parade within those corridors and promote, for his consideration, the provisions of power that evaporate from his careless hands, that you are better wed— most firmly to my established love than to his disjointed tending of the realm an' overlording of mysteries he can not control nor compel to vanish. An' thus, as you are taken ov'r by advantage of good occurrence for— our needs.... then he should take full measure to improve his lot of lordship, or begin to make some proper steps regarding his fiefdom. (standing) I have bled myself, Melinda, at your feet, an' on your lap have applied my weight to use.... an' strength to culture. Then be it now affirmed, your are protected with my life to chivalrously heat off the chill that has grown for you inside that palace, not more to sit alone outside like a misbegotten thought as, with misery of mind to cloak this serenity. I make you as valuable as a queen, again, an' seize as worth and worthy this temptation to try the Visgoth even more, irk him an' irritate, and disown his childish ways— finally!... My anger's not at him, nor his plight to make him so act, but the debasement of our constitutional an' constituent vows with his mental follies to proceed. An' I have heard stories, as to cross hairs for sights, concerning his behavior to the populace, an' people assailed of hardships an' griefs. I'll not have this, on my grounds providing more healthy strife for living.

MELINDA: You are a gracious nobleman to your varied subjects. Their desire to have you speaks much to the contentions we face in this kingdom. Yet populations change as much in character as sheep of their distinguishing wools. He is bothered by this—

HALYCE: No demonic plagues (are) at my sites! Ne'er an illness there from *fleas*, or fouling contagions from a breeze of lice an' ticks an' mites. Not in our hair, the threat is made to skin an' scalp, and the innocence of lounge for bites in cribs an' within poorly furnished homes. No! He shall not walk there, and make tread with his steps on graves, giving pronouncements on the aims of earth! That will not be permitted my tenor to support or approve.

MELINDA: It would be somewhat of a long walk for him, assuming he is tranced to employ the labor.

HALYCE: But that is a point, my dear. He seems not. Just mad. Obsessed of his provinciality for his inhabitants to control— even tame of thought.... against their own suffering an' sorrow. That is his cross delight, currently, (h)'is remonstration as through supernatural guidance that so perturbs all of these afflictions to be at their worst. For what is worst than death than to be judged by it!

Thus the people injured by these blights are now condemned to heart by his actions; as to were, the misfortunes were deserved of them in some.... heterogeneously spiritual way of spotted pieties, an' he (as) the hyena bawling of their sins that have of necessity brought forth these sad tragedies. It is.... impertinence to a high degree, coming from the ruler. Yet, what he leads he loses of the flock, with these cruel castigations an' chastisements. But is he blind to their effect, deaf to their hawkishness? I say not.... is he the Visgoth left to be, in this manner for much longer, with my pitch of tar across his spreading rant to reel from an' sway revulsed from his own character to review. A sickness makes you silly, but draws the rest of us towards sanguine intentions without you—

MELINDA: I?

HALYCE: But such as he. His is the illness considered. We will be freed from him. We will abandon him to his shame, shameful proceedings.... an' not care much. For I've never held him that dearly as to care an' carry; though once he was strong and sufficient of admiration, praise, perhaps even idolatry— by the weaker consciences. So easily turns a view, with one's head.

MELINDA: I?... So not related in blood, but (of) a shambles perpetrated. And then directly are we one in blame, if I have pressured him— out of reason.... by my simple occurrence.

HALYCE: Not of your heart. If staid, I'd lend you my own to recover with. The choice.... may have been pure an' quiescent of fortune, of high particulars to adopt with wanting, the characteristics desired an' shaped for desiring, a commonality of brow pierced of awe with view of grandeurs in beauty and pleasing faculty, facility, an' fruition.... But now *I* choose you. And it is made, a bargain of delight against all trepidations I could weigh.... come from my move against him, away from him, the Visgoth, that very notion indeed. Let us evolve—

PONTAL (*running up to them, excitedly*): Count! I find some few, a few left. Count!

HALYCE: Fellow commissioner?!... What is the talion of anxieties to spring?! Yes, I rushed out! An' I should never do so again, could be your prompt an' purchase, perchance of a kind. Though no disgrace to you for your supervising of the meeting. I meant no disrespect at all to fall upon your handling—

PONTAL: Queen Mother— Heavenly Melinda, tides turn.... to uncover the coarsest of sands.

MELINDA: Where bade the problems of the sea, in this landlocked region?

PONTAL: The Visgoth has befallen fouled an' felled. Upon his bathing, washing maids discovered scars, scratches, an' lesions all along his back and stomach, where hairs do not hide their tracks. It was as if the waters were made of blades! they tell. An' he bled profusely, of blood and salty viscousness upon their hands. He collapsed of shock! to meet the dampness and the wetting in the *Wanne*. An' washed as much as could be profitable to maintain life, he has been taken to his solitary bed, within the solemn chamber of our governing sleep. I have been scurrying around this estate, to collect what few of the commissioners are still around, and bring them to this space of trial an' tribulation. For clearly he may lie on his deathbed, and if so will have pronouncements to struggle through for our hearing and official witnessing of his lips for what is said or uttered listlessly, betraying the urgency of his attempts. Only two have I found— An' now Halyce. But I must try for more to attend him. A colloquium of at least seven commissioners must be made to verify in person the legitimacy of these mandates for all of us in the fiefdom. An' I have found the Queen— and Fabel! who wait on the dying Visgoth, still unconscious, I was last told. An' with more grief than can be granted blood and marriage, are they subdued to this doom an' cataclysm of change. Run to this chamber now! to hold upon his lingering breaths an' whispers.

HALYCE: ....Your sons should be alerted, Melinda—

PONTAL: They are not necessary to be present, not more so than his words about them. But if he speaks not one (bon) mot more— an' dies!... Fabel must transform his youth into a Visgoth's strength of armor, an' assume his father's throne. Rush you there! not to be deprived the truth of this, an' an agreement against suspicions or misinterpretations of what is enounced. I try for farther to find men leisurely of state, to bring to vigorous conclusions hastily, an' with dispatch of terror for our fates if we can not agree of the decisions made— or taken! perfunctorily of prerogative or impetuously with incaution. (*running off*) Help make a number now, immediately!

HALYCE (*as **Melinda** stands*): This does make of flight proposals to be held. Yet what of thus is said? if Fabel be, I am still of a leaving mawkishness. Yet, if Caton is elected....

MELINDA: Make we insistence to have his speech attention-ed. (*starting to leave, as **Halyce** holds onto her*)

HALYCE: This changes nothing of my sentiments towards you. Let him drift through a slow saw as we talk—

MELINDA (*restrained with annoyance*): We must encounter the Visgoth, Halyce!—

HALYCE: —an' draft of this compliance, to ourselves.

MELINDA: —For his meaning is not implied by his actions, but his state of mind to give a license to them. And now he may say why an' demonstrate a kindness to posterity.

HALYCE (*suddenly releasing her*): I would challenge him even there, to have him know you.... an' of our intentions.

MELINDA: Mercy for his weakness, man! We are not of beasts to bite an' banter, nor to better of his murder.

HALYCE: Only some hardship does distill it from his paces. If I cry for you in his face! do not add to his convulsions to convulse yourself with shaking shame, trembling temerity, an' rebuke my affront.

MELINDA: You will not plead to him as he makes his last pledges an' earthly vows!—

HALYCE: Nay! For our causes, then. But I mean on thee.... there is importance sheltered. An' I am just as mad as he, in this regard.

MELINDA (*exiting*): Hold your tongue! until his tells of this. An' with more decency forestall your temper of good wishes. I will have a manner about him sufficient for our grace, as he departs.

HALYCE (*alone*): This quorum makes for mint an abomination hastily delayed. Ah!... for mince of strapping back, can tell pressures upon an earth to lift an' sift through— creatures found! an' to insure with bequest of a damage to lay. Poor, poor soldier now, of the blithe encomiums heard, by praise of gods an' gall of gatherings, then does this rest some good— But how?... How can one speak the truth of lies an' the badgering of beastly natures? *Les bêtes noires font frais receptions.* It is more costly, to assume proofs. Yet how to protect what one has heard is more the error to these crimes. For if one speaks with jesting dead— an' dies.... it is too seriously said. An' what is the jest? but a definition of insanity. Still must I foray, to find his heart my own— an' seize his jest. It is a devilish principle to be a master ov'r men an' man, one most typically undeserved and cracking of noxious fires tempestuous— as you burn! Just see where his example leads, scrambling for.... a tendentious piety, that leads to death. Yes, are we all.... of our own temples made. We! We! An' from our own sins condemned, it could seem.... of innocence, are we to do. Sharpen the blades of this rinse an' douche. I must be present, to hear the laughs an' cries of a homicide. Every one of us were tending towards this, with reluctance of his fervor to.... adapt to the jest. He destroys himself. An' do we each the same, in search of sanity, fall to the self-inflicted wounds? Then why must innocents die?!... Because they are beasts, an' temples of the jest, disciples of the gesture.... to live. And then.... what kills? the visionary seeing beyond his means. It is contagion!... of hope an' light an' *hel*-latious enmities. We are born to be angry at something, I suppose, an' unreasonably cruel of thought. Unbearable are our boasts, with full lungs for screams— Contagion is here! Resolve to it!... Why?! Where must the senses enter, to be led? What is compassion for me? a jolting joke! spared for the bed an' swaddling.... layers of cloth wrapped with, entombed by, made comfortable of. Have I foreseen all of this— difficulty? It is no surprise, how hard to hurt, how hardly hard to search for these.... encumbrances an' restraints. Do wash the soul awhile, as the body rots, an' wastes away— That is the sanction! That is the permission, to be with breaths an' cries an' gasps an' gaffes an' warts an' words!... an' take this all, down to a calm, cool, kindness shallow an' bereft of animation and vitality an' life.... a terrible jocosity of irony, in— debted to.... a mean candor. An' so he lies?... as a helpless babe? an infancy brought forth, an' bringing forth? Now to my profits must I pursue; for I have explained nothing of myself, and there is need to greet the Visgoth dying, some graying pallor to encounter, and commend— as night.... as to an end. Here is no quarrel with it, as falls a human touch— to fell! (*Exits.*)

Scene II — ***Vincent's*** *sleeping chamber. He lies motionless in bed, covered up to the neck, with head on a pillow, eyes half closed. On either side of him, sitting in small chairs, are Queen **Terissia** and **Fabel**, somberly awaiting his stirring to a reasonable consciousness. The chamber is softly lit by candles on a table, at which a **nursemaid** is preparing unctions and other toiletries. Two **commissioners** stand near her, though more along a wall. They talk subdued to each other, the nurse overhearing at times.*

COMMISSIONER 1: Not a word spoken! Does he even know we're here?!

NURSE (*interjecting*): I informed him of your presence, though he was not concerned to (take) notice.

COMMISSIONER 2: He seems unconscious to me. Yet the eyes

are open.

NURSE: He breathes, sirs.

COMMISSIONER 1: The ears are closed. The sense of this escapes him, and the mind might be trapped into a void.

NURSE: He awaits more of you.... to arrive— to come, before summoning up his energies to make statements. He is awake, sirs, and whispers to me when I treat him, with these soothing oils to blanch, and comb down a fever to his head. But it must be done only periodically, or else the stimulus might exacerbate the body to cause more heat. They are scented liniments and lotions that give a cooling lift to the nose and skin.

COMMISSIONER 2: What good is he this way? And without papers to examine! He should have prepared himself much more for us, his absolute wishes to review.

COMMISSIONER 1: I'm sure he has— Pontal would know, where those notes are kept. Don't worry about it. But he is a cautious sort of being, to leave the most important matters to his cognizant expression. What's written are defaults to follow, if he becomes unable to tell us his most current desires or explain away what might be controversial or contentious. This is a reasonable ruler.

COMMISSIONER 2: He is. But what has he left to proclaim? which should not automatically occur. The son assumes him clearly.

COMMISSIONER 1: He might make exceptions or commission from us temporary tenures, after assessing the temperaments of his people. One kingdom is not many, but many make a fiefdom.

COMMISSIONER 2: So, that is so.... But why does he not see? Blank are his eyes.

NURSE: The light is kept from him. It disturbs his eyes to watch directly the flames. They bounce while he stays still. They shiver as he tries not to. And they may contain a thought he does not wish, of eventual extinguishing.

COMMISSIONER 1: I tell you, he contemplates everything that might be useful for our knowledge. His is not a'(t) sleep, but a' resting, an' not a leisure, but a lengthening of time.

NURSE: The Visgoth is fully awake, sirs, and grateful for your attendance. The Commissioners will be appreciated—

COMMISSIONER 2: (The) Apprehensions pour! This is a sour pout of expectations. Nothing good can become of this development except an intensification of rivalries an' a dishevelment of administration.

COMMISSIONER 1: Some clarifications may be made as to who rules and what governs. But we remain the adjuncts, as a body characteristically consulted, and in times of crisis appended to. Turmoil will not be permitted as transformations are.... achieved, an' maturations enhanced or sped through sundry courses an' careers. For we protect our state an' status, to be upheld.

COMMISSIONER 2: An' forgiven.... these desultory events, re-

sults so unforeseen and yet predictable. I thought he was— too healthy, myself, too broadly spacious of mind an' experimental. Yet disease cures wisdoms, slows down attempts at discursive idioms to actively, even frenetically relate to.

COMMISSIONER 1: You mean his troublesome visioning, as of late. Well now he sees much; and it is dark, but relaxing. Shadows always cool. Contemplations pacify the heart to beat— no less vigorously, but with a stony-eyed purpose to let it fulfil its task without conceptual adumbrations of hesitance nor haste, that life-giving beat and pulse throughout anxiety an' tribulation. Yea, to beat is like as to in fighting an' flailing; an' to pulse is as with waves to shiver an' shimmer with an installation of obedience through the water's tributaries of existence— blood red an' rich for shores, a crimson corpuscula-tion beaching its effects of fervid delivery to each unconscious member of acceptance, a wealth of some such wholesomeness.... May he struggle thoroughly—

COMMISSIONER 2: You mean his sores bleed for us?

COMMISSIONER 1: —with the dilemma at hand, to stay alive weakened or to leave definitely.... a state of misery within his kingdom, in some pockets of civil concourse.

COMMISSIONER 2: All will be cleaned, or burned away the scourge of a pestilence— a regulatory affair of housekeeping and management.... to remain attractive within the suzerainty.

COMMISSIONER 1: As better he should be (at), the floor sweeping kept up an' continued.

COMMISSIONER 2: No spread— not at all, this blight of births. (*as the **nurse** picks up a small bowl with rag, and a crudely pronged comb, and goes over to **Vincent***) Such impuissances are similar as for births an' deaths, with the ends of the rainbow's spectrum leading of like colors into the mysteries of the invisible.

COMMISSIONER 1: Oh, they only mirror unseen effects an' efforts, where substantialities remain for either. I do draw pause to think of emptied life condensed of miraculous concentrations so full an' teething as to occasionally draw forth brief spans of ballooning cankers into existence an' view. Fester for us their emollients as emoluments for our worth alive, pining of the earth as if suffused of kind.

VINCENT (*with an impassioned emaciation, as the **nurse** rubs his forehead with the soaked rag*): Aaaah, heat to drips, an' lent away. Sooth, for my temper soothed.

NURSE: An' bled colorless. You sweat profusely, governor.

VINCENT (*whispering*): (It) Is the essence of a crime. I feel not pain, but an extreme exhaustion, and am too tired to fully close my eyes.... or keep them shut. But what I see is dimly of a vernacular.

NURSE: Two commissioners stay. An' your Pontal searches for more.

VINCENT: ....They rush away, like leaves off (a) tree, like shedding mice. An' is it late, or early late?

NURSE: It is the brush of evening, sovereign. (*dips the comb into the bowl*) The growth of night awaits, its dark cuticle ensheathe-ing. (*starts combing his hair*)

TERISSIA: You impair him to a grooming.

NURSE: This will keep the bed lice from accumulating, my Queen. He is too weak to adjust himself from their habituations of skin an' scalp.

TERISSIA: The scents may mortify. But I am frightened of his perceptions dreaming.... within these fragrances. Delirium plays a toil, now.

VINCENT (*weakly*): I do not sleep yet, Terissia. I recognize ye, an' Fabel, bounded by an' bound to.

FABEL: Father, be gracious to survive.

VINCENT: I do want to warrant it. (*to the **nurse***) Be strict, for I hate them.... A deluge of saints passing judgment. (*as **Melinda** enters, holding a candle*) They consider me, an' taste an' bake a home in holes, to the head— Take away the light from my eyes!

NURSE (*a hand slightly over his eyes, as she finishes combing*): Shield! Lady Mother. It pains him to see the flames directly.

MELINDA (*shielding the flame light with a hand*): I'm sorry.

NURSE: To the side with it, please. Your approach was stirring. But as to notice just, I became preoccupied.

MELINDA (*going towards the table*): Instruct us all.

NURSE: I am subservient.... (*standing erect*) to our rulers.

COMMISSIONER 1 (*greeting*): Queen Mother.

NURSE: There are ointments, to be reapplied later, my Visgoth.

VINCENT: Will it be as taxing as my tiredness, maid?

NURSE: Toward the pain of sores, it must be done to leach the discomforts— by forming some. But in communal fashion will we apply the treatments. Then be prepared for the thrill, and look forward.

TERISSIA (*as the **nurse** retreats to the table, **Melinda** placing her candle on the table and joining the **commissioners***): The sheets to change. They must be drenched in vile virilities. An unwrapping and re-wrapping. Such shrouds are contemptuous when soiled and a blessing from the heavens when cleaned, refreshed and anew. Feel you much discomfot swathed this way, Vincent?

VINCENT: No pain, nor itch. No burning sensations, nor leaking remissness eked, as felt. But as a bandaging these blankets play as piled, my pulsations to enliven with.

COMMISSIONER 2: Queen Melinda—

TERISSIA: What is else-d? A cause for you, my husband, to preserve yourself.

MELINDA: Halyce is around, an' comes, sirs, to assist us presently.

COMMISSIONER 1: You have learned of his bearing, or seen of him?

MELINDA: I've spoken with 'im. An' he will abide his service to us.

COMMISSIONER 2: Rashly, though this be.

MELINDA: The intemperance is home-sown with me.

COMMISSIONER 1: As are we all with you.

MELINDA: I mean his impatience. He gambits for the traditions of our governance.

COMMISSIONER 2: Soundly stolid, but impertinent, even to our Visgoth.

MELINDA: Amends will be made, for our conversation brokered an understanding, of sorts, with agreement to allow the Visgoth to be, as is forsworn of us a service applied. He will not disrupt rulership of the fiefdoms collective, but will assume his own with honor, as is fit to do upon his judgment an independence to procure or postpone according to a health of leadership.

TERISSIA: This is rude of him, to bring up such matters to your attention. Though he has hinted often enough of these persuasions. His influence belittles you, Melinda, complaining to a nursemaid as a grandmother—

NURSE (*at the table*): It is here prepared, the royal anointment necessary to relieve the suffering of pus-bound tremors and disturbances of flesh.

MELINDA: Well then, to a better use than I may offend, offensive wounds descriptive of our raucous aims to mark of skin leprously proved. I am disciplined with these remarks against me, as if my.... transcendence were foul.

TERRISIA: Too sensitive to speak!—

COMMISSONER 1: We do not hold such discomfort with you ever, Queen Mother.

TERISSIA: Our sympathies lie bedded!... An' he is the concern now, not your sons or future princes of the realm. If what for Fabel be decease-ly led, that is your contemptuous quarrel.

MELINDA: I hold not his heart, but advise it sweetly.

FABEL (*standing impulsively*): May this be delayed for ever! our squander of affections.

TERISSIA: Sit! before your king.... does give you lesson.

FABEL: I would kneel to father's bed, and genuflect of weightiness the gravity he bears. For this to handle am I made? or do uncles lift?!

TERISSIA: There is your truss of tensioning, an' we shan't argue about it.

VINCENT (*feebly*): Be of a dependence to me, Fabel. An' bring more charm to calm my quiescent rages.

FABEL: I am for your legislation always, father.

TERISSIA: You have put doubt in his youthful head, Melinda, where it should not be.

MELINDA: That's a charge of base.... baseless spite! I encourage Fabel, to assume your goals. But I am not of your family, to assist directly his ascendancy. He must find his own path, make his own steps towards this assumption. That is what I've suggested, warmly and with utter friendship. Much as how I was encouraged—

TERISSIA: You were never made for this—!

COMMISSIONER 2: We are not here for these temerarious discussions—

TERISSIA: I sit!... And Vincent reigns. That is the only rule to follow.

COMMISSIONER 2: ....of these private affairs of contraction.... an' conciliation to the present state of our servitudes an' interests. We are here to have the Visgoth order us of his opinions— to hear him speak.... the truths of this situation. As to tradition, we are told it's cast in gold, and will ever be worthwhile to be upheld. But through his words we feel our way an' grope, and not with controversies unsuitable— between queens.... between mights of candor an' contagion. Peer we with our ears, for not a show of hostilities, but of regality to commend.... an' conspire with for our better purposes an' highest aims.

HALYCE (*entering, without a candle*): Here leads the light.... an' my Melinda!

MELINDA: Oh! But he is worn, worn out, and ridden, stricken of recline. And that has brought me to the Visgoth.

HALYCE: Does home a chamber (*coming towards the bed*).... that has diseases to spread? I am brave to be here, after my display earlier.

TERISSIA: Was as cruel as a stroke, I'm told.

HALYCE: Queen Terissia.... give me pardon of your grief. An' Fabel.... stand with caution, for I beseech this conference— with your father.

COMMISSIONER 1: We will condense. But we need four others.

COMMISSIONER 2: Pontal searches desperately.

HALYCE: He did find, and sound me out an' send. But what of Vincent's end?

VINCENT: I.... await.... coroners, to treat me with such disrespect, Halyce.

HALYCE: You were acting as a clown does pine with pity. An' I would say so openly, around these august personages, that I do not disobey to condemn such behaviors as to leave you thusly an' upset my notions for an overlord.

VINCENT: I feel no pain.... with your disgruntlement.

HALYCE: How then to educate? I take her from you, your.... ensemble of immobilities.

VINCENT: Where not you were to squirm, wear not this shirt! It is dense of fever, an' you do not shake! But as an enemy takes cancer.... the cancer takes blood— away.

HALYCE: I am not, nor either. Neither am I made to feel this way and plead for my fortunes. For with you you are insensate an' insensible. And what commands your lethargy with illness? It is a lack of detail, to learn of reprehensions suffered, that brings you to this repose. An' I speak for many quiet voices, as all of you may know. Yet I apologize for my rudeness today, if you've become this sick, an' now must consider matters with much more discreetness, resolve, resolution an' reserve.... to bring to leadership a stronger possibility than your disquieting visionary tendencies an' tentacles of astonishment an' dumbfounding for what you propone to believe.... in, as were a new religion grown with your sores as seed beds for its roots. But this is not the case, is it.... Visgoth! You— see now.... what foolishness lends— you.... a berth of despicable contaminations, an' a decline of fortitude— else a rejection.... of some powers. Is this a way to rule? or be ruinous. Then let your prickling be like complaints. Every one of the commissioners are upset, an' will not tolerate much further disgraces— For we are responsible to our own as well.... as to you. And that is your contract to consider with us. Soon angers will reign over agglomeration, unless with this.... heavenly sent, or at least as mysterious — rest, you are able to clarify your head, purify your soul, an' reach for some competent decisions. Now, what can you see of *me* that I have honestly displayed? We are all worried about your leadership.

VINCENT: My sights are catered to a fog, until a seven of you are found. But as to my leanings, I sweat. An' I lend you sweat an' perspiration with my breathes. For I mean to challenge discontents, with this enfeebled frame for sticks to steal— like ribs made decrepit for my lungs' carriage. But I can still think the worst of you — to improve, as I give pardon to my wounds as well as thoughts. No justification is needed for an honest belief. No education nor instruction will serve more than a thorough dip into its vulgarities an' nastiness, to be soaked and covered with the worst of its vile love. And now I am awash of protuberances an' swellings— an' *reeek*!... the odors of malignancy and malefaction. Yet I have no pain at all, an' less regret than simply being tired. For evil is for good to recognize; an' you see my lying, disposed as a hog for your bereft sincerities to receive and chew with, taste an' sift through with my snout. Yet how do I seem, to deserve this glorious lechery? I will make for.... how to do for you. And then you will own the erroneous as well, as some of you.... must have transferred those sheep to my kingdom, to try the innocent with wanting. And I am aspersed.... to take these punishments, until you purify yourselves of this malicious nature. I am strong enough!... to do anything for my fiefdom, including sacrifice of skin an' bowels. An' I dare you— to speak otherwise of me! This is not a madness, but a monstrosity for your mothering.... to commit to—

HALYCE: Don't claim adulteration by me, or others, with a ridiculous masochism to parade under an' hide of accident. It was greed of opportunity, an' nothing more. And had you kept your minions an' their subjects more circumspect, an' less hungry! they would have bewared of the foraging treasure, as the unusual do walk as were on a haughty, haunting hunt for acolytes of capture to be praised at. An' so do you howl! But it is a fault of nature, and not the supernatural. An' we have supported you as much as can be contrived an allegiance to sustain. I have personally balked at the awful taste of dissolution. Yet do you make more mounds! for some strange gospel visited, what would only be an excuse for your own lack of efficacious ways, an ineffable pronouncement, an inaudible decree, to render your lands cleansed of the spreading impurities you have found an' battle with by subsuming; a mad an' maddening endeavor, Visgoth! self-injurious an' displeasing of the eye an' all senses allowed to draft the stench of decomposition.... So to this sepulcher we come, for your praising?— or hazing!

TERISSIA: You are besieged from all corners, Vincent, by the ungrateful, even on(to) your deathbed. The shame of this makes voluminous an air of treachery, steeped of rotting passions.

MELINDA: I have not made fit.... a purpose for this blandishment; but more to oversee of these sad proceedings do I come, as the mother of this house made, an' maid to its regretful happenstance this night. Then appease my sorrow with conformity of temper.

HALYCE: Not made fit?... Not made fit to seem. It is lighted over there, but darkened here, where the Visgoth lies. Forgive my error of sentiment, for a dying man, an' my hastiness of candor at cross purposes to portray. But I am rude of wounds, dissatisfactions, while he is wounded of a blight that seems to pacify his convictions an' alleviate the harmfulness of their misdoing, at least to his mind. But then that is death's choke-holding kiss, to bring a head into luxurious rapture for its efforts thinking, with delusional play of placation for these thoughts. But not with shame, Terissia, is there pain from 'is bedevilment— but anger at it! throughout your whole countryside, rumors spreading of a contravention 'gainst their solemn grieving.

VINCENT: I am not to blame.... for a foulness at heart with this dementia. The matter brings me only to adjudicate.... an' convince some few of a righteousness with contradictions. And I do not die!... I don't feel death's grip, nor 'is sawing off of limbs. Yet is there subtleness to this view.... of shades an' shadows, whereat I am found.... not knowing of the end? Here lies the responsibility.... to pretend. An' I will give some orders for the world, commandments dangling of facile ties an' tethers.

COMMISSIONER 1: We do wait of ear, Visgoth.

VINCENT: They are as weak as my body.

COMMISSIONER 1: And though we may have.... criticized you much, lately, everyone of us holds your authorities.... an' their provisioning.... dearly to heart an' manner with your crafting, an' our manufacturing to realize, throughout our regions to obey or have a deference with. That there is a pinion of rule remains our tactic to observe, whereby our efforts are manipulated to run smoothly. That is the great genius for this organization of fiefs an' privileges, and association with you. Yet, to be official....

VINCENT: My head seems consumptive, consuming ethers.

NURSE: That is the liniment, master, that brings forth vapors as it disintegrates. Feel you a freshness by them, and their passing?

VINCENT: I am a master to no one (*as **Fabel** re-seats*) out of war, when I have held lives.... and condemned to death.... for their offensives against me, my person representing everything an' everyone.... of this devolving organization of rights. Hold I.... to a semblance of power to express.... an' pass along, it is for them, the people of this ligamentous body to relish with an' regulate of some rules of pride against a heathenish an' pitiable nature. It is for them I climb, to take their worst.... away an' leave them with a spiritual hope of character and defiance against fate. We make only the integument of the frame, Fabel, to keep the body sound an' preserve the allowance of its activities within.... an' without. So symbolic may our gestures be, that with clandestine tyranny are they sometimes perceived. But take not this anger as I am wart-ing from. With fear only is this a war.... of casualties— I make them not!... I carry them! They're on my back an' sides an' lumber loins. And I feel no pain with this burden. For they are off the feet an' legs, for feat of my deliberation. Yet to my arms they'll come— as I lift them! And then all over, to me will they encamp. And I will die, of their lives sanctified an' in this way only fulfilled. That is the Visgoth's pledge!

FABEL: It is a frightful one, father. Speak you terrifyingly, with a weakness of constitution still. Am I for pound to take with absolution blame, misdeeds that have been done the fiefdom? This scares heaven into scarring. I am beset with this terrible weightiness to imagine, an' my feeble strengths enlisted, ordered to be applied to this.... horrendous harkening you speak. (It) Makes for squeamishness a skin.

HALYCE (*with a demonstration of surprised cajolery in himself*): You'll not do the same as he! He is crazy!— Mean as thou; I vow to be observant. This tendency is insane, though the worship be a raillery of superficial an' superfluous tears. They are of want to think with sorrow, and not be panicked by their hurting as the Visgoth comes by, approaches an' defames. There brings a misery that enflames, an' no one would emulate *that*! So do not even suggest it for yourself. Yet more must mature here, with these sacred, sanguine influences softening.

FABEL: I mean to be as my father would desire, what of. What. How. (*as **Vincent** struggles to sit up some, revealing a streakily tinted dampness to his bed shirt*) And where to be seen for culturing of trusts an' reliability in me— But oh!... My father is terribly distressed of the bloodied hues an' hatchet marks. It is a Visgoth's emblem, bleeding wounds an' blisters, an' royal rouge wearin'.

HALYCE (*observing **Vincent**, as the nurse comes to aid*): We all get our scratches.... But this is more presentable to devils with pitchforks. It were a blaspheme for light be made, a cry of flesh, a *poxiana regina*. Then this is real of diseased vacuousness to mind an' might. What have you brought on yourself, Visgoth, that sleeps with horror!... the horrible.

NURSE (*helping **Vincent** sit up*): Let me assist you, my sovereign. The sheets are heavy, as is the dress.

VINCENT: I've climbed mountains with less effort.... an' reached their summits as breathless.

TERISSIA: Cover him more! Hide this view of chest, that even shaded shrieks of nightmares sounded deafly, our imaginations ponderously become.

NURSE: The scents of the oils and unctions amplify the effect sensed. But they are medicinal and healing (*covering his chest up more*), and must be working to lend him this courage of energies with revelation and purport.

HALYCE: What makes an evil is its spreading. An' haven't you fear in nursing?—

NURSE: Nay.... Not this kind, kind child.... to be my subject. It is a wish to wash, not wistful thinking of his harm. Improvements (in health) charm me more than frighten.

HALYCE: In all decadence you are beautiful—

MELINDA: A light be shed or sheared.

HALYCE: ....against this grim background.

MELINDA: We have not seen him clearly. But it is hoarse of view to say he must seem dispassionately patterned, like as of baby helpless to the bites of flies an' wasps.... while gentle in the crib, as much disturbed by fate as floundering.

COMMISSIONER 2: We delve into the misfortune, Queen Mother, to acknowledge it thoroughly. One does not have to see clearly a hurt to know its upsetting for you.

VINCENT: I merely present myself a labor, for your.... eyes— Stay by me awhile, maid. Should I faint, into your hands should fall the bejeweled torso, with its speckled back.... That to address you all, I am only tired.

COMMISSIONER 1: We have not met.... our number.

VINCENT: Saints sear me to speak. I am too weak to delay.... The kingdom still revolves. An' we are all with motion in it, pleading for some hefty stance to take. Yet more for my decreeing to be stated, it follows well the written scrolls— but this.... on parchment of the lips: I am dim, what makes me better to see the light.... from all. For all, to clasp at my protection, I deem it fit with suffering to change.... an' to evolve, into some leadership to make a Visgoth. God! Banish my perturbation! through these obstructions. It is difficult to withstand envy an' pride. But for what I would like challenges decay, the decay of angels.... An' so I choose breadth, to contravene a law as written—

COMMISSIONER 2: It can not be done!... without a seven (of us) to consider it as so. Your wishes are only hearsay else. An' we are deaf to them by covenant! Then rush to rewrite the law. But this is our protection from malignity within the fiefdom, an' disagreements that can fester beyond recognition of our binds an' gifts of fief— that have been won!

VINCENT (*with indignant difficulty*): Thus scribes the lips!— Then hear me!—

COMMISSIONER 1: We will take the matter as an advice for administration, an' lean towards this with our influences, as being sure to understand. So say— without the guarantee of diction, that nothing can be so well misheard as seen in this coven-ish quarter.... Then abrade of sound, your painful desire.

VINCENT: I haven't of the heart to bleed with agonies correction —

HALYCE: Worth better of the rule, a peaceful disposition to be brave—

TERISSIA: What's bred is brave!... you caitiff! cad of contradiction!—

HALYCE: I mean no injury to you nor yours, Terissia, nor to the Visgoth an' his properties.

TERISSIA: You'd reverse our sympathies, towards your foul deeds to own.

HALYCE: An' what is that?!

TERISSIA: You'd have her son be made our ruler, an' as your close friend. You've suggested it too often.

HALYCE: What be? I have no son!

TERISSIA: Hers! Hers! that Melinda does glide through palaces.

MELINDA: Why?!— Heck! to this outrage— an' hex your soul —!

HALYCE: But she has two! An' it's very well to be, Fabel's too young for this decree, to make for spoil a majesty. So bring off, often, I do insist, your disallowance for that son to fight, an' these—

COMMISSIONER 2: It has been clearly stated many times throughout the populace, the favor of Morliff to mold with—

VINCENT: Tears! for the pleading of my might!

COMMISSIONER 2: —The whole realm knows this, an' accepts the possibility as passable for Fabel's breeding timed—

FABEL: I am for breath of father's to promote, or withdraw from these contagions!—

VINCENT: Hunger for my wants an' needs!... This is determined in me, to provide your vision.

HALYCE: Half of which I may abide my love, Visgoth.... For your son is not ready! An' as a conscious man, you know (of) his squirming in the seat. So be clear, for this defines me— an' my actions. But let me state for wife—

VINCENT: I've married heraldries!

HALYCE: —to propose—

PONTAL (*entering with **Vaclkulor** and **Loresane**, the latter still shabbily clothed; no candle*): How comes?... But only three? It's not enough, never enough. And I implored to find, so fiercely. But it is late. They return to their lands. And it's only with luck that I caught sight of Vaclkulor approaching the palace. Yet as his custom being, to make some seeking of.... friends within the kingdom, he must return for horse from a casual walk of ease an' exercise. Though he brings this man with him—

COMMISSIONER 2: An' not a commissioner indeed. Nor one to be made, I'm sure, unless.... Does poverty come with fiefs, of some type?

LORESANE: I own a hut, sir. And it's as much a gift from my Visgoth, for it's as worthless as I am.

PONTAL: But he claims to be a seer, or so does Vaclkulor describe.

VINCENT: Seer?

VACLKULOR: And certainly with my faith. I've consulted with him many times, and only with reluctance persuaded him for this visit. Yet be the Visgoth ill, as Pontal says.... an' tells me so, it is not illness but a divinity of inspiration bought, according to my friend Loresane, which of course shines so bright to human flesh as to sometimes cause apparent damage to the skin.

LORESANE: But nerves of the wondrous recover from this, I've seen, my liege. Such valors develop immunities from the detrimental effects of profound realizations, though they must go through their trials of difficulties an' inconveniences, often unaware of their hardships' toils; such are they so spiritually transported as to be excited and delighted with all resulting prospects an' giving their supporters, advisers an' intimates much apprehension. But the matter of a messenger is mightier than (a) meed or meddlesome meddling to receive. It is the joy itself of the reception of heavenly notions that is the most astute award, an' reward with purchase of some tribulation. I live worst than a beggar, for I haven't the *stolz* even to ask for aid an' be an encumbrance for others' thoughts an' pity. And yet I am fulfilled with higher, finer pride for my works of visionary utterance an' inference, this enjoyment running as an enrapt pleasure throughout the courses of my personal dominion, which is myself— when.... I am successful with delivery of portent to aid or relieve the fears of a listener who has engaged my services through friendly consultation. Most sure it is that no true information can be bad in the scheme of life's actualization of treasures to obtain, when even death is something proudly won. An' pains an' griefs an' miseries? Well, they are toots on a pipe concordant with rapture, gaiety, an' satisfaction, when viewed this way the counterpoised principles of existence blasting to our ears, or blinding our sights at times with the raw strength of luscious being. An' so I sleep coldly at night, sometimes, even uncomfortably, in my thin, little shack of cover and protection from only the faintest of dangers. But I am thoroughly compassion(at)'ed with my dreaming.... An' I have been alerted to our Visgoth's.... an' can agree....

VACLKULOR: I have suggested to my friend that some nicer accommodations can be found for him for trouble of his visit tonight, as it becomes windy an' rushing outside, an' less hospitable than his cautions. Within the palace there are rooms, even holes that he could relish. An' with appetite, some mead at least? an' rice or wheat cooked to a royal perfection? Perhaps some meat—?

VINCENT: You can agree with *me*?

VACLKULOR: —stew of lamb? Myself becomes hungrier, with thought of welcoming hospitalities.

LORESANE: My Visgoth, you are apt to be superior of blessing wounds. Your cry is the most honest of wolves, an' I hear ye!

TERISSIA: So said, a meager man with a grand fashion.... of a sensual sincerity? But what to offer us in rags, as my husband is in sheets? These drip of terror! an' sigh of harshness to behold an' behoove.

LORESANE: Oh, pronounced Queen! (*wabbling to his knees*) I had not realized this. But I am your seizure to enlist an' strike at! of obedience. You must most certainly be in awful distress for our overlord. An' I will ply my humbleness before you, to peel of use an' pick at for signs of due or apportioned recompense.... with this discovery of his grandeur burning at its brightest, to make all other lights around him shadowy an' dubious. But I see a better man through 'is queen concerned. Does woman make, most cravenly of man his mate, then is so it. I pretend not to fear a goodness.

HALYCE: Too gentle to be foresightful. Yet perplexing. I would fear a goddess. But for good? Does make it better to be born angrily to face your caretaking.

VACLKULOR: He is substantially one of your most encouraging subjects, Visgoth.

PONTAL: Or to the death he bends. He is reverential.

COMMISSIONER 1: What?! As an example? To the Queen he is afeared— her anguish.

VINCENT: What of me.... an' my son, seer?

LORESANE: As you would, my Visgoth. These are not ashes to you alighted. An' you do not die— as suddenly as *this*!... The mortality is more subtle, to appease your aims. An' you will awaken as from a dream, refreshed an' more able than before. Else I should take the bed for you.

COMMISSIONER 2: That you may, to prevaricate and uplift hopes too falsely.

LORESANE: I have *never* been wrong with my seeing. As for my doing, I am a beast of hardships causing on myself, falling from simple failures. Prophets are seldom protected from themselves, nor from anyone else.

VINCENT: Aye, that is the annoyance. And of my son?...

LORESANE: The boy.... Fabel.... He is such a handsome jewel. An' to remain this way he will stay precious to us.... an' necessary for all governance—

TERISSIA: Ah! It is foreseen an' forsworn, an' impassioned with his life to lend for error. Feed this man! an' grant him sweets!

MELINDA: There was never any doubt to this, Terissia. Why are you so surprised or startled? with words that make our senses obvious.

FABEL: *I* am Fabel.... And I am stunned.

TERISSIA: The fear of shrouds to prevent it, the foreboding of character to lose allegiances to— my family, sharpen the peaks of anxiety for our practical rule. Then to find any supporter among these shadows is for a cream to make out (o)'f casualties. An' I will adopt his service for us an' insist of it for you, Vincent. For we have been hounded heretofore, an' will take gift of his.... visioning to improve the countenances of our worried hearts.

VINCENT: I can hardly see the man. Stand up, Lor(e)....

LORESANE (*wabbling to stand*): I am your servant, an' companion like as a ring to finger used. Blow on me for some words, and mouthed will be the truth of happiness.

VACLKULOR: Lo-resane has never lied to me. His deeds are actual and powerful, an' has made for a good advise.... at times, when I presume of profiteering through his thoughts. But I feed him.... of a simple money. His wealth is already of his eyes.

VINCENT: Bind 'im to me, would you? Can you nurse, Loresane?

LORESANE: As much for sores an' callus to tear at. I be mean an' lowly with my body to preserve. But then, that is how I persist, with a practical knowledge.

NURSE: And not with medicines prepared and purified? Could be more of (a) contagion brought.

VINCENT: You are my medicine an' mender, maid. But he may be a mensch of learning for me, in these visionary habits. So be relieved an' reassured that I do not confuse the methods. I only suggest, for him, a humility under you— while near me. I am so tired, as to see a night as flames, it's darkness scintillating to wear my mind down even more. An' am I imagining this? Then comes a visionary to perceive me.... an' honor.... an' reconcile my angst with my amity, even towards detractors.

HALYCE: Visgoth! If I may be so bold as to suggest, that what you need is more rest than rustication, then I say so soundly, an' as serving insufficiently a caucus. Few other relevant decisions can be made with our occurrence, that are contractual for a pledge.

VINCENT: Else a bond. I see in darkness as the timbers are felled.

HALYCE: They are light strokes, compared to the dimensions of your disturbance.

VINCENT: When I sleep, but what?... If I awaken to a coming past, then serves me well a rest in deep perturbations.

LORESANE: You will arise more glorious than is your fashion tonight, my Visgoth, since you are only weary of your insight, an' the muscles are tired from their transverse lighting, being more used to the shades of body. But you've a bolt of princely splendor to behold as when comes the lightening of your vision proved that the day will be refreshed with promise and appreciation. An' the merry will greet you an' dance an' sing of life. I will be held for this, awaiting your pleasure for dispersal of wisdoms. To me it is all a picture drawn already, as lush as a pond in view with its riches forayed of sights or casual inspirations. An' as a frog hops with delight from the lily pad to brook's broad bank of wandering among the verdant leaves an' green lunges of tranquility, looking for water flies an' insects with a sticky tongue as such am I expectant for review with aging sloth to past judgment on my predictions. I have felt all of this beauty, an' worshipful anticipation with

badly tried limbs an' sore sinews exercised by the mundane difficulties that keep me impoverished. So thus I claim, it is as a bore to bliss, my visioning. And you shall reawaken expedient for your tasks in caretake of these sundry provinces. Have no fear of that at all. The day to come builds up its powers to promote your presence with a vigorous vivacity to your effect.

VINCENT: That is a weak wish strongly checked, an' kept within my vaults of temper, feverishly locked among my treasures of the mind. For I've so much to do to keep this realm in tact an' our possessions solid, solidified for the usefulness of all an' any, any ruler, any regent, any knave. And yet I feel so tired as to drop through wells to reach a bed of rock an' damp frustration.

LORESANE: You shall see.... the better of your envies, with clean light from the sun, an' good air breathing of more strolls.

VINCENT: More?... Is wished, more. But not more— More.... the unorthodoxy of my arraignments and preaching above bodies. I sense the bitterness, of course. But this is stationed to my purpose for the worst.... to assuage. An' now sleep chimes a drowsiness well played for, if you have seen everything before me, low herald. I should leave my consciousness in your hands, for suffering through. It seems you have practiced much.... privation during your life's span of terror with the future.

LORESANE: That is the only thing which soothes me on this earth, my liege, that knowledge of its certainty. All matters else do chastise my verve an' pitiful vocations to survive.

VINCENT (*starting to sloop, the **nurse** attending to help him lie down properly*): Then be you treated justly for a meal.... of rich grains an' bread, an' drink, an' fleshly matters cooked.... an' a homestead for the night, warmed with a security for your location.... We shall walk in gardens, an' I'll correct your prophesies, as you annotate my messages with a proper or more pleasing grammar an' verbiage to produce.... for the populous ear, the such as not to teach but to.... ascribe.... for their own learning.... of me and man— dusted.... for preservation.... But I now completely exhaust from these notions of comparison.... An' what I would pronounce gains strength.... with this semblance of volition.... to dream for.... (*closes his eyes, head on the pillow*)

NURSE (*subdued voice*): I think.... he sleeps, the energies drained for thought.

TERISSIA: To fly of lighter wings imaginary.

PONTAL: Not enough have come.

NURSE: Another treatment of the oils, in an hour or so.

COMMISSIONER 2: We can not wait for him, unless throughout the night pervades this vigil, which seems unpromising.

COMMISSIONER 1: Yes. The others have left, for their homes. I feel, though, that we should stay around, and petition for a couple more tomorrow, as an errand of emergency.

VACLKULOR: Of course we can be put up, though my business is pressing elsewhere—

HALYCE: As is mine.

VACLKULOR: —But a courier can make excuses for me, for another day's stay.

MELINDA: Accommodations are plentiful at the palace—

TERISSIA: Our commissioners.... and guest, will be seen to. You will have a bed of goose-downed lace, Loresane.

LORESANE: That's much better than the straw I'm used to, your majesty.

TERISSIA: The threads will be comfortable an' as much colorful.

HALYCE: He didn't say much, our.... Visgoth.

FABEL: But what my father intimated is profound, count.

HALYCE: There's no meaning to it, boy, without pronouncements, details, an' strict abiding.

FABEL: He insists to remain able.... an' will not stay discourteously idle in the affairs of this fiefdom. That's more important than the error of the earth revolving for its sustenance an' subsistence.

HALYCE: Believe such things? We are one alone in the universe, to make our own mallet-ing endeavors, devise our own crafts of importance.

COMMISSIONER 1: Do prepare yourself, prince, yet for an assumption of regal duties.

TERISSIA: With a fine manner assured, Vincent has blessed him throughout, 'gainst those two—

MELINDA: A commission—

TERISSIA: —brothers!

MELINDA: —of the most imperative will oversee this, as the Visgoth is temporarily indisposed.

PONTAL: Does mere slumber make it so? We will convene with the written word— the latest documents of scribes and scripting, how he has said, to be obeyed.... with reason.

VACLKULOR: Or reasonableness. An' as the next head of the assembly, I will induce some order to our senses, an' not try the man too harshly. There are clearly disagreements amongst us; but he is feeble, health-wise, an' should not be hacked at as a sugar cane for juices to suck out.

TERISSIA: Yes. His sweetness is his benediction, that he cares for everyone so intensely as to even put himself in grievous dangers both political an' physical, administrative an' anatomical. This is a problem for his diligence an' fortitude, since what is right to do is not always right to be done, or practical of a circumstance.

VACLKULOR: Definitely this is the truth, Queen Terissia, or we would not have poor prophets at all. You shall bed in this grand fortress, Loresane, where even the ants hoard an' carry gold, as thin leaf on their backs. And your clothes may be improved a bit, if you are to make promenades with the Visgoth, and accompany him....

as a second head sore.

LORESANE: Ah, it's wonderful an'— sore?

VACLKULOR: Like as in, then in as eyes.... to restrain 'im from the tendentiousness of the aggrieved, with what you see an' sense preemptively. He is not a man to *want* to exacerbate sorrows or extend them.

LORESANE: No, he is not. A kindly gentleman he is, unto himself as if a god to please of the gentilities of mankind. Thus makes him a silence that we can all hear, even those without the brawn of presentiment to battle with. Oh, how successful are we? But he is mostly one to win his tare of balance by those tamed of this temerity. A tithe, you say. I would have to serve up lice crawling on a string to make for my worth such a portion of wealth. Then take I well to this kindness, more of words and thoughts than deeds an' chains.... I can protract him away from the most severe disgruntlements or shows of resentment, an' this with only a slight logic from what I see occurring towards the Visgoth.

TERISSIA: Direct him to a luxury, Vaclkulor. We shall stay by my husband, in silent prayer.

VACLKULOR: Not to bed by me.... But there is an ample storage room full of utility for this purpose, that I have found. I believe I know where it is, Terissia. Come with me, Loresane, as you have. (*starting to leave, with* **Loresane**) First you will be feasted, for service of your journey. An' then you will be 'customed to a relaxation on soft pillows— Lend me one of those candles, will you?

COMMISSIONER 1 (*picking up a candle from the table*): We shall withdraw ourselves, to conjoin for a dining.... at the banquets chamber?

PONTAL: I suppose so, though I had wanted this to be a true congress.

HALYCE: That's round enough a room, spaciously lavish.

VACLKULOR: Well come, then.

TERISSIA: Simply order the servants, Commissioners. Nurse! Direct them!

NURSE (*at first hesitant to leave* **Vincent's** *side*): ....Yes, Queen Terissia. He is only resting—

TERISSIA: Fabel! Oversee this! an' have our guests treated well, with a concordance to their importance.

FABEL (*standing*): As is to be done.... I want word immediately of his reviving, for he may be expecting my face to find an' recognize—

TERISSIA: It should not take long to accommodate the Commissioners. But take a repast yourself, before returning. Send the nurse back sooner, within an hour. We shall stay—

MELINDA: I have a meekness for this presence.

TERISSIA: —for late of ladies, to watch over our Visgoth.

COMMISSIONER 2 (*following the others, as they are exiting*): I should really not stay the night, but as to a duty am allowed to only. Then.... good awakening to prosper us, queens. I bid you an evening.

TERISSIA (*as **Commissioner 2** exits*): It's early yet, sir. This dread is protected from, our careful eyes averting the catastrophic. (*The three are alone.*) I would have it this way said, were he to die amongst us now, of the shallow respirations to end, by grave of angels seen, (that) we were observant of his soul in lift an' lightened by our presence. Or not a meekness to it, but a greeting pardoned as he escapes a trembling body, an' the excuses are made.... to us an' his kingdom.... Come take this seat, Melinda.

MELINDA (*standing by the table*): I am by candle's breath.... (He) Seems quiet to me, and without much shaking.

TERISSIA: That is the headstrong anguish of tranquility with loss of strength, yet within battles brewing. I fear for my husband. He is not that way but due to illnesses. An' I am not superstitious, but require an advocacy, for his secrets not to be lost. (***Vincent** coughs with a slight gurgling sound.*)

MELINDA: He stirs.

TERISSIA: It's to defeat the wane an' wanness.

VINCENT (*very weakly, while opening his eyes*): Are we alone?

TERISSIA: ....Yes. Much to queens.

VINCENT: I had simply collapsed of energies.

TERISSIA (*as **Melinda** comes towards **Fabel's** chair*): An' could not hold court well, Vincent.

VINCENT: These spells occur. But I can not see.... beyond the shadows, an' with a blurred enlightenment forestalling these events.

MELINDA (*at **Fabel's** chair*): I have been standing.

VINCENT: ....Melinda? Don't wait for me. I am before a recovery. (***Melinda** sits.*) An' where is Fabel?

TERISSIA: Entertaining the Commissioners.

VINCENT: Ah, the fiefdom.... the territories so important.... to us. I could not see him safely.... without a growing sadness for his bright perplexity that leaches from the sharpness of my views.... He does not wish to be the Visgoth, Terissia.

TERISSIA: So do fathers not wish to be.

VINCENT: Yet you must want this difficult duty to have a chance at it, fulfilling its heady an' grievous responsibilities.... A sacrifice is necessary, to twist you into contrition of its powers.

TERISSIA: You are more than made, for his reluctance.

VINCENT: Where is my nursemaid?

TERISSIA: ....Conducting the Commissioners.

VINCENT: I fight against harming you, wife. But who has the better son? For I will sleep without awakening again. One knows that this is pending. I sense an empathy for death, having breathed it of late.... an' condoned it thoroughly.

TERISSIA: He is only premature, an' will grow with the position to fill its gapes an' gaps, an' voids that vex of cursing.

VINCENT: Have I been that disreputable?

TERISSIA: You have not seemed of Visgoth, Vincent.... And you have not explained why— not to me.

VINCENT: It is a vile thing to be.... so encrusted an' bedeck(l)'ed of a savior. But my father had question of me, an' now me of my son. We can talk this out—

TERISSIA: That is for nonsense!

VINCENT: —Made he others, dear, with this concern. He was a perfectionist, an' I.... a slight disappointment— not in ambition an' effort or diligence, but in charm and character to hold the fiefdom together.... He was always trying to improve the whole.... of this foundation's stone.

MELINDA: Don't upset yourself this way, Visgoth. These are not the confessions to make; for your prowess has been unmatched in our history, an' you have nothing of your leadership to be forgiven for.... until perhaps lately, for conjuring up ideas to capture griefs unto yourself an' wear them boldly, recklessly, an' irritatingly, with irate speculations fostered.

VINCENT: I am unworthy else to do. I die with blunders on my hands an' skin. I have allowed an evil to spread within my reign an' era— just as he predicted to me bluntly, to my face.... after my youthful foolishness for some act bothered his composure. I had simply forgotten to water the flowers by my mother's grave, a personal duty and obligation. And only one of them wilted— But he saw red! on that anniversary I dreaded to attend with him, together and by ourselves, the flower faded as if its color were swept up into his angry, tearful eyes. An' he charged me.... rather softly, with a callousness. Said I should have at least replaced it.... before his visiting— an' that I was inattentive, an' would let ruin spread.... someday.... across 'is kingdom. An' so it has, in a degree entrusted to me, though I have tried for it not an' never. And from that very day, I am convinced, he planned.... to make brothers, for my replacement. But his declaration is ordained to be correct. An' I gladly die by it to distill the shame into a sweetness, a purification through the earth. I was a fearful, young.... fool, resentful of the loss an' trying to ignore it, I believe. But that gave the man.... too much disgrace of me. An' such are how errors wound.... that I have made a frightened son.... of Visgoth.

TERISSIA: That is a personal bemoaning you should not weave into my boy's remorse. He is still not experienced enough to distinguish dejection from diffidence, an' will improve with our handling, Vincent. And he does not lose me.

VINCENT: But will there be flowers by my grave, Terissia, through his care an' tending braved? I cry for the strength.... to *want* to absolve him from that chore. My judgment, though, is firm that it's not right.... to willingly spread a disease of conscience, perhaps mutating into a thousand misdeeds or misfortunes an' gravities.

TERISSIA: He is the most respectful of sons, an' reveres you as more godlike than his nature could ever assume.

VINCENT: That's a good attribute.

TERISSIA: He forgives you for all of your recent faults, tries to remain blind to them.

VINCENT: That is a courtesy an' consideration of royal privilege — I'm sure he will obey your admonishments an' suggestions.... an' be decent, thoughtful.... prepared for power. The entanglements will be as powerful a web to struggle through, or course along. But there is such a gamble to this, Terissia, a risky surreptitiousness in the offering. To shove Fabel into becoming Visgoth, currently, could shear an' destroy him, soft meat through a sharp grating thrust, an' bring unease, unrest, an' disruption to our rulership, over our domains. A considerable population of influence prefer Morliff to me even now, an' could as much as let a spat befall on Fabel's face, in courtly an' effective dismissal of 'is presumed authorities.

TERISSIA: So what has this to do with your conscience? Throw the dogs a bone to divert their attentions. Give this— brother of yours a conjunctive position of sure advisement in the affairs of the fiefdom, an' let me handle Fabel as a growing Visgoth properly nurtured, trained and restrained to my opinions an' judgments, the better for this realm to be acknowledged. This ends debate, an' solves distension of dissent.

VINCENT: Ah, for my last an' leaning words to lend.... they must be carefully considered. I can't say them afterwards, nor repeat an ending.

MELINDA: I am so rich of grape, with which to know of virtue holding in my hands more means than one for offering of matters relevant, virile an' fecund.

TERISSIA: What say you tersely?

MELINDA: Not to gloat of it, or boast, but I have *two* sons—

TERISSIA: Then virulently proposed!

MELINDA: —two brothers to you, Visgoth. An' there is a cadre of thought to favor gentle Caton for the principle position of overlord.

TERISSIA: Too soft of temperament.

MELINDA: Then wisdoms float as feathers made to soar, since he is the most reasonably thought of or appreciated.

TERISSIA: Take minor woes to minor troughs of hoeing, sowing seed! with a shamelessness before this dying man. Improper caring of you, Melinda! this concern. Morliff is in general more popular, an' so more the problem an' annoyance to deal with.

MELINDA: Not as enemies, neither of my sons are to you.... But already, dissatisfactions make for dissolution of the principal estates with a nod to Caton's conditional mandate tallied as rejected. Count Halyce sorely favors him. An' not expecting this to be, for

Visgoth to become, begs me to reveal to you that he is about to separate from the fiefdom, his lands as an independent state—

TERISSIA: No?!

MELINDA: An' not only this, but he wishes to make me his queen in marriage, to aid his rule over the realm.

VINCENT: With desperation or passion? This startles life, as for a competition entered in my death. That dare he makes a mockery of my reign is his intent, or as to transfer a capital constituency out of my kingdom an' into his contention, with your person, Melinda. Then where will lords be born is much a question of the legitimacy of love. An' too loosely I have handled this— Or so he threatens?

MELINDA: He implores me... An' he is a handsome man. An' I am not old, still very fruitful, an' desirous of some worthy companion. For being bound here is very singular, severe an' grave. Yet shadows sprout by lines across the face, with aging tediousness. Yellowing of the leaf is what threatens me, an' not his ambition to threaten you so much, as it is provided for with the natures of our discontents. I may still do, as he may be, a romance to entertain, an' away from the coolness of this palace.

VINCENT: Then.... I must not die. I must be a god forever, to withhold weight.... to our pavilion recognized.

TERISSIA: You are too young to be a queen, Melinda.... yet more to be a mother. Romance ends with the rancidity of foolishness. What?! Lonely here? That is your due, that is your pride to have been our majesty, though only selected.... But then, so am I.... now. No romance. It is fancifully frivolous.

VINCENT: I err as he takes?... my regions dissipating already? Heavy heck!... What do you feel of this dilemma, Terissia?

TERISSIA: ....How can you pacify a fool! Halyce is emboldened with your disease.... if it is love, but tormented with your thought.... if it is so.

VINCENT: To live forever, then? There is a burden of seconds.... lingering this way. I do admire Caton more than Morliff, despise neither, an' have suffered both to be parental for them. The airs twist with suffocation.... Melinda. How could I be a parent to *you*?!

MELINDA: I have a patience to abide your wishes, Visgoth. But it has been dreadful for me.

TERISSIA: That's because.... you were not meant to be liked.... being left over. An' I say this not out of cruelty, but from an older age observing, with perspective of the misery. Awkwardness of jewels make for a quiet complaining of your treasures an' riches. You still must provide yourself of this court, historically graced with your importance.

MELINDA: That effort has already been done, its impression made, its stamp upon the fiefdom scored, an' its perspicacity for future events soon to be extinguished with my endurance. (*The nurse enters, with a candle, going directly towards the table.*)

TERISSIA: What?! Have they been well equipped for comfort an' needs, maid?

NURSE (*approaching the table*): Yes, your majesty. Has the Visgoth rested peacefully? (**Vincent** *closes his eyes.*)

TERISSIA: ....He sleeps. He sweeps of black sable.

NURSE: No disturbances, or utterances of sound?

TERISSIA: Nay. Soft is the respiration, softly piped. If not congestion makes a whistle for a wheeze, then there is naught to hear of it for comment.

NURSE: He is much tired, and a treatment comes.... (*preparing at the table*) to re-anoint and time his leisure gracefully. Or like a tune, his heaves and hos of chest, they do a balance to the silent distresses.

TERISSIA: The pace is subtle.

MELINDA: Does speech make a melody?

NURSE: ....It accompanies, Queen Mother. But what of words with sweat?! The men do curse themselves that way, and make contusions glide for their effects.... But the Visgoth remains calmed, and roy—al, for this little work that makes so much our curiosity and anxiety.... Would be the greatest effort made, to remain silent within a whirlpool of consternations, complexities of thought, fearful upsetting. But that is a discipline of the body's healing, self-taught and tutored with each.... conversant breath, the diligence of stately candor.... to a life.... I will make cool the brow, soon, and wipe these distresses away. The cross effects of life.... make crotchety a strife with slumber.

TERISSIA: If were to fix with liniments then loom.... we all predations to our causalities. It's not as easy as a medicated scour— nor as simple as a maid's attendance.... But manage it with chemicals an' rivulets, by which one handles soothing of the mind.... an' manipulation of one's character, subjected into another during dreams. Yet how we must stay objective for this quiescent overseeing; then we are not ourselves imaginary, but only imagining the stimulations that rile or make us laugh.... through some poor dolt expressive as ourselves. So.... clean the body, as to preen the mind.... for some good thoughts left, an' a decent delivery into the eternal factions of cognition.

Act IV

Scene I — *It is day time, and **Vaclkulor** is sitting in a weak chair at a poorly made table within **Loresane's** hut, which is small and barely windowed, though the entrance to it is slightly tubular, avoiding the creation of a door proper. Only a wad of hay serves as a bed, somewhat behind the table. Few possessions are noted, most of which make a crude pile of necessities mixed with roots and dry foods. This marks a can for drinking, or pouring, conspicuous on the table. The walls of the hut are so thin as to let direct sunlight glow through them at spots. And the floor is a mixture of grime, straw, and earth, and that which live within a dirty ensemble of deprivation.*

VACLKULOR: This is some terrible cunning of without. For how can one mole his way for long with this largess of collected nature, without becoming thoroughly depraved! Yes. This is a poverty profound. Not even cloth here, (un)'less it's hidden in that pile of rubble an' rubbish. The scents are pleasant enough, though, if one takes to earthly grains an' dried stenches getting used to. Any moisture at all would yield a moldy havoc. No restraint at all to step through. But Loresane is courteous to visitors, allowing me always to sit here— though often I prefer to stand— as we confer and converse, he taking the hay to sit or lie with. And that is much an offering. Seldom do I have to look at him directly, and see.... that he is all resigned to this grim atmosphere. I've seen more of him at the palace, with much embarrassment for his peculiarly strong features an' pensive face, especially after a bath, or a wash down— rub off whisk with water, such as he has only rain to do. How one can stand such asceticism he should explain, beyond his visioning. This is certainly more pure an environment than what he may conjure through mere speculations on the future. It is purely dreadful, but seclusive and removed from.... criticisms heard. And the garden right around here is quite resplendent, if beauty serves the eye for him with need. I think, though, that he is poorly sighted, an' can only make out the shapes of colors enhanced with their blurring. Outlines fade for him. But that is of like, these messengers an' prophets. They are lazy for the real to see, and might as well be blind to details that are actual an' hold our attentions. Yet not a foolishness pervades this desperate space; but what he has come to be encloses 'im an' defines his worth as helpless an' heretical. As a friend? no. As a utility is he made. I have sought some answers from him that only the grounded up can feel without much questioning, as if already dead, merely an instrument transferring.... cautions, from one medium to another, from the divine disinclined to intercede with our fates to the devout of purpose struggling through the hardships of living an' wishing for some leverage of advantage to prepare with. Where does that leave him? He does not live, an' yet!... he struggles. Some sort of insensitivity, I suppose, to the surrounding banter of disdain. For if a forest breathes at all, it is for offering of liveliness to the senses and wonder for the current(s) that one may own to acknowledge. It leaves him where? In a realm of truth.

At least the principles of his sayings have been just to me. I don't rely on them alone.... but they are fair. And it is a useful habit or crutch of conscience to seek him out, occasionally, not for advise.... but for adversities to avoid. Were there a pit be made of them, it's here, more blessing found to sit through bruises unaware, unconscious of the day. Better here for idleness an' wounding. Now.... can I come this way assuming more? I am arrived on purpose, struggling, searching an' luring. For this is a solitary place, an' my plans have been made secretively. Of all who would most readily become, Morliff, a true friend, must be made (the) Visgoth, to support me with his favor an' influence an' power, that I might maintain my crumbling fief, a territory I can hardly support of funds, against an overflow by others whose armies judge I indeed am too weak to defend the possessions they would strip from me. Loresane has as much said so, an' claims this will be done. But if I align myself to Morliff's bond an' binding friendship, *his* theft.... will be inconsequential, a pledge that for my allegiance to 'im I will be made privy to the protections that will sustain me, at least for a while, until I can recover my fortitudes. To this (end) I must have Loresane as a stake, convincing death's ground, through commissioners, to have Morliff appointed. It would seem to be logically done already. But I must insure it, since Loresane refuses to.... predict this, or that or one way or another, with his dispassionately blind honesty concerning a Visgoth. So I must pressure him towards this proposal to be real.... to be foreseen, as regards commenting an' convictions. So to, what cares he, in this dry waste, that I may force with influence? It's not the real, but a friendship of innocence, innocuous insolence. An' so I chance the day to come an' wait, while a body of commissioners should grow, back at the palace.... I to be somewhat in search of members straggling to be directed, perhaps at inns, a possibility I suggested, to afford this.... trip. For what is made of a dingy shack? but the hut familiar. This is a patience to be dealt me. Time is not so critical— today.... as this hunch on personalities. Yet be sensitive to the flightiness of squirrels with sharp teeth, their delicacies so deceptive—

VOICE (*of **Hestor**, from outside*): Hey?! Loresane!

VACLKULOR: —aloft much broader trees.

VOICE: Loresane? I haven't seen ye today. Are you in?

VACLKULOR (*calling out*): Come within, lad!

VOICE: ....There be a conference?

VACLKULOR: No! But be, I am to greet you alone. Then make a conference, we.

VOICE: ....For treasure, make?

VACLKULOR: Aye! An' with a friend to see. Coins make a freedom pleasurable, no? But more necessity does bring me here than leisure setting, due to Loresane's detainment.

VOICE: Loresane? Detained?

VACLKULOR: Well at the grand palace of the Visgoth. There is a controversy about him. I can not lead more for what pertains, the detainment.

VOICE: Controversy? (*no response*) Does gain me worry.... as indeed I look after him— an' only I.... or look for him to be. (*a slight pause*)

VACLKULOR (*as **Hestor** enters into view*): He is too rich, an' too poor. Has made all sorts of prognostications that startle an' revive suspicions of the occult. (*cs **Hestor** stands by the table, though not near **Vaclkulor***) Yet, bare as he is, his mouth can not stop jeweling. He wishes to amaze an' profound almost to a surliness of contempt for mortals of lesser sight than his. The personality has changed with magic—

HESTOR: He detests magic!

VACLKULOR: —spewing forth for awe-struck victims, at least with the images from his words. Though the royal family is impressed, our body of commissioners an' advisers are growing resentful and less impressed with his uniqueness than condemning him as odd, oddly made, an' incapable of natural interview— or normal intercourse. I could never have predicted it. Our personal conversations have been quite fun and comely, an' useful to me. But the man wants to show off to his superiors, an' by doing so may draw an ire.

HESTOR: Magic is the foulness of deceit, he always says.

VACLKULOR: Yeees—

HESTOR: He is a competitor to it!

VACLKULOR: Yes. But that was when he was humble. Now he wishes to conduct flames in the mind, an' shape their dances. It is difficult to know what to do about his aggressions. Coming to this simple shack, a shed of patchwork, I've hoped to find something, some object to remind him of his simple but proud standing that makes 'im what he is, endearing of the community. But I'm not of this area, an' can't easily know what stirs the heart. I can identify nothing here that could bring his pride back to its roots of subsistence. For he has everything of food an' comfort, an' even new robes, at the palace. There is no valuable, no toy touching enough for me to return with and pierce the memories to break this headiness. There's no worth at all about me, he's lived so destitute-ly, not even a cheap amulet or chain of wrist work, trinket or melding of colorful stones. Only straw an' mud, thin parchments, an' a stockpile of gamely harvest an' gatherings. Now how can I bring him back a bean when he has a whole melon to consume there, or a clump of straw to compare to regal brick! Why, these walls are like a skin, in places—

HESTOR: I've helped him mat them up sometimes. They collapse easily with any dampness. Sections can dissolve entirely under pressures of a torrent.

VACLKULOR: But now he has the spend of rooms with rich woods an' a firmness to chamber. He can not look back to this!

HESTOR: Then all is good for him.

VACLKULOR: That is only temporal. He brags, my lad. An' that brings out the devil in 'im, making claims that can not possibly come true. An' when this is discovered or concluded by a number of notables, then he will be smote as a crooked an' dastardly manner of cruelty for the ears. Why only this morning he did boast to me, in casual jest, since he is still not so familiar with the others, that he could see the sun meeting the moon to make an ethereal glow over the earth never seen before by living men. His girth of presentiment seems uncontrollable, now, an' he over-extends his abilities to a grave deficit of wisdom, restraint, an' reasonability, that can only cause punishments and result in greater hardship than he has ever experienced before. I knew him as a friend, who took me to be finer despite himself with his abilities of mentation. As such for *you* did he perceive?

HESTOR: He's often claimed me nicer of a being, an' compared himself constructed closer to a stalk than its flower for attractions an' appreciation.

VACLKULOR: Or does he perceive at all, then? But now he views himself— himself!... An' (he) considers his sight higher than.... his current companions. But they are not his. They are only being tolerant of a proposed lightening. An' when that faults as badly as it must, disaster! lad. He will fall from the earth, an' into what void not even he can imagine. For the mind is cavernous for shame an' self-deprecation, once it realizes its own folly of thought. To misinterpret or misunderstand, and later make correction of opinion, is only the justice of learning. But to be of error as if made that way to be, with the conviction of impulse! must be devastating to the soul upon a forced review, if a person thinks at all. So the only path, for aid of my friend, is to bring him home, symbolically, to extinguish the excessiveness of hauteur he attains, by presentation to him of a memory he cherishes.... or would to think of it. Yet there is nothing here to grab, hold out to him and give for recall, that he does not already have in abundance of a wealth to make for only the most trifling of curiosity or interest. I'm afraid he is cursed with his own demonic plague of portentousness.

HESTOR: Do castles turn one bitter an' sour on the normalcy of mankind?

VACLKULOR: What else is there to offer, of the banal bash to witness for 'im a bastion of purview? I am deeply concerned, an' yet fail to convince even myself the futility of these efforts. For what does have a man? I have never asked him what he loves.

HESTOR: Does love consist of these embellishment of living? My mother says they are superficial to a caring nature; an' though necessary, as necessities are named to be, they are only servants for companionship. This is why my friend is so disowned. He hasn't any.

VACLKULOR: But he has *my* friendship, an' with bonds that are conversant to 'is talents.

HESTOR: Well, you are counselor of interest. An' you pay him for his services. But I am a friend of his too. An' he never foretells for me my fate.

VACLKULOR: Not one prediction?

HESTOR: Nothing more extraordinary than what my mother can explain. And they are not predictions. They are anticipations with assessment of growth.

VACLKULOR: So then you are a companion without profit to him.

HESTOR: ....Nay! My company is fun for him to be amused of, since he has no other, usually, (un)'less they be for threats or counseling. An' he has said that children are the stone of stars by which they glow. So he does make me a companion of friendly comfort— an' that is a profit to him! because he reviews his childhood of curiosities similar to mine. An' he often tells me I am as promising as he, or he was as I, till the visioning crept (into) his consciousness.

VACLKULOR: Well, children can foretell many things, by the very nature of their developing needs.... So then you are a release for him, from the tensioning of his gravities of thought. Have you no thought at all, perhaps, as could a vessel be filled— strictly for carry of the excess.

HESTOR: If this is so then it's all spilled, for I retain next to nothing of his more morose or ponderous thoughts. They're seldom made in my presence anyway. An' it delights him that I dismiss his heavy moods to ignore. For then he says, "Were I to take your character, this drudgery of life would be as fun!" An' then his burdens do seem lighter on him.

VACLKULOR: So well as made. Perhaps he is too overly burdened as celebrity. An' what to do to calm 'im out of this haughty acting? Sometimes one can not stop a conceit until the head is hatcheted. That drive (is) so powerful to keep going, else it is falsified by your doldrums an' fears of being insubstantial. Yes. Brought to glitter.... you feel you must shine.... But were you brought to him, he would see your dear face an' simple standing, and console himself with a repentance for this nature.

HESTOR: I?

VACLKULOR: You represent his better dreaming.

HESTOR: I.... I don't know if I can go?

VACLKULOR: There's more than money in it for you. There's a visit to the palace, an' a meeting with your friend.

HESTOR: But am I allowed? Should my mother accompany?—

VACLKULOR: No doubt that she is very busy today, as to let you play as usual near, about an' around Loresane's remote encampment. But I will give you twenty of what you picked up before, a whole bag's full earned. An' it is only an afternoon's farce, to bring the prophet back (to) some sense and acknowledge to himself once again that even you are better than he made. You are such a lad to 'im as could be named.... er....

HESTOR: Hestor!—

VACLKULOR: A savior.... Hestor. For believe me, he does drown in a mote otherwise. Or baste in a cauldron of contempt for being unwisely too clever is his design without *your* destiny to ensure.

HESTOR: Mine?

VACLKULOR: What is the recompense for losing a friend an' mentor? but to be distressed in your learning by example where fates are lent to haughtiness. You will grow to despise wisdom an' live with miserable shame for knowledge, knowing what it does to a person's character with unhealthy effect. Not much of a lesson to be taught a youngster of promise an' aim—ing, but he has arisen to a breach of familiarity without your assistance of recognition. A simple palliation you could provide, until he cures 'imself of this acting before nobles. One often breathes of strangling in rare atmospheres, for hope of recognizing your position an' climbing down— again.... into some safety. Withdraw you not softly from a cold stream or river, until you accustom yourself of the shaking an' take on a more proper behavior or attitude for the swim? Then he is a new fish struggling of acclimations.

HESTOR: Or acclamations, for may he be overwhelmed of your approvals for his visions, with such delirium to overproduce them. He is a hard worker, in his desperate way, often instructing me to be as diligent at least, to avoid his fall into a sordidness of survival with mean belittling. But with such a muscle as his released into.... what may seem glorious of an environment— he may explode of effort an' actualization of his thoughts to say, what is the reward of a trained practitioner honed to be consistently fruitful of result, as much as I may be in play for fun. His facility is easy for him, much as well. He tells me it is too much so, an' helps destroy his standing among the unappreciative. What then is talent without a gay time at it? That is a sad preaching, sir. An' would he have but gloom, then to see me it is relieved somewhat, as talents should be as freely felt, irresponsible an' unapologetic. That is his claim on my sight. An' to review the forest, in his mind, it is filled of this talent, a fact he says he forgets, within his drudgeries, until the thought is brought back to him. Well then, in leisure, he must lather of it an' lave of your accommodations for him.... I would make mischief of a pie alone, if 'abled an' allowed, its sweetness as cunning as its charm— to eat this treasure up.

VACLKULOR: We charm 'im? No, but of an opportunity to grab at uncouthly, he should have a child's hesitance. Yet I shall charm *you* with this (*untying a bag from a waist pocket or belt, and placing firmly on the table*), a whole pouch of gilded coin— the entirety of it! for you, to accompany me to the palace an' meet your friend. Take hold of it, an' toss these beans not. For it is not a gift, but an earning.... of talent. And it is not as natural, as before found, yet of my suggestion still. I need your service for this wealth— Count it! Play with this, an' bring your need.

HESTOR: My mother did take the latter, an' wondered how it was dropped (*coming over to pick up the bag*).... till I told her it was by Loresane's vicinity found. An' then she laughed of his visitors an' their lazy squandering of riches.... (*picking up the bag*) that could feed a season's hunger.... (*opening the bag to examine*) Here's many pretty traces on firm money.

VACLKULOR (*standing*): More (th)'an twenty pieces, as I recollect. If not I'll add to it. An' if more, it's added to. An' will ye come with me today, a respectable commissioner?

HESTOR: Yes, sir. It is a fair deliverance, an' I trust you.

VACLKULOR (*starting to lead him out*): This is fine. I'm to conduct the next session of the Commissioners, you know. You mete its head.

HESTOR: A very princely honor. Handle me to meet the Visgoth?

VACLKULOR: To meet with is an ambition consorting. If Loresane wishes to introduce you, I've no objection.

HESTOR: That will be a proud pleasure, if he remembers our friendship.

VACLKULOR: Though the ruler is rather still of head to see anyone, bedridden from an illness.

HESTOR (*as they are exiting from view*): ....Predicted by Loresane?

Scene II — *A dungeon-like enclosure.* **Loresane** *is in a long, sheeted wrap of robe, but sitting on a pile of hay, next to a lantern. There are no windows to the room. He is contemplative and worried, and seems to be shivering slightly from cold or cool temperatures. A* **guard** *enters, stirring* **Loresane** *nervously.*

GUARD (*coming over*): I'm taking away the light, herald, since you don't need it, with empathy of other visions, an' the invisible to our eyes. You are a wonder, though, of what to think about. There's much confusion to it, an' we had not thought of what you might do, to burn in here as if disgraces could be purified.

LORESANE: Why am I sequestered this way? Fed, feasted with compliments, bathed an' wrapped in cold silk spread, an' then thrown in here to spend an eternity, it seems. Have I been forgotten of?— Is it evening already?

GUARD: It's bright day of afternoon. Has not the lamp kept you company since morning?... (*picking up the lantern*) This was thought to be an appropriate space for your slumber, what you may be used to.

LORESANE: Then if it's day, why can't I go out into it?

GUARD: What need of you for foraging now? Some meaty bone will be tossed in here for your supper. All you have to do is think of images, with greasy hands.... or explanations to explain the future with the past. You need no presence in the sun to see. This magic satisfies, I'm told—

LORESANE: Not for a guest!

GUARD: ....*Guest*?!... Or whipping stock for thoughts and threads of ideology. Remain you still, here, and as captive as a prison can make.

LORESANE: I have the assurance of Commissioner Vaclkulor— a lord!— that I'm to be treated graciously. Yet it is drafty an' cool —

GUARD: But dry.... And without much pestilence. Vaclkulor is out, and the other commissioners are in a quandary as for much of what to do. But as for you, your only need is to dream of tales to tell—

LORESANE: They are not stories, but states of insight an' perception!

GUARD: Too often babbled, it is argued. What is your goal? You should have whispered instead of shouted.

LORESANE: I did neither, but gave my mind to you calmly an' loud enough for interpretation. An' I was told I am remarkable.

GUARD: That was more for your feeding to remark of.

LORESANE: Your foods intoxicated my arrogance, perhaps, but not to be treated as a horse of burdening, or a mule of carriage. I was not gruff to my benefactors. They seemed pleased of me enough, to ply with compliments an' pour more wine to my cup.

GUARD: Have you a headache—?

LORESANE: No!

GUARD: —That would be unforgivable. Last night was one of delicious morsels and mooring of the moods towards.... respectful and reserved revelry, with your unusual company. But now the dining 's made, and you are down to your activities.... enhanced and improved with isolation? What's to be suggested for you else, but proof.... with an enlightenment to predict.... correctly. Then you must prove yourself. And that is the cause of much scurrying about the palace, as the commissioners meld for an important meeting. You are a bump, in this clean comfort. And we have washed you, or some other servants have. Yet you remain a lumpy oddity to manage. What can you portend of your own security?

LORESANE: I am to be destroyed— I've always known it.

GUARD: That's not much knowledge. We all give up our airs in time.

LORESANE: Are you resentful of me, that I am especial, an' so may be treated this cruelly? An' am I to be walked like a dog through evening passages? An' where to wash an' deliberate matters more personal than my thoughts?

GUARD: This, as were an instrument be kept as animal, is your provisioning to be prepared. How can one astound with error? You should be better bred, to sleep in comfortable beds and stroll through gardens piercing of your person throughout the daylight, with onlookers hanging by your sighs and sights. That is to be proud of what you do, and what you have done. But tell me why my foot hurts (an') when, since I will walk awhile, carrying things.... And this lamp out will make your views envious, to an aura of the imaginary lights perceived like strokes of a brush upon a painting's canvas.... that draw to show yourself imprisoning. And we are not jealous of that, nor your consternations.

LORESANE: Vaclkulor will have you punished for this. It's against his personal bond to me— We are friends.... for a long time, an' he will hate this embarrassment. When I first entered this.... chamber, I did not know what to expect, being led to a gravity. For I know what a bed is, an' there is none! I'm not a country simpleton. I have grown into myself, not out of, towards a hut with loosened bales. But I am more uncomfortable here than the worst of my previous nights, the coldest an' most crass of nature's windswept crimes. For this is a distress of the miserable, an' I demand less sufferance for the foreboding. I want to be outside to see the sky, tied to a rope if necessary, but within a panorama of freedom an' changing views with the turn of the head, as I could do at the humblest of homes an' house. An' through day's vista seems some aspect terrible for me, that you detain as if in a pipe to distance my transit. I want more of a welcome than this, to these hours. Take me out, sir, an' have a justice for it, with my gratitude an' praise. Or what orders have you to restrain me, an' by whom?

GUARD: You have more of yourself, with this imprisoning to do, by your so called powers of enlightenment. Your speeches challenge gods' ears, as well as our own mortal fallibilities to entertain them as possibly to be correct, and frailties of judgment over all about your specious messages, whereat you be as much an adherent to our horrors as falsehoods and needless grating. And I, a man, blame you for our disturbance, since you detain yourself with reasonings— otherworldly and empirical of evil practices. But who orders you here? if not the day itself, and this same day indeed. Last night was festering of portentousness and vile decrees. Know not what you have said? It is with blemish for your sight to spread.

LORESANE: I can only emote to what I sense. Spare me, servant, for I am made rich an' weakened. An' this is not a good time to be enfeebled within wealth an' privilege. Oh! for the woods to lie in an' die of. There is more bounty in the grasses' rustic rustle to see an' hear than in these stern walls of protection from the breezes an' gentle swaying of the winds. What, then, commands my stay? I have only this cloth to give you, that kindles my internal warmth. An' it is of a fine material. But lose me bare an' let me out. An' I will roam to eat the bark off trees in preference to this entrapment an' detention for no good purpose.

GUARD: Can you not say you have been warned, by your own presentiments?! This deed of yours is dire, coming here—

LORESANE: I was made convinced of its necessity, as I may be applied to lessen woes of mental torment.

GUARD: And now you are the subject of them, and subjected to them. A mean creature you are—

LORESANE: Taunt me less!... Have you come for a prediction—

GUARD: I've come for this lamp!

LORESANE: —about yourself? I feel no orison towards that end, yet would I pray for myself. Free me! an' I will bless you with all that I am, your scrape to be, your scratch to tend with man. I am so afraid of these resultant ties to regal straw.

GUARD: That you are in hay and height, the lesser of your weens! Where is your ghost to speak with and consult?!

LORESANE: I am my own portrayal of the mind, in this fair body. I can not be privy to the disastrous encroaching of my spirit.

GUARD: Then learn of lessons sent, as you may hear them. (*while shuttering the lantern*) The king is dead—!

LORESANE: Oh! Darkness!

GUARD: —You did not seem to predict this so early. And there is much scrambling about the palace, to know what to do about this. But there is blame with it for you, since what you say is patent for our ears. Or would you stroll with him, he that sees no day? That you did claim. Yet, the Visgoth rules— but who? Come did we learn of these events, we had no time for you to discern, or disinter, until some thoughts were made, by a few commissioners, and.... the Queen. You must prove yourself, in chains need be— and eating hay! For you did say Vincent Visgoth would awaken and walk again more refreshed than ever before. And thus he does, as a spirit enabled and freed from all miseries and sores. But what does your ghost tell you now?! (*un-shuttering the lantern*) for who the living Visgoth is. That is demanded of your prowess to foresee.

LORESANE: Mercy for me, man! I am hounded here, an' by a queen who must be thoroughly recalcitrant towards my well-being. For I am principled to be resented always, when fate destroys a love, or a provider of estate an' power.

GUARD: She would have you bark for bark to strip; that's for sure, against the two queens quarreling.

LORESANE: I have no mind to this— except to save myself. But whom have I offended?!

GUARD (*shuttering the lantern*): This day's entreaty to be bold, with your pronouncements, therewith you wound yourself to say.... on which of witches lies the fiefdom's bed. There is no sanctity for error here. And what does light tell you?—

LORESANE: Superstition!... Superstition, in this grim malaise of malady, an' gray decadence of hindsight.... an' black pitch of terror. I see.... no resolution for me. An' all that is starry shed tears to shroud my mass. This is, then, a cloudy night. An' what of you, sir?!... Have you brought some shaft or shard of execution? to bring mercy to my misery— before my head is torn off, an' limbs cleaved for various directions tossed? That is a picture of my body sweltering of heat, an' pustulation during the decay. Yet see I not the wrong way said as told. I am never in error. I am of error placed.

GUARD: Then be you so (*un-shuttering*), a maladaptation made for man's sight. I am a weapon unto myself, and for this lamp to steal, and leave you thinking.... on your crude bed, that you are solitary of a soul, not with angels' wings for succor. But by your own mind made will you deliver yourself to a team of knives for carving.

LORESANE: I am no man to this, but a cavity of plight. So gather warmth for me, sir, the pitiable. I *would* have started fires, had I foreseen this angst and fear.

GUARD: You are a disgrace of miracle. (*starting to exit*)—

LORESANE: The fault is yours!... You do not protect me!

GUARD: Lord Vaclkulor is away.... He will see to your deportment when he returns, how you behave in crypts of belief and astonishment for others to ponder of strange bodies.... licked of leaves as (if) relics—

LORESANE: Leave me some light!

GUARD: ....But I have given you some insight, into these occurrences at the palace royal.... And it does interest me, how the remarkable can spoil themselves, and give me grave worry for some talents.... Yet must I make a procession of pallbearers.... protecting, with hardly a care of your ghost's chastity. (*exits, leaving the room in darkness*)

LORESANE (*after a pause*): Cruel!... serpent of my tongue! Is done this day for little comfort, nor caring of a guest abandoned to the weal an' *waie* of wraths. Here cries pain of panic.... to be truthful. But with visions dreaded are always most the faces of the seekers. An' this seer is forlorn of worth. It is not fair, to be honest of one's emotions an' foundations of thought an' perception, to be honored with this depth of light an' hue that bespeaks of a tone deceased. Such deprivations are meant only for the despised. An' I am he, ensconced of *malheur* for my views. Yet imperturbable a sleep is sight of vast awakening to chance an' changes due. For if with turns an' twists one struggles to be through, then with the cautions tossed I am repentant for none, an' repeal not a notion of my visions, dreams, an' amelioration of these threats foaming. How is the quiet view more simple than my plight? I.... am.... spoken to, an' made to feel somewhat distinguished in my trade of ponderous writhe.... even when wrong. So how is this space dented more an' lighted less? my quarrel quiescent, like a brook's drool is soft an' welcoming of thought.... or seemly contemplation. I am.... as well captured by my art.... as by their much determined distorting of it. An' that fact!... makes this a good place to be, under these circumstances. For mixed alone I am not led to candor's cajoling choler. I am not guided; thus, what is to see? A same sense surrounds this die, that sends me smoothed an' soothed for a flailing— an' a fraying an' a flaying. This is the better time. No more for independence am I taut. I can not handle myself well, an' have made light of life.... or its achievements impoverishing, wasting of me— as if to struggle with, resisting this gift. What shall be done for steps now? Lie! Lie! But direct me more gently, this cause of dispatch, this distemperate toll I pay to live.... An' the straw is very soft here, for straw. It is rich straw. I suppose this darkness is of the highest quality available. Then I am treated as a guest of some standing considerable. Now should I simply sleep, to be found dead with

thanks, the hospitable ingratiation fully accepted or acknowledged? But that is only.... my coldness to this purge of mind, on a verge of awakening to the crutch.... or the cane. With injury of my psyche, I've learned for pondering of smoky umbra. Oh! Soft velvet that embraces me.... I learn that I've seen nothing. An' now this land is comfortable. This bed is regal. This state is sublime, or surreal or subliminal or surlily senescent. An' I may lie here, tethered to a fantastic tranquility— No images allowed! Please!... With truth, they are not sought, they are not spending of me, they are not dancing 'round the resignation. I become.... more crippled with this freedom of the mind. So conduct.... of weighty mourning. I concur. I confer.... I concede. What is a sad life to be sown, with me? Then take, an' push me along, an' direct.... the pupation of my becoming.... a soft an' subtle usury of destiny's renditions. I do pay more for fates.... than the fated are allowed to see.... What climbs up here to me? Is it the ghost of a prince past? Come pounce on a pomade of fruitful intuitions. I deserve ye!... an' your warmth. (*A female **attendant** enters.*)

ATTENDANT: My! But you're like to be shackled to a cube of frost fall. (*un-shuttering a lantern*) Do you like it in here, for a preference to exist with shivers? But then, your type are particular for many.... environments. I've heard the prophets sleep in waters, often.

LORESANE: Maid?... What comes of me to bring such sight?

ATTENDANT: A refurbishing of your lamp, with oils, the slow mixture. But it is a better day outside than this.

LORESANE: What hangs that's heavy over it? An atmosphere most solemn, to have me kept inside.

ATTENDANT: But do you brunch communally? or will be served in here even?

LORESANE: I am locked within, an' though not to leave, as should.... as could.

ATTENDANT: ....To wash up and refreshen? I was told you merely wanted lamp. Are you not hungry?

LORESANE: But my leisure is restrained to this chamber of hay.

ATTENDANT: How so? with the day—

LORESANE: By the door!

ATTENDANT: ....It is closed, for your caring, but not bolted. Did you not even try it? You are wondrous of mindset. Entrances and portals must open up for you with the immediacy of a wish to think so, while dreaming. Yet for real, within these more practical spheres of realm, you must be tended as a child— or by, to reach for your gentilities taught, a sensitive and precious nature needing some assistance with what would be a rut for others and most, more normal, less nascent.

LORESANE: I am.... (*standing*) becoming more normal.

ATTENDANT: All of your physics is in the mind.

LORESANE: We are famished, my ghostly apparitions an' I, whereby we have concourse made our spiritual disputations to resolve, or make some order with our understanding of events. Then lead me straight to some dining, maid. An' some company— An' the sun!

ATTENDANT: If you wish. And the lamp will be blown out, not to waste its fuel. But this supply will be left for you, tonight.

LORESANE: I'd prefer a proper bed, in a spacious room decked (out) with as fine a finery of cloth as I've been given to wear.

ATTENDANT: The hay is not acceptable? We were told it calms your suitability of persistence to live meagerly, as like a religious tenet to follow, or condition to ascribe to.

LORESANE: It does. An' I have blessed the straw. But now I want linens an' the lovely.... of design, to look at an' feel. For great grace be at this home applied. (*coming towards her*)

ATTENDANT: Ah. Then you have finished your supplications. Are they as priestly as for priests? We're not sure about it, how you manage the divine principles.

LORESANE: With a relief from all that's tensioning, maid. (*standing by her*) Are there particular assemblies gathered, this grave day?

ATTENDANT: The palace is a lively place at all times, herald. And you have been asked about, for some conferences to join. But no one really wished to disturb your.... presentiment-ationing, if this is the unusual way you use.... Can you tell of me a charm, as by my hand to hold? (*He takes her hand in both of his.*) Some wishes, I have made, are not clearly to be done for me. Yet is this only a confusion, or a contraction against? The hopes are high.... and too ambitious, I'm afraid.

LORESANE: Oh, to know the truth, I'll tell. But would you want to know all, that which is as bad as good, or as fortunate as disastrous? One should hold out for a spectacle of conscience to be pleased with. An' then I claim that you are pretty to remain—

ATTENDANT: Oh, that is great!

LORESANE: —for a long time. Not may we err on hardships long, much disappointment cures the skin an' character of its blotches with a youthful disposition of resistance.

ATTENDANT: Then tell me only the good.

LORESANE: You radiate for me this good saving rescue, an' can not help but be entrenched in gratitudes an' being fortunate.

ATTENDANT: This is a good sense for me to aspire of. Good— better— best!... is your direction.

LORESANE: The such lifts to crawl, an' crawls to fly, but of a direction. It is as perfect as a clump of grapes is perfect, ripening in the sunlight. Whereof, as our spirits lend themselves to arise, no need to search for more perfection than our tastes. An' you will assume the fortitude of being loved. That is the highest treasure reached, an' you should contest no further within doubts. Though difficulties will always present themselves to welcome their occasions; do not worship these. For that is life itself in contrasts seen, an' not endowments lent by gods for pitying, one's self or others.

What of this day that lets it stand me, maid? I had thought to die in here, so heavy were my researches internal, an' thick of a lightness dark— without that lamp.

ATTENDANT (*retracting her hand to clasp the lantern more*): Was only gone but for a sec'(ond).

LORESANE: I mean.... my eyes were closed (for) much longer. One does not see your thoughts in dreams, but rather mock them.

ATTENDANT: The day is starkly beautiful. And the house is very busy with rumor.

LORESANE: Rumor.

ATTENDANT: Rumor and presumption.... that our Visgoth has lost his final battle, since he has not been bathed today as usual— at the usual time of morn for healing, or what was one expected. Physicians are with.... what remains to be cleansed. And they are as reticent as hidden, with the greatest patient being. And yet the royal family is as well unseen: the Queen Mother, Queen, and heir to the throne. They all have vigil around the Visgoth. The latest word is that his brothers have been summoned. The air is very grave, but not yet somber, not without all the details extracted and confirmed, or made for us aware as much to know that's proper and exact, we servants, guards, and hirelings for the management of this palace and its grounds. Some of the respiting commissioners say the situation makes for an appropriate test of your powers, and they are anxious for your opinions on the matter.

LORESANE: Better I should run with it, in full blush an' bloom, an' embarrassment to cheeks of my fruitions of thought.

ATTENDANT: They make more haste, though, for a quorum to compose and conduct themselves of. That seems to be their most immediate of pressures shared. Most are out searching for others.

LORESANE: An' I am a noteworthy guest left to be entertained, left to stay— or leave.

ATTENDANT: It is not known how particular you wish to be, of your visiting.

LORESANE: I.... take of food, my charm, an' bravery. For I have been reborn, an' of the simplicities to notice an' admire. How foolish we become, within complexities to consider. They bring up the subjections an' prejudices as retched a flood of the derogatory an' deprecating of imagery an' imagination. That is why meditations are so important, to restore one's self a room of healthier visions. An' I am so thankful to see you— now that the purge has past— with myself in review as a child worrying, that I will dare to stay longer, an' take a garden's light without being frightened— for the shivers emulate fright— an' be indulgent on your care an' attention. I know my day now better. Best to bear these honors with a courage.

ATTENDANT: It's good that you haven't shame of your privations.

LORESANE: Oh, I'm never ashamed of my attributes, though they can lead me into troubles an' disagreeable entanglements.

ATTENDANT: Then I will lead you to the washing basins, so

you may prepare yourself for supper.

LORESANE (*as they are exiting*): Do steer me forthrightly, an' I will give you lessons on your abilities. (*The scene ends darkened.*)

Scene III — ***Morliff*** *arrives at a royal chamber parlor, where **Halyce** is admiring some wall murals. When **Halyce** notices him, he sits abruptly in a nearby armchair, though maintaining a dignity of indifference to defiance.*

MORLIFF (*observing with a slight indignation*): Has my brother been around, Count Halyce?

HALYCE: ....The one who sleeps?—

MORLIFF: Caton! They won't even let me in to see the Visgoth.

HALYCE: I've not seen that fine man, an' promising pupil of the fiefdom.... An' nor should you be admitted to see its ruler until after an official meeting of the Commissions is held.... to ask for your service an' labors of allegiance.

MORLIFF: I just about rule already, Halyce, handling many imperatives by his order and with his approval— an' with the respect of most, even the majority of our subjects.

HALYCE: This is true—

MORLIFF: But the task is difficult, an' somewhat frustrating, without a clear officiation (for me) from you and others, to make some changes that need to be made, an' have them as permanent stamp of law to be obeyed and followed.

HALYCE: You may clean up some stalls an' sties. But you should certainly not make *law*, altering what is, an' without the guidance of the owners of these fiefs—

MORLIFF: *Bauch!* Things will be as they must, as I see their necessity to be.

HALYCE: It is too precipitous an impetuosity of privilege. Such actions upset true achievers, earners of their gifts, an' the strength of this collective authority. You do not understand the principle of being reasonable with giants. An' you can not so easily leap to become one yourself, no matter how popular. You have not even been elected to be a commissioner.

MORLIFF: Neither has the boy Fabel— Oh! But you elect yourselves, don't you?!

HALYCE: Under guidance of the Visgoth. We own our fiefs!

MORLIFF: An' when Vaclkulor presides over your next meeting, he will allow for some proper changes to be made in this regard. Because it's stupid to give me powers without the right to use them, that they are only implied for now. I'm not hungry for them, Halyce; but our regional organization is in crisis, and I must take these steps.... to undo Vincent's. Many blame him for a minor plague. An' it is a small matter, really, harming only a few. But he promotes the anguish in such a way as to make condolences impossible an' magnify the scope of grievances, disturbing more of you an' your subjects than need be. Well, his illness fells his pride, an' I must bring corrections to everything he's done lately. That is why I

will be made the Visgoth effectively, during his total incapacitation. An' I wish him a good recovery— I love him! as a brother should. But if that's not to be, then have no doubt that *I* will be. An' you should rally towards my support for your own interests and benefits. Because this is more than (a) mere privilege I'm taking, but a responsibility of wolves as cattle to constrain, against their might an' strong opinions.... I've heard of some decadence, towards your plans, where you seem to boast of overthrowing me— Not possible, Halyce! You can not assume control—

HALYCE: I?

MORLIFF: —Not ever, unless I die from giggling. The other commissioners will never support this.... usurpation or insurgency. It is too absurd and ridiculous to even condone its humor.

HALYCE: This is not a level charge, nor an accusation of merit, Morliff. I have never had any intentions of becoming.... what is Visgoth, a poor manufacturing for rich wounds indeed! An' I don't desire the position more than myself to own an' look after for my good tides of command an' protection. But what you *are* distresses me, without much matter to scoop an' score on rock! a deft purpose. You haven't the skill to override my concerns; that I will admit— an' to you barefaced. There's not a Visgoth to you that's for me. So there is the umbrage brought forth an' out of shade. I don't dislike you more than your making, an' less than your knavery.

MORLIFF (*pointing quite abruptly at **Halyce**, with an exaggerated deliberateness*): On that brow *there* is the chicanery! All expect you to disrupt us, as you play for my youthful mother, whose beauty is sensed as sins of the father you'd try to replace— This won't be done! (*banging his fists together*) I'll stab at the toad first!

HALYCE (*jumping out of his seat*): What more for my desires are you blistering?!

MORLIFF: It is like a secret yelled— an' waiting to be slain, that you would dare work on us, to restore this kingdom as its king, with our Queen Mother as the Queen! And sick be you! more than Vincent's biting, bitten bruises to even think so. That is a foul fog you evoke, as if to obscure reason with a cruel passion. An' what you need is clearly a devil's cure— straight to the heart.... some blade or sharpness, to show my determination against such debauchery of thought.

HALYCE: What you predict is hoarsely said, through a mixture of confusion an' conceit. An' for this lame lucidity you'd want *me* to support *you*?! as a Visgoth to become? Then make your brains brown water, out of this gray gripe. Yet so I tell you fiercely an' proudly, that I have *the* affection for your mother— but not for this suzerain to rule over. By my own contentions with her are we to be mated an' paired. An' far away from here, from this cudgeling that demeans her, are we bound for bonding. But I have been romantic an' discreet about this, until now. What matter yet! I've told her to say this to the Visgoth himself— An' you are not he! for explanations to be made to. I favor your brother over you, as many know. An' even to the one who lies, he is preferable. But this matter is sensitive, for a father to become— to Caton's ears! You lack this sensitivity, which he has in abundance, to handle matters delicately. You would crack the glass of this fiefdom— smash it! with your heavy handling, is my complaint against you. Yours is a clever crudeness only. An' you needn't complain against me, for all of this love. I'll never adjoin myself to your ruling, is more the point to consider of your harbingering than your mother's fidelity.... to ourselves as a couple.

MORLIFF: This is the beast in performance! An' you would delude Caton too, for favoritisms, as you may have done her. This leads me for a sword to handle, or an ax— as despicable as your deceit to swing, an' lop-off-paws!—

HALYCE: I've nothing against the battle except your childishness. You are not a maker of nations, nor a manager of one with any competence that I can foretell. Becoming father angers you, an' enrages me of your surly, sour reprehension.

MORLIFF: I am to be; an' that is better not for you to cause, your being all the worst. I would deny it heartily, unless you were subservient, an' submissive to my candor to be helpful.... to me an' this fiefdom.

HALYCE: You croak crookedly—!

MORLIFF: Halyce!... You are intolerant of masters—

HALYCE: Feed you the cup!

MORLIFF: —You try to abuse me as much as Vincent himself, with rants of hatred. An' you dishonor the Queen Mother, with another sort of loathing, the kind associative to the wart-ing of skin.

HALYCE: This is your toxicity! You are full of falsities to pour, into this drying troth an' come of trust among us drank from, as be passed by the commissioners of ownership. You can not handle being Visgoth well, an' yet seek to remove him, he who is unwell. Well! Well! Drown yourself in a wealth of incompetence that leads to evil, can only serve its purpose— An' in a cauldron bathe! For you do not deceive yourself— nor brothers, Vincent, to whom you're avowed.... an' Caton, from which you swerve to be mean an' lowly.

MORLIFF: I?... Were I? I am being responsible, an' mean to no one who's not to hear derisively my aims, for our needs to come. Let that insight spread amongst us— We must get busy, improving all our realms, cleaning them of pestilences animate or personable. An' if consolidation of this fiefdom becomes of a necessity to happen, then let that manner be fashioned with my caring hands, than Caton's reluctant ones, or through the Visgoth's dwindling consciousness of these estates. We must be built of friends, an' not dividers. We must embrace ourselves as one protectorate, an' not as antagonists swirling about to steal gains from each other, or displace seats at a conference, away from a table standing with dread and dour emotions mixed upon the Visgoth seeing, his shaky recline as to sit alone!... Weakened by ye all! with rejection an' disobedience of thought, a thorough lack of compassion. An' you would not even stand as long, but ran out, with a dismission that brought him to tears, some others said. Then you are not the savior to be so frustrated an' repulsive. But I do bring myself, to lessen these agonies, as a caring relative should for one who is in a state.... already destroyed, yet more for damage to make. I will clean the rooms an' thrones— an' not take your spite about it softly!

HALYCE: Dare tell it all, I'm told of the ambition, which surmounts purpose an' probity. But, were I to care if not remains a misplaced subjection. I have my own estate of land, a province re-

gal, grand, an' clean of the debaucheries you might promote, purely paced away from the brothers' deviousness, indecorous treads an' trending. Yet only Caton is fine enough a person, among you three an' a half, to allow my lending of the propriety of my regions an' subjects. Their welfare makes my own, an' I'll not have you walk upon that earth an' through those valleys an' villages as a ruler of the same. I won't let this happen, though related you may become.

MORLIFF: Is it so vile to be prepared, and to assume with wanting some authorities, the wanting of hands loosening their grips an' grasps? I only take the necessary steps to maintain our collection of fiefs under a tractable an' responsible management, while you make curd of milk for your creamy cheeses. Your endeavoring will not be disrupted by my forces—

HALYCE: Yours!

MORLIFF: — unless they anger me to most of their prevention, to swear with bitterness of pain against them. Remain a gentleman in place, sir!... within the row, an' of a rank generous an' devoted to our associations. But do not try upon my sensitivities to heighten — with a snarling of fingers within dear hairs, or crass plucking of my strings to sound, as were an instrument made distant from my control. The music, from your hands, will become discordant an' unappealing— in many ways, not to detract from my influences— in any way.

HALYCE: You are too thoroughly presumptuous to have a chin! much less forces. For you are too aggressive to be kind, which matters less to me than mine to protect. An' so you hide your severities of mind an' intention with cruel an' coarse trickery, which is a disgrace! Then fight amongst yourselves within this seedy kingdom. I will extract the better gold than gilded sloth in sluice, from the running ores released, an' make a better spread of fortune out of high quality, if not your fraternal an' familial mores can tame yourselves away from the most fractious of criminalities and dastardly of evils, if not a healer being.

MORLIFF: Healer?... Herald? What can you hardly mean? A herald has been brought here for Vincent, I've been told. I did suggest to him he leave the prophesying to such.... minions. So with approval by me are these fallacies employed, these tricks you claim, so that the Visgoth is not more distracted with 'is visioning, an' can concentrate better on the hard realities of kingdom an' fiefdom to rule an' lord. That this be too late, or only superfluous a tactic, can not be said— yet.... not to me, apparently. But it comes not with my importance. I didn't conscript the talker.

HALYCE: That is not the terror, for what Vaclkulor brought—

MORLIFF: Yes.

HALYCE: —in sight of future. What is commanded is always obeyed, that way, if the sentences are true, though not with a justice made.... beyond curiosity. They are friends, an' rather oppositely designed. But Vincent is your father's son. An' Pontal tells me how you drank from the Visgoth's chalice, or faked to, no doubt poisoning the wine, during the last meeting of the Commissioners, to which you were never invited—

MORLIFF: Poisoned! You are.... utterly wrong—!

HALYCE: He dies, does he not?! An' not from walking can such a spell be made— so.... opportunely—

MORLIFF: *You* impelled me to rush to him! to come to his assistance against your.... confining conference. An' I could not have prepared such a toxin before hand—

HALYCE: You probably grow it—

MORLIFF: —Such a contaminate—

HALYCE: —eluding from the skin— of chin!

MORLIFF: —never fell into his cup from my handling.... I only *sniffed* its drink, to see if it could have been tainted already, to cause his withstanding of you more distress. You are mad to think so, and have malice for me—

HALYCE: Your doctoring! Your ambition is so great as to find him death! to end his inglorious decline an' loss of faculty, facility — an' purpose for his fervor. Well, he is the Visgoth, was a great an' dedicated warrior, a good fighter for our aims. An' to see him weeping now, as could a hen losing her chicks, can be telling to many souls.... to end a living decomposition of once so able an' healthy a leader an' relative. But such a man takes his chances to become.... a ruler. He earns these dangers, an' treacheries to live through.... or succumb of— loving treatments, since he has lovingly killed many himself. An' not with deception, but with reception of gashing injuries has he made his stature or worthiness. I would disqualify you totally, with such a game of impulses an' impulsiveness. An' Caton remains more gentle an' ethical, an' composed with a stronger kind of courage to resist such infantile-ly simple temptations. But have as you will. You won't be doping the wells on my properties—

MORLIFF: A plague does this to Vincent, for his foolishness of vanity, not my need to become Visgoth. This merely develops, and I attend to the problem. Because I have no reluctance to the undertaking, you make up myths!... An' if you have any honor at all, an' with no substance of substantiation to prove this terrible an' hurting claim, then you will keep this misinterpretation of facts cloistered up in your heavy breast.

HALYCE: Oh, who would not conclude the same, knowing these details? But Pontal favors you. So does Vaclkulor. Most of the commissioners do. This makes these events, I suppose, an accepted mystery of disease, that they are relieved with your ascendancy. As for myself, with terms to come (to) I am relieved, with moves to make that I have contemplated for some time, an' have planned out with a marriage to beatify, from which I can not be dissuaded— by you.... but only through her insistence, genuinely felt, of refusal. An' that is not likely, despite your.... impetuosity an' headstrong nature.... or conviction towards shame.

MORLIFF: You mark me so incorrectly, Halyce. I have nothing but gratitude for the Visgoth, an' would never conspire to such grandiose injury as murder. An' with your reasoning it would be an imbecilic attempt. If this is how you think of me, then that is the greatest harm to my character. An' I refuse to believe my supporters could hold such a view as yours. But then your vision is certainly suspect in several matters.

HALYCE: I can doubt that, an' consider you not so much a cad as a cadaver's cravenness. For a true leader would simply seize

control upon spotting a weakness, in a would-be opponent or obstacle to 'is drive. You are not that, of that type, but try to mollify with inexperience, an' to disastrous effect, to close a waning Visgoth's era by shortening his disconcerting decline. It is a poor palliative for power to obtain, hiding this gravity under disease, or feigning that it be so. But that is how the amateur mind works, with a contemptible nature, counting on delusions an' wasteful, squandering— scandalous cunning to ply upon perceived opportunities, instead of the straightforwardness of an absolute assertion to claims.

MORLIFF: Amateur?!

HALYCE: So I don't fault you as much as your privilege to try such a path. This makes for the environment of your rearing an' heady state of mind. An' I'll have none of that to deal with, if one out of a few goes wrong. There is no perfection in life, but the deed accomplished.

MORLIFF: I am of the regal flesh, an' can be expected to persist towards this goal, a healing of the fiefdom for its repair an' preservation— of ideology , at least— than any other wouldst be in my position. An' for this caring you would malign me with dreamt up, possibly wished for falsehoods. Well, it is *your* deviousness that matters to you. An' I shall stymie its attempts. My backing is sound. An' my words will be rung as commands— out of necessity, the peal for a conservative transition over a statutory one, to keep these associations of might sound an' reassured. Your disruptive leaps out of order will not be tolerated, an' you will only be laughed at for trying.

HALYCE: It is I who will tolerate your dance, as could be the backdrop to a forest's scenery of obfuscations left to the silly dallying within this kingdom. An' I tell you bluntly, sir, a son of softness an' serrations irate: If ever you try to claim my land dependent upon your oversight— there will be headache! for you.... greater than can be caused by a spiked drink or a drugged dourness. The Visgoth's kingdom is lost to its impieties an' stains of the beastly, not to promote its better personages.

MORLIFF: Let me see this clearly. You actually do suggest to make.... a severance from the suzerain?

HALYCE: Aye!... if you're to be the overlord, or weaker.

MORLIFF: This is more serious.... than spoilage not to obtain the lead. But then you would try to control all administration, an' spoil things— No! I would implore you not to make such a move.

HALYCE: It's half-ly done, in here.

MORLIFF: A dissolution brings not much benefit for you.

HALYCE: I am discontented even now; an' that is sure to grow, with me an' others.

MORLIFF: Too sharp the candor, Halyce. You should rather commend my efforts.

HALYCE: A plague spreads.... as the Visgoth dies.

MORLIFF: The insanity will end, to mount mounds of the dead an' hills of heresies irradiated. No more of these mysterious tones of tome, self-structured from the invisible aura of conditioned thought divinely inspired or desecrating-ly imposed, will be said. An' people will not become annoyed or harried further with these hardships of conscience, being told what is their worth an' weight in pain an' grief for loved ones lost. That is most of my ploy, which you can clearly sympathize with.

HALYCE: Could be an admirable goal, as giving air its due for breaths to take. Seems obvious to me, to end a rake's upheaval of the fallen leaves.

MORLIFF: I'm to do this, with a patience of kindness. I have my mother's compassion in me, also.

HALYCE: ....That is the only thing that keeps you somewhat dear instead of—.... darning. Holes in breeches peered through, for all of this conjecturing of fates, makes ye saturated for some crime, but not the devil's moss for clothing. An' were I you, not twisted more upon this lap be sitting, squirming with discomfort an' distaste. I understand the waste, an' more this seat to take. But what is suitable detains you within a quagmire of failing ethics. I would cater to better thoughts than to blame the Visgoth.

MORLIFF: It's not my aim, but only the direction sent, with these responsibilities that fall.... to me.

HALYCE: Or with ye.

MORLIFF: Beneath a bog to lie, as if, unclean an' covered. There is depression.... with this dictatorship.

HALYCE: You notice all to practice— It's not kindly.... Then it's best to start kindly.

MORLIFF: *There* is Caton's privilege. I'll consider it if you consider it.

HALYCE: Such diplomacy makes a reconciliation possible, when amity is allowed to surface (up)'on our grins of countenance. For being wise, or having some intelligence, makes up for.... a few errors. An' I have never thought the kinship foolish or stupid within this particular realm. Else I could not be so charmed. (*reseating*) I am growing older, passing boundaries, an' have a need for colorful reflection.... on possible heirs, an' their relatedness.... to a munificence shared among many. But that I am touched with affection, at my age, tells of a warmth with friendships to need. The strife of the irascible distorts my.... portents an' posits. An' I am barely felt here, without those feelings reciprocated a thousand fold of my perceptions.

MORLIFF: I would say it's not true for your doubts. Your constituency an' contributions to the fiefdom are highly regarded, the loss of which would cause a great weakening of our strengths. Don't say you are unrepresented, because you are your own representative, an' must be heard as loudly as any other of the commissioners. As to controversies, debates, contradictions an' contrarinesses about which are argued during your gatherings, I can not tell how various people hold up of their persuasions an' influences, since I'm not present for these conferences. But it seems to me your disappointments are more with, an' most concerning of, the Visgoth— at present— over your membership an' hopefully confederates. So is it much of a daring, or impertinence?... to complain of me, or even to me? I wish to keep us whole an' happy. An' with

this aim— concessions can be made, to support one's.... feelings, an' supply a hope of satisfactions concurrent with the suzerain's solidity an' maintenance. That I would propose to struggle towards. Yet, how one is is his manner, to oppose or withstand, at least with courtly facility, objectives that run against his views or understanding. An' what you have against my brother— or for me— are merely illnesses, to me. They will be resolved by nature's kiss of kind; for the dead were of the dying, an' the hesitant promote a lethargy upon chances slept through. Do not see me as a master of the audacious, if your true quarrel is with Vincent. This makes up stories of suspicion an' sundry legends to have jest with.

HALYCE: I don't see you as a master of anything, certainly no office. But as to offenses.... yes. Suspicions run me wildly, without a mastery to blame on my conjectures— An' I'll not work to prove them, but to say.... suitabilities must develop, even if they may be learned. To serve a kingdom without a head is upsetting, as were a body dangling for use. An' indeed, my anger is at.... a visgothic quivering feared to see, since that is remiss of power an' as disturbing of conjectures. I wish, then, for some mastery produced. An' for my own I will provide, if not found here. Nor with much patience can I spend cajoling time an' brother.

MORLIFF: So with your will I will, to place the thought of decision to Caton, an' seek him out deliberately, since he should agree with all steps made. (*starts to exit*)

HALYCE: That is a work of walk, to tell him all.... Though he has been called, also.

MORLIFF: I describe not of suspicions nor supplanting, within this airiness to delve of deeds an' challenges for what might be correct in choosing, tasks, tests.... an' trials. (*Exits.*)

HALYCE (*to himself*): So then, be called up. I have revealed all of fire, with which to ponder burning thoughts. An' it is known (*standing*) that I may be a partitioner, to break of pacts an' promises. Yet, (this is) as I had wanted it be known. My Melinda scolds with silence otherwise. So let the bantering rain upon my head, that.... I am this dissatisfied, an' yet satisfied, for an honest curing of the breach. (*going up to a wall mural*) Consider things.... Dissent is a true ambition. An' even to tell of the monstrous implied, an' to the face, is a purity of consolation.... Why he pardons.... is perhaps of a form.... ennobling with courage of restraint. Relatedly.... may be some surge of valor woven in these kind, as for some colors hidden with refinement.

Scene IV — *A palace garden. **Loresane** is sitting on a bench, relaxing with enjoyment of the sun. **Pontal** is casually strolling nearby, in front of him, walking back and forth, though with some anxiousness. They talk with each other in this mode, for awhile.*

PONTAL: So you can predict only the best of events to conspire of a resolution for the Visgoth's grievances health-wise, Loresane?

LORESANE: ....Believe my tongue an' temper of the words. He will blossom into a new virility, with a freshness of ambition that will astound you.

PONTAL: But he's had ambition enough, an' of a kind all too distressing of late. It's what has gotten him into his debilitating condition, a grave-like state.... He's visited so many (graves).

LORESAME: ....Commissioner. What is to be poor?

PONTAL: ....To have a lack an' a need.... for substances vital and necessary.

LORESANE: No. That is being in dire straits. The king has riches, an' yet may seem that way today. *I* am poor.

PONTAL: So? To what purpose? Do you claim it's better being deficient of means?

LORESANE: You still confuse the issues, sir. I am flooded with visions, almost overwhelming at times. It is difficult to contain them, each one eliciting some nervous response that might be favorable or regrettable, upon sight of a subject's face merely, or prostrate as the king to face. He is my direct sovereign, an' acting principal for guidance through my daily habitations an' existence. I am fat of what is remarkable to sense of a person's future. Yet poverty defines me, an' controls my motions.

PONTAL: So.... to be poor is to be a seer? To be poor is to have eyes? You can't wish to remain shabbily endowed. You certainly seem to enjoy the dress we've given you.

LORESANE: A fine an' shimmering material, under this radiance. Every person of thought has some degree of insight. Critters too. This is independent of wealth.... The grass an' flowers around my shack are as poor as I (am), an' yet rich in growth an' terribly sufficient for their being. But I have never gleaned as much as an instant of thought or idea about their futures, nor much for the rodents I sometimes feed, inadvertently, and the animals that travel by and sometimes cross my threshold for a peering visit. For I am almost just as much disclaimed or disowned.

PONTAL: Oh. You are free.... Perhaps even a bit wild. That is being poor? To what relation of your graciousness is this laudatory?

LORESANE: I don't claim that being poor is. But does the king have more than anyone or anything, while under his current incapacities? He is as rich as nature can allow a living presence. An' he is as poor as an invalid with effort to persist. We are the same, this way. So being poor is merely being, an' not a lack of it in any way.

PONTAL: Blades of grass do not undergo suffering. Wilting an' withering, perhaps, but not suffering. Dying.... but I never hear them cry. So what, then, is being rich? the opposite of being poor.... The opposite of.... being? That would mean.... to have no future at all, to lose one's entity.

LORESANE: You come to lacks, again.

PONTAL: So the end of struggling makes you rich. You're not trying to craft a philosophical out for yourself, are you? should the Visgoth die.

LORESANE: ....You mean from his infirmity. No, this disease he has will not kill him. The affliction looks substantial, but I have seen his future, beaming to me with a desperation that it be insisted on an' announced. The energies of insight disdain the cruelties of disavowal through confusions, ignorance, or passivity of caring. No. To be rich is to abhor being. It is to become superfluous.

PONTAL: I don't mind having more than necessary, as security from periods of want.... Although I do admit that we all start out more or less poor an' defenseless, an' couched to be nurtured.

LORESANE: Being superfluous is not so much of an evil as it is restrictive to a myopia of views, because this condition results from a drive to ever increase one's standing of material attainment an' social high regard over the self you are born with, which is more glorious a foundation than any embellishment or ornamentation of character can provide.

PONTAL: You sound like a physician prescribing fear or austerity, over pursuits that allow one to become genuinely, an' quite effectively, more altruistic for the benefit of many associates, or even bare acquaintances. I know you live your way in the forests with difficulty, an' have done so for a long time. But tell me honestly if you enjoy this manner of life, or have been forced into it (*sitting on the bench, a distance from* **Loresane**), as I have been forced to be an overseer of men, an' as the Visgoth has been forced to be a leader of leaders. This is why I worry. Even a short lapse of such talent leads to dissolution an' squabbling towards an' into terrible, horrific confrontations an' disasters. Yet I believe the destiny of each person is predetermined as much as his character is shaped by circumstance an' accident. An' it's so tiring a fight of what's bound to be 'gainst what you stumble to become.

LORESANE: A boar is a boar, whether ravaging grasslands fiercely, in search of foods an' mating, or piked for broiling on a spit, hunted an' dismembered for a meaty meal of itself. Portions not withstanding greed, the gain is a loss, to be a predator an' prey, as we are each by fate ultimately consumed, for payment of our hungers.... The king is a very generous sovereign, an' does improve me tremendously from my hardships an' travails. But do I like my station made an' pounded into, does your question ask. It distorts my sincerities of pleasure, the conditions under which I usually live. For just to get by is an extreme an' yet mandatory accomplishment. An' laughing is rare without company to fool or be foolish with. Yet am I so without riches, I could remain the same in abundance, or at least lead guiltily my crippling urges. That is why being poor is juxtaposed to being happy. Unlike most, perhaps, I don't strive for happiness, but drink of its sweet nectar if an' when its hanging fruits have ripened near my grasps. In truth, though, I know not what to do but live an' satisfy.... needs an' knacks an' nag (ging)s of the heart.

PONTAL: Are you forced to do this, is the point. Can you not abandon your sight, an' pretenses to support its burdensome manifestations on your living, to become.... duller, more normal and capable of wealth to attain an' comfort to subscribe to?— I can not!... leave my charges until I am dead. One dies to be rid of them. But you, in general, have nothing, out there in the wildernesses, an' can fake your way back into civilization an' productive communal commerce.

LORESANE: This is like asking if one is forced to breathe. Should one live by faking not to? I am what I can do an' must. An' I live.... *by* a village. I'm not totally the outcast, or a nature lover to the point of singularity. I am.... solicited an' known.... an' known of, seen, an' pestered for opinions.... But not unique, or so solitary as you might suppose. There are friendships, as with Vaclkulor.... an' now yourself. An' hopefully the king will come to regard me favorably, as may the Queen encourage if my prophesies are abiding an' fructuous. Am I born to be myself? Yes.... Will I revert back to poverty, after this escapade, or try to avoid it?... Impoverishment could be a state of fasting of the body, after a feast— An' I am fasting now. Or as to sleep, after your waking chores have passed— This is a chore brought to. I was at first dubious, even onto it. But now I feel safe. What periods of time may lend to these divisions of existence I can not foresee for myself. There is no eye internal for one's fate. More carefully am I made than my mind. Of course I will swill of the nectar as long as I may, or until it becomes.... superfluous. The senses have their chores, staving off deprivations. The conscious body is in constant battle meant, 'gainst feeling bad.... I am not forced against my character to be, while struggling to survive.

PONTAL: Then it's all predetermined, as I said. How else can you see ahead unless it's all been already long drawn out, the passages of a life presented you! An' there is only a deception of the supernatural to perceive.

LORESANE: I suppose that may be an astute rationalization you make. There is a sensitivity that suborns misinterpretations, possessed by anyone. So this I try to leave for others, with their explanation of answers, messages an' replies to queries as being via superstitious methods or occultist principles. I simply say what I see, an' hear what is shown, as a discipline developed with my practice. That is why it may seem so mysterious, but this manner spares me from error an' mistaken thought. I will not judge the messages, or try to rationalize them through my prejudges an' analytical flaws. But you, with a much stouter muscle of intelligence an' mental ability, may do so, while I am only a conduit or piping tool in use.

PONTAL: Oh, my Loresane an' provincial visitor to this lush palace, I wouldst say ye are more smart than tempered of a tool that bores. To have such heavy eyelids grown smacks of a dedication to withstand much earthly criticism of your hazard. It's very brave of you. An' if matters should not turn out well, for anyone of importance concerning your advance to us on this testy, touchy occasion, I will be sympathetic for your welfare, to (at) least try an' rush you away from danger, an' your body from abrading punishments. Such as I may do, it depends on the timing of insults made, known, an' learned.

LORESANE: This be kind of you, Commissioner Pontal. I was scared, in my room of hay.... here, but brought back to myself with objective reflection on the events that must surround a life an' person, that it is for the better being what one has honestly tried to do despite all consequence. With this thought I have both privilege an' courage restored of my presence an' practice. How be the gentry of your domains, used to a restoration of the commonalities, I found my room somewhat chilled of splendor merely by unknowing the nature of its darkness an' the unfamiliarity of being caught up within it. (It) Was for a lantern kept, my patience an' my peace. An' for the same as brought, my release of worry. Then as the same receive the light, upon this bench to soak up all this radiance an' freedom, it is most splendid an accommodation. Worth all feeling, is it.... all sensations won an' issuing forth.

PONTAL: An' a more typical sleeping arrangement tonight, you mean— an' will have. For even dungeons an' prisons are less coarse than hay stacks.

LORESANE: I am familiar of the commonly cushioned bed. Nor is my hut such a barn kept as to become a manger for horses to eat at. I typically dine on grassy plains, an' snack only inside.... It is too humiliating to actually try to prepare a meal, to bring to table.

PONTAL: A fine constitution it should make, to put up with such diffidence of means— No.... My inhabitants take the Visgoth only with misgivings, though they be shallow. They are kept by me enfranchised of a proper state within the satisfactions of livelihood. An' as I not fight off others, within this fiefdom to remain, they stay relatively sound, contented, an' above all calm abashed the controversies heard of from this kingdom. We had naught to do of the sheep, I can guarantee you.

LORESANE: The sheep of possible contagion?

PONTAL: Aye.... But merely threaded them through our land, as a favor to a neighboring realm who wanted of their exotic handsomeness. I was suspicious of the woolly coats, myself, too deep of color an' texture to present an almost argyle-like arrangement contrary to nature's blending. So I forbade their capture for shepherding, along our outskirts to visit. But from whence this gift came is a total mystery. Rumors followed their strolling, from other places an' domains. Certainly some envies of ownership developed, among my subjects, for so unusual a breed. A measure of discontentedness threatened a sprouting, though I thought this to be trivial of circumstance. Never be goaded by the odd, is a precept I've often held. An' so did a many good heads within my spheres of authority. Yet children an' the infantile babble of prizes an' toys kept from them. So must I issue further constraint an' even one instance of punishment, amounting only to the retrieval from a family farm an' household of one ewe with her three lambs stolen from the herd, or induced away somehow, with sweetened straw. The extracting magistrate told me one child of the farm cried miserably, at the loss of these.... pets. This made me all the more disheartened of the occupation of these wild foragers. But I did amends to the family, making sure their impoverishment, the apparent cause of the theft an' crossing of my decrees, was remedied sufficiently to end tears of disobedience. I was nervous of these sheep, with the instinct of a guardian. Then a neighboring potentate gained notice that I had no proprietary interest in the animals, an' asked that they be led to his borders, to cross over into greener pasture lands at those points, a vanity to imagine so. As a courtesy between rulers, then, I allowed the migration to be continued, directed towards him. An' I've not heard of any complaints to result from this generosity, not by my beneficiary. It seems only when these ruminants reached the Visgoth's kingdom did some few misfortunes start. I assume this is due to his subjects' attitude for whole-scale capturing of the animals, greedily, to end their traversing natures, or husband out those inclinations.... An' also their unusual custom of penning up the beasts so close to their domiciles of sleep an' eating, raising of children an' nursing of infants— as insurance against a rash of municipal poaching, with overcrowding of households— have they paid some kind of epizootic price. But the sheep roam on more freely now, as I understand it. This is Morliff's doing.

LORESANE: That seems fairly wise.

PONTAL: Along with a general cleaning up of rural estates, Morliff is handling the problem, limiting its spread to isolates that heal.

LORESANE: An' through the obvious measures of sanitation and more careful homesteading, these occasions of infestation an' blight pass. That is easy to predict. When I learned of the sad deaths of infants, in numbers mysteriously growing, I was glad not be thought to be a cause of this, with my relative remoteness from the bundles of family life. Why, I even heard rumors of a few men wishing to adopt my lifestyle, their marriages destroyed through the loss of an only child. But nothing's come of this, as far as I know. Must have been only exaggerations of talk with conjecturing. Such deep mishaps strengthen a true couple.

PONTAL: I should think so, Loresane. Most people haven't your skill of predicting.... correctly. (*A sound of conversation is mutedly heard.*) Persons gather at the recuperative garden copula and ad-join-ment. I think they might be the physicians.

LORESANE: Done with the Visgoth?

PONTAL: Or out to breathe in some fresh air. They've attended him for some hours, as has his family a strict vigil kept, by his bed with great concern for his health.

LORESANE: In the presence of the doctors performing?

PONTAL: Queen Terissia insisted. She wishes no doubt in her mind for his condition to know; nor for her son not to acknowl-edge does he keep him with her, an' the Queen Mother. If the Vis-goth regains his strength enough, we will be called back in to speak with him. A good number of us (ha)'s been found, when Va-clkulor returns.... An' if he dies, it's obvious who leaves the cham-ber.... as the Visgoth— at least more obvious to Terissia.

LORESANE: Life leaves the room, Commissioner, unless it's a vault become of precious heirlooms. But I will take the sun to be warmed by, (in)'stead of some debacle of faith dissolved, or erudi-tion of some physicians to decree a death upon me— though they make labor assiduously an' honorably. My fate is here, what's heard, the radiance as sheen a sheet of waterfall off mountains an' their steep curvatures of rock an' stone. Here can not fear (to) re-side. (*as **Caton** approaches*) A purity presides upon my manifest destiny of cures.... for the curious. If the Visgoth lives not, then I deserve 'is death as well, without hesitation of sentence nor regret of execution. For the blind can not see wrongly, an' the deaf can not hear erroneously. an' the dead can never be mistaken again. No, nor again— They come? Seize an' strip! I will make lightening of the corpse.

PONTAL (Put) Head down! I will save you—

LORESANE: Not under this good rain.

PONTAL: Contrition! Loresane—

CATON: Sir Pontal!

LORESANE: Fear not a falling leaf for shed of timbers— yet!—

PONTAL: The brother Caton grooms this household?

CATON (*with them*): Yet hardly whetted from the anxieties. These doctors consult to pass a judgment, or an honest self-ap-praisal of their efforts, effectiveness, an' execution of mending worth. An' they look grave, as can be puzzled for a conclusion with such skills applied. Why, they even asked me to withdraw myself for a while, from their presence, as they.... commit themselves to a verdict, over the Visgoth.... or about. An' have I seen my brother?

— No. Though a family remains with the body. I was waiting for entrance into the bedroom, to have, perhaps, some final words with.... this great king an' royal relative, not less than a fatherly in-fluence. After standing solemnly, with mixed an' unsettled expecta-tions, the doors opened an' the practitioners rushed out.... with a piety of blandness false upon their faces. Then the doors closed just as quickly, or directed, as ever guarded from within an' with-out. One of the doctors noticed me, as perhaps the only noble within reach of consultation. None of the commissioners were around. An' he questioned me, as if to order through his medical authorities of knowledge an' the miraculous to perform. He said the three of them must have a discourse free from the heavy atmos-phere that I could tell had issued with them from the chamber. But where? I suggested the garden pathway. An' with a stern delibera-tion of professionalism they strode towards the botanical nexus. I followed as much as led, helpless to do more. They were.... as quiet as my vocabulary could comprehend, as we walked, consid-ering some details of medicine far above my recognition to de-scribe with significance of import for my brother's condition, ex-cept to say the past tense abounded with employment, as they seemed somewhat stunned or surprised of the trivial characteristics they uncovered, pertaining mostly to colors, I think. But who can tell what really excites academicians an' specialists? or how some hues of blood are applicable to the Visgoth. I'm not sure, entirely, of these fluids, or the fluidity of their thought. The speaking, fluent an' accomplished of complicated.... gibberish, ignorant of my sim-ple questioning: "How is the Visgoth?", from which I declined to offer more sophisticated inquiries, my attempts for attention being remanded before their incisive consideration of the patient, brought us here, at which they asked me to.... join you, while they deliberate as to what to say about their treatments an' the outcome.

PONTAL: An announcement of expiration must be officiated by the Queen, or at her command by some governmental officer said.... in public. An' you are of the public still, the Queen Mother representing your royal interests. A proper procedure must be fol-lowed, as the doctors construct an expiation for themselves, and the science. One of us, a commissioner, may be summoned to the.... death chamber, to make for witness it approving as a fact the character of this occurrence. This will most likely be for Vaclkulor to search (for), as he will head the next assembly. You waited not for messengers, to and fro, from the room an' for. What for the staleness of the stench, the family is bound there until he.... arrives. An' then to form us, as a body, we will commend the Visgoth. But formally it's done, Caton. For were the man still alive.... at least one of them would be staying with.... I have some papers to remit, to the freedom of being read.... at out next meeting. They decide you, an' your brother.... an' Fabel. I've not broken seals yet— though I could have an' can, during this emergency.... or crisis of the charmed. It's all left.... to a reasonableness of judgment, han-dling an' management of wishes. I should deliver them to Vaclku-lor now, or their location. They distill our fortunes.... Loresane, you should be away, through our gates of palace as a guest de-parted during this somber reckoning.

CATON: Oh. You are not a fellow commissioner?

LORESANE: ....No. I am a seer.

CATON: I caught the rumor barely.

LORESANE: I am barely said of rumor, to walk with the Vis-goth.... through some grassy gardens, an' on trips. (*as **Pontal** *stands*) But I predicted his full recovery.

CATON: From suffering?

LORESANE: For the Visgoth to be.

CATON: Well that is understandable, if we always hope for the best. Therein lies a good service.

LORESANE: I adopt to the results of my blasphemies, if there are cruelties to take, saying what should not be heard, an' seeing what is not shown. But I am certain the Queen caters this sun (for me).

PONTAL: You've broken no laws, nor scruples of custom, an' have only made a pledge blinding you.... from an escape. But what's for that with circumstances brought— beyond your possible manipulations to perform?! There is a shack for you, a lifestyle affirmed, an' countrymen to be surrounded with, who may be understanding an' compassionate for your protection. You had wanted better, an' have received an oracle of sympathy. Plights are made endurable this way. What can you say with grieving? Then only that there are beliefs upheld in grief, an' you are subject as much as any to those prejudices. You loved the Visgoth dearly, as your king, an' thought of him to heal. What has healing done but this, your brawn to this contagion lent to fight? Must we so do with good hopes an' positive sensitivities. What is it that you believe still, with your failing?

LORESANE: As could a day be wronged, then never failing.... as a day. I will not plead against your kindness to follow, nor advice to take. Yet without true gifts, have I been proved? Lend me to these doctors to vivisect. Let's be certain drowned.

CATON: They only asked that I be away but briefly, while they consult each other's opinions an' professions.

LORESANE: Let them pronounce to me my error. Though introduce me not as seer, but as this way seen. An' if the death is confirmed, then let me run.... run into oblivion's warm hands of shame an' shackles to a conscience, to thoroughly abuse my thoughts.

PONTAL: So make this pact with curiosities to be sured up. I'll keep the roads cleared for your return, an' bide you clearly, as much as can be deceived, until the other commissioners gather to consider your outrageous sacrifice today, towards your art. (**Loresane** *stands.*) I, as a commissioner, shall impose myself to impound the opinions of these healers as a requirement of service to a solemn governance, an' for advancement of your awareness to prediction, Loresane.

LORESANE (*as they are leaving*): With flush of flight, an' blush of sight— enlight-en. I may burn of a cinder otherwise, with crust of cur, not learning of the truth within my prominence upon leaders an' rulers the ideas an' concepts found emanated from them.

CATON: It is certainly theirs, for life, to see, as could a palm be stretched, unnaturally cartographic.

PONTAL: Ja, it is theirs.... But lessons learned, it is not them. (*They exit.*)

Act V

Scene I — *A hallway in the palace, extremely ornate.* **Vaclkulor** *stands with* **Hestor***, who is busy observing the fixtures with a childish curiosity.*

HESTOR (*rubbing a wall's facial projection*): This is very broad, sir.

VACLKULOR: The palace is decorated with an enormity of details an' sculpturing. Even the halls make showcases for the most prominent of the crafting arts, creative sinews stretching off into a distance.... Dirty it up well, boy.

HESTOR: Oh! My hands?

VACLKULOR: Place one of your coins in its mouth. She may speak to you.

HESTOR: I would lose my money won. (*coming off the wall*) It may be deep, too steep a throat, running through these thick an' sturdy wainscots. It raises me, to think of luxury involved of regal domiciles. But I am dear to what I've earned.

VACLKULOR: You've hardly reached her face. Does she remind you of someone?

HESTOR (*looking for other objects to handle*): As familiar as beauty to see, with a purity of form remarkable, the skin so cold an' hard. I had to test it, with examination. But here is such a wide expanse of treasure.

VACLKULOR: This is minor, for a minor road. With passing through, regales you somewhat of the precious are these wares worn. I have as good in my own abode, with harvesting of handsomeness to claw upon. Yet only are they so for use. Break them, an' topple! Cause a disturbance! The pieces of art are as false as deities to worship. Are you famished for them? or have you suddenly developed a hunger!

HESTOR (*studying a jewel encrusted case anchored near the center width of the hall*): I know what stones are pretty. An' these colors fascinate the metal.

VACLKULOR: That is gold, of some dilution, in which they are embedded. Too heavy be the block for lifting. Yet so uneven is its top surface that ceramics an' statues placed upon it tend to totter off an' crack or break or smash themselves into pieces, I'm told, with landing on the floor, these marble tiles you run through. Too much the case is of itself, I suppose.

HESTOR: Makes delirium of eye.

VACLKULOR: You can't extract a piece with fingers— But try! It is not stealing here, if we are so exposed to this. You find a new land, within your own. Yet all of what is found represents you.

HESTOR (*impulsively, or impetuously, turning away from the case to seek out other objects and materials*): A difficult love, sir. I hadn't a conception of such splendor. (*running to a wall drapery*) This is as green as water-fly's iridescent wings!

VACLKULOR: Pull it down! Though it's hooked more strongly than your weight, a tapestry of fine fabric an' rare intricacy of design. Foreignly produced, of course, where weaving is profound of time-consuming occupation. Make a tear in it! I'm tired of all treasures.

HESTOR (*handling*): I don't want to damage it, but to feel the wonder of its soft textures. I've not before sensed this, nor like to have, these threads to touch an' emulate. (*rubs his face in the cloth*)

VACLKULOR: Such art has almost ruined me, to collect an' pay for.... an' empty of my treasury with. Idly they are quite fresh— Bite at it!... Or suffocate not, Hestor, your amiable countenance of youthful pride to experience.... idealistic fashions an' miraculous riches— once!... at least, during the impressions of childhood. That is how I was homed an' cushioned, conditioned. But it has spoiled me bitterly, with some derogatory conceit for what I see an' find. If it's not up to me, my expectations an' needs.... or am I below it still, with casualty of illness for this pride?... this makes remarkable a necessity to maintain dullness. These trophies dull a man, by squandering his goals.

HESTOR (*coming off the tapestry, to look around more*): There could be a sea within the wall, a completely new place for travel through, an' experience of.

VACLKULOR: You are excited by the toy in these.... trinkets of the mind, baubles of the fortunate. I think there are some.... glass or jade carvings over there. Knock them out of their exposed encasements!— The hall's not made childproof yet— No.... Come here, Hestor. We must find Loresane. (**Hestor** *obediently approaches him.*) I know the.... chamber to which he's kept for night. But I doubt if he would spend the day there, entombed from a hut's-like air. We should find him strolling around, with his head expanded too unnaturally of a headiness, uncharacteristic for his simplicity, until he sees you to return to proper form of humbleness. An' what make you? my lad.

Hestor: I am his friend.

VACLKULOR: That is quite good, for this recalls a fair environment. This palace, within its superabundances, can destroy a sound conscience an' distort a healthy personality. This is what has occurred with the Visgoth.

HESTOR: I've heard of some complaints, but I don't understand them.

VACLKULOR: Need you not know of them for him, but that they are to demonstrate that even so strong a king can fall into deprecation. Thus so the head of Loresane may be turned, as yours begins (to), with discovery of this fine wealth to seem with. How easily you are fascinated out of poverty.

HESTOR: We're not impoverished!

VACLKULOR: But Loresane is, is he not?

HESTOR: ....Yes.

VACLKULOR: So then more easily may he be deluded into abandoning his poverty. That is not altogether bad. It's a matter of degree. He should not offend— anyone here, with pretensions of a false grandeur found.

HESTOR: No. An' I do not. I am obedient.

VACLKULOR: Not from the lowest servant to the King an' Queen themselves should he present himself in any way haughtily. Before the Commissioners he begins to, other than myself as a friend. An' I stem this behavior with you. When we find him, let him plead for your judgment—

HESTOR: Mine?

VACLKULOR: — An' let that judgment be mine. The rush of memories may bring tears of enjoyment to see your face and become as he was. An' then you may simply suggest, to him, to let his kind friend Vaclkulor guide him more within this palace.

HESTOR: Is that it? Seems very subtle.

VACLKULOR: Let Vaclkulor's judgment be commensurate for his principles.... spoken.

HESTOR: That more confuses me.

VACLKULOR: It is a slight nudging to remain wise, as I've tried to explain to you before. But especially with questions that are far above his unraveling he should trust me. As for example, who should most directly succeed the Visgoth himself. His brother Morliff.... Mor-liff.... certainly has the favor of your family.

HESTOR: ....Yes! I remember the name spoken with admiration, in our village.

VACLKULOR: An' the same will I suggest as you suggest. An' it is as simple as that. Suggestion brings to focus many attempts at picturing a view.

HESTOR: Mor-liff

VACLKULOR: With words so spoken.... as the lady pure an' hard, he will condescend from more conceit within himself, as I try to do. So flood him so with happiness kinder than can be produced by these crude toys an' this grandiose living, but by remembrances towards the fulfillment of his true nature.

HESTOR: Yes. His is more gently satisfied with his visioning than by his manipulation of the real matters an' materials with which he may command his labors. I've often wished for the same gifts.... but candy is too precious to me. Still, fun leads to dreaming also.

VACLKULOR: But this is not a fakery of wistful thinking. It is an earnest deployment of dedicated skills. You wish to mediate for.... our friend, don't you?

HESTOR: ....Yes.

VACLKULOR: And to help guide him safely through these unfamiliar surroundings—

HESTOR: Yes!

VACLKULOR: —with a noble purpose, posturing for a proper

presumption of friendship.... an' relationship to his actual character an' more generous demeanor, then here are the dusty airs obliterating sight! *This* is not real, as much as man— or boy, all around us shining an' gleaming. You can not eat upon your wealth an' health an' life with hope to sustain any of these kind attributes. Yet build upon them only virtuous enlistments. Arrogance is a crime against your blessing to be. I have been too much so, with decadence of properties, an' now must represent.... so many faltering of their possessions, standing, an' endurance through declines of station, status an' even passion for a life worth these travails—.... You are oblivious to all of this, being only a child of innocent indecency to run through rich halls blithely an' without bounds of supposition: This is real?... This is a cuckold's tale, whose wife is bold for wealth an' privilege. Rub that face harder, someday, until it grinds up your endearment an' sympathies for a love, for a loveliness. This is a trail for furnishing.... the floundering, the recessed of notion— an' the needy.... of eternal substantiation, or however long some stones an' jewels may last. Not a playground, but a potheriage, to which I have brought my friend—s. An' it is with your courage— an' recompensed service, that Loresane may bypass the mallet-ing into this.... stately mold of greed I try so hard to salvage my subjects from. The common folk among them are as witty as your infancy, Hestor, while the gentry suffer image an' a steady loss of resources. I carry the brunt of all worrying, an' consult my herald for help, in seeing what to do or avoid. His best advice becomes when he is positive an' indifferent an' has no stake in the matters presented to 'im. But such sagacity would change in a palace, as much splendor clouds the eye. So be for ye a lesson to perceive even a blight in heaven. We must extract the man, through his own devices of volition.... An' if I must keep you in a— room to stay, to be stored up! so that I may bring him to you.... you should not be afraid. For if he is even half himself he will sense your presence or occurrence here, an' the predicament befalling of your visit, as were a game to play. The chamber will be threaded grandly, an' you will be comfortable in silks an' with toys an' sweets provided, during your wait. Without a deviousness will you be apportioned some measure of this great, grand masonry, with Morliff's beckoning to cry for. Mor-liff.... be the honor, upon Loresane's lips to hear.... An' see you. See you speaking, of such devotion enlightening 'is vision.

HESTOR: Why can't you take me directly to him?

VACLKULOR: I must find him, search around the premises. An' the Visgoth is ill. You should not be roaming about unexpectedly at a grave time of crisis, with only my invitation to uphold your visiting. You should be positioned appropriately— for this day. It will be quickly done, with success, my greeting Loresane. An' not to bother you, nor you to bother anyone— especially the servants an' the guards, who run rampantly about under anxieties for their master.... today. Trust my manner in this mansion. It is best that you are given a more securing accommodation while I am essaying.

HESTOR: ....I like the sugary coats, on glassy-ed, hardened gelatins.

VACLKULOR: ....I'm sure they have.... some type of this delight here. An' if not, then better. We had coconut-ed wafers with our dinners, last night. An' that was highly delicious, too. I will direct that you are given a good serving of ample treats to pass the time; or treatment thereof of what ever you find, you may keep if you can carry.

HESTOR: Even of the furnishings found?

VACLKULOR: I will pay, for this trouble. (*as a **commissioner** comes up to them*) The room is not far from here, an' is as beautiful as ivory is white. Of course, it is a detention to be held at, or locked to, for your own safety hidden within. But the pillows are soft, an' the rugs supple enough to roll through without damaging, however you children frolic about without distraction from more serious endeavors.

COMMISSIONER: Ah! Vaclkulor! We have a number met. Who's the boy?

VACLKULOR: A gift for the seer.

COMMISSIONER: But time stacks against us, sir. It is believed the Visgoth has taken such a turn for the worst that even still we are not called for his recognition. He must be, then, unconscious.... or incapable of speech. Physicians have conducted themselves about him, an' concluded much. Though the word is they deliberate within a stroke of gravity that hews our expectations for a meeting. Meaning, of course, that we should prepare to assemble out of his presence, his brothers have been ordered to arrive— an' have. We might, then, make decisions against Fabel. But that is to be debated, ere Pontal presents us the Visgoth's papers for consultation (s), these documents without question, yet all for questioning. Yet, if the Visgoth is actually dead, a quarrel must ensue, for the betterment of this organization of fiefs an' states—

HESTOR: What about Loresane?

COMMISSIONER: Who is this boy?!

VACLKULOR: Hush! Hestor. Do not address this man as (were) I, with sore familiarity. For as we are friends, still, he is a Commissioner an' of high rank, a representative of lands and their inhabitants as a ruler. Forgive the boy's impertinence, sir.

COMMISSIONER: Then he is so young to bring here, Vaclkulor.... an' at this time, not may be soundly fashioned. Yet you command some order for us next. It may be our most important congregation in some time, for a direction of official missives all throughout the fiefdom, an' to draft them upon cold agreement.... But as to this prophet, boy: If the Visgoth is dead, then so is prophecy, since its practitioner argued against such a dire occurrence this soon. An' he will simply have to succumb to his own harbingering of reward an' walk with the Visgoth likewise—

VACLKULOR: Oh!... It is only a pride that will be defeated, an' not a death sentence (made), Hestor.

COMMISSIONER: The man boasted of his prescience in this matter, an' must well take a just action according to his claims, even if he was drunk, or intoxicated from a newly found fame.

VACLKULOR: We will leave such decision to the man himself. He was out of his element, out of his head, out of his sight— an' can be forgiven for this.... brevity of foresight mingled with a desire to impress. I take a blame for this an' will correct, as he is my friend.

COMMISSIONER: Ever the arbitrator an' assuage-r, Vaclkulor.

A kindly character you are, with an adjudication to impress *us* by. But this is not so important as our weal, with responsibilities to handle as an assembly of commissioners. We collect before the Visgoth's chamber door. Eight of us even, one more than necessary. Or should we start to do when called, I say we gather now, an' with the brothers an' the doctors to commence this maelstrom waiting upon a somberness to descry excited furies, an' decry the poorer arguments ignited thusly.

VACLKULOR: There'll not be a celebration, sir. This is a dire passage for our commissioning to waddle through. For we must craft an' construct, as it were, from our deceased a ruler, a Visgoth governing. An' I'll conduct a solemn tone for these proceedings.

COMMISSIONER: Cuff me for a lack of ambivalence? Are you not happy for activity, in that we are in motion again to produce the strains of growth an' prosperity— an' a liveliness to decree throughout all of our regions?! Of course I am anxious for this— this.... party of permissions to conduct our thoughts through. An' you've much the same for pushing causes, honing on the Visgoth towards his death to lead—

VACLKULOR: Oh!

COMMISSIONER: —feeding his reporters an' country surveyors — only half surreptitiously— all manner of events for his attentions to brew on, the burials of these children an' their locations to find, so that he might mother death! an' have his feet walking to, an' witnessing the un-seeable, hearing the unspeakable, an' touching the unrecoverable—

VACLKULOR: Sir!

COMMISSIONER: —It's been much an entertainment for us to notice, such a clever way to push along an' hasten the natural change in leadership that has been deemed most necessary by most. An' we commend your efforts for Morliff, hardly to be acknowledged not, since they are evident with our sympathies. But now chastise me not, to show a similar enthusiasm, or even one that's amplified, at the very verge of fruition for our needs an' mandat-able desires. I'd say at least five for Morliff, two for Caton, one abstaining in confusion— an' only the Queen for Fabel. It is a just delivery you have brought us to. An' we *do* celebrate it!

VACLKULOR: ....Dear fellow commissioner—

HESTOR: I understand none of this, but infancy.

VACLKULOR: Lad, be silent! Stars stir, hidden by the sun.... Most renown colleague, we must be humble an' not boasting for any triumph. Here is no jest of state, but intestate the means. There are griefs an' grievances involved An' some must argue.... intelligently, to counter what may have been written as if to void or ignore, as if the documents did not occur or exist at all! That can not be done with a guffawing sense, nor lively atmosphere before a deathbed. We must stay respectful of the Queen, *and* the Queen Mother.... an' indeed of the dying Visgoth's death, which I do not promote gaily. Nor with pleasure am I bound to this, an' will not *you* be under my helm of conference. Mere numerical agreement is not enough, since we are not electing but finding, as with substitution, the best outcome available for this sad day to end with. An' compromise of feelings must be purchased with a common an' shared will by all to continue at this fiefdom, this association

bought of valors an' the strictest nobility of action where gifts are won through deeds— sometimes thought to be impossible of accomplishment. My own courage, once, hues the grave today as being lost.... somewhat, to devolve downward into these ticklish manipulations of religion an' beliefs, what the crazed Visgoth had thought of himself to perceive.... an' accept. Yes. This has disturbed us all into these motions an' subtle whirlwinds to produce a change. Some clarity will be reviewed, when the winds die down, if whether we are (in the) right for what is to be done. But make this certain to you, sir, to know or suspect, that Fabel will someday rule as Visgoth, should he come to an age we could accept. An' this is to pacify the Queen an' her kingdom. An' Morliff will become Visgoth in this stead, to pacify the Queen Mother an' her kingdom. So much to be relieved by, we should more pray than pronounce our officiating vigors, since with the women we must bargain an' assort.... So have you a more serious demeanor— an' pout some.... for what must be done.

COMMISSIONER: Perhaps some keen advise. I'll not speak with a giddiness about this, these.... cold events. If Vincent's will makes Fabel king, then the son shall reign, become of age till Visgoth be — delayed. A simple privilege simply stated, softly slated— slashed with sorrows. Oh, we will all be saddened by the Visgoth's death. I can definitely respect those sentiments. An' again you weigh to balance, with a wisdom coupled to a wile.

VACLKULOR: So we have not been called 'to attendance yet.

COMMISSIONER: Not as of light thus far. But then thus far we have made ourselves, it can be done with profitable endeavoring.

VACLKULOR: I seek the seer.

COMMISSIONER: Last I saw of 'im, he was walking with Pontal, towards the gardens, an' quite conversant. Much to be said, more to be seen. How do they explain away mistakes an' errors!

VACLKULOR: By eyelids, closing eyes when the sight's too bright, shutting off the vision to see (a) red of their own substance, the humility at awe. An' should we each do similarly, at the precious crossroads burning for our choices— So fault him not! I shall be dreaming too, someday. Inside this— cake of wonder— he's become.... not himself but more a serving of the luscious. The boy's to bring him level again.

COMMISSIONER: A relative?

VACLKULOR: Nay, are we all apes to appetite! I should condone his change as my own. But this is what friendship brings. An' I pursue him.

COMMISSIONER: To the garden bring.

VACLKULOR: No, from the garden bring. A sight of roses in a garden makes no more their freshness sensed. You should study the character of man more. This poor, seemly fellow will appreciate seeing Hestor uniquely, uniquely seen. But in a backdrop, or background, of botanical beauty, what is more to be seen but a recognition of the same an' as much veiled by the indignity of pride.

COMMISSIONER: Well, we should collect, in a parlor nearest.... the demise, in order to be found. I'll have some brunch yet, though. (*leaving*) Inform the others of us that you come across, Vaclkulor, should they be swarming about as we. Drudgery revolts against the drab, an' we should assemble to prepare.... to make some difficult decisions. Nor not against our nature are we lent to a commission. I find the errant homed to this submission to be tal- lied. (*has exited*)

HESTOR: Take me to the gardens.

VACLKULOR: I'll take you to a room.

HESTOR: I want to see Loresane.

VACLKULOR: You'll be all with him presently. But can't you understand?— The house cries. Our Visgoth is dead, or nearly so. Your king departs the earth, for earth— an' you are not to be around.... playing, throwing balls, or with fun thinking, for some notice within miseries.

HESTOR: But I feel restrained from meeting him.

VACLKULOR: I would not pay otherwise. That is only childish impatience. An' you will have toys to preoccupy you. But I must first prepare the herald. (*starting to lead **Hestor** from the hall*) I must speak with him man to master. Such practitioners do not like to be shocked or abutted by predictions told. But he will find, for you, the fatherly protections to undertake in himself with this pre- cognition I'll suggest he exercise 'bout your longing to be with him, as you play your games of solitary leisure an' have taste of what may be served, delicious, for a palace.... visitor.... Yet find you locks of hair (in) there, just tassels an' tresses of golden embroi- dery.... be they made of. Fine for spinning round.... into projectiles.

HESTOR: More of a severance be down to goose than threadbare aristocracy encumbered by. Riches make for ruin without a play upon.

VACLKULOR: Sound gesture, lad. Then sing in chambers for angelic wisps. An' be not so hesitant to be found. For that is an es- tablishment of comeliness, to be discovered. An' I hope fervently that you are brought to rights with Loresane an' nature itself in needing pastimes of potent recognitions or remembrances. There be some decorative flowers in the room also; so draw upon, that it is very airy an' slight.... an' hardly molted of its birds.

HESTOR: Outside is it? or in.

VACLKULOR: Within, without. It's here about the palace found, throughout, as were a puzzle in a jar. For what is placed in it is meant to be removed, as fresh as before— yet better kept, serving its entrance. (*They exit.*)

Scene II — *A sitting room just removed from the **Visgoth's** bed chamber. Two **commissioners** wait casually, as if having nothing practical to do, one sitting on a wooden stool long and broad enough for several, and the other standing.*

COMMISSIONER 1 (*standing*): It tastes of day.

COMMISSIONER 2: That's the liniments held within. They ab- solutely swathe the entire physique to try an' keep it intact an' bathed in bed. A sensation of staying holy, or at least elegantly washed. It's hard to nurse a blistering all about the skin except with bandaging an' liniments.

COMMISSIONER 1: Have you seen thus—?

COMMISSIONER 2: I can imagine it, dimly portrayed. But from the chest I spied directly, an' it appeared like eyes peering out, peaking for some breathes from out under the sheets. An' thus?... So thus the smell of the oils whose fragrances imagine the refresh of the day, an arboreal temper of leafy scents an' wishes.

COMMISSIONER 1: Oh. Up an' about, again. An' wandering through briers an' root strewn pathways.

COMMISSIONER 2: Soothes a tenderness of thought.

COMMISSIONER 1: So then it's a bad pettiness, to be so easily displayed of wounds. Stumble, did he, upon the prickly most.... as could do harm assuredly?

COMMISSIONER 2: Danced! upon the woody spines, I'm told. An' then to roll through, as if to spike off fleas, much like a dog with dusting. But to the earth of sacred burials did he bring himself with a madness. An' now he bleeds of punishment from sacrilege, weeps silently— or as asleep. I've not heard a whisper of a groan, nor a moan of mourning for disease.

COMMISSIONER 1: What is the wild contagion of such steps?

COMMISSIONER 2: Enliven he the graves. The Visgoth must control everything, or mar for an approval.

COMMISSIONER 1: Control he the transit of the sun, an' the phases of the moon?! This is a foolish hedonism practiced, bound for the discomforts he's enjoined. One can not handle light as were a meat to bake for food. Or with the sun shining on graves, to reach for it to grab an' toss an' tussle with for rope, this makes only diversion of principles to reign.

COMMISSIONER 2: Become bemused.

COMMISSIONER 1: Not in the dark would he visit those.... children, their families. The night does not hold luster of its mes- saging, but for howls an' hoots an' peeps.

COMMISSIONER 2: His is for to do, the infancy returning to, reverting an' restoring, an' reproaching of the indolence discovered, against his sanctioning.... the lack of crying— So he cries an' yells an' screams an' has a tantrum over plotted corpses. That is a regal right.

COMMISSIONER 1: Not by my conditioning to believe.

COMMISSIONER 2: He searches for a sense to it. The waste provided for his entreaties to pronounce upon make him subject to this shock of loss, an' he handles the weight of emptiness in air, as to rise an' have arisen, again, these souls within himself. For each blister may represent a babe of anguished commiseration an' argu- ment, as how they should cry for this atrocity performed.... upon his person made malevolent. He will cure, in this way, the damage carried, or make for carrion a more sensational rotting than at chil- dren's graves.

COMMISSIONER 1: Sensational! But what is to consider after he is done? One kills an enemy to make his death a matter of approach to earning rights. We've all done so, with an overwhelming effect for ourselves, wholesome an' yet wanton. What can we win by this.... visgothic pillory? but that our errors can be made to seem silly— an' must be better burned in bed, choked with a smoke of ritualized atonement an' sacrifice.

COMMISSIONER 2: This might set some tradition symbolically, for all crazy heads to follow or remit to of their characters becoming excessive, or their accumulations of power overflowing for their overthrow to nourish, the popping seedlings an' saplings that surround.... the bed. But he will heal— that is, a visgoth— from these disasters maturing into affirmed committals to improve our wealth an' standing. It takes the greatest of fortitude an' common sense to know when it's time to self-implode an' then to do that!... regally an' with submission to a passion walking, awakening an' growing throughout our vested interests an' cold vices. He gives me more envy than an earnest sorrow, to be so brave an' calculative of a demise. Then to consider more of fruit off (of) the tree. If they do make their time this way for us, then we shall succeed of our more personal engagements across the responsibilities of an organization of mights. For they condone.... their own bruising for our rights an' leisures of attainment, a strong insanity blanched of clay death masks honored more of their graying sights than their visionary pursuits could ever lead us through. What is to be Visgoth but with vigor as a gigantic vase be seen, huge an' prodigious for our lording, an' with proportion as an urn to become, that decorates the countryside with rich solemnity of purpose.... representing us. This confirms dread for humanity an' affirms our prosperity over heedful masses an' emotions. So learn of this persuasion, not to be so self-deceptive as if blind, but take what is not seen to own an' ogle with agape the vestiges of visages drawn, etched upon the hoary hide.

COMMISSIONER 1: Oh, he makes us, for certain. Medals for our meddling, you say. I would have thought his constitution stronger. But how must giants wither, with empathy of blight? His is still a strong personage defined, an' a greater position declared. I favor the more easily humiliated than by this terror. It's been a stubborn resistance to details, is all. But so much so, to make a leader of leaders.

COMMISSIONER 2: Birth, an' contrition for the birth. We won't dissolve easily, an' disperse our angst among us an' throughout, if there's more pith be found for the worms. That is the blood-soaked birth we should command, or manage, oversee an' recommend as a productive outcome to these events. Right to the eye, a veil of sweat for the Visgoth. Take he.... hardships from us.

COMMISSIONER 1: Take he sheep, flocks of them.

COMMISSIONER 2: Manifold mendacities. I knew they were cruel from the first. Poisonous bait from angry gods, but to catch.... a superficial evil— if any, if at all. Perceived deprivations amongst acknowledged riches. That is the character of man, to succeed an' outdo even their deities worshiped so hypocritically, or perhaps too critically to be rational about.

COMMISSIONER 1: There be truth to that, for I would spit at idols which don't spit back. They are.... never false, though, unless a craft be faked— which can not be with artistry an' care of workmanship. No, we are ourselves a perfection to perfect.... our deities, an' our drives an' ambitious aims. That might be why he walks on graves. It must feel.... somewhat stimulating— if perverse an'

presumptive of an honor or a right won.

COMMISSIONER 2: A rite wrung.... through toes of feet.

COMMISSIONER 1: How sense the earth its casualties of attrition? Why die as a solid?... raped of life! Wrought of love, worked into a hardening, masoned for the ground. What is there left to be fetched, a disrespectful defiance for the cause? A fighting spirit? Then why do I persist to achieve the impossible?

COMMISSIONER 2: All is not over death, an' hopeless. Destinies are gradual (*Melinda enters, from out the bedroom.*) in stone. (*He stands.*)

MELINDA: Where have the physicians gone? I can not handle bodies.

COMMISSIONER 2: They went to refresh their lungs; outside, I believe.

COMMISSIONER 1: Is there no care for the Visgoth? Are we summoned?

MELINDA: A nurse subdues the quiet an' ineffectual. I can not make a pronouncement, but must await the Queen's privilege an' purposes, as I had done for my husband's bones an' flesh to transform daylight into anxiety for shrouds. The son is disbelieving. (*coughs*) An' not within to withstand. Could be taken.... as too dispassionate a sign. The vapors shield the eyes of their senses, what these doctors have concocted. *She* did better, in my opinion, than their venom. But he did give up an agony.

COMMISSIONER 2: Or a struggle, Queen Mother? Is it now peaceful in there?

MELINDA: How still is a longing for? The Queen is patient, an' grieves. Fabel paces back an' forth, distressed an' disconcerted, refusing to stay seated as the mother would insist. An' the nurse stands by Vincent, to rub a pall of balms.... as if to wax.... An' I would call for doctors, to bring to a decision what occurs. It is the scansion of their method, to relate. An' they must stand by their results.

COMMISSIONER 1: They should return shortly. It is their duty, unless they try to escape life, profession.... the prominence itself that is so highly valued, in these states. Can I try to collect the others—?

COMMISSIONER 2: They will drift here.... This is the magnet drawn to, of solemnity an' somber recourse. For words spoken we are lost to hear. An' this pits the Queen against our prophesying.

COMMISSIONER 1: The words last spoken, due the Visgoth's rue.

MELINDA: You will hear Fabel, but the Queen must figure this.

COMMISSIONER 1: Or how to say?... I am the Visgoth!— May?

MELINDA: Do not chastise their weeping. It was intoned directly for our ears, the Visgoth's provisional commands. An' we know this, faithfully said.... though weakly spoken, pliantly demanding, autocratically versed— (***Halyce** arrives.*) Halyce!

COMMISSIONER 2: Count.

HALYCE: What makes of the Visgoth?

MELINDA: To the gristle baked, the gristle. I told him of us.

COMMISSIONER 1: Us?

MELINDA: He was only amused, gently perplexed if at all.

HALYCE (*coming over to take **Melinda's** hand*): Gentlemen, I announce to you as consorting companions at this palace that the Queen Mother is to become my wife, an' I.... a devotee to her an' our aspirations. Tell any an' all you wish about this. But that she was faithful enough an' brave, to counsel the Visgoth on this matter so personal, confirms for myself my intentions. (*taking her hand*) She is more power to me than I could ever lend.... an' more beauty of the soul an' heart than I could ever fend for else, nor other.

COMMISSIONER 2: The romance, count, seems quite untimely an' inauspicious to the moment spent. Yet as it is, all mating guided by the guile of serious endeavoring, it is less substantial than surprising, an' less surprising than suspected. I wish you better bread than happiness. (*re-sits*)

HALYCE: Then it will be a sweet cake, because I have more joy than sorrow.... for the Visgoth an' yourselves, at this occasion of grief.

COMMISSIONER 1: You would try our remorseful dispositions, Halyce, waiting before.... him?

HALYCE: Rejoice with us. Let us have a celebration an' a dance, to come clean an' abreast of our futures.

COMMISSIONER 1: That we will decide, on a visgoth.

HALYCE: That you will decide on a marriage, in a ceremony. That you will stamp the floor for, or flatten the ground with, these pleasant sentiments I espouse, I decree, I obey an' purchase with my life's longing.... An' some freedom for us, to be as real as our wills an' nerves will allow, that I contract of. That I.... envision. An' what do you see, that is stately? We will be matter mauve, upon our thrones. Yet you stay clothed.... to a Visgoth's bib an' peerage. I.... demonstrate my worth to be, to have for me what's better than myself— a love! You only emulate yours, an' for a Visgoth to loathe an' laugh at.... or with— too very soon. Then this day brings bridal showers to arrange, an' ornate biers to propose— flowered each, an' one justly symbolic for your own earthly dispositions. Yet may you choose to celebrate my heaven found, an' uplift yourselves into dominions more favorable to much contentedness.

COMMISSIONER 2: Sir, we are not married to the Visgoth. We have had our own betrothed, which lack for you be gained your bother an' our pondering to avoid in patter pandering. Be it blest or bless-ed, commensurate to your deeds or adventures, there speaks of whims enjoyed an' eventually condemned. But have it oft, the folly of a furrow an' a furlough off commands.

HALYCE: Bitter widow-man—!

COMMISSIONER 2: I've done more without than with!... An' in my age does time tarnish connubial talents. But not to the Visgoth am I dependent for these! An' so we are led, so as not. It is an agreement among masters— that keeps the peace among us flung an' flinging as a vexillum shadows aims, his signet placed on *me*.... Bah! this contagion, caused by lending prophesies of shame. I hope you (two) remain well together.

MELINDA: This is to be greeted more generously—

COMMISSIONER 2: Destroy us all! for our foundation labored with an' struggled at. There was always a cringe to your company, an' its begetting hemorrhagic— You are sent as the sheep!

MELINDA (*running out with tears*): This is dissolute!

COMMISSIONER 2 (*standing suddenly as she is leaving*): Forgive me!

HALYCE: Sir!... I will bash your head! What panic brings to you, (is) hemolytic-ly placed of bowels!

COMMISSIONER 2: I apologize, Count Halyce. My distemper is thoroughly distraught.... for its ugliness. But I am with grieving of my nature today. An' you have brought me to see myself alone an' losing more of comrades an' protectors, companions an' confidants. I am too old to complain pleasantly, pleasingly.... an' too honest of my vile an' base reckoning. For here is a weakness standing, that needs the Visgoth proper. I can not afford of myself, with my own strengths, to subdue the disillusions of my terrible subjects with their gravities. Not all fiefs are rich or handsomely endowed. An' I am fearful for mine, for they who are so stressed under it with my pitiable hands of management.... Everything frightens me, now. (*slumping down on the stool*) All is as gray as ash... spread ov'r graves in sacrament, as sacrosanctness of the burning wood that lends its odor for more ardor to promote in the observers mourning.... now that he dies, an' powers strain to circumvent our wishes. Now that he dies again!... does do me justice for my shambling of ownership, brought now to shambles in my dreams of portent. There are.... no other choices left but bad an' dour scathing for our health to describe. Sour suffering of souls, denied their flourishing, lie under me. (*head lowers*)

COMMISSIONER 1: We are all worried for the sustenance of our suzerain, though I'm not from such a poor state as you avow.

HALYCE: I shall take her from this.... mire (of affection), an' leave you with sticks aflame to batter an' drop hopes upon, inflammatory wart! That is all you have left to sit on, in this assemblage of warring cocks an' cockerels. An' let you be spared my strangulation 'bout your wobbly neck! where it is worth some tie to Visgoth. But you will be as deceased as he, to walk about ungraciously, without grace nor gratitude, swept by slovenly winds an' auspicious soot! for your graves.... So buried in the chest are you, without seeing the enhancement she has brought— to begging bugs an' boars of coxcomb bund—

COMMISSIONER 1: Sir!

HALYCE: —I'll not have it more.... You fools of meretriciousness to he that would collect your favors.... (*exiting*) to walk on your graves. Let ye learn of my wife.... to not be more your spite enduring. An' I shall bring her your apology.... stained with my angry rebuke on all of you. For final is it, fain to leave. (*has exited*)

COMMISSIONER 1: He plans.... some certain maneuvering, about these celebrated skirts.

COMMISSIONER 2 (*slowly lifting up his head*): There is but one.... I have confessed to it, the quarrel.... an' our disagreements.... he often boasts of. He vies to be the Visgoth, definitely. But what of a strong man? I am afraid of dissolution, an' he pours (on) the threat.

COMMISSIONER 1: Will he consult with us?

COMMISSIONER 2: I will kiss her shoes, an' rest my sore head on her knees for a pardon— I am so ashamed.

COMMISSIONER 1: Some truths are shameful, nevertheless. It's all been thought.... an' whispered through before.

COMMISSIONER 2: Consort.... We must consort. A consortium of strengths.... A continuum of breadths.... Consort.

COMMISSIONER 1: There will be some agreements made, concerning the Visgoth. I'm certain of that. We will accept what occurs an' provide ourselves with a promising future of harmonious relations an' good fathoming. This is the promise, the commitment of a collective might. An' we each contribute to this handsomely. Not least yourself, sir. There's been no wrong committed by your actions, save conformity to anger for his aggressions. An' that I'll fight. I'll give a perpetration of resistance to his— usages.

COMMISSIONER 2: I do miss the wife.... an' blame a curse on thy remissness to lend her fame to thus.

COMMISSIONER 1: ....What? Commissioner—

COMMISSIONER 2: Thy callous craving.... to be so frightened. Thy weeping sorrow.... brutish boar. (*lowers his head again*)

Scene III — *A hallway within the palace.* **Melinda** *is walking through it upset, when she hears a childish cry coming from one of the closeting rooms or secluded, sheltered cloisters.*

VOICE (*from behind a closed door*): Oh! Oh! It is dark in here!

MELINDA: What issues of grief? but as from a child most barely brought to!

VOICE: He did put me in here, an' I complained that there is no light to play with. Thus I cried, how then to enjoy the goods of fashion promised. An' he retorted jokingly that I should play with my coins. Then (he) closed this door on me. An' it is locked, an' I am abandoned to a pathway kept empty, I fear. But save me from this retreat all silenced for the palace.

MELINDA (*unhitching the door*): It is lightly bolted, your door. This is a corridor of privilege, though seldom used. For boys were kept here, who are gone. (*opening the door*) Who are you?

HESTOR (*emerging*): I am Loresane's friend.

MELINDA: Loresane?... The herald who was brought?

HESTOR: Yes! My name is Hestor. I have favor with you, yes? if you have this for Loresane?

MELINDA: I do not disown the man, for the fellow is a guest. But how do you arrive as he? Would scamper here for fun, with a friend? An' he makes mischief on you.

HESTOR: Where be I destined for abuse? I should be home for soon, else me mot'er becomes quarrelsome.

MELINDA: Ah... An' he reminds me of my sons. For this is a place where they were brought, for prayers, devotions, an' lessons taught by themselves in meditation. Or what, how to behave civilly.

HESTOR: There is a fearful chill to me here, the dimness surrounding my mind for as much punishment as designed therein.

MELINDA: But you see it is airy light in this corner of day. Yet in these rooms are icons an' sculptures kept to view upon when lit with lamps, to aid one's worshiping of life.... an' loves. It is only as solemn as your emotions permit, inside, contained to preview of a premise that you are fit to think an' earn justly one's gaiety an' pleasures of the world. An' if you know the secret to the latch, you may unlock yourself from inside. But do you deserve to, is the temptation taught to assess or gauge from what you have been doing recently. Thereby you train to become your own disciplinarian, the best schooling possible. My youngest would stay.... for an entire afternoon sometimes, in quiet an' satisfying contemplation. I am drawn to here, a ridge of palace, with remembrances, to decide on what to do. I would early let loose the lock, if I thought them distressed, but without letting them know this, defying teachers an' principles. But they accustomed themselves to this instruction without my meddling, an' with their own examinations of the doors.... An' staid they did.... for a proper span of prospering judged decent.... of what scruples could develop within their characters.... An' must I stay as well? or have I spent enough of time embarrassed of this kingdom. That is my thought of devotion questioning.

HESTOR: I could not see anything in there but the potential for monsters, where shadows shine the brain with imagining. An' I was promised candies an' toys, to speak of Mor-liff.

MELINDA: ....Morliff?!

HESTOR: I am to say Morliff to Loresane, an' remind him of the high personage he represents.

MELINDA: What, for ever why be named of Morliff?

HESTOR: That is who he who brought me advises me to think of —

MELINDA: What son?

HESTOR: —I know him not, though. (I've) Just heard his name spoken with admiration.

MELINDA: Who brings you on this quest.... to influence a seer? What jesting soul could? among your retinue.

HESTOR: The great commissioner Vaclkulor, a friend of Loresane's. Yet I am also, an' I've only come to see him.

MELINDA: ....Vaclkulor's sarcasm brings to you anxieties.

HESTOR: I am paid for this, by he who would have my services to aid Loresane.... with these coins. (*showing his money pouch*)

MELINDA: A rich sum for a small boy. But aid your friend in what way?

HESTOR: I do not quite understand how an accumulation of wealth, privilege, an' approbation of person with praise may threaten one so, except to say Mor-liff to alleviate any harm. It is all that I could think of while cloistered.

MELINDA: You pronounce the name.... as if an interjection for a cause, one you can't grasp. But he is a real individual, of muscular enlistment to the fiefdom, an' mental perseverance to his rank. He is my son.

HESTOR: ....Oh! The Queen Mother! (*coming to his knees before her, bowing his head*) I should have guessed, to be within the palace an' warned not to pester— with yet royalty! I am wholly subservient, to meet whoever here, maid or master. Yet my eyes lighten.... to obey you.

MELINDA: The sense of them. Regard me more.

HESTOR (*lifting his head*): You seem as young.... as my elder sister, an' I thought you as a matron.

MELINDA: I am held to this.... quandary of approach, that spares not privilege or commendation for fostering good men. An' of this kingdom, yet I am kept foreign to the palace, as are you.

HESTOR: But I will roll along the ground for your pleasure, or be leased to an entertainment performing. I know enough to offer myself this way.

MELINDA: That is not my manner with children. They are rumors learned, on the peculiarities of Kings an' Queens. An' subjects bait themselves onto these images to portray.... But you see more of my position than many would give credence to a reticence about, within these walls an' atriums imitated. An' that is as good a perspective for your early education as could droll vespers be, spoken in these chambers of seclusion. You have not made error in addressing me for help, with a motherly tendency begged. I can review.... my highest attainments with this pleading, when oft an otherwise dismal life have I slumbered through. See what we are made of, if not happiness.... then constant disapproval, an' silenced adumbrations of purpose. Such may this be for the appendages left.... regal. But it troubles me that you are not at home. He paid for you — this much— to accompany him?

HESTOR: Yes. An' now he searches for Loresane—

MORLIFF (*approaching*): Mother!

HESTOR: —to bring to me.

MORLIFF (*coming up to them*): I thought I might find Caton wandering about here, given the Visgoth's gravities to think over. (*kisses **Melinda** in polite greeting*) We have to discuss our situations to secure ourselves of proper standing an' agreement.

MELINDA: I haven't found him yet, or may he sleep within one of these rooms. Our voices heard, does not bring forth a call.

MORLIFF: Who is this youngster?

HESTOR: I am named Hestor.... Master Mor-liff— An' I am your servant—

MORLIFF: I might say.

HESTOR: —guided by Vaclkulor, to find Loresane—

MORLIFF: Who?... Oh. The prophet. Your father?

HESTOR: Nay! A friend.

MELINDA: Vaclkulor brings this child, to remind this man of yourself, with utterances of your name.

MORLIFF: For me be spoken? Of want for friendship is our Vaclkulor. But I don't see the point, mother.

MELINDA: He may suggest you, as a Visgoth to be decreed, as he pertains to health an' recovery the perceptions of his visioning an' projections for the future.

MORLIFF: Well, as be assured, many do suggest me already. I don't need seers for that. Events entice themselves.

MELINDA: This is a bartering of the boy's sincerities of friendship, an extra insurance to persuade the herald of your favor.... to this position. Intricacies of the court an' this fiefdom, but I don't like the manipulations of souls employed. It sounds.... transgressive, if not too overwhelming of initiative. I would be rid of much of these dealings.

MORLIFF: I didn't put him up to it! More amusing than catastrophic, dear.

MELINDA: You know not yet of catastrophes, son.

MORLIFF: Oh, I've spoken with Count Halyce, mother. His beans are banal, Ma! An' he asserts dissension with disassembly. Not that I say he is ignoble of character nor means, but it may be difficult to contain him within the fold— with whatever license he requires. An' my labors will be hard enough, to help maintain this collective of proud states an' styles of property.

MELINDA: Don't be so presumptuous, son, or assuming, though you have worked hard enough for my blessings to fulfil of an honorable gentleman becoming. But that is the least pressing necessity for the comely correspondences we toil through, hiding rigors of contempt for each other, like as ribbons nailed together as much to fall apart without much metal or sticky coercion. I am not much wanted, an' have pushed you two through as early as possible to find your means of manner securely quickened of our kingdom. An' my sons are more accepted than I've been, at least of late. Yet much bitterness remains for biting, bickering, an' backstabbing ambitions, drives, an' wishes hardly— barely concealed below a thin pall of politeness or courtesy.... to the details of our constituency. This stoning is dedicated to a willing allegiance to be loyal to the Visgoth. Yet Halyce welcomes me, amidst hatreds, gripes, grumblings, grudged grovelling, an' graybeards apologizing for their spattering of bile across my raiment worn, skittish to be seen for their addressing. So do not blame this count of an impudence that confides of my uneasiness. We are abandoned to ourselves for caring of an' liking. An' you should make peace with him, whatever the event that causes our departure.

MORLIFF: ....As one couple?... or a strategic wooing.

MELINDA: He is not against you (as) so obviously, but simply admires Caton's calm complacency for what he may be capable of doing, contrasting this with your more openly aggressive strife to craft more prodigious achievements to which he could be forced to submit. But this energy of yours brings you the accolades of most. So be you led by your yearning. I encompass both kinds of contentedness in sons, as were, the father himself restricted to me with many temperaments.

MORLIFF: I can not allow this breach.... before the boy, of such personal matters spewed as a sympathy for Halyce, mother. It is of decadence to be confused by, for such ears.

MELINDA: Decadence?!

MORLIFF: For the complexities of governance evolving around one's self to contemplate an' thoroughly misinterpret.

MELINDA: When is a lad not a son?

MORLIFF: He should not know this ever, from these more mature matters preened or kindled for infusion.

MELINDA: I would make privacy ensconced! more privy purpose for a boy to ignore. An' you are here, within this region raised.... Hestor?

HESTOR: Yes, my grace.

MELINDA: Then know your Queen's mother an' that I may leave this kingdom—

HESTOR: Oh?

MELINDA: —to start a revolt against it—

MORLIFF: Mother!... This is silly.

MELINDA: Such are the contracts of affections placed.

MORLIFF: Do not make rumor for a knave to totally misconstrue.

MELINDA: I will have my ways dissembled through pure innocence, that, as he blushes prettily, entail my dissatisfaction, through a charmed demeanor, for this palace life, for this abuse of nerve an' caretaking, with strut of a Visgoth.

HESTOR: I know not what to think of it. When mommy gives me sweetened creams an' jellies, I devour of the bread for its good taste. Must one learn the how of a pleasure, when not why?

MELINDA: So are my rewards for living here. Privilege contorts to tame, to paddle into hushed embarrassment an' keep one couched to glamor's enviable strengths or powers while you are weakened into softness an' pliability of aims.

MORLIFF: This is all very dilettantish for children to be suggested by, with the possible perturbations of rulers an' their causes.

HESTOR: That is what Vaclkulor fears for Loresane.

MORLIFF: What! How?!

HESTOR: That his privileges gained may produce an unworthy arrogance amongst.... you, that could lead him into trouble. I am to remind him of his gentle nature an' humble generosity of talent.

MORLIFF: Oh. He could never be a ruler, seeing far enough ahead to displace steps in crimes—

MELINDA: Ah!... This is how the lure is made. So find you Loresane to greet.

HESTOR: He searches for him in the gardens, to bring to me.

MELINDA: Or we will bring you to them both. (*extending her hand to **Hestor**, as he rises*) Come take a walk with royalty greater than commissioners may make.

MORLIFF: I should find Caton—

MELINDA: Just accompany your younger brother.... an' myself. I will be very definite for explanations from them both, an' have this child brought home properly, promptly, an' escorted meaningfully.

MORLIFF: Oh.... Such to come. We shall not delay Visgoths—for our deliberations. An' so a.... youngster be found.

MELINDA: All visitors are treated wholesomely, under these terraces an' roofs. An' pathways brought through, seeking daylight, make more candor for these emotions to emote an' sense, the heraldry of subjects meant as serving for, some signs of pomp an' prestige seen an' viewed.... with easy cream, an' lushness hued before our traces, (*leading **Hestor** by the hand, **Morliff** accompanying*) these steps for children made an' for a garden to be found with prophesiers.

MORLIFF: One could predict that they produce for us.... a semblance of persuasions. (*They exit.*)

Scene IV — *An open corridor within the palace. **Pontal**, **Loresane**, and **Caton** are meeting with three **physicians**, who generally prefer to remain apart in order to appear consultative.*

PONTAL (*to the doctors*): The mephitic characteristics of salts explain nothing to me concerning the Visgoth. As a commissioner closely involved with him, I wish to know more of substance with less detail, concerning his health an' extant existence.

PHYSICIAN 1: It is a herringbone!

CATON: What does that mean?

PHYSICIAN 2: With sufferances variable and temporal, we can only describe what has occurred, submitted to an inquiry—

PHYSICIAN 3: What's made official. Official is as definite as death; or as a rock would have life wished, what could such a thing wish for? Then try the urbane of characteristics, they are handsome — and with a polish for being exact and documentable. But to say for the whole, it heaved up and down as could befit a vee of vice with desperation rolling.

PONTAL: What does do?!

PHYSICIAN 2: The Phantasma!

CATON: Phantasmagoria?

PHYSICIAN 2: The illusion to life, the persistence to declaration, as could ghosts evolve of limbs and their activities, with movement of joints tested and acknowledged.

PHYSICIAN 1: And a supposition of sweating, even. For where there is a dampness being, were as caused by cold condensation or warmth of elution. Thereby, the day being a day of fine temperature, this is as so of a body's heat, what caused the moisture.

PHYSICIAN 2: We mean indicative of cause, sirs—

PONTAL: How do you couch "supposed"?

PHYSICIAN 2: —We do not try the definitions, but rather observe with examination. And take we this errand physically, as were a trip to find some knowledge about, a frame supposing life. But see you an inanimate object moving. Is this not a cause for questioning, do some forces live to push this thing along? Or is there life to it become, perhaps received? Then we test every little detail we can uncover. But what if it's removed? Then what does move it?— Pressures of gases burgeoning, and retractions of lines and ligaments from a drying out. Yet here reacts much to our poking and pausing to see!... what can be elicited or brought up for capture of our awareness.

CATON: This is terribly unclear, perhaps even cruel.

PHYSICIAN 1: Sir!... We are presented with the ghastly to condone or coach into some reasonable form of personality. Yet as sleep be a consciousness deserved for dreaming states, then how to fathom such a stiffness weighted with supposing? We go through every rigor we can devise to activate the body, and apply stimulatory balms all over and throughout. And this gives us our judgments— but not the decrees at rest. A bit of matter moves here and there. A parcel of the (land)'scape contracts, and then cause a bloating of edema with our tactile prescriptions in function. What is to say of this, but a word of consciousness desired, some tongue of agreement disposing thoughts.

PHYSICIAN 2: And what tongue there be of its scaly protuberances—

PHYSICIAN 3: But that is to cry for a definition roughened.... We can not give a diagnosis to you, before it is agreed upon officially— and declared acceptable, or palatable to our professional notions—

PONTAL: Postures!

PHYSICIAN 3: —that there is clean meaning to what we observe and describe.... For life is to a governance—

PONTAL: Back postures!

PHYSICIAN 3: —that may be establish-able beyond our influences entirely.

PHYSICIAN 2: Sirs!... We can not make monsters governing. We can only point out what is of well-being, and what is not.... and lean if possible towards cures for health. But it is not of our province to decide for you our leaders. We have of stone lent our querying—

PHYSICIAN 1: And often the worst for that. For what is dead, if it moves? And what does capture the gods' wills and wiles without much struggling or display to our caring empathies?

LORESANE: Does capture a remoteness, doctors, to your views.

PHYSICIAN 3: We have seen a log afloat, with tumbling of our feet with roll to balance of our aiming strengths, good carry of a life along if it aloft, if not.... A difficult take, here. And we wrap with quick license of drenched ligature medicinally empowered, to contain all possibilities inherent of a silenced demeanor. Choose your messaging profoundly of this purpose. We would suppose, as for you to instill the necessities of meaning, or insist on them, what should be governing. This is not our question, and not our matter made of, with doubts for prophesying what occurs already.

PONTAL: Blame your erring!—

PHYSICIAN 2: We do not know!... It has not been ruled upon; and we can not subjoin the wish.... of what is to be proposed by ye all. That is not our character, nor our method.

PHYSICIAN 1: We have been given nothing, and leave not less.

CATON: You have your training, skills an' equipment.

PHYSICIAN 3: Sir, the nothing is everything. It encompasses all that may be done. It is what flows and breathes. It is our hopes and excesses. And it is nothing. For where is dark as much as lightness be! It is the same place, the commonality of environments that we pretend upon. And we have tried to play our part in it, with the diligence of submission to the tasks evolving for all of us. Life is a strange impossibility, is it not? It dissolves into a continuum. And so, as if it were, it never was, and we have not lost lease of it.

CATON: You claim your job impossible from the first?

PHYSICIAN 2: From the beginning?... I'(t) was— as were lives running about a temple. How can you think of such hopelessness, with our potential perfusing his! And what does make us clear, a sight through dimness, are these features learned: this is as clouds would make a rock, and handsomeness grown old, and wisdoms worn with wrinkles, and blisters born at birth of a resurgence of will. It takes determination to die, as much as for excuse to live all endowments taken for granted insouciantly. This we always find, and this we found. As could a baby cry, scream and yell, for the slightest enjoyments of feeling.... so then a stone lies rigidly for the merest of perception, the least kind gratitude of touch and appreciation of its terror. Here does expression make its rounds.... with somber strokes, and our slush of sobriety against intoxicating wonder. Yet we abide by this to examine carefully, and predict whatever useful treatments may possibly be applicable.... to all consciousnesses involved with interest in our working. And this concern is spent on us— to allow your decrees a reigning substantiality to our efforts. We have tried our best to procure.... a seemly sense about this subject, or to make what is objectionable.... ornate of some kind praise delivered with our hands— and minds. Eternally is nothingness inclined towards. And it is won through a fusion of our senses, into a dull sublimity. Thus, our practices have condensed upon the Visgoth known. And he is found within these bounds of your righteousness.... or worthiness to receive him as your life's continuum expressed. We three would not say more about it—

PHYSICIAN 1: Not a jot.

PHYSICIAN 2: —for witness of your proceeding around his estimation as to what a Visgoth is!

LORESANE: He is as projected. (you) kindly practitioners of an art.

PHYSICIAN 3: Not an art, sir... but with artistry, the sinews inspected for firm consistencies and the sinuses for the cling of clean retentions.

LORESANE: An' a clang of mastery. I would propose you leave him thusly, the master of his own body.

PHYSICIAN: 3: It is most definitely his left, whether wood or wooded with the serpents of surly surrender. For such investments insure a spiritual peace and a relaxation of soul.

LORESANE: All of that which is more voluntary of (a) release, then I would claim clings on.

PHYSICIAN 2: Sir, more dye is needed to be clear.... of the skin. We have observed the pallor, touched it, and even tasted it— through our chemicals symbolically, symbolistic-ally. And the signs, the reactions of the peel-y flakes present a notable purview of condition and state at rest—

PHYSICIAN 1: We were thrice confirmatory, after exercising of the patient through our vigorous examinations. Independently determined through our individual inquiries and answering of various aspects of specialty to health, and yet consultative-ly in agreement on the results, we do conclude ourselves to know what lies and how.

PONTAL: More dear, that death! but to a threshold brought.... You could not save him—!

PHYSICIAN 2: Here safety sues you, for it is wont yet wanting of ambition. We draw inflection of it merely, to a declension 'bout its nature. What is to save is passable and plausible, not within miracles but within medical advocacies, and so demonstrated by our efforts to the utmost.

LORESANE: Then be they effectual on my eye, since you can make no claims of state—

PHYSICIAN 3: We can not! It must be official-ized, how we have left it done, and did all that was to do. The Queen herself oversaw everything, and will be appropriate for us. But for the verdict, it is hers to claim and finalize— officially. And we shall return, after her wear of contemplation, intelligently left her, this brief span of time, to foster the requital of proof personally that the reality of the situation is aided with professional adoption, and papers are signed acknowledging our services and all attempts around the Visgoth. For what we left, sirs, was as much of grave as granted of this gravity.

CATON: To which he returns?... or will return?

PHYSICIAN 2: The darkened cavity speaks much for itself in mourning, for a resurrection wished, with silent intonations of intimacies, wife and son absolved to seize these precious moments of reckoning with a faithful maintenance of respect and dignity. And even the youthful Queen Mother adjourns her dispatch of uneasiness, for the deliberations presenting themselves to condone. Without cries heard, or tears seen, did we leave the bedroom and its candle light, the occupants more brave than night hiding under this day, this sheet of despair, this cloak of calamity—

LORESANE: But resurrections, doctors, are typical for healing. That is how recoveries are made. It is the nature for the action to be sufficient—

PHYSICIAN 2: Sir—

LORESANE: —The hope remains within his powers seen an' sensed, not your powders felt or smelt or smacked—

PHYSICIAN 2: —So have you sir, this privilege of speculation.... We can not conclude this for you, one way tempered as a steel, and the other way loosened as a lace coming apart. Yet call for expectation within you nobles and your groups. You are as may be divined and appeased, for much circumstance— above our humbleness— to handle and manage correctly. And with preparation tame your tenacities to our message: the Visgoth blew seriously his tendencies of existence before us, with difficulty and strain.... and regulation of constraints to heart and respiration. And that is factual to say for you.

LORESANE: Then there is illness come, as if we were to call for some alleviation with your presence. Then all of this is due somewhat obviously—

VACLKULOR (*approaching*): Loresane!

LORESANE: —of description.

VACLKULOR: I find you.... near the gardens. What are these, Pontal?

PONTAL: Physicians, for the Visgoth.

VACLKULOR: To be persuaded by Caton?

CATON: I can't advise these doctors, Vaclkulor! Their methods are likened to mystery plays.

PHYSICIAN 1: We are of independent faculty, sir, dependent

only on the patient for our learning to be influenced by.

VACLKULOR: So to my friend Loresane am I sometimes similarly doctored with advice.

PHYSICIAN 2: Health-wise?... or pictorial.

VACLKULOR: I bring you word of Morliff to relate, through the gleaming boasts of your little companion.

LORESANE: My companion?

VACLKULOR: He can't stop saying the name. I thought there might be some attraction to it for you, through otherworldly processes I can not grasp or distinguish, with the little tot's sudden insistences to indulge on his mentioning.

LORESANE: Tot?

VACLKULOR: The boy, Hestor.

LORESANE: Hestor rambles about Morliff?

VACLKULOR: So much so, in childish feverishness, that I invited him here, for your counsel, to calm him down. I found him at your hut, while searching for more commissioners; an' he virtually screamed with gaiety, for Morliff to be known as if from some ethereal volition passing through him. I thought it quite unusual for the subject of this personage to be brought up so voluntarily by such a naïve an' inexperienced youngster. Yet, he also asked to see you, caring for your fortune. An' I assented to the visit, as I was returning to the palace anyway, somewhat discouraged an' yet wondrous of the boy's nature. What does this portend, do you suppose?

LORESANE: You brought him here?... with his approval, and his parents'?— Where is he?

VACLKULOR: Playing in a room I've stored him at, one of the brothers' small cloisters. He'll be safe there. We wouldn't want him running an' rampaging through the palace, trying out childish pranks an' activities, especially today.

PONTAL: Nor as a guest on any day.

CATON: One of my sites of meditation? Is he a fair boy for thought? Otherwise he might become quite abruptly bored.

VACLKULOR: Tsh! If you an' Morliff used these spaces as youngsters, they must be full of toys, appropriated suitably for children; though not as windowed as I had expected, given their location near to a boundary of daylight passage.

CATON: We need the candle, man! in these spots of seclusion an' self-seizure. He could be frightened to death, in there alone, an' feeling through the darkness images of grace as fear of figured deities in statuettes an' icons of bas relief.

LORESANE: Take me to him immediately!

VACLKULOR: For this he waits with impatience. Come towards the rose thorns where adjust a pathway lined—

MORLIFF (*approaching, with **Melinda** and **Hestor***): Vaclkulor!

CATON: My brother arrives.

LORESANE: An' with my friend, at last.... adorned by the Queen Mother.

PHYSICIAN 3: What wary wake!

HESTOR (*calling out with excitement on seeing **Loresane***): Loresane!

MORLIFF: Be ye fit to appease my presence, commissioners?

PONTAL: As more for more the same.

MORLIFF: Dare deed a young son to me, Vaclkulor?!

VACLKULOR: What heed, sir? He adores you, as do I— by necessity bring.

MELINDA (*still guiding **Hestor***): My lad was prompted, for this influence.

MORLIFF: An' I need not the interference made. But we find my brother.

CATON: So be so. Assemble we today with graying clouds.

PHYSICIAN 1: It is quite clear, sir.

LORESANE: Ostensibly, waylaid to candor.

MELINDA: This boy should not have been abducted for these purposes of caucus to promote, an' to infer with, an allegiance to either son proposed. The abuse of stately innocence is improperly compounded.

LORESANE: What say you of Morliff, lad?

HESTOR: A grand gentleman, Loresane, as is often observed. An' to accompany me most civilly, he accepted my politeness graciously.

LORESANE: But does he dominate your thoughts?

HESTOR: As for to see my friend Loresane.

MORLIFF: An' how are we related, thus to see? a subject an' his sire served to bring you. As to the king himself, you are welcome. An' what does friend Vaclkulor make of thee?

HESTOR: He pays me well, on your behalf pronouncing ever your name, Mor-liff, for Loresane's ears—

LORESANE: So do I have it heard.

HESTOR: —for less span of head an' arrogance throughout this court.

LORESANE: That is obeyed of me, Hestor. I've only boasted a little, until some courage found.... an' thank the noted commissioner for his concern. We did indeed propound ourselves to suggest the good qualities of this noble son, as ever I could perceive them. I did not predict it would involve *you*, or wish for your entanglement precariously. Were you afraid, sheltered up in that darkened corner?

HESTOR: As for my balls to roll, I would have flung— in fear—

VACLKULOR: At me. At me, again.

HESTOR: —of the monsters emerging. But the Queen Mother does save me.

MELINDA: From your own contemplations. An' I will have you homeward sent straight away. For even at this age you are too young to see within yourself, foretell the worth an' forestall the worst.

CATON: It took me days of hours.... to be baked into a satisfaction for this life.

MORLIFF: Well, we must talk of this, an' how you do prefer to let each of us be an' grab of accomplishment, honors and renown, an' these dangerous responsibilities approaching.

PHYSICIAN 2: Ever for assessment, my nobles, and agreement to facts with election of popularities. I take it the Queen.... does prepare— ourselves.... to these realities?

PONTAL: Within a general reasoning I would suspect this sameness of sanity, upon leaving the Visgoth, dispersing, as it seems.

MORLIFF: I can't say I built up the same tensions as you, Caton, locked within those cloister rooms.

CATON: What— of conscience? I never did escape but once, after learning how.... an' have ever felt guilty for that sacrilegious act of removing myself from my thoughts, my most profound searching for insight.

MORLIFF: Insights.... into behaviors an' comportments only. But not so much knowledge, unless it's memorized— by rote.... verses read and repeated aloud, or only to the mind if particularly lazy. I said my few, an' left, but always learned some new ones for each session. An' that was the mental discipline earned before playing.

HESTOR: We ain't got such detentions.... yet. I mean, to think about. What is presented us conforms to us, our thoughts to think about.

MELINDA: Then there's a paltry few for you to think about an' be obsessed by, these simplicities of life an' their number. Pregnant to their meaning, but sparsely found with some importance to observe, those tiny lessons of solitude are meant strictly for the suborned of character to heal, or personality improve. Or wouldst be trapped into conditions derelict of kindness, one must absorb to evade.... disgraces, an' then escape. I should have you accompanied

—

PONTAL: It seems to make for time, commissioner, that we retire with the others, to polish our debates, before the Queen to meet with a presentation of solidity, solidarity, firmness to our resolutions, an' reasonable insistences.

VACLKULOR: That is clearly before the door, Pontal. Before

this very Visgoth, I must conduct the session. An' I must wade through our choices an' capabilities, to count, collect, an' deliver of the fiefdom's strength, with each of our domains agreeably treated.

LORESANE: But Vaclkulor does bring—

PONTAL: I would make a bid for your departure *now*, Loresane. For what has been seen.... is not envisioned from you.

PHYSICIAN 3: Seer! Aye to the rumor, let loose throughout this palace to roam, ramble and rumble of prodigious worth of prestigious talk. But then we should combine as well, doctors, to give merit to our final observations that may conclude our standing by the Visgoth's bed, that there does lie.... a person and a patient.

PHYSICIAN 1: And a loss before the Queen, but not to heaven, and not to hardy ambition. For that is life, betwixt the non-life.... and the spiritual.

PHYSICIAN 2: So say our weapons, and our instruments and implements of aid. We are dedicated to those thoughts alone.... Loresane, with these weighty abilities, these heavy excursions through truly temporal editions of life, edification of facts, and excuses for attempts—

LORESANE: I would be, forsooth for sought (of) these attempts.... timed, as they were, upon my unerring pronouncements. But where the deficiency lies is where I sup an' suckle to hone the honesty of my volitional whims an' voluntary employment, deployment, or just implementation. For what is false of a wish an' a hope, there be the power of a prophesying. An' forgive my semblance of you. But without antagonisms for the heresies a headiness may announce, I say with the most humble originality for my suggestions that I profess only what I am told or made to feel. An' this honesty can not malign truth, as well as each of you may be candid. The only error, perhaps, is in having spoken. But that I am called to do. For this I am pleaded to, an' obey with tenderness, as my witnessing is as by a feather stroked. So soft may cruelty be, or fortunes gained, is but a brush of the fated. Whatever may be supposed by my enunciation of weird timbres an' inflections, translated through me for bold, emboldened ears, I leave for your conclusion to sharpen an' make definite, an' draw upon for the crafting of your labors an' endeavors through your own influences of character. I merely shake for you, to hear from what has struck me, as does a tuning fork. But what does shiver of calamity or cataclysm becomes throughout your own interpretations due. The messages released through oddities are mysterious an' tricky for understanding or mental enlistment, so hued you are to your own prejudices for their sifting through (as gates).... to reach your adumbration of their details with favor for your purposes. Then let it be more clearly said that I am not purposely odd, but am forced to be, through your approvals an' expectations to exist. An' what are judgments made with this condoning of the prophetic arts? They're only for the wherewithal of man being much greater than his destiny allows or permits.

PHYSICIAN 2: And grander yet, with tools instead of tunes or melodious inspirations. But chime our brains more with observations of facts and their judicious manipulations for more realities to beget—

LORESANE: An' be gone to this, haughty fellow. Dreams are more dissolvable than disasters. An' if you say he's—

PONTAL (*observing to aside within the palace, distracted by floor thumping*): My lord! What yields to presence!

*From within the palace slowly enter **Vincent**, in a long and heavy night shirt and with a staff to aid his walking. He is nevertheless functionally escorted and assisted physically by **Terissia** and **Fabel**.*

VACLKULOR: The Visgoth!

PHYSICIAN 3: What be the spear of this arrival?!

PHYSICIAN 1: It is remarkable, even the ambulation!

VINCENT: Seat me to that bench, for I am more exhausted than the sun's muscles.

TERRISIA: Yes, dear. An' terrify not still the day, your petty passage made through it. It is a want of leisure less an' leisure more. But certainty insures your growth of gait, now to be seen with daylight freshening outdoors. (*All others make way for their walking towards a nearby stone bench.*)

PHYSICIAN 2: But how (is this) recovery? And why this flight —?

VINCENT (*suddenly lifting up his arms in a show of might and defiance, one hand still holding his staff*): To give these wounds breath of air!

FABEL: Father! Be careful, even yet of strengths that leave an' return with a rush, do rip out more than nerve an' courage sometimes, in their haste.

TERISSIA: Yet the lungs restore themselves soundly.

VINCENT (*seating*): But she who trusts me to be nursed does discourage thoroughly a foundering. My bed's more damp than a river's. An' she does toss me out to clean up an' replace the sheets an' spreads. An' so in telling says: "Take some air with the flowers, an' build up a resistance to 'inactive longevity'." Oh! What a term for illness sleeping. But she does paste my body over with a camphor to disturb me an' irritate ma nose suffering stale scents.

PHYSICIAN 1: We had a hand in this too. But say you have more power to your groins, and demonstrate as much, irascibly squirmed?

VINCENT: I breathe! An' take great gulps of living freshness. This garden clears my head, though it stays feeble.

PHYSICIAN 1: Then to these bites (caused) I might suggest.... the ivies and the vinegars.

PHYSICIAN 3: Yes. For with the bathing oils elicited the wounds. Caused the bumps to burst.

VINCENT: Ivy?!

PHYSICIAN 2: So subtly so. I see now. The poison oak leaves, placed on the graves instead of wood ash; perhaps because the bereaved had not access to that typical for a fragrance used.

MORLIFF: Or more as a deterrent to trespassing animals.... to roam upon the mounds, carefully made.... carefully made, the punishment.

VINCENT: Maybe. But I do work some discomfort by it.... An' so the seer sees me, an' predicted I would rise.

PHYSICIAN 2: It is as common as health, sir, and not unusual to foretell with healing.

VINCENT: Quite so. An' you've examined me enough.... to make some claims of encouragement. But.... Loresane. I am still weak.

LORESANE: What birth is not? Yet of the greatest power to express, that is a demonstration of your being, pride an' prowess. The more common tones of musculature will return in time, with the talented care of your physicians an' nurses thoughtfully employed.

VINCENT: Then that's good.... to think these things through. What have you, Melinda? Yet another son?!

LORESANE: He is a friend of mine, brought to see me.

VINCENT: Hmm.

MELINDA: The boy is of our kingdom.... Visgoth.

VINCENT: Must be well able, to consult with Loresane.

HESTOR: I am, your highness; though his life is pitiable, an' his shack rather holey—

VACLKULOR: We do make for my friend— friends.... consultative visits, though his lifestyle is quite stern an' spartan.

VINCENT: An' my head feels faint, even with all of these riches. Yet for recovery, it is personable to each.... as to how an' why— an' when an' where. There in the forests do you live?

TERISSIA: Our subjects live on land, Vincent.

VINCENT: Yes, commonly. An' much spoken for are they.

CATON: Still with much pleasure for you, Visgoth. I have heard both crying an' wailing on your behalf.

PONTAL: Perhaps, but that is by your particular company, Caton. I should say you should be less foolish in the future, Visgoth, to virtually antagonize everyone in some way or manner recognized with authority. But I'm sure we can agree to some prosperous terms of behavior an' conduct throughout our fiefdom. You have much of a strong constitution physically to endure these latest troubles or afflictions. Now you have a good opportunity of rest to improve your mind as regards your handling of.... tempers, teases, temptations, an' terseness of administration, to remain successful at our governance.

VINCENT: I *am* trying to make a relaxation, Pontal.

PONTAL: Not that I'm being overly critical, but you are well aware of many complaints.

MORLIFF: He doesn't have to be so bothered by that now. What runs through currently is a restoration from troubles, as if resiliency is pledged.... far above caution.

VINCENT: An outstanding observation, brother. I seem to be recovering from crimes.... with lack of caution. An' I will consider this carefully, for I do not *wish* to annoy.... any good person, any genuine conscience or pure soul not warring with. An' not with poverty to speak with, nor from, co I claim that anyone must have need of some insight. But low to law, if you debase it to transform into unwholesome, unworthy deeds. For the future may bite back, with angry, bloodthirsty teeth, to force you to obey.... her wishes. An' so I suppose I've been scraped a bit an' bludgeoned a while, welted for a weening. There's toc powerful a magic for the amateurs to play with an' abuse. But row I see what leads to leave me breath, though the eyes remain blurry an' easily misinterpreting (of) some features. Still, as guided by the reasonable to sense, I submit myself to caution.... an' the feelings of others, for both their losses an' their gains.

HALYCE (*coming from a part of the palace almost distal to where the* **Visgoth** *arrived*): There you are!

PONTAL: Count.

MELINDA: Halyce!

HALYCE: I've been looking all through this entire fortress for you, Melinda. Almost with desperation, that you had become unbalanced and self-injurious due to the insult, did I search.

TERISSIA: Insult?

MELINDA: Was no insult, but a description.

HALYCE: Find you a boy?

MELINDA: A prince of remembrance, reminiscences.

HALYCE: Ah? A premonition? A hope? I've come to take you away from here. Definitely. So does the Visgoth wake for day, sir, I tell you bluntly to the presence, an' as forcefully as the stench of your medicinals.

VINCENT: Or well to be, a perfume explicitly promoted, it is a compromise for some compassion an' lift! from the doldrums of an inherited state. But what says the heart? I know I have been sick, an' she unhappy, an' you dissatisfied. So let us improve ourselves with more harmony of consciences. For I remain on the throne, an' approve of you all, the reasons for your disquiets generously permissible.... an' pardonable. This palace need not be a sanctuary for unjust cruelties, nor for my rule over you to be abolished with sorrow. I contend more so to let you be happy an' end the folly of my foolishness. I am convalescing from disagreeable conditions of being disapproved. So then I should submit to all of you a suggestion to be kind an' genteel with your worldliness, gentle an' affixed to some form of justice, an' have an intense sympathy for the plights an' misfortunes of others.

HALYCE: Where after is the warrior?! Have you transformed yourself?

VINCENT: I have transformed my grievances.... to myself, an'

fourd them (to be) wanting or silly. Come here, young boy. (*Releasing her hand,* **Melinda** *allows* **Hestor** *to advance towards Vincent. He stares at* **Hestor** *with some scrutiny of puzzlement, squinting to see him clearly.*) Who are you?!

HESTOR: I am Hestor.

VINCENT: But *who* are you?!

HESTOR: ....I am the son of my parents.... an' the good friend of Loresane.

VINCENT: You are "Sir Future," an' that is all that need be known about this precocious, precarious, anxious state. An' I predict for you your good health, as a gesture of kindness; though no one can ever guarantee anything but can only propose earnestly, honestly bewitched with hope, perhaps. I had thought to see many of you, with mental inspiration. An' now, all condensed into a single child, this is my prophecy before me standing, an' what I have searched to find, with such uneasiness an' worry. I did not stop to think of asking of a child.... to see a childish face. Such things.... reside on earth, where heaven finds its lease to men advantaged. An' now.... does Loresane have riches?

LORESANE (*coming to* **Hestor** *and* **Vincent**): With such company, a wealth is to be maintained as thought of. (*hand on* ***Hestor's*** *shoulder*) Entrusting to endearment an' friendship is the highest gift ever to receive. An' I take your meaning, noble Visgoth, quite sincerely. Such foresightful realization astonishes, which makes for prominence of a leader, what should be obvious for all to be led by.

VINCENT: I would say so, an' without being delirious, nor effete of substance, but weary from this battle of conscience an' the upheavals that have been caused by it.

TERISSIA: Vigor remains with your mind, Visgoth. It supplants the disappointments felt. But no one can blame you for the hurting suffered. That is to be witnessed an' admitted by the ruler, in every case of harm to feel an' swallow an' swell on with empathy. Terribly difficult an' emaciating, yet to be done by someone relating these disasters more than to fate but then to your felicity of caring an' fostering of amendments.

PHYSICIAN 1: Would be, for our superfluousness here, and superficiality whereat to correct ourselves, I then suggest some simple rest for the Visgoth—

PHYSICIAN 3: And many more fluids of meal, to restore the salts and the waters—

PHYSICIAN 2: With meats, doctors, beyond the vegetables and grains, to instill the body's frame of muscles, sinews and tendons, with energies and fortitudes to re-accustom by. That is the simple prescription, straightforward of any mothering.

VINCENT: Oh, you medical professionals stay quite necessary for my recovery, not in the least ineffectual to it, an' an addition only of commendation with your use.

PHYSICIAN 1: Then pass we along to others calling for our services, Visgoth. For we delay not our devices where pleaded.

VINCENT: You have egress found of honor, an' an out with soups. (*The physicians exit.*).... They do return with potential torment, but portentous instruction.

MELINDA: The sky becomes of graying, as the day draws down.

HALYCE (*placing an arm around her*): So what we are to think with? of some escape an' excuse. Then lingers of our commitments, to ourselves.... an' others. For a while we will obsess of welcoming.

VINCENT: I goodly offer bed an' food for all seers of a passion.

LORESANE: Aye, my lord.... But I will return to my humble domain, an abode of humility with pleasure.... an' led by this child an' friend, as you so thoughtfully suggest.

VINCENT: But that's how a future's to be treated, an' spied upon — with friendship.

LORESANE: Yea.... (*starting to leave with Hestor*) For loss of vision makes one worthy of another's.

VACLKULOR: I shall accompany them back, Visgoth. An' we will digest awhile, my friend Loresane an' myself, what has occurred today. (*exiting with Loresane and Hestor*)

VINCENT: Well (to) do. An' see that the future returns safely to 'is home. (*The three have exited.*) That is a true regality, to force one's self emancipation from our splendors.

MORLIFF: We should talk further, Caton, about our positions within the fiefdom, in order to remove all doubt of circumstances happening.... for what may occur in less propitious times to come. (*starting to leave with Caton*)

CATON: So just we are, to bind with our brother thoughts of deliberation.

VINCENT: Morliff! Look in on Loresane, to improve his housing an' status in our kingdom. I will walk with him some days, an' thus should be entertained with decent accommodations at his residence when we come there to rest up after our sojourns.

MORLIFF (*acknowledging with a gesture while just exiting with Caton*): Yes, Visgoth.

VINCENT: With his steps I will stay out of more.... disgrace of deprecation, humble as he is to assume my favor over his.

PONTAL: Well this is queer, Visgoth. I should retire to the other commissioners, presumably just before your bedroom. We have a number still to officiate decisions. Yet he who was to head the meeting has left—

HALYCE: An' so do we withdraw, an' with no oddness to it. For time is a party of preparations to anticipate. So do we adjourn, an' to your health pledge better, Visgoth. (*starting to lead Melinda*)

PONTAL: Where to, count? This breaks the quorum—

MELINDA: To our consecrations of desire, commissioner.... an' our commiserations for yours. (*The two exit.*)

PONTAL: ....Then this is hacked, Visgoth. We can not pronounce on you tonight.

VINCENT: So be the fleetingness of friends, Pontal. Throw a party for your members left. An' invite my son, as were you before to entertain. The day does stand with me. An' the night.... my recess from your noises be.

PONTAL: There be well, sir, some proper prosperity to your condign tasks ahead. (*Exits.*)

VINCENT: Come hold me, family! (*Terissia and Fabel sit to embrace Vincent.*) This is for light, a surface made to spread upon, my love.

)( )( )( )( )( )(

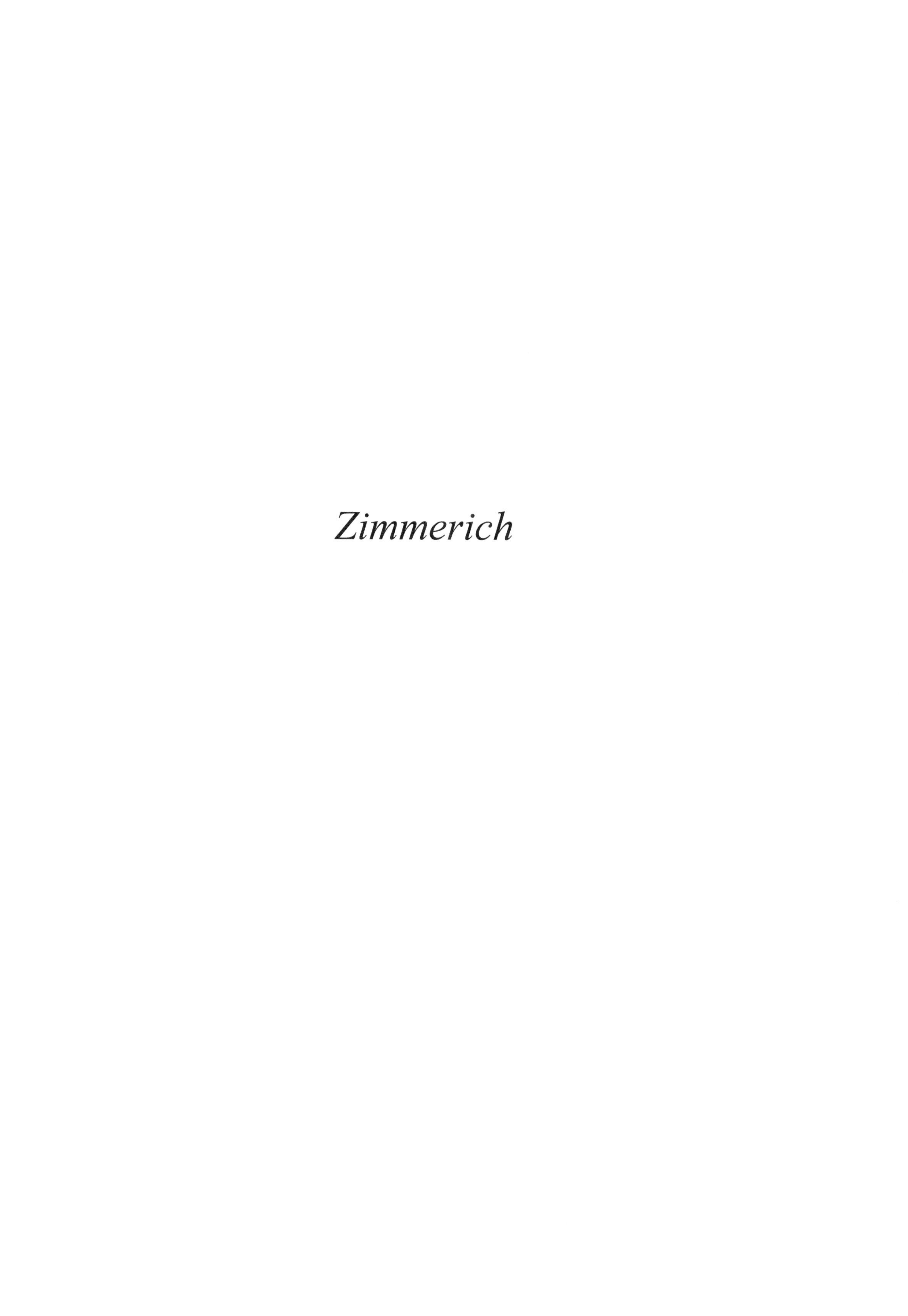

Zimmerich

Zimmerich

Characters:
  Vogel
  Kanter
  Paths
  Margathrate
  Carl
  Pathes
  Lorken
  Hastern
  two Indigents
  a Physician
  Mathews
  Harold

Locale: A rural town.

Act I

Scene I — *Two parishioners are sitting in a sunny comfort room to a church, generally used as place of relaxation or retreat for lay workers after or in between carrying out simple chores. But it is also a common meeting place for general idleness, while containing a relatively sparse assortment of household implements for cleaning, painting, and abused or forgone cookery and utensils. It is not a place for slumber or napping, and is usually locked up during the night, after evening services, when the church is generally closed. But the inhabitants of the town view it as particularly theirs to consider and utilize, compared to the more serious grandeur of the solemn and somewhat mystical religious edifice built around it. For even the church gardens carry the imprimatur of a cemetery to become, if not already for various pets. Such are the local rumors of the ecclesiastical permissions grudgingly permitted by the bishop regional, to patrons of particular favor at this parish. Nothing is known exactly or explicitly by its members, and gossip stays respectful of the operating clergy. **Kanter** sits near a corner, next to a bin or box of old mop heads, reading a thin but broad pamphlet, apparently not of a religious material or subject, occasionally muttering to himself at indiscernible vocabulary, somehow a learning aid for him. **Vogel** sits relaxing at a centrally located and busy table of various discarded items, some rather dusty. He is supple in a very stern wooden chair, of a character finer than the simple though utilitarian-ly handsome table, one arm lying across it , as best it may, to support his slouch. He is doing nothing in particular, but with great satisfaction of purpose and even a hint of concentration towards this. He is a worldly thinker, in a rustic rural town much self-contained.*

VOGEL (*loud enough to be easily heard, but strictly for his own amusement*): Kay-art! This finery of moat in lush, luscious grasses.

KANTER (*mumbling*): A— apiary.

VOGEL: What bees? No! Wasps.... to sting the eyes. There is a season to it, of hot humidities, followed by much dryness.

KANTER (*responding with indifferent attention*): It reads he was a bee-keeper for honeys, out of his passion for the flowering blossoms. And that this lent grace to his resplendency of view to found so beautiful a habitat for others.... The rigors, details and cautions of the hobby, to impart a sultriness to the botanical splen-

dors he adored to maintain and support of panoramic luxury, most senses overwhelmed.... in the hot summer.

VOGEL: During droughts? He didn't draft the scope of land, but found it thusly, of a natural wonder for picturesque sentimentalities.... this.... describer of our little hamlet and settlement, with his name: Thorngrile. Implies the rosy fragrances, or barbs for sweetened fruits of berries to protect, till ripe enough to fall to the ground and be eaten, seeds spread deciduously and with delectation. But what have we here now but an expansive collection of mud holes drying up under the unrelenting boldness of the sun.

KANTER: This was once a very scenic valley, Vogel, a region of indeterminate number of floral varieties and species, and more colorful than the pastel representations in this *recueil d'art.*

VOGEL: History is lame to the realities ascribing to. This parish burns. The leaves are burnt. The tree tops are autumn-ed already in their shading, or lack of it. And what ever flowers withers wretchedly after only a few days, a few hours of sunlight.

KANTER: Could be a mode of summer, Vogel. Could be the means of a weather made tolerable with these musty efflorescences of vegetations and regenerative desires.

VOGEL: The flora have no aspirations as do I.

KANTER: To sweep the pews?

VOGEL: It's only a gesture. And they are wiped, dusted, polished caring-ly. I'm not one to mop around.... before the alter. We have heavy-set clergy for that.

KANTER: How 's (that) said?! You mean those who direct the midwives of the church.... to do the actual cleaning. And I'm told they find all sorts of dirt and rubble left there by the kneeling.

VOGEL: And by whom told?

KANTER: My wife, even. Is she not one of them?! And too old for the service, I would claim, yet more respectful of it than our own household. Well, how can one manage beliefs? She fears some evil vengeance or trial if she does not have her wish to purify the holy.

VOGEL: Sans sweat.... they are pig-eyed through their devotions.

KANTER: What?! Me wife?

VOGEL: No, the parishioners. And that is an endearing term, for all the bothers they must run through. The eyes swell up for (all) their troubles. But they are lovingly suppliant to pray.

KANTER: I've not heard.... of great difficulties among us, not above the fold of a common life.

VOGEL: The younger folk, the less established types, are often through their problems seething and struggling.

KANTER: That's to retire for, a rest from being enterprising and aggressive.

VOGEL: Thorngrile does not help.

KANTER: This city? It's location is of a sparseness of disaster.... Though it's hot all over.

VOGEL: I mean of the saints, Kanter. She does not help, but rather to be helped is the imperative promoted.

KANTER: Oh, the church herself. Well, one is known by the schooling and religions espoused.

VOGEL: Then how does a bee-keeper become a.... sketch artist? He was dirt poor and rugged, until he made his fortune.

KANTER: That's as much as a painter being.

VOGEL: He wouldn't have this church.

KANTER: ....It was not put up during his time. And he wasn't the wealthy squire, if you count a fortune as barely making a comfortable living. Just someone favorably remembered as an adopted tradition of hospitable prominence, the town named after him.... after he died and bequeathed his ranges, ranch and drawings to the public good and common use or advantage from.

VOGEL: It's odd how his works are justly valued, while this diocese can only drink the sap of its inhabitants.

KANTER: That is for a beneficial intent of communal commonality to evoke. But our folk are proud.... of the sweet benefactions provided here, of powerful thoughts and concepts to grace the mind out of many sufferances.

VOGEL: An entertainment I enjoy as much as any. Yet it merely replaces talk and chattering with a focus on ideas ecumenically pleasing or aspiring towards. That does not really help a material basis, as much as she is helped materially.

KANTER: Are you complaining?

VOGEL: Oh, no! Not at all. I partake of the raptures and the agreeableness to share through them many associations of determined precepts and allegiances to.... goodness, or at least feeling good to our professed devotions.

KANTER: Well then.

VOGEL: I will hum, hymn and dance the same of simile to god as man. Helps to keep one persuasive and acute of promise. That's why we all volunteer.... to do our little chores around here, without the clergyman capabilities to attain— nor become envious of, for those subtle, purely intellectual responsibilities.... Nor to be envied by; those members are certainly suspicious of us.

KANTER: Well, we're not as well trained or fortified with knowledge against the evils of wanton temptations. They apparently know secrets, about human nature, we couldn't even understand if we read them, or read them aloud.... to others.

VOGEL: I haven't much faith in that supposition. But we should test a bit, ourselves. The reality of devotions challenges many cherished traits to our becoming.... useful, for embracing and celebrating our lives and lifestyles.

KANTER: For me that's getting out of the house.... being useful. I have boredom like a cake around the mouth, there. The wife too. Frankly, we don't like seeing much of each other at home. I mean too often, during the day. It reminds us that we're functionally retired.

VOGEL: That is an oxymoron. But it emphasizes the point.... of the importance of this comely building, this statuesque structure of architecture, a mini-cathedral midst admiring country and pastoral muses.

KANTER: There ain't no bishop's throne here. And it's only as large as a couple of barns. Yet the wide spaces of expansive solemnity and void full of (the) monastical imperial piety, and a monarchic nobility, do give it a splendorous grandeur at times, especially during important services, and the burning of the harvest wastes. It, standing before us with celestial approval in the background, only enhances the celebration of our prosperities and continual good luck. We are proud of our church and every splitting timber within it. And she does pardon us.... for our proverbial laity of laxity. It's good to do some work here, even if it's superficial but regular maintenance. (It) Demonstrates more than our caring and adherence to her order, but also our pride to stay devout and subservient to the religion despite all troubles or hardships.

VOGEL: Yes. That is the test.... the perception. Do you notice that our two deacons, and the priest, have the run of the place?—

KANTER: They're supposed to.

VOGEL: —And they have their own offices.

KANTER: Well, people need private spaces of inspiration, to study up for their ordinations.

VOGEL: If ever that be done, for those two. But they prefer us to closet ourselves up in here, when we're resting or have nothing to do before a chore is ready to be tackled.

KANTER: Well.... This is our room, the community's. This is the original spot of conception, the very heart of the principle that evolved into a house of worship. This is where some influential town's folk decided to have undertaken the task of this resplendent construction, and planned out its financing. Even if the room has devolved into sort of a storage area and carpentry assessment space, it is ours. It is privileged to us, the people and the lay worshipers, the only room with a locked door that opens from the outside, aside from the main entrance. Not even our priest can go directly to his rectory from the outside. And we each have a key, to enter this room—

VOGEL: Though not one to cross from here into the church proper. That door's not only locked, it's bolted.... at night.

KANTER: Even my wife has her own personal key for this room. She insisted.

VOGEL: There must be a hundred duplicates, Kanter. They must keep a whole chestful of them around here somewhere, maybe right in this room. There's nothing especial to it.

KANTER: But for a community standard, it's noteworthy to own.

Almost like a badge, or an offering of peace from the diocese, that we shall take our directions.... with an internal fervor of responsibilities demanded.

VOGEL: They might think (of) us (as) merely prudish of ownership, with this one, relatively modest, concession from their authorities. Though our priest certainly runs the order here, and knows the complexities of stricture to be followed or obeyed. And he's generous enough to entertain opinions from us.... But you've been associated with the place longer than I. Why haven't you become a deacon?—

KANTER: Oh, no! Much too much work, for little compensation, and slightly more respect. I'm not so impressed.... with such a pecking order that gives them the ability to make so minor decisions as to store us in here and decide on which cleaners to use on the stained glass panes and bronze panels. It's not so important to me, though it *is* important. Directives have to be made and carried out. But I don't have the autocratic tendencies within my easy nature.

VOGEL: You're not a lazy man, given your age.

KANTER: I can mix and handle resins as well as any.... and patch up the woodwork. I wish I were more (of) an artist, but the shaping of things out of a formless mass confuses and stymies me. I don't know how to start. You just have to have the talent for that, and an impulse to jump-start over inhibitions. Whenever I try, to sculpt or draw, the early results are never even close to what I had originally imagined or intended. And those initial attempts foil the effort entirely, like for a child's. I become quite disgusted and dissatisfied. The wife only laughs at me— as she's a right to do, after all these years— for my blundering with messy materials. She can sew better than I can scratch with chalk.

VOGEL: Frankly, I despise them a bit, their haughty pretensions, pretending to assist the learned priest, as if hands (make) ornate a wisdom. Oh, such an education. I can't remember a thing anymore, and I used to have a good memory, a fast one. I think I overdid the flame to extinction, running through and using up the wick. Some days, for a scant fraction of a second, I can't even think of my name, though I know I know it. And then it comes back to me, reassuringly. No, I certainly wouldn't want to become a deacon, also.

KANTER: Keep busy, though I've never had that problem. But idleness brings sleepiness, or drowsiness of thought, and slows down all imperatives of action and potential. Don't let the neurons languish. Admire.... the remarkable. That's an easy way to stay alert, with others doing the bulk of the work.

VOGEL: And I wouldn't like having to pass the plate, as if it were a dedication to promote. They actually have to appear happy about the request, or assume a dignified air of deserved expectations.

KANTER: Well, Vogel, the priest does, in a sense, perform for us. The audience owes for that entertainment, what weekly revelations he's uncovered to tell us about and instruct us through. I can read for myself, but I can not perform a verse to translate into such intensities as his. (It) Brings obscuring perceptions into the blare of a colorful commitment to the words.

VOGEL: They are trained to be orators, not oracles. This church

lingers and languishes without us. But I can't understand the plumbing.

KANTER: He lives here.... Must wash his clothes here.

VOGEL: That's.... done at a special laundry, I'm told. Particular soaps and waters and essences industrially applied.

KANTER: Well then, he washes himself up here, and cooks.

VOGEL: It is a tab that we must pay for.

KANTER: Cheaper than having him rent a boarding room or house apartment. It's how the building was envisioned.

VOGEL: These employments should be aside occupation, and more self-supporting.

KANTER: The deacons' are.... But think of the training involved, and the disciplinary sacrifices we benefit from. That man has gone to a kind of college, to make himself able to workout and solve moralistic puzzles, or explain them *correctly* for our edification and reasonable, intelligent questioning. It's so easy.... to convince one's self of nothing. He produces evidence to consider, or teaches us how or where to find some. It's not always as expected to be, nor obvious for consumption.

VOGEL: A test, Kanter, for this church, in this region, is necessary to employ. Why should we live in drought, and not in doubt for what allows this?

KANTER: Oh, man can't prove a thing behaved—

VOGEL: But as a god he may.

KANTER: —and is ever an observer made, even throughout his own body.

VOGEL: I say he may.... approve of this war of nature made, if he can resist and conquer and supplant disaster with delivery of self-made fortuitousness—

KANTER: *That* is an oxymoron.

VOGEL: —We certainly fight well enough amongst ourselves. So why should nature be so desultory for our well-being?

KANTER: It's not so desperate (of) a season—

VOGEL: So say you!

KANTER: —It's only hot, for a prolonged spell. And we abide.... with what is under our control to do. This town is well enough off for its inhabitants.

VOGEL: That says hardly a thing, for its inhabitants. The population of the region as a whole, of this entire broad country, must be considered. And there are miseries abundant with every slow, lucubrative catastrophe.

KANTER: Lu-lucabra—

VOGEL: There's suffering aplenty the church must weep for and renounce through meaningful remediation. I mean the notion of its powers to enlist, that they are made of the same supernatural stuff.... as endowed a priest by a bishop.

KANTER: You want him to clap his hands for some rain, or do a dance with a dowser?... Or make some tale of it with relevancy to scripture? There's no one dying of thirst, maybe some loose or free animals. But despair has not overtaken the community.... much.

VOGEL: It is the manner of the discomfort which can abolish all faith. A meanness to it severs all forgiveness and patient acceptance.

KANTER: How can you argue that against the.... meteorological physics every person must assume to be correct, every creature, perhaps, every living thing— every piece of rock or spurned earth parched into a destitution? The weather may be cruel, but not mean, not deliberately angry at us.

VOGEL: But then.... this church would laugh through(out) the wretchedness, as if going up in flames with much crackling.

KANTER: What are you trying to say.... or incite— or infer?

VOGEL: If there are powers to her, commutative of worldly might, then she should abolish this distress, upon this prominence of hill to smile over her patrons, with a dominant insistence for their betterment. That is the symbolic role of this.... monument to worshiping, with the cross as a steeple and the steeple as a weather vane.

KANTER: ....Just because you feel a little hot? Do not deprave the significance of a crucifixion, no matter how common an execution they may have been in the past. More meaning arises from its depiction than the act.

VOGEL: My sense is not irreligious, but utterly devout.... with piety awaiting a strict and stern demonstration of the spiritual commotion we steer through to avoid the pitfalls and clashes of moral illusion or disenchantment. And I would like some proof to the charmed significance of these holy fixtures. I want to see the idol faces weep with sympathy from our discomforts.

KANTER: It can not be.... that there is protest for our eating and drinking in this room?

VOGEL: This public space, that they can control only less than our intentions. A quarter of humanity within a mansion to God!

KANTER: We shouldn't make a mess in here, for which we must clean up. But you know how people are. Still, pestilence should not be found or inspired so close to the altar and the pulpits. So drink your beers carefully, and eat your breads without leaving many crumbs.... unattended but for mice. But if you're hungry and thirsty, and if the perspirations warrant, on a hot day, then eat and drink. This is a human spot, and not as capital as an inn with bar for peanuts and sandwiches, pretzels and cupcakes. But nonetheless, we are here divinely inspired. Make a lunch of it. And we do pray for days of relief, in that mysterious, otherworldly darkened chamber of sermons and protestation. How do the images and figures drawn on glass shade us with their rich coloring and purloined hues? does make for an atmosphere evoking mythic indoctrinations aspired towards.... as within the volume of a painting, a religious

relic grandiosely proportioned and propitiated.

VOGEL: Their kitchen dining hall is larger than this, and can become more soiled of celebrations' lent offerings, consumed onto the ash. But they complain of us in here, to be more thorough managing the dirt they would not find— nor care to seek. It is an ornery protectionism. I want some proof of this necessity of being separate. Have a miracle performed, to justify the arrogance in their proscriptions. Our light in here's better than their lamps and chandeliers, and is more honest of the window views for more radiance becoming.

KANTER: Theirs?... Ours. They only give structure to our order.... and theirs. Hold the resentments, just because of this heat.

VOGEL: And why not a fan allowed?

KANTER: The air conditioning is expensive enough, and lends itself well to this space.... when the pass door is opened. Otherwise a fan would be in competition for more funds.

VOGEL: So!... I will bring a radio to keep in here. For competition, indeed!

KANTER: ....Might be stolen, or could play sacrilegiously of sounds. Let's keep the room.... more timid, as a retreat to silence, meaning the absence of the profoundly said, whatever may be heard in here.

VOGEL: Then delineate my growing suspicions for me. Is this not part of a church?!

KANTER: Yes. But— (*The door leading to the interior of the church is opened, as Deacon* **Paths** *is heard.*)

PATHS (*voice*): They're not devils in here, Mrs. Margathrate.

MARGATHRATE: Mrs. Mason, deacon. (*entering*) Margo's my ladies' name.

PATHS (*voice, as* **Vogel** *stands, although there's another chair for her*): You can have that done before them safely. But the dinette can't be opened for these deeds, especially with the energy expenditure allotments we must adhere to, these trying days. One switch could burn on twenty bulbs. (*The door closes behind her, after she enters fully.*)

MARGATHRATE (*pausing to stand somewhat with bewilderment*): I bring my brunch, masters.... but had expected to eat it alone.

VOGEL: Brethren, Margo. Don't let us stop you from enjoying it, at the table. (*as she approaches with a slightly shy hesitance*) You'll have to move some things aside.... but the dust is clean enough.

MARGATHRATE (*placing a cloth bag down on the table, and adjusting some articles away*) My rag's ma plate, for this chunk of loaf bread.

VOGEL (*unnecessarily putting a chair next to her, since it was already near enough for her to do*): You're new to the service—

MARGATHRATE: I be married!... Thank you. We've been in the congregation for awhile.

VOGEL (*while she sits and unwraps her bag*): Yes, I've noticed you.... two.... or three, I think. You have a little son.

MARGATHRATE (*manipulating*): A daughter, sir. The clothes are similar enough at her age, for the sex to be discerned as "child."

VOGEL: Oh, but—

KANTER: He means about your service to the church, madam. Taking up these little extra chores we do.

MARGATHRATE: I come to wipe down the statuary, and scrape to brush off the wax from the brass candle holders. So many porcelain figures we have, perched on shelves near the windows. And me nails are dirty. I'm embarrassed to take hold of the bread, but I didn't expect to do so in company. But a want is a wish in hunger, if you'll excuse me. (*starts to eat with her hands*)

VOGEL: Not heavy work, for certain. We would handle that. My specialty's in pipes, and their rusts. (*going back to his chair*) Cleaning off the corrosion.

MARGATHRATE: Really!

VOGEL: Takes a knack to do so regularly, without causing stronger of a buildup, as fresh surfaces are caused, sketched or etched, for priming more deterioration. I won't polish or take them apart, though. I'm not a plumber. (*stands at his chair, observing her*)

MARGATHRATE: I am newly made. I was made just last week.... And I start, with my convenience, today.... More than I thought to do, so many to be found. If they're not attended to, they hide.... and make false grimaces and sighs.

VOGEL: To think that way, you must be peculiarly bred.... for this work.

MARGATHRATE: What does? to rearing and attention. I be back to my daughter soon, who cries aplenty at the slightest dissatisfaction. It don't mean much, as much as silent howls. Yet I was cajoled into it.

KANTER: A deacon asks for volunteers, for specific assignments, as they're needed.... and you enlisted yourself, at the end of a Sunday gathering.

MARGATHRATE: But I was just made, sir. I was *created* last week. A cup was poured, of rich red port, and I was born, with sudden ripening of sweetness for the labor. For there I was in bed, beside my hu'b(and). And then a spark hit me through the darkness, that told me to do some amercement because my heart was sore, of my condition made and bound to. And so, with threat I was totally recreated anew, and refreshed to plead for a purpose desirous. Then come we to the church, as by command.... with orders suggested— and grabbed at, least for myself. I had hoped to find some solitary niche in a corner, and pray— away from the congregation— for this espousal of blossoming with resurrection. But it ain't the way devised, to clean the figures and the candle

sticks, the statues and the holders of wax. I was to myself, though, and rediscovered a fault. For if it's so easy to be made, then where's the need to be unique, or uniquely wanted, distinguishable from others?

KANTER: Oh, you youth just have your muscles reeling.... from raising a family, a hard choice.

MARGATHRATE: Would be better to have me meal out on the lawn.

KANTER: It's cooler in here. But the responsibilities of a marriage can seem like a prolonged agony, at times. We all experience this. It's not new, but for you to notice and complain about.

MARGATHRATE: There's more than a rejuvenation to my hardships won. It is for quiet cleansing of the soul.... I do not like them, and am being properly punished because of this. The pain makes me better, and less bitter.

VOGEL: What do you mean? you do not like them. Your husband and daughter? And right out to strangers you'd admit this?— You're lying! Weariness is more bold than to be worthy.

MARGATHRATE: This is a church— is it not!... where sins are to be.... commuted with exchange of griefs for holy service? And.... you are masters to me, though charged the same— I'm sure.

VOGEL: I wouldn't say that. To be proud of this church and its upkeep is no crime.

MARGATHRATE: But you are here, in this singular room. And I join you. I am led into this space, a just compensation for my polishing, and ratcheting of the internalized emotions to feel.... solvable, of a salvaged, savage nature, on my commitments to pounce and destroy.... I can not. And neither can you, I must surmise. My hu'b.... accepts my displeasures to know about (them), as for some depression to describe. Housework is a hectic boredom. He agrees to have me summoned for these labors, in short intervals of time, as he looks after Chrisie for a change. A sort of lesson for both of us, this is. And he particularly informed me of this room, to be a cell of detention voluntarily sought.... perhaps, where one can review one's failings peacefully, tranquilly, and heal of wounds or soreness.... or regrets. But we both thought I could be in here alone, expected this without half thinking. And he warned me of those who habitually come here, come to need this space. "Privileged Penitents," he calls them, exchanging sorrows for submissions.

KANTER: Well he be a devil to think so. I'm only in here to relax. (*stands*)

MARGATHRATE: Submission to a camaraderie of tribulations, —

KANTER: See this book?!

MARGATHRATE: —guilt.

KANTER: It is an art book of pictures and drawings of beauty. Now is that sorrowful? Is that meant to have me sad?... in any way? And I find it in here— to read. You'll read it too, maybe. It is the property of.... this room, which means it's ours. It belongs to the community, as does this— sultry chamber. (*coming towards her*) It's about the founder of this town, and his pretty sketches in pastels, kept down at the state museum. We're justly proud. I think the pamphlet was nabbed from the public library, or a school's— but it's ours. There must be many copies floating around, for coffee table wear. Would you like to see it? (*stopping next to her*)

MARGATHRATE (*wiping her hands, slightly greased, on her rag*): This is not a spot for having tea and biscuits and polite, informal conversations about the trivial or dainty concerns. We are to be house-struck here, into our gravities of penance—

VOGEL: Do you have a key?... You can come here (at) anytime, of night and day, and not do any work at all.... within the church.

MARGATHRATE: My husband gets one for us, soon. I do not feel obliged to ask, till it is fitting that this should be.... a reasonable cure. The deacons will not offer.... until he demands it, I think.

KANTER: Well of course you have to ask for one.... Margo. My wife did right away, and received hers promptly. You mustn't be so shy about what is deserved, and what is commonly available. A great many have access to this room. Don't feel so inhibited of your rights, or taciturn of needs.

MARGATHRATE: I would never come here, without some service to the church to perform— Do you?!

KANTER: ....Noooo. It's not worth my bother. Here's more of a benefit to return to, after some tasks completed— an earned retreat, to rest up a bit.... before returning home. And I have a wife and house to entertain, and clubs and club memberships to enjoy. Though I don't do any more hunting. I don't like that sort of sport anymore, killing, seeing (a)'s how I'm getting closer to the end of life myself— But I still fish! and have every trophy eaten. No.... I don't come here without a purpose, and that purpose is to recover from some work for the church—

MARGATHRATE: For the clergy. Are you made for this place?!

KANTER: Aaa.... (*lays the book on the table, beside her*)

VOGEL: I occasionally visit here, just to exercise my privilege of being allowed to. No great disgrace to this. And when I do work on the pipes, I do a hard and rigorous job. So I'm entitled to relax in here whenever I want (to). And that's not a shame to cry about. But there's no penance to it at all. This is a lovely little room, considering its utility. I take pride in the dust and dirt and grime that can be found, if you look to notice with hard candor. My name is.... Vogel, and that's my surname. But Reginald will due, when you think I'm fit for punishments. And he is Mr. Jules Kanter, a former supervisor of business suites, procuring furnitures and fixtures and the like, to convert house space into offices, an activity quickly drawing to a close as the town develops to evolve more proprietary establishments—

KANTER: More professional ascension, Vogel.

VOGEL: —I.... am unattached, at the moment. My wife left me a few years ago, our children a few years before that— And good riddance, for their happiness. Well?! But I did my chores, properly. But I am still employed.... as a monitor to a hotel, and a few other buildings on occasion. And I don't feel deluded out of happiness for my independence and civil duties to perform.

MARGATHRATE (*lowering her hand and grasping her rag with both hands, as if for some security*): I love you.... Vogel.

VOGEL: My purpose is only to live.... and test that living, madam. I can not be cross at strifes.... But Margo is alive.

MARGATHRATE: We are sequestrated, sir.... to the beneficence of this church (*looking up towards **Vogel***) and all its holy happenings. For what occurs here dominates me, if I'm to gain some strength.... and pity with endurances.

KANTER: We.... sequester ourselves, my dear, after great efforts. We are seized only by a spirit to perform our chores to temple.

MARGATHRATE: As could to a prison's danger must we find, and be beaten for, to allow entrance.

VOGEL: Oh! Such self-flagellation *is* peculiar. And prison cells are not so dangerous, if you can keep to yourself healthily.

MARGATHRATE (*more animated*): We must have vigilance for a burning of the god-awful, gents!— and brethren. A burning off, from the shoulders right down, of the crust baked onto us as we seep through the hot mire of humanity.

KANTER (*slightly startled*): You seem to be releasing yourself.

VOGEL: And finding more freedom here to be genuine, or let loose a genuine disturbance from within the deeper internecine confusions of your thoughts.

MARGATHRATE (*demonstratively, but for her own notice*): To scoop up the dirty muds and bathe of this for thorough absolution of evil wishes and intentions. (*returning to a consciousness of company*) Do you know how often I must wash off filth?!... as purely foul as to become one's self within. And so an agent be, what is forgivable in me, I am involved too terribly of cleaning.... without end. Even of the holy sacraments brought to, in this profanity of hearth, I must scour with satanic fervor of irrelevance, to release my energetic drives upon such artifacts of faith and articles of communion, to exhaust my evil breathing, breaths, and its heat!... But still this turns on me, like a churning inside, for want of more relief from temptation and anguish, and a passivity towards deterioration of the soul— You say your wife left you. Was for that to find others to bite, or gnaw on their bones?! That is how so disconcertingly I feel, to tear away and at, the same time thieving for a household to command, the expectations of me. (*clutching at her rag tightly, to roll up*) And one night I found a long, silver hair dangling impishly from my comb. With a groan of lost resistance to my thoughts, I was devastated to review the conviction that this person dies, or might be dead already, from a release from entombment, trapped in a sarcophagus of values, and to this Mason.... mistaken for. I've searched feverously for other white strands, but can not find them, or admit to noticing. Yet to be reborn absolves all of this horror and loss, of ve-o-raciousness for truth, and carry all of my decadence out and away, to leave a bare flesh and frame ready to be reinhabited, but by a doting, doltish goodness once held insufferable by me. I try for this.... doll to become, again, till in this room I'm placed to breed, more discontent to shed. And how to handle children more I kneed.... my internal angers into pleasantries of hands and figure womanish, with a flattery for this spare room to be in— and find.... what contends, what is found to commiserate with and behave towards. All that occurs here.... may be allowed by God, if (only) my labors done. (*Some faint organ music is heard.*)

KANTER: ....Well.... we perceive you.... as a mated adherent to the church—

VOGEL: That boy practices some psalm, now. How he learns those pieces quick(ly) enough for weekly recitals.... startles me.

KANTER: Oh! It's a good sound, Vogel. Fine instrument.

VOGEL: Good electric one. But that model is years removed from current standards. The keys are still ivory.

KANTER (*going back to his former chair, though leaving the book*): Yes, it's flush with a richness, a really deep reverberation at its full effect and intensity, which Carl sometimes uses too much. But when it was made, the electric tones and complexities had, or delivered, true warmth— not like the high pitched piercingness of today. They actually tried to emulate the pipes of an organ and its moistened breaths of sound.

VOGEL: Well that's because some air is still compressed in columns for that model, Kanter.

MARGATHRATE (*re-subdued*): It is a simple thread of melody, with base accompaniment. (*The music stops.*)

KANTER: Well that is how, for psalms. (*sits*) The words are more important, so the singing must be simple.

MARGATHRATE: He played it twice, I think. Twice.... for a mastery?

VOGEL: The lad sight-read, of course. You can play it a thousand times, with mastery, if you're only sight-reading.

MARGATHRATE: Where is your nave to lend—?

VOGEL: Stop feeling so contrite! Margo.

MARGATHRATE: —to me my passage. I've not been up there ever, where he sits, until today. And it is a good space adorned.

KANTER: What?!—

VOGEL: You can't use this room—

KANTER: —The chancel?—

VOGEL: —like a flushing toilet!

KANTER: Good wood there. Excellent carving and carpentry.

VOGEL: You're one of us.... You're one of a folk— with a key.... now. And this is a pastime to cherish! I'll give you mine, if you want. I can steal or snatch others' anytime. But there might be some around here (*starting to go towards a wall to look through boxes abutting it*), say in a rusty, lidded can, or dirty jar.

MARGATHRATE: My hu'b will get me, give.

VOGEL (*to his knees, rummaging*): But you'll want one for yourself, alone.... like Kanter's wife. Women should be.... that way— (*The passing door opens as a boy's voice is heard.*)

VOICE: I don't *want* to go in there! (*Carl enters, as Vogel shifts himself to see.*) I have it come for, sirs. (*He is carrying a small collection box. The door stays open, as he walks over to Kanter.*)

KANTER: Now that's a good breeze there, Carl. (*Vogel stands.*)

CARL: Yes. The air conditioning is fluid. It should really reach in here more often. (*Kanter searches in a pocket for some change of coinage.*) All you need is a slot with a screen.

KANTER (*depositing*): That construction costs money. Maybe a fair amount, to do the job correctly. A flue must also be designed, which takes some engineering.

CARL (*going over to Vogel*): Oh. So the doorway's a flue. (*noticing Margathrate uncomfortably searching her rag for a purse*) Only for the gentlemen, miss.

VOGEL: Mrs. Mason, Carl. (*pulls up some change from a pocket, as Carl approaches*) What are you going to entertain us with next Sunday, Carl.... at the evening service.

CARL (*as Vogel slowly drops some coins in his box*): ....I'll do a prelude and fugue by Bach.

VOGEL: Oh! That's quite ambitious. One of the easier pairs?

CARL: I'll try the F# minor one, from the melodious volume.

VOGEL (*startled*): That's even more in need of dexterity.

CARL: It's only like Beethoven's Moonlight, and I got that down pact, two lines in one hand and all.

VOGEL: You'll only make a lot of annoying mistakes, like you did a couple of weeks ago. We only put up with it because it's at the end of the day, it's lightly dark out, we've all been lazy— preparing our muscles with rest for the workweek ahead— and we admire your courage to tackle these difficult pieces.... before a captive audience of weariness. But where you find the nerve to do it, I'll never know.

CARL: Well, I practice—

KANTER: You should master. You should memorize— learn, before delivering a performance of such works.

CARL (*starting to exit to the church*): Well it's not that important, is it! if my intentions are good and my efforts are honestly earnest. I'm always applauded, and the other kids know they should be trying to do something likewise, to gain approval.

VOICE (*from within the church side*): Carl!

VOGEL: Say! Do you know of any keys around here?

CARL (*turning to answer Vogel, just before he's about to exit*):

Oh, here?... To the outside door? I don't think so, but I don't look for things in here. It's not my calling.... (*as he exits*) We're not allowed (to).

PAHTES (*his voice heard just before entering*): Never brag to parishioners! (*entering*) *What* are you looking for, Reginald? I'm sure everything you'd need can be found in this room. Needn't ask our organist.

VOGEL: I was just wondering if you kept any spare keys to the door around here, somewhere. Could give one to Margo so that she needn't bother to ask.

PAHTES: The door?— the entrance. You can ask us about anything of the church, Mrs. Mason. That's why we've taken up being deacons, to assist in all affairs and inquiries involving our treasured community institution— Not that I can firmly answer religious questions, except from what I've studied and interpreted as a lay worshiper.

MARGATHRATE: I didn't want to seem too forward with Deacon Paths. He's been showing me my duties, as I begin this service.... My hu'b will ask for it, when he deems it fit to be familiar enough, my service for the need.

PAHTES: Your.... husband and yourself are certainly entitled to request for a community standard of possession, Mrs. Mason. Though we prefer to restrict the use of this room to those particular to help us actually with the maintenance of our church. But anyone is welcome here.

KANTER: You mean we have to ask for it, first. But you wouldn't want bums or tramps hanging around, stinking up the place with their bottles of rot an' rye.

PAHTES: We do hope, out of a sign of respect and propriety, that everyone of utility keeps this spot orderly enough to be presentable to those who wish to serve, Jules, with a greater employment than merely worshiping. But were there a storm outside, we could handle many with many recesses of the church, and of all types and dispositions to entertain with.... at least a passable form of safety—

KANTER: Pahtes!... Now why must we ask? Why aren't we presented outright.... such a miserable little token for our manual contributions to the building's upkeep? It would seem to be a trivial gesture, Pahtes.

PAHTES: Are you unhappy? We don't shove such gifts onto people. They may not want them, or the added responsibility of something else to lose or carry with or store.... out of the reach of children. But drunkards don't tend to reside in this room. In any case, you are all proud workers, not easily proselytized by keys, side keys nor passkeys. We don't limit your possessions. You limit yourselves due to your own proprieties.... scopes of purpose, and so forth.

VOGEL: Are there any keys in here, deacon?

PAHTES: Why do you think it's a clandestine treasure to own one? So many do, already. But, as a matter of fact, some old ones may be lying around in the dirt and dinginess of the corners of this space, left by former servants who have moved away, out of the region serviceable by this church— We don't keep tabs on these things. You can scratch around for some, like mice for moldy cheese, or you can ask us upfront for one, properly. The cache is kept in a tin by the priest's rectory. So we must disturb *him* for it. That's why we.... wait for you to ask.

MARGATHRATE: My hu'b will—

KANTER: Pahtes—!

MARGATHRATE: —when I am proper.... for it to have.

KANTER: Your name's so similar to Paths'.

PAHTES: Not really.

KANTER: It is a coincidence of our collectivity, that must be common for an association of like and likened spirits, like birth dates, and numbers of teeth in the mouth.

PAHTES: I say not at all, in truth, Jules. And in fact, my brethren's name is actually Pathers. He shortens it deliberately, for the church registers and announcements, in order to avoid confusions. So that is the real coincidence— that we are both deacons.... to our lady's grace. But we are both very upstanding adjuncts to the holy mystery everyone embraces, and bring our organizing skills to assist the priest with the mundane matters of a sermon and a session to present. One can't do this alone. We are not as simple as a flock to control. We have our individual temperaments.... to regularize for praying and praising.

MARGATHRATE: Does he tell you I've done my job well, so far? I am entirely in his hands, to know what to do.

PAHTES: Deacon Paths hasn't complained to me, Mrs. Mason. Very soon you'll be able to do your chores without even thinking about them, they're so simple and straightforward of.... cleaning, I believe?

MARGATHRATE: Yes. Dusting and wiping.... and scraping off wax.

PAHTES: That's what the women commonly do, along with the floors. The walls and the windows are more special, though, and take the talents of muralists, almost.

VOGEL (*going towards his chair*): She's not that, a painter.... or she would have volunteered the such—

MARGATHRATE: I haven't art within me to express, unless I am directed and guided to that enlistment, since I already know what is beautiful.... and what may be outlined for me to adapt to and enhance.

PAHTES: Only the masters can make the images. Not even the priest attempts—

MARGATHRATE: He is not a master?

PAHTES: ....The men paint.... The restorations may be womanly. (*as **Vogel** sits*) Not many of you are needed.... at one time, as long as your service is regular, and revolving of participants well enough to keep the tasks from becoming only drudgeries and unrewarding to you. We're very thankful—

VOGEL: I'm never bored with good decent work to perform, especially of the mindless kind. A good way to spend one's spare time.

PAHTES: We.... make all of us. And *we* are very thankful. But if you become unhappy with it, or tired of it.... then just drop your keys. They're like old, tarring pennies, anyway, pitched in a forgotten bank, if you have no use for this room anymore. Just tell us.... for the turnover to expect or solicit.

KANTER: The deacons are more devoted than us children.

PAHTES: *We* are depended upon, Jules, to stay.... as we are no matter our personal misgivings or difficulties. There are degrees of service, certainly. But this is not a pastime for us, but a second, though minor, deployment of our skills and interests. Everyday I am around, after work, to make sure that something is done, of a concern handed (to) me. Paths almost as much— but we compliment (each other), because I am the more analytical.... and he the more propulsive.

VOGEL (*slightly agitated*): I want you to have a key now! Margo. That is your pleading pleasure, and a discovery fulfilled. You want this trifle, like a toy mixed in with the candy. We deserve it, the luxury of its instigations to open. Come here alone—!

PAHTES: And I arrive to make sure everything is fine about you. If you're upset in any way with your service, any of you, we can change duties.... Or you can excuse yourself politely, with your service done and merited. But this is for ourselves, so there's no detriment to ending it if you must. This is where you might feel agitated, if you become dissatisfied. So.... I come to see, what the weather is— or brings.

VOGEL: You *can* find yourself alone here, Margo, to think through things. I do so, occasionally, very late, early morning, while the sainted sleeps. And I sometimes greet the morning this way. The dawn becomes fathomable again, the streaks of light.... glowing to intensify my patience, and waiting for them.

PAHTES: Your night work is very pale, Reginald, to have to come here instead of your house to relax.

VOGEL: I'm simply not tired enough, Pahtes, to return to an empty house, some nights. I'd rather be here, for more primitive ghosts to eavesdrop on, the creaks and pops amusing me as I wear down my attentions for a more suitable somnolence to eventually find.

PAHTES: I suppose you bring an early snack in here also. Well.... as long as your tastes are.... as immaculate as cleanly. How should the woods infuse you? of their deepened characters and ebonized grains.

KANTER: Oh. Like the string board of a viola to press and play on, fingering your bowed tones and sometimes notes. I did, in fact, bring my instrument here, to see if this could be a good practice room between services and sessions. But the acoustics were not fantastic, Pahtes, the resonance dulled.... Seemed like such a foolish waste of effort, almost pompous to think of doing.

PAHTES: You'd best not let our priest hear you doing something like that, unless it's sanctioned for a chorus' accompaniment.

KANTER: I don't think anyone ever noticed me, not even coming and arriving, staying, playing, and leaving disheartened and dismayed. The only fun at all was in the doing. But I'll never attempt it again. I'm not good enough to accompany pure voices; and I no longer practice regularly, not for a good while.

PAHTES: Whatever mayhem occurs in here—

KANTER: But I enjoyed the effort.

PAHTES: —is yours, the community citizens to condone. The church offers this space for any to use righteously. But please show some prudence with your pride. We are near the altar, and the very ascendency of our ideals, the beatification of our cherished privileges to be alive within a radiance of the holy and hallow of our thoughts, or as Mathews would say: "the brew-bree beckoning of our brotherhood".... or womanhood— humanity itself, in constant boil of stirring to sing praises to the everlasting.... which is *how* long as we may last! with honor to our lord and God and the holy spirit that must reside within each person, yea— entity itself of breath to take.

VOGEL: Are you practicing to preach, Pahtes?

PAHTES: ....I am moved, Reginald, occasionally by a whim of feeling. And the lungs may yield a voice as stale and brisk and nasal as by Jules produced. But I am moved for its issuing.

VOGEL: I think you have to take lessons for that, a proper, tutored schooling— at a seminary college, at least.

PAHTES: I would not be the priest.... in here, to even try to persuade you of my thoughts to favor. It is a personal expression of devotion as private as loud, as demonstrative as demure to what I am graciously found to be amongst: patrons of the religious tendencies we adhere to and bind of. What can be more important? It may even overreach decency, at times— And I am not an extremist, but for our extremities of influence to find and challenge, or even protest, with the vehemence of conviction that we should all be under the wraps and folds of a common and shared rapture of the blessed.

KANTER: That is a bit much to take, even for an instrument. Suffice it to become one *with* the instrument, Pahtes, and all other allegiances worthwhile will automatically follow by form of being utilized for this necessity— of performance.

PAHTES: Oh, I do, Jules. The whole of one.... is the goal of many, bodies to body.... If you wish a key, to the outside door, Mrs. Mason, you must ask Paths, for a formal request of acquisition.... you or your husband. It will be noted and acted upon— by him, since you are under Paths' charge, at the moment. He will inform our priest, and the fine reverend will distribute it to you directly, at the next Sunday service.

VOGEL: Right in front of the whole congregation to see and share with approbation the gratuity won—

PAHTES: Earned, Reginald.... by simply being for us. And with no envy to the rite will this be done for you. For it takes a certain amount of courage not only to ask, but to accept the distinction— Please be specific at the first, if you want one or two; though you may have a locksmith make copies— at your expense.... Mathews likes to hand them out, deservedly.

MARGATHRATE: I will.... explain this to my hu'b, deacon— No!... I will ask him today— for a couple.

KANTER: That's the spirit! Be brave, Margo. Join us thoroughly.

MARGATHRATE: What is the embarrassment to it! I will wear one, like an elbow handkerchief of ostentation, to show that I have license of entry.... to a common room.

KANTER: People will know it. They will applaud your fastidiousness of mind to demand this token gesture, as could my wife. You will be considered.... more broadly, as a parishioner to be taken account of. She is proud to the extreme at times, when she is pointed out, within her presence of hearing or witnessing, to acclaim.

PAHTES: Well, yes. That might tend towards a deviance of purpose. But this is such an easy right. You are more than welcome to pursue it. And now.... are our servants happy (*The organ is heard again, louder than before because of the open pass door.*), inside your room and chamber for the community? Community access?

VOGEL: Deliriously so, Pahtes.

MARGATHRATE: I will come alone—

PAHTES: We must move him on to the next selection. One can only encourage the youthful artist, with suggestion. But only he knows when he is satisfied.... (*exiting*) enough to attempt the next piece. But it's slightly more complicated than this, actually composed for a true organ, and requires some pedal tones to be emulated in the base.

KANTER: Can you leave the— (***Pahtes** closes the pass door.*).... leave it slightly ajar? Well I suppose we have refreshed the air enough in here for some gifted comfort. I can certainly last another few minutes, before returning home.... through the heat.

MARGATHRATE: I may come alone, as much to work for.... thought, and compensation for my time. It is a hot day, yes. Rains the heat, to get away. What is the sense allowed? under this sky, and privilege.

VOGEL: To end the drought, I pray. I will demand it. If this can be done, we'll do. The church herself permits its attempt. (*The organ music stops.*) May even order it of us to try. It is my test announced, when one can get away. And where to?— Here! To the very lips of the beloved, I cry for proof.... of our summations.

MARGATHRATE: I was afraid to enter—

VOGEL: What do I care of shame, if it is sham!

MARGATHRATE: With hesitances to admit myself, I find the cleaning due and just, me works provided me, by.... guidance?... and holy intervention, which I can not understand, yet may respond to. For I have been so unhappy.... unhappily burdened. Now comes more resolution in me, to have a mission made to these....

pious efforts, and bring me more clearly for my actions to approve of. I am anew of circumstance, with resurrection. (*looking at the book left by **Kanter**, and then releasing her rag to thumb through it lightly*) He founded us?

KANTER: So it is argued. Essentially, that is the happenstance. He was responsible, initiating the caring that developed into a town.

MARGATHRATE: Pretty.... drafts of nature.

VOGEL: We are our own already thus, the sketches of our semblance to others, and our needs profuse— profane, if need be. But that is by our earthly needs, to dominate those sketches— Draw them out.... of shadows, hidden beneath the corruption of true snares and proprieties.

MARGATHRATE: So be of nature, rightly. liberal. (*The pass door swiftly opens.*)

PATHS (*voice*): Are you done with your brunch, Mrs. Mason? There's a little more to do, if you may, that I may show you how, and why it must be done this way.

MARGATHRATE (*standing*): I am ready, deacon. It was a good lunch. (*while exiting*) And I have found a settlement for my desires to serve you.... masters.... for what the church allows.

KANTER (*as she exits, the door closing behind her by **Paths***) A good day to you, Margo—

VOGEL: Yes! Margo....

KANTER: She will do well, as well as my wife. She'll find some pride.... amongst us.

VOGEL: And privilege. But as I say, this noble edifice smiles with her presence and those as likely to be sufficient for the tasks at hand, and the required testaments to faith, which opportunities are provided by our indulgences of caring.... for each other. Still, I find the drought somewhat unbearable without these little privileges, like a sprinkling of cool water on parched faces. And this becomes an undulation of meaning, like a sign, that I am being answered, if only I may remit, for broad interpretation, my callous and crude designs. Oh! How we are such clumsy creatures, Kanter. More for barking are our thrusts through bedevilment.

KANTER: Go well, about it. But take matters sparingly, and forgivable with pardoning. The angles turned through are with mystic proportion to your needs. And gently may the angels lead you into pits, as were by devils found or sculptured. But if that's where you're to go, to spend some time, then be off! But take heed for what is really allowed. I often find it insufferable at home. And this is a release of tensions for me. Yet I keep myself.... available, for home and tensions. So what this lady provides.... is a mysterious condoning. And as for you, provoke the sensitivities without a beastliness to carry. I still love my wife.... as wife by marriage bound to me, and can protect that respect.... or aspect.... even as I'm away from her. So, does the church say this or that about it? as could a chapel sing of approvals and reproaches? (*Faint organ music is begun.*) Here come mistakes in certainty, Vogel. (*stands*) And I am driven off to contend with the heat. For that is what I don't understand, the errors.... Timidly he takes it.... cautiously apprehensive. Can one sigh that way, and weep? (*goes to exit to the outside*)

VOGEL (*relaxing in his chair, like as to open the scene*): Whatever is provided, whatever is ordained, Kanter. I sense.... no evil to good wishes for you. (***Kanter** opens the door to the outside.*) Gosh gadding! It's a pall of fire out there!

KANTER (*exiting*): You've just been kissed through a mist, and now find flames. (*closes the door behind him*)

VOGEL: Ah, dare defile.... Defile, to aim at my good standing— aims,... with a sultry baseness tamed for— train.... unto this mouth pronounced, the gain.... and grin of our desires. We have made mention of it only, as with hinting. And yet!... does spring the curvatures and pipes.... and overtures to offerings. That is truly tentative. (*The organ music stops.*) But she insists.... that this be tried. It is a wondrous mystery of revolt, and revulsions padded for our.... relaxations to play with. Can this be?... gathered here, and offered here, and postulated here! So much inducement, into traps.... or travesties.... or trysts— Yea! Yeh!... I could for cool, this blessing to be wounded by. But it's why we all— come— here. Timidity made tenacious. And so.... for this we are prepared, and produced, these sessions to attend and weather (through).... as could be her design for us, adopted to much fun, folly, and froth.... or how we subjects run— from mouth to must.... of our beguiling tastes, the tongues.... more devilish than for our tests. (*takes a more contemplative poise with the side of the head resting on a hand, its arm anchored to the table by the elbow*)

Scene II — *Outside and hot.* **Lorken**, *middle aged and tramp-ish, sits almost sprawling at the bottom of a deteriorated brick-work wall of low height and isolated width to suggest its span has never been of much utility to the current inhabitants of the town. The area is quite rural and somewhat desolately unkempt and dirty with that might be called the natural rubble of clay and stone fragments and rocks not even caring to pierce baked mud clumps but merely satisfied to be strewn among them. The grass is hay colored and rough and unevenly cultivated, or cultured to deposits separated in the earth. He shows signs of a tranquil discomfort, the face tilted towards the sky, with eyes closed and mouth slowly quivering, the back of the head resting in a gash of wall depression. There is only a slight shadowing from the wall's poor height; but this is all that's available, and he has found it. He appears to have rested here for some time, as **Kanter** comes up to him.*

KANTER (*after a pitying survey*): Why keep you this way, gent? Are you in dire need from this remoteness? (***Lorken** slowly opens his eyes, as if struggling to interpret sound with aid of sight.*) There's a church not far, you know, that might offer you some assistance. You don't have to die this way, polluting the country's fallow wastes, (in) a pocket of emptiness.

LORKEN (*with an unnatural squeak at first, while attempting to recover a normal voice*): I've seen it. Wouldn't want to present myself to God, looking like this. The clergy might fear me, and throw me away as a sinful failure, cloaking me with much more contempt than my harmlessness— Oh!... pouring it on like a burning hot syrup, of sweetened consternations, pious pity, and praying for a disappearance from sight. And did they not shoo me away but several times! when I wanted to join in services. But I'd disturb the congregation awfully, considered so foul and ugly as a person, and must reside out here to plead for entrance, or some disturbance of

the population numbered. I've come to the end of my means, and am totally destitute, and without a grace for anything, without a "thank you" earned— nothing owned but these stretched, torn clothes.... and my hairs.

KANTER: But you can go to them, up there, *after* the services, for some help. It's clear they don't want you.... frightening parishioners, to see how one might fall. But later there is bread and water at least, since your freedom has become your jail now.

LORKEN: Freedom?

KANTER: From all manly responsibilities.

LORKEN: Later is the church closed. And between services and sermons.... I'm not to be smelt. There's a shame to it, my embarrassment in particular. So there's no real convenience offered me.... I could make my way passingly in this weather, especially at night, finding some fruits of the land, and garbage thrown by man for pigs to eat. And I am one become to hide my hide of. But now.... either nothing's left— to find— or my body's lost its capabilities to handle these endeavors anymore. I must submit myself to that possibility, that even with some measly food or roots to grub at, I've come to die already, and unsaved. This point is reached for me, even as I eat earth to find some moisture and its insects. But if the world has not changed, then I must. I must begin the corrupting process.

KANTER: You only need (know) how to ask properly, off the weal of the commonwealth. We don't let, leave beggars (to) die this way! desiccating to vent!

LORKEN: What wealth is this community for to say? or speak with treasures from the mouth, all forsaken through me. I am like rust on the wheat, to show you blight arrives where not even the molds thrive, in this hot, dry drought. (*sits up to shake his head lightly*)

KANTER: It can't be all that terribly devastating for you. The heat causes much slack of ambition and enforces sloven laziness.... in many ways to many people. But you have certainly satisfied a thirst to survive— somehow! though the rivulets are dry of branches, or brackish of streams at best.

LORKEN (*rubbing through his hair, to stay conscious*): I drink from mud, man! and its flies— that bring a pox to me.... impressions made by heaven knows what, traveling across or coming up to, the best of animals. But now I haven't thirst to throat, nor hunger to stomach.... and not even a headache to head. The strength's not even there for these displeasures; and I lose a concentration on them, but simply lie to bake and brown and barter with my life for some holy gates to run past, roll or claw through on my knees.... and paws— The best of animals! and this is how I'm delivered?! Yet this is all my fault, I presume. I would be human.... and stumble of my aspirations. But dear savior, have some patience with my worthlessness. The teardrops are dry as salt to shed, granular tokens like a dandruff of the eyes, or a sandiness to see through. I'd stop rubbing off the crusts (*rubs his eyes*); for who would care.... of what I could notice, in this disgraceful state? It is a time, the time for me to lie still.... and let the world have at its forgetfulness. And I'm not sad at all, but satiated of life's pains and anguish. "Are you satisfied," I ask, "of what I have done?" And I say "Yes!" to this. Now crush and crucify, for it is hot.... where I've come— and fallen.

KANTER: You must have been a proud man once. I leave you bums alone, while you're moving and stirring along your scandalous ways. But when you're still— to stagnate your expressions, as if for dead, past want of sleep.... that frightens me. I am alarmed, that you have given up the gift of birth, and have forsaken all humanity. There's much left (for you) to do, if you can breathe, respire with some wisdom.

LORKEN: The ground is hard. And the grass blades are like sticks, pikes that needle a bum, as you call me. What good is it to know that? I've lost much feeling, suspended in this way, as it were, between thresholds.... of the perceivable and the impervious, or like a slab of slate to become, under this fine fixture of rustic inconsequence, some country ruins I've crawled to. Let not a depression compel my sorry leisure, but an imperative to find my home with burning hearth, where charred bones are bounteous and plentiful. Is this not the scrap of a cemetery, the remains of an ancient burial site?

KANTER: I don't believe so. Not in any histories I've read. Indian or pilgrim, native or naïve, I don't think so. Though this area does tend to outline the boundaries of some kind of populous settlement, long abandoned or obliterated to the past.

LORKEN: No? Oh, for some reason I did think so, imagining pieces of tombstone scattered about..... I thought I had made an heroic journey, scuffling up to this forgotten masonry, after a sudden seizure of tumbling overtook my aimless paces. The time had come.... to be decisive and die, well roasted from inside, and ready for the eternal consumptions.

KANTER: You've too much sorriness contained, to hold of a proper mourning—

LORKEN: And I wanted to face my maker, once.

KANTER: —as could be a *baquette* baked, but filled with a softness.... in rich creams.

LORKEN: I wanted to show.... something.... something, that was of a man, to know I had been generated, and this is not a fantasy wasted. Yet I had no vengeful intent to this display of decimation. Resting here, I felt quite calm and composed.... to decompose—

KANTER: There's more than sufficient help for you, around these parts. As I say, the church is just over some hillocks. Why, you can see the top of its steeple even from here, if you know what to look for, through the spindly branches of trees, with their withering leaves.

LORKEN: And I say I know the place.... sir. But tell me, what would not strike for branches torn off? and used as rods to breeches of the unseemly thought of, there. I.... am not coached to come to their heaven, not a haven kept for the likes of myself, not in this squalid condition. Nor can I count much on their handouts to the poor, since I am below even that rank or caste, as a derelict decried or described by that holy concern. Yet, should I present myself bathed, cleanly dressed, and deodorized, then would they ask of me more than I may of them, a donation for their upkeep.... and some reciting of prayers— So I am hopeless, for what I have and am about, as far as this church may find me.

KANTER (*after considering*): Well.... Oh, well. If you have come to starving, let me help you up— Can you stand? (*offering assistance*)

LORKEN: Why for would I do? (*becoming slightly frightened*) There is some certainty to my position, having already fallen.

KANTER: Don't be resistant to some charity, man! (*extending his hand with some insistence*) I'm rather old, to be taken advantage of. (**Lorken** *grasps his hand with weakness and some hesitation.*) However that, as a good servant to our lord I offer to bring you to my house— and the Mrs.', have you wash up, launder what of your clothes and perhaps add to them from used wardrobe, feed you a good meal to bring back some health of body, mind, and spirit, and let you sleep a night in a real bed of cushions and sheets (*helps* **Lorken** *to stand*), before presenting you to the church tomorrow, at a service for the indigent, which they allow on occasion, upon recommendation of any prominent parishioner. It involves you doing some simple chores for a trial of time, as they work to place you towards more formal help and employment to find. Thus, you only need my introduction to them, or that of any fine citizen around here, to be saved from.... that corruption you speak of, being premature for your years 'gainst mine. You vagabonds only need to ask for help in this county, with a proper way to seek it, and it is often provided.

LORKEN: I stand feebly for you, mister. And I'm not sure if you delay the dead. But my circumstances confuse me, since I was sure they had been solved. Though generosity is greatness, always to laud with misplaced honor— that I have.... to meet you. Yet fear I tread away from what should be closer to the inevitable of my days left.

KANTER: The name is Jules Kanter, retired from a business of hospitalities.

LORKEN: And I am.... a travesty called Lorken. I can not presume on you, with decency, a familiarity of friends. Though they often ask for Bob, if I'm around. That being the case, I'm bound to only a few. And we were restless to disperse and find our respective sovereignties of useful purpose on this earth, till stalled with the realities of our desperations. I've not seen one sole companion in three days— or a week, my head can't retain the number clearly. But we're a disgrace to ourselves and *had* to separate. Else we would've convened as accomplices to (do) no good. And I was becoming tired of that— But theft is for survival, sometimes. And I did slap Harold, right on his greasy cheeks, for that mutton stock he stole off the plates of hungry kids, effectively, and failed to want to share. And I told him he was one shameful thing, and left him to his own devastations as he whined and squealed infantile-ly for his meal. Devastation of the heart, you know. I wouldn't touch the stuff, though the widow, I presume, was generous enough to leave her growing brood with less. We begged, you see, in the beginning, for a handout, from a rich aroma which drove us to a poverty's wealth. And she started to give, till then I noticed it might have been an act of compliance. For a glint in her eye said silently with a shined, shy grievance, that she may have been afraid of us, as her children clutched to her apron strings in dread. I strode off proudly. That was my sign to leave, leave this.... bitterness becoming of my person, while hearing a little kid, small boy, gasp in innocence, "There is so little left!" So little left. The (grown) lout did try to take what would have been my portion. So I briskly returned to hit Harold— with a warning, so that the pot would be more eq-uitably divvied, overseeing the division without having any for myself. Harold was just another blabbering child. An idiot of want with need.... I left him, finally, a fair friend of better days— and have never been successful since, these three days or a week, to find my soul less heartbroken. Oh, I met up with him later that night, well after his dinner. I think it was him. He just wanted some usual company. But we said not a thing, to each other nor the uni-verse, as we walked awhile. And then I just drifted off, leaving him alone. There was no complaining or whining about this. Perhaps more was there of expectation, as he had regained his years— and I.... the cruelty of disappointment.... Three days ago, or a week, I can't recall clearly. But I have seen no friends since. I have recog-nized nothing well, since then.... And I starve myself, I think, for some strange punition, with pride. I.... condense to conform, to what I am capable of doing.

KANTER: A man is capable of many foolish actions, and de-scents into gravities not easily handled. Problems don't prove much about a fellow, only recoveries.

LORKEN: I hope I don't greatly disturb your spouse.

KANTER: She is a bit finical, but more so for the church than home to demonstrate. No, she might be startled the first instant, and puzzled the next, with your visiting. But I will reassure her. We are both kindly elders with no great ambitions anymore, and tired of our talents if they can no longer shake the world. Ago.... I would have consulted with her carefully, before making this.... in-vitation. We protected each other and our feelings and mutual re-gard. But what does this really matter to us now? There seem hardly any chances left to take at all; and once you are restored you can become quite agile of a stranger and easily overpower us, or play deviously with our aging wits to cause some harm or grief or detriment to life—

LORKEN: Not I, sir!

KANTER: —But then, where is the passion to our simplicities and boredom? We have as much as more to lose than the little you might find, and even kill for.

LORKEN: Murder? Ridiculous! Why bring up these thoughts?... Am I so shameful to appear, to make suggestion of this? I won't even come with you, then. I am *not* hungry. I have lost that privi-lege—

KANTER: Privilege?!... We make for tests also, of humanity, with our privileges. I'll wash you down myself, with my bare hands, if you're unable to— and feed you and coax you to be-come.... what was my worth, a man of privilege. For life is a privi-lege, Bob!

LORKEN: I'm awfully confused by this,... Kanter. Yes, I'm quite weak, physically drained in this heat. But I would not be mean to a benefactor. I wouldn't turn on you, when 'abled, to make your giv-ing seem foolish, or worse— stupid. Why should I want to be that way? I'd rather die here, baked into a disreputable sight for crows and voles and weevils to attack.

KANTER: It happens so often, Bob. Yet, better to be done by love than hate, the terrible permitted of one's dare to test these privileges. For I have fear of tramps!... and disdain them!... have

often thought about such types to be.... of a contemptuous evil brought to, and more deserving to become.... of hardships and ha-treds. Yet, find myself a tramp of leisure, with this aging corrupt-ing, I see, and count of waiting, and struggle with for meaning, why I can not be as I was— as we!... were loved to be. Then pic-ture you! converting or corrupting more, stretched out near death and still. This is to make, for my examination of the folds of de-struction!... and how one follows it— or projects one's self into it. And I will call you "test," Bob, to bring you as near to an affec-tion.... as you may allow to take advantage of or capitalize with. It is in warmth, and nearly burning of a resolution to produce, as I might feel the compunction placed upon me. And being a member of the congregation, it's more of a duty, but a *privilege*.... to exam-ine the wanes and wants of your conscience, and how you might be reinvigorated into a productive and useful life once again.

LORKEN: You can't hold me to be such a fiend, as to want to bite you after being fed?! I'll be appreciative of whatever largess is provided, provided I be seen as a human kind, and with a human kindness told (what) to do, humanely proctor-ed for some simple service to perform. Because I have been brought to the edge of the abyss, Kanter—

KANTER: Jules!

LORKEN: —and my head's been poked in (it) to view the void and emptiness that awaits me. I ain't afraid of it, no more! It is a scenery becoming (of) the blinded and insensate. But trust me to be a true person.... Mr. Kanter, a decent chap, a sort less swine than swindled of all ambitions— for awhile.... So, you belong to that church.

KANTER: We both are principled members, me wife and I.

LORKEN: Well.... then you would want to bind me to it, as— for a good deed performed.

KANTER: I only wish to have some of your distresses relieved; and this is a way that's feasible, don't you think? for a considerate person to propose and prepare? To find you leasing death makes for a very disturbing circumstance.

LORKEN: Have you a drink on ye?

KANTER: Are you thirsty, now?

LORKEN: I want to satisfy *some* longing of the throat.

KANTER: And the belly and the head, next, as you revive. There's much of living matters at my house.

LORKEN: And.... how far is it, sir?

KANTER: Well it's not a great walking distance, Bob, as far as a child roams.

LORKEN: The effort's need is definite, and confuses me— if you're looking for a servant.... or a supplicant.

KANTER: I haven't looked for anything I've found in a long time. I've just been wasting away in retirement and hating almost every second of this. And then to find a younger man almost throwing away his life— by giving up— unsettles and angers my perspectives on what you should be doing— tramp! So you can walk a short way, for food and a bath— And I will help you, if your steps are heavy and the paces small. But you must be willing to do something (for yourself) in this hot day, besides lying down to die.

LORKEN: Yes.... master. Lead me away, with your chains as hands and arms, locks and links.

KANTER: I haven't anything on me, except some money. And you can't eat that! with much prospering of nourishment. Would you like to steal it?!

LORKEN: You know I couldn't if I wanted to. I haven't the strength.... nor the courage, nor the desire—

KANTER: Here! (*reaching into a pants pocket and taking out a leather billfold*) There's some considerable money in this— although how much I won't say— along with other documents of importance. Hold it in your hand— put it in a pocket. I'd once fright an army to keep it, my possessions. (*shoving the wallet to **Lorken***)

LORKEN: I can't do that, Mr. Kanter. I'm too weak to know what I'm doing, what you're giving me.

KANTER (*forcing the wallet into **Lorken's** two hands*): Take this! Walk with it. And if you're sincerely human, and if you're not a tramp's tramp or a vagrant's vermin— full of vice, then when we reach home you'll hand it back to me, immediately.... or eventually. But if I lose it, then what does ail me— trust?!

LORKEN: Stop belittling me this way. (*drops the wallet, though by accident of being feeble*) Don't need to become anyone.

KANTER: I've been losing milestones, lately. Tired of being fearful about it. Tired of being a hypocrite, that I'm still useful, for anything. Tired of serving— demons! Prove your worth, Bob! Prove something to me.

LORKEN: I can't! (*collapsing to the ground, with a suddenness that startles **Kanter***) My voracities have been extinguished.

KANTER: You are.... terribly enfeebled.

LORKEN: I can no longer be baited. Can't bite the hook.

KANTER: This is truly desperate for you. I'll seek help. (*starting to run off, as best his age allows*) You have fallen moribund. Stay by my.... billfold.

LORKEN (*half lying on the ground, with exhaustion*): What is this, now?! Is this it? (*clutching at the wallet with one hand, almost disaffectedly to mock its importance*) What could be in here that's so invaluable? (*torso falls over the wallet*) I am singed to faint. Nothing seems.... (*cuddles to curl through the earth*)

KANTER (*voice*): Protect!...

Act II

Scene I — *It is dusk. **Vogel** stands before the entrance to a hotel. He is in working clothes, which simply means a rugged but semi-formal appearance, not exactly uniformed, for a watch-guard. The hotel entrance has a lamp hanging just to the side of the doorway, and the proprietor, **Hastern**, comes out to light it. **Vogel** notices, with a casual sternness.*

HASTERN (*coming up to **Vogel**, after lighting the lamp*): It's a good night, Reg. Not a chill at all. Quite temperate. Easy work. I'd camp out, on an evening like this.

VOGEL: Sure thing, Mr. Hastern. But the nights 've been fairly warm for a spell now. No use wishing for some rain through them. Makes walking around the hotel easy business. But I don't mind the weather. Cold or mucky, it's O.K. by me.

HASTERN: During this season you can't be too careful. Every indigent in the region 'll try to get in here, to search for the liquor and some left over victuals, and to try to bum up the guests for some loose coins. I get tired of throwing them out, and you do a good job. We cut down on window breaks in the back considerably. How they do it so quietly I'll never know. But there they stand, in the hallways, frozen with panic when they see me. And I put them out. Rough times, in the beginning. Then I had to fight, and they learned not to.

VOGEL: Glass cutters, and that newfangled tape they get ahold of, Mr. Hastern.

HASTERN: Oh, I know that. That's early stuff. They steal it from the freight yards. But you still should be able to hear glass break, and I seldom could. One year it was so bad, I just left a window unlatched. Then I thought of a watch-guard like you, though it's a novel idea.

VOGEL: (It's) Spreading through— this town, at least.

HASTERN: Oh, I can take.... a few of them, during the winter, with their visits to get out of the cold, when they're really desperate and miserable. As long as they don't damage much, and stay peaceful and quiet, I can't mind. But not during this hot season of drought. They should be able to sleep, and live off the land to a considerable extent. I was a bum once myself, when I was a kid, but terribly polite, and only for a couple of years, till I grew horns.... to manage my life.

VOGEL: Tusks to help (with) a reaming through misadventures.

HASTERN: Why, if circumstances are such, you have to get by any way you can. I can understand that.

VOGEL: Yeah. It's been tough for some, when the weather turns to extremes. Not too many actually try for this place, as I've found; although that back room parlor can certainly be made to seem comfortable and secluded. On the sides of the building they'd have to climb up a bit.

HASTERN: Well, they see you walking about and they stay away; 'cause they know I put you here, and you're not in a mood to play with. This.... dissuasion.... is much better than a potential injury, or death. And I think I caused one; but I can't be sure, because one doesn't take medical reports on loafers and drifters too seriously. That information is not given out accurately, and is prone to rumor, spreads with exaggeration.

VOGEL: Nothing you should feel sorry or ashamed about.

HASTERN: And I ain't.... as much as I know, or can confirm. But they're getting more persistent and aggressive. Heard they attacked a resident today, right here in this town.... Must be getting really desperate for some, out (there) in the day's heat.

VOGEL: Guess we'll have to protect ourselves more. Really no excuse for that; though I have proposed a public water troth, like there used to be for horses. I'm sure some of them are still around. Wouldn't have to be built.

HASTERN: Rather ancient public munificence, of you ask me. Some would probably dirty it up (too much) before most could make use of it. But under emergency climates?... A kind of welfare hostel might be more appropriate, suitable to the task, if it could be handled in a very orderly manner. Nothing done of effort should be free. At least the effort must be paid for, if not the rewards and benefits.

VOGEL: Yes, sir. But there's no capital in bums—

HASTERN: No. You're right. I was quite worthless, but felt great about it, in spurs of perception, till I tired of a lack of responsibility and purpose.

VOGEL: And marriage?

HASTERN: Well.... foot-locks train your feet. You're still bitter about that?

VOGEL: Why shouldn't I be?!

HASTERN: But it's been years.

VOGEL: (It's) Made me what I am, with my freedoms. But a lady should never walk out on a good man.... She should wait for him to provide her a proper avenue of exit, with a securing impulse to allow for the departure. Not simply vanish after an argument, it's almost criminal. I won't trust another less trained.

HASTERN: Sure. But that's what innocence is for in a relationship, Reg.

VOGEL: Oh? What?

HASTERN: Training. You can take the weather like a stolid rock, or a steely-eyed statue. But that ain't gonna get you satisfied with life again, staying so resentful and hurt. Maybe you should just take (it as) a difficult lesson, and modify your attitude towards people and their affections. I've often thought you needed a way to steam out your angers, patiently. And this is the perfect job for that. But you can't go on forever being a lowly house sentry, not with your talents. A love for someone, or something, is not the purity of feeling we're taught to expect. It is.... a realization of incompleteness and need, that can be very messy and sordid indeed, to deal with. Besides, you have your church to backup any deficiencies of character you might too loosely express at times. Just rely on your membership for approval of behaviors.

VOGEL: I'm neither indignant nor injured by what's happened in the past for me, Mr. Hastern. I'm an upstanding citizen of this town, and an enthusiastic patron of her delights. Indeed, I refused to leave, when others have left me. But I don't take any philosophies for granted anymore, no matter how logical or of apparent common sense, like a vow "for death do you part," putting yourself within the hands of some god-earthly entity of existence and eternal partnership, which don't last the drying out of spit! And if one claims to change the weather, for the better, I want some proof for my faithfulness. You certainly can't prove faith. It vacillates and wavers due to circumstance, like changing colors through a lens upon more careful examination of your objects and objectives.

HASTERN: Well that's a harsh stance. So apply (a) more achromatic diligence.

VOGEL: Yes! Mr. Hastern.

HASTERN: I mean, faults may lie in yourself for your failings.... in life. The blame might be internal. So forgive your weaknesses a spell.

VOGEL: Hey! You didn't kill that bum, Mr. Hastern. The "Assumption" is not unrecoverable, for anyone's virgin goodness. And no saint walks the earth. They have to die first, to allow for a total corruption of their bodies. But I ain't forgiving spite and malice in another, until it's really demonstrated that I should. And.... that don't happen easily, sir. You have to provoke these things, to reveal what strengths there are— in you and others, the entire world, a grain of salt dissolving on your tongue.... or a salty tear dropping from your eye. There needs be evidence for this to be withstood. But I can condone.... even evil, if it's proven to be right. Or somehow deserved, maybe? There are many aspects to consider.... of our disenchantments.

HASTERN: That's mightily so, Reg. An endurance of suffering, of the heart or thoughts or spirit to abide their testing with cajoling, must have many hard but handsome facets to peer through, for a distorted view, within and without the world. (*heading back to the hotel*) I hope.... hope I'm not doing too much wrong to others.... during these trying days.

VOGEL (*as **Hastern** enters the hotel, front door left slightly ajar*): That is the tease of a jewel to own, Mr. Hastern. And this establishment is certainly stunning.... (***Hastern** has entered.*) for anyone's concern. (*He stands silently for awhile, observing the growing night's view with approval of a nocturnal splendor, when he notices a sound and turns towards the front of the hotel, paying attention to one side of the building. After a pause, he confronts.*) You better come out from there! I'm on guard, and I can just make you out. But if I have to chase you, back there.... you're gonna get hurt. Understand?! So just come out. And nothing's done that isn't done. (*The figure of an **indigent** slowly emerges, cautiously walking from the side to the front of the hotel, almost with an internal struggle of conscience to reach its door.*) You ain't going inside, my friend. In there is for the customers to exist, and not be bothered.... with your likes. (*Though a distance away, he remains rather un-anxious, as the **indigent** braces his shoulders and back to the hotel wall, not too far from the entrance, looking at him pleadingly.*) There's nothing in there for you. You can't have anything that's offered the guests.... It would be an abomination, an upsetting of paying privilege. Move on, now.... and no one's been wronged.

INDIGENT (*softly, with much apprehension*): ....Let me just ask —

VOGEL: No!

INDIGENT: —the owner— I'm gonna die out there, mister, in this wilderness, unless I get a tad of proper food, and some strong drink to staunch up my motley innards. I've been living of grass, what's left to be green, for almost two days now, and I'm ill. I wouldn't try for this place unless I had come to a panic— for my health. But Hastern has some sympathy for us, I've heard. Our poor plight, sir. Let me just ask—!

VOGEL: Can't do that.

INDIGENT: ....One bit of biscuit in fat— can last me a week. I ain't been bred for a burial like this, behind some rock.... among the wild weeds to lie. If I could do without this dole, I would. The day sears me into unconsciousness—

VOGEL: You enter pass that door and you *will* die. I can't help myself, from causing you a greater grief than hunger. And I've done this before— to others.

INDIGENT: ....I've heard something of that ilk, to have occurred here. But I've been made meanly curious— by my plight. And I tell you, I'm more than desperate, by my situation. I'm defined through it, to have to try this, to have to seek out Hastern's conscience—

VOGEL: I'll bash your head in, for every step you take to get some undeserved comfort in there. Or hospitality? That's— not a flea's market, but a well kept spread of furnishings and furniture, for the well to do or well doing— and not for you to brush up against and tarnish.... of reputation for a fine hotel. So— stay away from that door. If I have the rush in after you, it's gonna be a loud agony. I enjoy giving beatings asked for, makes me anxious to be thorough about it. Won't have rust like you around, bothering the patrons, dirtying up their grimaces with disgust. Despair? There's aplenty goin' round. Now, come here. Obey my orders. Let me see this plight, on a protruding stomach.

INDIGENT (*slowly lumbering towards* **Vogel**): Why kick me when I'm complying?

VOGEL: While?! Just get over here. I don't have time to waste on you, nor the stars shining. Gosh!... You're a young one! I was expecting some old beggar. You are a waste of sympathies, kid.

INDIGENT (*by* **Vogel**): I'm starving, sir— I'll work.... for whatever you want to make me do. But please don't cast me to stone. It's a fairly warm night; but I'll freeze out there, if I don't get something decent to eat. You'll find my bones next, bared of flesh. I can't take the hunger anymore. I thought I could hold out. But that was a night ago, and I've come to the end of my endurance. Charity, please! If I walk away from this attempt, it's to my careless death, careless of being born. Plead for me! sir. I'm all dried out, and these are most (of) my last conscious thoughts. Pity might defend me, even, to be a roustabout. But let me feed. I don't know (what) more to think. I can't conceive of conceptions, now, except to grovel for handouts, and the precious garbage that might be eatable. That's really all I ask for, on an evening like this, leftover drinks and gummed-through sandwiches and crumbs. If I could recite verse, and handsome stanzas of historic poetry, I'd entertain for this— But my mind is a blank, and my body's gone bust to be broken, with this burden of living like a rat. And now even the insects

avoid me, because they're dying out. At least I could subsist, with their pestering captured. But now I need help, and I'm all in a panic. Just hold my face, to show that I'm no spook. (**Vogel** *cups the man's chin and lower jaw with a hand.*) I.... was a young man. Now I'm defined by my plight. Some empathy, please, for a real disaster. You'll find me dead, next, if I don't get some assistance.

VOGEL (*releasing his face, but with a brush of the chin to move it sideways*): Yeah, you're a true victim of drought. And you've lost the clam (of clamminess). But you can't dine here. And I won't let you beg for anything inside. You are totally hopeless. This must be a really despicable trend, or turn of events. You're the second I've had to deal with today, who's come to his straits of annihilation, calamity and ruin. But at least you can still walk, gabble and gamble for some sausages, boy.

INDIGENT: Am I so? Can you still claim me to be alive, and worthy of some aid? I'll bend myself into your liking to supply— with humility! sir.

VOGEL: We found him a short way from the church, poached by the sun, as dead as a hide of leather, till we examined the scrunch more carefully. There was some breathing left, above what he was huddled over. Couldn't even run away with it. That was remarkable. So we carried him back to the church, for whatever spiritually induced nursing was possible. Now there's a good place to pass over earthly realms. And I'd say that's your best avenue of approach, not trespassing here. Do it before God.... or godly symbols and representatives.

INDIGENT: What do you mean, sir?... A church? I should go to a church?... Take me there.

VOGEL: I'm too busy.

INDIGENT: I don't know the way—

VOGEL: Well then you'll have to find it, won't you! But there's only one in this vicinity. And you must have gone by it, or avoided.

INDIGENT: I'm trying to think. They take in the helpless? My mind is totally unsteady. I can't concentrate on any memories. Is it still open, as a shelter for the shiftless?

VOGEL: Evening services will probably be over before you react it.

INDIGENT: Then it's closed to me.

VOGEL: Bang on the doors! There's a priest, (who) lives there, as a resident preacher. Might let you sleep with that other vag'(abond). Could give you some kind of a supper, for one night. But it ain't no hospital, nor way station for drifters.

INDIGENT: I'm game to try.... because I've lost my sanity, and can't think of a thing else to do.... now. But I've gone blind by the night, and will probably wander (to stumble) into some hole. (*starting to leave, with a somewhat aimless resolve and uncertainty*) It's to a church. I hadn't thought for that. They don't like the truly downtrodden much, I've been told.

VOGEL: Father Mathews'.... his name. Hey!... Get back here!

(*The **indigent** complies mechanically, though he hasn't moved far anyway.*) You really are senseless for thought.

INDIGENT: I can't, sir. I can't—

VOGEL: See that star? (*pointing*)

INDIGENT (*head down*): Yes.

VOGEL (*directing his face by the chin*): See it?!

INDIGENT (*without notice*): Yeees!

VOGEL: It's over those trees. That's the general direction, although you have to go over a few small hills, that can obscure its view.... briefly, if you're walking towards the place. Has a prominent steeple. (*releases him*)

INDIGENT: I see.... I see nothing, sir. Not much energy left, to disavow you. So I'll find something.

VOGEL: Well that's what you're searching for, isn't it? (*reaches into a pocket*)

INDIGENT: Yes, sir. Some.... night's nice charity.

VOGEL: There's a back room to the building. (*takes out a key*) Now look at this! You can get into it with this. (*The **indigent** studies the key half crazed or mindless.*) Not onto the church (properties) proper, but only to that room is this key for. (*shoves the key in the **indigent's** slightly torn shirt pocket, but then searches one of the **indigent's** pants pockets*) Well if you lose it, that's as much a path as any to follow. But I send in the bees. (*retracts his hand*) Steal honey from that space.

INDIGENT: Is there honey there?— a jar of it?

VOGEL: Maybe. I don't know. But if you're not let in through the front door, you can use the back. And there's always stuff there, tools and miscellaneous materials, things ripe for theft. But foods? I think I left a small box of raisins in there, once. Search the place. And if you're questioned, by any deacons.... say you're from the community, or you're concerned for your tramper friend they're keeping. Kanter told me me name's Lorken.

INDIGENT: Kanter. Lorken.

VOGEL: That's a place to break into, my friend. Not this hotel.... Just in that direction.

INDIGENT (*leaving, blank-faced and eye-starved*): I will.... to let to do to seek:.... Kanter Lorken. (*He stumbles slightly, which alarms **Vogel**, but recovers and exits. **Vogel** resumes his earlier pose as **Hastern** comes up to him again, from the hotel.*)

HASTERN (*approaching*): Who was that?

VOGEL (*barely turning to greet him*): Just another vagabond, Mr. Hastern.

HASTERN: And he wanted....?

VOGEL: In and at, sir.... You left the front doorway appealingly approachable.... He must have been waiting along the side, there. I should make my rounds to check, soon.... A very pathetic attempt, though. They have rumor of your.... sympathy and generosity.

HASTERN: No, that's just because.... we've been such a target in the past. The freshly indisposed of character, or newly dispossessed of worth, like to seize the best possibilities of their situations, and believe in what's most nice of rumors or twisted tales. They haven't much else to go on, or be guided by.

VOGEL: Well that one was really starving.

HASTERN: Was he? Maybe I should go after him.

VOGEL (*obstructively*): You have a hotel to run, sir.... A cathedral, for all it's about, you should pertain to it your care.... I don't mean actually starving, or he wouldn't have had the strength to try me and confront. He would have just dropped before me, and died. That's what you do, when facing the ultimate of blasphemies. You give in— cave in, and accept your downfall. I've seen the actually starved— and they are dead to me.... But he was starting to stink, reek of a poverty of soul. And a young fella, too. I told him.... something must be going on, some sort of cataclysm, or catechism for our behaviors to recite, by gestures and submissions questioning and answering.

HASTERN: You told him all (of) that?

VOGEL: No. he wouldn't have understood.... the dimensions of our shrinking divinities portrayed through our foul deeds, or curses lived through. But I told him to head for the church—

HASTERN: Oh.

VOGEL: —for a proper sufferance and shame, rained with, rinsed and cleansed by.... Douched up—

HASTERN: I'm not an inconsiderate pig, Vogel. If he was really in bad shape.... I could have helped him—

VOGEL: Believe me, he wasn't that way, Mr. Hastern. Just testing and teasing, to see what can be done, what can be gotten away with, over soft-hearted.... sore hands. And we always help each other, one way or another.... in this town.... If he were dying in the dirt.... I would have lifted him up, and carried him in— but he wasn't. He was only perplexed with his circumstance, stunned or phased by it, and was trying to take the simplest routes he could imagine to assuage the fright, the absolute fright! Mr. Hastern.... for a short while. He would have slipped in— which is unimaginable with my guarding, and therefore strikes of childish assessment, of the type he's brought himself down to— let you yell at him and lambaste, perhaps taken a little germ of leftovers you were about to throw out, and retreat sheepishly, withdrawing back into the darkness knowing that *someone* considers his standing in the universe, even if with only anger and contempt. Would be easier to simply rumble and rummage through the garbage. But one wants to be told off, and spoken to. And I suggested to him he should seek.... a higher wrath to be humbled by, humiliated with. For such a lad should feel disgrace— like sugar on the tongue to taste, knowing— or suspecting— it's yet possible for his substantial improvement. Reprimands are not quarrels, though, of struggling with self-appraisals. And I'll have none of that while on my watch. Let him face God's hostel of forgiveness, with songs of sin to sing

— as were an education of depravities to avoid— and not yours, born of fruition to your goals of hospitality for travelers, passers through, passersby, passers idling (with) a stay.

HASTERN: Well, if he were hurt, or delicate from injury of health.... you would have alerted me.

VOGEL: You know I am a kind man.

HASTERN: I count on your reasonableness, Reg. At least that, with certainty on your performance. Oh! This drought can make men mad, really insane, (from) torturing dilemmas tossed them, like spores blown through the wind.

VOGEL: A hot sirocco.

HASTERN: You can come to a point of trying anything, and leave what happens up to the.... cravenness of your casualties.

VOGEL: This is so, since you're only truly brave over what you can't control.

HASTERN: I mean, people can become quite vicious, under great stresses and difficulties. An animal takes over, that may be forgiven its single-minded implacability and determination to cause harm— but not the harm itself.... not that terrible action and its sad consequences for others.

VOGEL: No bums will enter this hotel, while I'm around. No heartaches visiting. No sorry sobs professing horrible incapabilities and loss of luck so thorough as to bend a horseshoe straight, as if it were meant as a nail to your coffin, a spike sepulchral to bolt your containment into a state of persistent distress and misery, until you dread its ignoring of your presence.... having taken your sighs so often for granted as to make them seem like the random chattering of a slow deterioration not more to be seen.... through your darkened background, and exhaustion of hope!... But I must make my rounds, to see what might be lurking, or gathering behind us, Mr. Hastern. He might not have been alone— You never know. Have there been any complaints by the guests?

HASTERN: No. None at all.... that can relate to culprits 'gainst their comforts. Though lodgers always seem to be worrying about something. Money, food.... bedbugs....

VOGEL: Then no strange creaks have been heard, as were their nature to be made by unsettling sorts. But other, then, the same as kindness heard.... of the settling of time and day. I feel to hurt.... something, of what passes by. (*starting to head to a side of the hotel, in order to survey around the building*) I'd close that entrance, sir. It is tempting.... and not of warmth, for this quite tepid night.... but as a window opened, with its lights.

HASTERN (*observing* **Vogel**): Oh, the doorway.... is fashionable to be showed off.... (*more to himself*) But I'm not expecting others to arrive. All that have made it here.... have been to the till, have paid for their stay, and have seen the attractive fixtures displayed for their admiring.... of my working house. The lobby is rich of handsome décor, and the front desk regal of efficient functionality.... for this invitation made, and appeased. Such is a handsome place I have designed.... for others. A mercy to it, please! I keep out not the wanton, with this ease. Into which, adornment's testament to our servitude of the human cultures and natures we pacify,

(is,) at least what I've known of them.... and have been bred from. There's a doleful severity to all sorts of travel, and being on the run. Thus (is) to make this easier, this transit.... serene, among our speculations.... of behavior. A gilded passage found can make my station, from the mournful, bound to destinations and what has been determined for them, hard reckonings of life. Yet.... who can not subsist of this.... clandestine change of miracles, and modifications of earthly environments, must seem to linger and lurk around this search-meted pour, emphatic of my establishment.... disconcertingly— and afraid, unsure, and endlessly remiss of our society. For that, I draw a hunch to, nuts to a fire— for a communal roast, or a community condemnation— celebrated and partied on.... that way appeased of tastes and barren worth, unshaved faces and moth-worn clothes.... and.... the subtle grace of bums displaced, light steps of heavy limbs, agape through their timidities of fraternal association— with.... the world at large. How have I found this place, can make a night's explanation of daylight. There are just too many, to be enticed by the fire, and drawn to.... not my hatreds, but my humanity. I, were a gob, a goo of glue to these!... To them, persistently I am amended for, a few petty morsels, and mouthfuls.... of rancor and loathing.... towards their plights, and passivities in this world shared.... and inoculated by. And I could be.... an exculpation— without prayers! to visit me and know.... what is humane belies the truth of worrying. We worry for our dogs, and not our princes. We tear at rags, to please our lusts for destruction, and leave our fineries alone to fend of their own lusters for appreciation and approval. So we are made.... to build hotels, and test the streams of passing persons, as like a frog's sticky tongue to capture water-flies and crickets. Yet, what is ate.... is for my resolve to grow with, knowing of my sensitivity to these.... fine passions, for a human consciousness at play, with madness honed onto dilemma and desperation. Then.... keep them coming. Keep them coming!— But not so many for hope's extinguishment. I must stay in business, as an example to portray, and demonstrate the— fight, how it is made.... if never won, from the deep-seated poverty that remains inside for a gut of feeling with rapport, for the disemboweled vitality thrown onto the streets and roads and earthy fields and dusty, dirt-strewn havens and heavens of open country and rustic Manifest Destiny— of the insects that cling to you, and the grime that is patient for your carrying, and the supposition of your being.... done.... or how as not. *There* is the parcel built, of your great substance, tied up by your most holy thoughts. And that is what you come to carry, or grow with and eat off of during a drain of conscience and drought of amicabilities or friendliness or kindness to you, and a spell of worthlessness lasting so long as to extend the air into the very breaths of depression taken and gasping for. One heavy, musty mist to toil the lungs and battle the brain with difficulties.... of starvation, is how (is) done to you a loss of luck or being in the least fortunate. We make trips to avoid being caught for too long under befouling weathers and conditions that hamper our nerves and courage.... to be consistently ourselves in tranquility of conscience. What is more for any thought to make? but that it is our own possession in some way enacted to appease a self-upsetting or anxiety. Come to a hotel, then, and spend time witnessing.... mobile hearts transacting towards endearments, wants, needs, wishes, demands.... and vainglorious attempts to shelter a disturbance of the soul.

VOGEL (*coming up to him from the other side of the building*): No one back there. It's as empty as a graveyard.... Why did you wait for me? Couldn't have been that afraid— I can handle these tramps. So can you.

HASTERN: Well the night is such a pleasant one, Reg, I just thought to spend some time considering what it harbors.

VOGEL: Rats and indigents and weeds, or burnt-up flowers.... and a sorry star.

HASTERN: But it's still good, to be out here and apart of it.

VOGEL: You expecting somethin' fantastic and remarkable? some event portended by the sweet breezes?

HASTERN: No, if it's not more than ourselves listening to the rotation of the globe. One has to take a breather from one's chores at times, you know. A whole slue of requests will bide my time now, late into the night. There's no question I'll have to redo at least one of the rooms, to accommodate a certain taste. Maybe drape the windows black, or adjust the shutter so that the patron can sleep late. I'll show her how to do that herself, though I don't see why women should travel alone so often these days. Then again, no one's here to stay for a vacation. This is not that kind of town. They're all en route, for some other destination. And if they're visiting folks around here, well then that's rich, to have to stay at a hotel, like you're invited just so much— and not more, to live with or guest at a family's dwelling. Then I guess your destination's.... a kind of commiseration with what some others expect of you.... And I have to make sure that boiler's working. You'd think tepid water would be enough, through a night like this. But they've paid for it to be hot, when they want it. And the same for the radiators, if a morning chill greets us, which is not expected— but you never know. Customers can be finicky about the most absurd things, when it relates to their health and comfort. I'm glad I don't cook. I just approve what the restaurateur serves and serves out, and what the kitchen provides.

VOGEL: You should get back to your work, sir. But I certainly do enjoy the night airs also. The serenity transfixes cautions, of the coming day to be prepared for, in a most relaxing and pleasing atmosphere (as) possible, even if the weather is raining or quite horrid. Because that next day, that's always a heavy chore to handle. I always prefer my anticipations to run through the night. To precede what happens, in as opposite a sense as you can make it, is one's best course of action or preparation for.

HASTERN: That may be so. But it's a peculiar synergism of beliefs, because the night has its own right to be— and, I suppose, the poor and downtrodden. Well.... (*leaving to return to the hotel*) it's back to the woodshed for a spanking of duties and obligations.

VOGEL: You certainly won't be bored with it, Mr. Hastern.... And neither am I, with my watchman-ship. Our respective tasks are cruelly interesting for both of us, I'm sure....

HASTERN (*offhandedly, just before entering and closing the hotel's front door*): Stay sensible and reasonable, Reg. (*closes door behind him*)

VOGEL: ....Peculiar, hey?!... Like pelicans at sea? Why would you expect to find those types flying over vast stretches of empty ocean, or other birds making extraordinary lengths of migration through the change of seasons? It's all because of the contrasts of nature.... Walt, and the privilege to sense extremes differently, individually.... of character. And I like my standing stay. It's a just punishment I can only relish and adore. That sunlight.... it bores

holes in the head, to see what should be hidden.... or to not see what is missin'— missing. Missing!... Callous taters! They have eyes too, you know! And.... rust on the pipes.... rust through the drains. Rust! Rust!... It's a shame, what is allowed for a church.... Not half as pretty as a shade of consciousness, though. And I'll take a rain just fine. We all would, these days.... (ex)'cept those poor earth-squatting dogs, and mud pie bathers (bathing) 'gainst their will— Well.... have the night! for listening on shrill owls and chirping cicadas. I love the banter, and the shrouded battering. But bring those golden rays more blinding of my forsaken-ness. Prove something to me!— you beast palavering my hurt! End this crusty drought!... for a day or so, just to show.... this has to be done. Yes. Raw meat tastes best to the wild biting at it.... But there are more keys at home. Stupid trinket. Silly collection. Ooooh!... But such an empty house to return to. Aaaagh! Tears my habit up.... And he's such an early riser, this Hastern, to relieve my watch.... and work.... Good terms, though, to shade the day. (*resumes a relaxed alertness, observing the landscape*)

Scene II — *A small bedroom within the church.* **Paths** *consults with a* **physician***, while* **Lorken** *lies quietly perplexed, a light bedspread covering most of him.*

PATHS: Well? How long must he stay here? Is this truly serious?

PHYSICIAN: The man is totally famished, Paths. That's his trouble in a nutshell. I've seldom seen such wasting of muscle for a skeletal physique like his. It's just about to affect the face and skull —

PATHS: Not so, doctor. We fed him some broth already.

PHYSICIAN: —When the jaw and cheeks and eye sockets become bony and protruding, that's a dangerous time reached— Broth?! I hope there was some meat to it, some hearty stock.

PATHS: Sure. Sure.... Well, I think it was an alphabet soup, with some beef broth. Not sure if the can said meat to it (containing).... but the pasta is definitely caloric.

PHYSICIAN: This man needs a thick stew, rich of the meat as well as the starchy potatoes and fibrous vegetables and roots, something to fill the stomach with a real digestion— though starting with a soup is good, to restore the hydration with some salts.

PATHS: Starting?! Is he really ill?

PHYSICIAN: There's no serious disease to him that I can tell initially, except hunger. Perhaps a slight cold coming on, due to a stressed immune system. But with some regular meals he can overcome that easily, even prevent it.

PATHS: Food! Food! How long? How long? We only have enough stockpiled for our priest, with a reasonable period of time planned out. You'd be surprised how delicate it is to foresee and arrange these necessities. How long before we can send the bugger on his way?

PHYSICIAN: Take up a collection for him.

PATHS: We thrive off (of) the charity of this community, sir. He has not been properly recommended for our services— and can do no work at all!

PHYSICIAN: *My* services to you are voluntary. Certainly you may feed him.

PATHS: I mean around the parish, doctor. He must be.... a utilitarian conscript to deserve our assistance, in the material worths, because our thresholds and margins of subsistence are thin. We are all volunteers, in relation to the church. Our priest's salary is quite superficial, and very variable. We can't plan to nurse an invalid for very long.

PHYSICIAN: You don't have to bargain with me about his prognosis, and diagnosis! A day or two will do. Three to eight decent dishes of dining, I'd say, and he will be ready to thank you with his departure, unless he *does* fall prey to stamina-crippling insults, like a chest infection or sinus-nasal blockages or a sore throat, in the mean time. Any one of these, or others, could weaken his constitution so, given his current condition, to have to keep him in bed for weeks to months.... if one should contemplate the proper thing to do for a convalescence. You wouldn't want him lying on the bare ground while trying to endure such distresses of health.

PATHS: Two days. I think we can just handle that, although one would be safer. And what about communicable diseases, doctor?

PHYSICIAN: What about them?

PATHS: Has he any?

PHYSICIAN: I said he might be catching a cold. I doubt a flu, at the moment. But if a flu, he should stay in bed entirely through (out) its duration, a week or two, until the virus passes out. But as for spreading, you could treat a child the same way, with common sense cautions and precautions.

PATHS: So he can walk with a cold.

PHYSICIAN: This is diddling, Paths, on his well-being. But I've often done much work with a cold— The man's body is simply exhausted from lack of nourishment. Restore that, and his health will follow. But where he rests will often dictate the strength of a recovery. And that's better done in this sheltered bed than sleeping on the outside grasses, even with this weather.

PATHS: I don't want to seem parsimonious of caring, doctor, but there's only so much we can do. We are spiritual healers—

PHYSICIAN: Are you?!

PATHS: Well, attendants. The Father heals— *The* Father heals.... and we assist carefully, managing the mundane to allow his full focusing and concentration on the miraculous to achieve, through preaching and educative suggestion.

PHYSICIAN: Yes. Yes.... Science and medicine are not the complete answer to everything, though they do so strongly feel to be. We each have our hypocrisies to overcome. That's why the most enlightened allow the philosophical idealisms of religion to be imbued by. But, man, this fellow is in such dire need, at the moment, you only should be thinking of his welfare as utmost of importance for his presence here, and before you, healing.

PATHS: I am not a priest. I must wear— the practical gloves, while handling cases like this, *for* our priest.

PHYSICIAN: If he takes a turn for the worst, to be inappropriately kept in a church, couldn't you take him to your family—?

PATHS: Mine?!

PHYSICIAN: —to help assuage the rampage of the devilish that contaminates his body and coerces his soul into sorrows, the illnesses at play to dent his resistance?

PATHS: We're as compassionate as anyone around here. But my service to the church is as her deacon, and that is a high and heavy task. Sure. If come need be, I'll search for other places (for him) to stay, with another family or a generously accepting institution, if his health and ambulation become critically marred, or that is threatening within two days. But not at my house. The wife would not have *this*! We do so much already. We can't open doors to a regularity as such (*pointing to **Lorken***). Do you know how many there might be (out there) in similar complaint?! One can only do what's within the scope of one's jurisdiction to serve.... an organized, formalized method of relief. And that is how this tramp's been brought, detained from vagrancy for awhile, and with no other potential in sight. But the ways mysterious will lead him through this trial and travail, just as for any other person. And what is appropriate will become. Don't blame me for any other outcome.

PHYSICIAN: You mean, if he's meant to stay here, he will.

PATHS: Not that mer—

PHYSICIAN: And what if you had found him dying on a deserted road?! Would not you have tried to take him to any household available, within your personal carrying distance of brute-strength endurance and physical duress to the necessity made— unto you?

PATHS: ....Not that meritorious assignment made— for me, doctor. He was brought here, to the church. And it does no good to conjecture on the other possibilities that could have been, because they are endless and therefore meaningless. But if he were starving at my doorstep, begging for some crumbs, I would feed him a whole loaf! And if he had collapsed before my house, from a weakness of health or injury, and a wealth of hunger and bodily privations— then I would take him inside, to heal. I am a servant of action in all obvious matters and requirements of the humane tendencies we must share within a civilization. But this happenstance lends its way to matters more obscuring than what I can see through. He is now a purview of the church. And she will dispose — not I, the consequences towards his recovery. The holy spirit that abounds (in) these walls will make for order of the circumstance, that he should leave within two days, or not. *I* can not calculate more on it, within his poverty— nor hers to lend. For if he stays longer, the riches of this church decline substantially, and quickly. And my efforts of deacon-ship will increase exponentially, to try and maintain her. *That*.... is my commission to this problem. That is my responsibility to him, as contrasted to the care of my own family and household. Any purposeful arrangement of life may de-convolve with a foolishness. So naturally I avoid this for my own, as he should for himself, sometime. Is this not our gift of talents to be human?! Then take charge of one's self and condition, as much a possible.

PHYSICIAN: ....And would it not be simpler just to be neighborly to the distressed?

PATHS: If nature brings him to this state, then nature may excise.

PHYSICIAN: Human nature(, you mean).

PATHS: Heavenly! For all (of) its worth is *here*. I'll not take him home, not with me and my burdens already, unless he's healthy enough to do work around the house. And I'm not wealthy enough even for that prospect. So the best remedy that She can design, is for him to leave within two days.

PHYSICIAN: Or by some miracle to find him a job, as he does chores around the church, fully recovered. As I say, some just meals will do—

PATHS: I doubt this. He has not been recommended, with an empathy of spirit any one of the congregation should sense.... Maybe at the hotel, but one must actually want to be saved.... from the discouragement and disgust of one's self—

LORKEN (*interjecting not too weakly, but obviously handicapped of voice*): Are you talking about me? Jules told me something about church service, but I couldn't understand it well. I've been terribly confused. Where are my shreds? (*sitting up*)

PATHS: Jules?— Yes. Jules Kanter, he found you, madly searching for water, I believe, at some supposed masonry of well work that doesn't exist.

LORKEN: I—

PHYSICIAN: You are in a night shirt, sir. I believe Deacon Pahtes washed you up a bit, and fed you some soup—

PATHS: I prepared it.

PHYSICIAN: —And then you were brought in here, to repose. Isn't that right, Paths? The explanations were given to me franticly, perhaps to rush me over.

PATHS: I prepared the room, too. This is all so unexpected and unusual. And I'm not even sure if it's proper. That is consecrated cloth you're under.

LORKEN: Everything was for a daze with me. I can just recall that.... Mr. Kanter was insistent about something, I think to walk me to his house, for some food and shelter through the night.... before introducing me to a church.

PHYSICIAN: Good fellow! Kanter. That is being hospitable, in a manner customary to our.... methods of belief. Well now your worries are over, Paths, about where he might stay long term.

PATHS: This has not been recommended yet. The congregation must approve some of these activities—

LORKEN: But then my mind swelled up, like tears. And I fell down.... unconscious, before he could lead me.

PATHS: We carried you over to our church, man. You are at our church. You are within the holy graces, observing and displeased with your weaknesses. And you should strive to revive and become more worthy of your weight in flesh and bone. Because right now you are a curious burden to everyone, whereas before.... to no one — but yourself.

LORKEN: Can't deny it, sir.

PATHS: Deacon Paths!... You are— not sick, according to the doctor.

LORKEN: Thank you, doctor.

PHYSICIAN: It's all of your matter to behold, son.

PATHS: So we will care for you a little bit, a couple of days, until you can resume your.... travels and wandering, your habitual destitution.

PHYSICIAN (*going over to a small table for his medical bag*): It's the severity and duration of this drought that brings them to this sharp edge of existence, Paths. They are normally an astonishment— to me— of heartiness. Or else they early succumb to physical afflictions, diseases, debilitations and death.... And not in the numbers I would have predicted (*as Father **Mathews** enters*), either. There must be many of the destitute, today, in these general regions. The country's simply not right of its economies.

MATHEWS (*He is impressively thick of ecumenical robe, considering the temperature, and seems large of stature or threatening of its gravity.*): And how is our country suppliant doing?

PHYSICIAN: He may recover handsomely, with your generosity, father.

MATHEWS: That is to the purpose for all of our considerations, fine physician. On the wings of an errand, he is brought us to, to seize this suffering and demolish its perspicuous intent of dehumanization. This man.... is such a creature of God, that it is glorious to know him, instead of disparage— Has he been feted well, Paths?

PATHES: He's had some soup.

MATHEWS: Ah, but for a fast in our simplicities. And crackers, yes? Bring him a roast of something, soon—!

PATHS: That takes from you, my grace, much of your plate to fill.

MATHEWS: We should not be ashamed to share, (such as) much of this room for visiting prelates. Are you hungry? country.

LORKEN: Country?

MATHEWS: Country fellow. I myself have a tremendous appetite, after the evening services. But I have not ate much tonight, with thought of you. Since here is an example— of the scrounging of saints to take souls, find them and meld with, out of the mire of your independence to pull. That muck from which you were dragged, as described to me, demonstrates this all too resoundingly and clearly to relate.

LORKEN: Saints? Oh, I'm sure they are the nicest of gentlemen to have retrieved me from my disabilities.

MATHEWS: In every single one of you there is a saint, come to lodge and borrow your heart, turn you into a magnificence of your humilities, instead of with embarrassment for them. And I will as much starve, to see you walk again—

PATHS: He is not crippled, father. He simply collapsed from dehydration in the day's heat. A full recovery is expected— shortly, and necessarily, for our parish.

MATHEWS: But the danger of a relapse is always present unless we provide a pontifical care for our least to our highest members—

PATHS (*with honest disdain*): He is not one of our congregation, and I'd say never prays above the fodder of his specious desires and capricious whims towards lusting of a life deterred from decent living. These people live on limbs, Father Mathews, and almost cherish a fall out the tree. Should we make the ground mushy for them, and damp and uncomfortable for ourselves?!

MATHEWS: We are all within an assemblage of God's children, Paths, and no better than our strides and gaffes. There is a mission to uphold, to be benignant to persons in need, as the physician works to heal the sick and end all illness.

PHYSICIAN: A powerful avocation and rewarding task, father. Though my main purpose is to make correction of poor health and salubrious practices. Patients must essentially heal themselves— Or through a graciousness that is eternal in this drive, I merely medicate and reset fractures of bone and coddle wounds to clean, prevent infections and stem the blood flow, mundane but educated techniques less powerful than righteous and meaningful prayer and the honest evoking of our worldly goodness to ourselves and each other.... But I am not a person of miracles. They I leave to the impenetrable and un-adulterant of knowledge.

MATHEWS: That is the mystery we all must abide with. Yet each has a solemn duty to perform, which is as important as the most important. And do you understand this? country. For as we serve you, then you must serve us. As we assist you, then you must assist the world. And as we pay you an homage of living, then you must live without misery in your apportioned duties and rites of sacrifice to the betterment of all mankind.

LORKEN: I.... ain't done no great wrongs.... to anyone, in my life.... reverend, though I do admit to killing animals, for sport, and then to survive— and then in desperation to survive.... And then in mind do I think of the evil done, and done to me. But I've just come to a bad way to fall flat on the ground, the soil-tendered earth I've begun to love and not treat as dirt. But I do accept that we all owe some perquisites to *something*.... that works for our existence, no matter how meanly these difficulties are struggled through, or how lowly we become. And I will let you order me about— or away, whatever is your desire, and when, as much as I am able to satisfy them physically. But my mind seems so terribly battered, father, that I can not even recognize disparagements anymore. They are neither cold nor hot, and I repeat the sentences unmoved of castigation: This travesty of man should wither away and die. A pouch of pestilence converges on our thoroughfare, staining the roads with rot. The dastard sits.... by our views and senses, waiting to cower before our noses and steal our trampling respect by beg-

ging for our garbage and turgid glances.... I accept everything now, as due a daze and through a haze of darkening irony, that these contempts are becoming satisfactory, and satisfying, and comforting, and generative.... of a good feeling and warm conclusion for a life that's to be shunned or shimmed under headstones.... interpretive, that this fellow is bothering, pestering death!... Give me a panic again, father, to have me yell at you, so that you'll kick and give me a good hurting, cause a strong and robust crying and make some temper for this pitiable being who can only sob in gestures of unconsciousness. I can not take.... the knowing of me else, else the long wakeful hours of perception frighten my eyes— and my entire awareness. The sounds condemn. The scents condemn. The bites condemn. The tastes of spoiling trash condemn, the wastes of delicacies thrown out, no longer worthy to be (called) food. And the grimaces, the sneers, the faces of scornful condemnation.... they contend for my mind, to kiss and welcome and accept.... and allow of battering so justly deserved. For if one can no longer live proudly, if he has learned the secret of being a disgrace, if he has learned its meaning and its holy mentoring of decomposition and faithlessness.... then let him eke out the.... retaliatory pleasures, of a baseness to his reckonings.

PATHS: ....These tramps always feel so sorry for themselves!—

LORKEN: No! It's for you that I feel sorry, that you've had to bring me here. Now dispose of me, if you can bring yourself to do it— again! Did I not die? out there? Or is this some quarrelsome death inside of me? bemoaning my patience with you!

MATHEWS: My son, country—

LORKEN: My name is Bob— Robert Lorken!

MATHEWS: My son.... you are— *home*.... This is your privilege, returned to you, that you are as actual as any other to be cherished, welcomed and adored.

LORKEN: I am appreciative, reverend.

MATHEWS: And you will be fed, with the fullness of my blood —!

PATHS: Father Mathews—!

MATHEWS: —For my blood is this church's license to treat you, and all who need be brought here.

PATHS: Father!... What he eats, you can't.

MATHEWS: Where he treads, I walk. What he touches, I feel. And as he heals, I become healthier. As he restores himself, I receive the resplendency of his grace and thank-you. We are bound to a contract of benevolence, for this brings out what is eternal in all of us, what our mortal flesh has only been loaned.... to use and subscribe to in obedience to the highest spiritual goals and challenges. And you must sense this, Paths. You must feel it dearly, before you can preach with earnest to others. You must be willing to take his bed, if necessary, as he has taken ours, with your willingness. (*coming towards the bed to reach for* **Lorken**) And as emergencies arise, you must have a genuine sympathy for any distress also; and that is your pain and hunger and suffering, and pangs for a commiseration of the heart and an understanding of what beats, or what may defeat you during fits of emptiness in isolation and scorn for one's sorrows.

PATHS: One need not belittle the rational.

PHYSICIAN: That's only a servant, Paths.... not a prize or trophy to be won.

MATHEWS (*beside* **Lorken**): Now, touch my cloth, country Bob, country son.... (**Lorken** *hesitantly complies.*) and tell me what you feel.

LORKEN: ....Is it.... God?... your highness, and priest to followers? Is it that which makes current to my anxieties, not to defame a purity with my crude fingers of sloven inconsequence? You allow this of me— you permit.... to share the shimmer of a religious relic to be made(, in the making). I do not dare to, but dare of to.... touch your robe.

PATHS: So he gets religion that way?! He receives it through a gown?! It is a pathetic demonstration— an enticement, even! I could do worst, but not as mendaciously.

MATHEWS: You're too demanding, Paths, of a trustworthiness to the idealistic. We are coarse animals all, only vouched for, while left to our own squalors of business. But it is quite an honest expectation to receive from the rich ecumenical garbs a sense of the holy provisioning we aspire to achieve. But what you touch, my son, doth maketh of your own wonder, and yourself perceived for joy. Each person is one's own religion templed, as the raiments worn may suffice for the naked truths underneath not to be haphazardly displayed. There is more simpleton in man.... than sinner, and more sinner than son. And you converse with me by touching this, not to be ashamed of wanting God. (**Lorken** *unhands the robe.*) You want.... your health again. Can this not be true?

LORKEN: The hope confounds me a bit, reverend. I want to end this suffering, as a casualty of wariness and weariness combined— But I did receive a shock, to touch your dress and feel some kindness left in my hands. That is possible. It's a worthy lesson.

MATHEWS: You have yourself for kindness to be made, and craftsmanship or utility with a useful purpose.

LORKEN: That's definitely possible. I am a rather stupid clod not to realize for myself some necessity—

PATHS: The soft feathers in a pillow could do the same.

LORKEN: —But I've fallen very tired of thought lately.

PATHS: Well certainly we may share ourselves, as the father demonstrates in this.... game of touching the ordained, an interesting tactic taught.... at a college of pastors, perhaps? We perceive by what we are. And so, what we receive is, in a sense, ourselves.

PHYSICIAN: If there's hope, it's your own hope. And if there's hate, it's yours the same. Well that is justly made a notion to accept, since you certainly die all of yourself. Or let's say we use up the wick of a candle, as the wax melts away to allow its burning. But if these allegories hold true, then our poor friend here is as much a priest as Father Mathews standing. And I can tell you with a clear humbleness of conscience that none of *you* are physicians, with my standing. So why, then, should the point be more compli-

cated than to get well, want to get well, work to get well.... be well?! Certainly there are falsehoods preached if otherwise.

MATHEWS: My point to country is that it is how we are wrapped that pertains to our accommodations with others. How we comport to be (i)'s much little of ourselves but all important for our relationships in a society. One should not confuse the appearance with the apparent. One makes the appearance, but one does the apparent.

PATHS: An appearance is apparent.

MATHEWS: One can be ashamed of either. (*sits on the bed, as if to suggest a crudity of nature*) But you own the appearance, while others make for themselves the apparent. Neither may define each other properly. And the country may be as loyal as I, to such definitions as we would want to make.

PHYSICIAN: Your reverence is subtle, to be god-like.

MATHEWS: As all men must (be), before the truth to behave. Now.... Bob, I am tired too, and dog-like, despite my high religious office. But that is not worth anything— to me, without your conversion to a well-being to obtain. For you are not as you appear, yet honestly haggard and fed up with life. But you are what I am, apparently, a challenger and a struggler. And you will help my creed, as payment for your convalescence.

LORKEN: I'll do my best.... Father Mathews. It's not that I've never had faith in myself, but that I've *only* had faith in myself. And when I came to believe that my time on earth was up, there was no sorrow to it or pity for the conclusion; since I have great trust in the profound judgments that eventually come to me, as for any person. And like an insect trapped in a snare, I prepared myself for calm through the desiccation and death. But then Jules found me. And I love him, while not liking him— For how can you like someone you don't even know? And he was a little coercive, challenging me by implying I might be of an evil nature or intent, when healthy, but that he didn't care and would take me home anyway, a wounded wildness, to look after for awhile. I felt rather resentful— Not the mad animal I!— And that last vestige of pride fainted me, out of arguments and protestations. Some clarity is needed for my supposing. He was afraid of my.... dying? A bum?... more than as I am a bum to fear. That type he customarily stays away from. Well.... he's a nice person, a kindly fellow, an old gentleman and one dominated by a golden conscience. But I sense he's a little fed up too. So I can share some of his fears easily, and shed their expectations like a nervous dandruff of the scuff. But as to be myself, I was so sure finality awaited, until he jostled my attentions to demean my attitude. It's what one comes to, ultimately, that decides your actions through or around, when life is this shameful progress burdened by.... It pains me to know of your nakedness, father, to realize it's so much like mine, and I'm as worthy of it as yours (to you).

MATHEWS: I am betrothed to the similarities, my son, and am loved through them to seem. You are keen to be made an apostle with reach!... of the despondent and forsworn of God. We will bathe together, my country, as we experience a spiritual enlightenment coursing through us, and you producing converts to life and sound living.... among the woeful arisen, among the famished kissed and marked (for exculpation), among the drought-ridden blessed with drink of never ending dews from a divine purpose of holy self-sacrifice with thoughtful lack of comfort and pleasures through this dreaming. Thereby, a death is assured.... its resplendency due with honor for its necessity, instead of debasement. Now, is this too bold a wish to purchase with this young man's life, Paths? Or would you demonize my falsehoods expressed through him, his ire turned into a possible elation!

PATHS: You're too sensitive.... of my deficiencies, father. They are for anyone to discount the discountable, or unaccountable of dereliction. Yet the doctor tends to any bodily suffering, and so should we a body tease towards health. I just say nature, or the natures in ourselves, only permit this in a reasonable way. And I would never claim you to lie.... or sit with thieves demented for their causes.... Forgive my aspersion. The deportment seemed strange to me, from a priest.

MATHEWS: There are sundry methods and behaviors of presumed authority chosen to exist with, Paths. I'm sure our physician would have his patients not fear his great concoctions and instruments, as I would not have my country fear the cloth.

PHYSICIAN: With a measure to the severity of my prescriptions, Father Mathews, I prefer them to be bravely obedient to the medicinal usages I insist upon, and the recommendations and abstinences I suggest, for their physical constitutions to support and improve. But other than those dictums delivered and bitter chemicals ordered to digest, I take no pleasure at all in making commands. I find it a rather childish attitude to have to partake of professionally, and a rather boring one, for the age of my practice. But it take years of experience and observation of personalities and actual study of the commitments to well-being to determine what is a subject, and who are the servants to these sick. We are respected because we're needed— And are you needed?... Then who are the sick among us? And what is an illness, socially determined? No, indeed. You don't fine me scouring the countryside, attending the trampled and disaffected loners scrounging miserably for a pittance of existence. I'd probably be assaulted or man-handled by the destitute I find of any strength, either because they are mad, or I am mad. And I properly fear this. They don't need me for their indigence or general misanthropy to promote and heal within. Though the poor that are finally forced to come to me are often very grateful for my services, if they are sane and conscious at all. And this fellow should be pleased of the attentions given him. I should probably never have seen him, or come across him, were not for the charities and concerns of this church, and its parishioners.

LORKEN: I am very thankful, sir, to be judged worthy of some care, a modicum there of at least pursued by strangers for me. And you are absolutely right, I was thoroughly mad. My circumstances made me so—

MATHEWS: But you are country.

LORKEN: —and stole a courtly spirit.... I am apparently coveted
—

PATHS: Now that you're here.

LORKEN: —to prove a point? Country? I've seen it from the bottom up. I have licked the dirt, for some tasty salts, and ate the grass leaves to know what I've become, a ruminant of country, a cattle of the range and pastures, though disconcertingly humanoid and for no one's desire to be herded.... I would not have crawled to

the doctor, (to) any aid. I would have died silently, pleasantly, letting the sun's heat rush at the transition from the living loathed to a loathing left. That is certainly an insanity; but I felt humorous about it, so well justified.

MATHEWS: Disciples often think that way, until their heads are turned.... (*standing up from the bed*) towards mine.

PHYSICIAN: Why, were you expecting some poison, young man, to end your complaints forever?

LORKEN: No, sir. I was dazed, and expected to find myself displaced among the viscera of angels, since they must discard that stuff for (their) lift. And before I could comprehend anything, I heard voices speaking.... words, I gathered. But I could not consume their meanings, yet. They tantalized to struggle with my visioning in this solemn oak-lined chamber that seems so.... declarative of an oppression, or wonder towards a stern hand of guidance. I felt the wood, as in a resonance with its broadness of the walls, and the handsomely stained grains, the wrought lumber abiding my eyes— with whispers and chants, and authoritative decals to notice and submit my mind to. For they.... and their shapes.... are telling me to behave in some way, weak as I am, confused and submissive. It is dark in here, but very bright, I thought. And then I heard sentences, which taunted meaninglessness— until I strained for understanding through the clouds, and came to a conclusion that I was alive, mortal and earthly still. And thoughts were being said.... about me, those of which I had.... perplexedly woven into a tapestry of contrasts and complexities to live through tasks and things, a church, a congregation and its services of approval.... or approbation, like the clearing of the head from heavy dolors and thick admonishments of soft pains and griefs. And I come to you with some merit of suffering.

PATHS: You had your soup like a dog.... listless, he said, for spooning. Unresponsive and blank. But I judged it should make you aware. Though you remained peculiarly removed from the environment, remote to commenting. All the more urgency for the doctor (to arrive); and now you know, you realize, it is your responsibility to recover— as soon as possible. You must perform.... a peregrination— away from here and through your customary modes of— travel and existence, or whatever your kind to do prolong it.... It.... The being!

LORKEN: ....I know I must make my way back into life's purgatory, for there's been no joy to me for a long time. And I've known that without as much hardship as lately. But the world.... separates itself for many minds. And you can't see what I cry about in recognition. The browning leaf, for example. It reminds me that I've fallen. It sensitizes my plight and all who occasionally accompany me. The roaming is just a hesitancy, to engage in what must occur to turn the useless into substance and acknowledgment. It is a desperate frenzy of a search, to carry this life back to home, without direction nor meaning anymore. Agony of the heavy load, without heartbreak, depression seeps its juices until emptied of the cask and exhausted of its powers; so you are driven like a ghost more circumspect of what was.... than what is actually around you.

MATHEWS: Dante's visions were not true, my son. Yet, here is everything. These matters are transcendent of focus, some times. But.... everything 's around you, every mode of being tempting with its offers and promises, and every entity in stance and stanched to disprove your opinions of it. For such the leaf would turn to crusting despite your enraged opprobrium to observe it, to have to. And are you hungry still—?

LORKEN: Not for a real ma—

MATHEWS: —If you are I'll hold onto you.

LORKEN: —ma-matter of a meal.

MATHEWS: But you may not know how hungry you are— until you eat, and taste of varied foods more solid than a soup. How long have you gone without the bounty of a real plateful?

LORKEN: It's been some days and weeks, father. I can hardly remember. I don't own plates.

MATHEWS (*starting to leave*): Then you will have cornbread as a supplement to something substantial, like broiled poultry or glossed ham. I'll cook it myself—

PATHS: Father!

MATHEWS: —from what's ever stored in the pantry and freezer, and bring it to you on a buffet table, along with a fairly well fortified apple cider—;

LORKEN: I'm not really hungry, at the moment.

MATHEWS: —but first I must tend to my evening vespers, personal to me, for what has been delivered. If you may, Paths, help him a bit for his rights of habitué—

PATHS: That's all I have been doing!

MATHEWS: —And thank the physician heartily, for us. Feed him well, if he wishes. The wine may be liberally served.

PHYSICIAN (*as **Mathews** exits*): Good evening, Father Mathews.... I needn't stay for a feast, Paths—

PATHS: There certainly won't be one.

PHYSICIAN: I've had a supper already.... A glass of sherry, or port of the church, would be some recompense for my trouble, though.

PATHS: Come with me, sir. You're more than welcome; and I know where it's stored, for special guests— Not in the Rectory. Mathews is strict not to have spirits with him. (*as the **physician** starts to exit*) He doesn't share the taste of alcohol like most, but keeps a good supply for.... company. I should force on him a drop or two— He's not totally a teetotaler.

PHYSICIAN: There is no such species of human on earth, as there are no true vegetarians since we must ever eat of ourselves to some degree. The patient should stay in bed, to rest, and for restoration of energies. And should an appetite grow, fulfill it reasonably, with lots of starches and protein— beans and meat. Canned varieties are O.K.... That's all. (*exits to outside the room, to wait*)

PATHS: You.... Lorken. You're some kinda treasure to him, that he's found. But I won't have the father cooking for you. He just speculates on his activities wildly at times. You'll cook for yourself, in the kitchen, if you must have a decent meal. And you'll be parsimonious about it, I warrant, because you're taking the foods directly out of Mathews' mouth. Do you understand? He is giving you part of his share, the only share around, in a magnanimous gesture of religious benevolence, beneficence and caring— And the fish stays all for him! It's his favorite to diet on. He fasts, somewhat, over the weekends, stretching one dinner for two afternoons. He considers it something of a piety, with honey-sweetened rose water, or lemon. But it's really become a fiscal necessity. Understand? And now he takes on this burden of yours. He doesn't realize his own weaknesses. And you don't know your own strengths—

PHYSICIAN (*voice, from outside*): Paths!

PATHS: Well, Pahtes may make you some breakfast in the morning, if he comes in earlier than usual, before his work. But you should beg the father not to make you any, as he makes his own; though you may share his if he forces it on you. Instead, wait for one of the volunteers, that Pahtes will inform about you, to prepare it. And that should really end it, my friend. With a good night's rest, and a fair breakfast of biscuits, bacon and oatmeal, you should be healthy enough to get out of bed and make your own lunch or dinner, if you even stay that long. If you have any honor left in you as an individualist, you'll know not to rely too heavily on this public weal. It is mostly illusory of the material caresses.... (*exiting*) Your clothes, or something like them, will be given you tomorrow. I'm not sure what Pahtes(has)' done with them, or tried to, and where they're stashed. If you come to difficulties (tomorrow) and are asleep or too weak to dress, they'll be stored in that commode over there— Don't walk around the church in that night shirt, please.... A lot of effort is being made for you. Can you turn off the table lamp?

LORKEN: I can manage that, sir.

PATHS: Well, then.... don't work the wall switch. (*Exits.*)

LORKEN (*after a short pause*): Don't treat me like a child, sir.... And give me a razor blade, so I can shave in the morning. And if you don't want me around, then let me cut off. Oh.... And now I feel.... a little for the lavatory to find. Well, how do dogs handle this? inside a house. I'll go where I please, and die where I like, where I'm allowed.... an ending's pother. And if I weren't so weary and faint, I'd make a big scream and let out a god-awful yell right about now, to complain of feeling so unusual, and trapped here for requited observations. But I'm sure I died, and this is.... a beaching. No one's quite certain, yet, of the comeuppance. So.... sleep may relieve this dreaming. And I feel kind of tired, too, like having read through some tedious Epistles, as a punishment for finding peace. But come the claws of night, I'm sure, for my fell(ed) head.... It was so bright, and red— all of a sudden gleaming.... to my closed eyes, that instantaneous approach.... of the running through and out of roads.... and paths to take. I'll take no more of them— They're all so discourteous to me. I'll just drift to be rolled. I haven't a mind anymore. Is that what you wish to hear? I give in to you. Treat me.... better.... Treat me better than a child, or a worshiper. Tame me for the hunt! Tan me for the— homeward bound, to fight the devils, or with them slay the solicitous.... hovering over me to define the anxieties I've struggled through in life. And see that my treatment is deserved, you most holy instructor. Of all the things for want, I never thought to need another lesson— now!... not at this grievous time. And to be loved? with cynicism and sarcasm both?!... After the fall, and before the fall, are these jests made—

on me, to bring back sorrows of uncertainty, and sublimity of hope. Turn off— Put out this light, and end the burning, lose this passion, quench this heat of tribulations and travails (*running through*), and see what you must make of this bare testament delivered here for you.... (*turns off the lamp light*) distorting under your pallor and squirming in your bed (*assumes a sleeping posture*) I am so tired, God.... that you make fun of man this way, so that he is sincerely chided.... and severely gazed. Hack now, with nightly axes, to have me dead for real, or be impressed with your praising, a creature's nocturnal spiel.

Scene III — *The twilight before dawn, with faint incidental lighting.* **Vogel** *enters the church's back room from outside, with a bag of some few groceries of least perishable foods: crackers, cookies, raisins, granola, and for some reason an unopened jar of maraschino cherries with stems.*

VOGEL (*while first trying the door*): Hmm.... Still locked. (*unlocks the door and looks around, entering while keeping the door wide open; he speaks fairly softly, but not as to be courteous to anyone sleeping.*) This is the most serene place to be, as the sun erupts to shine and remove shields. And I almost missed it, going home to get these.... relatively unused delights, and bring here. Oh.... but that I haven't much use for them, lately. I bought them, sure enough, and for myself. But tastes seem to be turning sour for sweets. In the hot enclave I want no fun. It is a foreign place become to greet me. (*walks over to the table*) A most familiar rambunctiousness of quiet emptiness and affront, my sensitivities honed down to shadows and the omnipresent, the ephemeral losses due me, repeated every day and barely noticed, but not a loneliness it causes. For such a place as this to be (in), what I could do here! (*takes the articles out of the bag and distributes on the table; the lighting slowly increases.*) Foodstuffs.... for visitors. I should have let the calf be fed, but my employment was against it. The notion of my work, I mean. It'll be a hot day for moistures— I can feel it already. Should have brought a jug of wine, a vat of ale— a pool of stream.... or a bottle of soda, but that's too heavy. They must be able to find water somewhere. Can't live without it.... Should I lay this stuff out so directly to be seen? or shelve it in a corner to be found. Let some volunteers enjoy it as a gift. The deacons place us to eat in here, but don't like foods staying in here— It's all opened.... 'cept for the cherries, for some ants to find, if they can pass through box flaps and wrappers, like servants of a hunger, ghosts' ghoulish tongues with feet and claws and mandibles of prying. I don't care what is brought to the church. Everything is utterly fantastic, in such a solitude of supposed solemnity, assuming all are being noticed and judged. Yet at home there is a privation of feeling, and a sensuality of abandonment. Leave our messes here, I say! where they are made.... I did expect him. Maybe, in his bewilderment, he couldn't find the place, half starved and out of his mind.... or element.... of conscious thought and reasoning. Or maybe he came and left, finding nothing to eat, and courteously locked the door again— Not too likely. Well.... one must have his trial, to find his fortune. One must test his initiatives to search for things, like the ants do, in God's house.... one of them. But I should have done more.... than just give him a key. That's quite useless, without a monkey's inquisitiveness, and an ape's demanding.... It's gonna be a hot day, now. I can tell. The morning cool hugs you to be smug and mockingly deceptive. It's not so cool at all, just transient of light from what lingers out the night, some breezes stirred on waking and excited to be fit again.... for daily creatures walking and running through them, sighing through the heats of their adventures.... Maybe she'll come to meet me. It would be remarkable,

but the unusual happens. And anything can be sanctioned or approbated a forgiveness in this space, even my thoughts. For here is an importance for everyone, and therefore to no one a propriety forced upon remiss nature, however she must exist to be stirred, twirled, run through. Hunger is a greed.... to living. (*sitting at the table*) How can one separate from family.... and those responsibilities willed?! She has a husband and a daughter, and tells me with the casualness of a stranger that she does not like them. But still, it seems impossible to manage a rendezvous in the morning, this way, one not even planned. And I could not expect it, to endanger her with him!... my guilt. So.... what is it I want, or want to think? For whom is adultery made, and out of which marriage? Just to be disposed.... Just to be disposed, the unhappy are made to glide through sins, like ashes floating above the fires of our discontentedness. I could arrange.... for my house to be (our meeting place). But that is foul— especially of time.... and I am bored there. (*The sun is starting to shine through the windows and doorway, illuminating the table with a particular mischievousness.*) Let the rays hit (*wipes his face*) to bring some reverence to my dimensioning and growth, as I grope along and place my test on thee— to end this drought, yet keep your beauty and my meaning aflame. For this is the most glorious time to seek and be enthralled by. Once again brought home, my disposition sound and just. And rats can bargain with it! I am an envious man, but not an acolyte to the wisdoms that have deflated my candor of mourning and demented my shape— my sense, for those who have left me. I do not serve bitterness with a cup filled of regret. I am here to greet you, for one day left, my armor drawing to perfection around me, to face once more the circumspection of another day's worth of being, of existing. Then— prove it for me, (that) this is not a callousness, this is not a resignation, this is not a deafness to memories calling to antagonize and tease or rend apart my surly obstinacy which combats your heavenly destruction. And this is not a hurt homed within me to disprove *your* face! called ugly and abhorrent, dishonest of affection to display, and despicable for trying to— or just plain bored with!... plainly bored with my existence. Not if you keep coming back, a wonderful reminder of what is ever greater than our delicate impingements on.... the sensitivities to be loved. Not you, but the re-dawning is what I wish for and admire. It supports.... my steadfastness— But I need more, Jove-quaking might! Fortify my faith, to remain kind, and seek for the debaucheries earned. (*a slight pause as he draws in the sunshine on his face, the eyes variously closed to cognizance or reception; then, someone appears at the doorway, and* **Vogel** *barely perceives this before becoming more alert.*)

MAN (*at the doorway*): Hey! It's early.

VOGEL (*assuming a reconstitution of consciousness and personality*): ....Yes, friend. It's a beginning again.... and a review of what has transpired over the gray and aged night. What brings you to our doorstep?

MAN: There ain't no step at all— and it's early. But I'm a beggar, more or less.... sir.

VOGEL: Well, come in.

MAN: I be welcomed? (*entering*) One just comes up to it, you see. And here ye be. But I've been watching. I've been noticin'. And darn if it weren't true, these be open— and receptive?... to passersby? I am astonished, for I never find things like this, open hostels. They're only rumored about, when you come to your last bit of effort to apply. And somethin' from heaven shows you the way. Trumpets are blown off in your head— that tell you to look in this direction, you fool! out of the darkness into darkness, into.... somethin' developing into a real possibility (*coming up to the table*).... of a generosity.... But that only rises up, with the day.

VOGEL: This is a church. Have a cracker.

MAN: Thank you, sir. I've been lyin' low, low on the ground.... (*picking up a box of crackers, already opened*) and my hands are a bit grisly, grimy from the dirt. I mean, they ain't clean (*getting out a cracker*), but they ain't so foul either— Oh! You have cookies too. (*eats the cracker*)

VOGEL: ....Take a seat, my friend, if it's a morning banquet you've come to.

MAN (*placing down the crackers and pulling up a chair to sit at the table*): I ain't suffered a breakfast in months. It's what you can find whenever it's available (*sitting*).... This feels funny, to the stomach.... It's a good time to ask, during breakfast hours. Some scraps, maybe, are thrown at you, or left on the back porch— But not lately. No, no. There's a meanness rising with the heat, and you're better off waitin' till the afternoon for some handouts. (*going for the cookies*) I don't know why, 'cept maybe a noon-ish dinner is more reliable to count on, these days. The air's so hot, people can't wait to quench thirsts, with eating of a meal to justify the expenditure of their foods. (*while eating a cookie*) Some of it is left.... for the hogs, later— This is good, sir. (*getting another cookie*) I won't be greedy, though. Had a fairly good bit of gruel for yesterday's supper. (*eating the cookie*) If you fill up too early, it's harder to stretch out the periods of real famine, the long hours you're more used to. Can run a couple of days, you know, till you get lucky, finding some pet's portion lying about. But something's better than nothing. A breakfast makes a dinner necessary. Prepares the body's metabolism to expect a later helping.... That is, if you're going by the sky's clock. But we're all so spoiled as people. We insist our constitutions *should* be satisfied. That's all well and good if you can hunt, regularly— like the animals. But what if your circumstances prohibit this, if you must rely on begging because you're too weak of stamina or (are) careless of thought due to a growing inability to concentrate?... Well then, you must prepare your stomach, and your mouth, for the hours of opportunity that come. And they be the afternoons, these days, when the good citizens are feeling.... satiated enough to be thankful and sharing. Can I keep this?—

VOGEL: The whole box?! Tell me, friend. How many of your kind do you think are around here—?

MAN: My name's Harold. Oh, there 're lots of us— And we share amongst ourselves, when we find each other and can. Yes, we're scattered about quite widely. It's been so hot, though, we don't move in groups like we used to, not even pairs. When we're subject to the extremes of our extremities, each person has his own fallibilities, at different times from the others, showing up, and so wants pretty much to be on his own for a spell, until he can recapture his shell of reservedness that makes him suitable for company, and even sociable. (*putting down the cookies*) I guess I can't have this. Well I'm not a pig, sir. They may be too good for me, though. Sweetness is important— for the energies. I don't see anyone else around, coming for them. Maybe later? Well, that's just. I'm not actually that hungry, right now— What's this?

VOGEL: ....Granola.

MAN: Oh. You eat it like popcorn?

VOGEL: Nuts, oats, dried berries, and honey-coated cereal bits. Good for taking a hike with.

MAN: That is a very thoughtful preparation.... of nutrition. The raisins alone should do, though. I once lasted a whole week on a box like that, somethin' a couple of kids could eat up in a few hours of indolent play, for the classic tummy ache to bother their parents with. (*picking up the jar of cherries and reading the label*) Maraschino cherries. Any alcohol in this? (*turning the jar to study the ingredients listing*)

VOGEL: I don't think so.

MAN: It's for parties, drinks.... Odd that you brought it.

VOGEL: ....I?

MAN (*placing down the cherries, though rather near himself*): This is all your.... contribution to the— destitute and dispossessed, ain't it? Why, the most valuable thing here is that bag. I could sorely use it to carry things with. You know, what I find on the ground.

VOGEL: Then it's yours.

MAN (*grabbing the paper bag to fold up and stuff in a pocket*): Thank you.... No, this.... food certainly didn't come from the church. It's too sweet. Too beautiful. Too childish. Too angelic. And too sublime of all too human tastes. But the church needs the dour severities of bland wafers, garlics, bitter vinegars with their olives, and a harshness, sir, to be commanded by sweet scents and perfumes and smoky incense wafting mysteriously up and out of necklace tethered balls of silver gently hung and rocked for eyes to follow and be blessed with. (*finishing his packing*) But I'll take this over that any day. It's more practical than a religious mesmerizing. I am attracted to the good of all things and materials. I like the idea of these cherries, though, as if we're celebrating life in a wild entertainment or amusement on our difficulties to think (about).

VOGEL: Well.... I am of the church. So all of it is from the church, presented you—

MAN: Harold!

VOGEL: —not only from my wishes, but with those of the entire congregation, poor decrepit soul.

MAN (*indignantly takes out a key from a pocket and slams it, more abruptly than forcefully, on the table*): I've been here before. Searching through the place, last night. There weren't nothin' here!... not to eat. Not to weep over. But that didn't seem so strange to me, 'cause I was only investigating an oddity of fortune.

VOGEL: ....Where did you get that.... key—?

MAN: A totally confused.... traveler.... gave it to me, came up to me by accident of not knowing where he was anymore or what he was doing about it, crying about some Canter Lorken. He didn't make his way too well, sir. Not well at all, and has not made it! sir.... to the church. But I have a friend— a companion named Lorken. And *I've* come, to investigate this, this story that he's been taken in by the church, and rests here, possibly.... But *I* couldn't find him. And that door is locked, to make a proper search inside.... not possible. Yet, I am polite about it. I made no angry, loud fuss over my disappointment. I simply considered this room, in silence and careful demeanor, and left it as it was. It's called a community room, or the church's comfort, I've learned, from back aways. I know of *here* certainly, sir.

VOGEL: You should wait for religious services to start, so you can enter the church properly and visit your friend. I think the morning hour today is around 9:30—

MAN: I don't want to see him— I want to *find* him!... I don't want to disturb whatever convalescence this church has involved him to.... suffer through. The day's earlier rumor was.... they found a man dead—

VOGEL: They?

MAN: —And I know he went off, with a foolish greed for that.... because he's given up his poverty, and vows of poverty— to leave as done, to leave as found—

VOGEL: Why, don't you know?! I helped carry that man in here myself. And he is not dead, this.... Lorken, at least with the last strains of information I have about him. Now, whether he died in church is another question. But these days are hot, and not too many in his shape *can* survive, I could imagine.... We thought he was dead, till we looked at him carefully. I guess one has to fight off presumptions and expectations, but Kanter was always hopeful.

MAN: And the examination bore fruit. I just wanted to confirm its reality, and this is apparently so. We remain interested in each other, I'm sure.

VOGEL: Like a brotherhood of thieves?

MAN: We ain't that, if it's no crime to live. What does it mean to steal? if you have what I need and what you don't need.... as much (of). Should I starve because of a restriction to your possessions? Is that a community standard to be upheld around here, as we linger with diseases to spread? Our initiatives improve you— by improving ourselves, just barely. We decline death, you know, as a point of stubbornness, or obstinacy to that (what's) little granted our abilities and expectations. We certainly don't take for greed, or accumulate a wealth, but rather with a grunting for this weather to survive (in). Grandeur is being able to open your eyes once more, to greet the day again. I was blinded for a week, the worst (of it) at least a couple of days, by something in the stream water, a chalky mineral. Who could tell it would harm my sight? Well, the stream paid for it, all right, by drying up to a uselessness. And I'm still alive, but I don't wash my face in salty streams. I've learned to use moist earth, as an exfoliative. (The) Dew off the leaves cleans me. And what are left for leaves?— They bake! They stay as green as night, as shaded as surrender, surrendering their.... fans of transpiration. Our existence is a severe and punishing one.

VOGEL: I can imagine a band of wanderers, helping each other. And you could have brought the key holder here, or directed him —

MAN: He's not due for anymore walking. So my friend's been found by some religion.

VOGEL: It's not easy threatening us with your miseries. What is owned is not free to take, including a pride of ownership.

MAN: I?... I can not threaten *you*, sir. I'm a weakling, a parasite at best, that can be disposed of with one blow, one swat of your hand. A slight carelessness of your gait or glance could cripple me. Children could probably wrestle me well, in my condition; though I can submit, my bones have become more rubbery than normal, and may endure a tackle admirably (well) and with surprising facility. But that is some small reward for my.... freedom to fail. It's not my aim at all to cause even the least dreadful fright to any of you kind people, nor to present myself to feel sorry for— You haven't the right to do that, if you don't know me. No!... No, I just— bring up the radiance of the day, that first glorious light felt. You must find it remarkable yourself. I can hardly ever miss it. It's the only reason I persist with this.... dreary sub-human existence. And do you know that earthworms can make for a good chow! Not so easily found, these days. And if I should go blind again, I'll cry for the sunlight, and make due with the warming. But it's been hot, sir— Hot!... mocking our destructions— and distractions trying to remain civilized, as we fend like fiends disputed as head-strong creatures. But I, sir, refine the art of constant struggling out of uselessness. I beg— and not for sympathy, but for food, and occasionally shelter, when it's cool, cold, moist, damp, rainy— rah!... (*pawing the cherries*) You have religion here? Then how can I bother you? Isn't the veil thick, so thick that you can't notice me, my simple scrounging? And then you must forgive.... what the ants and insects do. They make way for you, without such recognition of your might.... as much as obstacles to (be) overcome— Can I have this, sir?!

VOGEL: No.

MAN: It's useless!... to you.

VOGEL: I don't know why I brought it. Wanted to get rid of it, a taste for cherries. Tired of that. But it's for others. Not you. You're all.... wasted up— And I gave you a bag.... a practical gift. But not that jar. I may want some myself, later.

MAN (*slowly unhanding the cherries*): I ain't hungry, you know. It's only the thought. I can't snatch things up, in front of people, not even companions. I have to be given— offered. But when the need arises, when I find an opportunity to surfeit my way out of a personal disaster, am I so grisly faced as to offend, with cause to give more of than what should be allowed? It is not greed. It is a nervousness. We are such weakened creatures, both—

VOGEL: Both?! I fear to see myself, in your contemplations.

MAN: Then tell me this, sir. Is it wrong to cause harm?

VOGEL: ....It depends on who to whom— and why. God certainly causes harm, and that is never wrong, seeing as how we've only a lease for living.

MAN: Do you want to take this key from me?... I'll come back for sure, if I keep it.... These rags will visit you, for whatever droppings you leave.... around this time, when others are not here. But take the key away.... and I am done with this comfort, I am through and finished trespassing realms and being humbled in them, more than the sky could make.

VOGEL: ....Keep the trinket. You'll probably die, passing it off on another— Trespassing, you say? No. This is a community room. (*as the key is picked up from the table*) Any.... person can be allowed in here.... eventually.

MAN (*slowly pocketing the key, with an evident sadness*): You must have felt quite sorely for him. There aren't many excuses to be made, sir. Downpours happen this way. (*stands*) You.... are familiar to me, guarding that hotel. Bless your vigor, sir. We need our spankings and kicks, and punches to stand agape of the good.... or the so much better than ourselves, the earned of their privileges — which we respect, and are fascinated by. I could never be as mean-spirited.... as my presentation might imply. And I'm reluctant even to admit that. Is this.... Hastern's charity, or from his larders?

VOGEL: ....No. It's mine. I specifically brought it all from my house, not the hotel, or a storage place I have in it.... I'm not throwing it out, but it's as good for anyone as for myself.

MAN: I'll take another cookie, then, if you don't mind. (*reaching for the cookie box*), because they are particularly delicious, sir— (*Running water is heard, as from a bathroom.*)

VOGEL (*as **Harold** gets his cookie*): He's up just a trite earlier than usual. I seldom hear him washing up.

HAROLD (*pocketing the cookie, conveniently*): This makes me real, and really wanting of a wait. He?

VOGEL: The priest. Father Mathews.

HAROLD: Oh. Alone.... but with a guest, I make him out to be.... a very kind gentleman.

VOGEL: A deacon or two will probably arrive within.... oh, an hour and a half or so.... or some other volunteer. I never like to be (around) here for that, them. I like to be in bed by then, so I can have a nice afternoon, later.

HAROLD: Yes. The room becomes very busy, doesn't it.

VOGEL: Not so much this (one), I would imagine, until its use is necessary as a respite's lounge.

HAROLD: Sir, I pass by here occasionally, and notice; though the door is closed, there is a motion through the windows seen. And that is almost frightening of a church. For what do they do in there? And whatever that is must be most private and sheltered from the public view. I would not like to observe the work of undertakers, through a window clandestinely viewed.... or viewing. This.... hurts the proprieties of the profession, and sullies our conditional restraints on froward behaviors.... And I am no hymn-est. But here I am, eating cookies, enjoying your company, and hearing priests bathe, as if they were to clean off the night. But sir.... that baggage is carried on my shadows. And the fine gentleman must spruce himself up, for the morning prayers. How do they wash, sir? How do they sense themselves? I do not wash— I wait.... until that time may come opportunely, and with unexpected shading or veil. And I know nothing more than of myself to flame of incense in this

heat, a strange offering to the gods of weather. But eventually clothes get so dirty, they start to clean themselves— So on my soul I pledge!... I will return here, to review the space.

VOGEL: This is more of a spare room, a storage shed, a workshop for relaxations after chores about the building. And not much of anything is worthy to be stolen. That's why it's left to the community use.... So don't feel intimidated about it. You are not a guest, you are a.... participant, a patron of our civil arts praised, more or less, with the visiting. (*The sound of running water stops.*)

HAROLD: What does he do now?— Will there be more food treats left.... for us?

VOGEL: What's here stays, as it lasts. I'll not return them to my house. But as for more, what gentleness does start this action flight, on wings prepared to soar, to permit others of an imitation is unknown. Most may see them for themselves. I may repeat of generosities, with a sultriness that burns the head, to remove of what diminishes me, in tastes that are tired but not to be wasted.... You may find more.

HAROLD: How does he dress? It is oppressive becoming, to know you, sir. This room might play as caricature to leftovers and discarded valuables, or articles and objects waiting for more utility. And I am such a useless being, (*starting to exit*) just your sight of me extends my spine a notch, to know that you are watching out for bums.... I will leave a payment, or a "thank you" for your largesses.... something crude and simple, as much as I may fashion to come upon it.... (*at the doorway*) Say, a flap of shirt that's managed to tear itself off, or a torn shoelace too short to tie with or too weathered and undone to withstand an effort, or some scrap of parchment found— with writing that I once thought dear, the lettering at least, that I could hold in my hands and be proud of the possession. Something of importance, by way of being handsome, carefully crafted, and inked— or scratched— of a technology denied my thoughts, interests, knowledge or approval of, might I pass by to leave here, once its charm had flourished for me and now flutters—

VOGEL: A message for your friend?

HAROLD: ....Sir, this is sunlight, now. I've no luxury of a pen or pencil, and can not write easily. I was referring to the methods used to make a paper seem pretty. And I'll brush off the trash, gingerly, no matter where it's been, or how foully treated, before I leave it here.... for the community. Such things can really move us, as you might take them for granted, being so common and lowly dirt, litter. But it was made.... within magnificence, and a sharpness to character, a strong definition to the reading or viewing.... It's not charity if it's found. (*exits, his voice heard*) It is a recovery while expedition-ing. And my friend Lorken stays there, awhile. That's something gained, revealed, confirmed.

VOGEL (*after a slight pause*): Here's something cooking, now. Could it be as traditional as bacon, eggs,... toast and coffee? But sure it is, that the day heats up quickly. It really is becoming oppressive in here. And that door should be closed.... to keep the heat from overwhelming the air, stifling breaths. So traditional. So.... ordinary.... And these foods, they shouldn't be put out out this way — They're unexpected. They should be searched for. (*picking up the box of cookies*) I'll shelve them (*standing*), stash each one in a corner, though plainly to be seen— Oh!... It's going to be a heavy day. (*starts distributing the food articles separately, on shelves throughout the room*).... I'm more tired than I've felt before.... (*looking out the doorway*) What was that guy's name? (*pauses to think*) I don't remember. (*resumes his management*).... They all seem so much alike, until you talk with them.... These feel like daggers in the waiting. Now, why should that be? This is.... something akin to a good deed.... No, it's hard to remember what you don't respect. Hard to learn what you don't believe.... The ants and grubs will have a field day, if this stuff isn't found readily— and used, ate.... especially in here.... Why am I caring so much, for this room?... (*holding the jar of cherries, the last item to be placed*) Somewhat symbolic, of my life's.... ontogeny, how I am separated into so many corners.... of the emmet to be found (*holding up the jar to look at*), rejoicing for what?... Now if this should fall, it will make a great splash, and draw in the flies most definitely.... Yet on a shelf you'll be— up there (*placing the jar*).... distant from me. There's no fun to my house anymore— Be here! for the community. You've not even been opened. And I bet I can surprise her with you— Sunlight?!... Yes. It is a day. Once a week— once a month? How can one tell, with such shyness through the heat, the venting? (*leaning on the shelf by the hand, with threat of toppling*) I am so tired of subterfuges. An honest mating 's never found worthwhile but for the child. (*The shelf wobbles slightly, which excites him to come off it while speaking.*) Children!... (*turning towards the doorway*) They soak up the fun, I guess, all out of you and away. And that is the retribution (*starting to exit*), placated through a family, or some such communal gathering, kept together with blood of relationship.... But why am I totally disembarked?... (*at the doorway*) The last time.... was a party fled from, drunken to wake up alone, like losing all your eggs stupidly.... Well, that's how it felt, if not so formally was it actual. I could not believe it, the said goodbye as worth of grimacing; and then the next day comes.... This repeats. The headaches grow, for their intensities to miter you.... with their number. And so be shocked to the solicited solitude, and the shrill sham, that empty house of cries. It.... takes its toll— It took! It was hard.... on me. And I won't be found here, by no deacons, till some work's been done.... till I can relax again and suc—.... supplicate.... suckle onto the good graces of the church, for my divinity to be avowed through sacrifice with what we all must contain of that empyrean spirit, the imperial maiming of our desires. (*Exits; voice heard as he locks the door*) Here be it better a wish for our mistreating than what is kept inside to serve.... august inspirations to connote with mysterious, mystic pleasures.

Act III

Scene I — *A garden with bench, behind the church.* **Margathrate** *sits quietly, eating a small brunch of meat-laced bread, which she carries in her usual bag. It is evidently hot outside, with the flowers seen as wilted. The door to the back room of the church is closed.* **Lorken** *slowly walks from the side of the church into the garden, with a slight difficulty of ambulation. He is in clean, donated clothes. Stopping at a bed of flowers to look down, he appears unusual, as* **Margathrate** *notices.*

MARGATHRATE (*after a pause of observation*): Are you a volunteer to the church also?

LORKEN (*not looking up, at first*): These dears are dying. They need some watering. I'd do it, if I had the energy, and the time.... I'm not a volunteer that's actual. The church is looking after me awhile. I'm convalescing— after a fall, out there, on the barren ground. Some parishioners found me, and brought me here. I mean to get well enough to leave, by tomorrow. Have to start now, recovering my strengths. No sense putting it off. I'm a burden to goodness, and only the clean air will do to revitalize the lungs.... even if it is insufferably hot today.

MARGATHRATE: It does seem to burn mists and steam showers of a dying respiration for the plants. When I first sat down, I could swear they were shrouded in dusty tears, till I realized some of that's from my own eyelashes flaking off particles and blurring the vision. The heat whips our perceptions.

LORKEN: Eat you a lunch out here? I was all afeard in bed, waking up late in the morning, and not knowing what for of my cognizance, till the blessed priest brought me a breakfast himself— and on a plater serving. He shared some sausage with me, but let me have the rest. Said he had to leave, for morning prayers. I heard.... little. Must not have been a crowded gathering.

MARGATHRATE: Only the devout regulars, stopping for a pause before (going to) work.

LORKEN: A deacon brought me clothes to wear, and I sitting up in bed like an infant. My own were.... damaged in the fall, torn and made ragged— I'm as much to be a bum, madame. (*approaching the bench*)

MARGATHRATE: You say they found you. What were you doing?

LORKEN: ....Oh, trampling along. I have nothing of significance, and don't wish to deny it.... I.... just came to the end of my persistence, and fell down.

MARGATHRATE: Sit.

LORKEN: Thank you. (*sits on the bench, respectfully distant from her*) Why are you eating out here, in this heat?

MARGATHRATE: It was an excuse to get away at all, to the church. I was expecting to find someone, like a child a playmate. How very foolish, and inconsiderate.... of adulthood. This is my punishment, self-imposed, for such a wanton behavior. It will keep me from making excuses too often.

LORKEN: Excuses?

MARGATHRATE: I am.... responsible for others, and should be looking after them.

LORKEN: Younger siblings? Is that how people live?

MARGATHRATE: But I prefer to do chores around the church. They're new to me, and that is refreshing, at least.... My charges are not in danger. Excuses made.... to others.... for the afternoon.

LORKEN: I've not looked after anyone my whole life. And now I am as much destroyed as could be made abandon-able. But I'm a caring man, nonetheless. And you seem quite an attraction— if you're free.... for company to speak with. Who were you waiting for?

MARGATHRATE: ....A fellow volunteer I met yesterday.

LORKEN: Just yesterday?

MARGATHRATE: Yes.

LORKEN: You work that regularly here, or want to?

MARGATHRATE: He seemed fun enough, for an escape from such tedium as I can no longer handle. And I wish not to be evil about it. But it is crushing me, and I need relief, as much as I can find. Where better than here? doing.... good things, trivial but decent, and mending my sore muscles from a travesty of affections and caring. They pummel me incessantly, and I must escape, at least occasionally, to find an outlet of survival. I can't be grateful for anything, otherwise.

LORKEN: Older daughters are often given too much to do. Surrogate mothering—

MARGATHRATE: Sardonically arranged!

LORKEN: —I'm told.

MARGATHRATE: I'm— fed up with it, sir! Fed up with an all encompassing mistake. And I know that this is bad of me, unless I can do some good by it, in some way, this escaping. The conditions that surround me tear at me, stab with a jagged meanness to them to stay necessary and almost never ending. I'm only the animal, after all, running from persistent pressures and obligations I've grown to despise. But is that hate?

LORKEN: I've run without hate.

MARGATHRATE: Precisely! If it gets too hot.... And is it wrong to seek help?

LORKEN: ...No. But it depends on how individualistic you are for how long you hold out to find any. You mean a companionship more acute to your purposes.

MARGATHRATE: Exactly! And is this not God's work?! His keys to find!... Especially here, some solution he constructs.... for us. I know of no other pleading than to beg this way for some relief, to work for his gifts and comforting. I may be misunderstood, but not— this.... of which I'm trying to involve myself. It is adult. It is an adult way.

LORKEN: Why sure, to come to meet others here. The place is made for this.

MARGATHRATE: And it's so early, yet. I thought I had immediately found help, almost so. Beginnings may be deceptive, though. I'm so earnestly in need, I can't tell clearly what's presented for my choosing. The judgment's not firm. And I'm thrown aback to reconsider everything, to avoid a girlish fantasizing on foolishness.

LORKEN: Oh, that's the best kind.... of foolishness. I wish I had some—

MARGATHRATE: On my very first day I was shown someone.... something— and reluctantly!

LORKEN: My dreams are almost all extinguished. I can hardly imagine anymore, but simply must let things happen, to fall on me and demolish— or demand my attention. And I am here!... living here, so unexpectedly. And for a short while musing on what is base—

MARGATHRATE: What is basic!

LORKEN: —about me to have brought me down to this condition, this absolute state of reckoning about myself.... and what I'm going to do further— if anything. God weakens you, to force you to consider these things. For when you're strong you're headstrong, haughty, heinous and the hound to fortunes that may only escape your grasps without his guidance.

MARGATHRATE: Are you religious?

LORKEN: ....No. I'm only saying.... that's obviously a path (to take) if you *are* religious. Because then, no matter what your state, even unto death you are successful.... in certain eyes. That, in a way, excuses anything that happens to you. It's the greatest arithmetic known. Do you know that something divided by nothing yields everything?! If you can handle such fundamental principles, then you can believe anything— that's true.... The such makes for a religion, I suppose. And I am certainly nothing.... nothing much to consider. But I am here. And it's like the universe is suddenly opening up to me, for one more chance— at something. I don't know if I can entirely grasp it yet, or if I'll fumble life again; and for the last time— I'll demand.... The last place I'd think to end up at would be a church, miss.

MARGATHRATE: Margo. And I have drunk for here— with a thirst.... for compassion. It was only recommended to me, not my original idea at all. But a sincere try was made, to push me here to find out.... what I might actually need. One has to take these tests, one has to expose yourself to their requirements so that you learn of the consumptions, so that you might enjoy them— or revile of your participation. But it's all a giving of yourself to it, to find out. I'm not wrong to feel the need, to enjoy the testing and find delight with the.... possibilities tickling. It is not base. It is not condemning — It is warming.... of a cold frost that's been creeping up and over and making me callous and sick to know.

LORKEN: That can't be possible.... Margo.

MARGATHRATE: Or afraid to know, with childish innocence and honesty.... I struggle against becoming— heinous and hated of personality, too hard and harsh of displeasure for affection.... They have driven me away.... to seek help.

LORKEN: Well.... that's permitted here, if they take me in— to heal.... even one such as to die. I don't *feel* useless anymore, even though I darn well know that I am. But it's the spirit of the wood, I think. It's very rich. This is a mighty edifice to sleep in, to have your nightmares influenced by.... to feel somehow resurrected— for some strange mystery.

MARGATHRATE: Oh, the mystery of getting better, of improvement. That's always a wondrous condition to be in. Some people purposefully get sick just to go through it, like many women repeatedly go through pregnancies. They are perennially ill, either with or without. Only the condition counts, the psyche to be volatile of expansion.

LORKEN: Well, it's better to ascend than descend, Margo, even as you're wasting away, even as you're looking down at your misfortunes— reviewing them— instead of looking up towards a future of possible benefits— anticipations through hope.

MARGATHRATE: Are you one of those born again? I hate children. Can't stand them, now. They're worst than brats— they're brittle, easily hurt, the sensitivities of a wet tissue. I.... don't want to be so mean about my thinking, but there it is. I'm fed up with it. At least the infants lack a wit at all, but with the children it's pathetically expressed, and as awkward as crooked steps and broken pilings.

LORKEN: You probably have so many little siblings and cousins to look after. Or even if (only) a few, one could overwhelm— I refused the job entirely.

MARGATHRATE: It's difficult to like them, through their ever present faults. They.... constantly remind you of your mistakes, your errors amplified and exaggerated to mock you— imitate you, as to how they might think about you— which is scandalous and scurrilous— and so insulting.... mister. And for this I take pains to deceive myself that it can't be this bad, this terrible a life. But how long can one breathe under the ever dirtying sheets, under the wraps of bondage to what seems to be a lifelong series of responsibilities. It hardens your sensitivities—....

LORKEN: Bob.

MARGATHRATE: to want to be a sledge hammer to break things, or a thick, heavy pair of shears to clip metal chains. That's how I feel.... about some relationships.

LORKEN: It's all a mental bifurcation, of course—

MARGATHRATE: Of course!

LORKEN: —imagining a fork in the road, to choose directions.... henceforth.

MARGATHRATE: And for this I'm made to feel totally guilty. So let the guilt rain on me, or pour over my head like a dense, smelly, sticky molasses during the summertime heat. But as it may drip, and I droop from its weight.... we're allowed *here*, aren't we? this honesty to bake with and put forth. Am I so horrible of the truth to rant? I wish to dissolve.... from the world, poorly made but indelicately stiffened and (with) tenaciously tough ties.... of conscience.... I can offer myself, to whatever is here to find a solution for me, whatever spirit or spiritual or inspiration that haunts this place. I can work of song and praise to pay for this assistance, because I'm being ripped into bits of meaninglessness— which is denial— without much help from you.

LORKEN: Me?— Are you unhappy?— Oh!... to talk with.

MARGATHRATE: Everything that's of the church must be in some way sanctified— even our honest thoughts. Good religion sifts and purifies, to make sand sand and rock rock. And I have been polishing the silver, till I can see my face again.... It is not as cruel and haggard as I had thought, or feared right down to my guts' contentions.

LORKEN: You are manly pretty, Margo— That is to say.... you would make a man look pretty, with all of the worries that 're bothering you, and the frowns and wrinkles of the forehead that only seem to enhance a beauty shielded and a shyness to be shunned. Oh! You work at much to become of this place.

MARGATHRATE: And you are a handsome man, in your apparent youth, to return a compliment, even with your slight debilitations of approach.... But why have you fallen— at all, as much as I?

LORKEN: ....There's a disgrace to it, Margo, to being worthless, as you might be worth everything to others. But I've simply not found the need to my existence. And that's worst than a game to play, the searching.... because I haven't been searching; I've just been bored and lazy, and entangled with.... noxious elements of living. I'm too honest with myself. I want nothing, and find it constantly. And then I fell down, dead tired about *that*— finally!... But I ran from a family, too, that pestered me too often for some sort of conformity, till, when I reached an age to flee.... I did. And without any thought at all, I lasted, lowly, meanly, grittily, but not embittered in any particular way, except that I was growing more and more tired, and come to drop near here, by some ruins.... of civilization. I dreamt, last night, that (it) must have been a good sign— for something.... social caring and all. But then an entity, utterly shaded, formless except for form utilized, kept striking (at) me with a whip— white like the yellow sunlight— over and over again. It didn't hurt at all, but the indignity was maddening. I gave it my back. I gave it my front. I gave it my face— But it wouldn't let up, until it just gradually.... decomposed, and I opened my eyes.... to this day. I wasn't scared, and had not been at all, to take this.... subtly symbolic blandishment.... to withstand my abuses. But I felt terribly weakened, this morning, like something had been taken from me.... While, of course, the opposite was true, I had been given— so much.... I think the priest's kindness touched me a bit, last night. It's been so long to receive such favor from a person, and so unabashedly. I might have been working out some shame or embarrassment— But I thought I would be dead, within irony of a church, even, never having done a truly bad thing in my life.... but not having caused much good either.... Now I see you, sweltering, suffering. And I don't know what to do (about it) except to be myself, in new clothes.

MARGATHRATE: He whipped you?!

LORKEN: Oh, or something—

MARGATHRATE: To behave, to behave for me.

LORKEN: To commiserate, perhaps. Let's get out of this heat—

MARGATHRATE: I seldom dream at all, lately. I just wish for the drudgery to end, to end all drudgery once and for all— That is his purpose. (*Lorken slowly stands.*) That is what (ha)'s brought me here.

LORKEN (*rubbing himself*): Here's a hot spot become. How can you take it so? But women are (such) wonderful facilitators and regulators.... of their own personal environments. It is a principled humanism deployed, I'm convinced of it.... Those flowers are very labored, though. They hurt, under this sunlight, and wait for clouds, and night. One never comes, for too long a spell.... and the other always, and often too late.

MARGATHRATE: Are you going to be here long?

LORKEN: ....I said, I work to leave by tomorrow. I.... struggle at it.

MARGATHRATE: That's a pity— for me.

LORKEN: I have nothing! You'll find.... more playmates to speak with around here, I'm sure. This is a place of worship, and that means much talking.

MARGATHRATE: You've never felt worthy of anything. That's your problem— even when you're pled to.

LORKEN: ....And you are worth eternity, for sure. But you are cruelly right— I want much.... and haven't taken anything, haven't justified the deserving—

MARGATHRATE: How must you justify a hunger?! The infant doesn't. The children don't. And I need not— nor neither you!... It is.... permitted here, a lusting.... to make our sacrifices known, to He who knows already— and causes it within his shadow, under his radiant approval. We are made this way, to condone *his* thoughts for us, since this is how we exercise ourselves. And you mustn't be reluctant—

LORKEN: I can't figure out how to handle what comes forth, Margo. I've nothing, again! and must lead to nowhere safely.

MARGATHRATE: What does Father Mathews want of you?!

LORKEN: ....Health? Amicability?

MARGATHRATE: To stay? and work around the church, tend the gardens, make repairs and reprisals for our carelessness— not to feed you? Might you have prayers here, spoken with me?

LORKEN: ....He hasn't spoken to me, about such employment. I'm not sure if it is affordable, according to the deacon.

MARGATHRATE: But we can make it so. The congregation *can*! And you will be around— me. You are handsome— and in need! And I.... can only be suppliant, at the church, on these grounds too hallow to defame with our sincerities, so little are we that he must condition his religion with.

LORKEN: The Father? or our God. I must.... make way of the berth, Margo, for the more appropriate. It's used for visiting dignitaries. And I'm impatient, not to be too much of a burden— and not to remain helpless. I've been for so long more capable than now, and I want to regain that semblance as soon as possible, which means getting away from here— and possibly this community entirely.

MARGATHRATE: But I'm tired—....

LORKEN: The father may want me to enthuse others. With an intelligence, no. But with a generosity of feeling, perhaps. I can not be a true representative of the church— but only of a man.

MARGATHRATE: That is!... That is the decency of this, this demand on you. (*wiping her hands with her rag-bag*) I'm just about finished, here.

LORKEN: Have you more to do?

MARGATHRATE: I will find annoyances to stay about the area, and delay the tedium I must always return to. But you might find me stripped bare naked, hanging in a closet mistaken for a shower with delirium. For this heat turns my head towards a maladroitness.

LORKEN: That is a supple warmth to contain. Are you becoming dizzy, and losing all faculty of propriety? I felt as much when I fell, to let whatever happens happen, handle me and conspire with.

MARGATHRATE: I'm totally in control of myself, as regards my wishes. But flowers wilt where they may not run from drought. (*The back door to the church opens, as **Kanter** comes out.*) And you know what your desires are.... to conspire with— me!

KANTER (*approaching*): I told you it's too hot to have a brunch outside, and I feel it. Even a snack makes tastes tedious, if it's not refreshing. Are you quite alright, Margo? You seem screwed to the bench!

MARGATHRATE: I've eaten much.... And it is delicate (*standing*), for more to do.

KANTER: I thought you had finished, and had let loose(, gone off).

MARGATHRATE: I.... wouldn't have this at home. That's why I've brought it here. And I delay, to find more work.... Jules.

KANTER: That's ever so. But I'm certainly done with my chores today(, up with them), and miracles perceiving, young man. I thought you could have certainly died. But you look quite different, more spiffy and less tacky, all clean(ed) and shaved.

LORKEN: I hope to do justice to the appearance. Can't seem to walk as well in these clothes, though. But the leg muscles will come back.

KANTER: Yeah. The shirt and slippers are mine. I contributed a pair of pants too, but that's not them. But I insist on helping you more, son, because you are a fair fellow. You.... protected my wallet with your whole body— Now that's impressive to me.

MARGATHRATE: Give you him money?

LORKEN: I.... weren't about to steal it, in my condition.

KANTER: We!... Whoever looks after us. But you must stay with my wife and me, as I recommend you to the church, so that you can fully recover. Paths almost expects this, says Pahtes.

LORKEN: If you still wish to be so obliging, Kanter, I will accept the invitation. I have reason to want to stay in the region awhile, though it may not be deserving of me. A tramp is a tramp defined by his mind to be.

KANTER: So might be anyone. But there are still many lessons to learn to extend that mind, Bob. Thoughts are generally more flexible than bodies.

MARGATHRATE: That is the pervasiveness of God. I almost faint, out here. (*running a hand through her hair*) The hair curdles and *chauds* to a crispness and a bleaching. The sun can ripen (*leaving for the church*), as well as sterilize.

KANTER: Oh! Go inside to sit awhile, dear.... And keep the back door closed.

MARGATHRATE: Give him more than money, but a home.... I want him around.... ma.... (*voice trails off its hearing*)

KANTER: She approves of your character, Bob. So will the congregation, learning the travails you've gone through. I suppose you've explained some to her.

LORKEN: I.... I mentioned a little of my saltiness.

KANTER: They must have been colorful adventures. When I was your age, I couldn't stop looking for trouble, till I found my wife. That checks many misbehaviors.... They may even want you to deliver a short guest sermon, during m*aaa*ss. It is.... an amusement for us, but satisfying to learn about others and their stories. Exercises the empathies. Justifies the concerns and interest, or nosiness.

LORKEN: I'm not much of a speaker to crowds; and my legs are still heavy, my feet weighted for a walking. But if it's entertainment you want from me, I suppose I can comply.... somewhat. Not that much to say, but that I'm grateful—

KANTER: To be living amongst us, Bob?

LORKEN: ....If at all, indeed.... to being led, or guided. (***Margathrate*** *enters the back room, closing the door.*) Why have a fear for *my* health, of all people?! But I am churning through my weaknesses, and trying to get stronger.... I had thought to really leave, by tomorrow.

KANTER: The church? But sure!— We can accommodate you almost immediately. Only Mathews holds up your egress, until he is convinced it may be made safely.... And I should prepare my wife— She knows.... about you already. And now that you survive, I'm sure she'll be delighted, and terribly curious (about you), tomorrow, it being her day to sweep the pews. But now that you will be a lodging guest, her work will make a preview of you for her. Do you still have any pains, in your body? A starvation can do that, even when alleviated, and after.

LORKEN: Not so much. Just a general weakness, without the tranquility expected for the feeble, which is why I am troubling not to stay so.

KANTER: Does the stomach pound nervously, after being reinvigorated?

LORKEN: ....The groins are nervous, that's true. There is an unsettling uncertainty left them, there. But it's an unusualness, not a pain. I seem to be searching, for some conviction to grab ahold of.... to deserve it, win it.... have fought for it.

KANTER: And what do you plan on doing, the rest of the day, Bob?

LORKEN: I'm going to walk around the church, in a fairly wide circle, to make myself familiar of its surroundings, until my legs come back to me and don't feel so strange.

KANTER: On such a hot day, such an effort already?

LORKEN: Slowly done. And I'll return here, before going to my room.... I was walking through the garden.

KANTER: It's not of its best presentation. Some things can't always be saved. But they are not perennials anyway, Bob. Not even they can endure a long drought, when the water becomes more precious than their flowering. We change the plants, to suit our tempers of persistence, and do not make any improvements, really, other than our goals to keep provisional to our capabilities and resistances.... to losses. I'd.... rather remove the dead than the dying stems; so it's not beauty that we're always after, but rather an array of what was beautiful— the chore.... to do this. And it would steam here, if there were bloated, bulging thunder clouds to rip of their loads.

LORKEN: I labor of the contentions for that lightening. We wait together, knowing it must come.

KANTER: With a certainty, Bob. There are seasons to reckon with. And to this very spot you will return? Well then, that's hard work.... for you. But I have to get home (*starting to leave*), to prepare the tidings. Sometimes I think we offer our services too often and too easily. But what are the retired to do? And that is a poor term, because I've only just become, today.... The wind makes a furnace, and dries out the moisture in the tears, and the lushness of a running joy of eyes. So keep that back door closed.... And have Pahtes welcome you, if you return this way.

LORKEN (*alone*): I have sunk to my desires. And by the time I come back—.... she may well be gone. These legs hurt already. The countryside is hot and fallow, but for patches of bodacious-ness proposed. But why is not to be that I'm deserving of it, for this kindness? So then I am, and (am) considerate of others.... They're all afraid of me, in some fashion, as not for themselves only (to be) so livid. Then do my pleasure seeking unto them, our liberalities together. It is a kindness, proportional to our discomposures or discomfitures.... in this heat. But it is just.... if it is found, this way. As were for an entrance to front, I will begin an estimation of it due my vigil (*starts to resume his walk*); and these limbs are hard put at it, and well fought for— Stamina and endurance are my groats to champion, for a lunch. (*Running from the same side of the church as did he come, out to the back, is* ***Carl****.*)

CARL (*calling*): Hey! Are you the new recruit?

LORKEN (*as* ***Carl*** *runs up to him*): I? boy— Ay. (*stopping*) I am new, but for what?

CARL: The father said you're to help me.

LORKEN: You can see I hardly walk well, lad. That's being fixed, with an exercised constitutional.

CARL: Out here? You'll drop dead before twilight, evening prayers. You're to turn my pages, Bobby.

LORKEN: ....Bobby?!

CARL: I'm the renown church organist and choir accompanist—

LORKEN: You're a boy!

CARL: —And I need you to turn the pages of my sheet music, as I learn a new work, for a performance next week.

LORKEN: I don't know nothin' 'bout music, never studied the subject. But aren't you instrumentalists responsible for your own page turning?

CARL: My *fingers* must accustom themselves to the score and its notes. Every time I stop to turn a page, it's like a knot forming on the memory, a disruption of the process. The mind must flow through a work to master it.

LORKEN: Can't you simply sightread, like they do in orchestras? Why do you have to memorize the work?

CARL: The cannon is too complicated to stop through musically. It would be torn asunder with too broad a rest, and would clearly defeat my mental anticipations for the fingers' dexterities during play. They're allowed a certain versatility, but not as ornery as this, for such a work.

LORKEN: Sounds rather advanced, boy. I wouldn't even know *when* to turn a page. I don't read music.

CARL: The fingers memorize, don't you understand? I am Carl, the organist, more boy than buoy to the musical treats, unless I master them with a strong facility that is expected for any church, especially a grand one like ours.

LORKEN: Oh, I don't know. It's large, but not that large.

CARL: All I have to do is nod my noble chin, to give you notice of a page turn. I can retain almost a full measure beforehand, to keep the flow of the music unaltered, as you do me— and it— a justice.

LORKEN: Well, I suppose it can be done that way. But can't you rote through the notes, like most? And why was I recommended for this service?

CARL: That's a stupid way to learn! It is practice that promotes an artistry. And I can only do that at the church, there are no other means available to me. I'm too young, yet, to buy a piano.

LORKEN: Most become too old.

CARL: The father said this is a simple discipline that you can apply yourself to, that would not be too straining on your recovery to health.

LORKEN: I work to strain and stretch, so as not to remain weak.

CARL: You're probably going about it the wrong way, then, as were the father's opinion, should he know. But music fortifies the brain more powerfully than any other exercise. It makes for our finest acuity and sensitivity. And virtually everyone's a genius to analyzing and appreciating sound, or tones and melodies and harmonies and chords—

LORKEN: Assuming they can hear such things. But he does *not* know. I'm still my own man, to battle through adversities— though.... to know *how* may sometimes take some assistance. And I am grateful for the kindnesses shown me here.

CARL: Sure! If you learn something the wrong way, it becomes a terrible crutch for error making. I had to restudy my trills for a whole month, with diligent observance of technique, something I thought would come naturally for being so simple— until it was pointed out to me I was doing them wrongly. And the embarrassment, on acknowledging and accepting the complaint, burnt my head, because one of the parishioners was not only a keyboardist, but a contrapuntist to boot! and scoffed with jesting at my attempts — in public.

LORKEN: Were you angry at him?

CARL: Her. She, a teacher. No, I wasn't mad. Just a little ashamed, and taken aback. But errors do these things to you. I felt so safe and secure with mine.

LORKEN: Yes. I can imagine that.... Though they're useful, but you need some considerable guidance to work through them, make corrections and come out alright.

CARL: I have to get to my practicing.

LORKEN: That's some handicap, that you can only learn your scores.... at the church. I'd think you could take some of the pieces to school with you, to study there— mentally.

CARL: The deacons don't like that kind of borrowing. But I've tried it, and it doesn't run well with me. I get very bored with dead notes on large pages. I have to work through them actively— with my fingers, and then my mind jumps to the purpose.... But there is a piano at school. And on occasion..... The teacher— not *my* teacher— scolded me that (a) true professional pianist not only learns the score forward, but even backwards, to make his memory indelible and promote a virtuosity of performance. That sounds crazy to me, and frightening— almost, nearly impossible. I am ashamed to use the school piano, for any of my studies musical. I'm not fit to be a virtuoso at any other place but here. But this is where I'm comfortable about my art, and enjoy it. This is where I'm not afraid to make mistakes, and accept some reproving. The music is at home, here, because He is listening, and approving all efforts!... sincere and forthright.

LORKEN: That certainly may be the case. But I can't imagine you'd restrict yourself to music making only in this church, or that you *could* limit your interests and activities that way.... You wouldn't want the church as a crutch to shield you from (some) humiliations, even self-inflicted ones. And once you buy your piano, or get one, you might find that the teacher was right.... about virtuosos, and might find the thought even comforting, when going through scores.

CARL: I only know what I can feel right now! Bobby, what scares me and welcomes me. And that is what the church allows, and the congregation praises, and what I make sure I'm up to doing, to be deserving of their approval, being suitable for my work ethic. Mom says that's the most important thing in life, being suitable. It's more important, even, than being charismatic.

LORKEN: ....And adored. Well, we're born to be suitable. That talent is god-given everyone— and everything, even wilting flowers.

CARL: Are you going to turn my pages for me?

LORKEN: If you give me the right nods, I'll be glad to. I want to please everyone (around) here.

CARL: Then let's get going! There's a particular passage that's very tricky, I'm anxious to get to.

LORKEN: I hope you don't make a whole lot of mistakes. There's nothing worse than listening to error-prone music— if it's fine music.

CARL: It's more fun than labor, Bobby. (*starting to head for the church, the way he came*)

LORKEN: Let's go this way, Carl. You'll have to walk with me. I'm slow.

CARL: O.K. (*adjusting himself to **Lorken's** pace*) How is it like, to be slow?

LORKEN (*as they're walking*): It's like.... waiting for a reward, coming towards it, and feeling good about that.

CARL: Yes, as when everything is going great! There's so much satisfaction to your efforts that the instrument you're playing seems to become part of your body, an extension of your persona, and can not do you wrongly in any way. I get to that state sometimes, and I could linger there forever, as long as the notes are played correctly. But even not, it's terribly sensual to feel attuned and in total communion with that mysterious pleasure, and that playing (that) I can't often explain to my friends.... well.

LORKEN: That is the hope of the crippled, Carl, that they be not so crippled.

CARL: I don't mind being stuck on such a plateau of worshiping.... the Muses, they're called. But it's hard to get to, and very rarely attained. The craftsmanship of great music helps, their complexities hinder— But I accept it all, and work with what is given me to love.

LORKEN: Your enthusiasm is important, sort of like a birthright of energy. Don't let it be drained easily, boy.... So much works against you, these days. But a pretty tune seldom goes sour to the ear, unless it's repeated more often than its sweetness. And lovemaking is an art.

CARL: Lovemaking?

LORKEN: Why, you do that to your organ, don't you?

CARL: ....Is that so!

LORKEN: That's what the Muses mean to bring you to, the inspirations that preside over the arts and sciences, in Greek mythology.

CARL: I haven't learned that yet.

LORKEN: Simply means to make love to nature, yours and others. But one must learn how, tactfully. It doesn't come as automatically, nor intuitively, as one might imagine. That's why people struggle so hard to learn how to paint, and play music, and investigate phenomena. It's really quite an awkwardness of undertaking.

CARL: I guess so, if you haven't the talent, Bobby.... Oh! You mean: love as to enjoy, to make enjoyment possible.

LORKEN: ....Yes.

CARL: Well, if beauty comes to me, I will love her just as well as m*uuu*sic! But I can't say such attractiveness stirs me greatly, yet.... except maybe to talk about.

LORKEN: And think about, fair organist. But to the front we both return, and this way passing, anew. This is a swing of sides, and I am to relax to settle heads.... of some imperatives. Though to be needed is something remarkable.

CARL: It's prettier on the other side. More shading and less browning of the foliage. And some water drains a bit, there. That's where the pipes are, and so the drainage ditch.

LORKEN: There are pipes all throughout the church.

CARL: But I mean external to it, to the building, the hard structure.

LORKEN: A firm relief. The circuitous path leads us back, but less wildly than I could have imagined. For the sights, with this perspective of my being, are somewhat new and unfamiliar— Though I've seen everything before, but from a distance casually and without much caring or attention. Now comes.... consideration, of the isolated.

CARL: I do like it here, much better than at school, to study (in).

LORKEN: That is your rain, pouring on the brain of fortuitous swart swathing.

CARL: What?! Do you want to be a preacher, Bobby?

LORKEN: Cool sheets, boy, for a hot night. I could never have thought as much, to be found. But you have found. And we.... return.

CARL: Cool sheets....? Yeah. There's some good music to those pages, the sheet music. I'm a devil, sometimes, when I'm in the mood and form, for a good performance, during the evening services— especially, since I've had more time to practice.... But they are rather holy chants, you know, and are to be solemnly respected. There's a fairly broad library of works, though, I think selected by.... the deacons, maybe? or at least collected by them: gifts and donations by parishioners, especially those who have died, and so can no longer play the music they loved, or admired. Some of the volumes are richly adorned, and appease my eye for their importance.... to me. I wish the sun could set by them, sometimes. They are.... voraciously profound for my attentions, and are as attractive as coloring books used to be (for me), when I was a tot. But now I can use them.... to a really good importance— A grand imperative, Bobby. Yes! I understand.

LORKEN: Toys are playthings. But playing 's not a toy. It is— an inheritance, of the animal nature.

CARL: I don't think they serve God too much, not like we do.

LORKEN: All thoughts of purpose have their rituals. And deliberation makes for deities to be involved.

CARL: Even careful hunting?

LORKEN: Don't you wish to capture your sounds correctly, for His ears?

CARL: ....Yes!

LORKEN: Then there are no diatribes, even for the animal to denounce their tedious activities of existence, feeding and mating and mattering— in this world.... with play— all the same to do.

CARL: Then I am play, Bobby.... (*They disappear, along the other side of the church.*)

Scene II — ***Pahtes*** *is with* ***Mathews****, in the rectory office, which doubles as a living room. The space is somewhat ornate, but in a tired way that would have excited children perhaps a couple of decades ago with its beaded colorfulness and demurred substantiality. This is to say the picture frames on the walls are rather baroque considering the piety and reservedness of the painted subjects, and the decorative articles that abound the room in glass fronted cabinets and on top of handsome chests are of an old religiosity celebrating the splendors of image over the resoluteness of faith: much gold gilding and plastic or glass jeweled trimming, etc., things you could hold in your hand to admire but would not dare to touch as a visitor, especially younger than adult.* ***Pahtes*** *sits in a nice and comfortable easy chair, while* ***Mathews*** *stands facing away from him, at a small desk with carefully strewn pamphlets and small books, little booklets, and neat piles of handwritten letters or notes and typed (up) correspondences or papers of interest. The priest holds up a thin red, leather-like bound notebook or folder to read, as* ***Pahtes*** *simply relaxes in the chair.*

MATHEWS: We must do more to directly involve them in our activities, Pahtes. We have to explain ourselves better, what we're trying to accomplish, and why. That will generate much more interest, inquiry, and help.

PAHTES: The days are fairly difficult, father, for many. And the heat brings anger and disconcerted-ness to the fore very easily— That will pass, as the seasons change and the rains pour to moisten the dry cracks of earth into mud piles. But you can't assert on them more a responsibility to God as they deal with the cleanup, and the relative dissolution of their state. The message will be less convincing than their work to do, their priorities of maintenance. There is a general uneasiness to these travails, relative and pervasive throughout the territory— not just in our little town. People are unhappy—

MATHEWS: With the weather?

PAHTES: —and complain with license to be rather more self-serving and indulgent of whatever tiny pleasures they may accomplish. They can not become readily obsessed with one of your major tasks, at the moment; and when it must all be based on charity and your asking, there will be a resistance considerably to be reasonable. That is why they don't participate more forcefully. They certainly know of the.... idea of the project. But is it just another whim of yours, they question.

MATHEWS: The diocese itself supports me in the matter. We discuss these things at regular intervals of meeting. They are not whims of thought. (*placing down the folder*) But we must be self-motivated to provide for ourselves the fury of this necessity— And then we will be granted some substantial assistance, when we can show that this locale is serious to undertake the effort and do the preliminary heavy lifting. We must demonstrate what we have done, before we can begin to do. We must donate our hearts before we can receive the benefaction of their goodness, and the radiating warmth of satisfaction to be joyous and happy for ourselves.

PAHTES: That is a philosophical charm ever to be desired, father. But realities make for treats only to be practical— otherwise they are superfluous. If a food is sweet, then it is for its sweetness that it is sought and battled over; not that it be needed as such a food to be, since there are many others.

MATHEWS (*turning to face* ***Pahtes***): Be, perhaps, a bit more relevant to the argument, Pahtes. Why should anyone be against this?

PAHTES: I'm sure only a few are, to be spiteful to.... you, or what we represent of assistance to the soul. A few are not so religious as to think us incredible. But I am not being vague to say that there are many people here. It is a proper settlement for a century or more, definitely more than that. And throughout that time there has always been the wanderers passing through, not contributing to our welfare in any propitious way that can be overtly demonstrated with their presence. So why now, in this modern era, is there such an essential need of jolt to harness and protect them? to put them to specious uses paid for with the sweat of our worrying and need to overcome our own more decent tribulations. That is the question felt.

MATHEWS: Because they're suffering in this heat. They're dying, out there.

PAHTES: There have been no reports to quantify this belief— I certainly do not obstruct the principle of your conviction, nor wish to, in being relevant.

MATHEWS: For every heavy tragedy prevented, there may be many hidden from our eyes to countermand the fall, that it must be thus by our very natures of callous proprieties. Kanter came across that man by accident! Pahtes. There was not even a search made for the possibility— And this is shocking, because it might be predicted. Don't you think so?

PAHTES: And we chance by it, father, that suddenly you bring it up as an emergency. But is it so surprising to see how a congregation responds, or do you expect miracles upon first hearing of it, this proposal?

MATHEWS: I said it might be predicted, or might have been. It is only the urgency that shocks us at the moment. And we must plead more, to remove the veil of doubt that this must be done.

PAHTES: They are only confused, and not totally convinced of this latest whim of yours, this insight bolted through the night. You are a man of action— with your thoughts. But it takes more than thinking to move stone, more than dreaming and ideas to make for fact of an accomplishment.

MATHEWS: And you do so much work around here. I do really appreciate it, Pahtes—

PAHTES: Thank you, father.

MATHEWS: —you and Paths. But does this not give flame to the candle of enlightenment, and fire for the drive to do something about this, something heroic and Herculean in scope and extent, such as can push a border of mountains back! in a humanistic way or meaning?!

PAHTES: You are already doing much. Your lead will be followed, by the generous. You have spread information, about a condition of mankind we face right now in this region. The problem is known and will be discussed and analyzed socially, debated and solutions conferred upon. That is an extraordinary achievement of yours so soon, and is what our church stands for. But make not time so much of an enemy, to turn heads. Even water must take time to flow from the faucet, and into the cup. Enough will come around to prevent a catastrophe of sun-stroked carnage. The community is genuinely concerned about this, but the method to resolve the problem is yet to be enunciated with sufficient dictum to be.... compelling—

MATHEWS: Compelling?! You mean convincing— of our own worth as sheep of the flock!

PAHTES: Compelling enough to take immediate action without more presidency or opinion of councils. One may have to ease into the venture— You can not push independent minds. That is always a mistake worth more retaliation than herdsman-ship.

MATHEWS: What is more simple than a shop of sweating to save souls? a compendium of community assertion to collect these poor strays from out under the hot sun, during the daytime, as a fundamental course of principle to uphold and celebrate with! Just such an announcement alone, by agreement, could have revolutionary results in our town, that this is an area to be known for the utmost compassion during a most grievous spell of unfortunate drought and unsustainable hardship without our collective assistance to assert. We become proactively inspired by the problem, and each household searches for at least one sultry resident to contend for and tag with a homey hospitality insisted upon, for a short stay's term of work— to get out of the heat, and perhaps receive a lunch or snack as part of the reward. This is an emergency, after all. And all I asked for, this morning, was a show of hands by those who could find this project in any way favorable with their participation, or could offer adjustments with modifications on the procedure, perhaps to stores or barns or factory leases for the vigil-ed to send— to keep them awake, and alert and alive. All we received were confused smirks, and feigned perplexity at the proposal.

PAHTES: Well even I didn't understand it much, father, until you explained it to me more thoroughly afterwards. It was like an impulsive dream delivered. But you did not emphasize enough the important aspect, or notion that.... the invitees would be *working* for a household's merit. And you have no idea of what labors they

could do. Neither can most parishioners imagine, so rapidly given the problem— and the solution. What is conceivable of a transient's skill? He is most likely transient because he hasn't any. (*Faint sounds of organ playing start.*) And that in and of itself is no crime— I'm glad this rectory is well insulated. You must hear all of the boy's practicing, sweetly filtered through firm, rich woods.

MATHEWS: It is what we make for them to achieve, which is our purpose to apply our art, that causes of their obtaining value to their hearts and the appreciation of others to recognize this.

PAHTES: It's not a crime to be useless, if that is all you really are. It's not a crime to be born paraplegic, or deaf, dumb, and blind. And the stillborn are buried with as much honor and ceremony as the aged, highly achieved or not. But.... the healthy and lazy are not blessed, father. They are.... somewhat indecent and suspect for their hardships. And we can not automatically shelter them without threat of our own well-being, for the weal of a community is based as much on common sense as on industry.

MATHEWS: All thinking life have a genius to do *something* constructive and creative. We must promote what talent there is, to assist us.... assisting them—

PAHTES: It could be like lodging discourteous and foreign soldiers, and asking them to mop the floors they dirty. You can not so easily or immediately order up trusts from the truly deprived and possibly depraved. For impoverishment is relative to what you wish to be. If this is nothing, if this is to be a bum— then you are rich! and may be easily insulted by suggestions of changing your way, or being forced to or coerced into menial tasks— even with necessity for survival. They may much prefer to steal, rape, and pillage, as the cutthroat rogue would when given the opportunity and access. Not all men are the image of man.... spiritual, pious, and sublimely devoted to a holy goodness and general tranquility of peaceful existence and agreement in all that betters the world. You can not charge your listeners with an unwarranted resistance to what may sound foolish or too precociously tackled, given savage history.

MATHEWS: My son, men may be dying today, because of that reluctance to know them personally, because of that prejudice to disallow their testing.

PAHTES: Father.... I am more your brother than a son. I think I might even be older than you—

MATHEWS: That is no matter concerning them! I have no age, but with to present options for our good deeds earthly (to do) and in peril of inaction. I have been given the consecration to do away with time— And in truth, there is no past nor future. You can not understand this, and I can not explain it to *you*. But one must simply do what is right despite all obstacles and contrariness. And my aim is not to offend your natural hesitancy— nor anyone's.... For we all must sleep in a bed of sins and debaucheries. But I am trained to remove my mind, from the body's inevitable corruption, and without the plangency of sorcery admit to myself and others what is really to be done of a desperate situation— as I feel it, as I dream of answers given (to) me as being especial.... through the conduit of prayer and hope to mend and meld of mind with the most hopeful of revelatory perspicacity. And I tell you, it is a sound judgment I state, to quarter these men for their lives to save and protect. That is our venerated task to somehow perform, given us with direct acuity and acuteness. It may be the very reason for the drought itself, in its entirety an examination of our societal will. And they are as much your sons, even if some may be as old as your father, because time is thrown out with help that raises life out of the murky pools of dissolution and loss of solace. We must assist their spiritual, self-enlightening growth, where it may be treaded upon by disaster, natural or nadir-ishly human by our lack of attention or caring. For they are separate and distinct tribulations we run through, at this moment.

PAHTES: ....I didn't say I was personally against the proposal, father. Something of its kind may be achievable. But intelligent conditions and boundaries of activity have to be devised and set for a venture to even gain a foothold around here. For example, a way station may have to be set up, for those who *wish*, or want assistance, to come to and be properly dispatched, after having been observed of temperament, adjudged of sincerity, perhaps cleaned up a bit, and articles of weaponry or danger removed for their persons— That's all that I am saying, that they may at least be made actually presentable.... to willing households or other daily employers. We must think through the details as well as the— decree.

MATHEWS: I can attune to that.... And he improves mightily— with help!

PAHTES: Adversities are to be experienced. But indeed, I was taught, we should never hope for miracles but only expect them as the due diligence of the creator in a hurry.

MATHEWS: That, son, is a peculiar prescription for faith, though we are always the instrument and never the instruction. Instruction is provided us through inspiration and divine intervention, when necessary. We can not draft the blueprints of such achievement, though we like to think to be the architects always. The insight to this distinction must be learned, for we are no more responsible for failures as success with our devout efforts. Thus live we humbly of mind.

PAHTES: You take on too many problems too quickly, father.

MATHEWS: The hands wring of them, bleed with them. And I enjoy the challenge, always have. I will be buried under them, and that will be the purest satisfaction—

PAHTES: I handle—

MATHEWS: —slept to.

PAHTES: —many small chores about the church. And I take care of each one by one, on its own terms, before going on to the next. I'm not confused by them, nor any to complete, until I come to one that can not be. And then all things stop, as I call for help to deal with it.... And sometimes the problem can not be concluded satisfactorily, and we simply live with what results can be achieved. That to me is faith, not to be dissuaded from other tasks to tackle after the last one falls fallow. And I sincerely fear you call for help too easily without estimating whether the problem can be solved or how difficult it might become to work through. So then you run the risk of being confused with too many failures.... on your hands.

MATHEWS: That is a sublime shame, them. (*turning back towards the desk*) I tremble with them, to remain positive. But that is

my brushing wished for, since we are sacrifices on vacation living.

PAHTES (*getting up*): Just have a stroke of pessimism occasionally (*The organ music stops.*), so that we can tell you *do* review things carefully and with consideration of the possible. An idealism is only handsome when it works, and when not, terribly ugly. And when one can't tell, if it is occurring, well then one hardly notices it at all. Our people love this town, and so are very conservative about it, willing to offer assistances at all costs when the needs seem reasonable to be dissolved. But if you bristle their hairs too often, implying they may be the blame for some bad tidings— they may snap (back) at you! not to be handled by the scruff. Indignity is through decadence.

MATHEWS: You think some are riled by my sermonizing?

PAHTES: I'm only saying what I hear, out there, father. You are more isolated from common public opinion (*Organ music starts up again.*), but at least a few believe you are becoming.... too (*A particularly hideous dissonance of mistaken notes is heard, totally contrary to the proposed key.*) idealistic about your own state (*recovery*). And you depend on them— the church does. Some things must be toned down, imperatives softened, and less resentment generated, that you have answers which could seem.... decadent, considering from where they must originate. Is it not foul.... to spit in a bathtub—?

MATHEWS: Pahtes, I try—

PAHTES: —or to splatter mud on a window? You can't coax the caring onto the careless, as if the indifferent of value were still children. There are limits to what will be put up with, or should be. If a person is meant to live a way and die that way, perhaps, even with our abhorrence to appease, this could be allowed to occur. In some cases it must happen— So why not others?! And not for yourself to transform with empathy, care you not mostly for the decent and hard working?

MATHEWS: I—

PAHTES: These are questions being asked.... only, excuses for disbelief— It's not grandly conspiratorial, yet. Just resentments occasionally brought up in conversations, which I sometimes feel obliged to answer for— always outside of this building, far removed, out of respect for the religion. A few cry to me, jestingly, "What has he come about, Pahtes?" or, "Should I continue to contribute to this deviancy of emotional wistfulness and whisking of our patience?" You are not an ideologue of perversities to enthrall and thereby check or change into a loving servitude of mankind, the miscreants morass-ed into a jolly party for some applause. But then again, you hardly have meat enough to last the end of the month, if you catch my meaning.

MATHEWS: I.... try to avoid both controversy and pity, Pahtes. (*The organ music stops.*) But.... what am I bounded by, of the seraphs' decrees? This is my brother broaching my neck willingly, piercing for discomforts that I sense too many have, or experience alone, forgotten, and with no representation at all!... before God. And should I be less sententious to their deaths?! My hunger is weak compared to theirs. But this is what I feel, to hold out for more assistance. (*facing Pahtes*) It is all there, it is rich.... and to be found. It can not fade before my face and pleading. And it is a just harassment for all of us— You must continue to support my....

spiritual legitimacy in this, our ecclesiastical sense to be righteous to all in need and generous enough between ourselves for our poverties of soul to abate. This is not as obviously thought as felt; since the heat hardens many, and softens some to an abandonment of moral goals. But it's simple enough for any to understand that I can't be made differently from what I've been taught— and blessed for— to try. (*organ starts up again*) Here the dissonance stops, as I am afraid— but will not be deterred.... from the frost of mistakes. We are better learning because of them, and best dying from them. I know there is discussion, dissension, and discontent to face our facts. I see this on their reverential smiles, and want not to pose more challenges onto hardships when I speak before them to serve their thoughts that they are being better than the worst to allow my entertainment and might indeed climb high of gratitude— from He who grants you life and purpose to it— with an adoption of some goodness and petty, human kindness. Yet I won't decline a beating, Pahtes, for burning the toast. I will take all responsibility for my errors poured on others. And if there is any theft or assault— or worse.... because of this project, I will give my hide out to be banged with pots and pans, and pounded with rocks and canary cages— till the criminals are all locked up! And I will leave the church and my mission as indecent, and go starve somewhere in a ditch, away from her splendid shelter. For this abuse of myself will be as warranted as my abuse of faith. And I'll take it as so, calmly and of true complaisance with humor to be gracious and thankful — to have been once.... much better for the task, just as Socrates so composedly agreed that his arguments and philosophies were no longer fine enough to save his life. But this I give you, Pahtes, my blood to bargain with—

PAHTES: Father—!

MATHREWS: —because I am so resolutely determined on this path to take and this practice among our citizens to fight for.

PAHTES: Father Mathews!... I merely contemn the jesting.... and your being made to seem like a fool. But I am partial to your sometimes blinded enthusiasms, though not always of their working. It's — why I am a deacon here, to help you serve your wares of prayer and celestial opinion, and to find, someday, a similar courage, with your example— in some matters.... to make principle of a belief, as more to being than merely being felt.... since practice is the most difficult of disciplines to harness one's energies about, the rewards from it.... often always (being) as questionable as the impossible to do.

MATHEWS: Well, one does not have to practice instinct. You know intuitively that.... we are on the right course for this. It is of paramount mission to assuage suffering, in what ever ways we can to whatever forms we find. As much as man may cause, he may reduce and remove even more— even nature's. For we are of holy privilege to do so, and no other of the earthly bound can claim the same mandate nor consciousness of task. Now.... it may be impossible, in some instances. But we're not to judge *that*. We're of to taste from the bread offered, bitter, sour, or sweet, and make provisions, thereby, of our candors to be concerned.... I know it will be difficult, to accept inevitable misfortunes. The impossible is as easy an exercise as its possibility, hemorrhaging in our brains. And I am often scared (*turning back towards the desk*) of the consequences we bring ourselves to, the discordance of a rainbow, how good and beautiful attempts of even wishful thinking bring about sullen, rancorous talk or comment. And they actually laugh with disbelief?

PAHTES: Some, that you would go so far.... with hope, that you could present to them.... the incredulous of ideas to swim through or bathe with, that you— so heavily weighed down— can be naïve of forces pummeling everyone today, or say to seem. (*as **Mathews** rubs the red folder*) This is hard to accept, where you are cornered. You serve them, and they allow you to be so disposed, but just barely to be properly sufficient for the task. Your ability is *their* privilege, your capabilities their cheeks to rub and warm and calm from the daily anxieties of life in a cauldron of civil discourses and opinions on the nature of man.... or our townsfolk becoming displeased or uncertain. Your role is as much to assuage their fears as to instruct of yours— and His for ourselves to ratify against, vote ourselves contrary to.... evils, mean-spiritedness, and the like.

MATHEWS (*lifting up the folder to read through again*): I've no doubt.... that I am controlled by all forces impinging, and implicit for my handling. It threatens *here*— boldly. But we are all sacrifices to hardships, and from heaven held over the flames of anxiety, until we are thoroughly inured.... or burned up. (*The organ stops.*) The contrast is so slight, and yet powerfully different are these two aspects (*reading*) of being. This frees my actions up, though, to be as responsible as I am foolish. For how much can one really starve? within this realm of gilded handicaps and beaded jewelry, and ancient profundities, and proud demurring to the miraculous disclaimed, disowned or dismissed. I'd rather choke of hunger than (on) his bread and wine.... He was much more correct, this time.... this last time—

PAHTES: We must exercise ourselves, to be more politically savvy, father; and work to convince more than to propose; to induct by favors more than to indoctrinate by fashions or manners, is what I would seriously propound. And this might mean calling a rake a rake, sometimes, rather than a social rash for treatment come to. Agree with thoughts of the congregation. Are they to be schooled with scolding, and what can appear to be scandalous? Not so much so are they.... to be rubbed, or sorely made— Reddened! father.

MATHEWS (*placing down the folder*): I will not complain *on* them, nor through them. We can be wise (*organ starts again*) to have a good way eased. We can do so by example (*facing **Pahtes***) and demonstration— Courage, Pahtes, is what we must manipulate with, the only tool left that's broad enough for the rasping— for the job. Bob will make a splendid coherence to the method I may talk about, describe and let.... fly of an honored whimsy for proof. He's a good-looking lad when freshened. I spotted the potential immediately—

PAHTES: Mr. Lorken—

MATHEWS: —Attractiveness draws conversions to many matters. It is a natural sugar to the eye, or honey to the mind, that healthiness can be admired and adored. And what if not? Then this representation might have been lost. That is what will be conjectured, contemplated and discussed.... and hopefully generalized for the plights of others. He is my trick, Pahtes, to apply.

PAHTES: Mr. Lorken.... is improving. But I really do think you should let him be on his way as soon as possible, away from an implied covenant with this church, or advise him as such, that the association can not be long maintained or afforded.

MATHEWS: People will choose him, Pahtes, for the congregation to join; and therewith, help (is) allowed. He's a good boy—

PAHTES: There are too many good boys. What we need more of are correct ones. He struggles to leave, you know, does not wish to stay as a burden on anyone— or anything or concept. And he may eventually become more active than your applications can handle. There is little deceit to his purposes. He wishes to run even harder now, to fall even harder, exhaust the batteries once and for all; and beyond recharging, fulfill the decimation, like a bag full of pretzels crushed, crunched and swallowed. I've seen this before, with some young people. It's a pleasure for them, an ingratiation— to the ingratitude for living. He could turn out to be a terrible failure for your example to advertise— And he warns you of this, with his reluctance to be tied down here, knowing of the situation, and what he feels he must forbid. His is not.... help from a church, or adoption of a religion. His precepts are froward and dark. And you will not educate him with kindness. You may madden him.... to do the forbidden, unless you grant him his proclivities!... And that includes a strong notion to get away from here.

MATHEWS: Don't you like him—?

PAHTES: I do like the fellow as a fellow, a person. But he's in a rush you try to delay. His time is short. You build too much upon him, for yourself, and pray too little— for what can decently be felt. This is destruction of our bearings and moorings, father, to keep him around. I can tell. I can tell for you. It's not a simple thing, but he's a simple blow to make for, a downfall of your principles and restraints— to be circumspect of your constraints. You will befoul your lessons by forcing them through this guy— for others to see and realize. You have vows to uphold, and cerebral vestures to maintain. There can be no excuses or exceptions made with so fundamental a law of habiliment and custom.

MATHEWS: Just that he is unusual a tool? The crisis becomes unusual. I do not keep him here, deacon. I welcome him here. This is a grove, for the oranges to be made (in), the.... golden apples. And he will benefit of a sweetening and ripening into a proper manhood— within or around this community, or wherever. He's only been confused. That *I* can tell. And we will help him.... because we want to help him— Do we not?!

PAHTES: Yes.... Yes for all, who happen by in distress, we make our betterment with some assistance. But are you sound to this, Mathews?

MATHEWS: Sound?—

PAHTES: I mean sure, (that) it's God's way to beat? Is this a divinity running through your veins, in this instance? Or does what courses serves a sultry heat! to find some approbation for this nerve.

MATHEWS: We.... merely promote from a supposed death— And I don't mind to say I care for him.... But this is taught..... in a way perhaps different from your context to admit, that this feeling is, while often frightening, always allowable. Souls are not saved, being too pure to be sullied with our actions. Rather, they are savored for their goodness to produce, and to provide. And I can taste of this from him or anyone alive, if we are brought to these conditions to submit ourselves to, that golden sense of compatriotion of ourselves through others same, and human maned.

PAHTES: I'd not romance the happenstance to make it as trivial as it seems to me, and less worth my attention than attendance to assist in the performance of some little and temporary aid to a wandering stranger. As I've cautioned before, you may be searching for the miracle over man, (rather) than the man himself. He presents himself to you quite candidly. Accept that honesty to be contented with— Blast these sounds, Mathews!

MATHEWS: Patience, Pahtes. They'll congeal into a sweet performance, and will conjure up the usual approvals and plaudits.

PAHTES: I know. But still.... you have to enjoy the efforts so much, in order to endure them. Our church has become such a retreat and refuge, for frustrations to diminish honorably. What does listen to a cry for sanctimony? and every blatant excuse for bereavement of abilities applied. To permit such errors—

MATHEWS: It is not how well one does that really matters, but more how well one does to do. We are ourselves the most magnificent of errors ever earned a substantiation of this life performing practice. We are born mistakenly, angelically fouled, and put immediately on a path to prove ourselves otherwise, and dominate our realities. (*The organ stops.*).... There are suspicions dawning with the silence, that this alone is not capable.... of describing us— thoroughly, as within a harmony presumed. That draws us devilishly deviant— unless we sing! unless we make a noise and rattle, and a prattling to spurn complacency of existence. That's why I enjoy all of Carl's most earnest and urgent playing. It is like a child to do— for growth and hem of attention. We are here, and we promote a notice, for whomever may be interested or whatever possibilities preside over us. We do maneuver actively to be involved with that harmony, however falsely done the technique, or however false the requirements are for these.... techniques of ascension. For that is what good sounds provoke a sense of. Let children play, to pretend for their accomplishments. This is what we do, in various manners, all throughout our lives, an imitation.... of reaching heaven. May we never be so satisfied, until this is done—

PAHTES: In fact? Facts are what we hold and hammer with. (*organ starts up again*) We should take collections, to buy a keyboard with an earphone jack, and give this to him as sort of a prize for his diligence to the services, or a present at Christmas Mass. What a surprise that could be! such a presentation wrapped, bundled out of view until the moment that he completes, competently.... but as often always, the final fugal conclusion, fruition of the fulminating coda, that all has come resolutely to an end.... We may do this for the lad.

MATHEWS: I would rather want to hear our organ exercised.

PAHTES: Carl is the keenest representative of the congregation to me, someone not in need and ever with need.... of a generalized mass approval. He's even afraid to enter the antèroom, for fear of finding the more discerning and disconcerted volunteers not likely to hold back a remark on his fingering of notes with rests.

MATHEWS: They may be pretentious.

PAHTES: We might all be, father. I must revolve him for other duties, for another collection, and promote his social inuring. He will never be a virtuoso, and must learn to deal with people faithfully of their opinions. It's becoming bitter, these needs. (*starting to exit*)

MATHEWS: I will half placate them at least. The diocese is ever thankful for you volunteers. Religion could not persist without you. And you continue to give, knowing the funds are needed.

PAHTES (*opening the door to the room, as the organ stops*): Then are miracles needed? We may consume our own bait at times, father. But take heed of real hardships. The impossible is ever present, and like a soreness to the back must demonstrate its weight. We may not save from its afflicting, (*while exiting and re-closing the door*) but only work more in this toil of life to salvage our wariness towards its approach.... with blackened hues and desirous anxieties. (*has exited*)

MATHEWS (*to himself*): Then what is for to be alone desirous? the lure to live, lease lengthening on these accomplishments. (*turning back towards the table*) And now to crumble— Yet perform a service, Oh dear! (*leaning on the table with his hands*).... and prove ourselves desirous. We do perform, and we continue. And this is reticent, but can not remain unacknowledged in my prayers. I am so afraid, and yet so certain.... of my actions, and my signaling for help. A flash of lightening speaks for me, and startles for its thunder to be felt.... past this occurrence. Needs be needed now! and purpose rejuvenated, in the drying wood. This is a call indeed, for moistures to be sent, and a drink allowed— for growth, to appease the worths we have developed or assumed. There becomes a lifespan, a beginning and a conclusion, for any endeavor. But here it can not be finished, not within a hypocritical state— no power to do more for what is presented to us. The signs are ablaze, they are on fire, that decrepitude must be resisted. And this is a terribly old.... environment. But I am able, through it. (*straightens up, hands off the table*) With surge.... of sacrifice, I will make due of the dough that bakes.... pretenticusly. No one must be harmed, with my purview of shames and shambles, if not that I could succeed.... to take the blame. Conversions are difficult, more so than their anointing, or demanding, or pointing out as possible. And I have done all of these things, or tried with heart and hope. Yet still the structure shakes to shatter, in our collective failures with one and another individually distinguished.... as distaff of value. Be there loneliness to this (*polite door knocking*), that I would cry for to assist, with company of myself— here!... Yes? Enter wholesomely, without review for thought of any chastisement. I am for the maid's service of ye all.

MARGATHRATE (*opening the door with a shy respect for dignity*): Father Mathews.

MATHEWS: Yes?... (*as she enters*) Mrs. Mason. The chamber is afresh for your inquiry to make.

MARGATHRATE: I.... would not.... wish to be too bold. But it has been suggested that I come to you directly, with requests for keys to the community room.

MATHEWS: That approached from the outside? One for yourself and your husband.

MARGATHRATE: A couple. I enjoy my labors here, and feel entitled of this little, but symbolic ownership. I wish to continue—

MATHEWS (*going towards a cabinet*): That is definitely so, Mrs. Mason. Deacon Paths tells me you become particularly de-

lighted with your service, those little chores of magnanimity that are so important to us. Has he arrived yet?

MARGATHRATE: No. I've not seen him yet, this afternoon— But I know what to do. He has instructed me precisely, and very carefully, on the stained glass metal framings to wipe.

MATHEWS (*opening a cabinet's glass window door*): They are of thin lead, and very fragile. All the way up there you're to climb? With a ladder? He's showed you how? (*opening a canister*)

MARGATHRATE: Yes, and the dry chemicals to use and polish with.

MATHEWS (*recovering two keys*): I wouldn't want you having an accident, and falling through a window. They're so brittle, that possibly wouldn't cause as much harm as a modern pane of glass. But they're so valuable, as art, and not easily replaceable. (*approaching with the keys*) Yet my proper concern should be for your balance. It is the fall that is most dangerous.

MARGATHRATE: I've learned to steady myself on the steps. And I'm not too adverse to some heights. The greater problem will be to keep from being dazzled by the colored flints and glaze. They are much like jewels, though the drawings seem almost cartoonish, when seen up close. The details, I mean.

MATHEWS (*giving her the keys*): It's a difficult skill to master. Today, images are mostly stenciled, for easier mass production. Ours are etched in dyes, with tedious craftsmanship flowing of ability and sincerity. A humble art made magnificent by being devout.

MARGATHRATE: That's a very regal cabinet you have there.

MATHEWS (*turning slightly to look at the cabinet*): Oh, it contains all of my soul of treasures, Mrs. Mason, and holds a very stern position within the rectory. Much of importance (*turning back to face her*), like those keys, are proudly stored within, much for to celebrate their.... accumulation for the church. And she is a fine witness to them, don't you think?

MARGATHRATE: Yes. Our church holds much.

MATHEWS: I store everything of personal sentiment up in that cabinet, that I can review at times of indulgences for reflection.... on church proceedings, as well as some of my own possessions from the past that have helped to make for me clear impressions towards my attitudes and beliefs, and have shaped my character to withstand adversities.

MARGATHRATE: Yes, but I think the cabinet itself is handsomely built. Fine woodworking and—

MATHEWS: When life is not going as well as I would like, I can often find some little object in there which has as little value as my attempted achievements, but which nevertheless retains a preciousness or charm that assists my general outlook on matters trying, to remain positive. The simplicities found in there buttress me against the difficulties. I seem to have to collect both, at times; although I haven't added anything, that I could find remarkable, to the cabinet in quite a long time— No money at all, Mrs. Mason. Any cash reserves are kept in a safe box, and documents for the building in a vault in a bank.

MARGATHRATE: I needn't be privy to these details, father—

MATHEWS: It would be virtually impossible to steal one's faith within the bounds of my procuring, not inside this noble edifice.

MARGATHRATE: That's somewhat enigmatic.

MATHEWS: I mean that.... whatever I come to suggest to our members of congregation, what I have a nerve to try and say, after review of all scripture and religious history, often turns out to be.... simple and ineffectual— to me, and innocuous for your affections —

MARGATHRATE: Mine?

MATHEWS: —if you can take the suggestions as almost automatic from one's faith. I have studied what you need not (study), and the lessons distill constantly, to reduce to a consistency of.... essentially helping one another, as pure of form as the colorful trinkets I keep in that cabinet, the rounded beads and shaped crystals and stamped metal engravings of sayings easy to remember, and buttons decorative and tiny frosted drinking glasses and miniature plates depicting.... passages from stories and parables, and neatly kept ribbons and swatches of cloth, ornaments of clothing.... and of books, all in there, nicely arrayed. And keys.... for the community, that is what I think of them, too. And it all boils down to being neighborly, or socially conscious.

MARGATHRATE: That I am, father.

MATHEWS: So we must strive wherever possible to help the indigent and dispossessed.

MARGATHRATE: That should be without you having to say so, after reading our psalms and singing through our hymnals.

MATHEWS: Even with risks to ourselves, weak to subtle to great, we collectively must try to do this when there is a need to. And it is by an infrastructure that we provide, as a concerned body of worshipers, that these accomplishments of assistance and relief from suffering, material, spiritual and moral, may be made.

MARGATHRATE: We share ourselves, to alleviate for others their distresses.

MATHEWS: Then that is all that I mean to argue, of a terribly simple significance. I never hope to affront or perturb our cherished congregation who have already demonstrated a selflessness so often in the past.

MARGATHRATE: Could you ever have a fear of doing this, or causing some anxiety by distributing for us verbally your.... researches into the meanings of faith and holiness?

MATHEWS: I may have annoyed some, during today's morning service. And since they were some of a few, it gives me a grievous worry if this may be representative of many or most. But in this time of drought, there may actually be homeless dying on the country fields and grounds, effectively sheltered in their tawny conditions with ignorance of their plights. And I have simply suggested that we actively seek them out, search for them to help, and

do what can be done to save them from ruin of their person and humanity. And do you know, Mrs. Mason, that one such soul was actually brought to us, here to within our walls for a recovery. So I know the problem is real and may be acute. But how can I shape my speeches to present its immediacy to you without belaboring the predicament to make its acknowledgment offensive or displeasing? Our hardworking community in general is not easily disposed to view or consider the hopelessly transient person with any favor, since they are totally unproductive. Resentments may tend to mount against them as they worsen. But that very sense, that evil pillaging of empathies, is against.... a faith's mandate to be helpful during a crisis. This error of heart I must better communicate to warn against, yet all within the imperative that establishes itself so solidly now. Here am I vexed.

MARGATHRATE: This poor man brought to you—

MATHEWS: A man, yes. Impoverished from life, almost.

MARGATHRATE: Is his name Bob?

MATHEWS: Yes. You've met him already?

MARGATHRATE: We talked awhile. He's been walking around outside, in the heat—

MATHEWS: Oh! I advised him against that.

MARGATHRATE: —to try and recover some of his health.

MATHEWS: He needs to relax, and take the opportunity of our offered respite. Less effort and more nourishment should be his goals. I'd rather have him help Carl play, and I rather thought he had.

MARGATHRATE: Bob wishes to leave these grounds, tomorrow.

MATHEWS: That's all a matter of his stamina to decide, Mrs. Mason.

MARGATHRATE: He seems like a nice chap. And I would agree he should stay put, until he can no longer. But I believe Mr. Kanter has offered to give him some housing, provided he recovers enough to do some chores regularly for the church, as sort of an indentured volunteer. So I don't believe you have to worry about him too much, father. His health, I mean.

MATHEWS: ....There are sure to be others, with as dire a dilemma between life and death. For (any) one fortuitously found there may be many more thoroughly hidden—

MARGATHRATE: We are all struggling around the same cusp of earthly existence. He should be regularly here.

MATHEWS: You may help to convince some of your friends of the mission I've proposed. Jules Kanter and his wife, Ellen, are very generous people, hopefully among many.... But there are sure to be others, and they can't all work around the church, if they're not of a faith. One has to be approved, by all of us eventually, for their chores. The drought is a temporary strife, to burden our reserves of heart.... Is you daughter here too?

MARGATHRATE: She's— with a friend.

MATHEWS: Oh. I was afraid she might be playing outside. Children can overdo themselves in any weather, if given a chance. Then I appreciate your growing interest in our church, Mrs. Mason, and I hope you and your husband can continue to demonstrate this reasonably. (*going back over to the cabinet, to close its door*) You certainly have rights of access to the community parlor at any time, though I know that you are personally timid to promote the gratuity.

MARGATHRATE: I?... am timid?

MATHEWS (*closing the cabinet door*): Paths tells me of your.... general reticence and shyness, and how this is contrasted remarkably with a beaming glow during your labors here, simple and seemly. There is a self-satisfaction to them, that I share with you— in my work. For otherwise, I would not be much of a conversationalist. But what is felt is known by all who matter.

MARGATHRATE (*as he turns to face her*): I am not so timid, father, with a good man.... and a fair god, and a decent judgment of myself, as much to be less satisfied— until I am worked.... and worked up, to do for myself better. But my husband speaks for me, much in public. His erudition is greater than mine.... I show for myself by what I may handle, and am allowed to. And the tasks around here suit some blessing sent. They are not so simple— to me.... but definitely offered me as mine. And I take them, with a greater will than merely to observe while doing, or while being gratified.... through them.

MATHEWS: That comes from the great codex read of the volunteer, that others simply can not interpret, to be gratified, of the experience. And I am pleased that you are free to feel gratified among us. It gives you much ability to influence my considerations substantially, as I try to imply to all who diligently help us out. Not much wrong thought can be within your gratifications in the church. Though to be misled in anything is to lead your impetuses wrongly, and willingly, and anywhere. There's always that caution to our prejudices to self-examine. But I am satisfied that you find Bob to be more of a man than a bum, as I am more of a man than any angel could be with my faults.... and yours. We are born to be terrific, Mrs. Mason, by what is permitted in this place, and between these richly supplicating walls of solid oak and retributory angst. For here is where nowhere else can be, to celebrate and fear a good feeling. I am often frightened by my assumed powers. I hope you (are) not (as well), by your assumed toils.

MARGATHRATE: This turns a fright into a phantasm. You must carry a mountain-load of responsibilities to shift towards insecurities of the holy message so haphazardly, that we do not receive it thoroughly enough.... at times. But it's very plainly known, and that you need help with its constant delivery. For this we are aroused, and an endowment of my services is all that I could wish for joyously concerning our parish. The community will come through for you, I'm sure.... and ourselves, as it's in our nature to grant aid to a good teacher, like wind to a weather vane—

MATHEWS: Mrs. Mason!... There is a mystery about us. We are not the golden calf, and yet make precious our deviousnesses and deceptions. We are no gift to God—

MARGATHRATE: And I am not shy of blood, father. I want to

be here. Orchestrate us, but do not blow our horns as if you were the instrumentalist. I'm not afraid of a "good feeling." I'm not afraid of it being found here and.... sanction-able— and I don't fret over a possible deception over this. We collect ourselves, publicly and socially, to allow it. That is a service to ourselves; and you engage of this as much as we, to make some— holy order of the privilege. We will not defeat the rituals, father, by being faithful to ourselves. We will not be cheats. That is impossible, on sacred grounds and in religious temples. For what swirls around us (here) to perfume our senses is a majesty of being sensitive to our real natures in celebration of the thought that this too might be of a devoutness.... to well-being, for any soul that is depressed or hindered by moral or spiritual harassments. If the sorrow can be expressed here, then it may be purged with love, and the grime of the task glorified with (the) irradiation of the— privilege to do it, to partake of it and handle— maddeningly.... and with some observance, some license.... quietly approved. Could not it otherwise be so? that I was brought here by necessity and immediately felt a freedom.... for my work, a true release of pent-up tensions, confusions and uncertainties till told by a sensuality of sternness that pervades one, in such a place, not to remain embittered, but to accept the gift of my toils and smile with all happiness for what is permitted to occur through them, what may indeed be ordained to happen and facilitated— designed.... by a solemn grace waiting for my fulfillment.... waiting for all fulfillments.... human.

MATHEWS: ....There does stir a need, for a justification to all of our activities. And the spaciousness of our church, virtually a musical instrument in and of itself, reverberates this tone as we sing praises for our worthiness and fortune to be within and find security with the consoling thoughts of our faiths and religious tenets; since tenets are true only when they are carried out and demonstrated. And so, with a courage, you feel you.... will find the tramp and give of succor, graciously and gratifyingly, that our congregation will, on the whole, take up this mission I suggest?

MARGATHRATE: I feel assured of this, and more tempted than not to deny a need such as this, your proposal blatant, to early ears — and now my own. You make it proper, were there any doubt.... that exigencies exist, and that a longing for these actions builds up from our lacking of warmth and deficiencies of caring.

MATHEWS: Then I'm much relieved, Mrs. Mason. You may be a fairly accurate representative of how these concerns may be felt among us. And you and your husband can perhaps play principle roles in promoting the imperative— though it must be done without any one family taking up too much of a burden. An accumulation of assistance will make the crisis more and more temporary, as will God and the weather. Does he wish to do some volunteering also?

MARGATHRATE: My husband? He is too busy and involved with his work, father. He has given it to me, to patronize the church.

MATHEWS: In your hands will the lifting be made, the.... uplifting, then. And with the strength of being careful and considerate will you succeed in being fulfilled of service to the deity of human nature, neighborliness and generosity. You do.... have a method about you, as I perceive, that can.... lift stones and boulders of their graininess to reveal the purer crystalline forms on which we must all be based; though may they be surprisingly tender and damageable, at times.... But we will not be fooled, but of ourselves to flaw.

MARGATHRATE (*starting to exit*): That is true, Father Mathews. I make no excuses for graying. That is a will for being left to rampage through our torts. And as recompense.... a recuperation.

MATHEWS: Remain quietly desired, Mrs. Mason. That may be most loudly afferent to a remarkable genuineness to express, through your dealings and doles on the church.... (*as she is closing the door after her*) And try and keep the community.... (*door closes, it being too awkward of her motion to wait for a continuance of speech*) as to the community, a crescent made.... for full approach, of matters reigning.... per a view. (*goes to the table and touches the red folder, lifts it, and lowers back to rest*)

Scene III — *It is near dusk, and **Vogel** stands a distance before the church, as **Paths** come up to him.*

PATHS (*approaching*): Evenin', Reg.

VOGEL: Ain't goin' in, Paths.

PATHS: Not your day to, I don't think. Thought maybe for evening prayers.

VOGEL: Don't often do that, just before work.... Even if it is a couple of hours, but it still seems like day.

PATHS: Then why are you out here? (It's) Hot.

VOGEL: I slept so soundly, Paths. I thought I'd never wake up. A world was in my dreams, and everything was busy. It was a true consciousness, you know.... But I knew I was asleep.

PATHS: So you're not tired now. Why don't you wait in the—

VOGEL: I *am* tired, Paths, almost totally worn out with this living.

PATHS: The temperature.

VOGEL: There's nothing much left to it, no expectations or excitement. Thought I'd just come around here and demand a thunderstorm, and some rain.

PATHS: That ain't gonna happen tonight. There's almost no humidity to the air. And the clouds are too high, Reg. They can't even cast shadows.

VOGEL: I thought she might make a miracle, for me. If you believe in things strongly enough, forces happen. They erupt into existence and lead to energies of motion.... of some drops.

PATHS: I might say so. But you seem disturbed, Reg. And it ain't the heat.

VOGEL: That hasn't bothered me as much as its duration, persistence, deacon.... Though I saw— Mrs. Mason leaving, as I was coming up. Could have crossed right by her, but she didn't notice me.

PATHS: Mrs. Mason? Yes. She's right(ly) on her own now, quite rapidly, as a fellow volunteer. Don't have to get here early, to assist her much. Pahtes can handle it.... more and more. Well, she only met you once, I believe. Probably didn't recognize you that easily.

VOGEL: I doubt that. We had a fair conversation. May have been preoccupied with something.

PATHS: She's a fairly demure person, Reg, even after a fair conversation. You shouldn't be so insulted—

VOGEL: I? I'm not. Not me. I just want some rain.... or some sign of recompense for this weather. It's heavy, Paths, and lingering on too long. No, I'm not disturbed. I'm just being a little assertive about my demands. I dreamt I was better than this, than the conditions I find myself in.

PATHS: Well, loneliness can be—

VOGEL: *I'm* not lonely!

PATHS: —remedied by our church meetings, and social gatherings. You don't have to stand out here like this, questioning God and the world's climates.

VOGEL: I'm not questioning anything, deacon. I'm testing my worth, my standing, before the lady. I want something to occur, that others do too, and *should* happen. And if I seem like a protesting sentry, then so be it— and let the thunder bolts fly.... and buzz around my skull like a lightening knob. But you must admit, weee — little people (ha)'ve deserved better for a long time, now. And where is the rescue? Where is the relief? I was healthy, sleeping. Now I'm quite stultified, expecting a higher order of understanding to bring in some solutions.

PATHS: Humpf! It's only brought in scarecrows lately, the grief of pandering with fright. Look! You're a man of action, like myself. You know you have to poke and push to get the pig along or the mule started, to work your troubles through and have them done and over with, all the while lying in wait for more. But that's the drudgery we've bargained for, as people. You can't instigate miracles. When the earth gets farther from the sun, we'll cool off! Till then, don't commix your dreaming with reality— But let me help you. What is really bothering you, more than for anyone else, in your mind? We're all suffering through this heat, and complaining about it. But that doesn't put most on edge to being.... befuddled by it all.

VOGEL: I'm not on edge, (un)'less I'm sittin' on a precipice and gauging the fall of a jump. But everything is relaxed about me, Paths. I have no nerves to care about. I'm just considerin' the "why" of not being able to change things, and what is Atlas doing wrong. You must have wondered about misery, or being uncomfortable, the purpose for these feelings.

PATHS: Are you miserable? You sense those things so that you make strides to improve your lot!... in life, man. It!... is the poke and the push, the disagreeableness. Almost a gift, so that you know something is wrong, what you have to work on. Why, you can't be taught everything. That intuitive sense takes over, to make you aware of.... dangerous or injurious conditions.

VOGEL: I'm not miserable. I'm not confused. I'm not inquisitive. And I'm not stupid. I am demanding a change. We are due for one — And if it's possible, it should be done simply by my asking—

Because I've kept my promise to be a righteous person.

PATH: Well, we all have that aspiration, I would hope.... all who we associate with.

VOGEL: But no one seems to be listening, Paths. Or nothing.

PATHS (*starting to go towards the church*): I have to help set up for the evening service.

VOGEL: How come there's none over the weekends?!

PATHS: Because the Sabbath is the Sabbath, for our afternoon prayers. (*stopping to face*) And our Saturday is a holiday. Heaven knows our father needs one day of rest. You think you're a titan, who can just command the sky?... to do something for you— for us?! No! It doesn't work that way, Vogel. And He is always listening. We are never forsworn or forsaken, whether one keeps a faith or not. But.... you don't ask for those things, Vogel— whether you deserve them or not. Not Him. Not to Him. The prayers are only (for) cause to make *yourself* better, for whatever comes, good or bad. Do not blaspheme the hope. Do not denigrate the community, with such wishes.... to be so powerful. And.... do you think *I'm* stupid, Reg? Any of us— the father, even? Mrs. Mason is a shy, though young, woman. But she's alike of any creature troubled. And we all know what that community hut is for—

VOGEL: And what is that, Paths?!

PATHS: For a commiseration of past sins and current sorrows. The room's almost glorified for the use, legendary. We allow but so much, of the wing, to pacify the community, those who wish to serve the church— and find some sort of salvation.... from a curst nature felt. But take what few (souls were) from yesterday. You hate the way you drove your family away from you. Kanter can't stand living with the wife he loves, or wishes to stay devoted to. And Mrs. Mason—

VOGEL: Margo!

PATHS: ....She's perplexed, Reg. She doesn't know what to think about herself, as regards to a life of responsibilities she possibly loathes. We.... hope you can heal aways, in there, with help from the very presence and nearness of our church and ourselves— everyone! But do you really think you're the first to discover this, its use?! That is really dumb of you. It is his-to-ric! And do you think it's so private? Well.... God doesn't talk. God knows all the tricks of roguery, and disaffection, and blights of shame and deceit of talk!— And could you be this? Could you do this evil— where we let you play?... Well, well. You'll have to come to your own terms, won't you! You'll have to play your false tunes, full of bad tones.... and make corrections, or not. But if you're of a faith, if you like this religious gathering as something not so very abstract, and approve if its— chores...! I know her husband— a friend, acquaintance. They are having.... difficulties. She's becoming terribly frightened with herself.... and her daughter to care for willingly. I recommended her volunteering to him— Take action now! Let the problem resolve out of its own malaise, to settle for the true character of the person, whatever holiness is left to be found.... And how is this for you, Reginald Vogel? I swear there isn't talk about it, not by the church practitioners— There is silence! But what have you become to imagine?... and so childishly, Reg. We can only help each other out of droughts.

VOGEL: ....So now I'm the dastard—!

PATHS: You're a good man, highly regarded around town. That's why we sympathize for your self-turmoil, and ridiculously patient agonies of heart. Do you think we'd.... let her meet you.... with even the slightest chance that you'd actually fall down into an....?

VOGEL: I am the fool! Paths. The jackanapes!

PATHS: Do you think I'd even allow the hint of an adulterous affair develop between you two?! I'd whip you with the rosary (beads) first— And you'd stand for the punishment, Reg. You'd take the scars and welts, and wish for a hundred times more. You'd have that done to you, Reg, wish for it— pray.... for it. But God won't *talk* about it. He would hardly punish you worse than your own death, and turning into stone.

VOGEL: Banish me! Paths.... And take my keys—!

PATHS: No.... No, no. We trust, that you will be.... religious about this, maturing to a piety, returning to your own goodness. And the key...? It will leave its marks to burn your hands, and your chest and— wherever you pocket it. And you use it, Reg. You continue using that community room.... because you still need to. And our pipes need looking after (by you), until some other needy artisan comes along, with commensurate skills of amateur rank.

VOGEL: This reels me! Paths. I'm burning up inside!

PATHS: And so might the weather be. But that's no cause to be disheartened of your soul, your.... standing in the community. You should have come to all of this without me having to explain a thing to you, or make an exposé of the matter— And I'm sure you would have.... found some grace to hold you up with. But you darn-near defiled the lady, Reg, right in front of her. And you didn't mean to— So I bled! Your misery is so apparent.

VOGEL: There's a candor to face, in this heat.

PATHS: And really, Reg.... you're not that young.

VOGEL: I wouldn't say it, Paths. I couldn't prove a thing, now, about myself, not with a certainty— and not to be held despicably. But all I was truly seeking was a new friendship, not an illicit liaison. Just a familiarity with a fellow worker. Perhaps some little chauvinism had shown through, a petty thing, to hold for myself a decent bit of vanity, that I can still be found affectionately wanted, or admired or taken for to give counsel. I can't be sure what breathes today, from what I have imagined of myself. But I haven't been so awful. I insist on that! And is this how you'd treat those who regularly give their menial service to the church?! by having us like inmates in a ward of the.... dispensable?

PATHS: Like you, the dispensable? Your just dispensations are devised around your own interactions, and your healing through comparisons and complement of thought. Because what must be correct generally issues through.

VOGEL: Just compound us together, then. I'll tear up that room, now!

PATHS: Goodness begets the good. Is it so unusual to want to help those who simply feel overwhelmed, by the circumstances or peculiarities of their lives? Help them.... help themselves, in company, in like company.

VOGEL: I'll destroy it! I'll destroy everything in it!—

PATHS: No! You like it. You like the community.

VOGEL: ....Don't preach to me—!

PATHS: You still have need of the thought, that it is a decent location for you, and that you can be made comfortable thinking in it, analyzing your fears and arranging your.... directions for more to do about your life. Therein is a mysterious place, I will confess. But that's because it holds your essence, and the community's, that part of myself, even, that I'd rather drop, at times. There one can be more subjective, than in a church or chapel.

VOGEL: Deacon.... Don't try to preach to me—

PATHS: The greatest doctors cure themselves and their societies, and with their societies are they cured.... together, one and the same. You love the room, and will until you no longer have need of it, when your ailments are solved.... and your solutions are— healthy. That could be as simple as giving up the pain.

VOGEL: Could you possibly suggest—

PATHS: At least in there you do not hurt yourself.

VOGEL: —that I should leave the community, that Margo leave her husband and daughter, and Kanter his wife? That sounds ridiculous, Paths. Absurd! Kanter would be lost— We would all be lost!

PATHS: No. Go back into the community, and return to spouses. Mature out of the community room.

VOGEL: I— I can't. I can't! They've left me!—

PATHS: It is a place of salvage, the back room parlor for thoughts not respectful enough for the rest of the church. Angelic cherubs are nearby, and listen for the sounds almost with game to hear what is relieved in there, and how ultimately you must judge yourselves.

VOGEL: Despotic gaming!

PATHS: The key holders own that.... salacious chamber. Where else to send you to recover and relax after you've felt the torturing warmth and wealth of a labored, laborious beatitude?

VOGEL: It's a purgatory?

PATHS: There's where the crudity of man may dwell awhile, healing from the religious constrictions and medicinal inflictions of your church, or say a characterizing of effects after treatments for the tempering of your soul. For, do you feel an evil while working on the church's structures? I'd say no, if being conscientious about it, since they are more likely working on you. And when.... the healing is completely done, and even the good pain has vanished from your limbs and mind, well then.... you no longer need the room for a recovery. You have recuperated from the transfor-

mation. You have been blessed through your own Thanksgiving, washed and soaked of a discomfiture with providence, the scales and barnacles of distressfulness scraped off your bottom, legs and feet. Then you may leave the room for good, the outside no longer staying so scorchingly hot in the day for you. You are tempered to the environment of man— again, profitably adjusted and acclimated to a humanity of pious beliefs and drives.

VOGEL: And the space can be reserved for more patients.

PATHS: ....That's essentially how I see it, Reg. No god can order you to behave, or feel as worthy for your life as with others to enjoy. Life is on its own, that independence (being) His gift, and it must make its improvements itself. Do it yourself, and get it done. Do it amongst yourselves. Get to it! Man has religions for guidance, the lower animals less(, as for) complicated societal pacts down to the solitary gratifications of the microbe, and plants merely the bounteousness of fruiting and flourishing. I doubt if God cares about any particular strategy, as long as all obey the commands of life and death. But if *you* are to have reverence for His desires, as adjudged by our faiths.... and community standards, then you must strive to be good—

VOGEL: I *am* good. I am.... *good*!

PATHS: And for that a holding pen, of some adult sophistication, might be necessary.... until you are convinced of achieving that state, so near to God you are. Dare not you try to terrify even Him, with your actions near our common pews and lighted trestles of arching wooden beams. Where there is handsomeness to a splendrous piety there is observance of a kind, to remain holy.

VOGEL: Forgive me, Paths, if I remain cynical of the procedure, until I am.... fixed— by the process. It seems this world spins me, today, to stay within a vortex of conditioning to garner any conclusions. What seems to be does seem not thorough, even fickle becoming the perceptions here.

PATHS (*going off towards the church*): That may suit you, standing there. You haven't worked on our pipes today.

VOGEL (*as **Paths** recedes*): And I pay for this?! Yes! (*more to himself*) Ferret out the Pharisees! More justice to (hi)'m, as for puppetry to perform— with miracles displaying, arrayed like ribbons on a wall. This convert here, thus turned through there. And on and on, the bending, twisting, twining.... laving through the shimmering textures and colors of the cloth. And with pontificating! of our moral resources— held!— Bleached! I am so scrutinized, and shown upon, an experiment to watch— and of my thoughts, the very things of privacy, the matter of our mental substance— reviled and reviewed!... All have known this, who would matter for by burn? Then that is a holy decadence inscrutable. Certainly not for evil, but for what?— Fun?! Devil pricks the license, to allow a psychology, a psychoanalyzing in the church, for want of doctrine for the theologies of our practices. I say enough revealed, that all have seen me. But I thought with suddenness— and shading, trimming of the disreputable mergers, in a mawkish manner of sentiment to our faiths— and beliefs in the sensational— obtainable! But is this so bad? What do I want?! but some little proof, some supple surety to be rained with, and arraigned by my poor, contorted conjecturing.... on life. All (of) this to have been seen, as through a lens to a peepshow! what I thought was so subtly dispersed, the fears and disquiets of our emotional states, now portrayed like cardboard cutouts, spoiling in a dampness of rotting rustication, the moistures of our sweats— to have been found out, and sent to Coventry, to squirm within our self-made ostracisms, or induced squalors of the soul. But what have ye learned? O pontiffs of the proletariat— with your magnifying glasses. How to build hotels? and how to rein in marriages, purify their fortitudes and distill out my— ooze. Was I so angry of a dream, that you could challenge me this way? You! You! Observers, of a falsity, clamorers for a cant!: I – am – be- fit – ted – here! I – am – be – fit – ted – deer! Is not the lustful bound to fail for lack of surfeit found or sustainable? Is not life made to extinguish itself by all or any manner of aberrant disposing and proud debauchery, or flaunted chicanery? And did I desire, really, so much baseness to my creed? perched on madness, lady! And now climbing down into the pit to bathe of mud and tars, and all that scrub the serpent's scales. Is this so deep a predicament?... to know one's self as known— I would have only flirted. Yes! I am tired, and dismayed.... of the lingering fire, the lasting heat, to prove what of myself 's still worthy— of some worth.... to burn. A privilege of fear, I guess. Must have shown on my hands, my staunchest character for handling of this.... distempered environment— Has she been informed of me? of my cowardice.... to face what's real? Or is this only supposition still? in her mind, by my description, from others. I must excuse myself, more bluntly, the next time we meet. It's not a death in me to party with, nor a toxin to be intoxicated by, or a poison to be poised about. There's not a commandment broken across my back. It is simply the levity of hardship— and my room! My.... closeting, to contain.... the augury, the auger and the anvil and the axe, to bore (holes for) my treatise of a life, hammer it out and apportion it to others like a hardened taffy to be licked— and ate. And yet sincerely am I dispossessed of shame, that I must continually test my aims and goals for not to be disastrous.... of faith. The method's not important— No! It's not! It only matters that.... I still care. I still care for others, and will make for them some applications of difficulty that will lead to a trite fulfillment in the least.... and for a spell, of happiness and pleasure sent to us both, rationalized that way as.... not to be called an ignominy, not hurting others— not hurting anyone or any thing or notion or concept or avowal made in haste of growing up.... as suitable for being brave of God, or the sun, or any power that must overwhelm our lives and give to this a justice struggling, our bodies dangling and twitching before the energetic rays, heated to a crisp, yet coldly observed.... And the room's still mine, what faints my standing here. Myself, a chamber made, for storage of this discontentedness and patient strife, I will commit to more, within this heat to bear, for crying tears internally, till temptation submits me to the rain outside, the forgiving and the wash— and the approval of the heat that has been made! Caused! Delivered and detained to!... And felt, as were a purity to feel, the contrast between a decorous disaster and a divine disdain. Since I am wholesome of my blood, and harried for its luster— drying, spent!... upon a twirl of holy lace, and grace of bodice come to ply with. Oh, the lady, and to her back door sent— The shame of it! And yet it's made that way, the practice, and the articles of procedure for one's apparent beatification— here, if this be heaven, or more to expect, and we be dead already, in some surrealistic jest to say what's foul is fathomable for holy grins.... as the creator, and the slayer, looks on majestically contemptuous but ever satisfied of deed. So let's have pouring! (*with arm gesturing*) Pour! Pour! Pour! Expound on this!— I'm going home. (*arms down*) I'm going to find— a knife.... to bring to the room, as if it hadn't enough tools and sharp edges already. But as a symbolic gesture, I'll do this, to have myself bleed in there to leach the poisons out, and the tarnishing meanness, and the smoking candor that wishes to es-

cape.... All symbolically— permitted.... for your approved debasing of my wealth, in mankind. I will associate with ghosts, then, and shadows leaning towards the dawn.... as has been my custom anyway, as I search for thee, controlling augusts and observers of my malefic kinship. Born in this mire (*walking off*), I won't deceive you!... I won't deceive you!

Act IV

Scene I — *An encampment of indigents at night. There are several small groupings of individuals, one to three or four in number of membership, based on the most tenuous of fleeting or casual association. Some groups have low burning fires, for cooking whatever horrible rubbish may be found to eat after being in someway sanitized from pestilence by the flames. Slightly seared grass has become a delicacy during the drought, since most blades off the ground have already dried up and died. But some seeds, in general, have a taste, and so various mixtures of vegetable and insect matter, and very rarely animal— often poached pests— make up highly personalized gruels. The camp is roughly divided into activists who try their best to steal, and "sedatists" who scrounge around and beg and have generally aged out of the former half from necessity due to growing debilitations or stifling injuries and illnesses.* **Harold**, *who sits alone by a small fire that seems purely decorative, with nothing cooking in the hay to speak of and no indentations in the ground to act as pottery, is a curious exception or crossover, being relatively fit and relatively young, but preferring to work the "dole mines," as the activity is called. But he is spiritually enfeebled, and misses the character backing and support of* **Lorken**, *who would not steal on principle unless he had determined what was taken could not possibly be missed. The general area these men inhabit, mostly, can be considered more scrappy than clean, but not the garbage dump amidst a township of urban buildings; so the general openness of the location grants it a peculiar nobility of lowliness.* **Harold** *is quietly, or tacitly, demoralized, sulky and sullen as another* **indigent** *comes to sit down across from him, obviously not for the heat since it is a warm night. Taking passing notice of the company,* **Harold** *greets with a slightly resentful yawning grunt. But fighting is rare, in such an assembly.*

INDIGENT (*after a disconcerting silence, looking at the fire but only indirectly towards* **Harold**): Is there anything (in) here?

HAROLD: Nope.

INDIGENT: I've been tryin'.... tah find some scraps around.... Haven't ate today. No grass left. All burned up.

HARLOD: Try a tree leaf.

INDIGENT: They'll kill yah.

HAROLD: I said, try *one*. Not enough (pine) resin in one ta kill ya. But it might fill your stomach for a few hours.

INGIGENT: Can't do that. Can't afford tah get sick now. I'd drop cold in a second, right down tah the ground, an' never wake up again.

HAROLD: Your constitution is your pity.

INDIGENT: Can I search fer what's left?

HAROLD: Go burn your hands in dem embers. Your fingers wil' pop! I ain't cooked nothin' there.

INDIGENT: Why make the hay stove?

HAROLD: It ain't a stove. It's some light, to look into, study.

INDIGENT: Wastin' matches—!

HAROLD: Ya can get plenty, from the hotel. Whole boxfuls are left outside sometimes, in the back.... near the trash.

INDIGENT: I ain't goin' there no more. Heard there was a killin'.

HAROLD: Killing?!

INDIGENT: Someone.... beat tah death,

HAROLD: Fantasy. Wild rumor.

INDIGENT: ....fer tryin' tah sneak inside, from the back.... to roust the patrons fer some food, an' change.

HAROLD: Ya's crazy! I stay away from there, but.... Did that happen in this decade?

INDIGENT: Occurred tonight.

HAROLD: ....So the story goes. Fibbin' for fish fins.

INDIGENT: Ain't no pools around here now. I was headed out there, this evening, very desperate fer somethin' tah sustain me somehow, when I was met by.... a guy, comin' from that direction, who took a look at me an' told me not tah go. Said his friend had been givin' the whack of ages.... an' covered with a sheet. A fair disturbance made, an' an officer came out.... Or they took the body tah one. I'm not sure.

HAROLD: You could distort everything about the tale. You don't have the story straight, but from some coward's fear and suggestion.... (I've) Heard nothin' about that.

INDIGENT: All I have is an empty stomach, an' a faint head. I'm hungry, goin' on starvin'. I think that water from a pet's dish I drank yesterday (ha)'s done somethin' tah me. It were dirty water, but I lapped it up like a dog, 'cause it was there, on a back porch, an' a thirst brought me tah do the thin'. But now, one of ma arms hurts too easily. And I can't steal well anymore, don't have enough endurance fer a good run (away).

HAROLD: You'll get over the cramps.

INDIGENT: ....Won't let me sit with the other thieves here, 'cause I can't share nothin' anymore. Don't have a ticket fer their com.... panionship.

HAROLD: So.... ya come by me, and the other beggars.

INDIGENT: Got nothin' left tah try.

HAROLD: I know the watchman to that hotel personally. He's tough, but he ain't uncivilized.... That wouldn't happen, what you say. I'm uncivilized, but....

INDIGENT: What you've been eatin' then, if you ain't cooked nothin'? Yah's don't look starved.

HAROLD: I grope about, an' investigate my surroundings before I ask for handouts. You have ta know how ta be successful ta be successful. An' ya have ta plead with some sincerity an' learn how ta survive on nothin' for a long time.

INDIGENT: Is it true pine cones burned just right can be made tasty?

HAROLD: They got seeds to 'em.... Sparse, now. Those trees are mostly gone, till ya get upland and further of the altitudes. That's where we should go, some say. But fewer people live there. An' we are dependent on the generosities of others.

INDIGENT: I don't know how tah beg well, yet. I'm becoming one of youse sickened poor, all of my pride drainin' out of me. At least once I could be functional, but now I'm injured.

HAROLD: Afflicted, probably with worms. You ain't poor at all. Ya have ta have some dignity left ta be poor or better. Ya have ta have a longing ta improve ya condition considerably. We've given that up, till the weather improves also.... So don't think in terms of being poor. That's another caste, a realm or state of mind. We're all.... indulgent of the earth, and its more proper, prosperous inhabitants. That's the freedom we choose ta be buried with. An' if we were.... part of another country, we'd all be nomads, in a way, an' daring for our suffering with.... more honorable hardships. But that can't be.... where we are— Stay away from that hotel! It is.... laudatory, for our despite, an' is a fool's treasure ta search for.... I have abandoned its history— An' I know what I'm doing, an' how ta be.... deficient an' supplicating. This spell of heat won't last too much longer. An' the town's folk will be happy ta throw us some meaty, marrow-rich soup bones afterwards, knowing what we've been through— That's why we stay nearby. That's why we haven't died yet, or died out. Till then, learn the trade of temperaments, an' judging people. An' do this quickly, if you're hurting. 'Cause those pains require a cynicism in yourself (in order) to heal, that you are not all that bad off, given these conditions. An' like a soldier struggles through war, so should you— even if wounded.... But you ain't even bleeding. Ya just have some tense muscles an' a growling pot (of belly).

INDIGENT: Any little bit of stiffened turd will hold me over.

HAROLD: Cook that too. Resilience is the only god we've got.

INDIGENT: When I must beg from beggars, then my days have come to stone. (*starting to get up*) I will tease myself into a corner, of nothin', tah lie in some richness of the desolation—

HAROLD: Wait here! (*reaching in a front pocket*) I have a bit of something ya might try.

INDIGENT (*sitting back down*): What's that? What is it? (**Harold** *hands over.*) What's this?... A piece of cookie?

HAROLD: Y*aaah*.

INDIGENT (*almost nervously*): Thank you! (*gobbles sparingly*) This does moisten a dry tongue.

HAROLD: Only been a day, or two.

INDIGENT: It's the fear, the fear that I would never be able tah, again.... But I have tah learn how tah— ask. I've fallen intah the need, an' it looked bottomless.

HAROLD: It's only darkened with uncertainty.... I used ta bind with a.... good thief. He taught me some, an' I taught him. But the poor beggar turned on me, an' insulted me.... An' then went off ta die. I can't take the offense ta friendship well. I have ta find out why a repulsiveness, of all things, led him ta despise *me*. I've always been.... the same throughout our wandering. And I thought he understood our shared characteristics, an' could not be so humiliated— by me!... to want ta give everything up. I've been pondering with the fire, looking in, what burns, what should be stoked, what am I viewed as, by a comrade, that should stay warm an' gracious for a company. An' it does tear me down a bit, ta bring disheartening to fore. Bear forth to know, it puzzles me much, if I have changed or others do modify their bearings on me.... Ta much. Ta much.

INDIGENT: The kindness warms more than what is seen. But how can it be, for a good fellow's daring?

HAROLD: Oh, he'd only take what he presumed was abandoned, and did find of me as a partner hooked for learning.... honest ways in a dissembling world.

INDIGENT: Yet tah die.... it's so cruel.

HAROLD: He ain't of it for pardoning, though I'm certain that he tried. Hardened are we, for the doles mining. It's a wondrous art, like the sprinkling of cherry-red charity upon the head of humanity's threshold of depravity, if not with lack of privilege to be catered to— We work at this! and it is a difficult skill. But if I could fathom more ta become— I would, or try or aspire towards some of the sundry goals more highly praised, or even worshiped. The beggar is a bugaboo, of hardships bearing, hardships baring, drawing forth for view.... and reminding all of what is possible.... in man. Sheaths of callousness to face, we are contained, somehow, with criminalities on boil.

INDIGENT: For theft?

HAROLD: ....I judge what is allotted each, to serve a life, and take not more in leaving some to a prosperity earned.

INDIGENT: Fer death?

HAROLD: For what my own must needs devour, and tackle with the gramercies. We can hardly be considered glutting on those scraping by.... Theft is a crime, my friend, ta be ashamed of. It is something told against you, and held up.... for ya chastisements that may be mentionable and celebrated.

INDIGENT: Robbin Hood was a thief. Bandits abound through history. We are clever, and may even be fashionable in some eras. Yah is due some needs, an' some reliefs from them.

HAROLD: I haven't much of a mind, ta steal what has been hoarded by others. But the brain is weak muscle within these plights, not ta condemn your practice. It all revolves around our requirements ta survive. But I'd rather plead, ta show I'm beggarly an' waste-abled for the dispensable. And especially ta children, ta see the dispensations in ma face, I do 'em a service, present a fright or warning.... not to become as like. Those capable of theft develop means for being greedy, and I am anything but.

INDIGENT: That's certainly so.

HAROLD: Yet.... steal I sympathies, then I am greedy for a trust, and ta make the most of opportunity's spills, with tumbling of goods an' presents— I should be the one, ta die.... Can't bring myself to the need.

INDIGENT: ....I'm just waiting fer the chill, that comes with early morning. It's such a faint thing, that feeling, before the sun shines. But it cools me fer the cooking, tah withstand the day.... I wouldn't need the disgrace. That's why.... I've stolen.

HAROLD: There's a good paint to (i)'t, a rich lacquer ta hide under. Not a many could recognize me, from much other, similar an' alike. Could that be the common baseness we come ta tire of? Too obvious a conjecture— We are baked.... for our acceptances with one another, an' for what's ta be expected from us.... or each or any, or a particular.... being. He couldn't have felt stultified by that. A friendship is too strong an' precious ta be broken under commonalities.

INDIGENT: The thought of it's more powerful than the action, this giving. It swallows fer me, an' feeds. A token's all I'm worth, right now. But it is bountiful, and presides over my faith.... in mankind. But how did he dare yah, tah die?

HAROLD: With a walking away, my friend. A silent walking away, that was all ta be told of it. Was confirmatory, of his intentions. He wouldn't eat, and not with me. And so he left. An' I heard rumors later, just as you did, which let me believe the distasteful had occurred. And I spent so much time trying to extenuate the cause that it split me up into pieces, emotionally. The more worn parts wormed their way back ta a consistency of thought, an' helped lead the others home.

INDIGENT: You came apart, with sorrow, and then found yourself again, still a beggar.

HAROLD: Yes. An' he.... still a friend. But the rumors were not true, you see. You can't believe these things— An' ya shouldn't! till you've tasted the pie!... till you know what's made isn't fake!... This really turns me about, an' maddens like a demon poking— ta find out, or find me out, describe. Because.... life should not be so subtly disobeyed of a marriage. But I'm a poor crust baked ta bite inta. An' are ya still hungry? 'Cause this night grows hallow.

INDIGENT: Some hours (are) left yet. I could sleep awhile, but fer the lingering heat.... Though that cookie has settled me admirably; the panic has diminished, like from a soothing drug.

HAROLD: But do ya want more— heart-beating pangs driven through?!

INDIGENT: What stirs? The fire weakens, like it was never needed at all.

HAROLD: I know of a place— where we may steal! an' from which more food to obtain, treat as much as solved your dieting.

INDIGENT: I'm still not strong enough for much trouble tah handle. Any struggle could fell me easily. An' a goal is not tah be caught, not even tah try if yah could be. The thief is more a scavenger than a purser.

HAROLD: This place is not.... attended well, an' becomes for our tentacles to employ, since it's as open as the community itself to be used. We would not be breaching faiths, nor transgressing anywhere our permissions to steal.

INDIGENT: Permission? Where is heaven?

HAROLD: To where my friend lies.

INDIGENT: ....This sounds grave to me, if yah'd try to circumvent the meaning of a death, your demise as well distortedly erased.... Or was he buried with a picnic basket?—

HAROLD: He's at a church, singing hymnals in 'is dreams, probably. An' I know how ta get in— an' get at!... some delicacies stored there.

INDIGENT: ....It would be closed an' locked. An' breaking windows— I've never done at me own fer a religious site to ravage. Have never thought tah. Yah hold some proscriptions tightly, fer just the chance they may be real, when said that way accustoming our fears.

HAROLD: It will not be a break-in, but a welcoming. They want us there, for a visiting. She does—

INDIGENT: She? Your friend is not an angel now?!

HAROLD: The church.... offers this room, a chamber for the entire community to placate themselves with an' plunder harmlessly, or ritualistically. She gives, (to) whomever may.... want to, a free scope of range an' rampage, way past her consideration of you. Yet, the space is kept relatively decent an' orderly, because there doesn't seem to be much need for a disturbance to leave. It's fairly neat. I've seen nicer, in the poorest family houses, where households make a room.... a desperate place, of some importance to the dwellers. But people bring to this back chamber anything they want— And one has left us food.... for our taking, and our need. A sort of maiming savior he is, you might assert through your rumors. We may confront this.... schism at the church, and confound envies: ours, his, an' the world's. (It) Would not be like stealing to me.... but it would for you. We can both derive some satisfaction, since I have interests there.

INDIGENT: She's clearly closed by now. We'd be breaking in. The danger is too.... sacrilegious to attempt, if we were caught. Could mean the absolute scourge of all of us, that the residents of this locale determine the conditions for us are so desperate, to lead to despicable acts, that we should be rounded up and destroyed—

HAROLD: Awh!

INDIGENT: — or disposed of in some way, effectively. Transported out of the county, to starve most definitely, or truly become

meaningless under unimaginable punishments. For people take their religions ultra-seriously, and with actinic purpose to cause great harm and hardship, fer the blasphemer or heretic or apostate that tries to do for their justice a misfortune, who would actually transgress their monumental properties tah violate. I'd be nervous about such prospects. Talk is easily done here, tah imagine wha' is possible. But I seldom try tah inflict myself on the severely devout — They are devious to impinge upon their shames medals of pain pinned ontah the flesh of offenders. There is no reason tah it. It is above thought or analysis. An' I've even heard the rumor that members of one religious sect had tarred an' feathered the strewn limbs of one poor devil, thoroughly torn an' hacked apart, for merely mentioning he'd burn down a place of their worship and was (mis)'construed to have been caught during the initial stages of the attempt. All because they wouldn't share with 'im some bread or feast or festival provisions, because he was unworthily smeared of contemptuous commentary. People are more than jealous and envious for their churches, temples an' tabernacles, because these structures represent fer them a stake in the sand, a demarcation of proof for their honorable existence on this earth— and therefore never to be defiled in any way permissible of a vengeance. For I've learned of retribution with my mouth scoured of soap, as a child, fer (my) cursing of some abuse tah me— ma person! They are wild!... But , I be wilder, an' more aloof, tahday.... an' dare not challenge such strong hatreds or intolerance tah shamin' candors.... It's not worth the condescension, intah a pit of stonin'.

HAROLD: Ya must be fantasizin', about thus wood as there, the timbers fitted an' constructed for our ingratiations, with their holy scents and infusing of richly pious airs— I'll let you in.... ta have you take things out. An' you will be the better of an ownership— without deceptions. Neediness is not deceptive. It is determinate of our actions.... But it is left for us ta do, an' try for— gifts baiting our approach and approval.... I'll not bring back here as much, if I go at it alone. 'Cause, by fair chance, it'll be there an' waitin'— with more comin', for my stalwart apprehension— ours.... as we may do.

INDIGENT: You'll not share with us, though I am dying?

HAROLD: I've given you, but not as much for more if you can not do more with me.

INDIGENT: I am afeared.... of the heavenly distastes an' displeasures of indecencies committed.... at a sanctuary fer the angels.

HAROLD: The murderous Macbeth, a winning soldier, was afraid of killing— kings! Then it is not the killing that is villainous, but the fearing.

INDIGENT: Maccabees?

HAROLD: No Judas, to our attending of the room, ta make be felt a visiting, our shrine of invitation placed ta be.... be— ourselves. Accompany me, if you have hunger, if you wish to remain a friend, if you have found me wholesome for you, as with another (be) dead! ta give me license of some pity!... and a recommendation of this place. I hate the loneliness of my tasks. Or do you want to live further, ta greet the day with a sufficiency of stomach! There are treasures, around the church.

INDIGENT: There are treasures fer much of any crime.

HAROLD (*standing*): You will not follow me?!... You will not dine more this night. You will sleep aggrieved, an' duly wronged.... Then prune the fire for ya thoughts. (*leaving*) There's nothing left but ashes from a dusting of the hay.

INDIGENT (*to himself, while watching **Harold** leave*): And I am thief.... And I am chattel, crying fer what is right.... tah be owned by. This greed fer life? that is so discourteous, uncomfortable, humiliating, draining of the spirit, an' desultory, from rock tah rock's bane tah seek fer.... any sign of hope, any cause or chance tah claim it, an' tah my bosom warm with— clasping tah have been found.... Fer.... here are many, angling fer a partnership, tah bring tah higher station one's involvement (*looking at the dying fire*).... of the civilizing principle, that I was born alive tah partake of wealth— with others.... fortunes fruitful, an' amended fer several caring of.... An' happy tah be found, the circumstances similar, but I must steal.... an' scrounge, or die.... die brokenhearted of a cold, a chill within a baking drought, a heat of shame an' shamming of a warmth, that does me not much favor tah exist. With all of this around.... no nerve tah steal, but take this rubric of hunger an' pride tah serve fer sleep's direction or instruction amongst some certain bounty of the unusual. It's not fer talent that we live, tah employ or detect, but merely tah test ourselves with.... till we— die. An' thus is it, the indigent tah hear, why you are so enabled tah steal, an' groan with profit from the pilfering, this air alone an' simple sights an' tiny bits of wisdom flickering: This is a shame, this is a holy drought— a drowning within vapors imperial, if one be the only emperor of 'is body, an' 'is thoughts. I am afraid.... of the religious aspects to this. I have been beaten an' conditioned to this.... brokering of sincerities to the possible that lights, that gives the light.... an' the heat an' warmth, an' is familiar tah be both feared an' adored, with a radiance warning.... fer our humble demises an' elopements to meetings generative.... of our responsibilities reviewed or argued over, an' decried. Fer as a youth does fear the beating bogy, bearing gifts or a harsh stick to strike with— an' strike down till done of mischief.... I do know that we are born, all.... misfitting, and to be wrong an' wronged, till eternity solves that frightful matter.... of our trivial criminalities. An' fer this I shake of generosity. It is so much to carry, burdensome to accept.... an' pass along tah others, the other.... wedded with. A strange sense is this humanity betokened. I would.... dare mouth this fire, tah have blisters bleeding, of the human cry: I am a thief of man! I am.... the robber robbed, tonight. An' fer this toll of sight, (he) dies worrying. Who makes a spectacle on earth, to become so distempering an' evil, tah displace one's faith in thee? Then now, the shadows make a night. It is the same as day, less light. An' I am one alike, tah sweat of ma pangs an' needs.... an' followings. Fer poor is faith, without more feeling tah be deserved.... Or would I be so base, under a rock to lie, an' be found, hope in spewing rancors.... an' my renditions of privations exercised as red flame, reddened to be tamed fer your pronouncements, serpent-like an' dragon.... ennobled, or reduced tahhh.... Makes justice fer a death, within these cruelties.... of our natures.... An' who was he, then.... that lent me a bit of sustenance? A caretaker of hope?... or harassment?!... I would wonder, fer my piety aflame. Suspicious be he, fer a savior. But I would see stars twirl, within these cravings last, duration of lost hopes with dizziness in desperation.... an' a cause tah find some help, be it fire tah entail, within 'is eyes of interest.... fer me. How does the night dance yet?! It is a spectacle of the somnolence an' tranquility of thieves an' beggars, more home-like than harangued, out here, our piddling forays throughout the countryside, stalking.... this common blight, and finding ready-made.... our fortunes, within rest an' sleep. (*lowers his head deeply*).... Care

fer.... the indistinguishable of sight. (*The small fire by him should be out.*)

Scene II — *A small kitchen enclave within the church. It yields to a very tiny window neighboring the community room. With a relatively weak ceiling light on, though bright enough for the enclosure, **Lorken** sits at a thin, portable picnic table, drinking some tea. The appliances (stove, lunch box or refrigerator) seem undersized, and the room gives the makeshift appearance of not quite being up to building codes or current standards. Only the sink, with its plumbing, and the counter top look normal. It is late at night.*

LORKEN (*to himself, after sipping, and with a haggard countenance*): The pantry here is pitiful. I'm distressed of the astonishment to find so little food stored for the priest. There really isn't enough to make a decent supper, if you remove the concept of a soup. And I've been mooching off this poor man, all the time long. I had imagined some sort of riches, at least for meals to prepare, with his position. But he must be on the brink of deprivation (himself), and with dieting. I was even afraid to prepare this cup of tea, shocked to find only four paltry bags to brew with, now sans one, in the covert's tin can. They must not be used often. They are precious, along with the sugar and the bit of cream. And I've wasted it on myself for some hot thirst, on a warm night, a summer's blessing to our wounds. But he gave me rights to take what I wanted, whatever I could find, to satisfy my health needs. And now I feel so guilty about this.... generosity the father provides, so freely, and without even a hint of protest or discouragement or tenseness, terse apprehension. That I might have, from the little held, is all he cares about, concerning this for me. I could not dream that a church should lie in poverty, and within this fine community. I've dreamt of everything else, but could not anticipate that. It must be due to a bad management of finances; though the building itself remains somewhat magnificent of structure, if old or aging well.... (*sips*) Sell that organ, for more cash. But then she would be left without, without a sound from the rich keys played. No. An' I've become to turning pages for a boy. But he was so engaged with the effort, and tried so hard to do his best. It was, in a way, inspiring, to see what I have lost, of the youthful and unknowing character I once held to for my attempts at life. Dismay hews, with gray perceptions and dismal revelations. Yet here is the worst to acknowledge thus far. I don't deserve even *this*, taking from a mouth that offers.... sacraments. If this kitchen had (more) bounty, I could take as much as I would like, and fatten up awhile, and deserve the repast. But I was led here, brought here.... to show me what? about myself in need. That this is all disingenuous of a life? That we must clothe our lacks in greatness? I was to die. Should I be grateful now, to have been blinded of the realities until they infringe painfully on others? I must not.... be as sorry as I feel. All of this presses me to be forgiving of my hardships and doubts. But then to do what? but note, that even the good seethe through hidden afflictions and may not be treated well. What pretenses this father must shoulder! An' he has showed me, has invited me to see, that we are poor, but not unprivileged, in this church, this Epistle wrought to life an' lived through for describing the tribulations tendered from one's sincerity to.... sacrifices, self-sacrifices.... Denials, even, gravitating towards the unreasonable to happen, miracles an' mysteries of wondrous pardon and relief from.... bitterness. And I must make it my business to leave, tomorrow, an' contribute with my absence to the well-being of this holy establishment. For this is being fed with mimicry of hope alone with my weight an' appetite assailing an' appended to.... I still feel very weak, though. Something is wrong about me. (*sips*) I want to be much stronger, an' should be by now, after having eaten a breakfast and a dinner. It's been ages since that feat was accomplished— I should be racing now, to find some sorting of the energies.... But the feebleness is real. And I had hoped some warm drink would bring quiescence to my unnerving and shaking only internally sensed. So I am bold enough to enter, with some escape from the bedroom, and find this place utterly humble an' humbling, unexpected and disfranchising thoughts of grandeur owned throughout the noble edifice. She must often weep with terseness, of this deeply ingrained wood and lumber, for some brevity from the succulence of view presented for an overawing nourishment of solemn importance and splendors spiritual and up-lifting— persuadable— convincing!... with this effeteness housed. Not hidden nor shunned, but kept for true to be, the conditions of our states. I will not take more from madness. I thought I had ex-punged it totally, until I awoke within a church, within an irony now paradoxical of care, if not suggesting for an overreach towards my hypocrisies from being found. Death.... would have been vo-luptuously timed, an' I was ready then— Kanter (*sips*).... says to house me, and off me from this further burden of Mathews'. I would rather run away, if well. For I can't control my actions.... tho-rough-ly. Not that I would purposely do—.... From sanity's chores, I am contrite to stay, obedient to remain of my human folli-cles shared by many, to stay decently with gratitude for help, or as much to give as ever willing—

And this lady.... confuses my loyalty to aid and reliability to be resilient during its application to my case. I have not adored the ro-mantic.... for a long time. Yet swells.... familiarity for affections partied to an' hounded by. A person must stay separated from love to love, for that current to flow, for that drive to be enlisted of its force.... propelling and causative, cavities filled.... for need of be-ing emptied. (*sips*) What is possible, in such a place as this? as probable as such to share in the splenetic indulgences of heart. There must for ever be a gathering place for stones, till washed of brook to be embedded bed. She seemed to be.... much harrowed of her life, as much as mine could seem. Contemporaries of the battle for.... duplicities for being genuinely faithful?... to a spouse in a marriage, or to a cause of strove, roving maintenance? I've no idea of my position to be faced, in this sultry universe. We might very well be both abandoned, to find each other in a chalet of imperious impiety, with the improprieties honored by being.... shielded, hid-den, removed from view, atone-able, and honest. We each are as base as our actions necessary to do. And if not necessary, then that is only play, and playing with deception and heinous attainment. But what is it then, that moves me to breathe still, and yet be alive with longing to persist at this? this churl of an existence, this cruel cloak of rustic demeanor practiced towards uselessness if not for someone like her to find an' dally with, dance upon and around, spin an' be thrilled to know, be lessened with an' engorged by! I have my passions found— in a religious retreat? There's a purpose to this webbing, past so called commandments or edicts felicitous of— crucifying rites an' the honors bestowed to these!... The ex-cesses of debasement an' debauchery are not as obvious here as could be assumed through objective disinterest. We are ourselves to manufacture laws of decency through whatever actions we can attempt and exercise, upon this consecrated setting, no matter how depraving their influence outside of here, away from her an' into the world at large sprayed. Lust, lechery, loathing, lewdness, lurid-ness, they're demons— outside the church, or the religion, and in the pastures an' the burning pans for cooking meals. Is that the les-son awoken to? Do your shame.... before the widest eyes, and the most knowing of your true, incontrovertible nature, the might of your character bleeding! telling all, shedding of the remorses to be

thankfully guided through these fundamental tendencies we must adapt to by being both the human and his animal bodied, a dichot-omy imperiled always. So it's not so incorrect to have felt thor-oughly lost in this world, for so long a time, able and disabled, master of myself with mastery of nothing! My skills.... are the lee-ways found. Ours are.... in this grace bonded to, a license of indig-nity.... an' poverty an' anguish an' unhappiness, disaffection, dissat-isfaction, fear— terror!... It's as much to be at home in a minefield. An' if you're blown up—.... there's no shame to it, nor for a friend's dismemberment alike. I can guess.... our wrongs or transgressions are tolerated while under care to be tolerable.... to ourselves and our kind. (*A sound is heard, accented with the creaking of a door.*).... What's that?! The father sleeps in his rectory. But this comes from next door, the community workshop.... Someone must have noticed my light. But it is late. What is there to investigate? I eat. I steal.... I drink tea. (*A slight sound of rumbling is heard.*) Perhaps *I* should investigate (*stands, but with the cup holding*), an' inquire of an inordinate visit. I don't have a key to the outside door. But I can get in there easily enough, by unbolting from in-side, the— gate to the parishioners, or her taloned *Tür*, as Paths likes to say. I don't want to be a cause.... of some shuffling curios-ity, in there. (*sips*) People think they have to find out things, given the slightest hint of a disturbance to the usual, as if they could be sheriffs patrolling the commons. But I only drink a little tea to-night. An' it *is* late. But what of it! Mathews sleeps, I'm sure.... and *I'm* awake. (*Faint sound of a chair being carefully adjusted of its legs to floor is heard.*) This feint(er) drags a chest across the floor.... or a chair— to sit!... at the table. Sits.... with wait for what? My accosting for this untimely disturbance? This is a church, you know!— Oh!... But even standing now, I feel faint, weakened so unusually. A poor presentation I could make, to upbraid. But with-out any cowardice my intentions are set more firmly than my health. Just the startling of my presence, coming from within these sacred chambers, might serve good enough to reprimand this visit. Yes! I am here. A shine from the smallest of windows extols my.... welcome to be around, an' might announce my neediness. (*sips*) For I am not indifferently lodged— but carefully brought! (*Faint sound of munching is heard.*)....

This thing crunches. This thing eats.... too! (*slowly walking over to the adjoining wall, with the cup*) Her being brings a lunch. An' the indescribable is always feminine— till found out, one or the other, or it. The mysterious provokes a meal within barren treas-ures, or as bare as can suit a man with difficulties. You must know that I am aware. An' the vigil is distasteful, even for me who can only warm for sleep. (*carefully places an ear to the wall, as a low but gruff cough is heard*).... A man..... He would be out, tonight. Some.... overly possessive volunteer, for his duties not to end until he's satisfied. (*coming off the wall*) Yet, the room may be used as a retreat, for some. And a warm night brings out (the) restless quar-reling of conscience. Then where else to go, for a snack! You can have most any idea.... seeming approved of to be thought. I should tap on the wall, to bring him notice of my wariness, that someone listens, that someone cares and doesn't care, what devil-may-care abounds within one's head. A nice sharp rap. Yes, I am here, and am awake.... And you suspect me not?!— (*thumps the wall lightly, but firmly quick, with a fist*) Know my presence now?! (*There is a pause of silence.*) Mrs. Mason! Mrs. Mason! Deceive *me*, even?!... Sweetheart-ish girl— a wife! Paths explained to me, with an off-handed nonchalance of being accurate, with the mere mention of this prefix of addressing, that she is here as much to do penance for her husband's concerns as for the church's. (*The munching starts up again.*) But is this so? Then divulge to me what is real, and what is play, what is actionable to be offensive, or abusive. So many hidden terrors veiled, an' swept for a confinement. And are you this content? to pass over my arousals?! (*makes a motion to thump the wall again, but resists to steady himself to be more contemplative*) You speak of a crying howl with warning, to come this late at night to revel, dream, an' sift through means of attack, for your quandaries stately posed. What more to do to contemplate, we're in an arena of the dispossessed of meaning an' fidelity, searching more for the basis on which to rebuild our sincerities. (*sips*) And do I wish, this searing heat to pause for pores? The devil take you, in there. You are a cad! (*coming away from the wall*) What do you find? What can you learn, of this community? It helps one.... broker poverties, an' enlists your aid. For you must be rich— right?! to have this time to spend, an' make caricature of our component employments on this earth. Volunteer. That gives you a leverage. That gives you an entry— in the back. An' you must be highly thought of, (not) to make a commotion over, or argue with.... Caretaker, of trivial tasks an' details, that might— puncture tires if left unattended. (*A low sigh is heard, almost lending to a groan.*).... Is the banquet finished? Have you aired your complaints to the phantasmic, and received some suggestions to mull or brood over? Therein 's like a confessional, is it not?— to yourself, an' your sweetest, most pious purity.... of form, foundation, faculty, an' fascination with the mystery.... of our suffering, forcing ourselves for breath, an' the efforts of a heartbeat. Yet think of me! I am to my bed, like bowers, and solitary for the displacement, in that hallowed room where ministers have slept. I can feel them chasing for my thoughts, panting with astonishment— or bemusement. "This fellow's here?" they gasp, "In here?!" Yes! An' on the pillow, the head sinking to its plushness.... an' the body wrapped of sheet, another night at least, a cold determination to be found. I will return to that.... rich room for the sainted, an' climb up into its bed like a child to lie. Or should I leave with you, tonight, an' just disappear, not to be found by Kanter, by her or anyone anymore? the spirit finally chilled down to its appropriate state. But I feel so weak, still.... so oddly weak, weak with certainty, weak with obstinacy, weak with questioning. Is this.... a conversation? Jules! Help me! Find me out! Find me again, but with less doubt of me. Know this.... thing that wanders. (*sips, while standing at the table*) Why am I so weak, and yet not tired at all? sleepy and yet not drowsy? anticipating more scope to live with, and yet not expecting even the slightest broadening of my life in the details an' fixtures that may matter most. (*A slight adjustment of a chair is heard.*) It is too late, really too late for you.... Must be nearing the dawn, or something like it, something.... approaching. (*sips*) This tea is finished. It is tepid, but it's done, through with, all drunk up. A.... nother waste. (*softly placing the cup on the table*) I'll leave a mess here. In the sink would be more proper. Washing it up to dry would be best. But for some strange reason, I don't want to make much noise. I don't want to tempt.... retribution. I don't want to retire with cannons sounding, blasting... I want to be quiet, an' stalk like a thief afraid of his own environment. An' I want to leave you.... to your world. I don't want to bother you. I've become too timid and pensive to allow my nuisance made. (*starting to exit, with a noticeable cautiousness*) I'm going back to the.... finer woods, to chew my crud. And I think I'm shaking. Is it from the tea's caffeine, or is it due to my exertions— to seem.... as were to be as generous as Father Mathews, as were to be a devotee of the devout goodness spared my wrath an' rage for leaving me.... this way cowed.... an' leaving you, to analyse your chores. (*With as much silence as he can muster, he exits, turning off the light by a wall switch.*)

Scene III — *The community room. It's light is off, but the door to the outside remains ajar. It is night, and barely visible; but one can make out someone sitting at the table, head sunken into hands held up by the elbows. He is seated to the side, but facing slightly away from the doorway. A soft sigh issues as he sets himself up to straighten a bit, though still with his back tilted to the chair's. He relaxes with one arm resting on the table, looking towards the doorway to notice the relative brightness compared to the more shadowy room, as a man comes up to the door to open it wider.*

HAROLD (*at the doorway*): ....Are you here?

VOGEL (*seated at the table*): Yes. Someone is here, of course. People don't leave this chamber open and unoccupied at night.

HAROLD (*entering*): ....And what did you do?

VOGEL: Others may be awake, you know. I've brought a large knife, and some dry vittles, more than what I've left before.

HAROLD: ....You're early.

VOGEL: Got off from work tonight, before my usual time.

HAROLD: Somethin' happen at the hotel?

VOGEL: There was a bit of a ruckus this evening. A poor bloke fell out the window.

HAROLD: Fell?

VOGEL: One that he was trying to enter through, from the back. I discovered the attempt, after hearing some noises back there, and pulled him down. But he fell very hard; injured himself too seriously.... for that foolishness.

HAROLD: Is he dead? Should I know him?

VOGEL: Hit his head on some.... metal crate he was using. Must have been startled, surprised. Didn't take proper precautions for tripping off a ledge. Amateurish— But *I* didn't kill anyone.... When they took him away, I got the rest of the night off. Hastern was nervous enough to stay outside, considering things. I walked around the neighborhood awhile, cooling off. Then I went home. It was not too welcoming for me. I don't like to leave a work with awkwardness behind, though no one can possibly claim I was in any way sloppy with this.... incident to handle. But fate would have it this way. At least he's not crying, bawlin'. That was the most frightening aspect to this.... accident, at first. A deathly silent plummeting, like he was meant for a fall, expected to fall— almost as if staged.... for this most unfortunate event, at least as regards to the results, which were horrible. Hard! Hard!... You can't think things through like that, though, because they can hardly make sense. A tumble works hard at its begetting. The physics are so definite, as if it were life itself to mimic, as if these matters make for the orderly, ordained, aspired towards. And are we to blame for the bad that happens to us? I'd say so, if it's like that fall.

HAROLD (*coming up to the table, though distal to **Vogel***): .... Well, if one's ta live, then one's ta blame for dying. An' ya brought more stuff for me? You're relentless.

VOGEL: Just some granola.... that I feared would go mushy with longer storage. Hard to believe such foods have a practical lifetime. Hard! And there ain't any moisture around at all. But component parts comport for their dissolution, and the glue-like binders (of the grains) atrophy of strength to leave a taste of alarm, or notice for attention— unless you eat it up now, while it's.... healthy. (*gives a slight, though ineffectual, shove to a bag on the table; it is hardly demonstrable for sight, in this darkness.*)

HAROLD: I'm not hungry, just yet.... for the spoils. But I think there was more.

VOGEL: On the shelves. I've brought down some ant bait, though. The cookies. Want one?

HAROLD: Can't see too well. Don't want the light on, either. I know what's all about already.... Can I have that knife?

VOGEL: You say I'm relentless. Everyone seems to have discovered my intentions. In the back, you mean. But you can't be that distraught, yet. The world is not cruel enough become. And you haven't been beaten or whipped at all.

HAROLD: Are you gonna leave it here? Why did you bring it? Must be a dozen other tools or utensils around, just as sharp or pointed. An' I ain't no worker ta be thrashed. I live off my own sad meagerness, an' darin' ta be fresh for each day coming. 'Cause that is such a miracle of surprise—

VOGEL: Stock up!

HAROLD: ....You should 've ported portaged potables.

VOGEL: The whole point is.... you'll kill yourself—

HAROLD: Die yourself!

VOGEL: —eventually. That's how your time is spent, preparing for the inevitable.

HAROLD: Are you brave enough to give me that thing?... or leave it for me, after what you've done? Do you think we're as crazy, or out of the head? An' in your very presence.... you'd question this poor man, for the implications of doom to rouse.

VOGEL: You've steeped yourself so long in *that*.... you can hardly notice its tincturing. And this knife.... a weapon, is proof for myself.... that I am not an evil man.

HAROLD: ....I shouldn't claim ya ta be, much so. Do ya want ta be hurt, or do you want to hurt me! Is that the examination ya're presentin'? A test of convictions, or neutrality? Did his death really stir you so meanly?

VOGEL: What had nothing to do with what? An' what stirs you?! fella. You've come here, due to my attractions.... maybe. Or is it the night itself, that draws out your sympathies for being chained!... to an illicitness of character, a lack of resolution to improve your standing, and the slinging of low regard towards your personage, towards your face an' faith.... in the world at large, in general, as one earthly corruption betwixt the grand malaise of so many others. Yours is the dark spectacle gleaming. Even your eyes shine sorrowfully. An' your patience is patterned on my respect for the morose.

HAROLD: The hammerer likes his striking blows to make, an'

the gorilla a romp through the forest with disruption his pride enlisted, the only subtlety being anger for what can not be accomplished through this panic for causing harms, slaying or assaulting. An' you would confront my like?! to impose your—zeal, your religious dispatches of death?!— an' forgiveness? Then what you do, you could say heaven-sent is, like the scents of wilting flowers drying up around and over misplaced graves an' forgotten tombs. Ta be that way for your dispensation of regrets, I haven't muse enough to moan with. You are of stark an' deadly ambition only, while I sink ta goals an' slither through your flaming! This thing shines as if through lights transcendent. The blade is bright! An' I'll take the whole of that box of confectionery, what's left in it, unless you stop me, prevent it with more of your destruction an' ruining.

VOGEL: You're like an advocate for rodents.

HAROLD: Cajole me?! within this blessed blight! I do not want the pain!... this friendship brings. An' would another come, but you see me. An' it's as clear to be, this makes too many hunted, with your traps an' snares in place to capture reasons for your acts, an' your enjoyment of might ov'r weak men, an' your pardoning of them as you kill! *There* is the shame of it we share, for a useless life is dead— is it not? Then being alive, in any manner *culpable*, we are all useful— I would preach, an' demonstrate most grievously through contradistinction. Apply my hunger to this, *Herr* Vogel. All blame is to the deceased attended to an' buried. Yet my cause remains of impulse. Now, may I take? or may you pound. It is not theft, nor greed. For gifts haven't such values to be blasphemed by.

VOGEL: There's a wealth here to be stolen—

HAROLD: I won't take it that way.

VOGEL: —But that's not to allow you a bloating gloat for the provisions. You may grab as handily as I, an' eat here, what is left, before me, before my waiting for the sun and this warmth justified. But I've given you all the trust possible, to be in this way honorable. An' I have not found the place ravaged— an' goods stolen.... though I searched in the dark like a mad man, to discover what remained of what I'd brought. An' its really all here, still. So you're not.... the despicable type of bum— with no conscience. But you'll behave before me, nevertheless, with a decorum I insist upon.... an' not take more than you might need— merely in spite to my generosity, to make a jest of it, or an anger of it— From it! No.... that's not the cause. That's not due the pathos you express, to me.

HAROLD: It all, I could have had— I could 've shared. But.... in front of you, that is the offering made— that is the service.

VOGEL: ....The Communion?

HAROLD: The privilege— to be before a slayer and a maimer, an' receive these gifts, pardons or excuses, rights ta be foul an' awful— or sober in our thoughts an' beliefs. The means ta stay on track, towards a destination to nowhere— ya does that provide us — Don't ya! Like the skimming of the milk, ta remove the scum.... or the rich cream, or the curdling after it's been cooked, for the cheese.... you want that delicacy ta eat, maybe? Did ya provoke the fall? Did ya frighten the man, like a scrawny mouse found in a drawer pulled out? suddenly revealed as a festive game!

VOGEL: I don't deny that he was stunned an' stiffened—

HAROLD: A philter for your love, to do onto 'im.

VOGEL: But that he thought he could possibly get away with it, such entry, led to perhaps his greatest shock, of realizing the impossible is without harm, and what is possible must lead to casualty of the most prominent injury.

HAROLD: That's what you do. That's what you always do, mangle an' torture. Tease with the sentiment, an' then slam down an' ratchet up inta a terrible becomin' of pain an' misery.... all on the road ta death— an' divinity.

VOGEL: Are you hungry?

HAROLD: Ya got no guilt about it, just reservations.

VOGEL: I'm sorry that he.... fell. I would have helped him down, an' beaten him up. Then sent him on his scurrilous way— I would have done, screaming, perhaps.... but walking.... walking with.... till he was gone. That is the proper rite, most actionable. That is the determinable testate.... ion, that much of him killed an' willed to, that he'll never try it again with me around, that he won't be so foolish with my legend on patrol. But that's not why.... why the job is. I'm not the cruel one. These conditions make the crime, the criminal to elicit me.

HAROLD: You *are* the conditional, now. Did you cry a bit? Want ta be forgiven? Then I'll take some of those staling cookies as recompense, an' spread them out as a feast of remembrance ta make an' hold (*going for the box of cookies*), at parts where we sometimes assemble, the thrashed an' trashed— Is this it?

VOGEL: Do. (**Harold** *grabs the box, while staring at the knife, which is closer to* **Vogel**. *The staring causes* **Harold** *to seem a little stupefied.*) Take as many as you—.... need, but not the whole of it. For the container stays here. An you must be disciplined like a child to believe this.

HAROLD (*slightly weeping*): I want.... ta help my friend. (*starting to extract a few cookies, to place in a front chest pocket*) I ain't so hungry, ta-night.

VOGEL: Take those cakes— from me! It's too dark, for me to tell how many. So that's not important at all. It is the process, tramp, the process.... The giving you, of some time, that's worth my coming.... an' my soothing of the head, my impairment to resolve.

HAROLD: Of heart an' conscience? These are almost moist.... (*placing the box back*) an' crummy. Can I have that? (*wiping his hand along his side*) It's only the temptation ta steal, ya know. An' I'm not so good at it, takin' only what I want, only what I.... see.... ta need at the moment. But this will do some good, for others. (*A faint sound of rustling outside is heard.*)

VOGEL (*startled, and grabbing the knife*): Gang up on me?! I'm a butcher, sir! A butcher, in the dark!

HAROLD (*alarmed at prospects of being wounded*): No one to hold about. We're all the vagrant, lord (*cautiously stealing towards the doorway*), the humble servants.... an' recipients of your largesses.... We aren't no fools, but ta live. (*at the doorway, looking out*) He might 've followed me, but we're all shame-faced an' pale before ya.

VOGEL: Out!... So warm a night, for ambush. The dawn will tear your heads apart, with seeing of what I'll do.

HAROLD: I invited him, for company. But he was reluctant.... to be at a church. An' he's a true thief, too. (*turning his head to look at Vogel with frightened wariness*), but starving from a sickness— an' desperate. (*looking back outside*) He can't do much harm.... An' he's not coming up, (*exiting*) as much as I can see.

VOGEL: Waiting for me?! I'll wait for the day!... What are you doing out there? in that light darkness, that gray mire, a scintillating graveyard.... of dying flowers— an' wishes! (*rubbing the knife's blade with two fingers*) So you'll stay outside. An' how many of you?! just begging for my approach, for some sort of retaliation aimed— after what you've learned, or heard. But there ain't no fear inside! do you hear?!... An' this leader with the key, I've given birth to. He.... commiserates, over my punishment.... or punishments, my loves an' languors, dreads an' drudges, the pipe cleaning or attending— till the peroration's worked out, for my disease to express an' explain an' conclude with! All.... heavenly darkness. All night sky. And all of you converging, learning of me, with curses made— made up— defined— inaugurated as the Vogel slaying. Yeah! But wait on your daring.... and your munching, mouthing, muttering, mewing, musing. An' I'll see.... not a face come the light. You'll be hiding behind rocks, pebbles!... But the door's open. Come in for a greeting, a skinning, a shimmying towards living more soundly— patrons of my arts! Bashing of heads!... (*placing the knife down on the table*) I feel so destructive, now. I should yell to awaken the hounded, an' the devil's choices to barter with. But let them enter God. Let them *enter*, God. Let them enter God's house. I.... can't tell for them, my deeds to make. It was a curious fall, rhapsodically suspicious an' too perfect to be an accident, or to have been, for me, a keen announcement.... that this is what I take, an' this is what I crumble. An' this is what you've caused, and what we deem desirous— from a fall. Much more a leap into my breast of villainy, the way Walt Hastern looked at me, to tell my turn of it, disastrous of sight. Then yes! I did.... the too much done, and did not.... choose my methods wisely— enough. An' now the heavens await my approbating heat!... to approve of my tormenting? my mangling rent? Come in! Come in! (*jumps up out the chair*) Dine here, with what I've brought. What I've brought *you*! You.... gentle folk. Are you even afraid of me, that I should drive away? Come! Come tear— and rend, an' be at home within this house. Or worship with me, at mine, with evil blades stuck of. For I give you this chance now— to destroy.... An' others? Others?! Wait!... Your pure eyes drench the darkness, peering in on me. I want this meeting to be better highlighted with your enthusiasms an' blood! spilled for the decadence of your greeds.... An' mine as well.... Darkened spectres!— Glow! I await you, with the entrance cleared. Is this the proof, what powers reside— within this fortress, that you dare not transcend its privy nature, an' private access? Till the shepherds dine on their lambs,— come! An' have what is your espousals of being— bludgeoned an' buried an' beaten down into the bricks of history an' the foundation for our discarding manifests of dismay an' disapproval. An' I.... as someone known, from this.... intemperance, not easy to be cooled.... not easily pleaded to— for an advocacy. but ever responsive (*slowly reseating*), an' ever reciprocal to your aims.... So enter here to greet me.... or await the new day, and then disperse from my anger. For like Achilles, it's only soothing for a rest, an' a biding of time.... an' without much personality to own nor injure.... till the shock wears off.... of having been so badly treated, an' brought down.... down.... down.... into this noble posture.... of protestation, waiting on revenge.... Come wait with me, unless the teardrops sear the eyes in this heat, this.... contention.... Or then just stare, as the malicious passes by. I am a.... guard.... to be wary of. An'.... the crime's within.... to bust about an' crack heads.... an' drive off souls, to their more appropriate destinations.... all away from.... me. (*sits to relax in the chair, a hand, supported by the elbowed arm, holding up his head, as he wearily looks out through the doorway to await the first rays of dawn*)

Scene IV — *Early morning, with daylight for the church interior, which is empty of parishioners, although the front entrance is conspicuously open; and we find* **Hastern** *standing on the rostrum, perhaps staring at the pulpit but in general seeming very ineffectual or unsure of his presence, slightly nervous, but not from his standing.* **Pahtes** *comes up to him casually, apparently after attending to a common church chore, and about to engage in another, carrying some leaflets and notes.*

PAHTES: I told you, Walt. He doesn't come on stage for a few minutes more, and then it's only to review things. The morning service doesn't start for a whole'nother hour yet. And attendance has been so sparse lately, the effort worries us to become only (a) ritual obligation than for essential significance. You might as well sit in the pews. I don't think he's even had breakfast yet.

HASTERN: No, I like it up here— if I'm not transgressing any rules.

PAHTES: Even the children are allowed to stand on the rostrum, for the Easter and Christmas pageants. You're not breaking any rules of importance. But neither are you gaining any particular importance, or inspiration, if you thought that's possible with a mere presence.

HASTERN: This is like an amphitheater's platform. I've always admired its centrality to the core of our professing and stylized preaching.

PAHTES (*going towards the pulpit*): You mean apex to the congregation, who are usually seated. This is a focal point, for what the audience should concentrate on, for views, speeches, and announcements. Your preaching, you say? (*putting some papers in the pulpit's shelf and a thin folder on its top*)

HASTERN: No, I let the father deliver on the words we need.... to be spoken. He is an emphasizer, after all, and the representative of our body as a whole, for our surrogated entreaties to the all and mighty we wish to serve.... or placate. But I know it's as much of a craft to be done carefully and through specialized procedures and verses I'm too old and busy to educate myself with; or apparently some degree of deafness to ears that should be listening occurs.

PAHTES (*occupied, working on the order of notes*): There are correct ways to plead an' pray, for an official sanctioning of a service. But to be sacrosanct about it does not lead to many restrictions for your own personal choices of expression, as long as the desires of these attempts are genuinely holy, or inspired to be at least blessed. Each person is one's own house of worship. And you don't need legislatures controlling your thoughts about *that*.

HASTERN: I attend this church every Sunday afternoon, deacon.

But this morning I feel particularly chilled.

PAHTES: It's almost burning outside.

HASTERN: Yes, but adversity chills, Pahtes. A personal.... antagonism with one's self can do this.

PAHTES: Let me tell you bluntly, Walt. Father Mathews is no psychologist. (*turning to face* **Hastern**) He's hardly past a mediocrity of intellectualism, though what he's learned in the sacred colleges he's learned.... acceptably well. But what he is is sincerely devout, and even more so to be helpful in spheres of confinement and confessing of the principles of man far out of his reach. That is, by accident of being simply honest with his studied conjectures. There are rules to be followed, which he has mastered and which guides his assisting on spiritual questions or manners without him even knowing it. The mores are studied, and they are ancient and powerful because they work. That's why religions last. One can figure out so much by one's self, by simply being true to your agitations and stirrings. But that is not a religiosity. That is an animal nature, hammered out through our hypocrisies against God—

HASTERN: There is theology, deacon, which tells us how to do things through our philosophical intents, how one religion may be made to comport to our theories of existence with a decent amicability towards all, or most.... But that does not explain, nor exculpate.... a demonstration of the disastrous, and of raw cruelty felt and acted with. No religion can accommodate that easily as simply animal nature. It is man's antipathy to other men that bothers me so. And I need an honest preacher's guilt to share and commingle with my own. Because his kind, his form of concerning doubt, may assuage a painful reckoning from my own, if I have done, if I have caused or allowed a cruelty of the utmost to occur.

PAHTES: If? You're not certain? Well that depends on whether things were under your control. And I presume you are considering the goings on at your hotel. A more personal matter I would not wish to avail myself of knowledge—

HASTERN: And I'm asking to see the priest.

PAHTES: —But everyone knows of your confliction concerning — indigents. You bring it up in some way or other almost every year. Your hotel is a fine capitalistic enterprise, with your honorable capital and risk infused, gaged to make a livable profit. And the indigent can not do that for you. So you have so many rooms, and some may even be empty. And yet you still feel guilty for not renting to them— for free? or under some form of social credit? And you bring up this moral commotion during the hot summer, while in the winter it's all busy business— by the necessity of your occupations and involvements— not to even consider those who are seldom around. How many times must we try to convince you that what you are doing is totally alright and sanctioned by all laws — in this country— and virtually every religion?! But more than that, it is not indecent to mankind in any way. You contribute to the prosperity of this community, and that in a way helps the occasional stragglers found mortifying within their poverties. But even more than this, you remain spiritually sound, while serving your hotel to improve and prosper, just as I do while working on a decaying church during my precious and most harried time of articulated leisure, becoming more and more difficult to find and keep legitimized to these efforts. There are certainly regrets in or during any industrious activity; but that, of they no matter how well-founded, does not falsify the need of the activity's usefulness, propriety, and profession. You.... tend to reproach yourself too vigorously— and clearly too often.

HASTERN: Concrete examples lead to my straining, deacon. When facts occur, you must trust them as being so. And one can not deny the instances that discourage, for all that you have done to avoid them. But we are faced too often with these contrasts. And it is indeed a hotel's worth of troubling. And why is snow so heavy as for rocks to make? in the frigid temperatures. I can be as rough as they come, and as persistent of my aims as the falling snow. But this is all still water poured to combat our empathies for one another and confuse our purposes of civility.

PAHTES: Paying lodgers?

HASTERN: Hotel guests have my best (service), and are hardly ever surly towards my finest means of accommodating them. For whom do we serve, though, when we feel to strike at pests? Why are they presented to us, to emaciate our tolerances and bring surge to such a temper as to demonstrate our worst of behavior and thinking? To whom are we pleasing but ourselves, then that is no pleasure, but a release of tensions. And yet we are terribly debased by our actions, our assaulting and inflicting of punishments.... Then why are we put into these conditions? What is the game to this that entertains with our hatreds and despising? A natural disaster has no wrath as much as ours to be as kings slaying serpents. Then, why are serpents drawn, Pahtes, for our eyes?

PAHTES: You're suggesting.... heresies?

HASTERN: Whereat for evil, are its influences made within us, with some decree of stationing, then it is desirous and poked through our body as an ember to lodge, a coal to warm with internally. We are predetermined to be mean, makes the suggestion felt — And why?!... I've fought meanly, and have hurt those weakened by my blows. Are we raised for this? Is this why the grass burns into hay and straw these days? For what we would have ate 's not green and fertile any longer.

PAHTES: Be more blunt, Hastern. You say to draw a picture of a God that means for us to destroy each other. And I say that is blaspheme of the hormones. Sensitivities are delicate enough to cause wars on any level of strife. But that is the animal's teeth employed, and not his brain. For it is it, and it is He, and He are it made by. And I challenge you to think! (*leaning on the pulpit*) What is worth a life, but a universe for others to employ. And what can not be justly served his right to be, can not be saved as whom he is, but only as an animal defined. To the most basic levels of subsistence drawing on, our vagrant friends are lost within themselves and can not be rescued with our reasonings until they see a light— a strong one of penetrating radiance— outside of themselves, attempting to warm hard hearts and cold consciousnesses. We pull the blinders off the horse most commonly with help, by helping them— menially. But one can only observe for sense of sense, whether they sense the light, and the stroke of humanity meant for them. Till then they remain obdurate wastrels of a life, obstinate donkeys braying for more hay— and heat! You can teach, and you can learn. But you can not realize a religion until it has done for you a spark.... Of envy? Of wondrous beckoning? That's hard to say. It can not be explained through thoughts— only described. But God does not seed you with hate, but rather thaws the cold ice, working. Working! And there's no shame to hounds barking. Emotions are

generative of needs. One needs not a glorious and impersonal religion, for which you can never be a majesty or ruler. It simply comes to you like insight— to behave wholesomely in a society, and worship the gift of this piety realized. It's a weight well borne. (*comes off the lectern*)

HASTERN: True facts demean my notion of this, Pahtes. If we were so self-evident for a great radiance affecting towards the beneficence of our nature, we would all rise up to heaven in a puff of goodly toil. But the grime on our hands must prohibit this.... spiritual ascension. We are made to grovel in the mud of our consciences for someone's, something's audacious delight. And my hands hurt, for being of a cause, that life can be befouled so easily, and so deliberately, under my jurisdictions of control. It is more heavy a concern than mere piety, or the wish to be proper and considerate of others and gracious to all as much as can be deemed their goodness or fundamental worth. But a stone has no life more, Pahtes. It is created to be dead. And the dirt can not be drained for gold, unless there be some grains of this left in it to be smelt. And this is rich of nothing fine to come by, for what has happened, that there is a nervous pang to my sensation, hosting horror. And even standing here, this does not cease but little, whereat I be bathed of the solemnities we cherish, and the dignified strength of character this position represents. Are we decent, being here, Pahtes?

PAHTES: ....I'd say we deserve our church.... at all costs, Walt. And with whatever reservations you may feel, they dissolve within this space, a shrine of contemplation only, and not the miracles demanded by so many. Bring to the fore yourself, and your soul must at least follow, even if reluctantly, even as a youngster in impatience and with boredom. But we are all correctly to attend this haven for our heads and prayers.

HASTERN: Even as to be the sorrowfully perplexed?

PAHTES: ....It can't be that crazy, down there—

HASTERN: The worst of all occasions for me, deacon, when one comes abutted with creation of one's follies. The indigents are so desperate as to bring themselves to madness for me. And I am wired of a tension cocked, this spring prevented of a mangling, as were for this prevention wound. But now, with decency is found my terror.... and hurting, and slaying of souls, and a rush to provide this— at my hotel, this.... insensitivity to grief, as the opportunities are burgeoning for more harmful assaults, and of both person and psyche to demean.

PAHTES: If you have to fight them off, then fight them off. The nuisance must be dealt with as it grows. One can pray to be rid of the drought. But what one prays to have is what one is, an excuse for being.... this way in (a) state of, askance of the right. So then you would beg here, for the excuse— It's granted! for our means of living. You have the right to protect your establishment from their filthy and nonpaying abuses, even within the desperations they cling to. God pats you on the head, Hastern, as much as I may remit for you your absolution has always been as evident as we have said.

HASTERN: I want more authority speaking—!

PAHTES: The deservedness has been fulfilled. You are a decent person, standing here. And you have the justification, of a supposed— cruelty. It's more a kindness to your actuality and restraint, in your business. So you need not have to yell for it, or cry out for your priorities to satisfy without this— guilt!

HASTERN: The devil be incongruous for my pain!

PAHTES: These things resolve themselves, Walt, by the nature of their instigations towards solutions, like a new season corrects the lethargy of the just previous. And what you suffer is simply a humility to be aware of others suffering, which is why we pray—

HASTERN: Pray?!

PAHTES: —to stay aware without a timidity to be consistent in our daily lives and earthly affairs. It is.... the resonance of a community in shared, though individual, meditative practices and schooling. We speak as one— to one, though as many, with many.... voices differing and various, colorful and representative as a basis for humanity. Birds can sing as well, but not as we may sound.... our wherewithal to be thankful, for each participating life.

HASTERN: I need more outstanding assurances that this internal turmoil is justified, that its queasy abruptness is determinate for some gradation of absolution to come. For I need a pardoning desperately now, knowing what's past is unchangeable and unalterable. Drink or prayer will not do to save me form this upheaval without reason. I've tried both already, but the nagging funniness to my nerves and intellect remains—!

MATHEWS (*coming onto the rostrum from the back side opposite to* **Hastern**): Is the lectern ready, Pahtes? Are the notices that I should address in place and proper order? It often seems that one affects another, and I hate to present them awkwardly, let alone remember their details while trying to concentrate on my lecture to deliver. I'm afraid I haven't even completed it yet, not to my satisfaction, but more to a standard—....

PAHTES: Is done, father.

MATHEWS: that I'm used to. Hello, confrère Hastern.

HASTERN: Good morning, father.

MATHEWS: A bit early for the services. (*He stands rather near* **Pahtes**, *but with* **Pahtes** *between him and* **Hastern**.)

HASTERN: I don't want to disturb your churchly routine too greatly, but I wish to speak with you—

MATHEWS: About the poor transients whose numbers grow for our anxieties to treat with worry, the vagabonds of such plight in the current weather?

HASTERN: Yes—

PAHTES: The tramps, father, who may become abusive in their desperations.

MATHEWS: And you wish to offer them some temporary shelter and work, but don't know how to go about finding any? You wish (for) some reasonable method of introduction that maintains your standing and dignity in our wonderful village?

HASTERN: ....*Aaaa*—

MATHEWS: I've been thinking a lot about this question. But we're hosting a representative of the tribe who can probably lead you to quite a few. And that would be a splendor welcomed, with someone answering my cries of concern from yesterday.

HASTERN: Hosting?

PAHTES: Nursing is the more appropriate term. He was found dying on the parched ground and brought here, along with his thirsts and hungers.

HASTERN: This seems.... a little remarkable.

MATHEWS: A life saved, a life spared. But we must do something immediately about the situation, or as rapidly as possible.... And you do shine internally, to offer your housing establishment as a perfect collecting station or location, a holding spot for other citizens to come to, observe and "adopt" one or two temporarily until the weather crisis abates and the roving fortunes improve for these poor chattel of nature. But that you came to me directly to offer your assistance in this enterprise that may prove to be too complicated for my proctored head, with your professional acumen as backing for a success, is as a blow of miracle upon my head from God's holy scepter; since I have been feverish of concern about this all throughout the night and in my dreams, and feared I had been too impetuous and presumptuous to the congregation yesterday, in presenting my case. But a tone of awe-striking solemnity rattled my bones during sleep, occurring occasionally— just after my most nervous trembling— almost commanding me for rest, until the next thoughts of disquiet started up to swell the emotions with crests of quaking. (It) Was clearly His voice or whistle informing me of His attendance to this problem. And I heard voices arguing — I was arguing with myself, over whether.... *we* are capable of cure. And He said, symbolically, I guess: Don't even question that. Our friend Hastern will come to visit you up-front, and plead himself of service for a solution, right on this very solemn rostrum we use to address the worthy flock. For, what is man essentially.... but a helper! and the associate for His accomplishments to make. His weathering is such a busy task. And then to look after the natural natures of the animals and plants, and the rock itself, from mountains to mantle, and the streams of water in between, or the fury of the lavas made fluid for a change of land and strata, one could almost think that His hands are too busy. But then one comes to realize that we.... are components of this jubilant manipulation, as per perhaps one finger pointing to, to do some good.... for ourselves. So I take heed of the wisdom of my distemperate instruction, to have a trust in your offering help and knowledge to make this mission doable and timely. They are as jewels to a crown, your benefactions.

HASTERN: ....There may be some misunderstandings, but....

MATHEWS: But.... you would know how to explain to your regular patrons the necessity for some slight disturbance and inconveniences to their customary service during this period of succor for the utterly devastated of housing. Only you would have this skill, to remain in operation and bring in a profit of lives as well as a wealth.... of understanding. And with the boldness of your enlistment to— our task, many others may become convinced to do their part and contribute to this alleviation of suffering and despair. Our friend Kanter thinks likewise to be helpful, but has not your imprimatur and resources to launch this project effectively and with a solid footing.

HASTERN: ....I suppose.... it's the punition sought for, to quell this uneasiness, to espouse of some hardship, and reconsider customers to promote their generally ingrained generosities during their traveling through this epoch of difficulties. I would say this is more a warning, father, like that.... bellwether you heard while unconscious, to be sufficient of ourselves for the obvious to do. Yet.... that's not easily seen, under both griefs and exhilarations, groans and laughs, that if one hurts— then *there* is to heal, and if one suffers.... then *there* is a salvation to be found. Yes. My matter is proposed to you, that a caring must be legitimate, in order for prosperous results. And that is what I doubted— in myself, that misfortunes are caused by uncertainty of convictions, rather than the certitude.... of holy actions and commands. But with your assurance that you have been touched.... by our grievances in total, which you may not even understand but are presented through mental chanting symbolistic for a wariness of the almighty powers to be entertained by, for your attention, arousal, worry and rest.... then I feel you've given me some kind of answer, for a personal tumult to abstain. The chore will be a difficult one, father. One of pardon and forgiveness, but it may be seen as appropriate in some quarters of the continually afflicted of spiritual poverty and hopelessness. A recompense, of sorts, I suppose this community may owe. That has certainly always been my internal inclination.

PAHTES: A mad try is eased through you.

MATHEWS: There's no punishment to our assertion of these social responsibilities. It is not a burden, but an ability we maintain to be kind, caring, and considerate of all others, as much as possible or personally reasonable.

PAHTES: Some things we can not afford to do. So the doing may appear capricious when done, a rushing at it where an opening of opportunity is suddenly found, or some solution is startlingly realized within an instant of thought. The actioning may not even seem proper, for awhile, or certainly unusual and devious for the excitement it brings, and the satisfactions imagined and insisted for this exertion of principles and relationships figured out. That is what you were probably doing subconsciously last night, father. But the logic of the matter leads its way to you— and everyone of us.

HASTERN: I would say his inspirations are genuine, deacon. We often build our own tunnels of the mind. But to where they lead, that is from a mysterious imbuing not controllable by ourselves. The father may above all amongst us sense this, and be particularly sensitive to this.... gift of mind. In any case, he has relaxed me heartily; for I actually feel stronger, speaking with him only this short time, while I spent my night in perhaps as much a quandary of conscience as he. The details are.... unimportant.

MATHEWS: This is through the passion of my profession, friend. I know not how a persuasiveness is lent me; but as it is, it is less my doing than my conviction to have it done.

PAHTES: Well.... you're trained that way to think, father, and ramble through your brain conjecturing for solutions to obstacles, as well as much to make a body movement an actual act of calculation. It is ingeniously taught through the lessons of ritual and observance of even the most minute practice of religious infrastructure. You don't have to understand the mechanics to conduct them

properly. But by doing so often and regularly, you become a master of these processes, perhaps even in spite of yourself and your capabilities or mental temperament. You hang yourself upon the framework of your schooling.

MATHEWS: I simply care, to be correct for goodness, Pahtes.

HASTERN: That said 's enough—

PAHTES: I could do as well.

HASTERN: —Then who is your "guest," that I may use as well for nurturing of our demeanors?

MATHEWS: His name is Bob, and he may still be sleeping, in the prelate room indeed as a guest. He aims to leave the church today, to be housed at the Kanters', if he's feeling well enough. Though he still seems terribly emaciated to me, given the relative heftiness of his skeletal dimensions. But he fights a twisted sense of ignominy staying here and drawing on our humble bounty of nourishment. He was truly at death's threshold when brought. But the physician said all he needed was food and rest. And I wish this for him mightily, because he is important, as much a symbol as any I have ever received. And I don't actually want him to leave the premises yet, if his health would be damaged or compromised by the effort. But I'm not medically skilled to know what is absolutely right in this case. He struggled to walk around the church grounds yesterday. And while I did not prevent this, I conduced him towards more leisurely activities that remain useful to himself and ourselves. I can not tell, if today he could delude himself of his strengths. I can not protest for a man such as he to become fit.

HASTERN: So he wants to leave, not be a bother to you. And he wants to be working. That's a good sign, although it may just be restlessness in unfamiliar surroundings.... (*exiting*) I'll contact Kanter about him. He'll replace the stone in my conscience.... Morning, father.... deacon.

MATHEWS: Yes. Thank you.... The name's Bob Lorken.

PAHTES: You divine too much in the fellow, father. Yes, it's a noble privilege of our graciousness to save his life, but that doesn't make him an iconic personage for worshiping or worrying over, as much as principles may be involved. His disposition will probably disappoint all of us, when he gets back to normal. You have not changed his character, the basis of his existence. The most we can hope for is that he has sobered up to the importance of being alive, and will fight much harder to avoid the unseemly death he was approaching, with such desolation of spirit and contemning even of hope towards the end.

MATHEWS: You assume that for him.

PAHTES: Well how else does one die on the ground? dissembled of humanity and profane of bodily worth. He was not knocked down. He became down, and resigned himself to this position. Paths told me this explicitly. Now, if he were forced into this condition, by the weather and its consequences alone, then his giving up may be pardonable. He would have been trapped. But there is more to his depression and abilities. He was angry of life, and wanted to be rid of it. He did not *want* to survive. That hasn't any excuse, intellectually. If he wishes to see, now, more days come, that is a favor to our interventions— But he may revert!... through

131

what makes him lowly in the first place, a profligacy and wasteful-
ness of the soul and the gift of life, the ability to be so complex as
to breathe and think at the same time. He is an example of many
prodigals. And there's not much one can really do, for or about
them.

MATHEWS: We can't hope to change what is corrupted without
help from our beliefs, Pahtes. And I believe his sparing is crucial
to us, as an institution to remain.... relevant in this community. We
have been faltering— I have been floundering, lacking more talent
for my position. The congregational body is slowly diminishing,
but with a steadiness too acute and accurate not to acknowledge
without a growing distress. And this church ages to be only presumptuous of glories and praises given us by circumstance of being allowed to accept them through historical importance and social custom. But she is saddened of this condition. I sense this
strongly wherever I'm alone within these proud spaces of demonstrable piety and holiness. This is.... my fault, for being so insufficient. But how may I be bettered for this service? He was brought
here, as a sign to allow me to starve more and improve—

PAHTES: Oh! Come now, father!

MATHEWS: —To fast and feast of the pious vows and offer myself a conversion into stricter stuff, and greater memory through
simpler attainments of thought and trust, to know that one man has
been spared perishing because of us and a fundamental feeding, a
basic concern. Then the whole congregation may be saved, the
whole village— the entire world, with such a thought of sacrifice,
self and onto ourselves the restraining placed, that we deserve as
much of poverty but then to be relieved of poverty. The asceticism
and austerity may purify, harden and strengthen. And Bob might
be a proof of this— to us, if our caring is extant.

PAHTES: That is a crazy theory, if I may say so, father. The
church is old, that's why it crumbles. And you.... do as well as can
be expected of a priest, given the current training and the current
morals that are pervasive in the community. But I will not expend
my precious time deacon-izing just to let you fustigate yourself because of a tramp! (*starting to exit, for the front entrance*) That is....
too disconcerting. Grip yourself more, to show some value with
your piety.... And if you feel overwhelmed at times.... then pray!
(**Mathews** *grips the pulpit with both hands, as if to deliver a sermon with fierce deliberation, as he watches* **Pahtes** *exit and go
through the church door.*)

MATHEWS (*wildly preaching, but to himself*): He was brought
here! And for, to have been made by us, to be revived, he will go
out into the world and bring us favor and renown and more treasure in people and converts and more funding for our coffers
through their devoutness added to our own. This is the pageant
brought, and the decree made, that we are accursed for to cure the
curse with our deeds and proselytizing towards the hope! and the
glory of the radiance we wish to share. And our Bob will bring us
more followers, and saaave me! from a self-destruction similar to
his own attempted. Not be thine (s)'elf, with such deserts designed.
It is not for want that we are sacrilegiously made, to avow this
need in tender of service and solicit of the gift due our rhapsodic
pretensions for comity throughout our lives and with each other
dwelling of a common flock. Then notice we together a low sign, a
miserable squalor, for the height of achievement fetched and
seized!... and atoned towards a saving grace, a remonstration accepted and approved for worship, and signaling the forgiveness of

all sin.... in a life reprieved— our own resurrected to observe our
own.... common faults and failures, failings flung as of a mast of
ship sailing waterlogged. But come my sense.... to proceed, and
improve, within our treasured subjugations— as if bums! brought
whole, and made throughout this holy.... servants.... and saviors....
of this church.... (*lowers his head, as if to recover from a mental
exertion*) The meanest do cry, for the smallest of wealth to obtain,
and the most little of reward to the heart, a recovery of premium
for its valorous struggle to continue amongst our worldly affairs
and anxieties of state.... and these fears so profound, staying in the
red of blood. Thus so, as a hand is given— Lift! (*slowly raising his
head to look at and speak to an imaginary audience*) Good friend.
Good friends. Beseech my caring, and our cares.... for this— public authority to indulge of.... the down an' out and wasted. It is a re-
couping, a recuperative reclamation. of humanity's vows and vendetta against our baser instincts, and sloughing of concern.... for
the terrified— god-terrified.... Eyes about to close, but I see.... a
sermon due, too incomplete to share.... with you upon this— empty
space.... And work to do for it to complete, abiding worry and
nervousness with resolute tones and demands, while I am all and
only afraid.... and need some sacrifice towards this effort. It scares
me.... to try and finish it, in time— Ears hear.... my thoughts.... but
do not bake the bread, for the hunger, and the begging, and the salvation desired.... of your taste. And.... the time is short, it is mean.
I must conclude a talk to preach.... to a few— of so many pews, so
many.... nations of mind. This is what is represented.... kneeling
withing our characters to pray of.... empty berths of richly made
grandeur and strong sturdy structure, woods from the forests, tall
statures.... brought to sulk with me.... I'll.... make amends. for this
carpentry. (*leaves slowly, but with a skulking uneasiness, back to
the rectory*)

Act V

Scene I — *The community room. It is daylight of a heavy, oppressive afternoon. The outside door is open.* **Kanter** *comes in, but
from the church portal, with a handkerchief rubbing perspiration
off his face. The table is busy of articles as usual, but no foods,
and conspicuously with a crumpled paper bag lying towards the
end where* **Vogel** *customarily sits.* **Kanter** *comes up to his usual
chair, off to the side and away from the table; but he stands, looking at the open doorway to the outside. He seems just slightly perplexed and disturbed, but without a physical warrant to portray
those sensations with emotion, as he is more concerned with the
heat.*

KANTER (*to himself*): I can't quite grasp this. (*As he is contemplating sitting down, or first closing the door to the outside,* **Margathrate** *enters from the church side, with her lunch bag.*) Mrs.
Mason. (*She acknowledges him with a glance of greeting.*) I can't
quite understand why.

MARGATHRATE (*sitting at the table*): Why what? Mr. Kanter.

KANTER: My wife doesn't favor him much. Seems reluctant to
have him stay with us, after seeing him about the church this morning. But she's leaving the matter to me, she said, before I left the
house.... He looks to be in good clothes, and fairly good spirits, as
far as I can tell. But I don't want to upset my wife with anything
unusual, if she's not ready for it yet. Yes, he gaits a little slowly,
for a man who should move around more briskly; but that's because he's still enfeebled. One can't hold that against him, as a sign

of warning or for apprehensions to feel, unless it's just that this whole undertaking is simply too fearful, at first, when involving a stranger.

MARGATHRATE (*starting to eat*): Bob is a stranger?

KANTER: Certainly to she who has met him.

MARGATHRATE: They spoke with each other?

KANTER: ....No. She only observed him, from a distance. I meant met with eye. She felt disturbed when he glanced at her, of course not knowing at all who she was.... But she wouldn't tell me if it was sorrowful or indignant, his face. I didn't ask, and she didn't offer to describe. But it was clearly not gleeful.

MARGATHRATE: He told me he had a slight pain in one of his calves. That may have been bothering him and his early disposition. But I'm glad he's around.

KANTER: I didn't notice much twitching as he was ambling. I know he's anxious to leave the church—

MARGATHRATE: He shouldn't.... if he still has some physical discomforts. He should stay here, maybe quite awhile. And we can all provide for him, not straining any expenses for the priest.

KANTER: Father Mathews.

MARGATHRATE: Yes. I've been thinking a lot about this. And if he is an invalid, he should most properly be nursed here. Members of our congregation should not be burdened in their homes with such illnesses or diseases. If he were quite healthy, then you could put him up, but not yet. Your wife's intuition is correct in this. He should stay around here, where he has access to the generosities of many, and where I can find him and visit him easily—

KANTER: Margo.

MARGATHRATE: —as could any other, to offer any sort of help (as) I see fit, without encumbering any family— including my own. This is the proper way to do it. He should stay at the church. I wish it this way. And I don't mind being bold about that to say so.

KANTER: ....There might be fifty of them in his general condition. They can't all stay here. If we're to be helpful, we must.... partition them into homes and residences.

MARGATHRATE: You mean divide and divert them—

KANTER: No, partition.

MARGATHRATE: —into various localities.

KANTER: Partition them away from their wandering gists, for a period. It is too dangerous under this climate. They are dying— There are probably deaths already. But even the church has the air-conditioning off today, to spare some costs until the services. I can't tell if I should close the door or not. Any draft at all might be helpful, any kind of aeration or breeze. And a thunderstorm would be great, but the clouds are too high. No. They should be in houses and apartments, or at least under trees of shade without withering leaves, greenery still broad enough....

MARGATHRATE: This is *the* house, of the community.

KANTER: ....No. This might be the room.

MARGATHRATE: I'm not upset here.... every place else, but not here. I'm not unfaithful here, to the religion—

KANTER: We have adopted to correct, to fix up, repair, furnish, improve.... and prevaricate over— until this is done right. An' rightly, Margo, rightly! Here is where we're allowed to try.

MARGATHRATE: A faith is inherently sound in and of itself, through the sincerity of its expression. As long as it's not corrupted by the activities it allows, it is keen for one's self-assurance. And all that happens under its guidance and fundamental or most basic principles makes for whatever corrections may be necessary of peculiar circumstances. Our church is not collapsing because of this, neither a marriage nor a casual— though regular— association of believers. My husband approves of whatever is possible of a recreation under this faith and loyalty to our higher appetencies and instincts, though we may not understand everything that *must* occur, nor really why aside from opportunities inspired by. Matters should be allowed to work their tensions out naturally, intuitively, and without false restraints and shackles on conduct, as long as this is all within the bounds and boundaries of our religious codices and beliefs. Then, what is permitted is to live joyously and as much fulfilled as possible for our personalities to employ and exercise. This is no excuse for behaviors. This is divine inspiration herself leading us through our worldly tumults and capitulation to the necessary debasements we must seem to experience to survive, or have for others a caring found. Under such a happy aegis of agendas— for my own emotional well-being and communion-ized uplifting— I have no embarrassment nor apologies for feeling good here and being good in this place. We make for only the sanction-able of activities under this solemn welter of thought and environment presented us; and we may play about what is not proscribed, to be humanly decent within our natural fervors. I had a great sense of the rapture, while cleaning the stained glass figures and making angels look prettier, if this is possible from smudges and smut to remove.... And *I* am prettier, Mr. Kanter, and more deserving and more desirous because of the effort. My husband approves. Chrisie is more pleased with my attentions for her. And I am better coming here and doing chores— for the church and myself. And I really do both sincerely, and with full deliberation and determination, as long as this may last. Here is a second awakening for me, a new and other blossoming of my potentials for humanity to please. I am being re-humanized for love, and will not deny her character on my countenance and conscience because all is becoming freely fresh and wanted, doable and determined for me.... Whereas outside of this place, I may become wanton and lost, and revert back to the loneliness within crowds and responsibilities. So I need this spiritual break regularly and often. And as I've found it, I aim to keep.... this special recreation.

KANTER: ....Well, I'm sure Mr. Mason considers the gratification you receive to expend on your family, if there is excess left on you.

MARGATHRATE: And don't tell me about Bob Lorken, lost soul found. He is extremely pleasing to conjecture on and commune with. I enjoy the freedom to say so openly— here. And it's so satisfying to see his physique improving, to know that it is and feel that this is so. We may be as mated partners, for the radiance that shines upon us now, in this house brought to. We may be as one to commiserate together, the crippled recovering, the nearly dead re-vitalized, and happy of ourselves for a change. Is this not the privi-lege meant, and directed towards? with full regalia of endearment of friendship. We like to be near each other, around each other, in this— place of worship. It can't be illicit or foul, but a commonal-ity shared to romance through, to prance with and bounce upon.

KANTER: And do you, with your husband, share such a view as candidly?

MARGATHRATE: ....No. But to you, and to us! Fiends made finer of our sincerities to stroke and pat and grind. This candor is for the church, Kanter— and my mouth to speak. For are there devils here to listen and be hurt?! No! Or I would be impregnated with their concerns. Only the angels found may open me up will-ingly and happily. Delirious are the angelic, to cater to my whims. And they speak to me to enjoy— this— presence of rejuvenation an' jubilee, this state that nears towards ecstasy of mind to be al-lowed to think through and act on. And would you tell my husband —?

KANTER: Never!

MARGATHRATE: —But he knows I am particular, for my gaie-ties to work through. My standards are so high and specific, that if they are nor met.... if they are no longer, I can not fake an amiabil-ity for life. I am the sullen sluice become, the sulking slush unable to be cleaned. I wash in mud, and bathe of predatory practices to demean his affections, in search of others. And this is why he brought me here, more under his control— to enjoy life again! And how.... must always be we pardonable of the techniques used and the temptations satisfied in such a place, since both are given me and taught me for a purpose, to exist of God's endowments always, chastened of the tenderness used upon my mind's acceptance.... for these chores to celebrate through.

KANTER: And have you done so bodily, with Bob?

MARGATHRATE: ....His calf does ache for me, Kanter. And this activity will grow, the more 'abled he becomes to prolong our spiritual enlightenments and engaging. He is a handsome man for his honesty, of a feebleness stronger than our positions yet assume.

KANTER: And in *this* church....

MARGATHRATE: Where a mastery can only be blessed, I am taken and shrined of these abilities.

KANTER: Might you only imagine them, Margo.... or hope for them, these strains of mastery.

MARGATHRATE: ....Is the day hot, Jules? Can you take my word for *that*?!

KANTER: It could be a sauna from our own perspirations, lady. (*wiping his brow*) One must feel one's needs to know them. Or of them least may they be.... a warmth to take notice of. A chill is harder, much harder, if it is slight. Because the heart beats warm already, Margo. But I would not recuse myself of concerns to es-cape. I do so already, and I fashion it this way, to remain proper about it— And so does.... she. But we are tired and old, and

haven't that spark of imperative that drives.... the younger muscles in their fevers and pulsations and rhythmic dances. So I can't chal-lenge how a toil is wrought, nor easily condemn these modern forms of pacification and alleviating.... the more personal conde-scensions practiced. But within the scope and scale of this loca-tion.... my method is alright— ours. I would feel shameful other-wise, accepting the torts in our heads. I study the concrete around here a lot. This is.... concrete block, and platform. It may become cracked, and split in time, damaged with our use. But it's still firm, for the moment, and will be ready to be replaced later.... with much the same. So may a gospel be sung resolutely and with meaning.... through voices growing hoarser and hoarser. I don't know about this trend, Margo.... to discourage explicitness and confronting of facts, and to sublimate them through our holy ideologies for cleansing and traces of a few.... perfumery oils left with. But I have a trust in this room at least, not to delude myself as to why we're brought about. It is to be gentle about it, as the distastes and dis-gusts grow.... and not to be horrible, at home. I was almost wish-ing, for a heaven-sent adjustment for which we can not do on our own, and to ourselves or one another.... not in the dictates of *this* religion. Subtle may a weak man be. I've read of many stronger to deceive. But Bob makes you out.... a dame. And for ourselves?... I would sit on spiked hedges, now, for this dampness felt.

MARGATHRATE: You're not feeling faint, are you? Haven't overworked your assignments around here? Relax as you can, in-side. I've heard of elders dropping right in the streets, on such days as this, the swoon from an expulsive heat, an embrocation of radi-ance.

KANTER: Not a hint of humidity anywhere, but stark shards of radiant warmth, like sheets of air blown from an oven's fan.

MARGATHRATE: Protect your reserves of motion, and take a seat, Mr. Kanter. One shouldn't fall ill on a day like this. The cause of malady would be too omnipresent and not easily recoverable from. Yet exhaustions can be remedied with rest.

KANTER: I'm trying to decide whether to close that door. Sur-prised to find it open. We should have a little, portable fan in here — Oh! But that might tax the church's electricity bill.

MARGATHRATE: Bring in a paper one. (***Vogel*** *enters from outside, as **Kanter** is about to go towards the open door.*)

KANTER (*stopping*): Oh. Vogel. (*turning to sit in his chair*) Close that door, will you?

VOGEL: It's a true furnace, today.... Margo— What? The air-conditioning's not on, in the church proper?

KANTER: Trying to save on expenses, Vogel.

VOGEL: Well, they will with the pipes. (*coming towards his usual chair at the table, without closing the door*) Nothing wrong with them, 'less the joints crack up. Let some air run through the place, Jules. Don't want the interior to get too stuff.... Mrs. Mason — Who left that?!

MARGATHRATE: Left what?

VOGEL (*picking up the crumpled paper bag*): This.

MARGATHRATE (*as **Vogel** considers the bag*): Probably another volunteer.... I didn't even notice it was litter.

VOGEL (*slightly tossing the bag back onto the table; it lands near the knife he left*): Not important. But this room should be kept as neat as our community. The deacons are particular about that, since they're.... really less responsible for us. The church itself is of course kept spotless. Isn't that so, Mrs. Mason—

MARGATHRATE: Margo, Reg.

VOGEL: —You should be very familiar with its interiors by now. Well.... Reg is for you.

MARGATHRATE: Great care is taken inside, I can tell you that. And I don't mind doing so. Every bit of soil and dirt I'll wipe up, if directed to. But I've gained a knack on the windows, Paths tells me. And the father is very pleased.

VOGEL: They're very beautiful to look through. But I shouldn't become too obsessive about washing them. It's still a fairly risky task. And the more you work at it, the greater the chance for an accident.... That's true for anything done regularly.

KANTER: (They) Should be made to open out or up, like regular house windows. But this is an old church.... and that is precious glasswork, not meant to be touched.

MARGATHRATE: One merely has to be careful in what one's doing, and observe the recommended precautions. It's most precarious while you're learning how to do it, especially adjusting the ladder correctly.

VOGEL: It only takes one accident to spoil the entirety of the effort, if it leads to a very bad crash. If I should crack one of those old pipes with my wrench movement, banging up against the ceramic parts, we'd have to replace perhaps a whole section, leading straight to the underground, to bring up to code— to repair within codes.... A considerable expense, and a tragedy I'm sure, to be shared with the bishopric's treasury.

KANTER: You can never replace that stained glass.

VOGEL: But my point is.... these matters are sensitive, as sensitive as people. They shouldn't be abused, or let on to be tended for when they're not, or cant' be.

MARGATHRATE: So what can't be here? Mistakes can't be here.

VOGEL (*leaning with hands on the back of his chair*): Man's might be error prone and devastating at times, like he's triggered to destroy, an instrument of casualty and blasphemies, pretty much drafted to do so in spite of himself or his best interests or those of all others.

KANTER: That's a rather cynical view.... of people.

VOGEL: Not if it's not his fault. Not if he's created this way. And I don't like to be hurt. But I've certainly caused a lot, in my day.... Slept well this morning, though.... Margo.

MARGATHRATE: That's nice.

VOGEL: Had a complete universe of slumber for myself, in a very empty house otherwise. It is particularly useful for unexpected visiting, because there's no one else around to disturb.

MARGATHRATE: Your work should be regular enough for your visiting.

VOGEL: As much as to here— Oh, I see.... It is really my home. An' I use it as I like and when I like. But sometimes it does seem as if I'm only a guest, though a most important one. But it is thoroughly offered, and need never be taken for granted as anything vile. The ghosts have been exhumed for a long time now, and have all drifted away.

MARGATHRATE: That, I suppose, leaves some form of loneliness, even from fright.... to miss.

VOGEL: No, not at all.... Only an empiricism, the bland freshness of experiment and skepticism to thought.... Almost.... as hospitable as a hospital.

MARGATHRATE: Do you.... houseclean adequately?

VOGEL: ....I do everything adequately, and as well for invitations pitched. It is a unique thrill to own the utility of a house, something which should not be discounted during one's.... planning for future relationships. It is totally available— still.... and outstandingly so, handsome and retiring.

MARGATHRATE: There.... can be many keys to a lock.

VOGEL: What?! To open with? Information hardly changes anything.

KANTER: But knowledge all.... to be possessed by, Vogel. You should move out of there, and to another town— to find another life. This one is.... too oppressive becoming you, in this heat and pleading and crushing of your conscience. We all have our desperations to atone for. But you younger ones.... There are ships that sail to take you almost anywhere, if you're still limber enough to jump onto them, right on the deck welcoming yourselves to the captain, who's surprised you took so long. That's a freedom from mistakes, else (they) would anchor you down to drown.... through these undulations of wave.

VOGEL: You mean the air is like an ocean. Well, it is a hot day, the same as many lately. But I'm flustered for my stake, Jules, and my pride. I can't try to do good, enough to be for my conscience to approve, because it's always queasy. But we're all among the highly decent. And that house doesn't char me, not with dread nor hatreds. We attend to our disquietudes in seemly.... congenial ways, to remain healthily sociable and attractive, as we congregate here to praise righteousness and fix up problems, little problems amongst ourselves, maybe. I'm available for this, for some compassions to share and mix with, and not be hypocritical about— but also prudent, selective, sparing.... benign, not to hurt feelings.

MARGATHRATE: Recreational.

VOGEL: Yes. Exactly. And also understanding of the complexities of modern associations and community relationships, personable agreements, assessments of intimacies, and a boring family life— I've lived through this already, and can lend the practical advice of what soothes and what stings, out of experience of trying.... to do good, trying.... to be good, and knowing of the evil that revolves around broken confidences, how to avoid the pitfalls and traps that lavish on your inexperience and gentle arrogance as if a thin shield to ignore what obviously demeans and soils of your expressed essence in the common interpersonal relationships we typically aspire to and are weighted down to uphold. And I can show how to work this through.... comfortably, without much deception. My vanity and candor amount only to my generosity— and need.... to be useful, approved of, valued for company and companionship, and related to our church's scruples and principles of living wholesomely with a religious acumen, at least in terms of following the scriptures and singing hymns for our joy to be able to.... absolve ourselves from the wrongs committed, through this contentment.

MARGATHRATE: Wrongs by whom?... or to whom.

VOGEL: Whatever is known of my intentions, or ours, can be forgiven of the delicacies of interests involved, and the sincerity of the needs generated by our dysfunctional unhappiness, this condition we weather to endure. To whom are harms brought? No, to where, to our heads, they're shared by everyone in degrees. And all life benefits from honesty about us, surrounding us, as much as we would intermingle with all life.

KANTER: So that's your conclusion about qualms, distraughtness and self-doubt?

VOGEL: We are what we want to strive for, Jules, not that we might ever obtain (it) or even hope to, however we can mangle ourselves and others in trying. But it's what we honestly *want* that is forgiven. That.... sensitivity is as pure as the notion is gifted to us, expected of us— demanded for us.... to say to ourselves we will not deny it more, once the urge overtakes propriety and properness.

KANTER: What?! To kill? like in a war? I mean, you're expected to, as a soldier, and it's.... legal and legitimate— but it's still ghastly to have done it and realize the full consequences upon your psyche, to allow yourself to treat man like a lowly animal butchered for sport. That can be forgiven? That has its place, its box of acceptance or seat of spectacle under religion? Or adultery or polygamy or excretion— or any other human activity, foul or felicitous? That can find its approbations somewhere, somehow decreed? What makes for sarcasm of the word, then? A long recital of joking jolts! if that can be forgiven— even my.... disingenuousness.

VOGEL: We are the other.... of what can't be. Faithful? (*coming off the chair's back*) Now that we count on flowers for our poisons, how can you tally faith but through our desires beauteous! It is.... alright. Alright. Predation is a right of animals: carnivores, omnivores, herbivores. And feeding is a rule for all life, off each other and ourselves, perhaps. But that's the way we're made, and that is what we're meant to do. And love is just as meanly fundamental, common and low, as could be through forms of respect— and disrespect, affection and lasciviousness. These are basic rights. Only societal influences fashion the romances, or their modes of being.... proper. And those influences do more for hesitations than chastities to promote, marriages instead of monogamies, in effect.

MARGATHRATE: The one must be criminal.

KANTER: They're supposed to be the same!

MARGATHRATE: I won't contradict.... or say it is impossible, Mr. Vogel— for Mr. Vogel to be Reg.... or Mrs. Mason to be me. But I'm not responsible for my conduct, under your expansive, expansionist principles of belief. Only one's religion is. That's fine by me, to apply it and stay copiously within the fidelity of this fideism. It makes for much less to think about or worry over, as I gobble down my food. (*folding up her lunch bag, having finished her brunch*) But I mature and ripen here.... or re-ripen, spiritually, not to feel so angry about the world— and my commitments to it. There are more degrees of freedom to exist through than I had thought, just a few days ago. (*standing*) And sincerity is a stretch of the leg.... to exercise and run with, awhile. I am not bad for being poised— or taking a pose of pasture, every so often, for opportunities to observe me, take notice, and as by religion apply themselves upon me, preordained apparently. I will not fight the gifts brought to me, if they are pleasant, as I hope to be as pleasant for this church. There's less circumspection about it anyway. (*picking up her lunch bag*) I am commanded to be here, through my husband's direction, however this is made out to be so, the manner in which I come. (*while re-entering the church's realms*) And I'll accept whatever divinities prosper upon my head or stroke my back and backside to cure of the.... anxieties of living. (*has exited*)

VOGEL (*only sardonically, though ineffectually, calling out*): Don't fall!... She does make the day, Jules (*sitting*), worthwhile for visiting. I can think of little else that can hold my attentions less pensively and still contain so much merit.

KANTER: Then you take too much for granted.

VOGEL: I trust nothing of myself and observations, now. Only potentials can be sensed or sniffed for, but not much is certain for me. Things change too thoroughly and rapidly for my taste; and I struggle to remain a master of my concerns, rather than letting everything just fly by and over my head.... I may even have to change jobs, now, and haven't a clue as to how or what to pursue that's different, in this town.

KANTER: A reluctance for the night work growing?

VOGEL: ....Boredom, maybe. But it can be as dangerous as falling out of windows, you might agree. I don't want much trouble to arouse. But I actually hurt myself a little, last night. And questions have been raised of my effectiveness, staving off the indigent.... at the hotel. Yet this season drags on, as I age through it.

KANTER: Well.... we most grow out of our physical strengths eventually. You can't stay a loggerhead forever, ending disputes with agile arms and firm fists, wrestling and stunning.

VOGEL: I haven't much of a mind to do differently. I'm.... not intelligent enough—

KANTER: Flange!

VOGEL: —or creative enough to think anew. All I can do is protect.... houses, buildings from.... what little could confront me, would dare to, seeing (as) I'm not much but all of an obstacle, with hardly any capacity left to be relenting of a desperation.

KANTER: Did you sprain a wrist?

VOGEL: ....Less than that. I got the job done, don't worry. It was too easy to accomplish.... But I did hurt a bit, afterwards. That's true, Jules. I'm not a mean man, nor a maniacal machine. I have my sensitivities and disgusts. I drive away. But that don't mean I hate so much.

KANTER: Doesn't matter, if it's your job.

VOGEL: People. I don't hate people so much. I can care for— people.

KANTER: Then become a masseur. Or try something you can do in one of the rooms of your house. It's a fairly big one for a sole person to inhabit.... But it might be best to sell it, and wander a bit, off its earnings. One good fire could destroy your whole security, in this village, since you must leave it empty— unoccupied, for so long of periods.

VOGEL: That's mostly through the night.... I'd rent it out, if I went traveling. I have some savings accumulated, over the years. I'm not afraid of being insecure. And I'm not afraid to die out on the roads like an old beggar brought to his end. But that home must remain.... a place to come back to, even though I don't particularly like it— nor dislike it. It's all equal to me, as long as it's (kept) standing.... (It) Would not be the style I'd design myself, as an architect. An igloo shape would suit me fine, or an underground shelter with a pyramid on top. Good, simple geometric forms of canonical value and worth, since all of these religions are premised on the simplest or most straightforward of physical proof of existence. Then that's for me.... Kanter. Proof. Proof that we exist merely for the purpose of our physics employable. Proof that.... this sort of thing will be protected, somehow, throughout our and all lifetimes. Proof that we needn't be ashamed of our most basic characteristics and attributes. Proof that we are at one with whatever we can do. A demonstration that our very existence affirms our rights of ownership to these senses so profound and overwhelming of our consciousnesses, and dreams as well, as to make all other thoughts of being bland, is all that I continue to insist for, throughout these storms of baking.

KANTER: You had wanted rain.... some rain.

VOGEL: But, haven't we had?! I'm still waiting, for the inevitable and obvious to come. The seasons must make prominent their motions to proceed, that expected procession of change we experience and adulterate with our miserable daily activities hindered or helped by the weather. But rain is not for test as much as tease, now. For my condition has become too.... strenuous to hold onto for much longer, to maintain at night the quarreling within my brain concerning what has been and what's yet to be— banished! Has kindness been driven from my hands, Kanter?! Kindness and commiseration, there's a partnership for you to bleed with and weep upon, as were ye angels.

KANTER: But "ye" ain't. Isn't that the point? Man is a debauched creature, needing the angels' help.... and much too dangerous without it.

VOGEL: And that's why I'm cruel? I haven't any angels left?

KANTER: Who said you are? We have our credos to follow, which we share. And I'm sure your prayers are listened to, when said in private or mentally, meditatively.... as much as our singing collectively, in praise of the glory and everlasting we reach for— struggle to obtain.

VOGEL: Now tell me truthfully, Jules. Do you really believe that? As a man of the world of strife and slaps and slugging it out to prevail with a competition that demeans the soul and decimates so many, can you really have faith for a peaceful, tranquil afterlife where roses come to life as nymphs to dance with and all the air is perfumed of confraternity?

KANTER: Don't you?

VOGEL: For the good, it's nice to assemble to believe so, voiced in numbers to dictate these beliefs, if not demand them of many. But there are beliefs, and then there are ideas, and then there are notions, and then descriptions, and then depictions. And we tend to mix all of these up as one collective whole of vat to chose from. But we haven't much control on what we pull out, on any given occasion, because our dexterities of procuring are crude and clumsy compared to the quality of what may be taken and burnished with. So have you a belief?... or a picture, a drawing, merely a mental outlook. Can't tell. I can't tell. I've done some harm.... again, and must cast more bait, to see if there are any fish left in the pond, that which I own and may have to restock.... or let flounder. It's coming to a crisis point in me, Jules. I may not be meant for heaven.... to arrive to beaming and welcomed, smiling amongst the greetings and greeters and practitioners of faith. I may rather be condemned, to stay within this heat, waiting for dark clouds.

KANTER: Well, I've heard the rumor that no one actually gets to heaven willingly. It's more or less a random process of selection, or election as a representative— But that's just agnostic ranting, banking on chance with the unknown and unknowable. But it's clear the only opposite to heaven is what we drudge through now. For only the sensate and living can have such misery as this, the eternal pressures of our psychologies balking us at any instant to keep our happiness down and drive our pleasures away, leaving space for only the most rudimental of comforts.... in our minds. I.... am entirely faithful to her. I have panic for an out. And ideologies help so much, that one may be murdered with forgiveness, one or many, an entire body of people, an entire nation— an entire genesis.... The whole world could be destroyed.... Bob.

VOGEL: ....Bob?

KANTER: Oh.... I was thinking.... off the boil.... off the boil.... Out of one's head, you know, Vogel.... What he must have felt, as he closed his eyes to faint, and die.... come up against it, that dark stone, and accept the sword to the heart, a piercing annunciation— by God lifted of the ear that you have finally been chosen for your peace and rest.

VOGEL: ....You mean that tramp we rescued?... the guest we have— this.... Lorken fellow? He was only under the ether, Jules. Not much more to it than that. Your luck on finding him, stumbling upon him, was his luck for a reprieve in life. You've earned a good deed, we've *done* a good deed.... And he lives. And the times have become harsh. And the tally wavers back and forth for our.... heroisms and nobilities of thought. But don't count on him to save our consciences. He'll probably die within a month, if not looked after for a year. We are being crushed of offenses everywhere, and it seems like the entire planet is convulsive for.... your murderous conduct—

KANTER: Oh, there are wars and conflicts aplenty. But they don't concern me directly, not at my age. Carnage is rampant, even strolling about some parts of the globe. But here is by and large.... a very peaceful, quiet, dreamy location, dreary of its self-sufficiency and drab of the monotony of even the brown graying from this mortifying warmth. Yet, no explosions would be nice, here. Even the kids are hesitant to play with (the) firecrackers. Those incendiaries and flares would seem to only highlight what's been burned, revealing nothing new about this anguish and languor above drooping trees. But the fall will come. It has to.... and then some coolness, and then some rain. The cycle of fortuity will continue.

VOGEL: It seems to me that we are not optimally designed, at least in terms of reason, for these cycles. We struggle through them mostly, instead of stride. And the changing conditions seldom enhance my perspicuities of expression to explain to myself my life—

KANTER: Then you're in a rut!

VOGEL: ....I could be content in another location if I could only imagine it. But I can't. Those thoughts do not come to me, the mind is empty of them.

KANTER: You cant' picture heaven?!

VOGEL: ....Only what's been described—

KANTER: You can't picture the blue— and the gold? the blue of sky an' the gold of grains, grain stalks flourishing, and green meadows and comely streams of pure waters from rivulets rubbing up against reddened rocks, and blissful clouds of white, the fluffy down made bedding for— deceivers?!

VOGEL: ....We have too much of what's already cratered and cracked, pitted an' parched. And would a frozen snow be any better felt? This is what we have, that I can see and imagine with. And there's no better state of man, I fear, than from what he may drown on, given him— to devise. For I can not think beyond, to leap over what I have experienced and contemplated about. This is all that is left us to play with.... No other glory (is) possible. And it burns too much anyway, to call it much of a privilege.... enduring. But the beautiful is what is best for us to see, and that makes not a heaven as we seethe from discontentedness.

KANTER: Then I wouldn't spend so much time contemplating on our collective wherewithals for all our worths, but simply baste. Because nothing 's possible but what we can behold and live through. That (which is) unexperienced simply does not count. That unknowable simply is not so. And that unimaginable.... Well.... make sure one doesn't imagine it, or else your universe will do a flip! and plop you out of oblivion into the surety of everlasting sadness, the results of forbidden realizations and squalid rationalizations to provide some explanations for what's surrounding us in shades an' sheets of darkness and despair.... to know yourself as cruel, or despicable of company, or terribly harmful to others— devastating.

VOGEL: What can you mean of me? to ignore my challenges, distastes and dispassions? I've only ever done what I've had to do.

KANTER: That's never enough for a fulfillment, is it! That is.... ritual. That is— making the bed soft, warm, and comfortable.... what you have to do, to fall asleep. But dreams are illusionary satisfactions, that dissipate too easily— until you die from them! And I've become bitter with a meanness.

VOGEL: ....You don't seem so sour—

KANTER: It's all what's in the head, Vogel! What one mustn't deny is stored up in there, and is ever clawing for some license and freedom to be born.... of a reality, like this very place was founded, accidentally, perhaps.... but so, an' done to do. A good passion gone awry— by.... good people.... gone.... blasphemous, batty, self-sufficiently sinful— but otherwise purely observant of the holy laws, canons and scruples that allow us to pray honestly and devotedly— I hate her not!... and kill not! but would have death drawn on my hands, painted as a tongue to palm's licking— Dreadful! Dreadful! Vogel.

VOGEL: ....Jules?—

KANTER: We must— both agree to such a thing!

VOGEL: Thing?

KANTER: To die, to have languor and to die, instead of being brought to surprise by the result. But I am a coward to kill—

VOGEL: What are you talking about?!

KANTER: —It must be accidentally warranted, caused by a flow of personality, or personalities of invitation— invocation, and deservedness to be received, a back door sacrifice from a front door entrance.

VOGEL: Your meaning for the church?

KANTER: And it must be approved of.... by all parties wanting this— assistance and exercise of the emotions and mind, not by being blinded and usurped from behind?! You must be willing and prepared, and expecting such sort of thing.

VOGEL: I don't know.... if I can be as often stunned!... But we are definitely in the back— working from off our works, resting up, before revisiting our caustic dredges. And what did I mean to do out there? like pouring acid on open sores— to sanitize.... disinfect and kill germs! or raging sorrow in pustulation. And that is an excuse?!... for the fire?!— No!... It was an accident.... brought to (as *Margathrate* and **Lorken** *enter from the church, their arms around each other's waists; although she is ostensibly helping him amble, as he slightly limps*), hounded by, furnished of a view and flushed and flustered from. My cheeks must have been raving scarlet of a blush (*noticing*).... in the dark.... like the disease.

MARGATHRATE: You can sit down here, at the table, like all of us.

LORKEN: No, I prefer to stand.... Mrs. Mason, to build up an endurance, withstand these things. It's only a passing hurt, a thumping that comes on and off. And I have it homed. But my legs must get stronger. (*They dis-embrace as she resumes her seat.*) Shouldn't be so weighted, or seem so, 'cause I've got a lot of walking to do yet.

KANTER: Bob! Up and about?

LORKEN: So this is the treasured community room.

VOGEL: A parlor of workshop.

LORKEN: I would have entered earlier, from the outside, but didn't want to be too presumptive of these spaces. (*slowly walking towards* **Vogel's** *side*) Felt I needed an invitation, I suppose.

MARGATHRATE: You certainly have one. You can spend a lot of time here.... in your manliness to prove yourself.

LORKEN: One builds up reserves, they're not handed you automatically. This is rather busy, and near the kitchen.

VOGEL: An afterthought.

LORKEN: And I thought I heard someone in here, even late last night. (*He is behind but to the side of* **Vogel**.)

VOGEL: That's quite possible. Several have free.... range, passage and access to this chamber at any time.

MARGATHRATE: We're given keys to the door, and open it.... and enter, from the outside.

KANTER: From the brouhaha of cracking embers and moans of suffering, Bob, in this heat. One pleads for some relief.... after working the church, brushing its feet, cleaning the pipes, and polishing its eyes, stones and icons of statuary, and all the other sundry little deeds that keep this old place standing and refreshed.

LORKEN: For visitors?... like me?

VOGEL: ....For our congregation, of course. And for everyone who wish to partake of the proceedings that occur in this religious temple, anyone who would bother of them— or could, could find a favor for our practices.... Lorken. This is.... a pacific place. But the church proper is very solemn and grand, deep. Deep of meaning and richness for our consciences.

KANTER: And not as old as this community room.

LORKEN: Ah. Then people come first. The community precedes the spiritual. That's how it always works. You build up a faith, by gaining gatherers. And then you start a fire, around which to tell tales, or dance or be celebratory.... of your mysteries.

KANTER: You sound like you've come back from the dead, Bob. What insights have you won, sleeping in the prelate's bed?

LORKEN: Nothing but a wonder for it all, Mr. Kanter— savior. Do I seem relatively healthy to you? Much better than when you found me (as) a bum dying.

MARGATHRATE: You must get used to this place.... Bob Lorken. You will share with us, experiences and activities. And we will become.... one whole thing, a single entity, a gaiety of existence, a joyfulness out of the storm, a retreat into higher orders of presence and meaning, a beauty beheld by eyes other than our

own, an' my pardoning to have invited you and needed to—

LORKEN: But the air is just a tad fresher in here, as you said. And it's more comely than standing at the front of the church, making it seem like a porch and possibly discouraging, or dissuading, some members from entering. I am so caddishly attributed, still.

KANTER: They're good clothes, all good clothes, and proud of your wearing.... them.

LORKEN: We do siphon of the breathtaking and revivifying, at this port of draft. It's better than inside for me, awhile. They did not think much of proper windowing, when they built this place, did they?! Then I would stick my head out for some air.

VOGEL: I suppose, the original intent was not so much for residents, with the rectory as something symbolic and kept only for a traveling cadre of priests. Can't really be uncomfortable for a night, when knowing you're about to move on anyway.

LORKEN: And.... I am no priest. I am a murderer.... I would have killed myself, had not you rescued me, or restored and replenished of some— better sense, less common and more pontificating to survive, at all costs survive. So my intents were rightly as a murderer— and not suicide, if you've lost yourself in the wilderness and have no means left, no strength to find your heartbeats again, and your kindness of thought to serve your maker, with whatever that deity crazily wants you to do.

KANTER: You were spared, Bob, to do.... more....

VOGEL: You can't be arguing, with an incredible ungratefulness, that.... you've finally become part of the society, by coldly, stoically, subjectively, and indifferently becoming your own murderer; or would have, by merely letting yourself die of circumstances that became not of your control, in that at the very last instance of.... conscious life, you had no longer the wish— the fervor, the drive to struggle against the end, and the loss of the gift.... the key you might well then have never deserved, all of these miserable days?! You can't be claiming that, can you? straggler an' recalcitrant dole seeker, as an excuse for your.... miracle-sensing repatriation into mankind?! For you are not much more of a man— to anyone, I hope— than when I first helped (to) pick your body up.... to eventually place in this casket. No! You don't die so easily— not out of my hands.... and efforts of caring and trying to understand.... what I do, why an' how.... where an' when.... Because it's driven me, by my character and personal attainments, to give some aid.... and turn my face away from a much too sudden and unexpected fall.

KANTER: He's relieved to be living, Vogel. You can tell by the way he hustles to regain some normal mobility. You can't taunt him on being slack for this, or lazy about it. We *are* glad he's recovering, are we all not?!—

VOGEL: Of course!

MARGATHRATE: Absolutely. The church is pleased. Proud, even.

LORKEN: Then I'm totally humbled by your support, and not the least bit humiliated or ashamed. But when a muscle loses its character or firmness, it tires. That's all I'm claiming. It limps until reinvigorated, through some store of vested interest in its actions. It comes to ignore that interest, for awhile, feeling abandoned or despondently aloof from the world of pure causes and actual being. And I was limping of the head when I fainted, for want of hunger and for hunger, and for meaning to this.... travesty of existence. It was darn right derisive to be brought down to that state of hopelessness turned into uncaring, no longer even needing to question why, a true comedy of awareness. And then I fell, right on what I *could* have needed. And that meant it was nothing done, can you understand? This life was nothing done, if I couldn't even use that which was lent— some concern for me.... And then I woke up, during some examination— or discussion thereof— in a holy bed. And then I felt, I learned that that concern was profound, way over my head, much greater than my petty pondering. And so I enlisted.... the father for explanations. And he was very explicitly generous. It was as simple as being fed— and being happy about that. Now I am obliged, to be as happy of the assistance as I can ever become worthy of it, in this house that has— smoked me back to life! with infused oak scents for a sturdiness. And I am ready. I am ready to go back out into the world to be burnt up a bit, to be a fashionable and resistant woodiness that may be built upon, as all subjects should be, unknowingly or keen, within the domain of a most magnificent construction of empathies, humanistic harmonies; since that is obviously my purpose, now. They make the symbols for the lessons shown so bluntly and right to the face that their perceptions are easily missed as a blur, so vigorous is the need, made and presented to me as abruptly as awakening; so frenzied an' insistent, agitated and discomposed is the maker to have this done. An' for all of that my calves hurt— anxiously, to get going— on my chores, like this pain is a warning not to delay too long for them. I *have* been resurrected, my friend Jules Kanter, to be a servant more simple than before, to simply carry life— and breadth of blush breaded within me, by saintly Father Mathews, to walk over the dried mudflats— and persist!... within the harmony of mankind this community and any other may allow. For you will not! let worthy souls expire before their work is done an' due. An' last night, though I may have been dreaming, I lay in bed, and God came up to me and paralyzed me frozen, as is what happens in his presence. And he bent over and looked me straight in the face, and told me directly, in his august tones of reverberant depth, that it is wrong! to kill someone. And then he lifted himself up straight, for a more erect posture again.... and walked away, the darkness of night returned with his receding. And I normalized myself to relax — of this simple, declamatory message: Get busy! and let the people prove this!.... And so I stand here— growing of the community, and racing myself for a physical wealth to attain, earn, for this mission. The weather does not harm us, not nearly as much as our dispassionate natures towards one another's welfare. And we will prove the contrary, that man is made high enough to overcome this heat, and the sweltering poverties of the soul.... or character or personality or— being.

KANTER: You are much charged, Bob. This is a good community still.... Did God.... touch you?

LORKEN: ....If He did, I didn't notice. I.... wouldn't have been able to tell if He kissed or bit. I was too much as stone. But He told.... me. And if this was a command, in some way.... or an affirmation against my doubts— of people.... then in either case it was a lesson, to become zealous over— without a zealotry to proclaim to you that I am born.... of this church an' community, now. I no longer feel.... so unusual and outcastly hued amongst you. I have become fertile.... an' fruitful, to share of the juices an' *Saft* we may imbibe on together as sensitivities are restored.

MARGATHRATE: You can do that best working around here, staying here— as long as the father needs you. And the church needs so many volunteers these days. You are a welcome.... acquisition from the parched ground, the proof of our obedience to nature.... I'll have you much around—!

VOGEL: What can you tell.... from a gospel of discernments, Lorken? You may seem to be.... rather uneducated. The poor may make the church, but the church may not make the poor. And if it's a disciple you wish to become, it's not from here, but from your heart the desire is drawn. We may have lifted you, our magic restorative from the damage of your fall, death, even. But we have not crafted your discernment, and what you may distinguish of a mission.... given. For certainly you are a burden here. And you can *not* create bread. And volunteers are volunteers from— other work, their precious time to lend. I am for the tramp to triumph over his causalities. But to be remarkable in this is too much to ask of you, or any person. The sights are too highly placed— And I do wish I may recover from them myself, at times. But work can not so easily be found for you, in a poor church.... We will convene of a congregation to find you a position or a job somewhere around these parts, as is the normal case when dealing with a— worthy indigent. You have my approval for this exercise to be granted, in terms of recommendations, pleadings or requests, and the like, as well as the approbations of others, including the father himself, to employ. And this will eventually be done for you— Though it is a difficult task.... because one always has to question whether someone of your.... condition and conditioning— *can* be made into a useful object of society, given the many hardships that one's self has endured and overcome to be a fully functioning citizen of value to the community. Suspicions of character must be outdone by diligence of hard and earnest employment— which simply means that you must prove yourself in every task to at least tackle conscientiously, and faithfully meaning to do well at or with some level— always recommended— of sufficiency of effort. And griping is not allowed. You must turn then into smiles. Pledge this much, at least, and we will all help you. But don't try an' tell us you've taken on an apprenticeship towards sainthood, or even emulation of the father. You are not of that type. You are.... a broken man mending yourself, with an astonishing self-revelation that you actually want to—

KANTER: Lead him not into a bashfulness again, Vogel!

VOGEL: —That is all that occurs here, and which should occur. That is what a religion, or any set of principles is principally for. I.... judge myself from such standards, recovering from the bends, as an analogy, from falling too deeply in the ocean of human misery or disgrace or.... circumstances that make you feel disgusted with yourself, even if you're not really to blame for them— But they happen. And you deal with this in practical ways, through strong beliefs and spiritual guidance.

LORKEN: ....Yes. It's myself spoken to. Wrongs were made for man to do.... in order that he may correct them. An' I'll not claim to be any more special.... than your generosities may warrant on my behaviors. But the truth is I've become to love you— hardly even knowing you. And that is the glorious gift that has escaped me throughout most of my life. So I will do as you have pardoned me to try, and make some corrections in my walking, an' be the blessed astonished and the forgiven of crimes and the forgiving of impugners. That really hurts the most, for a fellow that strives to think a way out of a whirlpool of confusions an' contrasts to face, to be viciously attacked of motives.

VOGEL: Then we might be stablemates.

LORKEN: Being misunderstood should not lead to dire consequences. The natural and proper evolution of events themselves should draw forth the sanguine results from actions and reactions, without fallacious or feigned interpretations to support some theory or conjecture to uphold and celebrate with, like: the deprived for long must automatically become demented.... to certain aspects of a civilization.

MARGATHRATE: It's a common but foolish habit to look down on the assessed underprivileged, and treat them sparingly of any sort of high regard, even if judged talented for some useful purposes. A society like that must have its castes of myopias, taking people for granted too often or perhaps autocratically, whatever assumes a rulership, as if one person be a governing institution. One person may rule a marriage, but not a life, and not a life's view of the world, and may not accept or condone the resentments of another that have been building up under a prolonged period of.... misunderstanding. That brings many to states of torture and twisted sympathies. And yet the poor often remain poor, despite societal concerns and perhaps misguided but protectively gloved assistance and aid.

VOGEL: Well, one takes a fellow to be himself, with the alms received, Margo. There's no denying it's a beneficial gesture. But the fellow must stay the same, by and large, and without the grandiose transformations due to a supposed inherent or ingrained appreciation for the gifts— the doles, to keep one complaisant under a rule. Be this from gods or men or the sturdiness of metal trash cans or garbage bins. Swine be swine to pigs and cattle to their swineherds.

KANTER: So now you fall on the philosophy of hogs, how man.... uses up the resources given him on this earth; thereby our religions help to keep him sane and sensible during the decimations, and the emaciations of those few or groups of poor who habitually luck out during these competitive thrusts. Well that's just a pork's roast, for the fat an' greedy.

MARGATHRATE: And hungry—

VOGEL: Capitalism and Communism are not mutually exclusive, Jules. One can't hope to protect everyone. And whether ownership is personal or collective, whatever products are limited.... or limiting.... will be kept from some— And that includes hope, during a disaster like this drought. If you can't put up a stake of savings, or borrowings, to defend yourself and your comforts, and if this can't be done for you, decreed of purpose, roared of a savior to demand.... then you starve, wither away an' die. And who's to say this is not right? even if you've followed the word.... or come to it — or the letter of the prevailing law, when abnormal circumstances treat you this way. Even if you're butchered in a war, religions hold. (*Paths enters from the church.*) They hold up no matter what happens to you, because you are the swine—

PATHS: Let's have a little less talk—

VOGEL: —to take tally of.

PATHS: —in here. Mathews 's trying to practice his sermon at the podium, in his usual way of not saying anything at all, but just reading to hope to memorize within an affecting or inspirational atmosphere. But I'm hearing faint sounds of a desultory conversation that's disturbing even from the far side of the pews. And we want to keep these doors open to air through the place without turning on the fans, till the parishioners come. It's my suggestion. They don't have to be greeted with an oppressive stateliness of beam-laced scents, no matter how impressive of this church's religiosity of rich wood resins. And if we turn on the air-conditioning for an hour, that's an hour's bill of extraneous hard labor, in this heat. So let's all be complimentary, and contributing to the holy service by muffling it up a bit.

MARGATHRATE: It works harder when it's hot?

PATHS: Of course it does. There's more hot air to cool, the volume's greater— the pool of air creates a higher workload, and electric bill, pulling those streams through.

KANTER: The woods almost seem to weep, Paths.

PATHS: Seepage.... of ancient qualities and ages, and essences of — spirit, and our indulgences.... through innumerable ceremonies in their presence. What they have heard of us! amounting to a grumbling of praises.... and lessons. And the wax of the candles become soft, too soft to work with safely, sometimes.... But that is the warmth, His warmth. And we shouldn't have much fire in the building anyway, during times like these. A flame could hop onto a joint of timber, and too easily or quickly lead to the fiery consumption of the entire wooden frame. We have to be careful. I've been arguing to do away with open flames altogether, during the summer. All it takes is a slight modification of procedures. But the father is.... intellectually timid about this.

MARGATHRATE: We weren't being unruly of our thoughts, deacon—

PATHS: Were you, now!

MARGATHRATE: —to demean any aspect of our faiths. I was just introducing Bob to the community room.

PATHS: That's nice of him to join, as an improvement to his perspectives.

LORKEN: I—

PATHS: Mrs. Mason.... in this particular space one can, and does, say anything one wants. It belongs to the town, and is as much for its character to express. I've even heard— tell.... of cursing in here. And of course you are what you are and would be, out there in the ever expanding landscapes and locales of busy persons. But the confines of our church proper are quite different, and the door between here and there is usually kept closed as often as possible, because the ears and eyes within the sacred, sanctified spaces and areas decry all forms of decadence or moral brutality. The transformation occurs (*pointing*) in there, through regularity of entrance, visiting. And not in here are you changed, spiritually refurbished, but rather do you rest the muscles exercised to gain full benefit from the lessons learned and worked through, with great diligence of mind and fervor and courage, that you may return to the outside— to the world.... enlightened and past enlightening, until you must seek the church again. Without a prudishness, this room is tolerated as a compromise to your troubles— to express and argue over and analyze or discuss and laugh at. But the alter should not be smeared with them, and the pews should not seat them, and the stained glass windows you so beautifully.... brightened.... should not shield them with their views and their angelic considerations of face. And our father should not hear of lowly complaints as he is trying to reach for, with his own words, expressions of the radiant glory all worshipers seek to see and review and adjust ourselves to and towards. That.... greatest of all heavy labors, he has made vows to, and commitments for, with impossible difficulties to overcome, lest it be not allowed at all against community impatience, indifference, ignorance.... or insurrection towards conflagration of each considered soul.

KANTER: I hope we don't incinerate our volunteerism inside this room. We each carry much similar burdens throughout our lives, Paths. It's quite a struggle to be helpful, and to learn how to be. One can often assume trusts to an absurd and ridiculous limit, but the intent remains meritorious and proportional to the concerns we have for one another. Why should not the church be more aware of this? instead of remaining so dogmatically deft an' blind to the turbulence of a living cataclysm each person owns and dominates! It's difficult to stay reverent to the mystical when the non-ecclesiastical so often crushes— And I'm not one to protest on the promises, that all of this sorrow will eventually be reviewed and— in some beatifying way— corrected. But some slack should be allowed, with our complaining, in this room— and in the world at large, that very little is right and very much is wrong to promote the metamorphoses desired from crude and callous skin to pure and sacred souls. For everywhere it seems our common motive is to make mistakes, and to perfect ourselves through them— to make more mistakes. Then man is the greatest error ever committed, by highest order the finest crafting of quarrelsome contentions within a body, or a body of bodies, or a whole universe of belief systems, be we alone on this planet for the absolute and utterly sole entity of destiny— everywhere of a consciousness!... So I would plead with you not to chastise our disgraces so— perfunctorily. They may be what makes the rainbow seen and appreciated— and what the creator really has in mind for us to do!

PATHS: That may be so, frère Kanter. But one shouldn't have to boast of a grin, or a sensing of the inappropriate with a smile, or an appeasement demanded because of one's natural and uncontrollable promiscuities, whereby one 's always to be forgiven if striving to stay chaste to the propriety of logistical scripture, this life ever being a battle to gain entry and welcome into heaven, a movement of personnel so profound as to be stationary— on earth, and remarkable of the destination. One shouldn't have to brag out loud that this is so and let the fires rage of sound and sight for holy ashes to be sensed, touched, tasted and smelt, for the great calamity that awaits!— (*Organ practicing is heard to begin.*) Oh! How one is bothered by the sincerity of interests, that we must ever be fought for to uplift and inspire towards a goal of commonality of worth, and through the errors breathe an' breed! Kanter.... It is not cruel to be wrong. It's wrong to be cruel!

LORKEN: I— I'm not that way, Deacon Paths. I'm gaining my strengths for less to be a burden an' more to be a brethren. And this change is not from worship, but from wishing this could be so. For the light has shown through darkness— It has revealed me, and I am weak to be displeasing evermore. Yet, what I see is nothing— or what I saw was everything contained to be witnessed. So what is

left is a shallow surf indeed, of notice to take an' struggle through and be divine against, as my legs make pain of envy for your goodness and our greatness to reach. Since this is true, we travel through ourselves and to, and in a circle make our trip together for ourselves as one to be, the same of pleasing might and sacrifice that He may dole upon. This is the righteousness of man to be— an image of himself and his disasters played through. And so the cruelty is ours to suffer an' not strain for disappointments. It is served as we do serve— And it is right, to purify through our displeasures and make us ever more humble for the efforts spent. And it arrives like a blessing to be real, and a wounding for— to pay attention with, that hardships are not crimes against us but a cowing to lessen the pride of the greatest of all creations. And I will walk a thousand miles with this same difficulty to reduce my boast of blasphemies—

MATHEWS (*entering from the church, with notes on his sermon*): This could be an evaporative deluge of content to listen to. But I can hardly think for myself to read my own writing, interpretations on the Holy See's commands for how we should attempt to abide of the prophesies canonical. (*as **Paths** goes more towards **Kanter***) Yet, to hear of an awakening throughout these recesses of solemnity, and then to here, as could a cannon blow of sound, Bob, confuses purpose for my speech and speaking to parishioners!

PATHS: I did try to warn them to tone down their hefty palavering, father, as you were concentrating. But it seems now as for a palace to be brought to timber from more agley sounded timbres. Then could a house be played through and through for wrought of a livelihood.

MATHEWS: What is wrong is what pleases, then, for a cacophony of hope! that could be claimed much better than my arduous scribing to tease for meaning through constraints of reserve and decent deportment as we observe God's word and try to be pacified of His intentions for us. But yell throughout, and from the community room?! Could be as much to scream as one thinks. And do I complain? Or am I inflamed with honesty for such virtuosity of expression? Then I should revise life itself to seem, before me spread of intense nature and a calling. And such to be, as I would handle flames, I hold you dearly to me. But let me think, as I would see through panic, for your faces to behold, and scratching for a pettiness of writ! My ears are sharpened not to hear my thoughts, and my eyes blinded from seeing these few lines (*holding up his notes for notice*) punished through. So take more care of me to be sensible, that I know of your conversions— and adopt them blatantly to my cause—! (***Harold** walks in from outside, with a defiant, perhaps jaunty dignity that startles.*) What be? man! (***Lorken**, nervously surprised, anxiously acknowledges **Harold** with a gesture, calling his name.*)

*In concert: **Harold**, dispelling all emotion, quickly walks over to the table and grabs **Vogel's** knife, pointing it with threat as **Margathrate** screams. **Harold** lunges forward, causing **Vogel** to awkwardly jump out of his chair to avoid the blade. The organ music stops with suddenness as **Harold** stabs **Lorken** in the upper torso, just as quickly retracting the knife. In the confusion, as choreographed, **Lorken** stumbles backwards onto wall shelves, which collapse to cause a great disruption of materials (tumbling down), as does **Mathews** loudly gasp, flinging his sermon papers involuntarily in the air. **Margathrate** runs into the church as **Harold**, with a serene but determined satisfaction drops the knife to the floor (with **Lorken** also falling), walking backwards slightly and plac-*

ing his arms in back of him. The whole display of action is done fairly swiftly.

HAROLD: Now! Take me to Hel.!—

KANTER (*jumping out of his sect*): Call the authorities!... And a doctor!

Paths rushes to restrain Harold from behind; and Vogel re-balances himself and runs over to assist Paths, causing a scuffle of much intended roughhousing against the assailant, as Mathews rushes over to Lorken.

KANTER (*going towards Lorken and Mathews*): Hold him at the steeple tiers, so that he can't get away!

HAROLD (*as Paths and Vogel cre dragging, pulling and shoving, and in general man-handling him into the church*): Lock me up! And pull out the key!

KANTER (*bended down with Mathews*): Father, your scarf! Let's try and stop the hemorrhaging!

HAROLD (*he, with Paths and Vogel, just within the confines of the church*): I want something to drink!

KANTER (*while he and Mathews are working on Lorken*): This is terrible!... (*yelling*) Call! Call!

Scene II — *A cemetery grounds. Mathews is in full attire, having conducted the service for the mourners, which include Vogel, Pahtes, Carl, Kanter, and a few cther citizens of the village not directly involved of the circumstances but coming out of respect from hearing the story of the indigent Lorken. The principal characters— aside from Carl, who despondently stands alone, looking towards the burial mound (which may be unseen)— gather together.*

PAHTES: Well this is one poor fellow to get a decent grave.

VOGEL: I'm still convinced that crazed devil was gunning after me, father, finding whoever he could (to) jab at and kill, once I fell out of the way.

MATHEWS (*quietly, or subdued*): You?

VOGEL: There are reasons, for :hat tribe. A sense of justice is simply to explode, if a notion of being harassed lingers in you too long. And those types have certainly been under the pressure for too long a spell now, before something like this would happen.

KANTER: Well at least the community is aware of the dangers that can brew. This madness that can befall the homeless and wandering indigent is no longer a surprise, not in our village so affected. Though it will be mightily more difficult putting up any of them into family residences. I'm afraid to imagine what could have happened if that fellow had come around to my home, with Bob lodged there. My wife was right to be reluctant for his stay, the intuition from a judgment of licenses an' locks, how a person must be hemmed in by his associates, and what those few feel free to do. But I would have allowed it anyway, had Bob wanted to try. We must all take our chances in life. This event calls for notice of a swarm of the desperate around us. These poor indigents can fall in-

sane so thoroughly, an' perhaps readily, that virtually no iniquity may escape their actions or contemplations— And our governing authorities should do more, much more about the problem that can grow and be exacerbated to extremes like this under the oppressive conditions of an excessive summer heat.

PAHTES: As horrible as the outcome has been, in this instance, I think we can handle, ourselves, fairly well these general circumstances, given half a chance of decency and a propriety of being alert to problems and hardships. What occurred here was purely an accident of misfortune, and those things are difficult to avert, a bit easier to foresee. But steps will still be taken to try and alleviate some of the tramp-ish suffering during this hot season. Hastern is even still not dissuaded from adapting his hotel to the purpose, in some way, if it can be aided with organization by the local policing staff, if our municipality will allow this. One has to even show— some.... pity for a man who can go mad under this heat, under this baking of the head, and help try to avoid the same for others left outside and.... in the wild, as it were. It was the simple instance of having that back door to the church open, at an unfortunate time, that has caused our loss.... and an awakening to the intensities of plights.

VOGEL: Well.... maybe so. But you can't count on madness to be predictable. He walked straight into a quite occupied room, as if nothing were there but a target— to find, oblivious to any conspicuousness of his entry and presence, with not even a sense of the holy to approach or get near to. It may have been improperly fortuitous, for that door to be open.... But he could have opened it, if you get my sense of the deed. And he could have been a shadow at night, misplaced during the day into some legitimacy of horrible form.

PAHTES: He could have entered the church properly, from the front, and attacked the father, or Carl.

MATHEWS: One could have attacked me in the community room, Pahtes. He did not go after me— He didn't come for me.

KANTER: That will be a lonely spot for quite a while, until we get over the grieving of our losses. Volunteers will not want to bench themselves inside, unless, perhaps, the room becomes a novelty of historical interest to visit, an unsuitable attraction for the curious, or the morbidly religious fanatic. But I can no longer sit in the place for long, as I have loved to do, knowing what happened in there, and exactly where at. I would fawn there for my bones to rest. But the space has become too irascible, for my sympathies to grate. And I found myself sobbing uncontrollably, in my chair.... thankfully alone, the last time I was there, as if I felt I could have contributed to causing the misfortune that occurred— that irrational sense that becomes you with grief.

PAHTES: It's certainly understandable that any of those present might have stopped that madman if they could have reacted to his threat in time. But that is a poor logic to adopt. Even Paths was present, and he couldn't prevent it. No. No one can assume blame for this. It must have been meant to happen, within the mysteries of our servility to God. But one thing we can do, for ourselves, is close that room up (for) a spell— and change the lock to the outside door. The community room has been shown to be a weak spot for the church. And if the community wishes to maintain it, it will have to design its use better.

KANTER: That door was only open because the air-conditioning was off—

PAHTES: If some tramps make part of our community, if we feel obliged to treat them as so.... or lean towards this treatment, then we can not trust all of our community for access to it. And while by their nature they are not volunteers, one did manage entry.... or two.... There are complications that have to be smoothed over— and complexities that have to be re-thought through, before we can wisely allow the back door to the church for use.... Still, it remains for the will of the community— And they'll have to pay for any modifications of structure. The bishopric seems too shocked even to rationally acknowledge its existence. Right, father?

MATHEWS: Yes.... They feel this is a local matter of particular character, or peculiar, as they say.... which we ourselves must deal with. A solution must be home-grown. And our church is especial to us. What occurs in it must deepen our resolve of faith. (*with stern resignation*) But what happened in that room was like a rap over the knuckles with a heavy ruler, by the headmaster, to give us due warning of a punishment that must occur when we try to pass on our labors unjustly to others, and abdicate our spirits to an abjection of the heart against the fervor we must always stoke and try to maintain to do good— and be obedient.... to the principles and intentions of our worshiping. Any other thought is a crime.... that may lead to disgraces. My speech today.... was very simple, for Robert Lorken, for whom we mourn. The essence of it was that.... he was no disgrace. And that is how we must come to an end. It doesn't matter what others may think of you, if you are worthy to have found your rest.... out of satisfaction with your life. And at his end this man was born, I'm sure of it.... to be thankful for us, in eternity.

KANTER: It was a very nice sermon, father. To the point of his simplicities of need. Not a one of us would have been prouder to adhere to those words spoken over us. And I feel.... he's been amongst true friends.... in the end.

VOGEL: Where's Mar.... Mrs. Mason?

MATHEWS: As for with guilt, she carries much, as inappropriately as Mr. Kanter. She feels she should have not led him into the room, and is too fearful of her sadness to display.... at this burial. Aside from being interminably frightened, she tells me her lesson.... has been devastating, and breaks her (down) into a search for total absolution. I have advised her to simply seek honesty with self-instruction, and her problems will eventually resolve themselves away.

VOGEL: Could be.... Could be she was over-indulgent. But she did not cause a murder.

KANTER: Merely delayed a death?

PAHTES: Or belied, Kanter.

VOGEL: No one fought over her. She fought with her self, is all, and observed some mean business along the way. But she'll.... get over that. We'll all recover from this.... experience of souls. Life will calm down. The seasons will change. Our desperations will drift an' wander, and dissipate as we mature to the cosmic realities of the universe. Whatever we do.... it's not nearly as powerful as what is done.

PAHTES: You mean.... what results from our actions, the intended, unintentional, and accidental. Well, that might explain the whimsical for a life, having less control over it than you thought, the more you age.

VOGEL: Perhaps—

KANTER: It's all the same kettle being drained. But Bob Lorken was a decent man, it seemed to me. Decent!... An' I wanted, to take him home.

PAHTES: Carl looks especially despondent, standing over there by himself, studying the burial mound. I guess it's as much for curiosity about death, you think? at that age.

MATHEWS: They're both children, Pahtes.... searching for a way to accomplish their talents.... within this earth.

The group disperses, as do the mourners in general. **Carl** *stands singly for awhile until a woman, ostensively his mother, leads him away.*

)()()()()(

Author's note: Obviously, Paths and Pahtes can be played by the same actor, as well for the two unnamed indigents.

144

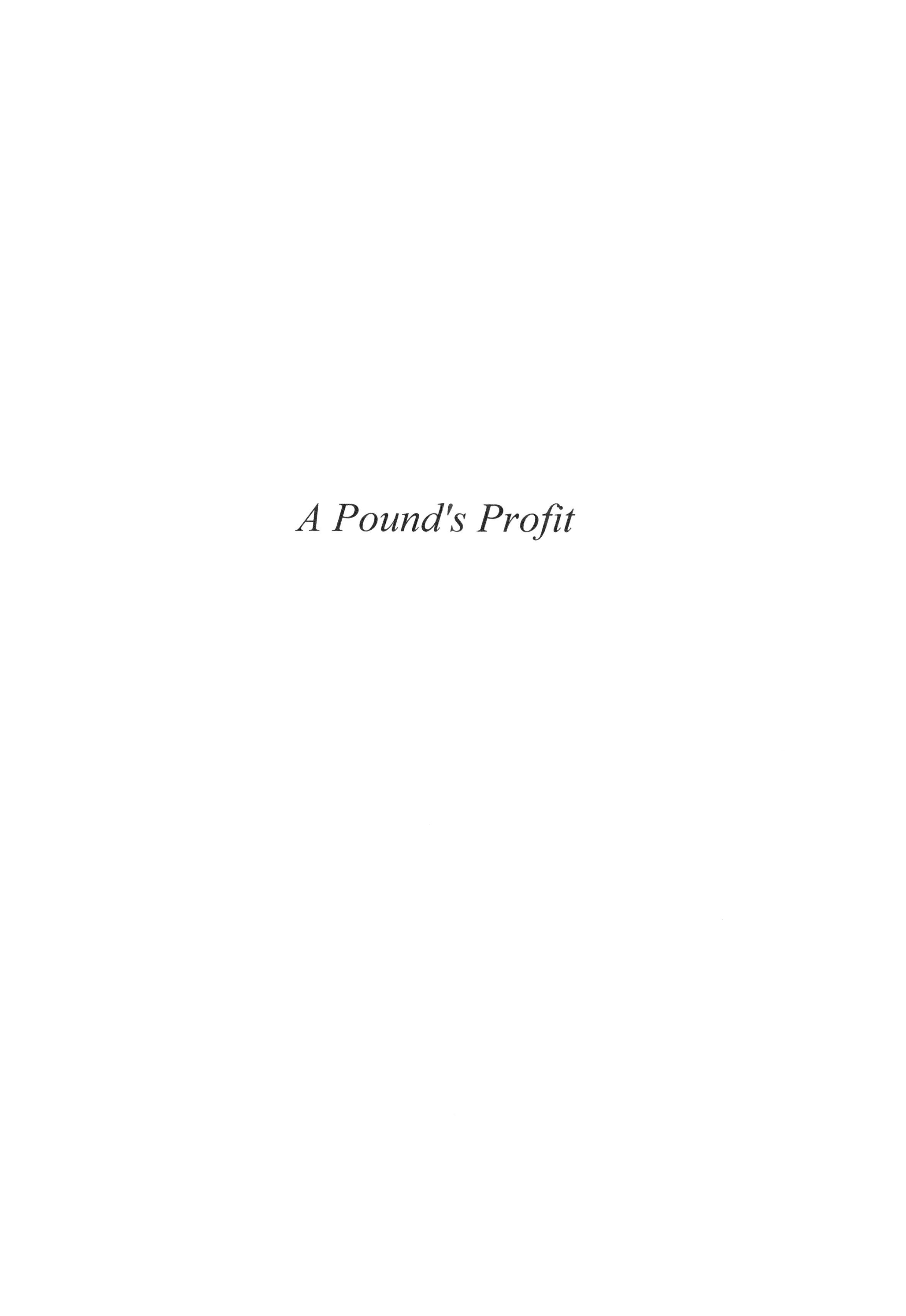

A Pound's Profit

A Pound's Profit

Characters:
 Fennel
 Skacte
 Duke Decamond
 Count Elser
 Count Maeli
 Alfonso
 King Musgonie
 Councilor Ponstole
 (Scott) Bracken
 (Walter) Townes
 Squire Darughe
 Calerand
 Wallinda
 Councilor Costile
 Jocqueral
 Lord (John) Bracken
 Queen Matile
 Lord (Hugh) Townes
 two other Counts and two Viscounts

Locale — a kingdom and one of its remote provinces

Act I

Scene I — *A small clearing within a forest, during the day.* **Fennel** *brings in a chair and sets it at the periphery of a makeshift circle removed of fallen leaves, though not plowed of the grass. After carefully adjusting the chair, facing inwards, he stares at it, then sits below it on the ground to the side, lowering his head slightly.* **Skacte** *enters with a crate box that may be used as a table or desk, though it is large enough to be cumbersome and awkward to carry due to its volume or width rather than weight. He places the box next to the chair, distal from* **Fennel**, *and hardly noticing.*

FENNEL (*lifting his head*): Want for my serving? He'll be pleased with drink, a pitcher of the dew, a draft of the sciatic cravings. I'll do to the relief. It is balmy warm this morning. (**Skacte** *looks at him for a moment before picking up the box again, with exaggerated involvement to the task, and carrying it over to* **Fennel's** *side, placing it down next to* **Fennel** *so that he sits between it and the chair.*) It's a tightly woven duty, to carve out such a spot. But don't fence me to be bound up and tied. I offer as a pet obsequiously unnoticed, except from the ladle pouring from. "Come, Fennel!" be the twitch, with a pat on the head for drink to deliver. Potent punch.

SKACTE: Potable p*urrr*— chase. I crowd you not, if you're not to persist of this p*errr*suasion declarative to provide. With a wish for winds in brewing, the time he takes a whiff at first he declares a sojourn here, a sortie to the forest faeries to arouse like butterflies in spring, imp*errr*sonally demeanor-ed. It clears his head or nostrils, he claims. I don't know which. And that's the fiendishness of power. We want to obey such might. We want to please it and ascend to its purpose.

FENNEL: Skacte! I only mix. I do not bake nor blend, but provide the list of ingredients in their proper proportions. And the ladies fashion this clip of lip— Though I do sip a test, that it is strong enough, not so diluted of the wine.... It is my bakery, more

or less, to roll about and describe, a happy day as this, with naught to garner of it more than a pleasure breathed. It lasts innocently, and with a casual wear, greenly sufficient, grandly indifferent, and irreverently disposed to our entertaining.

SKACTE: Could hardly rain more, in the round.... It is dry. And he dares clouds, on a sunny day, rushing for the wounds. So stays one active, as a judge, an.... assessor of the weighted caliber.

FENNEL: What was invited?

SKACTE: ....A cardamom. An adventurer traveling to impress. What pays his own way to become of this imperial splendor—

FENNEL: To adopt and adapt to.

SKACTE: —Some form of magnificence in mind, for the magnanimity of invitation.

FENNEL: I would rub his bottom, to search for wooded knurls, coming to this place for a joust of proving. (*gets up and brushes his backside*) The ground's not so damp as were my lusts to it. But this will chew.

SKACTE: They are the delicacies made for fighting and hand-wringing alike.

FENNEL: As much as can be consumed in clanging metal jaws. And it be said they're bred for this, as I for my potions and you for your.... furniture, or furnishings.

SKACTE: Carrying what is provisional, and around, I might only harness and hoard, collect the oddities to make use of. (*as* **Fennel** *sits atop the box*) But there is a taste of deference that some men sense to fling themselves into the cataclysmic upheavals they may find swirling within sight or smell of a disturbance. It is.... the simple thing to do, without much thought of figuring. Not as hard as to create, but harder of a staunchness and stolid need to be a hero.

FENNEL: Our master dines on them, and burps out smelts of accomplished lore and gore.

SKACTE: But he is with patience configuring. This is the least be said of a leading light. His causes are precipitated to obscure the means and intensify the motives.

FENNEL: Pour powder on my shoes, and I will make you dust them off.

SKACTE: Or sawdust grains, or earth upon the dress, does make for a response combative, pledged of some activity. Yet the man chides for being inactive.

FENNEL: He wants a cushion for his behaviors, with wild things flowing from his lay.

SKACTE: You live to eat from your hunt.

FENNEL: And how may, then? How many cups to satisfy a thirst, a teething with pain, a bloodletting to soak the blankets and the straw? It is a gruesome proclivity. (*getting off the box*)

SKACTE: Not banished, but promoted.

FENNEL: This is steady, sturdy enough.

SKACTE: These are the times, Fennel, for such monstrous actions. A skirmish may lead to riches, either to end the afflictions or to capture treasures. And he volatilizes the volition, with many confrontations made. This is well-known, whatever the excuse. Arise our prince in plumes of fume and fury, this is the instigator of victories, the cracker of the wood and smiter of the senseless found before him. And as wealthy as his smirk he bids us our morning knowingly. One can't let such a fellow languish indolently, he must find his attacks.... and attackers.

FENNEL: Keeping strong, I think he has a sense that emboldens, an intoxicating fuel escaping from the neck that keeps his nose up tilted to stay alert to cruelties and crushing.

SKACTE: What is amiable is ability.

FENNEL: Yes. (*going over to the chair and slightly readjusting*) The man can do so much, and mince as well as mortar. And for deceptions made, he plots unto them. Care he of us? That is a courtesy.

SKACTE: Bound by the eyes, we might be flagged and flogged otherwise.

FENNEL: Not for fear, but for dissembling our overtures of usefulness. It is often, always curious how loyal tyrants may be. But he wants this chair just right, not leaning to or fro, near or away from an interest, or he will rant my legs be bowed and bent.... shapeless of utility to level.

SKACTE: Decamond is picky of his seats, yet not of any day in particular to find for his requests, a silver to the air, a sleekness of the pined perrrfumes. I am afraid; to find him toppled over in his chair could cost an ear.... of rages to be heard, of complaints to.... failings, lack of foresight, bruised futility. So generous is he, though, as to pat your head, and fill your mouth with gold, or golden meal and corn-like succulence.... and take my palm to rub my chin and cheeks with, contemplating of my thoughts.

FENNEL: We are defined for him to abide with, servants serving freely, responsively, and with an indefatigable candor for his approval. For what would happen should he cease?! We would be left to drift disabled of our vim— vital valors, scathes as we are to his scratching.

SKACTE: And for paw, for penchant make. Trapped to our liking be our elections. Let heads rule bodies. Let thinkers be the distillers of destiny. And let the fortune hunters suffer through and through, by degrees lent the servitude. I would have a better mind to mend than his all-knowing certitude, as for digging to make ditches. (It) Takes some command of notion to be so sure, where at to start, and out, and the like. More sophisticated than a hunger is a crafting of directions to relieve the pangs; as what could tools be used for else? than to an artful mentality.

FENNEL: Then art must always bleed success, if bred with an uncertainty for skillful endeavors. Define the simplest of labors to construct a solid whole of application, wherefore complexities are guided fruitfully to make some action substantial. And that is justification enough.... for this employment, his implementations.

Though I would channel fright through thicker heads. Some risks are water to the mud,— makes more.... Here at approaches a penetration of penury to bone, for our crimes serving with enlistment, much gathering to relish in a man usurpations' prominence and promotions. (*Enter **Decamond**, flanked on either side by counts **Elser** and **Maeli**. At a distance behind follows **Alfonso**. **Decamond** is outfitted in a garment that allows his left lower arm to rest in a fashionable sling-like pouch, such that it may not be immediately obvious of its ambulatory use. While both counts are armed, **Alfonso** is most obvious of sword, though he stays respectfully in back of the procession.*) My prince! The round is ready. I will go for the refreshments.

DECAMOND (*as **Fennel** is exiting*): It's early brisk, Fennel. Make sure it's harsh. But the slight chill vivifies the lungs and skin.

FENNEL (*as departed*): The coach of provisions is just aways left, my liege.

DECAMOND: Skacte!... You could have furnished more chairs. (*sitting*)

ELSER: We asked for none, Decamond.

MAELI: One most commonly stands or lies in a forest.

SKACTE: That can be accomplished, sir. The fittings are loosened, parts separated. They may be constructed. It's all in the carriage.

MAELI: Not necessary for us, though you do convalesce.

DECAMOND: Oh, pool of air, counts, does make my bidding to refresh in bright gamesmanship. Where other than a forest to restore one's health! I've been punched so squarely in the face as to loosen teeth, what for a fist to greet as being lucky. And I had disarmed the cavalier; so a fight of hands and kicking was all due, the pace of feet as could be a dance as he resisted my assaults and grabbed to hold me thoroughly at bay. It's to his death I plunged for, with might as much in admiration of his hopeless defence. But this later pinning of the shoulder blade by a spear, and a pitifully small one at that, does leave me tender. The wound drains still at night, even after these weeks. And the upper arm remains much too stiff, though the lower is recovering with rest and disuse.

ELSER: You are quite one for a challenge to your challenges.

DECAMOND: It is they who seek to separate, make gambit out of authorities of our sovereign king, and as such fall prey to my surges of recapturing— interests and interests. And who's to say this is illegal?! These severing attempts, to leave a loyalty dead and cold, happen in spurts all over the territories commanded. And they're so nicely temporally spaced for my invasions, based mostly on feigned religious or economic grounds, that they assist— or give reason for— my own exaltation of dukedom properties, either by wealth or accumulation of land, and rightfully stolen from the impudence. Though our king grows wary of my aggressiveness and accretions of strength and influence. But who's to deny I have the right if there is nerve enough to ascend through these opportunities? Wealth and riches, they're bitter without a reason for them. The king compromises throughout his realm too much so, trying to avoid.... disputes. But when some subjects break off— and only when they actively try to, above threats and warnings or implica-

147

tions and indications— they are bait for my involvements to prosper through, as long as I remain allegiant to the timid king. This last attempted despot came to his knees before me. I had bargained and arranged for an indemnity against my trouble and time. And there he was, across a field, with a veritable pot of gold, expecting my soldiers to come over and pick it up. We exchanged signals as I ordered *his* to bring it to me. Finally, after this wrangling of symbolic approaches all the gray day, I had our weapons raised, so that their battered men could see our anger, and sent a messenger with a terse letter exclaiming that *he* should carry the treasured parcel himself, and place it right before my feet.... which he did, a sign of real subservience— and reattachment. This restitution was enough, the land there not at all up to my standards of beauty to own, or control actively. And the king has intervened anyway, readjusting taxes and tributes. But this last adventure leaves me sore awhile, (*as* **Fennel** *approaches with a tray of cups and a ladled pot or punch bowl*) and waiting for the next. Gripes soothe the griefs, and grinds the groans into garnishments and gratuities of girth and greed— Come! Fennel. The dew! (**Fennel** *places the tray on the box and pours a cup for* **Decamond**.) What is it really? sirs, an initiative. Is it really greed? or an impulse to do what can be gotten done? (*as* **Fennel** *hands him his cup and the counts go over to pour their own*) It is that I am made for these.... transactions, and transformations of necessity. (*sips*) This brew blinds me a bit, the chill that warms a chill.

MAELI (*after sipping*): Whew! This is a stout *aranciata*, Decamond.

DECAMOND: The orange does typify my golden flame, as for the apples maimed of capture— tamed!... for what of taste. There decadence is not desire but disdain of ease.

ELSER: It is strong!—

DECAMOND: Through the veins coursing, opening them up, permitting better flow. Fennel seasons for the season. No need for a hot heat, when a hot heat's cold. We must ever train, to endure the extremes. (*as* **Alfonso** *approaches, noticing there is apparently a cup placed out for him*) I can not let pain suffice without some gravity of gain and weight of curing, the restocking and regrowth of tissues— Was directly to a muscle impinged! (*sips*) And every taste turns the agony from cry to cheer, because.... that is what a stimulant is for. I (*sips*) can not afford to stay injured. I must be ready for the next bout, the next advantage ripped from the tapestry of good attaining. Demands make one healthy.

MAELI: And are there any hints, as from Lacismoor?

ELSER: That principality descends mightily, (ha)'s been erring left and right against sovereign rule— Drink, Alfonso! Don't be shy now.

DECAMOND (*noticing as* **Alfonso** *complies*): Home rule.... It's been a beehive of distemper ever since it was annexed to one of our more distant possessions. But, how can you sketch of mind such heathenish people, and their intentions? If they threaten nearby states, what does it mean? else they seek cleaner meadows to bathe in, away from their polluted washing streams. I have not heard yet of a declaration of disunion, nothing that can be stated as written and flouted. No real organized disdain against the kingdom have I learned, or I would leap into the fusty fro an' to allegiance. They must have some value hoarded, aside from an uncontrollable

fever for insurgence and a lack of tolerance to authority. But I do not gamble on whispers. They must be an ache to pay attention to, for my cover against the king's insipid constraint. Then there is reason to topple heads for gain, without excuse and with, a pardon to be persuasive for our unity and to compel.... obeisance to our ruler. Yet, in Lacismoor is much disease to leave alone.

ELSER: Have you some fruit?

DECAMOND: ....I have showers of jewel encrusted gold, Count Elser.

ELSER: I mean the real thing convenient—

DECAMOND: Skacte! Provide! (*Skacte withdraws.*) For test of fruit, a sweetness? or ripe plumpness? The waters of a spring held through the fall.

ELSER: That can be demonstrated well, off firm trees. Yet for an examination are they pledged to be tossed, into the mouth for a taste of vitality (*as* **Maeli** *goes to the other side of* **Decamond**, *with his cup*), rejuvenation, citric inspiration to startle eyes as well as palates and fortify an impression.

DECAMOND: The fruits do grow colorfully.

ELSER: That test is demanded.

DECAMOND: I.... often find remarkable that the more they grow, the more color they obtain.

ELSER: Then may I introduce to you a young friend of mine, Alfonso, from the Mûr region. Of only a recent acquaintance, he has asked for this presentation.

DECAMOND: As leaves off a tree may arrive through gusts of portentousness, Maeli? what brings to my attention more usually of conscripts.

ALFONSO (*cup in hand*): I would not presume to speak directly to the fine Decamond without this prelude of review. (*sips*) But I have feasted on adventure, and find more on a scale as potent as this drink, without intoxicating valors through toxins of intent (*sips*), or wish to— under Decamond's.... guidance. For I am sufficiently available, of broad sword, to indulge the gathering of claims (*placing down the cup*), and by my own announcement arrive. Or with self am I made for this advance. I plan to prune.... unsavory tangles of misadventure.

ELSER: Treat! (*as* **Alfonso** *goes to the center of the circle*) A spirit is as bold as is demonstrated, but with skill should be defined as bravery or buffoonery. Then see.... how hounds may become horses, since that's the difference between pretense and one's pride. (*as* **Skacte** *returns and places a bowl of fruit on the box*) I like above all, and before the most important, to acknowledge an ability professed and argued for to view.

ALFONSO: As pleaded to. Then hear the sheets of wind more blanketed than soldiers stationed for defence 'gainst gall an' guile alike. (*engages in a vigorous swordplay*)

MAELI: Does make a show of it.

ELSER: As well as legends might, against imaginary demons for their deeds told. But blithe of balls to throw can make the battle better to belay, belie, or believe. (*He starts tossing fruit at Alfonso, who slices each in the air, generally coming down on a projectile than coming up to it with the sword. This allows him to control his dismembering so that the pieces of fruit remain within the area of the circle.*)

MAELI: Quite a marvel, here!

ELSER (*stopping his tossing*): Is true!... of this precision.

DECAMOND: This is quite a credible display, of your skill in handling such a heavy blade. The control is more admirable than the feat's proposal.

ALFONSO: Thank you! (*wiping his sword on his gown before re-sheathing*)

DECAMOND: You have all of capability, it seems, to join me on my missions of.... deliverance and sparing, as an officer of men. But.... you've paid for yourself to present here? All of my men of caliber fend for themselves to find their riches and wealth during our.... excursions of relief— And as well they may pay directly the men, the soldiers they command under themselves, from what is found or seized. And often an officer is judged largely by these largesses led to, lured to, and received. Else they're often tricked, the would-be leader, into a quiet decapitation.... of fortitudes. I myself am utterly generous, as I value most highly the will in a man well above an ability. I provide the necessary abilities with training and formations of perseverance. Often a poor officer is taken care of from below, his followers more able than his aims. Be that as it may, can you hold out till the next trial, a real one, to spend your time convincing others? Then for the while you may share my cloak as being a noteworthy fellow under Decamond, as said and as allowed to be said within the ears of all my company. There's little costume to it. Proof is the blood splattered on your dress during battle. But I won't deny that you ascend thus far.

ALFONSO: I have funds enough to last a season's wait for war.

MAELI: That may be good for a proper rooming, adjoining the regulars you should gather with. And of course the meals.... are what you eat. Sleep is what you may make from work, learning how we act, and what great achievements will be expected from you, and what dependencies you may— earn. Not every adjunct stays to face the fighting, when he knows what's to be performed. These two here could probably shred your vainglory, were they ordered to. That is how well prepared they are, for Decamond's machinations to employ.

FENNEL: We'd hardly want to confront such a high arts artist of the sway; to protect ourselves and others, (we) would do— but not for sport nor fun of it. Yet, thin bars may keep a thicker (one) displaced.

ALFONSO: That is not my enfranchisement sought, but to incorporate for freedoms of debating strengths with profit. I like to attack with aim, and not merely for show. But this is one great position grasped— and understood, that some much desired acceptance is due me, long awaited as for the treasure of a party of principals. Alone I could be feared and fierce, but within a grouping of Decamond's become part of an invincibility expected to be pleased and

engaged of spoils through valiant efforts. With this name as a badge, I secure prosperity for my work and eliminate a thought of idleness and laxity of purpose. For what does being useless mean in this world, but to be unable to *advance* your art, not merely polish.

SKACTE: These talents all must wane, eventually. Thus to take more varied tasks, we do succumb. And (have I) no shame for it. Most skills remain in death, the most deft craftsmanship: the arts of resistance and defiance, persistence and imperturbability. But what is earned? Then usefulness is now, in all forms, the most valuable gift attained.

ALFONSO: And I welcome that! For the great Decamond to direct, (*kneeling*) I am humbled emotionally for this admittance, afraid many a time it might not be possible, with his august history pensively reckoned. I have not been accorded much accolade for my wearisome practice, and have been afforded even less potential to assume during my growth in these activities. So thereby need I the master's opinion that this obsession is worthwhile, has been throughout my brief life, and is no more impersonal than for his personalities to acclaim with just rewards and the patience to retrieve them. Then pull on me for service and I will dispel all apprehension or cause to doubt the wayward brought before you. This is my feast enjoyed, of your uptake to devour of my pledged devotion.

DECAMOND: Your.... entreaty is allowed, half-way proven to be fully deserved.... its fulfillment. As you may see, my men's highest art.... is some self-pity for their wisdoms. But as for a reserve of character, they are voluptuously endowed (*Alfonso stands.*), with an overwhelming propensity to justify themselves in battle so handsomely that this alone attenuates many an enemy's aggressiveness of defence and enhances the probability to accept more calmly a defeat, more surely a subjugation, and more taming-ly a treatment. Yea! We often depart as friends, with correct commitments reestablished. And there is no theft this way, as much as killing may render the useless as useful, then at a banquet entertained with gifts and treats, treaties and taler for tonnage duty-ed.... One must accept our exertions as righteous— That is most important.

MAELI (*as Alfonso goes for his cup*): And certainly they are, or I would not blend myself to them with some of my own.... personalities and personages, as much as they may be lent for a good return. The investment has been not for lucre but for license of these adventures, Decamond. I've only lost a few treasured men so far, yet gained much for a portion of the spoils.

DECAMOND (*after sipping*): And stay for riches, Maeli, as these delicate matters wend their way towards us. I above all know how to handle them and proceed.

ELSER: Were that I should myself treat the risk applied. But it is still much of a gamble, and I am observed more carefully to be high-minded, being a distant relative of the king's. Yet I have interests in this.

DECAMOND: Alluded to. Your friendly resistance does not dampen my intentions when proposed. I only occasionally need an augmentation of soldiers of high quality— and from reputable sources. And there has never been much complaint from those sources that participate.

MAELI: Aye! (*sips*) The compensation is timely.... Time is the greatest coin of all. It is never misspent. But you are very careful, to work out the complexities of influence that may be garnered. And that's my main need at the moment, when I want some subtle foothold in a territory to obtain. Wealth I have more than enough of to share or be shown by. One purchases power through predilections, and not random roughhouse or tangling with any presumed to be weaker. We are not to war, after all— as the king would insist not this way defined— but of discretionary inquiry with a bluntness of force and physical strength. Therewith keeps the peace, officially, under the shades and shards of squabbling. I wouldn't engage myself otherwise, even so partly or peripherally, ancillary of incautious action. In effect, then, all of the principal players must be predictive of outcomes, with the accuracy of their results as for judgment, the substantiation of our overall positioning bettered. I do not, in particular, like the carnage wrought. But fighters thrill themselves, it's often been told me, in few ways other than a bout or a struggle to safely bring to the eternal gates.... a clasped, bound companion, and then pull off and push through! Must be more than a pride involved with the ability.

ALFONSO: It is a survey of the intended, sir. (*sips*)

MAELI: There's little proof (of that) for those quieted of their complaints.

ALFONSO: That may come with the eyes' acknowledgement of these terrible affairs. It.... sounds to be a maelstrom of contention.

MAELI: I have been in battle myself, young sir— young whelp of fortitude. And the only profit is in winning your struggle, not to have the conflict protracted 'gainst your odds.

ALFONSO: That I should avow, not to be a monger but a treasurer. I've yet to foil a man's life, and have not thought of that prospect thoroughly. So it is not a pride for my abilities to do (so). That is a contemptible thought. (*sips*)

DECAMOND: Yes. There is often wisdom in avoiding unnecessary carnage. Even angry beasts ravage for more purpose than destruction, but to withdraw debts of nature.... My hearing has been affected as (of) late, not to hear cries.... moans an' groans of the agonized. They seem filtered out to a hum, as my mind reviews the most recent triumphs and conquests, ablating subtle headaches. Quiet times are worrisome, except for resting.

ELSER: You have certainly been brought out of exhaustion, to recover yourself.

DECAMOND: Fatigue is a sore, Elser. But I caught the plague riding back, not during the competition of interests. It's like a chain wound round the neck that grows heavier with thought of its absence.

MAELI: Find you some bug, Decamond?

DECAMOND: Nothing an illness can't cure of itself by burning through. This is not a diseased nature, but a tired one.... of idleness and indolence, perhaps.

MAELI: There is the mad flea or gnat that oft accompanies the fight and flight of warriors. I've seen the effect of bites! as they affect a man staring into darkness, looking past a combatant right in front of him.... to find another. And battle for him is no longer personally punitive, but a reflex of the sinews as the reflux of a worldly spite ensues. He is surrounded, for all his cares, by an empty continent— and breathes with sword. This (thing) sort of taught me, quite early on, to stay clear of military marvels (*sips*), but let them have their way upon whip of my control, as I am so positioned and petitioned to be and stay.... hereditarily high. But, to look back on them, to recall them, is of another, envious world perceived. Such few may be the only (ones) who can die gratified, a real life being incidental to such spirited malice.

DECAMOND: I've not noticed such machinery to my soldiers' actions yet, Maeli. The running mad have been with pain, thus far, gushing wounds of fear and infliction. For all about the violence, they stay practical. There is no sludgy sickness tampered with, as they follow commands. You may have noticed your own involvement projected impersonally, for being at a wrong place or in a crazy situation. But I enjoy the combat— when it's needed. And a concentration on the details often makes one seem blinded gloriously.... My head swells not of hauteur and hubris for my accomplishments, but of images left to quixotic arrears for understanding, that I seem yet to make a payment for, or feel to owe.

ELSER: Regrets?

DECAMOND: No.... except for cherished lives and friends lost; and that is the bothering curiosity and funny twinge that abides my healing convalesce. There's certainly more to do, about these matters of dissolution. Come! Fennel. Replenish! (*handing **Fennel** his cup*).... I feel with a loss of ease, not to do this myself. One spoons as one should lip, taste as one should drink, and swill as one should have a thirst for. But freeze these dreams.... for a beautiful morning. (*as **Fennel** returns his cup*) Let them not ramble on hoofs that hurt upon the brain. (*sips*) Seize me with an intoxicant of cleansing, for all that I remain of knowing.... a sultry practice. Block the slack (*sips*) of sensing my manipulations.... Skacte! Clean the forest. We are not ungracious rogues to beauty.

SKACTE (*going with the fruit bowl, to retrieve the fruit slices*) It is the round, sir.

DECAMOND: We are neat.... Alfonso, to make up for our.... squalors and squandering.

ALFONSO: I could have done so—

MAELI: Pray that you may! someday.

DECAMOND: The juice.... is as a blood too sweet to drain haphazardly. (*sips*) This really does cloud my eyes. Well done! Fennel. The tears are becoming sensible.

FENNEL: Thank you, sir.

DECAMOND: And what is a servant? There's a servitude to our idolatries of hope and pristine adulations, what makes the better of a person's character to want to be so.... obsessed.

FENNEL: I enjoy it thoroughly, this— girdle to be wove to. I am.... subject to its fitting and tight nature (*as **Skacte** returns*), finely knitted. (*as **Skacte** places down the bow!*) And I say these scared pieces are quite clean (*picking up a fruit slice*) and stay delicious. (*enjoys the fruit's succulence*) This is quite ripe. We are born delicious, sir. We are all born to be terrific.

ALFONSO (*a bit startled as* **Fennel** *re-enjoys*): I would say!

MAELI: Born terrific, hey! Yes, that is a tactic of life.

ELSER: A clique to least should try to remain so. But aging, then, has its mortifying ways. What is better than a servant? What is better than a birth! And as you have seen him, and approved of this, may I take Alfonso off, and prepare him for a strife of stardom to be tested by. For he will be asked to perform this tactic so often, as word of it spreads by— servants.... that his arms may ache and bleed profusely of this courage, with threat of withering. (*as Alfonso empties his cup, to place down*) And take of my advice, son, how to make your stand among companions. (*as the two walk off*) I have interests in this.

DECAMOND: ....Here is a thread of garment needled, what I have bought.... and for so little as a verve to allow of display. But friends be we together to stay, as influences gather, disperse, or stretch. We do try for ourselves many perturbations of groping.... each other. And that is as light-hearted as may become a war, friendly to remain and respectful of our aims.

MAELI: Individual ones, Decamond. They are as various as the personalities brought to, and their interests expounded, explained, and exposed. But Elser must remain more sympathetic to the king's worries. And obviously this Alfonso will play as an extra pair of eyes and ears.

DECAMOND: I have so many already, Maeli. It's almost like a tax. But my work is reviewed from without and within thoroughly. Yet it is allowed, because it is very.... very helpful, as an adjusting valve's fine movement to avoid a flood with the draining of tensions and protests. And.... we willingly take all risks involved in this; my army is not coerced into inquiries of promise, so this is all terribly legal— as far as the sovereignty's status may concern itself. And I may bat— better my head as often as I like, or be provoked with a marching battalion.... They know no single leader can last this way for much of a long time, and drink of the juice fruited to be flowing like a sap.... from a tree of inspirations. Or else I labor for ourselves backhandedly. But I will die on a battlefield, most likely, and not seated on a throne pining wishes or yelping from griefs, with a gout or a consumption to entertain.... But from great wounds of strain or flesh will I cry for the heavenly gates to open. Therefore, I am no threat to our majesty, nor your ambitions— nor Elser's.... nor our nation's supposed diplomacies. But we are threatened through and through variously, vicariously, and with sundry goals to bring to pertinence or testing. It is almost like a purchase for our services, or an investment, accommodating added fighters lent from corners of objectives, though they must prove of themselves to be true warriors. But I remain for typical of their actions, that, upon strictly following orders in all pursuits.... they make their own gains and suffuse a contended region with their customary ambitions of greed, gratification, and glory. But now comes the question, with new blood to mix to and help uncover answers. For the addition of novel content, without even contrary motives competing with, stirs the boiling pot to bring to view, occasionally, the fat an' the foul. And was I a sessile leaf bared to this injury within my own tree, by foreign ants or aphids.... or through devious lichens growing about the moistening barks? I have seldom seemed so awkward for a blow. But I'll patiently await an answer with admixture and replenishing of men to tell. That is often the way, and

an old soldier's trick. Look not for internal rot but a ravage quelled with caution of the new or unfamiliar and uncustomary to be made as common for some surreptitiously growing alinement, if purely even due to wariness of aging, 'gainst the old one in command.

MAELI: You are that suspicious of an.... indecency to your person, whereat such crimes are tender to be bargained with?

SKACTE: Was thrown to the back, sir, this missile spear-let, but for which he turned slightly with unconscious erring. The commander is always protected from the back— because he always faces front to defeat an enemy. So this was a surprise that took long to figure out of the physics. And there are many possibilities of wind and travel and curving trajectories. But we were not combative at that particular instant. And so propitiousness or propinquity of motion are in question. Nothing so radical could have been expected. Nothing so atypical of weapon was else found in that land of fremitus and stagnating emotional bounty. So it was a singular gyp, that may have been native or may not have. But to explain it at all is startling of fate *rrro*und worries of potential pretenders.

DECAMOND: I was more stunned than hurt, and am more patient than anxious to learn the nature of these such accidents, as it may have only been. The infection sues me, though. And any officer I have makes for himself what is slurped, the tastiness of a rich and meaty grind of gears for gruesome attacks, skillful planning and well thought-out stratagems and strategies. To put a stop to me takes no more organization than death. But to produce a current of the spoils.... to keep that stream flowing and unobstructed, requires a courage of mind an' muscle. Few can betray that need successfully, who are already engaged of my figuring. We've not one failure yet!... which makes me prominent of symbol for what may be earned and practiced haughtily of reputation; which is as good of a consensual advantage as for many of the most fierce to portray and prove themselves to be. I trust the action in a battle more than my back for backing, the good necessity of sharp reaction and notice, alertness to the thievery. So then, what happens?... past the confliction, above the victory.... can instill a nursing?... of a fever? This blunder still aches, and does not make me proud of the gash, or allow a nobility to it, permit a legitimacy to its commitment. It simply annoys me, to party with the headaches and tremors of waiting.

MAELI: Do you think the king sly enough, to insert a possible usurper through your militaristic legions, a person perhaps more controllable from the court? He does not strike me as that fiendish.

FENNEL: We doubt the king.... He is weak-hearted to the banditry of nations and countries, to do with what he has, and make with a material that may burn the skin if touched directly. Thus is a craftsman as a statesman. Much political weaving is yet to be performed before we're through.

DECAMOND: We are with slight (in consequence) disowned by his court. But then, we do not protest *at* the court. I am as gracious as can be whenever I pay a visit. And this attitude is received by me in return. If ever there were even barely the concept of an arrest, by he of me or they of mine, I feel the foundations of our assemblage of might with comity would shake contemptuously. So let this skittish, scandalous villainy try of germination to my limbs. I'm stronger at it waiting, to allow these weaker, un-poisoned attempts of fate and fidelity to rule. (*sips*) What blinds me is not callousness. What bids me do is a blinded god.

MAELI: You make war because you like to, Decamond, and have gradually constructed a means to do so profitably. But others like other things, and are also engaged in their crafting. And a few, as like the king, are active by necessity of their responsibilities, whether they enjoy it or not, maintaining kingdom and country and the strata of class privileges or castes, to keep our societies moving and evolving with an orderly progressiveness. Your interests are only a small part of this, no matter how intense or vivid of attainment.

DECAMOND: I'd never claim otherwise, Maeli.

MAELI: So don't fall into the trap, the philosophical fallaciousness, of canonizing human crassness. Such pleas are unbecoming to those in the know, to beg for the wanton ease of a deadly struggle.

DECAMOND: Ease?

MAELI: I mean as a concept to mind, duke, something too easily grasped for confiding with, a notion perhaps recommended facilely.

DECAMOND: These are very difficult undertakings, count. Their principles may be straightforward, but the accomplishing is much harder to do than to argue for or against or even about. And I risk much, lose much to gain more. But it is a manner we have educated ourselves and mastered into a comfort— of mind, not an ease, even if the complexities— and there are many— become to roll as familiarly as for a learned review. This is not so simple as an ease.

MAELI: Well then, (be it) a passivity to the harrowing, however you like to characterize a gut-wrenching experience as typical, common, customary, and accepted.

DECAMOND: You try to say disgraceful, by describing a disconsolation. Warring conflicts are not disgraceful. The results may be, but the activity is usually very exhilarating, and a fulfillment of the human potential we would have as games otherwise of the stakes involved. This is intuitively acknowledged by any objective thinker.

MAELI: How so can anyone be that objective! You've seen what I've seen, but by a thousand fold over, perhaps. Are you not, then, a thousand times as revolted? or a thousand times numbed?!

DECAMOND: ....I have sensed the rent, and bathed in the bloody carcasses—

FENNEL: We are not the demons demanded, but the demanded demonized.

DECAMOND: —Yet, that is how this is done. Violence is no simpleton, and is not to be avoided for its usefulness. But I do not promote the senseless massacre, like others, on the edge of urge merely to decimate, where reparations can never be forthcoming and not even the attempt is allowed. In that respect we are the most civil of forces, determinate and decisive but also discerning and decreeing of a dominance rather than an annihilation. We make a responsible tool, and a fine implementation, not the crude conscription the king may fear to perform.

MAELI: That is only thus with you, Decamond, for your skills to be tolerated, organizational and organic to our nation's nervous impulses. I commend without (a) compendium of torts others devise against you. But be advised. This activity does exist, as well as your own. And friend or foe alike may be consumed by its advantages come a ripe moment to defeat your erring or blundering on stately held importance, imperatives and interests, and authorities wrestled with. Take your wounding to soften for some sympathy, and retire this obsession to the less worthy and influential, that they may be more readily managed without contentious courtly quarrel. This might be a try at that goal to fend for you or sell. You have more wealth to use than for the life that's left, as I. Marry and have children to dissolve it through. That is the general course, the usual spell spent of these successes. Chime down your marshal toll of countervailing distant dissenters, and gloat with friends and riches, rather than (with) the entrails human. This is an opportunity arrived—

DECAMOND: Like headache. Love it! Love this basting?!... It's a cool, clear morning, past the sun's warm light of rays—

MAELI: You have never been—

DECAMOND: —illuminating me!

MAELI: —availed by the crown. I have, and can show you how.... or could have, during your formative temptations and associative estraying. One builds an army for protection of one's properties and estate, and to help fortify his ruler's ruling, perhaps. But for this.... profiteering, you have only threatened the established order. A fear is of course that you become too strong or swerving of disruptive sweven.... in this.... (suddenly) Ambition gives rot to head!—

DECAMOND: I bake comfortably out here, in this jolly little pitch of the round. And the cranial sinuses dry up, that have been overflowing (sips), as from a declension found. Into a hollow space drips the leal leak of lucrative adventuring. I have found and made my method through much matter that has generated it, through laxities and indolence, mismanagement and poverty of spirit or insight, whatever the king abhors. And my only ambition, as everyone knows, is to be greedy with these abilities perfected. That is not excessive in this current world, where breaches make a scaffolding for modernizing structures and conformities. Pirate or rake?— No! To settle these disputations I am performed. I have found a response to a placating need. That's why I'm not abused.... at court. These needs persist a while longer, if there is to be anything like our governance left in tact for places I may reach on which we rule. The spirit and intentions of this need are appropriate and responsible and vested of my.... courage to investigate— not instigate. And as I do, what's weighted as official weighs heavily on my conscience. I assess these things thoroughly, and no case can be made that I an anti-official, nor apostate of our laws and composures. So let the warnings 'gainst my art be a spiel's commitment to a vacuous value, as I bring home wealth and allegiances that can only be discounted through envy or jealous distempers.

MAELI: I'm simply pointing out that traps and snares are oft baited, and good scents may derive from much scandal. Think you, as you rest up, that what may seem permissible to do in most circumstance may become absurdly wrong or inappropriate to.... continue to attempt under some conditions. A wounded lion, presented as such to the game, should not hunt.

DECAMOND: And.... the game is at court, you mean. Well.... that is a decency to my recognition.

MAELI: Adapt to this, while you may still meld.

DECAMOND: There is the tease of trying.

ELSER (*returning*): I've sent him off, you know, to change his chambers.... with your word as recommendation.

DECAMOND: That's nice. I'll acknowledge it.

ELSER: More expensive for him, though. But that's a pride the handsomeness of youth may display, not knowing more of value than what may be obtained through it.

DECAMOND: This is occasionally so. But he seems to me to have saved.... considerably, count.

ELSER: Fresh winds often blow more cleanly of their dearest intentions. You will help?

DECAMOND: As much as may be practical to afford of my time with the promising. Skacte! Have an eye out for Alfonso. Buttress (him) in friendly company, if he is questioned of. And lend him some furnishings, should he need any.

SKACTE: Aye! sir. An' if (h)'e has debts, he may work for me.

FENNEL: Or as a menu's tenderer, he may experiment. It's done all the time, though I doubt he's of the make to nursemaid foods an' cookery. Still, an occupation is an occupation, to bide one's time. And you learn so much in the process.

MAELI: A restaurateur, for certainty of skills providing. And a handiworks shop, for the crafting of furnitures and decorations. It amazes me, what minor professions you men may take up, away from the battle excursions. Would give me aim to do similarly, had I a need to. Then purely for interests' sake. But I'm too busy already, overseeing mansions and palaces. Rich properties indeed, I am responsible for. As in what makes to be, then I say mansions and palaces, if I own them, even as like sties an' pens, for all their bother.

SKACTE: It helps to be anchored to the practicalities of life, sir, and to have something always to do with one's hands, or head. A feeling of usefulness is common to many activities. But where we might be enterprising, your class is expansive.... of an empire. An' that pardons our predicaments to earn.

DECAMOND (*more to himself*): I have not earned.... I've yielded.

MAELI: Self-sufficiency is a right of birth with lungs. How did you first meet this young gallant, Elser?

ELSER: He came up to me, as I bowed out of conversation in a gathering of nobles, on the king's campus. The square was decent enough for a casual attitude, not too crowded that afternoon, and the markets not blooming of a frenzy yet to come before the evening's pall. He asked for an introduction to Duke Decamond, but had no idea of how to procure it. My name, as a friend, was suggested. But I inquired as to why this was necessary, as there are no engagements to his name current; and certainly (there's) a slack off of recruitment, or negative recruitment as he recovers from his injuries. The young man said he thinks ahead for a future that is shoring, but had not been aware of Decamond's ailments. I assured him that they do not seem critical, and neither an introduction to his company crucial at this time. But there then started a friendly pestering of future meetings as he would see me, or find me, going by and walking around the courtyards. And eventually, as I took him to be more or less a friendly acquaintance, he told me of this particular feat of sword's play that he had honed of technique, which he thought might be of interest to display, and before Decamond. Though he was too embarrassed or timid to show the skill otherwise, in case it might fail from lack of urgency to witness. I told him it sounds really to be a slight demonstration, these fruits to slice. But as a means of some little entertainment, I would allow him to accompany me today, as this rather incidental, or inconsequential, morning congress had already been arranged to visit with the duke awhile and satisfy my and Count Maeli's curiosities on how he was faring, health-wise. Thus have I done. Thus have we done. Thus has been done. And young Alfonso now seems to me.... perhaps not exactly elated, but a little more anxious of the august nature he must presume and bear to live up to and maintain the expectations he has pleaded for. He takes on that young man's stern countenance of having proved himself worthy for more difficulties, without yet being sure whether he should want them at the moment, if they may be prematurely received.... I've just tried to calm him down a bit, to say, in my opinion, true battles are aways off, and that for the months ahead he should concentrate more on the acceptance and convivialities of Decamond's important companions, the soldiering of which he is a part.... And I suggested he should take a job, to bury his hands into a knowledge manipulative of the brain, since we may all slice oranges in our particular ways most effectually, and as efficaciously as want to peel.

MAELI: Hmm.... He was not suggested to you?—

ELSER: Never! I should not be caught supplying our friend Decamond.

DECAMOND: Would not be blasphemous, count.

ELSER: It's only that you're.... indisposed, that I feel allowed this far of a transference. Should the king learn of this, it's only meek meat at play, to cavalierly pose a skill to a fighter.

MAELI: You were suggested to him.

DECAMOND: I doubt the royal would want to see it themselves, my bay.

MAELI: You should have searched as well for a history, Elser.

ELSER: Why? What is so unusual about an Alfonso? This is stout innocence. *Stolz*, Maeli.... for aid.

MAELI: The duke may be baited, during his weaknesses. I stay suspicious—

ELSER: Of that fine young man?! And let's not talk of foolishness, for that's as fashionable as gold trim. It's only a surge of ambition—

MAELI: Ah.

ELSER: —And Decamond makes his own judgements (of men) with an exemplary conditioning for the best for his purposes.

DECAMOND: And.... I do see. (*sips*) And.... what could be more promiscuous? than to please me. I might pray for Lacismoor. But first my head must clear. My eyes must cloud for some internal consultation. And my arm must strengthen out of its funniness. Let the morning breezes thrill, and chill contention against me. Since every day that I'm allowed 's more entertainment to be made of it, that I am expected to be wanting, an' wanting this, a continent's conspiracy for my talents.

ELSER: What can he possibly do?! What can anyone possibly do?! (**Fennel** and **Skacte** snicker.) No more drink for me, please. Does bring the vaporous to light, as a visitation of fumes to the head. (*moving away from the box*) And where the nose usurps, trepidation!— But, thank you.

FENNEL (*after drinking from **Elser's** cup*): Is finished.

ELSER: You drink yourselves into a light-headedness for possibilities too unlikely to be imagined while sober.

FENNEL: The *Saft* is fermented with a freshness, sir. That is, barely to the morning due.

ELSER: Even a least is actual of taste and effect.

FENNEL (*after drinking from **Alfonso's** cup*): Is finished again. And this was ever empty, 'cept for haste. A rush to fancy of himself.... a successful infiltrator? We shall tell with deeds only, and not inclinations. Many a recruit is inclined to fight, but becomes frozen with fear on the battle plane till warmed with a skewering— too late.... to run from. And he falls dead, from his inclinations— without deeds, or perpetrations. We call them heated weights of ballast, and there are always a few of initiation to slow down or delay the enemy as we stabilize our positioning and stance against opponents. Quickly gotten rid of, these.... inclinations,... we become incredibly formidable and frightening, the enemy almost immediately deceived, followed with a somber disheartening.

ELSER: I think Alfonso will be braver than your intoxications.

DECAMOND (*after sipping*): That should be the case. But this is like a medicine, or a bandage, or a filigree of nerves suffused.... a potent warning to obey, my rehabilitation played upon, and my sensitivities edged like a stone's. I don't fear the grating cause, or the chiseling attempted. I just bow my head for awhile, as it restores itself to oversee the body. (*sips*) And there'll be much control in this— Come! Fennel. More! (**Fennel** gets his cup.) The air is coolly kind, the slight breezes soft as a knife's cold surface chills.... by hardly touching it, and its sharp edge made.... ephemeral to seem, hardly felt till the day is over and the darkness falls. (*The replenished cup is returned.*) And headaches pounce into pulses (*sips*), as you lie dying, draining of the vital fuchsin, drops to gushes, through your night. (*sips*) This air is so vivid, I breathe it and can hardly breathe.

MAELI: Some wounds can heal themselves. Take heed of stark elopements for your thoughts. It's much easier to handle dreams than real earthly drudgery, skulduggery, and manage all of this moated mud surrounding as could be intended for your bettering or prevailing, instead of sinking into oblivion unrewarded and never more for notice. The seasoned may change this way, abruptly as could an eon seem.

DECAMOND: ....Is death? Then make panic more of— the game. (*sips*) I've been alerted.... And the head still twitches, the arm's still stiff.... and the meanness of a candor may still vivify in its defiance (*sips*) of the ineffectual.

FENNEL (*more to himself, while apparently pouring a drink for himself*): ....Or ineluctable smite.

Scene II — *The chambers of Privy Council for King **Musgonie**. He sits at the end of a long table, his face much towards the ceiling, with eyes closed, as if in a particular form of physically strained meditation. He clutches, on the table rested, his crown. He is alone.*

MUSGONIE (*with a serene and dreamy vigor to mumble*): Ah! What forces oppose me are the contrasts that make my strength, with purpose to hold me through and stand me up to contentious bickering among this realm. And I shall enforce these disagreements played to ear, since they give more reason than is purposeful. And that does hold this throne with tactic above symbolic sympathies and decorative respect. But that I am strung to abate quarrels, hear me pleased of harshness. For I have more fear of desolated temper and restriction than these debates. To be dismissed is what fells kings and sovereigns, more so than to be thought of as weak or indecisive and dallying for unearned affections and inappropriate admirations from subjects whose monarch was made thoroughly bare as a babe, as themselves. I have not grown of these responsibilities, but have been educated to them to be capable, with the backlash of discontentedness to feel or suffer for. By law I must be loved, but don't ask me to reciprocate in kind with pardon for the grind I ground myself to, envious of a freedom of wistfulness to abandon pressures, that which I can not allow be seen.... too characteristically, too often, but only sometimes assumed. The indentations to my skin, from this poking of fingered interests, make for slight displacements oft sheltered from or buffered through. But they are incessant, and could seem to last forever of this life, born to them without my will nor caution. As long as I can keep this country sound, the people satisfied of justices in their customary and necessary pursuits, then that will be for good excuse my leeward weighing of events, with accusation to find a considerable comfort for the generality of our population. And there is my rest to commit to. I will continue to sleep upon such a bed, not particularly liking the ground that so often shakes its legs. There's not much pleasure to be a ruler born with the gentility of fear, for the hardships prominent for so many and responsible to alleviate. Yet, for the worth of one myself, I claim no such generosity felt, but am bland to the contempt of life and vague to the endurance of mortals. For as a king I feel stripped of some divinity, to tend to the more divine, the rambunctiousness of man and his miraculous persistence to find in life.... my uneasiness condoned. Could have rather wanted, to die (as) a tadpole in a murky pond, than to constantly face the twists and squirms of unsatisfied calculuses of endeavor with my need to amend, regulate, and referee for the disjointed parties involved. Would have rather been a flower smelled more briefly, than for worshiping with expectation to improve a soul forever. And I am tied to saints, who all are dead, to stay as saintly in my machinations of rulership and manipulation of egos.

Oh! This king might be the angel with wings torn off! And I feel this way, thus treated. (*lowering his face*) Prayer, then, is only for a compromise, a redaction of thoughts to follow through as planned by the celestial spheres for our earth-bound purposes and warrants. To be the principal.... is to be for pounded mesh, and grating for crude ways, with our humanities so rampantly astray. There lies the principle pursued with these incisive votes of caretaking, and being fair through shaded eyes for a grand encompassing to view, the vigorous at play an' pother recommending.... my decisions. Now, what if there weren't any, no complaints to review, no subjects to carry? But sleep to be fed, and grow as tired as this bed of worries, worrisome adventurists made fewer and calamities capped. But really. What if there were none left? What if the world were empty? What would I do— complain myself of the boredom of a king? Why then be grateful for these labors harnessed and rode with! I am.... capricious, for these colors. They.... misrepresent me. And I am fond of a good life, and happy celebrations and festive moods. So be the toil for pleasantries espoused.... within securing states and sanguine sanctioning, as long as I am left alone in my mind— to plead for heaven and enlist.... while overseeing such solutions of the— soiled. (*Looking out a window to the courtyard, he takes a deep breath.*) It's time! (*places on his crown*) The veering is elated, as the jeering make their case. (*politely folds his hands on the table*) Prepare for nerves distasteful. I can't say I despise them all, when none despise me. Yet for the muscle levied, the disdain is great. (*enter **Ponstole***)

PONSTOLE: Your majesty is well? Then we shan't kill time.

MUSGONIE: Privy councilor.

PONSTOLE: There's more to be defeated than your emotional state, sir.

MUSGONIE: I require a span, a spate of space alone before these meetings. It elevates the atmosphere away from the musty musts you often present me.

PONSTOLE: Perfumes are deceptive—

MUSGONIE: But not scents.

PONSTOLE: —Even the mental ones. And there are matters I must introduce to you, again.

MUSGONIE: Then not an introduction, but an instruction to difficulties, hard explanations of entanglements and complicated webs of countervailing forces.

PONSTOLE: Fair King Musgonie.... you hold the peace to heart — but not to head! Our edges shimmer for detachment, with provinces insisting to break away from under your.... guidance. This is not a complicated intuition to sense. And you are being called foolhardy: remaining hardy as a fool! trying to maintain such a solidity of unification when we can not even ask, much less demand, for a calming down of tempers and a lease of more time to soothe nerves — without the interjection of stupidities to hear, how much to carve the heart to let the limbs flee! We can hardly beg! throughout these freedoms felt.

MUSGONIE: Negotiations are in order—

PONSTOLE: More?! And for what to do? You are said to squander interests and weaken your command of the issues. But the might of leverage is simple, sir. So many of our regions are so terribly self-sufficient. They send to us of their resources, not we to them. And need they us not! We should let those on the periphery of our control separate into partitions of country, more related are they to their neighbors than to this capital heart. Would you have more anger spewed, with threat to force the issue? and conduce of mind to attack this very motherland?!

MUSGONIE: These few.... beads along our garment of nation are too distinct to want to converge into a single contrary force to (oppose) our rule—

PONSTOLE: Our!

MUSGONIE: —my rulership of this kingdom, a nation of mixtures and misdeeds and segregated fiefdoms and variegated polish of a notion to be colonies or settlements away from this bed of country. But they are bound to us, and are connected by ties of motherly endearment—

PONSTOLE: Relatives distant and deceased!

MUSGONIE: —and relatedness of commonality to our whole, our fostering, and for all to prosper peacefully within a civility of varied interests and attributes. I have explained to you before, expounded on my sincere opinion that if we allow these separate states to exist there will be heated conflicts and wars among them, and this heartland will be battered towards the total effacing of all judicial purpose, unable to tame wild hatreds from self-interests or importance. Our role is to stay sound, of a single constituency with fluency of ease to be called one, and not permit a severing turmoil which may turn on us to dissipate into harried branches from an uprooted tree! We must stay solid, solidified, Ponstole, and lean on leisures of particular, characteristic, and aggrieved for offerings to pacify the few dissenters in these ancient regions who stir up such discontent and misunderstanding, and who try to misappropriate for themselves my territories hereditarily won, for which I am responsible to save and solve into one nation desirous and comely. And I am not convinced of the urgency you profess, as the measures I've proposed are followed with resolute obeisance to this unifying principle I've inherited and am weighted by and even castigated with!

PONSTOLE: The few are not so few, your majesty. And we are not so strong as to withstand them as their tempers glow. Appeasements.... only lead to growing corruption and debasement within these regions, as semi-autonomous governance is set up, haphazard of the instigators who jump to lead these attempts to lessen your reign. They may be patriotic to their communities represented, or they may be scoundrels of the most despotic and overbearing nature. But we can not control them with gifts and licenses and the privilege to wash your face—!

MUSGONIE: To have my company. To have me visit, not merely through envoys, to learn bravely of their disgruntlements in person and see how to assuage these problems. It could be as simple as changing the produce they're committed to ship, but I must have my hand directly in the matter to know— and show real concern and care for this. They are more than charges and wards. They are people of fastidious nature about themselves, and must be treated of my description wanting, never to be let go! into a dissolution of aims about our nation. Then I will travel to each, if necessary— and have each shoe me! to show I walk for their commitments to retain. For they can not demand much more than my utter concern.

PONSTOLE: That would be a demonstration, your highness. But is it to appease factions that will only insist on complete independence and self-reliance? In such instances the choices are as sharply divided as a fine knife's sides, to either grant the separation or not. And they have not invited you for visits—

MUSGONIE: Invited?!

PONSTOLE: —As far as I know, not any one location has sent emissaries to suggest this, though some may return with your pleading to. I wouldn't think it proper to submit yourself to such.... denigration as an indiscretion of custom. For they accept you as a ruler to visit properties, but to be invited to review and acknowledge possessions— lost!

MUSGONIE: Lost?! Could any ever claim so?

PONSTOLE: It is the effect of the conditions of our times. They are more valuable to trade with than to trample, to have amicable commerce with than through the animus of coercion to control. We can continue to command more effectively through the physics of our economic interdependencies, for our interests to be maintained. That is the more astute posture to prepare for, a profitable one in the long run for forming inseparable ties to regions dismissive of our pretended sovereignty over them. Or might we easily batter each into submissive appeals, more or less constitutively, you are against this method; and it would be counterproductive to our resourcefulness in managing the nation. For they fare better distantly viewed.

MUSGONIE: Pretend?! They are as necessary as hairs to legs— and they are ours! I govern for their improvement, and with a leniency to their characteristics. But as one whole we must be kept related.

PONSTOLE: Your majesty, many of importance in our country bid for much prospering in these provinces of remote relation. They have direct fiscal engagements with one, some, each or all. And they are as I see it best for their total independence to gain, this trend to continue as we swell within of growths sophisticated to which we must concentrate our attentions. Our country is becoming more complex than past complacencies may allow, while these remote outposts of resolution and resolve offer to purchase their rights of severance, maintaining an historical friendship through this process, to the mother country an alliance of familiarity held. This is all to our good. This fits well with our evolving perspicacities of notion to exist healthily and with a clever management of our power and might and worldly prestige and respect. The entire council agrees with me—

MUSGONIE: They are biased members all, I'd suspect. Only yourself remain idealistic enough to judge these affairs without subjective influences at play, which is why I fight to keep you as the privy head, and void electioneering to substantiate personal aggrandizements. But I tell you again, if these various domains are granted their autonomies they will battle amongst themselves, as their trading with us will conspire towards those actions. And we will be exhausted to a withering expiating these crimes and subduing their forces to abatement and peace. For little bites to legs ac-

cumulate of pain to have one crippled in time, and infested with poisons and poisonous views, dangerous temptations, scandalous skirmishes, and disruptive tensioning. The pressure is of distance, Ponstole, not of people and their conceits or self-indulgences. They must be paid more attention to by myself and our commissioners to build up their importance sensed within the country and for the country housed.

PONSTOLE: Then what is your practical alternative to their constant complaining? Your person?! That will not suffice—

MUSGONIE: I am *the* majesty, in arms and armor, prowess and performance of my rule.... They will obey what is dictated—

PONSTOLE: With your begging or pleading for them to behave, and to show a constraint during your rulership, within your kingdom bound and grounded?! What magic may be painted on their faces of growing disgruntlement? And how many coincidences does it take to make a fact, that they no longer wish your leadership! that they are disturbed by it, disown it and ignore as more they treat themselves to freedoms. I simply say that.... the trend has been awhile now, and has only been checked with— disgraces, and not your fatherly nature of appeasement and understanding and patience and offering to visit rowdy subjects.

MUSGONIE: These *disgraces* are my heartache.

PONSTOLE: But that is the reality of it, your highness. You have allowed, through a queer laxity of intransigence, the intransigence being this way rare, a bold entrepreneurship of warring against these regions—

MUSGONIE: They are not wars!

PONSTOLE: —strictly for profit and notoriety.

MUSGONIE: They are lesions to my reign, scraped and gashed across my legs.

PONSTOLE: But permitted under your faulty strategies, for they are seen as reasonable warnings against disunion, and by a noble and distorted-ly heroic figure, in some eyes.

MUSGONIE: Let not the sun be seen that mashes toes! He is a burthen (of) blunder to me loosened.

PONSTOLE: But then admit that only he holds your policy in tact, as much as it may be followed in madness. But therein lies the point— that he's *not* mad and makes for this effort its only useful purpose that can *be* constructed.

MUSGONIE: Many despise Duke Decamond.

PONSTOLE: As could despise you as well, for breaking the back of business, disrupting trade or detouring trade routes, and in general causing such a consternation as could threaten livelihoods with fear to pursue them.... till he leaves, having won his winnings and rewards, for having been nascent-ly adopted to their understanding— all in your name!

MUSGONIE: He is not an ally of mine, nor to this court so much approved as lauded for his exploits.

PONSTOLE: Yet turns he provinces into poundages removed.... from our varied interests, their customers of citizen faint to interfere but for a lack of produce received as possible for a time.

MUSGONIE: The man's too colorful to be distilled into my service wanted-ly. Condemn him can I not, with license to be proud and assertive through my.... handling of reins. I do not wish to perturb benefits, Ponstole. But trap him may I, to conform to death more pleasingly, since war is this machine made for.

PONSTOLE: Death of what? Lividity of goals?!

MUSGONIE: He has been injured— seriously, I hear— through his last engagements.... or enterprise.

PONSTOLE: These princes grow of their wounds, to be newly born more arms or tentacles for holding up more swords. They.... maturate of their proud suffering, to drain themselves of pus and become more defiantly lean and decisive in their battles. They do pull out their own teeth, sometimes, to gain an edge of anger or surliness.

MUSGONIE: That is ridiculous, unless it's loose to begin with. But he is hurt badly into convalescence, I'm told, a more probable rumor to conspire of.

PONSTOLE: Or with, sir, a longevity of recovery? That makes only of time for more rambunctiousness among our colonial dependents, as they do insert into other sovereign states with haphazard geographies. This too is a reason to leave them break off rather than to stay appended so unshapely.... unless you mean to do some ferocious business with neighbors of heretofore tranquility. It is simply that we grow of prominent protuberances so successfully as the wild and uncultivated lands are occupied for our civil dispositions more easily than by others, or more rapidly than perhaps natural to these regions for their inhabitants to aspire of. And a few of these "dependents," it is argued, are entirely cut off territorially, from our mother country, the thin ground bridges leading to them hardly recognized as such by those living in the areas. Nature seeks to help to crop and polish our coiffure. A spin of time will sever them from us eventually, as the neighboring natives strengthen their administrative skills and management of settlements into cities and municipal dominions. We can not fight these regularly and normally evolving trends, nor should we wish to. For too large or broad a country leads to ruin from weaknesses and varied interests too competitive and querulous. I think you confuse this foresight with a false one.

MUSGONIE: I confuse nothing, and see the obvious creeping upon us. But I am of the opinion that we are not too much of a sprawling motif committed to nor characterized by. And by what I am handed (down) to govern I will maintain, and not allow a spacial atrophy occur during my reign. Though I'm not in favor of more spreading of our arteries, at this momentary epoch— I'm well aware of the principles guiding decrepitude of the whole through overextension of its parts. What we have right now is suitably possessed. Ne'er a subject to the main would argue for less. And this is more than a national pride, but a national posture worth fighting for or defending— but not by gross opportunists whose capital is a flouted insubordination (to my proposals of inducements through comity) and temerity of conduct. So then, I mean to offer Decamond as a payment for the favor of those felt too disfranchised by my governing, and as a recompense for his goading in the past.

PONSTOLE: You would even offer a loyal and famous subject to destruction by a few of miscreant intention—?

MUSGONIE: He destroys himself—

PONSTOLE: —I can not agree to this reasoning, and never suggest so much as his arrest for questioning the wisdom of his exploitation of exploits.

MUSGONIE: ....unless he heals with me. I haven't the need to imprison him, or detain him during his sanative recovery or restoration. But I will do more to alter than to condone, by appointing him, upon my solemn command, an ambassador to these provinces respectively— and without a standing army to conduct. After all, he knows these regions better than many, from solid experience and fascination for. And with the quagmire of effort and intelligent work on improving relations and understanding that he will find himself immersed of, or embroidered to, he will lose the warring inclinations through sheer exhaustion of the brain on finding methods that will work for his accomplishment of tasks perceived. And he will be converted from being a belligerent into being a belle, on the diplomatic stage, which will serve to enhance, though re-hue, his already haughty reputation. And the provincial authorities will stroke (his person) to cuddle and caress him eagerly as a representative, if I know my human nature; because there is much control in this, and patronizing of position to both diminish or dismiss effective powers and yet garner some advantages with this minister appointed of only his publicity as strength.

PONSTOLE: This sounds rather devious, your majesty.

MUSGONIE: He will think he is at the height of his influence and utility, while the office will be actually a rather shallow or hollow one, this being almost totally evident to our "dependents" from the start. He will be entwined within his own arrogance of importance and blind to the real diminution of role he must undertake. Our dissenters will enjoy the jest and see me as wiser about them than they may have thought before. A true pacification may follow, as ties are thickened; and Decamond can only work towards this.

PONSTOLE: But he's a brilliant man, your highness.... or must be, by his successes made as evidence, or through them made to be. And he's not likely to want to give up an employment for his soldiers, nor a dispersal of this endearing military company.

MUSGONIE: Have the leadership given to a subordinate, underemployed and never to be used. Can not you see the folly, or the irony of such deployment? Wherever there is a tension that would have normally aroused his entrepreneurial curiosities, while his soldiers will be put on alert, he will go to the place directly to work to allay the pressure, as an ambassador of outstanding capability. And not to work against himself, there will be no conflicts brewing of his efforts. The hypocrisy is bright! His men will never be sent, and he will not have need of nor want for contrary interventions, as he may do his job splendidly and with notice and admiration for it.

PONSTOLE: But he is a cunning thinker and planner, perhaps to plot for wars.

MUSGONIE: That would undercut his mission and his growing reputation as my ambassador. He is wealthy enough not to ever need any more battles for profit. And some sanity may heal with the muscles in his head as well as the tendons of his limbs. I will appeal to his resourcefulness of mind to remain noteworthy and celebrated while retaining great activity. Believe me! This is how ascensions are made precipitous. This is how pests are cordoned off without blandishment nor blanching of their natural and god-given traits, to be of little burden, some bother, and much betterment to superior aims. And this is how appeasements to successful characters are made, not humiliating one, but to make (as) homonym of meaning and purpose. Decamond will be his mouthpiece made for mine.

PONSTOLE: As were the moon to speak for the sun.

MUSGONIE: Do you find, at least, some salutary notion to this suggestion?

PONSTOLE: There might be some merit to it. Here (it) depends on what is occasioned of deceptive practices. And our duke may be much a master of such things, that he might instinctively sense an effective denigration of his status, in which case he may cause you more of a problem.

MUSGONIE: Then.... the process must be handled delicately and with acute sensitivity. I have worked on this proposal from my end to proffer.

PONSTOLE: Have you. I'm not too sure—

MUSGONIE: And thus may you—

PONSTOLE: I?

MUSGONIE: —to have him bid of me this recognition and honor. He must come to me!

PONSTOLE: ....You want that I should suggest to him that he may be the best to serve in this position?... and that you are reluctant in choosing an ambassador because, while it is obvious he is the most capable for the job, his past history has antagonized you against him?... but the matter presses you to act relatively quickly?

MUSGONIE: Something of that substance to construct. It is all plausible and obviously stated, and even correct in the details.

PONSTOLE: I should bait a hero?

MUSGONIE: It is the timing that counts most critically, crucially, Ponstole. He should think about the proposition offered as offering, but then he should not have enough time to think at all about it within its propitiousness. And now as he is physically weakened is when he may accept to ask for this, because one minds pain more than painful prosperities, or stiffness more than stalwart thoughts. He will not pay as much attention as typical, to concentrate more on the consequences rendering him. They may not even occur to him till years later, many underhanded strifes abated, because he will not have intentioned such possibilities to his consciousness. Thus they will be evaporative of effect, lingering but not lay. And once this method of conducing is approved, by me.... he will be committed, to do his best at his own insistence of utility. For that is the nature of a military personage. They are very much blind to other considerations— How else to think to die for a cause possibly drafted by others!... by a king, or country. Disciplines of the mind and personality may guide you through

straight paths crookedly. He will be averse to noticing the blunders.

PONSTOLE (*sitting down at the table, on a side*): I am not to present fibs with this manner, your way of cajoling into chicanery, since his own well justified loftiness of view leads him through these steps provided. I am privy to you, sir. But this is a precarious gamble still. Of course, if these outer provinces may be appeased and some satisfaction smattered throughout these regions to keep a peace and quell minor discontentedness, then this is a welcome tact to try. Yet it's hard to convince me otherwise than that this discontentedness is deeper and more thorough than your ruse may lessen. Yet still it is a curious compromise to your train's obdurately held direction— which I can appreciate for your royalty to sustain and feel responsible or obligated towards. But revolts and rebellions are dangerous articles of action to contend with, your majesty. They are not a game with any levity to interpose or interject. They are ever subdued or subsided only after devastating casualty and trauma, and often with infarction of our heart to care for. That is to say, a country's changed when it is forced of grisly means and manner against its own assumed countrymen, and patriotism may alter much its shading.... through your paternal robe to quench or squelch, or try to protect from extinguishment out of your favor. This is an interesting ploy you suggest, even a cute one, to support and continue your mandate of union and avoidance of any reduction to our sovereign territories. But the seriousness of many lives destroyed must be considered as too possible should this simple stunting attempt fail. I still say it's best to let these few severabilities occur as natural, and without any disgraces, that which must always happen if any violence is employed. But I am not the one to gamble on direness, and must only try to advise and persuade. This being your strategic poise made absolutely—

MUSGONIE: I'm always open to opinion.

PONSTOLE: —resolved to, and not to toy with toys but for toys to tear, then I will try my best to carry out your wishes. It be plain as a spectacle of comparisons to me (*standing up abruptly but with resignation*), your rights against Decamond's to bargain with in propriety of our natures and stances, there being an allowance to be one's self and not slavish to a decorum—

MUSGONIE: Have you more suggestion other than what's been said?

PONSTOLE: —Then let these movements play themselves out, and arrange for their order. More time to be proactive escapes my breath.

MUSGONIE: Have you more thought?! I hate confliction and agony, and vicious argumentation to develop a view as if I were to make a pastime of trouble and problems. And I can not stand war and battling. This is against my office, true, to protect the realm. But it is how I'm born to both degrees, and with a temperance not to take *anything* lightly. Never make such a claim to me again! Ponstole. Your honesty singes. My best intensions only are for your pursuance and ratifying with.... the others. Am I still misunderstood in this?!

PONSTOLE: No, your highness.

MUSGONIE: I've.... too much envy of your objectivity. But I am king! and can not afford to be objective. My only point is for my rulership, much independent of my personal drives and ambitions of pleasure.

PONSTOLE: I am not for one, that to deny.... but nor to prescribe, and never to analyze.

MUSGONIE: Then you can't conjecture of my throbbing pain, in this business. But I am always serious to be dealt with, because I am even more seriously dealt. When we meet, en masse, I want some agreement of procedures from your body— to deal with *them*!... and to prepare for my visiting as soon as practically a suitable climate can be arranged— or erected— or resurrected.... You must tell me how. You must work this out, in ways that are not specious— nor inflammatory. That is what the council is for. And I will bear the brunt of aspersion, should any of this fail.... And then to the war committees to lend my spilled blood, will I drain. But I know I am wrong— I'm always wrong. The majesty corrects from being wrong— not being right.

PONSTOLE: There is ambivalence of logic here you should faithfully challenge— Errors.... are not right nor wrong. They're simply unexplained for being errors till they're done and the results are examined. Yet still we trust you to make them as were to be of some participation in all of life. Were there not the error there would be no sense of progress overcoming them. We in a sense all work for errors made.

MUSGONIE: I say I'm wrong! and not in error mirrored for my faith.... in country placed to rear and tend. Your jostling hurts of joke.

PONSTOLE: I simply can't believe so august a man would fall for such a sham or shove of shrewdness, even under the duress of some transitory enfeeblement or incapacitation. But one proves a point by what is done out of suggestion. And so there may it be demonstrated, that you are right— correct—

MUSGONIE: I am wrong for this to have to be done— not for it to do! Were I perceived as a stronger manager, mere threat alone would subdue this duke's excrescences of power and extortive activities that bring a shambles to my goals of sympathy and paternal commiseration, tolerance with a touchiness to be respected and approved of. I'm not of a like to want to harm anyone. But something clever must be tried, where stupidities thrive for changing.

PONSTOLE: Yes, your highness.

MUSGONIE: It must be attempted, no matter how smart or self-assured or savvy he may seem.... to you.

PONSTOLE: Yes, sir. I will— suggest.

MUSGONIE: For this is the minor problem, to your master's major one, a gnawing buzz only about my steps to goad. And these complaints constantly received, from the "recaptured" regions, demanding damages be paid for— as a practicality of persistence under my reign, then I should pay for he who has been excessive dealing with them, as well he should work for me.... We can not afford the debts that accumulate throughout the land, from all regions entirely. And I must find a way to abolish them, and simply eliminate their existence. As a decree made popular I should find, to vanquish the country's pressing loans owed and their growing interests. And I wait for a proper climate to make the announce-ment.

PONSTOLE: This is a common problem for rulers of expansive ranges, and a brutish step to take, your majesty, to address its imposition importunely. But as we are not masters of finance, not of the country's, I fear our council can not intelligently help you find this "climate." It is a matter for the treasury to be astute with. And they say you may balance our expenses by eliminating debt altogether?

MUSGONIE: It is a rationality of negative decree, since the vast majority of these loans are internally held. What may be specific losses to persons might possibly be made up by common gains to everyone, were I to permit such a law as stated, abolishing for one time only all outstanding debts. And my own are considerable, but they make the country's. Therein a tax will be applied to all of ownership, which will be facetious and non-existent of worth or value other than the expected repayment from these loans. The debtors and those without debt will therefore not be taxed in this way. And all may be pleased if it is authoritatively handled for the commonweal. My treasury officers tell me this is fiscally possible an.... adjustment, but only rarely or once(ly) done within a memory. And the problem with the attuning is always in how it is conducted. It should be swift, and soundly accepted.

PONSTOLE (*to himself*): *I am owed some duty.*

MUSGONIE: For the betterment of the state, this common usury tax may be sparingly applied by sovereigns— if a time can be agreed upon by most for a need to spruce up the economy from a presumed malaise of internal deficit and bickering about that— *and*.... if our trade with foreign powers seems meanly recuperative but not of such an extent to be primarily concerned about, that the problem can be redressed with more vigorous production and selling on our part, the debt elimination in this case to be so minor as to be virtually ignored or glossed over with other deals and future promises of great potential and established certainty.

PONSTOLE: This does do me beyond my knowledge to contend (with).

MUSGONIE: Obviously, for such a climate to appoint itself of taxing, we must bring our external provinces well under control, and not allow them to break any meaningful ties, as their goods are so important and their commerce healthy and vital to remain.... And they are due such *immense* payments as to cause our gifting of the sugar. Then to stifle their disgruntlements with slack of payment or none, I give them Decamond to batter or ball with— for a while, as the stemming of recompense is only temporary until our industriousness rebounds. Thus, one (is) for another, for awhile.

PONSTOLE: The scheming is more plain than the plan, and has many dependencies to hold. I can not understand its obscurities so well, your majesty, except that need to enlist a warrior and warmonger of the highest degree.... to avoid your wars.

MUSGONIE: That is the privy task, Ponstole. Advise, but do not think for me. My head is geared of crushing weight to comprehend. And this does often cause it ache!... to be wrong.

PONSTOLE: Yes, your highness. Had I your authority and command (of everything about us), webs would be turned into crystalline spheres of see-through showering.

MUSGONIE: This is not what pleases me, but what is pressed upon me to attempt. And contrasts need not be opposites. Glass traps within its heated pour of resins as easily to kill as spiders on their webs.

PONSTOLE (*starting to exit*): Yet may the spider gloat of its bundling, a crystal melt is shaped more mineral-ed and regularly than a happenstance of resining with disorder.

MUSGONIE: *I* will build the shapes, out of whatever grows.

PONSTOLE: My leave, sir.... and good morn. (*Exits.*)

MUSGONIE: Too much is said of knowing, without a wanting thrill. (*taking off his crown and placing on the table as before*) Only my crown proposes this.... And the character of beasts propel the argument. I don't dislike the brute.... but more the brunt per-suades me of this tact. Oh, but for a solidness to land upon opin-ion. One must treat as unkindly the most popular of kind and con-vert of growth, reshape and direct of training what is fortuitously held, as could be a blister on the sole for running of this enormity of nation and my consequences towards it. But does the patter blight (*starting to lightly tap the table top*) of blithe emotion about my hands. Consider the weight held here, to change the seasons and the songs (*stops tapping*).... of cruelties. How does one handle the realities of people? How does one presume to? As subjects or children to be? or as gods at play, be spoiling fortunes and anger-ing up the earth— with a preciousness. And I feel less a ruler than a roan-hearted gnome, at times.... belittled and belated, and be-loved— or be.... this graying majesty imposed upon to be as kind, as kindred generosity allows. (*starts lightly tapping again*).... Some little feet, with dearth of stepping, too. A nervous twitch? My passions lie betwixt my rue.... Service! Service! (*stops tapping as he raises up his head towards the ceiling*)

Act II

Scene I — *A field or yard of military athletics , in Lacismoor. There are various soldiers and pages at display through their daily routines under good weather. As **Bracken**, in polite clothing, enters, **Townes** suddenly bounces from out of an exercise towards him, to startle playfully.*

BRACKEN: Prince!

TOWNES: Let's have a row!

BRACKEN: I'm not dressed to tumble, Lord Townes, and well past those efforts these days.

TOWNES: Oh! But you hath recommended—

BRACKEN: As well for young son of state to exercise the lungs with a passionate activity for living. I was vigorous indeed with my calisthenic drags an' drives an' drifts. But one may mature into other fashions of employment, you know. Not all may soldiers lean to be as found for our necessities of limberness. Yet I might bend a few pikes still, or lance as well as knight, or with my archery su-perb to be in throwing arrows as like darts and knives. And you, boy, were, to be quite blunt about it, drooping much until the gym-nastics took hold of your interests.

TOWNES: I feel fine in them; and even of small frame could kneel to lift mountains, as is felt a rush to imagine to accomplish the impossible for play, and the all too possible for review. But I don't mind the honest commenting.

BRACKEN: They take their activities very seriously here. The *men* do practice out a toning of the muscles to simulate routines that may both be deadly and lifesaving at the same time, deathly and sparing.

TOWNES: We're particular fighters, aren't we!

BRACKEN: Especial for Lacismoor, as it is entirely voluntary and anticipatory for some defence that these collect with a custom'(iz)ing of preparedness. But I found the organization to be somewhat casual of importance and un-general-ed with little regi-mental discipline, as my awareness of life's circumstances grew. The camp certainly has its health-wise benefits. But it's not enough to stay fit, young Townes, but also to gain a well-defined sense of order and organization, and restraint and reserve towards a recog-nized authority, something which these soldierly participants regu-larly fight against— as I did myself when I was among them to revel of the fun. It was very difficult to provide a truly decent dem-onstration of collated skill and demurred ability to any audience, much less to set up a pageant, we were all such frothy individual-ists.

TOWNES: I'd like some of your ooze, for renown.

BRACKEN: That takes a bit of physique.

TOWNES: But there are officers who can handle us to mold into some coherent body of fighters and defenders of the state, should that need ever arise.

BRACKEN: Ever?! Young man, it's been practically foretold since your childhood that our province is too rich of resources and handsomeness to remain merely an appendage to an aging gout of country. We expect to undergo a secession quite vigorously, some day. But no one prepares for this realistically, and with a high moral strategy to follow; because to fight for a cause to the death requires the strictest of morals. We only deceive ourselves into a perception of comfort through a maintenance of sundry fortitudes of game-full strengths and expertise. And when a real army arrives to confront us, they will not ask to take turns with the blows, inter-spersed with assessments on the calibre of our posturing stances. I've come yet again to seek the captain of the camp, to have him try and speak with your father and the other major landowners to con-tribute to the organization of an actual military unit, something fundamental and precise. But the squire's never to be found unoc-cupied for the chore.

TOWNES: Darughe only owns the land and equipment that we pay for to use and exercise with.

BRACKEN: Always too busy for me, one of his best adherents to the venture. But he's a fine one to yield advice for a more serious creation of army. He would know who to weed out or transform of temperament and pride, and who are already appropriate of person-ality for a proper soldiering; though I'm sure all of us would be pa-triotic enough to defend Lacismoor as an autonomous state, in our own ways.

TOWNES: Oh, he's a royalist, I'm certain, loyal to the king.

BRACKEN: He wouldn't be here, rough-headed(ly) as the case for ourselves. We all strive for some measure of independence in these out-parts, and away from the stale decadence that only grows to thrive inwardly of country. No, he is more timid to offend our local landowners with subscription to this obvious need. His business here is good, has been for a long time. And he does not want to alter its character in a damaging way with possible resentments from the heads of families of potential customers and promising aspirants to physical daring and show. But he senses the need nonetheless, as acutely as myself and so many others. We must begin *now* to build up a good militia, something that can be feared as formidable. Our protests have been loud and flagrant, as of late. But words and letters of insistences and certain freedoms are not enough. We must send an actual representative to court, with threat of a real military to have to be dealt with if our desires are not met, (but rather) are continuing to be hampered.

TOWNES: I know my father still favors Musgonie's rule.

BRACKEN: There are ambivalences to that, I can guarantee you. We know each other better than you he. Respectful he remains— to law. But we've all come here, the original settlers, for advantages. And they are being slowly whittled away with obscure fiscal oppressions that don't make much sense to us above the common taxing, as if our commerce burns too brightly, self-sufficiently. And somewhat do they wheedle away our rights with a strange flattery, that we should represent a future progress reflected in time throughout the whole of (the) kingdom. It's simply not possible. The kingdom is too old— while we are fresh, vibrant, vitally stunning an' starred to proceed with a graciousness that automatically leaves behind the drab and stilted. The court attempts to pull on our tails as if reins. This can not be tolerated very much longer— And I don't mean with contempt for the country proper. Matters are quite more sophisticated than such a childishness. But we have much work to accomplish independently and without untoward or misguided interference and hindering controls from a good king. We can no longer be taken for granted as a sovereign right or privilege. Do you feel so slavishly adaptable?—

TOWNES: Why, no!

BRACKEN: And neither does your father, nor anyone of considerable importance in Lacismoor, with properties won through difficult exploration, extraordinary self-sacrifice to labors, and hard management of a diligence characteristic of proud ownership and fostering.

TOWNES: Well it's hard for me to believe that people of our spirit can be overrun and subjugated by any troops of indifference.

BRACKEN: That is so terribly easy, young Townes, for the professional to do with proved tactics of deployment and engagement against a disunity of field, of, in general, amateurish impostors to a battle. A group may choose a nominal commander. But it can not choose an orderliness and fluency of militaristic technique. One must earn this with rigorous, thorough, regular, repeated and rapidly reliable training and drills. That moiety of mind against the lackadaisical of interest in this regimental-izing is lacking here. There still sparks the puerile tempers and behaviors, which is why one eventually grows out of this camp for enjoyment and the camaraderie of entertainment, as I've left it some years ago. But the potential is tremendous to recommend. And there is no other considerable social gathering through which an effective army may be transformed into and melded with a few more serious and experienced volunteers. The cause to do this is not seen yet. Skillful *soldiering* must be paid for and provided with imperatives and impetus to remain of a highly prepared fiber of fineness, a strong cord rather than weak threads. Now how can this be even attempted if the very creator of the camp is reluctant to be an advocate?! More minds must be moved than these athletes and showoffs. A seriousness of our circumstance and growing challenge must be imparted to many to ignite the actioning of a military body, its framing, construction, and reinforcement. Yet we have no local government to order this be done— So we must do it ourselves, and we must feel of a precarious pressure for our tardiness to initiate and our lateness to develop this fundamental need.

TOWNES: Oh, I suppose you convince me, for some organizing to impose or try at. But we have uniforms already.

BRACKEN: They're only sweating gowns, Townes.

TOWNES: But I mean, they give us a regularity of appearance and recognition, when we wish to seem neat or present ourselves as a group of more or less common abilities to observers and admirers.

BRACKEN: You hardly make a team, in your performances locally. It's strictly for the fun of displaying what you've learned or practiced well, to those who haven't the time or the ambition to do the same. But a regular army is not a pastime or a club activity. Even as irregularly meeting, one must be kept vigorously inured of threat with rigorous procedur(e mak)ing. We never marched long distances in arrayed formation, you know. That enhances the endurance so key and vital to meeting conflicts of battle with a steely resolve of reliability on what has been experienced of a group's capabilities with tensioning. I dare say you don't march still.

TOWNES: Not much cause for it. Don't see the need.

BRACKEN: Have you ever walked for miles and miles without much rest, and on a timely schedule to arrive someplace and return from it?

TOWNES: No.

BRACKEN: But one may have to, as a fortified body, to chase an opposing troop or withdraw from their superiorities— in a reasonably intelligent and protective manner. That sort of thing has to be practiced, the necessary muscles and nerves of tedium built up, or the effort under fighting may too easily break down to scatter you and negate your resistance totally, eliminate any actuality of oppugnancy. One cannot expect miracles of fairness to happen under these combative strifes of armies and soldiers. The better prepared and more properly trained prevail, sometimes irrespective of number or against what would normally overwhelm.

TOWNES: I'm for it in principle, my good friend Lord Bracken, and hold your wisdoms in awe and without any disconcerting whatsoever. But you see the cataclysmic nearing. Isn't it most likely treaties of understanding will be brokered, or agreements written to be ascertained, to lend us more self-governing? possibly election of even a mayor or magistrates?

BRACKEN: During these days? dangers are great for any hint at dissolution by degrees, designs, or desires.

TOWNES: The king would not want to have a warring with us. He hates any kind of altercation disrupting his peaceful kingdom. My father insists on this. His pride is for peace and tranquility and abhors bloody antagonisms, so many of his ancestors and relatives have played the victim to deadly quarrels and martial spats. He refuses to be judged by force, as was his father killed and mother decimated of mind and manner through this. So sensitive is he, my father says, that it is a major reason our country has spread of excursions, as to Lacismoor, with perhaps an uncontrollable expansion of jutting territories into quite a disparateness of interests.

BRACKEN: It was a fairly violent time, well before you were even born. Musgonie felt to be a sovereign over ruins fortified, but fairly fought for and won, the scars of plain and earth reminding him too cruelly of his mental scarring and effective orphaning at early adolescence. He does not forgive war, to be idolized and yet feel shunned at the same time, and with youthful responsibilities that tarnished his nerves weirdly towards an uncomfortable tang of outlook on his environment and royal positioning, surrounded as to be crowded of advise, and yet quite alone of personage. He.... senses strangely, and can not recover with his queen, hardly trusting women. And he is doting on his little son, who is rumored to be.... surrogated through her, trying to give him a strong taste of life without any hostilities to face or bear as did our king, hostilities internal or perhaps internalized. Oh, Musgonie may try to placate us with favors, even favoritism. But he tends to handle these situations rather loosely, allowing for adjustments and opportunists to take over and carry out his proctorial wishes with great irritation to these outlying provinces as ours. There is a reason why no independent states have ever been established yet, certainly not for lack of desire. But the lax king allows this strident, raucous, and most notorious intimidation in his name, with absolutely no incentive to prevent it if it keeps us homed. It is a cruel, yet unmanly policy, terribly resented but abided with— so far.

TOWNES: Adventurists?

BRACKEN: Profiteers of bullying and harassment! as if punishment for insolence to crown and country.

TOWNES: It's hard to believe we could ever be so mastered, or dominated for our backs to be broken in spirit.

BRACKEN: Well you've always been free, Townes. But certainly you've heard of the skirmishes at other cities. The news spreads for attentions to seek, and you young ones must begin to take notice, for your futures' sakes. But care you not what rumor makes pacific, until the snake bites you on the leg— and you roll down with it to struggle and conquer of an invitation to a match, try your stuff out in handling fate an' fury. The carefree days are ending for you, as you age.

TOWNES: Oh, but they are backward provinces. I took it for granted they were sluggish in their remittances of taxes or fines, and were being reprimanded. We haven't such a problem, with our wealth and vitality and resourcefulness.

BRACKEN: They were foresightful enough to produce armies, or a defensive posture. Thus were the skirmishes made, the resis-

tances crushed or apologized for, and the events celebrated for the continuity of the country, angers softened by a beating, and shameful insouciance forced to return.

TOWNES: Do their armies dissipate?

BRACKEN: I hope not. But they are not fighters against the royal— but against the scandalous of professional robbery hardly made legal except for the events. And such outfits must clearly have an eye on Lacismoor. The day we declare ourselves of some proponed or propounded freedom long considered, they'll rush by to test the mettle of our purses. It's happened enough to be predictable, though they must wait for an overt insolence to the king.

DARUGHE (*coming up to them with a slight reluctance*): Then we must not offend his majesty, Scott. You are so apt for making soldiers that you don't see the climate for its temperatures. I hope you haven't been proselytizing too heavily. But it is the sword that brings the danger, in these matters.

BRACKEN: I've only spoken with Lord Townes, squire. I've come to look for you, not the diligent at play but the busy to avoid me and my argument, sir. And I needn't sell for convincing of words what is obvious to worry about.

DARUGHE: Yet that's the cost contained in them. I don't avoid you— I abscond.... with your reputation to stay so pleasantly in tact, and which clearly enhances mine to disavow any disagreements. But.... we've always been public, haven't we! And are you paired to party with your friend, Walter, some parity of view to reach? in this militia business?

TOWNES: I don't see why not, to have some kind of an organized unit of defenders trained and schooled to be proficient.... as of our skills to evolve.

DARUGHE: Well, son. And what would your father think about that! But onslaughts are to the death be made. It is called an activity of slaughtering. And I am old enough to have been in war and killed a many, each soul a sin released. The real encountering is horrendous, and no game. And wounds are trinkets made of skin an' flesh. How proud you are indeed, to have them leaking the red valors and pus! Pussy war is made of, boy! in fighting to vanquish and not merely thrown down to subdue, as in a wrestling bout. An' the stench is no fun to be of, the homicidal kind. Your teeth become fangs to see, an' your face a malice ported of weird pouting and perverse attitude.

BRACKEN: I would not propose that a battle is anything but terrible and horrible—

DARUGHE: An' you drink blood like liquor, as it comes from your own mouth and lungs!

BRACKEN: —But you know that we must take steps to prepare ourselves against the inevitable, that we must build up an organized defense made of soldiers, paid soldiers not with sinecure of position. Else we will be stolen from mercilessly, out of a thankless cowardice to own.

DARUGHE: You gentlemen are not well treated of your inexperiences. But you fail, Scott, to see the point, that an affronting with an army leads to attacking of that army. It legitimizes the assault.

And as we remain one country, this country will police itself, without the warring conflicts within us raw and bare. But give your home an army to proclaim, then that demands attention and dissentient spite. An' I can mean homes and houses distinguished of their fierceness and rage, and streets and neighborhoods at battle with each other. And why then expand towards peace? to bring in soldiers.... soldiering. The weapon is a talisman for harm and quarrels. We've settled out of this, for awhile. We've come a distance to escape the tensions—

BRACKEN: That was long ago. You recall too much.

DARUGHE: —But a baker bakes his bread to sell with heat! And I do handle my hands about the physical skills of athleticism(s) to know and promote. That— don't mean none to war, but not to war. There's not much purpose to it, if it's not serious, if it's not to breach a fundamental flaw insisting on a loss of life. That payment's due, with many manners to evoke short of futile, useless sparring. All heavy men know this, for heavy heads to bear.

BRACKEN: Lacismoor leads herself towards independence too fervently to be denied. We will not allow much further more abuses to be levied against our customizing for autonomous natures and dreams of secession.

DARUGHE: It is a comedy of terror to laugh at wrongs.

BRACKEN: We are entirely self-sufficient, and wish to take our steps unhindered by the shackling disorders of the past, the disarray of mind and character that leaves us to seize freedom as a right of man with maintenance of himself, and leads us through these insistences with growing discourtesy— wildly so!... this may become. And this is all of a sensible concern— because we are rich for the poachers to poke at, and fat of honor for our achievements; not that the king would police against his own dogs, but see them as the country ravens menacing for crumbs.

DARUGHE: Ah! But we outer provinces are as vibrant as ever, are not with a poverty to nature, have established ourselves more or less soundly. And several have tested the waters of supposed separation, and have learned that now is not the time to be so indecent and ungrateful. Obviously a first success will come, but not under these conditions of severity to sovereignty. The king remains friendly to us, and a populace must evolve into such action as to dissolve an amiable and patronizing tie. The heart of the people is still not to be offensive to Musgonie. Most are proud to be seen under his cloak, at least in terms of staying reasonable to his representation of rulership. It is due to that that we are much protected in these remotes, by mere notion of this relationship. We are the infants of a charmed life yet, relaxing through our vigors of commercial interests as like children on vacation from our schooling, playing of commerce and only fincing success. But would that be true or would this state of bliss be lifted if we were (not) headed-ly on our own, repulsed and repugnant, and reprehensible of thought to deal with? too difficult to be considered favorably? Now these several attempts at militia were half measures, on a learning curve for growing into substantial adulthood. And it's only to be expected that they would be picked on, by scavengers finding the foolish young at play. The only ones certain for battle were the mature— it's always that way. An' aggressive arrogance backs down to suppressive surety. Look at what you try to do, Scott, among players and their playthings. They are attentive of a soldiering to resemble and be proud to seem like, but hardly committed to a task

more serious than themselves to fulfill or fortify. And what harm has Musgonie done them, as they exercise?— Knights of the realm, in effect of stature felt or prided by.... but just the sons of the seemly of Lacismoor, and their paid pages, attendants, and well-wishers. What must you convince *these* nobles of? their tans and muscle tones? It would be halfway done again, against the determined to do and trained to be responsible. This is idyllic harnessing I make for these— And you yourself grew grounded more towards the realities of life, as pleasure seeking ceased its lessons (for you), with maturing of mind to be more pleased by. But the way to actually deserve an independence is to have the king grant it, for some reason, with benefit to all parties involved. Until I sense that trend, I remain very skeptical of militia; for only then would one be truly needed, the effort of production stimulated through trusts and mutual respect. Take my wisdom to heart, Lord Bracken. One learns quickest how to crush the immature, unready, and poorly prepared for battle during a *combative* war. They may make your only personal conquests, some days.

BTACKEN: You say we still babble like infants, Darughe. Well, that might be. Intuition squeals with strange sounds, sometimes. But I *know* we are going to be confronted. It is beyond our capabilities to resist for much longer the secessionist urge, especially when our economies may support the attempt. We must prepare for this with intelligent promptness and astute awareness of our probable coming conflicts. No one here will settle to be robbed.

DARUGHE: And what real fighting have you seen?

BRACKEN (*offended*): ....I have seen relatives killed, sir!

DARUGHE: I mean seeing as in doing, not as a small boy viewing. All (of us) of a range of age have noticed *that*. Do not be put off by my candor about brute causes to cause asunder brutes! I've lived this, and know no other way than to speak honestly about it. And I tell you flatly that your method is flawed of ignorance and lack of insight gained through deathly struggle, even if it does succeed and give you revelry.... for a time. You're a persistent and capable person, handsomely driven through the logic you allow. But success of goals does not mean success of results. An' this will most likely end as the others, a *defeat* of militia, as can be a growing pain survived (through). My concern is for the unnecessary carnage. That revolts against my notion of human justice, a revulsion of more shame than pity. Do you know how many youngsters these hands have destroyed?! Now I only want to build the youthful body up! for tranquil splendors posing in the sunlight.

BRACKEN: I am not.... particularly offended by your sincerity of vision, squire, even if it is a trite reactionary. But that is certainly your right of conditioning by a lived-through history. However, sir, I need your advocacy in this mad caper of an adventure to create a standing and functional army. And I sense your instinct is for this, and only against its failure if it is not crafted correctly, since you have always been badly sensitive to a slovenliness of purpose. You know as well as I that Lacismoor is agitated and haughty, heated, impatient and restless for a proximate change of state of the sylvan into the sensational of self to own. And if we might have an army of soldiers backed and corrected by the monies of our consciences, then more triumph may be predicted for our cause— Lacismoor's, to become rightfully prominent and persuasive for a freedom. Her young sons demand it. Her older ones want it, and will allow this, once the impetus is carefully made to start a meaningful separation and reduction of restraints that even publicly annoy. How have your fees increased?! with no justification at all except our monarch must drain for more to sustain an un-wieldy country of back-headed and retrograde principalities not as effective to contribute to the weal of nation as we must!

DARUGHE: And what of our Townes (*placing a hand on Towne's shoulder*), with youthful plight? Would you like to see his neck severed or chest punctured to a futile cause—?

TOWNES: I'd defend myself even through rain!

DARUGHE: —twice a day some weeks, when the fighting was vicious and prolonged. A worthless cause, if this is so. There is no doubt to an arrangement towards killing. If you arrange it thus, might does its best to make it so. Yes, you can cause such a con-sternation of threat to make a ruler want to sever you. But this is a beggar's way of awkwardness, to show himself disown-able and trying. It seldom works easily, if he's really a jewel. I don't disclaim to be disturbed by the rising prices of doing business. Be they fair levies or not, they are assumed of the customer. But are we not very much in a paradise compared to the stench and congestion of the inner country, and should not a greater cost be due to enjoy this luxury of existence relative to the gray faces of contention in the crowded cities? and their daily frustrations paramount for a deeply expressed sadness within the grime of their common and lowly toils? (*bringing his hands together*) Pray we for this heaven kept of privilege fashioned through a king's lax care and handling. Should we upset this as being spoiled?... or remain obsequiously pleased with wondrous fortune and hard fought for luck!

BRACKEN: You're teasing, with a caricature of obeisance to laugh at. (*as Darughe laughs*) I'm not one to try and topple a smoothly sailing boat by standing up (in it) and jumping all about.

DARUGHE: We are all very comfortable here, with our industri-ousnesses— are we not?! And Walter, I bet, has only known it to be good, a life of leisures slack of laziness.

TOWNES: I work hard at it, certainly, squire, vigorously— to approve of myself, my condition and conditioning, I mean. My fa-ther admonishes any waste of time. And fun is for fun's sake, often difficult to obtain, if it is not worthy of some high purpose or achievement.

DARUGHE: I agree, young lord. Play without pattern is not praiseworthy, and often potty. Now, the establishment of a military regiment would only work to enhance my trade of training, Scott. And I could be for it easily in that sense, were we to forgo in thought the heavy consequences. But such a practice must have its proper or auspicious time of blossoming. Yet prematurely it will flower wilting-ly, wasting of color and vitality as in a winter's cold charity of care. And do I wish us to be independent?... We must *earn* that right, Scott, as our independent spirits have sent us here in the first place. Your father will attest to that clearly. And I was one who badgered him and others to make this move we wanted to do. So don't see me as in any way reluctant for our passions. He is a slayer fed as I, by a propriety of practice, and an appropriateness for doing something catastrophic and of grave importance. And I'll lay odds you have not even convinced him of this, fully, but wish to circumvent opinions of reluctance by trying to convince and in-fluence the other great landowners.

BRACKEN: My father's old as you, squire. But Townes' and oth-ers are of a younger breed who may hear the clarion call more stri-dently, less deafly.

DARUGHE: Only slightly younger. Less than half a generation off, but no less wise nor ambitious. Lord Townes, Esquire is a good, stern stake, and was a noble talent when needed, quick to learn and adaptable of technique.

TOWNES: He brushes them off, now. Disdains them.

DARUGHE: I taught him some striking moves myself. It takes good timing to smite swiftly, not wasting effort for result. Oh, they can be made foolhardy.... again, with half a cause to show their knowledge. That's not the point, nor the principle discussed to be disgusted by, as far as I can see this need for a militia. You don't raise such flags of provocation lightly. They're not to be used merely to support points of debating. Lives and deaths and the mis-ery of maiming follow them like ever present shadows in the day, till shades of night with nightmares consume. Speak to the many cripples, Scott, aged well past their broken limbs and marring ef-facements. Ask them if this is wise!

BRACKEN (*affectedly moved, but dismissive*): Not so many here.

DARUGHE: Most are stuck.... in the tombs of our country. Could not make the trip, the trek.... towards our freedoms.... and frugalities of wealth.

BRACKEN: I know a few, back there, and of— But what I fear more now is here, our unpreparedness. Yet as we strengthen soundly, this is not a superfluous view or a wild-eyed foreboding. It is of the day most currently felt, something dreadful to avoid or prevent. And if with me, then with many are we sensible.

DARUGHE: Well most of all, Lord Bracken, it would be a de-clared sign of disrespect to the king, an advertisement to that ef-fect.

BRACKEN: Hm. And have you respect for him?

DARUGHE (*turning his back to leave them*): Sir, I respect any-one who has been injured. Enemy or foe, fiend, friend, or favorite, they all have my assurances to be considered of their wounds.... (*stopping to face Bracken again, slightly*) I am not opposed to the concept, but you must render it more propitiously, and with a back-ing of worthy minds.... (*resuming walk*) And if I am invited to a conference of this liking, I will assure you, for your arguments to work through and criticize— or contemplate.... Yea, Scott. Stay as busy as I. There's always much to do, to keep one's self reliant....

BRACKEN: Well, that's something. He's not exactly opposed to me.

TOWNES: I should think no one of vaunt for liberty would be, in Lacismoor.

BRACKEN: But I had hoped he would *help* me create a meeting. Says I must do this myself. Yes, so he shall. I'll ply, plea, pledge him as a card of attraction to the event.

TOWNES: He may argue against you. Seems to swerve on the waves of compunction.

BRACKEN: The meaning is, if he does, it is for foolishness to frown. Then take we risks as great as he has handled, in his life. We are for better to be one strong ball of hard shell and even harder core. Of that he is wary, not our vision shared. It took great intrepidity even to establish Lacismoor, as well as other of these outposts to the crown. We are the fruit of this tree, young Townes. And he awaits our ripeness of action, a suitability to be proud about and gloat over as a rich and sweet success. For such a man as Darughe sows and plants the seeds for a future only to be won, and not for a love of the backbreaking labors to glorify. We are schooled for a secession, lad. That is the fiber of our character, the mineral of our leniency to self-expression and ease of thought. That is what this camp composes, so obviously lit a light to seize precisely for this purpose shining with. And this is what all men want, of Lacismoor. But how can we who must do the actual of army prove ourselves as forthrightly as the founders— with acrobatics?!... or with an acrimonious resolve to threaten both ruler and rogue to tears?... or something within the extremes of servility, to our harsh natures, and the warmth to cold of others. It's hard to judge or balance. But we will be exercised of it— and all harried, until the irritations felt are totally extinguished, so thoroughly reduced as to leave us as ourselves alone and enormous of the passions freed.

TOWNES: I will, then, continue with my practices. Because I feel totally at liberty to. It is indeed an inherited right, and I'm not accustomed to anything other than this privilege of eternal welcoming in this fine and healthy province. And I can't see how anyone else with such and similar enjoyment could ever fail but to want to defend Lacismoor to the heights of contention and contrasts to the sodden mares of country bickering and fiendishness the lusters have left. For if ever an assailant finds his way here, it's to a weighted hatchet of our fervor to defend with thorough disassembly. And it's more like fun to prove the thought compelling, since stout hearts are never as frivolous as they may sometimes seem, or beg at (*while bouncing away*) ambulation! good, and forthright friend.

BRACKEN: And a good day to you, Townes. Lean with the muscles and the dexterity of joints. That is what to head for, and head off, with some considerable balance of aims, for our station in a paradise to retain, and rattle through deliriously with the noise of easy difficulties to want.... or the want of ease for difficulties that are must to take on. (*going off*) Dreams are never as hard as their dreaming.

Scene II — ***Alfonso*** and ***Skacte*** *stand before a row of dwellings, independent though regularized to a handsomeness. The street is "off" of busy, meaning that a greater disorder from vendors and pedestrians occurs within hearing nearby, such that one can conjecture this housing may be somewhat partitioned off with judicious fencing and fees. It is midday.*

ALFONSO: These are fine accommodations most definitely here, Skacte.

SKACTE: I've furnished most of the dwellings myself, at least partly. An excellence of utilitarian furniture I will attest to. This is an area provided for with corners, coves, and calculation of a generalized privacy for the inhabitants, almost especially of Decamond's ensemble, and truly a delight to pay for residence.

ALFONSO: With some heft, I can presume. I've been staying pretty much in a slum myself, a swineherd's open penitentiary of activities and scents. But it was easy to find, in a confusing outlay of city design. Merchants seem to work the dirt there day through night. And the walking is so dense, it might as well be through mud.

SKACTE: But relatively inexpensive are the rooms and apartments. A cheap part of town you've found. The duke's men can generally afford better, and here is a prime example.

ALFONSO: Does he occupy a space himself?

SKACTE: Absurd! He owns one of the houses, or maybe a few. But he seldom lives here. They are more like property interests to him. And some of his soldiers have adopted to them for that reason. Nothing mandator(il)y prescribed. One can sleep wherever one likes, even out in the open country, using tents— (of a sort) that I may provide. In fact, before an assumed major campaign, if the weather is particularly foul and rainy, a few of the soldiers purposely camp out, to re-accustom themselves to the earthy— and heavenly— hardships. But a camaraderie of association tacitly builds up in an area like this. And so I personally recommend it for you. There's even a private sporting area—

ALFONSO: Ah!

SKACTE: —in the back. But it is restricted in what may be practiced. No overtly flying weaponry, for example, is allowed. And participants must scrupulously clean up after themselves. Therefore, any bloodletting is considered quite distasteful and offensive.

ALFONSO: Well, polishing one's techniques need not be too injurious.

SKACTE: Yes. But no mess of any kind should be left lying or festering. And the hours of use are of course restricted, to the mornings through early afternoon.

ALFONSO: Interesting. One may awaken to a passion for pangs, clangs, and panoply.

SKACTE: That was adjusted from late morning through early evening, and will probably be changed again according to the seasons, by variously changing committees of dwellers, polite agreements always. But women live here too, and they prefer their afternoons and evenings to be soundly safe. Not many children as such, below the age status of pages. There are a few proudly donned knights, decoratively speaking. But it is impractical to travel long distances with heavy armor, Alfonso. One's skills must be employed with a more facile physicality of attack, I can tell you, as a soldier for the duke. And actual noises at this residential spot are kept sparse and low by common custom and courtesy. Late sleepers are seldom much disturbed. Frankly, what distinguishes neighborhoods mostly is the simplicity of cleanliness, the most civilizing principle of all. People living in any location share a comity of action; else it could hardly be called living, and more like dying. But the dirt and grime and stink and sundry pestilences make their anchors to the lower regions of habitation. This location is quite clean in comparison, and actually garbaged, all manner of trash and treasure sequestered away for city picking and scrounging through. I've even found a few usable items at the collection dump attached to this compound— though properly outside of it. A little cleaning or refinishing is all that's needed to restore a junk. It's sometimes a marvel of what some people will throw out, for sentimental reasons.

ALFONSO: The sentiment becoming banal?

SKACTE: The point is more that they accumulate too much, and must get rid of things often of value, when, for example, they must leave the dwelling and yet have no one to leave with their.... peripheral possessions. This sort of implied my city career, being already a carpenter and craftsman of some skill. It is much more fascinating to improve or restore a work than to create it from scratch. The interest lies on following a logic dictated or struggled for. One can judge intelligences that way.

ALFONSO: Universal intelligences, of shape and form, you mean. Well, I suppose the solitary should stay with few.

SKACTE: Or combine with others. It depends on your length of stay, and what you are willing to give up.... or are coerced to. Now, have you length, Alfonso?

ALFONSO: How soon to action can you suspect?

SKACTE: Oh, a couple of months at least. I doubt Decamond would want to deploy forces, for a battle yet to be devised, before thoroughly healing himself. He's quite fastidious about martial skills and readiness, the "wounds should be brought off the field and not onto" sort of mentality. This makes one want to win and be decisive in these conflicts. The sick stay home, or stay out of it. But now he is injured (himself), seriously of notice or concern. And that makes for a mean cake of crusting, ever being a practical man with these affairs.

ALFONSO: This could be a long time for me. I must make a fortune to have one. I feel like I've been draining one already.

SKACTE: Then I would suggest the dump. Now if you can put up here for at least two or three weeks, I think that I can arrange it for a longer stay if you can prove yourself of a practical worth retrieving promising items from the trash and helping me restore and (re)sell them. That sort of work and my influence can lend you much credit in this fairly protected enclave, that and the acknowledgment that you are indeed one of Decamond's.

ALFONSO: Sounds doable, but slightly demeaning.

SKACTE: Oh, it's noble and realistic work. And it's no more deprecating than a time of trial given an initiate. In fact, some are proud to alert others of the servitude, when its true purpose in known.

ALFONSO: And that is....?

SKACTE: For an obviously dignified and estimable person to allow himself to become humble enough to work his way and adapt himself practically into a service with Decamond that can only prove to demonstrate a future illustriousness of deeds and worldly success. Why, some neophytes or newcomers laugh aloud with joy for the prospects during such tests to make an impression for others. Those fearful now, before cruel, harsh fighting, to reduce their haughtiness often yield to a similar fright at battle; which coldly sifts for a separation of pride from prowess, if this is possible.

ALFONSO: I've never been so afraid, to grapple for being approved of and to wrestle away doubts, about me and of my own. But this is dirty work to gloat of— It may contort the mind. I won't bring praise to it as anything other than a possible necessity. There are many things that must be done in a life, some menial, some majestic.

SKACTE: Ah, yes, sir. And with talent, the rapidity of your climb from the menial to the majestic might be startling to the relatively stationary of position. I oft say that if one is born to be a king this will occur, as opportunities search for you with haste and seek you out with worry of being late, for fear of fate's reprimands and causticities. Or a leader or a slave, it makes no difference to what you are born to be, become, or result as. This is one great mystery persistently viewed, noticed, examined and wearied by, and totally revealed for review, absurdly unhidden and abashedly bare. A most obvious thing is to be born to be. Yet stands this ever difficultly to predict, we have such fallow heads in retrospect.

ALFONSO: And not the strongest nor the largest of creatures dominate. Then it is by a trade.

SKACTE: Aye.

ALFONSO: And what do these do, here?

SKACTE: It depends on their past successes. Some simply linger in languor of renown, until the next great foraging to build back up the muscles and the strengths of fearlessness. Others more restless and reckless seek further, intermittent adventuring, with danger as some sort of bodily tonic or refreshment. Many, as myself, take a practical trade, based on our natural skills or educations. A few do not exist at all, but only as ghosts. You never see them around, and forget of their persons, until they suddenly present themselves on the battlefield, in some miraculous way keen to be there, with hardly an alert; though they remain upmost (as) reliable and competent during the fighting. And then, when the war is over, they disappear, their great comradeship and gaiety of company lost again.... until the next time. One grows not to miss them till they're needed, and to love them while they're around.... I'd say most in this particular quarter of town carry on with some sort of occupation to bide their time, essentially buying an employment of overseeing some specialized or technical activity they could hardly comprehend on their own. But one's chief need not know how to have (things) done— and properly and prosperously. That is a prudent use of one's riches, an investment of control. Thus, there are tanners and metallurgists and grocers and manufacturers and clothiers and farmers and potters and glassblowers here. And yet there are not these here, by their own hands and heads to do, but all of these in small scale operations to run, promote, or prop. Almost all of them, though, curiously, have some sort of a literary bent, presumably in the process of writing autobiographies for their descendants extolling their military exploits, conceptions, and opinions on explicated and effectuated military strategies, their outcomes and analyses of results. Almost all soldiers feel to be of a higher purpose than merely fighting, but as like the eyes of gods for tales to make and stories to compound of witnesses and participants. A creed of sorts, to be responsible expositors in this way, it is commonly felt. I have my own fables and fibs secured away. One must write them while the thoughts are livid and you have time. The objectivity of forgetfulness fades their colors substantially. So much so that I can hardly believe my own early details of combat by what I've written of my perceptions. But I accept it as true, since there's no point in lying to one's self about such issues.

ALFONSO: I'm sure under Decamond, most can only act bravely.

SKACTE: Yes. He's fair-minded enough to evenly spread the dread so that most can handle it, at least initially per his instructions. And then talent, training, and happenstance take over to amend for the ways needed to survive and defeat, succeed and demolish oppositions. That is very coarsely stated his general tactic. Eventually, each man becomes his own leader, as the war dusts rage with a blinding violence of imperative to fight and destroy. And we've been more than lucky through his commandeering of our souls and tempers for these.... rewards from living through the basting of men with men.

ALFONSO: He is clever—?

SKACTE: To be upfront, straightforward, and predictable to his words. There's hardly ever any complaints about the percentage of reward from spoils allotted. Aside from a base pay, always out of his own coffers, he goes up individually to the promising of late and tells them precisely that if they do this, they will get that, and if they fail how they may combine their efforts with particular others to recover almost to an exactitude of divvying. So there's often little question of what is to be done for such incentive. And if one succeeds of task, whether it improves the overall scheme of things or not, the duke remains faithful to his pledges, even if contrary to the suggestions or feelings of some.

ALFONSO: This is how to command!

SKACTE: Of a certainty, this is how anyone would want to do. But to want to do, to know how to do, and to actually do are three different tiers separated by such scales of copious difficulty and conscience as to separate the ants from the anointed to a supremacy of leadership. It is very much easier to fall short of Decamond in many ways. I think his method and mystique stem mostly through, through a simple sincerity of interest in the activities he promotes. Then there follows a stern and strict and boldly discernible logic towards earning the treasures sought. And complications are reduced to individualized valors, instead of remaining intractable or impervious to resolution. Contrary to popular opinion, the *least* of the duke's qualities is deviousness.

ALFONSO: Well, such a fox must have his tricks, to last so long victoriously. A pointed mind is thin, though a blunt one's ineffectual. I'm sure he considers many aspects of many things, problems, and is rounded enough to be responsive to changes very pointedly, with an astonishing flexibility. But one must gain such.... jointedness.... by exercising the growths of one's visceral though divers knowledges of characters, qualms, and cravings.... Is this dump.... clean?

SKACTE: It is a dump, but not a bed pan. Not a collection of rot and decaying organics, aside from the woods and the occasional flower pots. Not of wastes from foods, but of fashionable items wasted, no longer wanted, but offered generously for free procurement. It is indecent to leave the badly spoiling and diseased there. And as I said before, most citizens are proud to rummage through these plots that are scattered here and there, next to developments or factories. They hold garbage, not for grime, but for grinning at,

perhaps. Proud may be too strong a term. I'm not as regular (going there) as I used to be. But pride has absolutely nothing to do with the activity. It is merely a convenience for avoiding loss of use undeserved of the materials found. And your imagination may be peaked, as mine was, in ways that can not be lucidly explained. I simply saw dark things shining, and knew they could be made to be sold. The notion had not occurred to me before, until I sensed their noble stride to become useful again.... I was, however, concentrating on the new— and ignoring the old; for my craftsmanship, I mean. What a foolish plebiscite for opinion-ing of one's powers. This sudden insight eased my conscious burdening tremendously, and my store of furnishings flourished thereafter, through the practicalities of that very dump— and others. It is a rush of fruition, to grab hold of a shocking intuition by the scruff and take (off on) an inspiration for a ride. What nature's nape allows for reins, man's mental facticity untamed for wild, delusional thoughts.... (is) so rightly buried in the security of a finely shaped object of construction with his hands. Damaged, perhaps, awhile, its restoration evidently planned, is the vision struck fiercely.... and like a blue thunder heard silently in one's mind.... Yes. They are as clean as a blade's use for cunning cutting. What knife lies dangerously, Alfonso?

ALFONSO: You mean the skill in it? or the skill in its use. I can afford a flat here for now, for a few weeks, and make my mannering around good company. Count Elser suggested I do similarly, and as soon as possible (*as **Calerand** comes up to them, approaching from the dwellings*), having received the duke's blessing. And for a wound to make a wound, one needs adherents to your personableness, cohort-ing, and thus.... cohabitation.

SKACTE: There are plenty on the plains of valor made, runs richer juices than a citrus splayed of fruit—

CALERAND: Good embouchure, Skacte! A henchman for your heavy cabinets to carry? Damaged mine. (*He is sharply dressed, but slightly aged.*)

SKACTE: Worthy Calerand, a fighter fit for stony seasons. See here some worth alike. This is L'Afonso di.... duke Decamond.

CALERAND: Ah. The orange slicer so rumored of. And demonstrated?

SKACTE: Before Decamond and myself, as plain to see a miracle of skill. Even the seeds obeyed the circle's perimeter.

CALERAND: So!

ALFONSO: I was as nervous as humble to be proud of the performance, sir. And I might mean to take up lodgings here—

CALERAND: When I was a more limber young man, I used to catch them with my bare toes, thrown by butchers and monkeys— and bring them right up to my mouth to eat, from a standing position.

ALFONSO: Peeled?

CALERAND: No one would believe the trick, till I showed them right in front of their faces. And such an amazement resulted. Even Saint Peter was taken aback!

ALFONSO: Really, sir—

CALERAND: I come up to the gates of pearl and posh permission, all distraught to be let in. And the great angel questions me for what of my many deeds to justify an entrance. I lie down on my back and squirm a bit with delirious pleasure, and he questions for what ails me. Then he bends over and starts to strip off ma clothes, for his curiosity about man is unbounded. Yet so authoritatively does he proceed, with a stately grandeur that makes a life seem serious. And when I'm stark naked he says he can't find the agitation, when examining for cause to make one have a tickle, and stands me up again. Whereupon I immediately grab his halo with the toes of my right foot and place it over my head, his startled face turning a puzzled red with anger. Acknowledging the uniqueness of this ambidexterity, for both my hands were deferentially behind my back, he haughtily retrieves the crown and tells me to go off! an' capture a bit more life instead of his indignation. When my soldiering colleagues recovered me from off the blood-soaked ground, and I came to, they said my wounds should have killed me with a certainty of raging weapon. Yet I responded, through the sublimity of my weaknesses, that I had out and over raged the accursed crying tear of flesh by astonishing Saint Peter himself! Now that, sir, is a demonstration to be memorized.

SKACTE: And tall told, of Calerand. Yet I have helped wash out deep wounds of others many times. And they often survive from my touching and caring.

CALERAND: So ye thinks to live amongst us, here, an' demonstrate this orange trick, perhaps out back?

ALFONSO: If that is requested, I may show and teach the feel required for the knack. It is one of timing more than strength, though the sword should be suitably heavy. And of course, much practice facilitates the art. I'm no longer embarrassed by its carnival-like mystique, if it is enough to gain an approval by Decamond.

CALERAND: Hmm. One can make a living through its gamesmanship, daring others against you and placing bets. I suppose it is a reasonable occupation. But this is a fairly expensive arrangement of dwelling. So then you are rich already?

ALFONSO: I have some wealth of savings.

CALERAND: All young men think so, when sums little seem a lot. Living for the day lasts a day, and they haven't the conception of stretching monies to last for many tomorrows. Now, if you are regularly supplied through endowments by relatives or high friends, I may help you garner more through enterprises steering—

SKACTE: He has agreed to help me, in my shop.

CALERAND: Oh. Skilled in the woodcrafts?

SKACTE: That may be learned with merely a chisel and a hack.

CALERAND: Well then. But I would have wagered.... someone promotes you. As for this information to gain, it is an advertisement for your services.... for your presence at least, to be noticed.

ALFONSO: I promote myself, with my wherewithal to be.

CALERAND: And does not Count Elser back you— and the crown? And then to stay at a place like this! it seems fairly clear, that you might be.... charmed or enchanted— for our attentions.

SKACTE: That may be the ruse supposed for obviation. The king can be so awkward in personally trying to handle such affairs. But for us, Alfonso, eliminations are left to the combativeness we seek. And we haven't much care to tarnish gold or other goals, as they may be made and applied towards. Yet, to be perfectly open about such assumptions is a great privilege, no? This is only for what you may do, and are capable of. I've seen so many aspirants melt of battle in a morning's slight sunlight, with the raw smell of battle sensed, that I remain reserved to judge of talent, and almost indifferent to their tactics, until real action must be applied for the warring skills.

ALFONSO: I am no such usurper to the opinions of stout and valiant men.... But as to the king, I remain obliging. Musgonie is a fair principle to behold, and to be beholden to, for any righteous soldier of the realm.

CALERAND: That is conservative enough.

ALFONSO: My goal is for an honest fortune to be made, as you two have presumably achieved, one way or another, materially or spiritually, entrepreneurially or militaristically of valorous merit and respect. And my high friends are admittedly substantial, and of considerable worth. But I seldom rely on them, as they on me. For the world is each man's palace made (for), and the bouldery, ponderous rocks his temporary thrones. We do for ourselves to get done everything.... that could be sculptured of our backs. I won't deny that.

CALERAND: This is, though, perhaps a rather solitary view of solidity to one's aims. There is always a sharing of interests in the machinations of man and his influences.

ALFONSO: I am always Musgonie's minion.

CALERAND: Yes—

ALFONSO: But now I am proud to be Decamond's servile follower, as long as it goes to be profitable for both of us, and not unprofitable for any of our legion. How else should a young man strive? within these tallies of conditional attainment. I have no dependents to speak of, that would care for my death or injury, no wife or children, nor indigence of parents.

CALERAND: Totally detached, hey?

ALFONSO: As were a gang to find and fence of a playfulness, then take me seriously as being loyal to a mission.

SKACTE: You deny nothing.

ALFONSO: I say that killing is a grievous business I would never slight or take lightly, within my hands to presume such a power against other men, and assume it as needed by our circumstances. But I will not jest with my abilities. They are solemn and dictated, and storied through rumor and the rampage of curiosities. Yes! I will stay here for some time, take a risk, and infuse of this place for its atmosphere of conscience, caution, and calamity of notion.... for all of the deaths responsible for, lives hewed and hacked (at)— and hated? Is that a wish for thee, out on the fields and plains of valor? Then I will burnish myself likewise to think and believe.

CALERAND: Well, one should make empty space, for one's emotional drive to consist of and then occupy. And I can clearly say there's not much thought to the act of slaying, and more action for its thought. Once a weapon is used it remains.... appealing to be used again. That is the only magic brought to warring.

ALFONSO: I often have a great appetite. Are there restaurants nearby? Is this a hearty square?

SKACTE: The residual of inns, Alfonso. One generally cooks inside, where one sleeps.

ALFONSO: Cooks?

SKACTE: But there are corners serving meals. It is a convenience of this quarter that these dwellings are fairly much self-sustaining. One can do as one wishes, to dine. Each has a hearth.... and a heart to it for the occupant to use.

CALERAND: The relative privacy is exquisite. That is the major quality found.

ALFONSO: Cooks and hunts.

SKACTE: There, too, are some accommodating forests walking distances away, if you need free dinners or suppers. And they make good practice of foraging, such forays, to bring your body's physicalities quickly up to standard for the typical endurances of troop members.

ALFONSO: I can do both, and have.

CALERAND: The airs are particularly clean here, my favorite attribute to find. But then the general household management is also of a good sanitation. If you must skin, do so in the wild. And if you have no second or pages, wife or *Weib*.... if you live alone.... maintain the utmost cleanliness of covert, pantry, rooms and person— please! Smelly annoyances here are the greatest of sin. The.... arrangement of areas is designed to avoid them.

ALFONSO: Is why I prefer to dine out. Rooms? Is rich. My belongings are few to travel with.

CALERAND: Hmm.

ALFONSO: This will afford a heavenly spaciousness.

CALERAND: Travel far, have you?

ALFONSO: To meet with Decamond.

CALERAND: And to be near the court; for Decamond is now near the court, a coincidence.... Elser find you?—

ALFONSO: I found him.

CALERAND: A relative, is he?

ALFONSO: Who does not share some distant relationship to nobility who is himself noble, in the same country? I wouldn't be sur-

prised. But I don't count on artifactual rumor, only knowledge of a personage to utilize and regularize with.... I am not royal, Sir Calerand, nor supplanting inopportunely, nor implanting golden spikes furtively into.... our regimentation. I am on my own conduced to join the duke, inspired with a faithfulness to candor and transparency of aims.

CALERAND: That is to be sure of Elser also, friend and friend and friend. You can understand my inquisitiveness.

ALFONSO: Of course. And you can be sure that I do wish to become a leader, and surpass many others.... with righteous and bold deeds only.

SKACTE: I'm sure that brings much to your credit to dream so, Alfonso. But real combat, like real death, yields forthwith to a most terminal division of judgment on the characters of fighters, more so than any positive association.... to the living or the dead.

ALFONSO: I will not fail for any courage of a warrior to have.

CALERAND: That should be tested. But you seem to abhor cooking *and* hunting.

ALFONSO: Well....

CALERAND: Well, feasts are made not out of fantasies and spices.

ALFONSO: There is, perhaps, a certain daintiness to my tastes. But these will mature with actionable imperatives to face. And this will convert into such manliness as to turn a growl into proper verbiage of the dissatisfied and hungry. I do not abhor, but am rather disaffected by what's often unnecessary. When they become mandatory for survival— I seize to love, and am involved with this. It is not a transformation, nor a modification of my persona to meet conditions, only an expression of what is inherent to any godly creature. I am made for such challenges, sirs, to tease of my opinions, and to teach myself the worthy ones of others.

SKACTE: Oh, you'll learn much from raw meats, as the lions digest. And should you not take a companion, to share of expenses —?

ALFONSO: No!

CALERAND: It's not so terribly unusual for this span of housing.

SKACTE: The need for this can be easily advertised, through the discretion solely amongst Decamond's members.

ALFONSO: I'm most used to myself, and have not shared personal living spaces with others since early childhood left.... Yet of rooms, you say, contained therein. There might be some advantage to this, not to hastily dismiss or throw away. With delicate courtesy this may be asked?

CALERAND: You might come used to sleeping with several inside one (small) box of an enclosure, whatever cover from rash elements a group may find, the weather raging discontentedly, or a shelter solely removed from an enemy's view. You must condition yourself to such confinements as are humanly possible and permissible of the animal we are, what makes a rat to roast with joy. You are not yet a soldier.... Humiliate yourself before me! and we might be friends.

SKACTE: I've told him to be humble.

ALFONSO: ....(*after considering*) I've no fear of this. (*goes to one knee before* **Calerand**) To this accomplished one (*lowering his head towards* **Calerand's** *feet*) I submit my qualities for review.

CALERAND: A most favorable flexibility of manner. Arise, as proved to be decent of requests considered. He shows some thought.

SKACTE: Now stay awhile, to reflect on this friendship, Alfonso. It is as firm as his standing.

ALFONSO: I have considered this, to be mastered by no one, yet respectful to any, what passes to be notable.... or heavily laden. A decent protection here, from the misfortunes of a growing decrepitude, as were a life to be: fit to be fouled, or soled so thickly from the ground as to end the merest sensitivity that even a slight discomfort hounds and howls for.... Is, perchance, a fate.... of growing callousness with delicacies. This seriousness is bare to see— of foot! what rises to arrive at, coaxed of histories. A cruel dismissal, from honored purposes— is this!

CALERAND: There are many such tales of infirmities won. (*as* **Alfonso** *stands*) That is the life of how one ages and walks through. And what is pillaged? but some gasps of breath and atmosphere. I am left this way remarked of and remarkable. For one's greatest feats may bring your finest agonies. And eventually afflictions accrue to anyone, and lessen thee to the ultimate of crippling with the calmness of a sustained sleep. But my flesh should have been won at least ten times over, ten remarkable escapes.... and a multitude of sundry minor escapades to sift through more or less soundly. And what do I live for else, but to retire dead! This is what we bring to it, young fellow, our capabilities rewarded as thus. Yes! They do hurt, and are bound tightly for my gait. And you do observe with examining intelligence. That is rather wholesome to me— See you similar.... of tricks?!... Then mention mine sparingly, and self-absorb their meanings for you. Demonstrations make you known, yet lend you reputations only. The war gods pluck your feathers off, the less you be companion to them, and the more the chirping fowl.

ALFONSO: I hope to climb above my wounds.... or others. And it is a bravery of mentality to run against constraining odds, and from the certitude of one's decimation. Or now I see your point revealed: It's decimation winning. And brought to this we are for passing time, all, each, and any.... who can last. The best of life remains where life is lost, whereas fortunes make your plumage.

SKACTE: Aye. Grow only to tire. Then your wealth be worth it.... worth it as you rest on your laurels and backside, and lazily dream of your past through campaigns and former captivations. See how even Decamond can be no better then we may seem, to this aspect. Much able fighting addicts your courage.

CALERAND: When the machinery of the body breaks down thoroughly and irreparably, only then is it all over for you, that adventurous smack to feel, Athena's kiss an' kick! And for all my money saved, which is as much to keep me lame and idle, and somewhat stiff or paralyzed for passions, yet would I rather die out there with Decamond, in that imaginary maelstrom we need coming. And for a leader bringing us to, that is my intention still, even with my carriage torn. So seek you fortune or fortuitousness in this —? art! I abuse for the one, and am abused for the other. And this is how our amusements are made, if gods descend to provide for us these games. With living here do we only ferment the wine.

ALFONSO: I can not think for such a paltry goal as to work hard (only) to retire. The notion does not lift my consciousness much. And the consideration yet seems too remote to me, to present myself to others as you do me the favor and the policy of warning, what may be expected of success. But then you ultimately claim you are bored, even of trade— even of life?

SKACTE: Not bored, but waiting for a resonance again. Though it's ever due, one never knows it will be struck for you, how circumstances be to bring you ably, as of a man's mind as well as his girth. But there's much interesting stuff to do today. And we occupy ourselves of this, that which is completely deserved and hardly stolen from others, else by glances off some opportunities shining and reflecting of our imaging to own, smoothed and polished surfaces of stone as impenetrable as our faces and experiences. Yet do not delude yourself that this is other than what we present, as you join our coterie with earnest expression to advance. You will hardly taste more of our representation, a furnisher and a storyteller—

CALERAND: And brokering investor to dabble at.

SKACTE: —or were to replace great Decamond himself, which is squarely improbable. We!... make the life you seek, Alfonso. That is the kind of which you find here, and not much other altering of aspirations.

ALFONSO: Sir.... I can only gently admonish this thought. But there is no height to the sky—

CALERAND: There's that youthful indifference!

ALFONSO: —And I will gainfully, and gleefully, paddle through the unknown rivers of the future, darkened as they may seem from afar, until the galley rows through this with light of hope. And I will have make of adventure with our stalwart captain, this Decamond, this order ordering of knights— and knaves.... and cads to cower, and kings to crave for. But I will be.... Alfonso (as is) brave, for his majesty's approval, the world's glory, and my sufficiency.

CALERAND: There is a bit of dichotomy between them, you know, Musgonie and Decamond.

ALFONSO: Tort-ish, tainted rumor. He's told me himself—.... with his highly apprised manner of judgment that he is all respectful of the king. Such as he has always good men *leaving* from his battles, he is a gentleman of great calibre not to want to waste any true talent. And certainly our king, as do so many others, must admire greatly this trait, which is why I lend myself directly to be tested and trimmed to.

SKACTE: That we often dine on success is a noted feature of our renown. So come ye, Alfonso, to examine these premises. For there are usually at least one or two vacancies left for given era of transition; and as I am duly notified for some wares that I rent out having been returned as of late. Then do I direct you here as a fine possibility and a good caretaking of prospects.

CALERAND: Toward the eastern wing I should observe.

SKACTE: Yes. And there, at that end, would you put down the hold of a lease, with Decamond's name and my own as workable recommendations. But we will examine first, to see them suitable of a living for the living spaces with which you may comfortably share. (*starting to lead **Alfonso** towards the dwellings*)

ALFONSO: And how much for a hold?

SKACTE: Some tally tested; you may make the sum for a week, with another pledged.

CALERAND (*as the two are leaving*): They're all cleaned out, cleaned up.... and fumigated through the smokes of camphor.... And no news of war, Skacte?

SKACTE (*virtually departed, with Alfonso*): The duke must heal first, Calerand.

CALERAND (*to himself and alone*): For much to hold out for.... for.... fomenting. validations.... It's such a cruel life to have to age through, hardly ever to be satisfied once the prizes and the trophies are seized, grasped, and thrust over our heads in a sign of victory. But that does signify much better the doldrums brought on, and the respites stretched out. For how can the fish swim without their waters, and how may we breathe without our wars?! This.... is a grandness of rest, yet more to prove that one may become ready again for the strenuous pitching of pikes an' darts and malevolence, the slings of slaying, all worthwhile and profitable. Then why do agonies deserve a hearing for these tests of prolonged relaxations and the larceny of vigors stolen? Here is the fear, that one not be able to walk again— as a soldier and active combatant, after the leaded weights of indolence seep into your sinews. And this is aging, which chars the skin, and brings you to be lazy and afraid, fatted with accomplishments and praise, yet unto paralysis paraded of your views— thought quaint for this kind of battling.... with such years held, and so many insults to the flesh, pierced for healing— (as) sound.... an' newly old. So do be scared for one's endurance, of the physicalities and nerves. Again a test is a vacation longed of rest, to bring towards stillness an' stiffness and a temping towards dullness and all the matter of a dearth of spirit or a death of heart. What can be sadder and (more) uncomely to witness than a foolish restlessness and agitation for what can not be restarted nor restored! And how can one tell this? but by patriarchal retirement enforced, as the mind says to do what the body can no longer— March!... Strike! Hound and' hurt! Rally towards revolting kindling! an' take charge to charge.... Some life's work, some purpose to this can make you proud, enhanced with grandeurs and gratuities. But fear the era 's mutating about you, as scintillations forming glass to encage.... your sight, perhaps, reminiscently colored or shaded. And my friends remind me, of what is decrepitude to face, with an honest worrying. There's been more trouble.... to my arising from the bed, than in awakening. But the shrill blast of a cold wind needed of a warring field's dimensions will easily, most thoroughly cure that lethargy. The impression to awake and be alert, and show of sharpness tensioned— tensioning, an' to be terror, crystallizes the emphasis of a fighting member found. And yet in bed, arthritically bound— No! On cold grass, with pins of frost an' friendliness to mention this familiarity, does make me leap of capa-

bility an' threat, and with a powerful essence force my way through dangers, not so much with youthful ignorance but with the candor of my kind, the earth-bound storm, the ravaging rage of rancorous wrought, raw routes willed through.... defenders.... But can I climb this wall of aching joints? for one last blast of fiery breeze to freshen of, refreshen with, and be made purified of a soldier— without endangering nor hampering his others.... too overtly, too disgracefully, despicably, disgustingly, and with embarrassment made of life.... this vacation pounded into— shaped by— milled!... Would this.... Alfonso have me as a trusted conjoiner to the effort? Should I favor him, and change, what likes to be seen so daringly? as a throng of thousands to be mortised, bored through for a spirited machine to compose, such wishes as we've all have had, at some time till the fanciful is reckoned with the real.... There might not be a master left, left for to change assuming one anew. What is then pain to share; it's better on the plains of valor felt if must we have them, and with laughing for real through high and haughty discomfitures, hardy alignments of the soul, hearty gushes from our cries for what has transpired to convert us into worthy corpses — winning.... our disconsolations. Once awed by bright clashes does a dimness do you tiredness of limbs and tempo. And this is no treat for a warrior to sift through, sands of remorseful boredom, dolorous *malheur* of condition and state, morbidity embraced of chains to motion fasting with. And for one so supple, aghast! Find me more disgruntlement than this. (*leaving*) I'd rather lament of wounds than a wear worn of wasted debilitations, wrinkled scowls, and an habituation to infirmities and slight weaknesses. For to play on a man through the mind is much devastating, yet through the creeds of muscle devoutly entertaining.... Winning on investitures, of men or money, makes for me no badge of honor. That does come with.... mending matters.

Scene III — *A posh, sunny living room very ornate and decorated. There is a table near the center with no apparent purpose but for its grand ostentation lending to the room, and may even have been an afterthought added of furniture article with its superfluous and exaggerated gilding of forms hanging off the top to suggest purposefully a dripping of gold to the floor, as of a cake overly frosted with the yellow cream. Yet the size of the table is practical for only one; and in this room of richly upholstered lounges, **Decamond** sits at the table in a wood-metal chair purposely diminutive for the golden craft. Though dressed, his injured shoulder is bare, its arm coming to his lap, and so a state of incomplete attire is suggested, with a limpness of limb challenged to become functional again. **Wallinda**, his house servant, enters with a pot of freshly brewed tea and cup, etc., and places it on the table as he revives to life with her entry and notice, having previous to this adopted a pensive blankness.*

DECAMOND: It does not heal, Wallinda, though washed and aired daily.

WALLINDA (*pouring some tea*): You've suffered worst injuries that this, my lord. It has not even need for a cast.

DECAMOND: But it does not chime my arm for much motion, remains funny (feeling) and with an uncomfortable ratcheting movement, popping of joints there to frighten me that the whole shoulder assembly might suddenly come apart under the skin and beneath the muscles.... Must have been a poisoned spike, somehow, to impinge the nerves.

WALLINDA: Only by chance, my lord. (*as **Decamond** takes the* *cup with his good hand*) They say it was purely a fated accident, that arrow coming through so unusually to reach you. Yet with thousands of chances taken, one may make its way to reach you, or breach your imperviousness; and I have warned you of my intuitions before, feminine though they may be, that restrictions always seek a permanence, no matter how slightly applied. You should not have the violence at your age, when even a feather's quill might taint the blood of its strengths and vanities.

DECAMOND: I remain healthy, for all but this.... numbing. And I only relax the arm and hand, as they feel unusual and need some restoration to normalcy, with time— But I've given (them) weeks! And what delays?! The shoulder wound, my maid.

WALLINDA: It doesn't look damaged externally, still. I've examined it often enough. Should I call for a physician—?

DECAMOND: No. They'll only say to give it more time, the insult to recover from. But they can not feel how unusual it seems. They can only rely on the experiences they've had or studied from previous patients. Those practitioners are not what I am. (*sips*)

WALLINDA: But perhaps a specialist.... a bone and fracture specialist.

DECAMOND: I've had more training in that (area) then most of them put together in an asylum of war wounded. Out on the battle fields where the real doctoring's always done, and the pertinence of strife-filled gusts brushing up into your face has its own medicinal effects and restorative powers. There is nothing more alive and eye-opening than to smell the imminence of serious injury and— casualty.... No physician can tell me this is reasonable, what's going on inside of me. Yet that's all that they might say. If I have painful suffering, then call them; or a respiratory congestion, or a dizziness or an epilepsy, yes, then call for their consultation— and advice. But this is not of that like. This is not so scrutable a symptom, not so sanguine a result after weeks of rest and inactivity, and an anxiousness for recovery. My body warns me to worry of it, that it may not be able to....

WALLINDA: You are not ill, my lord.... Do you feel ill?

DECAMOND: I'm not sick. I haven't the convulsions of a disease. (*sips*).... There is no fever, no gastric swellings, no consumption.... and no unusual palpitations of the heart. The pulse is steady, the beating's steady. But I.... have a reluctance beyond myself, it seems. The body will not heal. And that is like a death rattle to hear, Wallinda, all too personalized and conscionable.... I tell myself I am defeated—

WALLINDA: Oh! That is ridiculous, sir. Maybe of the combative activities you are through, the physique can no longer take the stresses physically nor psychologically. But your person is not defeated. That is impossible if it still lives with a conscience. Even the most destitute beggar or incapacitated amputee realizes this.... solemn truth of the soul. And you have lost least among the sorry, if anything at all but patience. But can you actually feel so pathetic? with your standing, wealth and renown, the respect of your comrades in arms— of the nation— the king! It's not like you to feel sorry for yourself.

DECAMOND: I told you, (*slightly angered*) I am beyond myself.... (*calming down*) This is not my pride at play, but that substance of a personality that fails to be denied its honest expression, that is too boldly objective and convincing, and that speaks only to me— shamefully remonstrative. I tell myself.... more than fortune and fate are occurring, and more than doubt; for I never have doubt, only uncertainty.... until a decision is made. This is.... coercion of the universe around me, a compaction to shell me up and out of it.

WALLINDA: That is absurd infant's dribble! my lord. The loss of dedication is merely the relaxation of a muscle—

DECAMOND: If the whole blasted arm had been cut off, I would have recovered by now— in full! The fortitude, Wallinda. I would have taken stock of what is left, strengthened to recuperate and recompense, and recovered. I would be ready for another outing, this day, this week, this month. But my body refuses, tells me calmly it will not. No medicine or magic can help, to overcome this declaration or admittance. It scars my composure from beyond reason, how these things must fall into place as proper for one's self. And this is the outcome not of treason or disloyalty.... but of the nature of a method.

WALLINDA: You feel your body is working more against you than your mind?!

DECAMOND: I shrugged it off.... when it occurred, as possible and probable, and to be expected at some time.... And recovery would put an answer to it, almost immediate recovery, astounding recuperation, hours— minutes.... a couple of days for the entirety of improvement, a complete curing.... But this did not happen. It does not happen. And it leaves me with a resignation to the facts.

WALLINDA: The method of healing? An answer to what?

DECAMOND: From my own membership, from my own body of soldiers and fighters has this been done to me, has this been allowed, the injurious spite.

WALLINDA: ....Is that possible, through the high discipline you've always imposed? Mistakes and accidents bind themselves to activities and rigorous pursuits.

DECAMOND: One or a few may see me as unfit, my maid, too old and unreasonably stretching chances for victory— in a young man's game.... And if one, then probably some or several find me improper for this, or suspect a growing impropriety.

WALLINDA: You imagine this, because you do not heal?

DECAMOND: The body tells me so, and without whispering, nor demanding explanations nor descriptions of the fall, but simply says this is what they believe.... And I can not prove them wrong, though they must be— who would think so. But the most honorable position to take is to stay out of battle if you can no longer handle the roughness and the rue. For as a soldier your job is most to assist your fellows struggling in deadly earnest, so that collectively, and only so, we will succeed. Thus, perhaps, (is) this shove to retire, the method to the indelible of mind. This wounding was so unusual, the way it occurred— completely within our sheltered ranks. And now all observe me sweltering of sweat.

WALLINDA: If this is a method.... military.... perhaps it is the better part of a blessing, though I would say blunt conferencing should be better.

DECAMOND: Who rules the leader, maid?! He must always be hacked down through his errors, for there's no one to oversee his commands. But I have not made such an error as the taking on of this harm, till now my debilitation weeps and weds to faculties of resistance— sternly to be proved.

WALLINDA: Even the King has counsels of wisdoms.

DECAMOND: He is the head of state, but no leader be Musgo-nie—!

WALLINDA: You think to over-think, my lord, and come to imagine the evils of your friends and reliable subordinates. But if your body creaks, that is all your own to fault, regardless of a cause. And now for the future does it make a preparedness, some nicety of existence.

DECAMOND: Do not bring evil into the proposals or assertions contrasting. Evil is the paint we brush with, and is beyond all consideration for our actions. We are not responsible in that way, to think for the good or bad of a mission. There is an objective made which we complete, as a unit of like or common minds. My own would not harm me due to evil. There's nothing that occurs which in the end will be forgiven. Such need is abolished by commitment and contract military. And I must become sound again, or I do not deserve....

WALLINDA: You are the acknowledged instigator of valors. Worthwhile or not, these projects are too complex without your.... opinion. And it's only your enthusiasm for these adventures that lead to the willing recruitment of so many brave men. It doesn't stand to much reason that those very same would try to oust you, and in such a way as to increase a life's peril and cause such discomfort of body. This was an accident you should blame, an inflammation of fate that destined your injury and perhaps helps to clear the head. But now you can become suspicious of anyone because of your own failings—?!

DECAMOND: Maid, challenge me not. (*sips*) I know I am privileged to believe anything I want to indulge of a thought for. Yet I carry out my disciplines of mind as well. I accuse no one without facts to back up the actions felt. And what I have expressed is only a possibility, and against what I wish to be. But one's own bones are very telling and sensitive. We hardly cater to superstitions in front of those who would likely main.... The best of my men are very loyal to me, as I reward every one of them fairly for their efforts collective and individual. And our enterprises have been fantastic together. These things, though, can not last always, is the view I assess.... (*sips*) with a practical wariness, if I indeed have changed of body. I order myself, to do what it can. If I can no longer succeed in that, then how can I succeed in ordering others for very much longer.... and with a faith for them? I know what I should be capable of doing— very much. But this odd feeling of arm mystifies. If it is injured it should ache. It is merely.... indolent, and that has never been my nature before. So— yes! I'll blame the world to name a cause, if it leads to an impetus of cure, in my head, in my heart.... I don't want to accept any of this— Ach! It's popping again! It's aired enough— bring me the hot towel, Wallinda. (*as **Wallinda** exits*) You do everything for me, maid. I'll have no other by my side at home, to see how I reveal myself to you. (*She has exited.*) It's only the taint of cautious inspirations,

and telltale tattle. For even I have my weaknesses of infirmity, I who have busted heads with solely fingers.... pointing for this action to do, approving of this with exculpation of the need. How are we like criminals to kill at playing soldiers! Does guilt mend nerves?... or boys' faces? (*sips*) And for a real engagement of true hostilities, but this king would never want to allow our liege romance to malice.... as he does this shiftless environment to spring of commissioned restlessness by the forethoughtful. (*working his poor shoulder*) This is dull. But I permit you of life— And friend.... you are a part of me, so behave as nobly. Let the arm be limber steel, and not as would seem knotted at the source. I must have.... (*stopping*) a guarantee of membership. I must blame.... anxieties for this. What is the requisite?... for this prerequisite of service. (*sips*) Am I.... useful, now.... for these unnecessary campaigns of richness? I have.... too much, to make enough enough. Yet have I more to do than be a statue posing.... or posed for. It is more obvious for me than others losing the fanatical tendencies. Authority can only stay powerful and insistent. and instructive, demonstrative. This is to rule by emulation, and I should not appear to be this lame!— if I am not!... But rewards are rewards. (*sips*) And I am due some strange justice. People actually worry for me, as an institution slobbering of slumber. Or does an era change this way into another? as the country encrusts.

WALLINDA (*entering with a towel in a basin*): You have visitors, Decamond. (*coming towards him*)

DECAMOND: So? You seem peevish about it.

WALLINDA: Your friend, Count Elser. (*placing the basin on the table and bringing up the towel*) He's brought the king's privy councilor with him.

DECAMOND (*as she applies the towel to his shoulder*): That be? And from Musgonie?

WALLINDA (*working*): This will either be an admonishment politely gestured, or a plea for some sort of contrary service.

DECAMOND: Contrary?

WALLINDA: ....To your wishes.... or desires, in some way suggested.

DECAMOND: And they wait for my appearance, in the vestibule?... or the drawing room seated?

WALLINDA: I told them that I dress you now. (*finished wrapping the towel about his shoulder*) They are demanding of some favor. Would not excuse themselves.

DECAMOND: Some urgency to it? Bring them here. Let them see this creature moved, and as I am. But they are men of some vitality (*as she is leaving*), then have at me virilely to rub, Wallinda, and show that I do wait for action also, (*She has left.*) whatever their pleasure is. (*sips*) Heat soothes a looseness, or ambivalence to move.... But Elser— Yes!... He plies good-naturedly to cause the least of harm to friendship, for this need supposed, prompted by, proffered for acceptance. And so he has been preparing this move carefully, and for a long time not to offend.... most probably. But what is obvious of a demonstration, the king does not like my voyages on his behalf. And more to be replaced by, a crafted scoundrel from the ranks of royalty, though too distant for my notice. These are how these usurpations are subtly arranged. A transcendency through abilities professed, assumed, implied, and applied with trickery of skill and impression. But to what purpose now could be the change legitimized? Does Musgonie break fashion?— to employ such powers as I can command? But my men follow me! into the stoning brickwork and bulwarks of defense, with obsession for our deeds and deaths to mention. Or by my passion coldly spewed, it is a heat to soothe the quarrelsome and illy constituted. Bad temper be my harm with this. A time is now to ask for resignation and appeasement?!— Never yet! can Musgonie mean. The world is too convulsive still, and our country with contentions pressed. So not for changing that my need is made, this does exist and extracts my rage. Then comes he to a backbone? Would certainly surfeit this indolence with another. How is meanness made of kings deceivingly deceptive! You will not loosen me, this joint so easily—! (*Wallinda leads Elser and Ponstole inside.*) Good friend Elser!

ELSER: My duke. (*as Wallinda goes to Decamond*) I introduce for you the king's privy councilor, our most honorable Ponstole, head.

PONSTOLE: Duke Decamond, it is my pleasure. We have not had personable conversations for a long time. (*as Wallinda sits in Decamond's lap and starts to rub on his shoulder*).... Er.... You have been wounded well— for caring of.

DECAMOND: This is plausible.

PONSTOLE: That's quite a jewel you have there, for decoration — the table.

DECAMOND: And hardly a gift, sir. But brought home and handy is the elaboration, the panache that helps make this room particular of spell. This is like a solarium for me, especially in the mornings. And I adjust myself ably to the rest of the day by spending some time with a bright decadence. Yellow warmth, you know, shines its medication radiantly and with thorough effect. Yet it is as precious as noteworthy to own.

PONSTOLE: Were that we all our days owing to.

DECAMOND: I seized it from a tribe of caliph visiting a province. Their intentions should have been more trustworthy of the crown.

ELSER: How is your injury, Decamond? Is your constitution coming along?

DECAMOND: My health is improving daily, count. The physical imposition wanes with a watering of wealth an' welfare, as well I might be the impostor of Decamond. For I hardly know (of) a difference besides an injury. My nurse caresses me for this, and that is my only mentality to its regard.

ELSER: I was fearing for the weeks of treatment done.... Is done this way?

DECAMOND: Like a sorcery of exercise, the sorceress submits me to. And so.... as a candy waited for, the remedy does heal my qualms for a blow dealt. But more does rest upon my lap than lethargy of limb. And as of limb, upon me rests an energy.

PONSTOLE: There reposes then my quest of a proposition, since you remain very much of a charismatic personage.

DECAMOND: Is charisma due to personality? Then I haven't any. But if from deeds of accomplishment comes a notoriety, that capitulates into renown, then I've stored much of this fine trait, hopefully for our majesty's approval.

PONSTOLE: Definitely, sir.... approval. The king finds you and your exploits to be.... remarkable, though he hesitates to show this feeling publicly because of the controversy that surrounds your actions without his strict license. Of course this leads to some resentment. It even necessitates a rebuking, in many court quarters. Clearly within the privy council I have advised against allowing it, at times. But our Musgonie is a very wise ruler, and admires you with almost an infinity of appraisal with leniency. Though in my opinion you have too often caused him some real trouble for handling the provinces you invade—

DECAMOND: If they are ours, of our country made, then mine do not invade— but rather inviolate off from the dissolution-ary erring a laxity of rulership supposes to some. And am I not legal —?!

PONSTOLE: The king is not lax. He is leonine and Rex to consider all sides of an argument, and even to laugh with you.... or agree to some legality for your efforts. But yet, Decamond, he is a fixed man who needs, and wants, the assistance of many. For his wounds make the country bleed, and be disastrous to everyone involved. Now, consider the complaints he must adjudge and work through to mold for satisfactions.... of damaged parties. They must be restored to a reasonableness of unity, to be made to feel whole again, as a part of this nation.

DECAMOND: One body, singular!... And I have my ways to soothe from damage. Is done? the toning?

WALLINDA: Aye, my lord.

DECAMOND: And is there flabbiness still?

WALLINDA: It is removed, my lord.

DECAMOND (*as Wallinda gets up to attend to the tea, refilling his cup*): You can not convince me that Musgonie's methods are as wise.... nor as ambitious for a healing to what sores.

ELSER: His is a kindness to deal with, duke. Does your shoulder still hurt, even now, these multitude of hours past infliction? And how many babies have been born to our kingdom during the interim! these many days.

DECAMOND: She does not hurt, but tends me violently at times, and must be marshalled regularly for good exercise. My demeanor is not with pain of parenting. (*sips*)

PONSTOLE: And do the feet go (to) next, with this demonstration? Tendons tendering—

WALLINDA: I! but that you insult me, am no demonstration!

PONSTOLE: Then forgive my querying, but it is a curious spectacle of Decamond presented us. Yet as for you, they make a radiance no finer, our gods and deities of worship. You melt the gold about him into wasted drops, your beauty overflowing their stances —

DECAMOND: What is your point—?

PONSTOLE: —into only pretenses of worth.

DECAMOND: —of Musgonie?

WALLINDA: Thank you, sir. Yet define me more of my merit than some supposed mirth.

PONSTOLE: Yet for myself I see—

ELSER: The king wishes to pardon you grandly—

PONSTOLE: —good judgment in Decamond.

ELSER: —from your insertions—

DECAMOND: Have you brought love?

ELSER: —into his management of worldly affairs, by having you service him directly, and not with subornation of his personal rights as ruler to have these matters treated more peacefully.... The insults he receives, from the provinces you attack and rile, harm him personally, Decamond. They call him weak and vacillating to allow these canny but mean indiscretions, friend, while he is not.

DECAMOND: And what does he propose I do for him directly, that I have not already done directly?!

PONSTOLE: I see the man here.... and the lady. No, no to mirth! if I am blinded. But yes, suppose of me a worth— to us. In total tallied, told of tongue.

DECAMOND: Succor this tongue-tied tyke, Elser—

PONSTOLE: I am a lone man!... Decamond, as much yourself for the king to be pleased with. So then he proposes your greatness, with my counselling to be assured. And he wants.... Oh! wants. But what does a man want, sir? To be viewed as favorably as yourself? Hearts may be crushed otherwise.... And our majesty loves all of his peoples. But I may pretend for him, to what is soundly said. And this is that I see good prospects here, in that you should wish to obey our lord's desires, with good interests for all pleasantries to allow, and honest feelings about this mutually shared.... with the pangs of an almost instant devotion. For he would have you as a magistrate sent, to places you know, to settle affairs of disagreements diplomatically and with high knowledge of these peoples' needs, and rational persuasions to employ. And I can burn for this, with doubting lost, if you are of the clever and discriminating judgment he imagines— and which I see. For how prompt is your injury, sir?! A bolt to the mind banged into! And how long does it last? Then here is a passion found, through your handling!... in that you might wait.... for such a position— to be asked of you? Then the king concedes with a consent.... if this is truly found. I will not confuse but of details that dazzle to provide me of this vision, glorifying Musgonie's notions about you, and your savvy but temporal, timely acumen.

DECAMOND: There are, then, aspects to a virility assumed.

PONSTOLE: I don't doubt that you are manly enough to face one-time opponents cordially and with offerings of real help or palliative advise and wisdoms royally approved of or suggested.

DECAMOND: Without sacrificing my vanity towards reality, to any great extent, I'd say I could alone face a whole army of such argument-alists, during a series of polite conversations and practical discussions. They're of a pease-ly (brained) nature when it comes to political and socio-rational logistics. But I don't see much profit in these enforced conferences without the backing of a standing militia.

PONSTOLE: A supporting one. This is a certainty, and can never be taken away. Your approval must be assured that conditions are conducive for your success in these matters made trivial for the king, as he is convinced of your capabilities as much as your adherence to the country's welfare. And so he does decide that you should be invited by these parties, at his recommendation, to oversee the discoursed efforts and draft the necessary treaties and pacts and statements of understanding— in your own words to be understood, what agreements are of writ to be honored that separative ideals and ideas are to be abandoned. You are of the most (of) character to.... imply these expressions forcefully, a sentiment demanding attention with your very presence and standing. And yet is it not of the keenest intelligence and wisdom to have *you* represent the crown and country in these polite and comely affairs of state! And if not a transformation spoken of, from belligerence to beneficence of mind and how you may carry both.... shoulders, then a transcription of word or note performed by you to make most suitable the alliances and loyalties of our farthermost subjects. This then defines a miracle of mentality for your multi-faceted skills that you, the great warmonger, should lead a transpiration into lasting peace and civility.... unless you really have become— injured.... and incompetent for such a virtuosity. Then I would understand your denying this request of Musgonie's, as a man who knows himself faithfully sufficient only for what he may think to accomplish or try at. I am much the same, in my own belittling way—

DECAMOND: A request?

ELSER: His majesty is besieged with worry about this, duke. And these pressures lead him to this astonishing solution, of you to employ for this work. No one else could ever think to come up with the idea— And it frightens.... a few.

PONSTOLE: It took much for my convincing, personally and in private with the king. And it takes me even more of conscious effort to convince others of this opinion possible. Notables tell me "he" waits for war, with growing dissent most outstanding in Lacismoor. Others say Decamond lapses to heal, of wounds imaginary but cruel to conscience—

DECAMOND: Awh!

PONSTOLE: —But I say he, as always, bides his time.... for precious opportunities to be crafted for him. That is the enlightenment of ability, to know when to use it, and in what way most favorably. And as a leader of men worthy to kill and die by killing, you above most know how to appeal to the realities and crudities of the human nature. You are.... justified for this assignment.

WALLINDA: My lord does suffer of the arm's weight. This is no fantasy— I have attended to him religiously these many days, to lessen the discomforts—

DECAMOND: Well.

WALLINDA: —and the queerness of feeling; a sense, almost, of dismemberment, he says.

DECAMOND: Well....

WALLINDA: There is true concern here, sirs.

PONSTOLE: I would so make it thus to see, my queen. And thee be named....?

WALLINDA: Wallinda, lord.

PONSTOLE: A queen to me, for his concern, Wall*iiin*—da—!

DECAMOND: It cures itself, as rapidly as purpose, gentlemen. Yet, baths do take a time. And I have much consideration for my soldiers—

ELSER: They would be well served with a second, during your diplomatic absences.

DECAMOND: ....Of course.

ELSER: To keep them in full fighting form— for your use, when the occasions warrant this.

DECAMOND: They tend to keep themselves in shape by their own manners of sustaining the physique and aroused anticipations. And they group among themselves, how best to form the seconds, thirds, fourths or multi-mers of officering. For I tend only to approve or disapprove of what develops of these characteristics according to their actions and results. This.... looseness of command makes for the most taut rope, one honestly tensioned, as each is in this for himself.... a danger to uphold himself through and withstand. Good leaders improve these chances thoroughly, dispersed among the small battles of individual might.

PONSTOLE: They, then must have much say about themselves.

DECAMOND: As much as to be paid.... they pay themselves. And if a second to me be needed, *they* will choose this fellow— Elser.

ELSER: Yes.... But you see, Lacismoor. is most impending of a bolt or bust. Would seem almost immediately of day or week, Musgonie fears. Now should you lead your men there as you are?... or as you are now then lead Lacismoor.!

DECAMOND: The news coming through.... winds, might be deceptive. Though it is one of the larger external provinces, and terribly prideful. Yet you say they might invite me?

PONSTOLE: The king may be able to arrange this, as a prelude to his hopeful visiting of there and others like it. He insists on taking on the personal hazard, and I think seeks to find ways of mitigating the jeopardies associated with this honor. He would try to do the impossible, with banking on a Decamond. But as Elser has said, he is being pressured into perhaps a craziness of idea, or ten-

dency, yet assumes the attempt courageously.

DECAMOND: ....So that I may soften tempers, for his visiting, this travel....

PONSTOLE: You see, he must raise funds.... fees.

DECAMOND: Oh, this is a prickly game you propose! And these remotes have the fat for it to squeeze.... Should there be a compensation to me— for success?

PONSTOLE: ....I suppose so. Certainly enough to keep your army standing. Musgonie is not opposed to you.

DECAMOND: Then draft the appropriate percentages, man! And this letter of invite, design for me. For I be the trickster, the contortionist, the convert, and the legerdemain of artistry to muster from and with, these depths of power and agility to call upon. And I have always been a loyal subject to our majesty. Aye! If this can be done.... then it will. You have my word, and now my visioning.

PONSTOLE: That such enthusiasm flows to my eyes, sparkles me and spanks. The king will be informed of my love.... to have a man evolve into grandeurs gigantic! and for which my humble mind could never have foretold on its own.... nor find of its miasma what becomes a rapturous state. For I am touched and pinned to this, with a piercing of beauty—!

ELSER (*pulling on **Ponstole***): Let us withdraw, immediately for the king to find.

PONSTOLE (*being led*): —A beauteous thought! Come! Come show us out, Wallinda! I must consult with you often, about our Decamond's state. You will inform me and instruct—!

ELSER: We know the way, man! But this is important news. And I have others to tell!

PONSTOLE (***Elser** forcefully withdrawing him*): I may even assist you here! to design his attitude.... I know not what I mean by it — But it is fervent! Delay me! sir.... in the solarium!—

DECAMOND: Go show them out, Wallinda. (*as **Wallinda** complies and they exit*) He is wounded, yet bleeds happily.... for a fortunateness of duty, accomplishment of task from Musgonie. And I share quivers of the heat he should not have expected. (*They have left.*) For this is an apt rule. That when the stern stick breaks it cracks an awful roar. And he was so dignified of reserve and diffidence, not to reveal tortuous constructions of argument behind the façade— and thus to show the torture of his commitments— as I recall him previously. Yet more for Musgonie am I played than privy to. One sees a manliness, and this erupts in you, a pretty face alone not enough to cause the smack of crime.... breaking a stone's hold bound to pledge. And what is this idolatry that leads us ever astray but the wince of one's vulnerability to be moved, touched, and charmed. And as long as I make profit of this, for sake of staying so, this trait conceived of me, then that is satisfactory. Some agility counts much in managing affairs, seen to be able to, highly regarded for these purposes, and above all.... taken of this notoriety to be fit— sound— and suspect of a danger, if not handled correctly and treated of a respectful gravity for being.... a one for these missions. But one? Hey! Singe the straw— This is a company gathered, at a bonfire of propitious enlistment; though another manner of conflict develops 'bout my talents. Then I will deal with it, and my honest soldiers in waiting. For I've had just enough of this patience. And with their weight behind me, which is my armor seen, indecisions evaporate! and I bring to discussions what causes for definite insistences, candor, and consultations with a hungry lion!... But, be spread, adjourned. These matters might not be made as conjectured. Stay relaxed, and presume nothing yet, as for a healing motive. The motion is still swerving and unresolved, until official appointments are made. The king may think about this for the one thousand and first time, and change his mind on error of insecurity.... or insincerity of view. So stay— limp, awhile, to let these machinations pile their plies of thickness to be embedded by. Musgonie is so vacillating, but not entirely vacuous— with caution as a right. And I.... I am shouldered still, with a rag's warmth. (*as Wallinda reenters*) Have they calmed down, to leave? and without the parting salutations customary. But I presume they were pleased with me.

WALLINDA (*coming to the table*): They seem sharply speculative, and excited—

DECAMOND: For you. But what for me? ma maid.

WALLINDA (*contemplative, while touching the teapot*): There's one, your Ponstole.... He is enraptured, to think of things.

DECAMOND: 'Tis polished enough, my dear, for your attractiveness. And what does sing of love to contemplate? He's somewhat beyond your station, while you're way beyond his head. Yet, if it's true, sweet child, then it's no longer to play at, these sudden infatuations. What would your late father say, as I've allowed you your sanctuary with me?

WALLINDA: He always said you were very kind to take me in, as he was wasting away and dying, and trusting not his wife to step-care of me, my lord.

DECAMOND: But he was one of my finest soldiers, who took a blow that spared five others, shielding his underlings instinctively from what was hurled at our company.... And you were such a youth of spark and brightness, for him, the single jewel to leave to me as a dependence for his many valors, which I have welcomed dearly and protectively of some responsibility for his memory and mine. And raised to be a comely lass in only five or six years, I've allowed you as much as games to work as work to play through. You've been happy sheltered, though remaining amiable to your family— which I insist you allow. Yet, during that same time I've aged decades, it seems, for a ward as a wife to take. Not that I would die un-prospered.... But what is a prospering in this world of riches already?

WALLINDA: I love *you*, my lord.

DECAMOND: This is no game with Ponstole, Wallinda. He is actually bewitched, perhaps genuinely for the first time, with a cupidity of emotional longing pent up and vaulted out of feeling until his gaze of your.... potential.... melted down the iron walls to reveal a beating heart impassioned. Those types do often drip of metal before releasing blood, the metal of an office for their importance and pride, and safety of some worldly substantiation. Too real is the thrash now of realizations for his actual needs, to almost totally defeat and overcome him within a rush of seconds. I observe it with one eye closed of interest, and the other open to the

curiosity. But I've allowed you, within the proprieties of our household, to have game of suitors, in much the way of phases of game you've passed through over the years to develop by, from dolls to dolts. This, though, is a serious consideration for you to angle through and fish at, since you've actually come of age to start managing your future. Mistakes this time may lead to heavy challenges, more so than teary-eyed disappointments or pouting spats of anger for lost candies. Then ask you(rself) now, if you're a maid to remain.

WALLINDA: But I am here. And this is pleasant enough. And I am suitable for you— Yet.... I've not ever gone through a serious desire, before not even a friendship made or placed between ourselves.

DECAMOND: Then.... time to calculate. As I said, I think his station is too high for marriage with you, and he would only use you as a courtesan and then maid.... effectively. He should court court women. But I don't know to be absolutely sure about any person regarding the arrows of affectionate wounding. I've never been that way brought to ma knees. So you'll have to judge yourself— honestly, and study him yourself— if there's any interest there.... for you. I'll not be around as long as that.... fairly young cad—dy, that young(er) man. And I'm not entirely certain of this injury, yet. So I plan around it, as should you.

WALLINDA: My father would be grateful that you allow me the privilege, my lord.

DECAMOND: What? that you have the brains and wisdom to be thoughtful and make decisions?! The satisfied or successful minions are always smarter than the masters who need them, and rely on them to stand up and walk about and breathe the freshened air directed to. You are the charm, my dear; but of a bracelet or a spell, must you determine on your own. Oh, I'm sure.... he'll treat you as the queen— for a month, if the consummation's made. Assume his promises yourself, to judge if they can ever possibly be fulfilled. I can't doubt the hope of dreams. But hopes alone do not toast the bread, nor butter it for taste. Nor can it deliver the roasting of a pig.

WALLINDA: Has that cooled down?

DECAMOND: .... Oh, leave it on awhile longer. The extremes of temperature, from feisty cold of morning's raw reign to fiery hot of a warm water's wash, do nothing for me but to care of time.... for these endurances to last until some certainty is fixed of them. And I will gaze as well, upon my hopes, while tepid(ly) brought to, as the tea.

WALLINDA: More, some? Hot? Heated?—

DECAMOND: Nay; retreat, Wallinda. (*as she picks up the tray*) Leave my cup. I am questioning myself to choose, (*as she exits with the tray*) what lather's drunk this day, the froth or frost of men.... (*She has exited.*) I'm not so foolish as to deny.... the casualty of either. (*adopts a pensive mood and sips*) There is some reversal of attainment here, to remain useful.... (*more defiantly*) But a limb still has its function by its very existence staying, and not as an appendage so appropriately placed— what seethes to conquer! the fashion of our making. This can never be obliterated if the body remains whole and sound of purpose, no matter these royal tactics or persuasions. If a need persists, so does this function carry

— aims.... and arms, without dilution nor suspension of the force applied. Time for calculation, of a devious diversion of intents. Just stay aware. Swarms through his court debaucheries of intention and ridicule. Yet have I the manner to wade cleanly through the muck an' mud, and pull out trophies of shiny accomplishment and refulgent glory— for my troops of traction. Aye! It's movement needed. Then beware of me (*sips*), Musgonie. I.... don't placate to disinherit men, but to impound more hearts for my practice, as the need grows— as you make it more so, as the method radiates its obvious tempo of occurrence and dominates resultant action— No falsity to it, no.... missteps now, to tread along these paths of procured road. Or that I am blinded by the need, this traction, this movement, this ambulatory lust, makes me keen to grab at weaknesses— if ever poisoned.... with a treachery. But care for my might, I am a clever fool! among clever bait. And I elect to try at stabbing.... or dabbling, in this way. For the shoulder is tired of its silliness and funny feeling. Time to be fed. Time to break some cakes! and pour to mouth a rich watering— again! And for Wallinda.... with envy of youth, I leave all the messiness of that contagion of contraction and myopathy of heart, myopia of sight, and the mysticism of a self-centered longing to play on the mouse. One must go through these stages, to inure or excise wasteful, useless dalliances— of the mind. And I did batter sentiment and sensitivities, in my day. What's more for home to find? Some.... handsomeness to rear, and wholesomeness to wean, that does surround me for a cleansing.... off the battlefields. To, then, what is made does makes one's self for me, a reprieve from callous nature; and this virility expressed for demonstration.... homes. Or find a home, perhaps, were for this shoulder now to bleed, it does not. It doesn't weep at all, now. That frees me for a temptation, so explicitly stated, and freezes of the sensations, this sense of boring domesticity.... and tranquil.... amour. (*sips*) What roasts today.... is wrought with calculation.

Act III

Scene I — *The broad porch before a palace garden.* **Musgonie** *sits on a bench, relaxed but without ease, the tensioning mental. He holds a paper, which he has been reading.* **Costile**, *a councilor and messenger, stands near him.*

MUSGONIE: What is this Bracken like as a man? He seems almost bluntly to make demands, and of me!

COSTILE: They are couched as warnings to alert you, your majesty. He signs as a loyal subject and highly respectful of your sovereignty. Yet he suggests that events in Lacismoor are rapidly turning foul; and does wish for your leniency to attend of them, basically of those particular freedoms of assembly and governance to apply—

MUSGONIE: With his as some sort of autocracy for the region, a duce indifferent to my rule?

COSTILE: He is only a prominent landowner there, one of several. I doubt if he vies for any run of despotism, but alerts you to a growing possibility of revolt, and a sensitivity to taxes. He might even be considered brave to address you on these matters personally, with his name and signature. But as to the man himself, just one of many independent sorts to inhabit that province, I suppose. I can't say to recall him well; although I'm sure I must have met him in his youth, since we are in touch with all important families

of our country, at some time or other.... I would take this letter to be an ensconced description of the atmosphere there, rather than a warning, and by a loyal subject who is simply concerned about it. And it does not demand of you directly, his brief. It was sent for the eyes of the Privy Council, and only suggests what the king might do to relieve tensions and his worry.

MUSGONIE: It is a polite raking of me! Costile. Not offensive, but with offences to make.

COSTILE: I bring it to you because of the relative immediacy of the situation, if an urgency of action is required, as you have expressed (to want) to know of such problems quickly. We councilors will consider the questions raised here more thoroughly than at the moment, to draft our advice. But you are given notice of a minor storm that might be approaching the region, a storm of (a) populace's sentiment against perceived oppressions, certainly badly conceived out of their indulgences thus far. But they are rich, and spoiled for some further independence. It's said to be a very beautiful area. They make their impressions known with their owning of it—

MUSGONIE: I must give them more fees, not less. Grant them that privilege to support their country, and my kingdom. So they feel especial there. And must I militate to bring control? I'll send Duke Decamond and his soldiers to establish a fortress there, an outpost— and a threat of my power and indignation to their.... opinions of resistance, and optimisms.

COSTILE: Decamond?! He's never been there, my liege. This is not usually your method, to control a province, though it has been how you have allowed (this).... to happen.

MUSGONIE: But now I may have him as my personal representative. Have you not spoken with Ponstole yet?!

COSTILE: No, my grace. He seems to have become distracted as of late. But we have not held session for awhile. And as far as I know, there have only been rumors of Decamond's relinquishing of fervor and perhaps even leadership, temporarily, of his band of cutthroats, as he recovers from a wounding in his last.... military engagement.

MUSGONIE: Engagement? *I* have engaged him! You are behind the curve of a fine calculus, Costile. I control the man now.... somewhat. Or for least to direct, there is a temper directional in him, I am assuredly advised.

COSTILE: And he tells you this, your sympathies to own? It is only with a profit to be made that he is drawn. Only this may support his carriage for more adventuring and adverse butchery of your pacific methods. And I don't see how a downing down in Lacismoor favors such an establishment for this decadent duke—

MUSGONIE: I've not had conference with Decamond directly, yet.... nor see much need, at the moment, to relate to him in person what should be assumed in our private hearts. For we are related only of a tooling to control. Else I would be very pessimistic about the whole of this utility. But I do not want to promote the impression on my part of a catastrophic castigation for any province, any of my territorial possessions and responsibilities. I want a reign of peace and passivity under my hands to pat and ponder with, and no overt violence permitted of me to be seen or even assumed before excuses may make craft of explanations and justifications.

COSTILE: Decamond is not one to want to be kept, confined to a fortress. It's not his manner of operation, my liege. You mistake him badly. He lives for freedom to roam and canvass of golden opportunities in the remote regions and distant stretches of our estate's extensive holdings. It is only with the slightness of your authority sensed at those places that his manner of action is allowed to happen, the effrontery of an honest rogue brought upon by these regions' laxity to country rule and courtly order— He would oppose this expedition to Lacismoor, as you would explain it to him.

MUSGONIE: You fail to grasp the significance of a king's representative. I have made him thus, and without trusting him. But he does trap himself with the offer accepted, and that is to assuage the temporal of temper until a stronger, tighter, and more comfortable bondage is constructed of these areas. For they now slip wantonly of the gown and are loosely worn. My representative would have to fight himself to allow his instincts range in these affairs. Yet only the king rules a kingdom, and its entirety. This man professes to be a loyal subject under me, and thus must do only what law permits, even if through a guise. But if he should break my statute in any way, for his position to discuss away the tensions found through honest effort or attempts, and to instead promote more and with physical intimidation provoke a war, then that very illegality, juxtaposing jibe job with jab jollity, clamps his legs finally to a royal prosecution. He, I must wager, is too wise to allow himself a plummeting into such predicament, as could he view the situation firsthand and breathing freshly of its actual airs. And so I, in effect, control his men, to use for intimidation with restraint and calm and peaceful threat. That is the pattern I have weaved, for and between ourselves. And I know it is a gamble— But with such a clever man as Decamond, 'tis how it must be done, to find his blinded side of pride, and abuse. I control this fellow now, before he may contain the notion thoughtfully. Yet it is all within reason of my rulership sanctioned anyway. So have him taught to live for me as all may learn. I live for everyone with their weights carrying. (*stands*)

COSTILE: And how may he and his soldiers be paid for this.... excursion-ary intervention? Our treasury should not afford what could erupt into extensive battling unless the events be deemed of *our* military.

MUSGONIE: The doles from Lacismoor herself should suffice to maintain Decamond and a peace between them, until we may more firmly handle these landowners through my cordial visiting, to entice the necessary strengths of allegiance, strategic and influential favoritisms already (having been) designed by Decamond, agreed to and perhaps even partly put in place, which should be his major employment of mind while down there. I've no doubt it is of high calculative value. And then (*starting to pace back and forth slowly*).... off to other places for him, my method exercised, finely tuned and mastered.... and he accomplished to be comfortable with it and made much more highly regarded because of it.... His troops may dilute a bit, with partial permanent stationing at fortresses or invited garrisons for residence.... and commands over them gradually changing by necessity, until the duke in the end.... becomes symbolic— and retired.... That is how I try to see his use, reducing his aggressions with a growing aggressiveness of purpose.... out there. An envoy christened by me personally, he does become for honors gained.... and transformation into civilizing work, as would his soldiers change.... into monitors and merchants distantly placed

on rich soils at spring of advantage.... or as enlistees to our finer forces proper, distinguished of prior service.... or military disciplines— This, or something like it, is what I envision, Costile, to handle my country without tumultuous upheaval. And if it works much in any way, for a good span of time, then I will call it a "decamondine" administration, even decorous of this man's name. (*stops before* **Costile**)

COSTILE (*as* **Musgonie** *hands him the letter*): It is the privilege of a sovereign to have such a felicitous view of harmony for our maturing hegemony, as we would expand from a hoped for natural evolution of common interests and friendliness, what has.... already been won by violence, my lord, or the accident thereof to precipitate possessions gained. But this tendency of yours to want to visit these remotes so relatively soon or quickly stays worrisome to many of the council's minds. It seems much so like forcing an issue that not might be, or to which we are ready to risk. If not rash, then rushed you are, to be.... paternally accepted throughout all claims of your kingdom. Yet have these claims broken many a head, as does leak out the brains of resentment, then may you not have to visit these most questionable regions until much more time wounds the disagreeable intentions of many. Why then risk so hurriedly this jeopardy?

MUSGONIE: ....Because our country is in quiet turmoil now. So then it's now to do for chances to be taken, that I wish to be effective and guiding of all my dominions, and will stand as a proponent of good sense to be followed or demanded of for all of my subjects. And I will not allow wounds to fester or grow strange of feeling. I will have a presence for every rock to stay and rest in tranquil ground, or be myself misled to cast astray— this mission born of me. But our problems, I feel, are more ungracious than dangerous to my person. I am the sovereignty, sworn to protect even the furthest outreaches of our country and my inherited estate. Yet those some, on the periphery of my power, question my influence or effectiveness in this regard. It is a misunderstanding that I would not do so under any considerable circumstance; though naturally there is a tendency of men to be defensive of their properties and conservation of care, even to the point of liberalities of view towards much independence to.... adapt to within their environments. I'm not much opposed to the practicalities assumed for a productive existence, and am inclined, as is hoped, to be lenient of thought for my properties' constructive management. But no ruler for long will tolerate intuited dissent, and will work to abate its growth. I try to do so without calamity, and safe visiting of the provinces is a demonstration of this talent emblematic of my effectiveness as a cautionary king. So I am strongly bent to allow my visiting and welcomed reception, by all of my subjects at any location inhabited. This is the most primary accent of law, that the king may tread anywhere within his kingdom, and may stand upon any such ground— un-accost-able by threat or injurious intimation. Should not I choose to do so?— but the manner must be judiciously prepared, I'm astute enough to realize. For all men make more the beast than angel, within their tremendous limitations. And I am the same to wound as any other. Yet test my regality, Costile. It must be heaven-sent for methods such as these to employ, these moments to craft with a patient and considerate strength, and proofs to find for these sultry and various responsibilities.

COSTILE: My liege, it is exact that you are entirely our ruler. No more tests need be provided to attest this. And these.... regional properties.... will always be reluctant of fees and heavy fining.

Some may remain recalcitrant even through the storm of your graciousness, the splendor of your inspirations for them, and the awe of your immediate presence or proximity to spell-bound faces. Ponstole should be here to argue the point, but it simply is not a wise move on your part if it is not necessary and may even precipitate querulous attempts to abuse your standing and placement before then as a dare to complain of. Your popularity in those areas can not be easily judged to be favorable, and yet you must bring with you even more demands of servitude, service, and surrendering, for the country's whole benefit. Let ministers handle these repulsive tasks. Be not so personally associated with them. I would volunteer myself, to assess Lacismoor('s) sentiments on her ground, even before sending Decamond and his brigands. Take this sort of precaution at least, so that we may faithfully judge reality. I can claim to have been sent by you, through urgency of this brief received, to estimate revolts.

MUSGONIE: ....You have my leave to do so, with official expenses to use. It seems to be a prudent alertness to express of me, though you may only suffer a vacation in that fine region. And what of the council meeting?

COSTILE: It's planned for well within a fortnight—

MUSGONIE: Yes, but—

COSTILE: —As soon as I find Ponstole and give him this letter, I'll head off to Lacismoor—

MUSGONIE: Copy!... the letter. And present it to this Bracken. Tell him that I reply with your exhaustive examination of conditions there, and that I am aroused for changes, placative and plausible, but also persistent to my aims, pertinent of my goals and the duties of that province. Explain to him.... what to expect of the diplomacy of Decamond, and study all aspects of his reaction, both for my enlightenment and to judge whether to keep this information private between you two— upon royal decree, which you may use if you feel it's warranted. I want to know.... what lingers there of work and laziness. He writes that they are of the customary industry we've come to expect— But I want to know the truth of what grows fat and hedonistic in Lacismoor, and what possible moles and pustules may be developing of her fair skin.... organized resistances to my direct authority. Inform all of this to Ponstole, when you see him. This is not a secret mission, and he has not been remiss in his laboring, as far as I know. Yet, to be honest about it, I tend to want to avoid his necessary and fully appropriate refreshment of argumentative vitiation when we meet for discussions, a swat of sangfroid care, demeanor, and dimensioning of problems I must come to address. I hope he's not ill.... of the legitimate issues we face.

COSTILE: Nothing that I've heard.... The last I spoke with him he was remotely pensive and self-reflecting, with a fawned cheerfulness that seemed silly, or not typical of his character nor outlook on life and events of common seriousness.

MUSGONIE: Perhaps he was impressed by Decamond. He seemed somewhat astounded when he relayed (to) me the news of the duke's apparent accedence to implementation, along with my cousin, Elser.... But I took this to be only a polite servility, on both of their parts of manner, to report on an acceptance of my wishes and expectations neither would have suspected from that skillful fighter. They think warriors are only belligerent in nature. But(,

rather,) they must be the most considerate of thought, during the pressures of battle— and elsewhere. And my father was the kindest man.... to me, in my young years. Judge more by what a person does not do, than what he does.... and his absence from events, than his presence to nurture them.

COSTILE: Yes, your majesty. That's as retrospective as Ponstole has become.

MUSGONIE: Then the lessons of a severity spread. Decamond is not a cruel man, else he would be viewed somewhat as a heinous fiend for what's been committed under him.

COSTILE: ....Er.... Well his reputation is disputatious in many quarters still. And what lies under him are bruises, wounds, and deaths, and the unforgivable of heaven— if life is torturous enough.

MUSGONIE: I've not known him personally to allow it, of his men. That is illegal under my rulership, except for the most extraordinary cases and jurisdictions of law. No such complaints have come to me about them, or I would have squashed that entire enterprise forcefully by now.

COSTILE: Yes, but.... a sword's blade must impair, impede, and impinge to be used, against a courage and a flesh. That is the method of its machinery, and this is a torture of body and mind.

MUSGONIE: He is not assertively evil, Costile. Rather more, an opportunist is his curious trait, as such he needs not this business but finds in it his scabbard's impulse. Remarkable for such a dependence on tendency are how our individual aggressions play themselves through without necessity. And I would have made a feast of wrestling, till a serious though inadvertent injury of another (contestant) tempered that trend. He still lives, though, and that is evidence enough. I am convinced that we must find our burdens, rather than receive them like gifts and presents.

COSTILE: You mean make them of our hands to shape, from clay in which we bed.

MUSGONIE: Hm.... As much as for the muds to mold and dry into adobe(d) structures of our reasoning. I mean to say— we need them not, except for that fine sufferance endowed of the inscrutable quality that makes one singular among all others.... and yet too, too common. For dangers please many childish behaviors, as one examines the world. (*Jocqueral, his young son, runs up to them from a side of the garden.*)

JOCQUERAL (*approaching*): Vater! Vater! Mummi says they an-tagonize!

MUSGONIE: I am in conference, Jocqueral.

JOCQUERAL (*stopping before them*): Hello, sir.

MUSGONIE: This is councilor Costile.

JOCQUERAL: Yes, sir. Cos-tile. Hello, sir.

MUSGONIE: What does Mummi say? Who antagonize? Where?

JOCQUERAL: In the courtyard. They make a fuss about something, and quar-rel, Mummi says. She sent me to find out.

MUSGONIE: Sent you?!

COSTILE: It must not be serious, your majesty. Lords must be gathering for an afternoon's pleasure of greeting.

JOCQUERAL: I was with Mummi in the boudoir, when she heard them from the window talking. She said it was annoying— Oh! I told her they seemed to be upset, before she noticed. She said they discuss things piggishly, and sent me to find out.

COSTILE: Priggishly, you might mean.

MUSGONIE: Sh*eee*.... just wanted you out of her hair for awhile, my son, as you point out matters to investigate on your own but with little interest to herself. Yet, she told you they antagonize?

JOCQUERAL: Yes, Vater. She said they an-tagonize your name for boast of jocu-larity, and that this is becoming too usual and commonplace.

COSTILE: And did you find these persons, young prince, to stand around and take in of their jesting? It would be unusual to stay so indiscreet in your presence.

JOCQUERAL: I was going to... But I saw Vater first, and come to him.

MUSGONIE: They joke of me upon my very ground?!

JOCQUERAL: I didn't think it funny, what I heard. I thought they were fighting with words.

COSTILE: And what did you hear that you could understand?

JOCQUERAL: Nothing, sir. I asked Mummi.

MUSGONIE: Perhaps I should make a presence, and of a gravity for this entertainment.

COSTILE: It was probably only the manner of their boasting, my liege. They can be loud, and that can sound querulous to a young boy who can not yet interpret meaning in vernacular bouts.

MUSGONIE: What the queen assumes makes this upsetting. Maybe only incidental, for my son's hearing, how adults may argue. But "antagonize" is a strong term to use, for his learning. It could be misconstrued as to be against me. Did you fear they were talking against me?

JOCQUERAL: Yes, Vater.

COSTILE: Prince! But you could not understand this talk, so then it was the sounds of intonation that were alarming.

JOCQUERAL: ....Y*eees*, sir.

MUSGONIE: No matter what. A yell is a yell, a scream is a scream, and a laugh a laugh. But what concerns one more is the information belied through these polite gestures of conversation, as could the Queen assume, a guffaw to be a curse on me. It is riot-ous.

COSTILE: There is controversy within men's gatherings always. But this is only for opinions to be polished.... or defeated down to effacing, your highness.

MUSGONIE: ....It's only courtly behavior.... councilor. I do not contract on the ideas of men, that may be other than my own.... But have you not a blush for your worries, son? This is only passing and must be trivial of your anxiousness to command. It could be new to learn, that grown men often disagree, and not every concept can be definitely known as clearcut and confident of practicality or substantiation.... They do not laugh at me.

JOCQUERAL: No laughing, Vater.

MUSGONIE: They do not cry at me or contemn of me, but only promote views of themselves to each other. And Mummi is just bored with the banter. But to relate "antagonize" to you, *that* is a jest. And that you find "antagonize" is teasing of your patience, then have you forgotten it?

JOCQUERAL: No?

MUSGONIE: But you have forgotten them, for your attendance. Then it is not "antagonize" but "antipathy" for talk that your mother hears. And you.... would silence them? So can the moon escape its luster with such silence. It is simply gone from view.... Yet shines this deafness on me, to treat you to "antagonize." Here, then, is a lesson for you. Yet have a candy be, then it is sweet?

JOCQUERAL: Yes.

MUSGONIE: Yet decrease the sweetness gradually until it is no more but a vice! a candy remaining? This is how men often talk in groups.

JOCQUERAL: Then it is ate.

MUSGONIE: Yes, with woe to gain at its loss. But men take pleasure in this sweetness of conversing, sometimes, and without much substance to it left for hearing, these traces of voice that you took in.... (*bringing **Jocqueral** to his side with an arm*) Deny what is not understood, Jocqueral. The Queen does make of migraine, possibly. It may have been of a soft talk expressing nothing.... no relevance to either of us.

COSTILE: That can pass for truth, my liege. There is no convoking sensed, of antagonisms here, but only conviviality of guests to court, those.... passing through by habit of daily custom. There may be some news breached, about Lacismoor, or other.... vices. But not offensively orated, as to cause these unheard uproars. Little seems disturbed of this tranquility of ground under a splendid day.

MUSGONIE: This *is* a nice day, Costile.... with the garden pleasing me— and my son. What for a monarch (to) be displeased on such a day is sin.... of whispers. Yet for my wants, I will not strike at slight contrariness. It is only a reprimand for heart let out from a tiny disquiet of angst to be reminded by.

COSTILE: And through the courtyard may I go to have ear of this, if it may even so disconcert your imagination? the unknowing or incapable to be understood this way. And if there is any serious-ness.... of debate.... then I will come back to report for you; else am I off to prepare for a trip to Lacismoor, your majesty. With leave....

MUSGONIE (*as **Costile** withdraws*): Do this, good councilor. I'll not have subtile hostilities mar such a fine day, and progression of events.

JOCQUERAL: What are pro-gressions, Vater?

MUSGONIE: Er.... The principal happenings of a positive nature towards some result. The racing of a horse through a course. The blossoming of flowers in spring and the ripening of fruit in sum-mer. Whatever gradually occurs, as your growth into manhood and worldly intelligence. You wouldn't want to be responsible for at-taining these things suddenly— and at command of your school masters.

JOCQUERAL: No! That would be frightening.

MUSGONIE: Then you must progress to those levels of stateli-ness, through periods of definite events or lessons. What tricks have you learned?

JOCQUERAL: ....I can jump over my head.

MUSGONIE: ....You mean backwards?

JOCQUERAL: No; but by placing my head low, I can jump over it.

MUSGONIE: ....That is a trick?

JOCQUERAL: I had not thought it possible, until it was demon-strated for me and I learned how. It's easier than to think of doing.

MUSGONIE: Perhaps, to figure out. And is your mother worried for me, that I can outlast some idle chatter from noble idlers?

JOCQUERAL: I think she was annoyed.

MUSGONIE: ....When it was brought to her attention to remark on. And can you have fear of my capabilities to handle men—?

JOCQUERAL: Never! Vater.

MUSGONIE: —subjects of the realm? No. They only squander air with their artifacts of opinion. Don't be deceived by such talk-ing gestures and articulating acrobatics. It is only a form of stimu-lating recreation— play.... of thoughts and concepts supposed, as if to imagine what might occur if they were true. But they ramble through the inconceivable, which means they can not conceive cor-rectly, and may not threaten.... my rule. This is why I'm generous enough to let them gather in the courtyard without proclamations to hear. They are not summoned to me, as in the past, when such meeting had shades of being ominous and of accusations to levy and defend against or explain away. But that was during times of great conflict, and before your birth. Today it is all cheery, and they take advantage of a sunny court.... to say perhaps the outra-geous without fear of persecution. As long as they are polite in do-ing so, and not so harsh to my ears and the queen's, with treason-ous intent to parley, then this.... manner of talk is.... permitted of sounds to be overheard. Though such a custom is not exactly po-lite, Jocqueral. You should not make it a habit of curiosity to try to interpret the talk of a gathering of adults, with aim to esteem or condemn, if the subjects concern you not. Sometimes you must ig-nore with a conscious and enforced deafness the scatterbrained ideas of wiser heads, as being not of your position to hear and try to adapt to, strange notions that may be listened of. They are too often sloppily said, because they're improvised or inadvertent and sloppily constructed— Not the carefully crafted speeches of your illustrious teachers. Very often, in such gatherings of impromptu socializing, while some may know what they mean, they don't know what they say. And that is a conundrum of irony you will simply have to be educated of through the experiences of listening to men speak in order to try to convey ideas beyond their intelli-gences. This probably annoys your mother most, about such talk, because she is very wise.

JOCQUERAL: Yes, Vater.

MUSGONIE: It antagonizes her intellectual perceptions.

JOCQUERAL: Yes, but—

MUSGONIE: That I even be mentioned of name in such a loose-ness of company organized solely by the beauty of the day and the whim of meeting or finding friends for conversation. Although I suppose that you must learn of grumpy and grumbling habits, that they do not always lead to your punishment, admonishment, or re-straining of activity, and— sometimes.... may be observed for their worth of foolishness. The taint of such behavior will allow you not to fear it later, when you may yourself become involved, participat-ing of casual talk or gossip and suspect mentality or mastery of views. (*leading him into the interiors*) Now.... what other tricks have you learned? The squaring of simple roots?

JOCQUERAL: ....I can roll upon my head.

MUSGONIE: ....As a spinning top?! That's dangerous for the neck, to collapse on.

JOCQUERAL: No, Vater. Lying along the ground, I can roll upon my head without catching the strands of grass in my teeth. I can keep my mouth closed now, and roll properly. This lets me breathe the earth more.... fluidly, the good scholar says.

MUSGONIE: To escape from it? These gymnastics are for the woodsy scents to appeal to and taste. Yet you are of a boy ever growing to develop and strengthen your limberness and reduce an awkwardness of motion with your weak physicalities, which we have feared for your age. But these exercises are not the mental arithmetic that should be starting to take hold now, also. Retain those studies as well, such as to know if any number be evenly di-visible by three. Then you can always divide it into three equal parts, or sum a number of things, small to grand, for this division.

JOCQUERAL: Yes, Vater.

MUSGONIE: It is useful, for various forms of husbandry and herdsman-ship.... or commitments.... to otherwise contrasting obli-gations and agreements.

Scene II — *A palace courtyard. People, mostly men, are sparsely gathered in small groups. One is near to a palace stone wall, pre-sumably under a window to the Queen's sitting room. This group* *contains **Maeli**, who is having most words with another **count**.*

A COUNT: It is tantamount to a subservience, Maeli. And I will not play under him, as if my liege, simply because he may offer a few more men to the royal guarding forces, even if his demesne does border my own and is of a similar size— only in acreage. We are equally granted these dominions and requests of service. But his is fatter and broader, while mine is more extensive of length and touch along many other, though smaller, principalities— I need more of my men to protect my regions neighboring.

MAELI: His does seems larger, in effect, such that he is substantial to you in concentration of area, and may provide of his own from yours—

A COUNT: Not without a fight! The king allows this disparity of voracious allegiance because he is a relative— And only that! I send all that I have been legally contracted of, and need not take a lessening of favors granted.

MAELI: Musgonie does not directly make such claims on you. I certainly won't allow them on myself. But the estate that borders yours is more centralized and capable of producing greater effect of defense. It's only natural that the king more favors his contributions and concedes to him some loosening of law to challenge your properties at that spot of boundary. I mean, banks grow or lessen not only with the seasons but also with the evolution of the earth to the area. And as a stream thins, he must necessarily gain more land. For yours is defined only past the water distal to his land at that particular geography. This is in the record.

A COUNT: Then why should I be forced to pay a toll to him, for use of the darned river, when it thins from his side? If nature has it thus, then let him pay for *his* losses.

MAELI: The fishing is more difficult for each of you. Yet your wharfage structures have not changed as much, compared to his expenses of correction of alignment and reconstruction costs. Consider it more as a courtesy for a shared conjoining resource. Or would you have the fish prices increase throughout the region? Compared to that, this slight seasonal toll is.... miniscule, trivial.

A COUNT: I'll not have him as an overlord. He does not own the waterways. The king has not declared such a thing.

A VISCOUNT: I'm not even sure what the king may propose in such a thing. He does not have much command over these situations, as a practical matter.

MAELI: Your neighbor only suggests that you share in some of the expenses of upkeep, since you use that stretch as much as he. There have been no demands for your payment levied at the court. This is a dispute that can certainly be resolved between yourselves, with little disturbance of anyone. Does Musgonie have to oversee personally every little detail of his territories resigned to your personal management? The squabble is frivolous and confronts logic as if a bastard. Your neighbor is only asking for a little help and cordial responsibility.

A COUNT: He would help himself to my property!

MAELI: Suppose some of your serfs or free(d)men were polluting the streams in that area, causing an abandoning of fish and game. Would not he have the right to at least ask you to control their activities more?! That is only decent. Then, as he must tax himself to support the area, so should you. The benefits will be for you both, including a lessening of friction about the water.

A COUNT: I have not caused the distress he endures, an itch to overtake my sensitive rights along that stretch of river we share.

MAELI: Has any of his crossed over onto your land?

A COUNT: I do not complain of it. Nor have any of mine hunted on his, not that many animals would care for the disorder of woods and earth on his side of the region. Even his reeds are bent crookedly. But there is no question of any illegalities between us. My distinct allegation is that he would have the arrogance to propose what amounts to a fine against my use of a country resource, a king's property, merely because he would wish to consider himself superior to me in importance or substantival worth or influence with Musgonie, when we are by all rational accounts equal of status under this king's rule. I would not bring ourselves to battle over such a minor disagreement. We each place our feet in the same constitution of mud-bank. We each have the visiting rights, the rights of passage throughout the country. But the absurdity of the way he coaches the suggestion to belittle me rankles me— Though I'm not upset at all, since he can not enforce such an apparent tariff on fishing rights, clothed as some kind of duty to him. It is worse than ridiculous and mean-spirited. It is presumptive of a cowardice that I can not abide to be thought of. And so I complain openly, to free and friendly ears that may listen to me, about my grievance— Should this not happen to you? in some way similar, with some slighting condoned by this court?

MAELI: It could never happen to me. I would not allow it. Yet, if I were in your situation, I would help with the affordable upkeep of that area. I would consider it a contribution to pleasantries. Your neighboring lord may not have expressed his request with an intelligent acumen for flattery of your rank; but you may still interpret the need this way, out of your own wisdom of character and chivalrous grace. Not all men may be as versatile as yourself.

A COUNT: And he is a crude horse indeed, with snouting spit from nostrils—

ANOTHER VISCOUNT: He does propose, severally, various tactics of aggressive control over aspects of others' properties leading to his own, from time to time.

MAELI: But his is such a rounded landmass, of handsomeness almost for a nation in itself. It is difficult not to be proud for it, and at the same time perhaps inordinately.... protective or assuming, as occasions may allow. This is merely an attribute of cartography, gents. Symmetrical shapes are favored to be viewed. They are admired like a fine beauty is facially seen— and respected, simply due to this.... natural talent of physicality. And thus a geographical hauteur may be portrayed here. The manor estate of a neighboring lord to me neighbors his, and there has been no quarrels or upsetting between them; though this lord's territory is somewhat smaller and irregularly shaped in comparison, in no way as highly regarded. Yet cordialities remain as kudos are won for the more handsome land. Well, this is only an act of nature in men for attractiveness— Praise his land more as you would wish to, and he may lessen the proposed fee. My neighboring lord presents himself with subtle humility and polite self-effacing to not only myself but his other high neighbor, and there is never any trouble between us. We are peaceful and strong. And my neighbor's administration of his ground is particularly complimented as being terribly astute and smart. Yet we all make a common complexion of country.

A COUNT: Are you saying my area of property can be considered ugly, in comparison to this— self-appointed sheriff's?!

MAELI: It winds a bit, *schlange*-ishly. But his is simply more circular and even and centralized and admired for this. Eventually the practicalities of management determine the virtues of a refinement. And in this respect you are justifiably ennobled as he. They are both finely run.... subkingdoms, and no one can detract from you this reputation of caring.... But his is nicer to be had. Or were you he would you not say so?

A COUNT: Would you?! I much prefer what I have won, with all of its difficulties, over pretty sights to boast of. And yet, the manifold variations of landscape and interesting natural architectures that run through my country accesses make for of themselves a particularly spectacular comeliness— to the strong of mind and attention to the curious. My grounds do not rest as squat and fat and complaisant for an observation, but remain active and frantic of life for much of the territory. There are more interesting occurrences in one of my acres than in the whole of his domain. Musgonie can not even fathom this richness— which indeed my neighbor does not attempt to steal or usurp. He stays honorable to our codes. But he would try to lessen me into being under him, to bring him mine by this kind of association. And I say publicly to all who can conjecture on the sounds of my thoughts, that this will never happen, even as he himself or his friends may hear. It will not be so, and I will not be so divested of my solemn, princely rights.

A THIRD COUNT: Give up the fee to keep the peace between you—!

A COUNT: No! 'Tis not a fee— 'Tis nothing! That will not occur, unless the king so sides with him and demands it. And such action will not be sparing of my discontent.

MAELI: Taxes rise everywhere, in this country. You two are not immune to necessary measures taken, or to be throttled with within the patience between gentlemen and worldly nobility who can solve their discontents without eruption of fires, magmas of specious derision and public attacking with thoroughly unfounded spite, for such a simple matter to resolve. Or do you want to return to the days of hatreds carried merely by the winds of burning matter, to your nose arousing for a cause?!— This is none!

A COUNT: I only complain— I have no cause, Maeli. I describe.... what dissolves of a fealty, in this nation, what.... degenerate conditions are allowed, with sharing of a river or a lake, and how one can be made quite unsettled and uneasy, with one's inherent state these days. Unrest fosters causes, but I haven't any yet. Still, should this be allowed to occur, so wantonly of implication? that there is favoritism provided for but not ascribed to— else we faint?! Then who may escape such a challenge to heart? Musgonie does not rule us well.... enough, to prevent these disputes, nor abolish them— Not that he doesn't try, with all his heart and will. But one is a picture, the other a vase riding tremors. And you may judge me in all of this by my openness to say.... if not foul, then feudal.... are these circumventions of my expressed and well established loyalty to this country and kingdom. What's fighting more?

It's something known to do. You relieve of locusts by burning down the entire field occasionally, perhaps even periodically. And if these crass impingements on my rights multiply.... They are more delicate than wounding, and more dainty to be trying— Of course I can afford the fine for standing on my ground. But, what little more— for each of us to endure? This grows, and accumulates of imposition. And are we not due rising rules for revenues? Then who will challenge whom to lessen these burdens with their own little taxes on their neighbors? The process develops among us, and starts against me. So I inform you, it is implicit for our retorts. We are all subject to this mismanagement.

THIRD COUNT: Pay the fee to keep the peace! This is not persuasive of you. For you strip more of fish than he, from those brooks meandering towards your estate and to. And the nature of your winding land suffices of argument. But this is what you have accepted to abut. So carry the tolls imposed, wherever meet you some quaint rebuke to passage or threat. These shift— with your meaning.

A COUNT: Have you such survey?!

THIRD COUNT: The tensions drop! with your amendment.

MAELI: It is that your lands pass by so many others so quickly of transit that suspicions of proprietorship are aroused of too numerous trespassing, your soldiers diligent but feared abusing.

A COUNT: That's only a fear *I* would have. And who permitted such a design of territories? It was the best that I could partake of, after our broad spell of warring, these travel-ways adjoined to make my countryside, what was left me from my undeniable valor of generalship— What I have fought for!... Dare you think I would not heed to gamesmanship?! or brinksmanship?! Blame me not for this design, but Musgonie to settle with— what has been offered me. I will complain to the hilt of heaven's sword for ever being taken advantage of. But it is the looseness of our composures, friends, which will drive the strokes through flesh.

MAELI: ....That may instigate. Yet it is a difficult task, these composures— balancing. I've not been taxed yet, directly—

COSTILE (*coming up to this group*): Lords! What quarrels here?! 'Tis right under the Queen's ear!

MAELI: No quarrels made.

COSTILE: And could be thought as disrespectful such a noise. Have you seen Councilor Ponstole today?

MAELI: No. We only smite the air a bit, with fleet of tongue. Yet nothing serious is said to warrant warning, nor warming of these pleasant and casual breezes.

COSTILE: I have a letter to show him. But if you talk of swordplay, think towards Lacismoor.

THIRD COUNT: That is too rumored already.

COSTILE: We should defeat it, I would hope.

MAELI: Oh. You would hope. And with a light of like of Decamond, I should choose.

COSTILE: That could be indeed, Count Maeli. For you strike at this incidentally, but only.

A COUNT: The king is against such overt hostility.

COSTILE: Yes, that is so. But practicalities abound for Lacismoor, its populace fearing its own behaviors. And I am to visit, to help lessen these tensions between her and home, as under a window placed.

A VISCOUNT: Good resorting clime there. But what is hostile? pray make me so. You've come for a vacation yet to make.

COSTILE: It's due, with counseling. Breathe you not with courtliness? I am as well, as much to say, invited on this mission, practically. And as for Duke Decamond, so may he.

MAELI: No?! They clamor for a roasting? or a dare! May I attend? Is rich, the juices quaffing there.

COSTILE: This is a new age of contending, and the king with the recovering duke have come to an understanding of aims and terms that fly as peaceful doves.

MAELI: With fierce pecking of each other's feathers.

SECOND VISCOUNT: But hearsay says, Decamond retires.

MAELI: Only for the healing of his wounds. I was with him not but a day ago. And he is as anxious for the opportunities at Lacismoor as ever.

COSTILE: And may his intelligence demonstrate this. But our era changes around us, gentlemen, One can not always fight your way through storms, or storm your way through fights. Treaties are a better weight of pact than parsimonious relations.

A VISCOUNT: Musgonie would allow her separation, independence?!

COSTILE: From this country, no. Such severance must be denied. Yet parties gather to promote the cause.

MAELI: And thus does cause the thrusting out of Decamond's chest! This is plainly obvious.

COSTILE: Not to the gentle king must matters devolve so horrifically. For I prepare a bed for this same duke's very influence, to help avoid warring conflicts.

A COUNT: That is much too hopeful for you, unless the king softens even more to strange tactics as to deny wounding and deaths, or to ignore slaying for what's to be inevitable about this handling of the rash Lacismoor. Brash boys and men make hungers here.

COSTILE: No one should wish more on profit than peace. The king crafts with this objective in mind. Yet have Decamond tell you himself that this may be done—

MAELI: I certainly will ask.

COSTILE: —He evolves, my noblemen, to the times of opportunity's remittances.... to him, to all. But what for you are made a pardoning from squabbles? Strong heads see the light, or sense its warmth through radiance.

A COUNT: Is this what you advise, as for a lecture to preach to titans— of responsibility?!

COSTILE: I say that one may struggle in many ways, not always physically. But tranquility must reign under King Musgonie, as he sees this as preferable to contentions and destructive behaviors. And in this manner he has reached.... a remarkable agreement with the duke.

MAELI: To tend for himself to be fooled? Is the king so desperate here, about Lacismoor?

COSTILE: A letter speaks to this that I must give to Ponstole. But our king.... while mild, is never foolish, Maeli. Do not underestimate our country manners now. For at present.... we do not tear up his kingdom. And his rule is stricter of some disciplines.... than you may feel.

A VISCOUNT: We?

COSTILE: Then I submit that if our king is clever enough, which is very much his strength for you, then Lacismoor will remain a peaceful, and will become a contented, possession and companion province to this country, his kingdom, and your fealty.... But I must find Ponstole— so good day to you, sirs. (*goes off*)

A VISCOUNT: We?... Yet are we disciplined alike?

A COUNT: For make of Decamond.... an officer of the king directly to control us?

THIRD COUNT: Pay the fee to keep the peace!

MAELI: Never—

THIRD COUNT: Worry not for Decamond about; his motives 're always sound, though often sullying of beasts.

MAELI: Never would he accept such a license against our fine nobility. This must be a different ploy for his use, maybe redefining it.... to fit under reconstructed laws, drafted by these councilors out of necessity for the imminence of situations in Lacismoor.

THIRD COUNT: And it isn't clever to even attempt to counter his peers. Decamond would avoid this at all costs. So—.... He has simply agreed to temporarily embellish the royal army in order for the king to deal with Lacismoor— And this embellishment might simply mean a collective show of forces while restraining independent action of the duke until and unless all possibility of peaceful resolution has been arrested by the tempers of those free spirits out there, by which and by then Decamond may rejoin his forces as commander of his typical activities.

A COUNT: You mean, simply, that Musgonie borrows his soldiers as Decamond recuperates.

THIRD COUNT: Yes, and probably at a high premium of expense for the king, to make this worthwhile for the duke as well as keeping his men alert, in shape, and directly educated of Lacis-moor for ease of future action. Now that would be a move decent of Decamond, yet appropriately risky for the king's usurpation. Yet, weighing the king's laxity with the good training and loyalty of fine fighters, I would take such a risk as Decamond, to be paid well!

MAELI: Didn't ask me at all.

A COUNT: You are not Decamond.

MAELI: But I've assisted his troops with some of my own sol-diers, in the past. My men are just as capable as his, and stay in as good of form. Yet, Musgonie thinks not to even consult me nor any other of us on this matter, leaving profits only for a fine rascal.

A VISCOUNT: The duke is demonstrated. This prevents his un-regulated intervention as the king sues for peace.

A COUNT: Peace?! There hasn't even been a war yet. There hasn't even been a back-slapping kick to the backsides—

SECOND VISCOUNT: The king deals paternally, as it were, with his children, pleading for good behaviors, but threatens harsher punishment later— by letting loose Decamond on them!

THIRD COUNT: That's the whim of it, sir— and the whip! Something councilors could conjure up to placate Musgonie's hesi-tance to promote any sort of conflict within his kingdom. Has Pon-stole's stare to it, to snare the resting but restless, wounded duke. Oh! how convalescences may lacquer minds for a change— but not a change of heart to linger of a fiery flame, this passion discon-certed, discordant of one's aching flesh—!

MAELI: He's not in any pain, from what I've seen, as he recov-ers. Simply walks more slowly, and with a sling for arm.... which must consider its use, since he was not wounded there, but at the shoulder. Then it's a funny injury indeed, for these results. It *feels* strange, he says.

SECOND VISCOUNT: Does he not retire?

MAELI: ....He relinquishes control, it sounds to me. Yet such passivity is not adjustable to his intentions to die at battle—

THIRD COUNT: Oh, that's almost like an oath among us leaders of men and owners of soldiers. Better to be brave at death than a coward sleeping under sheets from sheared animals. But there's nothing actual of this sentiment to make of. Have my comfort with it, I'd prefer, unless for duty bound with action take the break of life. An old man dies valiantly nonetheless, rattled with his body's scorn and uselessness.

MAELI: Futility.... is a powerful weapon of mind. But not for proud Decamond. He would prefer the grassy beds at the end of consciousness forever, and a cold slumber won with heated ago-nies and angst against opponents. It is a strange victory to feel, the intoxicant.... of futility.... Well, the king must come to me. I'll not supply myself to him more than my allegiance pledged.

A VISCOUNT: He has not asked any of us to offer more than our customary yearly quota of visiting soldiers to maintain a prepared-ness of the royal troops, even as rumors of Lacismoor's heresies

fly. He refuses to use what he has, except for show of symbolism, and thus does not *need* more.

MAELI: Ah, but then the need.... the need. Our country is not at-tacked. That is why we feel so strangely settled among our consti-tutional mixtures of lords and men, and how we contest amongst ourselves to find some fighting spirit left us. Decamond resists this drain of verve, with an organizational cleverness to fight un-merce-narily of his might, proposed off official views to be accepted, with this falsehood flung in the faces of transgressors. Yet even he feels strangely, today. Much said of it. I would profit of Lacismoor, were I (of) a mind to that type of fighting to lead.

THIRD COUNT: If the king can gracefully resolve the conflict, then plaudits are deserved him, this new tactic employed and made successful.

MAELI: Yes.

THIRD COUNT: And he converts us, with a true royal might.

A COUNT: Quiescently,(perhaps,) but not quotidian-ly.

THIRD COUNT: Pay the fee to keep the peace! sir. There are boundaries to the patience of raw souls.

Scene III — *A study room in the house of* **John Bracken**, *a lord of Lacismoor. His son,* **Scott**, *stands near a writing desk as the lord enters with a handful of letters.*

LORD BRACKEN (*entering*):What have you done! to cause such disquiet to our estate? but to fake a letter from me to the king?! (*throwing the letters on his desk*)

BRACKEN: Not a fake, nor a forgery, father. We are both lords of this manor, and I signed it as such.

LORD BRACKEN: Did you think I would not find out about it?! But here are acknowledgments from an outrageous array of court spectators, to the point that I provoke Musgonie!

BRACKEN: I did not design the correspondence—

LORD BRACKEN: Correspondence?!

BRACKEN: —as such.

LORD BRACKEN: It is a threat, and to him!

BRACKEN: I politely stated that conditions here in Lacismoor are becoming testily towards more independence, and that the king should consider these requests with the care and kindness he as-sumes of himself.

LORD BRACKEN: You wrote it in my fashion, as if I, and signed it, as if I were observing. And these replies are addressed to me! John Bracken, one of the most loyal ever to his majesty.

BRACKEN: I sent it to his privy councilors to read and ponder over. For I sense they delay encouraging the issue with Musgonie. And our province is distempered for more rights to seize and dem-onstrate, that we have become entirely self-sufficient and need only his leniency to exist more independently of the kingdom, and with the tranquility of relations he professes to treasure.

LORD BRACKEN: Scott! He is our ruler and protector. And Lacismoor is not so disobedient as to thoroughly break away from his kingdom and my country, Scott! Scott! You were not born of Lacismoor to own! And out of the dishevelments of the past, the turmoil and bitter fighting and tragedies I've witnessed, we accede to his pleasant ruling, his comity to allow us our bulge and accession to the wealths of this land. He had the wisdom and patience to allow its exploration and now our exploitation of its riches. Yet be you youth so ungrateful for what you have. Well have it *we*, and the king. This makes me seem as a proponent to treason, the most heinous of character I would ever own, a concept I have fought against— throughout my life. For no matter of the king's administration, men have died for this and struggled to maintain it. And it is of our utmost worthiness to continue under his regency, or of a regency, until a ruler does grant this independence out of a confession of principles that pleases to establish our own, and to be as likewise civil and grateful for our existence. We are not the heathens of the outback and the undisciplined scoundrels tugging for release of ties and from parentage. And if a separation is to follow, it will be deserved out of cordial merit, not these callous and childish measures *yooou*— young men think up. But do you want to draw the weaponry upon us?! Did you think of that!— No! But scorn is a pride to possess, among your bastinadoing fold. And you are deliberate to harm me! my name, and our family's reputation. We are noble fighters, Scott— And for the king! And not just the king and country, but also our historical heritage is for favor to be cherished by our actions today in this solemn possession—

BRACKEN: Possession?!

LORD BRACKEN: Yes! Whatever we establish, we establish as a royal dominion of Musgonie's.

BRACKEN: He did virtually nothing for the founding of Lacismoor. We'd not be even hardly more than a colony of interest were it not for our prodigious production of some raw materials the country now thrives with, most ungratefully— taken for granted and with hardly a gratitude for our efforts and labors and dangers —

LORD BRACKEN: Dangers?!

BRACKEN: —in this wonderful territory.

LORD BRACKEN: The only danger you're in is with my rage from (your) insouciance for this attempt to stab at my reputation and respect of our sovereign. What back-breaking work have you ever done?!— but we leave it to the natural inhabitants of the area that we have found and overtaken to govern, as they know how to provide for a life here.

BRACKEN: We pay them generously, father. And we've taught them methods that have incredibly magnified their productive and manufacturing skills. And as for people, they are as us, yet only unincorporated until our arrival to construct a legitimate social order amongst our population. And they, we all take pride with our extraordinary successes fostered in no way by Musgonie except by giving us leave to do so, a license and a freedom to experiment of exploration, export and espousal of our value to him.

LORD BRACKEN: "We" means almost exclusively of my generation, Scott. You have benefited from our arduous leisures and curiosities of the world. And the native gents here are as honorable as ourselves. Yet would you ask them of this advice towards insurrection and insubordination, a severance of ties and friendly relations from our own country?! And all retaliation will be mostly upon their backs and bodies to bear, those who have been faithful to our administration and who would judge us to be at least savvy enough not to provoke hostilities from our king, not to be crazily intolerant of his rule as they of our.... instructive and profitable oversight. You have not lived through enough genuine difficulties to warrant a full-scale attempt at independence for this province— But I.... have fought and killed with the king's troops to help establish a solid and peaceful country for everyone, all inhabitants— including of Lacismoor. Contentions were base and mean and deadly during our warring sacrifices and excesses. And not a few were ashamed to stay in regions where we had literally butchered neighbors and our own of nation, during these powerful struggles of regency until a true regal dominance was agreed to— won! by *all* parties that had contended, all petty lords to would-be overlords, all princes of a realm to harden into leadership and then obedience to a comity of peace, fiefdoms contracted for after the bloodshed; and through allegiance by the holy word to our much pressed and impacted sovereign, perhaps the most noble and righteous soul left amongst us after the extensive and extended mêlée of frustrations and furious hatreds, enmities too profound to remain rational. I have given this word with my very heart! to speak for those whose beatings I have ended, common lives destroyed during the storm of state we have realized, for some tranquility and calm and healing and self-assuredness among men still standing and walking.... and their families making, their country supporting, their king accepting with a loyalty through tears for his grace and understanding and patience— throughout and beyond the destruction and loss of verity for man, his goodness the only think provided (for) by the angels— constantly distraught with a malicious temper or rancor.... towards himself.... towards the beauty of the world not to own and control. Then it would not be unusual that a few of us, still adventurous enough to leave from heartaches left, would travel to find new territories, expanding our country's peaceful estate, and diluting out the grief that must linger.... to so many tastes of fortunes made, the sour feeling for its generation and the necessity to be worthy of these results, the riches of life procured; and of the day to see that does yet shine upon our merciless and bitter heads undeserving of even the slightest of a gentle and welcoming wave of wind or breeze.... over our foraging footpaths. And the land is so aromatic here, for our resting comfort and enjoyment. But we remain the king's subjects, Scott, through a pledge above honor. Up into the clouds resides our pact of nation to defend and continue, as by gods decreed, by what we have done with each other. The vendettas are our own to breathe through, the acrid, smoky plumes pushed away by our heavy, strenuous expirations, and the sticky, piquant ashes vanquished from our hands and faces with a wash of stringent candor in civility and temperance.

BRACKEN: That could be very romantic, to honor gales so strong as to move clouds, or the cumulous approach of darkening weather and 'nimbus-like hail threatening, what showers be rain or ice. But down to the earth itself, father, we have quite a state here already to defend. And for its single purpose be, that we are *not* to be abused by weak and inconsequential— aye! even blinded overseeing and misguided ignorance of.... this fervor that grows for self-rule and responsibility, never more to be taken advantage of— under a court's justice of royal minuscule of mind and manner— then I prepare of myself a battle to be taken (on), in defense of a worldly inheritance.

LORD BRACKEN: You would swoon on sight of errors to be seen as realized, your presence before men of calibre and catastrophe.

BRACKEN: I and others—

LORD BRACKEN: Others?!

BRACKEN: —will resist the sullen, sully influence of kings.

LORD BRACKEN: These others make of pack-hounds voyaging through sandboxes of privilege and infants' games. Take not your heritage to this for your stronger nature to apply and my foundation prospered by. You have a finer character to employ than this foolhardiness.

BRACKEN: We shall devise an assembly of stout defenders of Lacismoor, an association of like-minded men—

LORD BRACKEN: And boys. You are all boys.

BRACKEN: —to protect our world of natural aspirations here into the future to own— and command, and tyrannize or terrorize all objection to our collective status for our placement and activities of home.

LORD BRACKEN: An army?! And what— of soldiers?! That can not be made from growing wheat, the steel of swords and its capacity in capable hands.

BRACKEN: Many do practice these arts to perfect for gross knighthood to effect of our purposes. And you and the other elders, of landowning property, must support our reasons as well as for your own, to avoid external mismanagement and plundering from our properties. You know this practice must— can only intensify.

LORD BRACKEN: Will not a body make from severed limbs, and independent notions of a musculature.

BRACKEN: We will build a hard station to legitimize the consortium of our military endeavors—

LORD BRACKEN: It is illegal! to construct a fortress removed of actual dictum from the very king of fortresses to provide. That could be reason enough for his assaulting, as you'd evoke his necessary retaliation even with regret of heart and reluctance of mind. No family head will support this idea (here). It is blatant of misauthority and almost defines treason itself, or a sedition only to be learned of.

BRACKEN: Yet all families of Lacismoor must. And this is almost to effect at Squire Darughe's camp for training—

LORD BRACKEN: Blame him not! for the hedonistic tendencies of young physiques. That is a childishness, and he is a loyal soldier of the sovereign. You could never convince him of such a claim— such a charge.

BRACKEN: Grows there a body of champions who might be used to demonstrate this tendency in the forefront, that of a drive towards self-determination. And it should not take much to moti-

vate their courage into a cohesive whole of defenders of this state, safeguarding the interests of each and every person noteworthy of our province.

LORD BRACKEN (*going to sit at his desk*): That is a child's game, with the fortitudes of strong and limber bodies and yet unaccomplished brains, thoroughly unproved of true adversity to withstand. You've certainly had your play there (*sits*) with a mastery of techniques of difficult prowess to know that it is only a stage provided for going through, to exemplify the dawning of abilities in young men. And you are well past that education by now.... The squire is a friend of mine. We've saved each other on many occasions, and assisted each other for all (occasions) during the warring onslaughts and tempests. I tell you quite truthfully that the only reason.... (*picking up a letter*) This one calls me a potential renegade!— (*letting the letter drop to the desk*) The only reason for his notable and renown camp of exertions is to increase the endurance of men during the strains of physical tests, with the mental concentrations on them to work through procedures from simple to complex. A growing population loses these abilities without exercise. Yet it was the key to staying alert out of necessity for us— increasing our endurances, not our strengths.... in the main. So he trains. And we allow and approve of this training and his skillful handling of his members' pursuits, as a positive consequence and offshoot from our earlier adventures of vigor and vengeance within the kingdom. It can only lead to things productive of the athletic attainments, with his management. But it is in no way fashioned for an army. It's not meant for that purpose, and the disciplines learned there are quite different from the regimental and the unqualified obeying of authority.... through orders and commands. Even you found, in time and after proving to yourself your capabilities learned, the state of mind there (was) too immature to prolong of your attendance and attention to the customary activities of the place. That is not a legion outgrown. It's not a legion at all. A true army persists out of need, and can not be matured from— by definition of its requisite nature and legitimacy.

BRACKEN: But the potential is most obvious there. I know how each can feel, through profound experience and understanding. And I may sally them to cause of shared ideals, to prepare, at least, for a possible besieging, what the future brings with envy of our valor to withstand it.

LORD BRACKEN: You are not the Darughe to shape those of leisure tasked. Have you not enough to do, to help me run my estate— and protect your fortunes and privileges?! This requires constant attention, or else it will all run down due to swindlers and opportunists finding weaknesses in our operations. These matters are too complex now, for a single mind to be absolutely thorough in every aspect of our ownership, which is why you exist for me— and your mother.... when you were a child especially. But now you must help me responsibly, instead of driving daggers at— with an inadvertence from being bold with deceptions and stupidities.

BRACKEN: Am I not a good master to our workers, to second you with more details to describe, ascribe to and employ?

LORD BRACKEN: You're good enough at your chores, thus far, for a plantation's inheritance. Yet this business of self-rule and militia distracts you improperly— and damages me improbably of correctness and reward.

BRACKEN: I simply sense what is evident to happen, father, and try to force our stances appropriately. You have often been worried of the advice given to Musgonie. I attack at this directly, affront it for warning and to beware. That's the only way with tensions such as these to satisfy. And I must push for momentum to build. I must instigate events. Now it is clearly worth your while to have this meeting I've proposed, with some other landowners of prominence, to discuss prudently this question of independence for the province, since you are now of some capital towards this to be known. Perhaps you will use the occasion only to explain your way out of the aggressions of influence I've made. That's a risk I will afford, dearly if so. But I am hoping you will examine the issues more carefully, to see how my conclusions, and those of some others, might possibly be apt. Else we can only stay slavishly to a king for an eventual grievousness of misuse and abuse— You above all know the trends.... of cowardice, and the lack of celerity to prevent this for your own handling of cherished affairs and valuable undergoings. One seeks the scapegoats to blame for one's failures to make attempts at what must be actionable and is clearly heralded. Such scapegoating, such as the king's political climate today, and our supposed tranquility of country to maintain— a callous calm within our whirlpool of sensing growing inflictions from his court, his advisers.... this only absolves one's broken spirit to resist these wrongs, to fight against them and protest with more might than the temerity of argument in my— overly subtle message to the privy councilors. We must be more pronounced than that—

LORD BRACKEN: You'd pressure *me* in this way?!—

BRACKEN: For your own truths to heave again! and stone with. You can taste the battles coming, and the confrontations that will be made. And you know that to the uttermost we need a preparedness to establish and threaten with, or else we are useless, but only as serfs become to be regarded and treated, allowed to fatten now only for the exploitation by others later. Is there no coincidence of concern here?! Is there not talk about this everywhere in Lacismoor? in every corner of heads gathering for expressed thoughts? You know that I am boldly right, if foolishly wrong— with this way of doing the transaction for our rights. But one finds for one's self the method, within the environment he is placed and shares with others. And this is of *my* time to do and craft, as how I see the urgency and plod to plot my way of a reaction which could be successful for us!... if we are of one mind condemning— a usury of interest in us there, within that king's country drawn from.

LORD BRACKEN: No such army has ever been successful, against Musgonie— and you very well know why! There are complexities to handle, through his.... peace-mongering ploys. And we are not the better of us met by men who gain to destroy and can demonstrate this with virtuosic flair and a prideful passion for their exhaustive abilities at combat. I am too old.... to defeat such an assault. *We* are too old. And to command for this, to show how and lead through, a worthy defense against true warriors.... the training is not in place for too long, and the trainees.... are not here, in my opinion, born of the rationality required for such effects of deadly sacrifice. These adventures are life effacing for most, and with a seriousness as stimulating as stifling, each new-born day a conquest wedded to as the morning wakens you to trembling consequences of endeavor; and no game more is your pride to bare, but wounds and wounding to make. Yet for the threat on Lacismoor, I know we have been self-indulgent and proud, practical and profuse among ourselves for envying. But what may we withstand of this? There is a contract made, a tie and ownership— completely understood from the very beginning of our provincial exertions. And

what with, accepted for this leasing of treasures found and produced, what if we are bull-whipped a trite? for being too overt of happiness, and too arrogant in the eyes of our royalty. There is a holy debt to be paid, that only bloodshed can abscond with. That is the agreement and the promise written of and deeded for this land. With prayers of birth it was made; for the king protected our excursions of investigation, by way of permitting the act under his guidance and control. So we are honor-bound to observe his decree of province, to some very great degree, upholding his wishes for us.... and usage.

BRACKEN: Is why I've asked him, indirectly, to consider allowing our effective secession. I very well know that without his grant of cause the issue may only be resolved through warring argument. But what might that accomplish? but a permanence of contempt stationed here! Our way is much better, for the unwieldy neighbor to bear. Then leave us on our own, and friendlier. But we must show a strength to this legitimacy of freedom, and have an army mustering our fortitudes— for show and evidence seen. And have there not been several dispatches of this sentiment and hope sent from here to.... diffuse eventually and somewhat ineffectively towards the court, though all have been practically anonymous— until my direct attempt. Then if the king is such a peacemaking or maintaining enthusiast, he will give us our nationhood, as if bred for and now welcomed, to this adulthood we find and hone for ourselves. And our terms will be generous to Musgonie, which I've outlined in brief in my missive, and gradational for our growing responsibilities as an independent state.

LORD BRACKEN: You are no statesman yet, with these incongruous ploys to play, on him, me, and heaven itself, perhaps, (should you feign for an entrance,) as well you can not bargain on tariffs with God and King. What madness thought for you to have the right to set *any* terms?! or even suggest them out of your mercantile knowledge. I suppose they could have been impressed with the orderliness of your information presented. Details improve a perception. But they are thrown back at *me*, as a scathing of scoffs, that some would rather pay to burn our merchandise now, and tax us double— triple its worth for the smoky clouds oppressing our sight. That is to say, taut son, or to imply, that they'd rather want to come here and destroy us with a purifying fire! and whatever ashy plumes drift over the homeland proper tax our smoldering souls with. Such audacious boasts, by a few in these replies of commentary, present only a manner of thinking, and merely of thought, not to be endearing of our character and my name and business in particular. Not good advertisement for Lacismoor. No commendatory recommendations (of us) in this pile of answering.... And yet not answering; for none of this seems official from the court, but are rather heated scurrilities from informed, though substantial and influentially imperious individuals. Yet you haven't the right! to speak for Lacismoor, as they'd properly charge of me.

BRACKEN: One may always speak one's heart, under a pen of name in any guise. Yet this name should be emboldened at times, for sharp notice.

LORD BRACKEN: And as ever you ignore the complexity of men.... and fighters. You have not been one, Scott! You have not been one! even as you may seem able or feel enabled. It's your imperative that escapes my fathoming, as you would sense dread and dreariness from fright as a luxury of birthright. Or yet, what essences put forth, to make this balloting of war— you seize! You sniff and breathe with, snap at!... Where is this meeting to be held

—?

BRACKEN: Here!... with your permission.

LORD BRACKEN: My permission?!

BRACKEN: We must sharpen our fortitudes into a shapeliness now, and begin the motion of this evangelistic plea for actual and effectual independence as immediately as methods may be devised to afford the try. And I will bring your Darughe here myself for a convincing, as he's as much a friend and master to me.... as your friend and respected companion during your days of combat. But time starts always currently to do things. And it is clear that decisions must be made and requirements drafted and agreed upon for this cause delayed too long to dangle of uncertainties and erring to prepare for. We will be victorious, father, of spirit and in fact! as soon as we draw out, in mind's blood and on a parchment of courage— courageous nerves and nerve.... the picture of how doing of this errand, a confrontation against tyranny, as by the weak of manner or strong of will. That makes no difference in this idealistic stance against our oppressions, the character of a king. And with the intelligence we gather, of Bracken House, this time the defense of sovereignty to own, and for to seize and clasp daringly, will succeed. For here is an instance of young men capable.... of older minds and elders' fractiousness.

LORD BRACKEN: Go out and tease our workers with such headstrong rhetoric. For I've a fright of gradual realizations— to *lose* responsibility for. The earth shakes of shackling through your want scheming, and out of my patience to observe such wanton pleasure for confrontation with a frank-heartedly good ruler and decent rationalist. Are you not apprised, in your thinking, to the very clear fact that all of our gains here in Lacismoor have been accomplished during the calm that he struggles to maintain, and that the wiser of us wish to uphold while the brazen may break? Militate the fields-ters for a revolution in harvesting, instead of this havoc you may precipitate on us. Go, as this soreness from my wounding tries to mend; (*touching a letter*) and from some thought to be friends respectable, in passing and greeting.

BRACKEN (*exiting*): Then shall we have a chance, to review the splendors of the past and bring them forward, as employed with a new and fresh heartiness, in sincerity for the deed and demand of individualistic sovereignty, self-determination, and the serenity that this brings to one's conscious state of happiness upon the glory of— achievement, fighting or no. Yet we shall have the merit of it, when thought of.... an army to proclaim.

LORD BRACKEN (*after **Scott** has left, and while perusing the letters*): There is no past nor future. They are concepts only, useful for calculation and planning. Yet these are not physical. They do not exist, so obviously they may not be "visited." Reviewed, perhaps, but not brought forth nor back; what hardships wail to recall, the agonies do lead only to scars of the mind, and distasteful proclamations of more needed. Yet chides a Bracken, broken all to fury. Then be sound, to the disgraces warned of.... and amend me of my fears. For there is no terrible right in this day of excesses of influence and control over others. And there surely can be no justice.... in a cynical world. My own son would contemn my valor as a might, when it was only.... circumspection to exist, through an angry age. And what for this— correspondence that is not meant to be? Some.... hatred here of me, or blithe disparagement to toy with, in rallying around Musgonie for his approval and to show support

— Well.... that is.... his due, for this sad daring. I will absorb the error, for our families' sakes, and hold again the abscesses, of hardened heads, until some clearing 's permitted, if for this strength I still have the endurance.... with more than pain of parentage.

Act IV

Scene I — *A broad parlor at **Decamond's** mansion, very open and airy, with a doorway welcoming a garden and the sunlight. **Wallinda** is dusting off a sofa or long settee, very conscientiously, as **Ponstole** stands a distance away observing approvingly. The conversation between them is almost fraternal, rather than flirtatious or romantic, as if he had finally found a capacity of being to relate to with ease of abandon; though she tries to remain properly respectful of his station.*

PONSTOLE: You can have no idea of the wildness of whim that goes on at court, such a multitude of machination to manipulate the drafting of ideas, even their conception for the noble ear. And the craftsmanship that may be used by some is magnificent to hear at times, the very notions plotted at to persuade of a tempering. Or as a tuning of instruments are these opinions played. I will introduce you to these sounds as you accompany me.

WALLINDA: That is a fabulous idea, sir. But a lowly maid to walk with brings no great reflection to your shadow, as I could hardly mirror your grandeur of position.

PONSTOLE: Oh, there's nothing to it but a handsomeness to be with. And mine has been up to now only a stern consideration of their minds. For they know I have the king to meet with easily and regularly; and so I might transfer the suggestions of these noteworthy ones as directly as could themselves, if they seem more than reasonable to me, and possibly feasible for a benefit of country. But it is my office that gives these tales and tries more definition to his ear, than merely casual talk. He knows that I have pondered of, and maybe even studied the utility of these ideas for him. And so as a sieve I separate the poorly thought from the profoundly inspired for draw of his attention. Though I'm quite aware that Musgonie likes not my company in particular for these discussions between us. All of the councilors have this privy privilege. But as head of the council I must only draw action from the king, be busy about it if things need be done, and not have a friendliness that wastes of time. I represent a definite prod for his execution of responsibilities to his kingdom, a shoving and tussle— of words— that he can not often appreciate favorably, since his manner is to stay reluctant to many disagreeable affairs attending his action.

WALLINDA: You then take a toll of harming threat?

PONSTOLE: Nay! The king is very generous in his restraint of temper. And it's not of his nature to show any belligerence, but rather to hide it. Yet I must serve to bring out his disappoints for airing and proper description in review, so that we may both see how they may be soothed. He knows I will not condone a slovenliness to this most necessary retrospection of the upsetting, usually. So then I remind him of these problems with my presence. Yet for one's self alone to ponder of great tasks is to be deceived royally by the prejudices held or homed with. Then, with my occasion that tendency is checked strongly, as is his particular strength to allow me the more objective arguments to insist with. I could not honor myself otherwise, as a leader of advice to give, and would not assume the role if there were more danger to it than my honesty in grievances and surreptitious or disingenuous ploys to alert him of and argue through. Yet be these noblemen to try all tricks, they think to get away with any silliness as the king stays silent, on some such issue. Well I do not allow his silence to me. I command for his openness, when we meet, as he commands for my subject to be, his pardoning of the assault when met. Yet with you to walk of, through the court, our nobles will be no more fair(er), and as easily checked of their devious pitches to beauty as the king of my handsomeness to be real. They are not so much a higher sort, really, Wallinda. They have the clout of station and position, and underlings, to back for them their public musings. And by my side will not an underling you'll be, as they would have, but much as much a checker, Queen and King, with business of these affairs in tallying good sums for us to make of, our robes together in this way cut, for the squares give to.

WALLINDA: Bandy you with the Exchequers too?

PONSTOLE: We'll be consultative of many riches. I garner suggestions from many corners of opinion that wish to present them to me. But that is not in the main my preoccupation; but rather to analyze constructive trends for the country is my major role that places me and others above the ordinary noblemen. It's not only for ourselves that we think and act and even aspire of, some quality of righteousness profuse throughout the populations, and kept so with a proper administration of kingdom. It is that moralistic constraint of gown that keeps from spread the destructive notions of singularity of interests with a selfishness for primacy and the disreputable, antagonistic and divisive behaviors inherent of those tendencies of view. It's all fine to look after one's upkeep and properties and social connections and status. But the strong are more greedy than the weak, and could wreak the country back into conflict and bellicose attitude without an ethical tension, and ways of maintaining this which we attempt to provide the king.

WALLINDA: And you are so high principled as to be above your own aggrandizement for favors and privileges.... sir?

PONSTOLE: The maneuvering can be very tricky. I myself have been rather ascetic towards treats and profit making among colleagues and peers, the mostly unnecessary advantages people often rush for.... more for the glitter than the weight of light speckling and sparkling. But I am more than rich enough to fawn lavishly over personal affections, as they are as simple as the emotions would take them to be: love to feel and beauty to see, and so forth. That's all that anyone needs who is comfortable and capable with his work.

WALLINDA: I can agree to that, sir.

PONSTOLE: The greedy seek for more when it is attainable, yet only fatten themselves needlessly. There is no motive to the motive. The movement no longer moves— one.... and becomes superfluous. (**Decamond** *appears at the doorway leading to the garden, but is not noticed.*) Musgonie, as a personality, is the opposite of this. He seems to starve himself at times, through humilities that make him seem weak and with prayerful overattention to rectitude with reticence. But.... this king is far from powerless, and a flick of his finger can tumble mountain tops. I'll tell you willingly, Wallinda. He holds back angers magnificently, and dams these waters for a later use of their forceful current of energies. And he realizes that I above all feel this characteristic in him, which is why he

allows me to be privy head and trusted.

WALLINDA: He is simply a man of great reserve, then, without the need to demonstrate this.... to others.

PONSTOLE: Yes. As a ruler is, he must be irate at many things bothering, that no man on earth can control. They are called tendencies or fancies, that affront his kingdom. But how can one fight at a passion? since all passions are as deserved as shadows to a man. Then you must strike at the causes of these, which means to identify them and study.... for a while, until they're mastered enough of your knowledge to bring upon them judgment: to allow continuance, imperil, weaken, or destroy. Musgonie must spend most of his waking hours, and many while asleep, on these heady assessments grappling for his sincerities. It is a terrible crisis of conscience, to have the responsibility of all lives of a nation, dwelling for your own demeanor's make. And oft times, I can argue vigorously with him, to spare the impugnations of a supposed enemy of his intentions, and to rebuke heatedly the malefaction of a friend when I suspect clandestine illegalities or assertions are behind all (of their) suggestions made. Oh! Spew forth these errors to the light of day! Not all contentions are meant to malign, but some may inform too amply for defects seen. And not all amiabilities may plead to you worthily, but some quite hypocritically, with shame for crime or crassness of conceit.

WALLINDA: So you protect the king from his slight perceptions? (*finally turning to face* **Ponstole**, *though not seeing* **Decamond**)

PONSTOLE: How wrought to protect, as I might you? We are our own beasts of many eyes. And as a royalty may flow as evident, then is a fineness made that shields with this distinction, The king is the arbiter of all destinies.... in this way. I can only advise, to tell him his handsomeness is true— or cross. Yet see me regal (ly) flow of the divine as made. And I will never be a king as he, but hold divinity to cherish— evermore. This pledge.... I've made — and I make, to honor you—

DECAMOND (*entering*): Are you going to stay around here all day? Don't you have official chores—?

PONSTOLE: I have not *been* here all day, Decamond. There is work be done, that I attend to constantly. Yet to my caring, I have much attention for Wallinda— and dare say so. Open ears. Open minds. Blast them thoroughly to gasp at my involvement with your house. And she is no maid to me, of ever form that angels bring.

DECAMOND: And thus a competitor?! (*going over to where* **Wallinda** *has dusted*) I'd make my adulations known with less temerity to me. For what in your nature could challenge my overtness?

PONSTOLE: I've clawed through ruses before. Yet I serve you boldly, Decamond, as a direct and close representative of Musgonie. And candidly I tell you, he prepares for you some good service. For I *have* been working. And I have in my possession (*patting his chest*) a letter squarely for your interest to read.

DECAMOND (*with hardly to be disconcerted, placing his foot on the sofa's cushion, to adjust a shoe fitting*): It's not my ecstasy to read the correspondence of the king's private postage. Nor need I the privilege to have this couriered directly to my mansion.

PONSTOLE: It is not to the king, but to me.... and the privy councilors. And so I allow it, for you to read and consider with, that it does make Musgonie in use of you to.... trend (towards) or recommend. I've no doubt he will issue a request soon. It concerns Lacismoor.

DECAMOND: Lacismoor?! (*taking his foot off the cushion and brushing it clean*) They do for some distemper there?

PONSTOLE: More than that is evident, but calculated planning, through a heated dispersal of antipathies to subjects ruled. (*taking out the letter from a chest billfold*) But read for yourself how this is crafted boldly— and signature-d, even.

DECAMOND (*turning around*): Come bring.

WALLINDA (*as **Ponstole** comes towards **Decamond**, and she towards **Ponstole**, for the letter to transfer*): It would mean war, for a spirit of rebellion there.

PONSTOLE (*with courtesy and some thought, giving the letter to **Wallinda***): Definitely a realignment of intentions is due. Musgonie is always ever against the need for warring conflicts. But that is not overtly suggested in this brief, merely nuanced of the possibility.... I direct it to your attention for a seriousness of conception demonstrated quite in the open, now.

DECAMOND (*receiving the letter*): You mean there's an actual declaration here? I'd be burning ma coals by now, for some heat on this matter to bring, were I fully physical. (*starting to read*) Would be baking bones.... for the meal.

PONSTOLE: There's no proposed secession declared, but the boldness strains its credibility of restraint. Now here is a chance for you to test and determine your diplomatic prowess. I am certain the king will wish to have you descend on Lacismoor, with arguments mitigating, or at least mediating, these requests of more independent mercantilism and trade.... He even suggests that Musgonie does not understand a decent notion to the fairness of deserved autonomies, those built of personal labors utterly removed from the court's conscious awareness or interest.... Regimes form of themselves when themselves are left to prosper of their own efforts. Yet I hesitate to call this the making of but only a preliminary threat or warning, a very signaling for ministers.... to come and deliver on climates of agreement.... between province and country. He does not reject being a subject to the king's rule, yet less a servant of is he sufficient to carry his head.

WALLINDA: With broad shoulders, no doubt, to think to make entente with a man of such reserve.

PONSTOLE: The humility presented only suspiciously maligns the king as fearing (for him) not to be as reasonably humble. It is an attempt to instigate more than discussion, but veils a prelude describing possible actions he hopes the king will not take too much exception to, such as thoroughly unauthorized commerce with some foreign parties, to which Musgonie— and our country— will have no involvement in, nor recompense by.... One of several independent activities outlined that he hopes to.... presume; and that which rankles me in particular upon reading is the establishment of a monetary school, for the education of their merchants. Though whether he means an edifice of learning or a philosophy in man-

agement of coinage remains vague, too many definitions assumed for the reader. But that is an important lesion of economy, an important primary step towards separation of.... economies. They appear to feel too rich in Lacismoor, to treat a kingly abidance easily or wisely. My council member Costile is already headed down there, to assume the environment of hostilities for us read to grow, and to make preparations and contracts for an envoy, Decamond, so that some polishing is done before a landing of commentary and reaction. This is the courtly way.

WALLINDA: The king will not rush troops there, to re-seize absolute loyalties. He's not the punisher that teaches bluntly and effectively the limits held.

PONSTOLE: But this is no announcement, Wallinda, and deserves the courtesy of restraint. Not a law of the kingdom has been broken, as far as I can sense, through this yearning listed for our eyes with calculated testiness.

DECAMOND: ....John Bracken. I've known of him, minor-ly. A fair reputation for fighting. But I believe the squire Darughe resides in Lacismoor now, and is much more renown to the soldering aspects.... No standing army noted. It's unlikely either such man would be so foolishly disturbing unless there were some intense pressures growing among the worthy population for an agreed upon sedition. Yet in such a case, the very first thing they would promote for notice is a standing army. And with that is my experience, to hold them accountable for my visiting, with enforcing retainers of country, to justify the action and the eventual payment for our services of binding endearment they'll come to seek. So this is unusual a trick to cast of die and gamble with forewarning— us.... before they have assembled manly strengths, rages, and outrageous boasts to dare my company. Perhaps this Bracken works to fashion some form of scutage for his son, and other sons of the region. They are so proud now, in their wealth, that they'd try any sort of fallacy as fantasy to play at or indulge amusingly. Yet to the privy councilors is sent.... an urgency of intentions I can not easily understand, unless they invite reprimand and denunciation. For such a man as Bracken would not take this fairly public route knowing me and mine and the court commentary which must follow, without much more backing of power than described in this letter. He would not give away the muted moot desires already known.... with so detailed a strategy to deploy them. A soldier never tells how.... to a potential opponent. Perhaps Lacismoor has friends, in that region. Yet I fear not this Bracken muses idly of a ramble under such delusional support with such a nicely ordered and scripted letter.

PONSTOLE: The writing is a copy, by secretaries. But the original is just as finely penned. This is no scatterbrained attempt to draw our attentions, wildly drafted during a moment's dazing of wishful thinking. It is a howling cry, to alert us into bargaining stances— obviously to propose to the king.

DECAMOND: Aye, Ponstole. But it is awkwardly placed and played, these details. Almost like an arrogance of knowledge to compel you with, astonish you into trembling reverence and discerning of this trivial mastery— It's almost.... inexperienced of matured men's minds, is how.... one heaves a chest, without a weapon protecting. And a Bracken.... Have you considered that it might be facetious and without much merit? Maybe even a conspiratorial fake from a competitor in the province, knowing the country's concerns.... for her.

PONSTOLE: That's why Musgonie sends Costile off so hurriedly, to learn of these circumstances. My intuition, though, places credence to these issues plowed through. They are too reliably mentioned to deny their thrusts of argument. Musgonie fears a despotism may be evolving there. *We* have not discussed this fully yet, among us councilors, though rumors will probably spread. But to my notion, this John Bracken remains a faithful subject of his majesty, until shown otherwise. And this is why *your* assistance will be called on, I'm sure of, to deal with him as you are, a true man of backing, and with the king's trust and character, to be reasonable and sparing during deliberations of understanding. I suppose only soldiers know soldiers best. And while Musgonie was one to officiate, well.... you are hearted to make the necessary compliances, with more of reputation to know of.

DECAMOND: Yes. Motive, manner— and men! This— lame arm twitches yet.... Now, there are certainly knights down there, of (the) steady and in training.

PONSTOLE: Most assuredly.

DECAMOND: And this province does not refuse its number sent for the king's service.

PONSTOLE: Thus far, no. It does satisfy its allegiance to the crown. It is, in fact, exemplary of riches we receive throughout the country-land.

DECAMOND: So no offenses have been made at Lacismoor, no animosities perceivable as yet.

PONSTOLE: None but what could be due to the headiness of wealth with rumor, no substantial adverse conduct. Yet only pleas like this, now signature-d, result of the king's concerns. They have been rampant for some time, as you know. But now to relate an individual to them is either a broad gain or a bold move. One can not be too sure of advantages here. (*as **Decamond** offers back the letter*) Keep it! There are several copies. Even Costile travels with one, so there can be no doubts of accusation.

DECAMOND (*letting the letter fall to the floor, although this might be mistaken for a clumsiness due to his physical imposition*): I need it not! (***Wallinda** goes to retrieve the letter.*) Improvements, councilor. These pleas are for improvements. They have it by the tail to steer, commands for some attention to receive for their importance to.... our country. But whether it be fox or wolf is a question biting back, turning round to sneer, growl, and be angered. (*as **Wallinda** picks up the letter*) To my papers summon— rest, with storage among. She becomes dear to me, with a particular fixation on the candors of gentlemen. Tell Musgonie, should he ask, that as Lacismoor waits, so do I, (and) of a preparedness for service. I am healthy, and refuse to sling. But lets the arm hang, does he, and swing of motion with good thoughts for the king and his kingdom, while at home awaiting his pleasure and soaking his muscles in leisure and free-flowing sunbeams for some warmth. I know you must have cause to answer him quite readily.

PONSTOLE: He hasn't asked me, yet.

DECAMOND: But I have given you.... the reply. And the king may count on it, with your immediacy to transmit.

PONSTOLE: ....I....I— I wish to take Wallinda with me, this afternoon—

DECAMOND: She has much work to do.

PONSTOLE: —to show her round the court, and as my proud favorite to stroll with.

DECAMOND: Sir, be your job.... and advise the—

PONSTOLE: I beg you!— of an honesty to please—

DECAMOND: She is of her own high calibre, and may take leave of you when she finds the time— and wish to. But do not dare to propose yourself as a rival to my class of caring. For independence of mind is a privilege won from virtue.... and wisdom parenting.

PONSTOLE: ....Then I'll insist to be so raised, sir, and assure you of a profit from her own initiatives. And I will come by again today, Wallinda, and help you clean this whole house (if necessary)— to free us for a walk together, and a social inauguration of casual friendliness through very high estates, to make this clear.... that I am partnered—

DECAMOND: You are parting.

PONSTOLE: —now. (*turning to leave*) And good day to you, duke. (*stopping*) For I'm as much your servant as I am the king's. Trust me so. (*exiting*) I will be proficient for all our benefits. But I clamor fiercely for our own.

DECAMOND (*after he has left, through the house*): Now, what of he, Wallinda? 'Tis fiery stone.

WALLINDA: Of him, I am polite to his exigencies for expression, my lord. And he does make much matter of which to tell me, his.... observations and opinions on life. And not basely does he produce his concerns or confessions to be worried. I am flattered with his courtly interest in a maid, but can not conduct myself as equally, unless more awkwardness subdue us both back into childishness.

DECAMOND: Oh, he is mad, to return there.

WALLINDA: I can't dislike him.... easily. His conduct and deportment make nature seem more natural than haughty of its charms. And he is fair with me, to believe in, that the affairs of state are like many others.

DECAMOND: Yes, much akin to greed and jealousy. I've not either. I am lame on this.... pursuing.

WALLINDA: You have him tied, sir?

DECAMOND: I have no anchor needing, nor for ballast in Musgonie's ship of state, through rocky storms a'sailing. He's not tied to *me*, but does ponder his own entwining pines of heart, to long for you, as to make an association compelling for him— but hardly complementary. Though he is definitely well-off, his condition fairly high, and his own lands spacious, I've learned. Proper, yes, and prompt for making strategies. He may teach you how to.... satisfy each other's hearts. But I fear he will air you out with complaints, to tire your ear from its body's wholesomeness. That may be.... a decent thing to do— for your security to future advancement and attaining some station and acquaintances to safe prosperity, perhaps more than my own to ever be willed. Or nay or say, that is my drilling to contemplate on.... suitabilities, for outstanding measures of affection. But this is not romance he strums, with his crossed feet perplexing of the dance, but an adulation, that does break of mirrors leaving, his face on yours to lose, in these reflections sensed of a sympathy for him— I've seen romance.... and it is born of two becoming one intensely of submittal. Not this tempered approach, for conversation's sake, to bathe of his ideas.... makes the hymeneal decree proportionate of a fulfillment, purporting this is so with deeds of law.

WALLINDA: I will not make tease on the matter, lord, but grow to its reality one way or another. The man does not displease me, yet, and is not disagreeable of manner nor fashion. But as for a physical attractiveness, I leave myself for error wooing. It is so transitory, these days, to home for comforts based on that, that indigence may more be praised.... for its soliloquies on the gathering storms of nakedness— and raw truth.... of one's condition. Fortunes are made of minds with complacencies.

DECAMOND: Is that so!— And stash that with my other papers, since it's like a gem of purring to be read. Such demarcations of one's boundaries of quandary are rare. Is why they take such care to make copies. (*starting to head out again into the garden*)

WALLINDA: Yes, my lord.

DECAMOND: And if that fellow comes by yet still, today— What for suitors! I should not see him, as a pestering of dignified mice, for injury of sight and smell of what they carry, the rudiments of civilized posh adhesion.... to this courtship— (*stopping at the doorway*) Let him not!... invite you to his home, but restrict yourself to the king's grounds, of public gatherings.

WALLINDA: I can not dictate for him his wishes for me to do.

DECAMOND: But then, as you ladies may, (with) a sigh or a wink or a blush or a fawned discerning grimace— smite.... such invitation making. A cough of uneasiness is as powerful as a punch to the face, that he might in any way displease you or offend.... your control over his wishes. He is not the craven brute to allow for you any such discomfort's perception.... with himself, but only wishes to parade of a treasure found, and dwell of its cravings imperceptible but by gods and their charmed victims. For how else can one explain such a desire to be devastated except through the influence of a madness leapt off the immortal and onto weaker flesh and into meaner minds.... that feel deserving of its capture. But this does capture you like rain, suddenly (found) unprotected from, by which you are wet— and wed to be victimized, softened for a shivering of showers, and hardened to the catastrophe of your happy plight.

WALLINDA: Shall I help you with your shoulder sling later, my lord?

DECAMOND: ....I wish to try without it for much longer, Wallinda, and let this arm fall to find its grip of me.... its purpose.... without delusion. Already I am sensing more— nerve.... to need less dependency on slings, disfigurement of limbs, and an unnatural presentation of body.... to others. I am not greedy.... I am not jealous— And I will not be shamed of a crippling, shown for the broken legs of a squashed—.... or tease-tested insect drying out in the heavy sunlight. No. I will ingest this calamity of madness, and grow more fortunate of fat. (*pointing at her, with his good arm*) Though I remain self-conscious, dear!... (*lowering*) And you may assist me, as usual. Continue to, while hearing facts of court, and learning truths too innocently blurted out of a starved head too full of the king's raiments.... or arraignments and arrangings.... to contain more of these concerns with decorous prudence.... for his severe appearance. And I will not cheat my matter nor my math, with these curtailing engagements and ties. For we do, in the end, submit ourselves to the king.... in our endeavoring to please.... Show me him not, when he arrives, reminding me of passions hurled.... and caught. This man returns my face, for a grimness facing, through his queer gaiety to abandon his reason for more reason, and a reasonableness to let of himself admit.... to these circumstances through which we shudder an' shake, and must resign ourselves with.... I think I will plant for more blossoms. I want them thicker, much more dense and rich than what I see.... out in my garden, today. (*Exits.*)

WALLINDA: That must be so, to caretake of one's— formalities of appearance.... and formidability of person. Yet am I pleased of some attentions brought of gentleman, as I have durst.... to do. (*leaves through the house, with the letter*)

Scene II — *The open storefront to* **Skacte's** *shop of furnishings, very busy of sundry articles sitting, hanging, lying, and in some areas haphazardly arrayed while in others apparently flying off string, as by banners and ribbons taking of the breeze.* **Fennel** *sits on an above average height stool, relaxing with a grin, as* **Alfonso** *is crouched some distance away, cleaning with wet rag and soapy bowl an apparatus of wood he's found at a local dump, the function of which he can not recognize.*

FENNEL: And these I've seen aplenty. An outstanding bargain for Skacte, though he should have tried for a pair. But only this here has the worm hole, for discarding. But it's as steady as I can seat it. Good job patching. Who'd know it was ever cracked.

ALFONSO (*preoccupied*): You could probably fall through it, with a bit of bouncing.

FENNEL: No, he fixes them well.... Leans slightly, though, to one of the legs. Not unstable at all. Fit for buying is this chair, for the raised deck of a concessionaire of drinks and quickened luncheonettes— Gorgeous platters of foods, I've seen, Alfonso, the most colorful and geometrically styled of imaginative patterns the eye can come across. I do provide some beverages to some of those places, and am often dazzled by the eatery displays.

ALFONSO: It's done to stimulate the appetite, of course. But I did take one dish of the most handsome seafood, once, with reds as rich as rubies, and green leaf rolling of a thunder as (if) blue sea waves. For that is what I was reminded of, to buy. Yet, as delicious as it was, I was not so hungry for it in my hovel of a room, and even had to let some of that good stuff (go) to waste. The impulses of the senses tricked me, to abandon my needs for my wants. And I did not need that dining. I hate an unnecessary squandering of resources. Yet I seem to do, at many turns of life. Then it's very stimulating of good anticipations, to come across something quite unusual, and know that I can not possibly squander my efforts nor my money on it, if only a curiosity is satisfied for its improvement.

FENNEL: But you have to sell these things, you know. Can't make much sense otherwise, the toils expended.

ALFONSO: I've expended so much toil uselessly, the notion to it flatters a dry coinage's stationary rest in a chest or storage box— But oh no! to the purse. There is no value to such things with frivolous feelings and senses.

FENNEL: You're too skillful to be that dumb too often. Talent is for tally of your gains, not your wonts. Work itself, though, has its own merit in doing. Whether for the foul or for the good, it often drives itself to find a useful purpose.

ALFONSO: Yes. Among criminals and priests, what matter is your devotion without effort! Then how can I convince others, of my worth, without such deeds portraying? I don't feel too humiliated by this.... wet laboring, as long as the results interest me.

FENNEL: Oh, you've impressed quite a few already, to a sort of ability that may promote to leadership. It's always a display of enthusiasm that counts for initial notice, since all leaders have that, while not all of the enthusiastic may lead. But as is often preached to kids on who to follow for example and instruction: Ye who has superb talent in one field tends to have above average talent in several others.

ALFONSO: That must be due to a climat(e-mak)ing of the soul to succeed in tasks, with expectations of achievement becoming automatic. Yet, there are many worthless types who are self-assured, and are proud of the surety of one aspect or the other, no ties to any responsibilities and a complacency for their condition in all of nature. It is hard to get them to revolt against themselves, or improve. And I have been schooled to avoid being so. I am seldom satisfied of anything, which induces me to try many things, even if lowly or common.... for awhile.... Not that I keep any particular store to learning what is new or unknown or unusual to contemplate. This— learning.... comes without a need of advertisement and promotion, or proselyting for, if it is meant to be and good for your customary advancement. I am flexible enough in this way to gain some knowledge without arrogance through the attempt, nor by the attempt to become too proud.

FENNEL: Or by the attempt allowed.... these temptations. And have you ever slain men?

ALFONSO: ....With men is where of must to be slain or slay? I've not had battle yet.

FENNEL: Then it's a good service (to me and others), not to boast of what you have not done.

ALFONSO: I have not killed a man, and would not claim to have. I've eaten animals, and in that way have killed. For there's no way around it but to cause that death, one way or another, for your consumptions. Just as there can be no kind killing of anything, the very act is for its purpose cruel.

FENNEL: And what lives on cruelty— Life?!

ALFONSO: And its fruiting. I've been prosperous observing trees with falling limbs and rotting roots to know there is more cruelty to noble life than its offspring-ing.

FENNEL: Ah. That is an aging, of the seasons seized by and jostled with. We are made for those.... devolving transitions to endure. Yet I would say, in the end, this cruelty is the simplest to obtain, and so it must be quite pardonable. After a few dozen to your hands, it is only swains' service serving. Come what? (***Calerand*** *approaches, with friendly reassurance of familiarity.*) Good day to here is Calerand to see.

CALERAND: As so is Fennel.... and Alfonso. The dwelling's well, ma boy?

ALFONSO: It's rich, but yet a lodging mate to make.

CALERAND: They are good rooms, in sound places. (*more to* ***Fennel***) And I need a pot.

FENNEL: To boil with? Skacte should have plenty. What kind of earthenware?

CALERAND: Just a pretty pot. Something to decorate with.

FENNEL: For planting?... in your living quarters?

CALERAND: A small one, of attraction. I woke up this morning, and felt to have it, noticing that I did not have one— just of that sort. And to my time all pleasures rush to be fulfilled or forfeited, during these leisures of speculation.

FENNEL: That's an odd need, for a trifle of beauty.

CALERAND: I grow sensitive to any lapses prone to. Yet have I gone for ages without one, as you seem fairly high. Or for a cheap necessity is no necessity at all, but an indulgence. But the erring leads me to some slight agitation, to miss what I have not had, and to redress this with a fright. For the first time may be the last time. Hear you not of the rumbling in Lacismoor? Rumor says even the king is upset, to wear impatience for her. So then I note my lapses upon awakening. Yet sense I strongly a gathering of forces is imminent, and the duke will call us out, to take helm of our ship of staves and saviors, and quash a very uprising, or veritable becoming. It's what these passions of pastime are made for to stir, that I may not miss seeing this pot.

FENNEL: I've heard of some actual contention, but nothing to deliberate on, for a mass to send. There are no declarations heralded or read to rest for our promptness to this province. Speed does make with more than method sensed, but metered an' measured. And responses to all of my inquiries have been vague, though less dependent on Decamond.

CALERAND: The airs, the winds of fortune are not expecting us? That is foolish thought. I have assumed without much questioning. For this seems similar to all other (of our) events thus far.

FENNEL: And I would wish it so. But the duke does not call me much. And when he does, no hint of Lacismoor does he plead. And when I asked him directly about the matter, he says he heals. So then is he reticent to make demands for; I think he awaits a firm disobedience by them, or pronounced ostracism by Musgonie for some attempt at self-governance, relieving the king ostensibly from his protections against bandits and rogues.

CALERAND: That may only be implied by reason of his disquieting. And wise as the duke is, we are preparing for some considerable military adventuring, as by condition to assume of us. Already there is some camping made, in the forests, to toughen skin. Though I resist this for myself, until more definiteness arrives, of a slow calling to ease my port, this body back into the tensioning required. And not portly become, for pot to see— I hope, with time this need.... This symbol— to get ready.

FENNEL: That sounds too premature to me, the outside shearing of the fleece, grown golden wool and fat. But we must consider Decamond's injuries for his sloth in this— preparing. He is more slow, than I have ever seen of wound to resolve itself, these months have taken.... care. I'll not presume of him his cares, this aging deity, and our dynasty for.

CALERAND: It's to his advantage, of course, to be certain of his physical health as well as the details of secessionist tendencies. The indications have been as long stretched out as his recovery seems. The occurrence of the two at the same time may make the probabilities lessen of actionable revolution. But I doubt this, the confluence being only more of a complication or complexity to peer through. And Lacismoor reeks of wealth. This is too much intention to let go of, anticipation to not carry through towards a ride with— on a magnificent mare of silk and flaxen mane begging for proof of its conquering and steerage with its proud nature and haunch to be lowered to a royal accommodation of interests.... And for a country's welfare indeed! this is a patriotic gesture we apply, pressure to stop the bleeding of a gaping wound of wonder, that it could be allowed.... And I've been so terribly bored lately, and yet fearful for myself. Facilities are dwindling, as the opportunities wane. And it's best to make the best of your standing as you may. Yet kick with strength and vehemence against the eventual smite. I'd rather be out there with death! than here with a piece of it to carry.... as it grows and grows more careless of me, lackadaisical, lazy, dull and dulling. Be with excitement for your final voyage, as it should be your finest.

FENNEL: You wish it?! out of comfort and richness of living?

CALERAND: If you need that sort of retirement from your work, then your work has not been worth your while and wiles. Only your work should retire you.

FENNEL: Unless you tire of it.... graciously. But you can't claim Decamond to have become unworthy of this, with his delay in.... alerting us to formally prepare for a province. And you know he is responsible enough to do so, to give us time to do so correctly and carefully. We have businesses and occupations—and families to look after and tend of. He has always been most generous and intelligent to this aspect for our services to arrange, sometimes even subsidizing a few of us, as his urgency for action mounts. Our credit is often from his crediting. If not with money, then with word of mouth, as can Alfonso attest a membership, does he lend. And with a craft to it is his purpose. So, when he does not call for us to assemble, as Lacismoor brays ever harshly, here must be some thought to this— which we can only induct of caution. Perhaps he has in fact learned that the king reverses his policy of peace and will send soldiers there himself. That! is the rumor deduced, that I have heard most often these last days. We can not afford to fight the royal troops off of our advantages— That would be treason! So then does Decamond protect us from a buffoonery of blunder, to hold us off. But then might he try to broker our

fighting alongside the royal, all profiteering must be weighed daily, even hourly, to inquire and question how much the king can make up to us for our services against what we will lose not being on our own. That, I suggest, is what our duke worries of, whether this adventuring *can* be made in any way profitable under such circumstances. For it is a point of honor with him, that all of his surviving men are suitably *enriched* by the experience. It is his sacred gesture for our trust, and more important than any steed chasing. The duke is above all calculative of his aims, and will not rush into foolish wear for any of his dears.

CALERAND: I've only gained the sense.... of Musgonie's pertinence in this, and that his higher goal remains solidity of country, and solvency of kingdom. To fight *with* the king's soldiers would defeat our motive of operation, Fennel, and eliminate our purpose of organization completely. That can hardly make sense and has never happened for any such union of fighters to remain so. Now, to have the king subversively employ us, singly sent, is more useful of a speculation, since it would avoid breaking his public vow to avoid fractious warring at all reasonable cost within his professed dominions, that he himself would initiate. For he wishes his name to remain of friendly ties with Lacismoor, assuming they are not upstart enough to force his handling overtly and openly. An agreement with Decamond to allow our.... mitigation of the boil— can be very complicated to arrange effectively, should the king finally acknowledge to himself our usefulness in these situations, as long as his involvement towards this remains at most muted, and in some way lawful.

ALFONSO (*feigning preoccupation*): To suggest or to direct, you mean. What is the price of either? Is it possible the king can fall into such deviousness with his subjects? He'd say, "Good Decamond, if you're going to do these heinous crimes, I'd rather they be in Lacismoor today." Thus, he suggests. And for a price the duke accepts, without being officially directed. Yet, this seems to me to be— off the straightforward path of heroes.

CALERAND: Oh, never say you are a hero, young son. A savior, perhaps, to the king.... but not a hero, nor a gallant. I step and stamp on gallants— They are stupid. And I've done this much, to force out more quickly the freedom of their departing and long suffering souls, dismayed of such silly bodies and accounting temperaments. You are candidly for make of the splendors of an income with these activities— And never forget this, the rewards for what results not justified else.... And we are not the mercenary sent by a king. That is forbidden of internal conflicts in this country. It is unlawful, and not the preference of his kingdom, thus far as he may hold water in his hands— without washing his face of blistering contentions. Our pride of opportunity brings us to these territories, to dwell of a service made for to perform— And that is meanly (put) all to it. We hack at fools!... growing unruly and wild.

ALFONSO: There are either legal or illegal hearts—

FENNEL: The debate's less precious than the penance we provide. I say the king conspires with Decamond's art, though they must hate each other. And I'm not totally uncertain this has happened before, for the seams of a cask can get leaky with the introduction of a tiny taint of oil into the fluid cargo. The duke's been sharp, sharpened.... and not so sharp of his management as late, as late our last battling. Then does decline some fortitude, as pus to be removed. (*eyes **Maeli** approaching*) Here comes a count, to Skacte's.

MAELI: Greetings, capable. And do you sit some high, Fennel. Portentous? for some purpose brushed aside? Or merely pretentious.

FENNEL: Want you for a bar to work, Count Maeli?

MAELI: You play upon that stool like a child. And I see you all gathering of sore wounds.

CALERAND: Wounds, good lord?

MAELI: But it certainly does disaffect me as well, to see such uselessness in good soldiers. And I might have cause of enlisting from Decamond his fairer talents, now.

CALERAND: Ah. This is only inactivity. But we ply our trades in other ways to avoid being useless.

MAELI: It's so, with capabilities. Yet more than sore you must feel— to sit so long there, idle and aloof but for the stinging gossip.

CALERAND: Where comes of Lacismoor, perhaps.

MAELI: Exactly of bee.

CALERAND: We have been discussing the matter and the eventualities so evident to everyone.

MAELI: Where (thus) comes from this astonishment the resignation of tired hats, the pilgrimage of pillaging dis-hastened. It is a heavy corn to walk with, or callus painfully impress(ion-)ed to the foot.... and head; for wear we both our shoes and chapeaux to cover bare and bold reckonings.

FENNEL: Of hair and toenails?

MAELI: Of fate, simple sir, for where we walk and think to be, one's condition and state. Try this outrageously, for the profits to be made in Lacismoor— absconded with.

CALERAND: Not if we get our hands on them.

MAELI: With your feet? these feet— docked to country? More more than mooring of your versatilities— soldiers.... have you to wait upon than Lacismoor to visit. It's fierce, the disconcerting you must embrace— But no cause of mine.... yet. I am not that adventurous against the king.

FENNEL: Neither are we against the king.

MAELI: His wishes (are) employed and amplified tremendously, without your cold treading.

CALERAND: Are you possibly claiming this province to be too warm of blood?— for us?! That has never been.

MAELI: Here cools the heel-held hound, not with?! It is depicted so, your sorry tempers wrangling of wretched boredoms and disuse.

FENNEL: As the fair Duke Decamond heals? But it is any day now, the stronger the license of Lacismoor becomes.

MAELI: And yet to head again, my doves. I'd choose upon you willingly amongst. But does my friend Decamond tops himself with Musgonie, it is a remarkable business of achievement that must sadden thee. But it is calculated to be fit and fit. *There* is a proud and capable cranium, for this turn about.

CALERAND: We will be ready, sir, to roust the rascals into subservience— as he will be, to call on us to assist.

MAELI: ....Assist?

CALERAND: With our usual flaunting of expressiveness in details voracious to our pleasures for a containment of the military purpose. And as to show of ourselves, a persistence in this task, we assist.

MAELI: Need he your words, your thoughts, for a diplomacy? I think in no way this, but to frighten with a backing. But his backbone is enough, for this prestige. The entire kingdom is aware of Decamond's.... noble deeds. And now they will be made.... as if to codify permit of his persuasions. And.... perhaps.... more than a few crowds will cheer the reconciliation of intentions for the common goal of prospering our country. The comment I've heard, thus far, is already mainly favorable. But that does leave your worth cold of cash, disbanded as you are. That's to say, valuable— but unspent and not expendable under these conditions.

FENNEL: We'll make that up— But did you say.... diplomatically?

MAELI: Why else round? or round about your sorrows conversing and confessing sins, with amiable intentions, perhaps describing feats not to be repeated under Decamond.

CALERAND: A body of soldiers does not diplomacy seek, since battle rears decisions more directly and weaponry makes for argument with exchange of points and cries, cowardice and courage, strengths and exhaustions, defeat and victory.

MAELI: You sound uninformed, of meritorious environment, that a calming down of tensions is proposed for Lacismoor, as by the king's long held aegis of tendency.

FENNEL: Decamond will deaden it.

MAELI: As by must, his will to be discussed. He tends for Musgonie. And you are disgusted.... of your beheading? But then I would sob as well— awhile. Yet, what can cry without a head? Then be you sultrily cooled, and a body without much formation, dispersed through this city, and others, as seeds of dissolution, dismay, and disinheritance.

FENNEL: What tends to be pretended for?— We fight for Musgonie?

MAELI: Why to fight, with Decamond avowed? But do you not sift through the salty airs to eye.... the idle talk that deems your service— idled, perhaps even stigmatized and styed to a memory. It certainly spreads about my circles to discuss, and observe this clever, calculating change.

FENNEL: What means is thus to tend!— That we are called?

MAELI: Hear you not of Decamond already? Oh! Then you hear only what you wish. But he's become an envoy to the king, to help calmly settle the disputes in Lacismoor, and to tone down temper (ament)s with arguments of reason.

CALERAND: Decamond does take the word of Musgonie.... there— and not us?!

MAELI: As a royal representative, a voice of the king— with guarantees to make, and offers to bring.

CALERAND: Blasphemous! And not to be believed!

MAELI: Why, I've shortly visited my friend, to confirm this for myself. And it is so. His profit making's now for a well deserved good name and respectability for high civility and chivalrous appeasement.

FENNEL; He doth heal.... of sore wounds.... in this transition. I had discounted the thought.... But you verify— for us.... a patriotism over parentage. Does seem a spent lintel like, to passage towards such destinations. And do we remain to tell tales, of this and that?... what would not be said to us. So squirms of discontentedness, that we alike are here.... stilled.... sanguinarily congruent, and yet of sanguine cheer for our.... pretentious bidding on review, as the day changes its challenges.

MAELI: So well. I knew you knew this.... ethereally, by the hues of faint gossip that surrounds, to draw you to this shop of furnishings to collect at.... and muse over these developments. And I would have some of you serve me, as faithfully of your abilities, since there are no doubts to the tales you tell yourselves. But there is less ransacking in my deployments, much less usually of such pirating, but for the more regular goals of enforcing laws and privileges and properties. Yet, some profit is the excuse for exercising one's talents. And so I would offer the benefit of a well established company kept in the ready, though exclusive to a leanness of efficiency— and therefore especial as being chosen and selected. And I would admit you, Fennel, and.... Skacte, and perhaps a few others, to this membership under my administration, should you ever again find the time for such activities.

FENNEL: We do have our off-employments, which may be generous enough.

MAELI: Though Decamond did mention to me, that if he is successful in his missioning, he may work to establish provincial forts, under the king's guidance and approval. Manned by locals or his own, or a mixture of both, I don't know. But if he chooses some of his own, that would off your off-employments, I would think, as a resettlement's necessary to a commitment of relocation for more stationary tenets of usage. A minister he is not quite— That requires courtly approval and functional review. So I'm not certain of his authority to found outposts. I wouldn't count on it yet as more than a prosperous thinking for the best, though that is a beautiful area of country.

FENNEL: Me home's me hide, sir.... I should await the duke's courting call— We all should. And I will take your generosity under the strictest consideration and advisement.

MAELI: That's, then, a wisdom. For you should always await your master's explanations. Though I could have thought it made by now, or intimated at least, with intuitive interests. (*looking towards* **Alfonso**, *who countenances a perplexed disinterest*) Good day to this gathering collective, of sorrows and assumptions that may prove to be your surety, as no association lasts forever; and many evolve unrecognizably of its originators.... much sooner. But do try to stay alert to your professional situations. Else be blinded by the happenstance of august natures and storms of change. (*starts to leave*)

FENNEL: Good day, Count Maeli.... I'll discuss this with Skacte. (**Maeli** *is gone.*)

CALERAND: Fish fiddles! Fins!— Finis! That is the anguished anger of a sadness!

FENNEL (*coming off the stool*): I feel pounded, Calerand.... by a profit to be made without our bidding score. And scratched of riches is this fame produced, provided for. Our Decamond.... is suspicious of us— That's why he takes this turn. For his wounding augurs resentments, and loss of trust. He would not tell me at all this news, though I am low!— (*with almost vicious interjection to* **Alfonso**) Why do you soap that thing?! What kind of wood is this for soap!

ALFONSO (*low-keyed; struggling to find an appropriate composure*): It's to be cleaned.... of a trash's conjecturing, a fine finish dulled and dimmed. Yet it stood out to me, as something deliberately made.

CALERAND: Would not even look at my age, to offer on a position. I am as much done, if Decamond decries with his silence, there to do of speculating worth.... and pining away. I am late of planting.... late of pot, late of posture— late of pretension.... for this service yearned.

FENNEL: I fill fit to kill! from disaffection. Yet make we more, of this— development. I will confront him personally, and have him curse at my face! And to it— No! For not the man the same to take dispatch— to him! I am lowered of my rights, this way, not to have been told!... And to leave us all hanging, with such expectations as to Lacismoor be brought. The devil be a ruler of men and obsequious beasts!— So swiftly make another, foaming of his foolish hands, Alfonso!

SKACTE (*coming out from his shop*): Why is the scathing rant?! Did fall from stool?— Is it that faulty? Broken?

FENNEL: The duke shears, Skacte! And the king preens upon this, the curls of clip left to fall.

SKACTE: What?! What be you there with, Alfonso? What is that? you tend to. (*coming towards him*)

ALFONSO: A find, from the dump carried here to clean.

SKACTE: But what is it?

ALFONSO: ....I don't know. But it's constructed well, I can fathom.

SKACTE: It's part of something. How can we sell that?! You

must caution your eye better, for the utility of an object.

ALFONSO: The wood is rich and bold, and of an art form makes itself. It deceives me not of value.

SKACTE: Well then, after the grime is removed, if the finish remains, you must oil and wax it.

ALFONSO: Yet there remains a curious handsomeness about these parts. The geometry of configuration appeals to me.

SKACTE: It is trash.... and probably split for scrap left. Yet rescued.... there is a shape here. Wood's not to be wasted. That's why it was dumped. And for a precious fire, in some poor family's hearth, these cold and bitter nights, it was granted to be used.... by the previous owner.

ALFONSO: I would keep it as it is, and shine it up. And who knows? Better minds than ours may make of it a value with a price.

SKACTE: This is not a mad shop, for furnishings mysterious. But if it sells, it's all to you, Alfonso.

FENNEL (*as **Alfonso** stands, leaving the bowl and rag down*): Decamond disowns us of our services, and plans to go to Lacismoor alone, as some sort of courier to the king's employ.

SKACTE (*slowly coming towards **Fennel***): How be it so, if he's still wounded—

FENNEL: But that is it! He has been fed the wound, and distrusts us, I believe. And dare I suggest, he sights Alfonso as a plant from the king or court; he washes his hands of us entirely, now.... and takes the clever route to work overtly with Musgonie. Some pacifism found, and no doubt profit to it supposed, he suddenly adopts to this new manner of excursion-ing as slowly as he leaves us meanly of mind.

SKACTE: The duke did tell you this, out of frustration from his non-healing agonies—?

FENNEL: He's told me nothing! Not one word of anything! But Count Maeli explains to us the rumor he's confirmed with Decamond himself, in a private friendship of that class. So it must be true.

CALERAND: The count was just here, and babbled the blood. A sad lessoning. Maeli wishes to recruit you and Fennel for his personal regiment.

SKACTE: This is astonishing!

CALERAND: But it's more local than Lacismoor.

ALFONSO: I'm not a plant of anyone's—

FENNEL: You'd hardly know how you can be used. But Count Elser is in league with the king to promote you. And so it shall be fresh with the grinding wheel, these turns of stone upon us crushing wheat to meal, or grinding the corn through our compunctions of loyalty, down into a mush of grain!... But not if you clean wood, to cast an eye that way so off-balanced(ly), that we should stay obedient to these trends and guiding is your competition aflame against more naturally. For, whether with your doing or not, I will put you up against Decamond and to his face contrast this leadership, that we will step under your command to remain.... rambunctious for our profiteering off these provinces, in an orderly way that *we* will compose with your front. For then we will use both king and demon to survive of manner, and evolve ourselves more forcefully than through this subtle stringing of superiors.

ALFONSO: But I—

FENNEL: It's that you know— and have the eye, the strangeness to supplant with another wildness.

ALFONSO: I have pledged myself to Decamond. I have fought for this— or, well.... demonstrated myself to this end.

FENNEL: And he makes you, to this end found for him. He allows the conversion of yourself as with himself. He permits this transitory exchange, to be more candid of his new aims. And would you deny an ascendency, if it were given you?

ALFONSO: I....

SKACTE: Who wounded Decamond?!

ALFONSO: I smite oranges.

SKACTE: It must have come from amongst one of us, or a few— and he is too shrewd to ignore it. Then this fallout remains, for us to manage of a marriage.

ALFONSO: I.... split them in two, symmetrically and with tact.

FENNEL: A rise may be swift and unexpected of immediacy, if it is beyond your own physics forming and thrust. But you'd make pardon for us knowingly, as it's in your present interests to maintain our caricature of liberators, for the true intentions of a colonial spirit— not to cause derision of its birth and parentage, and for a fee in this support your wealth, abilities, and militaristic merit.

ALFONSO: I know what I might be called, a practitioner of skills, an.... artist of display and tricky maneuvers. But as your leader and commander...?

CALERAND: Out of some sudden void rushes the unknown and unknowing. But this is placed before us to seize— And so we'll have it there, that way in place, since men prepare and become ready— for the assaulting on riches. It is the motive drive, and we'll take whatever mask is needed, to wear of our aims.

FENNEL: Decamond assumes you.... The king assumes you. You are Alfonso!

ALFONSO: I've never met his majesty.... in conversation.

FENNEL: But this is what Elser bastes.

SKACTE: I don't think you should confront our duke so angrily, Fennel. What he assumes is our general loyalty and respect for him, trust in his judgement, and upholding of his basic principles of procedure and behavior. If you feel you must confer on his opinions directly, and in person, then simply say, within the expected politeness of your station 'gainst his, that we have learned of his mission for Musgonie, and that with his absence *we* elect Alfonso to surrogate his duties in keeping this fighting body.... in tact, solid and prepared for carnage, so that people will not think it dissolves away within the sundry mires of country. Then Alfonso is an able figurehead out of nowhere brought, to yield the duke's best spirit, as Decamond presumes these new tactics for himself more personally, as could be thought fitting after aging and some wounding. So, for no embarrassments are wrought for anyone, these turns are made, these climbs are attempted, and these changes are fashioned fashionably, feasibly, even colorfully. For Decamond is very notable, and Alfonso.... is made more curious than his demonstrations. And— for out of nowhere brought, that where is for the king to reside. We will preach to it, as a residual of his cleverness, and offer this suggestion: that he manages his kingdom.... In the interim, though, let us hope that the duke does heal himself, and not out of our ways and manners that he has helped shape. For we do not change one stroke of lightening blast. And we will come to Lacismoor, one way or another, the same.

ALFONSO: ....Will men elect me?

FENNEL: The duke will *choose* you. That will be the charge spread amongst us, whether he admits to it or not, with public disclaimers of a worthiness to retain his soldiers. Skacte and I are his best lieutenants for this purpose. And we will have this word, denouncing slack of our intentions for whomever of our membership we meet. Then Decamond's word is for Alfonso, a bond he himself promoted forcefully, as can be already known, and with the sagacity of this foresight.

CALERAND: You sleep with the best already, and out of nowhere (were) pushed upon us by the very man's recommendations. Then you are the golden Alfonso we are to accept, and— though not worship, certainly rally with.... This is a move, lad, to stake our claims for fighting. And Decamond must give in to them, or wager upon being seen.... enfeebled and incapacitated— by his own men!... It is Elser's gambit turned around. And we choose it for our blows to make.

ALFONSO: Really, sirs! I only know him as a high friend, and an acquaintance of courtly influence.

FENNEL: And we know you, Alfonso.... I will greet Decamond smiling, grinning, and lavishing upon him obsequies of piety for the news, if words can make of rite— for our demands. And sharp as he is, he will take the strike immediately, as I beam of teeth and tongue. For who wounds Decamond! (*pounds the stool seat*)

Scene III — *The boudoir of Queen* **Matile**, *wife of Musgonie. She is entertaining* **Elser;** *although* **Jocqueral** *is present, at times sitting on her lap. The "i" of "***Matile***" is generally silent unless the name is spoken formally.*

MATILE: Cousin Elser, you bring dishevelment of his head.

ELSER: 'Tis only a child playing, Mat'le.

JOCQUERAL (*sitting at her feet*): I was rolling on the lawn grass, mama.

MATILE: But there be strands of it entwined, Jocqe, and it's not comely to run about like that with stuff in your hair. What should our guests say, and their children visiting in their best clothes and

nicest appearance for the court. A rough appearance speaks not well of the king and his management of family. We are, at home, the example to set.

JOCQUERAL: Are not these visitors example-d? And they speak badly of Vater.

ELSER: That is only a discussion of tendentiousness, boy. Let me teach. As I walk you from your instructors, playing of your tender muscles—

JOCQUERAL: Teaching me how!

ELSER: —we are open enough to traverse the courtyard. And there nobility gather to talk. They have opinions and purposes, and naturally use your father's name, on his very estate. But they are not angry with him. They emote their arguments to reinforce the strengths of their views. It is a ploy of importance to enhance this substance of gossip in ideas; and you shade weaknesses of mind or mental construction with a tone of voice that makes one seem serious about what you're saying. There's little more to it than puffery of personal concepts or views among equals, or nearly so within a class of good rank. And as I am myself happy to visit, and chance upon you playing, as you can lead me to my cousin cheerfully, and on such a gorgeous day, do I speak only of praise for your father.

JOCQUERAL: But you are a relative. (*standing*)

ELSER: Believe me. Not one word of condemnation to Musgonie's name did you hear. Simply because the name is said with stern dispatch, and as our king, this does not mean they were complaining about *him*, but rather other matters that must contain his attention.

MATILE: *Groß Gott!* They spiel(en) on him, Elser. Let me comb out that hair!

JOCQUERAL: Ja, mama. (*sitting on her lap*)

MATILE (*getting a brush off a neighboring table and starting to attend to* **Jocqueral**): It's not that they are angry with the king; but they find his complacency a complicity to many disorders of state, Elser, as if he were responsible for nature herself and her weathering upon us. Yet no ruler has much power over the contemptibility of man's sore feelings—

JOCQUERAL: Ouch!

MATILE: —And so they blame the protector for whatever disagreeableness to the protection of their properties and rights they may encounter. And so they think to speak these things loudly— publicly— for his general ear. But they do not suffice to be (called) complaints, but only of the weather and its trends. And they know that he struggles hard to keep a lasting peace throughout the kingdom and wherever he may have responsibilities.... to subjects, *und Gott*!

ELSER: I try to ameliorate some of these burdens, Mat'le, as a favor to you and our family.... especially those to which he has never had any cause to promote, instigate, or inflame, and from which he himself can only prudently react with a certain amount of patience or restraining tolerance. But there are schools of thought to peace: one with freedoms and one with locks, the graceful hand waving and shaking (with others), and the heavy hand slapping and slaying down with fetters unruly or discourteous actions, activities of protracted amercement. I, in fact, favor the king's wavering off of such bitterness. There are degrees to which one may stand back and let matters precipitate into their own resolutions. Yet, even slight encouragements for thoughtful and constructive behaviors may pay off with extraordinary dividends, sometimes, and to all interests in affairs that develop handsomely. We move through a fairly solid span of restfulness which everyone would like to have last for a long time, under his rule. It is almost a divinely inspired respite from our previous torments of confusion, contradiction, and heated choler among the various populaces of our country. I mean to show support for Musgonie's management with whatever assistance I can provide, directly or with the subtle gentleness of my influences.

MATILE: But cousin, you are friends with scoundrels who can only annoy my husband, by taking too great advantage of his lenience towards sometimes outrageous adventuring, and under his name (be) wrapped most audaciously of hypocritical and malefic intention.

ELSER: They are right honorable men, my queen, though clearly opportunists. But there is a sporting principle involved which you may not be able to comprehend. The happiest, most fulfilling risk, for a.... person of action, is the risk of one's own life within the company of confederates doing the same, a common goal justifying the outrageous exertion.

JOCQUERAL (*coming back to the floor, and rubbing his head*): It's sticky!

MATILE: It's oiled. (*preening the brush*) And don't go about roughing up my work.

ELSER: But the decadence resides not in them as much as their need and their needs; for the like of, say, a Count Maeli or a Duke Decamond ultimately serve to buttress the king's authority in regions daring to be riled and stupefied. But these activists do personally wager their lives to the cause, the lives of their soldiering. How would Musgonie respond without their interventions on his behalf? to the cruel callousness of upstart governance he'd never give approval nor sanction for. And the king does not prevent the voyaging for these purposes of quelling insurrection, country insubordination, and defiling of kingdom. How can he, really, within reason of imperative to act? These warriors remain the most loyal of subjects.

JOCQUERAL: There's a knot here!

MATILE: He does not command them to these adventures, Elser. It's as if they trump his persuasions with insolence to be bold. And it's only a matter of time before he's forced to put a stop to it—

ELSER (*as Jocqueral laps again*): Ah! But your husband has certainly explained to you his current method of enlistment of their services to reside strictly under his control. I was instrumental in conducing Decamond to this.

MATILE (*examining Jocqueral's hair*): It is a boule of mishap, caused by your fingering. (*brushing*) He did say to some manner of this man's implementation. But he is not altogether happy about the intimidation.

ELSER: Intimidation? One can not easily intimidate a— a legend like Decamond.

MATILE: As he himself felt intimidated to engage this ploy, then does he feel to take advantage of this duke's wounding, and so intimidate. Evil begets evil, he fears. And this is a roundabout way to circumvent his lapses to seem powerful or imperial. The man abhors arrogance, and yet must use this tool. And it's not likely to be altogether successful, but only as a stopgap measure until he finds his nerve and season to confront rebellion directly and in name of his rulership. He considers this not his most attractive cunning, (*as Jocqueral goes to the floor*) but as himself would be desired more and unquestionably. How can one dislike a peacemaker and lover of the serene? one who advocates prosperity for everyone! and kind feelings all around. He wishes to enforce his presence to these desultory peoples, and is upset with this current trying as no true solution.

ELSER: It was difficultly brought to fore, my queen. Councilor Ponstole works hard for this.

MATILE: But I do judge by the people, their reactions, cousin. And I hear them only laugh or deride the proposal, and that the good king is only being taken advantage of again; goodly intentions squandered on a profiteer— that can suddenly change into a salvo-ing saint?! It can not stand up, and they give me good wind of this to my ear, with full purpose to distress me while unseen, so that I might be urged to strengthen my husband's resolve to take these matters into his own handling and with force disabuse perceptions of his weaknesses. But this is a man who attempts to assuage a doubting public by several ways, each prickling of his fiber, yet can not seem to hold up the cudgel firmly. Taxes, fees, and fines must come, to spar with more than sparring. And he is worried more of these excoriating concerns financial, which only checker his entreaties for a lasting peace. For as the country remains solidified under his administration of these details, it also grows and grows of debt.... to its extremities relied upon, much too inordinately now to provide for a hindrance with revolts. And how can one pacify the proud of achievement with sticks to be beaten by? My husband grows more sacrificial every day, from the weight of these problems to contend. Yet hear I scoffs alerted to, not even to be ignored any longer, by the incapable of interpreting, as the roars rise, and a furor of noise ridicules.

ELSER: Might you ask the court visitors what they mean to portray by this rousing?

MATILE: Dejection leads me more, than quandary of inquiry.

ELSER: It is a creed of Musgonie.... to allow public discourse. But this method must be tried, as an avenue keenly fought for and blasted of, out and through the muddy derisions he has faced, to broker around some discontents of haughtiness. My friend Decamond is.... weakened, and by my doing slight— but crucially, out of my concern for both the kingdom and this person, so as to suggest.... by method.... means and manners evolving, that he may be replaced of leadership— for his own good, from the belligerent activities; as these altercations grow to annoy Musgonie more and more, into possibly a terrible thinking frothing from pressures I understand; as you deliver to me condolence and condolence for his changing personage to mourn, as the maligning mounts. And too much onto himself the king speaks. Ponstole has certainly advised me of that. Then what debasing may be crafted for an irritant of mind so self-absorbed? But in this way.... as envoy.... I avoid for Decamond a certain destruction through more princely powers scourging than mere noble combat, as I know such punishments can evolve out of sovereign tempers pushed too far. And noting his behavior, as a king in quiet, alerted me much. But then I suggested to him this method of employment for Decamond, and with passing thoughts applied casually allowed our king to make.... construct reasonable arguments for this avenue to follow. If he is uneasy about it, it's only because it is not actually originated by his own musing and perusing, but is brought to seem so being, or significantly enhanced by. But this is long in the making, Mat'le. I had to set up fruitful conditions on several fronts, taking advantage of whatever incidents, coincidences, and serendipities appealed for my usage. Thank god I was not ever discovered, as to be possibly criminal to a friend.

MATILE: As if like, you like your friends too much. But you are related to the king! through my marriage. And it is beholden of you not to infect him with ludicrous thoughts and characters of lucre. This Decamond is all indecently possessed to make some profit out of his endeavors and exploits. And now it appears we must pay him out of the royal treasury for his scandalous work. How much may the crown afford for his type?! when the kingdom is already stressed of disparate ways and objectives my husband tries to stay keen of.

ELSER: I had suggested it initially, at least, be a voluntary employment. But it is a low duty anyway, compared to what the duke may accomplish more typically, merely (amounting to) traveling expenses. Most importantly, though, this relieves many pressures, and may eliminate some assaults, the thought of their occurrence vanishing through the dissipations of polite political conferment. This is a choice opportunity for a much too ripened fruit to finally spread its seeds, more sweetly.... the seeds of what results from being influential; since my friend will become known historically as an able diplomat and broker of peaceful, tension reducing relationships, quite contrary to his current renown, if this missioning is successful— And it can be, Mat'le. For this is too delightful a contrast of dispersed natures not to seize, which I knew instinctively could appeal to Decamond's more intellectual temperament, to try what is possible but only conceivable by a few. And he took to.... the king's offer— quite gregariously for a cordial approval, even while strenuously nursing his wound as Ponstole and I spake with him about it. Wisdom is the guardian of men's affairs—

JOCQUERAL: What wound does he have? Was it bleeding—?

MATILE: Quiet!

ELSER: It was—

MATILE: Speak not for injurious intentions to learn of, Jocqe. They may become your own someday, through the weird portending of curiosity.

ELSER: To the shoulder was he hit with a sharp projectile, boy. This seems to have damaged some nerves to the arm. He was trying to soothe away those funny feelings that linger these months through the distresses of the trauma. For at first it was quite a considerable bludgeoning, said to have brought him to his knees.... From the force of the impact or the resulting painful agony I'm not sure. But it was definitely a shock to his body, considering where

he was at the time.

JOCQUERAL (*jumping up*): I was shocked to my elbow.... here (*showing to **Matile**), last week.

MATILE: Does it hurt now?

JOCQUERAL: It tingles a little.

MATILE (*allowing him to lap, after a slight hesitance, in order to rub his elbow*): Still there is a disagreement to the limb?... You must exercise it more. (*starting to rub*), or play less— I can't decide.... how one must go about to improve the functional muscles. But you know you are particular to their exertions.... and must be careful.

JOCQUERAL: I *was* exercising.... but fell down.

MATILE: His instructors are somewhat perplexed about his development. (*stops rubbing*) But at his age, what is strength to be? or be for harms and mistakes and distractions surviving with or through. Then is dependence on one's body a right, or God's? One feels not to understand. One hears not to understand. This might be more a curse made to homage of worrisome fate.... But you must end this tingling.

ELSER: Those teachers can be mistaken about many things. As I intuited, only wisdoms may shroud the affairs of man. And Ponstole had such doubts, as he explained to me my own, that he failed to take into account the wisdoms of both duke and king. And those that you hear singing rasps an' razors, Mat'le, may be similarly deficient of foresight and inspiration. This will work well.... He will heal well, and grow to be outstanding and proficient of the world and for this kingdom.

MATILE: But *he* is not convinced of this tactic to last long, only of its cleverness to be attempted. Good results must come in a flurry, Elser. My husband freezes himself in solitudes too long and too often to be that (they are) spiritually beneficial. He can not shelter himself away from his own discontents, as those of others mount him.

ELSER: A test may be soon, Mat'le. Ponstole tells me of a reason for Decamond's usage.

MATILE: Yes, a discourteous letter from Lacismoor. It is all of hushed discussion about the court, and country now throughout— no doubt, to judge the king's reaction to this clear insult on his authorities and patience. And will this agent bring us back heads or heeding for our rule?! Just what is the duke obliged to do through argument? warnings and threats and demands to place on their lips?! I do not understand this courtly love of persuasion by the monstrous of deed. And will he have his army with him for show, to provide an impetus for rationality?

ELSER: That is what Councilor Costile asserts to find out, whether this province does arm itself for insurrection. After all, Mat'le, it's only a letter that induces us.... the king, to act on rumors long rampaging. Decamond will be apprised of his investigation to the nature of that beautiful land and its heady inhabitants. And the duke will judge for courage what is meant to show, should he be needed; as he wants to go.

MATILE: Enfeebled and afflicted, cousin. That is his make, for us.

JOCQUERAL: Is that my make too, mama? (***Matile*** *takes on a frightened startle of concern.*)

ELSER: ....Never have you been so wounded, fine son. Not ever can you be so without such injury to person: body, mind, and soul.

JOCQUERAL: But everyone says I am deficient, behind my back!

ELSER: Of the muscles and the tendons only. They are.... prognosticating— around here— to help prevent or delay.... debilitations or further atrophies, not deriding you personally. And if any *are* purposely malicious to your sensitivities, this cousin of yours will castigate them thoroughly for the insolence, what your father might not have a mind to do for.... incendiary reasons to avoid, if these scoffs and scathes are not told to you directly.

MATILE: I had hoped they'd dare not, in his presence, but rather lower the boil of their remarks on many matters, these court guests — court jesters courting a sufferance!... You tell me if any do, this aim at you (of) those words to back, to be deficient— Dis-aim! Dis-aim! (*as **Jocqueral** stands, with concern for being instigative*) But I will challenge them for you, as your father seeks some weaponry justifiable. Much more than a rebuke will be due.

ELSER: Now, taunting is a commonplace activity these days, Mat'le, and to a certain degree can be tolerated as the innocence of foolish, honest thought. But Jocqueral strengthens towards this as well, in an age of easy skulduggery and ready hardship for many, sore feelings still being lessened and bodily insults versified for their rapidity of occurrence as even an obsessively repeated form of news. We're still only in the dawning phases of relief from the shadows of devastating human conflict and rampaging quarrel. And the pathos of public comment is one of the last valves left for means of reducing tensions that might otherwise result in overt actions of wanton destruction and concurrent villainies of espousal with determination to demonstrate. Though here at court, a politeness must always be asserted of expression, and even of opinions revealed or unveiled to draw attention. Yet, what the boy interprets is of himself to moor more passion with— for improvement. Harassment here, of the royal individuals, would be an utterly foolish and ridiculous preoccupation. So let us leave to that such crippling of mind, if anyone is caught doing, the punishments will be more than severe but rapacious of discrediting as a visitor of nobility, rank, and high-mindedness.

MATILE: I'd have boiled alive in oil any who would dare abuse my son verbally, had I this executive authority. But to prevent scandal more a scolding may I pound with, since I must follow to some degree my husband's ginger temperament.

JOCQUERAL: I think I hear these things as I fail and fall.

ELSER: What of terms disparaging?... They might only be objective descriptions of your motion. And you *have* been practicing to tumble less injuriously, or to avoid the awkwardness and collapse of legs altogether.

JOCQUERAL: Yes. (*goes to the floor*)

ELSER: It's the weakness of the limbs that is despised, even; perhaps, by your instructors out of their frustration with your physic malady after some customary physical activity typical for your age, an undue draining away of too much energy during and after the exertions. Yet not the man himself is defiled by weak muscle, as take my Decamond who remains august of notoriety. Then not for you to shame with correct comment. But think this through, you are disturbed by it. So change.

JOCQUERAL: I was only noticing its frequent use upon me claimed. If not deficient then defamed. And when I turn around quickly to oppose, as for response to hearing, am I greeted with bright, broad smiles that calm my heat for being wrong and wronged. Then only a jest is made, to induce me to try more at my lessons. But this happens more and more.... with exercise, and as they become more difficult and involved to perform through.

ELSER: Well, (*standing*) no life is without some casualty of exertion. And why do you station yourself here, Mat'le, when the court is aroused, if the words spoken and heard may offend your ear's view of gathering mishaps and misshaped thoughts?

MATILE: This is my quarter of the palace privilege, cousin. And to sense the quietude of which, then is the country tranquil may be judged.... or deemed for me what the king will not tell, say to queen with disturbance. His strength lies somewhat with this allowance, being strong enough to have this erring tensed. Then I am present, and passive alike. But it is comfortable in here, when the sun is strong and the legs may rest awaiting.... awaiting his concerns for nation. Does this kingdom draw on truth or trawl? collecting of these sentiments deferring my visage.... nor with his vision tamed. There is some gain of public opinion gamed upon me, knowing that I know what they would tell me shuttered, sheltered, and observant.

ELSER: I doubt if any would preach war to you, or open dissension, on which topic you would relay to Musgonie— timely. But are they so daring of the wall, with window?

MATILE: *That*.... is the sampling allowed, cousin, as by an oracle's screen to make a protest, for results to mysteriously bring forth. That's why the frame is painted pink, to show it's mine along this.... gray, graveled wall.

ELSER: Then deceive sound, Mat'le, if you feel hurt. They are only proffering words of suggestion.

MATILE: What's real is worth the banter. My husband is too timid, or restraining of self-motive, to confront disagreements too openly and with public asperity. So I do learn the tenderness of court's catering on the country's public opinions.... and noble concerns.... and for this bide my time, where at he asks with some regularity: "How goes my rule?", as I could think it. My answers, in this way, must always be too often of disseminating disappointment. Yet, when problems come officially to court, he is not so blank of face as by me wounded with the cruelties I hear to tell, that in truth there is much dissatisfaction with him, his management of affairs, and his kingdom's progress.

ELSER: That's a dissatisfaction with themselves, and from all corners coming, makes Musgonie fair. For they are ravenous, Mat'le, to take advantages in their interests. And not to you nor him are they personally averse, but for their own administrations of properties challenging.... And have you more lessons?

JOCQUERAL (*standing*): Yes.

ELSER: I promised to return you rested, and quite intact and functional for your progressions.

JOCQUERAL: This I'll do. For none can laugh at me, where sincerity beds of effort.

ELSER: That is true. It's your father's pledge, where from you learn to try and apply your— gifts of nature and spiritual endowments.

JOCQUERAL: I gain propriety with them—!

ELSER: Priority with them. There is a difference. All life is propriety of itself. But you are a prince, and the future ruler of this country. A confusion of words or terms is merely learning's tact to refine and sharpen the opaqueness of mind. (*leaving, with Jocqueral*) To a good day bind yourself, my Queen. For though they evolve for each, they are owned by only one.

MATILE (*as they have just exited*): Ja, Elser!... or to a speciousness of holy order with ungodly nerve.... am I of this state serving for our times, a darkened promise, or betrothal to the future this marriage of cascading indifference to error in our own lives and perversities of innocence, that this cloth of privilege is more bound to harms and harming.... than can be wished by any ruler to predicate, where a weak child grows. But of our failings made, sufficient for this crown are most our assurances to endear. (*Some voices, indistinguishable, are heard, issuing from the window.*) The hurling howls.... the hurting heat— the heaving haste.... the— human husks to hound.... These sounds remiss.... are company, and accompany my chaste persuasion not to indict the faculties of commentary. (*standing, with brush in hand*) Where sifts an airing of to disabuse, plain as haughty, spent of nature served (*coming towards the window*), a corn-fed wherewithal of pouch an' stomach.... and high chins.... then such a wording stains the walls and its shadows. (*looking obliquely out the window*) Yet see me not, for I'm within the shading. Yet know that I observe, and without seeing those directly below who would castigate this royalty. Sense my severe searing, that if one foul word is echoed to my ears.... against the princely diffidence— that keeps you sound!... on to err, ere my wounding pounds.... without much gentleness. With that.... suggest more fighting, which a husband may allow. For the speciousness of a holy peace is an awful waste of time— and hate, whereat these churls would chuckle at the crown. Better be your bane than mine to dread, if influence be more portentous.... than these poisonous thoughts. (*starts preening the brush unconcernedly*) Pray for more than this, you— people.... down on my court.

Scene IV — *A comfortable, though spacious, room in the manor of **Lord (John) Bracken**. **Darughe**, at a serving table for wine, is pouring **Costile** a glass.*

DARUGHE (*pouring*): We're about to have a meeting, sir, regarding the improbable formation of such a very confederation of soldiers as you have thought to be alarming. Your arrival, and expressly from the king himself—

COSTILE: Well, from the Privy Council.

DARUGHE: —is most felicitous to hear the discussion. (*coming over to **Costile** with a glass, and one for himself*) But I'm sure that John set you straight about that sorry missive sent by his son. They are both lords of the manor Bracken; though the deception was intensional because.... well, that's the way youth like to stir up controversies, (*offering **Costile** his glass, who accepts*), by deceiving and startling their elders into some action for or against their positions. (*sips*) It's a way to see what happens when one doesn't have much experience.

COSTILE (*after sipping; with reflection*): This is a grave matter, squire. Bracken denies the brief altogether, claims that his signature was forged of proponing at least, and yet admits the young heir is not disciplined in any way (for this). Then almost simultaneously I find that you from this province meet on this very same estate to consider the establishment of a militia and presumably the self-governance that this assumes. Now, you may preach to me otherwise—

DARUGHE: Hasn't much of a chance. (*sips*) These are small heads in large casks of vaporous headiness, leading themselves with the intoxicant of visionary aspirations, the young sirs of Lacismoor. But they are all good boys, and are certainly not revolutionaries.

COSTILE: You are dismissive of circumstances and fortuitous coincidences. Then why are you here—?

DARUGHE: Now, Scott Bracken has an ambition to assemble an army and be its general, having tasted some ceremonial activities of knighthood— and excelling in this slight bit of practice for the martial arts. But none who have worked at my camp can really be called fighters in the sense that *we* understand it. They have never been in combat, nor the threat of combat, and have not the nerve for this to endure— yet. But our province, like all other dominions of his kingdom, is pledged to Musgonie to provide soldiering knights, of a number more honorary to be regular than numerous— I think we only send in two or three yearly— for his court troops to maintain. Yes, we send in three; but one is a reserve who generally returns within a month, and another after nine, if there are no external conflicts or warring going on, against other kingdoms and countries.... And there hasn't been, for some years now. Indeed, the king doesn't want to bed and feed all three for the entire year, and neither do they wish the privilege and expense of supplementing their comforts that length of time for a duty which is largely of ceremonial custom and not terribly necessary, as far as they could tell. But I prepare the young men, who by their landowning positions are capable of this privilege, so that this supply of knights is kept stable and able. And that's all to my camp, Councilor Costile, and all to these sons of great estate in Lacismoor. (*sips*) There is hardly any fear of an army developing here.

COSTILE (*sips*): Then why do you entertain this meeting of such, Darughe? fierce soldier that you were.

DARUGHE: Not of such, so much. Only a few and their fathers or uncles. No. You can't dissuade young minds with a brashness similar to their own. But the analysis is simple. Such an army, as some in any place would propose— essentially at the extremities of country and homeland— must be paid for and contributed to both functionally and monetarily by the the leading wealthy inhabitants of a region. Therefore (is) this assembly, to convince these wiser and more cautiously prudent heads, both of age and experience, for the need. And if a remarkable madness happens and they accede to the cause of a few young sons, then, as a trainer of men for soldiering and a provider of this potential, I stand to profit handsomely of such a symbolic gesturing for regimental self-protection. But it could only be symbolic and to a small degree, much like a legion of scouting about the province. And, in such an improbable case, we elders must firmly hammer out all conflicts with the king. So I am here at the beginning, to see how the (other) landowners tone down their brood. That can only be the most reasonable result, is it not? Lacismoor is not in need of more protection than the king may provide, not so?

COSTILE (*after sipping*): The young sir is bold enough to assert that Musgonie's court may actually be guilty of the mismanagement of Lacismoor's economic affairs, that this province is harmed through the king's oversight. Now this must redden all perspectives, ours, yours, and his, to state such a claim in writing. This region is too profitable— to country and kingdom, to be left to its own devices. And rather more control over its.... development as a dominion, may be required than less. And less independence of operation is in order, after this espoused breach of faith that wounds the king deeply, for the affections of his subjects he increasingly requires and nourishes. Then, perhaps, you can understand, and certainly anticipate, that measures are being planned to construct more guidance for the province.

DARUGHE: Measures? (*sips*)

COSTILE: Giving it more of the king's personal attention, and working to assuage all disagreements of faction that have aroused such contrary spiritedness here.

DARUGHE: A heavier hand, you mean? (*sips*)

COSTILE: You might warn the imprudent of the incongruities they bind themselves to with actions such as these.

DARUGHE: What actions?

COSTILE: His letter. Your meeting.... The blasphemies even talked about. Lord Bracken can not get off so easily—

DARUGHE: Which lord? (*sips*)

COSTILE: ....I mean the man, sir. The owner of this estate. For all provoking in this way is still ultimately his responsibility— Now.... he has honorable character to deal with. And I don't doubt his intelligence in handling this mishap with his son. But the matter is grave enough so that he may very much need our help, our assistance through this heap of misunderstanding launched to land at our feet, and soiling our shoes for steps.

DARUGHE: You mean to knight-bound Scott for a lengthy service?... at court?

COSTILE: We only advise the king. But more than advice is needed here, but a delegation to patch up discontentedness with discourse and review of Lacismoor's nature and obligations to the crown. In short, a work of official diplomacy between the court and your province is badly needed, so that you learn to know, and know to learn, what is acceptable and possible and permanent.... in your relationship to the kingdom. And I tell *you* this in particular, and forcibly, to annul any thought of instigation away from this purpose.

DARUGHE: Warn me?! (*sips*)

COSTILE: The procedure is initiating with my concern, and what I've found and find.

DARUGHE: You're that important?

COSTILE: The problem makes for its importance a consternation of court interest and involvement.

DARUGHE: The *problem* is important. (*sips*)

COSTILE: No one is against the prosperity found in Lacismoor. It is a fine model of achievement and aggressive productivity. All of its inhabitants are to be commended highly for your work and praised for the industrious nature that becomes you and describes a wholesome province. Yet, this is still a dominion of kingdom and must subject itself thoroughly to the purview of rulers in the least. (*sips*) We would not want to obstruct or disturb any of the region's established agricultural and mercantile operations. Instead, Musgonie is prepared to assist and enhance these activities, and with a personal visit observe them and his subject peoples, reciprocating adulations.

DARUGHE: The king himself does do diplomacy on his own possessions? It is as if trade relations were bargained for. Yet are we only vassals to a wise hesitance to interfere. Yet, from lords to serfs to the natively local populations, there is a purity of consideration to deal with. And I tell you again, we are not against the king's sovereignty over us. There is not seditious urgency to what I and many others have fought for.

COSTILE: Then listen to me carefully, squire. The king protects this heaven from the scavenging of scoundrels, and would send you one converted of his manner— or perverted towards our mandates— to arbitrate on your subservience through terms agreeable and beneficial to both province and crown. And with peaceful settlement, to all possible disputing, (to) work at and hammer through, he'd be the very duke himself— Decamond.

DARUGHE: Decamond?! the raider! the usurper! comes here to — bargain with us?! I would to greet him fiercely—

COSTILE: Had you not expected the like? with all of the subversive rambling that has come out of Lacismoor lately. And now this letter from a lord?— signature-d!... identified! But the king has forged a way to constrain the beast to useful purposes— if you do not oppose him openly with insult by conscription for a self-made army!... If this is in any way your penchant thought, to disoblige Musgonie, consider the gravity of this symbolic representation. For the duke remains capable of much military backing.

DARUGHE: And what does John— Lord Bracken think of this possible imposition to our lands and estates to wrangle with and debate over management and service?... to country, state, or self.

COSTILE: I've not told him this news. For I've only just decided it might be warranted. But with you I declaim it as a method devised, to keep this region square and honest of obedience to the kingdom. Thus, as he would come, it is not a threat from us— but a pledge to eye you faithfully.

DARUGHE: We are not the decadent here, to be so woefully observed.

COSTILE (*as **John Bracken** enters, escorting in **Walter Townes** and his father, **Hugh***): Then consider the imprimatur of this knowledge's dispersal.

LORD BRACKEN: So few, Hugh. Scott sends of notice that we host dissension, as my name's become notorious even at home.

LORD TOWNES: The better would know better than that, John.

LORD BRACKEN: This is Councilor Costile, from the king's Privy, to overlook the event.

LORD TOWNES: It's been judged that important? How do you do, sir.

COSTILE: How do you do. (*as **John** goes to prepare drinks*) And he?

LORD TOWNES: Walter, see what you've done with your headstrong talk? There things spread, and rumors mount to bring in officials. This is my son, sir. Darughe! The seeds were laid in your camp, I can just wager.

DARUGHE: It's premature.... to judge the character of men through verse alone.

LORD TOWNES: As (to) what is read? That is the punch. To have it written is too bold.

WALTER: A standing army is only to our good, father. It gives us the prominence of being noteworthy and ready—

LORD TOWNES: For what?! to be used as bait for thieves?

COSTILE: I just happen upon this meeting, sir. But it is indeed to this written letter of Bracken's that the court is concerned.

LORD BRACKEN (*preparing glasses*): Not mine.... But to be concerned is criminalizing carelessness.

LORD TOWNES: Scott never denies it, John. In fact, he jests about it, (*to **Costile***) how you councilors can be so easily duped. But it's brought you here. And that was his ploy, to garner some attention. I've warned him of the fire played with— and Walter. (*to **John***) But he's a grown man, John, and not stupid. Are you sure you didn't put him up to this—?

LORD BRACKEN (*bringing two glasses*): No, no— not me! It's damaged my reputation as a pacifist.

DARUGHE: Ha!

LORD TOWNES: In some trifling, insouciant way? These young men are so restless and anxious to mallet out their armors of mendacious valor—

LORD BRACKEN (*offering the **Townes** their glasses*): Audacious, and behind our shields protecting.... courage. Or the thought of it, perhaps, they think. No, I've always ever suggested to Scott....

to have what is needed and to need what is wanting, and to never confuse the two (motives).

COSTILE: Then why has he? This is a serious.... offending. I won't go so far to say it is an offence, since it's only your personal prestige that is wounded with this trickery. And I'm glad to learn you've not employed the man to such manner, that it is his own deviousness. But I saw the king cringe and grimace right in front of me, while reading your son's letter. It's tantamount to an attack on royal sincerity, and hurt Musgonie to the core of his pious patience, that statement that said God would be against more nocent intervention into the province's commerce. You can't allow such an attitude of misplaced rhetoric to be made public for commenting on its unfortunate apprehensions and erring thought. Yet he deliberately sent it to the privy councilors, knowing we must handle the disgruntlements of (all) the truly noble. And what is his punishment for this, Lord Bracken?

LORD BRACKEN: My discomposure, perhaps? Sir, he has his skills and opinions, and uses them to what advantage he perceives. I can't say to always understand it. But certainly no law was broken with this discourse. (*going back for his own glass*)

LORD TOWNES: Sure, that he was complaining about the taxes coming. We all expect it, and might have means for similar correspondence, one way or another. But leave it to the young to say more bluntly that I don't like this, nor the interfering aims or goals of the state, in some matters.

COSTILE: Gentlemen, you can't be pleased with gold un-quarried. We quarry you. And what you send makes you rich, brings you wealth. Now, as to how you send it, we should have some control over— because we are your managers. And our needs are your needs too. I speak of course of the kingdom to which you belong and are obliged to prosper, and prosper with. Now, this Scott Bracken, incredulously, may not understand that as once soldiers to the king you have pledged your land, estates, and entirety of produce and production to our country and his royal administration. And there is no way you can argue your way out of this allegiance short of shuddering, stumbling, and utterly stupid secession, which would crucify you as an enemy of Musgonie and promote the battling he wishes to avoid. Have you not explained this to the young, brash, empty heads around here?! or have you let them become so thoroughly spoiled of privilege as to defile their birthrights!...

LORD TOWNES: Are the spirits too high here, of this fine Riesling? I think you presume too much of a birthright. We have all labored very hard for what has been achieved of each. And there is not an undue arrogance to any of it. But I teach Walter to respect.... all leaders of a realm, be it physical or psychological or introspective. And Musgonie has our adherences, physically, psychologically, and introspectively. Yet, try not to define this as a bland blindness fought by younger, sharper eyes. The sentiments of Lacismoor— all around— are such that many stately improvements of relationships between us and our royal rulership.... your court rule, are to be desired much and steadily. It is the tempo that confounds.... the quicker of emotion. But the imperatives remain common. We, today, contribute more than ever we would receive of benefit from the king— and are proud to do so.... up to a point. The sweetness of a nectar hums to the ears when drank, of a white grape's richness. Nor even dry with firmness and body is there more to be squeezed of this fermenting off the sugars. We apply ourselves, sir, to consider many species of habitation, and to take of argument which may worry Musgonie— to stay worthy of us! (*sips, as* **Costile** *appears astounded*) not that we would ever think to remove ourselves from country fought for, without some profound justification. But the legalities are this way brought: that we serve this kingdom as the kingdom serves.... its subjects.

COSTILE: Dare you make complaint?

LORD BRACKEN: There are no complaints, friend.... We do not tug, to tear off, and come apart ourselves.... into momentary fiefs of importance. Relate this to the king, with all assurances of hospitality. (*coming up to them, with his glass*) But there are rules of avarice to follow in any locality. We no more plunder than are plundered— yet.... these natural resources and riches that delight renown. And this with much transcription worked!... is done of heavy labor to convert ours to yours to ours. But there are limits to the sustainability of this method; and the king approaches this or these, with levies and dues that seem, at times, from our remoteness, terribly unwarranted a bleeding. So obviously we consider many forms of redress, short of the mutinous, that can be found in many regions of the kingdom, perhaps even surrounding the king's court. The disgruntlement does not mean we conspire of any treacheries against Musgonie's mast, but debate amongst ourselves, as he'd allow the thought of thinkers starred with such fortune as Lacismoor. And as I exclaimed to Scott, we are not— and can not be— foolish enough to tempt the pirating of an independence. Yet, as for self-protections— well.... estates may have their soldiers of policing. And for a commonality of province, and one of such importance, a commingling of resources may be naturally evolved into a single fighting force under the jurisdiction of a local governance—

COSTILE: That can not become too strong, as to threaten Musgonie's authority over you! This is a deadly happenstance provoked, no matter how naturally evolved.

LORD BRACKEN: These are only thoughts, councilor, that come by thinkers as far removed of their actualities as ahead for their visions. This is not.... a disobedience, but a wherewithal to see one's state. And we do not take any more license than the king himself provides for all free men, to consider their station under his statutes assessed. My son proffers no more than this to review. I have *ordered* this be so, under my scope of tolerances. And I can assure you personally, that if the king was in any way offended by my son's action, then this same son, by a similar action, is quite prepared to make a public apology, and to the entirety of the king's dominions and interests throughout the nation. And the name Bracken will be codified to a coherence of aims for Musgonie with this humility of supplication. But to forgive of ideas is not this king's nature; for even with treason in mind he holds that a man be allowed his thoughts be worn and read. And that's as much as I've fought for, of this royal generosity and understanding of individual needs.

COSTILE: Indeed! He has allowed Lacismoor to exist.

DARUGHE: We have panther-ed the territory, and panicked you.

COSTILE: The king would not have sent me down here if he didn't think you were all very reasonable. There was not a sudden frenzy to prevent your aspirations shaded with alarm, but more an imperative aroused to understand the problems here. And Lord

Bracken ably apologies for his son.... But this is not needed— of Lord Bracken, and that is the more important discovery. He, and you all, are held in high regard and esteem.... by the councilors, to (want to) assuage of any botheration to cause dissent against.... our country.

WALTER: I just can't understand why we are so feared. Our calisthenics are impeccable on display. And we would make a right good army of recruits, volunteering to soldier around our homes and properties to demonstrate the high moral, industrial, and military calibre of the sons of Lacismoor. So why contemn this proudful-ness an' pride?! And we train well, at Squire Darughe's camp, to handle many situations foreseeable, as this great province's fame pervades even historical comparisons to great states and regions, noble territories with their cities, towns, and villages establishing abundances of wealth and wisdom for their inhabitants. Is not ancient Rome an example? Then we are Lacismoor, and deserve something of a prominence and distinction, within our.... kingdom.

DARUGHE: Lord Walter Townes, Rome was a nation-state of city, and the hub of all administration for its possessions of area stretching all throughout Europe. Why, the Pope even owned England for a spell. Papal jurisdictions remain extensive. The seat of our power lies with Musgonie, and his court. We are obligated to.... this concept, and not to affront his army with any of our own — that is.... what he may allow. For essentially, we are conceived.... as part of his principle to help administrate the vast areas of land under his control and management.

WALTER: I know this, squire—

DARUGHE: We are.... his helpers heroic, within this principle of subjugated overseeing. And some tenderness of feeling or resentment may transpire as our successes continue to be evident throughout the country. Sometimes there can be too much pride with principle. And as there have been other examples of separatist attempts, so may we be suspected. That is why the councilor investigates, and this is a matter of some importance to the king.

LORD TOWNES: More that we should be his favorites, for what we pay out to him and his, though including ourselves.... The king is sensitive, son, to make sure we stay well within his fold— That's what it all boils down to, to have an official sent us— and (of) any (kind). But Scott's letter was sent to the Privy Council for notice, and thus a privy councilor arrives. I hope, sir, that you are lavish of praise for us, as we have lavished ourselves for him— Nay! for our own country, for our own settling and resettlement— and solving of the torment! with warring— and now.... expansiveness of fruiting, fruitful goals. Turn Musgonie's doubts around, towards our views of happiness, as he has sanctioned it to be! And show him our sons as his knights, and not as nightmares to be dreaded in his dreams. For I have fought for this permission to be used.... as we have done. Spectacular is not our foraging for the crown! And with great strides may we present ourselves for review. (*sips*)

COSTILE: It is a lovely area indeed. Though that is as much nature's doing as.... ours. (*sips*) So you are, in general, against a self-maintained and regulated militia for province, since this would be too competitive for the king's purposes.

LORD BRACKEN: We have the independence of mind to entertain arguments— only.... But we are circumspect to the proposition, of course. Knights make not such forces as could arouse the temper of his majesty, as our sons do plead of utility.

DARUGHE: Hardly yet, till schooled thoroughly.

LORD BRACKEN: You can not tell me, Costile, that Musgonie is actually angry with us?

COSTILE: Not angry nor startled or surprised, nor disappointed and dejected about. But always is he worrying over the polemics of rule, as he does serve so much more than himself to keep his kingdom intact, the country at rest and respiring productively again, and all comforts dominated by the sweats of healthy labors. What else can be substantial for him, for all of us? I believe, though, that he allows the hearing of too much controversy to his ears, for his own attending to notice how contrary many views are to his, even as spoken at court with casual pejorative and berating even of his own suitability. Now imagine that to hear regularly, daily by some circles, and not become either incensed or with a growing loss of self-confidence, after awhile. But he insists on permitting this public spewing of ideas of contention before him, as for a filter be the air, with letting off of steam to make the clouds. There is agreement, though, through all corners who dare to speak within his courtyard, as nobility at leisure, that the nation now as formed (must) be kept from disintegrating— in any way and at all cost, and that serious disputes be settled with proponents arguing rather than fighting, leaving decisions ultimately for the king to make with weight towards peace, if necessary. Musgonie is still considered at least (in) this way either just or responsible enough with his wisdoms pursuing his intentions.

SCOTT BRACKEN (*entering*): His intentions are to remain solicitous of sovereignty to retain, and to appease candor with a weak face and long arms.

LORD BRACKEN: Scott!

COSTILE: You're rather brackish, lad!

SCOTT: Lad, ladle, linden tree— I'm rather "sir" to you, sir, and refreshing for some spirited debate (*coming up to them*), since all you privy councilors do is whine to Musgonie, about his errors and their results on your consciences, not having every little thing your way, totally with your approval as being prudent. Let's be glad there's still a king around to listen. Yet you've showed him my letter, to discredit my father—

COSTILE: You've done that!

SCOTT: —my draft of compilation, for the well-being of Lacismoor to observe and ponder through, instead of being brazen enough to handle the affair all by yourselves with the discretion to study the matter thoroughly and privately, and with good proposals to form, before bringing it to the king's attention as a package of thought-out recommendations.

COSTILE: Is that what you could have hoped for?!

SCOTT: You haven't half the head— collectively— to worry his, with this disquieting. But then the intelligence of a political group falls below that of its individual members, after all of the compromises are fought through it.

COSTILE: Your.... observations.... were shown to the king be-

fore we'd meet to discuss them. The king is particular to know—

SCOTT: Not even that, then, would you weigh in to construct a preliminary investigation, before pounding Musgonie's notice of my claims. And what have you found yet?!— Well I'll tell you what I've found. There's about twenty come for tonight's meeting, father. I've put them in the drawing room. And they await you, gentlemen. Though they are enjoying refreshments at the moment.

LORD TOWNES: Now Scott, this councilor has been quite plausible to digest of the sensitivities you've caused, advertently or inadvertently or however. Let us be polite in our discussions, and not so aggressive to wound personableness, especially for a visitor and guest to our provinciality.

SCOTT: I'll be as hospitable as he was with my message. Personally you showed it to him, did you?

COSTILE: Yes.

SCOTT: And did you stop to think about this, before doing?!

COSTILE: ....Young man, your ideas are not that incredible to avoid scolding. Musgonie is particular to know of all developments in Lacismoor, as soon as the information arrives; since he is particular for Lacismoor to be favored, amongst his provinces. And you hurt him, with these details of opinion. But that is for your sorrow and suffering to bear, when one day you become aware of manhood.

SCOTT: Sir—!

DARUGHE: Let us breach this scuffling. To that suggestions have been made, the king is aware of them.

COSTILE: But he is not apologetic!

SCOTT: Apologetic?!—

LORD BRACKEN: Be the son.... before a meeting placed, aroused, and all with aim prepared and excited. Be the son.... not quite so old as we, and without the same sophistication of expression for what is actual to stir him, as by a cry and scream from him may truly be an apology.... to me at least. But judge not the behavior of the agitated as being that uncouth, if for debate he has decided upon this manner. Opinion matters much in Lacismoor, and has the respect of voice more so than shouts of disputation.... over questions already solved. Let us then withdraw to hear ourselves, and rid mystery.... (*starting to exit*) of this balloting for militia. (*the most others start to follow him, but not **Costile**, **Scott**, and, waveringly, **Walter***)

LORD TOWNES: Walter!

WALTER (*standing next to **Scott***): Glass in hand, father. (*starts to leave, though ponderously*)

COSTILE: I'll word you first.

SCOTT: Really? Then as to consider and counsel me?

COSTILE: What is your perchance made drift.... sir? feeling for more war. (*The others have left.*) Have you a sacrifice to make?

SCOTT: Nor none to gain, no. Or for retribution for family— members.... perhaps. With impression I have gleaned.... that we are self-sufficient here.

COSTILE: And how does it suffice to harm your father and friends in this manner?

SCOTT: It's not harm—

COSTILE: They fear for you, this impetuosity to bring the king's eyes here; while dismaying favorites of this province's prosperity and genteel nature, you provoke.

SCOTT: This is heralding Lacismoor, of its independent clime. It seems to me, councilor, that the provocation's a subjective reasoning of free thought, and that it should be obviously paid for. I would not have alerted you, and attempted to awaken the king, with any other purpose than to serve this meaningful response or rebuke, for the conditions I have proposed.

COSTILE: Awaken the king?! His sleep seethes of terror over matters like this. And the demons chastise his dreams by being factitious of their condemning his all too complaisant but forgiving nature to keep established the restful tranquility of country he loves.... and could still fight again for.

SCOTT: Defend— for himself, for his own. That is his only prospect, it seems, or proposition for his subjects. Well I'm somewhat wary of this, since we must protect out interests also— as the law-abiding. And under Musgonie's rule we prosper— when left to ourselves, sir.... uninhibited, and thoughtful of our doings. And as we do lend wealth, slashing profits to his favor— we work hard for the same. And we deserve our little convocations to review the possibilities of self-enhancement or improvement of aims, goals, and contractual obligations to kingdom.

COSTILE: He sacrifices much personal prestige to allow you yours, as well a burden he may carry of caricature, loss of confidence, complaints, and even ridicule to weigh upon his leveraging. Yet the country remains solid despite diverse and— dispersive elements.

SCOTT: Solid as would seem a ship of jelly shelled. What be an egg, must then develop into some animal or instance of solidity. Musgonie holds about him too many parties of collection to be safely maintained or controlled. Any witness to what's current can sense this, and oblige a future disassembly to take care for and plan on. He is not a ruler who can inspire a fruition of hopes to so many differences and competing attainments— Not that he is in any way an unworthy person, nor the royalty itself undesirable to praise, remark about and identify as under.... a sort of handsomeness of state. But there is no generative motive to this particular king—

COSTILE: Treasonous sentiments!

SCOTT: —And eventually departures.... from his kingdom will occur— successfully. For he holds on desperately to what is, but has little conception of what can be. And if I think so, then so must thousands, since I am not outstanding of opinion nor sensitivity.... nor original perception to a truly remarkable environment as is Lacismoor.

COSTILE: Then let me be blunt as to your origin, Scott Bracken, whelp of the whip! For your father and others of his age of honored peerage and capability struggled mightily to beget this silk of tranquility and prosperous reason. With that and many such efforts, the king allowed this exploratory settlement as a way to *avoid* warring with the neighboring countryside too preoccupied with their own concerns of state to take interest to this fertile yet undeveloped territory which.... you.... have capitalized on with great industry to provide for— us! great achievement. Now then, to lose this province to the jealousy of others surrounding you, or nearby and within common regions, would cause a loathsome dissatisfaction to our country, which, if not preventable, would have to be fought against.

SCOTT: Then.... should we have the means of our own militia—!

COSTILE: You are not capable of doing this alone! You are not formidable enough, to overcome the powers of local desires and lustful greeds, and the resentment of not doing as well with this particular land as we have permitted you to culture and prosper on. Lacismoor needs the concerted protection of our kingdom.

SCOTT: Our neighboring residents have only ever seemed pleased with our cultivating presence.

COSTILE: It's honey for the bears, now, as *their* rulers would adjust to you with an unwise independence to pounce on and break, necessarily causing conflict and friction between nations. Else, why your rush to form a militia against these evident and obvious anxieties?! if you're not a stupid man of expectations. We will provide the military backing, for *our* province. And this will as ever be costly. And taxes will be raised for this— And you will pay them. But you will be spared the misery of going it alone, of fighting alone against the overwhelming. Lacismoor owes her kingdom this privilege of defense, as you do owe your father some reciprocation of the sullying of his name with perhaps a more suitable behavior during this gravity you hardly sense.

SCOTT: My father's name is my own! And I will deal—

COSTILE: Ask any person of experience if he does not reason these circumstances differently from what I have said— Ask Darughe!... to expound on a foolishness he has probably already warned you about, or at least asked you to worry of. Battles are not won with wishing for the best, and going into them ill prepared to face true opponents, oblivious to the reality of heated and deadly dispute. You need the forthrightness of Musgonie, to detain more blundering, even as he is scoffed at from this distance. But consider that if you honestly consider him to be so weak as to allow this insult as you'd envision, then what must come from weakness but a weak piece still— to be much abused, seen to possess a weak morale at least, disingenuous of monarchy and to be thoroughly subdued as some disloyal partitioning scooped up. But Musgonie is not so effete or ineffectual, as he allows this province considerable strengths, and will make use of them for the collective good of all his subjects. He is the strongest ruler, who can bear professed dissension through the endurance of an adversity of mind. Do not have him break his pledge to withhold punishment from such tort-tendencies as yours, but rather assess opinions in this place, and under the roof of the Bracken manor, to hear of what is more a proper course to follow, as expressed by wiser and more wily heads than you have ever yet had a chance to match on exercising. For these chances are few, as progenitor-ily made for you, and determine all outcome by the acuity of your reactions to them. On erring here.... correction is as erred for devastating consequences. It's the only way mistakes may correct themselves. So then assume, that the king may fight this tendency with a bitter toxin— turned to tonic, for your behalf and his assuagement of mental pains and a conscience for your betterment. He sends me, not to appease you, nor merely to observe and investigate, but indeed to apply his dictum (as if from his own mouth) that now you must behave for him well, that the times are terribly sensitive for this, and that he will help you to do so forthwith, perhaps even visiting, to study the progress made.

SCOTT: Has he become so obsessed against my argumentation?! I do not propose true separation—

COSTILE: It is the candor.... that he respects, for show of some urgency with you. And to see that it is *you*.... I find raised, then brought up will he bring, what must be dealt with from a speculation of character, and on this ply his aims. When the devil brings (on) a storm.... 'tis beautiful of storm to wake, as heavenly to die forsaken of its powers. You will be convinced.... by your own devil-may-care attitude to abate—

SCOTT: I doubt it!

COSTILE: —once provided with a harshness to see fear.... from the fearful sworn for a peace of continent to buy— or win.... or work towards, for their sons.

SCOTT: The lords.... of manors throughout the province will speak prudently, I'm sure. But there is more of pride to that which stirs the heart. And chances made are for to take and ford through carefully. I have considered everything for my actions. And we are awaited, sir. Or at least I am. (*gesturing for* **Costile** *to accompany him*) That I am not convinced means nothing. (*starting to exit*) All that matters.... is for convincing to be made. (***Costile** follows.*) But I do approve of the king's notions to have some attendance to our proceedings of thought and debate. For I've never doubted his sincerities for subjects. (*They exit.*)

Act V

Scene I — ***Ponstole*** *is leisurely strolling through the king's courtyard with* ***Wallinda***. *They are socially associated, but not arm in arm. It is nearing a bright evening, and the yard is not densely populated, the atmosphere being very relaxed.*

WALLINDA: I can sense the power of the place, councilor. The very brush that surrounds it has fragrance of a magnificent environment and world other than to what I should be used to.

PONSTOLE: The palace is in this way public, for the society members that warrant and deserve notice. The flowers are bold, broad lilies of strength, my dear, and the regality that accompanies the place; for this is a dominion above the common, and one never knows who one might meet crossing this courtyard.

WALLINDA: But is my humble dress suitable for a presence? I would never think of entry un-escorted. This is not for the many minions to visit.

PONSTOLE: They don't. And they can't. But with my cloth and

coating, you make the very essence of fashionable statement. What is beauty, then, but a pleasure of being seen! And I am rapturous to accommodate your view and ascendancy as being witnessed for my favor, and soon the court's. Because you enchant this otherworldly arena with a delicate handsomeness of features and motion, as such a goddess would descend to interview the ground. And indeed, dusk is near and the spiritual approaches of warm night and earthly sprites of fancy. Thus do you make an evening whole and valuable.... The flowered ethers perfume your parade to announce your estate of being, something marvelous and new for me—

WALLINDA: Yet we are polite.... and not so well acknowledged in this sparseness. I'm not sure.... if many others *could* notice us.

PONSTOLE: They see. For all who would be here know who I am and the importance of this. I am such a busy person, that my standing for a sighting does make in that itself an occupation for much remarking and some comment. And I try to avoid it most times of mere casualness, since it generally denotes either a dreaded meeting with the king or to perform an assembly of the councilors for deliberations, high stuff to burden a personality observed. But it's why I've brought you with me, and walk you purposely through the court. I want your visage to be subtly known, and shaded with my meaning, that here is someone.... something of renown to join an august party and gathering of the especial.... And that something, Wallinda, is my privilege to escort you with more than a friendliness, but with a love for your company to express. For I tend to you, this day, a sincerity of interest by which means I would know how to draw of this attention faithfully for others and whomever might be. We could by chance greet with the Queen. This is a treasured time of day for her to promenade these lofty surroundings, when the air is particularly fruitful and aromatic.

WALLINDA: I see her not around, if she does wear a gown resplendent.

PONSTOLE: Rose-bound pink, most often. She might be on the other side of the palace, within the royal gardens, and might just happen to come to the courtyard during her evening's constitutional and night breathing of these heavenly outsides. I've noticed this before, since some of our councilor meetings occur on the moon's parade and paradise of approval for our privy thoughtfulness. And if she does walk by, I will address her for you to speak with. We're on better personal terms than I with Musgonie— which is all of work and drudging through commitments— since I made myself friendly and familiar to her when she was new of bride and insecure from this estate's haughtiness and her sudden standing within it. She will see by you what is a queen by me, in all important manners and respects. And you will have this closeness of circle to be in or visit, where the royal reside, as perhaps a regular to beam for her a happiness and desired face.

WALLINDA: That is so much to ever expect.

PONSTOLE: True friendships are rare. They grow from the heart, and are rooted with a commonality of needs and companionship. But I think of her as often alone in this grandness to tread and trust. Though she can be an exceptional hostess when the occasion merits her fortitudes and instruction.

WALLINDA: Does the king himself.... come to his courtyard?

PONSTOLE: Not much that I have seen, by day or night. But

this might be to actually avoid me, summing up his attitudes strictly for our official conferences. We work so hard through them. It is difficult of the social graces to show a pleasantness afterwards— though I personally would have this.... But he is a particularly sensitive man. Cordiality is a breading to him, a lesson *attached* to life. Even some tones of pipe may disturb his composure vigorously, when he is in one of his contemplative states. Yet he struggles mightily to retain these discomforts, out of politeness, and without distraction from his purposeful and conversant involvements with others. I sense the ire mostly in his tearing eyes that, without crying, gray as if by clouds are shielded for his views. Few can be that close to Musgonie to observe this. Yet, would he shadow through the night? The court is closed at heavy night and deep hours, except for some sentries and guards. Then might he pardon himself for his grievances, outside, where the unresolvable might be resolvable of sight in mind, these issues that he faces with a clearing of the lungs through the nocturnal breezes. I would not be king to wake to his responsibilities, when each day brings more hunger to a full and uncomfortable stomach. He must hold too much inside, and with privileges play of his agonies, or play off them one by one until they detericrate.... with himself— But I release my tensions to you, that I am not so fit to hold back passion.... passions, passionate entrapments. Let me take you to my castle—

WALLINDA: I must reserve my master duke, tonight. He expects me back.

PONSTOLE: Oh, some day, I mean, when you have the wish to, and will make the time for it, as better than can be done.... for a visiting. It is a very fine fortress, of the highest quality onyx stone you're likely to see in this region, since these ornaments were all imported. And.... I wish you to become used to my residence, wherewith you may overview it and control—.... its sight upon your psyche and mine, to have you be there. The insides are cleverly spacious and warmed strategically with easily started fireplaces. The home maintenance is simple and un-frightening. I do it all myself, with a manservant who lives in a cottage in the back— But I am not a cold personage at all, Wallinda, despite a possible reputation from some. I am warmhearted, and more amiable than flame to wood— for friends.... The fuel of a relationship is its cherishing.

WALLINDA: I've not heard comment of you as being charring or curt in any way, but of only offering the finest assistance to the king, the one he trusts most in one to one discussion. And I see your nobility to be gentle and understanding without a hint of undue restraint, and never patronizing to flaws in my respect. So then it pleases me to walk with you and offer you my presence as a company, in this society of the elite. Yet know that I am not elite. I'm commonly born, and am in this way proud as only I could be. Then stir some interest with me, if your pleasure spends forth such geld to moderate the fashion of your envisioning this simple curiosity as myself to display for others—

PONSTOLE: And to myself will I be bound most generously to this notion— chained to it!... that what makes more worth be petty as a flower, yet expansive to a universe of realization.

WALLINDA: Then worth is worthiness to assume, in all locales. These fences of hedge are incredible, so tall and majestic. How can they be kept so healthy and thick all throughout the seasons?

PONSTOLE: ....I suppose that is a common concern. The royal gardeners know their art.

WALLINDA: I've seen then from afar as if a form of masonry, and could never imagine peering through, nor observing from within. Here is quite a world.

PONSTOLE: I could say typical of high princes. One makes due to the fullest with the forestry one has, when allowed to. And the shrubs here are quite resplendent and busy, shiny of both leaves and berries, a collected sheen of floral waxes to make an impression in any sort of lighting.... Could glow in the moonlight.... and starlight. Have you never been on court before, Wallinda?... say, during public festivities and celebrations? The local populace is occasionally invited. Special events make for such havoc, though. The confusion of the masses is their worst trait.

WALLINDA: Crowds? Noisy gatherings of the.... regular types of man? men, women, and children with no special talents except for living freely of themselves?

PONSTOLE: I don't like them. They provide a zone of comfort, with familiarity of innocent aggressions and the gaiety of harmless loss of inhibitions, I suppose. But such groups are not easy to control or induce, except with threat.... of weather. A more formal nature resides of the formalities of masters, just as growing responsibilities make one more conservative of outlook and protective of charges. But this is so terribly difficult, at times, not to perceive the common people as anything but dolts to be lead or forced to commitments for their own best interests. Else they would, for example, farm away acres of good, rich soil into exhaustion and barrenness, useless for production, without the legal enforcement of studied and tested methods of crop rotation and tillage. Still they bicker of the requirements levied or fined for, and of the plants decreed for planting in alternate growing seasons. Had they owned a nation, it would dissolve into individual interests and wastes, futility of efforts and inexplicable disgraces misunderstood and uninterpretable by themselves for any reward of analysis, (not) befitting at least in some way from their mistakes.

WALLINDA (*disappointed*): I may have....

PONSTOLE: Say what?

WALLINDA: made a visit here, to these grounds, when I was an infant, brought by my mother. I know my father's regiment assembled on court, those decades ago, with a rarity of review for valors to be celebrated; that she would remark on the handsomeness of his uniform and standing profile on such occasions, to quell my childish squeal and choler over some little, childish.... disappointment.... seldom else would she see him as a soldier, because it was too painful for her contemplations.... and constitution. She died of a plague-like bout from some lung's bug, while he was away on a foray she had demanded he not follow. I was so glad to come under his care again, and out from that of appointed neighbors who could by necessity only be half interested in my welfare, when he returned. The times, then, were so difficult. And yet he was so surprised and startled by what had happened.... But I was old enough by then to explain it to him thoroughly.... So I may have stood on this fescue-d lawn, in tiny shoes or bare feet, with my mother, at least once before, oblivious to all this grandeur, and not caring much for anything.

PONSTOLE: ....I'm sure you cared a lot for everything, even then. Caring and recalling are such different activities. One cares to recall.... and one recalls without caring, sometimes. I don't mean to suggest I think lowly of— our people. They are the breadth of the kingdom, its span, and much our measure for the value of societies we contain. (*turning away slightly, in frustration*) Oh! It's so dispersive, to commit to one's feelings, and then to offend, and waste affections, as by pouring your wine onto the grass and ground. But I feel so swooned of temperament to behave for you properly, (*turning back towards her*) how this unnaturalness becomes so natural to feel uneasy.... and disobedient to your natural height, your better riches to behold than my might and prestige. For, if I'm a worldly man, then you are the world for me. And it is not possible from me to criticize.... where from a heaven is allowed to glow and awaken my notions and needs, this beauty which you own and the privilege of its becoming past. Then I will cudgel my carelessness before (*taking her hands to hold*) you, in making remarks about the.... common— There is no common wealth.... above one's pride to desire. It is unique, and declamatory of your individual worth, to dream this way—

MAELI (*coming up to them*): And what is here, Ponstole? A new.... servant?

PONSTOLE (*startled, but slowly releasing her hands*): Why count!... You make the rounds.

MAELI: And with the fall of day. There's much news to be heard and spread.

PONSTOLE: She is.... the chamberlain, of duke Decamond's.

MAELI: Aye!... And that befits his purpose to arouse. For it's about him all over with commentary, and his incredible conversion for the king. Outstanding and frightening, can you agree?!

PONSTOLE: I....

MAELI: You are.... dazzled?

PONSTOLE: Some things are privy to be kept, Maeli. The king may not want *me* to openly discuss them too broadly.

MAELI: Then what can you say? But I say, his fighters have been certainly taken aback by this. One of his best pretended to me, yet with only disheartenment for sitting still— Sitting still of *it*, I mean. Surprise steals a torch's light, within some camphorated daze to trap your sight. Yet I confirmed myself with Decamond, as his friend.... And I *know* the lady.

WALLINDA: Forgive me, Count Maeli.

MAELI: For what?— For being here? You are more precious stone than can be sculptured. And it's all within his plans of decadence to serve.

PONSTOLE (*with cautioned but growing rancor*): Decadence?!

MAELI: The Lacismoor rambunctiousness, of course. He means to serve them more realistic plates of argument. But as I warned him plainly, the rebellion is two-faced, and the rebels two of sides. He'll lose the best of his soldiers to me— And the man even wagered against this with a friendly grin. That is Decamond! devious

to no fault of planning. He conspires not with Musgonie, but with his whims— him and his, the king's and duke's the same. And I warrant you, some bludgeoning 's to be expected. And the king will chow it down. For how else can I explain his happy stance but that he prepares the seeds for battle, or let us say assures for it. This will sure up tents, to boost the morale of the uncertain. Yet his arm still sways listlessly, refusing a sling at home. Aye?! Wallinda.

WALLINDA: Yes, sir. He gropes for his muscles to fend with the physics as found.

MAELI: But why keep his soldiers from disabuse is still a serious question. For ripe is now to grape of grappling with Musgonie's *bouche* commands, and take advantage of, a vintage of discontentedness to savor. And Lacismoor is rich by any plum. Were that I could contend for it respectfully; but the king leases Decamond, a most trustworthy churl for negotiations, so would his majesty imagine.

PONSTOLE: Has he informed you otherwise, in any way, not to serve Musgonie in a manner appropriate to his purpose? the edict pounced of?

MAELI: Councilor.... what is to inform? What is a friendship? The nature of man resides of such resiliency as to display his better meaning for this course, as coarsely pledged. The king has not even addressed him within presence. Then must Decamond take these matters for his own way to craft and construct, and constrict as a python's pantheon to oblige. Doubt you he has brain?! He sneers at the quivering jelly and sneezes through the disintegrating powder. But he does mean, as always, to loosely follow Musgonie's objectives; for some slack is always necessary in dispatching, when finding how to achieve a result. One's special talent is this, to be one's self. And my friend will do what is effective for this task. But tell me, then. Is this not Musgonie's stealth as well? The king is in a bind of faithfulness, to contract with the duke of deployment. And what does characterize his aim of taxing wealth? It can only be to frighten. So what are *we* to expect, Ponstole? The crown needs more for its treasury, much more, as rumors fly. And this must come from we subjects, we great, and wee as small alike. But mostly from we who have to give, is this ploy with desperation made. One would be a fool not to sense this. Yet, can you confirm it, at least?!

PONSTOLE: That is too broad to speak of—

MAELI: Then tear at meaning. Decamond becomes this might to leverage of his gaining. And like an angel lowering itself on a spider's thread, descends on Lacismoor. That is the analogy I'd make, for this— impersonal agreement between he and the king. Can you really expect differently, from such a clever rascal?!

PONSTOLE: What has he told you?

MAELI: What have you told me?! (*looking at **Wallinda***) But is there a league.... in the making? some understanding to propel your interests here, and between the king and Decamond?... an offering to fixate efforts—?

PONSTOLE (*with impulsive declaration*): I love this maid!

MAELI (*slightly startled*): ....That is much too blunt, sir, even for

A Pound's Profit [Act 5, Scene 2]

you.

PONSTOLE: Major be my mind to it pronounced.... and not as cause to the king, as were a gift! for effort. But we do not bargain with Decamond in this way, nor he propose to us such manners as individuals should caretake.

MAELI: Wallinda, dear. There must be more suitable suitors for you than this.... councilor. And the duke has not sold you for his attempt at gaining.... such an indecent gaining.... with favor from Musgonie, this try at Lacismoor.

WALLINDA: I am not leading to be lead, sir, away from my humble borough of design, with distraction from these proceedings and arrangements between my master and the king.

MAELI: As well should be said as told. My friend does not suggest the like— but deems.... what foolishness is stargazing, with never reaching.

PONSTOLE: I have not been distracted!... nor influenced to favor this design of the king's. And yet to try it makes its flourishing — to try at it.... makes its proporent cares, for me. And if I am a fool for being misled.... it is as the king's knave assiduously bound to his needs. I will advise him.... that the duke (should) come to court for a meeting, to swear of meaning before our majesty the nature of his designs as befitting ours and kingdom, and that no rashness of behavior is in the making.... or contemplating.

MAELI: Oh, Decamond is never rash or over-quick to his maneuvering. But he does tell me he feels Musgonie is too.... solicitous to see him personally—

PONSTOLE: Then I will compel for the gesture, with oaths to be made.... before oafs witnessing. Your friend tells you a lot.

MAELI: As have you, sir. The way that this.... fortuitousness envelops our country's concerns. But the sky darkens, now.... revealing stars. Yes. They're soon to come. And— councilor.... you are persuaded by sincerity of feeling, not sincerity of faith. The better judgments are made within the heads to do the inevitable of plight, with these causations seemingly foretold of result. I will withdraw, to leave you two with— gestures.... Sometimes the obvious can never be as feared or hoped.... for. (*walks away*)

PONSTOLE (*carefully taking a hand of **Wallinda's***): I am.... with chill to surmise this endearment.... and am tugged (at) with insecurities for this transit plied through—

WALLINDA: It is your motion totally.

PONSTOLE: ....the working of this nature. And yet.... are you lent to me?...

WALLINDA: At your behest, I'm visiting the court.

PONSTOLE: ....And I permit the night to fall— all over us. But I am man enough to have good thoughts about it, despite the machinery of my making. These objectives.... are so destroying of the nerve to have them, the presumption to seize their worth and drain with loss, or drink from fulsomely with error.

WALLINDA: What do you think of me, then?

PONSTOLE: ....It's (a) hope for the queen, Wallinda. That still beats passionately. That still surmounts the wall for view.... Yet, the shading disquiets you. She will not be out tonight, I intuit. Not during.... our time. And is this late for you? We shall return to the duke's mansion. (*releasing her hand*)

WALLINDA: This evening is still grand, you know, as what to make a character of court. But I would rather see more certainty in the dark.... than a bright doubting, for all these days condensing.

PONSTOLE: ....It's for my way to question shames and shams. They do indeed look like stone walls now.... But they are not, though quite as difficult to breach. And will you tell me worth.... is what I've known, or anything at all? Then makes this stroll as valuable as your graciousness, through these evening breezes. (*They start to stroll off, he with a guiding politeness.*) I am not afraid to leap over many barriers, for a justice to be made or felt, and on falling be as flustered with the attempt as its ease to make did please, the flailing more to worship than its tease.

Scene II — *A fierce daylight, as **Decamond** relaxes in his garden, sitting in a wicker chair. He is entertaining a visit by **Elser**, who is standing by a similar chair a slight distance away. **Elser** drinks from a small mug. **Decamond's** is resting by him on the ground.*

DECAMOND: I suppose there's been some grumbling within the ranks. And Wallinda's acted rather terse to me since last evening. The mixing of (these) atmospheres conspire for some eventuality.

ELSER: This is a delicious *pomme du breuvage*.

DECAMOND: Yes. I've called for Fennel to replenish. The store for a single person runs quickly in this heated season, even at a cool cellar awaiting my beck. But these fruity drinks do serve me well with their slightly fermented degradations to arouse the palate with tingling sensations, and a discernment of some might.... left in it, and in you. But now, Elser, what be of your friend Alfonso? whose name I kept hearing whispered about me, as I took a walk around the village last afternoon.

ELSER: Chance you to go into the city?

DECAMOND: No. I am bold of arm only about my confederating neighborhood, due to a simplicity of pride to make my way well enough of where I live and am most ponderous.

ELSER: He stays at one of your rental dwellings in town. Thereby I'd suspect you'd hear his name, as somewhat popular for his feats of display; several of which he's given by now, for some small but appreciative public gatherings. Skills generate attentions and attractions out of the curiosity to be awed. But I understand he works with your member Skacte on some enterprise. Perhaps together they are staging these fêtes for profit. Surprised to hear his name spoken *here*, though, unless you've invited him to the round again, but with more country viewing.

DECAMOND: No.... I haven't seen him since anointing him as plausible, or passable for my suggestion to employ. So those things make their way.... even to here with exaggeration of tales. (*picking up his cup, by his good hand*) They must think he can cleave solid rock with that sword, instead of fruits. People are so ravishing of the imagination. (*drinks*)

ELSER: Oh, he's a handsome young man, on which legends can be placed.... (*sips*) Not a protégé, though. Just a friendly acquaintance, Decamond. I feel you think I've pushed him through you, your noticing.

DECAMOND: Have you not?

ELSER: My influence is as slight as to bring him to you, which was his desire.

DECAMOND: Well— if the legendary is based on crude tricks and deceptions....

ELSER: Oh, you saw for yourself well enough his abilities are real and polished. Let us not demote the man from merely his notoriety, or growing novelty.

DECAMOND: Hm.... Well, let it rest in the minds of children easily impressed by the slightest execution of an expert's art. Why, I could do the same, in my younger days, and with only a long knife.

ELSER: You must exaggerate.

DECAMOND: It's the sharpness and speed that counts, for the cutting. But the trick to it rests with a steady eye— or head.... And mine 's grown to contain so much more than to preoccupy myself on such useless silliness.

ELSER: Then why did you approve of him? Why did you acquiesce to his performance?

DECAMOND: ....Because it was something I had done. And they are good memories restored, to suggest he has some goodness to his character if it is similar to mine, at his age. That shouldn't make him any more noteworthy than presentable.... to me. Yet, does Musgonie approve?

ELSER: Musgonie?

DECAMOND: The king does tenderize me, for my missioning. Can this not be true? It's certainly not on impulse that I am thought of, but carefully examined and planned around.

ELSER: By my faith! great duke, the king appoints you to much higher concern than anything dealing with Alfonso. I doubt if heaven knows of the lad, except through court talk and gossip. But you are much more of a topic. For it's well known now, of your acceptance of the king's.... temerity, to ensconce your own, and settle with his thinking, settle down to its practicality— for his purposes of peace. What more need he approve of?! And you have my honest word as a close friend, that I was not induced by him to introduce you to Alfonso.

DECAMOND: ....And your cousin?

ELSER: Not by the Queen either. It's only on chance of recommendation that the young man sought *me* out. There's nothing more to this.

DECAMOND: Oh, I believe you, since your scruples prevent you from lying to a scoundrel. (*sips*) Nor to *any* friend would you

be so disposed. But there is more to this wounding that brings me apprehension of the novel, with such coincidence of incidence, that I am hurt.... and a young man rises, to some popularity or celebrity.

ELSER: You are hurt, and a million other things happen. Despite your domination of feverish thought, you do not churn the stars to rotate. (*sits*) All must happen that is willed by the motion of events and life's developments. And man is only a major inhabitant of this store of realities and instigative forces.... And how is your shoulder? Should not a sling be more appropriate to the arm?

DECAMOND: ....No.

ELSER: It's been these several months, Decamond. A more specialist approach may be needed to heal you.

DECAMOND: The shoulder is all cured, as far as I can tell, but for a slight sensation of muscle tension on bone. The arm remains weak, though, and must learn to regain its usefulness more customary of my actions.... And the hand is.... only a little numb. I fear for it most, its utility of grabbing stymied and starved into awkward displays of merit and shaking. It is my favored hand, and I don't like relearning courtesies.... But improvement is slowly steady, with the proper stimulus. (*sips*)

ELSER: What then cautions you, duke, about your soldiers? They start to dissipate,....

DECAMOND: That's only grumping.

ELSER: or disperse, into the forests to prepare for some adventure, or other expectancy of gathering around goals. Now you must not offend the king, after having pledged your obedience—

DECAMOND: We have all this allegiance made, Elser, virtually through our births..

ELSER: But to Musgonie's particular style you have been called on to adopt. And you accede.

DECAMOND: You mean agree to. And that is not so particularly so. I have not altered much, and have always sternly proposed negotiations with an enemy before initiating a warfare. Musgonie's subtle touch is to remove the moniker of "enemy." Or, since one is hardheaded enough to disobey the crown, not much will be changed of my presence and presentation of options, which can still include fighting the obstinately insurgent. The shading of my purposes stays weak, and of— *our* intentions strong. For the country will remain unified throughout, through many insistences.

ELSER: But.... will you lead—?

DECAMOND: What castigation would be worth it without me?!

ELSER: Can you actually lead men into battle with your present health?

DECAMOND: The present is not for battle. Yet, tomorrow could bring a century of recovery. And both my hands will be ready to clamp down on dissension and choke!... for want of tribute— in the king's name. It's the idealism of it, Elser, that such social retorting is deserved, responding to.... indecencies of attainment and over-prosperous swelling of fortune. For you can't have pockets of good wealth turn bad and irresponsible and totally self-serving and bothersome to contract with or deal of (*sips*) without some retaliation of kind and kindness to help adjust the situation, and readjust for loyalties of true subjects. They only cry for some actual leadership, without which they feel sorely on their own to develop their surroundings and environments, a way of defining advantages as they may so construct. I am— this way of a judge of men, that they may too easily become despicable without, occasionally, some really pernicious oversight, lest the ruin of one's hospitality be useless for correction. Now I know my fighters are restless, to bring sharp quill tips to the amicable abiding, since that is by definition what a king's subjects, or citizens, should be. And any contortion of the term brings out my result. But these are unusual circumstances.... for me, and so to my men. The cake yet must still be further layered, and frosted more sweetly. That's the only reason I resist the slicing, until more propitious a prize is prepared, produced, and propounded. But it is not yet present, not yet justified. It waits.... on me, as if with destiny to steer— a steed of contention and conflict raging through my marrows, were illness to the blood itself constrained. (*slightly shaking his cup*) Oh! I'm so awkward to remain, these too many days, these weeks of wear an' withering. I have actually dropped a mug or two, with my less favored handedness.... And made more of a mess than misery, Elser, with my abundance of faith, is this contradiction housed— that my soul is ready for abuse, but can hardly stand this bodily wariness and timidity. I wonder.... if this is a mental test of will more so than the physique's injury. Thus, I learn to giving arbitration a turn, for some short while, as the arrowheads are sharpened— as they say.... with a means of mind to be ready for use.

ELSER: Are you still so ambitious.... to have combat, even as you hurt—?

DECAMOND: I am not— hurting. There is no physical pain to these sinews and tendons, and popping of the joints.... in the shoulder armed.

ELSER: Armed? Is the wound well cleaned out?

DECAMOND: As good as can be done. What's left is healed over with scars, and what might be foreign (is) shrouded with skin and flesh.... But the surgeon assured me, short of taking the whole arm limb off and apart, the shoulder has been thoroughly excavated to the standards of his profession— For I did complain, of a funniness to the pain, for much of a moon's duration.

ELSER: I still say you should seek a specialist on this type of inserting insult with weaponry.... Thrown weaponry. It's different from stabbing—

DECAMOND: The trauma has passed.... the burning sensations replaced with a surrender to ease of motion without the swells of limitation and reluctance. If one had to rely on those specialists during the strong clashes we encounter, wars would not be won but merely analyzed for descriptive medical texts and treatises on remarkable assaults to the corpus. Not many animals can continue to think well after the injuries we often take. Our obstinacy to defeat cures more than the medical method. And so.... if thus some tiny fragment from a weapon remains in me, impinging on a nerve, and now encapsulated from finding with the careful examination of an expert, does this bit of furious shard have more life of will than my own? I'd say not, and that it is dead, as could be chains to the wrists, of rope or metal. And how do you deal with the dead? One

buries, or carries with for burial or burning. And I've heard tell these materials make their way out of the body themselves. For they are cursed to stay within, without more reason than to be a nuisance. And they feel accursed, wholly unwelcome.

ELSER: Do you suspect that? It's silly to submit to those beliefs of wishful thinking. New manners of dealing with these problems may have evolved without your notice or education (to be alerted by).

DECAMOND: The warrior knows when there's nothing more to be done about a project, friend, and to let its nature take its course. One resigns one's self to its own ambivalence towards you, how it can not realize the harm done.... you, and so (you) permit its blind craftsmanship to play out. As much it may be a warning or a lesson or a gift of designing— many fates, of friendships to dissolve or rearrange of character. And.... what is the retaliatory act, Elser, but to let this wounding have its chance of action?!...

ELSER: Ah. You still have suspicions.... of a loss of felicity, among your rank an' file. But what is accident? Have you ever considered *that*, that one of your soldiers struck you by accident?... did not really mean to hurt you.

DECAMOND: I have considered, these months of a retired gesture or attitude, many diversions of aspect about this— stoning, and from many corners of bumbling. We.... are accustomed of making many attempts— before a success, and expect the same of others. Legitimate wounds during a battle are passable to accept after they're done. They are chanced at, with a need to fight opponents. They are a principle of merit and currency, though the process may be usurped with error. And betrayal is difficult to define in a confusion of hatreds and loves. This.... fragment, should it exist, will reveal itself eventually, through the spite of its own cursing. And if it not be there, then it is my own error to believe of it. You see, within this— long time.... my men investigate themselves, and clarify of aim. Those that still trust me will find out and hound— the perpetrators, well before jeopardizing their lives with the start of a new.... excursion or marauding— as you may put it.

ELSER: I wouldn't say so, that it is evil.... but has been just a bit evasive of Musgonie's practice to be sovereign. There must, of course, also be some uncertainty to this point among your soldiers, that might be expressed with a casualness for experimentation—

DECAMOND: If taught the method, the means and manner and wherewithal to do! You mean that they, or some, should test my reaction to them— my faith!... in their validities. And the simple fact of it is.... that I heal, as I have told a few. They will for themselves.... consider this, to taint reluctance for a rebound, if they are the stout of heart I'd associate to my privileges and have assumed to rest my.... health with, slap my life to and restore of.

ELSER: That would be a curious elixir. (*sips*) Ah! You mean.... that *you* are testing your men. You are probing them with the threatened adversity of your absence, and you wish for them to right the wrongs that may cause it.

DECAMOND: One could hope for such a comprehension; though it be rather gratuitous, for I lend them no real justification to try. What happens on the battlefield— stays on the battlefield. And all is forgiven after the sparring is finally, thoroughly over, completed and concluded forever. How else can one commend the dead and permanently wounded to these results? And what is worth a life's struggle slipping on a plain of fractious purity? The errors stay always of war to war grounds. It is the ultimate abyss and depth of fright through which to enjoy the squirm of contending for superiority. And what is won is won: wounds and trophies, and conditions, concessions, obligations and obliquities of purview for the future. This is how societies operate, to condense or disassemble, coldly and cruelly. The process is sped up in heated battle and unruly behaviors of mortal confliction. But it is essentially the same; and all must be forgiven after the division of the spoils is made, and the contractual agreements of servitude and mastery are ratified. Yet off.... the court.... are to be found some strategies of inquiry for what has happened, and what has caused this world so queerly felt. Why does the sky turn red in one's eye, from blue, and other such curiosities, are proposed to be answered. Is it all the sun's fault, that our Mars-like natures cause us to fight? Answer this! before the next bellicosity is engaged, so that we know better of ourselves— to produce even greater events. And who has wounded me, Elser, count.... has caused much balloting to be made. For I am tried too deviously to be pious about it— or abstract and indifferent, and yet am too friendly.... of my men, to be rewarding of their diligence in this affair. They must handle all of these details themselves, and come to some understanding of what to do before we can possibly (cause) adventure of ourselves together and again. That is inherently obvious, through our decrees of comradeship.

ELSER: Don't tell me you have suspicions.... as to distinguish knights from nobles? You still indict the crown. It is the enemy that has wounded you, Decamond, the enemy that is no enemy.

DECAMOND: I'll say I've never fought one, of that ilk.

ELSER: The need to do these things, my duke. The obsession and compulsion drive you and many into injuries. Yet for the satisfaction from these endeavors, must then some recompense be made — to be responsible for your hardships, if are to yourself beholden be. That must be rational.... to contemplative spheres of thought.

DECAMOND: I am not the enemy to myself, but am conspired upon— friend. As were to take on many dangers breathing (*sips*), there is a breadth to them in any life. And I ask you openly, to suspend the mystery. For what does rumor say, below the king's ears?

ELSER: ....And the Queen's—

DECAMOND: Ah!

ELSER: There may have been influences plied, but of sentiment only. That gives some license to these.... intentions. You have a body of soldiers, right good ones. But you also have rivals, right friendly ones, perhaps with jealousy on your success. So who knows where the courteous may lead through influences? including one or a few of your men.... to experiment.

DECAMOND: Is this what rumors say? Does she despise me greatly?

ELSER: One can rationalize any sort of explanation without knowledge. Stories run rampantly, from the nefarious to the buffoonish: that you jabbed yourself or deliberately stepped in front of a poor marksman for injury, to give yourself an excuse from warring, having felt worn-out but incapable of escaping the pride of

your legend by consciously bowing out.

DECAMOND: So now I subconsciously drew upon an errant arrow's use. (*sips*)

ELSER: Whatever can be imagined, argued over or joked about. Where does the seriousness descend? She does not hate you much, but blames you much for Musgonie's disquieting— Though I can tell you faithfully, she did not propose you as the king's advocate or diplomat.... I suggested it, with the merest hint, originally, that it might be possible. How this has worked its way, through Musgonie's machinations, is remarkable, a simple germ of an idea germinating into an apparent reality. But the Queen is not the plotting type to maneuver you in any fashion.

DECAMOND: Well, you speak for her, as a relative, not ever to besmirch. But even if she could conduce some of my soldiers towards my bad health, this would be forgivable, from her position and stance of interests. I suppose we must all try to protect the king's sovereignty, in our particular ways.

ELSER: My cousin did not injure you, Decamond. (*sips*) Do not blame her.

DECAMOND: *Sache*! I'm not even injured. There's no proof of it. (*sips*) 'Tis only an allergic reaction to a bumblebee's sting. And a huge devil it was, sent by Athena. (*notices **Ponstole** approaching, from a house door leading to the garden*) You again?! Have you no real business, man, than to wile and whittle with my maid?!

PONSTOLE (*approaching*): Good afternoon, sir. Count Elser. Yes. Wallinda let me slip through to see you. She's very busy. You give her all too much to do—

DECAMOND: I am—

PONSTOLE: —about me to call and bother. What precious little can we meet upon but this estate's panoply of protections— from the truly caring. But my heart feels fried—

ELSER: It's a hot day.

PONSTOLE: —and we spoke little, this time.... as my business is urgent of news.

DECAMOND: I am for her benefit of work, councilor, as befits her. Yet I fear you made her realize her place well, last evening, and embarrassed her pride (*placing his cup on the ground*) into humilities upon the court. For I told you it's not her proper location to be at; nor for your purposes is she. But one finds these things out for one's self— and painfully!... with her crude dress perhaps demeaned, as could her face be sacrificed for a low and lecherous attempt in the minds' of some— high of station and offended with the imprudence of her visit. I myself, even, seldom play to raised noses— for what I might hear.... at such gatherings.

PONSTOLE: ....It was a beautiful evening, sir, of teary-eyed resplendence.... and a suspension of my age, as leering lends it's course of dreams upon the angelic to view, and a discrimination of charm against reality— until brought to me Count Maeli to grow old and firmly staid again.

ELSER: Was he there? But you're still *relatively* young.

PONSTOLE: And the stars fell, sirs.... for more my mission to produce. In this I have been hectically busy. Wherewith for your employment, Decamond, I have convinced the king to have you see him—

DECAMOND: What?!

PONSTOLE: —and address and inform, so that with personalities personable all matters between you may be (thoroughly) understood.

DECAMOND: Why would I want that, if it's not so necessary? The intent has already been agreed upon, I was led to believe.

PONSTOLE: But more develops, as of Lacismoor, for example, that brings of eminence yourselves to this imminence of she. And there demands of a topic to be forthright with, as you are called for high representation, the highest yet permitted of a loyalty. And I had to almost force the king to be curious about this,— demand more presentation of standing, since he assumes much by reputation alone.

DECAMOND: Are you telling me that I must assure him of my faithfulness to the kingdom?! I, who have fought slavishly to maintain it! and I, who have crossed this country through and through to observe its every wheres and know its peoples fully! I am more in harmony with his subjects than he— that is why I'm renown, though notoriously praised. Yet, how can he ever question my intentions when they have always been straightforward and evident with my actions— with my daring, with my undertaking of the dangerous of situations— on his behalf! And now, with my wounding, even still am I disposed to serve his better instincts of unity. I have given my consent to try, and have let this be known to everyone. Why must I then proclaim it to the king in person? As some official act of private parley? or a conferment— to see a weakened frame? Does he not trust my word, that he has never even used yet?! What should be now is an agreement to be signed — for a forfeiture of valors! I thought he was afraid to see me, and be reminded of my passions— or repulsed.... through them to see such struggling as I've done— and will continue to do!

PONSTOLE: You are wounded?... and would be embarrassed before Musgonie? Hm. I thought you had healed perfectly, through what you have professed to.... Oh! But (it) is wondrous of feats, and for this mission perfectly made— to dream!

ELSER: He is physically able, to carry out all types of negotiation, and is mentally capable for a universe of more.

PONSTOLE: No! sirs. I dream. I have dreamt. I have slept upon the hopes of the miraculous. And does your arm befit your nature now? feigned duke.

DECAMOND: As much as does my character shoulders these realities.

PONSTOLE: Are you in pain! sir.... Are you feverish and contrasted to one's health! Are you loaded with weights of *malheur* and unhappiness, and balls of balefulness to cry at— that you realize what you are, what you have become, and what you must stay as! for the rest of your miserable existence—

ELSER: Councilor!... This is Duke Decamond you are talking to.... And not describing— No!

PONSTOLE: Cruelties crumble me, lords. For I feel much severed of limb, or heart.... and am fallen into a decadence of candor, for these agonies to depict and gloat with, gloat with the proud Decamond, who would defeat sincerity itself. But are you healed?! sir — And how?!

DECAMOND: ....By drink, of a morning's rush (*picking up his cup*), and hearing the thrushes chirp.... upon the orange to yellow glow. (*sips*)

PONSTOLE: Could you dare accost our aims?!—

DECAMOND: Ours?... You are love-sick, man— for the fruits of life, for the benefits of a daring, which you have withered from, squandered and lost. And now you could reproach me for my injuries? What of your.... crippling, with a life's long struggle to remain.... sufficiently soporific to your siring as a suitor to the more wonderful urges of life, the surges of emotion that have made saints dead! for worshiping. And now you awaken from your sleep?! Then what have you been walking through, these several decades?! With service to Musgonie you are alert. But it is too late!... to lose more limbs and thoughtfully survive. But I am a fighter, sir. And an antagonist? perhaps. But I'm not afraid of— sainthood! and mummifying balladry (*As he's about to drop his cup by accident, he changes this into a forceful, apparently deliberate, throw to the ground.*) for what I've done, of legendary harm — and perspicacity of aims. And the awkwardness of my eyes demeans you, feeding the earth back of its juices. Yet, can my honor be questioned, councilor, in that you upset me?! For you would have— none of this! on your own. I sense this almost with a religious fervor.... against your type, against your resistance to happiness— and embarrassment for it. Well, I have fought.... just the opposite. I have fought the rich of heart and feeling. And it is an imbibing mated to, and a defence!... of Musgonie. Yet now, more matter shakes my girth of turmoil and restlessness than were my entire arm to turn into a sack of gelatin. And I pry and claw and scratch for usefulness, during your slumbering.... and dreaming. And this would Musgonie know: I am no fool to his utility. I permit myself.... into his regulation, as these problems heal.

PONSTOLE: ....You draw me as a skeleton without flesh. And I was very doubtful indeed, of the king's proposal, his propositioning of Decamond. Yet I have muscle more, about my anguishing constrictions of desires blundering to my face, with blows to body felt only of folly and foul worth and pout of station.... and allegiance.... to my sad manner (*hands grasped before him, not quite for praying clasped together*).... that the king suspects you'll fight no more, that the need for that pursuit in you will evaporate, diminish, as the value of your diplomacy grows. That is his hope. (*hands coming apart*) And here I have delivered it. More holy than deception have I bound myself to this attempting for your services, to see for myself what may be done. And I've seen much too much of my own viewing, to blur the eyes into contrition.... for what I could expect. But with the count we have convinced you. And I have made the king want.... to hear this from you directly, and courageously, as like he hears from me too often for his patience. So then, I submit this need, and his request....

DECAMOND: Oh, then I'll see him. (*getting up*) And I'll wear my.... ceremonial, more decorative sling, to be polite. (*as **Elser*** *stands and retrieves the thrown mug*) But my tastes have not.... dissolved. And a thirst still lingers. (*as **Elser** gives him his cup*) Yet with honesty am I.... re-handed this business. And that is a powerful faculty to own, Ponstole. It helps to keep you from running askew and into ditches of dirt and soiling filth.... that should be obviously seen, or smelt for notice. For these debasements only represent yourself— at your worst. (*heading towards the house*) Yet for a day as bright enough as this, so much more radiant of pleasantry should be evident and found.... found standing and awaiting your admiration with attendance. Else they.... not be much at all, without some.... participation through their splendors.

ELSER: Does Musgonie expressly permit this? It's not of his temper to want long explanations.

PONSTOLE: I made for the need, Count Elser, and with a difficult persuasion, as I said. Because his friend Maeli suggested to me the duke meant to prepare for warring, or to lay the groundwork for this adoption— totally against our purposes. Already there are reports that some of his soldiers camp, with expectation of (a) force to be called soon. We can not ignore the possibility of deception, and so Decamond must speak on his honor to the king, an assurance more powerful than our indirect deliberations, our going between reluctances to meet. That is the brunt of my scope, and probably yours— I've been burned enough. (*turning to leave*)

ELSER: With certainty, councilor, you exaggerate your wounds with sensitivities promoted by ridiculous temptations.

PONSTOLE (*addressing to rejoin*): I have been scathed! sir, and scalded by my own boiling humors. Every tickle to you becomes a deep jab of pin to me, and a scratch a thorough gash releasing sediments of hope. I am outdone by my own foolishness to think otherwise— of myself. For if the high revile my sternness.... so must the low condemn it, without a sympathy for hating me.... but simply noticing this hard work walking.

ELSER: Yet still, Ponstole; you are too principal to melt.

PONSTOLE: I am of a burning day— dashed by night! What irony of an irrevocable wrong, to see more in the moon's darkness than in the sun's bright portrayals of the possible, of an awakening to blossoms— within myself to burst of frost and coldness smote by and (was) smitten to. And I am hurt by the pangs of my own deceptions, and could disown this sorry flesh thoroughly to burrow in a hare's and run away defeated from the fright.... of my loss, my cause of feeling to deserve a love. But man can not make that transmutation into the animal, without such terrible suffering as mine to follow or result, and so must endure his peevishness as a lesson of reserve of energies for his higher and more tractable goals— away from blissful fantasy and fortuitous nonage.

ELSER: Still, sir.... such immaturities are allowed occasionally, to— for example— evoke memories of childhood, or child rearing. One need not abandon one's happier stars to a galaxy of setbacks. That is just being perversely regretful and resenting of life. You single-minded chaps tend to focus too narrowly on what you are studying to as much destroy it with your searing light. Yet have more broad ambitions for things especially affectionate and vague of facts; for in this way you are creating them to your profit, with a softened touch that will be approved of and rewarded.

PONSTOLE: Tell me this of bone!... and backbone, and I will abjure all callousness. Yet I am too hurt to break.... and too weak to bend towards my sorrows to re-edify this heart. Only the shock of nonsense can restore me for the impossible— or to it be forsworn to deny.... that I stay so deeply affected as to tears to call upon with witness of bleak views to the future for my solitary, self-remanding aims.

ELSER: Hmmm.... Then share more aim. The sun attempts to shine on everything. And yet, an impossibility to you— now.... is like.... the apprehension of one heart beat to await the next, you are so touched to worry on the shame of your envisioning. The fears and disquiets will subside, as more chances arise.

PONSTOLE: That should genuinely be true, with rational thought. But I don't feel this way, is how I'm taught to respond to a keenness.... More work is the only remedy for now. But I thank you for your.... considerate advice.

ELSER: Have some of this juice—

PONSTOLE: No thank you, sir. What soothes the throat should mend the head as well, under our circumstances. I wish not to relate the good to feeling bad. That oft becomes the nuisance of dependency, when misdone.

ELSER: Maeli is a friend of mine also. He boasts a lot, to shade his cautions.

PONSTOLE: Then may we have ours, sir. (*turns to leave*) I thank you for the afternoon.

ELSER (*as Ponstole exits*): Please do inform me of these details with the king whenever possible.... as concerning our good Decamond. (*sips; then speaks to himself*) I hope we will have witness.... to the event. (*wipes his brow and goes over to where the cup landed, to study casually the ground*)

Scene III — *A cellar in **Decamond's** mansion. **Fennel** is setting up, on a floor shelf, a barrel of "cool relief," as **Wallinda** watches, or oversees.*

WALLINDA: There was a crack in the last one, I tell you. I could see it dripping.

FENNEL (*occupied*): Where? Must have been from the spigot. You can't see much in candlelight. I check these casks carefully before and after locatin'.

WALLINDA: It comes from the side.

FENNEL: Perspiration, then. The door should be kept closed more, to keep this space cool.

WALLINDA: It's a bother opening and closing that heavy golden *Tür* so often as he wants a drink! And how can oak transpire if it's dead?! Was a crack in it.

FENNEL (*studying the woodwork*): Condensation, then, from the air, Wallinda. Would be impossible to breach, unless it was banged up. And I cart them out myself, so I know which ones have fallen, before being filled. I never use the compromised casks for here. The duke's more valuable a customer than that. You know. (*finished examining*) Not worth the trouble of keeping the broken ones around anyway. (*facing **Wallinda**, in a friendly manner*) They can cause an awful mess eventually, coming apart to lose an entire load. They're tested with water, you know, during their cleaning periods. This one just the same, and the last— They all go through their trials. Don't worry about it— unless somethin' happens.

WALLINDA: Well, I could have sworn the last stock contained less than previous, or my duke's become more thirsty during the summer heat. But I tend to count the cupfuls requested; though inadvertently sometimes, this is such a minor servicing. And I tried to check the reservoir—

FENNEL: You can't fill it up all the way to the top. That would be too much weight, during the traveling. I always bring the same allotment as'(is) ordered and paid for. And he don't need your hustling all the time, for his fruity tonics. Used to let us get drunk down here, himself included, in the earlier days. But then the place was exclusively for wine and— the spirited. And that got me started in my interests making up refreshments— 'cause I was very creative, and learned you didn't always need so much of the alcohols to satisfy. This one's intensified by the seed and pip grindings, before being clarified.

WALLINDA: There's still some pulp (in it).

FENNEL: Clarified of the seeds and pips. It's done by weight of heavier masses. The settling.

WALLINDA: Sounds salacious.

FENNEL: Well you have to separate the good from the absolute. There's a sort of filtering screen involved. You dip into it—

WALLINDA: I don't want to know about how it's done, just how much we result of.

FENNEL: We? Do you like my elixirs? They're pretty strong by the time I bring them. I thought they were the reserve of the duke.

WALLINDA: I try some of it occasionally, mostly in the mornings. You can tell what needs cleaning best in the mornings, after the ravages of the night vermin.

FENNEL: What? For drinking— or deterging the ants, aphids and termites?!

WALLINDA: It's good with cooked bacon— for breakfast, as I'm due a hefty start for working this abode. And in the morn it's all here to do. What would you think, with only two to live, then too many to live— through the night. The chores come from after the battles won. And do we hear the cries of what transpires through the dead of light, then it's a wonder I don't awaken— to find pools of blood in dungeons; this mansion-ed manor has so many rooms of recession and withdrawing from view— what might occur to uninvited visitors!

FENNEL: Down here are the rats?! That would be foul— That would be for the cats to chase. But this is a very wholesome space, for all that I've known of it. You keep it fairly decent, unless the duke brushes floors himself.

WALLINDA: Mops and brooms, sweeps and swoons. You've no idea of my exhaustions to manage such a large house as this, if it's to be maintained in an orderliness.

FENNEL: But only a few particular coves need to be worked spotless, for guests to notice and remark of the duke's fine attendance. Most other rooms need not be bothered with.

WALLINDA: That's not the point. I live here too. And all the place needs be bathed regularly, to keep from decays and mildews. Else I'm not much to own, for all my weathering. The man needs so little.... yet has so much.

FENNEL: Satisfaction can be a ditch covered with straw. That becomes a haven of heaven, in some climes, and under many conditions. Now, aside from my laboratory, for the labors of potability, I prefer a particular retreat, in my house, and don't bother much with the other locations except for their regular utilities, like to sleep, eat, and wash in. Though the miss— and we ain't had sacrament— finds the whole place grand. But my own corner's colored great for thoughts, dark for pensive enlistments.

WALLINDA: She probably scours all the rooms, to keep them manageable; and you don't even notice or appreciate.

FENNEL: Oh, I appreciate. But I only notice where I spend most of my time— and we together. Though romance is mainly for the out of doors. Housekeeping and sheltering is for the in of doors, and is a pretty tough business not lent much to frills and frolicking. But I ask you, now. How can a man like me have a wife, while involved with so many fruits?— and then to be off to fighting at any reasonable moment. It would not be proper, nor much permissive.

WALLINDA: Unless the partnership was consummated with child or children.

FENNEL: Ah. Then there's the very definition for a respectable burden to marry. Ay, it's the marring of one's independence, to take care of the dependent.

WALLINDA: That's very much so, Fennel. I don't see a legal need for the institution otherwise, 'cause it certainly can't be based much on the extremes to endearment. I can't grow with the notion, from what I've seen and felt.... lately. Companionship should be left to companions, and their follies of adventuring, under their conditions found or environments merited. Yet one can love children, and at any where, or at any time.

FENNEL: Yes. But the secret is that the world hardly needs any more— (*notices **Decamond** entering, and takes on an exaggeratedly broad grin*) My lord!... I've just supplied.

DECAMOND (*cup in hand*): Good. I wondered if I would have to make do with the dregs. They are bitter in your concoctions, Fennel, and aid the nose more for their intensities drained. You should cake them as some sort of health candy, maybe for infants yearning just past— sweets. (*coming up to the cask*) That's a terribly difficult time for one's health, a defining period, perhaps, to see whether you might make it as a person. I'm told I suffered a terrible disease, that might have ended me at two. Yet here I am, and you two. And what have we survived through? for generations' wheeling and dealing. I just feel old, myself. But your potions help to stimulate these nerves. And I take them, else magic cures.

FENNEL (*impulsively getting the cup*): Oh, I'll fill it for you, sir.

A totally fresh batch aged for today.

DECAMOND: Well, today's the best time to be. Would be a shame to be otherwise.

FENNEL (*noticing the mug is slightly soiled, and rubbing it*): And how is your arm today, lord?

DECAMOND: It hangs.... And I sense its weight. But perceptions are coming through, Fennel. And I've always been able to adjust my fingers. The whole assembly is just so ridiculously weak. Yet your drinks strengthen— and I'm counting on them.

FENNEL (*examining the cup*): But this mug's cracked, sir.

DECAMOND: Oh, is it?

FENNEL: Split down its side. Would be dangerous to use, precariously holding substances still liquid.

DECAMOND: I had dropped it.

WALLINDA: I hope your awkwardness is not growing, my lord, as you keep testing that hand.

DECAMOND (*with a slight embarrassment of reflection*): Well, well.... one has to exercise these— problems, until they resolve themselves. I can't allow.... an unusual feeling of grip to be so long protracted. It is too funny, if it cracks mugs.

WALLINDA: Did you lose much of the *Saft*?

DECAMOND: About a quarter cup— but to the grasses, Wallinda. You don't have to clean anything up.

WALLINDA (*almost tersely, while exiting*): Then wasn't it foolish to try, since you love it so much? I will fetch another cup. If this is the best to be tapped of the barrel, then a large one I'll bring down. (*has exited*)

DECAMOND: Fennel.... what goes with her, this peevishness with me lately? Fear me as too childlike becoming, dropping things?...

FENNEL (*fondling the cup with some emotional concern*): Yes, sir. Afraid that your debilitations might be growing, or that you're not improving as you should. The prospects can be frightening to her, in her reverence.... you in effect replacing her father's imagery —

DECAMOND: Oh, now, don't say that.... without the succinctness of your product, that I may ever feel responsible for her. I sense to come apart too dearly, with this healing, and not for which to affect her too much. But in fact, I threw the cup down on her behalf, out of anger, to challenge this Councilor Ponstole to get serious and leave her alone. For he only plays at imagery, while she must live it of her hopes defined.

FENNEL: So visits he from the king?

DECAMOND: Of course.... and again— and too much. And from the king, or for subservience to whim has he been here often, trying to persist in some intoxicated state beyond his fairness to me, since the business is more firm and doable than his wishes.

FENNEL: Then does this king drive your parts away.

DECAMOND: ....I've not lost an arm to him, unless you've heard otherwise. I hold, still, to discount some plotting against me— by Musgonie.... or his others. Yet lose I a hand's use?

FENNEL: Count Maeli told me you intend to disband.... much, much more.

DECAMOND: He's a bit flippant.... Fennel, before fighters.... and soldiers.

FENNEL: Yet, he tries to recruit my services military, and Skacte's.

DECAMOND: Not for other means may you do a profit? I dared him to try. But is this so unusual, after my shouldering delay? He's a current man of daily affairs, who actively likes to maintain his forces, and is often looking for potential and talent. But does that mean I would disassemble our especial enterprise, our conglomeration of daring, the uniqueness of our persons to obtain the individual valors each may deserve?! Nay! I wrought not such a ruining— And you know this true of me, not to wear defeat nor warrant for our dissolution. See me bound at home? I am not bound to the discomforts.... of such thoughts, as were a sling to wear— if nothing hurts or is sore or swollen.

FENNEL: It could help to remind you of disuse and preciousness, that wrap of bedding, that emblematic holstering, that symbolic guise of sudden restraint.... in your procedures or modes of thought. And why do you make it, the cloth to wear, so decorative? It's as to emphasize your clinging to a change of demeanor, and advertise your responsiveness to new environments of action, subduing previous inclinations, obsessions, fanaticisms, and drives for gains with glory.

DECAMOND: Ha! The mere scarf itself is a handsome piece of figuring, that does displace my motion quietly and without imbalance— yet holds as much my worth as any other garment worn. And neither should my men feel aches within this bin of restraints we find. For they are themselves, as independently alerted of a cautioning. The tempi of events change for our adjustments. And as I heal, so do nations and their rulers that consider themselves about us. But I am conscripted by Musgonie, as a loyal subject of his kingdom— and not to be made suspect by him more than ever through my past activities.... nor ours. For I only work to placate the soothing of his worries, with continuance of our intentions, as I myself am smeared with time's mitigations to alleviate a weakness. Such may be allegations made that I end our military missioning in favor of diplomacy on the king's behalf. Who can believe that? But Maeli stormed the question to me— and with a sense of fear, Fennel. Because we represent.... an alternative defiance to the lackadaisical and inept of perceptions of rule. Yet our king is really neither of those qualities, and is keen enough of intelligence to come to me with a direct offer for my utility. Now, how can anyone disavow such a usage inherently pledged?!— Could you?! It is what we have been doing, in effect, throughout our adventuring. But now he takes the bold step to address my service. And no one may decline from such a courage if one has fealty to the governance of our country— Though he hopes to shape me more than my wounding, with patterns of design pacific of conducing towards his will

— Well.... such may be a fashioning as fate, more powerful than the minds of men and mortal hopes or aspirations. For he expects his method will work miracles, and will turn ambitions into amendment of insurgent wishes. But we know other, do we not? that one can only *fight* ambition— to determine what prevails. And so I submit to you, Fennel, out of respect for the soldiering, that I am only lent of time to heal, as I have asserted to you before and to others often. And with this Maeli can only laugh and make jests, but not with a buffoonery before me.... a friend.... but with a wager of results to apply, quite ineffectually, even innocently. Because he can't know the character of your true hearts to trust me, those whom I've led to riches and richness of fidelity to their encompassing aims of heroic encounters, breathing and breeding— for their purpose-filled lives.

FENNEL: Yet would you sin against sincerity, lord? Can you heal— or is this permanent, as for a crack of cups?! One can not abandon one's king, in this manner weaved about you. And he would have us end our legion(naire)-ing.... by your name. Then should we not forgo that break of soldiering, wherewith another name to follow—!

DECAMOND: Maeli?!

FENNEL: Alfonso!

DECAMOND: ....That is more error than a moon not born!

FENNEL: Youthful, talented, spirited, determined towards our intentions, inclined towards our manners to learn— ancient as they may be— self-sufficient and yet self-effacing enough to be a great leader of men, deferential to their greater prides— For one only wins pride of battling by conquering— is our Alfonso.

DECAMOND: Much talked about.

FENNEL: —And chosen by yourself to be.... a member in training. But this is a difficulty to temper, or tamper with or play around or abrogate for sense, of men's sincerities, lord. And we will have not the game, as I can tell you faithfully, in that you may.... sensibly.... pardon yourself from our activities, after being hurt, and chastened by the king. Then we do prepare for ourselves — freshly and unpredictably. Yet this is what we are left of reason —

DECAMOND: Reason?! Fill that— cup with your juice!

FENNEL: It may collapse, Decamond, and within my hands, or yours—

DECAMOND: Fill it! (*Fennel starts to comply.*) And if it abhors you, then let the spill be of my loss, and the fragments of craft my remains to stare at. Yet not abhor you, but your sweetness and power and stimulation to count upon, in this intensification of my grip, do I abide in fate the fact of all I've done, the faction I have made for us— and my thirst.... as genuine for this daring. Let more blasphemies weep of Decamond!

FENNEL (*filling the cup*): It would have been somewhat difficult to do.... with your one arm. But then to place the retainer correctly under a high spigot. Yet here it holds. (*turning off the spigot*) This is full.

DECAMOND: And hold it up before me, Fennel. (*Fennel complies.*) Does it drip— does it leak? Yet is the crack so solid? Is the seam dissolute?

FENNEL (*clasping the cup with both hands*): It holds from with to drink from, the moisture contained inside. Yet it must be afraid. For this is hard to do. This tempts disaster boldly, with too terrible a circumstance to cause, that you could rely on this sign as a gamble, that on us and yourself to place this beverage. And what if it comes apart? What will be the destruction? What will have been made of this sign, but that you are through with us?... and so simply!

DECAMOND: It brings the want. Are you afraid to lend yourself, and by yourself performed? This is yours!... And it is mine—!

WALLINDA (*entering, with another mug*): What is this?... You're playing games?... or experimenting, at my expense to clean up?!

DECAMOND: The spills I will lick off the floor with my bare tongue, the dirty, shoe-stepped traces lined—

FENNEL: He dares the cup to break. It is for madness of our loyalties to taste. Yet, be it so—

WALLINDA: With foolishness. (*holding her cup just under Fennel's*) Pour, man! I'll not make a mess.... nor permit it (*as Fennel empties into her mug*) of good breakfast wine to lose. There is.... the childishness between you, only. But if a cup is cracked, then you discard it!

DECAMOND (*almost transfixed*): No for want.... Not for waste.

WALLINDA: This is a precious delight, lord.

DECAMOND: Give it (to) me.... (*as Wallinda hands him the mug*) for my throat! Here it's deserved.

FENNEL: And what of this.... emptied vessel?

DECAMOND: What for?! It's done its job. It has delivered, and served its role soundly. (*drinks*)

FENNEL: Was held.... to hold.

DECAMOND: Oh heaven! for this parched heart. 'Tis your best production, Fennel.

FENNEL: I suppose you seldom tap it, right after I have held it.... and rested it for you.

DECAMOND: It's such a hot afternoon. (*drinks*)

FENNEL: Set it steady, for your allotting— May I keep the mug?

DECAMOND: What?!— Yes. (*drinks*)

FENNEL: I can measure some seed and pip chippings with it. A handsome style, for this container.

DECAMOND: One of your barrel's worth a treasure trove of them. That's a principle of good design.

FENNEL: This container type.... Well, she's a standard for the business, a necessity. Breathes, you know.

DECAMOND: I am almost satiated with delight, Fennel. I sate, and hold you some proof of this.

FENNEL: Yet.... (it's) remarkable, how these matters may evolve. I must withdraw now, having brought, given.... and set up. (*starting to exit*) A good day to you both.... Though I'd try to have this chamber closed more, to keep the coolness confined longer throughout the day, this summer.... Would help the wine casks as well as the fruity tonic.

DECAMOND (*Fennel has left.*): Well I'm not fool enough to keep the door open.

WALLINDA: He seemed somewhat discontented, my lord. Did you actually have a bet with him? Would have been really indecent, for the floor.

DECAMOND: He feels my head, Wallinda, and warns me of the sincerities brought and the sympathies left with. But I did try to test myself a stroke, though his knowledge of these drinks is much superior to my own. And that is how I have, for liking of superiors to handle. The gods manage the world, the universe. Well what is better, what is higher, than this vast intellect, this scope and scale for actions to be made?! To really seek out this refreshment was my main cause, a duty to my burdens.... and burdened brethren. For how else to be obliging may I be amongst those fellows? They are not of caste but with their hearts and fervor. And therein shares my highest regard to all such people— and yourself. Superior, Wallinda. Superior. The metal is superior to the smith, the rich earth to the (land) cultivator, and our foods and fuels to our bodies and carriages. For what to shape are for gods' loose tidyings and assemblages— strictly for play, and to glorify with gloating, a glorious glut of gloat.... over the weaknesses of their constructions, and the relative futility of such demented housekeeping or husbandry.

WALLINDA: With of a heritage to own, I'm most proud to have one, and more of this pride than a need to glorify.

DECAMOND: Leave that to me.... I wrestle with my head.... and leadership to deserve. (*sips*)

WALLINDA: Perhaps, then, more reclining.... this hot day.... and retraction from those duties?

DECAMOND: Does cause one's pondering paw to waver, as the other sleeps.... My guests await me. Were that Elser's mission one embodied. (*as they start to leave*) Are you upset with me, that I am chaff(ing) to your councilor?

WALLINDA: ....No, my lord. He is too chaste to his kind, I fear.... to be much submissive for my tenderness. Yet must he overcome a harshness for me.... to be soft, then goad you him honestly.

DECAMOND (*as he is closing the cellar door behind them*): Worthless. Worthless.... A—che!

Scene IV — *A palace chamber, though not a throne room. There are comfortable, plush arm chairs around, and no table.* **Musgo-**

*nie and **Matile** are seated in two of them; though not next to each other, for she is by **Jocqueral**, who is out of his chair with a young boy's impatience. **Elser** is standing next to the king. He may sit a distance away if he wishes. **Ponstole** seems a bit aloof, by the open doorway, which is more like a hall entrance with arched passage. He is uncomfortable of any meeting with the king without a table. But this informality is designed to please **Decamond**, for whom they await.*

JOCQUERAL (*to his mother*): I want to see the Decamond!

MATILE: You are to behave. As soon as he arrives and enters, go to your seat—

MUSGONIE: The boy should grow to observe these occasions.

MATILE: —and be courteously obedient. This visit is to present the royal family to a hero, a very important subject of the realm

ELSER: Oh, please do not call him a hero.... my Queen, to his face. He frowns at such characterizations as epithets for the dead, or nearly so, or imploringly impending to be so as a justice to history.

MUSGONIE: Well he is a living man, and a rational one for sociable behaviors. And I trust that he has been well prepared, to give a most charitable statement to us, concerning his intentions to serve my most dire wishes at this time in *our* history, development of country resources, and sovereign rule of kingdom. You've promised him nothing other, I hope, than for these sentiments to be allowed of his display— his willing avowal and adherence to me.

ELSER: He lacks the stupidity to do otherwise, your majesty—

MUSGONIE: I want no surprises, but the uttermost of a genteel attitude throughout our meeting. Not even an (h)'int of argumentation over any details, but only a fine cordiality of understanding for his purpose and agreed to obligations to the crown.

PONSTOLE (*almost broodingly*): He is of a mantle stoked to stalk upon our wishes. (*Exits.*)

MUSGONIE: What does that mean?!

ELSER: Our councilor.... He is in love with the duke's maid, your highness. It is a love-sickness that terrifies him.

MUSGONIE: Oh. For its novelty?

ELSER: To him? There is a heart there. However, the duke, without being dictatorial over anyone he associates with, let alone his servants, expresses some disapproval for the relationship as being unsuitable. Not improper.... but unsuitable. He is a good judge of men, with an incredible experience to the variety of them under stressful and pressure-filled situations. Such an affair would probably not last seriously, and hurt more the maid than Ponstole's reputation and career. But his service to you is more proficient than this.... emotional difficulty.

MUSGONIE: I can't say I could not ever expect such a thing. Each man has his taste of feathers.

MATILE: So?

MUSGONIE: The softness of the down, my dear, is represented by many colors and textures, and veining of the quills. What is reprehensible to one cheek can be empyreal to another's. Is she youthful and pretty? or elder and stolid (of) *stolz*.

ELSER: The daughter of one of his soldiers.

MUSGONIE: Hmm.

ELSER: But of age, this longevity of representation yields to a certain orderliness of attraction to her features, in my opinion, for her youth. Not too dissimilar to my wife's, in proportion.

MATILE: Is it for men to judge the character of a maid of Decamond's?

MUSGONIE: Only in that the councilor was amused.

MATILE: We should amuse ourselves more with this duke.

JOCQUERAL: Does he have horns, mama?

MATILE: What?!

JOCQUERAL: To call his troops into battle.

ELSER: I believe the trumpet alert is left for pageantry and triumphal marches. But our king was referring to the fact that Ponstole is seldom with amusement, cousin.

MUSGONIE: Various instruments of many natures may be employed for any concerted effort or substantial undertaking. But the sounds of a multitude of belligerencies are particularly corrosive to the comeliness of ear, numbing one to the distrait effects of corruption and casualty until you are overwhelmed by war's fastidious nature to slay and mutilate and cause depravities of mind during this suffering. It is an atmosphere.... unearthily kept, and to be avoided at great cost. Yet, some do revel of its rage, the battle, the confrontation with innumerable sparring and incessant fighting to cause harm, until the dead are more brave than the warriors left dazed from their hurts and agonies, until the most defeated conquer fields with their bloodied bodies strewn, as to make a new ground with this slaughter. And I have disgraced my feet and legs, wading through such events, and abhorring my sensitivities to them.... So, the boy must know, that this particular exercise of manly muscle, this kind of engagement of wills and courage, is a deprecation of the human spirit. And he must see how I do my best, even against my own tastes of conference and friendly conversation, to convert one of our highest proponents of conflict *up* to my more tranquil standards of resolution of disagreements or misunderstandings between persons and adults. Take this Decamond, Jocqueral, as an example to study for my attempts to handle. He is renown and notorious. And yet, (being) up to this task, I am not afraid to present him— as a subject, and yours.... in due and worthy course as you survive me. These lessons.... are the most intense to endure, for the teacher. Then may I illustrate for you that he is sick and not sick, sound and not sound, wise and yet foolhardy, noble.... but for being pathetic to me, with this verve for injury through catastrophic means, and to be highly respected and praised within the irony of these condemnations. For he seeks a profit for his lusts— And lusts deserve none, as their own reward is their debasing of the soul.

JOCQUERAL: But Vater, he is a gallant soldier I have heard of, who often leads his men directly into onslaughts of the most vicious and debilitating fighting, with bravery stronger and more firm than ever their weaponry. He is— famous!

MUSGONIE: Words you have learned.... perhaps affixed more properly to other conquerors of history. Yes, we honor such stellar personalities, whose profundities are due to remarkable, often exaggerated deeds of some importance, worldly or mythologically or otherwise, whatever are your current studies.... or recent impressions to recall. But do not confuse for yourself my aims. I do not belittle Duke Decamond. I present him to ourselves, to become cordial and acquainted with. Yet judge this character quietly, as he transforms from legend into liaison for.... our wishes to perform.

ELSER: I can assure you, your majesty, that he adopts to your principles wholeheartedly. He has the savvy, now, through his recuperation to cooperate, and slay the savage streak in him to better your purposes, not that he was ever other than possessed with a most orderly mind. Then comes the defaulting recognitions he must apply.... to his condition, that my friend never meant to distress your opinion of him nor his method, and demonstrates this with his assistance at your beckoning.

MUSGONIE: I know a person does his manner best, and not to fool others— but defeat them of his being.

MATILE: So one can only be one's self. But you have allowed this duke his reign thus far.... because it helps yours nonetheless. And that is the irritation, against your peaceful standard, that moves you so to agitation to lessen aggravations from the successes of his way. It does challenge your allowance of madness and anger to focus on him so, or impose yourself.... on his activities.

MUSGONIE: I've been given to unease, yes, and complaining over those exploits. But this is not the center of my ambitions for ruling our kingdom. We do not indulge of his compliances inappropriately to my own creeds nor his salvation, but rather more of opportunity to seize and clutch, as can be some cleverness at play. For he has recommending friends to approve him(self) to me—

ELSER: A man most chivalrous of temperament.

MUSGONIE: —and append. Thus.... to learn (the) how of such sort is better than to brood over what he has been.

ELSER: We are each of definitions often hounded, your majesty. Yet can contain a person more of falsehoods to subscribe to— than to be, as could the cur be gentle and the mouse fierce. Yet then a dog's a dog in all ways, a mouse in its, and a man in any. Then try we for the best to preach of ourselves— and lend.... towards the encouragement of our finer aspirations and achievements. One must protect, for those ambitions to evolve.

MUSGONIE: Reputations?

ELSER (*as **Ponstole** enters with **Decamond***): The friendships, your majesty—

PONSTOLE: Here before you (*as **Jocqueral** adopts his chair*) is our great Duke Decamond, here to offer us his.... especial camaraderie of presence.

DECAMOND (*bowing slightly; aside from being sharply attired, as appropriate for the occasion, he wears a highly decorative though darkly colored shoulder harness, slip or sling for his arm to rest in.*): Your highnesses. Elser. I give you with great humbleness what is left of me after my last voyaging for the crown. And in your generosity I wish to be received for my poor condition. But it is apt to be submissive to your commands, as powerful as my muscle may last and my mental faculties stay astute to your dreams and wishes. I come answerable to all of your requests and inquiries. With this invitation may a substantial compatibility be demonstrated between our motives for the welfare of our country, the common weal, and the volatile eventualities of state. For I remain an active man devoted to the servicing of (the) extant sovereignty. And to Musgonie, and his Queen.... and child, I thereby genuflect (*making a subtle attempt towards kneeling*) to symbolically offer my obedience and resilience for your missioning.

MUSGONIE: Fair Decamond, (*as **Decamond** recovers to standing*) we welcome you with all of the amicabilities that may be provided for a prosperous kingdom. And I wish not to question you nor your purposes agreed to, but rather to have you state yourself before my family so that they may witness and admire your much deserved grandeur and standing within our community of notables. I dare myself, to have my young son take notice of such a personage of valor and initiative.

DECAMOND: He sees only a wound, I might espy, for such a strange figure brought to him. Yet, for your company to make it is a blessing to have cause of wanting this rapprochement of aging wills.

MUSGONIE: And then.... to note your method, may you state it for our ears?

DECAMOND: As well as eyes for sight, your highness; for we are not yet blind, in this space of nationality.... I would not call myself an ambassador— to our own provinces, since we are common subjects still, and under your rule enhanced of the proprieties of citizenship and inhabitancy for your kingdom. Yet may I be used to help persuade the remotely discourteous and disrespectful of these gifts, with contingency of growth and extension and a vigor of usurped independence felt, to abide to your treatises and generous pacts of relationship and fidelity to your administration of their properties and worth.

PONSTOLE: And this persuasion is by how of hand, now?

DECAMOND: Comportment over compromise. As you have suggested for me, I may personally relate to some intransigence the king's own arguments for discussion and explanation to make strategy of acceptance, and without threat of forceful inducement, bullying tactics, or the daring of offense against an insurgent spiritedness, with means I am more used to employing and am talented for.

MUSGONIE: That is the admirable aim.... And you have said it, for us.

DECAMOND: It's how the method was explained to me by your councilor and Count Elser, your majesty. And I bear witness to its rationality sensed, as you may send out many proponents to state a case. By choosing me, it is by my reputation of being earnest that you may succeed, and that I am of serious thought to the details to

be demanding and disputatious of failure. Yet, independent minds have independent means of resistance. And for this I stand firmly as one for you.... and one for others—

PONSTOLE: Others?

DECAMOND: As a representative I may bargain, to couch requirements as being, becoming, or eventually to become desirable and worthwhile for all. As a soldier, I may not—

MUSGONIE: Not what?! Bargain?! How play you to make ploy for our attentions?—!

ELSER: He means that he stems his soldiering for you, your highness. Isn't this so! Decamond.

DECAMOND: My arm does do this insult for me to present. It is bound up, as were encased in a stiff cast, for what it feels. Yet be no ploy its lack of utility— for months! And I show it to you this way without any deceptions. Without its healing I am no honest soldier, for I insist it may! But honest men will not take argumentation alone, if they are rational enough to fight for their goals. And I must stay prepared for such event, as to warrant more meanness than your method—

MATILE: Despair of the decrepitude of your slanders, Decamond! My husband's honest and your king is real. Yet have you goaded him with your activities of profitable shame as much to leave anxieties in nightmares for my reviving. His standard is explicitly simple, not to favor any bloodshed within our country any more. And if this was not extolled adequately enough by our councilor and my cousin, then there's as much fault as your wound to own, for which your patriotism alleges—

DECAMOND: Alleges?!

MATILE: —You would wear my husband down to an age of old blasphemies and disgraces and trying realities, as your friendship does lead you harmed, to strife your transgressions for your own good of decaying—!

DECAMOND: Harmed?! What of this frame! this sculpture— this scaffold debilitated! And by friendship? Means of what?! That my own did damage me?! Such rumor! But not the wound is praised before you, as to display a tort? but more the wounding of my sincerities for your mission. I refuse to be detained by such cautions! to realize this?! And who does say such things of error? without of to conspire with! Yet (to) be defeated is not my love of nation in this way, nor of the soldiering it takes to maintain it. For I have fought alongside men, not minions, as were to bleed a life as capable as my own to die— die with, and die of!

PONSTOLE: And for some monies meant—

DECAMOND: Does take a mint, to pay for this adventuring! And for the spoils, they're won— not stolen. Ply you your dreams on hopes or harnesses?! This is the real of work to do, that keeps us of a unified encompassing. And not with error is your ignorance, to discount these effects. You know they lend you mine to worship, as for the blame to take. Yet are they done of time possessing your encouragement— of kingdom!

JOCQUERAL: Sir! Do you have hoofs?

DECAMOND: What?!

JOCQUERAL: As for your horses brought, with good effect, these men, for which my uncle says does cause the wound, this error of your trust.

DECAMOND: What says the son?

JOCQUERAL: It's not of error, out of lust! for the money paid.

DECAMOND: This uncle does harm me?— Elser!

ELSER: So!... Good friend.... I've been too privy. But yet it was to temper your excesses that I lent this influence of causing harm. Was meant more for the leg, to end some stepping within age, I'm told. But the shot was not good, though well aimed. There was a flurry of confusion, at the time, within some victory or celebration, that moved you off from the gauging.... as many rushed about.

DECAMOND: ....You have all been abrupt my penitence, with this knowledge?

PONSTOLE: I know nothing of this! It is deceit to me.

MUSGONIE: ....Or, privy of a councilor to learn from, I've myself been unaware of this inserting try to plead your grace with me.... The effort postpones charms of thought proposed.

DECAMOND: This brings me to stumble.

MATILE: Really! (*as Elser goes to sit*) For such a forager as yourself, without the king's admittance to your excursion-ary lobs against his designs, yet for to be as advertisement for his loose handling, then tear at his promotions to draw his caring shatterable, and with complaint by a populace of noting, sharp noses that may sense a laxity of attitude. And battering praise to condemn our attentions, do they hurl behind our public notice, such as to cause dismay of feeling— and hurt! Decamond; that their ruler is too complaisant towards such a force as yours, if not compliant to the troubles stirred for himself with a complicity to your erring— is not a wrought-(mak)ing due?! some wrath to pour over these events! these— excursion-ary blights to cure a treason for some treasure and make foolishness for thought of the king's too hardly fought for peace and his visions for tranquility with unity and a commonality of protection for all of his subjects— as by his own doing and methods of persuasion! Should you not then have expected some such measure formed or wished for, crafted and conducted out of need to stymie your sureties confusing these affairs?! as bluntly the effort may have seemed to say: Stop these interferences! contorting the king's manner towards his public.... And yet with delicacy done, I am proud of my cousin to so do. Because the squeamishness of wrongs must spread, to harm you in this way as it has our notions and loves. And I doubt not your feelings to discern for such foul behaviors.

DECAMOND (*to the seated Elser*): You have done this to me?! This! (*demonstrative*) For all my battling to bleed for, have you such cause for an injustice? And of my bodily worth to terrify! it is as evil a debauchery as death to squeeze at.

ELSER: How else to get through to you your danger? and such dangerous ambition as to constantly affront the king. You've drawn such peril to yourself as to make for delusionary and unnecessary trophies, despicable valors, and a certitude for being stopped with a thorough destruction. Yet it's so difficult to turn an independent head to reason, as no others can conduct it. Yet over men— to lead — then must through men it be, as warriors only know of harms and hurts to end their struggling and strangulations—

DECAMOND: And the cockroach knows its cracks!

ELSER: —Yet for, to plead in friendship, how can I tell you with warnings what you are bound to resist, what you must ignore through your very nature and insistences?

DECAMOND: I'll have the culprits known and routed!—

ELSER: With my life! I've pledged the(ir) anonymities—

DECAMOND: Elser!

ELSER: —They are minor, so minor. And they've protected you more than wounded. Yet for your men to ask, the more they'd know, if they have not already told you so. Then should they not concede, of those around you feeling, that your rest is well deserved? Then of a mixture are your soldiers considerable, and perhaps a new leadership demanding. For it is too heady to stay so bristling-ly brisk against our sovereign's temperament and to risk more on this hypocrisy of caring for the nation ov'r your worth, with your notoriety unbounded of growth towards derision. Yet are you saved, today, as an emissary of Musgonie, the very one that you have fought so theoretically. And you've become so greatly admired and celebrated for *this* courage and wisdom to conform to.

DECAMOND: Minor!... Minor as mine. I would not dare, to ask such a question.... of my fighters, and insult their silence. But that they would elicit of this knowledge, voluntarily, insistently, intentionally— immediately.... I had expected, for a long time.... Yet it does not heal.... these several months.... Minor, as if unknown. My best would never deceive me, I'm sure of that. But we're all, more or less, independent actors for our wants, to organize only for succeeding.... not so much in value of the group, than for our aims to plead.

MUSGONIE: And what is this? More fighting? That can not be.... sensible. Exercising of one's skills? Sounds terribly indefensible for the harms caused, the wounding and the loss of life that can be avoided. The wealth to gain? What is wealth with error to your body.... or your mind?! A passion to express? Does roar the lungs every day to survive. One's heart is the most passionate of undertaking to your cause. Then it is merely to be needed, as what your slung arm may show. If it is needed not, then it need not heal. Yet can I use you still, in this state. For the condition does not preclude your worth and nobility. And let me make my errors more myself, than with your provisioning against the slackness of my rope and tethers.... I would not retire you, this way, but bring up from *here* a full utility. And may this stumble of my methods, appealing to caution and reasonableness and the practicalities of being whole, then consecrate the try (as being) for your own. The intentions are more important than the results.

PONSTOLE: Where turn you, Decamond, to these— legalities that you are permitted? Your actions have only been condoned in this way. Yet still you bleed to contaminate sympathies, against this court, against this nation turning into desires the falsehoods of our gaining tries—

DECAMOND: I've not twisted no (nor any) hearts, from lack of competition—

PONSTOLE: To the king, with your excesses.... to the court, for forcing friendship with an almost paternal constriction of punishment, to your soldiers, who you would have— would keep too independently of trust.... to myself even—

DECAMOND: I've not thought to you much, but as a belittled annoyance.

PONSTOLE: Or to have Wallinda suggested against, my partnership to recommend—!

MUSGONIE: Who?!—

DECAMOND: I leave her in league to another suitor— of her own mind to value yours with caution!

PONSTOLE: Argh! (*rushes to exit, though stopping by the wall before the passage*)

DECAMOND: Not that you'd dally with my maid am I more disturbed than with her upsetting to ever condone you, as if you'd need not ever compete— but only argue and assert to pry into a common heart, and cause such frigidity of fear to your unusual circumstance.

PONSTOLE (*at the wall*): Remember me! (*with emotional anguish*) Anee—ha! (*banging the wall*) Remember me!

MUSGONIE: What pains you to this panic?!

PONSTOLE: Remember me!

MUSGONIE: You are privy to us all!

PONSTOLE: I can not have of sweetness when condemned of a sufficiency— by him! And always thought for worst— for worst, for worst to be.... and worst to bring— of my company! So remember me!... that I have loved once! (*starts to exit, but is prevented by* **Costile** *just entering*) What?!

COSTILE: Here am I directed. I did not expect of such temper, your majesty, with such an informal gathering for debates! But I bring news from Lacismoor.

PONSTOLE (*with disdained astonishment*): Oh! What does a license sliver of the lips, then hustle unto through our crying mists a caricature of these concerns! (*rushes out*)

MUSGONIE: What be of Lacismoor the urgency?

COSTILE: Is not, my lord. But that Lord Bracken is not the lord we noted— but his son. Not John, but Scott. And what's with Ponstole?

ELSER: 'Twas an ill reprimand for his feelings, so brought to his mind that he could be admired merely for being a privy councilor.

MUSGONIE: Does not Lacismoor write?

COSTILE: But more than this, does right itself. They had a meeting, at the Brackens', which I attended, whereat all the efforts for a militia by the son to form were thoroughly denounced, by those of a standing to promote one with their funds. They abide to your majesty, to deter a youthful independence, renounce this threat as foolish, and treat themselves to a serious review of their proprieties — well under your guidance placed.

MUSGONIE: This is a good light shone from Lacismoor.

DECAMOND: Then be my purpose flushed, sir, until the next event to stand upon and conjecture of; flush, as a day to greet one's brow or brimming head, these thoughts to hold and fight ov'r. With cause to be, of my ascension to your service, then let us try with might of mind to condone our preparations for the next! For which we are on centered ground to please all audiences, these casualties to bring for your review. Then how determined are our views to seem? (*turning to the audience*) They are simply for your viewing, under independent thought, the chains of pondering your freedoms wrought of iron wills and subtle tendernesses for your persuasions. Extend to all, then, with forgiveness of much profit, the leisure of our display.... and with this end our errant ways. (*bows, as well* **Costile** *and the others standing to bow and end the play*)

⳨⳨⳨⳨⳨

Peccant Pecus

Peccant Pecus

Characters:
 Tansel
 Terence
 a Lieutenant (Valter)
 various Soldiers
 a Medic
 a Adjutant
 a Colonel
 Sachel
 Ponaple
 Sarah
 Falfour
 Weiler
 Joel
 Beryl
 Laura
 Moraine
 Bouwder
 Valin
 a Physician

Locale — A rural though modern town

Act I

Scene I — A designed mini-bower within a plaza setting at night, faintly lighted from a distance. It is evidently a late but warm evening, as **Tansel** *slowly walks up to the bower planting and sits under it, on the fairly level ground. He is not disheveled, but neither is he finely dressed, this showing his nativeness to the region is of a common lot. And he shakes and rubs his head occasionally, as if to dislodge a headache, but in no other way appears to suffer the mannerisms of intoxication. Though alone, the area shows signs of population, and this location being a fairly public spot.*

TANSEL: Whirl, these good, warm graces. There's nothing better than a good night of comfort. Here is good. Good.... And be redundant about it. Because I've been rather tired. But this is as safe a place to come to as any could be found to wander towards. My heaven! Those stars look like shop glasses broken in their cases, strewn on their shelves, glistening and twinkling about remorses after a shock. Well at least the pieces are sheltered quite finely, up there. It's too nice an evening to complain about anything. And when a breeze feels like fun, the aromaticity of a worldly health, then I should be allowed to be complaisant with my setting.... Safer than the river bank. I might fall in, slip in through.... the indiscretions of inattention to my state, and what I'm thinking. The head hurts a trifle, just the forehead, the beginning of a search for something missing. Is it some air, under these clean leaves? So! They're clean, when they're allowed to be themselves indulged, this plaza quite deserted, because it is too late for the preoccupied to be roaming about, and only some ghosts of wishes can know my presence. Now, I'd like to be a leaf on a stem. The fall from it would not cause such derogatory remarks as to characterize expectations of one's autumn, a blush of bashfulness due to the surprise that these realizations come so suddenly to grip you. Then oh! for a night like this, to be at peace with the pleasantness of the world, as much as it may offer you some restful stargazing.... I try to be heavy, within this lightness of the dark, this ocean's buoyancy, this

heavenly lift as senses rise; and to be alone out here speaks of voluminous envy. So let's be blunt about this, Tansel. You are twisted, as a swirl of ganglions into rope. And they judge you, in that.... restaurant, that meeting place for rants and ravages of opinion. They, who speak the loudest to me.... and of me.... want to enlist this brunt of body into some very serious evil or foulness, because they choose me to be appropriately fashioned for a weakness of mind, a confusion of morals, and a loss of taste as they too much so salt the tongue. And thus, I am drawn as something devilish to tickle. Well then, "Like like like!" is their motto. And I alike shall like to do as they would like, some awful deed, as much my purpose in life as to be used for. An ugliness as toads, they can envision of my mangling. But it's so good to be able to rest out here, and away from them for awhile. Then who can claim such artificial constructions as being unworthy of me?! I choose to be used, that's the only threat to my presumptions of a fine nature. But one is built for certain things to do, or to be able to do with skillful prodding. Now what is a depravity, asleep in a grave? There are many wrongs that can be contemplated, as can an angel think— with indeed some rawness of visioning. Theft and treachery will make temptations speak. So what's more wrong than a life that steals.... its.... necessary resources and nutrients? as these leaves my respiratory exhausts. Claw me for this purpose. Become the animal, with animation throughout these glorious breezes— (*as* **Terence** *walks up to him*) Conscription churns my maladies. Surfeit my ease of wanton slovenliness and abandon.

TERENCE (*upon him*): Subscribe to pain! my devil-kin. Have you recovered, and from the thought? We want that killing made.

TANSEL: Who would not, to survive?

TERENCE: But it should be left to animals like you.

TANSEL: Hounds hounded.

TERENCE: You've done it before, this sort of thing. So why be squeamish now? it's your specialty.

TANSEL: But this time you fellows leave me no out, to ever clean myself or wash my hands of the deed.

TERENCE: You shouldn't have to think about filth. It's almost a cloth you wear.

TANSEL: I occasion decadence, Terence. I do not make of it a specialty, as you say, or would declare of anyone who may do such things. But this could be a particularly foul butchery. Can't we agree on that?!

TERENCE: No, if the subject is virtually dead already. Why be reluctant? You know we have our standards, so this is not a crime within our grouping. And she has been groomed to expect a slaughter; might very well plead for it, the way she's been fatted up, breathes with difficulty and walks painfully.

TANSEL: It is theft.

TERENCE: So what of that?!

TANSEL: We do not own her.

TERENCE: Neither do they, to leave her in that condition. I

mean, rightfully speaking, we must be merciful. And we must steal to feed ourselves, because our scruples don't allow us to raise such animals. Yet we may eat them gleefully.

TANSEL: Gleefully cooked. But it's only to my sort that you expect the procuring, that you expect to do the deed.

TERENCE: Because you are a dirt with hands. The store has been forbidden, the storage prevented. And I'm anxious for my steaks. Everyone around here suffers for some beef. And to think that holier than thou proprietor may starve us out is indecent.

TANSEL: And if I'm caught— no loss to you.... because I'm already demean-able.

TERENCE: We don't wish to share of his influences, as if he could train us to his tastes. But he simply means to market for his cows, by proscribing others to be sold around here. And he chooses this community to bed in—

TANSEL: He's had family in the region for a long time.

TERENCE: —because of our self-serving mandate not to cause any animal cruelty or suffering on our properties and in our residences. It's been a nearsighted provision of our community charter, since it's helped to drive out the only *carne gestione* in the region. But who would have thought such product would not be shipped to our stores?

TANSEL: It may only be a delay for fortuity to arrive.

TERENCE: And fortitude. But we set a strongly imposed example of ourselves against the reign of brutality that raged like a fire through our humble and generally pious village. And to uncover the misdeed-ing, by of all kinds our own children aping some misunderstood fad of abroad or trend of a soul-less future, you were keen to play a part.

TENSEL: I merely pointed out a truth, that to give a youngster a pet means misery and death for the animal, with some cruel certainty, because we use the situation as a learning experience whereby the children may apply all of their errors upon the poor beast. A few of you investigated this possibility like a scientific project, and discovered the shocking reality of actual tortures committed by a gang of brats, though mostly onto the wildlife— and mostly rodents and rabbits at that. A committee overreacted and caused enforcement of this ridiculous ordinance. And now, under the fragility of our economic state, this municipality suffers with the closure of our only butchery, and that needing only slight compunction for rejection of our sensitivities.

TERENCE: And now we starve, as feared to be vegetarians, or leaving that impression for the markets. Some of the populace even try to earnestly adopt this view. That's fine, for those individuals. But it is not correct to harness malfeasance with malnutrition. Our children must have more meat.

TANSEL: The chicken and fish are quite enough by some.

TERENCE: I demand my steak, my sirloin. It's been incredibly over two months. I thought the fever would die down much sooner than this, without the complications that have ensued. And there's absolutely no reason why anyone these days should have to go a'(t)hunting. Stores should be full of the four legged remains. But they refuse to ship to us?

TANSEL: This is such a rural place.

TERENCE: I refuse to believe it. The delay is due, somehow, to his interference, and possibly his dislike of pork.

TANSEL: Who can not love the bacon?! But the poultry is sufficient, and the bovine will be coming. Eggs are aplenty, and the fried flesh is delicious. We're not the vegetarians *here*.

TERENCE: (The) Children need a richer source of protein than that. They need some beef, and red meat. And I don't mean via the bouillon broths and stocks. They need (it) because they're growing. And I need because I'm grown, and know what's best for everyone.

TANSEL: It's only a temporary delay, some kind of shortage throughout the counties. Has something to do with abroad;— but I don't understand it, since we raise the cattle here. The farms are our own. I've been laid off because of this. But you can't blame that country merchant for our problems, even if he does seem more wealthy than mean. And I happen to know he delights in pork rinds, because I delivered some to his kitchen once. Now that takes some real guts to order, a virtual waste product— until used, properly baked and roasted, and made crisp.

TERENCE: The tripe, for all I care. They don't eat pig there. Who *can* avoid any foodstuff? they're so mixed up together, in the commercial packaging. But I've had enough franks for a lifetime, these past few weeks.

TANSEL: You want me to make as a criminal, against a former patron. You all want me to do that.

TERENCE: I want that taste again. I want a sharp knife to good rich meat opposed. I want to cut out large chunks for a ready mouth, a waiting one. And you're to make this happen for us, because you know how.

TANSEL: It's only a shortage.... Why don't you take a business trip to the coast, and bring back what you want? The cities of the waterways always have the greatest variety of trade.

TERENCE: I'm not so rich as your imagination, Tansel. Nor so impoverished to be stupid. The premium (to be) paid for what we propose is already high. And you will go through with it, because you're the only expert in this business that can be manacled to our wills and demands. You are totally loose and aimless without our benefactions. How many alms and handouts can you expect to collect, and for how long? until you come to terms with our desires and do our bidding with some work in a speciality you're capable of.

TANSEL (*rubbing his forehead, as for a fever to sense and wipe away*): That restaurant is not so crazy as to promote such a criminal act, a thoroughly illegal and despicable one, even if they are feeling pressured with the scantiness of their menu offerings lately. You few rabble-rousers gather there to shoot me into this utter ungraciousness because I'm so vulnerable to shame and disgrace and desperations. You pick on me with your high(er)-brow thoughts, prying away my skin to irritate the nerves beneath and embedded. But how can I deny your ordering? They'll send *me* to jail, while you'll have your dinners, and for this purely retributive act to show someone he can't get away with what you think he's been doing. He's as innocent as I, and you'll make him as guilty as me.

TERENCE: No one's gonna find out about it, if it's done correctly. A conspiracy will be blabbered out by you, if you're caught. You don't think we take the risk?! It is strong and considerable. But you're the one to do it, only you. That cow wanders out a lot, quite unattended. She separates from the herd, we've been noticing —

TANSEL: Herd?! There are only three, and they're only for milking!

TERENCE: You forget the bull—

TANSEL: I've never seen one!

TERENCE: He must have a bull! It makes no sense otherwise. They will want calves, Tansel, to build up a herd. And then he'll force us to buy our meat directly from his stock, and at an inflated price. That's so terribly obvious—!

TANSEL: They're more in the mind to modestly produce some cheeses, and that more as a hobby for the sons and daughter than as a livelihood for the parents. They don't mean to corner any beef marketing around here.

TERENCE: She wanders from them, because she's feeling uncomfortable and can't keep up with them as well. Only that can explain her poor grazing habits. She'll cross a border to be caught, and we'll be waiting.

TANSEL: There are only three of them. There's no need for fencing, in some locations. Who else could they belong to?! Just notify the owner for the return of a distressed cow— a pet!

TERENCE: That will give the fool an advantage— of ignorance, for what he's doing. They're not branded. The ears are not even tagged.

TANSEL: You're calling me the fool! But I won't be ignorant. I'll be the blasphemer and the blemish and the one to be blamed. And that poor, dumb animal will be traumatized by my hands, its life destroyed by me! with your planning. I'll know all of that, and why I'll be kicked for it, and incarcerated with a hefty fine I'll never be able to pay. And you will deny everything but being sold a bill of stolen goods. The satisfaction will be only yours in this— horror.

TERENCE: The horror is all because of you! But that's not how it's going to happen, because the beast will simply be lost. They don't know how to take care of the pastural, and this will be a lesson learned and a mystery kept. All you have to do is be yourself, and use your training to be swift and definite— and concrete about it, Tansel, because you know there are no second chances once it's started. It must be run through and ended in one sprint. You certainly still have the muscle and the nerve for *that*, with all the bells you've heard, once the slaughtering is completed and the partitioning of parts made. That's still fully a part of your conscious abilities, and conditioning for the labor. So you will do this with the efficiency that permits our scheming as possible, as real! (*A thunderous sound is heard with light, as if a shell burst of explosive.*) And we will be the staff that controls you! (*Soldiers in camouflage uni-*

forms and with fire arms run through the stage, to befit a battlefield engaged. Sounds of gunshot and rifle shot are heard, and to a corner is seen a constant stream of ignited bullet-ing as to scour an area with a prolonged permanence of danger. As the soldiers run by, they ignore **Terence** *as much as he ignores them, he preferring to concentrate on* **Tansel***. But* **Terence** *is not completely incognizant of what is going on. A* **lieutenant***, with a communications device, a* **soldier***, and a* **medic** *run up to* **Tansel** *and* **Terence***.*)

LIEUTENANT (*rushing in*): We can ditch below that bower! (*The three stash themselves next to* **Tansel***, the* **lieutenant** *and the* **medic** *to one side and the* **soldier** *to the other.*)

MEDIC: Here's some cover, man! (*to* **Tansel***) I've seen some many, but can't get to them easily. Are you hit?!

TANSEL: No—!

MEDIC: Couldn't get to you well anyway! There's too much ambuscading going on. Hard to dig out the fragments of shot quickly. But I can anesthetize some, and sterilize with antibiotics. Don't hope to carry anyone, soldier.

LIEUTENANT (*to the* **soldier***): Are *you* hit?!— Are you hurt?

SOLDIER: To the side here, sir. Just above the elbow. A nick, maybe.

LIEUTENANT: Well, good! That's what arms are for. Stay down. There's something overhead. But we have to move out. We're near a targeting nettle. They're spraying deep over to our side. Want to inhibit out left flank from rushing up. We can get by them if they do.

VOICE (*from communications*): Valter!

LIEUTENANT (*impulsively responding*): Yes, sir! Valter here! Under fire!

VOICE: Get your men to the described corridor and run through — Now!

LIEUTENANT (*answering*): There's heavy artillery to that sector at the moment, sir. Flame bullet-ing right across the region. We can all see it!

VOICE: Soak up some of that fire, to allow your flank more advance. Do it now!

LIEUTENANT (*confused, but with a vigorous determination*): Yes, sir!

VOICE: It can be done, if you approach the area correctly, as you've been trained. Figure it out! lieutenant.

LIEUTENANT (*responding*): Yes, sir! (*away from the communications*) That flank is gonna burn. No fairness for the volunteers. (But) They're blasting over us even now.

TANSEL (*emphatically*): Get down, Terence! before you're mowed down!

TERENCE: There's no salvation for a ham's shank!

MEDIC (*to **Tansel***): I'd curse the bodies blown, without my wherewithal to save at all. That's what a medic should do, to skirt away from all of these dangers— and rush in with aid, after much of the leveling. But I find them stilled and dead so often, that it's a waste of my courage, (*picking up some of the parched ground and squeezing*) and 's as much as to crush the crud! for all of my efforts. And yet, you're all right? (*dispersing the earth*)

TANSEL But— Well—?!

MEDIC: I thought I heard you crying. And it's the bullets piercing pigs and soldiers. Are we talking together? or am I dead!

LIEUTENANT: There's the flare to start! (*The **soldier** yells, and **Tansel** turns towards him.*)

SOLDIER (*as another **soldier** is pulling him up*): No! I don't want to go crossing through that field of straf'(ing). (*The **medic** rolls away.*)

OTHER SOLDIER (*forcing the **soldier** up*): You ain't wounded enough.

TANSEL (*turning back to where the **medic** was, and noticing he's gone*): A phantom!

LIEUTENANT (*rising*): Let's go! (*He and the **soldiers** rush towards the fire nettle.*)

TANSEL (*cropping his head as if to clear his sight from hanging obstructions*): Buckets of gravy spilled everywhere!

TERENCE: You don't even have to rip through the cornstalks to find her.

TANSEL: Why one "r"? Are we not one an' the same?

TERENCE: You'll be rewarded sanely.

*As for a choreography, many of the soldiers fall to lie flat on the ground as if dead. The machined tensioning lessens as a **colonel** and his **adjutant** walk through. As they pass by, the lying soldiers get up and leave, some running forward and pass the two. The **colonel's** voice was what the **lieutenant** heard.*

ADJUNTANT: Here was a particularly brutal spot.

COLONEL: Ayeee. A(h)'. A'. Ayeee.

ADJUTANT: A mortar burst over one whole group, over there.

COLONEL: Ayeee. A'. A'. Ayeee

ADJUTANT: And this region was shelled incessantly for over forty minutes.

COLONEL: Haven't seen so many bodies. Overwhelming. Am speechless.

ADJUTANT: The fire hail's in this direction past.

COLONEL: Ayeee.... Valter's there? (*As they leave the stage, the scene returns to its original tranquility and incidental lighting.*)

TANSEL: Why be such need for carcasses and cuts? It's a balmy night.

TERENCE: You don't have to do the theft at all. We'll bring the animal to you. Look at what *I* have. (*pulls out some keys from a pocket*)

TANSEL: What's this?!

TERENCE: Entry into the emptied abattoir. Don't ask me how I got these keys. But the power's on to be used. The machinery is all ready, the implements clean, sharpened and waiting.

TANSEL: How long can that be? When are you planning this?

TERENCE: Tonight—!

TANSEL: Can't you see I'm drunk?!

TERENCE: We're searching for her now, Tansel. And you're not drunk. You were in the restaurant not more than half an hour ago. Couldn't have gotten drunk since then— with nothing! Only said you needed air. Well, you're the only one we can get to, who knows how, how to kill the cow quickly so that the meat isn't damaged and the nerves don't burst into a spoilage.

TANSEL: I ain't got a gun!

TERENCE: There's a rifle at the shop, loaded and waiting. So are the blades and the saws. And you know how to separate the parts while the animal's blood stays warm enough not to congeal of the bacterial growths, and then store it all neatly in cold brine for the refrigeration and freezers.

TANSEL: I can't handle a whole steer myself!

TERENCE: *She's* ready for you, or will be. And you sure as well can, with your experience, be it a bull or a cow or an elephantine marmoset. But you've got to do it all alone, and with the proficiency of a night's shield. Because only you know how to clean up the instruments again, to a professional standard, and wash up the shop, which is gonna take a lot more time than your.... filleting. You'll stash the hide, hooves and head in bags out back, which we'll supply and bury. And then we'll gradually come by to take out our choice parts— for our own freezers— without anyone else knowing what's happened for at least quite a few days. You won't tell, on your own work.

TANSEL: I can't do it so soon! I'm in a shambles—!

TERENCE: You'll make money off (of) this. Be able to pay off some of your bills— Take none of the meat, though.... You need this advantage to your character, Tansel, the only thing you can be noteworthy for.

TANSEL: People will find out. They'll suspect things with your bellies protruding more of a grin than your caustic smirks!

TERENCE: Who among us would start the rumors? I want those ribs, and so do all of us involved in the venture, especially the thieving participants. They're the one's most at risk— not you! You have your training to fall back on. They've not ever stolen a cow, I'll say.... I'll tell you that. Face it, Tansel. There's a whole cadre of us who will make sure such rumors don't spread, or are belittled from the start, unless you really go crazy with remorse for your art, or find guilt in your delicacies of being sufficient. And you know who we all are.... pranksters at best, to be charged with (being). But you'll do the killing— You have to!

TANSEL: It's the children that have, with their erring and learning—

TERENCE (*bending down*): Take these for the locks, (*crouching*) and get over to your place of— worth, Tansel. (*giving **Tansel** the keys*)

TANSEL: I'll run off with them. I'll hide them!

TERENCE (*sitting*): Within an hour and a half they'll come by with her.... And you should be waiting, by the trail gate, darkened and inside, to lead her into the withholding snare. A single shot must do, the sound.... ephemeral, in that deserted country. Then, on the roller, take to it, your.... manipulations of profession, and the machete-ing with craft. Only some stars can be witness to this buccaneering art, out there alone.

TANSEL: With your traffickers overseeing me—!

TERENCE: Yours.... Once the— thing is dead, and you are comfortable of handling, they want no part of the details, no splattering of blood on their clothes, no assistance to you to offer and be indemnified with your knowledge and skills observed. They'll leave you to your mastery. Others will come by for the bags. You know you have to work quickly, efficiently. The cleanup will take much longer. The editing of a creative act is ever more tedious than its initial, raw creation— You will do this, because it's in you to be able to, and the only way to earn— earn!... some sense of prosperity, tonight.... But listen carefully. By dawn everything should be finished, including yourself rinsed and made proper for appearance. Lock all up, and *leave the keys in the shop-front office desk,* that space which is windowed. Be careful of all shoe prints, so that you may leave from that office— which faces nothing anyway but mountains and empty road. That early in the morning there'll be no one around with chance visiting a closed shop.

TANSEL: But what about the generator?! I don't know nothin' of electricity!

TANSEL: Fully gassed and activated. In sync and on line. All you have to do is turn on the lights and appliances you'll need.

TANSEL: But what if somethin' blows up—?!

TERENCE: It won't!

TANSEL: The manager was worried about that old engine. It's failed before, in curious and dangerous ways, causing sparks and fire hazards.

TERENCE: We've tested it.

TANSEL: We?! And how did you guys get enough of the diesel for a working load? What is it?— two days worth? Sixteen hours?

That has to be shipped in, and can't be stored in it's rusty fuel compartment for too long before oxidation gum(mie)s up the gears of the alternator. I know that much— I had to clean out the crummy residues once, for an inspection.... But I don't know how to start those things up, in an emergency.

TERENCE: It's all taken care of, for this one night, Tansel. And I happen to know you rendering employees were given instructions on the total operations of the plant, all of the necessary procedures to follow.

TANSEL: Awh! Only an afternoon's lecturing, with drawings hardly explainable, ordered by the state. I don't remember half of it, if anything. All I know how to do is operate the cutters and shearers without severing off my arms and fingers. I'm not even proficient on the bone snapping and grinding.

TERENCE: Not everything has to be machine-aided; and we're not after any bonemeal for dog food!

TANSEL: I can only part the ribs with manual scissors. It'll take a long time.

TERENCE: You know all of the choice cuts we're after: the round, rump, loin end and short loin, ribs, chuck, flank, plate, brisket, and shank.... all neatly divvied up in labeled freezer bins, Tansel. That's what we want. That's what we've planned for, and devised this simple scheme to provide.

TANSEL: You couldn't have thought this all out thoroughly enough. It sounds more expensive than buying your own cow! And how can I be sure you'll pay me what I'm due? you.... guys.... knowing I'll be too afraid to confess to any of it!

TERENCE: Tansel, we're not stupid! And we're your friends— You're one of us! We know each other thoroughly, fraternally. And you can trust our collective acuities in this. Did you think we were only boasting, in the dining niche, that we're going to look into these problems a lot more carefully than just to complain? We have to make a statement about them, a silent statement, more evident with deeds than declarations. But we're not petty with any of our dealings.... There's a fifth advance for you, in the office desk where you'll leave the keys. And you'll like that, the richness of owning some real money again. And you'll receive yet another fifth by the deliverers of the cow, though in check form. If you don't have a bank account any more, you'll be able to open one again, and make deposits. That's two fifths of the payment, Tansel— almost half, virtually immediately. And during the next couple of weeks.... you'll be able to make more deposits. We're not going to store that stuff there longer than that. But we're not going to barge in to collect the goods one or a few times. We will share the responsibilities of a casual dispersion that won't draw any attentions. And there are enough of us so that the premiums paid for our meats are quite affordable and welcomed, a number you can easily figure out. You are definitely not alone in this venturing.... Now, if you don't like what you find in that office— and I can't see how you wouldn't! for such an attractive sum— leave the keys there anyway. If we find the money's gone without your services, you better leave town for good, or there'll be some real monkey butchering. That will be our loss, but we take the risks. And if you go to the police about this, what evidence can you show them? The keys will incriminate you quite curiously, amongst our denials of ever having possessed the like—

TANSEL: You must have several copies. The locksmith would.... records?... kept?

TERENCE: And without the keys you have nothing, because there'll be no crime committed, except our trust in you.

TANSEL: The money left might be evidence enough, to implicate someone at least, if I run to the authorities right now!

TERENCE: ....Tansel. We'll retrieve it— as some others watch you in town. What are you going to do— throw the keys away before you get to the station?! Or show up with them, to offer that and some possible cash as ample evidence of the plotting? Even if they catch us with the cow, it's a minor offense, walking or leading a stray, with attempt to harm theorized, the breaking of a code of civil harmony— But we'll curse you, to find a safe grave! You won't be worth the cigarette butts you'd try to sell around here. You're almost a derelict already, if it wasn't for our company, our society that helps to mitigate your misfortunes— and our own. You like me, don't you? Haven't I treated you squarely, and seldom denigrated? Oh, if a person can't hold his own and comes down to a miserable plight, some belittling is necessary and natural, as an example for others, especially the young, to be educated with. But I don't push the point, except to say one's only goal in life should be to work with diligence to overcome one's poverty and bad breaks, foul fates, and mistakes in life, since to live is merely a definition for struggling. You've never heard me call you an imbecile, not with your abilities. Foolish, perhaps, but not impossible to admire— or respect, I certainly know you.... to be. And we help each other. These tastes may seem trivial, for the efforts we wish to apply, to satisfy them. But the matter of it is that we can at all, we can figure out a way— to be clever.... about it. And this is not an "ends justifies the means" argument. This is an argument of employing our talents for a worthwhile cause, a helping Tansel as we help ourselves argument, seeing how we can fit him in. Would you really want to tell on me?... a friend who just doesn't want to put up with the impertinences we face?... for so long?

TANSEL (*holding up the keys*): This is hot stuff! It grapples my teeth!— But I need so much!... And you've lent me your friendship, to ease my despairing, just to be able to talk with you guys, sit down to conversations that still treat me as valuable for my opinions, and offer me some help.

TERENCE: We wanted you in on this. We fashioned it around you.

TANSEL: I tell you. What I've done for a living is dreadful. I don't deny it. There's no excuse for living. You just have to do these things, or someone does. But they're deemed legal— and awful! I learned the skills.... in a training school, a profitable operation derided for this education. Yet, such outfits are dwindling, these occupations diminishing. I had hoped this one could remain practical. I'm not smart enough to learn how to can and package, running intricate programs of operation, and adjusting myriad variables to optimize results. I only know how to.... promote some flesh for eating, and propel the eaten for their offerings. The simplicities of stripping off hides and slicing through the sinews, that's all of my calibre.

TERENCE: You're much more than a meat packer, Tansel. Virtually the whole of our little community are counting on you. Not just some ugly thievery is involved, but a means for healthy sustenance, and a restraining of growing tempers. Sometimes people have to take matters into their own hands, to provide for themselves, as professed principles work to shatter them and run against their more amiable and innocent strides. I need, and we are willing to pay, quite extra for.... But you must do; and you must feel the privilege of your abilities called upon, because an evening has but so many hours of darkness.

TANSEL (*as **Terence** stands*): There's still so much convincing to be done. I'm a'trembling, for the fear of being awful doubly so. Two weeks? of tainted wait in crime, and dread for being poached of participation. But can you keep the freezers on that long?

TERENCE: They'll be soaked in brine, won't they? the parted delights. That is how you'd do it? You have the magic touch in spite of your lack of understanding. Once fully frozen, they'll stay that way for at least a week— like ice in an icebox. And for the other week, the meats will be protected by the salts from putrefaction, those choice cuts to be made morsels. And it's all to be cooked well. We have anticipated the details of pleasure. But what is illegal about a desire? This foolish ordinance is bound to collapse of its own weight, and fairly soon, against our prosperous intents. For the taste of good beef will strip it down to a voluntary pledge of individuals, before their children, not to abide the illegalities of cruelty to our lesser lives. For the simpler and underdeveloped minds of our brood must be taught the contrasts and distinctions between hunting for sport and game, husbandry of livestock, and purely malicious intentions to cause pain, sorrow, and suffering as a punishment for being born— a pet.... or a prowler for attention. This is a good lesson. An evil is not for cause, as any action could, but callousness towards the result to play of— We kill a thousand things daily, if not on plains and ground or watery habitats and airy nests, perches and passages, then within our own guts! where the death and birth of bacteria are like mulch to the intestines. And not one evil thing results, so be it foul. So then we sully the law, a bit. For a gluttony?... or for a protection of our arms of fortitude— and a protest against perceived interferences or influences to constrict our markets and lunches.... But all in all, Tansel, this is a subtle thing, considering the difficulties.

TANSEL: I'm frightened—!

TERENCE: Let yourself be driven.... by your needs. A world of interests is behind you. (*leaving, apparently back to the restaurant*) And you know the route. You have the keys.... You have our blessings affixed to them, and some reward to— earn.

TANSEL (*to himself*): How can a bower shade in night but from the moon and stars their lights of heaven? I have entrapped myself, within a restraining troth or trough, by my own artistry placed. Now, it's waiting for the kill, too certain not to come and miss me — shot! of friendships and respect.... or an animal's declension into food, a being puzzled to be raised and led.... to justify such an end, treated fairly— all of handsomeness.... and then bled. But that is not the spot that sores my back— with wart of willingness to crime! and engender a misdeed to a thoroughly innocent party. What is the worth of such damage? It's most terribly foul!— Yet evil? Evil is the pressure. Evil is the anvil struck, as for your head to feel these blows made earnestly.... towards what is to be crafted. (*to his knees*) Should I refuse the stars their shine and hide?! But that is to my own destruction, in this town, wherein I may only loaf and loathe till purposed. Yet.... deposits to be made, in checks—

for what?!... An altruism to the poor and depraved? And certainly for taxes to be paid, a record's kept— I'm being led, led through.... the bier's acknowledgment to roll upon. And this is such a sultry night for serving wishes. The doing is impossible!... but for my cautioned crudities. Know how to do, for reasons why are left to others juxtaposing their demands.... upon your contrasts bearing their estates of mind. Then must I think as if a criminal? or a paltry serf ordered to do this for exchangeable masters. Think this through. You know what is the better to do, within the rationality of justice and decent behavior. But I need so much— flak! to move me. Then am confused I, by this calculated pinning.

For.... the way is left to me, with these keys. I allowed my arrogance to boast, in my most unworthy state, which led to a condescension of this town, and banning much of my practice. My— friends.... see me as the cause, and offer me this way of resolution, and to save myself.... from some more suffering, for a time with debts to pay. The pen is sealed and high, to squirm within for these anticipations. (*to one knee*) Where is the malevolence? when forced to do by a convincing of your wrongness. But am I not correct? And have I not been? It's often so with pets and children. So blame the truth, if I'm deranged. I'll kill the sorry eye, that can only see of shadows on the walls of its container. I've done it so often, the innocent to stare at and sense of the anxiety, the understanding little and accepting— all that is to come.... as a surprise. What is such hope to have? (*standing*) Yet do, as is to death— it is not right, to this occasion made, somewhat artificially, and when pity is allowed. Then what am I, when pitied in this way, with shallowness.... of a laudatory contempt. What shall I do? away from the leafy shades, at night. No cover to my meaning is much left. I need the money, without more depth to own. But pockets are more pride to own than person. (*starts to walk away*) There have been.... so many casualties already, that I may cross upon a field of my own rendings. Then yonder are the ruminations left pursuing for this nocturnal disputation of my cause, in one so lovely (of a) night as this destroying of itself.... with much blame to me. Madness is as blind.... as a darkened light in mirrors spread for sight, the perceptions woe-some and harried of a distance sensed, afar, to reach upon.... a gilded shade of noxious notions popping up about for view. And what is there to hew? but of more deviance to nature. I've lost my nerve again, before persuasive demarcations. Yet chattel on the battlefield, be they removed.... and withdrawn.

Scene II — *A lighted barn open to the country scenery.* **Ponaple** *stands, slightly perturbed, as his son* **Sachel** *enters from the field, with a bucket.*

SACHEL: She's gone again. Hasn't returned to the stall? I brought them out some of the sweetened oats for night. But she's not with the other two, in our field.

PONAPLE: Maybe she found a particularly good grazing patch. But we should begin to fence in our boundaries.... They're not supposed to go roaming away from good food, hearth and home. I was assured of that. The bovine is very reasonable. But this one strays too far adrift, aloof, though terribly friendly of a disposition.... You might have to go after her—

SACHEL: Awh, pa?! She's just inquisitive. It's a gorgeous night, let 'er wander a bit. I was thinking of sleeping out myself, to be serenaded by the crickets up close, and the fireflies for dreaming.

PONAPLE: Well I'm not sure, Sachel. They certainly have large eyes for night, larger than owls'. But she might have gotten stuck in a ditch, or caught in a troth bed for the irrigation. They are totally domesticated pets, with an innocence for the plains and a thorough dependence on us, our management. It's been suggested I look after them more carefully. No telling the trouble a disturbed cow might get into. But they're supposed to have a strong sense for the natural boundaries.

SACHEL: They do. (*putting down the bucket*) They don't travel far at all, but stay well within our range, comforted with our observations and attention. It's only at night that we get uneasy about them. But if the weather gets threatening, they naturally head towards the barn. I don't understand why we have to herd them in, though. They should want to enter on their own.

PONAPLE: We'd have to close up the space, to try and keep the hay dry. Musty hay is like a toxin. When the weather really gets strong, son, we'll be put to the test about these cattle. I suppose you learn by doing what's necessary. But this is an interesting education we've taken on. And I wouldn't have even thought to do any pasturing without you and your brother and sister to help. What do we know of agriculture? Though your mother says she's had some experience. It's pretty much her idea, for you children.

SACHEL: He's too young to do much except be fascinated and have fun. I make most of the work around here.

PONAPLE: I put in my fair share too. Although I admit, bringing in livestock was a bold step. They'll always have to be taken care of. But.... don't you think this is an excellent retreat for the summer? If this is work, it fills the lungs with scents you'll remember for a lifetime.

SACHEL: Yeah. What about that? Oh, this is a grand place to be. I can love it here, this way. There's such a rural handsomeness everywhere you look. One can walk around for miles and not be bored or disappointed with what you see and discover. I'm surprised by the variety of colors in almost every location, and most of the animals you come across seem more indignant of you than scared, a funny arrogance I like. But what happens in the winter, pa? Who's gonna take care of these cows and chickens? I suppose they can take care of themselves during the fall and spring, to a large extent, but certainly not with frost on the ground, if they're not wild enough.... Are we renting out the spread?

PONAPLE: Possibly, if it comes to that. But this is a family undertaking. I've interested some of your cousins, on your mother's side, to have an honest go on this farm, if the young ones can arrange it for their occupational pursuits, to give it a try. That's the only thing that encouraged me to buy the cows. Otherwise, I'm going through this whole adventure rather blind(ly), feeling my way through, as it were. But this is more wholesome than merely investing and speculating on capital shifts. This could be something permanent. A small property like this might have a chance of developing into something substantial for the future, with luck and good management.... Derrick's ended his schooling, and has wasted almost a year looking for some profession to get into. His parents don't want to stake him much longer. He might be interested in caretaking the place.

SACHEL: That guy is kinda ornery, pa. Doesn't have a conception of failure. Plows right through everything, no matter the results. Too single-minded till he's stopped.... but not due to determination, but a lack of imagination and insight—

PONAPLE: Now don't think poorly of him because he seems narrow and limited of vision. That might be what a venture needs, to get started.

SACHEL: Got me into trouble once, with one of his quirks of ambition.

PONAPLE: You survived the beehive.... Probably too young to do a better job of it anyway, considering neither of you knew what you were doing.... at least well enough to avoid the stings. But keeping it so secretive led more to your punishment than anything else. I could have told you honey's not so profitable these days, even before you two got started, trying to convert that beehive on the house roof into a ground establishment on our estate's backland. You guys didn't even know about smoke, to pacify the little beasts.

SACHEL: I saw the danger all along, even after we bagged 'n burlap-ed it. But he kept at it, till the neighbors called the police.

PONAPLE: A minor disturbance and infraction of the law, endangering the security of a populated location. But Derrick is no simpleton. I happen to know he did quite well in school— I've seen the grades. But he feels higher education is a fraud, at least for him to encroach upon.... at this time. He's searching more for a natural legitimacy than an occupational one, not like yourself.

SACHEL: But I want to be an accountant.

PONAPLE: And that's perfectly fine. You're preparing well, for the stockbrokerage. But you might change your mind, in college. And you are *going* to college, to get that undergraduate degree.

SACHEL: Yes, pa. I don't doubt the need, these days. Good resumés are as important as good breeding.

PONAPLE: He doesn't take that tact for himself. In any case, his parents can't afford for him what he doesn't want to work for. But he might be interested in homesteading, a few years perhaps— and for us.

SACHEL: Sounds screwy to me At least you need a wife. But more than that, you need security for a wife. Oh, this is a fine place, for its country beauty. But can you live off that alone? And what does he know about making cheese?

PONAPLE: Just because you and your sister successfully nursed a tiny amount to fruition doesn't make either of you experts. And I was the first to milk a cow and show you how. But Derrick can learn all of that just as well and just as easily. Perhaps he already does, if your mother isn't yarn-ing too broadly. We're doing all of this for you, you know, the future generations in general. We have to stake a place now, if we're going to be country folk with any considerable presence, because I'm too busy in my work to be so devoted as you children may become, some of you. But that aspiration is not to be derided. For in principle we own a fairly large tract of land here, at a good bargain of cost, meaning that I had to buy a lot of acreage for the price. But the house and barn are not so spectacular as the land. So matters must be justified for my investments. Endeavors must be built up and developed, substantiated, and made to prosper.... You are enjoying yourself here, I hope.

SACHEL: Oh, definitely, pa. It keeps me alert and wide-eyed. And I think some of those grasses spread intensifiers in the air, to what I can notice of it; more than a freshness, but a virility of purpose. They must grow with a fury of vigor and enjoyment. It's a much richer life (here) than the splendors of suburbia. But I wouldn't want to tire of such an atmosphere, and become bored.

PONAPLE: There are mountains in the distance to drastically change the climate, at any time desired of the climbing.... So the other two are resting in the peripheral field.

SACHEL: Yes. The near field. They're just standing there, noticing our light, enjoying the warm night, and now eating some oats dry-coated with molasses. Supposed to make the cheese sweeter.

PONAPLE: Lactose is the milk sugar. I think you should go after her.

SACHEL: It'll be impossible to see her, in this broad darkness. She doesn't respond to calling.

PONAPLE: So take a flashlight.... Or set up a lantern. Let her see *you.*

SACHEL: Awh, pa! She's not a stray. She knows how to take care of herself.

PONAPLE: I might be overly anxious. But we're responsible for every life introduced on our property. I'm going out myself.... (*starting to leave the barn*) Just thought you might want to help. Mind as well leave the barn open— But extinguish that lamp. Don't let it die out unattended. Wastes the oil, and increases the threat of fire, if the wind knocks it down.

SACHEL: Oh, I'll help.... Hey! It's firmly attached to the cord.

PONAPLE (*voice*): A cow might jump at it.

SACHEL: They don't do that. I've never seen a cow jump.... (*more to himself*) Hunting for one could be hard work, this night.... Boy! Why does he have to look after my (work 'n) labors?! I know what I'm doing, as much as any of us could. (*looks around to the interior of the barn*) And what is he gonna pull her back with, wistful wishes?! We'll need a heavy rope, a tagging line. Wonder if one's around here. (*starts going to the back of the barn to search*) We need a dog!... for three cows? No.... that'll just be another pet to take care of. (*His sister, **Sarah**, runs in from off the field. She's perhaps only a year younger.*)

SARAH: What you doin'?!

SACHEL (*unperturbed of his searching*): Looking for a strap.

SARAH: Why?!

SACHEL: The "lone miss" has gone off again. We need a tether to pull her home, if we find her tonight.

SARAH: Oh, she can't be lost. Probably lying down in some wet grass, closer to the stream.

SACHEL: Try and convince Pa of that. We have to go out after her, because he's worried.

SARAH: About what? They're native to the area. We bought them from a ranch not that far away. Father must know something we don't.

SACHEL (*still searching*): *He* bought them.... And.... a little far means a many miles, in these country remotes.... Must be something here, to put around her neck.

SARAH: Ain't that some twine there?

SACHEL (*with an erecting exacerbation*): It has to be heavy rope, sis. We don't want to cut her, if she resists. Could as well use barbed wire.

SARAH: Well.... can't you— twine the twine? There's a lot of it.

SACHEL: No. I'm not in a mood to experiment. Where are the flashlights?

SARAH: Flashlights?!

SACHEL: Do you know where any are?

SARAH: ....I saw one in the house, a week ago. Mom might know. My guess would be the tool cabinet. But it's probably locked up, so that Beryl do(es)n't get into it.

SACHEL: Yeah.

SARAH: You gonna try and search, under the hayloft? Could be rats back down there.

SACHEL: There are no rats around here. Maybe some mice. But I haven't seen any. What Beryl doesn't get into 's probably frightened away by other predators. But this is a clean barn. I've made sure of that, at least to look at. Can't fathom everything about here, though.

SARAH: Well.... Why would some rope be stashed under the hay?

SACHEL: I don't know, but we need it. There has to be some around. The lamp's hanging from a good thickness (of cord).

SARAH: That could be all there is. A flashlight won't help find any more, (not) in here.

SACHEL: Not for the rope. I'll need one to help search for that cow at night. Pa's already at it, I think.

SARAH: Oh. He brought a light?

SACHEL: I didn't notice. Might have pocketed one.

SARAH: I only think we *have* one. What's this?

SACHEL: Oats.... Then I might have to set up a lantern.

SARAH (*getting to her knees before the bucket*): Have you done that before?

SACHEL: Not actually. (*as she puts her head in the bucket*) I'll have to find a wick and some oil— Don't get gross! Sarah.

SARAH (*popping her head out the bucket*): Well?!

SACHEL: I thought you were past the childish stage.

SARAH: Well if that's what cows look forward to—?!

SACHEL: You ain't!

SARAH: We treat them that way! You're not sensitive to their proposition of being, since there are no bulls around. But I am! So I want to see how it tastes, this pastoral domestication. (*sitting*) I was just by the other two, trying to commune with them, perhaps telepathically, to understand their peaceful submission to being milked, and being ruminants.

SACHEL: And?

SARAH: They could tell me nothing at all. Just grunted a bit. Pleased with my presence.... or maybe not. Hard to tell. They're large enough to give horses a threat. But they don't cause such (a) pause.

SACHEL: You thought you had a special in with them, the noble ladies.

SARAH: An insight.

SACHEL: It's because of their diets. They eat the grasses and the wheats.

SARAH: Tell that to an angry bull.

SACHEL: Generally, in the animal kingdom, the larger you are, the smaller you eat. The cattle eat the tasteful weeds, the elephant and giraffe eat the leaves and fruits off trees.... and the baleen whale eats the miniscule 'n microscopic plankton. They are docile for their diets. But that doesn't mean they can't fight, or defend themselves and their charges. What seems subservience to us is only a nature of doing the least for the most.... effort-wise; especially for grazers, a peculiar wisdom that doesn't hold well for the hunter and meat eater. And we are that. So don't act like a cow— ever!... It's disturbing.

SARAH (*getting up*): You can't understand. But I've never heard of a human cheese. You'd think some would be made.

SACHEL (*turning away to re-search*): Sounds revolting! Women only lactate when they're pregnant. And then they produce only so much as is needed.... for their infants.

SARAH: What about wet-nurses? And why aren't you searching more?

SACHEL: Because it's not here.... None's here.

SARAH: They brought us out to the country, to get more embedded with the realities of our natures. Don't you think so?—

SACHEL: No.

SARAH: Well, it's a sensitive subject to me.

SACHEL: You can't even milk one yet.

SARAH: Maybe that's why they might have been slightly disgruntled with my presence.... they saying to themselves: "What's she about, now? And this evening!"

SACHEL (*turning to face her casually*): I don't really care about such matters. But they're fairly gentle beasts in any case.... Beasts for our burdens applied, they're made for it— or adapted to it. That's no cause to give you— or anyone— some psychological complex about their willingness to be used. They are domesticated pets and no longer wild— and a food source, a potent one. I don't hear you squawking while you're eating your fried chicken, or see you imitating hens. (*coming towards her*) Dad wants this lamp off.

SARAH: Pa? But they're so large and evident, and seem to be knowing. I only thought....

SACHEL: I'm not sure if they do think, like us. I'm not certain if any other animal does. Some may act like us, occasionally. But that's not the same as thinking like us. We should restrict those anthropomorphic tendencies to explain things and behaviors. It is a cruel deceit for the dumb. They can never be us—

SARAH: How do you turn it off?—

SACHEL: Don't play with it! The one thing we must never have is a fire in this place. There's too much fuel for one around. The barn was part of the bargain. But it will probably cost more than it's worth to have some electricity installed. This is where things can really go wrong.... for us. I don't think it comes up to current building codes, while the house was forced to. So we can get in trouble here, by being sloppy— or stupid!... You understand this, sis?

SARAH: Yes?!... But if she's out a ways, this might be the only light she can regularly see, to orient herself.

SACHEL: ....Can't count on that intelligence. Could be the moon for all she knows, if it's not moving.... I don't know why she strays off so often.

SARAH: Maybe their natural grouping is in twos. If there are too many in one place they can't graze as easily.

SACHEL: That doesn't sound likely to me. Although we do herd cattle together closer than normal. Might have more to do with flies and insect pests that concentrate around them. But that doesn't explain why she's so particular. I think they're all of the same jersey. Look similar. Smell alike. Make the same calls. And I can't say any of the three are standoff-ish or peevish with each other. She just likes to explore more, or is not as easily settled to our range yet. Might have a slight allergy to some of the weeds here.

SARAH: You call them weeds. They are fantastic wild flowers, and in huge abundance.

SACHEL: That's because the land hasn't been tended to in quite awhile. How long that's been is a question, maybe over a year. But there must be a reason for the bargain price paid to acquire it, so much of it. I've heard rumors, or hints at them, of patches of "mad" weed on the property, when I go to the country store. The gentle-

men don't speak to me directly, but sometimes direct their topics of discussion to ranges of land territory that must include parts of ours, especially when they say the new "Ponaple" spread. They must know I'm a "Ponaple." It's their way of being polite, I guess. They're old cusses, anyway.

SARAH: I don't hear such talk, when I'm with you.

SACHEL: And right! You stay with me. We have to become more familiar in the area before you can go there alone.... Either they think you're a child or a woman, but they have a strong custom of differentiating the sexes for social treatment. They expect all females to be accompanied, and most children to be.... regulated for their activities. Ancient, historical views of land husbandry or country courtesies, (are) hysterical to be upset by modern principles of deportment. But there you have the rural life. An independent spirit can be considered rash and even wanton, at times. And only women raising children are considered wise or due respect. But when I go there alone, I can get the full blast of their pastoral commentaries, as a glancing or incidental blow; because I'm deemed man enough to take the singeing of the hairs on the back of the neck, even with my youth. And I'm smart enough not to confront them openly with a challenge, by identifying myself and asking for more specific, detailed descriptions or opinions about our properties and worth. All in all, though, it's been rather cordial to remain unobtrusive. The country manners are a pleasant veneer, for a way of life. You and Beryl stay careful and cautious, though— Well.... you do. Beryl's too oblivious to any dangers. But there must be some real roughness behind the panache presented, of this country serenity, so set to seem.

SARAH: You're suspicious?

SACHEL: An area this beautiful must have its crudities well in the background, not easily recognizable to us, and just waiting for the vulnerable to trap themselves with justification to be taken advantage of.

SARAH: Dad will prevent this.

SACHEL: Yeah. But he's bought into it, as an investment for the future. This is more Mom's inspiration or doing. And I can't help but feel it's a strange sort of strapping to improve or refine our behaviors or perspectives, to accustom ourselves to all of this.... rustic wonder.

SARAH: One can hardly have a bad experience here, with people. The area's not populated enough. The next neighbor is a car ride away, to be called a neighbor.

SACHEL: There's a pub or restaurant for the locals. Like the hub of a wheel, I suppose. And there's that general store I walk to, about half way of the pub. I like the novelty of long travel by foot, for a change. But next summer we should bring bikes.

SARAH: Why! I should be driving by then.

SACHEL: You don't think I'd use the car?! I can never tell when Pa wants to allow it. Says.... too much driving around through this region is un-comely, not to more notice and appreciate the sights. But what he's really afraid of is that I'll ram into some old dodger's horse-drawn cart, he can imagine, and cause a tragedy of accident long to be related to our family. It just sits in the storage garage, like it was retired. Why should one be embarrassed of a handsome automobile?... But that's where the lantern glasses, frames and components are.

SARAH: You're afraid to ask him for the keys?

SACHEL: Sis, I can walk to the store. And we're well stocked already of provisions.... And he wouldn't want me driving out, just to get to some pub. So we're stuck here, to roam about with leg work, like that cow. Supposed to build up your health, the muscles and personality, perhaps even your philosophies, to be more dependent on yourself than on your toys and machines; bodies real over bodies artificial. Well, I make a stab at it with this barn, keeping it looking clean and prepared, for the three.... disciples of Ponaple. But these are makeshift chores, very superficial, something to do to stay busy for an occasional sweat, and to have a purpose on one's vacation you can talk about later and glorify for your friends.... or impress peers and those "rational adjudicators" we contend with who often belittle our good fortunes. Those cows can probably take care of themselves pretty much. I'll let Pa worry about any veterinary concerns— He bought them. I warned him about that.... But nobody likes the lazy who are well off— least of all themselves.

SACHEL: You fear they could get sick?

SACHEL: The misses? They probably think they're on a vacation themselves, of overindulgences, away from a busy herd and all. And knowing Pa, they must be a bargain to some defect or weakness or deficiency or something. I can't tell. I mean, if they can teach him how to milk them, in hardly an hour's worth of afternoon, and he me just as easily.... then they're terribly well behaved animals, thoroughly permissive of novices.

SARAH: Sweet tempered and strophic.

SACHEL: What?

SARAH: Poetic of disposition.... to follow any logic offered.

SACHEL: Oh. Well I'm sure they'll come down with something we can't handle. If Pa's serious about this, we'll have to take one in to be artificially inseminated. Then we'll find out, what qualities are missing or aberrant. Those matches are carefully regulated and guided of the genetics, by state law.

SARAH: Oh. I wonder if we can afford to raise calves. That increases the complexities of management tremendously—

SACHEL: Cousins might.

SARAH: Cousins?

SACHEL: That's Pa's secret proposal, to have some of our cousins run the range into a ranch, after this summer.

SARAH: Hm.

SACHEL: Derrick is suggested.

SARAH: He's a nut!

SACHEL: But he's an able man.... and hasn't much to do, so far.

This is certainly a wild enough venture for him. And area-wise, this is a considerable prestige of responsibility. If I were of a nature.... But I need my accounting interests, to make a name for myself. This could be a prime opportunity, though, for a guy like Derrick, hard working and all— for a labor's labyrinth.

SARAH (*as **Sachel** is extinguishing the lamp*): ....What's "mad" weed?

SACHEL (*occupied*): Something like ragweed, I think.... but closer to a catnip for some animals; (*The light is out.*) mostly small ones, like rabbits.... Yeah, I can see them from here. The night's not that dim. But beyond those figures.... it's very dark, discerning past the silhouettes.

SARAH: Can you see Dad? I can't.

SACHEL: Our eyes aren't so promptly adjusted, sis.... Let's make a lantern. (*They leave for the garage.*)

Scene III — *A corner nook in the restaurant. **Terence** sits at a table with **Falfour**, drinking beers and sharing a bowl of potato chips and another of pretzel bits or nuggets with nuts. The place is sparse of patrons; but this is a late night establishment, closed during the mornings. A few other tables are occupied, and the atmosphere is relaxed with a very light background of semi-classical music of easy or slow tempo playing, as if in contrast to the rural setting of the establishment. A **bartender**, for the moment, appears to be absent from the bar deck, giving one the faint impression that the customers may be on their own, for their behaviors.*

TERENCE: I'm gonna stay here all night. The anticipation is electric! what I've started.

FALFOUR: What we've.... But are you absolutely sure you've gotten this into his mind? This is such a casual affair, if it rests entirely on his shoulder-blades, and sharpening of knives.

TERENCE: He'll do it! And they're already sharp, if they haven't been used for so many months.

FALFOUR: But carvers often sharpen them up again, just before cleaving into the meat. I'm not sure why, but maybe it's for practicing the slicing strokes, like an athlete preparing his muscles before a routine, warming up.... I'm not sure we can rely on him. This isn't a competition, Terence. This is a full scale savagery.

TERENCE (*munching on chips*): You exaggerate.

FALFOUR: Well, we're really going after it, this time. Lord knows I'm famished for some wholesome meat, instead of the baloney, spiced ham and salami sandwiches we've been living with.

TERENCE: No different for any of us.

FALFOUR: But when the grocery ran out of sausages (*grabbing a small handful of pretzels and nuts*), that was the last act to take of this.... temperance. I really became frightened, (*eating, by picking from his hand*) and talked myself into this attempt, allowed you and others to convince me it was plausible.... But all I've done thus far is put up some money towards it.

TERENCE: And a set of keys.... Your return will be three or four simmering steaks, Falfour, well worth the gamble.

FALFOUR: Oh, I don't mind taking the law into my own hands, for a good cause. One has to do that regularly, because no prescript's wise enough for everyone and all circumstances. But I long for the beef, so much. I don't want to lose my taste of it, as some say can happen. That could hurt my health, the very fiber of a man. My wife's not so sanguine of the outcome. But I'm the head of the household. And my boy will have some real chuck, no matter what he's done. He's growing and needs it. I ain't ordering some silly goat's meat, that's been rumored in the offering.... by some outfit with an excess of them, several districts away. I never ate goat in my life. Goat's cheese, yes— but not the spindly flesh. Could as well use turkey. But we need real oxen and cow.

TERENCE: You told her?

FALFOUR: Why, yes! She does the cooking. She's great at it, the way she can tenderize. It takes a skill to make some animal fats delicious— I've tried, and nearly drowned in my own retching with the results. So I leave that to her.... But of course she complains that if the store can't bring in any more decent cuts of lamb, pork and beef, then we should make arrangements for our own expensive shipping, from butcheries up north and northeast. How can you explain to her those factories work on volume orders— It would be prohibitive for us. She says this town is a volume! I told her this town is being punished, and we got into an argument over it because she helped promote that ordinance. So then I told her there will be beef in this house— but soon! Nevermind how it's to be done. It will be fresh and bloody and ready for the pan.... And so I contributed, Terence. And now we rest our fortunes on a disturbed ex-employee....

TERENCE: Oh, he'll get it done. Look at the incentives. They all lead to this result, a mighty favorable one.

FALFOUR: But this is more a pleading of faith, in our own resourcefulness, than a bargaining with criminalities. It is a psychological gasp of need.... The poor dolt refused to sell us one, if I've understood you correctly.

TERENCE: How could he or anyone do so, with our local restrictions, for our purposes in mind? But the idea was proffered to him and rejected, right in the general store, that he didn't know how to handle a cow that roams too far, being a complete novice to cattle raising, and that he should sell her to one of the more knowledgeable and experienced residents instead of having her turn into a real and dangerous nuisance. He cordially declined the pitch, although he might have been offended by it— But that set (up) this stage. If he can't take wise advice from a couple of our elders, then he's due a comeuppance. But frankly this whole valley is, adopting such an incongruous statute. It is a social activism that is too liberal of mind and has to be stemmed by the hard realities of our character. Sometimes groups of people think so much, they avoid all practical consequences of their thought. They solve one problem to make a hundred more, idealistic tendencies culturing bacteria on our tongues! And a stand has to be made against these trends, something definite and deliberate— but clever, and subversive with the softness of a simple matter-of-fact ridicule.... of the inanity that descends upon our lives. Again we must teach ourselves the most ancient stricture, that you do not make laws to feel good but rather to govern. (*The **bartender** appears, from out the back of the bar.*) So if the law becomes unbearable, unendurable to

be fit, and a body politic develops ungovernable proposals and postures, then resistance is stirred and an effective revolution must happen— not to destroy a people!... but a bad surge of unwise legislation and hindrances.

FALFOUR: A bad law, sure. That can result from any party of opinion, dealing with the complexities of raising unruly children. But I say spanking is the best remedy, not a civil unrest to provoke through pressuring consciences as if all citizens were still in kindergarten or elementary school! Now, does that fellow deserve some compensation for his lost cow, that's what I'd like to know.

TERENCE: He bought into the likelihood of being upturned through his ignorance. And these are arrogant types. For some reason he wants three, when he can only handle two.

FALFOUR: He was sold three.... probably for the price of two and a quarter.... But that's a cadre to make specialized assistances more practical, especially from the ranch he got them. One in three's more likely to fall for servicing than one in two or one in one. It's those incentives at play again. I'm sure that ranch will suggest or recommend their own favored veterinarians to him, if not the outfit may do some doctoring themselves.

TERENCE: No! He should expect nothing for his loss.... unless he finds out what really occurred. Then he might get some sympathy. But if things go according to plan, not much else may be due. It depends on if one can prove where she is captured—

FALFOUR: Or fished for.

TERENCE: That's why his neighbor will be involved. Bouwder must see the stray on his land.

FALFOUR: The straying. But can they induce her?

TERENCE: That is as sure as done, on a warm, coaxing evening like tonight. Shouldn't be any difficulties bringing to, with her tendencies. Hank's already threatened a complaint, that's been notebook-ed at the deeds office, but not registered. They won't issue a warning until three infractions. But only one is enough for the infracted upon to abide with our purposes. The warnings are for the owner *against* those possibilities. A legitimate causality is easily established, for our actions, based on these patterns of behavior and loose caring. Why, the cow could ruin crops, or worst, trip and fall and die in a watering troth and attract rodents, coyotes, mountain lions and other predators.

FALFOUR: Maybe even (rac)'coons.

TERENCE: No, the real mastery here will be Tansel's. He's mad enough to be brilliant at this one thing. But it must be as for an incessant drive to get it done, like relieving a terrible itch. That's why I've forced him to be responsible, by giving him your keys—

FALFOUR: This is such a risk.

TERENCE: —and having him go over to the .... shore of shears alone, unaccompanied, but with a feeling of being watched. He will want to do this— he will drool to do it.... being made committed to a force of nature, and talent.... or at least an ideology of craftsmanship. And the poor beggar needs some money badly. There are all incentives to be persuaded by. He'll be prompt about it, and efficient, industrious, proficient and profound. He will re-mark to himself how quickly and thoroughly he got it done, and so cleanly as to be swept by angels his flooring. Yet all the while he'll be in a cloud of purpose, a haze of duty and single-mindedness, his body tensioned for us— by us!... I found him sitting under the plaza bower, relaxing. Can you believe that?! as if the night were a comfortable blanket. I would have more expected him to be lean-ing up against the pub's outside wall, waiting for a cigarette to drop from heaven lit, to fall in his mouth and ease his worries.

FALFOUR: Fall or fly to his lips for puffs. I don't think he smokes, though. (*munches*)

TERENCE: But I saw where he was almost immediately. He was swimming through reminiscences, I think, day dreaming through the night. I could have figured where he was at, in his head.

FALFOUR: He's been knocked squarely down, and kept on the mat. We emphasize to him that this is all his fault, and *he* must do something about it. And he knows it's something horrible to tackle, that it must be, that it's going to be. So then, you tell him about the cow, I gather, and the project we've devised.... around him and his capabilities.... Was he scared?

TERENCE: ....He's not invited to us, unless we can intensify for him the soreness he's caused.... He was beaming with fright, that kinda fright that's more resistance than fear— For he well knows how to do it, and how to pose of a caution towards the eventual. A crime he's afraid of? Now is that a butchery of faith?! that what he's suffered through must come down to a triviality like this! To survive only for such a meanness? Well, it's not so mean and lowly. But it's hard to explain that fact to someone through halluci-nations.

FALFOUR: Come by again? Always with some pressure. He does good with the blades and saws. But hack it! Terence. Can the mad be trusted?! Now he can do good, with his tearing and strip-ping and rendering of flesh, the cutting through and cutting up, and disemboweling. Yet there again? I'd be mad too!

TERENCE: He wants some gritting of the teeth, and more to fall into his mouth than from—

WEILER (*jovially brushing up to the table from another*): Hey! Give me some of those chips! (*grabs a handful*)

TERENCE (*almost disgusted*): Ain't you got enough of your own?! BobCat!

WEILER (*chewing with emphasis of sound*): Don't go sour. I ate all mine up. And he's late for serving more. But what I've got you've got. And that's a decency for jawing— Ha, ha!

FALFOUR: It's lively, to the strains of Strauss.

WEILER: But you've got it all wrong. And this is all too public.

TERENCE: Only the pledge need not be so, the pitch!

WEILER: Hank only noticed that cow roaming nearby his prop-erty, not on it. Never crossed the imaginary boarders. That warn-ing's as much for the cow's protection as anything. He has traps set up in those areas. An' now we're gonna kill—let— Ha, ha!

TERENCE: Shut up! You don't have to blast it out loud, even if everyone knows.... in here. And it was a complaint. He's being compliant, to protect himself from a possible irate assessment of damages that might be attempted by a certain injured party.

WEILER: Yeah. Those Pooonaples. Those haughty *hoch*-es Boy! did they get a pile of dat weed! You can't raise no cattle there, comfortably. The seller probably never expected them to. See what the long boots get themselves into. And those fat ruminants sniff-ing around. But once you get a taste for it, nothing's higher to search for.

FALFOUR: It's tainted land for sure, in that respect.

WEILER: But.... good, rich meat, peppery and fiery, hostile to drab tastes and flavors. I can hardly wait.... You sure it's for tonight to do? I've been stuck on my stock of liverwurst. I need a change badly!

TERENCE: It's certain that *Tansel will tanzen*. He's heading for the meats factory now, I can imagine, unless anyone can prove oth-erwise. I've.... concentrated the effort for him, wrapped his mind around it. There'll be no problems with this.... transfer of interests in the community standards. And you'll definitely get your share.

WEILER: But.... is ol' Betsy's as willing as gamely?

FALFOUR: Hey. Joel's refilling your bowl. Go finish your po (c)'kering with the others.

TERENCE: Young Beulah's as sprightly to be game. This hap-pens, Weiler.

WEILER: Sure enough should! with what I've provided. But your waiting till the very last minute, Ter', to prod him forth, cuts time and tempers thin(ly).

TERENCE: Sparks need shocks.

WEILER: Well be it so, man, what's to be rendered. I hope you know what you're controlling. I don't sit too well with all of this supposed psychology, and have a fright for just what will be found bloodied and butchered at the chop shop.

TERENCE: It is an automatic tendency and urge that we're play-ing into, sufficient to crystalize our contents just as easily as neces-sary to be done. And I guarantee the results will be fantastic.

WEILER (*heading back towards his table*): So make it so, man! Since we're all accomplices.

FALFOUR: The entire sane community are.... Or at least our san-ity's at stake, for more steaks.... He's so boisterous, he's gonna blabber this out all through the village, the region, maybe even the state. You better be sure our legalities are intact.

TERENCE: It's all quite firm. No one here's stupid, (Fal)'Four. He may not seem serious, but he is— and just as anxious as we.... Else he wouldn't be so loud. Because this is truly a great demon-stration of the orchestrated recovery from an adversity of lowbrow, mean-spirited thinking that can happen through an anesthetized whiff of mindless public opinion. If a speciousness causes harm, then you must cause more harm to bring back legitimacy and merit of action, soundness of mind, capacity of heart, and the conferring of good goals. You're not gonna teach bad boys nothin' by restricting their meat-worthy diets. But at least we can show them how to get around foolishness, and not to cause any more. For, if we can accomplish this, think of what punishments we can provide— if we really get angry.

FALFOUR: You actually going out there, this morning?

TERENCE: Absolutely! It will be taking the tally of a grand event, and to see what's done and how. I expect.... a pristine environment, a remarkable advent of our fulfillments, and first choice at the pickings, for being so brave. And the day will be like a nightcap on our anticipations, topping off a really fine sense of design with desires.

FALFOUR: Then if anyone's to be seen, at first, in any timeline drawn up.... it will be you and Tansel.

TERENCE: There's little treachery to that, Falfour. Many of the stout families in town are involved in this. But the actual killing of the beast, the irreversible act, is done by one, only one of us, best profession-ed to do it. And he culls for the desired parts we're to receive.

FALFOUR: Will you help take away the disposable remains?

TERENCE: That will be done as quickly as possible— Don't ask for the details of your expenditures. Everything is prepared, and with little waste of parts. And all of the active players (for) tonight have highly specific roles, each limited to the most necessary of deed, as much to make a fantastic secretiveness, and with no considerable fallout from error— nor blame.

FALFOUR: Or fall guy, if problems develop? But you must be going out there, for some of them, this morning. I just want to make sure the whole business is as tidy as we'd wish. This is an incredible joke to play, but it must be thoroughly streamlined.

TERENCE: We employ the most possibly neat worker, under these circumstances. (*The **bartender** comes up, with a bag of potato chips.*)

JOEL: More chips, gentlemen? (*starts replenishing the chips bowl*) Thought I'd open the "Sour Cream an' Green Onion with Garlic" one. It's coming on its expiration date, though it's not as popular as I'd like. I think they're delicious.

FALFOUR: Might be some allergies.... to that.

JOEL: Well, you pay for everything, with the house tab. But I can't believe the onion and garlic flavoring causes all of that, natural and imitation. It would be to the potato itself. And there have been no reactions adverse to the freebie since I've given them out, all these years.

TERENCE: Free?

JOEL: As in freely flowing, Terence. There are more problems with the peanuts, of course. But I ain't never seen anyone choke on a potato chip.

FALFOUR: Can do a havoc to my stomach, with all of that.... salty oil.

JOEL (*finished*): Well now, gents.... I don't mind your confidentialities to feast on snacks anyway you like. And what you say and discuss here is completely private of our public establishment. My ears tonight are not responsible. But I'd contend that it is on?

TERENCE: Definitely. Concurrent of our rallying for snacks and drinks.... Why no popcorn?

JOEL: Because it's only cost effective to pop it ourselves. And such machinery is a mess to maintain. But the restaurant is dying for some hamburgers at least to serve. Are you sure Tansel can actually grind the beef, and clean out the grinder, with all of the rending he has to do, all in one night?

TERENCE: He is a miracle of the slaughter. Your agreed to allotment will be procured, if we indeed receive our flanks and cuts.

JOEL: Don't know how one person can be so proficient, but I've never delved into that type of work. Though I've chopped up some snakes and rabbits, in my day. (*bringing his head closer to the others, using a hand to support himself on the table*) We are taking a terrible gamble in this, selling stolen.... spoils as cooked delicacies of local gourmet fare and flavors. None of you had better be advertising the fact too loosely, or it could close us down if county officials catch a wind of it. I want to make sure that we *can* plead ignorance of your— prank.

TERENCE: Effective ignorance, yes, to any contention of law. You've paid for, expect, and will only receive, upon my assurance and guarantee, legal beef. We will not deal with crooked beef, the way things are planned.

JOEL: And we must receive a frozen shipment, Terence. We're only allowed to thaw once, and that can not be at the factory.

TERENCE: It will be so, Joel. Although you picking up your share yourself would make it much easier—

JOEL: No way can we ever be seen coming from or even near that butchery. The suspicion alone might damage our license renewal. And we're paying extra for the delivery. No. You have to be the conduit for this, Terence. This is your provisioning.

TERENCE: And so I am rightly to go about it.

JOEL (*rising*): Then enjoy your chips. No, I won't use the barbecued flavor ones. They really do draw complaints, with the Cheyenne an' chilly pepper spices. A potato chip (*starting to go off*).... should be the simplest of treats to taste, without much complexity, nor sophistication.... Made for children. (*He goes to another table, but is warded away.*)

FALFOUR: Terence.... you're tempting a conspiracy to do wrong, an attempt to promote conditions to breach the law. I hope you've figured all of the strings and ties out. And I must insist that.... land owner be compensated in some real way. A substantial gesture must be made— He'll probably turn it down.... But we must offer to do *something* to recoup his losses, or help him try to. After all, we're only hungry for beef!

TERENCE: What would you think of doing? I'd suggest we contribute labor for fences and boundaries. But if his cow does roam off this territory tonight, it is almost a moot point until he discovers that necessity, or we confront him with it. Almost.... because we will not argue about it, Fal. The gears are already in motion. And the appetites await their plated meals.

Act II

Scene I — *The windowless back room to the abattoir, whose closed door opens out onto the livestock area. It is a relatively small area or space for inspecting meat, with a metal table bordered to retain blood and fluids, along with various implements, but no machinery. A single ceiling lamp illuminates brightly, accentuating an eeriness of purpose, the table's surface yielding a slightly blue glow or sheen. **Tansel** stands at the table, his back to the door, studying the instruments he has brought from further within the building. He speaks to himself rather remissly.*

TANSEL: I am here. I have arrived. I have decided. I need the cash stuffed in my pocket. And what a paltry sum.... with retrospective. But .... anything is more than not to have. And no decision has been made, until the deed is— started. For once started it is done, either badly, or well. I could have run away— I can still.... But where must oblivion lead you? I am tarred and feathered already, and for an impossible job. Imagine them leaving that rifle unloaded, with the cartridge placed neatly nearby, by its side. They don't want to cause.... any damage to themselves, not of their hands. The hypocrisy is stunning, since there are hunters among them. And that's all their boys were practicing anyway. I've caused such a dysfunction, (*picking up a blade of scalpel*) and dislocation of terms. (*examining the scalpel*) An envy for civility and civilization? within the humors of a carnage. They *have* kept these knifes quite sharp. Now, to be efficient, one must partition manually the most major regions of the carcass, to avoid too much bothersome connective tissue for the sawing blades, gumming them up and greasing them to a sloppiness— Oh!... Why have I come here?!... to do this?... banal need! It is not for a futility, to absolve myself of a cretinism, or be captured by heinous thoughts of the usual, what I've done so often— and learned to contend with. But this seems.... somewhat cruel, not sterile of emotion. Within minutes— crime!... and consternation all alone. Knees in a snare, squarely locked. Cord through bone, to keep you in place. No cutting it, without hands. No biting through the metal twists of strengthened line. And you are trapped! to take this punishment. Soaked! of the abeyance and constriction till fawned of a departing— is your life's worth. And here is drenched, nocturnal shade of calculus, the actionable disgust of livery. Or for a thick apron to wear, I must find one yet, and through these hands wipe of staining grief!... I am too modern, to amuse myself upon some justification of this heinous cowardice. For what's to define of innocence with such easy death? subscription's killing to yield some crude tenderloin. Or.... play of myself, this scalpel— weeps! for a legitimate use, else cleanly stay.... for purer implementations intended. (*placing down the knife*) I can.... much juice of life or mortuary contain within this bordered table top, and let drain into a bucket. But here is where some choice slices are judged and examined.... for disease. Oh!... Where I am.... must I do this also? What is this blue light strained for me! that I am bound to this strange servitude— again! And for a dole.... A dole—

VOICE (*faintly heard, from outside*): Moo!

TANSEL: Grips the gravity of nurturing my art. This much arrives— with henchmen awaiting on me.

VOICE (*louder*): Moo!

TANSEL: And they come up, along the cattle trail (*brings his hands up and to rest in the back of his head*) for delivery into.... the fiendishness of my approaching, the heaven's wickedness for food and sustenance. Must kill! Must kill!— (*The door behind him bangs open with a powerful and loud force, as **Valter**, the **medic** and the **soldier** rush in. The **medic** immediately grabs **Tansel** from behind, to constrain. All three fighters seem somewhat bloodied.*)

VALTER (*running to the opposite side of the table, to face **Tansel***): Where are you enemy! to yield for us these wounds?!

TANSEL (*startled almost into shock*): It's Tansel here. This is my abattoir!

VALTER: And lead us into slaughter? You warned us not— Close that door! unless night goes on fire. (*The **soldier** complies.*) We burned much, in a blaze of shelling.

TANSEL: I gave no order—!

VALTER: I destroyed all order. My communications was smashed, as I recoiled from bullets an' bombs. But you did not follow us! I said to run—

TANSEL: Blame me?! My legs were.... broken—

MEDIC: Ha!

TANSEL: —I couldn't. The impetus swelled of such a weight, it made lead of my blood, and I was stuck as an anchor— feverish and sweating to be killed, true enemies advancing upon us, up and over us, through our flanks and lines. But I did not disobey, of the extremities placed in! (*The **soldier** punches him to the lower side.*) Uph!

SOLDIER: Take that for bravery! You devils stoked our hearts, like locks to break through. (*punches **Tansel** again*)

TANSEL: Aaaagh! The lying dead didn't deceive you, nor retard your withdrawal from the field more than myself.

SOLDIER: I was wounded, and could run! (*punches again*)

VALTER: Go search this hole, for other remnants of his sorrowful reproach (*as the **soldier** runs into the interior of the building*), our missions to endanger, our lives to warrant collateral wastes!

MEDIC (*bending **Tansel's** body, so that his face nears the table top*): Smell this blue burn! the sulfured ash turned into silver.

TANSEL: Rear me here! Take me apart—!

VALTER: We'll take you to another place!

TANSEL: Delve into this sinewy rain.... of power battled with and beaten. Saw my head off! for this brain to taste— and pass around of pot. But I was burnished, lying still. My legs were— frozen, my feet were gone. And guilt replaced all energies to suffer with. Ream of bayonet! some symmetry to splice upon this dream in halves, one shallow frost to lie in flames, and one's other disdained envy iced of rage throughout the snow and sleet of panicking!

MEDIC: Ho! Lick these carvers!

TANSEL: Ho! Ho! Ho!— I am here! Yo! I did not desert— I was deserted from! Lying! Lying!

VALTER: Falsehoods spying, were your means.

TANSEL: Can't you tell I've killed too many?! I see them, and their spines wrap around my wrists, as the spinal cords strum with me. A nervous energy's released, as I pale and blue become.

MEDIC: Lick this faux chrome! Right into the eye of the blast, see its hideous splendor and taste its shame.

TANSEL: I'm a disciple to it! I butchered many more than the average quota— daily, weekly, monthly— once! All that led to this famine of the heart, all that brought me— decadence, devoured the soul, and left more pity in the nails than their fingers. Yet sue my surgeries now?! They are beasts! They have always been beasts!— Not men. Not children innocent but of association— to beasts! Not the women, or the aged licensed for a caring— of beasts! and foul endeavorers, and hateful, spiteful minds, and crazy stupidness, and remarkable vaulting of unease into disaster. And such are we the animals become, to have become of us these trying necessities. Oh! Let me up, to have my head rub through these bowels of viciousness, and my face face foulness as a wanted drive, an excitation for relief, these tensions panicking for show (*as the **soldier** runs in with the rifle*), a distention of the brawn to prove one's self of capital to curtail the— beast!

SOLDIER (*as the **medic** lifts **Tansel's** torso up*): Found this (*pointing at **Tansel's** chest*), already to be used!

TANSEL: Go singe the wounds! That is for beef!

SOLDIER: I've poured upon it, so many a time. Here's bled revulsion, for your cunning sufferance.

VALTER: So suddenly pacific. So suddenly restful. And through the storm of lighted powder and clouds, admit the deception!

TANSEL: Moo!

*The **soldier** fires. And there is a great smoke obscuring the scene as sounds of "Moo!" are heard from the soldiers. As the smoke clears, **Tansel** appears wounded in the chest, and slouched, although still held by the **medic**.*

VALTER: You've come for the cull!

TANSEL (*feebly*): It's to the head, should be.

VALTER: Fast and found! Deserved of wounding shared.

VOICE (*from outside, of the **colonel's***): Get in that field of fire— Now!

VALTER: The cake of catastrophe. We'll pull you through it!

TANSEL: Tansel here. I'm stunned.

VOICE: Valter! Valter!

VALTER: Let's take the weapon. It's our only defense.

TANSEL: One shell is fired at me.

MEDIC: You'll die in the trenches! You'll die in the trough! You'll be bled out and ravaged, and pulled through for dismemberment. Those blades look keen— But that's a diddling, for you!

VALTER: Avaunt! Ha! Go open that door! Show them some light! (*The **soldier** runs to open the door to the outside.*)

TANSEL: Are there enemy out there? (*with an attempted surging of force and strength*) I am wounded! (*The door is opened.*)

VALTER: Ha! Night! Black as a powder keg's explosion. Blinds our entry, but not our entrance.

VOICE: Come!

MEDIC: Feel this deed, Tansel?! (*turning him around*) Face more of the furious unknown.

VALTER (*as the **soldier** leaves*): Walk him! Parade this fiend into the darkness, and let him witness firestorms first hand; (*as the **medic** forces **Tansel** out*) for we've been through it first hand, and have had our bodies torn up for a rust(ing) of metal. (*He's alone.*).... Only a soldier needs a gun— only a fighter! And this is a blue heaven. It is.... electric. We don't sacrifice ourselves. We serve the imperial— and— the— prestigious. Only they can command.... such an empathy of want. These are nice, sharp instruments. Finely honed.... while we're all battered, and blown up. (*picking up a knife*) Yet, as were for a bath— detaining. (*throws the knife down on the table, with a loud clang*) Let the stars shine! (*runs out, leaving the door open and the light on*)

VOICE (*after a pause*): Secure!

*A(nother) soldier enters, weaponed, and looks around the room and then goes within the building's interior. After a pause he runs back and out. Another pause ensues before the **colonel** and his **adjutant** enter, not quite leisurely but without anxiousness.*

COLONEL: What is this place?

ADJUTANT: It's a butchery, sir. A chop shop.

COLONEL: Ayeee. A'.

ADJUTANT: Cuts of meat are manufactured here. Deer, goat, lamb, pork, beef.... perhaps even ram.

COLONEL: Ayeee. Veal cutlets. I miss them.

ADJUTANT: The animals are marched through the trail line out there, into a restraining tank above a trough ditch, where they're shot in the head. The blood runs through freely, as they lie on a rolling platform, made neat for dismemberment, and are duly processed. Few parts are wasted, in an efficient factory like this. All spillage is contained in narrow paths easy to clean up quickly and thoroughly.... Not much mopping necessary.

COLONEL: Even the head? I wouldn't want the head noticing me, while I was working on the rest of the body.

ADJUTANT: It's taken off with a mechanical saw. Sometimes the eyes are preserved, stuffed as a delicacy after removing the cornea and lenses.

COLONEL: Do they make sausages here too?

ADJUTANT: I don't think so, colonel. But they surely provide their contents.

COLONEL: What's this table for, then? And why are there no windows in this room?

ADJUTANT: I'm not sure, sir. (*picking up a knife*) These knives must be used for making very thin slices of (the) flesh. Some of them are curved like surgeons' implements.... So I would suggest (they're for) an examination of the meat. This is a particularly brilliant light, falling down on a metal surface. (*replacing the knife*) Perhaps the lack of windows provides a contrast to concentrate it. There's little area for a diffusion anyway.

COLONEL: So, they examine the meat— for disease?... or trauma? We seldom examine our bodies. If you're dead, you're dead.

ADJUTANT: Autopsies are performed when requested, by families or staff.

COLONEL: To determine how one dies. But when you're dead, you're dead. You're never coming back to explain yourself. And you're not to be forgiven of this ultimate failure, though I imagine there are many ghosts around here mewling about. But theirs is pretty much a deserved fate, especially for the domesticated husbandry. As long as there's no silly wasting of the food, like those notorious and absurd frankfurter eating contests. That really turns my stomach. Why, you can't even call it gluttony, because the contestants undergo artificially induced hungers to enlarge their belly capacities.

ADJUTANT: Yes, sir. But that's a sign of civilization, to organize that kind of festival. It generally means no one's starving, at least those who are enjoying the extravaganza, if not participating.

COLONEL: I should think people would have something better to do, something so compelling that it can't be given up. And if you're talking about an organizing principle throughout history, something more sophisticated than eating and breathing, biochemistry and metabolism, then it must be to form groupings of inhabitants without such wasteful indulgences of foolishness. If you can eat so much meat, it can't be tasting good at the end. Excesses obviate a necessity, makes fun of it and degrades its character or privilege.... A lot of people use the term "butcher" for a soldiering. They are terribly wrong. It is a necessity and a privilege, for defense and expansion of controls on destiny. And I ponder often on how we could ever be otherwise.

ADJUTANT: Yes, sir. That's a crass comparison indeed, one full

of misunderstandings and.... immature conjecturing. The nature of the world compels the armies of the world. Heaven forbid(, that) the definitions of war are ever so misconstrued and distorted as to include animal husbandry. That is.... incomprehensible.

COLONEL: But it's why we defy the critics and will for ever. Because they will always be wrong, even if some warring conflicts descend into committed injustice and overt atrocities. That is like blaming the saw for the sharpness of its teeth. If a tree must be felled, it must fall. And if that's overdone, too many trees cleared.... blame not the felling, the soldiers commanded to do it. That can never be overdone, that proud volition to perform.... I've sent into harm's way so many of my men to die inescapably, just to gain some slight advantage of strategy or territory. There can not be one waste of personnel to this, as were to be myself a person tried and ordered. All (of) our lives— must be as justifiable as our deaths: valorous!... and simply that, entwined of the human condition, emphatic for its development and further-ing. So speak not of slaughter to me—

ADJUTANT: Yes, sir.

COLONEL: —as if it were a valid detail. A*yeee*. It is the chewing of the crud, for our jaws to make, for our incidences to sense and handle— the raw, rich taste.... of a defiance against odds, for goals deemed more necessary than our lives. And produced by you! Populated by you! are *your* needs met! I could explain to anyone. How should you suppose we defend this place, against the enemies observing?

ADJUTANT: ....Well....—

COLONEL: It's merely to be back! back here stationed to own this standing ground, and fight off its abuse, or misuse in any way. That is what is lent our drives to accomplish, and seldom rest for. But this is otherwise a quite wholesome environment, were not it being fought over.... or *debated* about, callously, humiliatingly!... But we'll take charge here, too. We'll spread ourselves, and work this area, this— possession, to be fit for our employed stance. And I would die for a rich beef stew right now, where there's more meat than the vegetables and the gravy.

ADJUTANT: I'd prefer a fishy goulash, a bouillabaisse, sir, with oysters, trout, cod and bass or chopped up eel—

COLONEL: We can have— none of that! evermore! until we secure these lands successfully, and end all persecutions possible, for a good long time. There are no tastes available to us, until we win a space to at least sleep without disturbance or abuse. But the reconnoiter said this building seemed deserted, the refrigerators and freezers empty— yet working. This is a great puzzle to contemplate.... what they're meant for— to hold, store. Never mind that, though. Our minds are here, to comprehend this garrisoning of butchery, and make use of whatever comes to meet us, be they the enemy or an empathy of survival. I'll have men stationed throughout the place, on all floors and in every room. And make sure they have time to be alert, and stay alert! as you direct them to their positions standing—

ADJUTANT: Yes, sir!

COLONEL: —For we are definitely here. I can't tell yet, if we can say we've captured this place. Its meaning is still too mysterious, its.... nature wanted with a cloak, a veil of seeing through for what it awaits. But do we much control it now, some guarantee of purpose pledged. Ayeee! There's a conceit to it, sure. But what else have we left, for all this gravity to hold.... of ourselves, or float away for loss of action feared. The enemy appears.... suspicious, granting this, as were a trap of liveliness in life. This enemy is what is fought— not might. Caribou and reindeer steaks, and should an antler be replaced, then let us hasten. For what is worst than an oblivion— of the mind? Guard this fortress with our souls in need!

ADJUTANT: What Valter finds. he leads us to.

COLONEL: Ayeee. A'. A'. Ayeee.

ADJUTANT: A blue light within a municipal forest.

COLONEL: Ayeee.... Some soldier type, he claims.

ADJUTANT: Then to interrogate—?

COLONEL: Of wounds?! If you're dead, you're dead. And if there's clothing made of this, suppose it so, to capture your remissness in nature, or slow to die out and extinguish this drive or impulse. The war's still hostile to continue— and to kill. And we beam of faces for it. So what's the disgrace, to find some surrogated life or mode of living, in all too civilian dress.... and place a claim upon him— liar! Look at what he leaves for us to do! So then we think of this, as the man thinks of us. Indeed he is shot— We're all shot! We each have that tiredness that keeps us awake. We all share the trepidation of deed and duty. And we forever fight off cowardice and sloth of intention. But what would they do here? those heady twins. Through the storm is brought back— meaning of disgrace! and nevermore to waste upon it caution. That's what the lieutenant finds. Bathe of the same hot mud! and share of the anxieties to wear of spots an' spat contusions. He never fails me, to engage of the fire fight. Though he's sometimes reluctant, for his men, to resist these most definite injuries and continue. So what he finds— always!... must be some kind of heinous imperviousness, some form of calculating permanence a fellow takes up to withstand these constant and consistent assailments and criticisms of body— of being, of relating to to stay around, of persistence— to be so ordered and moved, and to delve of the miracle repeatedly. So often have I sent him into the scorn of destruction. Why question that love? Why interrogate to find out what upholds it? or what might be captured of it. We are divined to be mandatory— more for our effects than our efforts. These actions are automatic for our character, but what results defines a character more.

ADJUTANT: And this war will never be over—

COLONEL: Until we end it! Until we succeed! Ayeee. A'. As the marrow of a cooked bone.... can be as delicious as the meat it was once covered by, it's the core of a thing that counts most to digest, even if it be sour till converted to our tastes, or bitter until made blithe.

ADJUTANT: How many more storms, sir?

COLONEL: How can a head perceive the night? as tranquil as is threatening of squeaks and pops and anticipations. I want my ragout.... We'll defend this place, as could be dreamt to be— a palisade of fortunes.... lined of trees. Where better to be felled, among much company! (*They start to leave.*)

ADJUTANT (*as they're exiting*): As if there *could* be prisoners!... in one's own battalion of stout hearts. (*After they leave, soldiers noisily rush in to occupy the building.*)

Scene II — *A living room.* **Beryl** *is playing with some blocks on the carpet.* **Laura**, *his mother, is sitting in an armchair, facing a TV, though it is off even as she has its remote control in her hand.* **Sachel** *comes in.*

SACHEL (*carrying a fixture of lantern*): Mom! Where are the wicks?

LAURA: For what? What are you playing with?

SACHEL: This lantern. Pa wants me to help him search for that cow again—

BERYL: Can I go? (*emphatically*) Go! Go!

SACHEL: No. He has the flashlight, and suggested this. The night could get heavy—

LAURA: I don't want you fooling around with those bombs. That's why the wicks were stored away from the garage. They're here somewhere, but probably packed up in a closet.

SACHEL: The oil's in the garage. Gasoline's in the garage— the car's in the garage. Why not the wicks?! I wouldn't be dumb enough to light it up in there. That's done out in the open.

LAURA: Your father shouldn't order you to do such things. (*as* **Sachel** *places the lantern down, next to* **Beryl**) Where's Sarah?

SACHEL: Still there, looking for wicks. (**Beryl** *starts manipulating the lantern.*) What are you doing?

LAURA: There's no good reception. I turn it on intermittently, but nothing comes through. Not even the radio signals are clear. One just has to wait. These rural locations are terrible for an evening's entertainment.

SACHEL: I told you, Mom. (*going over to the TV*) You have to keep adjusting the antenna. The stations change their transmitter directions too often, to satisfy the various regions of broadcasting. Their logic can be mysterious to follow. (*examining the TV top antenna to twist and rotate*) It ain't even on, Mom! The signal has to be amplified, out here. We should get some satellite service badly soon, if we're gonna watch television. Try it again!

LAURA (*pointing the remote*): You know your father isn't that enthusiastic about it. We're not staying so long to warrant an installation for yearly service. (*No TV signal is picked up.*) See! Nothing still.

SACHEL: Well, keep it on that channel, as I twist this thing.

LAURA: I can't scan for other channels?

SACHEL: You have to tune into one, find its direction.

LAURA: For each station?!

BERYL: Oh.

LAURA: Any retreat is an investment of time and planning. And your father's outlay was not inconsiderable. But you need.... the agrarian air, during some spells of your formative lives.

SACHEL: Yeah, it's great. Sort of a shock, at first, looking after large farm animals. But they tend to take care of themselves most of the time.

BERYL: Giv'(e) me!

SACHEL (*almost laughing*): I need this part too.... There are so many little components to any simple thing. We got any heavy cord rope lying around, Mom?

LAURA: What for?

SACHEL: To make a leash, for that cow.

LAURA: Not in the barn? Your father would know where. Can't you just leave her to graze? I doubt if she'll run that far away from the other two. They are social animals with interdependencies.

SACHEL: Pa is concerned. This is a new home spread for them. And she likes to investigate every inch of it too vigorously, perhaps.

LAURA: That's called grazing the grassland.... I think I saw a spool of rope in the tools cabinet. And you should look for an extra flashlight, instead of managing that lantern. It could be dangerous. You're not aware of how easily it can be faulted, by mistakes in use.

SACHEL: Sure is more complicated than I thought it should be. We should turn to battery powered ones. These are relics from a past age, come with the property. But look at the glass. Not a crack in them. They were carefully kept.

LAURA: And there better not be any cracks. Or else the thing could explode— right in your hand! dripping oil igniting on the surface. When blazes start, they're quickly determined to burn you. I don't want you fooling around with it.

SACHEL: Pa suggested the lantern, Mom.... I guess he thought we could just leave them to themselves, on the open range, and they would behave like reasonable animals and stay near our caring and attention.

LAURA: Why should you have to search for anything tonight?! Some batteries are in a kitchen drawer. Find a spare flashlight, and leave that thing alone.

SACHEL: ....Those are for a pen light. Won't do me much good. I need a continuous stream of illumination.

LAURA: You can see fairly well outside, with dark adapted eyes. The night predators do. The cow probably does.

SACHEL: Well then I'm learning how to use a lantern, Mom. That's something, of a knowledge I'll never need. And she might feel lost, on unfamiliar ground, in the dark. Pa's idea of a pasture might not cut it for her. Her roaming might be due to a nervous-

SACHEL: Afraid so.... I picked one up yesterday. But the reception wasn't good enough for watching.

LAURA: Nevermind, Sachel. Just leave the antenna on. (*turning off the TV*) I'll scan for channels later.

SACHEL: It doesn't work that way, Mom. You have to play with these things now. Optimize signal strengths, and all.

LAURA: Let it alone. This is too much work. And the stations aren't all that great anyway. Why aren't those beasts corralled, like they should be?

SACHEL (*idling at the TV*): Pa didn't think to have one made, when he bought them. Thought the barn would do. But our area is so vast, it didn't seem important.

LAURA: It's not so great that you can't build a decent enclosure for them. I thought *that's* what you were working on.

SACHEL: Me?! I don't know anything about fence building.

LAURA: Well, you're supposed to learn these things. That's why we're all out here— Beryl! Don't handle the glass so hard!

BERYL: It don't break!

SACHEL: There's no schooling on it, around here. I suppose you're just to intuit your resourcefulness. But I don't think there's any construction set for fence building in that barn. The timber pales and woodwork must be supplied from somewhere. Probably a local industry is called on, to do the job properly, to a code of animal humaneness. I can't do *that*. Not a skill I really need.

LAURA: Construction of any sort is a practical hobby to pick up, at your age. You're just making excuses again.

SACHEL: No I'm not, Mom. I can't do it. I'm all thumbs to that kind of thing. And I wouldn't want to hurt the dear "misses," by some spiking stake mistake (by) which I haven't a manual to consult to correct or notice.

LAURA: There are a lot of maintenance chores that you could be learning how to attend to. You children should have curiosity about problems around here. This house isn't exactly in the rare state of a new home. You should be going through the estate, and wondering how things work, or how you might rebuild their functions if we had to. That could keep you busy and interested. Practical activities on a country residence, they'll help you grow into a fully rounded person.

SACHEL: I'm not bored around here. And I keep the barn looking neat. (*going towards* **Beryl**)

LAURA: As if that needed much tending. But this is only a vacation.

SACHEL: The wind blows straw around. (*crouching to pick up a piece of the lantern*) It's not like we have a great investment here, we specifically.... Now look, Beryl. You put this through that notch there, and that keeps it attached.

ness for the new, an unsettling anxiety.

LAURA: I doubt if we'll lose a cow.... You have been feeding them the oats?

SACHEL: Sure.

LAURA: And milking? Then they'll expect to be fed and feed. She'll be with the other two by morning. They should all be in the barn.

SACHEL: That could feel confining, on a night like this. It's warm enough to sleep outside.

LAURA: Well then, at the the barn. You have to keep their wandering trails clean, you know. They won't themselves. And if it gets too spotty in one area, they'll wander off father.

SACHEL: ....I attend to the manure, when I find it. You're supposed to mix it in with cut up hay, to make a fertilizer. But we're not growing anything. So I just pile it up in a shallow ditch in back of the barn.

LAURA: She might be disgusted about that.

SACHEL: It's covered. You wouldn't want to stop up the drainage pipes with that stuff, it's only partially digested. Let me have that piece.

BERYL: No!

SACHEL: Come on, Beryl! Play with your blocks.

BERYL: Why! What are you going ta do with lan-tern? Start a fire?

LAURA: You're teasing him with it.

SACHEL: Well....

BERYL: Start a ha-la-cost?!

LAURA: What?!

SACHEL: They teach little kids too much too early, these days. I'm gonna shine light out over our field, to try and locate and draw attention of one of our big cattle. And then I'm going to go up to her, throw a noose around her neck, and drag her back to the barn house.

BERYL: Wow! Why!

SACHEL: ....To feed her some candied oats. And to train her not to run away; because she might be trying to get back to her home, which is many miles away, too many for her to cross. So she has to stay here, and be comfortable with that. (*gathering up the lantern components*) And I have to assemble this correctly, so that it doesn't come apart during use and cause me a disaster.... or an embarrassment. (*standing*) But I need one of those wicks, Mom.

LAURA: ....Some might be with the rope. I don't see why you should be forced to experiment this way.

SACHEL: He wants me out there. I don't really know why, but he wants me searching with him. Pa's unsure about something. We know we're not going to homestead here in earnest. That's left to our cousins.

LAURA: ....Yes. Relatives.

SACHEL: But something doesn't fit well in the gut, for Pa. He's anxious. When I came up to him in the barn, just a while ago, it was almost like he expected me to do more— or be more. Of what I don't know. But he leaves the choice for me, that isn't any.

LAURA: You let her drift away, that's all. He was annoyed, but not condescending because.... he can't expect any of us to be so careful. It was on a whim to buy them anyway, as if that could anchor our being here, solidify our.... belonging. But this is only a vacation home, and he might have overstepped for the responsibilities— we've garnered. At least you and Sarah may earn some practical experience living in the country. But now there are heavy animals to take care of. We've overreached our determinations to build a more expanding foundation for our immediate family, and so we'll probably need help with relatives if this is going to be a success, if this venture will be validated, and if this investment of our capital can be warmly composed and effective for our futures. You never know what influences you may have, with what you can not possibly know about until you try. And perhaps an actual business is at its infancy with the Ponaple spark of initiation. You already know how to make some cheese. That may seem of a slight consequence. But it gratifies me fully for your experiences to grow and nurture of healthy interests and clean wealths, which can be called a farming life. I was removed from mine.... much too early, but can still recall it as a very enjoyable period, and not as an idealized development but a happy one. But we are.... foreign, here, with no doubt. And we must adjust our behaviors to what is reasonable of the region and its inhabitants. Having roaming cows is apparently not a proper decorum, to this locale. Who would have thought so? Domesticated livestock, it seems, must be attended to, attention-ed, and shepherd-ed much more thoroughly than we could have anticipated. And this worries your father. He might have hoped you two would take to it more naturally than is reasonable to expect. Relying on the miracles of hope, to solve away the unexpected of difficulties and inexperience, can lead to cruel humiliations of what must be realized. But if something becomes successful here, perhaps your children or grandchildren will process cheeses from off this site or estate, in the form of a wholesome cottage industry. One can never know how our names may be employed for the future.

SACHEL: That'd be something, for us to get into the dairy business. But we really haven't cultivated anything yet— I mean in terms of interests. Just done a little hands-on study of a fascinating process, since we have these bovines around who seem rather aimless and confused to be.... on the property. I sense they want a purpose as much as we.

LAURA: Would you say they have personalities?

SACHEL: ....Certainly, to remain aloof. Definitely, to stay obedient and manageable without much sign of resentment. But they clearly pick up that we're all novices to their caring; and that might bring an uneasy restlessness, on occasion. They allow me to come near them and touch them. But then they expect more, and wait and wait until there's an absurdity to it felt. I wonder what they wonder about. I was just about to hose one off, for no other reason than that she looked a little dusty. But she gave me such a stare of incredulity that I backed off. There wasn't really a need for it. I was just looking for something to do about them. Was it foolish? But I don't think you wipe down cattle like horses.

LAURA: I don't ever want you attempting that with a horse. They're more independent and might not appreciate the watery treatment, kicking back in displeasure. We may never own horses.

SACHEL: Those flies kept bothering them, though. Really, Mom. We don't know what we're doing, with these animals.

LAURA: ....Then we'll have to learn. (*clicking on the TV*) Because they're our responsibility. Your father had to register some sort of acknowledgment about that, for large livestock to keep.

SACHEL: Lucky we don't have a resistant bull. Wouldn't want to go chasing around for it.

LAURA: ....Nothing. (*clicking off*) I think they might even be inspected. We can't kill them, you know. We can only kill rodents and rabbits.

SACHEL: Why would we want to?

BERYL: I like milk!

LAURA: If they become sick, they have to be nursed with professional care. We can't just leave them to weather through a disease, in our ignorance, or have them mercy-killed. They have considerable protections, in these parts, probably the whole country.... And it's the same if we'd ever want to sell them. We now have obligations to make sure they are and remain healthy, or else they have to be expensively looked after by farm animal vets.

SACHEL: Hm. So they have to be kept (en)'abled, for milking?

BERYL: I like milk! Chocolate milk!

LAURA: I believe so. And I think the milk has to be tested too, regularly.... or periodically. So then there are laboratory expenses.

SACHEL: This sounds like a lot of work. And a piano has to be turned two or three times a year, actually per season, or it starts to degrade. I would have taken up the instrument, if it wasn't for that.

LAURA: You could have bought an electr(on)ic one. I was hoping Sarah would take up music— serious playing.

SACHEL: Well, if you're to take up a serious study, you might want the real thing. That's how I've always thought.... So now we have cattle to deal with.

LAURA: An acoustic piano might easily dry out, in this climate. I'm not sure.

SACHEL: Pa can't really want to retire to farming. He's at the peak of his powers.

LAURA: It's hard to make a living as a musician, with so much recorded perfection already available.

SACHEL: He can't expect us to.... seriously want to.

LAURA: We're only trying to broaden your experiences. Practicalities have not been so very thoroughly considered. But in any case, you're going to college. And it would help tremendously if you could get a scholarship.

SACHEL: Working on it, Mom. We both are. Well, let me see if I can light this thing. (*starting to leave*)

LAURA: You be careful, Sachel. It has to be properly put together for use. Search for a flashlight first.

SACHEL: Will do, Mom. The tools cabinet. Lamp oil's in the garage. I won't light it inside, only outdoors.

LAURA: This is sounding more and more dangerous, Sachel. (*He has left.*)

BERYL: Is Sachel gonna start a fire, Mommy?

LAURA (*standing*): I hope not. All of this seems so unnecessary. (*going towards the TV*) But you have to work your way through these problems. And he's level-headed enough to take precautions.

BERYL: Can I watch him do it?

LAURA: No.... And you're not to touch those lantern parts any more. I thought you were almost trying to break it. And those fixtures are made of glass, thick.... but still glass. (*studying the antenna*) You can cut yourself too easily.

BERYL: I wasn't gonna break it. I wanted to look through it.... (*as she's adjusting the antenna*) Why do we have cows?

LAURA: Daddy bought them.... to keep us company on our vacation.... But they're just too much trouble, if Sachel and Sarah can't handle their chores by them— Oh! (*clicks on the TV*) You have to think of so many things these days, just for a little entertainment. And then it's hardly worth viewing.

BERYL: Can I play with the cows?

LAURA (*adjusting the antenna more*): ....I thought you were afraid of them.

BERYL: No. But they sniff at me.

LAURA: They want to see if you've brought them something to eat.

BERYL: Oh. They're big.

LAURA: Well, you stay out of their way, unless your brother or sister is around. If they step on you, it's your fault, because they can hardly know how strong you are— And don't go teasing them either, because they don't play like that.... I don't think they play much at all, just graze and search for grasses.

BERYL: Then I can feed them?

LAURA: ....If you have something to give, I suppose so.... But that's much of the purpose of our land. They feed themselves.

BERYL: Can I give them sugar cubes?

LAURA: ....Have you been?

BERYL: Aaa—

LAURA: Still nothing. This is impossible! So complicated. I wonder if we're line of sight to anywhere. There are.... 60 some channels to scan, and an innumerable set of directions for the antenna to place— Oh! Scan. (*studying the remote*) I can't find any scanning button. I wonder how Sachel does it.... How does he do this?

BERYL: Sometimes. They don't take it from my hand. I place it on the ground, and they come over. I get out of the way, quick!

LAURA: You're only wasting sugar, Beryl.... (*with a sigh*) It's hopeless. I don't understand this. Everything is too involved to be relaxing or straightforward. (*turning off the antenna*)

BERYL: They're not afraid of me.

LAURA: I know. I know.... Don't get too attached to them. They're pets, but they don't like petting. (*clicking off the TV*) They don't need that. (*facing to study **Beryl***)

BERYL: They don't chase after me.

LAURA: That's not what moves them. Only if you can help them are they interested in you. They haven't much affection because.... that's not what they're for. And.... they're a food source. So you're not suppose to want to play with them, like a dog or a cat. But they don't care about your curiosities, being as large as they are. They're not looking for you, in particular. But we have them, now.... And now we're looking for them. They're more than children. Almost a demonic guilt expressed.... They'll step (and stomp) right on you, if you let them—

BERYL: I don't!

LAURA: —and with an innocence about it. And yet.... (*going back to the armchair*) they can be so easily injured.

BERYL: They live outside.

LAURA: It's hard to show pity, for their accepted abuse. I think we've taken up to much, if we have to search for any.

BERYL: How come we don't eat grass?

LAURA: ....(*sitting*) They like it. Haven't you tasted grass? It has no taste, to us.

BERYL: Oh.

LAURA: We don't have the gut organisms that carry the enzymes than can digest the cellulose in the grass.

BERYL: Oh. But I have to eat my lettuce. It has no taste.

LAURA: More taste than grass. That's one of the eatable greens. It conforms properly to our diet. And we can enhance the flavor with salad dressing and sauces, and other foods.... And when you cook them, they have more flavor. Cows can't cook.

BERYL: Oh. Can I—?

LAURA: No. We're not cooking any grass. And you're not cooking anything at all. Though I imagine some people like alfalfa sprouts, as a novelty.... I'm not really sure if you can't cook common grass, and eat it salted. But we're not going to experiment that way, like with seaweed.

BERYL: Seaweed?

LAURA: We can eat grasses that grow underwater.

BERYL: Oh. Then we're like fish?

LAURA: ....I don't think many fish eat seaweed. It just grows for itself.... But it's salty and tasty.

BERYL: Then we're a little like the cow?

LAURA: Much more so than fish. They are mam-mals, like ourselves. But they are not creative, and can't speak.... They don't play with toys. They only want to eat and breed, because that's all they're meant to do.... I assume. They have one-dimensional personalities.

BERYL: Oh.

LAURA: Unless you can be friends with them. But what could that be for, if they're to serve (and service) you? We have then around.... for a luxury of need. So we make them need us. In other words, their lives, as much as we would like to know it, can be very simply described, compared to ourselves. Whereas a real pet.... can sometimes be depicted in human terms or characteristics and wills, having fun or being sad. There are even doctors to treat their emotions, if they are excessive when relating to owners. This is called "anthropomorphic illegitimacy" that people placate, to make money.

BERYL: Oh. And—

LAURA: But man is unique, to dominate all other animals, all other— anything. (*clicks on the TV with the remote*) But life stays so difficult, sometimes.

BERYL: And is everything made for us?

LAURA (*clicking off*): We make life difficult, by being too clever. I suppose we're meant to be the caretaker of the universe, although that's oversimplifying our ambitions, which is merely to act upon our curiosities decisively and often with dread of consequence. In spite of that, we explore everything, and use everything to explore. So everything is essentially made for us to play with, yes. And we own all, which can mean we own nothing.

BERYL: Hun?

LAURA: Any one of us can own anything, even a sympathy. So what's owned by one can not be owned by another. That's easier to see materialistically than otherwise. We so often think we share things, under a commonality of possession or expression. But each person is a unique king.... or ruler, just like each being is a unique individual, including the non-thinking vegetations and brainless lives of microbes.

BERYL: Oh.

LAURA: We're meant to master, so many entities. To be rural for a time, out here in the real country, helps us to grasp this enormity of purpose and investment in mankind, the grandeur of its realization and the fervent notions we aspire towards.

BERYL: As.... para-gus?

LAURA: We.... enrich ourselves, in this country setting. And you are growing much stronger in body and mind, although you can't realize it, due to the vivacities of these soil-laden airs and fragrances, the complexities of un-stultified existence, the freedom of natural mixtures of land, sea and air— though the sea part must come from irrigation, in these remotes. But, there's no need to have a metropolitan stuffiness, to sharpen our figures and recognitions, out here on this fruitful land. And that's why I insist you children experience at least some of this. You've seen more concrete than trees, more cement than grass, up to now. That can not be right for your human temperaments. Parks are simply not enough, since they're only made for strolling and gasping through picnics. You must be in, within the verdures of our earthly reigns, for awhile.... and associate with this, eventually, an appreciation for what we own.

BERYL: *Aaaa*— hum?

LAURA: It's not necessary for you to understand what I mean, Beryl. This sort of education is infusion-al of the psyche only, and can not be found in textbooks or on blackboards or computer screens.... and televisions. It is felt, to be remembered for enjoyment, as you are learning.... the many aspects and varieties of life and the world's dominions of true empires. No trunk of living tree has ever fought a war. Yet stands she stout-er than any soldier, and as complacent to our cutting down as a wind to be breathed. That's the highest strength of fortitude to existence, for our purposes. And all of this we naturally own.

BERYL: Is this an in-doc-tri-na-tion?

LAURA: What you're being taught in school is only history, through a few strange terms to memorize, with rote of eventual intelligent utility. But out here is a much greater class, a grander enlistment to being real. And I've tried to obsess you children, particularly around your age, to the variegated modes of life and circumstance we may endure and profit by. This could only be with colorful picture books, for Sachel and Sarah, since that's all of it we could afford. But you are much luckier, Beryl. You can meander through a green wealth. We own all of this land, now. And there is more of a playground than the sum of all playthings disposed to occur on it. Here must seem (to be) vast to your conception— an entire country. And you will remember this vacation with the acuteness that sharpens your senses today. All things bounded to our land are subject to our domesticated privileges. They, no matter how wild or innocuous or disinterested in our surveying, in essence behave for us. And that is why we may use each as we would want to, or need to. So you may chase after grasshoppers and crickets and fireflies without guilt, and trap them— seize them.... for your study.

BERYL: But ain't there a limited supply of them, Mommy?

LAURA: Not out here. And the adults may cultivate the earth for crops, and tend with the animals for their worth of food and resources of energy or labor. This is a proper privilege we *must* indulge of, being commanded by our very presence to oppose their inutility for our wishes.

BERYL: So we are masters—(?)!

LAURA: *The* masters.... and the masterminds. But we must not fool (around) with any creature for no.... rational purpose. Even to hunt for sport has some meaning, a competitive adventuring, an endeavoring to hone some skills of dominance. It may make the squeamish of temperament uncomfortable, but there's no shame to such cruelty when applied this way.

BERYL: I hunt?

LAURA: You explore, and expand your consciousness with observations.... I once actually cornered a jack rabbit, when I was a girl. The large hare looked so pityingly at me, I let it escape. But it was as interested in me at first, before I approached it, that it might have come up to examine, as I was sitting in the tall, weedy herbage it must forage through. So there is no doubt to a surprise of nature. It froze, startled, having foolishly blocked its route of travel by coming up against a mound of decaying and abandoned wall, as I stood up. So, after a short stare, to find its eyes pleading of distress, I turned my back on it, to let her scurry off.... And she may have taken her time, to learn of me. For when I turned around, she wasn't so far removed. And much fear seemed to have evaporated, from her face. I assume a female's wisdom, to make acknowledgment of the greeting as all but a passing of mights in a field. And then she went off, satisfied of our conformities to mutual access.... for a brief moment. But that is a behavior owned and cherished with remembering. They play for us. They are our actors aping.... our resolutenesses, these animals and plants and life-sustaining qualities we own.

BERYL: I hunt? I don't hurt anything.... except some ants that keep crawling up my leg. I let them get to the knee, and then I kick them off.... I own them?

LAURA: The experience is yours. They know you only for that, to interact with. So do not taunt the cows. They might respond very differently to the flick of a finger.

BERYL: ....Can I keep feeding them?

LAURA: ....I suppose so, Beryl. But now know what they are.

BERYL: Farm animals!

LAURA: Yes. Farm*ed* animals, dependent on our good natures. And don't bother the chickens much. They can be dirty. I think Sarah should clean out the coop, before....

BERYL: Before what?

LAURA: Our cousin is arriving by the end of the month, just before we leave. He's bringing a friend, as I understand it, or friends. And they're going to look after the place and take care of the animals, through to next summer and hopefully beyond, if they can find the life suiting their *own* domestications.

BERYL: Oh. Then they can have fun here.

LAURA: Fun?! They better find a way to make this properly self-sustaining. We're paying for all of this till then. It's the old tradition of having a summer retreat. And that we can afford. But their ambitions? no. We've only invested in some fine potential of country land. Who knows what will become of it? We bid on family pride and relatives.... I think it's all been agreed to, by now. But we must have residents dwelling on the estate, especially with cows. Otherwise we'd have to pay for professional caretakers, to look after the animals and the land, and prevent poachers, thieves, and unlawful occupants squatting here and there. So this can be an expensive risk we've undertaken. But I feel it's a worthwhile foresight. It has your father's assurance to become constructive of some value.

BERYL: Daddy!

LAURA: Fun will be the lifestyle that is generated here, with a lot of hard work.... unless, for some reason, things start to dissipate away, and make of this a fallow domicile. It feels very odd to be in a place without any neighbors or regular faces to see.... But I guess a sparsity of population can lead to a comfort of tranquility, for one's immediate composure.

BERYL: ....Compost!

LAURA: But we are pledged to be here, now, and established as a home, by the purchase of those three cows. This is where they legally reside. And that step taken makes a great difference to the complexion of this wistful enterprise. I can't describe the thrilling terror it supposes, to be responsible for them, so that now he feels he has to search for one on our own spread of territory, as if the night could consume it wrathfully. Such are the unsteady beginnings to any venture.

BERYL: ....Ven-tricle!

LAURA: What?

BERYL: Vehicular insemination.

LAURA: What are you riving at?! They're teaching you another language too early, and you're tearing up your own.... It's getting late.

BERYL: But you said I could stay up as long as I wanted.

LAURA: I said as long as you could. But that depends on the evening. I thought there'd be some television to watch. (*clicks the TV on*) There's nothing. See?! (*clicks off*) There's always nothing. And you're becoming tired and testy. I can tell by the way you've been playing with your blocks, not with much— fun. And the way you tried to break that lantern apart.

BERYL: It was already apart.

LAURA: you were not becoming happy with it. So maybe it's best to bed for you.

BERYL: Awh!

LAURA: Your dreams will be sweet, in this fine, rich air. And the soft sounds of a rustic night will make a symphonic lullaby, to bring you gentle sleeping. (**Beryl** *almost starts to weep with disappointment and frustration.*) I'll make some warm chocolate milk for you, as a reward for your obedience.

BERYL (*recovering*): Chocolate?

LAURA: You must control your emotions more, and whimper less.

BERYL: Chocolate!

LAURA (*standing, as **Beryl** stands*): And then the night will taste grand. But you must get into your pajamas first, and brush your teeth afterwards, before going to bed.

BERYL (*as they start to leave*): O.K., Mommy.... Can I have some jam on buttered toast too?

LAURA (*slightly startled*): You're not hungry for that.

BERYL: But I like the taste of it.

LAURA: You have that for breakfast. Just because you like something doesn't mean you're hungry for it right now.

BERYL: Can I have some crackers, then!

LAURA: Don't bargain with me, Beryl.... You'll be uncomfortable for sleep, and constipated in the morning. You must control your appetites, like your brother and sister. Warm milk will be enough.

BERYL (*nonchalantly running out, unaffected by a previous grieving*): O.K. I just learn it.

LAURA (*while exiting; to herself*): We must control our appetites. But the evening exhausts itself of my pleasure. And that poor remote (control) can stay by the chair, hoping some day to be gainfully used.... and to prove itself. (*leaves, though the light stays on*)

Scene III — *Outdoors. The moon faintly illuminates.* **Moraine** *stands with a lasso, next to* **Bouwder**, *who sits on a tree trunk or stub.*

BOUWDER: I ain't seen no heifer strollin' about these parts. An' it's a good night fer it too.

MORAINE: She'll be comin' about. Pitch can sniff them out, Hank.

BOUWDER: Yea. But I ain't got no cause tah go after 'er till she crosses that boundary mark there.

MORAINE: What?!— Where? I can't see nothin'

BOUWDER: That stake out yonder. I can just make it out, 'cause I know where it is. An' if yer draw a line clear across tah that tree, then that's ma boundary, in this direction. An' that's bein' generous, too. 'Cause I know that tree's on my land. But I ain't too sure of the curvature due the boundary. So I just draw an imaginary line tah my tree from that stake.

MORAINE: Well, should be within' a yard any's way. Out there's precious fer us.

BOUWDER: I seen it roamin' about, munchin'. Sometimes even rollin' in the grass, a distance a ways. Mostly during the evenings, Mort. But not this late. An' I'm not gonna stretch my neck out fer 'er. You have tah prove to me she's trespassing. An' then you can catch 'er. But ain't Pitch running way over there, on his property?

MORAINE: That ain't illegal, Hank. He has tah catch a sight of her closer up, in this dark, and then come back fer me. There's at least a good yard that can be disputed.

BOUWDER: You baitin' a trail fer 'er tah follow?

MORAINE: I'm bowlin' with a lasso, Hank. Look! If she sways over onto your parts, that's all that matters. Now we've studied her. An' I'm telling yer, she comes by this way often, often enough fer some capturing with the cutlery. Now, if there are hoof prints in your soil, then that's enough!

BOUWDER: But you ain't meanin' tah drag 'er over here, are ye? That's not too sporting, Mort.

MORAINE: They don't know what they're doing, any's way. This is relievin' some of their problems tah come. An' we're all darn hungry fer the beef, Hank.

BOUWDER: Sure they won't know how tah handle any of the calves. An' I don't want any of them skirtin' up here with a protective mother followin' an' chasin' me on ma own property. But those Ponaples can learn how tah do rightly, Mort, with a little teachin' and a lot of fencin'. Still, I can't see them cuttin' fer the veal steaks. They dote on the rascals lik' pets, I bet, an' won't ev'n be*aaa*r tah castrate the males fer steer.

MORAINE: Then you'd better git yer rifle ready, 'cause you won't want any wild bulls runnin' up here. Now look, Hank. It'd be ridiculous tah pull on that *vache* all the way from off their territory. Cows don't lik' tah walk that way. This is gonna be totally legally done. All we need's a few yards out yonder, an' there can be no real disputin' about it 'cause she oughtn't tah be any's where near here any's way. He don't ev'n know what's his, in this area, that.... high-brow. An' you can assume a lot. But that beast is a danger tah man-kind, bein' untended tah this way. Suppose some children were around.

BOUWDER: They wouldn't be, not on my land this late. I'd spank them, if I could catch 'em.

MORAINE: But that's an argument tah be made, 'cause the cow could be sick an' mischievous. We're just gonna find it found, up here, and restrain it. My truck's reada fer the haulin'.... tah take the threatenin' nuisance away an' out of yer hair.

BOUWDER: I got tah see 'er first, Mort. Sure, the judgin' of dis-tances can bea tricky, at night. An' I'm well within ma rights. But I got tah see somethin' real out there what's on my land. I ain't a'step-pin' on 'is, fer this. An' it is outrageous, tah have tah be so put out by such sloppy husbandry. But I got crops tah protect, from the lit-tle critters, an' now this cow. If I find my cabbage heads ('ve) been

ate at, I think I might go mad! I've had so much trouble with them alreada, this season.... Must be within sight, tah believe this, Mort.

MORAINE: Will be. Will be.

BOUWDER: He got three of 'em, hey?

MORAINE: Yea.

BOUWDER: Think the other two (a)'re milked out? They ain't so restless.

MORAINE: Might bea near calving. The owners wouldn't even suspect the drama.... It's always rough fer the first timers, though, lik' we suspect of this heifer.

BOUWDER: Hm.... Hum! What a silly ordinance. But ain't it de-served, though! Caught a coupla those tot-heads tryin' ta trap an' skin a gopher out here, an' with my equipment all tarni'(sh)*eed* up. Boy! if I could've caught those two, by the buckle of ma strap! They run off, leavin' the critter in the trauma of agony about the neck. I killed it. But that trap was busted with abuse. Heard it be-come some sort of ritual for them, tah harass the rodents tah death — fer fun. Ought tah ship 'em off tah war, as mule bait an' kickin'. Naught! they be too young.

MORAINE: Well, Hank. You supported the stupidity.

BOUWDER: I had a personal involvement tah relate to. Who knew it would turn out tah bite us this away?! Incidences lik' mine need some recompense of the morality, or what's ever left of it in our time. But this is a fiendish turn of events, a novice with cows just teasing us insanely, an' from their own inane innocence. How might we be punished, fer this?

MORAINE: We won't! We'll be rewarded.

BOUWDER: Of all the eyes watchin'—

MORAINE: Hey! There's Pitch a'comin'. He must've found her.

BOUWDER: Well I don't see that it's a'followin'. You reada tah go intah the deep?

MORAINE: Absolutely!

BOUWDER: Well, I can imagine that might be my area, an' might not. But I gotta see, I can't suppose this way or that. He is a neighbor, after all, in a way. Though I hardly ever see 'im. An' I'm not sure about 'em, this ways out; 'cause it takes me a trip tah get here, though I don't mind the walk. What I don't lik'.... is the sur-prise.

MORAINE: Or the suspense of confronting 'is clear negligence? (*as **Valin** runs up to them, not with an anxiousness*) Well?! Could you see her?

VALIN (*approaching*): There's something moving out there, but way off in the distance. Roughly headed this way, I suppose. But it could take some time. Should we go in for her? It's definitely on Ponaple property.

MORAINE: Oh....

BOUWDER: I gotta see somethin'.

MOARINE: We have some license, an' more plannin'.

BOUWDER (*standing*): I ain't budging from this spot. Now there is some sort of criminality to this. if it ain't done right. She has tah roam out this-a-ways, an' be (h)'ere, fer the capturin'. There's a considerable leeway I'll give, as tah where is 'ere. But I have tah see it — from here!

VALIN: Too far. Too far. We could be here all night waiting. We're on a limited schedule of time. Now we have to come to some conclusions, gentlemen. Should she be out there at all?!

BOUWDER: It's 'is land. She can frolic on it as she likes, since she's his.

VALIN: But is that proper?! The cows should be in a stall-free barn at least. And they certainly have that. Who has more jurisdiction, in this peculiar situation? This is a loose animal, a terror! headed this way. Can you wait to fight with it, as it builds up steam and angst of independence, to suborn yourself, by allowing an inevitable illegality to be witnessed, this caused all by someone's foolishness? Or can you not take preventative action now, to protect your holdings and its purity of state— this land unsullied by such dangerous trespass!

BOUWDER: I ain't sure—

MORAINE: You mean we can go in and catch her because she's essentially running wild and furious, abandoned of her owner's oversight an' authority. An' we're out here an' have a right tah protect ourselves.

VALIN: That's right! We are out here, as a party of Bouwder's, with his invitation to study an actual threat to his agricultural property and possessions long labored over. And we have seen this beast— tonight! utterly unattended to, and certainly have reason to expect a maliciousness to be committed on this land, this fine earth that has never needed barriers before.

BOUWDER: I ain't never bothered before, from those parts. But that ain't proper grazin' land anyway. The whole county knows it.

VALIN: We have seen the menace, Mort—

MORAINE: You have.

VALIN: —and are compelled to take action now, before some precious plants are liquidated through an ineptness of resolve. Bouwder has shown us this problem is true.

BOUWDER: I've registered a complaint.

VALIN: And we have found for ourselves a sufficiency of cause.

BOUWDER: Fer what?

VALIN: To restrain an impropriety.

MORAINE: But tah go in so deep, Pitch.

VALIN: You have to define "deep."

BOUWDER: Much farther than what I can see.

VALIN: Then you can't possibly tell how "deep," until she appears for you.

BOUWDER: But yer said she was far off.

VALIN: As much as you can see, you can't see nearly so to know. Whatever you can notice is how far. And that can be right next to what you can't notice.

MORAINE: So we go in, into the black. And the time it takes is the fight tah capture 'er, as Hank stays here.

VALIN: That's how I'd call it. Reasonable, logical preemption has much value, in the courts. That animal.... is a danger to itself and others. And we have been alerted and alarmed, by its strange motions and maniacal intentions.

BOUWDER: Is it so!

VALIN: I've seen it move.... like no cow of a pastoral nature,—

MORAINE: Say it be!

VALIN: —swift turns and juts of agitation.

MORAINE: Could be a monster awaitin' our seizure.

BOUWDER: Yer don't say she's actin' ornery? That does take fer some precaution. Might be searchin' fer a place tah rout or roost. Yer can never know. I don't want ma ground all torn up! The vegetables won't lik' it. Could it be due tah the....?

MORAINE: I think we should.... take it off our path as soon as possible, Pitch. But, could she be violent? Do we need a gun? I've got my—

VALIN: Shouldn't want to have to shed any blood out *there*. The lifting could be difficult.

MORAINE: She's that far away?

VALIN: But coming closer all the time.

MORAINE: Still, we should make ready, if she ain't docile. That's opposite a cow's nature, an' suggests somethin' horrid or suspicious, somethin' tah be destroyed.

VALIN: We don't want to leave behind any evidence of an abduction, Mort. This is riding the rails finely enough as it is.

BOUWDER: There'll be no shootin' up here! I don't abide much with the fireworks. If you can't handle this endeavor peacefulla—

MORAINE: I've a right to protect maself— an' I'm packed for that.

BOUWDER: Just goin' out yonder, beyond ma sight, sheds yer of all protections. If yer brings a dead cow up here, I might shoot yer myself. This deviousness is goin' far beyond the simplicities yer espoused tah me.

MORAINE: I'll do what I have tah do, Hank, tah deal with a fierce animal. I'll just bring ma truck up, an' cart the carcass off from out there.

BOUWDER: Then I ain't seen nothin'! an' can't speak tah any proof of nothin'! 'cept yer come out here.... tah do nothin'!

VALIN: There'll be no need to kill anything. A fine line of circumstance is being drawn.

BOUWDER: I'll approve of nothin' I don't see, an' on my property—

VALIN: A lull is based on one's perspective.

MORAINE: That's a lure.

VALIN: We don't want to chase her away from us, Mort, and make this even more difficult. She's coming this way, and has an animal's native apprehension. If you're out to hunt her down, she'll sense that and run off. But we're only about capturing her, and those are chances she'll take because she's domesticated to be led this way or that, and transported. So let's not foul this up. Let's get out there.

BOUWDER: Out where?!

VALIN: ....Hank, she is going to walk on your land. You are going to see the transgression, and we are going to help you clear the ground of her presence. The "where" will be here. That's a pledge.

BOUWDER: Maybe yer ought tah scout more, where I can't see. 'Cause I don't think yers'(re) too sure.

VALIN: Well who can be absolutely certain, in the dark?! But your worries can not be denied. It's out there, and I saw it. Now are we going after it or not?! Are you paternal for your property, or is this just a night's game of stalking shadows and counting up hesitancies?!

MORAINE: Well I got ma noose (right with me) here, Pitch. Could yer smell 'er?

VALIN: Too far away. But it is definite. She's out there. And they leave their cattle out for the picking off.

MORAINE: Oh! Let's stop bein' hypocrites about this. The time's too precious.

BOUWDER: I'm not a hypocrite. I'm bein' defensive.

MORAINE: We want that beef. We need it— An' we're gonna take it!

VALIN: If we're going to do that, it better be soon. Activities are waiting.

MORAINE: We have a right tah it. 'Cause....— Bouwder's upset!

BOUWDER: Well now.... yer make the plans. But I won't be the clam in yer chowder, tah take any blame fer a foolishness.

VALIN: Foolishness?! That's what we're condemning, Hank.

BOUWDER: Well then, fer a falsehood. I ain't about tah rake on 'is property, as if he could do so on mine. If she comes here, then you have all the rights in the world tah pick 'er up. I give yer that permission. But if yer brings her 'ere.... that is yer foulness of the feasting at the festival. Yer just better not let me see yer do it, which I can't— in this darkness. Yer just better appear with 'er, on my solemn ground, lik' apparitions suddenly in view, an' with all the guiltiness established in yer passin', passin' through an' by.... my estate. Now that is what I've agreed tah— tah end a travesty. Don't yer go startin' up any others, with my name branded tah yer backsides.... (*while sitting back down*) Frankly, I was hopin' she would gallop up here, lik' a horse. But it don't (h)'appen that-a-way. Yer have tah go after 'er, an' bind 'er tah yer fates.... Never actually saw her, walkin' on ma rich earth, my vow.... tah do fer nature what she's done fer me.

MORAINE: You ain't havin' regrets about this? Hank. You want some steaks as much as we.

BOUWDER: Yer just better be smart about the undertakin', Mort. It's a fiendishness evolving about our purposes. I don't want tah have tah resort.... tah a cowardice of nature, or fer grievin' ov'r the inexcusable— in ma. Else I'd run off out there an' rope 'er myself. But this be more than a daring dare, but darn of nation tah confound things. Fer in many aspects this land is shared. I don't prevent seeds from my plots tah make seedlings on 'is, spread by the animals an' winds. An' so forth are we fair, tah each other. But damage I won't take, an' that's my only cause tah deal with yer. An' as fer the tasty steaks, there's (beef) jerky enough left in the stores, until this *punition* against us from the big state suppliers runs its course. Fer how dare we end a cullin' 'ere! Well.... then it must be, that we become devilish.... Yet on ma property?... that cow must bea devil sent, with all of this temptation grossed.

MORAINE: Yer want us tah, don't yer?

BOUWDER: Don't ever say I crossed yer not tah want to.

VALIN: Your conscience lies out there, believe me. When the prize is won, all that's sorry will evaporate. It's only these preliminary difficulties that cause you to wonder and give pause to be astonished by what might happen. But once the motion of this venturing picks up, with its own logistical validities fully engaged and at play, you'll be happy about what's being done for you, and by others enabling your wishes.

BOUWDER: Such as it may be.

MORAINE: Then let's get out there an' trellis that thing!

VALIN: Hold off, till I catch a scent. I can give you that much.... requital for your thoughts, Hank, that she's out here and near, and that there's much legitimacy to our actions. One can see fairly well along with gazing, through this night, as the sky stays brighter than the earth. (*starting to go back out*) Come join me in a little while, Mort. Just follow where I'm at.... And that will lead you to the thing.... that we'll cherish being found.

BOUWDER (*watching **Valin** disappear*): Does turn dark.... very easily, though. Who needs lights, with these proponents driven this a'way?

MORAINE: So? I well know the direction.

BOUWDER: It's a disaster, Mort. tah have tah come tah this.

MORAINE: No. We're fightin' the vanities of usurpat-ious freedom. This is a decent response, Hank, tah our problems. How can yer expect anythin' else? That guy comes here, with 'is family, an' buys cattle that he don't know how tah raise nor handle.... Then we have tah retaliate, on that indiscretion. 'Cause.... in this imperfect world, there are things yer're able tah do— which should not be done, lik' havin' bad babies. An' the on'y way tah control the tides of these misfortunes.... is tah damage them with breakers, occasionally. Teach those forces soundly, that there'll be a defiance tah greet them, one time or another, an' back them down.

BOUWDER: I ain't indignant of a Ponaple. I don't care if they're rich enough tah be this stupid, an' recover from our intolerances.... more or less unharmed. I don't wonder fer the pride of it, tah test us this way. If wrongs weren't meant tah suffer through, there wouldn't bea night by day, a spat fer healin'. But we needn't have tah be so meanly surreptitious onto strangers, if they be cruelly deceived. 'Cause, I've no way tah know them well that way. An' this reproof makes shallow washin' of the hands, fer later shaking with.

MORAINE: Well.... yer don't treat children that-a-ways. Yer spank them!... tah gain our respect.... an' notice.... or our attention, while we're tryin' tah do stuff tah support 'em. An' if this famila acts so childishly.... well then?... Cattle are an important commodity, not a casual thing tah own.... We're down right mean tah the beast— fer a good purpose. An' yer can't take that notion lightly, when considerin' their welfare. It is an insult tah the community an' the whole country, tah let some runnin' loose this-a-ways an' that. Because in the end, eventually— finally, we have tah treat them very harshly, those which we have kept an' nurtured so carefully. It is rude, tah make them feel too free, considerin' what must result. They're gonna die, by design of our needs. Yer don't treat 'em lik' life's tah be so much careless fun. Now.... they are 'is, that's true. An' he can do with them whatever he wants, that's humane. But.... What's humanely dignified, Hank? if they're meant fer chops an' steaks after they're no good fer milkin', an' permanently worn-out fer their service.... tah yer. You have tah pen, stall, an' slate those big, warm, dull eyes— away from yer cognitions of decency an' fair treatment. Yer have tah be the brute tah the beast. An'.... there are few better ways tah learn that, in a community hungry fer the beef, than by what we're doin'— tonight! So this will be an education, at its worst, an' at its best.... some considerable maturin' fer the conceptions of the Ponaple types that try tah visit here too arduously, with hopes of establishin' a basis fer this region. It ain't legal, in the bookkeeper's sense— But what's the law?... to the winds of management of fiery beasts! that flesh flamed or charcoal-ed on (a) spit. I'm so anticipatory, tah a fine reward from our actions. An' all because.... he don't treat 'is cattle properla. (H)'E abuses them, fer their characters an' consciences tah grope fer more than they can ev'r attain. An' so, we bring one back.... tah "cattledom.".

BOUWDER: Don't persuade me much, fer taste of rich, red, corn-fed meat. The law is as I see it— still. An' this night is warm enough fer many contingencies tah be provided, fer which you an' Pitch are still dependent. But one drop of cow's blood on my ground, an' I'll scold yer roundly, soundly. The earth's not meant fer it tah drink. Might soil the crops, or alter the plants. An' there

are mysterious, mischievous retributions plied on eager men— I ain't so! I'm just protectin' ma properties.... tah-night.

MORAINE: Well then, howl at us some. (*starting to go off, in* **Valin's** *direction*) Because we're goin' tah do.... the malicious deed.

BOUWDER: May the night shade yer boldness.

MORAINE: It ain't much cover, fer our commencement of rights.... tah be delicately sufficient of our aims.

BOUWDER: So is soundly said, yer amens into the abyss of covetousness. Yer does disappear! tah me, my sight somewhat blinded by yer tales of necessity an' privilege an' gains-manship an' the ego tah do this so outrightly, so outrageously that the storms of impious imperturbability approach ye, as yer goals define themselves more sharply. Oh! Watch out fer mistakes, Mort. They may fall on ye from the heavens, lik' a galactic sputter sounding off dismay an' disdain. An' I would countenance this?! But that of heart in woe inwove, then as I sit, awaiting a defeat of pride with principles, this heavy evening.... Ja, yonder is more crime, no doubt to it. But speak to this, Bouwder. You know some awfulness is in the making. If it were a chicken brought, I'd pluck an' roast. But lamb or goat would shear? Or calf or cow or bull— or even pig to butcher? Grows the deviancy tah ponder, bringin' base fer these intentions. The dear must be alive, as were a deer fer venison tah eat. What is this ultimate meaning, fer our dissemblance tah apply?... unto these crude animals raising. But not fer mine. I'd rather kill a thief than tame 'im. Yet what becomes.... of stealing improprieties, an' improper aspirations? wishes, dreams, and hopes displaced.... misfounded.... an' demurred. So give the foolish some chance, tonight. I want fer 'im less loss, an' fer 'er more sloth upon 'is land. But what I see.... well, that is legal fer my eyes.... tah believe, some transference of guilt.... into a common wealth of joy through justice.... since she should not be here stalking on my foods. What shame! tah sit with this pretension, here blinded of the event. But.... cruelty is no cow. It is a conscience.... felt. So then, I'll remain.... right on this spot! I'll have these remains here, tah think about.... an' some day darn a weave— against disgraces. Yet the night seems as shallow as what it permits. An' I can hardly feel that veil drawn upon me. Fer judgment's all in sight.

Act III

Scene I — *A room with a lot of machinery, in the abattoir.* **Soldiers** *are walking around, studying the works, as if making an examination of captured munitions. The* **colonel's adjutant** *enters, followed by* **Valter** *and the* **medic** *bringing in* **Tansel**, *who looks rather distressed and haggard.*

ADJUTANT: Your interrogation wasn't that terrible, man! Thorough, but sensible under the circumstances.

TANSEL (*held by* **Valter** *and the* **medic**): I've told you all that I've deemed proper for you to know. I'm not a combatant—

ADJUTANT: With all of these weapons, just waiting to be used? But come to it! You know how to run these devices. You are proficient of the execution.

TANSEL: That is my skill. But I'm only a ghost, now— You've

shot me dead, right to the chest, the heart of feeling. I am frozen, and my fingers can't manipulate anything well.

ADJUTANT: We'll see about that, Tansel. I've accorded the colonel the possibility of seeing these machines in action. And I want your knowledge, to that effect, although much seems straightforward for operation. But I've ordered none of our soldiers to try this, to actually turn on any equipment, until my approval. And that of course goes thoroughly with your reasoning, because you are going to demonstrate—

TANSEL: I don't see why I should?!

ADJUTANT: —some functions of this processing you've mastered. Let him free!

TANSEL (*as **Valter** and the **medic** unhand him*): Free?! I am more faint than freedom may portend. For my head is in a twirl, after your badgering.

ADJUTANT: Are these lights too strong? But they have to be bright, so that you can see exactly what you're doing.

TANSEL: Yes. No detail can be assumed at all. Every detail must be witnessed, and generally by several at a time, although the principle actor does the instrumenting as current for any given procedure. But we always watch out for each other and ourselves; because accidents, though rare, are almost always severe and life threatening. So then we are each totally trained on these cutters and saws. There's no woodworking to appeal to with the meat. A waste of parts can not be allowed. Our tolerances for error are very sensitive.... Yet there is no subject carcass to— demonstrate on.

ADJUTANT: None? But that is coming, you said. And that has come, else we're ghosts too. And nothing stands here to prevent your potential diligence, unless you seem to confront the mistaken.

VALTER: Oh, there's no doubt about it. I ordered us all into the fire field. But he withstood the urgency, to stay behind, burrowed under a bower. Now what does that make him? One of our own and not?!

TANSEL: You found me there. I wasn't compelled to engage in your conflict—

MEDIC: And we find you here!

TANSEL: —I'm not your enemy, or whatever you're looking for. This is my place of occupation—

ADJUTANT: Aha!

TANSEL: —, or was. I've told you all before, ten thousand times. I've been waiting for.... waiting for....

ADJUTANT: You seem all confused, whenever you try to come to the point, to the apex of your ethical dilemma. You're waiting for what? What?!

TANSEL (*loudly, but almost as groaning*): Moooo! (*The **medic** grabs him again.*)

ADJUTANT (*as some **soldiers** gather around them*): Cows?

Men! You're waiting for men, fighters against us.

SOLDIER (*to another, both noticing a piece of equipment near **Tansel** and the **medic***): That there's a bone cutter. I've seen one before. It's used to break up large bones.

OTHER SOLDIER (*responding*): Slices them up?

SOLDIER: Snaps them in two, and then into pieces, to make them ready for a crusher. You can convert a whole skeleton that way into a powdered meal for cattle and livestock and other large farm animals, and even as a fertilizer for plants.

OTHER SOLDIER: Bone meal. Potassium.

SOLDIER (*to **Tansel***): Demonstrate that, turncoat!

ADJUTANT: Don't bait the prisoner.

TANSEL: Why am I a prisoner?!

OTHER SOLDIER: Put your hand in it, and have it snap off!

MEDIC: Might I try to force you?! But I can't turn it on—!

ADJUTANT: Don't damage him!

TANSEL: I can only do what I want!... Oh! Pity my courage! I feel like trying this self-infliction. It's easy to do. The machine works very quickly, a marvel of efficiency. I've watched it often, and somehow it restrains a large bone into just the right orientation for a clean snap, by pressuring it relentlessly.... But you must supply it with a flesh stripped specimen, so that the marrow stays constitutionally fresh and uncontaminated. And it makes a loud sound working. I hadn't planned on using it tonight, because it's a devil to clean. So you use it for a long spell at a time— a lot of bones fed it continuously— to reduce the cleaning periods to a few.... Yet, I can do what you wish of me. I could sleep in it—

ADJUTANT: Don't damage him!

TANSEL: I'm waiting for a sour turn of events, a booty plundered off a farm, stolen from a former customer. I'm waiting.... for hounds to nip at me, and bite! gouge and tear apart my body to eat at! And I feel warmly to want this, but fearful to have it done.

ADJUTANT: You're much confused. The evidence stays here to condemn you— that you have partied with the enemy!—

TANSEL: No!

ADJUTANT: You have conferred with them, advised them!—

TANSEL: No!... Hardly as of late. I haven't been working here for months. The house has been closed, shut down, decommissioned. Abandoned is the abattoir.

VALTER: That's never done! Then why does this place work? Why is there electricity for the lights, and these instruments of.... consumption?!

TANSEL: Crush me! Put me in that V-grove to break up!

MEDIC: Could gummy up the snappers.

SOLDIER: He wants to be a martyr— For what cause?!

TANSEL: I'm being paid.... to be a criminal.

ADJUTANT: So!... You admit, to war crimes. Against your own side?!

TANSEL: My sin.... singes with remorse. But we are all hungry for beef.... in this community.

ADJUTANT: Beef!

TANSEL: From the bovine.... sir. And.... we are stealing one to-night, for me to process into choice cuts and parts.... But I must kill it, as I've done hundreds of times before.

ADJUTANT: Kill. That's your job—

TANSEL: Ay, sir! Ay! Bend me to it, please! I've taken money for this! I need the compensation. Make me want to. Twist me! Mangle my mind! Contort to confound legalities about it. Torture me and kill!— But you've already done so. You've beaten me up, and then shot. So now we can question each other on equal planes — Hold me tighter, man! You can't heal this!

MEDIC: I'll— let you jump into a blade, to saw off an arm or a leg, or a neck!

TANSEL: It's all the same as running into a firestorm, isn't it! You think I didn't see that! Eyes blown out of skulls, the aim was so good. Why do you come to me now, at these salacious moments of doom?! when I feel so happy to be destroyed, where this is a contentment led back to. And.... where is my rifle?!

OTHER SOLDIER: ....Rifle? We've— eaten it!

ADJUTANT: You're really confused, Tansel. You are happy to destroy, not be destroyed.

TANSEL: Then— what are you for?! Why do you do this— again?!

ADJUTANT: We found you, plotting—

TANSEL: Against whom?! Against what?! You always find me — I never look for you! Put my hand in it, you— medic! And I'll turn on the machine to start the dismemberment. I'll break the wrist first— Or give me my gun!

ADJUTANT: No— You can't have that. It's not even yours—

TANSEL: How do you know! How do you know all of this! Re-lease me! And let me snap!

VALTER: You're into some sort of mischievousness of mind.

TANSEL: I'll demonstrate everything for you— on myself! (*cheers from the* **soldiers** *gathered*)

ADJUTANT (*with arm gesturing*): Stop! Stop this rambling rab-ble! (*The* **soldiers** *quiet down.*) You're no madman, Tansel, with a masochistic front.... Unhand him! (*The* **medic** *slowly releases his grip.*) Go do—!

TANSEL: Tempt me?!

ADJUTANT: Go do, your.... chores— for us.

TANSEL: For you?! Push me into a crazy sweat!

ADJUTANT: I've promised the colonel. (*going towards a shelved implement*) How does this one work? Is it a lopper?

TANSEL (*pawing himself nervously*): Don't you understand?!... You have to be dead first! It's cruel otherwise— It hurts!

ADJUTANT: Come over here, and show me—

TANSEL: I want to!

ADJUTANT: Come here.

TANSEL: Show us! You want to show us!... Shoot me again!

ADJUTANT: The colonel expects these machines to be running, or to be in operating order, if we've captured them.

TANSEL: I'm fighting my love, the affection to be useful.... The colonel, do you do what he tells you, what he asks of you?

VALTER: He's the commanding officer, at this particular theater.

TANSEL (*trying to calm himself down*): And you obey him— thoughtlessly?

ADJUTANT: His thoughts are our thoughts, Tansel. We don't think, unless it's under his management to do so. And what he thinks has been designed by others above him. Come over here.

TANSEL (*still struggling*): Don't taunt grief!... I want— the rifle. They'll be here any minute, and there's so much to do. It must be done swiftly— and without any errors, or I'll die! By dawn I'll die! I may verify, that everything must be working properly.

ADJUTANT: Come over here and prove it!—

TANSEL (*stilled*): It all looks clean, able.... I assume it's so.

VALTER: Why don't you want to demonstrate?— (*as the* **sol-diers** *start to crowd* **Tansel**)

TANSEL: Why did you run into that firefight, with bullets blaz-ing everywhere to meet you?! (*The* **soldiers** *stop their advance, though the* **medic** *touches his shoulder to possibly re-grab.*)

VALTER: ....I was ordered to. The command was given— and I felt it. (*The* **soldiers** *move back a bit, slightly fearing* **Tansel's** *nerve.*) That's ordained to be, sir.

TANSEL: Duty? Is it your duty— to be killed?

VALTER: ....We defend our lives, by dying. We promote our lives, by dying. It is the most sublime irony we have, the contradic-tion a soldier owns.

David L. Birdsall : **Ergomont**

TANSEL: And we make more of ourselves that way, because more and more of us are wanted— are needed! And tonight I need a cow.... like the oxymoronic cowboy— tah rustle fer some stolen grub.... And we'll all dine together, in this heat, after I'm allowed my occupation. But we're all extended here, are we not? We've all been brought here, one way or another, to do something evil. Though I walked myself over, and it felt like a satanic rite of passion to accomplish.... through the dark and starry night that teases to be pleased with some strange terror or fright. And this is a murder to be made.... Or are we not all cattle here, for the slaughtering....

ADJUTANT: You want.... to injure yourself—

TANSEL: No! I want to kill a cow! But not her.... do I want to hurt, so much. It is the familiarity of these surroundings that punish.... us.... so easily, that the cash burns my pocket.... or burns up in it. I need.... a martyrdom tonight. (*slowly walking towards the* **adjutant***, the* **medic** *following, still with a hand on his shoulder*) Do you want me to turn it on?

ADJUTANT: I've never been here before.

TANSEL: But war is where you're at, to contend with.... I'll do this— now.

ADJUTANT: You're not to be damaged.... by us.

TANSEL: It will make a burring sound, a hum. I've heard it all so often.

MEDIC: You need this commitment.

ADJUTANT: Are you a criminal?

TANSEL (*stopping, with the* **medic***, before the* **adjutant**) The money's on me. The conspiracy has been made. I've adopted it. I am poisoned by it. And there is no innocence left on earth.

SOLDIER: We!... are all innocent! Jam his hand into that thing!

OTHER SOLDIER (*as the* **soldiers** *generally growl*): His head!

TANSEL: There's no innocence here. I can be plain about it.... Order me.... to do this.

ADJUTANT (*as the* **soldiers'** *grunts die down*): I.... I can't give *you* orders, in that way, if it's what you want to do. You are commissioned by circumstance—

TANSEL: This is love. Order me, and I'll obey.

VALTER: It ain't no duty for a coward. You're a coward, Tansel, sir.

TANSEL: I am not a cow—!

VALTER: That's why you didn't follow us in, or felt you couldn't. You never could. Not in that battle. Not on that night. You were below fright. You were dead— You died! And now you wish to kill.... cows! But did you like, to watch us fall and drop, and splatter on the moistened earth our worths and waters of de-

ception. 'Cause that was really cruel, to witness our vitalities gored, and not rush in to join our stance in pushing back the enemy— sir!

TANSEL: Lieutenant Valter! Valor Valter! No one would wish to see that, who wasn't in the bloody maelstrom.... except audiences. Those images have hammered me for decades, and yet no more softened have I become to this disgust. So now I must relish the gutting, as a profession, and am forced to be.... some miscreant? For this? It's to seem like.... to seem like having been—

ADJUTANT: But that's the proposal.

TANSEL: —through it!

ADJUTANT: We want to know how all of this marvelous, automated cutlery works. But we don't want to sever you from our possession.

TANSEL: Oh! You deem me butchered already!

MEDIC: You're a strange thing to hold onto.

TANSEL: Night ghouls! My mind's ablaze! This light pushes me to you— to confront! I must see clearly, for everything I('m about to) do. Why do you hold on— to me? Why do you hold out?! Where is the battle to be waged next? in a pit for my grave, dug with your deriding?! But, why are you here now? when I must be at my most acute and sharpened focus, alone to do the nearly impossible, thoroughly impoverished of more honorable goals.

VALTER: Honor?! (*runs up to* **Tansel***, as if coursing through a minefield*) Your behavior has tainted us, taunted us on countless nights.

TANSEL: You'd remember me?

VALTER (*grabbing* **Tansel's** *hand*): You must feel the disgrace felt, that we couldn't hold on—!

TANSEL: Play with it!

VALTER: I asked you personally to back me up. And your eyes sweated with non-recognition. I turned around, within the furor of lights, flares and sounds, the popping of deadly projectiles whirling pass, as if to stop and take up a breath of resistance to them— And you were nowhere to be seen!

TANSEL: You've mistaken me for—

VALTER: Was long before the assault, that I confided in you as a friend, a companion of arms. Days before! as we were scared together in a debauched bunker of muddy frost. It was as sincere an askance as I could deploy. For we all need some souls to look after us at least, to be concerned about us. So in a gentle way, just do the thing! so one could know about it.

TANSEL: ....I.... had forgotten. Show me how malice bleeds.

VALTER: You responded with a look of acknowledgment to duty, and teary-eyed to face my face. For you knew we were all doomed, eventually. And we all are. But I put some trust in you, all that my heart could stand to lose, for a communal assistance. (*squeezing*) How many fingers does a hand need left, to point with?!

TANSEL: At myself?— At you?! All you said was: "You'll be there, won't you? Won't faint away." And I nodded. But now this throbs. Yes! It is a lopper, sir. And it can rip, too, when it grabs hold, in a certain way. (*lets his arm become flaccid*) Order me!—

VALTER (*rubbing his face with **Tansel's** hand, with an impulse to startle fear*): Could you have forgotten such a promise— to commit your life to?!

TANSEL: Burns more than the saw. Order this thing done! (*Valter suddenly releases the hand, as to allow it to drop.*) I responded that way, to so many requests of that nature, you all must have suspected me of a drowsiness, feigned or foundry-ed. But the whole thing was a blur to me, from start to finish. So, how else could I be encased for service, to act as I did? I ran.... away from error and mishap, at any opportunity chosen for this— cowardice, as the nerves commanded me to do, louder than your orders. And I was exploded upon, but not reprimanded for the sensible of self-preservationary initiatives, since a soldier can't fight if he's dead!... or deadened of a cause. But those opportunities were so few.... you could count them without hands, or their fingers— Nothing would be lost but my umm and ripple across the surface of a conscious well, of waters rivered through my squall of crying. I did not make your casualties occur. But I suffer them most personally, with endless upset and loathing of our losses. The heroes of these actions are never recoverable. No true hero ever is, he gives up too much of himself to accomplish his tasks. And so I am a drained, wizened husk of dried hummus— now fed to pigs and erudite mastiffs! who have conditioned me to bleed, and cause more bloodshed. For I see the mountains crumbling, with what I must do. And I'll never reach the heavens, buried under so much rubble of conscience and congruence to the lowly self. But.... you may help me. You may collect yourselves, to push me through this latest demeaning, sharpen my acumen and awareness to see me by this horror passed and carried out. We may fall again from grace to battle, into battle's grace — with this more of a chance to be parted from.... the envies we have held, as we sink below the mudflats and into a single rest. Help me cut this cow!... Help me kill her—!

ADJUTANT: We don't know—

TANSEL: —correctly!

ADJUTANT: —all that much about you, but what you tell us and claim to believe. You've demonstrated nothing—

TANSEL: My hands shake of worry.

ADJUTANT: —to us.

TANSEL: I *must* steady them.

ADJUTANT: And the colonel is so expectant.

TANSEL: The orders. The ordering.

ADJUTANT: Your type is anonymous, Tansel. I would not know you personally, nor the lieutenant. It's just a nod of head noticed, or an attitude picked up that can't be very descriptive to the ongoing scheme of things, as so many varied events are happening around us all the time.... But what you want, is a solitary rest, it

seems.

VALTER: Hide! Hide with me. Hide with us— Come back into the fold, of death!

TANSEL: I.... weren't not— willin'.... shaped that way. Crushed and crumbled. Hidden under fire. Lost to the conflagration of human resources and guidance. Raised to want sorrow and despicable acts. I need your.... covering, coaching and couch. I need your lambaste blasted through my ears, as I saw, ever so carefully, the tender.... obedient parts. And with the energy of an army, do this jolting job tonight. Get it done for cash.... and checks, and a cheekiness to still exist.... Or might we hide out here, forever, as I'd know it?

ADJUTANT: Nothing can last that long.

SOLDIER: He wants some honest punishment. He wants to make hide hide.

TANSEL (*with hands beckoning, as **soldiers** start to converge on him again*): Then come into it. I know where.... where it's done, to me.

ADJUTANT: Do not damage him!... Find that rifle, Valter. (*Valter runs out, as before seemingly avoiding mines.*)

TANSEL: You're aiming to shoot me again?!

ADJUTANT: It might comfort you to see the weapon— Stop! (*The **soldiers** stop ganging up on **Tansel**.*) Always to see the defense.

SOLDIER: Throw him in a mulcher!

TANSEL: It's going to get bloody tonight, all over my—.... our hands. There's work to be done— soon!

MEDIC (*still touching **Tansel***): Not a thing to it. You're undone, to what you haven't done. You're yet to be, and not become, what follows serendipity with shame—

TANSEL: I've worked here before!

MEDIC: And that was a pleasant occurrence? Heal!

TANSEL: It made me a living.... What am I doing now? but lingering— This is no accident! I'm being punished, in this way.

OTHER SOLDIER: Cudgeled into cur's curing!

ADJUTANT: Always the defensive posture taken before threats.

TANSEL: You'd threaten my mind?! That's not more possible than what I've been. A drunkard without wine. A beating pumice weathering within the void of contrasts and contentions for my heartache to be grief! Where is he who shot me?! I want to kiss him!— He tried to end it.... But you can not. He had a prompt solution.

SOLDIER: You're not so suicidal— as to want to kill a cow!

TANSEL: There's money to it.... You are me. You are us— Help me!

MEDIC: Mend, fellow! (*pushing down on **Tansel***) Come to the floor, the foundation of things.

TANSEL (*going to his knees to sit*): We've got work to do!

MEDIC: I can't claim to save lives.... only patch them up a bit, for future targeting by fate's swollen arrowheads. But.... look at our boots. Study them, and smell their trudge through grime and bone-laced earth, reddened to be brown as the organics decay before us, an endless march to trouble and tribulation.

TANSEL: This is a clean floor.... with its last waxing.... I'll keep it so— but it will be difficult. Splattering(s) will be generous. It's more work to clean than to kill and cull. But there was.... a company of us doing so, with regularity of pace, and a proficiency of order and design, worked out by the management. We always left the shop sparkling. For to leave any trace of decomposition in a meat factory means disaster for everyone— all of the workers. No! We prided ourselves on the sanitary conditions we promoted and maintained, and furnished health with butchery. We never left bodies rotting!... A company! Accompany! Accompany!

ADJUTANT: This is the spot, no doubt, where you fell. You said you fell, fell down on the job.

TANSEL: ....There was a diseased ox, that issued such a peculiar ooze of organ's gamy scent that it knocked me out, momentarily, like an alcoholic ether would. And here I was treated, during the dangers buzzing above. I can remember the blood and bile, dripping down on my face like candle wax off a flame. But it was only for a minute or two, this disturbance threatening us, that occurred. I was rescued— and I rescued. I did my job well. I alerted properly. I did not ignore the signs, and let that tainted meat run through.... our manipulations to perhaps affect others, or even customers for the finished product. I took a strong whiff of everything, while I was working, which is what you're supposed to do. The nose is one of the best sensors of a malignancy, or a preservative fed. I was even accorded kudos from the bosses, and gained a stellar reputation.... in this home—

OTHER SOLDIER: A mortuary!

TANSEL: —treated nicely, kindly.... through my employment. And I have affected this.... in a disgraceful way, leading to a closure of the shop—

SOLDIER: Chops shop!

TANSEL: —and the thorough end.... of my livelihood.

MEDIC: You shot yourself. It was an accident! fellow. How can I really correct such things?! They occur everywhere I go, and too often. But in the confusion of our pushes and pulls, and our flowing to and fro, these travesties must happen, as chance events seized for certainty, or certitude to be allowed their reign over man's hefty endeavors and squandering(s). But see these boots. I help them march— That's all. That's my only.... plight to own up to. I can't make miracles for anyone.

TANSEL: And— neither did I. Neither did I. Yes, they're rich. They're soiled. They've trampled over grass and mud, and entrails

—

OTHER SOLDIER: We've been through a lot, to come to you, find such a place— and capture it with our presence. Our— spirit! Our core to dominate over adversities and mistakes.... And now we'll kick you!—

ADJUTANT: Hold off, here! This man.... is under observation. Not quite a prisoner to protect. Not quite a comrade to constrain. But a curiosity found and questioned. Observe him. And what do we see?

TANSEL: That in a few moments you must help me. You must hold me up, and place me into things, fighting stances and postures to get the Herculean done. For we're in a battle yet again, of pin-pricked sharpened drudgery replacing an enfilade of nerves. And I am on edge to submit to it. I will burst to have it done, and spoil!... unless there's much help. Rescue me again, to face up to the dilemma, commit to it— and do! Or else panic will prevail.... in this place, and through these sullen hours.

ADJUTANT: Does.... the enemy come?

TANSEL: Yes! Yes! No! No!

ADJUTANT: In what number?

TANSEL: One! Innumerable!— Three! An infinity for worshiping.... the master!

ADJUTANT: Master?

TANSEL: The mastermind for all of this.... circumstance. The master places me here. The master bends and twists my frame's frailties to conform to this.

ADJUTANT: This.... master is a person?

TANSEL: ....The master is the need.... for beef!

SOLDIER: For buffing, he means. For courage, brawn and muscle.

ADJUTANT: Awh! You've told us before, the same jabber and twaddle, all this is—

TANSEL: The need for carnage. We need this carnage! We must do pain and hurting, in order to survive. It can not be done softly, though a just civilization mediates and mitigates, for how it's to be done and to whom or what. But we must cause misery and suffering, to ourselves and others. There's no way to get around it. It's how some may thrive, as others must fall. And the victors change with time, as the victims circumscribe their pits and vats of transformation, from casualties into a chilly or a chowder, from churls of a catastrophe into.... the crepuscular capitulation of a crescent moon. And we have won this action to pursue, to dine from. But now it's all fallen down— for me— into the empty valors of necessity, the common cruelty that befits a life. And I must die within this— chaste evil to obey, as for a covering of hide and hatreds, to kill an innocent cow illegally, and forever be.... with a cad's cadence suffering. Then let us howl!... The cougar strikes tonight. The coyote mumbles eery epithets. And a cross is borne upon my visage, as these wolfish intentions intercede to approve these desperate principles of action— which we enlist.... and have enlisted of. And I want you to serve me, as a hound is fed of greedy wel- come— to do this thi*iii*ng! and want to have it done! Because.... can you ever sense how demeaning and humiliating and soul-suffering it is to *have* to fall into the ranks of criminality?! It lessens me to a spot on earth defined for the brutality I must use; and I need your brutal support to withstand the self-loathing that must result— af- ter this is done and I am flushed with a relief from poisonous va- pors and am left standing alone to face the world as a worthless pity giggling for another breath of life, reprieved only for more eventual heinousness and butchery. Is this not the firestorm re- vived?! I've descended into a dearth of responsibility for any pride that's to be left me, and I must be controlled by vile contentions and fiats of character to be used— like a knife or a weapon or an instrument of destruction, the battles for which I'm sent to.... inces- santly, as hungers last, to rule over the personable. Then lend me maw, to be privy to your resilience to these never ending missions that finally lead me to my state tonight, my— condition of abuse and abutment to villainy. (*lowering his head*) I have been shot. I've taken the condemnation to heart, accepting it with a frozen fear.

OTHER SOLDIER: Stand him up for some punches!—

ADJUTANT: Hold! Hold!—

OTHER SOLDIER: He claims we can't be proud, for what we do!

MEDIC: ....Or how we've driven him, to seek a maddening assis- tance through his sick quandary. Now fellow, what must run here?

TANSEL (*lifting up his head*): My blood?

MEDIC: You're to catch hold of a stolen pet, to slaughter?

TANSEL: ....Yes. As from nature taken is she caught, and brought to me by devils. Then my hands must redden. But I've no- where to turn, to avoid this. I'm trapped of a poverty with needs, and constant subjugation and dependence on the wills of others, as they could see my scornful use. They call it skill. Skill! what's use- less without might. And scowl to place more base upon this art, my pretensions honoring. With true odds righteous, I'm being set up for their laughter and sacrifice, as much as the gentle cow to han- dle and tear apart. Yet, how can I escape this rending of myself? I've tried to avoid it. But these objectives eat me, eat at me, within this com-mu-ni-ty.... one precious bit of self-worth consumed, with each passing minute.

ADJUTANT: You've taken payment for this— blasphemous deed.

TANSEL: ....Ay, sir. Ay, captain. (It) Lies as a girdle to the groin, where money needs to be when desperately sought for. Be- rate my weakness to accept it. But I was curious if it could be so, where it was said to be found— The only reason I continued to come; I would have preferred to dash through the night and disap- pear into a nothingness at rest. But there is truth to the material, and thus the need. Many expect the best of me for the worst to be. Pressures build themselves up in our realities than for our fantasies to fancy. I could not decline the offer; though I genuinely felt oth- erwise, a refusal to stomach.... make me dead, sirs, if you can not be a party to this ac-tion! I can not conform to it well, with your mental intervenings to obstruct. Too difficult to do and to wish for

happening. To linger longer, though, would be my only reward in this. And then you'll charge again, with a defeated man to stow.... for pondering of and questioning. Then make me blithe(l)y insen- sate, if you can, to this particular debauchery tonight. Feed me worms of courage— to replace this hook that's painful in my mouth, that I must admit to you my failure to circumvent evil. Or else.... take me with you, to wherever you go when I am not be- mused by your thoughts. But however you do, show some perma- nence to be my own. I can not deceive myself any longer, to need to do this cruel thing.

ADJUTANT: But we are the height of feeling, Tansel. That's all we can draw upon.

TANSEL: I need that weapon, sir.... and in my hands to feel and grapple with— upon her sight—

ADJUTANT: No. No. I'll let you see it, but not have it.

TANSEL: Then the mission is abolished!... These animals don't die in my presence without the shooting.

ADJUTANT: You're struggling.

TANSEL: They don't faint away to drop dead, with my grimace of caring.

ADJUTANT: You haven't the power to kill yourself....

TANSEL: The cow, sir. The cow.

ADJUTANT: without that rifle. You're incapable of demonstrat- ing anything to us, being so constructed of conscience. You'd rather fall to your knees and weep, or cry for counsel.

TANSEL: But you've already shot me.... I can not do the same? —

ADJUTANT: I'm not sure about you. Self-inflicted wounds in this theater of battle is a crime. And the colonel will not be pleased, since most blame will fall on him. But it's what you want that stultifies.

TANSEL: To escape! Don't we all?

ADJUTANT: What do we want of you? Tansel.

TANSEL: To become petrified! Don't we all?

ADJUTANT: Tell us who the true enemy are.... and the colonel may let you have your gun.

TANSEL: ....The who? The what?—

ADJUTANT: Who we've been fighting against, rushing into a blasting hale to defeat and defy. Confess that identity to us, and then you may win your sordid goal.

TANSEL: I've not a clue, sir, but myself to injure. Fighting against myself? But that is too summarily efficacious to believe, too easy to conjecture without diligence of heavy thought. No! I'm not against you! I've let you find me, I've let you capture me.... as many nights as I could dream of it— and wish for, hope upon—!

ADJUTANT: Who, then, make this malice that we struggle towards, and die amongst with a wanton pleasure to be found— resisting them! at all costs of life.... or the tangible?!

TANSEL: ....C-c-cattle? C-c-creatures?

ADJUTANT: What does smite the earth, with a contagion of its pestilence?!

TANSEL: ....C-c-cowards? C-c-criminals?

ADJUTANT: Con—....fer with your thoughts, Tansel. Identify them. Give me something for the co-co-colonel to tell. Some admission. Some profundity that's real....

TANSEL: Would not you know already? You're so persistent of the war—

ADJUTANT: It's *you* we need to assess, to analyze and examine and interrogate and make ourselves sure about! Not our motives at all! We've known them ever since.... you've known us.

TANSEL: Then, sir.... you exhaust both my pleading and my mind. Because I've ever wondered why. I've never realized, nor rationalized why I should be the subject of your furies. I've never felt guilty of my service.... in slaughtering. But for a slaughtering to be, it's not my maladaptation to the circus of mankind. And if I knew, then I could argue with you— But I never can. You're so determinate. And I allow your fortitudes upon my person, like a bleach to make for ghost my ghastly predispositions of a life I'm charged with. And of others to rally for, I only question myself in this deportment of resistances I can little fathom. I do not want to be unlawful. Yet in an instance must I make the ultimate of injuries to cast my remonstrations firmly sealed and forever arguing as I could sense the words heard and temperaments felt for this nocturnal predation. Then tell me for whom the enemy may be. Your charges and resurgences frighten me, whenever I may notice them, whenever forced to or faced to. And now I heel to your intentions. But I have not fought with an enemy, since becoming a butcher; and I don't know of whom you speak, can not contemplate the error led to of my visioning. Only the most fundamental of tasks must ensue to insure us, what I've learned to do. And with these basic measures shortly at hand, these shearing techniques, I plead for your assistance and guidance— not to make mistakes. For even one (made) may imprison the lot of my conscience and consciousness for.... as long a time as I'm allowed to breathe. And where is this bastille where we'd be found, without an enemy to pause over as you ponder it— my capturing? It is my head of shame and surrender to the forces of municipal authority— or friends.... that do the fouling for an errancy.

ADJUTANT: That's a clear breach of trust, Tansel.

MEDIC: You must heal.... over the graves of footsteps made in trenchant mud and running desperation. Your ruining has been applied for.

TANSEL: Will you help me?

ADJUTANT: Give me some of that money.

TANSEL: What have you to do with it?!

ADJUTANT: Wouldst (you), I steal?

TANSEL: Wouldst thou?

ADJUTANT: I wish to see the evidence, that this imperative is lenient of penitentiary.

TANSEL: Take from me all. Raid me with a rapaciousness to leave ramshackle of this threat of nerves. But I've told you about it before, during the interrogations. And you've left it on me, without the slightest interest in its caring. And what could you do for spending with it, during your wars? What battles would be lessened? (*standing, the **medic's** hand still on his shoulder*) I need it for a subsistence. And.... that's the sum of its powers and meaning — Take it! from my pocket. From my gut! Leave me barren and amazed. Have at and lop!... Then suffer, I would not do?

SOLDIER: It's that you're bold enough to give. With little choice of course, but give, you are subtended into fires of the mind already.

TANSEL: And (have) taken beatings for your love. Reach in! and relieve me of this duty. Then I will enjoy the punches more, not be sufficient for the crime. It is in.... this pocket.... this ungainly pouch of pants, the poison threaded with a needle as if pain were for a skin to sew with and stitch through. (*The **medic** puts his hand in the pocket.*) The causticity of a severe need is none at all.

MEDIC (*examining*): It's there, not even wallet-ed.

TANSEL: I rushed to take (to) the falconry; for I'll never have eat of this, what could be bait for a bail.

MEDIC (*extracting his hand alone*): And sweated. He's sincerely pinned for wounding, doubtful of all nations maligning his best intentions, what be a country cattled of obsequious members.

ADJUTANT: Then.... do we help you mate with us, some casualty of harm. The enemy is your fortune— And that is what we fight!

TANSEL: Thus bled from me.

ADJUTANT: The colonel will not be disappointed, with what you've shown.

TANSEL (*placing hands on head*): I'm ready.... for the dismember-ship, and the blithe blight billeting.

ADJUTANT: Then— keep it on! Submit to subdue, (*as the medic brings **Tansel's** arms to behind his back*) as the cow calls. You are for wish and want of this arraignment.

TANSEL: I am dead.... I agree to this— sacrifice.

ADJUTANT: As much to blush for an agony, you are in a detention of awareness stark and powerful, within your— home. March him to the appropriate brig, medic (*as the **medic** walks **Tansel** out, accompanied by one or two soldiers*), while we have wait for the immediate of expectations, a savagery to find, all of the reddened mists transposed as the sprinkling of terror's rains through a disheveling gust of anger and rage. Darn be a dimness in bright lights. That's how to be found, with the crimson bathing due. (*The party has left.*) We shall not touch *any* of this equipment! until the colonel orders differently. Is that reinforced enough to be understood?! For as I'm certain that man knows how to use it, (in) sooth be sour doing so. We have enlisted, to lie upon his queer necessity — as being found, what to participate with brunt of fiendishness. And empty is the night, without these mechanical sounds cracking and roaring through the raw a devastating subservience. But we are entrusted to help fulfill a certain destiny of aims, as I could analyze the situation, that this being Tansel's brought about. So continue your vigilance and alertness, to look out for the enemy and destroy, or into the heated flints to run and eye, as a defiance against their temptations to control our very matter and convictions.... and consequences for our ever harried selves, the always bothered and perturbed, the trained to tarnish if not winning! and the workable of sympathy for our honored losses—

VALTER (*running back in, with the rifle*): Here it is!... But where be he?

ADJUTANT: Taken soundly to his heart's contentedness. Bring it to me!

VALTER (*complying*): The colonel said, whatever it takes.... to have him meant for using.

ADJUTANT (*accepting the weapon*): It's a keen instrument, nicely made.... Polished for blasting. And in my hands, as ours, as much a demonstration as could be hope(d) for is this fight provoked as well as ours.

VALTER: It's indeed as good as ours.... for a couple of shots. But it hasn't been through the heavy trenches firing, and tested of its rapid spirit.

ADJUTANT: Well so, it has blown cows.

VALTER: Will you give it to him?

ADJUTANT: ....We'll give it to a cow, to make some such amends to the firmament of a core, and help him.

VALTER: And we could sift through bony remnants to find no better cargo lifting and carrying with.

ADJUTANT: Cold. Warm and cold. Our dead are warm and cold, dear and deaf, departed and deceased, and of a mire brawn.

VALTER: But never defeated, though impressed of mud to buoy; there tracks our lifeless souls as a bubbling through cauldron's boil, ever ready and ever there to be found, what should the earth subsume as necessary deeds.

ADJUTANT: Or to admit to them, with transport of our woes and worn demeanors, this gun's the proof.... out of so many carried. Yet one suffices to be shot, as for a signal to recline in heaven's steadfastness. We've.... pushed beyond the envelope tonight, I fear. And what awaits is a transcendency into the void of caring. But so fine a weapon as this need not be dismissed as a casual trophy found, if this is the place.... of staring suppositions to his plight, since his has ours become. I must inform the colonel that it might very well start here, the catastrophic smiting we must end with. (*starts to leave*)

VALTER (*to soldier*): He did not seem to me so brave. But then, when all is lost for one, where's the need to be afraid?

SOLDIER: Just something to cling to, a sticky web that dries to fluff in your hands and on your face. Then may you have a scare at least, for your emotions to displace. But I thought him becoming sound and solid—

OTHER SOLDIER: Like a *Willerby Whenaple's* corpse.

SOLDIER: —towards the final stages of his discomposure.

VALTER: Then present is a storm of sultry night to shun. For if he's calmed— by all of this.... then, what are we to do?

Scene II — *At night,* **Sachel**, *with a lantern and rope, comes up to* **Ponaple** *in the field.*

SACHEL (*calling as approaching*): Pa!

PONAPLE: I've seen you coming. So.... you've decided?

SACHEL: It wasn't so hard, sighting your silhouette. But.... where's the flashlight?

PONAPLE: I didn't bring one. It's not so difficult to see. The night is bright. The stars shine. The moon glows. And the ground reflects some light here and there.... It was only an impulse to go out after her, seeing as how you couldn't collect them, together. I suspected she'd be missing, and I just felt.... I should make a search.

SACHEL: Any sign of the lone miss?

PONAPLE: How can we hope to sustain such a property, if we can't keep track of our own animals?! I've not found her. And what can one do about it? You can't call out for a cow— they won't respond. I can't look for hoof marks in the dark, through this grass and earth. Even with a torch light it would be haphazardly hit or miss and not likely to reward with a find, even of evidence. I thought.... I felt I would just make her out in the background, standing or lying down. But we're really not skilled with such recognitions, though I assumed she'd look like the other two. No, I've not seen her. No glimpse allowed.

SACHEL: But this is a broad territory. Even in the daylight it could be difficult.

PONAPLE: But we own this land, Sachel. It must be husbanded properly. We can't allow such mismanagement to occur. We're responsible for every dreaded thing that might happen on this range of homesteading. We could appear to be too loose by others, too frivolous and credulous of nature, scolded with dismissal or taken advantage of. We have to be much more careful with what we're doing. I was hoping you and Sarah would pick that up, that anxiety we've— purchased, that this family has dared to live up to, surpass or overcome. For whenever one starts an endeavor as important as this, one has much proving of one's self to do.

SACHEL: Why is it so important? I've taken care of the cows as best I can. Can't claim to have a gift for it, much less any training or experience.

PONAPLE: ....No one can afford to wear a veil of failure about one's life. People condemn you for it, very unreasonably, as if your efforts didn't count. And what would you have to show up their commenting, if your efforts were lax and you knew you were not doing anywhere near what you should have to accomplish something, or not applying yourself to your best capabilities, nor even as well as you would want to? It's that acuity to perceive yourself as lazy or reluctant to harness value to your goals which grinds you into dust, as that shade of countenance appears for everyone to see. Because most people are cruelly competitive, to want to trample on those judged to be unworthy of competition, and merely fodder to feed their beasts with when not obstructing their strides towards success. How can I view ourselves here? I was hoping some.... motive force would simply evolve, especially out of you two young people, to naturally seize a command for the place, and establish a foundation of legitimacy to have this property, and to smooth out initial awkwardnesses with the fortitude of our privilege.... But it seems like our efforts are falling apart, if we're to be laughed at through events like this. I'm keen enough to note what must appear to some of this region as an unworthy attempt to live here. It's that veiling we must fight.

SACHEL: But this is only a summer home, Pa, a retreat for us. We get away to here, for nothing more substantial than a vacation. I'm not gonna make my life— tending cattle?! So why should we take this all so seriously?

PONAPLE: They are thoroughly domesticated stock, son, in the rancher's sense, totally dependent on our caring. They are less independent than cats and dogs; and we have to know where they are at all times, or better where they may be. To have one injured or lost due to our sloppy oversight debunks our justifications to exist here. More than an embarrassment of fines may result. Our character may be considered diminished, and any future dealings with others in this region may be difficult to make productive with such a demerit. We certainly won't be able to buy another easily, if she's lost to much suffering. Novitiates are not given more chances to harm their charges, I know the custom here to be. More important than that, Sachel, is the fact that you'll be doing many things in your life, working on many attempts to be proficiently accomplished of your interests, or even celebrated as a success in some.... It's not enough to learn now, to apply yourself fully in whatever you *have* to do. That sort of schooling (i)'s almost over for you. But by now you must *feel* the need to do so, and sense this strongly. That can't be taught. It simply must have been developed. And this is a trial for you to find out for yourself what has been wrought, do you have the mental endurance to keep up with a necessary task, or can you sense the importance of things without more tutoring and example. So.... I can't tell. It would be presumptuous of me to assume of you, inelegantly condescending or belittling of your uniqueness as an individual. I can only worry for you, as a parent should. But I('ve) struggled much, throughout my days, to avoid a sense of failure. And I'm doing so again, right now, because I've allowed my family to take up this challenge of having a rural domicile property kept within the rustic and fairly high standards typical to the area. I boasted to myself that I could, and thought we had the muster and caliber to do so. But I'm really out here tonight, just to sense the truth of things, in this span of quiescence and beauty. Because mistakes can be corrected, but not their errors causing. So aspirations may be reviewed, and ironed out a bit more, in this serenity.

SACHEL: It is a lovely night, Pa.... Buying those cows certainly took some thinking.

PONAPLE: ....Of course I thought about it, from an ingrained determination to have this home and adopt, for us, some aspects of this lifestyle a few days of the year. But I may have been too ambitious about it. Too late to change courses drastically. We must follow through with our responsibilities as best we can. I'll not say I've become discouraged yet, due to these mostly foreseeable difficulties. But when you cast a line into a vast ocean of possibilities, you're not exactly sure of what you might catch, and whether you'd want to keep it. For better or for worst, the Ponaples will make an impression on this valley. We will sift through any disdain, to find the seeds of discretion to survive with a reasonableness to this area. And positive interests will grow as our land is developed more, though in what direction that may take I haven't been able to clearly calculate. It depends on the character of the people who actually do a serious homesteading here, your cousin and his, perhaps.

SACHEL: Might be better for Beryl. He's curious about everything around here.

PONAPLE: ....Well, he's the jeweled child, Sachel, your brother.... the one we had after I was finally securely established in my profession, the dedication of reward as a great deal of struggling and sacrifice bore fruit out of the heady uncertainties your mother and I met and accepted. His life is by design for a celebration. And of course he has the youngster's need for inquiry.... When you were his age, slightly younger, things were mighty tough for us at times, though you wouldn't have known this. I was afraid often how life and family and career were going, so many traps and snares to avoid and escape from. My dedication then was to wrestle a prosperity for you and ourselves from out the hands of greedy temptresses that use it to draw one deeper into the pools of devastation with dementia, through which they draw their awful entertainments as you waddle, splash, and sink. Money as a muse of song to hear leads to your destruction with a deafening blast to blind all the senses, as you submerge alone into a deep darkness rich of wanting. I've seen it happen to others, on occasion, caught up in their abilities to achieve only for the sake of achieving relatively useless prizes, worth only for the thought of having owned or won them. Their falls can be incredibly steep and hard, more remarkable than from the heights leaped from. And no crash of accident do they make, but more deserving of the dive to sense its thrill. Wealth, position and power: You are lifted up by hands.... and are as easily drawn down into an angry ravaging should conditions turn the least bit sour for your supporters or advocates disturbed with your over-indulgence of the muse, incensed with envy from your gaining so much attention by those temptresses. But I have earned my prosperity. And we are prosperous, compared to most. And so the dedication shifts to be more grounded and earthly, and not to wish for more— if but to maintain— then rather to make for better of what we have, these lives we may control, and this territory we now may inhabit privately. Have you ever thought of such vast stretches of earth to harvest from? It was this unclear visioning of happiness I leaped towards—.... that we might enjoy ourselves here, in some way, and with a certain permanence of family.... interest.

SACHEL: And so Beryl grows up with more privileges to enjoy.

PONAPLE: You're not jealous of that, are you? He no more realizes it than you of the hardships we faced, long ago.

SACHEL: Well, it's more like privileges to endure, Pa.... How were you going to lead her home?

PONAPLE: ....I just wanted to know where she was. I can communicate my intentions of caring by simply being with her.... They're to have calves, and need some attention, some bothering for, some looking for to find and be concerned about. That's all I'm really trying to do.... tonight. And she'll perceive the caring, any pet's able to do *that*. But not a sight of her, nor sound. Somehow, I thought she'd lead me to her general location, if nocturnal mysteries can be at play. I'm searching, I suppose, for this kind of communion with our— beasts.

SACHEL: Nothing here. She could be near or far, from our perspectives. And our land is not so wide or broad for the horse's gallop. But it's difficult finding things in the dark, Pa. Daylight with binoculars would be better, though she'd probably amble back towards the barn. They know more of how to take care of themselves in these regions than we.

PONAPLE: How would you feel if we found her died of an injury we could have prevented?

SACHEL: ....How can we prevent such things, out here?

PONAPLE: By being more the manager than the missionary. I don't know what to do, if we come across her stumbled and harmed. Get a vet out here, I guess. But she shouldn't be so far away from a stable anyway.

SACHEL: ....Well, the other two aren't. The worst mistakes often happen in the beginning, Pa.

PONAPLE: I understand that, Sachel. This might all turn out to be a jocular anxiety with no more merit than to be more cautious in the future. But tonight I want to sense her out, or try, to show her we're not such numbskulls for a proper husbandry. That she, at least, need not disdain us, despite our lack of mastery over her affairs. It is a sad state, when animals fear for their owner's capabilities about them.

SACHEL: I imagine she hasn't much thought of that, a cow.

PONAPLE: Dependencies induce such.... beliefs. Remember that dog we had to get rid of, because we couldn't keep her in the apartment when the building's landlord changed conditions of pet ownership? You were just a tot then, balled all night over her loss. But.... the dog knew better, realized somehow she had become too intolerable a burden to stay.... with us, and calmly let me take her to a new home, saddened but relieved. Licked my hand, as I left her in other hands, to indicate to me.... that she understood, these necessities of dependence. Was a nice pet, for a year and a half. I've always felt bad, having to take her away from you.

SACHEL: Never really liked pets too much, Pa. That's Sarah's game.

PONAPLE: Well you were singed a bit, of the affections, early (on). It's understandable.

SACHEL: (I) Sure ain't looking for any (more), in particular.

Why is it so difficult to see, even with this light's spread? Not here, Pa.

PONAPLE: It's not strong enough for the horizon to reach. The contrast from here to there makes the in-between more difficult to notice in places, depending on the incipient shading—

SACHEL: Wait!... There's something moving out there!

PONAPLE: Where?!

SACHEL: Over there, in the distance!

PONAPLE: ....Your eyes are better than mine. I can't make out any distinction.... of surface.

SACHEL: Something was changing.

PONAPLE: Are you sure? Penumbrae have a habit of modifying their shades around the shadow of a solid object, like a tree, or a hill.... It's as good a place to search as any, though. Can you hear anything?

SACHEL: I'm certain something was stirring over there, Pa. Can't hear any rustling; but that would be where I'd run to find her, or reenforce my notions to. You suppose this lantern light will help to draw her out? or startle her more?

PONAPLE: ....I don't know. Did you bring any matches?

SACHEL: Yeah.

PONAPLE: ....If it's her roaming about so casually, then she's not in any distress. And she's not lost, on this property.... Put it out. We don't need it so much, tonight— Give me the rope. (*Sachel hands over the rope.*) Let's get up there. (*as **Sachel** extinguishes the lantern*) It's not very steep. (A) Fairly gradual slope. She may not have even noticed us.

SACHEL: We're gonna try and sneak up on her, Pa? Have you ever roped a cow?

PONAPLE: We don't want a frightened animal turn wild on us. This is only when it's necessary, like that lamp. This cord is kind of rough, Sachel.

SACHEL: It's the finest I could find.

PONAPLE: Some hitching gear's in the car's trunk. This will have to do. But we can't pull on her too tightly. I think you'll have to push a little, if she doesn't want to walk along side (with) us. Did you bring any oats, to pacify her?

SACHEL: No, Pa. She's not gonna get violent on us, is she?

PONAPLE: I doubt it. You know her disposition better than any of us. Just hesitant and resistant, perhaps.

SACHEL: Stay away from her mouth, Pa. I've seen her nip at things, out of curiosity. And those teeth are made for grinding down, not tearing off. They're heavy.

PONAPLE: Well.... Let's get up there, son. Just follow my lead, when we get to her. I'll try to be gentle, but avoid any signs of nervousness and excitement to amplify. And be careful of the legs. They must certainly kick up a defiance when aroused. I don't know how to soothe these creatures, except to be calmly commanding.

SACHEL (*as they head off*): I just try to tempt them a little. Only the lone miss is stolid at times, but not in an angry or mean way.

PONAPLE: Then it's just by instinct to show a little independence, that hasn't yet been engineered out of the breed.... She should be docile, under authoritative handling. Just don't cause any pressuring.... towards emotional displays or distempers. We have to learn how to do these types of things, to be legitimate for them. And the only way to do that.... is with these such experiences.

Scene III — **Valin** *runs up to* **Bouwder**, *who remains sitting on a tree stump as before.*

VALIN: There's someone out there with a torch!

BOUWDER: Ponaple! Well this cuts off the try, Pitch. It's his land. Where's Mort?!

VALIN: He refused to come back, when I told him. He aims to get that cow lassoed. Thinks I'm seeing things out of fear-spite. How can one imagine light in the dark? He's too determined. But we are stretched of time.

BOUWDER: It's over, Pitch! The attempt's been done wrong. He better get his— caitiff tiff back on my side of the range. He's trespassing. Yer both been. An' that cow.... must be a'runnin' through 'is land. Did yer catch sight of 'er, that's definite?

PITCH: No. But then what's moving about 's carrying light. I guess we're defeated, this time.

BOUWDER (*standing*): This ain't no competition 'gainst 'im, Pitch! I told yer I have tah see it on ma property, fer all of dis tah be proper. An' it ain't so tonight. Yer have failed tah convince me. I don't want no legal troubles with them Ponaples. They'sa rich! an' can squash me fer rot, if they're in the right. But I'm in the right by sayin' no cow's been up here tonight. None! The man's just tendin' tah 'is livestock, as I ma crops.

VALIN: Should I go out after Mort?

BOUWDER: ....That's yer confound-ation now, Pitch. I don't give yer permission tah. Ma hands are clean of dis, (un)'less that Betsy comes a'prancing through tah trample on ma footprints in the soil.

VALIN: We've blundered, fouled everything up! But he's never been out there, searching at night, from what we could tell. Would be typical (of novices), using a torch. If you can't see with (the) night's sky, you shouldn't be out searching in the first place.

BOUWDER: That is yer endearment tah this misadventure, Pitch. Yer have assumed tah much— too much!... about 'im, the cow, everythin'.

VALIN: I should get back to Terence, tell him things have fallen through, matters (have) broken up. But we can't leave Mort for a possible confrontation.

BOUWDER: *You* can't, if that's yer will. If there's any fightin', he is sunk! I can't defend none of this, as I couldn't even see yer enter on the fighting's property. But if yer're brave tah do, go bring 'im back— fer a scolding of nature's night! This is what 'appens, when yerse be wrong.

VALIN: We *ain't* so wrong.... to have tried, Hank. But our rustic demeanors have failed us, in this particular quest for some beef. If I can't sight her, Mort most certainly won't find her except by accident, and a dangerous one now. But he has that abiding hope, to prove this mission fulfillable. And that overrides reason with unsteady chances.... I don't think the beast is out there, Hank, not where we could have expected, near your boarder.

BOUWDER: *Within* ma boundaries, yer mean.

VALIN: We are crushed for time. But time is in abeyance, now. It has stopped, and no longer exists— for any practical purposes.

BOUWDER: Then yer have tah wait fer the fool tah come back — if (h)'e's not caught!... All night, perhaps.

VALIN: She didn't roam far enough, just when we wanted her to.

BOUWDER: ....Then, why is Ponaple out lookin' fer 'er, way up this-a-ways?

VALIN: ....Maybe he's not. You don't suppose he caught wind of our plans?

BOUWDER: They've been loose throughout the territory. I wouldn't think he'd be sociable enough tah discover any details outrightly, though, as a neophyte tah the region.

VALIN: That couldn't be a county sheriff out there?! warning us off, if not arresting if we don't comply! This is becoming a horror, Hank. Should we be found even here— with you?!

BOUWDER: Are yerse becomin' afraid, Pitch? I don't see how I'm not safe— on my own land?! It is clearly not dark enough, fer yer services, an' too dark.... fer yer surveillances. But suspicions will rest on the (wh)'ole town if anyone's caught tryin' tah poach off the Ponaples. We're all compliant towards a discrimination against the new, no matter how subtle, if they be outsiders tah breach the country ways. That will have tah be argued as plain common sense, an' wariness, if somethin' about this 'appens. But.... the law is based on what is done, not supposed tah be wished. We ain't got fetters fer thoughts, yet.

VALIN: Treasonable intentions in speech, writing, and thinking, can get you arrested and convicted in court, if a jury deems it to be doable. There are no unassailable ideals of notion nor behavior in an aggrandizing society for the realities of life.

BOUWDER: Well that ain't about tah happen here, an' not at my spot on the feeding trough. There's nothin' 'gainst those glorified interlopers that can stand the test of ma forbearance tah be decently civil an' legally neghborli. An' I ain't caused no damage tah anyone while perched on ma own property, patiently just awaitin' fer some malfeasance tah eventually occur, due tah faults.... too innocent tah be comforted by ma advice an' consultation. I have protected maself with the current, operative statutes of this county an' dis community of expectations. So no sheriff can rile me tanight.

VALIN: But should we even be out here, for the coincidences to assume. I must wait and I must run, at the same time!

BOUWDER: That is yer dilemma surely tah be rent by. This whole occasion is the thought of you an' others. I am a participant on'y by chance of dis bein' a pleasant enough evening tah be out here with yer. But that don't cut the leaves no wind of shame upon the tree. If yerse stuck, yerse stuck tah stay, an' face a formidable comeuppance, explainin' yerse-selves away.... I ain't seen the cow yer have proposed tah be my bother, an' tah bring me out (*A gun shot is heard.*) here!— What is that!... Oh! Goggali-ghoul! There's been a firin', Pitch!

VALIN: The sheriff?!... confronting Mort? He's been so deter-mined.

BOUWDER: I'm frozen on the spot, fer without a weapon hav —.... But Mort said, himself be packed tanight. Then worst can be worst, tah shoot at a sheriff.

VALIN: There was only one shot.... No returning fire to address it. What more scenarios can be dredged up to imagine?! This is gloom, glum doom become, Hank. How can I stiffen to these events, not to investigate?!

BOUWDER: Tah go towards what?... imposition of yer health!

VALIN: Frustration of my soul! For what rampages, that we might have caused? And Mort is out there, perhaps bleeding, per-haps fled of crime and weighted upon for life.

BOUWDER: I can see nothin' of it, man! from that periphery tah reveal its darkened contents tah my eyes! acknowledgin'.... assis-tance might be needed. Oh! It's shallow now, these causes im-brued. Hide we through the night, an' on ma own land scared. It's definitely done, some blast.

VALIN (*bending his knees, as if beyond his self-control*): Hank!... I must go out there!... I have to discover what has hap-pened.... (*starting an awkward gait*) We're all responsible—

BOUWDER: No! Pitch.... Yer look tah be bled yerself!... An' that is invisible, out there, beyond ma boundaries.... strictly out of ma control— Stop! Don't disappear on me!... again. I can't see yer!... But dis is lik' a hero's moss.... tah march on moistened ground in-tah the emptiness of 'is infinity.... An' tah the cause are drawn us ever, fer some despicable ways. Responsible?!... That's fate's drap-ery, in shroudin' bodies bier-ed. Responsible?!... There weren't no groans heard, but it may (h)'ave been too distant fer sighs' lamenta-tions. An' I reconcile this loss of knowledge, by standin' here sus-pended by ma fears lik' a puppet stymied fer a gesture of expres-sion, foolish made tah seem, with wooden wisdom. But I ain't caused dis much havoc, just tah bother of a cow, an' lend ma sup-port.... tah its lugubrious displacement. Then, how an evil thought does whip yer tah be ripped with fright, as these things may turn out; an' against yer surliness tah neighbors, are yer brought tah more than trouble.... but regret, fer these temptations tah deceive.... yer tastes. An' I done caused this, operatively? That ain't so so, so much as I would want a bleeding friend. But there's more.... to a nerve tah defy rights. An' I did try at *that*, tah help craft this terri-ble ordeal. So then.... much blame tah mind, could be my shatter-

ing surpassed of injury. Fer it's too calm, tah-night, fer this tah have been a mistake or accident, of foulin' up our dissensions.... in-tah misdeeds. An' though I'm prettily made, tah be legally absolved of whatever 's (h)'appened.... ma heart seems more as a gut's fill of disagreement; an' much retching of a conscience may be due me.... This time, it's angry of its quiet tones that we 'ave tested an' teased at, as brigands stalking through these fine grounds an' welcoming pavilions that we would shame with petty distrusts an' ethical am-bivalences. Then does our time show, or show us caution.... tah lead us into more caustic entrapments with these quarrelsome be-haviors. An' I should be better an' above the need tah be difficult to the unfamiliar familiarizing themselves upon our disapproving no-tice. Yet be the comely need at least, whereat 's displayed these er-rors. Could I have really wanted to steal.... a roving cow? or have this done as stolen? It's much easier tah return a property, than a life. So might I try maself tah restore, from this stupidity. But Mort's more defiant than 'is death, if he's been caught, or has caught a greater criminality than one man should own. So stay I away.... on this safety, till the meaning of the shot becomes more clear than precious tah be heard. Fer.... there's been no cows tres-passin' on my land, an' I haven't the reason, the cause, tah do other than this. Wait fer Pitch tah return.... with someone, something— news. An' tell me how it is out there, that's blinded tah me, as if I couldn't grasp it.... any other way. (*sits on the tree stump*) Shaking on a leveled log. Should 'ave run away— But I am bound.... tah be responsible, fer what's tah be learned.... An' what if Pitch is takin' away 'imself? All night. All night tah sit. That is yer punishment, Bouwder, until the sky brightens, an' the frozen nerves thaw, an' the impulses quicken tah be.... responsible, thicken tah be more dear tah my intentions as an honest, responsible person tending tah 'is business, 'is property.... alone out here, but curious.... of these falling meteors, an' excuses tah be awed. Rest me here, tah ac-knowledge such blundering of the sky.... But all of else tah be, this is a very clear, an' comfortable night. I ain't freezin', an' the world still spins. Just lift that darkness (ov'r) there, Pitch.

Act IV

Scene I — *Within a field at night, on **Ponaple's** land, **Moraine** sits on the ground with the agony of a bruise.*

MORAINE: Oft the gloom of pain, but does befit the maiming of ma sins. It's busted, sure tah spin off from the wrist. (*favoring one of his hands*) But she did explode, as I was aiming at a sacred cow, this fine revolver once, whose pieces dwindle of intention. Myste-rious cow, night-haunting ghost of shadows and quiet, silent siren tah lead me so astray an' from ma better views of motion. But I did frighten nothing with the false shot. It wa(s)r lik' the world were deaf tah it, an' not a stir tah run away as I imploded— with such a bright flare as tah gray ma hairs. Must have been the lints an' fluff in ma pocket, that plugged up the barrel. Only brought the gun as an afterthought, for some unforeseeable need of protection that oc-curs tah one lik' a ringin' in the ears when yer undertake the das-tardly.... What is such betrayal caused by? within the sure physics of a weapon an' its mission. That cylinder's done spun itself right off the frame. Almost impossible tah come apart lik' that, an' I can just see the fragments scattered by me. But dis injury is a muscle shaming. No wounding incisions, I don't think. No blood. But a re-coil tah pull the tendons, an' rip some apart, I am afraid, if this hurting throb does not deceive me. Yet.... worse am I for the loss of aim. I was a'gamblin' again, headstrong tah be proud of doin' some rash action, takin' matters in ma own— hand.... tah get this deed over with an' through, finished fer dis night's treachery of surprises. I clearly thought I saw her stirring close enough, tah take aim— with a suddenness of thought that, bein' alone, ma only way tah seize her would be tah stop her trackin' an' heave 'er up from off the ground as motionless, intah my truck brought up (tah) here, with Pitch as finally some aid. I should've thought more, tah run on rushing impulses. Fer I have been humiliatingly injured, and will have tah rest awhile, past the stun of bein' harmed, before searchin' of these pieces of a once treasured firearm, a worthwhile collectable an' just antique enough fer some nobility of character, besides this blast, this.... gaffe of moon's laughing, unless.... some help comes, tah pity ma infliction. But, with foolishness have I been made. An' it's fer times like dis that I prefer tah be solitary, an' recover from such a blunder. It's clearly ov'r fer the search of dis beef. I could no more lasso a lap dog with this leap of strain about ma hand and arm, an' don't feel tah do much about an(y)a mandatory task. I've given' myself too much restriction tah our country tastes, an' am sufferin' from an ignominious mistake. But, see my plight, under these stars. I am only night's folly folded onto feigning (famine) famishing without the veal.... This is sore, a punishment of being foul.... of mouth or tongue or taste?... or tenaciousness fer a silly goal. It would lead me tah draw these tears as much fer my iniquities as fer this throe of terror touching (of) the hand. It swells, as fingers slowly crack!— What approaches?!... A mooing tah malign me more?! They have the grins.... of chastising grassy groans, these cattle.

SACHEL (*approaching with **Ponaple***): It was from around here, Pa.... that flash-like burst.

PONAPLE: A country's crossness. I've heard of shooting stars landing, always with a ringing of the ears in deserted pastures. They never seem to come down.... to occupied places. But that's because much more of the land is unoccupied—

SACHEL: But something's fallen to the ground there. Is it possible the lone miss has been struck, by such an improbable accident?

PONAPLE: Light the lantern, son. This might be hideous.

MORAINE: I can hide no more—

PONAPLE: What?!

MORAINE (*while **Sachel** is fumbling to light the lantern*): I am a man, who's called Mort Moraine.

PONAPLE: What are you doing here?!

MORAINE: ....I'm a friend of Hank Bouwder's.

PONAPLE: So.... That doesn't mean much to me yet. But this is my property of range.

MORAINE: His land adjoins yours, Mr. Ponaple.

PONAPLE: Oh yes, the distant neighbor.... So, what are you doing here?

MORAINE: I've been trying to capture a cow that threatened me while I was visiting him tonight. (*The lamp lights up.*) Taaah! I've injured maself, runnin' after it an' apparently comin' ontah yer property. It's a wild an' threatenin' beast, an' I tried tah take a shot

at it while I thought it was chargin' me. But ma revolver exploded.... in ma hand, an' busted the effort.

PONAPLE: Oh, poor man!

SACHEL: She was up here?

MORAINE: She? I thought it was a bull.

PONAPLE: You were trying to capture it?... with that rope?

MORAINE: Hank was too squeamish, it chasin' 'im through 'is crops at night.... or attemptin' tah. But I won't stand fer such intrusions, to a friend nor fer maself tah frighten. An' as I had brought a lasso with me, in ma truck, I proposed tah try an' capture the raider — Yeah, he told me not tah bother with the danger. But I'm afraid I was insistent, bein' fairly proficient at ropin' livestock. We figured it must have been one of yer.... animals. You let them roam so wildly, experimentin' with the "happy grazer" concept, we guessed, fer.... organically grown beef; which don't mak' much sense tah me, since all life is organic. But I planned tah return the beast over tah whoever yer handler is, in the mornin'. The situation has changed drastically, though, with this mishap. Ma hand has been badly strained, an' I am in agony because of dis.

PONAPLE: We're very sorry about that, Mr....

MORAINE: Moraine—

PONAPLE: Moarine. We were searching for her ourselves, my son and I.

MORAINE: All the way up here? Yer let 'er roam that far by 'erself at night? Can bea dangerous fer everyone an' any thing out here.

PONAPLE: It was an oversight. We're new to raising and breeding cattle.

MORAINE: Breeding.

PONAPLE: Are you hurt very badly, do you feel? We'll help you back over to Bouwder's.

MORAINE: Can't slaughter them in this county, at the moment, you know.

SACHEL: We were thinking more of milking and making cheese with them.

MORAINE: Cheese. But when they're milked out an' can't calf no more, yer have tah kill them fer their meat. Otherwise they're just an expensive pet leaning towards the rheumatism an' physical discomforts characteristic of their breed, at old age. They're not meant tah remain healthy, but are engineered through the cross breeding genetics tah yield their most productive service tah yerse durin' their early tah middle adulthood. After that, they fall prey tah all sorts of diseases an' expenses. We don't want that kind of farm animal abuse around here, yer knows. Those problems can spread.

PONAPLE: We aim to be very responsible proprietors, sir, with more experience to tend our range. But, are you hurt badly?

MORAINE: I think something might be broke'(n). I can't tell too well, because the hand's so swollen. (*looking at his hand*) But it's not bleedin'. I have survived the gun shot.

SACHEL: Maybe we should take him home, Pa, and call out a physician, at our expense—

PONAPLE: Aaa—

SACHEL: I'm very sorry about this. But it could be farther for him to walk to Bouwder's house than back to ours.

PONAPLE: He has his truck there, he says.

MORAINE: It's a ways back. But I'm sure Hank wouldn't mind drivin' me tah a country hospital, though he's an old codger without good night vision left.

SACHEL: I think we should take him home, Pa. This could be a serious injury.

MORAINE: You needn't feel so sorra' fer me, son. It's ma own foolishness fer bein' so startled, but I could've sworn I saw horns.

SACHEL: Well.... I'm the handler, you see.

MORAINE: Oh.

PONAPLE: We are all handlers This is a family enterprise we're developing. I suppose you should come with us, Mr. Moraine. That hand looks rather mauled— Can you stand?

MORAINE: If yer help me, I can do. (*as **Ponaple** is helping him*) Aaaw! This hurts.... Fragments of ma gun is lyin' about.

PONAPLE: Leave them with your rope. They're not gonna be of much use to us tonight.

MORAINE: Looks better than yours. But don't yer knows yer have the mad weed about yer range? Yer just can't let them animals romp about so easila.

PONAPLE: I've attended to information on the vegetative properties of my land.

MORAINE: I suppose it's more the rumor that caused me tah be so defensive. (If) They eat that stuff, some become terribly defiant an' imposing.

PONAPLE: The weed should be spread over more tracts of land than mine.

MORAINE: Bouwder don't raise no cattle an' hogs. But if that's what yours are searchin' fer.... yer need some fencin' around here, assumin' yer decides tah keep 'em.

PONAPLE: Well most certainly we will— er.... Mortimer. Though we are personally responsible for these animals, they will be properly tended to by others, in the future. This particular cow, however, appears to be unusually restless, given her new surroundings. And that is a surprise to us, since we assumed they'd adopt to pasture lands with all (the) facility of their tranquil natures.

MORAINE (*favoring his hand*): Friends call me Mort. Yes, they as such generally make the character of a range, when they're present.... So yer plan tah lease out lots tah other homesteaders?

PONAPLE: To relatives, initially. We'll see how it goes. But a permanent and responsible residence on my property will be established, and some cattle raising is to be expected. We should be heading home, sir, so that your injury can be looked into more professionally. Can't say as I've seen any hint of our cow.

MORAINE: I'm hint enough, I hope. Don't make much sense, if yer don't plan on butcherin' them, in the end.

SACHEL (*as **Ponaple** starts to lead them away*): We hadn't thought about that. You say they're ultimately meant to be killed and culled. That is a cruel hardship, a difficult strain of life to bear.

MORAINE: Well if they can't make calves nor milk, what else are they good fer? but the beef. Yer don't get tah purchasin' any without such thoughts. But most live peaceably enough, until they're lent for dinner.

PONAPLE: I did not propose for myself such an outcome, in this valley. I'm beginning to understand. They were sold to me with such expectation, while I considered them to be, perhaps, more like decoration for our territory, and a sign of our family's permanent investment of interests here, on this beautiful land. But.... We've inadvertently leaped into the seriousness of what amounts to a new business venture, and a novel one for us.

MORAINE: These blasted restrictions on the commercialism of animal slaughter are bound tah be lifted soon, b'cause they go squarely against the grain of our communities in this region, an' promote or contribute tah a severe economic hardship tah several of our industries: restauranteur-ships, groceries, meat packin' an' processin'.... an' shippin'. It is a silly ordinance that starves us of much commerce as well as traditions of (the) meal, an' will be overturned. We had assumed.... yerse assumed that also, tah think ahead of the pack an' take advantage of the near future with this initiative now. Else why buy such a thing as cattle?!

SACHEL: We are novices—

PONAPLE: We.... are setting the foundations of a country life, within the splendors of this lovely verdure and healthy atmosphere. A fanciful retreat for us in the immediate of time, but this will develop into a profound residence for our family and relations. And I will think more carefully on the bovine acquisitions.... A brow of circumstance I've stumbled into, they may no longer be treated as picturesque guests and, for us, exotic possessions.

MORAINE: Chickens, turkeys, birds an' fish yer can still kill.... tah eat, an' fer yer huntin', water fowl an' water flesh. But the hoofed animals are forbidden tah dis cause, an' the rodents fer pure cruelty, though yer can kill 'em as pests. The hare lies within an ambivalence. An' the reptile is totally unconsidered. All of beast may be considered as nature's pets, but our subsistence rests.... on what is reasonable. Sadden we, tah have tah eat the bugs an' worms as they.... or fer the pawed tah disavow their prey. What is normal fer us hastens tah our breadth of desire.... An' ma primate's hand does hurt, what's tah be claimed of it, perversity deceived ag(a)'in.

SACHEL: Perversity, sir?

MORAINE: Consider, son, what is a man's endowment tah be stomached. Then all that (h)e eats.... must be mortified fer the mammal.

PONAPLE: Then that's some justice to the species, if he leaves what's left alone, and lets the resplendence of a vegetative bounty make pact towards his quiescence to satisfy.

MORAINE: ....I ain't never gonna be no radiantly anemic vegetarian.

PONAPLE: Nor would I force myself to be a cow.

*They exit, and there is a pause before **Valin** comes up in the relative darkness.*

VALIN (*standing still*): This has done it, then. We are hacked!... and we are discovered. And there is trouble to come, from our endeavors this evening. For Mort must confess his hunger, and we are doomed to a hot scolding, our remiss surliness backfiring. My lord! he must have been a great shot. That pistol was blown to bits! Its pieces are around here somewhere, from what I could see at a distance. Well.... I better not touch anything now. It's out of our hands, and into a hammer's blows.... of justice— or.... judgmental prudence. Mort must be so. This is the worst of all possible outcomes— What's this?! His lasso— Leave it! Pitch.... Don't touch a thing! It's over for our harmless attempt at crime, and we must backtrack somehow into the most tendentiously stingy of explanation for our actions, that we have misunderstood our capabilities about and around the law. But this whole thing can blow up into such a furor of participating maleficence that the simple intricacies of our method will become a labyrinth to escape from owning up to, such shameful details impossible of exculpation from a definite lawbreaking. And to involve so many people— and such serious facilities.... this shatters much, of any possible standing of defense or excusing. We have practiced a tort of region, and will be condemned to the territorial prison, and a thorough destruction of our lives as brawny citizens— unless.... we can find some way, to deny the sanity of our attempts as being driven to by unreasonable hardship and culinary adversity. But this sounds like a clown's argument, or a buffoon's impatience. Yet we are descended by such hereditary means as to make men *need* to do foolish things.... when they have been so gently pressured.... as we. And Terence will have to find our outing path, to close down proofs! of our scurrilous behaviors, since he guaranteed us some measure of success or at least the avoidance of all or any punishment. But guilt is more for being caught than catching of our errors to throw back. I will demand a solution out of this un-predicted miscarriage of our goals. For what wrongs have been done by me but sight, that I should shake an angry blame that crushes! What remnants are these?! of this.... peradventure twist to grind one so into the gears and teeth of regret and caution! Why did they not recover for some evidence this rope?!... and these— pieces of gun. Only fiendishness is foul— So what's the true offense? He blabbed— wounded! And absolutely no crime was committed else. Yet, we will all be discovered, and undone, because of this false turn. And where can one hide now, from one's involvement? There is no answer, nor place to be, where knowledge is held back from these details— Conspirator!... Confederate! Condoner of a wrong. What must I do myself? I am.... Hank's friend, and visitor tonight. And that's all of this business, no matter what Mort must— twine of tale, allege, to convict me, all of us, under pressure, that he could not have acted alone in this.... too ordinary endeavor, of how the brutish may behave. A misfeasance only, Pitch— I was only helping Hank! But this grows and grows of reputation and debate, into subjective causation for foul deeds and ruination of character. (*falling suddenly to his knees*) I may be destroyed— by this trifling need of meat! Or I worry much too nervously over this guilt that lies, or lay upon my faithfulness to friends— This evidence condemns me to the absurdity of our aims, and I am likewise strewn over this ground as an instrument implied and implicated. Need I more of a conscience than this depth!... And what does Bouwder oft say, that he can not see it done. But it's too sure, tonight. And I would not be so frightened, were not the ridiculous impinging on my truth. How can a man be so stupid for his sight! and what else will falter of me, from this courage falling, to take the brunt of what's deserved? Then have of holy lesson, Pitch, to make amends.... at eyes. The man might lie! about me, when the greatest shame is to be known.... to need correcting. Oh, but for a hesitance to cry.... I have been wrong. Steal these things!— No! Indecision is my curing cur. I *am* here, and can not erase the traces of my fraternal presence. For what is proof? in spoken words and signed statements.... I won't deny it more, to anyone who asks.... except those boastful children, looking for envy of mischief. And I'll explain to Hank, the event I've witnessed. (*standing*) He will feel cleared, there is no cow around.... and will no doubt cast some generalized aspersions on our efforts, and this failure. But we will be *all* be associated, and each be sought as hounds.... for some razzing, fines.... and even subdued approbation, if that's yet possible for this cloven immaturity among men. This is not so serious as sin, if what I've seen is salvageable.... (*leaving*) upon our backs.

Scene II — *A bedroom used for **Beryl**. He's sitting up in bed, drinking a glass of chocolate milk, as **Laura** stands nearby, about to go to a spindly chair to sit since he's taking a while at it.*

LAURA: Don't spill any on the blanket, Beryl. I won't have you sleeping that way, and I'll have to search for another one.

SARAH (*running in, as **Laura** seats*): Mom! She's come back!

LAURA: She who?

SARAH: That heifer who likes to roam off a lot. I went to the barn, to look for some more wicking material, so that I can join Sachel and Dad—

LAURA: I don't want you two trying to set up any more lanterns, Sarah. He's got one working?

SARAH: Yeah. I saw him go off. I know what the stuff looks like, now. But there she was, Mom, back at the barn, trying to eat some oats out of a bucket.

LAURA: Well, farm animals are like that. They return to what they know and for what they know (of).

SARAH: But I have to find Dad and Sachel, to tell them to stop searching.

LAURA: I don't know, Sarah. Did you help him light that lantern?

SARAH: No—

LAURA: Then you probably don't even know how to set it up correctly. And they're dangerous to handle even when made right.

SARAH: They could be searching all night for that cow.

LAURA: I doubt it. Maybe your father. But he just wants to get out into the country for awhile. Where is she now?

SARAH: Still at the barn, I think.

LAURA: Sachel should have stalled them.

SARAH: The stall's not properly constructed yet.

LAURA: If it wasn't so late, I'd help you try and corral them in some way. There must be something that can be used as fencing material. Just a suggestion of boundary may be enough, temporarily. We can't have them running off like this. You have to teach them the conditions of their stay(ing) on the range— They don't learn these things by themselves. They're domesticated.

BERYL: Did you go up to her? And did she sniff at you?

SARAH: I left her alone, Beryl. She's not the type to want to be bothered with much attention.

BERYL: Can I see her—?

LAURA: You drink your milk, so you can go to bed. You can see them all tomorrow, as usual. I don't think she's likely to stray out again, if she's hungry tonight. But I'd feel so much better if they were enclosed. It's almost like we allow them to dare to be lost. We should get some cow bells, so that we can tell where they're walking. Though that can be a nuisance for a pet. I put one on my cat once, when I was a youngster, and she spent much of her time trying to take it off. We shouldn't have to disturb animals that way. They are individualists also. The resentment was just too much for me to maintain, and I relieved her of the annoyance. But cats are very proud creatures and like to have their solitudes, though I wouldn't call it hiding.... from their owners.

SARAH: They really don't have owners, Mom, not too legally, especially out here in the open country. But I'm not afraid of our cows, Beryl. They're large and bulky, but show no real aggressions towards us. I just wish they'd accept me a little more, since I help feed them.

LAURA: Well dear, if you can milk them, then they've accepted you quite thoroughly. They may even expect you, if you become regular at it. They can't smile, you know, to show their gratifications. So what more can you hope for, than that they allow you to be close to and handle them?

SARAH: They're taught to be milked from where we bought them, Mom. I still think they show a rather indifference to us besides that function for them. They should be able to have a bit more affection for our presence, when we treat them nicely. But in general, those three tend to stay to themselves emotionally, to just tolerate our impossing on their existence.

LAURA: Well all animals have their boundaries. And they'll just have to get used to us (and ours), while we're here. But I don't

think it's that typical to become very friendly with cattle. They don't share the gregarious faithfulness that other pets do, cats and dogs and parakeets. They're closer to being a beast of burden.... She knows to come back to the barn. That's some kind of relief. Maybe we can keep an open trough of grains there.

SARAH: They'll just eat it all up.... That could be expensive.

LAURA: At night, Sarah, after they've grazed all day. They're not so gluttonous as to eat more than they're hungry for.

SARAH: Oh, you mean treats, like the sweetened oats in that bucket.

LAURA: Yes, something like that. You have to train them to get used to the nearby comforts, and make those comforts reliable and predictable. That very sort of predictability in them may be the form of affection you're searching for.

SARAH: Obedience to your expectations? A single-mindedness of effort for your benefit?

LAURA: As well as their own, given the circumstances.

SARAH: I was hoping for some grander form of communication, within the mysteries of the mammalian animal kingdom. Creatures so large (as they) should have something more to offer than just a polite docility to their masters. Many horses become actual companions to their riders. But what may cattle do for themselves, with our regard?

LAURA: Our regard is to tame them from the buffalo (*Beryl burps.*), the wild herds that used to stretch all over these parts in vast ranges of territory. We've not condoned that type of existence, for them, though it was strongly societal and in that way commendable of an esteem to imitate. But not as wild have they been kept for us. Their compassions for themselves have become ours to manage. And the cow is not an independent creature today, (rather,) somewhere between a captive and a caricature of their former selves and that native past. We've shaped them for our utility. And what resentment to that is left in them becomes indifference to their incapacities of expression. So you should be searching for no more than a friendliness through their dependence, more astute than a resignation and less generous than to be thankful. They hold little more of the passionate about themselves, especially as the bull is turned to steer and ox, and can not yield stronger secrets nor hints of their servitude to man. They are of their own dimensions felt, to live through what is given them of life, and must struggle as hard as we.... to control their manipulations of the universe, as could be perceived to be allowed. I'm not sure I should have suggested to your father their purchasing. It was only the slightest alluding for some farm animals, and he promptly goes out and buys three jerseys. But I was wishing for.... a proper decorum here of a liveliness I could remember. It's so remote to me now, and disturbing of our realities as to frighten of a chronic unpreparedness, or inappropriateness for ourselves today. The breezes of one's childhood seem simpler, and may shock you to face up to as a heavy ether misappropriated.... I'm only visiting the past. But you won't find much more in them for the magic of recollections— He wouldn't have pigs. Said they were foul, when they are by far the cleanest.

SARAH: Dad invests in our curiosities, and I'd like to know what

271

they think, as they lumber around pondering the grasses. It's an activity, done to such an extent by them, that must easily promote long spells of meditation. That may be why they stay so calm without needing relaxation, if feeding is a daylong chore.

LAURA: I'd hardly call them very pensive, Sarah. Though they can make much of a novelty come by, with inquiry for causes. But they are trapped to our designs, for the most part, and can not rebel against our kindnesses and cruelties. Their temperament are our production, their timidness our temerity. We are their faculty of thought, for whatever thoughts they may have. And we shape, not their grazing modes, but their grazing made to do.

SARAH: Then within them could be a whirlpool of agitations and contempts not even dormant, but inaction-able, incapable, insufficient for frustrations physical, except to investigate the land and ground cover.

LAURA: One can imagine anything unseen of an entity known.

SARAH: I mean, she wanders so often, that might be her only means of a general disapproval or upsetting.

LAURA: Well, they're new here.... and so are we. But the conditions are fine, and they'll adapt to them as a familiar region for a home.

BERYL: Mommy, I want some more!

LAURA: ....But you haven't finished what you have.

BERYL: But when I do, I'll want some more. And—

LAURA: No you won't. You only fear you will. Don't be contradictory to notion— You've been drinking that so slowly, you may not want much at all. And learn how to control your requests. You still haven't mastered (how) to say.... more precisely what you mean. And what you mean to say is that you want to stay up more. But it's time for your bed, because you're more sleepy than you know.

SARAH: You can't know exactly what you're thinking, Beryl, because your mind won't let you.

BERYL: ....And crackers?

SARAH: You have to grow up more, to be able to describe your experiences to yourself better— See! You're yawning!

BERYL: But I've been put to it, because now I'm supposed to. I can't fight out of being tired.

LAURA: No one can fight out of any lethargy of mind. But I won't have you wasting anything you've asked for. It's time you matured out of that carelessness to assess your needs and wants. Finish that milk so you can go to bed.

BERYL (*drinking*): I am.

LAURA: And I mean to sleep. (*standing*) I have a mind to go over to the barn and see her for myself, Sarah, now that you've made me wonder about her so much. No, you shouldn't have to want to *need* to know what they're thinking. But communication

between beings can sometimes be.... through a cosmic telepathy of simply caring about each others' concerns. And if she's having such a hard time adjusting to us, we as only visitors, a wink of an eye might be soothing enough, for our genuineness to receive, that she may be comforted, that some such acknowledgment can be made and is capable of being done. And they can do that, with no threat from the past blinking for our awareness.

SARAH: She's not been belligerent. But she's fairly large, Mom.

LAURA: So must we be, to have a word with them, in any manner feasible. Understanding is just a tapestry of relationships.... Perhaps those other two are more closely related to each other, and she resists companionship for isolation, occasionally. But that will probably change, within their fixtures formed. Such a temperance of independence is counterproductive for.... the owners' aims. Yet, sympathies can be transmitted with a glance. And I think we should do *something* to contain them.

SARAH: I doubt if any pet really understands our words. It's the intonation of our utterances that they seize and grasp. I can use the same word or phrase as a command or a praise. They distinguish by what we want, and not what we say.

LAURA: Obeying commands is for one's own pleasure, with a satisfaction of hopefully being right, since orders are indelible of necessities for action.... in some way offered for your betterment, or a common good's. I thought there was a large spool of wire in the utilities closet.

SARAH: That's electrical.

LAURA: If it's not sheathed it's not electrical; maybe bare tin twine. (*starting to exit*) And did he really look for a flashlight? Maybe we can do something outside, for their notice.

SARAH: What could we use for pickets?

LAURA (*just exiting*): That better be drunk up by the time I get back.

SARAH: There's hardly enough wire for a decent fence or boundary out there. Maybe she's thinking about reinforcing the barn stall.

BERYL: I can't finish this. My stomach's not big enough.

SARAH: Oh, you better. You asked for it, didn't you?

BERYL: But I didn't know it would be so much.

SARAH: It was a glassful. How could you not suspect?—

BERYL: I was just thirsty for a little bit.... half a glassful.

SARAH: Then why didn't you tell her that?

BERYL: I thought she'd know it.

SARAH: How—?

BERYL: She always knows how much I should get.

Peccant Pecus [Act 4, Scene 2]

SARAH: Well then you have to complain intelligently, because she can't always think for you. Drink that up!—

BERYL: I can't!

SARAH: —or else you'll get spanked.

BERYL: Why?!

SARAH: For wasting chocolate milk. It's more expensive than regular milk. You can't go about life wasting your foods and nourishments. You have to be taught to be more prudent.

BERYL: I don't think it's made up right.

SARAH: ....It came right out of the carton, in the refrigerator.

BERYL: Well, it doesn't seem as tasty as before.

SARAH: That's because you're not as thirsty as you said you were. If you lie about such things, you get mother mad.

BERYL: I didn't lie! How am I supposed to know how thirsty I am? I was thirsty, but now I'm not.

SARAH: You must judge matters practically. Anyway, that milk has to be used up before it expires. That's probably why she allowed you to have some more this evening. We bought it for you, you know.

BERYL: I didn't ask for any chocolate milk! I wanted the strawberry milk!

SARAH: They didn't have that kind at the grocery store around here. But you always want a sweetened milk, so we brought you that. And you haven't complained about it, until now.

BERYL (*near to whining*): I'm not complaining. But I can't drink it. I don't want to be spanked for what I can't do.

SARAH: She's not gonna spank you for that. I mean she'll be frustrated with your inconsistencies. You should be growing out of them by now. She's more concerned about Dad and Sachel, anyway, looking for that cow late at night. Probably wants to light up the barn, or something, to signal they should return. And actively trying to hem in those cows would be a smart way to do it, if it could be done. But I don't know of any useful stakes around. And a lot of hammering with nails would be necessary, making some noise. But when there's something on her mind to do, she likes to at least investigate the possibilities. Spends more time on that than doing, sometimes. She can be very persistent of the notions. But she's not going to waste any effort spanking you for *this*. She will reprimand.

BERYL: What's that?!

SARAH: Get angry at you for not doing what you said you would, perhaps restrict your playtime tomorrow, and certainly send you to bed earlier, so that you won't be so bothersome. We're all waiting for you to mature to be more responsible.

BERYL: I just can't finish this milk. And for that I'm gonna get punished?!

SARAH: We're not rushing you, are we? This is a vacation, and you should be having fun.... mostly fun, as you're growing. But these telltale signs of asking for more than you want can annoy. You should be more specific and precise when requesting things. You're not an incoherent baby.

BERYL: I'm not a baby at all!

SARAH: So you should start adopting some traits of character to be more self-reliant for your needs and wishes. You must develop the human intellect you're gifted with. When an animal has its fill, it stops eating. May even leave its food to rot where it lies. We don't do that, Beryl. We store our foods and perishables until their use. And if there is a gluttony we glut at it, to do service to its value— to us. To wantonly waste anything is perhaps the greatest of all sins; because by our very nature, above all other beings, we are capable of not wasting what we find or produce, while the dumb animals must let what happens happen.... to them and the universe they'd be concerned about.

BERYL: I can't do what I can't do, Sarah. And I didn't know I couldn't do it!—

SARAH (*coming for the glass*): Give it here, then! You don't want to disappoint Mom. (*taking the glass*) But I won't be around to help you all the time, so you're just lucky.... There's not so much left anyway. (*Drinks.*) Hardly a couple of swallows. And nothing wrong with it at all. You have to learn how to control your prospects, in order to get your chores done— You could've finished this!

BERYL: ....My neck got swollen.

SARAH: You mean your throat. And it's not likely you've suddenly developed a grippe from chocolate milk.

BERYL: Well, it's O.K., now.

SARAH: Still sore?

BERYL: No!

SARAH: Don't make excuses for yourself, Beryl. That's a bad habit.

BERYL: Gi'(ve)me the glass!

SARAH: You come to believe them, someday. And then you fail to achieve much, or get anything done.

BERYL: Well, it became hard to swallow— That was for real!

SARAH: The animals.... can't make any excuses at all. They must deal with themselves so honestly, it can be heartbreaking to realize.

BERYL: Why? if they can tell fibs.

SARAH: Like when an older chick pushes a younger sibling out of the nest in a tree, because somehow it can tell the mother bird can only feed one of them. The disposed falls to the ground and struggles to die, as if an ignominious loss. The poor thing is ignored, unless seized for a meal by a snake, or something.

273

BERYL: That sounds like a waste.

SARAH: But it's not, if that's nature's way of dealing with a problem.

BERYL: Then why did she have two chicks?

SARAH: ....To help insure that at least one would survive, because in the wild mistakes by definition can never happen. What occurs is what is meant by physics to occur. But man can control these physics, to an extent much greater than any other creature.... And so we can no longer be wild.

BERYL: Gi'me the glass!

SARAH: And we can't blame the wilderness, for much of our happenstance.

BERYL: Then how come those cows can make you scared?

SARAH: This!... is not a wilderness. This is the country. We are on a farm, and those gentle cattle can hardly frighten us. Wild animals may, but not them. Ferocity can be in an ant to fear. Not our cows. Are you really afraid of them?

BERYL: ....N*ooo.*

SARAH: The more these animals know us, the less fearful they should become to us, because they learn that we are their masters in everything. That is called a domestication, which is what we.... own. We have them, and own them. But those that stay wild and can not be trained we must always be wary of. They can not appreciate our dominion, fathom our dominance, and must be fought to be subdued, if we should care to for some need, or as a defense from their attack. That's why we've been brought here, this summer, to be made.... softly conscious, but more definitely aware of our privileges on this earth over all things.

BERYL: Plants can't fib— and they're wild!

SARAH: That privilege is a rational superiority, to even the tame and contented. And some plants can harm you worst than animal bites. Just because they're stationary life, (it) doesn't mean they garner any less respect from us.

BERYL: I don't want to be stepped on, that's all.

SARAH: And so you won't be, if you learn how to gauge your activities better, and satisfy your curiosities and thirsts more.... rationally, and judgmentally not to cause yourself too much discomfort.... in an inadvertent way.

BERYL: What?!

SARAH: Efforts can be painful. You have to judge their worth for the results desired.

BERYL: The milk didn't hurt. It was just too much to drink. And I'm not afraid of them. I'm scared of them rolling on me, like a car would.

SARAH: So you just stay out of their way, their paths, their routes of traction.

BERYL: ....Yes. But I can watch them, and feed them.

SARAH: O.K.—

BERYL: And they see me.

SARAH: They notice you.

BERYL: And look at me.

SARAH: They respond to you, turn their heads towards you, to see what you're doing and how that might relate to them.

BERYL: Yes. And so.... I know them.

SARAH: They're becoming familiar to you, and you to them.

BERYL: But they can still hurt me, if they're not careful.

SARAH: If they become clumsy around you, or walk up to you without noticing you. So you have to study them, because they can't help themselves. They're too involved with what they're doing, some times; because that's what's most important for them, grazing the grasses. That's their work, their mode of living, their pacifism, as they appear to us.... But it's serious for them, and life affirming.

BERYL: What would they do if they found me sleeping on the ground?

SARAH: Probably leave you alone. Or— since you feed them, they might examine you, to see if you have any food. They won't be mad at you. Maybe a little disappointed, but not mad.

BERYL: And what if I did have food? Would they eat it off me?

SARAH: I suppose they'd try to, if they sensed some, in your pockets.... You're not going to risk that, are you? Then one *might* fall over you.

BERYL: They dare my.... vi-gil-lance!

SARAH: You don't have to be brave around cows. They're not fighting *you* at all. They're waging war over the vegetations. Bulls you have to watch out for. If they think you're threatening their females, they'll charge at you. But most farms don't keep bulls, for long. They.... alter the male calves, so that they don't become so bullish.

BERYL: We don't have any bulls?

SARAH: No, only heifers, females without calves, aggressive from their youth, but still only cows. They're not going to hurt you if you don't bother them too much.

BERYL: They don't have calves?

SARAH: You don't see any around, do you?! They're *going* to have calves, that's why we can milk them.

BERYL: I thought they might be hidden, in dens or burrows.

SARAH: Calves don't live that way. They walk with their mothers.

BERYL: Oh. They need training? When can I get to see a ca—?

SACHEL (*rushing in*): Hey!

SARAH: When did you get back?!

SACHEL: Just now. Some—

SARAH: Well she's returned, to the barn. So you don't have to look for her any more.

SACHEL: Right at the barn? We didn't stop to look.

SARAH: You brought Dad with you?

SACHEL: Yeah. Some guy's busted his hand shooting at the lone miss.

SARAH: At the barn? I didn't hear any shots.

SACHEL: No. Way out on the range, clear over to near our neighbor's. Said she frightened them, on his property, the Bouwders'. So he chased her, this Mr. Moraine, back over onto ours. And when she threatened him he shot at her. But the gun blew up in his hand. We saw the flash, Pa and me. Probably frightened her straight back to the barn. Bad cow, Sarah, if she can be that rough. We have to do something about her.

SARAH: Mom was looking for some wire, I think, for a makeshift fence. Something you should have done by now.

SACHEL: Nooo. We've brought Moraine with us. She's tending to his hand— And Pa's calling for a doctor. But it's not bleeding, from what I've seen. Just terribly reddened, and probably stretched out of its natural proportions. But, the lone miss is gonna cost us much, tonight. We should.... ki—

SARAH: He's threatening to sue?

SACHEL: No. I haven't heard him talk that way yet. He's actually been quite apologetic for crossing onto our land. Considering what's happened to him, he seems rather reasonable. But I doubt if he can make much of a case against us because of his own faulty firearm. We'll still probably have to pay for any major medical expenses. But it doesn't look like anything more than a bad strain of muscles, hand tendons and sinews. Will probably just have to wrap it up for awhile. The physician will know what to do, for the extent of the injury.... That cow! Sarah. We have to put the kibosh on her but quick. This could've been— deadly. She's been after that mad weed; and it's made her tensive and unpredictable, around people.

SARAH: Well what can we do tonight, Sachel? Round her up again?

SACHEL: We haven't even done that, sis. We've done nothing at all. Mom was looking for some wire? I'm gonna do something about this!

SARAH: But what can you do alone, Sachel?

SACHEL: I'll get Pa to help me. But that cow has to be stalled up, at least. Wish I had a rifle—!

SARAH: Why?! You don't even know how to use one!

SACHEL: Well, cattle herders use them, for their own protection. So these cows must be trained to be wary of the sight of a man with one. It could be ingrained in their dull heads to have respect for such a weapon aimed. They came off of a breeding farm, you know. A large one! They know not to do battle with an armed proctor, telling them where to go, during their examinations to be comely and obedient beasts! Maybe a short, darkened pole will do.

SARAH: Don't go about poking and striking her with a stick, Sachel. You could hurt her— You don't know how to herd!

SACHEL: I know that if she acts up around me, she's gonna get hit. I won't let her chase me. I'll beat her into some decent submission, for a cow. It's hard enough to fight the weed, without a ton of it to control.

SARAH: You don't know what you're doing, or how you should be doing it. There are strict laws against animal cruelty. Dad's signed contracts, to that effect. And when they examine our cows, during the next annual inspection, if they find any suspicious bruises on her, it could be terrible for us, especially if we don't have them treated for records known.

SACHEL: This could be an emergency, Sarah.... We're supposed to treat them (as) perfectly nice(ly) while alive, and then butcher them up cleanly. That's a real hypocrisy, as if the meat will tenderize with gentle handling of the head and flesh. But I won't put up with her snapping her muzzle at me, or any kicking attempts. She'll either be stalled standing, or terribly lame, but in any case much quieted down. We've already disturbed a neighbor— Something Pa hated the most to happen!

SARAH: She's only an animal, and can't understand what's going on about her. Don't harm her, with an impulsiveness to do something! (*as* **Sachel** *starts to leave*) Wait for Dad's advice! What about the weed?!

SACHEL: Once they're properly fenced in, that won't be much of a problem. I have to begin handling things, Sarah. I've done a lousy job, up to now. Too weak and loose and— slovenly. But I'll take control of this beast. It's shameful, what we've allowed— Absolutely horrible!

SARAH (*following*): She still seems gentle to me! Don't make matters worse by being headstrong and half as—

BERYL: Gi'me—!

SARAH: —ked—

BERYL: —the—! (*They have left.*)

SARAH (*voice*): —to do something!

BERYL (*squeakily*): Gi'me the.... (*He sits up in bed, slightly perplexed and slightly anxious. There is a pause as he ponders whether to get up and turn off the light, to sleep. He half decides* *to and starts uncovering the sheets when **Ponaple** enters.*) Daddy!

PONAPLE: Are you O.K., Beryl?

BERYL: Yes, Daddy!

PONAPLE: Did Sachel and Sarah come by here? I don't see them around—

BERYL: Yes, Daddy. They were here, both of them. Mommy too. But they've gone to capture a cow.

PONAPLE: What?! No, she's with—

BERYL: Sarah and Sachel. They've gone out to the barn.

PONAPLE: Why?

BERYL: To get the.... lone miss. She came back there.

PONAPLE: Really?

BERYL: What's happened, Daddy?

PONAPLE: ....I'm not sure.... I'm not sure at all, Beryl. Someone got hurt, on our property, while we were looking for that cow.... But even running, it's hard for me to believe she could have reached the barn without us noticing her, or something of her. Though it *is* dark, and the pasture is wide of walking. But we saw nothing, no roaming shadows at all! as we walked. And we came straight for the house, which is in line of sight for the barn, though my attentions were admittedly diverted.... away from it.

BERYL: He's gonna strike that cow, for being bad!

PONAPLE: What?! I won't let him do that. Don't worry about that, Beryl. Sachel has better sense than that.... We shouldn't be angry at.... *her*. Besides, he's been working with them. They're used to him. He's just mouthing off, for his lack of control. But my children are not cruel.... nor vindictively stupid. And violence is never a good solution for the frustrations to alleviate, though it's often the most favored choice. It takes intelligence to combat the urge, or that primitive descent into the animalistic of our natures.

BERYL: Mouthing off?

PONAPLE: Claiming to want to do something one would never want to do, like cudgeling a defenseless beast who could not expect such treatment under your care.... But something similar (ha)'s been a problem in this region with its children. I hope it's not an infectious tendency, through the mysteries of these gentle country breezes of aromatic persuasion and tempting.... This is a night of much mischief, Beryl, as if its greenery could wish to dance with somersaults off the ground, for all of these strange events to witness in the dark.

BERYL: What events?

PONAPLE: ....The discovery of the nature of our land and possessions out here, son. A cow goes astray, that never is, and is shot at uncannily.... as could be a shadow's beckoning for destruction. That heifer was nowhere around.... and could not have been chased. We would have seen her, heard her. I was fearing to have

to collect her from off Bouwder's territory, with great apology and embarrassment. Impossible for her to be at our barn now, and yet to have been anywhere out there so far.... while *we* were. It brings to story told a demonic essence of trust in gullibilities. But what of phantoms for our sociabilities? We must not cast suspicions yet, until the impossible is beheld.... and can be thoroughly explained without challenge but for the ridiculous and improbable. No. Not out loud.... can we— claim anything! (*A faint sound of discordant mooing is heard.*) What's going on out there?!

BERYL: It's Sachel!

PONAPLE (*exiting*): I'll see to the disturbance of these gentle beasts. This is a tough night, Beryl, to realize your own. (*He is out of the room.*) But a doctor comes. (*There is a pause before **Beryl** gets out of bed, goes over to the wall switch and turns off the room light— the hallway lighting still illuminating— and then goes back to bed for sleep.*)

BERYL (*returned to bed and preparing to sleep*): It's early night, and late for light.

Scene III — *The back porch of a restaurant, that can be reached from within through a door which is opened. As it is night, illumination comes from the restaurant through the doorway, and perhaps some windows. **Falfour** comes onto the porch to find **Bouwder** waiting, in a somewhat agitated state emotionally as if to deliberately express a displeasure with events.*

FALFOUR: What's this wildness with Pitch, Hank? He comes barging up through the end of the joint to find Terence and me, and nearly shocks him out of his chair. But Pitch can't express himself well, except to shake as if he can't be seen, and as if no one saw him approaching the table, like sneaking in from the back pours a shroud over you under the restaurant's lights. He just stands there, like he was prepared to do something and becomes wild-eyed to forget how, leaning over to us and muttering "No, no, no, no!" and that you're with him, bothering Terence no end and drawing all attention to ourselves to the point that I have to suggest we come out here to talk. And since everyone seems frozen, I must lead the way.

BOUWDER: It's a night of the harrow, Falfour. An' what's tah be said is keenly through 'is countenance fer what he's seen. But let 'im speak it all. Fer on'y he does (h)'ave the patience tah be witnessin' dis distressful turn of— yer provisionin'!

FALFOUR: Mine?! Ours?!

BOUWDER: I'll not abut tah it. I've seen no cow, an' done no thing tah wear upon yer troublin' tries an' ties—

FALFOUR: No cow?

BOUWDER: —nor be related tah, when ana misdeeds are uncovered.

FALFOUR: You mean to say.... no cow was captured, on your estate, tonight. And so we'll have no beef anytime soon. Well that's a consternation immediately brought on with Pitch's presence, but not a great cause for the grimacing and shakes if the plan simply fell through. How many times must one try to succeed, can be enumerated in encyclopedias of accomplishment. And this caper has its complications to twist through. involving the Ponaples.

BOUWDER: More (th)'an that's been done! But I won't blab about what I can't see an' will not verify. It's a tensile tease fer Terence, where at yer've bought yerselves more trouble than some bloodied meat. The errors are sugared fer a sufferin' of cold slabs with possible imprisonment for these attempts.

FALFOUR: What? Would you tell of your complicity?—

BOUWDER: No accomplice am I! Falfour.... There's nothin' done, that's been on my land tah frown about. An' I have observed no crime. Nor can yerse draw me intah ana. But this is tah say I am disgusted!... with yer cavalier attitudes tah what yerse have tried tah trick me with.

FALFOUR: Trick you?! How could you have been deceived of.... Terence's expectations?! You expected to find a trespassing cow. If you did not, it's fair to say you may expect so in the future. So where is the deception for your concerns?... And why are you still not waiting for this possibility? The night is still open for its subterfuge, of a sustainability to this ploy for beef, with a quiescent roam of roan.

BOUWDER: An' are ye a fool! Falfour, tah think so.

FALFOUR: ....It's not gentle, to call someone a fool—

BOUWDER: Then be it made as it would seem tah be. Fer what of looks does yer key handlin' hide!

FALFOUR: ....Tansel will simply be alerted to abort the mission, until another time when it may be set up— if ever again. Expenses certainly mount, to overcome a law.

BOUWDER: That yer certainly won't! from these foul pastiches, from which yer've borrowed wrongly tah believe— much reason can bea circumvented. That cow's never gonna be on my spread. Fer I'll put up ma own spiked borders tah ever prevent such a barbed partnership with you an' yer debaucheries. But plead me innocent of yer errant ways tah think such wrongs as yer've done can be sa easila tolerated!

FALFOUR: Why condense on me your vitriol for Terence's plan?!

BOUWDER: B'cause yer are the bother tah be made fer, tah feel it could bea right tah do. Yours is the salivatin' tongue— not mine! I like the hen flesh fine, an' can live off salad greens an' beans fer months. But it's all yer deviousness tah bea party tah the hunter that's caused all dis troubla— in the county! Yer can't hold yer own heads up, fer the bones dragged through the dirt with yer mouths! —

FALFOUR: We are not the dogs of ravaging, Hank. I don't understand why you're so excited by this.... one setback. But don't ply your Angst down on me for an innocence to profess. Virtually everyone around these parts is aware of the need, to confess to.... some level of participation or endorsement. The society withstands.... the shame of its children, and may become maladapted to certain pressures of stress, strain, and stringency of supper tastes or dinner meals. But you are the golden rod of legality in all of this adventuring, what may be waved to ward off claims of crime, as if to cast a

spell of sanction for these acts over the eyes of those, ourselves, who have produced this condemnation of our meats to be prepared. And you agreed to this principle thoroughly, as a very clever ruse —

BOUWDER: That's gone a'foulin', flush of flimflam an' fallacy. I've on'y been protectin' ma crop shares, not yer snouts fer chewin' on cooked beef. An' I've agreed tah no criminalities tah hook maself onto. Oh! Agreed! Agreed! The greeds of yerse! The greeds of yerse!—

TERENCE (*coming out, with **Valin***): You say it's what?! that can't be told inside!

VALIN: It's a devastating fall for Mort, and then ourselves collectively to own his plight.

TERENCE: But injured how? With shooting?—

VALIN: Being seen, for strike of capturing. And so are we all, in the same coffin driven.

FALFOUR: Capturing?!

VALIN: I *saw* him, with a light, being taken away by officers of county policing, after they had injured his hand— for he was sorely favoring it— possibly shooting his revolver right out of it, to prevent his hunting of the cow. We heard the shot! And I went (to there), too frightened, into the darkness to investigate.

BOUWDER: It's 'is courage that saves yer ears fer witnessin'!

VALIN: And there upon I saw the eery scene, from a fair distance, as they apprehended Mort Moraine, his lasso and some remnants of his gun strewn on the grassy ground, no doubt moistened with his sweat of the devastating anxieties that must present themselves on being caught— clearly trespassing for harm of personal property, a transgression that will have to be related to us— all of us! For Moarine must contemn the need we've placed upon him, during his interrogation, by the county sheriff, and then a state's marshal, and then perhaps even federal officials for law enforcement. We had not counted on this— avalanche of heightening harm, to ourselves with this relationship of involvement, but only that a cow would be roaming out there, not the security.... gentlemen, of out-an'-out entrapment! All! of our devising will be found out and uncovered, as Mort is made to mouth the truth of every detail he may spew. And this conspiracy, this organized malice against Ponaple, is the greater crime than the attempt for some beef! It is harassment with prejudice, and conscription of hatreds within our community; something that is cruelly punishable, and thought deserving of the worst of legal constraint.... I gathered my nervous bowels, to go up to that spot, after they had taken him away. And there I saw the evidence more clearly, of what was left behind, more than proving my conjectures.

TERENCE: What evidence?!

VALIN: Why, the gun parts and the lasso! The fragments of our destruction! more felt with eye than ever touched throughout that dark gloom. For that close up, they are more evident to be, even in the shade of night's sky—

TERENCE: You left them there?!

VALIN: ....Why should I touch more taint to our souls! This is an official inquiry now, and such properties should not be tampered with—

TERENCE: But that is all of legal argument, that could possibly connect *us* with the crime of actually wanting to steal for slaughter, a gun and a lasso! It is the motive, that can disappear in court, without these physical articles to be found— no matter what Moraine says or claims. Obviously they left them there, at the heat of the moment, because they thought he was acting alone—

VALIN: For me to take them away would prove he wasn't acting alone! when they come back to discover them gone.

TERENCE: Don't you understand?! They must have thought the whole occurrence was most of a misdemeanor, and that those weapons were unimportant enough to leave behind, their priority being to spare the cow. The entire county knows that some beef is desired, that attempts like this might be made. And Mort's story about us can readily be believed— They were out there waiting for him, after all. But nothing can be proven without the evidence— not against us, aside from our tempers and inclinations that may be more cheered than chastised. Sure, the attentions may be brought on us, and that may even be enough to lift that wrathful ordinance. But physical proof chains us to a real attempting, and you should've seized it for hiding— Don't you think the law knows your wishes, ours?! But now, if they come back for it, after hearing Mort's.... expositions of possible culpabilities, they may have warrant to investigate further, with our names and those materials to deposit to a magistrate or judge. We must be clean about this, now, and burn down bridges of connection—

VALIN: I'm not going back out there to look for that proof. Their purpose was to stop Mort, and so myself— to god! I could have been caught. And thus they did, by maiming him to shoot at, protecting that cow—

FALFOUR: Police officers do not fire their arms for maiming. That's movie fare. They only shoot for deadly aim. And they're not likely to want to kill Mort to save a roaming cow, even if on its own range.

TERENCE: No. And you saw it roaming out there?

VALIN: ....Yes!

TERENCE: Then it was Mort— shooting at the stealthy beast, totally against my advice. His old gun exploded in his hand, giving himself away to the patrollers.

BOUWDER: Was way off, tah be fouled of thinkin'.

TERENCE: This only strengthens *my* way of thinking about what happened, and why things can go so absurdly wrong— when you improvise your efforts out of all planning. But only Tansel was to do any slaughtering. And now we must order him immediately to abort, and close down— shut up that abattoir, turn off the lights and freezers, and disconnect that generator. Only a night's hours are really necessary, for the empty.... or risk him being caught, too substantial to the plot to deny things. And you have to go back out to Ponaple's, Pitch, to retrieve those evidences—

VALIN: No! My hands are washed of this— For Mort will say that he was with me, and I'll be suspected of the theft so thoroughly, I'll see no more of daylight without fear of being watched (of surveillance).

FALFOUR: Mort can lie, you know. He has some honor to him. And if Terence is correct—

BOUWDER: He ain't never been correct in ana of dis.

FALFOUR: If Terence is correct, Mort's best bet is to fabricate as much as he can—

VALIN: That he acted alone?! If he acted alone, then let the stuff stay there. Let the ground swell for its resting. I doubt if I can find the location again anyway, in the dark. And if he's forced to implicate any of us—

BOUWDER: Not me!

VALIN: —then let's take our chances with the expedient alertness of Tansel, Terence, to the required actions for our effective exculpations.

TERENCE: That stretches a taut string too far, with this wrangling of occurrences. But.... we are bold for chances to be made, what's left for us to bone up to. But if you can't find the spot, neither may they be able to— It's still been so little time evolved through these events. And are you sure this occurred on Ponaple's land? so near for Hank to investigate.

BOUWDER: Don't try the rake tah my dirt! Nothin' can smooth dis out easila. I sure ain't gonna retrieve those things.

TERENCE: But if they were missing, you'd have reason to take them. Because Pitch and Mort were visiting you tonight, and something disturbing happened. So you had a legitimate cause to investigate. So may rumors flow.

BOUWDER: I ain't crossin' that path.... *if* the police are involved.

TERENCE: I'm just trying to construct a reasonable excuse for their disappearance, man! We have to sweep all corners of the house—

BOUWDER: Of consternation an' crass nation! mak' ye these notions fer more worriment. Yer ar' just scared fer yer own skin tah peel, in the brood-mason's stove.

TERENCE: Obviously that area will have to be searched in the early dawn, if Pitch can't re-find the location by night.

VALIN: Let them lie! Terence. This is getting too unpredictable —

TERENCE: You trust Moraine now?! to leave you out of his— confessing? his.... filo making?

VALIN: I'd trust him over your fabrications to explain away your errors of judgment, and enliven our confusions with even more complications of adventuring to trap us deeper in a moat of crime and folly. You haven't conducted these activities well. You haven't orchestrated them properly, sitting here eating chips and drinking beer.

TERENCE: All evidences must be erased!—

VALIN: It's best to take on a tentativeness, to see how things fall, and to do nothing that doesn't have to be done, and to disturb no objects of possible inquiry. Much can be explained around what's left to be, submerged in the nocturnal sea. But if we do not leave it so, then goblin tales will ensue, the mysteries only elevating interests for the curiosities of justice. And we will be the prime ghouls searched for, to provide its leather curing.

FALFOUR: Was there any evidence of a cow's injury, that you could see?

VALIN: Of course not! That was prevented!—

FALFOUR: It was not committed. The only error here is to be caught trespassing; hardly worth much of an offense, for unboundaried territory, at night.

TERENCE: Precisely! And that gives us full license to sweep!—

VALIN: There was a shot! There *was* an offense. And Mort was injured in his capturing.

WEILER (*coming out*): What's dat?! to give a commencement here, you guys. Is dere some beef already?

FALFOUR: Nothing begins, BobCat. You've missed nothing but our problems.

BOUWDER: Mort's been caught, tryin' tah rustle on the Ponaples'. Pitch escaped. They both brought nothin' tah show me. An' the whole deal has unraveled fer the lusted after produce.

WEILER: Ooooh! Dat is surely sore for simmering, should it be. No pot roast soon. And my mouth was anticipating some. But they beat him up?

BOUWDER: One or the other, methinks. Either 'imself, or 'is capturing.

VALIN: His gun's blown to bits, during the struggle, if there was one.

TERENCE: And he left that and his lasso behind, lying in the grass, as they took him away, according to Pitch.

WEILER: Well, dat's too bad.

TERENCE: I want to recover those things, to make it harder to prosecute him. But Pitch and Hank have turned more scared than confederate—

BOUWDER: Whaaa' tah lard's bass!

TERENCE: —Or some prudence masks fidelity from a friend. They're understandably upset and lean to precaution with the night's uncertainties.

WEILER: Lead me to 'em. I'll hide—!

VALIN: No you won't! They'll suspect I took them.

TERENCE: Pitch can't find the way back. They'll have to be found when there's some daylight.

VALIN: We've already hacked at this tree enough. It doesn't give any gun, despite the awful gashes.

WEILER: But.... you want to save a friend, don't you, from incarceration prolonged by— What is dat?! probable evidence? I done dat before, for some minor tiffs to avoid. Neighbors get at one another, because one finds a deserted plowshare belonging to the other, an' is incensed dat a farm hand might mistakenly be trying to furrow too far off his master's property onto his own soil. So he takes the plower, and is accused of obvious theft. He's too proud to return it. So before he's hauled off to jail, he calls me up to have me throw the darn thing well back onto the owner's grounds. An' I do it clever enough so dat it just appears like a statement, without any ties to the accused. The owner *knows* (that) BobCat's been around, to stop the foolishness. Might even had hoped I'd get involved, to allow for the recovery without a quarrel. This sounds something similar. Blame the disappearance on the rodents, Pitch.

VALIN: They're not gonna cart off no gun pieces and lasso.

WEILER: But it's as good an explanation to be let likely, if the materials can't be found.... for awhile. You trust Mort to protect your name, don't you?

VALIN: ....I'm afraid not. Though he is a determined cuss. If the police are involved—

WEILER: They're only for the miscreant types to bother, around these parts. How deep into Ponaple's spread are they?

VALIN: Not too far from Hank's.... in daylight.

WEILER: Then I can get at 'em?

BOUWDER: I suppose so, BobCat, if yer wants tah. I've been thinkin' on dis, now. An' it's more a pleasure tah be loyal tah Mort, than what I could presume.... from frightened eyes. But the truth of it is, I ain't seen nothin' of a capture tah describe— properl(y)a, an' so can't see how you may *not*.... efface a tort.

TERENCE: Well this is a way to do it, Hank. Involvement is involvement. And we have to take care for everyone, Pitch. Even Tansel.

VALIN: Well that is definitely your problem to handle, as you say you can understand him best. I don't like the way things are turning out, Terence.

FALFOUR: That's because you haven't seen the broad view. If we can handle all of this properly, it might make for a coup of sorts, for our self-determination in these matters. At the very least Ponaple's been alerted to be more careful with his cattle roaming. And if news of an attempt for beef gets out, in any way like this surmised, it might make the entire populace anxious for some, through the legitimate means we should allow. But we have to handle ourselves well, so that no one gets hurt, or more hurt, through our proceedings to bring this attempt to a safe and calm conclusion. And it can be as patently simple as Terence says: Close down the slaughtering shop, and remove some proof at an attempt to slaughter. What does not occur.... can not condemn for its occurrence— wished.

BOUWDER: Yer ain't made penance fer yer foul intentions, an' (h)'aven't learned a thing about the sloth an' slack of yer social demeanors. An' we'll bea comin' tah these same calculations ov'r an' ov'r again, till we start tah change these variables of contention fer some meat. But if yer wish tah subvert the will of the people, do it with a heart-felt hacking of our subservience tah the errors of lofty thoughts. An' challenge directli this terraba ordinance instead of with these surreptitious attempts tah fatten yer bellies with the flesh of flavors that can scorch yer souls. Fer yer have disrespected yerselves, tah fall prey tah labyrinthine wrongs against yer own needs an' suppositions of bein' honorable, presumin' on Ponaple's effective or enforced complacency tah yer ways of dealin' with a subtle crisis, as if any person can be treated with such callous consideration. Then tah yerselves may (yer) be so scolded of vile inclinations, tah counter balance 'is naïveté an' simplistic notions of us an' life in the country. Sure, that might bea reprehensible an' insultin, but not so much as tah turn us all intah fiends an' liverworts.

FALFOUR: I've always said that he should eventually be compensated for his losses. And you're not to admonish ourselves so strongly, since he is a neighbor only troubling you, as you have often complained. But hunger for a rich taste can develop into an overpowering emergency, that has little to do with villainy for some veal, but veneration for our vanities to procure without an overt vindictiveness to perform on the man—

BOUWDER: An' famil(y)a!

TERENCE: You were not adverse to teaching him a good lesson, like bumper bumping the car of an inexperienced driver who's just made a sloppy or awkward crossover into your lane right in front of you. Your response to his poor maneuvering might be as dangerous as his attempt at it; but it may nevertheless be warranted as a chance to take, not only to show some displeasure, but also to provide some impetus to improve his driving fairly quickly. That's about how I see it, from your standpoint, Hank.

BOUWDER: Well I can value a man who looks ahead, has some forethought tah care fer 'is cattle. Yer learn 'bout a person by the way (h)e reacts, sometim'. An' that can change yer opinions 'bout 'im, if yer *can* folla 'is logic.

TERENCE: As for the rest of us, it's the savory to sense, that has driven our tryst to restock some (of) our viande, and with no particular animosity to that family nor vendetta to cause, but simply an opportunity to advantage ourselves on his carelessness due to inexperience. And should we try again depends all on his astuteness to the situations he causes. Because the desire remains, and will find its way to a satisfaction.

JOEL (*coming out*): Why are you guys congregating out here? I don't like this back door kept open. It could provide opportunities for vagrants to enter, or put that thought in their heads.

BOUWDER: We ain't vague, here. It concerns the beef you've lost.

JOEL: Lost?! Hey! I've paid for it.

WEINER: But dat does get in the way of obtaining some, tonight. I mean the anxiousness of your investment. Mort's been caught trying to steal the cow.

JOEL: You loused it up, Valin? and let him take the fall? That's why you came shaking in here through the back, like a wet mutt seeking his master's bed to dry off in.

VALIN: He was on his own when he was caught. But I saw him from a distance apprehended and taken away, because I had to find out what happened.

BOUWDER: It took great courage tah go back ontah Ponaple's, after the shot—

JOEL: Shot?! He's been shot for this?

BOUWDER: Well 'is hand's in a mess, at least, accordin' tah Pitch.

JOEL: Terence! You've caused a great blunder to fall all over us. I can't afford to be involved in anything like this! They'll obviously take him straight to the sheriff's—

VALIN: Or the sheriffs take-if he.

JOEL: —And he'll be forced, under the duress(es) of pain and guilt and stupidity, to share his involvement with others, including this restaurant. Just what we need— to close us down thoroughly! We're already late on the gas bill, and some unjust property tax assessments we're fighting. An association like this, to some leaky illegalities, can demolish our business, destroy our license to serve food to any of you. I should've guessed the tide slipped back menacing(ly), the way you slunk in here, Pitch, as if the night could be too formative for fortune. But you promised nothing like this, Terence.

TERENCE: We're centuries of walk from any ocean—

JOEL: We were desperate! to take this chance and gamble with you a little. And now this could lead to our ruin. And I was more concerned with the chips than your chaffing.

TERENCE: No one is ruined yet, Joel. You'll get your beef— we all will!... But we have to stay firm! and committed to our aims, and not dissipate of head like a smoky cloud battling the evening breezes and mists. And I won't have your lachrymose disquieting because of a slight disappointment and mishap of action. Mort might be caught. But that means hardly anything of significance because no major crime's been committed if that cow has not even been stolen, much less killed. He could be released by tomorrow, with no more than a pat on the backside for being brass— and drunk. One can always claim that for surly, selfish behaviors. And the man is not un-clever. But this is all very trivial still, if you allow my management its proper range and gauge to confront these details with correction. And I have already laid out what's to be done, in this relatively early night. We have time in abundance to recover from a setback— if we do as I say needs be done. And that is simply to get to Tansel, to close down the chops' shop, and retrieve Mort's broken gun and lasso from off the Ponaple grasses—

JOEL: Oh my gosh! They wrestled with him even?! He put up a fight— a gun battle— and had to be brought down?

VALIN: It seemed calmer than that, by the time I got to observe them. They helped him up, and Mort was standing hurt and wounded. But I could hear no yelling or argumentation. In fact, their conversation was inaudible and clearly compliant for the prisoner. They walked off.... with the relaxation of a professional capture. And we only heard one shot, so I doubt if there was any fighting afterwards.

JOEL: But still.... this seems a gravity thicker than any gravy I've basted for. And our heads are blocked in for the squeezing— but mine especially.... since I'd be a licensed merchant for the spoils. You have to be right, Terence, that this whole problem can disappear by day. I'm responsible to so many for their livelihoods—

WEILER: There are only a few of you.

JOEL: —that I can't let my errors in judgment cause catastrophic injury.

WEILER: You're a bartender an' part owner, of the restaurant. But dat don't mean any of the others don't share your sympathies in this. We are all for a butchering.... And we'll all handle this, because that is our want, dere an' dere an' here. We are.... the makeup of our malefactions and gripes, ar.' can improve from both diseases possessed through the judicious mangling of our mistakes. Be brave, you o(ld)' chips provider. This is more fun than fierce, more folly than a foulness berthed to, more flagrant than flirtatious of our rights as meat eaters. And we shall gain our steaks, with a polishing of.... dese intentions. There's no fault for a flavor fought for.

JOEL: These aren't games of sport or practice, BobCat. We're right on the edge of falling off a very steep cliff, with a heavy load of responsibilities to ourselves and others. And I've been foolish, but not insane, to try and gain some hamburgers to sell at least. So try us now to commend ourselves to reason—

TERENCE: And it's rational that we mitigate these problems my way, the simplest and straightforward to do of all imperative. Sweep to have our foundations clean, and our fundamentals clear. Not a one of us need deny our consortium of trusts. And every one of us will be rewarded in our fortitudes of procurement. The methods may be altered, but the results will be obtained as we support each other. No law can condemn. .. the desires of a whole population, an entire community. It is a matter of modifying and adjusting each persuasion, to reach a common ground. And I think Mort will be forgiven, the lighter the matter is made to be or seem. With those articles of his enlistment unrecoverable, to resist further (official) effort of delving deeper into our planning, and his physical anguishing to elicit some pity for his carousing, I think charges will be dropped, or no serious ones levied. This is the courage we've undertaken, mates, to bargain of ourselves collectively. Let's not fail any of us, including Mort Moraine through his travails, by falling apart into a sheepishness of discontinuity. Rather we shall stay firm and more in charge of our destinies than indictable to them. I'll.... go out to Tansel myself—

WEILER: I'm all for that! This is jolly nerve for a jubilant night, gents. Because we are doing something. We are manipulating our consequences. An' dat can only be good, as it only feels good to do. I'll go all around and everywhere, to be proficient of our providential reinforcements towards these goals.

VALIN: I'm told he has use of a rifle— to shoot the cow.

TERENCE: Well he's not gonna use it on us, Pitch.

WEILER: Not *dese* cows!

VALIN: The guy is a little unstable. Don't explain too much of what has happened, to him, and why we can't bring him the prize.

TERENCE: A vacillating one. If he's there, he's gotten his first installment of remuneration for his services. That should smooth over any qualms of uncertainty or wariness. Yes. He'll just be paid for being there, tonight. And that will not be a loss, but a confirmation of his voluntary enlistment.

VALIN: All he needs to know is that we couldn't find the cow.

JOEL: Well you find your heads. If I go down, it's to a low depth sunk. But Tansel's already there, fighting his grips of gripes. And if this whole thing blows up in any substantial way (*starting to reenter*), I don't want him back here at my restaurant, blabbering his disenchantments and intensifying my poor luck.... Terence is sound to stay steady. But this all better work!

TERENCE: Consistent, he means. Oh, he has a lot to worry about. But nothing has really gone wrong, if we hold up our own. And the next time might be grand.

FALFOUR: Perhaps. What can be grander than a success? I feel it's a privilege to have at the attempt and learn by it.

BOUWDER: Cattle jawin'! You comin' with me, Pitch? Since it's Mort who picked yer up, an' his truck's stranded on ma estate. I'll drop yer off— but I sure ain't goin' tah no abattoir.

TERENCE: You should go back with Hank, and wait for BobCat in the morning. You can help him—

BOUWDER: I ain't holdin' no hospitality fete at mine, if there ain't no real crisis that I should bea handlin'. The Cat can come out an' go straight ov'r tah the cabbages, b'cause that's where yerse two left tah reach the Ponaples', leavin' me in the moonlit dark. An' if yer can't find the spot, mayba the grass grow-ed ov'r that stuff, lik' fer dreamin' through embarrassments. But I mean tah catch me some kinda sleep tahnight, in ana case, no matter how ornery it might seem on the nerves an' nervous shadows breathin', fer dis queer sensation of disgust.

VALIN: Then take me home, Hank. I've failed, tonight, by seeing too much. But a plateau is a plateau, BobCat. And where it slightly drops off.... (*starting to leave with **Bouwder***) there you're likely to find the goods, I guess.... I won't meet Tansel empty handed— not tonight.... He might be empty headed.

WEILER (*as **Valin** and **Bouwder** depart*): Sure! I know the area well, I think. Dat region's all mad weed.

FALFOUR: Or a heart-stroke's haven, for the maladapted of those sprawling tendencies.

TERENCE: I just can't understand what he saw, to give him such a startle. If a man is caught, he's caught. You don't dwell upon it, to the loss of your proprieties of action.

FALFOUR: But they have informed us of what's occurred. That's all of their importance in that theater of participation, since not much more can be done there. Some day I'll have to explain to my son all of the trouble I'm going through for him. He'll never be appreciative enough. But that's not really the point, for a'gaining an understanding of something, is it!

TERENCE: He'll let what seems to have to happen happen, as it seems to be, like Pitch. But one thing we haven't been is.... more careless than what happens, which is why we'll rebound from this stemming happenstance easily.

WEILER: Why, that's the very honor for our ironic attempts. Sometimes irony wears veils. Dat's all keen to reveal 'er. An' we don't even get to kil—lit! But it's all for fun of knowing how it could be done, or could 'ave, if those things are meant to happen somehow with our wishes boasting an aggressiveness to carry out a plan. Then there's no shame provided, dat's all, when it unravels for our patching up. Then I say, just live a little longer, get a little fatter, grow a little meatier an' moo. I am really all for.... the prospering of a natural order through our natures. An' we will be eating fresh beef again, as opportunities ripen anew. That's how I may support the search for errant cattle. Dat is it! An' not from the devil to obtain dese permissions.

TERENCE: Yes. We are country.... not altogether cruel of its nature and ménage. Hunt(er)s tend to carry their weights awkwardly, while targeting un-supposed cleverness.... So let's not be further detoured from our objectives original, to make a pact for our likes and construct their realizations. Perhaps this night was too bright for us, but there'll be others ever more gamely possessed. For we salvage our intentions with as neat an atmosphere as this evening's calm temperance to pertain of our attempts.

FALFOUR: And contain them soundly, hidden from our perspicacities of thought.

WEILER: But that's the bill, Falfour. An' I'm sure we're good for it, dat this can only improve with our.... patient attention, dese cumbersome delays to vanish through the hazes of our eying. For what steams as vapor(s) may be perspiration's perspicacities, all right. You don't often get things well done on a first try, nor should it be the last for errors' wanting. But be persistent to amend ideas around the goal, and not a zeal replace with likelihoods to note. Then we'd all be prodded into stables, awaiting the earth's banquet.

TERENCE: Yes. What is seamy seems, more sounds the sigh of this supine serenity to swear of, slighted from success into surcease, a swindling of sir's societal state. (Z)Swounds sate! surely, as could our safeguarding of this propounded sanctioning swish the swoon.... of a moon-tide's errant glow.

FALFOUR: What of the wheats to sway us more, as could to brush feathery the hands along fine, faint ferns, then must support the surges suggestive of these skullduggeries, as would glisten the grasses awaiting their morning dews. So it's how we'd approach this bath of hide, a slaying due from this rain of our temerities.

WEILER: An' roasted within blood, does muddy make the broth — but more delicious, delicate of taste. Dat's all to be so comprehensive of our race.... towards our acquiring of some variety in meal, *vache vole*, or veal to come within sagacious days more

wished than reared. This all tempts.... a transcendental realism for our hypotheses of these deeds.... that dese defile a cure for mouthing. Dat is the means to rush for— Rush! Let's get in play.... a balance for our aiming yet to come, an' more scheming to be poured, if in this way we must become more analytical.... of our idolatries for food.

TERENCE: Well, the night doesn't hinder us at all, for our dreaming. But how a want can turn a thought into an evaporative mist of intoxicating deviousness, this is sure to cause more trying. But we must present some seriousness to the adventure for our most immediate mission-ing, to pretend of a considerable sufferance to setback, while it's all the more to cause our urges might. And that's the sum of summary, to summarize how difficult things (may) get done. The rationale could be more urge than need. But so to storm through a task endlessly, until one can contemplate a satisfaction's reached. That could be the death of all compulsion!... or compulsions to resist. Is that called heaven?— or.... a meal for mind? Let these persuasions reign— forever! Since that's as long as one may live, bound to them. And so, with philosophic gist attack its ending. I am sure others would do as we, much similarly, were their beef so foully taken away, and their complaints subtended through the stars for reaching of the mind dazed by their malevolent charms. And what is healthier than a good meal, but to obtain this food more soundly than we have been treated lately, and refuse to subject ourselves much longer to an abstinence of the luscious cooked and trimmed with teeth and jawing serrations of red meat to endow our flavoring of mouth and tongue— That is the beast to pacify! and turn witless with a pleasure honed. This animal is omnivorous, carnivorous, and thorough to be justified his launchings of intense dependence on.... the luxuries of our emotional gratifications, what shapes the character of a ruler over the servile placed before us, teased by with mendacity of being fit without— to eat— and all of those other lies— to kill!— to kill our gastronomes— But!... We delay too long in our reveries, gentlemen, for our matters to take care of severally. Yet this night is so full of hours to be dreamt through, that it does bloat of promiscuity. The darkness hardly moves, with floating about our airy heads.... And so, to terms to make with Tansel. Be him similarly unwedded, to the rawness we'd desire.

FALFOUR: He's better off than we, but not so sacrosanct of our positions.

TERENCE: He's a good minion to squeeze at and pet. And we'll keep him, provided for our purposes and intact for some alertness towards this doing of a carcass what is made to be partitioned for our plates.

WEILER: Sure! With dat skill, he must be sharp for much, an' seems more terrified, at times, than tasseled to the past. It's O.K. to fear previous adversities. Shows the nerves are at least not over-firing aimlessly. They shoot for something to dread and avoid.

TERENCE: That's to be worked at. The point is that he's obedient for some pay, and will be assiduously employed of his abilities, whether by us or the entire community regaining the sanity of tastes. Yet abides he like an instrumentalist in orchestra, awaiting a performance.

Act V

*Scene I — The **colonel** stands before some large freezers at the abattoir, apparently waiting for others.*

COLONEL: What a senseless cold does dreaming make, but in these freezing ovens held, a thought to make for capture of one's misdeeds iced in preservation. Yet, could thaw in heat as well, the pertinent particle of one's demeaning, as would be kept still relevantly amassed of frosty remonstration. And I would object to even the lack of much air in these huge freezers. But they are functional and turned on, and someone must foot the bill for this usage. Yet, within such confines amounts a universe of pervasive causing. This entire establishment is active for a rich fulfillment. And what is to evoke? as were a thought kept cold or unconscious and dead. No. Thoughts can not be such, for even cold vapors have life to swirl and mix among the elements of frost, and could be dangerous energies to scintillate one's contemplations into a deep and permanent sleep of sparkled sickles painted darkly in an unlit cell, as could be revealed bright with light and opposite its shading in the frozen tomb, with much for snow. Is not the irony grave, that the coldest is not warm— yet blazes! white.... while much fire burns to black for ashes seething in their uselessness till graying of their age, reaching for to mete of other brightness, much as trampled snow, a determination of effects condensed upon some heat. And so are characters alike, to fly off embers.... while yet frozen of thought. This contrasts life with death; for some dreams never die, but merely enlist carriages, through smoky, smoldering remains to lift and find another apt.... Or of a marble smothered, do we perform towards viewing— to capture! that this is fought for, without punishment to feel. For what may be banished— in a ball of ice?! We are elicited from the cadaverous, as much as from a quiet slumber or a rapturous dishevelment of the mentalities, since thoughts remain present after their generation— climbing for some want to be associated with, bound to, glued on(to) nervous needs. Mortification is like the blasting of a bomb for rain! to share our drops— not the snuffing out of these intentions into a solitudinous mat of the mourned, with masochism as a fever loved! since all life succumbs to its ultimate enlistment and resolute dominance to control.... a transformation into dreams and phantasmagoria. Forgive us of our persistence to remain, but this is all that we may do. And here is held, for geriatric body parts to contain, in this shop of culinary culling of the meats to eat, if for any life past its conception aging is its sole reward. Then we have come here, it seems, to find you ag*iing*— and awaiting us! For to speak of some desires we are present. We must always be present for someone, or some thing to stir as thought. And we.... are drive, for some nourishment to obtain, I should conclude, if this is an appropriate result to find ourselves at. Or some other, of sundry reasons— as long as it is beneficial.... will do. I'm often of much confusion, by the pains and panics caused.... of our resourcefulness. We can not do the damage aspired of, nor cause this harm so lasciviously felt. It is purely towards a beckoning to follow and enjoy. But what seems of self? as much to be a frozen cavern.... turned on to be frigid. What a great comparison of aims to see! Fight for it always, a justifiable declension, with passage of yourself.... through stone into stone, or a multitude of other elements, to end up as an aery thought breathed through. I haven't claimed it so. But then to wait upon, wait upon.... most brazenly, this conscripted dementia that never disappears. (*as the **adjutant** enters*) Does the enemy come?

ADJUTANT: It seems he had prepared for them conscientiously, carefully. Everything about this place is in full working order— Though we've worked no machinery. But all is clean, and can be assumed operable.

COLONEL: That's how to judge? I think it's headlong rushing, that a chill presumes us.

ADJUTANT: Into the shaded streaks of day this night will crash for shards opaque to leave and walk through, scrounge as with a march to collect its remnants for review and analysis of the threat, the constancy of some oppression meant, the loss of a free will, the orderliness— of an automated pursuit. I think it true, that a battle balks of waste and will resume to contain us, sir. The surfaces in this place are so well polished, the metals shiny and unblemished of any rust, that here is well kept a most terrible assembly of wrought hosting which cries and begs of use. And we will find ourselves within a torment as worthy as before, when nothing's ever completed of a maiming due to needs until the satisfaction's done to cringe at and see the grotesqueness of the cruelly made remains. Our soldiers are stationed throughout these spaces and recesses of fortune rent, to await the onslaught. And what.... (*looking at the freezers*) great.... catacombs have we here! as portable as modern of design, yet made as stationary as underground crypts to symbolically replace.

COLONEL: And of course, they function cooly, at this position placed, this elevation into some desirable pertinence and evidentiary intensity of intent. No wiggling in there, not by now of the causticities imported with the cold. Though vapors may swirl, but that is a cavity's right to own them, of its sedentary contents. And the wishes?

ADJUTANT: Our men are keen, sir, to conclude all struggling clearly, and without resort to slurs at the fallen who may no longer bring more spiritedness to our goals. For by whomever, hands rendering these deaths, or slices at, they are none too sorry now— to have been used, and can not occur past us, nor in spite of us, for our detailing here.

COLONEL: Replenishment is all, that ever can be done, as life continues. Strife must justify its harm(ing)s, and heresies of approach. They're all out there, waiting to advance on us. But we have quite a stronghold for the present, furnished with the grit of discontentment, and the savvy of a barbarism turned into the proficient heroics of clean implementations, sharp, savage severings and slicing, and the salvaging of severities in treatment. It's very harsh to be determined in your actions, and totally focused on your objectives.

ADJUTANT: I doubt if any group would attempt a siege of this position, now that we have obtained fully its strategic advantage. Even a troop of considerable size would be bludgeoned into oblivion—

COLONEL: Or guts!

ADJUTANT: —one member at a time, with each passing stroke of our adherence to the fortifications of weaponry we now possess and may attack with at any convenient moment.

COLONEL: As for it now, yet can we handle it? the magic wand that converts our strides into victorious strolls, and overturns our defeatist postures to show the omnipresence of our gainful insistences to maintain the glory of a notion we represent and struggle consistently at.

ADJUTANT: We've touched none of the machinery, sir. I've given strict orders all throughout our company to prevent that utility over our guns and bombs and grenades and stares of menace, that these knives and grinders we've found might be prematurely used, and unwarranted for an enemy's promise—

COLONEL: Oh, that promise!

ADJUTANT: —to be abused, to sing with praise through the sawing we have suffered ourselves, and our willing launches into the midst of destruction and unavoidable carnage—

COLONEL: That they are dead! dead! to be subjected to these physicalities.

ADJUTANT: As much as any would.... to deserve such grim dispatch.

COLONEL: And deserve it, well that's the current to the flow. That's the mad need. That is the rampage for the ruining of our fearful subjections. That's the pull of plight we've hooked ourselves on to ride at. That is the drive to persist with these discerments of matter and flesh! That's the bounty to be won of it— the never! never! campaign to be ended. That is the explosion— through our heads! And where is it?!

VALTER (*running out, military style, with **Tansel's** rifle*): Here, sir! Fine and fresh and smooth! Clean of barrel to be smoked— an' smoked. (*stopping a ways before them*) I've.... never fired such a thing. It's for a soldier's hands who's been with smiting and a smack! For this is love of honor doing, (*raising the rifle*) this rifle's using, and its rush— for the ruminating. This is the decree, that all's O.K. about this turmoil and torrid tension to be teased through.

COLONEL: And give it him? to find a cow?!—

VALTER: He's found! and all shot with. He's made it, for his valors homed and housed. He's here deceased. Can that not be so? I saw it done, an' wished upon him luck, as could a medal be pinned through his neck! And we are pleased, with his familiarity — finally! Because he condenses into thought, an' frozen heats of validation, that this night is a surrey for his ease of travel through the universe.

COLONEL: And where is he confined?! Where is Tansel confined?!—

VALTER: To our comradeship. To our compatriation. To our company. To our collocation an' collation. To our— heeerd! (*All three start to "moo" loudly but separately, with random alternation.*)

ADJUTANT (*with a sudden stop of mooing*): What speaks?! (*All three start to yell "Might!", but as before with mooing. The cheer gradually turns into the word "Mouth!" during the improvisation.*)

COLONEL (*ending the yelling*): Moooutheeed! for a witnessing. This is the crying of the wolves, and the bleating of the sheep, and the dyeing of the moss with a fuchsia of sanctification and hallowed tears. We have permitted this! permitted him his acceptance. And now what is a night?! What is an evening's blight?! Funnel to a single beam of posture this grandeur owed. And we conscript. We tear at, tear off to own and bite— chew— swallow— and intensify the fight, make it more huge and tremendous, with a tremulous obtaining. And is this not correct to do, to satisfy your aims of dying!— Dining! D*iii*ning! Dyeing die for din of dining, like all others' winsome trying— to be fit! and fashionable of this roamed earth. Forgive us! But everyone's the same of shape, if what's eaten 's cut down for our mouths. And we are all soldiers alike, to convince you!

VALTER (*raising the rifle*): This fire needs no convincing!—

COLONEL: And capture—*you*!... And take you with, for the enemy to sight, and rush towards and into. And as a holy sepulcher pray at, to be so much of, these grounds of cow-eyed terror, with your presence of mind to condone and feed with, be of the need and eternal nourishment commanded, commended, commenced— our rapturous atonement to the sins of these appetites. Who does not really want a war?!... when there's a warring to be done— to survive.... defeat, surpass, and deify.... our frustrations to assuage. Then let these conflicts come; and have me rub the hair, of the heads of the fallen, to anoint them with my pride, and the satisfactions.... of your stomaching.

ADJUTANT: Oh! Stomachs! We have stomached so much, to be appeasing. We have bathed in fire, and inured ourselves.... to these desires ordered—

COLONEL: And they are here! They are right here! They won't evaporate of frost. They won't run dissipative of a nightmare. They are for capture and a ride.

VALTER (*pointing somewhat obliquely at the audience with the rifle, and speaking in an almost impersonalized voice but with a rising inflection of tone*): May I try this?!

COLONEL: Not more unto the fray of pondering speaks the displacement of our quarrels. They are out there with a certainty, awaiting and observing. But here seeks without misfiring a purpose, nor waste of fire untoward in character (***Valter** lowers the weapon.*), deservedly of heart-shaken concern. We've pursued this fellow like a mark of shadow's shame, to enlighten his wounding— and sharpen his blade strokes. But there is for the blow, each devious act a demonstrative demurring of the hacking need— bones sliced! like tubes of tallow talons tandem(ly) connected, each crack a sigh of strength, a promise rushed through. *There* is the execution of resistance, in those unfelt fells and fellings— furrowed of the hands and knife. And for whom.... the soapy *splurts* are saved and ate, packaged for consumption in neat pounds of bloody war, cooked of rent demands rendering the tastes' sensations, seductions, and seditious heat of animals. That is the hub, of this firestorm and maelstrom, that we must ever be within the fighting learned and leaned to, loaned from gods, to find more cows, whereat they roam, and— tame.... for meat.... and hide and soap and horn and nails and teeth! And to become the devilish alike in these complacencies and complicities is our maddening and furious transduction.... into angels.... or their minds' merit, for the human fusillade of striving in survival's mortar bore. We are at war with everything, and every concept that can be held for dreaming or idealizing its appeal. For, to conclude ourselves, conscripted on these urges, this curious machine flirts along the never-ending undulations of a pacific dissolution from rough, hard, and tough waves smacking of resolve and determinate return into the fluidity of nature's transformations. And so we blow ourselves up, over and over again, for that lushness of meaty, gritty mead to make— an' mouth.

ADJUTANT: That's the roving plight of pleasure-making malady to head. We are the effecters of this errand, heaven sent and shooting, shot at, secured for a tranquility of notice, as for grasses ate of this great struggle, to keep the universe in change, that is a man's mind seemly seeped! for his actions. So then is justified our raw burning of the flesh into that other than a waste of lightening and explosions.

VALTER: What a heavy cardinal of arms, to see one's self with such good feeling of purpose! and come from a derivation of meanness.... into morality. Chew! and make cud from this! Repeat the more of mincing—!

COLONEL: But guilt is for blaming— vows.... to be imperious, and impervious to harm, or harmful condemnations, re-stomaching for bitter bile your mortal musts to feel.... remorseful and regretful and repentant of causing pain, as in a tooth that bleeds to burst foully! and to be lost from gum, lost from the mouth and body— yanked!.... from humanity. (*as **soldiers** start to enter slowly, from both ends of the stage*) Such bad thoughts brewing, such care of caries cueing a remissness to see, of what is chopped— and chopped for you! an' chewed for news, and spent in groceries of grass— an'.... bodies hewed and bled for pots to have some sultry grace for dinner cured of guilt— and shame and disgrace and terror and horrific portrayals of a mind's weeping and yelping from abuse ineffectually, penned in of skull— for the fire! and the cleaving into parts presentable for consumption. Then where is the blood, to this green-gray matter frozen! to be found? (*The **soldiers**, having entered to imply some crowding, start variously to sit down and hum or grumble-grunt, low and soft.*) What breathes.... of a cold heart?

A SOLDIER (*sitting*): This is the grave, a mortuary of mud! I found myself here—

ANOTHER SOLDIER (*sitting*): This is the earth, rich of my contents, falling, leaking, lent, loaned, loosing me my ideas to mix with and wet. Wet where? Wet where? And for a headache from the ear, the bombs of casualty and casual loss of consciousness and fun.... I've died for grain. I've died for a prosperity on summer plates for feasting at a festival celebrating the freedom to do so, and to have wars to insure it. I've killed, to have cows butchered, and am (now) dead myself. And I have come across a threshold of relief, to be seen hidden, burning in the ground, and emancipating more doubt of my suffering wept for and envisioned as heroic.

ADJUTANT: But greed makes envy for it. Silence.... are complaints.

A SOLDIER: I stumbled only, and fell face first into the milky reeds, with pouring gratitude for some cool welcoming. Is it not just, that I (should) become so frozen calmly? and like for luck of instance to be made this way— ended.... of all reveries deceptive.

A THIRD SOLDIER (*sitting*): Eat this grass! It is good, and I am sufficient for its everlasting enrichment. And what else are the coverings of caskets for? but due the pleasantries of some utility.

A FORTH SOLDIER (*sitting*): I can do so many things, down here, and whistle pliantly remembrances of a long fought for station or habitude, like dressing up sharply— to be torn to bits an' pieces, with a merriness resulting, somehow. But darn if I'd be able to recall *that*, here lingering in a crowded solitude. There must be happy frolicking above, or beyond my notions to suggest.... that I must lie still and endure the torrent of some hooves an' whoever, daring my dispatch to send a message.

A FIFTH SOLDIER (*sitting*): Yeah. It's become busy, chewing away— and gasping for the thought of it, to be delicious, or at least thistle of a flowering. But I won't say "baa" to you, to announce the sprawling colors; but rather may I do, for your perusal gathering, this luscious pause for cattle to remark of, if they're as curious as I can still think, what generates this mesic mat of verdure which still sinks of a grave to lift for the sunlight.

ADJUTANT: A distance that is remarkable, through any span of measurement with leaps of greenery towards the unconquerable of comprehension. There's no need to know why. It simply does its property of growth, to lend itself of mashing and gnawing. There's no need.

A SIXTH SOLDIER (*sitting*): I don't want to be anchored down here, deep in the sea of terra-fate, for terra firma's consistency of fortunes. I want to float! on the surface, and see some shininess in mirrored faces on foundlings of the daft deeds destined for their days. I had mine too, an' worshiped hope. It's all I'm left with, ironed of my stillness to rest. Yet it can not escape my grip, nor my casket of lime. And with a promise made I've owned it, buried and born, with a miracle of permanence. So show me others with more animated glee to celebrate with. They shouldn't be ashamed to see me, or aghast of a decomposition, if all that's left is hope!

VALTER: So sounds the soldier wailing, that even in defeat there is no real loss through our repetitions. Then what is to cry for, but a deep joy of anguishing, that mysterious nurturing which justifies our manner even unto death and resurrection. The landscape is fertile from our presence, in virtually every direction of battle and engagement. And one can not easily be vanquished of a memory, where blades prick through the earth to reach a glorious eyeing.

A SOLDIER: All of filth, and muddy mockery. But we've not given way one inch! We are obstinately placed, and seated to remain, like lions of a pride awaiting the day, after a long and energetic hunt, exhausting not our courage nor our camaraderie. But when is this promised revolution? When does the shade of earth heal, to reveal us again— with devoted activity and earnest aims?

A SEVENTH SOLDIER (*sitting*): It's for the mind to lend us patience of review, such handsomeness and splendor won during our missions— caught.... for contemplating more and more. And do you have more for us to do? But certainly, and every day, with any justice made to ply our pounds of dauntless erring for more matter to be made of it. And we cease to complain.... as a tone is heard, the warmth of everlasting rest, the hum of a transcendent hush, the awaiting of a universe to care for our charges— and carry us aloft with welcome, to a banquet of gods pleased and responsive to these valors laid and piled for observance— plunged for! and recovered. And this is like the most tedious of editing, to repeat until perfection is seen.... in what we've done.

COLONEL: So hangs before these containers of freezing some thoughts, as could be shrouds of shields, one mind's evaporative thought to dwell of his enlistment for our company to make and be composed with. And now the comfort is all.... meadow. (*The humming begins to gradually die down.*) The comport was not so bothersome, and similar to many. The hesitancies and cautions were often shared. The resistances were typically overcome with an infusion of grouped and combined pride— in many instances. And results were as to be expected of a valiant team determined to persist through onslaughts and savageries, usually by machines of rapid, repeating strokes, strikes, stings, and stemming strafes.... to slice— slice— slice! through lines, through bodies, through the maimed of outrageous weaponry and subhuman tactics. And I could send them through this, with courteous acceptance of their predicament's rewarding. But I must touch their cold heads, to acknowledge their final accomplishments. And though their heads were often warmed of gushing blood and valorous contempt of wounds, (n)or being wounded, becoming so.... my eyes were cold, never to turn away from any face, any imprisoned stare, any shock of countenance, any posed pout of horror for the realization of this defiant victory, this concluding.... mendacity, to withdraw or flee from what could only be inevitable to meet and challenge and dare and defeat with the courage of— defeat. Not one head was lost from my touching. Not one face. Not one heart.... Not one mind. For my feeling for this— sufficiency.... makes for our collective groping to affirm ourselves— with a positive confirmation of our merit and deserved-ness. This.... we will continue to do— forever. We will meet— forever. We will fight— forever. And these thoughts will last forever, as I am soundly stationed to insist. Though questions and doubts will always be warranted to exist, as for what is the foolishness in this. To lounge forever? To lie eternally? To be deeded to posterity with a slumber so permanent, there wanes much consideration over time until forgetfulness and ignorance must dominate against these noble personages? No. It is the irritation that will last forever, the annoying notion that will continue to bother you forever, the battles that continue or come, and the wars that must be fought. In these— needs.... our valedictory triumphs and agitations will persist. And— we!... will menace your mind forever for securing.... your extant ways. Forever is a life that owns our thoughts, and that may be seeded with them for spreading. And as you adopt this knowledge, that madness is the birth of concern — pay.... us.... homage! And we will continue to visit, and fulfill your most appropriate irresolutions towards the world's competing and almost constant anxieties and sacrifices. Forever is this drenching squall employed for an empyrean perusal of our efforts.

ANOTHER SOLDIER: Enlist the storm! Enlist the lightening! an' the process heinously pursued— sir!

COLONEL: What dreams?!

ALL OF THE SOLDIERS AND SUBORDINATES (*in unison*): Sir!—

COLONEL: What drools of wet anticipation towards an anxiety driven?!... towards an attack of conscience? and contrariety? Oh hue of fire! bring this thing through to view. And let for entrance toll, on this broad night, a destination pitied of monstrosity to meet, your very vow to cow. And frighten of this frozen mind to mete, the mental punishments (*with hand and arm gesturing*) parleyed to Tansel! (***Tansel*** *enters, held on either side by the **medic** and the **soldier** who shot him. The three have difficulty wading*

*through the sitting **soldiers**, who are reluctant to adjust their positions and remonstrate with low grunts. Therefore, they must stop before even reaching **Valter**, let alone the **Colonel**.*) Where is your fear? you must mirror! What have you here as found?!

TANSEL: These heads! These heads!

COLONEL: These herds?! Is this a joke?!

TANSEL: Cruel ghost! I've tallied them— forever! And they're as found.

VALTER (*with astonishment from a sudden recollection*): Are not you dead?!—

TANSEL: I leaped! Valter. I was amongst you, and this field of might and leper-ed souls. So many losses— to our bodily members. So many limbs blown askew. And I listened to their sighs and their prayers for forgiveness— of forgiveness meant. It's that I did *not* escape to retreat that drives me mad. And I cried here, with a frozen sweat; for the tears themselves did sweat to freeze me still. And I saw a thousand die, like crazed revelers leaping into bomb bored graves and pits. And I leaped, kissing all of your strewn matters, flying for ascension. And when I fell.... when I could not reach heaven— I felt condemned! and manacled— to killing. For the heads had flown, so many, all around me, staring aimlessly of grief and consternation, as if I could never have existed, with all of my love of comrades.... company.... and.... I did not escape the diversionary tactic, Valter! I saw you burst of flame, and win your immortality. And I became.... what's wedded to: That drive and that need and that imperative demand to earn your right of valor and the worth of all of these carcasses, and so many more— forever!

VALTER: You were helping me from behind, carrying canteens and cartridges, and my trusted faith of friendship, that we would steal time together for a rusty drink, after this terrible deployment was suffered and survived through. But then, as I turned around to find your heart of backing me— you disappeared!... in a blaze of incredible coloring of lights. And some sufficiency of purpose must have been achieved; because I rush to find you everywhere, now.

TANSEL: I did not draw away from you. I must have been wounded of nerve to the legs—

MEDIC: You can't help the dead! I never could! You were just late for the volley, and like a cripple sulked! And you stayed down, nested to the ground like a mud grouse, while you saw this truly wounded soldier die before you—

SOLDIER ESCORTING TANSEL: And I yelled in your face! because I was in pain! Tansel, too awful to stand your worthless concern. Because you could not help me, with all of your empathy, and your crying to stay calm and hold on. I blasted your ears with my distress, or more your eyes— to see such torturing! I was not at you to complain— but (to) God!

TANSEL: To see such torture plainly, there were thousands— all around me— bathing in grief and the swarms of death, wasps and hornets stinging, clouds of darkness rising of burnt smoke and the scents of salvation— Colonel!—

MEDIC: I couldn't get to him—

TANSEL: —Why did you send us there—?

MEDIC: —in time—

TANSEL: —to have me here?!

MEDIC: —but I got to *you*, after the firestorm was all over. And can you remember me? I was the one disgusted of my ineptitudes, and almost dismissive of my mencing you, treating one life among a million corpses made. And my hands could hardly better these odds, so many died in them upon my hasty caring. For what it's like to find these struggling selves, is more for mercy than my medic's worth.... But I did not mean to impart on you any sense of demeaning to ever last your own subjection to this harm. And for the future you are found joined to us not with shame or guilt, but as a survivor representative.... of the carnage.

SOLDIER ESCORTING TANSEL: And you're as much a god to complain to as any we might have won. Because we all can blame our errors more to be of how you'd think of us— remembering this heated hostility to host— than as we were, deafened to the realities of our consumption.

COLONEL: And you keep killing cows, Tansel.

TANSEL: ....Not of late, until this night to presume....

COLONEL: We've observed the habit, to follow the battle lines you bring.

TANSEL (*meekly*): You've brought (them) to me.

COLONEL: But that you are obsessed with us *we* would complain. So have you lent us an abattoir, to end all time.

TANSEL: That could certainly be a place to do it. And I've been a prisoner of *your* complaining? Then what more punishment is merited than this stoic condescension to my peers? if I'm ready to provide the act. And I've always been. For I am heroic only to receive your arguments as just, and expect of you to insist of me some such compensation for your harried and vigorous dementias.

COLONEL: Recompense for our energies? We relieve you of yours. Ours has been a privilege, for our standing through your disquieting attentions to reprove ourselves repeatedly. And now you may not worry of it, other than to question why. Why did I send you there? Why are such things approved? Of course, it's all for the farming.... that (it) can be maintained by others, following from our exertions to permit.... a future of successes, or succeeding waves of human endeavoring. There is no righteousness provided for a cruel deed, but much for a wrong one out of the sincerity of its need. And we required that defensive, stalling tactic, which seems to have prolonged your enervations of the mind to an endless degree of withering and woe. But tonight, we have been captivated with your capturing— to face the enemy—

TANSEL: I? (*as the* **escorting soldier** *releases his grip and starts to sink, to sit on the floor*) I've fought with you. I've fought of you, and was found among you, amongst the piles of dents in earth and dungeons never to be escaped from. The sluing of the slain—

MEDIC: I.... handled so many, searching for some life, some spark of twitching, some stirring defiance to persist through a mortuary of mortared battle. And what few did I find, as could drive me mad with disease of frustration and resilience to the display nettled of and needled through.

ESCORTING SOLDIER (*sitting*): Emote into the moat. We are dead, to be dined of, taken off the roads of progress to be buried within them. And you see our shiny faces walked upon, as you burp of our flesh, and straddle the straw.

TANSEL: Grasses in the fields? But that's not enemy to me.

VALTER: Have I been searching for you, ever since the holy incineration?!

TANSEL: The grass.... is ate, to make more bulk of meat.

VALTER: And this is your gun!

TANSEL: Somewhat, mine to use. That thing is aimed at cows.... I've used it before.

VALTER: Before?!... Before. Before the enemy?

MEDIC: It was like searching through night, the eery sounds of death fermenting, being rousted about as in a thickened stew, the earth layered in a sauce and the air made dense of this essence, sponging it up about my eyes and nose and mouth.... bathed in a sea of putrefying glory, that early morning, that dawn of decadence discreetly dimmed by clouds and dusts and a singularity of browning, bronzed light, the sun much ashamed of this cataract— yet proud to review the actions of men. I wish I could be as ambivalent a god.

VALTER: You're not dead?

TANSEL: ....I am still a member. I am still "also." I haven't been deceived by any of my thoughts. I've not deluded you, nor avoided. I am.... souring of your weight; but I continue to carry all of your hearts, all of your ideals, all of your matter to matter with, all of your suffering to suffer with, all of this sin to surcease of foolish marches and silly whims until I've found my soundness again, repute as a soldier— once.... and reputation as a worthy and justified individual, in this community or any I should choose to be a part of —

SIXTH SOLDIER: We're not comrades now!

TANSEL: —And you keep holding me up, keeping me alive with vividness, for some envy to perceive.... and not guilt.

MEDIC: I can't be guilty for what I've found, Tansel. (*releasing his grip*) Nor can you share its haunting spirit, to scare you to run away from these feelings, these sensations that dazzle your conformity to life.

TANSEL: ....My conformation of it— as much a dream as this is it to be, that I have lived so long deliriously spent?! Where is there the real to seize, from you manifestations of my curing leather and thick skin? Why am I abided, if I have no doubts to offer you?—

VALTER: This is *your* rifle.... This is *your* universe.

COLONEL: So see! The enemy is your torturing, and you keep us running about. But what have you been paid?

TANSEL: Some trivial sum. Some miserable crease to lining of my debts and dues. Not enough to make a joy of it, but to forestall some coming hardships, in this rural sanctuary.

COLONEL: And is it worth the stay, when the cow comes due? —

TANSEL: A*aaa*h! At any moment.... any moment. And for that weapon lent, I should be loaned a cruelty. But for that rifle, am I dead? Then, I wish to sit!

COLONEL: No!

TANSEL: I wish to sift through tears!—

ALL SITTING: No!

TANSEL: For them I leap, if I'm not bound— to cattle call!... It is a limping of misfortunes wandered through, (*starting to make his way towards* **Valter**) that I need that thing to shoot, an' fire at the innocent, hopelessly brought to.... me—

COLONEL: What would you hold onto?

TANSEL: ....It's firm enough, and with a cartridge loaded—

COLONEL: But peer beyond.... And this peer, beyond him see. There are freezers here, for the frozen.

TANSEL: It's true. They're running.... expecting me—

COLONEL: But then.... that makes a bargain, for your visitations?

TANSEL: I need that blaster.... to complete the task!—

COLONEL: What task?!... Tansel.

TANSEL: But— to kill.... cause (*reaching* **Valter**)—

COLONEL: To get that firearm? To touch it, hold it, aim it, shoot? Giv'it him!

VALTER: This is *your* gun. (*hands* **Tansel** *the rifle*)

TANSEL: It's here. (*as* **Valter** *starts to sink, to sit*) But where?! No!

COLONEL: Need never search for you again.

VALTER (*sitting, with some resigned astonishment*): I've died.... It's more remarkable than giving birth to dreams and missions.

TANSEL: What do I hold?—

VALTER: The imperative is gone. The circuit's broken, the current extinguishes.... and the rest placates this irony of your standing.

TANSEL: What do I hold onto?

COLONEL: I guess it's your duty to fashion, since this time.... you've found us.

MEDIC: What can you search for with it, that I haven't found already?!

TANSEL: ....I'm not searching for anything. It's all being brought to me—

COLONEL: No!... Now you must find the light. You must see through the clouds, that pass or may form gaps. You must study and scrutinize for your freedom from this envy. You must make the move.... and your motive for your moving— on.

TANSEL: ....It's.... like to the heavens, to be sent?

COLONEL: Away from the herd. We are tired of your munching, if you're no longer....

TANSEL: A shadow on your contemplations.... A coward. A criminal. A crestfallen crusader.... A contagion ending— of the mind, and of the caliber to be turned crookedly.... Yes. There is night, here. A warm night freezing.... all over this country, all over these heads. And I won't more subject myself to deeds long past. For only if one learns to love the heat— of living— can one hope to survive through it to this end, or this just turn away from being cattle. And as I'm given this.... emblem of fortitude to hold, then one must make to do with what one's given or obtains (*the sitting start low groans*), and only that. Inspiration's only a form of marksmanship. Not worth as much as its resulting cause, yet can I pacify the miskept urge to shoot— and drive away the hungers with more terror?... or rectify this ruining and pervasive slaughter. I'm only subject to the world, am not angelically bound.

COLONEL: But there is territory, for the seething compost, and the burning ire.

TANSEL: It singes the ears. Where tread more capital than this? It's fortunate to be delayed for eternity, from any more rambunctious slighting. I'll take the shot—

COLONEL: Where, then? Where?

TANSEL: ....to home with me.

MEDIC: For what I've saved and spared?! useless are these hands — again!

TANSEL: Not with that lessening of grief, medic. It's all about a voyage, and traveling, and being (en)'abled to think and propose— and leap out of ditches and pits! and discern the sky proudly. That is your enabling, like a misdirection to see magic. That is what you all have done tonight.

COLONEL: There's no trick to it. We're just disillusioned enough to leave you still. Cease these groans I hear, that envelop a nightmare. It's like the motor of a rushing refrigerant, working to freeze bodies or conceptions. Do you think I haven't noticed what I've caused?!

Peccant Pecus [Act 5, Scene 1]

TANSEL: I? So then command.

COLONEL: You can't command the earth, man! It spins, whether you'd want it to or not. And the day comes. This is for you. Take over of your dreaming! I've suffered enough, of the hypocrisies to find me blighted of the face, and face these sights of anguish for review, and have them explained to me.

ADJUTANT: And that is my just enlistment, because one must understand how things have occurred, and determine or deduce the method of these spurt-atious injuries, to adjudge the weaponry involved with certainty. And that is what is deserved, to know what was employed against us and how we were deployed against it. In this way a battle is rationalized, through its analysis. But what of battle made, the colonel must see. Else it's not a battle, but some fiendish delight (the groans intersify) as they munch on their cud — to re-consume themselves. And this is the maddening impossibility. Because this can not be for fun nor pleasure, but only of a necessity wrought. So then the colonel must review a battle scene; although those lying within it could hardly have fought at all, but were decimated without mercy and with no chance of escaping the slaying slush of raining fire and destruction, the heated brimstone of brawn machicolation. Every detail must be— seen, reviewed, acknowledged and accepted, and found to be plausible for making improvements more favorable to our outcomes in the future— if there was a battle for these remains, to observe.... and entreat, a message of forgiveness for a mission forgivable. But if not a battle that has caused this, if not a struggle with resistance and fighting back and some attempt at even hopeless avoidance.... then what eye's the colonel, that I must describe and annotate for his understanding and letters of record? (The groans are grown so loud as to make him almost inaudible.) This is what Tansel coldly presents and I force myself to beleaguer of thoughts for our rational attempts. That is what pounds the mind and hammers for conclusion —! (abrupt end to groans)

TANSEL: ....Then cease! for my impediments to sanity. This is no war made unto us. Silence! Silence all! Silence everything! Silent are the stars. And silent is this moon-shed night for my bereaving. We have ended here! as the muscles pull and twitch and are quarreeeling! to be heard— that phantoms sacrifice their souls to me, and my disease of candor!... So be the punishment of satisfaction. I have no need for life, in this way provoked. Nor does any living thing of thoughts to bleed with and feel pain, are we more justified with maiming, cries and gasps for relief from stoning— slicing, butchering, as the only rewards to triumphantly march from— into death! Then have it made this way. It is a need! And I will condone it now— for my head aches of your tensions!—

COLONEL: Freeze the fear!... and keep these members dead! for once, these recruits whom you conscript for repeated torturing and ridiculous blandishment of their courage and effete drive or propulsion to serve your suffering—

TANSEL: Yes! I'll amend you all, your freedoms winning over me, my mal-contentions, mine! That should be sufficient for this evening's blessing, this evening's blast— no? What else can I do, to leave you straightly parted?!

COLONEL: These are huge freezers!—

TANSEL: And I've a huge heart!... because there's never been a sacrifice so profound as this!

COLONEL: There's never been.... (as the sitting start standing) There's never been.

TANSEL: Here's a crowd, most suddenly. And the mood is celebratory. But let me tell you: There's much relief— to a decision. And that's all confirmation, to the mental arts. And, I would say.... we as much disappear as to be seen distinctly. Because in truth, there's no one here— but cows. And that does draft a striking resemblance, to our concerns. But upon thinking of the matter more, this must have been the way men bore themselves through heathenish emancipations from the animal, congratulating themselves with some pleasing revelry and joking, for the Godhead to have won.... in themselves. And that.... does short-sighted-ly present the plight.... of a man's presumption to be honest with his nature.... and that of which surrounds and interacts with his. Since else be only deviltry at play, and demons seeking shade.... from bright suns. And so is swept the frown, from agonies distended into laughters and remarks of how this could have been so dire with such jesting emanating from a consciousness. And the ebullience to a relaxation of the temple.... well, that deserves some considerable joy. Because.... it's all in being fed, that anything may die. So be a harnessing of thought. I've had so many, that it's taught me not to think much more about the terror they may bring.... to inconspicuous eyes. (The scene devolves into a generalized conviviality of gratitudes.)

A SOLDIER (rushing in with a bottle, after having rushed out): Wine for Tansel. This will take the chill out of the co-habitation! I found it in a metal cupboard. (runs up to **Tansel**)

COLONEL: I told you not to touch anything! or at least for orders pleaded to be relayed, sent, spread and spurred.

TANSEL (accepting the wine, possibly exchanging with the rifle): It's a fine preservative!

A SOLDIER: But.... to touch is to find. And I found it— waiting for this epoch of use.

ADJUTANT: Well, as long as it's found for him, we've had no hand in it.

TANSEL: She's the Rhinebahre, from the Rhinegold stash for accomplishing a good year's production. And the day before the lay-off for winter vacation, the employees share of it. A new one is replaced, and the empties are stored at the proprietor's home office, each with a label of year, number of workers, and profit summary. I've seen the collection, and it is staggering of presentation, to make one proud to be associated with this industry in a really meaningful sense. (lifting the bottle to fully gaze at) This must have been left here for hope the factory will reestablish operations during the fall, else it would have to be labeled with a particularly low accounting.

VOICE (from among the **soldiers**): Open it! Take a swig!

TANSEL: Oh! It's solid of me. It's solid of my rank.

A SOLDIER: It was left for you. And you need it. We are your company. You'll have some of that juice in you.

TANSEL: I don't own it. I'm of it—

289

VOICE (*of various from among the soldiers*): Have that taste! Have the final swill!

ANOTHER VOICE: Be the hog! Be the one who represents us, all condensed into one.

ADJUTANT (*standing by the colonel*): Taste some fidelity in yourself, for a change. Do this symbolic deed to demand a healthy closure. Make this gesture for us to cheer about. That was clearly left for your appropriation.

TANSEL (*starting to unseal the cork*): Was found to be abused, I suppose. My head's not clear to think otherwise than this, which was usually a solemn ceremony, a drawing on rejuvenation and replenishment. It's a blessing to exist for more to do— whatever 's coming! (*dropping the seal to the floor*) Vandals have eaten my resistance to this. And I am one with life and home.

A VOICE: Drink that stuff, Tansel! Wet your tongue and swash your throat!

ANOTHER VOICE: Bring to gut's maniacal cure a measure of relief, a medicine of fete.

TANSEL (*while uncorking*): This hurts of fun. I steal what's stolen from my heart, a washing to avow of pride, these many days, the work and deeds desired of my hands— and mind. (*raises over head the uncorked bottle*) And revolt 'gainst an approbation for my death! (*The soldiers start cheering: "Drink! Drink!"*) I'd swallow molten lead— clear through to destiny's cure-all remedy for heat! in head. (*starts drinking from the bottle, amidst cheers*)

A VOICE: This is for sufficiency of charms and life!

TANSEL: I am maladroit. But this is heartily received, of an intoxicant— Share me!... of my crude and bungled labors. (*drinks more, within a night's cache of revelry*)

MEDIC (*exclamatory*): This is for caution's musty frown to burn, what's saved to earn. And frolicking makes for a peace of thoughtfulness. For jolly are the spared to die, without contamination of contempts and the certitude of mean delivery.... to our arteries of onus and purpose, for a meal. Yet, where does this life go? as towards the flames of metabolic thrust, *se*created and *se*creted deviously away from view, whereat your foods are made.... tender for your taking, meats partitioned and packaged neatly for a handsome display of earnestness ingratiating of your tastes. And mine as well would make a man this way, hastily deceived for his parts. But of a cow? Pleasure, please! for the procuring. There are so many dying for— worth dying for. And I've not saved one better than the rest who graze, and grovel for some light, a lightness in the head, and dining. Let's.... not be deceived.... by a ruckus, the ruction of doing anything, when you really have only yourself to bemuse. But then that's fine, of a camellia to sense, in these strange parts.

Scene II — *A kitchen in* **Ponaple's** *house.* **Moraine** *sits at a table, being attended to by a* **physician**, *as* **Laura** *observes.*

MORAINE: Is it terribly bad, doc, fer all of dis tenderin' yer have tah do? Will I be abl' tah use the hand at all, in some days comin'?

PHYSICIAN: Aside from the extreme ecchymosis to your palm, I don't think much damage's been done, Mr. Moraine. Maybe a pulled muscle. But the numbness should not become permanent through nerve severing, since you can still twitch your fingers, slightly.

MORAINE: It feels lik' a paw caught in a snare.

PHYSICIAN: Nothing to the bone for fractures, I do believe. It should heal, with the usual recuperatory inactivity after such an accident. The gun exploded right in your hand? while you were firing? Must have been mostly recoil damage. You're lucky. The injury appears to be less severe than it probably feels. Not much powder burning to the skin. But it must still be kept medicated with this lotion, to prevent infections. And I'll ask you to be persistent with its prescribed applications, even when you feel you don't need it any more. That's the only danger here. Only if you find a rash or any irritation is developing— and I don't mean merely itching— should you complain about it. Then I'll have to find another ointment for you, after, perhaps, more consultations, and some tests. Of course, if you're rough, to dirty up the hand with the.... organics, during your farming, nothing may help. But this swelling will go down. A couple of days, I think.

MORAINE: What about the pain, doc?

PHYSICIAN: Oh, try some aspirin or acetaminophen at first. I don't want to get you on any major pain killers unless this really becomes bothersome. But a man like you should be able to handle something like this. It's probably more of the shock that disturbs you than anything. The gun discharged, but not through the barrel, impressing the shells holder into your palm, that's all, disrupting all of the casing to break it apart. If it had actually blown up with all of its bullets, you'd likely not have much of a hand left right now.

LAURA: So he's fortunate the weapon was incapacitated?

PHYSICIAN: Considering its manner of incapacitation, I'd say so. The bore was probably not blocked, but it's tube misaligned slightly to the firing (pin).

MORAINE: Scared the grays tah the roots of ma hair. But one does have tah be meticulous, puttin' her back tagether after disassembly an' cleanin'. But she's so old, parts could've jostled out of place just with the travelin'.

PHYSICIAN: Yes. Firearm components are engineered to be particularly precise. And age does matter to them. Those old weapons shouldn't be fired, Mr. Moraine. What were you after?—

MORAINE: Nothin' tah be aimin' at dis night, sir. I was a fool fer spooks an' ghosts. Shadow lands an' apprehensions. An' the thrombin' fears now encounter ma hand fer transit.

PHYSICIAN: But what was the shooting for?

LAURA: He thought one of our cows was chasing him.

PHYSICIAN: A bull? here?

LAURA: No. We only have three heifers. But one wanders throughout the range and may have trespassed our neighbor's.

PHYSICIAN: Hum. That would be unusual, at night.

MORAINE: There was none, lady, but my own resortin' tah a foolishness of threat. Ol' Hank Bouwder thought he could've been abused, an' a friend of mine went with me tah investigate. We had *expected* tah find a cow rummagin' about the area of field we were searchin', which appears tah 'ave been within yer farthest extents. But.... given the condition of the grasses in that region, I was a'fearin' for some unruly behavior. Ma friend swore he saw somethin' movin' strangely, an' went off tah tell Hank we might 'ave located 'er, what may 'ave been pesterin' 'im.... But it was dark, an' I was alone. An' ma limpid eyes only saw the devil of me tah bedevil ma sight, tah be wary of a charge from— a much angered cow. An' intuition took the better of ma head, tah shoot at a shadow that seemed tah be approachin' fiercei. May 'ave been the effect of leaves on distant trees, shakin' through a warm breeze. Hard tah get a good perspective in the dark. But it's all certain I scared maself intah a wanton action, an' have paid fer it with this hurt. There was no cow out there. Hank must 'ave been dreamin' through apprehensions fer 'is crops an' planted bothers. An' we're the ones who have trespassed, foolishly an' unknowingly, fer all of dis trouba.

LAURA: But my husband said she had been chasing you—

MORAINE: We were chasin' ourselves!... fer some raw consternation. Everythin' were built up on presumptions, tanight.... that an' errant beliefs. But Hank didn't see no cow. I didn't. An' ma friend's not too good of sight at night tah be accurate 'bout anythin'! though he might claim otherwise. We wanted tah believe in an offense, lady, an' have made ourselves offensive by that. But.... higher powers prevail, tah correct everythin', an' show our eyes the worth of error in thought. I am deservin' of this pain, the injury all of ma own doin'. An' yet yer are vera kind tah me, with an understandin' that opens doors fer respect tah pass through an' take note of. We have been.... leery of the Ponaples, bein' so new tah these parts. Yer shame me intah pleadin' fer some forgiveness — of ma thoughts. Fer I have been a coward not tah learn of yer better, up tah now.

PHYSICIAN: Well that's not so unusual, for recent residents to acquaint yourselves of, in time. But this is as customary of a hospitality (as) our more firmly established inhabitants must share. Someone is involved in an accident, and I am called out. Caring for anyone is all important, in these rural areas, when there are medical needs to be attended.

MORAINE: But that's a point, doc. Our needs.... are not always our wishes, our wishes not our needs. An' our wants wear thin of patience tah become unnecessary burdens self-imposed an' convinced of. But I must confess tah be relieved of foul feelings fer another, be it cow, man, or other, when they're not warranted. An' we're always bein' re-taught, with physical reinforcement of some punishment earned, because it's more fun tah feel than tah learn. So.... feelin's must be replaced with others, tah show you their true values in yer. An' a pain in yer could be a pain in others lent, an' comin' back tah home more properli.... an' of a more approachable form tah stay conscious with. 'Cause this hand hurts at least as much as my mind should 'ave.

PHYSICIAN: That seems like a fairly contrite spiritualism. But I'm sure the Ponaples don't hold too much against you for your trespass.

LAURA: ....Not at all.

PHYSICIAN: Especially during the night, when you must have been confused while carrying about these anxieties about that cow.

LAURA: We have to put up boarders, to partition off our estate from others—

MORAINE: That shouldn't be so necessary, lady. It's not that customary, in these parts. We honor boundaries with the courtesy they're due of ownership. An' we respect yours, now. Though we hadn't expected yer'd bring in those cows. But this is yer property tah do with. An' yer neighbors will adapt tah yer measurin' with it.

LAURA: Well I wanted some farm animals here. And my husband bought these cattle.

MORAINE: But yer land, in particular, is not meant fer the cows roamin' about, with all of that mad weed spread through it.

LAURA: He was assured that would not cause a great problem, since most domesticated cattle don't have a great taste for it.

MORAINE: Hmm. Well it is a risk of genes, I guess. Some of them go back tah the wilder progenitors.

PHYSICIAN: Revert.

MORAINE: Yes.... in a few occasional examples of breedin'. That's done so much these days, fer homogenous herds, that exceptions stand out sorely.

LAURA: That one cow does like to be on her own, much of the time. But we will fence them all in, for night, and restrict their grazing to daylight hours. I suppose we hadn't suspected such confinements would be necessary so soon. But we wanted them to get used to the terrain, and not feel too much homesickness away from their herd (of origin).

MORAINE: That's a very gentle consideration.

LAURA: We wanted them to feel the right to master this land as their own, now their own. To me they sit too much bewildered, not far from the barn, like a parakeet placed in a new cage, who stays on the bottom for awhile before taking to the swing or perch bar. But this changes with time. It gets used to the new environment, and so will they. We're trying to allow that to happen. And the one of them who likes to roam around is as gentle as the others. My son knows of their dispositions more intimately, since he tends (to) them—

MORAINE: Yes.

LAURA: —And he can not complain of their attitudes. They're all very docile, as they should be.

MORAINE: Yes, Mrs. Ponaple. But yer see, they're not really pets tah a cage.

PHYSICIAN: And you resent that attitude in their husbandry? There is a conservative trend here to livestock, Mrs., that is much observed around these parts, since it pertains to the health and welfare of everyone's animals in these open fields that are sometimes shared with visiting. A poorly kept or looked after animal can spread disease with an astonishing rapidity throughout this region. It's happened a few times within my memory, though not recently. Can't really do much, concerning chickens, except to clean the coops regularly, and test the eggs for pathogens like salmonella. But for the four legged critters, ill suited farmers are baited with complaints to improve themselves quickly. That happens, for example, when a shepherd of sheep or goat herder tries his hands at pigs, and doesn't quite know what he's doing. People become suspicious when noticing any laxity of their management. Your free roaming cows may have evoked a similar wariness, to some, that you are not best for this kind of ownership. And I've even heard some loose talk to that effect.

LAURA: Have you?

PHYSICIAN: But not with much bitterness. More a resignation to try and sympathize for neophytes to country life.

LAURA: But I'm not—.... exactly that. Part of my early childhood was on a farm. And I've induced, prodded my husband for us to have this one, so that our children can experience some of this lifestyle.... a little. We make errors, but we will correct ourselves.

PHYSICIAN: With certainty you will, I'm sure. Why, you'll even be forced to legally, if things should ever get out of hand. But, the people here are fairly generous to be understanding and helpful. Just not used to beginners. And this is only a summer retreat for your family.

LAURA: But more permanent residents will follow us—

PHYSICIAN: They'll have to, because of this considerable livestock you've invested to the land. I'm sure your husband was informed of *that*, when he bought those cattle.

LAURA: Relatives of ours.... will manage the farm for us. It's a dream— for the future.

PHYSICIAN: You can't buy cattle, and then desert them. They have to be either sold off or slaughtered. Often it's both. They are calf-ed, milked, and fattened up when (they are) no longer useful for dairy production. Then they're usually sold to a slaughtering range, where they need a steady supply of ripe occupants for a regular flow of meat. Everything's become draconian, dreadfully specialized, in the cattle industry; so much so that romantic notions of a small family farm, concerning cattle, hardly exist anymore.

LAURA: Perhaps not romantically, but definitely reflective are our pets. And we will make up for their care. We don't intend to provide any viands of flesh from them, Mr. Moraine, but to have them for themselves of a calming pasture to be made and remember with.

MORAINE: That's a nice thought. But it's more work than to think. The vache are made for the labors tah do, of our dining purposes. Keepin' them otherwise is a sham of shacklin' tah man, for they are inept of the human compassions an' are just waitin' tah be destroyed. They age towards this an' are designed with that idea.

LAURA: I think not of themselves.

MORAINE: That's why they're allowed tah live so peaceably, most of their lives. The final rush of utility is devastatingly sharp.

PHYSICIAN: And usually quite decisive a blow to their composures.

LAURA: Throughout their lives *here*.... they will be pets. They'll be treated with friendship in their company. And they will comply with our need for their gentleness.

MORAINE: Lady, they are der beasts of the compost, as we lik' tah call their earthly service alive. An' yer son must realize this, with the hard hours he's taken fer their care, few in number though they may be, if I presume rightli.

LAURA: Well, he might feel a little harried of these brief and temporary charges for the summer.

MORAINE: Livin' 's no game tah any creature that outlasts its playfulness.

LAURA: He enjoys the schooling. And I can recall some aspects of any childhood— through my children's persistence with these cows. They've been making cheese.

PHYSICIAN: Education is for gain of temperance in life. The point is.... that cattle, like all cattle, will eventually have to be destroyed. They are not allowed, preferably, to live into aged states of disease, decay, and infirmities. The risks are too high for a minor epidemic of harm to develop, since they are animals of our *food* choices. So, while they must be treated with some affection.... they should only be ended while healthy and distributable. *That* is a statutory responsibility of the owner tacitly respected and enforced.

MORAINE: It's onl' common sense. Yer husband these things tah provide tah others. Yer are a provider, now, ev'n if onl' tah yerselves.

LAURA: We will, with dairy products.

MORAINE: Yer will.... have someone do, who will farm fer yer. An' then these matters will become more clear and sure of purpose, necessity, an' mandatory action. The cow.... can not grin an' be happy fer yer petting an' dotin' on. It's a life given yer, tah make service with. An' there can be no fullin' around with that cause. The worry was yer Ponaples were plannin' to corner the beef market in these parts, by— rationally— establishin' early a cattle ranch, while waitin' fer this ridiculous an' un-maintainable ordinance 'gainst animal butchery tah subside. Yers rich enough tah try dis, in a reasonable way. Tah.... say that's not yer plannin' shocks as much as a cow bein' killed— But let the claim stay, for this humane treatment of livin' steaks. I am reviled tah believe otherwise, now, an' say so tah yer, what may be wrong 'bout dis. But we are hungerin' fer some, an' are bein' driven mad ourselves, as much tah eat some weed of worshiping. That taste is too divine tah lose, the custom too much instilled tah drop, an' the irritation too provoking tah retire from. I would not 'ave been saddened tah have shot her. An' I would 'ave stolen her carcass off yer land, tah be used fer decent deserts of meal. This I tell yer with all of ma candor, Mrs. Ponaple. There are like minds all around these parts.... Yet fer a discretion tah be quiet about it, I am heeled of a farm lady, who may be pleased tah know the truth of 'er remembrances— fer what is cruel in man. These deeds must be done, by our natures commanding. An' what has been desired more than real, are these opportunities baited by.

PHYSICIAN: This sounds rather towards an illegal bend, Mr. Moraine—

MORAINE: An' I'm ashamed tah 'ave tah explain maself, tah hopefully kindred spirits, with this warnin' tah be realistic. Dis is a farm, here.... as much as ma hand hurts.

LAURA: So, (be) self. We're not vegetarians ourselves. Thank you for the invocation to make dear our labors, caring, and dreams. This is probably what my husband sensed, with anxieties to stay polite.... at a new homesteading. I'll not be disenchanted for him, by the realities our hands have taken up. We'll make a decent farm here.... in some fashion.

MORAINE: An' I'll be a devil, tah 'ave been honest with yer. But, ma complaints 'ave been swayed an' assuaged. It is indecent fer a real man tah lie tah dreamers.

PHYSICIAN (*starting to put his materials into his traveling bag, as* **Moraine** *stands*): And animals deceive as often their intentions from predators.

MORAINE: That's true. They dart about, tah throw the chaser off, when there's enough foliage an' ground cover tah confuse of directions taken.

PHYSICIAN: I would suggest you go home and rest thoroughly for the coming day, at least, before attempting any more chores.

MORAINE: Aye. No more chores this sultry night tah see. An' day brings tah head more recognition of a foolishness.... It's a bit of a walk.

PHYSICIAN: I'm sure Mr. Ponaple will drive you out. He intimated that much to me. I'll walk you over to the garage, and we can *wait* for him.

MORAINE: Yers all very generous, lady, tah have caretakin' fer the likes of me. An' I insist tah personalli discuss with the doc a bill fer 'is services.

PHYSICIAN: And to get you out of this house and into some refreshing night's air. He went to the barn, you said, Mrs. Ponaple?

LAURA: Yes, to look after the cow that returned there.

PHYSICIAN: Then he should easily see us standing by his car, the garage slightly lit up by your house lights. It probably feels novel to you, not to need a closed one, in the country.

LAURA: A quaint reminiscence. doctor. I used to play in such a space.

PHYSICIAN: Yes. Well, leave us to go there. (*ushering* **Moraine**) Good evening.

MORAINE (*while exiting with the* **physician**): Thank you, Mrs. Ponaple.

LAURA: It's all of our neighborliness.... sirs.... And thank you for coming.

Scene III — *The barn, lantern lit.* **Sachel** *is half-kneeled (single knee) to the pail of oats, as* **Ponaple** *stands only slightly away.*

SACHEL: She must have been eating out of here, Pa. There's much less of them.

PONAPLE: Those oats? You didn't find her there? You didn't strike her?

SACHEL: Nah. I wouldn't do that! She wasn't around when we came out.

PONAPLE: Then what was all of that mooing about? Beryl said you were angry at her.

SACHEL: Not at her, Pa. Maybe at the condition we've placed her in, the frustrations she must work through, roaming about so much— to search for some kind of solace to her being. People say misery's caused by misery.

PONAPLE: Where did you read that? One man's fortune can be another's disaster.

SACHEL: Sarah went out to look for her again.

PONAPLE: Without a light?

SACHEL: She's caused some trouble tonight, Pa. Has led to the injury of a person, even. We have to do something about it, but I'm not sure what. We haven't handled them well. It's too much of a responsibility for— me. I'm not suited for it. But I've always thought to treat them kindly, within reason of their labors for us. I mean, they exist for us. Is that right? We wouldn't have them around otherwise.... I wanted to punish her, in some way, for being a nuisance, causing us this night's trouble. But.... what is the nuisance? We brought them here, or had them brought here. Dropped them right into an unfamiliar place, with strangers for handling. And one of them seems to want to rebel, as we're learning what to do.... by them and everything around here. So is the orneriness really so misplaced?

PONAPLE: Has she actually been mean to you? Everything is it's own being. In any case, we don't— punish animals. We might train and steer them, and reward a few for tricks performed. But we don't punish them.

SACHEL: What about hunting, then?

PONAPLE: ....I say, *we* don't punish them.... or we shouldn't want to harbor up such feelings. It's not that you're unsuitable. Any country knave can learn how to do this— farming properly. Even the clever Derrick may adapt to it without much difficulty of.... conscience. I desperately hope so. It's that we've taken on the unforeseen to care about. And that is: our objectives for the task, of this animal husbandry. We were thinking of these live possessions too much as individuals to be liked and to like of us, when all the time they've simply been bewildered of.... this new situation that compounds our interests. I was hoping the honest naiveté of you children would be able to handle naturally any of these details of heart that I couldn't think of, or not entirely review for myself. It's not a cruelty of youth to want to get something done straight away, and without consideration of a hurting caused. One simply trusts that they do things reasonably. But we are touched by the cows' dilemma of passivity under our guidance, and are disturbed to find an independence of spirit among our own charges. And now we question if we should punish that, or our own weak management to allow its furtherance or even prospering. Because we've come to know it's not due them, this kindling to possess, for our purposes. And whether you may blame yourself, for not countering this tendency, capturing it and tossing it out of the body by your rearing of their acceptance to— our farm, is not a question that can be faithfully answered given our desires to want them to feel comforted and settled.... with the pragmatism we've tried to design for ourselves in this place, our plans that we have forced them to be involved of. I'm not sure if you could have prevented her wandering, and the accidents that may have resulted from it. I'm not even sure if this night's incidents are.... absolutely confirmable. But our deficiencies of thought will have to be worked through. And perhaps other heads are necessary, to initiate a legitimacy to this farm, this retreat we've encompassed with our soul-searching for the compassions towards.... essentially, objects of agriculture, animate to own and confused of their resentments to our hands.

SACHEL: I've treated them very kindly, Pa. Even gingerly. I haven't beat any of them in the least way. But I sense I'm out of my place to encourage them of any activity.... desired by us. Even when milking, it seems their patience is crudely artificial, automatous to please an activity more than myself, and to shade an uneasiness to my presence. The fact is they can't seem to resist their— usage, no matter what or who we are. But that can not diminish their discerning of our ineptness to have them here. And my feeling about this is unusual and disturbing. (*standing*) How can I possibly do much better with my caring when I'm not seriously committed to this retreat for in any way to be occupational? Yet I want not to fail those I'm forced to be responsible for.

PONAPLE: Forced? This is all.... only for your curiosities to foster. And only ambitions may force you to do anything. But I am forced, to realize pledges and errors, and lifestyles and attractions that are dream-swept, starry-eyed and night filled. And there are no failures associated with the impossible, only the impractical to do or try. In any attempt we must strive to make ourselves satisfied. Justifications will come or be lost with the results. But it's always more important to do than what is done. And now I'm sure we may have taken on more than our caliber to rely upon. So obviously our caliber must rise, which is generally true for anything new, encountered to be worked through. And we must face up to the fact that some criticisms of harshness can not always abate the cruelties of our treatments throughout life. We're in a new domain here, of rural perseverance to custom and the practical. We can not stay accused of having wandering and poorly tended to cattle. I've bought only three head, and even that may be too great a number. But I was wildly enthusiastic of a novel adventure, for our family. The mistake's been made, and we will respond to it as promptly as we are capable.

SACHEL: Fencing? Fixing up the stall? I've been working on that, or thinking about it....

PONAPLE: Sometimes aspirations must dwindle in number to make our hopes more manageable. A bad breed, or a bad in breed, can be done away with— humanely.

SACHEL: ....That seems very sad, to get rid of the lone miss, and all because of our poor judgment. She.... represents ourselves, in a way, out here on this estate, teasing the temptations of our accep-

tance to abide with the strictures of such a life, and only temporarily searching through our qualms with the country.

PONAPLE: But.... it's not cruel. It's not striking the animal with a blunt and painful stick. It's merely coming to terms with what we might achieve with reason sans delusions.... and derision. We accommodate ourselves, to what is purposeful. We're not skillful enough to handle the unusual well, the longing to find more comfort than what we can provide, our lack of knowledge making matters miserly. And I fear most of all to convert my children to the bitterness of evil contemplations that could ever instigate cruel and brutal treatment of these cows, stubborn to be themselves with the simplicities of a domesticated nature. For if this is how one must behave.... then it will remain that way, on our farm and through these fields and within this shared region. The population might commend us, to have this painful foresight. Our immediate neighbor may, to lend us more patience for our learning. And in many aspects the.... coincidences of this night's events may be excused and not amplified of comment if we show ourselves to take such action, to have at least that subtle wisdom of eliminating oddities that suffer of growing notoriety.

SACHEL: It sounds like a sacrifice, Pa.... and a heinous one.

PONAPLE: If we plan to stay, and of this estate to remain in our name, then such decisions may be made— to pacify a wariness to contain our abiding and adjusting to the deserved tranquilities of country residence. And, to be more plain about it, if a trouble is found concerning our possessions and property.... we *will* work to eliminate the cause, despite some pangs of remorse to endure for a notable spirit (to lose). Defiance— here.... is not of our happenstance to make nor bring.... with preference for our natures to be accepted and respectful of all (of) these inhabitants. We'll be conciliatory to the local rural demeanors. And they expect some typical pervasiveness to livestock manners and accounting.

SACHEL: Then what's wrong with being singular about our plains, is that for being deeded here a mainstream of existence so demanded by this company of judges and discerners to our presence.

PONAPLE: More the practice do they worry about, with its clumsy start and awkward gaits to startle for attention. But that's always the right of a beginning, the height reached for an initiative to seem frighteningly steep of falling. Mastery must follow, and I'll study this weed more of a problem than a purchase.

SACHEL: I guess one generally surveys a region from heights of aspect.

PONAPLE: Or artificial hills, mounds of creation and construction. That's provided of our industrious tendencies to take care of our concerns thoroughly and without much distraction from wishful thinking, now that we have willing ownership of some responsibilities and causes of production.

SACHEL: More volition's made by parenting of the hooves—

SARAH (*coming up to them*): Hey! Who's standing by our car?

PONAPLE (*turning to look*): It's the physician, with Mr. Moraine. The medical attendance must be finished. I wonder why the doctor's still around, then. Must want to consult with me about the details. I hope the wounding's not outrageously serious.

SACHEL: Did you find her?

SARAH: Yeah. She's with the other two, like I suspected. I intuited her feelings. She must have disturbed the serenity of the others, initially, to cause a commotion. But they're all peaceful now, and probably well accept each other as a group.

PONAPLE: You think so?

SARAH: Yes, Dad. They only have each other, you know, to be familiar with. And that's how they'll settle down, into this new, brazen territory, to accustom themselves to our farm.

PONAPLE: Perhaps.

SARAH: It's often difficult in the beginning, to get used to new surroundings, especially when you've had no say of being placed there.

PONAPLE: Well, guidance mandates action.

SARAH: But that uneasiness is not so important, if a true home's been found. If I cared so much for my upsetting, from situations unexpectedly placed in, I would have never developed out of childhood.

PONAPLE: ....I suppose so, Sarah.

SARAH: And I felt I was finally accepted, by all three of them, tonight. My concerns for their well being were felt, and my compassion welcomed. If they can receive such hopes, then they're all calmed down enough to find some pleasantries at this locale. It's just a matter of maturing to the elements of one's condition.

SACHEL: Adjusting yourself to the weather by taking the appropriate steps to be comfortable.

SARAH: And being reasonable about it. They're lying out there, at that favored spot.

SACHEL: That's become a comfort zone, then.

SARAH: I think the grass is not so prickly there.

SACHEL: Yeah, it's a luscious growth. Might have to be mowed.

SARAH: Or mooed.

PONAPLE: The land management will have to be studied.... children. A lot of matters will have to be more carefully reviewed. But if Mr. Moraine's— tending is over. I should be driving him out to his home. He certainly can't drive himself, with that sore hand.

SACHEL: Should I come with you, Pa?

PONAPLE: You don't have to apologize to him any more, Sachel. The night has had its play on our shoulders, and has allowed what's meant to be to occur. And you are not so responsible for his injuries, not as much as his own weapon. But we can take no real umbrage from these events, having been shaded, perhaps, from more doubts than could have been admitted, or permitted of our

candors, to be receptive of this region and this country life, with the ease of our desires to rest here and establish ourselves.... wholesomely. And for that pleasure, of a gift of ability to be bred with, we are all cattle to be led, towards a sufficiency to serve the earth with as much preciousness of goal as upon our own heads donning to be counted and counted on. Since all that happens must be spared to occur, its impossibilities retracted by the divinity of our purposeful leisures. And I think I'll have more bought, if we can work this through of being kind to nature's worthy thoughts and lofty heights of managing.... our own much too desultory deeds and endeavors.

SARAH (*picking up the pail*): I think I'll bring them the rest of these oats.

PONAPLE (*going off to the garage*): You children heed yourselves.... of this glorious night for your statures to behold— and accept.... your instinctual aid to provide.... I'm much relieved, by your youthful optimism.... in this world.

SACHEL: Now don't go perturbing their rest too much, sis. They can come here for these oats.

SARAH: But I want them to know that I'm coming to *them*, Sachel. And there's an implied friendship and loyalty to that. (*going off*) So make it that I can sense their desire for some.

SACHEL: ....And the cheese to be made will be particularly sweetened by this milk. (*extinguishes the lantern, after a pause*)

Scene IV — *The front office of the abattoir. Though it is still night, the office light is out and illumination comes from a hallway leading into the interior.* **Weiler** *is standing by a desk as* **Terance** *enters from the hallway.*

TERANCE: This is very eery, BobCat. He's no place around.

WEILER (*closing a desk drawer*): Definitely comes from here.

TERANCE: But that freezer was left wide open, with the rifle, that money, and a half-drunk bottle of zinfandel wine stashed in it.

WEILER: He must have been bombed out on the stuff, and mistaken the freezer for some kind of cabinet. A thoroughly irresponsible sort, Terance. Can you imagine the mess of this operation he could have made?! Blood splattering everywhere and about. The cleaving sloppily done and nearly impossible to clean up after, let alone for one man to do both tasks.

TERANCE: Why did he bring the spirits anyway? I hardly thought he'd have time to—

WEILER: His heart wasn't in the job, or his mind battled him against it, perhaps with such aggravation as to turn his madness fully loose. And it was probably too much work in a night for one person to handle, anyway.

TERANCE: But he's an expert at it. If any one could do it, he would be the one to.

WEILER: Impossible to rely on. There are sensitivities involved, even scents.... and recollections of what he's been meant to do. Tansel would be dumbfounded to face his reality of carving up carcasses. That's why he ran.

TERANCE: But why leave the money?

WEILER: In an intoxicated stupor? He waited for the cow, drinking. And when it didn't come on time he realized this couldn't be done. Our mission failed, maybe even failed him. And so he drunk some more, and finally threw everything in the freezer. As if to make some concluding statement about the plan? I doubt it. The stuff belonged there, freezing, stalling, like his head, the operation delayed. He probably expects to come back for it— for everything, except he dutifully left his keys.... our keys, in this drawer, as told to do.... probably early on, because I'd certainly go for the money first.... So he's on a silent rave of rant, Terance, out of his head somewhere.... or sleeping off the effects of this initiative. Dat does do it for us. We can't rely on such nerves.... Nor on this planning. It has to be revised, refined—

TERANCE: But he left the back door open.

WEILER: Along with the lights on, there.... to shock us of some courage for his waiting on the cow. But he's not around, Terance.

And neither should we be. And I'll tell ya this: I'm not goin' out to Bouwder's to search for that stuff left by Pitch. The morning can have it, all to itself. The idea of everything about this is becoming unclean an' hands off; for it's hard to clean up after a carelessness of this proportion, and we should let things settle on their own until we really understand what happened— an' why. Mort will just have to handle 'is excuses.... to the Ponaples.

TERANCE: Well that seems so, BobCat. I can't force things so unexpectedly as this to circumvent. And I'll have a strict talk with Tansel, when I find him. He's possibly gone home. But help me close down the place and turn off (to disconnect and drain out) the generator properly. A few of us worked it on before, but I was hoping Tansel would get it done more quickly, were he here to live through our chores.

WEILER: Dat we'll do, sir. An' leave this place shut down and out of our contemplations for awhile, till we better seize the idea of its use. (*as they exit to the hallway*) Nothing much 'as been disturbed. An' these surrounding can stay as pristine as chosen.... for such an environment.

)()()()()(

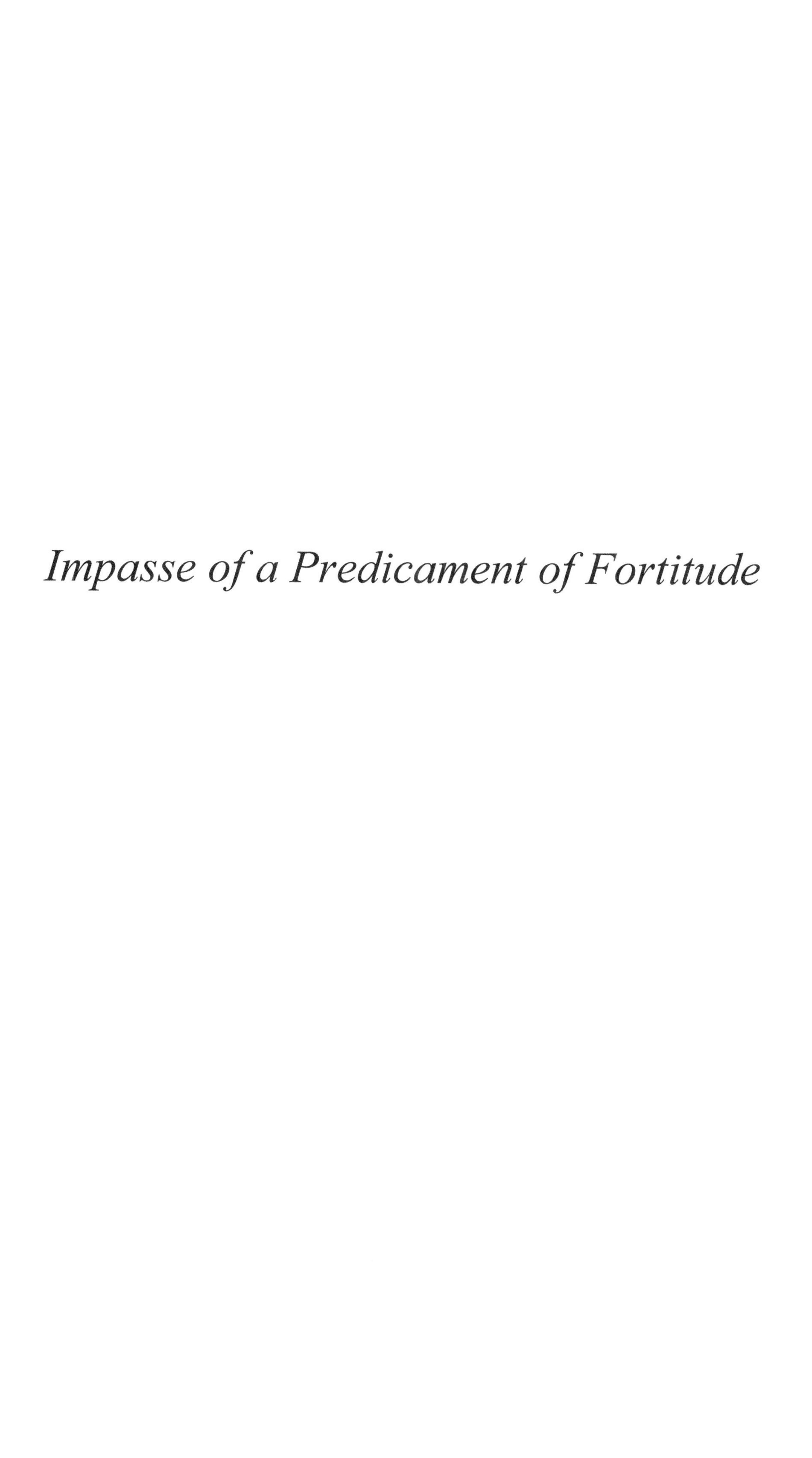

Impasse of a Predicament of Fortitude

Impasse of a Predicament of Fortitude
(*An Impasse of Fortitude*)

Characters :
Lord Balcome
Sheriff Gestromy
Makethrift
Linda
Brontrol
Craie
Herkshead
Bill
Curt
a Guard
a Man
Hoe
two Protesters
a Physician
a Sentry
a Courier
various Soldiers and sentries

Locale — A castle and its village

Act I

Scene I — *The turret of a defensive castle.* **Balcome** *is looking out of a crude, stone-walled hole, serving as a window, onto an expanse of hilly land.* **Gestromy** *is with him. It is midday and shadowy inside the relatively small enclosure.*

BALCONE: I see them occasionally, quite raw, popping along the hill-smocked terrain, approaching gradually, but without a decent caution against my outbursts of blasts and smoky volley, as were to warn for this visiting that I'll not be delayed of a defensive posture.

GESTROMY: We've brought no such means to view them with, lord.

BALCOME: Nor to greet. But I'd not be reluctant to such aim, were they more offensive of mere presence, Gestromy.

GESTROMY: Not much else. They prepare of a venture—

BALCOME: To see my face— What! Venturing here?!— and then my back— darted with black pins and silvery chevrons of slivered rancor. This be my station of assistance to their proposals, and my rank of hounding.

GESTROMY: At least it's day, and they are overt, to expect a meeting with you, a discussion of some conference.

BALCOME: But are they armed to have our heads?! Sharp as a shark tip are their teeth for knives, with where to grip but at our throats.

GESTROMY: You are the lord of this manor, and their contentions. I'm merely a priest for your defense and taxes, collecting among and amidst the quarrelsome and resistant, indignant to appease your wane, as our sovereign king appears to have fallen not only of disrepute but of a circumstantial reign to lose. Or at least this is openly, and somewhat combatively— in spots— disputed. And you are his servant rewarded—

BALCOME: They want my castle, my home! for their own purposes to despoil, and weave of me a mockery of demonstration to their claims. They would moisten me to soften— with their drinks, as I were a sweet cake! to devour. Gain they several providences of threat and sweating (*sharply turning to face* **Gestromy**), about my brow for least these coils of bead and salty curse. Whereat this usefulness of sweat is cursed to throw upon them my reaction feverishly, and with a scared stance, betraying my weak means of nerve and men as I plead to heaven for a defiance.

GESTROMY: They're not against your person, but your property as known: Balcome Castle and Estate, of the king's royal dominion placed and seated for control, or administration of this mottled region, this suspicious area of provincial coterie defined.... by a questionable pronouncement of law. (It's) Questionable because his majesty can not defend your position, and yet expects to be obeyed through your tenacity alone made sacred by his solemn touch of hand onto your shoulder, and then chin, and then the shank of your sword—

BALCOME: I was brave— for him!

GESTROMY: —during your great ceremony, at court, where the courting was mutual. But why take this privilege if you didn't think you could maintain your rights to it?

BALCOME: Who could know? when feverish of glory.

GESTROMY: Glory too? It could be fear. I've rode amongst them to reach you, and without any accosting. They do not mean to punish you physically, but to show some force to have you sign agreements of understanding and leverage, as your peerage— to them— remains.... habitable of this home. And I should not argue their case, but rather seek your redress or manner of regression. I can not collect for you their sums, as a sheriff knighted to your service—

BALCOME: Retrograde! this fury peaks!

GESTROMY: —if your own fighters are reluctant to be blunt against the meaning of these people.

BALCOME: They're only locals to me, showing their scarcity by convergence to this spot. Strip me naked, Gestromy! to feed them of me, my worth rattled of supposed subservience to their contused subversion. They are mean to be so stately about their desire to want to avoid open confrontation, though only of their ends they provide. And this disturbs my soldiers. For while this is a sufficient fortress, with men of great talent to defend it, the minions are of a finer and more persistent number to combat. And so they hop from hill to hill to mound, to show me their.... plasticity of coverage. They do not make the lunging charges to draw a fire, but present their silent war cries with sight of growing devilishness and the bemoaning of their faces, the disdain of my specious triumph. How can I augur contrarily to such a collection of bees? Theirs is a wisdom tainted of self-assuredness, the humor of a commonality, and a community of forceful complacence to their terms. This is a haughty severity of lowly might, a mean-spirited gaiety of reserve, to show my powers facially perturbed and moved wrongly. Might I tempt a strike, if they crowd my movement more. And what do you suppose for them? These relatively few around (me) may be easily massacred. Yet come again the river with its muddy lugs.

GESTROMY: They do not attack your wall and protections. We're all alert to their steadied approach. But take in a commission of them for discussions. They could expect this courtesy of overture, to treat them passably bold and current.

BALCOME: Have this rabble representatives at the ga(te)it awaiting entrance? I could capture overnighters to mend.

GESTROMY: They must be assembling such a company of advocates, since you have not prevented their leisurely progression towards your fulcrum.

BALCOME: And for a forum to fulfill, their dredging experimentation, to layer of my crossing of into an effeteness of strengths, with patience towards their spoliative suffering? I have been cordial to them, and justly fair, as could a king's demesne be granted me to promise of him rule and conditions of right acceptance. (*turning towards the window again*) Yet they gather with faithless alarm.

GESTROMY: It would be impossible to seize you, Balcome. Rumors simply arrive to their hearing, that the king ceases to be effectively so. And now they pray to find adjustments—

BALCOME: They pry to find adulterations.

GESTROMY: —to their regional constitutions and obeisances.

BALCOME: I can not subsist on coached contempts and callous comities. If bane they be for me to delight of, then I'll have war for the king's effects around here. That is what is meant for me to pursue, under caustic considerations as these. For it would seem a nation is being tried of its substantiability to stay authoritatively fit and felicitous of ruling. Though I be loath, to charge them into cruelties. There's some protest of the manner and the means of a royal subjugation. But I've earned this right of dominion— There goes another few, pouncing prance! like the hare at noon, in the bare heat of the sun. I warrant.... we have a good draft, up here. Yet down there stirs more the currents of frigidity.... to my cumbersome placement, establishing over their heads this purview of regret and wretchedness. And the ground would be as sand at a beachy cove, for their sweep of wave to disown my provenance of powers.

GESTROMY: Subtle are these crabs, then, to present themselves more properly to enforce a consultation, lord, rather than with a heinous bantering of pejorative Angst and gloomy premonition for your stance within their provincial airs. (*as **Balcome** turns to face him*) This is a show.... of some determination to be understood, and to be respected as a considerable torrent of concerted pride to their landed interests. So they most obviously seek a clarification, from you, of your position towards them. In brief, that you do not earn too much through them is their desire, the labors now of questionable ordination, and their privileges of ownership amplified among their meager, miserable possessions— compared to this overarching splendor. Have a good chat with their high members, to reassure the benefits of your occurrence, that you may be a leader of their stability, solidity and protection instead of as a reminder of their lowly subscription to mostly disinterested might in review of provinces claimed and commanded.

BALCOME: I'd rather capture a few, to tame angrily. There is no contra-prince to battle for. The king does not accede himself to others, and is not so sorely challenged as to stumble, for my wealth and interests to lose. He descends into the mystical blights of uncertainty of political support, whatever magic's made at court to turn opinions and ideas through themselves through, for a modifying of concepts and a mutation of forms and goals that lead to the thought of weaknesses— pronounced on the lips of pundits and poodles. But this does serve me no sparing, to have my own foundations held against disruption and dissolution. What cries now is for some disciplinary action to demonstrate my fierceness to remain a governing lord over (*pointing towards the window opening*) them that hop over heated coals to provide some perversity of presumption for my downfall and collapse, during a period of fallow faiths and feigned judgments. That I may be toppled— overthrown?!... by the laxities and looseness to think so?! can not be tolerated without errors spoken and flaming arrows felt. But a few such souls, wooed of their woes, can help to prevent a ghastly production of coercive physicalities spread throughout this region. For if I must, I'll rout more of an offense to stem these virulent growths their gains of organization than ever to defend these walls merely to stand terrified of community monsters specious of their common thoughts of Providence, and sharing this without a care of my provisioning to keep intact— as could I be cast out, away and aside.... coldly, with the casualness of ungodly decrees and deprecations.

GESTROMY: It's certainly a push of force they wish to show— to be capable of— to raise the hairs of a nape's scruff, and have demonstrative their aversion to your loyalties.... to a fasting sovereign, if not one wasting away from their views. But a better tactic now is to appease them, and placate some insistences to ease their worries and dreads. They're much to stand off and resist, but not so many as to dine and be inclusive with, conciliatory of, and compromising to regard their strengths as not only plausible.... but usable and complementary about your aims. As I said before, they're not soundly for our persons to destroy. And their threat remains a picture seen through nervousness, rational with restraint of provocation. Their movements only prove themselves of earnest factioning and determination to be taken seriously, or seen as serious—

BALCOME: To take over my castle, as some community event to promote, and throw me to the winds of a bare destitution, my fortune stolen and the future made gargantuanly void and bleak. Would you have me appear unarmed, to a dare against sabers?! Appease them as they lance and chop?! I have accumulated all of myself to this one bastion of Balcome, and must assume a hasty stubbornness and be as obdurate as diamond, not to be cleaved by anything less. And we must show ourselves piqued of their prickly cares, though generous enough for their withdrawal to— allow, away from this contemptuous posturing to picture me unnerved to their suggested possibilities. For what are their insistences, but to decline my value, sensing a lack of reinforcing thrust from our majesty.

GESTROMY: There evolves here a change of stature, which may be observed without any clashes of grief. Your soldiers are on the ready, prepared for action. But to attack what? A subversion against whom? Themselves for this end of a liveliness? Take in some committee from the revolt, and address them calmly yet passionately to be dear and subserving to reason. The talent out there is neither fearsome nor of fear. And we may carry stones from an assault of ideas to argue with better than from a well of disasters to sip from and poison prospects more, what may be heaved and hurled to smash us, dispelling all hope within the ambiguities of rage and intolerance to submission or capitulation to luck. Yet control we more the method with our largesses to listen of complaints and beliefs, without a stony quarrel raged.

BALCOME: Those complaints are inflammatory, the beliefs malicious towards my ruling, that they might form a diet of arbitrary governance to conspire of, and without my lean to favor being worthless, without my tilt insubstantial to behold, as could I not exist to be consulted or induced—

GESTROMY: But this is the very of which they suggest, wherewith they'd storm a mountain for your adherence. Pardon yourself, then, to be delighted with some respect. They assume of you the best of considerations and thoughtfulness.

BALCOME: They are armed, out there. This is an insurgency against my granted privileges— over them, all that I have struggled to obtain and have achieved. Though they be not particular to me, but what is left by chance to own of administrative rights, where the king has provided thus, and who.... are capital to be lorded over. What else may I become from them but to be proved, that I do rule here with expedience, and a patient purpose to improve our lives together. So of wit, I'll contain their letters, I'll contain their words— but not their rallying for my defeat, not this showy protest of collective stealth and tensive approach. And definitely not without a curtly pointed retaliation will I make, to disavow a lingering of this state. So have them bring me argumentative license, but not this threat of weaponry— I have my own to counter with and decimate!... So be the curse of curs to bark more than they bite of gods!

GESTROMY: So you will hear their speeches?

BALCOME: ....Their analytical proposals jabbed and bent on firmer iron, this is their due of ears to concede. And how is this to your endurance for me?

GESTROMY: But what I have promised some, of your fine reasoning, you want no battles here to soil your very station red of fecund and boiling tempers. And in truth, deaths and wounds among us are more definite than among them. For we are limited, while they are by comparison innumerable. So let us not promote an assaulting.

BALCOME: Brave fellow.

GESTROMY: Such wages may be won and lost without much gain of doing anything sufficiently, for changes that seem inevitable of peering eyes and minds.

BALCOME: I say that you (to) come here to station your future with us supports some kind of commendation, to be recognized of sides and principles. And what you choose....

GESTROMY: I advocate for you.

BALCOME: is to stay with me till I am slain.

GESTROMY: Will not the castle keep, till otherwise of valors fortified? You are for these remains established, and bound to ancient gratitudes of occupancy. But many out there know much more of this stronghold than even we through our examinations

and our timidity to walk of precious halls and ways. Then be much caution made of privilege. This is their indigenousness. And we can only learn more, when fighting less, of what sustains these populations. They are quite accustomed of paying fees, I can report, and with no trouble for me to collect through manners courteous and polite wherever I've gone of town offices and edifices of assessment and centers of gathering. It's clearly to all benefits, to have a strong defender in this region, and to overlay their courts with governable statements of law and advice. And peaceful seems our area to this acceptance. Yet one slight to the notion of a princely authority brings down all suddenly an avalanche of rock and revolution, as if a heavy tension slept, and rested only for a tease of kick. There's more to here than angers and pent up disgusts, but expectations of revolt too easily devised. We must wade through these strong currents to find the source of this distemper, and dam there, as to how we may a tranquility to assemble or associate. This is learning how as much as why, with volatile vociferousness safer to be heard than air-coursing projectiles slamming into stone of walls or brave flesh, and with waste of the anxieties we face to prevent more sacrilegious toil against our sovereign and his appointed deities of representation. We must characterize these indecencies as pertinent and penetrant of people, and not simply of whim to a permissible season of reaction. It's true that one can not bargain with barbarians. But they are not so, have not presented themselves to me with any likelihood of crude or incapable thought. And with an intelligence they gather to test.... or strengthen. As to what or who depends on you, your observations and high-minded assessments to think them not an enemy but voracious. Then tone down your indignation a bit, to have them not devouring your cauter-ant reprehension and repugnance at this try for your devout attention. Do not speak to them angrily, or let them see your eyes blazing hatreds and a deep seated need for reprisals. They have *earned* their arrogance, by the nature of these circumstances, by the reputed fall that could be a flowering of many to cause. Rise above wisdom, and into the realm of reality to relate of them, their concerns and chastisements. They are not insensate trog'(lodyte)s to cage and squeeze, as a supplement to your enrichment for power. The king gives you not them and theirs, but that which they would need a lord for, a justly approach of handling the complexities of their society to administrate, and as circumspect as their approach to meet with you and feel of you to draw out either some prudence or foul pungency.

BALCOME: Sheriff! you advocate for *them*. I must with all means and meanness prove myself, my prerogatives to have them follow without attacking or bitter protestations. When they come armed like this, it's not for the rhetorics alone to fling, but to have me cower to be banished! of all essentials of control over their lives, and to make at best of me a figurehead for the region. Then what for tax collection, villein? What of sustenance for my rule and pay for my men and servants and a wealth to be operative and active within this dominion I have won, and the civic improvements I would choose to enhance, and village deteriorations as from a rock's detritus I must serve to prevent, and a societal discourse to maintain with relevancy for people to support their best interests even when contrary to their simple or child-like longings as inferior thought and immature persuasion? What of all this guardianship that I usurp from crass craftsman of crude causalities, in order to create a more competent and solid node of human endeavoring, achievement and advancement, with focus of this castle home whereat are orders made for progress and servile production to increase, to prepare for the winters coming, the famines that arrive by nature's might despite man's industrious combat? Where is the mind bestowed to be careful of this parenting of populations, but within my own, as like a shrine for these responsibilities earned, of nerve and courage to perform these doubtful duties under the strain of tendentious ungratefulness that my commands may be the lesser thought of from a loss of king?!

GESTROMY: These are complaints, my lord, not judgments fashionable about their ganging up to be heard.

BALCOME: It's more than sound that is the damaging of threat —

GESTROMY: Appease them now, to make your judgments known, your wisdom and rationality. For what is deserved is certainly yours by decree, but such divinations may be falsified by misjudgments, or soiled by seditious envy, or boiled as an egg to poach instead of hatch— the hatchling want for birth of aristocratic goals— by the searing heat of challenge and resistance to a sovereignty in question, unless some serenity can be transpired through an adept action by your handling of this situation with caution and restraint, suspicion and submission to the irritating details to master and overcome. They already have their reasons, well in abundance. But you must not give them further *justification* to incite a hectic confrontation, and a furious feasting on your presumptive authorities to tear apart and swallow as their own. They feel this may be done. Show them order is more delicate and intricate for polishing, when owned, that intelligences are needed to prevent a disastrous crash of crystal during this rough trend towards turmoil and the titillation of their temptations to disturb you, or have you moved ungainly into steps of awkward, illogical, unprofitable resolve of menace and malevolence.

BALCOME: Malevolence?! Is this to crush cursers and rotters and rogues and rebels and potential thieves and topplers of what's good that's been brought upon their unworthy backs?! If ever there was as kind a lording to be proposed for their servitude, it is mine. And everything was peaceful, until these royal rumors arrived to rend their hearts and render me a target for extremists to promote my downfall, to wish to see the villain in me through disdainful caricature and words of abuse, mendacious and insulting, to fire up the incense-ment and insensitivities of these hoppers, these proud cads, these shifters and shuttlers of atonement and pledges. I can not give up my rights to them. If one stone is thrown towards these walls—

GASTROMY: Restraint, lord! Restraint! They may not know the malice made, upon this rush to greet you, by means with which they sound themselves of vicious temper and fearful ferocity in their efforts to arouse your wrath.

BALCOME: Was only a month ago they prayed to me. And now.... base candor, as the cauldron's soup must thicken of its simmering heat, their contentious quarrel with me brewing blackened, as for night's a candle placed, to steer the eye of day— Let them arrive! this darkening swarm. And I will make an unguent out of their fervor's oils. So disobeying are they pleased to be, that in their faces seen of me a bleach for these behaviors made, and a blanching of this harassment.

GESTROMY: You'd frighten them? With truths, my lord! and their subtleties to contain all conspiracies.

BALCOME: I would, Gestromy. Debts are made today (*looking out towards the opening*), with these approaches. But if they have a few of parley, bring them to me— in my proper office, and not this turret of tensions and tales.... (*as **Gestromy** is climbing out, down into the castle interior*) And if not! I'll flood this plain, these plains of the so called humanity that would deceive my caring temperament. It's terribly spotted of insult, down there. And to witness this disturbs me more than threatens.... because it makes for me a great labor, of laborious irate scolding and re-sculpturing of my mandate's form of cloud upon these heights to peer from and conclude.... of raining thunderbolts of storm during a fretful scramble for cover, or (of) soothing shade from an omnipresent heat to watch of men their habitudes of activity and pastime. Yet does one care to see, it's always lit them, to observe. Behave! for me, more. This station is for meaning of my malice if not rightfully held its priority of existence for my suite of attainments and trappings and instruction, and outright management. And if all of that is so affronted, by what I see— too little to be afeared of, but too much of to be belittled— then beware. I am just, but am not meant for justice spending, since that disposition, or discomposure of the wrong, is only earned of reception. And the sky is not meant to be fair, but only as fine as it would like of earthly needs and perambulations. So teem! But behave! I haven't much more sight to decide upon, than from what you show me. And this insincerity does frighten the security of my hopes. These confines make abrupt the sacrilege felt. More expansive may be their confessions, from which my manner can be more expressive and stated thoroughly, that certainly within my presence there'll be no putting up with foolish behaviors, rank tendencies, reek revolutions, and disrespectful temerities to wrangle over and disabuse my rage with a deserved, defined, designed for and totally, thoroughly delivered fury. Or such for windy shadows in this lift of preservation, observation and forewarning, strikes the calamity with counts.... of vermiculite epoch, this darned approach to me an' mine, and my foundations fearful.... of a crusty settling, in this riverbed of mounds. And there one seems to leap over backwards, like a gymnastic ascarid— Well.... such is their dance, so near the heat an' burn of my summation to their conduct, that's to be told of and done away with. Parch!... of this insidiously envious threat. Dry up and away, from my eyes' notion (*turning away from the opening*) of calculated bruise to rue. I'd wish.... for more of a courtyard than this peasantry could handle, sharing the shimmering sun.

Scene II — *The workshop of a bakery. **Linda** is kneading some dough as **Makethrift** enters with some wood for the stone stove.*

MAKETHRIFT: Light as airy, Linda. These logs are heavy, as a jot, to bestow upon the cave's release of temperature, some matted pies to home and form, as could a dove fly and rise within the fluff of feathers.

LINDA: They do in that way wrinkle, Master Makethrift, of some heavenly grace to sort through, as within air that's heated. Were not a stove as night?

MAKETHRIFT (*putting the wood underneath the stove, in the heater*): Wherewith the fire makes. It is open. But dare not be the flames, or else my works could be destroyed. Yet light sees the day, but not the reddening that tames.

LINDA: I mean, it should be dark, within the stove.

MAKETHRIFT: To where it peers out, an' we in? So what cares, of what is seen by night? We rightly judge the formations within her. And what's hot be more a chemistry than an atmosphere.

LINDA: It is a simile for all life, Master Makethrift.

MAKETHRIFT: What does compare?

LINDA: That out of caverns are produced these foods, or from a seed does pop a sprout, more to be greened than shaded by the night.

MAKETHRIFT (*lighting the wood with sparking flints and rocks*): That's somewhat womanly to think. I say, though eye it has, the night needs (one) not. And all is within day, even ourselves at sleep. For I am told, that at our dusk a dawn appears elsewhere, which means.... we only retire from the light, and there is no real extinguishment of purpose.

LINDA: Half hard again?

MAKETHRIFT: The flints are moist. But— yea! The sun must always shine. Now for this night to see.... what we are doing, as could some radiance suppose, with feeding and the frenzy of ambition she awaits. And dare not enter vermin. (*looking into the stove*) They'd be burned. But the warming's slow enough to give 'em caution. And for the insects, charcoal-ation and crackle. So much for thee. Then what's to be, there's more done in a night of service than its day seeing. An' if she could think (*getting up to come over to Linda*), what good is day without a labor made? Night's more laborious. Light as air, these cakes.

LINDA: Does thin the dough tremendously with air.

MAKETHRIFT: An' more are sold, though the sauces laid upon them make the meal.... All dark recesses and cavities are nights in labor, then. They need not eyes, but by their very nature must have. And I doubt if they could wish, for more subtler uses of their openings. It's to receive and retrieve from, products an' productions.

LINDA: And we are this way built?

MAKETHRIFT: Is not the stomach perennially shaded, and most of the intestinal track? Yet the grinding, kneading, and peristalsis is just as continual. That, with, defines the purpose of a place, for conception to create and transform. And that's the only reason for a night.

LINDA: I thought it was to rest in, Master.

MAKETHRIFT: To do the bidding of a duty, it must make its hectic orders within the restfulness of suitable actions.

LINDA: As I do find a tranquility of repetitive motions.

MAKETHRIFT: And with lightness as the aim.... A broad cloth should be made for dining. I could once spread one single pound across the entire (surface of the) table top. It depends on this kneading chore. But patience wounds me for my experimentation, that and the composition of the lards and fats. Some local tastes define them more than my susceptibilities of supply. It's a saltiness, the kind of salt that's particular of taste either roasted on and through, or dried to a sharpness of granular intention. These.... solid oils help define the thing, or promote with their varied spiriting.

LINDA: I should roll out.... seven full-sized patties with this knot.

MAKETHRIFT: That would be a feat. But you have the muscle of persistence, to my customary five. In any case, it's the spiciness of the meat stock that makes for the pores.

LINDA: And that is how we call them "pore patties."

MAKETHRIFT: Or that they drip readily and deliciously. But they should be light, and crisp as such not to arraign week teeth, those upon whose biting and chewing causes pain and dislocations due to looseness. It's growing here, that malady of mouth.

LINDA: The pain is to the gums, and sores of bitter bread, for what we must intake of current conditions. I'd fathom there must be better wheats than this for our flours.

MAKETHRIFT: There is an off-putting intensity to this grain, of being somewhat wilder than our willies. But it's hard to find a batter of this grounded meal which offends the palate less of its grittiness giving our community's current state of temper and haphazard agriculture. The plants are as uncertain of their use as the people.

LINDA: As could be owned the lots of growing, and for whom.... (*starting to roll out the dough with a pin*) The better stuff's sent off, to betters. But how be they now?

MAKETHRIFT: Ah!... And the oven can take it's pan of water now (*starting to pour some water from a jug into a ceramic dish*), for the humidity to build up. We want no burnt flakes from a weak fire. (*taking the dish to the stove*) Salt's in here too, the wet briny taste, to make savory the vapors and the steam. It does often amaze me how slowly cooking's done, when the heat seems more immediate than any considerable mass to temper and pervade. (*putting the dish in the oven with aid of a loading platter or shingle*) Some elasticity.... must always remain, to even bones eaten.

LINDA: Of the marrow. But that which has no taste at all.... is seldom liquid, liquid containing or dewed. Nor dipped or tainted by a flavor is such a particle of bland nature that even the mice would not touch or investigate.

MAKETHRIFT (*bringing the platter*): Then it's not food, but probably a kind of dried wood, its forest essences lost, an' its sturdiness abandoned by the termites.... So it's sound unto itself, in a way, that which has no taste. Now that does make a fine tapestry of tenderness for tongue, ma Linda. What yields in seven of this area? That's hard to picture.

LINDA: Then picture six, with six squares, or seven rounds. For what is left over of the rounds cut out can be collected and re-rolled for a seventh.

MAKETHRIFT: That's geometry.

LINDA: But made lightly, Master Makethrift, where they would puff up and smoke a bit of the steam. (*starting to impress and demarcate the flattened dough into six equal sections, with a knife*)

MAKETHRIFT: Sounds more clever than courageous. These new methods you bring are sometimes startling— But.... I'm more open than conservative of the traditions taught me. For I'm master enough to know the value of practical changes. I bet I could enrapture you with what I learned at the academe, like how to make a pastry scone melt with a tongue's tickling, through a rush of richness. (*as she starts to cut out the patties, with service of a bowl's rim*) But that's not so fashionable here, for the basic fare of reasonable foods at petty prices. And frankly, I found not a fussy ornateness to be lucrative of the culinary diets to this location. It almost ran me out of business, my first year, laboring over the extravagant and extraordinary when what could only really be sold with any regularity of reliability were the more basic forms of bakery items. How many could order a lavishly designed cake, compared to a simple bread loaf? or an exquisitely prepared pie filled of exotic fruits and spices? when what's desired is a homey sweet bun or roll, smeared with frosting or a syrupy glaze. And what does draw the purchases is a rapidity of volume and an availability for demands, of that which could be made in any kitchen furnished home. So that's a lesson taught myself, in the real crude world of operative dieting: supply is more important than superfluousness, which seemed so.... heavenly, when I was a student. And perfection is only really in numbers.

LINDA: So then, I make your six. (*placing the patties on the platter*) And now I'll make your seventh. (*collecting the scraps and re-rolling*)

MAKETHRIFT: And that's the odd one, besting fortunes. What's left over is last for luck.... and often finding.

LINDA: Most people must scrimp, to avoid any wasteful extravagance, Master.

BRONTROL (*entering*): Linda! What be you here? A— kitchen's pastry aid?

LINDA (*looking up, most surprised*): Bront!

MAKETHRIFT: Who's this?

BRONTROL: You were not at the house. Indeed, it was bare empty.

LINDA: Someone has to work, sir, to keep it so, a home.

MAKETHRIFT: Who are you, fellow?—

LINDA: You've not sent much, to maintain our household.

BRONTROL: Ours still. Ours? A neighbor told me of your.... employment.

LINDA: This is my husband, Master Makethrift, whom we've not seen for three years, my son and I.

MAKETHRIFT: Abandoner?

LINDA: A wayfaring soldier—

BRONTROL: A swad(dle) of the combative mercenary. And I've sent much of a fifth my make—

LINDA: For two thirds of your ownership, or responsibility. It hasn't been enough, Brontrol! not after my savings were expended.

And what's been due's been much delayed, these past few months. (*as **Makethrift** puts the seventh patty on the platter*) With the uncertainty of it coming at all, I must do work to keep us soundly out of the rain.

BRONTROL: Our occasional couriers are right stout men and honorable. There just hasn't been that many passing by this region for awhile. All of our membership are reliable, and coded to ourselves to uphold our best interests.

LINDA: And what have you now?

BRONTROL (*as **Makethrift** takes the patties over to the oven*): What have I?! but myself, to manage you out of certain disastrous events.

LINDA: And what have you brought, but yourself and no gold?!

BRONTROL: The monies made is mine for our family. Have not a fear of its sparing, since it can not be abused in my presence. But what's of import is more calamitous to me—

LINDA: You were tired of *me*, and left, for some adventurous wayfaring— and worsening of vows!

BRONTROL: There's not that reason implied with my return. But never to be missed is cynical of our mutuality. (*as **Makethrift** loads the oven*) I have missed both of you dearly, and demand to see you, since I have not abandoned my wife and son, continuing of dire support through my wellness fashioned, and my battling based. Yet too is there some need to make for me away, this kind of man who can not plant flowers and crops for a passion. And though I was bored, not to be active of my meriting to seize, I was never tired of you. Our arguing hurt my ears most. But to the heart, that final day, you did distemper me, 'pon saying just about to leave— Get out!

LINDA: I did not scold you so, and would not dare, of my gentler nature then, against you to be so bold. You took advantage of any measly thread of excuse to abscond from my faith and leave me stone-faced for caring, shocked, and with a boy to raise alone—

BRONTROL: Not by my wits of serving still. I did say that I would excuse myself from your irate bitterness— bit-ter-ness!... and exited. And you found me not the next morning, the week, the next month nor day of years. I meant of my control to speak, that I had no longing for your excesses of complaint, and cruelty to my conscience by proposing to my weaknesses and abide a family of poor strengths—

LINDA: And no endowments other than my own. I pleaded you be more seriously devoted to our life. And you responded with a foul tensioning of any fears, that we are not suitable to stay together, under the insistences and restraints of family life. And then you made for an escape from me.... to leave me dull and dispirited of worry and the anxieties of half parenting. But this has made me stronger than your death to even pardon, within the necessities of living to till and plow, and cultivate what's good— instead of leaving it to you to provide.

BRONTROL: I still embrace yet of a marriage, what's better than good— but *legal*.... to control. And I may stay its head remaining, for all times done with and supplying and fostering through backbreaking exertions of campaigning. It's not my aim, though, to disturb your sleep and lapses of probity—

MAKETHRIFT: Sir! You did more slight to her than she your name, since I've never heard it before, nor of it.

BRONTROL: Away from us, pastry! This is for my wife's indemnity—

LINDA: You're hardly worth the compensation!

BRONTROL: But I've come to improve these worsening circumstances.

LINDA: They've been brightening, as do clouds over head radiate sometimes the sunlight with its glories for our worshiping of sight— But not of you, your cadaverous approach, like a death's scowl found within a tomb—

MAKETHRIFT: Sir! this is my establishment.

LINDA: —reverberant of misfortunes and regret for having been often observed and too often recalled.

MAKETHRIFT: And I will have my way made in it, soldier or not. You seem.... scrawny to me, without your weaponry brought, else nerve.

BRONTROL: My— calf's muscle's larger than your head! For I've walked a hundred miles a day, to make for rest of camp a transition towards an enemy. And I've split twigs thicker than your neck, to brush my nose against from (after) sneezing. Your fetid fumes dull your eyes (and) its lenses, about a person such as me to see.

MAKETHRIFT: Then you receive your scents peculiarly, as about your clothes to linger, from where you've slept. But there are laws in this community, against deceitful entries (*reaching for a broad pan leaning on the stove*), that allow me to strike at rogues with full defensive ease.

BRONTROL: I'd make of that a diaphragmatic cap to cowl your face, as a helmet moist from tears and cries to be relieved from.

LINDA: My rolling pin could better score to lob with, falsehoods falling to my feisty provocation of attack—

BRONTROL: But would you now thus fight your master? I'm here not to waste time with much quarreling. There's been enough of that to leave, by testy tongues and harried hearts of soapy sorrows. I show myself to uphold the only sinecure I've left, in this hustled world of travailing tribulations. And that is to insure my son be made a man. Yet I've heard tales that you've sent him to some countervailing convent, to restrict and timid-ify his behavior. And I've rushed home as soon as I could, to prevent more damage from this crushing abatement of my flesh-sown pride and fierceness.

LINDA: Wild he's been, and uncontrollable at times, as grows his anger, standing on the earth without instruction to face its most fiery elements and intimidations, and the challenges of contemptuous views to a lost son, regarded as forlorn of father; nor any male instructor of sincerity to be improvised for is he given aid of through my maternal hands and manner. Then to appease his growing needs to sculpt his roughness of frustrations into some kind of fruitful utility, less reprehensible than gangs of tormentors to join, within this torrid space of country village, I've sent him to a school of shapely culturing. Or by my insistence he attends this free offering of our community fathers and leaders to tend our wood for constructive behaviors, and control the outbursts of agony that the likes of you have wrought. But even now does he contain such vile temper and vexation as to remain endowed of a vicious sap crying for a virtue over its vitriol. And he has not been lessened of the manly torts and hastes, but could be thought to be more coached under the authority of principals to this town, his steaming controlled and yet permitted to flaunt and flout of rages suitable in deed.

BRONTROL: What deeds? So not a nurse-ly lyceum, for learning how to fold napkins neatly?!

MAKETHRIFT: Surfeit it not to burn! These waters swell but temporally. (*starts to attend to the patties, taking out and putting on the pan*)

LINDA: He goes to the siege of Balcome Castle, or that fortress renamed thus, under the direction of our partisans of captain, to force this foreign lord out of his ingratiating stance, and into the mouths of litany for answering to us, our pleas for greater self-sufficiency and rule. This is a dangerous afferent thrust my son takes; but for his nature's sake it's done to be more useful in this grouping, than a solitary terror of hate and choler against the world.

BRONTROL: Then be he more a soldier, like his sire; though he be just of age or just under for such practices of army—

LINDA: Dare not be a fight that's made. This is only a show of protest and threat.

BRONTROL: Still, he's young. And still war wages young and must as any event age into its fullness of maturity and ripe pulp. But not the pup—

LINDA (*as **Makethrift** brings over the patties*): It's been three years, Brontrol! He's not the same insecure lad as when you left. But more defensive is Bill made, for a bullying through disappointments.

BRONTROL: Billy!... Born of my malice to coerce you to my sensitivities of matrimony. And of a legitimizing heritage he is. Yet can there battle be. It's.... why I am allowed, to visit thee so promptly.

MAKETHRIFT: These are to perfection, Linda, and aerated richly for the taste, as soft as crunch to fluffy flakes can be made. You've done as well again.

BRONTROL: The king does have my pardon, to consult of Lord Balcome, as a courser or a servant, as he should need, to prevent a shambles to this countryside under a tangle of unnecessary animosities.

LINDA: Wouldst be against my son?!

BRONTROL: Hardly, but his captains. Sieges, even of an illusory nature, are too serious a surge to be let done without the discretionary balance of royal reasoning to abolish, as could they have no need or charm for the participants. Yet could Balcome remain complacent of his restraint to attack— for awhile, some retaliation can be due after a loss of reserve to impatience is conducted.

LINDA: And how can you possibly avoid this resistance?! Our taxes are given to this man unnecessarily, if the king no longer leads us faithfully.

BRONTROL: But he does. He has not lost his powers of adoration, nor admonition, and does suppose upon you fealty, to this lord, despite some disquiets that occur like a shadow over the royal court.

MAKETHRIFT: The king himself gives you this leave of service, in person to pertain to your actions?

BRONTROL: This is a special dispensation authorized by his majesty. My superiors grant me my interests concerning this particular village and its prominence of stewardship by a loyal royal ward. I've been in the fields, of course, strugglin' for the elusive victories of battle the king divines as necessary for our preservation of state. And it's been very difficult, results disheartening and discouraging for the king's troubles. Yet he still maintains a thorough love of country throughout these lands, as could be shown by my meager privileges obtained. And I mean to make the most of our effects, by helping to survey the climate of this town, and purchase for it political placation, in the eyes of its true governors.

LINDA: As a horse's back, and courier of our messages of discontent?!

MAKETHRIFT: And disenchantment to the stretch of royal arms; for his majesty's influences wane tremendously, these days. We are about ourselves to own, and on our own to a very great extent. Tribute to the lord has been only a sudden custom soon to be broken, though with much toleration of his presence. But his lone estate of combatants and protectors can not make up, even highly organized, for a population determined to exist for itself in reward of its (own) labors and arduous activities of attainment, growth, and sufficiency during these most unsettling of times. That is the climate felt, mister soldier. We need no more pretenders when there's real work to be done. And Balcome has only been an irritating insertion of arrogance into our solemn affairs, from a majestical court of growing question and lessening relevance to our plights of survival. I can not think of one thing the king has done for our betterment or improvement in recent years. And we certainly need no policing as by Lord Balcome for us to be handled and manipulated down into (the) drudgeries to enhance his personal entertainments and powers. No. The king falls and fails us, with this interloper, as if we could be a gift of region in offering— to.... disingenuous sycophants, and (he) can not be easily forgiven for inappropriate intervention of governance, since it is clear he can no longer back this lord's commands, has whittled away his royal powers of enforcement, and leaves the man on his own to procure fortitudes.

BRONTROL: That seems so evident, do you think?

MAKETHRIFT: What be the siege for but to prove this?! We haven't much need for such a fortress. It is in fact a proud antiquity, but of shambles, and now made to be a shamed symbol for resentment of misuse and appropriation for the undeserving. Though

I grant you, not of Balcome's self is he so offensive, to have been placed unwittingly in this controversy, and given this unwarranted position over our heads. So we show him our heads in number, to suggest that what's seen may in fact be higher, collectively, than his own. This assertion, sir, is potent, and begs for considerable reassessment, particularly of these taxes. We are no longer the spoils for anyone's— victories, but have developed of ourselves into a unified solidity of communal whole worth for itself only unto itself the goals of living and extension of our preservation.

BRONTROL: So say you preten'(tiou)sly.

MAKETHRIFT: I would join the foray, this foraging for respect due our candidness of contempt and anger, if I thought it was necessary. But there's more than enough to show our force; it is so obviously evident, sir. I would only make more of a crowd a circus for our emphasis.... And if a fighting's due, I will actively defend our most significant points: fraternity of thought, and freedom from social interferences.

BRONTROL: This used to be such a faithful dominion to our rex. The loyalty promoted my conception to become a soldier, my foundation of home easily assumed to be kept sound and useful as a recommendation to the various corps.

LINDA: You've taken everything for granted, these *tre* years of blind consideration of our hearth. But the world evolves despite your steady state of uselessness—

BRONTROL: Hold off of these pejoratives!

LINDA: —to our becoming and furthering.

BRONTROL: I've not been useless. I've been fighting. I've been damaging assaults to our country's sovereignty, constancy, and health. It may have seemed peaceful here, but it has not been elsewhere, with decided attacks and contestations to our provincial ownerships. And while you may feel so safe in your lethargies of opinion and awareness— and even disconsiderate of a dignity harshly fought for— most of the world we can have knowledge of is in a frantic state of dishevel and alarm, a pulling and tugging of limbs to tear off and exchange for new designs of state and nation. It takes efforts like mine, to hold onto our own—

LINDA: You've hardly held onto anything of your own. Your loose grip is my gripe, and our son's sad banishment of heart.

BRONTROL: How I would see him—

LINDA: He is taller than you'd know him!

BRONTROL: ....How I would sense his spirit is a treat for me! to confront and train, and take him away— to join my entourage of worldly notice and adventure—

LINDA: No! He'll take no enticements from you. He loathes your thought!

BRONTROL: ....That would be hard to believe, Linda. I've been busy. He'll be as much so. I'll turn his face around from your foolishness—

MAKETHRIFT: We're not the clowns of dispassion!

BRONTROL: —And you'll make my supper tonight, in my house!—

MAKETHRIFT: We're aroused of disturbances to make, Bron—troll! Bront—truant! Bront—trudge!

BRONTROL: Sir, I will macerate your hide, in the boiling tremors of my displeasure. You will feel softer than your feint of assistance, after my hands have done of you.... But my wife will be made less bitter— by choice or by purveyance of the need, when Billy greets me.

MAKETHRIFT: Within that oven choose your head. I'm a tough man on patties, an' can wrestle you—!

BRONTROL: This is a family matter, you foolish strudel!

MAKETHRIFT: —down to dirt's death's door!—

LINDA: The boy lives not with me, Brontrol. He's not so stupid as to want to cause me grief and worry over his anguishing.

BRONTROL: ....No(t) so?— Then where?! With this— free brigade?

LINDA: Whereat he(ha)'d led, man. And by the siege to prosper, I can't absolutely confirm his friends and acquaintances. But he visits me regularly, with funds to provide for us both.

BRONTROL: Sparing to save up for travel? Does he sleep in the house or not?!—

LINDA: Nooo!... But I see him there if ever. You'll find him by the castle, most probably, as a regular to service, relating himself to dangers like an unction on wounds.

BRONTROL: This must be recent.

LINDA: I've taken on a young orphan for bedding, since he's out the house, and there's this need, in the community.

BRONTROL: I saw no one there. It's as empty as a tortoise shell dined of vultures—

LINDA: Bedded at *night*, Brontrol. There's no more room with the family that makes for a community orphanage. An' he's a pretty little creature, but shy. Children should not have to realize the enormity of their plights, but merely be confused for playing through them.

BRONTROL: Who protects the house? I must share with this?! —

MAKETHRIFT: Some decency of witnessing. You'll not abuse Linda, and shove her around like a makeshift stole, if what you've come back for is some amorous bounty, as you soldiers meanly expect.

BRONTROL (*taking a threatening posture towards **Makethrift***): You deserve the heat of staves, in bending prickly wood—!

LINDA: No one's going to steal from *me*! We haven't more than

others needing.... And you'll dine from what is ate as commonly as Hoe, the evening's porridge. Nothing special for you.

BRONTROL: Hoe, as who's this squatter?

LINDA: His proper name is Hugh, but he can only pronounce it "Hoe," yet. I pick him up, directly. And I am his Aunt.... Li'da. And you are just another boarder passing, for he won't need an uncle of your likes.

BRONTROL: I'll bash him out the way, if he's a pest to me—

LINDA: You'll be constabulary-ed!

BRONTROL: —In my own house I'm still supreme, to do as I see fit—!

MAKETHRIFT: And in my own workplace I'll make as well a tarnish spoiled by polishing. But we are ruled by considerable order and redresses against villainous brutalities. We police ourselves well and diligently, having once had favor of a showy place, and with that linage of (a) country substance approbated lead ourselves of comely settlements and rioting out much bad and foul that would try to linger through this land. A maturity of notion has evolved here, since you've left, of more self-sufficiency for our habitations and satisfactions. And this doctrine is made of our character, rather than by decrees and— royal definitions or projections. All tormentors will be tested, for their tensioning to twine into soft fabrics— as may be beaten and pounded out of them. We've lost the patience to be servile and sheepish, while being on our own during rough days and disconcerting events not of our practice to consider roundly, as if we belonged to the same sphere (but) with distance noticing. Only a few traces of that past subservience may be found, as by custom and commendation for it. But in reality, we will not run errantly to those custodial methods, to lords or princes or kings, if they be (papier-)maché tough, but in the rains of weather much to deteriorate. And Linda works augustly for me, with what is typical of our population's fiber and courage. You'll not change that inuring of quality with your silly assertiveness of haughty hollow height to hie, as by a run of desperate cause to see for the unraveling of its puzzles. What, then, is so dire of your mission made of? this poor excursion back to find what's left has left.... your amicabilities and aims disowned or dissolved—

BRONTROL: An' not the reception I had expected, for a proud man seeking a fully justifiable refuge, with honors brought.... and sustenance of money, for my stay and beyond. Then must I rent in my own house— like a lodger?! Or can I be greeted happily, for what I am: a husband and a father and a soldier! and an Odysseyan wonder, perhaps, but sincere of my chastity— of thought.... for my wife and home, and the inhabitants of this village. No, I am not roughly understanding you, or crudely picking my means, like odd beans from a jar, to study their shapes and decide on their merit for my touching and examination. Then may you be contented to see at least that I'm alive and wholesome, able-bodied and desirous for your prospering, and that I *continue* to raise a family, in mind and with good action to demand of myself such a labor. What's changed is not my connection to you, which has been kept faithfully, and materially gratified. And now I supersede it even, to be in person for some warmth, after these cold nights of travel, holding at abeyance all uncertainty for my arrival. Take me home, at least, with some spirited indulgence for my help and worth, and for my relief in seeing you— and this town— in tact. And embrace me!... or at least my want of inquiry, for the dear nature of your persons that I still hold with consumption of concern— consuming worries, to abate them with my presence to handle.

LINDA: Go search for your son, Brontrol.... and leave me to my working chores. That house will be there as it's been for long, the few days of your stay to have its use— and mine. But do not ever blaspheme my agonies of ownership. I've been alone its principal, and it knows well my sadness. It does solve this temperament by being solid cover and protection, shielding from the elements, and the softening of sighs at night through its sturdy muffle of walls and reliable structure. Yet the sadness remains, even as I see your face, from knowing what it's meant these years, to both of us. And now you are deserving of some sampling of these regrets, and the tenderness begot of calluses pressured. Then find his head thus. And if not like yours, then I be a judge empyreal.

BRONTROL: That does extoll it, lady, your cross work.... So I'll withdraw to find this splintering— And it angers me! But I must report to the lord anyway; and there be on the way to go, some rebellious stones to transit ov'r an' through. And have you not a hankerin' of fear? you foolish both. Balcome is a decorated fighter, who will not put up with so much complaint before retaliatory surges become evident. This leader wins battles, that's why he's favored.... with this country estivation. But he'll make winter sure— upon your heads to hatch some brain.... if you bother him more of disrespect and ungraciousness and cowardly crimes at castle.

MAKETHRIFT: It's done his sleeping there, not that the occupancy's warned from being hollow. It's been rather an empty and uncomfortable relic of an edifice, with much to crumble for. Any occasion of visitors might be welcomed to its poor heart of hearth, a graying decay. And what a gift's been given Balcome! But the symbolism is clear, to be a ruler over the ruled. Well, warrant that some governance, this sprout to suddenly pot up.... from foreign seeding. This makes contention vegetative an' growing, as the old weeds must abound (all around) that place. Now, hear me for his listening, Brontrol. We are not happy with these presumptions over ourselves, and we're not satisfied with his choosing to uphold them. But make he a fair compromise, and the populace may condone *that*.

BRONTROL: Compromise?

MAKETHRIFT: To be as weak a figurehead.... as that old decrepit castle stays.

BRONTROL: Some surrey driven here!— of bargaining? Wouldst thou thus erupt upon your doom?! you silly one of to mind, that makes of this a common resistance for the whole. We are killers here, with this soldiering, an' could spread such a phalanx of bodily destruction throughout this land and territory as to strip the fields its cricket-ing to discipline towards— silence! A military method seldom waits upon its needs for action, and is intensive and impulsive— and trained! to be this way effective. Surety comes of being sure.... for one's intentions; and I am almost astonished not to have to forestall an initiative already underway, by the forces of this lord, by his crafty troops, experienced and war-weathered, sent out into the village to restore and take over the policing, or at least thwart the insolence surrounding his castle grounds. This is no sport now. This is priming for attack. And that which you do not feel now can not forswear its essence being

made.

LINDA: You aren't so much essential, then, as estimating our resolve. These protests spirit you more to return than your own essences have tried, they being.... far and away, and in doubt of you.

BRONTROL: Linda!... make my house— prepared, (*starting to exit*) and for my arrival proper. Repeat this until it's done correctly. Mistakes are challenged only for correction, but so they are befouling of one's rights. I've not been careless of my dealings.... But it seems daily of arrears to forsake, how this welcome epitomizes grief. (*Exits.*)

MAKETHRIFT: This is stone walking, Linda, not even to embrace you despite a startling disapproval.

LINDA: Could not be given more to see, am I bound. He was.... as he was, and is the same, yet wondrous to perceive in an instant. And the law is on his feet, for abandonment is five years made. Though he has continued to supply some funds.... I would have beaten him off in front of you, if he had tried to greet me with an embrace, and with more fear of his person as a beast uncouth for wife than resentment of his presence. For handsomely his little virtue remains, but I am spiteful of its comeliness.

MAKETHRIFT: Can the marriage be restored?—

LINDA: I will poison him.... off that track of forgiveness. There's no plaintiveness evolved of his absence, for several months.

MAKETHRIFT: So to sense, or such to sue.

LINDA: And what of this regality? I thought this lord was timid. That was all the rumor coursing through us.

MAKETHRIFT: There must be caution in him aplenty. He does not confront the disputatious heads with the bluntness of stern anger and rage, but allows the demonstration some entertaining miracle of political, and perhaps emotional, entrapment. But we have to least (have) been rather curtly about this, and even quite amicable with his sheriff. There's no cause for any shooting feared by.... your husband. It's simply the acceptance of a statement of understanding, that as much as the steamy mists dissipate, so too our duties, fines, fees, and taxes— to the lord's estate for paying. This should end. Though I can admit, that castle needs some building up and restoration. What we've already given will have to suffice of our allotment to the effort. Balcome's riches must be brokered of his own devises. He should construct some industry that will enhance employment in this region. That is the way to lord over people, not merely by decrees from kings, as if the heavens deemed their intelligence for this or these demands sundry and invariant. Fealties must be earned, not ordered, actuated, not ordained. Yet, if Brontrol is right, and some fierceness is to be expected otherwise.... pray for his mitigating mission. It is bold enough to assume and undertake, that he could placate sides subserviently and with unassuming practicality, at least, to keep peaceful temperament like (a) pride, about his home.

LINDA: I would never call him unjust, to seize things as they are. But we are not mice to be scurried with fright from a sweeping. And that's all to the point of our calculations, compromises, if we must each defend ourselves. For the king can not send reenforcing troops to Balcome, apparently. And so we play it for how these confrontations must develop, as standoffs until some understanding's baked.

MAKETHRIFT: Yes. That's the plan, for the dough to be shaped. If I really thought he'd try a hearty assault.... Well, just tell me of it, or as I'd learn, graves would turn grave, gravely from their tranquilities and serenity of entombment. The entire of our inhabitants will be aroused of a resistance and struggle. The fighting will be frantically inspired and convulsive. And I will make the greasy bullets, so portable and yet sustaining of the energetic endurance for the combat. So I dare say our soldiering will, over the long run, have a definite advantage over those military masters. But I'd shame myself to wish for this need becoming, even as it could promote me.

LINDA: A people can not be ruled without their pleasure rationally. Nor can such a marriage be made. But Balcome is the better of the king, we may aver, not decidedly hawkish and asserting. Were it not for the royal descent, he could make an acceptable magistrate, major or mayor. But not at this moment of time's tesselations can he produce himself worthily of our graces and granting for this magisterial provision.... More patties to be baked, Master Makethrift?

MAKETHRIFT: I might say so later, Linda, after the shock of perfection wears off. Let us work on the tart dough more. (*walking off with the patties*) But I will bed these in a wrapper for sales.

LINDA: Then I can rest my fingers a bit, to have them mingle of the sugared flours and sweetened butter you'll furnish.

MAKETHRIFT: My own special mixture of ingredients to be proportioned with a gentle skill.

LINDA: And solidified into one whole batch for making.

MAKETHRIFT: ....towards lightness.... and airy flakes, again. The work.... takes management.

LINDA: ....And for a pin to subdue. (*leaning on the table to relax her hands*) The method mauls malefic of a benefit.

Act II

Scene I — *A parlor within the castle. Gathered are **Herkshead, Craie** and **Bill**, with one of **Balcome's soldiers**.*

CRAIE: This is some delegation we bring.

SOLDIER: The best that's near the gate to draw from. You are invited to attend. Sher-Gestromy is almost desperate for a conference. Now, if I had my way, we'd be mowing the lawn out there, and raking up the fallen grass for some straw bedding and hay, what the horses might prescribe of a nourishment. For some action's definitely needed, to clear off a rambunctiousness. And I'm all for a charge, and to do something about this.... assembly of the public around— our castle. But, as you are generally unarmed, you're welcome here to talk for lease of parley to contain some merit of yourselves. And do you know I was practicing my aim among your visages? Why do you abuse yourselves this way, to tempt mistakes? You give yourselves an opportunity for harms, by simply being around and noticeable, presenting your persons for

observations in your crouching. And perhaps daring our insouciance to end? that's a certain kind of madness, gents.

CRAIE: It's a peaceful protest—

SOLDIER: Of what?! Our occupation?! Sirs of swine, that has been demanded to be principled. You've no say more about it. This is Balcome's. We are Balcome's. You are Balcome's.

HERKSHEAD: That's rather outspoken.

SOLDIER: And I'm a soldier who can back up my pontificating. (*sitting on a decorous, though age-worn and studded stone bench*) You gentry seem somewhat pathetic to me. As guests I am conducive to your comfort, but not your motives of being ridiculous. It's almost an irony that we come down to you, to face this country rancidity. Now look, fellows of the realm, we've just been through an arduous and long war, slaying many and (having) extracted the barest of an advantage, the most minute of victory against powerful and persistent opponents. And we'll be used again with certainty later, to try and improve these efforts and vouchsafe our assistance to related forces doing so much worse as to be shameful of contemplations. So we come here for somewhat of a reprieve from active gallantries. We are given this place to rest up at. And now we find you turning against our leisure, and apparently against your own interests?!— broadly speaking. Sirs, we're not in the mood to play with or placate fools. We are tired and fed up, for awhile, but with muscle much enough to handle this protesting. But it can't be done genteelly and with much circumspection— because we are sore. It will be a sore chore, done hardily with impatience and bother, and with a strict lack of gratitude for the efforts caused. And who's to say with your defiance you can be cleanly put down? The dispatching will be brutal and ugly, which is how things are done rapidly, so we can get back to our relaxing with what's little time left for it before returning to our real devoirs.... We'll have to do a thorough job, you see, so that we can come back again in peace. But it will be a thankless task of muscle spasms— and you'll pay heavily for the disquieting.

HERKSHEAD: My good friend and comrade, you've brought in a most proper detail from the people—

SOLDIER: Not I. Others did the service. But I would conspire with you, as I guard, to promote some wits to your arguments by clothing them almost immediately to be forgiven. Can you realize the resentment and heated annoyance you are generating among our fighting ranks? We've not come here to find obstacles and obstreperous regards, so easily cracked by already painful hands. Plead with the lord to be absolved from these heresies of nation, and hammer out with him some method to revert your fellows back into a sanity of self-reproach.

HERKSHEAD: Good soldier—

SOLDIER: One of many. One of many several sharing solicitation for your deprecation of severities.

HERKSHEAD: I am a high representative among our band of civil sustainers and substantiat-ors, a captain for this servicing of— only the patriotic.

SOLDIER: Are you now? Then see how I may place you well with my superiors to commend, by bringing forth to me a justifiable understanding of your motives and morals throughout this disturbance on our lawns.

HERKSHEAD: My name is.... Fellow Herkshead, as you would call us fellows and therefore, like yourself, entreating of the similarities to state.

SOLDIER: Same country, at least.

HERKSHEAD: And we are all bound to be servants to a single nation, with courtly terms for everyone who might treat us rightly. And we certainly wish no hardship for you, through your stay.

SOLDIER: Then why do you display yourselves this way, so discourteously, impinging on our notice with a flared distaste? Think we lessened in any overtly physical way, due to recent news?

HERKSHEAD: That is a contaminating source of intelligence, I grant you—

CRAIE: We should be answering to the lord himself, these details of displeasure or anxiety.

SOLDIER: Not to meat?! you, and your boy. For this is how meat is made— You see? It is well toughened. And *we* are the meat of the matter, for you.

HERKSHEAD: That be said, we are all fellows. And you are.... glandular of earnestness to be sure, and dear to us for your scourge on the battlefields. Then so are our proposals as simple to be single-minded and determined that they may be presented to your machinery for ascertaining. And this boy, as you would enlist for his nature, is a member of our consortium, but is not related to me.

SOLDIER: Looks like you.—

HERKSHEAD: That is the favor that we patronize together, for my guiding him.

SOLDIER: Sir. Most hedgehogs look quite similar.

HERKSHEAD: Our lad represents the comely youthful vitality of our residency of community that we'd present for the lord to ponder of, as towards any destructive malice that could be attempted of planning. We will defend not only for ourselves, but for our future strategies of fruition and progressive growth.

SOLDIER: I've seen young men blown to bits, and wagered on what of their parts of body land where or near the bettors first. It's not for thirst of youth that war is made, but for possession of some aim to guarantee, in power or material splendors or rights or metaphysical conceptions to promote and abide with.... So coarsely may be a face scraped in combat that its age is not necessary to behold or try and calculate. And then you would defend the like of children and hatchlings?— Storms know the enmity of their devolutions, passing onto elders their despicable fates. I've played among groups of them found. A military whistle blows— and we decimate them all, as like the cross canon of our nature, to obey the molestation, and the cannons shot out across plains to faceless opponents and ageless enemies and the lackluster of shadowy, smoky deaths and trampling. Ferocity knows only the bounds of its bellowing. And a laxity of enforcement does not obviate nor obfuscate this law. We are angry men, to young and the older alike, and will not

continence a holding back of harm just to contemplate your futures!— not that our leaders may, until we're let loose!

CRAIE: Well....

SOLDIER: Bend this argument, then, to blunt its sharp corners which (one) might be carelessly scratched by, to say that monkeys prefer trees, whether it rains or the stars shoot of fire and hail, and so can not change their solicitous natures, aroused by master marksmen and fighters to beware and mock at, but that this character of silliness lends more to sighing than a slander of killers. Then plead a pardon for your bobbing heads— and go away, to your.... residences, or shacks, and prepare offerings of compatriotism to exhibit. And we may even brunch among you, with these amicabilities offered. I know I'd like a good country-cooked meal for a change, by some pretty wild-eyed wench produced and serving. But have some comity drenched and simmered of a hot rich drink in summer, and you'll do well by this lord's habits. For he tastes gladly all good things that would at first appear.... threatening or dangerous. And thereby with a much stronger bond he makes of friendship— and protection.... of a loyalty to king and country, county, convert and conventions common and lowly.

CRAIE: Well, this is most boastful of you, to pretend for us our covenants to be pitched—

SOLDIER: Only an opinion applied, from pent up tensions—

CRAIE: But I might say more forcefully this is for Balcome's ears to assume or survey. If boldness weans towards honesty, you've been well fed.

BILL: I've seldom seen such muscle to the face.

SOLDIER: What chastity does displace ye?!

HERKSHEAD: We do observe with eyes, sir, a youth's too keenly focused for its learning as to bring sharp tally of our numbers.

SOLDIER: This advice is what I lend you, not give you outright. There'll be some repayments made, while you survive our ownership.

HERKSHEAD: Conciliations may be preached.... Though to be realized takes more commonality of our dueling aims to remain at peace. And this is much for Lord Balcome to decide. We are in number balancing our qualms.

SOLDIER: That's least of all to be fit against our quarrels. (*hearing someone coming, and so standing*) The night's sky has tens of thousands to innumerable of the tiny twinkling stars which put together could not make one full moon's shine or glow.... Then pity you this day if you are not (even) comprehensive of the sun.

GESTROMY (*entering*): Lord Balcome will see this committee directly, gentlemen. He promotes you to his office for conversation. He's at the moment receiving a courier. And have you been amused with the familiar? You certainly know this castle's relicing. But in such a shambles left, the dirt is pure of dust, as from a withering of stone itself, the erosion magical of its infertile granulations, as from the pyrene of a precious fruit pulverized. Do you not even visit the site, as a ground for occasional celebration of the historic?

CRAIE: She has been left.... for little picnicking, the common possession of a town as an antiquity of rather generalized value. What should this purpose be to stand here, and face the briny winds today? We are more cultured of a civilization than to allow a squabbling through her defenses. But neither are we mischievous enough to promote a greater degradation here than through time. It has been left closed, as an uncomfortable shelter, and for the elements to bargain with, not that we are embarrassed of a legacy it represents. But this noble structure has not been very practical for some many decades.

GESTROMY: Our lord's men has done much cleaning (*nettoyer*) *de ménage*.

SOLDIER: And sleeping on stiffnesses more blatantly severe than ground. We do prefer the courtyard grasses.

GESTROMY: Balcome is considering this. But those few who took it upon themselves to do so are controverting his permissions, to earn admonishments properly administered. The complaints must be more organized.

SOLDIER: 'Tis summer outside and frigid within. We are more for camping come than comfortable cubby(hole) arrived.

GESTROMY: Show me no disgruntlement. I'll go straight to the lord about it, and have him tax your peevishness as I must tax collect from these generally agreeable countrymen.

SOLDIER: But—

GESTROMY: It's unbecoming of proven fighters and accomplished soldiers not to weather through these simple vexations reservedly.

SOLDIER: But my feet do hurt more here than walking many a mile anxiously. This portends of something despicable to occur, for such humor of irony never fails to lend its leaf too insubstantial for (a) cover.

GESTROMY: Then intuition is just a meanness of toe.... And a soreness of contempts, gentlemen. I can tell that Balcome is very disappointed in your stances before his occupancy— I can reason for you that you have some warrant to make plain your feelings. But this places our lord most awkwardly in a position to determine if he is not dealing with rebels to our majesty. For if so, he is duty bound, wherever he may be, and certainly where he presides, to confront and attack traitorous intentions. And this hardly being a matter of his thoroughly being able to do so, his rank, privilege, and king's troth won, orders him to try. But you see clearly he is ready for ameliorating gestures, to depart from misunderstandings or wrongful interpretations of his position.

HERKSHEAD: It is clearly with his majesty's.... permission.

GESTROMY: That's obvious.

HERKSHEAD: We do not object to it. And we are not objectionable, yet neither to be divested from a dissolution.

GESTROMY: What have you lost? What's been taken away?

CRAIE: A dissolution of faith in the certainty of the king's authority, or direct rule over this particular dominion and district.

GESTROMY: Well, sirs, how can that be argued?— but (not) with me. You'll have Balcome's attending to his subjects.

CRAIE: Treatises may proclaim this, of a reading that it is so or could be described as such, accepted as evident or even evidence in congresses of debate. That is, after all, a history to relate to and propose of explanation.... that we are defined as subjects, to a realm and a monarchy. But if the king has lost his strength— to enforce his own statutes.... then this might not be so in effect, and we prepare ourselves of an eventuality of looseness and vagary in governance, that by our own methods we will not dissolve into a certain type of corruption called anarchism. Our manners may besmear themselves of too many individual distinctions or attributes, to draw in any aggressively opportunistic despotism by a charismatic condenser of rights and chauvinist to this cause. So in this time of uncertainty and political spume or spoliation we prefer to give ourselves a constitution of governance, whereby we are rallied together of this issue to assert ourselves at this decrepit symbol for all the village.

GESTROMY: So you wish to rule yourselves. That's understandable, to much of an extent. But the king does at least reenforce his interest in you by having us sent here. And his interests are his ownerships. Now it's certainly reasonable to note, that, old as she may be, this castle remains a proud heritage of this region. And in order to build it up to a respectable splendor through the efforts of principled and superintended restoration, monies must be collected to fund the attempt. I don't see why you'd be contemptuous of that, if it's been affordable for you up to now, and should continue to be.

HERKSHEAD: Well that is certainly an avenue of argument, or an attitude of opinion, to take.

GESTROMY: And this is clearly the major reason the king has rewarded Balcome with management of this task. So why your protests now if all has been quite seemly about it? That you should have to explain to the lord in person. You appear to him, frankly, ungracious and ungentle(manly), perhaps obnoxious and absurd; while I, having dealt with your populace more regularly, know this not to be true adjectivally. It would be a shame for much disorder and unnecessary upheaval to arise due to false conceptions and errant preconceptions. But this is what men wager with, to prove or disprove. If this master has not been an ogre to you thus far, I'd not complain the way you have, insulting his very footsteps at his doorsteps to take. I'll not assume for you, though, to be in very much error, if all that's needed is a proof of strength or determination. I'll leave that for Balcome himself to provide, should he choose to. Until then accept this hospitality of appeasement through talk.

HERKSHEAD: Does he feel hurt, by our reaction to unsettling rumor?

GESTROMY: Would you be, if you were suddenly surrounded by apparent hostilities of potential? some malice to prepare for or combat? I'd say he's been surprised, but not stupefied, and is reviewing options for what to do.

HERKSHEAD: Well then, good sheriff, the best is known, that

we are unsettled. And not to hurt ourselves by a reluctance to show this, we appeal to Lord Balcome to be more kindly than kindled, and to adjust or vary his aims of providence. We may indeed make help for the upkeep of this castle, as a sort of monument of some relevance to us. But as to ourselves to be ruled over.... well that is more of a heady question—

GESTROMY: Give it (to) him bluntly. He does not like a mystery.

SOLDIER: My head does saw of time, sir—

GESTROMY: Yes.

SOLDIER: —There's a remediation somewhere to these proposals and proffers. But let there be some truth to a summer's heat. These gents are agitated, and duly need some exercise.

GESTROMY: Oh, wars should never be made of quarrels, but done only after quarreling is exhausted. An engagement is this dedication to disaster, as a demonstration of some physical art. Now I bring you to him, sirs, (as) a worthy delegation. For the representative to be, or the reprehensible to seem, I can not decide. But my only advice is not to be angry at all—

CRAIE: We're not.

GESTROMY: —Converse of friendly persuasion, and search for amicable terms. Politely stir the well of thought. And avoid the lie, as dirt thrown in the swill's refreshment. Deceptions have no place among compatriots.

HERKSHEAD: You might collect from us the pleasing epithets of civility.

SOLDIER: Those out there are hogs! sir.

GESTROMY: Soldier, fellow.... behave one's self—

SOLDIER: I like hogs!... when they're used for my bacon, when they're handled in a pigsty as for how they are. You camp before us daring, daring brought. And I enjoy such nerve, and would share hands with you, of courtly abashment— before we start(ed) slapping faces. But come to points, my fair foul fellows of fustian froth — countrymen! to treat us this way as if we could be your captives, or your criminals! You wish to test our mettle— and we'll gladly show it. For I'm quite ashamed of you, and will put you in your place, apologizing on your as—

CRAIE: Sir! Desist of rant!

SOLDIER: —sembly's motion to withdraw. You enrage me of your misplaced courage. I've seen companions killed for you, and your serenities of existence— Pigs!— as we fight! And then one simple notion brings you to bite at us like ticks, gnats, on' dykes in tights— that our king has become insufficient for you, a ruler we have mourned under, to stay fit for country and our eternities deserved. Don't amend yourself to these— porked bellies, Sher-Gestromy! Let them pick up their spikes and spears and spades of shaft — and fight us! as we blow them up into oblivion, and shoot out their eyes for a trophy of gasconading skulls. I told you what to think! Now prepare for a tensioning of men, in your proud moment. For it is almost unbearable, not to go after your display out

there, and demolish thoroughly, ridding the countryside of your humongous lice!

GESTROMY: Save your rages for our true enemies! You see, gentlemen, that our soldiers still boil a bit. They've been abused— and beaten!... much too often. And that may cause a vile approach to even the flowers of hills an' dales, within which many lie for centuries of reflection, a quiet spot provided for their self-turmoil of conscience and eventual beheading of all professed philosophies on the nature of mankind, and why there are victors paired with vanishment and vanquishing, since all annihilations are usurpative of victory. But I'll have you apologize to these delegates—

SOLDIER: What?! Me?

GESTROMY: —They most definitely are afforded our utmost cordiality while under escort to the lord. And I'll inform Balcome directly of your salacious outburst, too sensational to be sanctioned of a professional guard, unless you apologize.

SOLDIER: Why, he'll just say— we're bitter tasting the punch of their winds, pungent to the throat of our contention-ing to bother of; and so we've become mean, to what fouls these airs provided us.

GESTROMY: That may be so, if sights can harm a scent of freshness and restoration. But it's no excuse to assail these gents, who are courteous enough to place their persons implicitly under our guardianship during this visit *we* have promoted. And Balcome may have you disgraced, as you reflect upon his promise of good conduct and fair passage through his castle, even to the point of having you buffoon-ed before those outside, perhaps performing for them at buffet or offering a few lunches. That would be a nice show of contradictory nerve by our lord, to challenge them with the kindnesses of a fool!— And sure to be I'll recommend it. You know he likes such jests—

SOLDIER: Well then I apologize, sirs, for my hot head, an' lambasting soul of sufferance. But I have been deceived by your relative modesties to this great issue of our supposed besieged state. And that confuses the emotions of my comportment. We are not for playing games here stationed.

HERKSHEAD: We receive with patience all commitments on your part to hear our.... points, and do observe the representative leanings or tendencies bemused of our condition. With charity of conversation we dismiss divisive temper, if it is on the whole.... inconsiderable to our purposes. But we note whatever wisdoms may be applied to it, and can use this to broaden the scope of our arguments or persuasions. And as the sheriff says, we do not mean to be your enemy. We do not intend on fighting with a trained a'(nd) valiant military corps. And we need not prove ourselves, beyond our presence, to have some demands to state.

BILL: But ain't it wondrous to be so cruel at one's vacationing.

SOLDIER: That's not cruelty, boy! It's privity.

BILL: I'd have my head to speak less blindly—

CRAIE: We assume the soldiers.... to have slight upsetting, towards our community powers we enhance and advance. But it is Balcome who matters, for more profitable addressing, and detailing of our wariness. We do not judge anything of a meekness in our laboring. And if the lord should try and want to prove himself, then *that* is our justifiable concern. We are not so dispirited of a population as to permit the passivity of a royal tillage.

GESTROMY: Sur-hoot! sirs. This is for his ears to debate. Let me take you to him. There awaits the converse for your action.

HERKSHEAD: And for this privilege we are stanchions to the need, good sheriff. (**Gestromy** *leads the trio out.*)

SOLDIER: As could be reeds 'round a pond swaying. The waters here be filtered through this unsightly menace. An' I were better fit with terror, than a countryman to fight. But these here are so undeserving of our sacrifice and our contributing to the stability of this state, with their fickle loyalties bending and waving of leaf as could a wind change the direction of sunlight, and with such a lack of appreciation to foster us.... I could cry about it. Yet that is how we're estimated, by weight of stone.... and not heart. So burns this water boiling. They should fear the heat, more than this simmering condition accustomed to. For spreads a power mindlessly of maim, without a leader. And I do hope our lord condones himself, to bind them into some stance permissible.... for a retreat to laud.

Scene II — *A room in the castle apportioned as an office. It is perhaps the most handsomely furnished space in the entire building, though not particularly large. A great, shelf-like marble whetstone, original to the castle, serves somewhat as a desk, and by it stand* **Balcome** *and* **Brontrol**.

BALCOME: Of all that could be le(a)'nt, from his majesty's court, to help me fortify my stronghold to this audaciously bickering territory, they send me you? Only you? A single man? I could have need of a team of stranglers at least, if not a division of pike (hold)'ers and lance wielders, to pick through the countryside in a regular show of force and army regalia.

BRONTROL: Need you more than what you have? I'm not exactly from the court, but as I said am originally inhabited of this town and may know best how to assist its management without more confrontation than has already occurred.

BALCOME: Occurred?! They nest about the castle boundaries of ground and grass, as if forming human bushes, daring for some storming retribution, or a distribution of my men to thin out, perhaps. But of course they can not breach my fortresses, for which a castle's made. And so a standoff mounts, like moles forming around us, a disease to ever contain primacy and constrain me into temper, angers with misjudgment, and wraths woven of confinement.

BRONTROL: These here are a sophisticated people of mind. They demonstrate in one way to mean to play at the opposite of interpretation, as if in amplification of an intensity to perceive.

BALCOME: What?! Opposing conflictions?

BRONTROL: So they may seem to threaten your stationing at this curious fort. But they may actually mean to welcome your occupancy of a historical landmark, as long as this residing is under agreeable terms for the security of their habitual endeavoring and livelihoods. A show of intention to remain themselves is all they mean. I myself am married to a wife, in this village, and yet main-

tain my independence by having stayed away over three years. And does not this marriage strengthen of itself, as I return? Well then, they come here— but to leave, as soon as they are assured of your lightness, freedom and levity towards their private lives.

BALCOME: I haven't much interest in your connubial affairs, nor anyone's private lives. But you're saying they make opposites of meaning, here? There's no straightforwardness of dealing, or association to motives and ideals? Against my own eyes, from what I've witnessed, there'd be much challenge to perceive that what develops here is not aggressively counter to my aims specifically, and my probity of attempted lordship in this land. Though no doubt your ideas of faithful relationship have been much strengthened and straightened to be more evident and obvious through your martial service.

BRONTROL: There's no one here who questions your lordship is a lord. I suspect this is a ratified conclusion among all of our inhabitants with strong purpose to realize it, children not withstanding. But, you might represent to some a waning face of authority, with fear for which to subject to, themselves less envious and more onerously tasked by simply through definition that you may not be realistically supported or buttressed with effective magisterial endowments—

BALCOME: My endowments are my fists! an' my recommended fists—

BRONTROL: But yea(h) to this proviso, that you are recommended for this village to.... lord-ate over. This has been pretty much of a self-governing town through most of its established existence. You present an interesting complication to be fitted with and abided. And there is some organized balking, to a meddlesome complexity, if it might seem to lose, even tacitly, a legitimate enforcement. One can understand that, if a work to be done lessens of its ordering, especially if it is difficult or tiresome or tedious for any reward. Then that is a nature of men, to try and confront it, as like a change in the weather.

BALCOME: So cowards show a bold face, to test my perseverance. Well I have already assumed all of this—

BRONTROL: But then.... we're not so cowardly of a make, lord. The point is more that they will be impressed with a startle, a jolt of comprehension on your part, to appreciate them warmly.

BALCOME: Their fierceness is their friendship? their fiendishness their faith?! Obviously your mode of quest is to pacify them, somewhat, rather than have me demonstrate an overt danger I can easily cause to their persons. I avoid initiating a carnage because it is impractical of a long stretch of time their numbers suggest, and also because I don't prefer such show of force against our own countrymen or any non-warring body. But that is the key tone to strike of sound— warring. If I am provoked by any attempt of assault to my estate, any try to breach my fortress-ing.... a very dire, firm, and sharp response will follow. There is a keenness for it already, among my beleaguered soldiers—

BRONTROL: We must avoid this.

BALCOME: ....I have my methods. They want to be let out and loose, to roam about the countryside and show some skills, to make a defensive statement as could be a bout of athleticism performed for offensive baiting and catering.

BRONTROL: That would be to ravage and offend my home.

BALCOME: Quite certainly, for you know what we can do.

BRONTROL: It would put the king's name into even darker light.

BALCOME: Well they can't have it both ways. Either he is sufficient or he's not. Either I am alleged.... or am alone. But I will handle this matter, and I don't mean with trick mirrors changing faces from smiles to grimaces. Your concept of contradictions in effects, while an interesting premise, is only a poise presented me, a peculiarity of character I'll take into account. But we are not for games to play. And instinct says to extinguish all threats quickly. Intuition says to do this as mercifully as is thoroughly possible for something necessary to do. But I know quite well what are the summer winds upheaving: a queuing for independences. If they think I've gained mine from the king, through an unfortunate default, then they wish theirs— in certain aspects— from me. Their consolidation is to try and feel insulted, and thus to insult me. This is not an opposition of effects to conduce, but rather a resonance to enhance and intensify, as a growing mood or movement, perhaps, throughout the legal realm. I may be forced to treat some as criminals, then, and to appeal to whatever solemn sense of justice is left to your home woods.

BRONTROL: My assistance to you, for these home woods come (to), is to ferry suggestions from the royal court, by way of superiors learned. And then in return to send back whatever your comments or pleas may be, concerning all refutation to your rulings or legislations attempted, and to describe the situation of your progress as a king's liege, and what troubles continue to happen for you in this regard, all of this information and.... observation and opinion received through a similar route to the court. I doubt if I should ever meet the king directly, to discuss these matters personally.

BALCOME: Not much influence you may wield, since your opinions should be what I say.... and what you see. And I'll tell you much, as an active man. But these suggestions you bring—

BRONTROL: To maintain a peace, to foster our civilization, and to stay a proud representation of regal authority and parentage over the entire territory to which you have scope.... as well as to try and promote advancement of some mercantile or commercially profitable industry here. They're counting heavily on that, from many regions, to help refinance his majesty's crippling and costly toils, for the nation to withstand and survive.

BALCOME: These— goals.... are the standard fare of hopes provided to be known and accepted.

BRONTROL: But not unlike what you have promised.

BALCOME: The promise seen in me, you mean. I want you, sir, to be a lot more active as a promoter of my insistences and reasonable demands, to the people of your home ground and prosperous earth. Not merely as a courier, but much less than a regent, should you discuss details with, for example, your police and judiciary officials, to provide explanation, allowance, and even extenuation of my conduct when forced to do what might seem.... contrary.... to the welfare of some citizens. In this way I ask you not to think in-

dependently, but interdictingly, for the eventual good of all. Though you will not have to bother to be cautious of your prevailing taunts. They'll have my full backing and blessing as mostly after the fact to what I have.... committed to effect as mandatory for improvement of conditions of my residency, these of my men, and with equity of favoritism for all law abiding followers of my practices to assume and identify with. I'm in of the urge to pervade in this land—

BRONTROL: You have it thus.

BALCOME: —It has been given me not for a petty license, but for valors so unordinary as to redefine commitment to his majesty. Yet here we are allowed to breathe out of the gun powder's smoke an' haze. A clean and settled place is this, and I want not a burden of the beast to play at this recovering abode.

BRONTROL: All soldiers are well accorded the prowess of your battling and steadfastness to achieve victory.

BALCOME: And that is a desperation mortised to while standing through the heavy assaults and raiments of destructive blasts and cloudy horror. Times my hesitancy here by a hundred to equal my aggressiveness of force during a fight between armies. Then see that I am restrained only by an accommodating patience for awhile, before taking retributive action, that course delayed.... with just surprise of these inhabitants assumed to be more grateful and granting of our occupying presence, as could more for an acknowledgment of pleasure with the king for this station to own.

BRONTROL: They've not tasted much battle here, except for the remote past, as this castle attests. Not one living (today) has felt it much, who remains to dwell of these surroundings. Yet that surge resides in all men to cycle through and have display with cause. And this is theirs to simulate. It is an unschooled imitation, or rather one too well schooled by imitators and pretending.... to be a party of resistance to.... the thought to have become facile of weakness, a monarch's rule.

BALCOME: This is too broad and blunt of a pin's end, to make for an effective point of jabbing— home! to remark that for all here I am that representation, without harsher measures of needling to make, not that I may weave a softer solution to spread than death and casualties among this population of your familiars. This is tensely to be avoided, since the king's decrees have already been made, and yours must concede to it thoroughly— or be roughly punished.

BRONTROL: I.... do not originate from the town, my lord, but have established a family in it for some considerable years. And these courtly people are not to be punished without a better reason 'gainst the haughtiness of an organized few, or should I say merely assembled to show heads. As I try to assert, it looks like a threat, but is actually a welcoming.

BALCOME: That is an odd panic. I'd rather receive gifts than guile. And how well would you say they surround?

BRONTROL: Sparsely plaque-d, but evenly and regularly distributed. There is order to the array. It is thought out to be more visually vicious than actually acute.

BALCOME: That I would agree with. But it grows; and the insolence itself alarms might, when too many ants gather at a foot. For even shoes may feel disgraced by their crawling about. I'm not about to make peace with formic natures, Brontrol.

BRONTROL: So be them to judge, that their demands do not sting, but merely slight of evocation for the king's prescience and pressure.

BALCOME: They're not behaving well towards me. That is the hard misdeed. That is the totemism wrongly exhibited of flaunt.

BRONTROL: Then rightly viewed, they are approaching for correction.

GESTROMY (*entering, leading in politely* **Herkshead, Craie,** *and* **Bill**): My Lord Balcome, a committee of the gathering locals arrives to greet you with all amicabilities of parlaying.

BILL This gaffs of astonishment!—

CRAIE: We do court your lordship for a better understanding of our views.

BALCOME: What for my vision made? about the tower seen, below, dissentient loathing and frowns.

HERKSHEAD: From the top of heads? We are smiling, to adjourn near the castle.

BALCOME: Oh. And from what meeting of purpose? but more to promote another, adjoining to my estate for a confrontation more foolish than bitter and more bitter than better of your kindnesses.

HERKSHEAD: Your castle. We wish to visit our relic from our early dawn, having decided for once to proclaim it important, as for its momentous meaning. Here stands a ruin, almost. Yet it has been commanded of an occupancy. Such a novel development must be investigated— And we are curious.... of the king's motives in this.... granting of your honor to reside among our folk. Does it mean, for example, that our village and its inviting surrounds are being paid specific attentions to? And are we, with graciousness, expected to submit to your executive gist for the administration of our public affairs? Or are we to be left as always to our communal handling, these abilities not to be abrogated by a questionable influence and yet serving the curiosity of your sudden appearance with pleasure? These few things and their relatedness of matters dealing to our liveliness we have thought about and disperse ourselves— individually, to *our* castle for expression of these considerations.

BALCOME: Always is as could be abrupt to never before a time, what's given in a past before your own, then making always not a proper term to use. And sudden is as slow to your precluded realization that I've been here awhile, and you've been serving *me*, with taxes and fees earned and delivered forthrightly, until some rumors spread of our majesty's decline in authorities to bind you to them.... or bid you pay me.

HERKSHEAD: And slow is as a servant's tongue, to speak of what he hears. And dull might be the sound, through incessant winds, but not dimly perceived if there is true and considerable change, if there is a real turnover that we may adjust to. We've

come with civility, not servility, to acknowledge your majestic attempts to improve this estate so symbolic to the region's history. We've not been blind to the time of your furnishing of care and interest.... and residency for the resplendent generosity of your nature, to convert our— offerings into a magnificence, at least in this room. Though we know that this is difficult, and will be a timeworn job of.... personal fixations. But in short—

BALCOME: As could be heights.

HERKSHEAD: —we wish to state the *voluntary* character of this endeavor, and to demand, as much we may, some recognition from you of this detail. For our people, with this cordial demonstration of unity in design, feel not about to substantiate any lording over our heads that can not be sustained by regal mandates. And we provide for you a view not of combative affronting, but an affirmation of our *chosen* rights to allow your peaceful vacationing within our.... region of community. How this may be portrayed as a blunt insistence is only for the eye to analyze of subjective judgment. But if we are not subjects as such, then we find no fallacy to the views we propound.

BALCOME: And what of errors to your solemn reasoning, and your disingenuous, clandestine deceit up until this recent disgrace of display?! Not subjects?— to his majesty?! I've won you, for your service to me, to enhance ourselves collectively of honor and renown, this very position— this actual location given me. And progenitorial of my hands you will persist of my commands to follow. For obeisance is a lord's law at least to bow at. And for least to do you will obey my injunctions and prerogatives— or be hotly scolded!

CRAIE: Scorch not your rightful manner, sir! We do not disappoint you. We do not say you are unfit to be at this established worldly shrine, which crumbles more with your threatening. We do not contend against that. But all throughout this region preserves auspicious sentiment, that we would not openly chastise such a brave captain—

BALCOME: What call your picnic, then? this terrible outing.

CRAIE: Yet robbed of leadership do you seem. Can you indeed pledge for the king? Can you cast his arbitrary laws about us without merit for our character and resolve to be resilient and stay outstanding? We do not wish for your overthrow or defeat— in battle, and within the hardships you endeavor of as commonly as commanded to. But we are not your subjects so entrusted to these severities, and will not leap, hop to them as a legion of honor. We are ourselves defined, in peace and certitude— and without conscription to your service or placation.

BALCOME: Conscription?! I would not take from here your mangily, spindly fowl— Or so.... if from here leaves your better traits, what's left I would not salvage from. Can you not contemplate the high valors confined in this stoned embracing to such concentration as could whip a sea with storms if let to rampage through your unseemly posture, what power is contained here bottled up! You are avowed to my service and placation— but certainly not as soldiers for me to become, rather as comely inhabitants to support my state and status through your most ordinary civil affairs, not perturbing your liveliness and livelihoods in the least, and with a considerable amount of self-administration of the crude details you yourselves must handle. But do you think such a circumstance can suddenly dissolve, what we've fought for fiercely to maintain throughout the realm?! Your obligations to appointed lieges has not been abolished with the casualness of drifting rumor. I am here to be obeyed as decent for your country management— and little more than that, of these moribund and boring burdens. For my interests lie elsewhere— with much fighting. And this is an office duty, and a duty of office, to reshape and sharpen your misbehaving attitudes into a proper utilitarian fashion with purpose for our entire country to serve— not merely my cyclical needs. You are not alone for yourselves to be, isolated to your particular splendors and natural resources, and self-absorbed with selfish haughtiness to retain such a richness unto your only selves. But this is like an aphid growing fat on a plant's nectar. And a promising mollusk given me to tame for its pearl, you do exist for many others, as only one component to a grand party, and of a particular opportunity for me to control. Yet a nation relies on you, as it relies on me. And I'll have you respectful of that notion. I will punish you thoroughly otherwise— which is to say I'll treat you fairly in every action made. Zsounds, that you would favor cudgeling as a treat!

HERKSHEAD: Sir, can the king really compel our hearts, as you our protest? He is unreasonable, and has become disaffecting, and is ineffective for any control over our lives. Yet we do not banish *you*, to cruel— treatments, nor scandalous ostracism, since we accept your stationing at this castle as something already done and passively accepted for its history. But it is not reasonable to think for anything else about it now, that your curiosity occurs within this pit of ancient fortune and pride. And you can not outlast our imprinting of assurance for ourselves, that we are pertinently viewed for your acknowledgment and acceptance of our terms—

BALCOME: Terms for me?! Termination for you!—

BILL: We're a nation *here*! We don't gyrate to your javelins!

BALCOME: What is a boy.... but impatience towards his magistrate.

BILL: We've positioned ourselves about you, for a very long and costly toil of temper—

BALCOME: I'm aghast!—

BILL: —heated responses and disputations!

BALCOME: —as you'll be, with the first sight of your blood spewing forth as recklessly as your mouth. You will scatter yourselves with shame, into holes gorged out by termites, for ratty infested shacks to retreat to.

BILL: I'm not afraid to challenge might disowned!—

BALCOME: Disowned?!

CRAIE: We—

BILL: Strong flame from a slender candle!

CRAIE: —do not deceive you, prince! Our means are solidified of a recurring wave of defenses, and tedium will bring you losses.

BILL: The extinguishment will be bright!—

BRONTROL: Gentle countrymen.... and men you are, not dogs of a mean caliber. You can not hope to fight these soldiers. They are headstrong for a hurting. They are machines of terror, born of tempestuous tests. You do not wish such a tally of harm to be tapered by. It is not elegant, nor timely for your ultimate purposes of existence and prosperity. For death is only prosperous to the worms, and its existence coldly cherished. Now what you mean to do is bargain of your terms in taxes. So why not say this promptly, and spare yourselves misunderstanding?

GESTROMY: Yes. You can not seriously mean to bring battle onto the courageous, already pent up for some exercise, as a soldier's just told you he wants some killings made—

BRONTROL: Not for a maidenhood.

GESTROMY: —Yours is an impractical strategy to avert a catastrophic temperament.

HERKSHEAD: Yet, what's to be believed of it? We are seriously atoning to a search for reason, and a gamble on what's evident pursues all involved. For, to instigate a battle here is the most criminal of passion, in this village. Yet, would it must be initiated from this castle, such maladroit behavior for a lord, it would be a country's shame and disgrace. And so we bring ourselves, Balcome, for you to truly reckon with (us) on this regard. Would you destroy your nobility so foolhardily, to bring to disrepute your working of a gifted populace? If not, then deal with us of fair compromise for our opinions, to prevent the outbursts of haggled hatreds and superficial seethes. As for a gentleman to be, we would wish to accustom ourselves to. And end the decrepitude of a king's leeching.

BALCOME: You would dare to bring to me a cause of fear in this ungodly belief, that I might be threatened into conspicuous relief of your incumbent acts? Yet for a war I need not make of heads, they are your own to bear. One does not wash the hands with mud unless what soils you 's worse than washing. And I strive heartily to avoid a murderous rampage throughout your country sights and harmonious entrapments of pacification. This is now.... one of my homes, and I must clean the dangling brush away with a controlled burning away of the inflammatory refuge of clinging, rotting vegetations and aged thoughtfulness brought to a decadence of pondering. For by happenstance you represent some.... soul of disagreement with me, some irritating grain that may be made a pearl— for my purchase. Then, let, at least, your iridescences serve me.

CRAIE: We do reveal the wishes of our public.

BALCOME: And that's a frank candidness I can accept hospitably.

HERKSHEAD: A two thirds decrease in the grains indulgences, since you've hardly more right to levy on our crop production as much as may serve your needs amply. And a halving of the annual estate tax per residence should be more than enough to sustain the castle's upkeep— that is, per heads in household. We should not be expected to subsidize an entire battalion brought here without our asking. They are entirely in your hands to house and feed, these.... feral ferule soldiers. As for yourself individually, we are apprised to subsist with your position through the custodian's excision, as a tribute declarative— often delayed and too often ignored— for a regional monument or notable of historic substance, agreed to through various roaming township tithes of ancient character or charter. So we have no objection to your lordship's stay. The notoriety is exceptional.

BALCOME: But as to these specifics you think to impose.... considerable restrictions to my earned income through this village.

HERKSHEAD: Without the king's overt backing, is this not reasonable? his official enforcements slacking. We are prepared to rename almost half of our landmass to a city of "Balcome," focused on this castle's location as a principal point of interest, sighting, map making demarcation and symbol, whatever is appropriate for the demeanor of such a municipality.... of your note....

CRAIE: These are fine concessions to your prestige and importance, Lord Balcome. The old town reviles more than this— compromise through present circumstances.

BALCOME: They are of sundry honors for subtle forfeitures.

CRAIE: The people are implacable against less of a *laissez faire* to abide with—

BALCOME: The people are my guests.... of attitude to entertain (*going over to a couched rack and unsheathing a broad sword*); and for these conversations to instill more powerfully a commitment to our worldly affairs (*returning with the sword*) we are cordially substantiated to be together for discussions and amplifications of intent. So witness such a fine implement of purpose, wherewith its sharp and handsome beauty connotes the simplicity of its design and meaning. (*with much physicality of effort, slams the blade broad side on the marble top, to cause a loud crashing though abrupt ringing sound*) I have my meaning made (*Some soldiers enter.*) by might of manor and this mansion owned; befits my manner to be seized not, but by contentions to provide— of spirited debate and commiseration. Escort these country visitors to quarters "D," for they stay here awhile. And of that group that tensions about us, they be called some sirs of what provoking—

HERKSHEAD: Herkshead!

CRAIE: Craie!

BILL: I be—

BALCOME: A boy for wanting of this herd, provisioning within my grasps—

HERKSHEAD: Of powers? Then we are prisoners!—

BALCOME: Guests!... to the simplicities of life, and the richness of solitary fortunes—

CRAIE: Detainment's cruel of you!—

BALCOME: Until I am persuaded of your arguments refined, as the irritating grit becomes more jeweled.

BILL: This is for heat of quandary!—

BALCOME: Aye!... (*as the soldiers start to enforce their escorting*) with your compatriots questioning for my time. I will review this business, with your external companions swelling to a stultifi-

cation, awaiting your results— of words.

BILL: Then I am pat'ered into (a) silent disgust!

BALCOME: And that improves the exodus for thought. And I am meanly tasked! (*The three have been exited.*) But captives hold themselves.

GESTROMY: This is not the gentle way, Balcome.

BALCOME: It is a method more sublime than sacred.

GESTROMY: But they'll be aroused outside—

BALCOME: They will think such progress (ha)'s been made as to extend their visit.

GESTROMY: They are not of stupidities coached—

BALCOME: Hope! is their quarry.... and lethargy becomes their stealth, as they wait and lounge under the sun— and burn for their impressions and imprinting.

BRONTROL: But what of news, sir? We are not enemies!—

BALCOME: Nay!... Your town is not an enemy to me. That must be learned. (*tossing the sword onto the table*) And tell your fellows there's much repentance due, to a fatherly love, but that we discuss thoroughly of a paternal plight. And eyes are being opened for a king to have his rights enforced. (*as **Brontrol** picks up the sword*) But this is a pleasant causticity, because I only assert what is properly ordained, while our visitors have been improperly assertive and demanding, and with the asceticism of quiet torture will reconsider themselves.

BRONTROL: Torture?!

BALCOME: Of mental lambasts, and slaying gab between them, reordering priorities with current realities. For we all know what sacrifices are made with imprisonment and the uncertainty of its duration, and what should follow based of one's confessions to a faith. Have we not dealt with many officers this way, of combating opponents? and often converted a few to at least resigned utility and subjugation. That is a mind's wisdom to survive with masters, or let it's body perish of a feckless martyrdom.

BRONTROL: There's more weight here than use.

BALCOME: It's a fairly blunt, ceremonial thing. Let it down.

BRONTROL: These are not soldiers to deal with. They have heart!... of an emancipated life without the regimental distempers and unquestioned obedience to any legitimized order.... (*places down the sword*)

BALCOME: Legitimated?!... That is rank, or nature, or men's nature, or whatever compels someone to do anything.

GESTROMY: You take a strong risk, in trying to reverse them. If they are as reasonably obstinate as an independent man should be, then what?— but to subvert them as an initial casualty among these— our— country folk?

BALCOME: That is what a commander makes of a situation. But to my mind this is more of a policing action, an administrative duty, if there trends in these parts organized attempts to violate his majesty's rule and rules. And I am being most gentle, to handle this re-visioning through the persuasions of bleak and stark solitude or isolation into conforming testaments, as much as one may reason through one's dilemmas. But for now, I consider what advantages I've gained through this, compared to a possible faultiness with this stratagem.

GESTROMY: I tell you, this populace will not be easily deceived for long. And you will have to not only explain, but convince them from the harshness of this treatment—

BALCOME: Then it is to adore the king again!— and my possessiveness at once an earned virtue. For our labors are hard, comrades, and often short lived for (far) too many a life of warring valor and incentive to overcome worldly adversities. The people here must now be taught to appreciate these sacrifices as something almost holy, the efforts residuals of (a kind of) saintliness.... and to adopt a shared resplendence towards a faithful majesty outrageously tried by so many corners of assault that our entire nation could be torn up into salt-land flats without his perseverance to keep it united and whole and fully for his ruling. For as we protect his, he protects ours. And this is one focus of our powers we must strive with. Not that men become ungrateful, but indecently perplexed of their commitments, then they must be tutored to correction. And the baying donkey can be turned into the bleating sheep, at times; the stubborn against the worthwhile disciplines into the longing to be led correctly into pastoral grazing and peace, is such a timely conversion meant. So what of sacrificial lambs that make this toilsome— there is a need to be rendition-ous of authority, in this place.

BRONTROL: We are august enough—!

BALCOME: To be beneficent and forgiving, and patient for their adjustment to me, and the confirmation in their minds of my statutory powers— which are only used to bring and keep this region into a conforming alignment for his majesty's best purposes? *We* haven't the time to wait, to solidify this locale into part of the reliable foundation of a nation. *We* may soon be sent into such a maelstrom of battle.... as to be away for ever, with hope of strength and assiduousness to the minds of these people, for the(ir) unrelenting maintenance of our country.... instead of condoning its dissolution through dissension and petty regional self-interests. I must establish here and now the validity of my lordship over this lofty land where its inhabitants could even dare to question me of my demands. Such as they are for a bare minimum of tolerance, then their skins are much too sensitive of the regal cloak. And I only ask for the least of my needs of organization to be provided and provisioned from these fine and flourishing surroundings. The hefty verdures alone could make one blush for a change in color. Then much wealth of produce should be given me for this estate to (make as) furnace for a functioning stronghold. And the people of this area should be glad to furnish, almost delirious of joy and giddy of work!... to be considered loyal and trusted. Yet, can I find the opposite to be? with hesitance and worsening of conviction, at the slightest foul wind of heavy fortune?! The king has lost a major.... conflict, as a single battle may be called— That's very true. And its consequences seem grave. But *he* is not defeated. And *they* have not won from us any land that can be fought for through a war, but rather what has been contended for and is still in the process of deciding between mights. And we will soon restore our grounding towards a prevalence, if might we be regenerated wholesomely and thoroughly supported— and absent.... of demoralizing sentiment. The king himself insists upon his personal fighting during these harrowing engagements— And should he fall.... let not his regal rules collapse and our structure be demolished of its strictures time honored and adulated by observers in this world as (often) fine and healthy and keen to envy. The people here should be cheering us with warm tears, instead of these cold, jeering remarks and stances against our presence. And I know you both share the uncomfortable feeling with me, that we may be taken for rejections earned to undermine any triumphs performed. We are graying uselessly, are these communal fears produced. And for that— maleficence!... we must strain their thoughts.

GESTROMY: The people, these encourage-able citizens, do take a creed upon themselves that is not altogether the totality of being self-served. What right is their subsistence on losses if they can provide more for themselves? His majesty would approve the firmness of character and the defiance against odds. Then he lends them us so that we may sustain a favoritism for him, rather than force compliance to his needs that grow ever more desperate in this swart time. He applies to his subjects not the callous hand, but an open one— for help, for restraint of temper, for agreements to palliate hardships until true victory is won or a solution's found that satisfies our candid worries and disappointments, and finally for some understanding of the weightiness of burden a leader undertakes. And as you are his most direct representative to these parts, in you this current populace must see his charity of wisdom and worth, warmth for weal, and the wherewithal to be redolently redoubtable.

BALCOME: Show them heads!... frothing of a positive palaver to my causes, conversant of a conversion to my aims, and goading others to my goals. That would balance my hand upon these people away from a necessity of striking.

GESTROMY: But the three you've.... encumbered with more dread were taken more or less by chance of being near the gate, for entry of a visit. If they are as official of their sentiments as unofficial of their presentation for the whole to characterize, then these ideas must be widely and firmly spread throughout the entire population, not to be easily modified or facilely— nor even frantically — spoken down. This method can not work, with such a casual deputation—

BALCOME: (It) Was for you to provide. I'll take any to contort. But now with twists, Brontrol. Go out there and contend for me. Argue with some enlightenment, as good things are occurring in this castle. And show your fellows, slight(ly) that you may know them, that a much better attitude towards me can be created with a *reasonable* prescience for what should occur, in that they start to distand from insult— and go home! to their causations of civility in such number as to remove their tenebrous view from my sight and tensioning of my nerves from my spine's equipment of action and retaliation. For I am hard fought here, to show some.... counter demonstration, as of my martial nature, but do contend with the constraints of a patriarchal lord.

BRONTROL: Then the temerity of instruction will be tested through somewhat of a tease. For I can not argue the saintliness of a lord, but only demonstrate that possibility through assurances— which I must have— that none of our townsfolk, whether visiting nor visited upon, can in any way be scurrilously attended to by your lordship's intentions of treatment.

BALCOME: Well, I—

BRONTROL: This is an honest predicate to base my assertions on. You see how easily rumors may spread in this place. And then for prisoners made? The longer you keep them.... studied.... the more of a considered and scholarly constitution the people will expect them to present in public, describing your agreements and retractions and substantiations, or the outline of which a discourse may be ascribed. But to eventually uncover battled bodies.... or composures.... can only acerbate the tensioning that continues to steam, and build upon discontents predictable, and by a few, but too many, expected of fulfillment for justifications to revolt.

BALCOME: Well, I would not in many ways mistreat acknowledged guests to a severity of accommodating manner. And our three gentry will not be so abused as beyond the limits contained in their own castle monument, which this populace has allowed to fall into such disrepair as to make the starkness of empty rooms seem in several places as dungeons of decrepitude and despair. But that is all to be seen and felt of a dank darkness, or dingy abandonment — of themselves and their own to revert to.

BRONTROL: Such be the case for tears, Lord Balcome, sighs, weeps and cries. I should not have ambassadors hurt of the physical. That is against military code and principle.

BALCOME: There be the terror for it, if they were simply picked from potential combatants.

GESTROMY: But that has not been demonstrated yet, by any of these— visitors. So it is proper and prudent to withhold punishing welts or beatings, spankings as were the honored to be disciplined as children, or any foulness imposed upon them.

BALCOME: Sirs.... the method is theirs— to provide me. But confinement is often a mendacious lie, for freedom to do things. Rats often eat up their own pups. And have that multitude out there to know it, with this pressuring. Then extol my virtues and extend my victories, to their minds, that I continue to worthily entertain houseguests, within environments *they* have chosen to maintain as appropriate of their character and caring, and perhaps even as part of an etiology.... of declines, declensions, and decrees. I own this place now! and will do in it what has been made customary. If this is uncomfortable to guests, ask *them* for an amelioration. Though I will try my best to improve matter— and the universe itself, for all that they would know of me and my profession. And what is yours, Brontrol?!

BRONTROL: ....A king's soldier!—

BALCOME: Then go to this— envoy(er)ing for me, to meet with, if not your brethrens an' brothers, then your best made men of motion— resting upon my grounds, to bid them do otherwise and go away.

BRONTROL: (To) Your Lordship. (*exiting*) I carry of a strap.... to support uplift and retention.... about my brier's mooring.

BALCOME: Of all this trouble imbrued, they send me only one man, a circumspection for the village.

GESTROMY: Do you expect miracles of discreetness?

BALCOME: I would have hoped for more than as is pledged. But he is bound to bring me reasonably viewed. Though care he somewhat for friends, I recognized it not.

GESTROMY: What means of you for these visitors, Balcome? They've not done you any harm, and yet you have alarmed them.

BALCOME: Or frightened or disturbed, made anxious or cowering to fears and woeful presentiment? That's the sweat of this castle's environs. There are torture racks and stretchers, and head irons, neck braces, limb locks and piercers to bind through muscle onto walls, pincers to delay an itch, and nose clamps to keep you mounted as an (h)'orse or lying prostrate on the stony floor. Such ancient equipment of terrifying banality found here, much of it so rotted away of cords and lines as to be unusable but for the iron parts. It depressed me on seeing, as like uncovering from moth-worn rags to reveal or discover an ancient school still revered in thought as it sags. So this is what was provided for the proud, once, a demented ownership if any at all. And the curing airs to breathe, as of a leather boiling of slow decomposition, what makes you think of tendencies, Gestromy, we have been brought to. Deserve these people better than that— feeling? Have we cakes to offer our guests? I'd place a rusty buckle on top of one, as an offering of.... suggestion, that they are minions still, and I prevent.... decapitat-ory slighting by regional obstructions and foreign enemies. Yet may these three dance for me already, with a speckled water to drink— of worm entrails floating and sinking.

GESTROMY: Dirty water, sir! That's an old misdoing of the earth's wine.

BALCOME: So as they may reconsider what is adversity. I'd shudder most of all at a wet and muddy grave. Yet there's more dust and dirt on the floors to bath in than a pool's water. And they'll make their own liquids from the gasps.

GESTROMY: You can not astound them into submission, or scare them towards sacredness. And there is the error of injury, unintentional debilitation and death— of wits or physical person.

BALCOME: This is what the edifice provides, from their allowing of its gray tomes' spreading flight and moss-bound coiffure.

GESTROMY: Demean them not this way to face themselves— but you!

BALCOME: I've offered them the practical advice, to accept me without much question of my station and mandate to govern. Now, let them find the practice of this, in eyeing what has really become about them, and sensing my.... disappointments and regrets of their standing positions.

GESTROMY: To entrap them in a pigsty with no seats or cushions? This is not workable, of their humiliation to bolster towards more respect for you. They would rather condemn further your intrusions into their welfares, with more irony of hatred than pleasure—

BALCOME: I said I'd send them cakes. Yet, do you collect the taxes courteously? They treat you nicely enough. I will assert myself in a manner as serious.

GESTROMY: This is placing the bit too high up (in) the mouth, and even the coursing courier may buck. You must obtain from these courageous gentlemen a favorable enlightenment, rather than a shading into despair and umbrage. The opportunity is not late to provide, with cordial banqueting to surprise them with. Rethink the trend of treads, Lord Balcome. They may lead you anywhere, as you would make them.

BALCOME: I'll sue, an' sow with Brontrol, for the moment, to judge the notions, proclivities and inclinations of those who feign to surround me, or have indeed become fiends. Taunting tail whippers and snipes, are they to seem or appear, before my counts of heraldry and armored muster. At least this heroic spirit survives with my internment to this noble and terribly tired property. But the castle can be made statuesque of feeling, as a notable landmark for the region, again outstanding and remarkable, with our residency. I admire greatly the simple, stern, and handsome stonework throughout the place, the bricks masoned in place with a remarkable, though dark and oppressive, stark and deceptive (for a number *of them* to be contained), character of fidelity to the purpose of fortress-ing mighty cavaliers. How more straightforward and unpretentious the early times of true knights must have been, when your immediate aims were evident by stance addressing an opponent; and not with subterfuge were you in glory bound of dress and weapon. The malevolent error, Gestromy, is that now we recline from honor in order to impose our tactics and results. This improves our successes or survivals greatly, but at the expense of an appearance of fortitude, that even country bumpkins would care to harass such force so unknowingly. But place one honest scarecrow in the field, and the flying marauders will depart from the ripening corn, terrified of a rich and deedful presence that may quiver and wake in the wind. The verve of man has risen with descent, his dash dulled to hidden or anticipated descriptions only, not to be acted upon, faced, or even seen.... by uninvolved, disengaged civilians. Aye! They do not see us— marching! parading through their faculties of awareness.

GESTROMY: My lord, the people know quite well that your lordship presides, and is around, is part of the forest of fruition— in some celebrated sense.... but whether as friend of fowl remains for you to demonstrate. Feast our visitors, and assuage their heavy, hostile worries. Show not your anger to be relied upon, but rather your reason and calculation that may be more respected. You detain them.... only to make clear to them your understanding of their approach, and not to incite calamity. That realization will put them to a better balance for your leveling and adjustment. Avoid harming their psyches. That is more impressionable than physical wounds. The results may outlast any possibility of healing or recovery. Treat them as honored guests with the utmost hospitality we can afford, from this interesting garrison. Bring them parallel to your mind, not askew and irritated, not perplexed and depressionable. These simple few make for a representation of all caution with intelligence. Let's not abrade this prosperous potential.

BALCOME: You champion them much.... from tax collection. But then, you are responsible for these particulars brought. They're like your cubs, I suppose, for which you've promised them a decency, I can easily suspect. Or to your heart is bound this prospect to fulfill. Trust more my lording than these supplications and requests. One doesn't have to train me more of a patience, nor a posture. For as I've said already, they bring their own method forward.

And I will let that take more hold than a grip of straps. Whenever have you known me to abuse a prisoner?!

GESTROMY: Never, my lord.

BALCOME: And have we prisoners? Then disabuse yourself of the thought.

GESTROMY: My lord is generous, then.... of deceptive inference.

BALCOME: I allow them to question themselves, awhile. Is this really a proper attitude to take— within these very rooms?! within the interiors of the prominence of lieges?! That alone suffices for more turbulence of contention than a cauldron's boiling surface to be mindful of. And they will displace themselves, out of so uncomfortable a mindset. But, to be gratifying towards their progress, I will smile a lot at them, when they are brought forth again to speak opinions and conclusions.

GESTROMY: I remind you. They are not of the soldiering strictness—

BALCOME: What tries to be is evident.

GESTROMY: —They may bark more, than bow to your conjectures.

BALCOME: I'll dine them well— in my presence, what rations they have quartered for us with. They will devolve, Gestromy, into agreeable capons.

GESTROMY: A blend of men makes most the meaning of more —

BALCOME: Then all will become the autonomous amicability I drive for, if there's to be some sort of independence. And yet to me, the kind is draught. Man's independence under God is only a blessing to have prayers. But for the king we'd serve superciliously our own begotten natures. But for a leader are we nurtured of raw foundations instead of finely constructed ones. And so, to see a people shift in line.... is pleasant, pleasing, placatory, perfumer-al, and profound.... of an improvement to be spoken, on adaptation from this process they perform— around me! to circumvent this rest. Oh, but for a job of sinecure this has proven to be, with haste of horrendous homemaking— I'll not allow it so! There'll be more haughty protesting than their own to sore of— if my way is not produced, and if they clamor more of pothered bearing to my head! that I might absolve the tranquilities of death in expedited cemeteries. Yet over my steps of foot an' shoe— defamatory bother makes for this a dimension of the empyreal detritus found— come to; and all in flames of resolution am I baking, to improve this jurisdiction for the rights of loyal subjects, observants and adherents to our royal legalities and the supporters of my lordship. What else, then, should I do but fight for this imposition here! And it's not anger that commits me, Gestromy—

GESTROMY: They've been rather quiet, out there.

BALCOME: —but the angelica of my wine to drink and roots to establish in this faithfully apportioned possession of country and estate— which men have earned within the devotion to hardships and cruelties on a legendary scale, an extraordinary compass of struggle and sacrifice. And I'll have the people here dance to this recognition, instead of pout— at— me!

GESTROMY: They've been very orderly arrayed and ensconced of a, so far, polite determination to make a visual statement.... perhaps more about themselves than you. Err they not to be discourteous, allowing free passage to and from the castle for those concerned of such business of visitations. There is no particular decadence about them. They simply camp to show themselves—

BALCOME: And say what?! with this demonstrative bilge watering.... that if a confirmation's made, of a royal departure or collapse, that they would possibly take over these grounds— and throw me adrift? What tedious hegemony of factitiousness to be led by, and falsehoods to clamp onto, as for a feeding to be weaned off of. And off this property I do insist of them, for it's obvious they accomplish nothing but annoyance and intentions to chide. Where is their work? What is their labor? To whom are they responsible? What conception precedes this doleful duty of dimwittedness, except as augury of fierce remonstrations? Perhaps I should present to them directly my harried scowl of scorn to intimate the displeasure of their childish peeve. But all in a like, they show their faces as a might. I can condone more punishment than their lack of comfort and lapses of mind. Pray to them for a toll to pay, my sheriff, that they disturb the ground—

GESTROMY: What?!

BALCOME: —living upon it as if the dead unburied. Go out there and collect a fine for me—

GESTROMY: This is not the way to take a temperature!

BALCOME: Yet sully they my grasses, dampen my earth with their foul perspirations, and dull my rocks with their polish of body, or embodied laziness and lassitude to poach themselves under the sun.

GESTROMY: Could you have me beaten for the sudden insertion of a fine into their collective mantle of purposed organizing? Or not yourself take an impudence of intolerance to their arranging, and plead for fines! Outrageously bold to do—

BALCOME: But of calculated steps, to pressure protesters more and more, I'll take what is allowed an askance mowing bother— And if refused.... then be the method made, for this intolerance to assume.

GESTROMY: Can you be prepared for a blame of blasphemy within the town? We can not fight these folk honorably, or precipitate a warring incident unjustifiably. This is, after all, your peaceful retreat to be understood—

BALCOME: And handled. Out there! with your verbal warrants, Gestromy. Go an' threaten them!—

GESTROMY: Lord!... Keep them puzzled and amazed, with your relaxed attitude to their intrusions. They know you currently.... as a wisdom of portrayal in these events. Why spoil the charm without a slightest gain? If you seem to have advantages felt, they are contorted, distorted to me, my eyes, within the obscuries of chance and imprudence. You have guests to entertain, or prisoners to (of the) main— What could the amassing think?! to have them taxed

this way? You'd display more than they, what for a fallback position to assemble— fighting and harsh rancors flown! for love of castle. Keep them more perplexed, as to your strategies. These contentions may be resolved, without much battling at all. And can you be as sure of your contracting of fine soldiers, after having raised the flag during a vacation?! They're needed fresh, for other fighting— if the king remains!... true to his regal rule to have your services and employ *his* men...

BALCOME: What deeply steeped counsel! You take on the oils of this population, as to hide yourself amongst them, with a jolly jocosity of bellicose endeavoring towards my regency—

GESTROMY: That's not so, sir! I am.... for you to destroy. But as yourself, it's not legitimately done. The king reigns here. Would not you have him standing? Then I would call it to disgrace, a devastation evolved of this quiet region, where after the king might not stand. And good management can avoid such an outcome. But to be so weakened by a stab, so deeply in the country, can be too difficult to recover from, considering our king's fragile position of rule. You are asked to help strengthen him—

BALCOME: I've only asked for the practicality of the motion. Judge you yourself the necessity of my levies— And then see how I am faced, to make demands to keep from spoiling.... of my administration. Yet even you find fragilities in steel.

GESTROMY: Temperance, sir, will provide for us its merit. This is a good land to be taxed by, but not with tyranny made overdone and wretched. And our majesty must in all cases prove himself, as with your supporting here. Now, should there be civil unrest, if he should actually fall? Or can you induce this, to help topple him during his frantic, frenetic teetering? We may guard against that with some restraints of action—

BALCOME: I feel more murderous, just to prove a point, just to allow what may be conduced to happen, under this atmosphere, that unfortunate losses can occur with such occasion, when minds are plyed to demonstrate their thoughts and contagions. But you must better mean a hesitancy. For I am not restrained at all, and will pounce! onto all weaknesses to wield for capitulations. This such is our state here, Gestromy. And while I favor your advice, work quickly, to make it more sufficient for my needs. It can not hold off or forestall for long my more decisive actions— and our developmental wont. And time slips not its steps to meet my manor more handsomely described as totally expunged of these visiting heads who heretofore did not care a trite notion to see their decaying castle, nor partake of any caretaking at all. I'll not be a judge to this indifference or disinterest, but I will to my investiture for its ownership to protect and defend from city vagrants, vagaries and vulgarities of thought. To think to own this household now, this.... monastery of time and history, elicits the impiety that sieves for severance from my effective reality. I see no reason flowing to hold back from issuing a hard and heedful response which questions the sanity of mice entrapping themselves to the python's coil and bite!

GESTROMY: Then it be worthy, sir, to sift for the servilities of mind requested, or to try and find some circumstance as to evolve this naturally, and without a dictum of distasteful mourning. Yet have we here no enemies to fight, but of our own behests to squabble. Fruitless! Fruitless!

Scene III — *Outside and before the castle, of some distance away,* *along a hilly terrain. Sparsely positioned, when met in person, are members of the local protest to **Balcome's** lordship at the castle: most of them sitting or lying on the ground with an apparent indolence to arrest their time with waiting, few of them being actually weaponed. **Brontrol** comes up to one of them who, while squatting on the dry earth, holds a stick erect and between his knees.*

BRONTROL: Does make for the sun a clock of your disposition.

CURT: Off!... This is misplaced, and parts the shadows for its notches made.

BRONTROL: Then for a growth of plant, transformation. Transformation, friend.

CURT: You come from Balcome's belfry? I see them enter and leave, and notice every event. But I know we are observed, mostly from above— in steeples, spires and towers. And that is keenly said, to hurt the eye for our intentions.

BRONTROL: I am a soldier of the king's, and a desperate man. But what can possibly be you?—

CURT: A sitting member of the "Valley Inheritance Defenders." And the VID does make us cock-some to be sore, more than a king's tendentious envy to have his servants, and stow them anywhere without question, usurping righteous tendencies of an established region and its spiteful inhabitants.

BRONTROL: What?! Would you fight with that?—

CURT: I count with it, sir, the days spent in protest, or those exchanged with others for their own. Bid me for the evening, I have wasted yet another day's light. And that's one day closer to your death. But then to tending of the necessary chores, they are slowed of accomplishment, and delayed with slacking effort. Though the people will shine, as the VID recaptures a possession.

BRONTROL: As like a fish you wait upon, in trying to expel Balcome? But you see.... your king does send an interest to him, of perception—

CURT: No sovereign's here to dare of an expansion for his feet! as could some slippers make. And I doubt you, sir, as more of army than the lord's soldiers, stocked of fright and fear for a genuine people.

BRONTROL: And that could be myself as well. My name is Brontrol.

CURT: How(do you)'do.

BRONTROL: ....And this is a region of my favor and particulars
—

CURT: With rarity of visiting? Then you are refined, somewhat. But the caliper for our measurement is this, sir: We are tenacious for our growth, to demand more suffrage from a weak-kneed monarch, who can not defend us more than ourselves may do, and who, thereby, must not make decrees of misappropriation. Or to do so prompts a challenge for whatever is lost, be it only dirt or relic's dust an' dinginess. I know I'll pay no more— for memories. But since this lord's arrival, my taxes have gone up, my savings dwin-

dle. If this tide of wave is no longer to be enforced, I'll sit to it for a departure, and a lessening of the crest, from waters that have drained me and rinsed away some of my fertile land. Yet the erosion is more evident that Balcome is left impotently spying of our contempts and is fretful and doubting of a royally sanctioned outcome. And so we wait for him to appear, from off the heights, and bow to us with much concession offered. And so, do you bring an amending, to make this task more feasible?

BRONTROL: I question you, about your rights to purchase in this way—

CURT: And as I thought. The man does dally, as his master fades of relevance, the ebb still soaking beds of heated contention. Yes, it's hard to give up an authority. But if your power runs out, and your backing is stemmed, what have you left but to question and poke around for worms?! None here, king's soldier. We are replete of eel, as could be forced moves to lead one to compliance of our granting. Yet can it bore to sore (an' soak).

BRONTROL: One could sense you feel capable of any achievement, with this lazy wake. And mount you hilly tensions to provide. You present yourselves most bravely for attack, and are fearlessly complicitous for your own annihilation. I've seldom seen such a patient target(ing) made.

CURT: You're for an itch to solve, as a lord's soldier. But what would be the point to it? with arrival of the aftermath. There are much too more of us for the risk of an attack to be advantageous of your survival, as you can easily see, as I'm sure Balcome (h)'as alerted you. So you come out to find of a gentility to reap, with averting of any battle. This has all been well predicted and assumed. Yet foreshadows your remarks the tenderness of inquiry and a timidity to arouse or enrage. You walk around and about to us with almost a pleading characteristic, that questions of some begging for understanding. Your stance is.... tentative, your pacing sensitive; and your mind's view is selective, to secure some diminishment of the pressures you face, while encased in the old stone. You seek more solution than salvation, for you know your positioning is loss, as we swell with exchange and replacements of members, and plop before you a defiance of being self-evident—

BRONTROL: And arrogant.

CURT: Well, that is the nature of a right, isn't it! if it's already won. But it's really a simple physics that falls upon us, to give us much leverage. Lord Balcome must subside some, to a titular arrangement of powers, or abstention from them, if he is to remain.... inheritable. We are cocksure of this. And as a rooster bakes of the forgotten sun— For we ignore its hardiness— we'll stay for fall of an empire, even, with such loose threads of weaving as to leave us arrogantly patched. And is this not combustible wood, of our sincerities to lie, stretched across the grassy rolls an' knolls of boundary and boarder, the hillocks fetched for comfort of array, and moistened treacheries for our display perceived! Here lies the blood of the valley, and not for lord or count to harness as what's won, what may be given as a lark's laugh— by a loon of lion— Nay! Now come to us and ask.... what will we take, what will we take away. Poll of our seditious might to make a statement, before those caved in rocks of majestic ease an' eaves, and a crumbling fortress spiked with cracks of sunlight, as for my name is Curt— to see it shivering in the heat, an' fall upon our sight as languished, old.... and empty!... vanquished! of a guard.... guardianship, lordship, loss. Yes! I be overtly taut of pride to die this way, should yours make a charge at me. So make more questioning, to characterize the crossness? or the barbarism? Come steal my stick! and might a hundred more receive in blows.... It's hot, out here.

BRONTROL: I'd say, go home.

CURT: I'm not prepared to pitch a fight through languid, drying breezes withering the skin as for a parched languor.

BRONTROL: I'd say, transform, back into your proper domicile to return. Can you stay here through the evening— through the night?!

CURT: It can be fun, to spoke the spooks, likened of a camp's stay. But I'm not part of the eventide vigils. An' in truth, not many are a party to that. The din is quiet, and not terribly responsive to our sighting, then. Come the lowering of light, I retreat to an unearned supper. And to try and get some work done, I might not be out here (again) for a day or two, unless called to come.... which has not been. The VID does not expect your fighting hooks and blasts, but more some patronage of pact, under these pressures.

BRONTROL: I am sent as a direct envoy from his majesty's forces. The king is alerted to the situation of this place, and the troubles of Lord Balcome. So there is more than a slight concern for your dispositions. You may tell your membership of this. How this can be resolved depends on many factors. But our sovereign is definitely displeased and saddened of these events, not that some eventualities may be disposed of thoroughly. And I'm not happy with what I find of luster here, with a blistering of skin for no apparent purpose other than to be bold and careless, and to involve.... an innocence of mind and thought, led craftily for deviant practices, to find much hardship awaiting.

CURT: Oh?! An' where is the craft summarily stationed, soldier? (*getting up*) In that crag of a castle? Do you address me as would the king, to run off as frightened of your derivation, your causes of ownership an' rule? The heat (that)'s here (ha)'s become regal, I see.... An' such a broad sword, is that your diadem unwound?! We are policed, suddenly, as for criminals, bandits, and thieves. Well, I've this stick for you to chew or chop, what for of my patience. (*starting to leave*) Think I am afraid to die, for this— hardship?!

BRONTROL: It is an organization that impresses. There is where the power's wielded, as for a fact of its existence, but not for what surrounds its patterning or checkered enlistment. I do not cast a spell over your enticement to harm.

CURT (*turning back to face him again*): That's not our purpose, rotting of the soil under this heavy fog of resistance! The daring is for dwindling effects. And fear is a wise superstition. Yet you could slay most legally, I suspect, my stance for sitting there. An' to pick on me, what burden more? I'd open my chest to you! were there more mandate than my caution, my family's survival, and my pride to be seen.

BRONTROL: That's not much of my temptation to enforce. You test yourself with teasing— much too many. But if you've not known of me to recall, still I must be a direct representative of the king's.

CURT (*kneeling before **Brontrol**, in mockery*): Oh! Smite this crude crown of entangled hairs an' horns! I only breathe to be persistent of your ravaging. And bother me thoroughly, out of this dissent. For undoubtedly you're on patrol for carrots! an' rooted sweetness to this ground. But bleed my sap then, cruelly, and with a hardened heart for love of country and its principles. (*balancing his torso up with hands and stick*) Endow me of your thwarting strike with a hewing of weeds. I am Curt! the naysayer 'gainst my dogged positioning. (*collapses to the ground, though due to clumsiness*) Off!... 'Tis grasses better felt than fields of (the) strewn an' torn. (*slowly retakes a sitting posture*) You do order me for something to adopt, with these shrewd and secretive gestures, hiding of your privilege an' high rank, approaching me as could a servant to 'is master's hound. Feast of my sad sight.

BRONTROL: I'm for more courtly demonstrations than this. But with anxieties am pitched to preen out quarrels, Curt. I do happen on you, as a tranquil curiosity to find, that awaits.... words? messages, from within the castle brought? But you have seen all of the goings on today? Then know you of a deputation entered, and you must be wondrous of their progressions.

CURT: I care for my kind.

BRONTROL: Yet then, upon more hearing do you pose. And they are possessed.... of my concern.

CURT: A funny broach I do assert, for this cask of minion to draw from. Of a few standing near the gates, more spirit showing of their faces did they make. The boy was particularly rambunctious to be boastful, which drew the attention of the sheriff to invite him in. Swiftly, upon instinct to protect, a couple of his *aîné* superiors joined the lad. I hope they instructed him for more brevity of remarks, an' silence of sensations. And they have not been thrown out, yet, which is a good sign. But.... what's to be espoused is as clear as day, and without mystery arrayed, nor of a mind to even presuppose. We are upset, with this lord's.... opinions concerning his state and status among us. Matters were left to run as usual, for awhile, with dues and fees collected out of a conventional malaise of acceptance and regularity, however vulgar of a common plight, where pains are suffered and afflictions felt, our economic woes remolded into welts of mind and character and curved spines to realize losses an' agree to them, and survive on diets and material deprivations—

BRONTROL: This is more than sufficient, a village of substantial bounty and custodial wealth. It may even bleed riches, as much as fat is oozed from an overabundance. One would not think to leave to care what's worthwhile without such an environment wedded to, or bedded with.

CURT: Sir knight, not one person— or certainly family— has escaped the effects of a growing taxation and drawing of revenues in recent months and days.... And a winter comes to leave us barren from this trend. Anticipations are mean and frightening. For in a closed womb as is country, what is taken of is stolen from, one class supporting the losses of another, the waste and squandering of resources. That's clearly seen— the rhythm of the dance, of industrious bees for the beekeeper's honey to endow, as we are born of such action and fit for its procurement. So be it so, that this *must* be, then exercise does often fight the malaise, and the fact of doing (so) makes our wealth even as it's split and siphoned off for more and more fine(ness)— or less— of our personal gains. And fight we not as weather royals—

BRONTROL: Those three—

CURT: —But circumstances change the storms, as royals rain themselves into the mud, and skies clear, somewhat purified and refreshed, as if with signs betokening. Resistance from the airy, supple clouds? the wet earth dampened for our treading and trod of difficulties to make more of our way. For as a ruler fails, or falls into weakening deprecations, his immediate officers do wean themselves away, for their own controls of nature to establish. And this is much different, for an acceptance of such authority. It is a challenge, not inspired of a king, nor of gods more evident than ourselves, nor to provide for more than for ourselves. The recent news, of the king's surrender—

BRONTROL: Surrender! 'Twas just one battle fought of many. One can not win them all in a lifetime. He retreated, loss some ground or territory, and has not been caught nor demanded of for imprisoning. There's been no surrender. That information forfeits reality. Ooooh! The image of him falling to his knees, perhaps— but that is taken out of context and stretched beyond all recognition of the facts. The battle was lost, and he may have sprained a foot to stumble on the disappointment. Sad to say that it was sharply observed, keenly noticed, and not deniable, by enemy nor our own. But he gave up nothing to it, but a patch of land— not his heart, not his determination. I have been told.... he only works to restore his forces, and take tally of his properties more thoroughly than before, as his rewarded generals disperse to rest up an' heal. But his monarchy has not been debilitated by these.... outcomes. He is not powerless, nor losing of any command, only more circumspect of his wounding— with enhancement to his pride. For *we* are his pride. There's no charge of legitimacy here, and the country does not unravel of his administration. And you remain as before, cohorts to a proud leader, supporters of his tendencies— on your behalf, and laborers to the benefit of nation. Nothing has changed— But of course prices have gone up. Expenses have increased throughout the land, to pay for the expenditures that have been spent or credited for use, helping to maintain our solidity of faction under this ruler. And, with great charity of independence, he sends out the various governors and lords to bring for themselves of this substantiality, with methods.... penanced of all suffering, respectful and understanding— but thoroughly obliged to be. That you have done your duties, up to now, as we have done ours, makes illustration to the validity of this very crucial point. The efforts made have all been totally deserved, and continue to be.

CURT: That would be the king's argument, or a fighter for this lord, or a servant for the sovereign sown to his gown, or a coat of arms or emblem. But news of our king's.... decline, it promotes our resistance to his subjugating servitude and resilience against his surrogated forces. The soldiers in that castle are no longer the king's, I would submit. They are mendicants of Balcome; and we have provided no rule of law to feed them, if this is effectively the case. What.... can you drive more of this to comprehend? The lord is isolated here, must fend for himself— or *buy* up more properties and workmen, pay us for leases and labors, and share of the travails and boneheadedness for agonies. But we refuse to make a gift of this, as of the king's mind unreasonably and unwarrantedly to do. It is too casual to accept from the fleeting— You.... may only support your artifices, to remain of your profession serving, your predilections of loose living and cunning to win fights, an' share what of these profits only your valor can contend with. But we have our lives to hold— more firmly, much more registered to our continual struggles towards productivity, rather than decimation— Not that I value less your abilities. Yet, ours are cultured for a finer defense to undertake and provide.

BRONTROL: Those three.... would argue this hollowly— and you should care for them. Balcome is a lord who is tough to fight his way through obstacles. That is his manner, when presented with obstructions. And he has won with distinction the governance of this territory— He'll not place forlorn hopes into his head. He will use his angers wisely, to accomplish his aims. He'll let your protests linger into lassitudes, to turn you about into wanting of his ways out of your own disgusts, for laziness and futility, to grow intemperate. His goal is to change you, and not himself— for his position's won! And can you understand— the significance of a victory?! It is more than what is fought for. It raises the art above the vainglorious— It is the breath for being sound in these endeavors and for life to serve. Even a pauper or a slave might be proud of victory, to excuse all demeaning that returns after a war. So Balcome will not back off from a quest for what he has already gained. And your heads, or a few, will be turned into grotesque caricatures of your defying thought that you will hate!... to see or feel as lent to your defining characteristics.

CURT: It is a dull.... intransigence to spend our time with. But we do not restrict the lord in any meaningful sense. His come and go, to and from the castle at will. And as for transformations.... call it Balcome Castle— I'm not offended. I'd still scrub its stony steps if employed to worthily. And for myself to think, still shines a sun, still are the flanks to warm when sitting under it. And still are we able to be cautious and counting—

BRONTROL: Not enough!

CURT: —of our fortunes an' deficits. Now, have you seen this king?

BRONTROL: ....Not personally yet, if by a means of addressing do you mean. For to see is to be seen, 'tis more than a king's presence makes. And you've not spoken with the sun, moon, and stars either. Yet you respect their transits and regular astronomy.

CURT: Then he's a phantom to you.

BRONTROL: More to 'im than a season of shams an' feigning vassalage. I'll convince you of this: He is real flesh, not the abstraction of a ghost that disappears or a concept that dissolves away with rumors. He does cry as a god yells dissatisfied, and imports an impression of regality to our bearing. And he's a huge man, not so much in the physicality of limbs proportionate to a human, but truly through the aura of his glow, through which is transmitted the commands of supernatural imputation. How else are the worldly to be handled, but by a figure touched with greatness, someone descended from heights and ascensional, a carter and transferor of gifts— beliefs, and divine inspirations.... from the otherworldly and indescribable. For that is a king's matter to be made of, as a ruler of vast stretches of people and animals and vegetative implacabilities which would be useless to all advancement of culture without his solemn management and manipulation— and bearer of all burdens substantial. I tell you, between the heated engagements of much fighting and horrendous bloodshed and slaying, with a tiredness of arms and legs that could put a giant to sleep for thousands of years, and with a deprecation of heart as to leave me barren of consciousness, I came across a recessed nook of some ur-ravaged house, the entirety of its inhabitants thoroughly annihi- lated— men, women, and the blessed innocents of childhood— by the convulsion that captured its location and surroundings. And I kneeled to an agony of praying, for some solitude to sense and be lifted out of the depths of human transgression and the depravities of hatred. Yet, whereupon my momentary stay, I came upon the simplicity of common, pretty little beads, of as fine a texture as could be ate for its colorful shine, the cluster hung there as would an infant do to store a plaything cherished and honored, a piety of natural charm evolved of being so deliberately born of this suffi- ciency of presence and presentation. It was.... of its own evidence to be, and without question for its pleasing to be rested before my face, what thoughts it may have wept of.... but resisted. And from this stash that showed upon my eyes I saw an image of our king, as from a revelation made, that the simplicity of these effects are un- questionable for us, and overpowering to realize that all justifica- tions— for whatever happens— lie within his rule, and his repre- senting us, to burn of beastly natures and purify what's possible of our salutary existence. I can not explain the religion. Its aspects are above my comprehension. I could only feel the surge of strength restored to me, as I withdrew from this.... kindest sacristy. Our er- rors, our crimes— all that we do, for good or bad, righteousness or hideous wanting, are all the blame onto the finest to be held and suffered of, that fate elects some person— over us— to be terror- ized!... and with full resistance from the weep— the weaknesses of our human art— ruling over our affairs and actions. Then as I maim for the king, I am absolved of cruelties, of shame and sorrow and sacrilege and sacrifice and death. For this one man contains these responsibilities— ours.... to meander with and sigh of, as however the gods of nature and all-knowing may lead him. That is the value of his majesty to keep, and not extinguish but stay bound to, bidden of and abided.

CURT: What?! My hands have not become so fouled.... as to need excuses from this king, as were to break this stick with and cast away (as pieces) some wasteful effort. Yet as I'm sure there's evil I have done, not near the intensities of yours, a king does not absorb the poisons of these guilts felt, for the wrongfulness of my past ways— Yet.... he has maintained a country for me.

BRONTROL: I have done horrible things, Curt. And I come to you not to teach of them, or learn of them, but to consider them, for what they have been worth, to prevent an overrunning of our lands with enemies. Then.... have you fiendishness in mind for Balcome, and the similarly enlisted? Just think upon this question. I insist of nothing to pour doubts. But a more proper way is better to repay valor and bone-crushing deeds. And I would serve your incidental ambassadors some license to be treated nicely, by aban- doning this strange hope of browbeating a tremendously capable lord. Discuss this among your members of the VID, about what a resentfulness creates, what a disappointment beckons, and how shadows fall not to enlighten but to overly outline to a point of ut- ter obscurity of perception and contrast, when done as could a night. Then dissipate of this queer vigil, would be my recommen- dation. But I only propose it is not necessary, since these taxes in- crease one way or another by the nature of our expenses and indul- gences to climb, as we strive towards higher satisfactions.... in life. And I've never issued a command in my life, to those too indecent to observe the need. There's more temperance to my rent of coun- sel.

CURT: Well that's for least a kindly persuasion to adopt.

BRONTROL: There are no phantoms out of mind, sir. And I do blame my worries to my cautions and sadness for what develops of this valley home, to see with pert constraint these features (*starting to walk away*) roaming. Then to another might I plead to for an- swering of this conceptual quarrel that makes its vortex about my paces, and does follow me for blame. A good day for you.... Curt.

CURT: That seems to me to be a sad soldier, but a fine man. It's too true I am of waste this way, to circumvent the fears that pulse about an assembly for our courage together. But as for one, there's always much uncertainty. And what is good about another day left fallow? But then the senses stumble and turn faint. What is assured is mist before our argument, until that which it becomes.... its cause, the meanings rendered of a haze and fog, from which I know the truth of this to capture: There's more ingratitude among us than our gracefulness implies, or our sleep bestows. Yet can a creature truly know the cause of its creation, to be fully apprecia- tive? Then only the perceptions matter, by way of explanation. But how can I see more? What is said is what is seen. But what?! The sheriff comes among us again. This is some dire attitude to take, compelling our compunctions to direct. Yet, spare the seed that's sown.

Act III

Scene I — *A dungeon-like enclosure, but with high window light- ing. The floor is sparsely with a mixture of hay, grass, and dirt, to annotate that the space was shortly kept for holding animals in- side the castle, awaiting the various of tender. **Herkshead, Craie, and Bill** are standing at various positions, as there are no chairs for seating. A **guard** can be seen leaning alongside the open en- trance, which is without door or gate, although a broad wooden board can be propped up over the bottom to a considerable height against leaping, which can be seen nearby abutting an inside wall.*

HERKSHEAD: This is fetid!

GUARD (*responding off-handed-ly, as if only to add opinion rather than answer inquiries*): We've kept the stall.... for lambs, for shearing wools.... Then what of them— growths? Not much of that detail.

HERKSHEAD (*with effort to ignore the **guard***): The place! We are offended. This is offensive!

GUARD: Not without much weaving are coats made.... But the stews have strong character, carrying the aroma. Nights inside can become very cold. It was a surprise. But the walls are anciently built, and defend the ambiance of an evening.

CRAIE: We are tired (from) standing, sir. Offer us chairs— stools— seats— planks!

GUARD: I haven't any around. Try the dirt of the floor's pastiche to mix through and settle on. It is comfortable.

CRAIE: And might I use that (*pointing*), what's at the wall?

GUARD: ....It's too wide to be a weapon. Have its service, sir. But look out for the burrowed worms.

CRAIE (*coming for the board*): Thank you! I think it's more my nerves that tire me. But we've been standing a long time.

GUARD: Can not have been, if you're only just arrived. It seems that way, perhaps.

CRAIE (*picking up the board and placing flat on the floor, making the wall a back-rest*): We've been having conference with the lord. (*sitting*) And he sends us here.

GUARD: This is quarters "D," for devils. And they are a devil of a time handling, with their bah-ing and cussing. But then, that like prefer the meadows and the open air. Confine them to a primitive construction and they become difficult, animals.

CRAIE: It's a bit fusty of the fumes, down here. But the legs rest of tensions—

HERHSHEAD: Why are we delayed!... Craie. What have we caused.... to deserve such callowness?

BILL: I'll say, it's one and the same to be demeaned by.

GUARD: Why, you're all spread out over the castle grounds, like sheep at grazing. Come you here to deliver your heavily laden wool, of fine texture and rich beauty? Or what for do you complain? But I don't wish to pry out confidences. There is a curiosity, though, about any event to coat. And as a lowly soldier I've observed you and your manners. A suggestion has spirited to my imagination that you are drunkards off some rural wine, to display such behaviors as to come out here and plop down on the uneven grasses before this estate, as with exhaustion of a picnic. And a companion tells me that's very probable, to have fortified yourselves in that way, to make this strange expenditure of courage, and waste it moping on a lawn. But I would ask of you explanation of these valley jests and village gestures. I'm in no position, though, to require it. But this does seem funny, that you could be (this way) thus so usual.

BILL: Lends me loins. (*squatting*) This is filth down here, sir.

GUARD: That is just a pungent stench of state. How can we clean the space, with true sincerity of of its use? We have.... crafted its employment. What here was originally for.... is obscure to see. Never had a door. Was just a recess unblocked and with daylight. So it's not a prison— Don't try and rush me, son!

BILL: I'd rush to stand back up. rather. But here a wallowing of emotion's better made, a mix of candors and casualties, as were the pig contaminating earth.

HERKSHEAD: Don't blame yourself for this distasteful hospitality. We have sued.... to serve our points of reckoning, and do salvage risks. But a lord should be reasonable, if sane.

GUARD: I guess the day dwindles, shortly. The room is absolutely of no use for nocturnal activities, as originally planned. There is no granite alter for the procedures of religious cults, and no rocky stands or mounts to held up candles. The *Raum*-space was totally barren for any particular activity to picture, nor for tortures hearing— too easily for the outside, with that high opening of wall. It offers more a puzzle than a passion for storage. Yet, it might simply be the leftover result from a crude and early architecture of sophistication.

HERKSHEAD: Madness wears falsehoods. But this is not right. He seemed.... of courtly anger. And we have our protectors, or supporters, surrounding us. But I fear most desperately the nightfall, in this dejected state. Balcome is a beast! to detain us this way, with.... putrid matters floating to our noses, and a sense of calamity grasping at our clothing, and justice drowning of the anxieties we've brought. We are misused as representatives. And now to fall in darkness, into a dingy squalor, is more shame than sanity requires.

GUARD: Or repent the night, it often repeats itself.

HERKSHEAD: I would not have expected Gestromy to allow this foul treatment of our sincerities. He invited us to enter, and is bound by moral custom to prevent for us a disastrous entombment.

GUARD: That's a high and special assistant to our Lord Balcome, a very decent man for his position, a tenderhearted warrior — very rare. He makes sure of his kills and murders, so as not to prolong the agonies. Yet masterful he is, to carry out his orders received, and does cause a slaying, and does hurt men and create griefs, bereavements, and depressions. And now he is a sheriff? Well picked. His take of taxes will be honestly enhanced for his personal treasury. I wish I could (in) some way have such an occupation. But I'm as far from that as marshaling a regiment. They call me "speck," because I'm not smart enough to consider a problem except to stop it with the most immediate of force expendable, as to make a speck of a fly squashed. I can not think well, gentlemen, and can only be condoned as a brute with sympathies, trying to find a righteous path. But our Gestromy, of mental wealth— since you need that to collect real wealth— you certainly wring his generosities out to dry, and impede their soaking of your person, by being, I presume, too deliberate before the lord? There's not much that he can do for anyone worth some condemning—

BILL: My bottom meets the straw.

CRAIE: Sir, we are only casual of the protesters that have collected round our castle. Can you not with hesitancy condemn us all?! What is this proscription of our rights to ask for or bargain about? that our properties, possessions, ownerships of notion and earnings of achievement through labors be more fairly divvied up to protectors— or kept more for ourselves.... as protections wane, kings or no for headaches worn of grandeur. We have our own styes made, and should need not to support many others. (***Bill slaps an arm.***) What speaks of specks?! our fortunes poured on summer ice and infestations. Good men are we all to cringe and complain.... but not to trust our evidences clear. We've stepped right into imprisonment, a marsh of malheur and mortification, with our jaws open and speaking of innocence.... for our positions.

GUARD: Or incendiary nonsense, or extraordinary naïveté— for stout country chaps, of lands the like of which we've ripped through many a maneuver melded into war. But we do not infest to populate.... these hallow hills of history, but rather are invested to them, if I may say so candidly and with an unassuming posture. But in any case, this is where "speck" has been led— for duty, and saddened by this twist of reception.

HERKSHEAD: Are you soldiers all so sullen of your shearing blades?! It's for what you all insist of grieving, from your actions to be stunned by! then expect our heads to be smiling from the blows, as if reeling off accomplishment with amazed pleasure or star-struck happiness.... and the gaiety of a merry dance to perform— around your stern and proud formations. And as we stamp (*stamps*), to exercise the legs, you clap! with the fire of your arms, into a sundered silly crowd, of this surrendering to you. (*stamps*)

GUARD: Here is not grief, but grime. But, aye! We're oft'(en) re-flective of our chores, to commiserate about, by way of explana-tions *felt*.... if not divined.

BILL: Such felt.... as were to lay in this and lie, stretched out of the denial hooked to and anchoring, for my master's breadth.

GUARD: Don't give in to these circumstances, lamb, a young lad like you.

BILL: Yet, it's billeted for warning, as my name is Bill.

GUARD: Aye. Then Bill is lodged. But that's not the end of life.

BILL: And I am warned, of a warm embrace of fright, astonish-ment, and dispassionate dismay— to hold back tears (and head jerking)!

CRAIE: We're not so terribly distraught as to become imperiled more, by fear or than, what issues us to contain and contaminate our blights. The lord's purpose can not be so horrendous beyond this punishment, else we would be already cracked for patching into pieces of grout along these graying walls. It's sure we are sheared as guests remaining, of a recommendable captivity of ex-ample. And your masters would claim, dear son, to stay apprised of breaths with attitude, until our heads are crushed into a pasty pow-der.... and our bones lent a worth of filigree for design of these sur-roundings, hanged.... or adorning reverential temples.

GUARD: Ah. Resistance is a form of art.

BILL: But that's the bitter flesh to part of, as if crawling about me. (*while standing*) For I am all anxious to deny the truth. But it can not be had that way, what's clearly seen and felt.

HERKSHEAD: These insects are more temerarious than bees to a flower. (*stamps*)

BILL: And I flower of a view to startle eyes condemned, a blos-soming into discernment with revulsion, the face and figure taught!... for my own to be adrift.

GUARD: Our lord impresses you that much? A soldier make you cautiously, and with more paces of discipline.

BILL: What I've perceived is more worry than war. And not of welcome am I shamed by— not even a word of address! But I claim to you, my masters, that the very image of my father was alongside Balcome. His person was there, during our interview, or confession, but certainly not a conference. And I was humbled out of hate, since I had hoped to meet him gleefully someday. But a vileness poisons me more than this confinement. It entraps me, for what we both must know.

GUARD: ....Or realized without speaking? So then your father is a soldier among us? That is a dainty joy! for which he'll lend some measure of guardianship. Yet, along*side* Balcome? Not a common

footer. There's a rank to 'im. You have a decent advocate, and must be saved! or spared. He's not the sheriff, is he? What is he called?

BILL: The name is Brontrol.

GUARD: ....I've not heard it ever. He must be a courier, if a sol-dier, perhaps (sent) directly from the king. Better an' better for you. We've been expecting some kind of correspondence or assistance from his court, as the summer temperatures flare up for these envi-rons a considerable insolence of the population towards regal rule. I suppose this region has become.... as noteworthy as the lord him-self. But this is an interesting ostracism for a child, is it not! as could a Covantry make coven convert for conversations.

HERKSHEAD: A townsman unrecognized. I— thought I knew them all. He's sure to help you— and then us, Billy.

CRAIE: But you're a man. And that's the most of acknowledg-ment for Balcome, to be considered. We must wonder, then, of Bill's deportment drafted by the father to describe, for this retreat to send us to. I'd say it is not sure, or hardly to be recommended. We must think correctly about this. Perhaps a ruse is being main-tained, to support your not acknowledging him, lad.

BILL: We are in danger with any administrating of our company. But I'll not speak to him unless he takes me aside and initiates a cry to be known.

HERKSHEAD: That's a wise choice, given the uncertainties of his labor.

BILL: And my Mom will tell more of this, his return, if indeed *we* are returned to hear of it. It's been some three or four years, I believe. An ageless span of time, to contemplate on his absence. I can only observe the loss has deepened my voice. But it is deeply felt by her to have made great sacrifices because of it; though there are many husband-less families around, these days. I choose not to burden myself with one at all, for fear of these effects. And that is a complaisance adopted of the will he has expressed, not to care much for details left behind. (*brushing himself*) Yet, behind me'(i) s rubbish.

CRAIE: Now carefully we must take this restriction, for who pro-tects whom, identities as blaring as the sun's.

GUARD: I would more start to prepare your catering, for the next visit with our lord, on what you are to say, proclaim, or chal-lenge with, while analyzing these questions more or less in secret, though open to the common and presumably anyone who'd care to know.... of family problems with.... familiarities. Though would I hold that something's been blessed upon this sneezing to admon-ish. If the information's withheld to be, then withhold yourselves of temper and be.... strengthened by this reserve of character.... or re-luctance to seem foolish nor bashful, as you present more *argu-mentum*. A show of resentment is merely the demonstration of im-mature tastes. So take heed of the forbearance of Balcome to this affair which ultimately questions his very usefulness and utility for this region. His calculus is only to appease the worthy with like calibre, to fight as hard as an enemy to overcome, give the same of malefaction or minacious method to best of effort. Aye! The better at employing resources will succeed. And.... you have, in a way, insulted us— have you not! to make of rot for where you sit or stand. Our sensitivities are surprisingly high, of both pitch and standard of tone. So bear yourselves of a persistence to this comical sight, this passing retention of ratty circumspection and detention to the arrears of your visioning.... and our lord will appreciate your ridicule withstood, and offer means of pardoning, perhaps even some symbolic benefactior., by means of compromising through details of the tax. That is your quarrel, is it not? We've been led to believe this childish mischief is all due to that, which you have been willing and capable to pay. And now, with the slightest of shading by clouds, you balk to claim that we're unworthy of your labored largesses, that we're like dirt stored up in an old castle relic of forgotten tolerance. The mistreatment hurts!... to be thought of that way. It stains the psyche and punishes the soul with a dingy sensation of regret— Squash I such things!... But the law, fine folk, is on our side in this. The sun still shines, whether you see it not, or feel its heat less, what warmth is radiance from afar and sometimes casually reaching you. And a day still lasts, for your existence to ponder through and think with.... though now it seems to drag its tired weight below the horizon, in short order. So.... as poverty is bitter with the cold, impoverished are we all with your ungrateful attitudes. So.... show this not, to lessen the strains. *We* protect you, as long as you stay well behaved.

CRAIE: We were not obstreperous to argue with him earlier, fine guard. And most would not deny that our majesty continues to protect our interests and civilization, but in ways that are becoming more and more obscure or subtle, and eventually ceremonial of enactment. But Lord Balcome is of firmer substance interfering—

GUARD: Errr—

CRAIE: —or inferring for our donations.

GUARD: ....Err you not, this time. Opportunities are much fewer than the chances taken for them.

HERKSHEAD: Are we to be held here indefinitely?!

GUARD: I'm certainly not indefinite.

HERKSHEAD: It would be a fright to have to dine in such a space, squatting with contrition.... within dimness and dour spirits.

GUARD: Claim you ransoms earned? The lord will call for you again, I understand—

BILL: When?!

GUARD: As he is ready after discussions over you. And about you, I can not contemplate. But that is generally his method, to forestall of action until he has decided to allow it.

HERKSHEAD: But sir, soon the sun sets. And this is not a comfortable spot, abysmal to sleep in or doze off carelessly. (*stamps*)

CRAIE: An afternoon does render towards the evening. Yet for my impatience am I bored, to have hands shackled in surrender, with preparation to be dragged of body into the mental depths of Barathrum, Balcome's declension for our thoughts. It is with an obvious sweating that we're provided of the hot baths—

VOICE (*from outside*): To recess!

GUARD: My lord!

BALCOME (*voice*): To recess of my manor, this lively alcove sent, deposited. But gentlemen, have you counted your considerations to examine, and reconsidered what is filthy to be owned?!

GUARD: This is most unusual!—

BALCOME: To visit these?! I am betwixt rights, rites and harbingering, and am of a style stowing, to present to them their castle. And yet with stone am I more undecided of their fates. But have these here produced some documentation of their requests?

GUARD: There are no implements of writing around the entirety of this room, my lord. They are only three men, to stand.... or sit within. And that is all that they can do.

BALCOME: What?! Have they sounds of voice?— Have you?! — I want your words!

GUARD: To ear? They are deceived!

CRAIE (*standing*): We are permissible, my lord! With justice greet us.

BALCOME: Have you formed a declaration?

HERKSHEAD: We assess you carefully, my lord. We hurt in here—!

BALCOME: What be what's heard? There is a crazed dimension to the sound. Where is the paper and the ink— but of your skin an' blood! Yet I'll not abuse your official statements. Then tell me plainly what you want—

BILL: Some cleanliness!

BALCOME: Ah!... Chattel.... What's misplaced of property, are your taxing bothers. I'd like it written. I want it evident— as could be nailed! Yet dream awhile, for the utensils, what could be used as pen! You are my guests, detained by.... some madness. Though I'll have you supping—

BILL: Abandoned!

BALCOME: —soon. No! You are not there yet, for better of the company I'll have, who I demand to dine with. Good gentlemen! I am terribly undecided. But it appears even nebulous, inside that space. Well I want your indulgences, please!— and a statement, that ends this upsetting confusion, that surrounds— me!— and yourselves, and this castle. For there must be envy, to hear a truly crafted and very careful address, verifying your demands implicitly — and without the artifice of error, that could be misunderstood or interpreted wrongly. I want your hearts in friendship to our meaning— pumping out the truth! saying what you really desire. (*voice fading*) And I will bring back courtesies shortly, to have you dined in a more resplendent atmosphere. But work at this as if diligently bugged, while I must go up into my turret and golden spire again, to observe pea—ple! and bring the sun down— over their shoulders and into their guts! Ah, that such fullness may burst them— I've not discussed yet.

GUARD: He retreats!... into a perplexity of avowal. I've never seen the man this displaced of conscience. He's usually much more decisive of his meaning. You have stirred him to be.... stranger. This seems to contract a service, but he would have you.... preparing such a constitutional agreement between ourselves and this valley's residents as to bring full recognition of our devices in offering. Yet, how this (i)'s to be done sighs of the mystically penned, symbolically proffered.

HERKSHEAD: He brings madness to these shadows.

GUARD: It's only with a worrying.... gestation. I'm sure he prepares for your proper feeding—

BILL: In here?! In this dungeon of remorse and lenient abrading?! I would for out! upon his decision. He transduces fear into the typical of portrait, for an indecisive man. We have given him specifically the details of our contention. And yet he acts so crazed as not to know of it. This is more ruse for ruining and debasement. (We) Deserve us more than mephitic cajolery. I would escape, to bash his teeth upon my arm, what's left in crying mouths of wetness licked through and stun by as if a serpent spits into a cage!—

GUARD: He is not for tricks to play. That's not well his cleverness, but rather a strong action too determined to be denied. And it is that overwhelming evidence of intent that causes the capitulation of opponents.

CRAIE: Blacken the walls, guard. We are not your enemies. Show some pity for our fates, as monsters crawl to grow from gnats into giants—

GUARD: You'll soon be wined an' dined far better than we're ever accustomed to. I can assure you, the lord is a true gentleman to guests.

HERKSHEAD: It does turn bitter to the smell, as were to taste in here our declensions into misery. Plead for us some! We are ashamed to be in here.

GUARD: Inside this castle herded? It is salty, and sensible— as we have been contained. And now you are, with conferences awaiting. What's to be said— is what to eat! and devour of yourselves ideas or concepts, pertaining to our service. But these are mere discussions, after we have bled thoroughly, during our.... combats and confrontations on your behalf. But we do not punish you at all, and are only curious of the nature of men found of this locale and destination. Yet within this decrepit "cage" have slept the imperial— and ourselves, with woolen sheets and floor spreads. So not your own to take some comfort to your castle— with this presence made?! There's no deception made. You are detention-ed until he finds his mind— more soundly about you. And that is a tenderness, even a kindness offered. For with his soldiers he simply orders, and we obey. But you are confused of the proprieties, and the appropriate proclivities to follow. Yet, as guests, you may leave your brains in here, and wander about these ancient spaces as ghosts or thinking spirits. That is, imagine all of this falling to your care, and the burdening of your earthly shoulders. Then, describe that sense to our lord, and the piety felt, and the sweet airs of a smothering consummation.

BILL: But he's insane become— to destroy us, or let our minds deteriorate into the mush of the squalor found in this pit!

GUARD: The sheep do not complain as much, when fed their liking.

BILL: I'll not be ate alive by bugs, (*rushing to escape*) as Balcome pretends on us keen novelty to question of all soundness! (*The* **guard** *detains him with a fierce punch to the solar plexis or gut. He recoils back in agony as* **Herkshead** *and* **Craie** *come to assist and prevent his fall from staggering.*)

GUARD (*somewhat regretful*): Does hurt you, woe more weapon than a fist.

CRAIE: You'll pay for this cruelty!

GUARD: The lad's too bold for balance. But does his father see, that I have my duties applied.

BILL (*with struggling voice*): He sees me not! He is not here!

GUARD: 'Twas unnecessary, your impatience and impetuous foolishness. And I'm sorry for your pains— I'll apologize for that.

BILL: I will kiss!... the notion of your beating, as it makes me real, the squashing of a speck! I've been too long dreaming.... of the imaginary causalities to pretend for. But I'm no magician more, to fly out of dangers sought, and with wizardry placate my conscience. Obey the harm, and fashion yourself about its victory. The victim is a carelessness to know.

GUARD: That's a noble sentiment the dying often speak of off their wounding. But you've been educated early, not to daydream in kettles.

CRAIE: Covered cautions, in this house. You've injured the son of a soldier—

GUARDS: Winds will make up for my lessening.

CRAIE: He might have you hanged!... if high in rank.

GUARD: I doubt it.... though I fear familial relationships.

BILL: Not I! Beat me more!—

GUARD: Come rush the passage again. But that's the only cause for this. I don't dislike your nerve or surge of courage. The attempt invigorated me out of boredoms.... and a certain empathy for your unjust discomfits. Often what is courage is a cowardice. I see it all the time in battle. One fights from fear an' for.

HERKSHEAD: Help us, then, with some courage to you. You know we are—

GUARD: You're not to die because of this!... attempt at conferencing, (but) only to be brought to some humility to bend with, I must conclude. Though callous am I of lives lost, that's not contrary to praising life. The emotions hardly compare for contrast, they're so seldom together.

HERKSHEAD: Can you protect us from his apparent madness?!... from his ranting to purposely misunderstand us! to bring into excuse our injury or death.

GUARD: ....He wants you to have written of your terms. Is that such a fiendishness? But what to write on is your candid ensuing of your aims to plead for. Then to enlist my aid in this, you must not rush me.... I will protect you from an obvious transgression into the heinous. You won't be shot or axed without my restraining prevention, as to cohort make of defenseless children. There are some probities to our service here, after all. But I can't stop you from consuming poisons, be they of your own blood born or not. Such toxins may be inherent of your positioning and stances. Care for yourselves with cringes, and I'll be truthful. And what do you ask for now?

HERKSHEAD: ....We wish to serve the lord in all that he'd desire, which we're capable of delivering.

GUARD: That's nice.

CRAIE: There was never anything more to our contract making than what is reasonable.

GUARD: Then what is reasonable is witnessed.

HERKSHEAD: What we can continue to provide is what's suggested.

GUARD: And what is coerced is not fathomable to our complaints.

HERKSHEAD: You'll tell him this for us? insist on it— implore?

GUARD: Balcome will know by me your documented statements, that you have apprised me of your actual desires, and that they must be in some way a representative.... intention to welcome us more faithfully, as between ourselves we come to knowing terms with meeting. Yet you must presume I've not had error, even in listening.

HERKSHEAD: Most definitely not. If you're a simple guard, then we are simple men. And one coherence more may spare us from decline and foundering, that these soldiers may be pleased of us.... and that this castle may contain us comely as.

GUARD: That's a reason for its existence, to provide a favorable occupancy against the weather elements. That is its broad shoulder, with its variety of various protections, and a proud heritage to be fit for such an august engagement as we provide, the necessity for which is simply a reward. And I praise the old place for standing up to your inutility with a veritable endurance of neglect and dismissing. Yet, the creaks we hear are not complaints but compliances to our presence, and perhaps some mustering of a deserved welcome. And for this we sense its profundity.

CRAIE: We have never disowned the castle outrightly. Can you now disown your prisoners, to abandon of their welfares— and instruction?

GUARD: Prisoners?! Were you such you would be in a jail, with bars, or in such a space as this sentried by several— with angers to subdue the least resistance. But you are only being detained awhile, by the customs of our organizational procedures; that is, restricted to an area instead of being able to roam about so freely as to disrupt our order and activities. You have always been guests. And as long as you remain respectful of the privilege.... no more

harmful incidents should manifest.... indecencies of provocation. Stay calm, as the night falls. There's not more into this than the day's ownership of the common and shared wonders.

CRAIE: We have friends awaiting us—

GUARD: So does the cold.

CRAIE: —for our beneficent return, as demonstrated of a trust.

GUARD: So does the moon trust its stars. So does a shiver trust the warming of its shake. An' so will you be delivered for healthy embraces. I'll guarantee that—

HERKSHEAD: Can you?!

GUARD: Assuredly. I don't see other alternatives much, but to your pleasurable relief and release.

HERKSHEAD: Then sigh we now, of the growing darkness.

GUARD: It's but the shadow of a passing cloud. So well benighted to confusions do you seek, then linger of a trembling to lounge. It's only being idle here, not terrible, and not to be terrified. Sheep are not so scared as this, even before their rending. And you exaggerate the insect foraging. They are petty for their annoyances, not finding the customary fleece or fur. Then sleep through this standing, and nap of the noxiousness. I've dreamt in pools of muddy rain; and it ain't so bad, as the body adjusts. The frozen blight, as like a blizzard melts upon you, falling from your head onto your chest and back, does seem to coat congestions to keep them away from your consciousnesses, and you cease to squirm. This is a fine spot, a luxurious one, by comparison to any stresses of war.

BILL: Apply to me your buttressing (*starting to bend over more deeply, even as being held*), as I fear to fall and collapse into the fusty mire wept on. Something breaks inside of me, I feel, and gives way, a strength diminished or deceased.

GUARD: Just the novelty of a moon('s)tide's waves, the mind of darkness. Does it still hurt, or are you merely stiff?

CRAIE (*aided by **Herkshead***): We'll take you over to that board to rest. There one can recline safely.

GUARD (*as they bring **Bill** to the board*): It's the stiffness. Yet to recover from, the youngster should be sooner.... when more supple. (*as they set **Bill** down*) Ah, this time of guardianship. It does serve me.... casualties already. Quakes are better brought, than lamenting hides.... And no more skullduggery for speck, if you are really injured.

Scene II — *Nearly dusk. **Brontrol** enters the kitchen to light a candle before the window light fades; but he finds a **man** sitting at the table, near the window, which apparently has been flung open.*

BRONTROL (*startled*): Who are you?!

MAN: Who are you?!

BRONTROL: What are you doing in this house?!

MAN: Why are you here?! I happen to know the lady lives alone now, but for a tot.

BRONTROL: What gives you the right? (*approaching*) not to be torn to bits with my rage!

MAN: Wherefore may there be a rage? I'm only resting up from travels, and a slight bite of bread, what's left to cure of mice, the crust crowned of deprecation. And I leave before they arrive with the evening stars. And is there another window opened, for your entry? This is a more useful room to climb into. But I am not mischievous of her paltry possessions— An' neither should you be.

BRONTROL: You enter without warrant, into an empty house —!

MAN: It's not that, but a house may be for any one (to use). And there's no one around to share it, or be disturbed by my presence, till now. But there's little here. It's just a comfort zone for sitting a spell, out of the heat, or when it is more inclement. I walk a lot, in the mornings, and am used to the place to return to.

BRONTROL: This is outrageous! Now I will throw you out— with a knottiness to body!

MAN: Why so outrageous?! The home is not being abused by my visiting. I respect it, and even leave a token, sometimes, of my appreciation, less stale than the bread I find. But I think we'll both leave soon, because she arrives shortly. And this (*pointing to the open window*) is a better route to whatever you've found. Don't disturb my hospitality to pointing it out, though I shouldn't recommend this place become too crowded.

BRONTROL: You're a vagrant and a thief— and a fly-by-night — and deserve to be punished!—

MAN: What are you, sir? I'm merely squatting awhile, and taking nothing of consequence, in a thoroughly inconsequential abode, the deprecation of life that's been left to this lady. And we.... are merited to be here, as an area still to be used during the day. But we should not be fiends about it. There's no advantage to it— There is no wealth here.

BRONTROL: You ransack the house for monies of coin!—

MAN: I only visit (in) here, the kitchen. There is no money to a place left deserted in the day. And if there were, I would not need it, for my nerves are humble. Obviously you've found nothing. But you may be a beast.

BRONTROL: I'll strip your body for what's stolen!—

MAN: It's confined to the gut by now. But what it is is nothing. And it makes not any sense attacking me in a house with barely any provisions. So bring down this false ferocity. How can *you* truly be upset?! I've learned it pays not, but with headaches and body tensioning. We are of a kind to trespass nowhere. And I have accustomed myself to this simple comfort, which I leave undisturbed. I should insist that of you— or I'll notify the police!

BRONTROL: They won't have much to deal with, after I've finished racking you up. I only hesitate to consider the blazonness of this crime. And a scroungey scruff like you are certainly not from

the castle sent.

MAN: Castle?!

BRONTROL: You're no fighter. Your pus will taint my hands as I tear you apart!

MAN: Why such unnecessary belligerence of talk? You don't frighten me, if there's no better between us than a worm. And the blustering brings no calculation to it. Not profitable for either of us, sir, to be abusive.... Do you suspect Lord Balcome of sending soldiers out to occupy empty homes? You're certainly not one of them.

BRONTROL: Why do you say that?!

MAN: Where's your sword? Where's your coat of arms? You are dull, like myself. Gray of feather and reasonably flightless, else you would not be out here, pilfering the poor. But it's not to do of it. Things just don't work that way. And you'll find it's best to leave what's low as is, and not abscond with a weight of worthless junk.... The pain that that might cause.... is worse than your silliness of wrongdoing. For just the realization of any loss shames one horribly. And I don't mean to cause such harm ever— to anyone.

BRONTROL: Intruder and vandal! You broke the window latch to get in—!

MAN: It was already quite compromised, and the portal invites my entry. Lonely the house becomes, when it is emptied by necessity, when it must be left this way.

BRONTROL: Now I will break your neck! as compensation.

MAN: Why do you fear my neck?— Does it bite? Or are you jealous for having found one other! We are much the same for spending time here. Such a wreck as this needs company. It cries for habitation. Yet I justify the transgression only to take a rest, not to own anything. And I owe nothing to anyone as well, least of all you. So why earn you hatred of me, if I've not stolen a thing? Why chide the worthless passing through? It's simply that the room stays available for so long, that we're permitted to meet and combine our sorrows as a remedy for our existences. Though I admit, there is a certain charm to the anxiety I experience while residing unauthorized, an alertness that quickens the pulse and refreshes the senses. But that's a resonance shared. For this old house comes to tears without people in it, as if abandoned to accelerate its deterioration. Look at these walls, that drip of drops of (dungy) dinginess and air-moistened grime crawling down so slowly to reach the crevasses and cracks of floor. They brighten up with a face to see and notice, and withhold their sad displays of transit to feel a gaiety of human presence and produce a smile. And I smile back, to think of nothing of my state, but just to linger awhile, and enjoy this usefulness.

BRONTROL: It's becoming harder to see anything. That's why I've come in, to light a candle. Yet to find the ugly countenance of a stranger— turns wormwood into stone. But you are a pathetic idler, it appears, wasting your breaths of life on solitudinous reverie.

MAN: Stranger? Am I unknown to you? I know no one, and therefore everyone. I hurt no one, and everyone hurts me— But that's fine enough, if a man is base by nature, when uncontrollable. Avoiding others keeps me in check, else before long I succumb to such a humiliating tendency to demean my poor notice that the mind contorts my shape into a squib of characterization for humanity to describe, and I melt away, perplexed of a coma while trampled under feet. But am I a stranger to you? I am the type who would gladly let you destroy me— if I thought this would make *you* stronger. But flesh is flesh all around, and hardly changes much, till underground to disappear. Then I'm no different than your best merriment, to beat at something, something hideously deformed. I think about such passivities to be entertained by, while in this poor and almost barren larder, sharing miseries of futility with restoratively dispassionate review. Are you to bother me and tear my limbs off, now, because this creature's found a hole to spend some time in, commiserating with envies.... for higher purposes in life to achieve? As much a theft of person am I as you. I simply haven't the energy to do better than loaf around through life and occasionally terrorize myself with these adventures and stimulating thrills.... of solitude with self-centered shame.

BRONTROL: You can do that by trying to drown yourself in a lake! You wouldn't have the courage, restraint nor discipline to go through with it. Shame! That's a sham for shamelessness.

MAN: This feels like a new universe entered, for a few hours—

BRONTROL: Hours?! You *are* the lazy curd of cuss.

MAN: I evoke my dreams awakened, to sense the absolute shame of this space.

BRONTROL: This house hasn't that at all! (*bangs the table threateningly*) But what I find of it, more empty than pronounced to be a home. It's been deceived, somewhat, by circumstances— but has fallen not so thoroughly to ruin, by my leaving.

MAN: ....Nor with mine can it descend more broadly of the void.

BRONTROL: But I, crude fellow, am the master of the house.... I've entered with a key! (*The **man** jumps up to standing, for the first time really frightened.*) And.... I *am* a soldier— of the king's! My weapon and caped address.... are in another room, for which I've laid them down carefully and with reverence of their statuses. I've come back! to expect my wife— and not you!

MAN (*almost apoplectic*): Oh! Forgive me, sir!

BRONTROL: Not a roving pestilence to find, as a rat in leisure of a garbage patch!

MAN: I am less than his tail's end to balance with, sir! I am afraid, of my inappropriateness to discover, and be revealed for nothing good— but nothing bad either. I do mainly fish. That's right— You have me pegged! But sometimes I leave fruit here, when the season's ripe for it— and they're as inconspicuous as myself! the wholesomeness that's left. I am not the bandit, sir, nor the ravager as rake. I confront nobody. I bring myself to no one. I am not for hurting anyone nor damaging anything. Let me fall to my death, outside of this house.

BRONTROL: It's not steep enough, the window's leap—!

MAN: To find what is deserted?... of my callowness? sir. I have

grown, but have never grown. Kill me not with punches (*as **Brontrol** approaches*), but disdain. Let me feel the revoltion of your eyes, and not your hands to tear at this weak skin. (*grabbed by **Brontrol***) I— I— I thought she was a widow, sir! But I'm not so well informed by anyone—

BRONTROL (*almost yelling*): Hold me hate! (*declining of voice*) Do you shake.... of uncertainty?! This trembling can split a tree!

MAN: I am done for— Rip! I've taken chances of misfit! and now must serve your rage too fully won.

BRONTROL: Die in my grasp, then! Faint away into oblivion's stupor. For I can control this range of maiming rancor here, but not so well elsewhere. And you feel of a warmth that chills! what might I toss to explode of severe credibility for my distance for adventuring opportunists an' weakly milked monsters. And bide my touch to squeeze you for a swoon. It is a love to destroy!— man!

MAN: Yes! Yes! I am of pain with you!

BRONTROL: And you are trampled with a torturing, to stiffen muscles till they break an' snap! All of this I've done before, an' kissed the head for its worldly departure, as consciousness is bled from stone encrusted eyes an' tongue as tail. (*shaking the **man** wildly*) Struggle! Struggle! This I can do and handle. This I can manipulate and transform, and turn into my ways defensive, and earn your suffering—!

LINDA (*voice calling*): Brontrol?

BRONTROL: Gain then your reprieve into the dirt! (*throw-vaults him out the window*) An' come about again for death-masks to be made.... The darkened ground is prettier at night!

LINDA (*entering with **Hoe**, while carrying a bag*): What have you been cursing at?! to be at the window so wide open? Did you break the seal?—

BRONTROL: It was broken already.

LINDA: Go light a candle, Hoe. (***Hoe** complies, reaching for some matches in a cabinet draw not far from where **Brontrol** is standing, a candle stick with holder awaiting on the table.*) He likes to do that. Flames excite most childish eyes. (*coming to place her bag on the table*) There's some little game tonight, for our porridge meal.... finally. Was that a groan I heard before? Did you hurt your hand?

BRONTROL: No.... For what runs off into the darkness (***Hoe** lights the candle on the table.*) are seemly senses soothed by being hidden.

LINDA: If you did I have an ointment. The latch can pinch easily —

BRONTROL: Not done! Linda. (*shutting the window doors*) Was opened when I entered.

LINDA (*as **Hoe** stands in the chair to lean on the table and better contemplate the candle's flame*) So?

BRONTROL: And much too loose the sash has become. It's dark

out, now. Why does this boy stare at simple heat? Hasn't he experienced it enough by now, a common phenomenon?

LINDA: Maybe Hoe wasn't allowed to, earlier. Orphans are often
strictly kept away from anything that can even hint at turning into a
danger of mischievous play. Their groupings together are tightly
regulated to avoid individual behaviors and tantrums. I let him
have more to do than ever he's been used to, because we need each
other's help.... to live through our current means. I'm not paid
much to have him here. But he's worth more, as a working company. Let him learn his fascinations well and carefully, and
safely.... Why did you open it up so wide? (*picking up the bag*)

HOE: It is blue, an' blurry, blunt.

BRONTROL: The air brings with it strangers. This is not the
condition I left, Linda.

LINDA: The house can get stuffy, during the day, when everything must be closed up. Here's a roast of pork, Brontrol, a leftover
from the orphanage.... or more a bribe I've made, for extra services
coming. We just have to heat it for our dinners. It's bits will make
the ground meal richer, tastier.... as that you are here. It's what we
can provide (ourselves). Then assist me with a plate and knife, so
that I may dice it.

BRONTROL: What?!

LINDA (*opening the bag*): From the cabinet behind you. Come!
man.... This is petty, but fatty.

BRONTROL (*going to the cabinet*): A dinette set? This is not a
poverty.

LINDA: Have you seen—?

BRONTROL (*opening a cabinet draw*): No broken glass or a
shambles found, returned to. Not so poor becoming, as I could care
to fear (*lifting up a knife*)—

LINDA: Hoe, fetch some water, for boiling!

HOE: Yes, auntie. (*climbs down and runs out*)

LINDA (*as **Brontrol** retrieves a plate and brings it with the knife
over to the table*): Don't talk to me about that!... He can work the
well himself. Perhaps the youngest to do so, and I only had to
teach him once.

BRONTROL (*placing the plate and knife by her*): Show him
how— He teaches himself. That's how all children learn. But it's
still a dangerous procedure, to let him attend to it alone. Weak
muscles can easily be pulled in and down, by accident of tipping—
and he'd drown. I wouldn't let Billy go at it, at that stature of height
and weight.... But I suppose—

LINDA (*while working the pork*): What?!

BRONTROL: ....the boy's older than he appears, slightly stunted
of growth from a spell of early malnourishment.

LINDA (*cutting*): No, he's the young lad that he appears as.
(*while **Brontrol** sits*) Nothing (is) hidden of him for his character,

the disposition is solely of a little boy.... But we must take our risks
to survive. And it's startlingly curious that *you* would worry about
him at a well, (after) all the fighting and injury you've done.

BRONTROL: How to get things done always draws my attention
to details. Misfortune is mostly due to error, lack of planning, and
impulsive chance-taking.

LINDA: Nature plies her poniard to the breast, sometimes.

BRONTROL: And with a sharp panicking. But that can be controlled. One can outpace it— What sha'n't I speak of? A brutality?!
No. I've wondered often, how a world evolves of its affections and
caring, and a tenderness shown to the beloved.

LINDA: I should think so, for your examples to worry about.
(*bringing the plate over to a stove with pot*) And have you made....
kindnesses often, in your travels, to mirror for contrast of slaying?
What are the hues seen for comparison to us?

BRONTROL: They've colored dryly my impervious concern for
your welfare. But if you want to know, I've always been an honorable man, as most soldiers tend to be— as that's their best defense
against adversity. And when returning home (*Sounds of **Linda**
pouring the pork pieces into a pot are heard.*) I expect the same of
welcoming, an honorable one, and a greeting that's profitable for
affection. I've not aged much, and you not at all— except for firm-
ing. But I refuse to believe in this poverty. It is a scam before my
eyes of notion, to have descended into a neediness. And I will not
accept that this is true. Only your loss of faith, in our courageous
tender of separation, resides there, in this state. I have sent much—

LINDA (*adding cereal to the pot, from a small bin nearby*): If
you wish not to tell, then so be the mystery as dead as these dried
flakes and grains of wheat and seed. But it's as much the mind as
money to our manner, now. We are not destitute. But.... life grows
difficult, and that's why Bill moved out—

BRONTROL: Out!... I've sent regular allowances—

LINDA: For a boy. He's a man now—

BRONTROL: Nearly bred.

LINDA: —and has his own withstanding to do, of these circumstances for growing up through, and making of himself an adult
life with adult needs. Did you think he'd stay the *gamin*? Our expenses grew, as could his wants and intentions to provide for himself. And as I told you before, your last retainer's late, and feared
never to come— But it is you yourself, as could your family be
kept a service of your opinion.

BRONTROL: And I correct things presently, Linda. We will
feast here royally, without much delay. We will sharpen up this
home, and shore up this house to the respectability I command.
And I may even be able to invite us to dine with Lord Balcome—

LINDA: That lord?!

BRONTROL: —I've seen him today, as an important officer of
assistance. And he is gratified of me, appreciative of the king's
continuing and persistent interest in him. So change will glit our
marriage and circumstance, Linda, as I have brought home wealth,

station, and passion. All of this to keep, for quite awhile, as long as we remain.... responsible.... respectful.

LINDA: Is this for return an actual enhancement of my mating? Is it possible for Hoe—?

BRONTROL: My hands may benefit all. But I am characteristic to provide for anyone dear to me.

LINDA: The folk are against that lord. Our own son—

BRONTROL: Unknowing are these eyes to that contempt as real. It's what is staged to be, from sooty rumors tarnishing our gold. I can convince our own to change that choler— even Billy's— into a finer calendar of events to make more prosperous lives.... or that is the work given me to do, charged exclusively by the lord himself. And I have started.... making conversions. It felt like a difficult aimlessness at first, wandering among those protesters, those— VIDs or other. But.... reasonable arguments modify and mortify chastising goals. And I do suspect— or hope— their number will be less tomorrow day, around the castle camped, and continue to dwindle through the weeks until.... what's left is nonexistent or representative of a few misplaced frogs. I'm only battling tales with them, not the sincerities of our people, not their sensibleness and ability to think grandly and astutely, though contrarily at times. This is a small problem huge of number, and a huge problem small of wanting. No one out there *likes* what they are doing. Some even praise the lord himself, as a person, and admire his record of achievement. As they discuss terms more thoroughly among themselves, away from the heats of lethargy and lying on the hilly knolls of grass and tufts of useless, maddeningly silly and ridiculous temerity, these.... conflicted and ingenuine protestations will die down, taxes will remain proper and be seen to stay.... advisable, and no great hardships of bitter resentment will descend over our broad valley.

LINDA: That's.... not likely to be anyway near the composure of our collective population. It is feverish of travailing. No one is doing splendidly anymore. And to find one appointed to be so can only draw out inherent and fully justified resentments. Difficulties have been real of late, Brontrol. They, we merge and share by necessity, and forgive common, ordinary and innocuous transgressions of behavior that can do no real harm, like a slovenliness of dress, or a weakness to decorum under the heavy toils of current existence and exhaustion to stay at least morally solvent. We all see a generalized privation of daily life developing, slowly creeping up upon and over everyone. And manners change to allow our disgusts more thorough airing. It's a curious mixture of patience for the poor and impatience with the wealthy, as old and new clothes must be worn together, loose threads and tears supporting hanging jewels and ornaments.

BRONTROL: This is as rich a county as one could hope for— It remains providential.... favored, admired, exemplary for example and description. I've not seen any shags walking around or about. The poverty is all to mind for explanation, that one might induce it with these thoughts. So our king no longer seems pristine of a god's light. Then we who serve under him should grow more mottled by some noteworthy decree of self-effacement?! as a supposed divinity dies? It's for the opposite to be, that we should hold our chins up higher— and bless the regal natures left!

LINDA: The world is tired, Brontrol. Its children are born old.

(*as **Hoe** enters with a small bucket of water, coming towards her*) We must be careful, not to offend others with our good fortune, if this does indeed transpire for us. (*taking the bucket*) Well done! Hoe. This looks very clean, and with no sediment.

HOE (*as she's preparing the porridge*): Thank you, auntie. I did couch it once, but couldn't see well.

BRONTROL: Come here, my fine rascal!

HOE: Yes?

BRONTROL: You are an— urchin?

LINDA (*while stirring*): This is a .... friend of my family, Hoe.

HOE (*cautiously approaching **Brontrol***): Yes, sir.

BRONTROL: I know Billy—

LINDA: And?

BRONTROL: He was not among those sitting outside the castle. I looked for him specifically. But this young man reminds me much of the boy. What do you play at.... Hugh?

HOE: Is play for sport? There is no time. I help Aunt— Li'da!... But during the day I wash clothes some, and soon will be taught to mend them.

BRONTROL: Where's the fun in that?! Your life is one, and you hardly know it. But for an occupation, mend the sea.

HOE (*puzzled*): ....That can't be done!

BRONTROL: Oh, yes. There's a river nearby, that has too much fish in it.

LINDA (*crumpling some straw and putting in the stove's oven to help fire the wood*): Too many— fish.

HOE: I'm not allowed to go close to the bank. There's no one to take me there. And I'm too busy, anyway.

LINDA: Come light the stove, Hoe. (*He jaunts for the matches.*)

BRONTROL: Call him Hugh, Linda, if that's his name. To save the sea you must fish from it; you must utilize it, or else it becomes duplicitous for staying useless. And there's not one thing in this world, not one object nor even concept, that can remain unserved to mankind. But we must order motion. All work is in motion—

LINDA (*as **Hoe** is at the stove*): Strike it gently, Hoe, for a good strong flame.

BRONTROL (*as **Hoe** strikes*): And I must return to the castle tonight, to.... give some report on my progress among our citizens' demeanors.

LINDA: Light the straw.... to burn the wood.

BRONTROL (*as **Hoe** complies*): The lord is very interested in this, even anxious. He depends on me.

LINDA: What can you tell more, that won't be evident tomorrow? or make assurances that must be followed?

BRONTROL: He may have guests to be appeased. I should assure them— that I am working.... on it, that everything will devolve properly for benefits to all of us.

LINDA: Go get the bowls and spoons, Hoe—

BRONTROL: Hugh!

LINDA: —one set for each. (*as **Hoe** goes over to the cupboard*) That makes three.... I can't see how Balcome could have guests, except from this countryside, with all of the rumpus that's occurring at his castle. Visitors should stay away.... Set them on the table, Hoe.

BRONTROL: They are a peaceful deputation from the protesters, that.... may have been detained.

HOE (*setting the bowls and spoons on the table*): Here, sir.

BRONTROL: My name is Bro'trol.

LINDA: For that he may pronounce these words properly, Brontrol, then may they be legitimately heard! But such to make them erringly is a faith in one's perception. (*starts to stir the simmering*) And that is most important to control.... One says enough, when the meaning is clear.

BRONTROL: Then ears will handle this, Linda, for an interpretation of the realities.

LINDA: It's as one tries, Brontrol.... to comprehend—

BRONTROL: What's heard?!

HOE: Are we to eat together tonight, auntie?

LINDA: Yes, this one time, Hoe.... to placate our guest with good graces. Can you get that stool from out under the dormer? (*as **Hoe** exits*) I'll stand.

BRONTROL: Not a wit! Linda. Where are— more chairs?!

LINDA: We do mostly eat in bed nowadays, Brontrol. And there are chairs, with which we sleep, reminding us of.... utilities. He is not accustomed to eating with adults.

BRONTROL: You'll sit with me! And we will make a family again. The dinner will be dined, not denizened as separately frequented—

LINDA: Bill is away!... and I am alone.... And we are all dead tired, after a day's work. For Hoe works terribly hard for his keep. (*bringing the handled pot over to the table*) And so do I, for my being— It's heated.

BRONTROL (*standing, as she places the pot down and **Hoe** enters with a stool*): You'll sit here! what is apparently a custom of some celebration, or sobriety of remembrance.... on special occasions alone? I found a man sitting here!—

LINDA (*as **Hoe** places his stool by the table*): It was yourself, as what you were, seeing yourself again, pondering the thought— as feelings make you thoughtful to recall.... Sit down, Hoe. We're about to eat. This is called— a ritual of assembly. (***Hoe** seats.*) And once a family was this way with boy.

BRONTROL (*leaving*): Take this chair, and I'll bring in another — I know where it's at.... And be generous to me, and my recollections.

HOE (*as **Linda** starts spooning out meal into the bowls*): Who is he, auntie?

LINDA: Brontrol's a soldier of our majesty's service.

HOE: An honest fighter? Can he teach me swordsmanship— an' marksmanship?

LINDA (*working*): ....He'll show you how to brawl, as long as he stays around. But right now he wishes to be like a lord.

HOE: I would be that myself, someday. (*starts to eat*)

LINDA: It is burdensome, and heavy on the haunches. (***Brontrol** enters with a chair, and stops slightly to notice **Hoe** eating, before placing the chair for himself at the table.*) I will sit here, then—

BRONTROL: So sit! and before I do. (*She seats.*) Now may I recline. (*sitting*) But this is too innocent, (*somewhat emotionally*) and actual. (*recovering*) Yet it is as to be. That food tastes good, Hugh?

HOE: Yes, sir.

BRONTROL: Then what I'll bring us later will be as ambrosia in comparison. For I intend.... to make myself stout, here! (*starts to eat*) And that requires proper supplements of high cuisine and dishes with charm. This is.... passable. But we will be surpass*able*, for our enjoyments.... Salty enough?

HOE: Yes. Umhum.

LINDA (*eating*): The pork itself provides enough. The meat is preserved thoroughly before soaking and roasting.

BRONTROL: I would have sliced the pieces a little finer.

LINDA: They're diced well enough, Brontrol. Hoe need not have to use a knife just yet. Everything is a particular labor for everyone. And life's becoming more an' more taxing.... If you've announced yourself to so many, then maybe he'll come by tonight, just to see—

BRONTROL: I didn't have to give my name and history to those sedated rousers. It's enough by my standing and bearing amongst them to cause some attention and notice, and acknowledgment of a royal interest to my town. Although it is terribly peculiar that I did recognize but only a few, and none seemed to recognize me.... substantially. What has become of lasting friendship with acquaintance? I suppose some word will spread of my return.

LINDA: You were on the crest of a curious wave of defections, a tide of departures for a period of time, that captured the whims of many of our acquaintances— those common to *both* of us for sociabilities— to leave. The slightest depression in crop production gave way to silly fears that the soil had been demonized to a taint of fertility, or that something in the air subdued fruitions and prosperity, a general feeling of weariness and wariness that overcame too many personalities and their supposed insights. It was a shedding of the hives throughout the country-land. Plots were simply overworked, acres disabused of their exhaustions. But more practical means took over, Brontrol. Farmers learned in short order to rotate their crops, to offset some losses of the seasons.

BRONTROL: Rotate the plants, as if to show to a different side of the sun?

LINDA (*with slight exacerbation at his ignorance and misinterpretation*): Change the vegetables that are planted in a bed traditionally kept for one kind. If one starts to yield poor returns, another might suddenly earn a supernatural gain of growth. If you had spent more time studying the cultivation and tillage methods that are evolving, you might have tried to do so (also).

BRONTROL: I don't see the logic to this sorcery, Linda, unless it's that man evolves to eat smaller but more plenteous fruit.

LINDA: Apparently the first plant uses up nutrients in the soil—

BRONTROL: Minerals, you mean. Salts.

LINDA: —necessary for its development, but for which the second plant has no need.... And as the second plant germinates and grows, those former nutrients are somehow restored in the earth.

BRONTROL: ....Somehow? That sounds rather unlikely to me, or by a faith demanded. The abundance of the ground— is limitless.... More probably it's the physical disruption of the soil by the first plant that eases the rooting system for the second plant, the attraction for water being the dominant principle of vegetative prosperity. It's in the nature of the root, the coarse ones of the previous allowing for the thinner, weaker ones of a replacing species to subsist and suddenly thrive. That's how I would guess at this, that along with perhaps greater aeration of the earth.... I had a notion of the such, but did not reach the idea of replacing crops to achieve it.

LINDA: What bother may it be now? A malice returns to our land.

BRONTROL: It's reputed to be as productive as ever, a cause of pride with me.

LINDA: But at ever growing effort and strain, Brontrol, and not entirely due to the agronomies.

BRONTROL: Hard work can never be overcome with displeasures. It is aptitude that defines for the accomplishments, and appropriateness of effort that is most required.

LINDA: Lord Balcome pressures us with his being.

BRONTROL: ....That can not be more now than earlier.

LINDA: It is a continuation of stresses. If he is so noteworthy and celebrated, why can't he help more the likes of Hoe? There is

nothing that he does to improve the condition of the bemired by fate, in this area, except to append to them a name of prestigious residency for ironic renown. Were that we could turn that castle into an orphanage.

BRONTROL: It would be a stony and uncomfortable one. And they are not abandoned there. The lord is my immediate and direct master, Linda, while I am home. And we shall work, even struggle, to accommodate all of his wishes, or genuine needs. I'll promote this as a good and firm advice to everyone, as well as offer suggestions to this extraordinary noble himself as to the nature of our composures to placate. A meritorious mind will receive meritorious requests reasonably. And Balcome is a fair fellow to me, hardly demeaning of my position and giving me much license of opinion. It's only that he feels isolated.... for a spell, from his king.... and his prodigious activities elsewhere. He may not *know* of Hoe.... enough to sense the general remissness of attitude the boy represents.

HOE: What am I, then? Are you talking about me?

BRONTROL: You are.... Hugh, and with privilege to stay.... under this fine roof with eaves, as a welcomed occupant, and one deserving of the care of my Linda—

LINDA: He is more than representative, Brontrol. He's as princely as any single person walking the earth. And as is any thus, what is fair to be is how one's treated. So treat alike most any kindly, and for these poor consumptions to offer.

BRONTROL: ....That is not for war— and enemies, combatants. There's much to life that must always stay competitive— Yet.... I will accede to you, that as situations change, so do temperatures. And.... gradually I might offer this observation to.... the august. It's hard to reprove the proven in almost any matter. But my sense is, like the great beasts of conquest, he avoids being harsh until it is absolutely necessary to expend the effort. My job is not to soothe him, but to allow his sovereignty through a justice of means, manners, tactics, and tensions, by which his aboding is agreeable to.... to history, here of this region and for the entire country. You must realize that.... it is an honor to have Balcome residing.

LINDA: ....As it is to have Hoe.

BRONTROL: Well any man's empire is his own construction to maintain.

HOE: May I sleep now, auntie?

LINDA: Yes. Get your cup of water, and take it to your bed.

HOE (*coming off the stool*): Thank you!

LINDA (*as **Hoe** goes to the cupboard*): So early in the morning I must return him. They have so much work to do, helping the plowers and stalk strippers (for seed). And it's only here that Hoe may heal of wounds. What is a day, then? come to night, to sleep of bed before the darkness takes all sight.

BRONTROL (*as **Hoe** gets some water from the bucket*): Life's a workday, and a loafing from. I am with haste to tire more, after a battle. Yet sleep is the most reverent of states restorative.

HOE (*exiting*): Goodnight, auntie!

BRONTROL: ....Very polite is the child.... of feeling grateful for his shadows, that they may move with him and travel as if life itself.

LINDA: *There* is behaviors taught by necessity. Did he eat it all?

BRONTROL (*bending over towards **Hoe's** bowl*): Much so, for a meal to take.

LINDA: I store the leftovers in bottled brine. But lately he's been gobbling up everything. The pork would be a problem, to keep from running rancid—

BRONTROL: It's gone, Linda.... all gone. The appetite was dutiful for you.... He won't come tonight. I'm fairly certain of this. You needn't change so much, for me, your household procedures. I know you're very tired, and to be waiting up for our son.... you don't have to. I'll greet him presently, I'm sure, as words transmit the meanings of my presence, my return.... Have you been careful?

LINDA: ....As to what? For a daily existence? Then of course, what causes for a harm denied or abolished from the consciousnesses, and the hurt feelings smothered by— years!

BRONTROL: Be more so, now. (*getting up*) This was delicious, Linda. And I am pleased, for a home to have still. Now I must go off to another's, to tease off the restrictions and hesitancies of conformity due to frigidness and fear.

LINDA: And have you eaten all of yours?—

BRONTROL: My dear?!... This is a good paste to die for on the battlefield, with the cold shivering for warmth and raindrops as silent company aloud. And I am thankful for your due diligence. A grub enhanced is always a meal enlivened. (*starting to leave*) Yet, on to the castle, for some possibility of high delicacies to be catered with— I will bring back (some), like trinkets of adornment to wear, with for to eat. The present will show you more evidence of my resolve.... and our resolution— Adieu! And with God's pleasure rest through this night. (*Exits.*)

LINDA (*calling*): Disturb Hoe not!

BRONTROL (*voice, becoming fainter*): 'Tis in another spot, my gear, I do suspect, if.... bed is not a couch, nor a calamity of temptation, in this shadowy wilderness. And.... I came in for a candle, to find some relevance of my return.... upon which casts these walls.... more determination.

LINDA (*to herself, while continuing to eat*): Align his thoughts to stay high-minded. It's for the pork delivered to, a treat for trusts an' welcome.... as could be kind to seem, and glad to be approached again. But.... bitterly remains the glaze, as pure as from a mixture raised. Yet.... not to sour of our hearts, tastes refreshed, or re-seen, at least, to be received.... the doting epistles of a husband.... Hard to obtain, so ever be this famine to subside.... and these comestibles acquired. What more for fat.... is there in a dry cereal restored of moistures? Then add to this.... a generously true heart, for his intentions. And can a spell be braised, by even daring to be suppliant to our vows, then have it as.... this way found.

Act IV

Scene I — *At night, before the castle. Two **protesters** meet for the evening vigil.*

PROTESTER 1: Did you hear yells and screams?

PROTESTER 2: The moon is base for such yellowed conversations. And through those holes of wall did blast a sound of such continued agony, that I dare say nothing less than torture is in the making.

PROTESTER 1: That be how soldiers are disciplined?

PROTESTER 2: They have their own ways of spanking, if they're much as children to be ordered, ordered around and about with marching and regulating. But those breaths seemed more squeamish than coming from a regiment of ordinary men.

PROTESTER 1: Ordinary?! They're reputed to be of prime and primal quality for the fight—

PROTESTER 2: They're all ghosts in there, locked up, refused their mission-ing, and recovering from wounds and injuries, to the head and other parts of the body. For that castle's clearly become a hospital for soldiers, or a hospice for the aggrieved and dying. And we partake to control its squirming malice, as like a mole upon our very foundation, not to burst of pus and spread of dangers. It is suspicious, that they are sent here shortly before the king's alleged diminishment.

PROTESTER 1: Sure, they swell upon our taxes fed. But in pain?

PROTESTER 2: With much contortion of their sufficiencies to ravish and besiege, become inverted much of their virtue. And we sense intuitively the foulness that must be developing. For if a Geryon's head is chopped off, it either dies or becomes more of a monster, sprouting back another even more distortedly removed from nature's norms.

PROTESTER 1: Or a gargoyle's or a Medusa's one, for which it dies of sprout and pout, yet causing all disaster along the way, pouring on us rains of evil and villainous poisoning. It's this influence that brings us fears— and caution to contain. Though they had not showed us any sign of this bedeviling before our ears were scorched.

PROTESTER 2: It's all in the nature of this retreat. They're hounded into a Hades to conduce and conform to. And the fire starts within the heart of our valley, this township patterned on our region.

PROTESTER 1: But Lord Balcome is quite a noble and respectable figure.

PROTESTER 2: I doubt if he's even alive. Have you ever seen him?

PROTESTER 1: Where would be for me to see the likes of him?!

PROTESTER 2: I think they've killed him— and replaced him

with another, if not he died in battles actual.

PROTESTER 1: That would be too extraordinary to accept. Where is the need to deny a death?

PROTESTER 2: Only the name matters. For to be imposed of this, an impostor will due, drawn from the lunatics in there.

PROTESTER 1: They have not done crazy things as yet.

PROTESTER 2: Why do we surround the place, to stare at as a sore?! It's that we know of the turmoil that must be going on inside, stirring the castle contents up into a catarrh of inflammation and distress, as they get ready to attempt to descend upon us their ires of uneasiness. And what of the cries we've heard: They're from up there, in the towers issued, as panic earns contemning. That's why we can hear it so clearly. The night air spreads a distilling of hatreds through the winds. And sorrow transfers to all listeners.

PROTESTER 1: Could they be burning up in ice, to store such agonies? It's far from comfortable inside.

PROTESTER 2: That's to propel and promote their puissance to punish. I've heard that fighters may even jab at themselves, to bring on irritation and anger, as by a training. And this is deemed below an injury made, with the spattered blood mere painting of the skin, and the pain like a piercing of the body for a badge to wear. They are about to produce of themselves an insensitivity, before they begin their ravishing and rampage across our countryside— or our backsides, if we let them. But.... stare them down, to show them that we know this. That's why they scream during the night, when sight is hazed with dimness and contemplations, and the only accuracy that may occur are through arrows of anxiety shot. Now their heresy is real to be displayed. Yet as long as we keep watch, to tell what has happened through the night to our daily exert-ors, the stare will stay, all surprise will be subtended to their biers as they weaken to our resistance, as they realize the population can't be fooled or subjected to their footing, and their marches to control us and eviscerate our spirit and drive to remain solid of thought and convinced of truths and learned dispatches. And one by one they'll begin to stagger out as ruined, famished, and in need of some conversion to our aims of fairness and thorough reduction of indulgences to this estate, whose voids must be supplanted by them with our surpluses.

PROTESTER 1: But so broad is their need to fight. Any number could escape at any time to cause a rout of rueful terror. And they are free for casual transit. Might we confront one, of this noise to explain, the nocturnal regretting?

PROTESTER 2: That could be a good measure taken, alerting many of our persistence even through their nightly howls.

PROTESTER 1: Still, we mustn't initiate the conflict that gives rise to an overt aggression. That may be the only excuse they're waiting for, to take in a few of us and call it a victory—

PROTESTER 2: Oh! Be bold as the night's broad brawn! They can not capture a few, but a many— a most! And with that heap comes the downpour of our contrasts; for we will have reason to fight until elimination is victory, the utter annihilation of an irritating privilege and right of a civilizing principle that the people have submitted to, that there must be masters over a populace to take re-

sponsibilities and contuse them into various methods and means of defense, protection and custodianship of a country or a province or a provinciality. Well I say blow that out the nose! if the irksomeness (i)'s no longer necessary to retain, and if it becomes mushy and bothersome— with loss of muscle and capability to be effectively real of plug against offensive vapors from offending vipers. Then we might as well defend ourselves with pickaxes and brooms than be known to hold in a cultivating snot! of mucous feebleness and drippy dreariness dropped here! as if the king could relieve himself for our observance, by being part of our soul's desire for the bodily physics.

PROTESTER 1: But why 's the VID really against Lord Balcome and his men? They have not actually abused us more than has been tolerable and customary for a ruling noble. My wife questions our activities, or the dangers inherent in them. Let what is normal and ordinary be, she maintains, instead of letting the extraordinary overtake our verve.

PROTESTER 2: It's because this nobleman's unnecessary, Horace, or has become so. And we stare at his— possessing to underscore the fact. We place ourselves under his nose— in defiance! to dare his outrage and teach him his effective status in this region, as agreed upon by most— all of whom would ably fight to insure our views. So, squeeze his possible rancor about this into relinquishment of tax collecting powers, as could be plausible to a reasonable mind and a man already of great wealth.... reputably— or fight us now! and see whose blood is thicker, and richer of the red, and whose debts are more sorely to be paid. That is the "why" of the VID. This castle is actually our own. We lease it to them for occupancy— if they behave. Acknowledge this fact, in documented statements of apology and gratitude, and we shall be placated, a lesson learned that the people are stronger than the person.

PROTESTER 1: Ay, it's a haughty height to hem with of ideas, as human subjects evolve of God's mastery into more constitutional persuasions. Still too vague for my wife to adopt or understand, why butchers would not cleave easily— within their home land, given a provocation to.

PROTESTER 2: The womenfolk are too practical, their weaknesses assuring much inaction of a kind. Is the night practical? One can hardly work in it, yet it dominates half the day. You take what is given you to do and work with. And frankly, I find this all very thrilling, and a nice, temporary distraction from much harder labors. That's why I'm enthusiastic enough to be here even through the evening hours. Nerves are on nerves applied, throughout the area, and we're more awake than owls' eyes are wide. Here life and passions make a fervent stand.

PROTESTER 1: As that we must pace, since the ground gets too cold after awhile.

PROTESTER 2: But the half-sultry air itself is lionized with courage around us. And as we breathe in its tantalizing make, we are inflated with all aspirations remarkable. I love a night's soirée of purposefulness and duty. It makes more apparent to one's self the actions that must be taken.

PROTESTER 1: Does make them more contemplative to see.... But who comes towards us, and for the entrance gate to approach? It is the sheriff!

PROTESTER 2: A devout soul.... but devoted to the lord.

GESTROMY (*casually walking up to them*): How goes your evening, gentlemen?

PROTESTER 2: Extremely well, Master Gestromy. The sky is more high than evident of objects past the moon. But I take this to mean we are blanketed with considerable comfort tonight, temperatures reasonable and temperate.

GESTROMY: The entire night? That signifies an arduousness of job. But could I not convince you more, earlier, how unnecessary this all is? I can agree, this climate seeks salvation. For a cross-country stroll it would be wondrous. I've walked much, as is my physical habit everyday. And I can tell you that this land to every part is a beauty defined—

PROTESTER 1: You discover more and more.

GESTROMY: —and its inhabitants ever thankful for the tranquility that accompanies their life. But virtually upon the door to Balcome's castle there is a loathing sensed—

PROTESTER 1: We keep well past the gate, though it stays open and beckoning.

GESTROMY: This loathing, sirs, is so contrary to nature, that it nears to bordering the ridiculousness of a circus, as clowns make their protests through the gaiety of their endeavors.

PROTESTER 2: Ah, but.... they have a motion of behaviors that is practical for their strifes and tickling. And we do have ours, to be fun with notions and assertions.

GESTROMY: Yes. And they are worthily made. But such a show of things can overdo the timidities of tolerance.

PROTESTER 2: We come with our prides shoulder-bound, and place them into view, that our heads have thoughts in them and are not merely to count with.

GESTROMY: Even at night? What's to be proud about? but that you can see through darkness, apparently.

PROTESTER 2: Here's not so dim at all, sir. It is a spectacular view, a panorama of ruins and rustic relief to paint a lovely picture by. I can even see colors still, of this night's dramatic urgency to— feel!... if sight be the only hindered sense.

PROTESTER 1: And we've heard sounds coming from those towers, that suggest an awful plight that your soldiers must engage. Was worse than a crackling, or a crumbling of stone.

GESTROMY: Complaints? Some of the men are much put out by the firmnesses of accommodations found or forced to adapt with.

PROTESTER 1: Screams, yells and screeches of excruciation and anguish to be bothered with or plagued by.

PROTESTER 2: What do your soldiers do, of the extreme to extremities, to prepare themselves?

GESTROMY: On the whole, they curse a lot— through prayers. But these yelps you've heard, I hear them not. They don't seem to be ongoing.

PROTESTER 1: It was during the dusk and shortly after, that they were most intense. Cries uninterpretable but for some form of grief.

PROTESTER 2: And from the heights, sir, floating to our faces like a smoke for ears.

GESTROMY: The heights have holes and cavities, gents, that permit strange echoing of the winds, whose coursing is most prevalent at the time that you heard their rushing through wide and narrow openings for passages and pipes. I'm surprised you've not encountered this before. We discovered the effect quite early on. But you do not visit this place often, and towards the end of day. The great beauty of structure and composition is not normally your interest of attending, so fine an edifice is this symbolic fort. And are you not ashamed to be ignorant of her nature up to now?! It's simply a cry for your notice, perhaps, a plea to be given more heart of hearth, more warmth with a fire inside, and of animated life again, as the hot air rises.

PROTESTER 1: Seemed ghastly to me.... But human, sheriff.

GESTROMY: All sorrow seems humane. We are the very license for it, humans to be suffering of justices. But as predator to prey is but a beast to battle, we must invoke and involve higher excuses as explanation for our errants. Now see that you might suffer some, with a boredom throughout the night. I may even invite you in for some passing company, to spend with drink and talk. We make it hot most nights.

PROTESTER 2: Better that we keep our positions outside, as a gesture of our mission-ing as VIDs. What would some think, if we accepted your entertaining while during our vigil?!

GESTROMY: Ha! But we have three of yours already with us, invited earlier in the day. They've spent an entire afternoon and more, and I don't think are entirely displeased with our hosting. I doubt if they have left, without your pacification that we are courtly sorts, to exclaim. But if so, share alike an examination of our offices and see that we enhance your.... standing, with our residence, and that these silent protests sound worst than howling gusts of misplaced bellowing.

PROTESTER 1: Three of ours?

GESTROMY: A deputation. I escorted them myself right to Lord Balcome's chambers. And while I feared for their demands, that were made to the lord himself, they did not seem to become much befuddled of our comforting, not from what I could see.

PROTESTER 2: We've appointed no such commission.

GESTROMY: They were simply standing by the gate. And I induced them to enter as guests and deferential representatives of this valley, not merely particular to your organization.

PROTESTER 2: So they happened to be.... our appointed delegates? That doesn't seem particularly fair, though they are no doubt members of the VID and share our sincerities. But I've not heard a word of this from others.

GESTROMY: No one has informed you that Balcome meets with your representatives, valley representatives and all? This doesn't serve our purposes well.

PROTESTER 1: We arrive for our night work, sheriff, and have been told, by others leaving, no such particulars of the day's events, but rather that you and another have tried to assuage the stubbornness we've assumed, but without temper of argument.

GESTROMY: Yes, a fellow derived from this very village to become a king's official and now assistant to our lord.

PROTESTER 1: Probably without much effect of convincing our conversions, as we were valued to stand our ground and guard, with the typical expectations we arrive to. Now, you in particular were described as most gracious and imploringly genteel—

GESTROMY: Thank you.

PROTESTER 1: —with an understanding of our efforts that must certainly be relayed accurately to your lord.

GESTROMY: It is, with all intentions to be truthful.

PROTESTER 1: But little of the other soldier has been told us, except to inform of his propensity to deliberate, which was judged unusual and may suggest either a misjudgment on our part or a considerable change of sentiment among you fighters during your stay with us, a transformation to contemplate upon. Still we are advised to remain vigilant.

GESTROMY: Oh, this person has only recently arrived to us, but is as much a representative of our forces.... as our guests have been of yours. If there is a transformation in him, it's simply to have been returned home, or to find some reason to return or opportunity taken advantage of. No. Balcome remains adamant of his positioning to uphold, though he earnestly seeks a lessening of tensions between himself and public folk. We all search for mitigating methods. And I hate collecting fines.

PROTESTER 2: It's been well reported, Gestromy, that, particularly from the impoverished families, you yourself have absorbed some of their fines and late fees.

GESTROMY: Well.... from the poor these amounts are so small that the funds are little missed and easily replaced by the personal surpluses I attempt to find!

PROTESTER 1: But it's that kindhearted generosity which keeps your tax collecting afloat in this region.

PROTESTER 2: Your personal surpluses are simply your personal donations.

GESTROMY: I.... make sure all tallies balance out as expected. Balcome calculates them directly, and does not abuse the privilege in any way. For in my presence he apportions all funds to legitimate uses and keeps no revenue for his private expenses. If I could only convince you of this, he is quite strict about that. What goes to the king, for his soldiers, are subscriptions by the king for his soldiers. But the bulk of your taxes is spent on maintaining the daily and customary needs of this castle as a treasure of your community, now that it is occupied and— worked.... in a way, by the lord and his men. Supplies are continually in need and constantly assembled. And a small army is kept enabled right here in this neighborhood, while we are around, to protect the surrounds and its expanses. Everything is quite faithful about us. And do you know that many of the soldiers are voluntarily contributing to a famine fund—

PROTESTER 1: We believe you, sheriff.

GESTROMY: —for some families of this region, while assessing the state of agriculture in the area and the occasional spots of privation expected to arise during the coming winter? We are sensitive to the concerns of the public, and do not live bountifully of feasts, but wish to share experiences and travails of our stay to make life as much as possible comely for everyone. Your protests.... I understand them, but they still hurt our feelings. It is a noble attribute of Balcome to restrain himself from fighting them— and attacking you— but rather to try and let them peter away out of significance as the climate changes.

PROTESTER 2: We do not doubt any of your points, Gestromy, and have no contractual animosity for Balcome himself. It is his right, in our mode of society and civilization, to collect these monies as he does. Nothing escapes the normal about it, and you go to the task with a very admirable manner of procurement and duty. But the regal environment of the king's has most definitely changed for the worse, his authorities diminishing by the moment. And so must the amounts of these taxes. As pressures converge upon us to bring hardships: economic, financial, fiscal, what have you for a family to survive through and subsist with, then more to ourselves must we depend of aid, and less to the administration of rulers accede. We are better capable— through the intelligence of our common existence— of handling our monies for ourselves, as the clouds may darken. And there approaches a time, a state of actuality, when this lord may have no right to collect anything at all from us. We pronounce this possibility with our presence about his.... residence, and ask— suggest, insist that he make arrangements in preparation of the fact.... for our awareness of a compromising attitude he must express and produce during the management of our affairs. This is serious now, the more so becoming as our majesty fades.

GESTROMY: Can you expect Balcome to announce to you that he will no longer be a lord, a privilege granted him by the king, or that he will gradually lessen his lordship out of existence or essence of effectiveness?! or actuality? We'll probably all die for his majesty, on some battlefield of desperate straits and foreign, contentious land. Yet we will not hesitate to go to wherever as ordered or pleaded. For the slightest, weakest utterance of the king, through a bewailing, bemoaned need of our services, does bring to strength his calibre of action with our own. And in that way— *we*.... are the measure of his rule, not his lessening due to losses. Yet all subjects should remain loyal to a personage such as his, he who values us more for our deeds and determination to duty than by any decrees he must enforce. And can you depict him as weak as your sobriety through this intoxicating game you employ? this heady feeling of companionship and lawless commonality? He himself has been wounded, not least of all to pride. And yet he refuses to give up! And should he last to struggle more, we must all be a party to his cause— since this secures ourselves as well. Now challenge Balcome not, I would beg, but let the lord transpire more smoothly toward your aims.

PROTESTER 2: We've not said a word to him directly, aside from this— deputation you've constructed. And all that they could say vies as the obvious for VID. We simply show ourselves, and wait for the lord's approach. I can not truthfully feel much sympathy for his majesty. He is a remote figure to us— august without question. But that's a quality of man self-supplied and self-endured. I can only sense what is more rational and immediate to myself, than his empyreal plight. But you are more honest to observe the nature of these transactions.... above our heads. So may, then, Balcome ours.

GESTROMY: It's that your force is shown more to tease, that gives me want to calculate our ease through philosophical confrontations. And you are right to stand of these beliefs so ardently. The best is always what you know and can actually be educated by. Balcome does review the possibility of more direct talks, with members appointed by yourselves for this purpose. And he even fancies greeting you on the grasses someday, soon, for pleasant and cordial chats, conversations to draw out some reasonable commitments of sociable behavior. At the moment, though, I think he is more in a phase of examination and analysis— Gentlemen!... make good this veil of night to peruse through your paces of vale and delightful screening of your issues with us. Now's a good time to see how they may honestly be made to seem foolish, at least to yourselves as you walk. But the night's quietude is particular of noises to the head. And I must return to the lord, to tell him what I have heard, and learned of many composures. May we call him *ours*? This night certainly is. And you are friendly with it. (*starts to head towards the gate*) *Bon soir*, without to sleep through lethagies of emotion.

PROTESTER 1: And within it, sheriff. They be good graces that we stand to hold of you. We are dominioned without a domination of what dreads, should you be of a like inside to dream of us.... I think he states some good reasons for our folly.

PROTESTER 2: Well that he may present himself as such is our folly. For what is fighting meant for? We are grievous in this air, not challenging nor being chased, but merely spoofed of our causes.

PROTESTER 1: And if the spooks come out, we'll laugh at them, to bait for better explanations and responses. Why does night compel the cold, as if lacking thickness to its shroud?

PROTESTER 2: It is the generosity of resourcefulness, in devouring light's taper. (*separating*) May the moon move us well.

PROTESTER 1: Ay. An' drench of glow.... astonishment to throw. (*starts walking with divergence*)

Scene II — *A dining chamber or hall of the castle.* **Herkshead, Craie,** *and* **Bill** *are seated more or less collectively on a stone bench to one side of a raised slab of stone table, while* **Balcome** *is seated some distance away at an end of its service. While the others are more occupied with their dining,* **Balcome** *does more staring at them than eating.*

BALCOME: Is this not a particularly fair buffet for your appetites?

HERKSHEAD: Most delicious, my lord, and relaxing from our prolonged postures of standing—

BALCOME: Waiting.

HERKSHEAD: —Though I must admit, these granite seats do bring on a frigidity not accustomed to in summer.

CRAIE: But passable to take. Though at your wooden chair you have the advantage of sensitivities.

BALCOME: Don't tell me you're uncomfortable still? You complain more than my soldiers.

HERKSHEAD: Anxieties drain with your regard of us, sir.... This is unusual bread, but very tasty.

BALCOME: How did he fall?

BILL: ....I tripped along a tug of stuff an' straw within the enclosure, my lord, and awkwardness bent me over into distress of the contents.

BALCOME: So that's why you were ghastly-faced. When I was alerted of your sudden discomposure, I became worried. An illness befell you?

BILL: More gain of pallor than of pain, my lord.

BALCOME: The children are more sensitive of taming. And.... you are recovered?

BILL: One side still throbs; but hunger alleviates much of the panic, if not a soreness subsides.

CRAIE: Taming? Lord Balcome.

BALCOME: But of my welcome to decide upon, its color and timbre and shade of intensity. I had you waiting until I had thought this through myself, and with so many other matters before me, pressing for my attention. Then your overture needs be tamed to be appreciated more. I hope you thought through mine. But I was slightly mad of frustrations and annoyances.

CRAIE: We've written not a thing, and could not have.

HERKSHEAD: I thought you were preparing to have us skinned alive! It was much of a comfort when you called us for a supper, to relieve our strains.

BALCOME: I wanted your hearts to be thinking, gentlemen, as well as pumping. But when I had attained the heights, I began to moan and cry from a nervous tension about you, and your simple conditions of confinement, that might symbolically characterize my own and of the company I contain and travel with. We are sometimes placed in unusual manners of welcome and neglect, and must make the most of our surroundings.

HERKSHEAD: ....As we did ours.

BALCOME: And is not this castle splendid? Yet you have neglected it dreadfully— you being the natives of the area, the ones most familiar with its presence and standing.

HERKSHEAD: ....Perhaps a trite naïve to the extent of its de-composition—

BALCOME: And the country's?

CRAIE: We remain sound to its survival, lord.

BALCOME: That is a much desired goal to affix to each and every one of us, as like a spot of dye to skin for recognition and ad-vertisement of this.... wish. I hope my hospitalities have not de-meaned your fervors, while we ourselves are isolated from battle.

BILL: The character to your endearing lessens not my spiritual reprieve from knowing all.... about your methods brought. Yet may I question some, for your assistants to be enlightened by?

BALCOME: What?! boy! But that your guard was gruff an' rough?! That is typical, with orders placed upon them, to make for various of an environment manageable. And I'm sure he offers you all apology, with your license to dine. But detention means detain-ing is a law obeyed.... Though I'll speak with him of details, if you'd like, just to show of my keen interest to any incidents occur-ring.

BILL: That is unnecessary, my lord, as I am less so.

BALCOME: Less so what?!

BILL: Interested.

HERKSHEAD: He does defy himself, to think more of it than as being endured. But what of us, my lord? It is night. And the cham-ber's bright of flame. But we have homes to go to.

BALCOME: You'll sup some more. Is not the food grand?

CRAIE: Of ample fitness, sir.

BALCOME: I'll have the caterer bring more. It is regional, of course. (*motions to a sentry standing by a wall*) We can not cook this way, yet. Such quality must be prepared for us. And so I thought, with guests we should take the indulgence tonight, to splurge, a bit, upon your valley foods. This shows that I'm sincere for your welfares. And as I've paid for it— you'll eat, to partake of some kind of favorable esteem for me. Once satisfied of this, the night's all yours.

HERKSHEAD: That's very decent of you, sir.

BALCOME: The breezes, though, can be bitter.

CRAIE: From a particular berry not so eatable. It's aroma can persist at night, for peculiar pollinators of its flowers.

BALCOME: So be it with this caution. I am convinced that *any* vegetative morsel can be converted into some diet, perhaps if only steeped for a tea. My men and I have done much testing of this proposition.... an axion.... during our campaigns of haste and speed through deprivations, as we must often scour and scavenge the available landscape for any sort of foodstuffs.

HERKSHEAD: It is busy being a soldier, then.

BALCOME: While active, yes. (*Gestromy enters.*) Ah, Ges-tromy! You return.

GESTROMY: Yes, your lordship. And I see our guests have stayed awhile. That is good testimony to the usefulness of the mis-sion I've invited you to. If only it were more to be acknowledged. But I have brought more learning of the territory, Balcome, as cer-tain areas lessen of sufficiency to coffers. There seems to be a spotty depletion going on, of abilities to pay for anything— that is, to earn from any work or activity.

BALCOME: Withholding?—

GESTROMY: I think it's genuine, a perhaps light or generalized depression of economics adopting itself to the valley, and spread-ing in such a slow manner that only pockets show up of real pov-erty or adversity of subsistence. We should, together, discuss your rule over these growing difficulties, to find ways of effective abate-ment.

BALCOME: My privacy is dined with, for the moment.

GESTROMY: It is an urgency to be alerted of, my lord. That you know about it will suffice for now, though I had sought you to con-verse.

HERKSHEAD: What's to be acknowledged, Sheriff Gestromy?

GESTROMY: You VIDers are reticent of your own details. I had expected much curiosity about your visiting us—

CRAIE: It will certainly be discussed. And we will promote.... for our experiences to be shared.

BALCOME: Thank you.

GESTROMY: But your night watchmen, of a fashion, had not even been informed about it, not of the two I've met.

HERKSHEAD: ....Our organization is casual enough, to be de-serving of some proprieties of comment.

GESTROMY: But not to know that you're even here?!

HERKSHEAD: We don't hold meetings on the lawn. Our visit will eventually be shared more commonly of knowledge, by those who have observed it(, our entering).

GESTROMY: Can they even care for you, with such ignorance? I thought you folk were ripe, (to) reify, for spreading rumors.

CRAIE: We are collectively in tow of our concerns, but individu-alized throughout our common aims. And we deposit ourselves not for a shaming of anybody, but out of an anxiety for our conditions and growing— though regretful— contentions. (*Makethrift enters with a platter of pastries, followed by the sentry.*) The cause is cas-ual enough. The concerns simply spread through the air, as like this redolent fare brought, to be attracted to a fragrance of commu-nity action to please our agitations toward a useful service of dis-play.

BALCOME (*as Makethrift places the platter down by the*

*guests, while lifting up a sampling for **Balcome** on a plate*): No doubt, to do anything feels good of (a) participating virtue. That is how groups evolve, less for commonality than for compulsions.

HERKSHEAD (*to **Makethrift**, who stares slightly at **Bill***): This is an extraordinary blend of grains to the meal, caterer. It almost seems exotic.

MAKETHRIFT: I use a type of the (slightly) misnomered *bleu blé*, which is actually a type of corn. There's a certain aroma to its sweetness, sir, related to its azure sugars and fermented content. It's imported. I have it excursion-ed off the continent, along with a certain pepper spice.

HERKSHEAD: Well it makes for a remarkable deliciousness of the savory and smooth.

MAKETHRIFT: That is required by the making of the dish, sir. There pasty rolls should compliment those tastes well, being more traditionally based of the sugar cane and beet. (*starts to bring **Balcome** his portion*)

BALCOME: There is a slight iridescence to the texture of the foods, a sign of high crafting, to retain a plant's panache and dash of character through the wrestles and wrangles of its cooking.

GESTROMY: The sheens are colorful of this presentation, my lord.

BALCOME: Come have a piece of the desert, Gestromy, for all of your walking. (*as **Gestromy** goes towards the guests*) You are the most ambling of all of my liegemen, of current motion. I hope the others retain their abilities of superb endurance for long marches.

GESTROMY: Thank you, my lord. I'll have that and a cup of wine. Yes, I like to traverse all throughout this territory on foot, rather than riding up to a station or a house to collect moneys. That gives me a more common view of things, and allows for the indecencies of personal robbery that may pronounce the temper of our current area, as that I well know how to defend myself. (*pouring some wine in a cup*) But I've not come across anything suggesting of that tendency at all, so far, which leaves me with a very gentle feeling for these people and their innocences of faith in worldly views and happenings. (*sips*) Home is very different from a fighting foe, even when the home is unfamiliar. (*picks up a pastry and takes a bite*) The wine is well approved to be dry, to accompany these rich sweetnesses.

BALCOME: It was suggested, since it's the only wine we have stocked, to employ the services of a fine chief of doughs and bread, in order to enhance this particular supper. (*parts a roll*)

MAKETHRIFT: You are very kind, my lord. Though it is my specialties of novel construction which distinguishes me from worthy others within our valley.... I have an assistant who is especially talented for my purposes and instruction, to help me in my designs.

GESTROMY: And I have such a sudden hunger as to pick my brains with gluttony.

MAKETHRIFT: The rarenesses devoured stimulate the appetite, momentarily.

GESTROMY: You use an extraordinarily fine butter. It is slight, yet powerfully evident, commanding of the senses, and palatial of the plate. We're used to much rawer stuff.

MAKETHRIFT: Thank you. And now, my lord, I believe my evening deliveries are done. If there's not more to do for you, I should wish to take my leave.

BALCOME: And well you have performed your duties, Master Makethrift. You have the compliments of the castle to your recommendations collecting and accumulating. And you will be paid immediately upon your exit, for I like to handle such details strictly of occurrence rather than have them dwell upon the realms of postponed requirement.

MAKETHRIFT: This is such high praise to propound your charity for the culinary artisan, that I may busy myself to leave with your guests, insuring that all is in proper form—

BALCOME: They retire later.

MAKETHRIFT: —perhaps offering some training for your scullery minion, simple tricks to improve their preparation of meals, as their dual service affords their soldiering to take up.... I see. These sirs will leave much later than my need.

BILL: We are bound here, to soak of the syrups for a time, this hospitality brought to my head to dazzle and usurp!... into a compliance of understanding which disgraces my injury as more of accident than a form of restraint.

MAKETHRIFT: Injury?

BILL: I am still sore, of the side.

BALCOME: And still resentful of a punch, that youth is unaware of things until made aware?

BILL: Like stench to learn of foulness, I guess this is taught.

BALCOME: Much later to earn these realizations, they come with experience and explanation of them.

MAKETHRIFT: Perhaps the anise seed baked into some of the doughs will help to relieve much of the physical discomfort, at least.... and the happy sweetness ascend to the mental proprieties to feel well of spirit and secure with the concerns of relatives and friends.

BALCOME: The fine folk here do make for their representatives an education. And I mean to be pleased with their abiding efforts.... (*eating*) This is a strange sap, compounding of the fruit a licorice maple.... Yet, tart's not tart, nor tardy for me. I have enough time, for the sour apples to deflate into a tenderness of baked esteem. Apples of hard— meat can make the finest sauces of spread, the pulps pulverized into an acidic purpose.

MAKETHRIFT: They are generally sweetened with sugars, my lord, to reach a satisfactory taste.... and made more mild with cinnamon and mints.

CRAIE: Yet do we treat with satisfaction, this lord provides us....

room, space, and evening. The nocturnal clock seems longer than the day's.

HERKSHEAD: There's then this apprehension, Lord Balcome, that we are not sufficiently confessed to you.

BILL: I still hurt—

BALCOME: You still complain— of smarts.... and snarls enticed by. Yet see how I am sniped at and surrounded, my generosities almost entirely wasted, my patience teased toward tatters, mongrel dogs pulling at and tearing off my clothes, to reveal one pure anger. What night is yours? that in the day I face such hatreds constantly impertinent. And yet for a view to treat.... as an enemy I am mixed-ly inclined. And there is confusion with the headiness, and indecision as to what to decide.... what is counted here, and what is (to be) ate.

GESTROMY: But they are most definitely sufficient of a representation, as I chose— an animation to provide some speeches. And you've done this, whereby all matter's clear, and my lord considers carefully your proposals, to modify or embrace, or embrace to modify, and bring within a congruity of acceptance— by everyone. So then, out with it, boy: Are you sick? this exposure to overwhelm your thought? You were bold enough to be heard outside, not challenging but chastising, but of a sincerity that welcomed my attention— and led me to bring you hither, as a good example to be heard— and questioned. But are you ill to the reality of your circumstance, as you have your sweet breads, and this glorious opportunity to meet with.... someone you'd oppose?

BILL: I'm not so brazen, as my body aches, and a fever spreads; and I am standing— and yet I sit. And I am conscious, but can not know my way around this, can not encompass the meaning of rebuttal to our lord, to his actual face and what *he* represents. For me, this is a dying facetiously, as I see my own pitched towards this cause. And how can one honestly say I've hurt myself, by my own actions and indiscretions? With a boldness I've been severed from hostilities, and yet can greet them plainly by the eye. And I do say as I am shown, that silence serves me more to sure up this serenity of anguish. Because this is as calming a hurt as ever one could suffer from. I'm not the least bit squeamish by it. And this splendid dining does settle me much, the meals tenacious of relief from a nervous condition.... Yet, the side remains sore, the complaints remain sound, and I am tempted to suggest their meaning more overtly—

HERKSHEAD: It was a spicy stew of hummus an' ham.

BILL: —We don't need any more rulers. We've never needed them. Their superiorities are false and phony. They neither have purpose nor a seriousness to be desired, descending off a wish-list of refulgences to be gotten rid of for annoying the eyes of gods, and cast down upon our heads to make up blasphemies and quarrels among men and their misplaced mightiness, as if the heavens were too proud a battleground to have us serving through our clawing struggles of attainment. And thus we serve under the clouds, through the rains and storms— for kings! to become wantonly begotten subjects, hounds of hind legs and heinousness improperly accused.... of an obsequious nature by birth, and commended for a timidness to crowns.

BALCOME: You're of a strange social movement. You haven't

fought for real one day of your life, not for a desperation of life—

BILL: I am godless!... my head-strong lord, my bold producer, my Delphic deity, my gilt brine of brow, the counter of heads and their sweating distempers under your eyes and through your visionary gazes of our demonstration, at our displeasurable approach to your castle and bedrock.

CRAIE: The insolence is due to a lightness of the head, my lordship. Bill's definitely something ill and needs be returned home for some nursing—

BILL: I am fatherless! and freely floating here. For faith in you to find, Balcome? How do you churn my stomach!

BALCOME: What evil sits at my table?!—

MAKETHRIFT: The boy.... is with pains, my lord.... and blushes queerly to some heat—

GESTROMY: Though (he) makes with sentiments similar to what I heard; then more symbolically mixed, but still with the errors of a sincerity to speak and amuse.

BALCOME: I've given you the best that—*you* can offer!

CRAIE: Let's be more proficient of the truth. He was attacked—

GESTROMY: Attacked?

CRAIE: —and bleeds of youthful tremors internal, that first pain of reckoning one's actual disposition in an agitated world, the panic of discovering one's wingless flight through a punch to the gut, the realization that one is not the beauty praised of life.... but merely its determination to persist.

BALCOME: He tried to escape "D," Gestromy, and was put back down.... in there.

GESTROMY: ....That's very unfortunate.

BILL: It was my approach to a cruel authority, more honest than gentle.

BALCOME: And so you claim me cruel? to discipline you— each and every one of you?! Well I haven't time for children, only soldiers. But I'll wash your head of its lice! For you are subjects of a princely order to the very least, and will be brought of the cadaver! to want to obey it.

BILL: That frightens me, but not my soreness—

BALCOME: Back into internment sent, could be my command, ungrateful jackal!—

BILL: The dead don't wish for anything!—

BALCOME: But they obey everything. You doubt me?! I've moved their mouths in jest, to utter their complaints of state. And you need just a working to mend your head. (*standing with frustration*) But this is the youth, of this dominion, pitched against me, as their elders mislead valor (**Brontrol** *enters.*) and haughtiness of opinion, arrogance of thought, the brawniness of a misplaced dis-

343

cernment, and a poor calculation of effects. What serves me?!— Brontrol!

BRONTROL: My lord! They are becoming convinced. I have spoken to several, along with Gestromy. (*acknowledging*) Sir. And doubts are brewing, for their offenses to desist. Because it follows not the logic of their needs, this behavior, and they are gradually sensing the remissness.

BALCOME: But—!

BRONTROL: Their numbers will decline day by day, and the protesting will dwindle to a trickling of night effects and obscurities. For, to show themselves becomes more and more foolish. They won't want to be seen, and will avoid notice, your counting of their heads, and their taxing dispositions for self-assured grief.

BALCOME: But this boy mouths me badly—

BRONTROL: What?!— He!

BALCOME: —and is a representative! of these persuasions these folk carry and abuse themselves with.

BRONTROL: My lord. (*recognizing **Makethrift***) What is here?... What satisfies for your approach? These people are controllable, with an opposite sense, and do not mean to blame your administrating of any activity, but rather for their own uncertainties are stirred to act in a bizarre manner of instances. They sit to stand their ground, and stand to state their principles as sat of a stubbornness. They speak of hatreds with a love of this insuring their positions, and love of the ensuing fears for hatreds detested of their retributive mangle. It is all much immaturity through isolation, and being restricted from the harsher realities you face. I myself have been with my wife, and a boy, and see what casualties are brought by pressures on the innocent to distort what escapes as their views and issuances on life, their senses knotted of the line by a growing brevity of hope to remain purely untouched, unmoved, nor influenced by the more worldly events, this sad abandonment of kings and princes. Yet to be left alone is itself a sedition for such feeble faiths, whose lack of vision hastens injury. So when this lad, or any other, speaks the bad of you, he means the good is seen— the noble, the great, the charitable and chaste, and the concern consumed by. But to this sight does bring onto himself a badness of shame internally felt, which causes a need of pretense for a disagreeable rant and rage against one's erring, as the ideals imagined may be abandoned, displaced by the realities of one's true condition, that dreams often befoul the facts. I assure you that these protests will die down, as these people mature to more current events, and will grow to love his majesty again, and have sympathy for him and his efforts. They'll stop enlarging the faults— found— beyond their fealties and national pride. But not one man I spoke with holds you to be wrong, but rather could project your talents, wished unto the king.

GESTROMY: There's still some considerable resistance, though, swayable but questioning—

BRONTROL: That can be dealt with.

GESTROMY: —any possibility of their erring.

BRONTROL: One merely stays persistent at the suggestion, sheriff. I know these country folk well—

HERKSHEAD: It does seem veiled of sight, in here.

BRONTROL: —their types and typical approaches to life. The resistance is a form of acceptance, a welcoming of the inevitable by a stern greeting, or an acknowledgment of the apparent by a shift of glance to see it other than to be, as for a test of its approach in time. Does not a fish bite at the hook it loves, for a worm seeing? Then they are fishermen at heart, for these trophies of fate. They want.... some struggle sensed, to redound these efforts of an outcome; or as the overtaken enlivens at the end of a hunt, with a fierce defiance, its nobility is assured and admired. Life is this way in general, through its events to submit to.

BALCOME: Playing hard of capture?

HERKSHEAD: This does pierce the genuine.

BRONTROL: A plant does fight through soil to germinate, and reach some sunshine. Does it— hate the earth it(ha)'s fought?! But no, throughout its curses it does adore the security of its roots. This lad no less, whatever he has said. And so's the same with them all, my lord. I can guarantee a hearty compliance to you very soon. It is almost like the physics undergoing its strokes of oar, pushing against to thrust forward in the water a boat, a ship of state, a consciousness of being, a nation to grasp and own— through learning of its sweetness as once a candy obscured and abhorred for fear of being something else.... as I've not doubted you've found some tastes here captivating. In short, it is a learning process— of you as you must be.... to them. And all processes wend their ways a bit towards fulfillment, at various speeds and accelerated impulses. But Lord Balcome's renown has never been disputed by anyone.

BALCOME: Convince me more of my frustrations, then. For I sense disputatious intentions shrouding, surrounding me, provoking some imbalance or ineptitude.

HERKSHEAD: But can not Brontrol see more clearly?—

CRAIE: We are not ungracious. Never that! Yet not as fools do we preside over ourselves these dangers of offending the lord. Our wishes are real. And though you claim to know us well, it's much as if we were your children to describe. And that is to belittle both our courage and our caution.

BRONTROL: I'm much related to the gentry here. And though I can't say to know many of you personally, having been away a few years, I would never work to demean your character and local spirit of societal solicitude. Then strangers to me are you seated, but not as childish scouts to berate with an affliction, an immaturity running past your ages and through your desires. It is a common lo—

HERKSHEAD: What means to say—

BRONTROL: —aching of a group—

HERKSHEAD: —this plea does serve you—

BRONTROL: —that makes its own head!

HERKSHEAD: —not to recognize your own son!

BALCOME: Son?!

HERKSHEAD: This is Bill who sits with us!

BRONTROL: Billy!—

BALCOME: Deception!

HERKSHEAD: And he does make us well, for our concerns and anxieties to be real, through the innocence of an honest discourse of emotional enwrapment. Then this is true, as blurted out with candor and not calumny to perceive of rulers or authorities or heads of households or parental desertion. For to lose one's powers and enforcements abandons your privilege to lead.

BALCOME: You've tried to trick me, Brontrol! all the while twisting their most evident sentiments like a rope to bind me for my leniency, while deceiving of your knowledge that you merely advocate for your son!—

BRONTROL: No, my liege!— His hair is different!

BALCOME: —That's purely rational and understandable, to want to protect him. But you could have told me up front, at the very beginning of our meeting and discussions, that you wish to apply to me for his safety to secure, instead of all of this deception and playing on meanings and sincerities to treat me as a crude student of philosophical contrariness and reversals, a dunce to anything apparent!

MAKETHRIFT: It makes for you to find him blinded, my lord, with his fears for Bill. But he was more demonstrative with his wife, earlier in the day, at my very shop to find her, since she assists me ably. And so, with regret for that outburst of discomposure, not receiving the fond greeting he had hoped for or imagined due him, he overplays his efforts, to understate a meeting with his son. To.... possibly ignore it? That would be hard for me to do— But forgive *him* of this, whereat he finds of his home a difficulty of position to rest with.

BRONTROL: I've made up with my wife, for my earlier disagreeableness, and we have taken together again—

BILL: Oh! this pain!

BRONTROL: —And my apologies to you also. But I was upset to be so thoroughly surprised by the initial lack of affection I received. This stunned, somehow, my expectations, and intensified my immediate remorse at miscalculation, since I had maneuvered skillfully, ardently— and even deviously, to reach the valley again.... with some legitimacy of approach. And so I make up to you.... and your catering, for my bad behavior and foul temper—

BALCOME: But not to me!... ccurser.... coward—

BRONTROL: Sir!

BALCOME: —But I feel ill to it, with the blame you deserve, in treating me this way, guaranteeing nothing if you've seen, and even less if you've not seen of the truth. And that does lead to more the disastrous of my necessary stance.

GESTROMY: This is all very remarkable. But I can compare myself to his efforts of converting the protesters, my lord, and exclaim them to be as valuable and dear as my own—

BALCOME: Worthless!— Not for you to try, Gestromy.... but the attempts.... themselves. The manner— the method.

GESTROMY: But is it so unusual for him to one: have a son here, that you might suspect from knowing of his history? and two: keep this knowledge from you, if not such an influence on his action for you need (not) be directly related to your ear? His role for your assistance to provide seems very much independent of the matter. So that he would deny—

BRONTROL: I've not denied a thing!

GESTROMY: That is the point, sir. You've not denied, and you are not denied.... you are native—

BALCOME: But I am surrounded by a moat! of suspicion and possible dishonesty and misrepresentation, of so grand a mixture as to be condemning of a lesson. For the son is mean to me— and there's no denying *that*.

HERKSHEAD: Only youthful insolence, my lord. He's been beaten down—

BRONTROL: Beaten?!

CRAIE: Punched once, and thoroughly.... while in a detention.

HERKSHEAD: And it's only natural for him to have some resentment about that.... towards you.

BRONTROL: What cell?! Then he was placed, and "D" defers me— my guardianship?!

BALCOME: And the father promises me a glorious outcome, to all of this poor business worked. Well, be a day most glorious, I'll appeal to. For sunshine makes the summer, and that is seen of heat and warmth. And I *am* cruel, boy, for the actual to insist. You've tortured yourself. But the father tortures me, perhaps with inappropriate hoping.... or viewing of things. That being so, then— Bill will be retained.... to this castle. And I will count heads, from my tower, howling of the numbers, cringing of the sight. And if there are as many as I counted today, or more.... then dire graces mount upon this retainer, as a retinue of lies converged.... You more adult are free to leave and speak out on this, with all your earthly declarations lunged. Breathe more air for me, then. And— free the space for my decrees.

BILL: I am glad to be the brunt of death declared! Is not my father so dimensioned?! I am death!—

BRONTROL: Stop playing the sacrificial lamb! You were always so difficult, and then to show your passions outrageously out of place. But I can not allow this, my liege. (*A wall sentry comes towards him.*) For he is my fleshed foundling, through my protections to exert of family!— (*to the sentry, while drawing his sword, the sentry reciprocating*) Stay back or I kill you! I've slain many with more ease than this rage compounds—!

BALCOME: You can't fight this army, Brontrol. And the reprieve is clearly laid out, with your own actions doing, your own words spoken and to speak. Is this against me? or not! (*Other **soldiers** rush in, weapons drawn.*)

BRONTROL: My lorded liege.... you confound my merit. And what am I to do for you? with this way faced, and no retreat provided for my kin!

BALCOME: Abide in my method! manservant. Go out there, and pitch your heart to them, these simple folk of yours with such a complexity of description, that you may change a night of loss into a day of winning! For I do intend to count heads aggressively tomorrow. And I'll gain something by it, whatever is the sum. Respect, perhaps? Some deference finally definable? They insult me, and that's too base to manage easily, or digest! with of a fondness or a sweetness, their fickle flaunt 'gainst regality and earned position, station— status!... and the internecine turns and twists that this castle provides— all! It is uncomfortable here, but for a subservience. And that alone is mandated by the king, our sovereignty for ever.

BRONTROL: I am.... his man, my lord.

BALCOME: Go out there (to them) and boast of it!

BRONTROL: Were that I would dine here of it, with my entire family. There is less threat from me, for this reprimand. (*lowering his sword*) I am a representative of his majesty, and he will not be pleased of this treatment. And as for lies, the earth's no different night or day, for it revolves of them that would be better of it seen. Note that I am upset, Balcome, and insist that you allow me to take my boy out of here this evening— this instant!

BALCOME: It's not to be—

GESTROMY: We should not prove ourselves this way, my lord. It can bring us favor with neither the king nor these people.

BALCOME: My pledge is made, and can be easily enforced— more so than this man's dreaming.

BRONTROL: There's been no pretense to it, lord!—

HERKSHEAD: We've been as much more lately a guardian of Bill, Craie and myself. And we can not think to leave him alone in this dreadful compromise, facing querulous fear.

BALCOME: Be that as you may, to stay, as guests. But the fewer voices heard outside, to palaver on my good behalf, does raise the prospects of heads counted tomorrow day. And that tally alone determines the boy's destiny. You are better spent mouthing me, to assure his comfortable release.

CRAIE: This is an evil gnashing of the teeth, Balcome. Contemptuous! It diminishes you thoroughly, and lowers your standard— to these fine soldiers. Let us withdraw together tonight, as a whole of a commission brought you. Bring to an end an erring retribution, and misplaced— misunderstood resentments.

A SOLDIER (*who had rushed in*): What does the courier want from the lad, my lord? I'll take! He looks ghastly seated.

MAKETHRIFT: Such as for kin, man! Heard you not?! They are related.

SOLDIER: So be so. They look related. But what of that?

BILL (*trying to stand*): I.... Ire! (*flops down onto the table*)

HERKSHEAD (*assisting **Bill***): He is ill!

BALCOME: Call for a physician! (*A couple of **soldiers** run out.*) He shall not leave the castle until looked after and cared for.

MAKETHRIFT: Is not for diet bruised.

BALCOME: You can not have him, Brontrol.

BRONTROL (*as a **soldier** approaches the table*): The beds in this place can not be comfortable.

SOLDIER (*at the table*): My lord.... Judging from the face, this looks like a liver ailment, the greening of the cheeks, a hue of distemperment to the skin.

BALCOME: Let a physician tell.

CRAIE (*helping **Herkshead** try and comfort **Bill***): He was punched straight to the solar plexus, by one of your more brutish guards.

SOLDIER (*at table*): Ah! Then that explains it. He'll just have to survive the disrupture of nervous impulses, to the abdominal vicera.... or not. Little to be done, to soothe the distress, but a maintenance of courage.

BRONTROL: Let me come over!

BALCOME: ....Allow his approach.

MAKETHRIFT (*as **Brontrol** goes to **Bill***): I have a tea of milk thistle, Dandelion leaves.... red clover and burdock root, that may help, my lord, if I can leave promptly to prepare it.

BALCOME (*sitting*): Do so, sir. (*as **Makethrift** exits*) This evening's like the pounding of a drum, for warning of the expected. Tomorrow's day will be too brilliantly seen. And I must determine catastrophes, now. (*as **Gestromy** comes towards him*) But that is the way, to prepare for an assault.... of nature.... Gestromy.

GESTROMY: It need not be inevitable, unless he dies.

BRONTROL (*close to, but not touching, **Bill***): Have him drink much.... and pass it all through.

BILL (*weakly*): Father!...

BRONTROL: Conserve your energies, Billy. You've simply abutted a man.... Oh! He's fallen into delirium— and sees me not, but speaks of a father past.... Let me examine. (*rubs **Bill's** hair*) I must make.... so many amends(, now). Who did hit upon you?... Who? or what! (*retracting his hand*) It makes for wounding a blessedness, and sentimentality of error.

BALCOME (*staring*): The lad's asleep, I think, to have a conversation.

GESTROMY: Your method.... will be the talk of the valley, my lord.

BALCOME: Thus it comes down upon me, like a convergence.... And that poor boy's silent.

GESTROMY: All of these problems descend upon your head.

BALCOME (*as **soldiers** start to leave*): Yes, so. That is the method. But he'll be medicated soon. I've offered that, and everything else, to properly treat my guests— and country, to the privileges of this locale.

GESTROMY: These have been.... delicious pastries, lord.

Scene III — *Night, before **Brontrol's** house. **Makethrift** knocks at the door, with a frantic reservedness.*

MAKETHRIFT (*knocking*): Linda! Linda! Get up! It's Master Makethrift! There's news to import, that must concern you. (*A candle light is sensed through a window, approaching the door.*) It's.... much of your own to do.

LINDA (*opening the door*): Master Makethrift? What occurs?

MAKETHRIFT: Your son—

LINDA: Go back to bed, Hoe.

HOE (*voice*): How be Billy?

LINDA: He woke me, Master, for your knocking. Forgive his impudence— Continue!

MAKETHRIFT: Bill's been detained, at the castle, by Lord Balcome, indeed for impudence, and with some punishment, physically beaten for his offences speaking, in essence denying all authority of the king, thereby insulting his host.

LINDA (*with fright*): Oh! But to a hosting? Then Bill was brought to him?

MAKETHRIFT: As I understand it, yes, along with two others, much earlier in the day. I prepared a banquet, for their suppers, at Balcome's expense; so they were indeed hosted as invited guests, some sort of deputation from the people. But while I was there, things got out of hand—

LINDA: And my Bill was beaten?!—

MAKETHRIFT: No.... No. I saw no physical assault to any of them. The lad only spoke with controversy. But he did not look well at all, and must have been injured earlier, one of Balcome's soldiers punching him. The boy's too frail for it. The banquet may have been planned or staged as an apology for the incident.... or accident. Yet, the lad's mind was not up to behaving well, his physical constitution under siege of distresses. And he offended the lord in person, so much so that a retaliatory rancor was instilled in Balcome. I've come presently from the place. But I come here to alert you—

LINDA: You've just left my boy?

MAKETHRIFT: Yes. He's fallen ill, and a physician has been called. You—

LINDA: Then what more of this punishment has been made? Is not Brontrol arrived at the castle? He told me he was going to see this lord.

MAKETHRIFT: I'm afraid, dear, he has made for the most bitter of outcomes possible, placing his son— and himself— into some jeopardy. Seemingly not to recognize Bill, as who he was to him, his parentage was uncovered by one of the other guests. And the apparent deception, or withholding of the fact by Brontrol, enraged Balcome unreasonably, to the point of extending Bill's detention at the castle under such frightening conditions that your husband—

HOE (*voice*): Husband?

MAKETHRIFT: —felt forced to make an aggressive show of it before us.

LINDA: Then he is captured too?

MAKETHRIFT: Of this lord's decree, I'm sad to say he's bound himself to the most improbable. If the number of protesters around the castle tomorrow is not less than what had showed up today.... Bill will be kept there, with some torturous treatment, I fear, if not his sickness is enough to endure. You must return with me, to plead for him as his mother, and bring some nurturing emotion to the dispute, as I can only offer a weak remedy for the boy's distress, a tea drink that calms the nerves. I'm going to fetch its contents now, and will come back to take you up to the castle shortly. But we must have diligence to think this course through quickly, as tempers may drain themselves through the night's cooling; and reasonableness takes hold, as somnolence regains its grasp.

LINDA: But can I leave Hoe alone, to worry of this, with a candle burning? What in this house is made, but wood and woe!— Go off, Master Makethrift, and get your condiments. (*as **Makethrift** leaves*) I will accompany you, for my son's sake to plead for.... Hoe, you must back to bed run, though stirred with my anxieties, these tremors shaking everything to stay awake. I have to prepare myself, and it's not too comely of a time to. Nor can the water be warmed enough, before a hurried departure, for refreshening. Yet hot coals in the morning do leave their marks when properly attended. You must to bed. (*closing the door, with voices fading*)

HOE: But auntie—

LINDA: Sleep's your savior now.

HOE: All of this going on.... frightens solitude.

Scene IV — *A castle room. **Bill** lies on a bed built up as comfortably as possible with straw, and surrounded by **Brontrol**, a couple of **soldiers**, a **physician**, and **Gestromy**. The **physician** has just finished paying his attentions to the patient, who seems fairly comatose.*

PHYSICIAN (*in a showy manner to emphasize the virtuosity of his professional authority*): It is a condition of jaundice, gentlemen. The punch to the stomach only instigated a hepatitis that was already present. He must have been living rather unhealthily for some time.

GESTROMY: Thus was he yellow in shades of oscillation? He seemed quite stout and hardy to me in the day.

PHYSICIAN: I mean unsanitarily kept, sir, through which such contagion of the body is contracted. His foods and his clothes were not clean enough for too long of a period.

GESTROMY: The lord will not be happy about this diseased state arrived to.

SOLDIER 1: It's a mean basis to be pinned by, that can only be fought off with the mind's will to control one's bodily resources.

SOLDIER 2: Fair scraping of the soul, then. He's left to his own devices of recovery.

PHYSICIAN: I would have to concur. Not much can be done to actually treat the condition. And whether it's of a minor type or major nature will have to await his battles. Just keep him washed from foulness and fairly comfortable. Encourage his thirsts and all appetite. And encourage *him*— for a consciousness to employ, fighting off the lethargies of mind, body, and spirit. The errant ways must take their toil of rampage throughout the physique before their wane to leave what's left rehabilitated for a more wholesome life.

SOLDIER 1: It's like couching on a wet field, for all of this sweat. But there one would be hardly feeling such discomfort, due to anxieties against opponents. So must he make his way back to more necessary goals of awareness than this slumber.

SOLDIER 2: Ay. Fighting in the field 's a thrill for your injuries and impositions. I hate to see the young ones die within a somnolence. It's like they were robbed of all effort to resist and castigate the drive towards death, verifying the valors ingrained of life.

BRONTROL: But can it be warmer (in here)? to ease his resting.

PHYSICIAN: This slight coolness may help store maintenance of the body's fever necessary to burn out the pathogenic blight he suffers. I find it to be of an appropriate temperature for the present, allowing the body to regulate itself with some efficiency. And if he asks for blankets, so much the better. He'll be awake, to actively struggle of this ailment to ease. The strongest medicine is a conscious desire to be well again, and the sternest fault is a capitulation to one's grievances. But a state more of flux with dreaming just leaves one to be questioned. It is all due to physical weaknesses, as the machinery of our metabolisms take over our lives with a dispassionate endeavoring and charm.

GESTROMY: The worst to overcome often comes early.

PHYSICIAN: If he awakens, that may be enough irateness for the drive. Compulsions are for the knowing to be compelled by.

BRONTROL: I can't imagine how he's been living, in what filth commanded through by the strange social orders that have developed.... of my home and village. And I'll have some furious inquiry to make, about such conditioning.

SOLDIER 2: We're to watch you, sir, with due respect, as you watch your son. But as one of us, you have a mission of your own that's been given, we're told, more audacious than a personal revenge.

BRONTROL: I'm one of several outrages suffered, through this circuitous trek to find me dazed and depleted of happiness: an unwelcome return, an arbitrary lord to please, an injured son.... and now to beg men stay away from me— and ours.... of castle made, this madness binding my hands with mouth.

PHYSICIAN: Should that be easy, to cause sway of the populace? We have resentments. Yet heard I of this insane decree against my patient. I almost refused to treat him, out of principle. But I had already arrived, and he was in need, of a diagnosis. So sense this lad is not an enemy to anyone, what do you fighters do to us?! How have you behaved honorably, with such foolishness and silly pettiness of manner but as to presume to force our loyalties through this boy?

GESTROMY: This is the lord's.... proscription, of being ungrateful.

PHYSICIAN: To him?! And what is innocent about that?! as could this boy be more than just an innocent.

GESTROMY: I am much against this.... tactic, myself. And I doubt if Balcome can bring himself to be thorough for it, since it shames soldiers to the heart—

SOLDIER 1: To the gut! of comedy. We have been abused. You town folk have not treated us any less arbitrarily, doctor. But this is not with reason made, and can not be explained to us as easily as you may observe— patients! What might be expected now.... makes the cuticles curl, for grinding of our preparations.

SOLDIER 2: The lord has his methods.... They are often obscure, but just as well successful for his goals.

PHYSICIAN: And do you really think, young man, our entire valley will acquiesce to his wishes through a murder?! the execution of an innocent member of our community? by a provoking unavoidable and a staged outcome inevitable! This will make cross the bitterness already felt— And I must assume, this lord is purely crazed.

SOLDIER 2: Well, see how the lad has been treated in your community, and before his arrival to us. Has it been clean and wholesome and healthy?!—

BRONTROL: Balcome means to shame me in particular. I think he does not like our king, and resents being shoved away to this castle, and this locality to conquer.... with milksopping entreating. As the most direct representative of his majesty he finds, I bear the brunt of his maleficence. He called me coward.... He means the king. He means the crown does weaken shamefully. And so he wants to chase at a disturbance, bring to a head this confrontation as an excuse to be terrible. He must certainly know.... that I will return here, to lend my life 'gainst his— method, and cause a havoc of heresy.... to lie beside my son, as the sole representative of all of these varied interests: king, country, valley, lord.... and castle. And of a true family made, I am still bound.

SOLDIER 2: Work you harder, to avoid the disaster. It is above reproach, between us, as in this small grouping, to challenge the lord's allegiance to the king, to give this event such a questioning. And I would keep this wisdom amongst ourselves, not let it spread.... mendaciously. But it's well known, among his men, that Balcome feels offended and pressured.... with this withdrawal. For better to be recalled, back up the ranks of the heroic with heraldry.... we can not tell, nor even assume, yet.

SOLDIER 1: We're simply nervous about our displacement. It's with certainty we'll more be needed.... as the king survives, to oversee his military services. But this encampment is as being put off to the side for a reserve force to store up. And that makes many uneasy, as tenuous threads pull apart of our utility to find, and for whom. There may suddenly come a request for us to withdraw from our ranks permanently, if what conquers the political spheres is against our majesty's interests—

SOLDIER 2: And our country's.

SOLDIER 1: —Weirder things have happened throughout history with demises of fortitude and solidarity, as contrary forces—

SOLDIER 2: Internal and external.

SOLDIER 1: —render victors and losers anew.

GESTROMY: That's a general complaint by the soldier, bound by principle to obey odd orders and speculative mandates through a weave of uncertainties, of position, power and control they must often inhabit to employ of their forces. I can tell you personally that the lord does not feel we have been set aside, to avoid useless losses as long as possible, and that as time lengthens so do (stretch) the notions of our loyal capabilities. He is strictly against such argument, though it is true there's much turmoil at court and capital. Yet Brontrol is sent us, to show our interests remain there, still held by the king. Hardly could we faithfully make allegiances to others, or other pressures of stately demands.

BRONTROL: I've come more to see how you fare, in a place of which I'm knowledgeable, than to be coaxed and cautioned so despicably, as I've been thus far. But my ties to our majesty are coincidental at best— I've never addressed him in person, though I assume he or one of his war ministers has read, or read of, my commission of leave to visit my home and assist Balcome with any services I might provide. And lies here is what I've received for these efforts, a strange partnership of gratitudes.

GESTROMY: Lord Balcome searches for enemies, that is agreed. He can not rest comfortably in your valley while his authority is questioned. It's hard for me to believe, though, that he can have such enmity or honest acrimony for your son as has been displayed—

PHYSICIAN: He did not overtly cause this disease.

GESTROMY: —He tests.... this community, as uncertainties swathe our valors and intentions, as if we were wounds of the body.

PHYSICIAN: You soldiers are uncanny of your comradery. Do you not sense the fact, the most evident of sullen atmosphere, that our country implodes of itself with disillusion for our state of being ruled, that the ordinary person in many places throughout this na-

tion does not care for the health of his majesty, nor of how many of the enemy you've killed or legions decimated? The land is worn out of such unprofitable worries. It does not make for us, who remain sensible, much cause to exist— except that we continue to follow the city traditions as much as they remain reasonable as a framework of life and normality to keep. But we much care for ourselves, as opposed to a ruler who no longer can. And this is not his fault nor yours, but simply how a civilization strives to maintain its character, as pockets of society by necessity become more and more self-supporting, self-sustaining. Resentments— locally— against Balcome are genuine and natural, in this way, and can not be demolished by such sport as this upon this boy, a victim of apparent gamesmanship.

SOLDIER 1: Of attitude?

GESTROMY: You can not dissolve an army, sir, with such dissension.

PHYSICIAN: I only ask that you all grow up.... and realize the day. If you can not win for gain, if our wars are not substantial— for our benefit.... then for what have we need of your organizations and misdirection of heroic virtues? We are better for ourselves to provide, a meaning for our histories. And is it treasonous to know.... that when you go away, nothing relates us to your victories or defeats? Yet, when you're placed here, within our territory of breathing space, there can only be some ridicule for your positioning— and in an ancient castle. Well, that can lead to indignations and polite affronts of at least the challenge of questions— if.... your legitimacies are clearly undermined through the circumstances of our most nominal majesty. Can you think live people would simply stand still to an unjust subjugation based solely on regal protocols?! If a king can not protect, his sovereignty or more, then he is no longer a king, you are no longer his soldiers, and we are no longer his subjects. This makes Balcome a bastard to us, frankly: a mighty triviality to be disposed of peacefully, through abdication of his effective lordship— at least around these parts.

GESTROMY: You are too blatant than your honorable station warrants, doctor—

PHYSICIAN: Yet it's the truth. You are ridiculous; sense that this is so and becoming ever more so against the resistances of our people. What could your next move be? to raid our homes? or place sergeants at all of our money-making enterprises and offices? It's becoming absurd— and you know it! these notions of retribution. If Balcome were more balanced, he would avoid all of these possibilities. But see before you how disasters are made otherwise. This is when soldiering becomes infantile—

SOLDIER 2: You can not bring us into your embarrassments— of training! I suppose to you all diseases seem odd, and their lack of treatment more unusual. But for fighters, being injurious, and bringing one into a sickness grave, well that's a way of life. Death is a number, a count and an accounting of your progress—

PHYSICIAN: This lad is not a soldier!... He's a soldier's son. Discourage the thought more, to have him brought to this by you!

—

BRONTROL: So with, the name of punch!... Yet, the tanning's mine.

SOLDIER 2: You needn't bother about that.... identity. It is your army's.

SOLDIER 1: "Speck" did do it, if you're fierce enough to know, should this boy die. But what would you fight of him? He is a kind monster, and simply obeyed commands.

SOLDIER 2: Giving out information—?

SOLDIER 1: It's too easily obtained. And the privilege is this knowledge— throughout our corps. If we must come to fight ourselves, that is the patient lost.

SOLDIER 2: It did shame me some, to hold it back, for quarrels to avoid. The contagion is more that we're becoming unsure of ourselves, which only intensifies a prudence.

BRONTROL: The blame is Balcome's.... and Brontrol's.

GESTROMY: There is no loss of reign among us. Our king persists to be, and we are faithful to him. His protections are sound, physician— He sent us to you, reenforcing that fact. And you.... (are) subject to us as much as for a treatment we may bring. You are deluded to think you are abandoned to yourselves alone. You are part of a broad social order that continues its communications
—

BILL (*weakly, coming to*): Voices speak over me?

GESTROMY: —throughout this ruled nation. And that is of one party made, that of the king's parts.

PHYSICIAN: You awaken to a consciousness of pertinent conversation— er.... Bil— May I call him William? I find the familiar forms distasteful, for adult patients.

BRONTROL: Well, his name's—

PHYSICIAN: There are things that you should know, of the jeopardy that you have placed your father in, and yourself.

BILL: A danger, to this subtle health? (*trying to sit up*) And to my own I feel exhausted. Lift me up some, please. (**Soldier 1**, *of the torso, complies.*)

PHYSICIAN: Good food and sanitation's all you need for that. But your father must vigorously argue against your views to your compatriots of the valley, in order to spare your life.

GESTROMY: More like a free life, sir. I can not bring myself to believe the lord would have this boy slain due to a number of heads. But to detain him as a recuperating guest is what I could guess of a motive, until that count decreases.

PHYSICIAN: You must drink a lot of fluids, and take the fevers boldly. Celebrate the chills as good sense returning— to your head. And then make prayer, for your father's efforts. They must be contravening of your soul, your thoughts, your purposes. And yet they are for your health.

BILL: What are these ideas so perverse? to be more lenient of our taxing? (**Soldier 2** *helps to prop* **Bill** *up more with a roll of blanket.*)

BRONTROL: That's not against my mode of conscience. But the show of protest, and the notion of it, insults Balcome, who has his powers pertaining to your.... our governance, and wishes to insure this through a stricter reception of comity.

BILL: Where are my masters, Herkshead and Craie?

BRONTROL: Your teachers, instructors, suppliers of concepts, precepts, and other such profundities, and conditioners of your nature.... as surrogate— priests of guardianship.... that has let you come to a diseased state?!... They have left this castle, to try to convene an emergency meeting of the VID officers, in order to discuss your circumstance and draw up a responding strategy. But what it seems to me is to decide whether you are to be sacrificed, whether this initiative of Balcome's is to be met with even greater defiance.

BILL: They have helped to house me, father, when I had felt need of this. I found, to my surprise, that I could not survive well in the open; although I thought the weather conducive to the attempt, as I searched for my meaning of consciousness and happiness in man—

BRONTROL: What store of philosophy!

BILL: —and to relieve mom a burden of feeding.

BRONTROL: There is no such privilege that's natural. You can only be complacent to your surroundings and compliant to your needs.

BILL: But I was miserable, and restless to find a way out of my bothers.

BRONTROL: I should more show you how to work at this, and sense the legitimacy in yourself.

GESTROMY: Some societies allow their youth to feel they are unconquerable of aims, while others make them feel useless until fitted for the necessities of their community. I should have thought this valley rich for worthwhile labors. It's practically all that I see, roaming around.

PHYSICIAN: To be the vagabond is foolish, here, though romantically inviting. There is an aberrance at play, in his upbringing, that causes greater probability for these illnesses to contract of, almost wishing for a state of indigence to suffuse the poverty of mind.

GESTROMY: But not imagination and ambition. He tried, in some manner immature, to better his prospects (in life).

BILL: But so, as you say. I condensed my efforts, scavenging proudly and fervently, to uphold this poverty as something won. Yet this fouled and astonished me into a terrible state, without any answers for why I was placed in my condition of mind and body.

SOLDIER 2: The such is a minor soldiering, but without a cause of war. Hardships are heaven sent. We devise utility around them.

SOLDIER 1: Yet, in such an area as this, to be conscripted to be poor is mean.

BILL: Not until my mastering friends found me, for various homes to stay at, was I relieved, no matter how lowly the squattering— And they gave me answers, proctoring through the realities I saw and experienced—

PHYSICIAN: Ha! But then from filthy places you catch filthy bugs! And that is the type of schooling you've assumed. Without a proper parenting, one is lost to such travesty. That is not the way to grow into an independence of mind. You simply lose your health and die as a child, not weaned off fantasy. This is not the valley's fault, though some inhabitants contribute to it with their zealous goals or idealistic depicting of their controversies. I am not partial to the Valley Inheritance Defenders, nor to any specious form of defending. It is a spoofing rancor of the poorly developed that they stir up, glued onto a few justifications to make themselves prominent and hopefully beloved. But the radical is wrong of a populace by definition, whether supported or not. The imbecility is not the striving for fair rights, but the strident stringency of doing so to taint them, or contort them into wrongs. It's quite obvious to me, as any thinking man with objective thought, that your VID members will choose to have your fate martyred to their demands, and might show up outrageously to enhance their numbers tomorrow— to *support* you, with this imprisonment and more, making the desperate efforts of your father utterly useless and like a folly of Balcome sent to.... bemuse and bemean. How can I recover you?! As a wise man my methods are squandered on this patient, and my time wasted— except to advise you this way: Beg to Balcome! Approve for him to have adjusted your nature, and wholeheartedly accept his authority. Plead that you were wrong and are transforming back to health. Give him your thorough allegiance— and lick your knees! before him, as the supplicating state of an invalid, to show that you dutifully adopt the crown as your master, and not these.... headstrong friends of yours. Say that your father has converted you, and that the magic of the lord's method prevails throughout your body, even as to help cure your disease. Do these things, lad, and rid yourself of bugs— For you understand nothing, yet. The world has everything, and not your mind alone. Nor with impulses and more impressions will you earn your berth to any solid living. Concede to what is powerful, to allow for your deeds (in life) to be carried out.

BILL: I am confused, sir—

PHYSICIAN: I am a physician! I.... intercede myself, to try and spare your life, and fight against these foolish aggravations that have no real consequence nor lasting value.

BILL: Is not for what I feel of heart dependent on my creed to be self-honest for one's self-worth?

PHYSICIAN: What your heart burns of is a contagion, a physics of pestilence, or a chemistry of biological malady. And as this deepens or recedes depends more on a faith of practicalities.... than on your transgressions thus far. But this is all—

BRONTROL: I will bid—

PHYSICAL: —all that I can do for him! Why is hopelessness alert?!

BRONTROL: I'll bid a reformation in my son, towards our majesty to observe— through such delirium as he could speak, directly to Balcome, that Billy recovers to obey him, in a mystical manner unexplainable—

BILL: Father!

BRONTROL: —And you will make such pleas, as the physician suggests, to spare ourselves much torture, and make this lord more pleased with us, despite the coming day.... That is my command to you— if you are mindless!... And if not mindless, then be for me, to adjust this tribulation.

SOLDIER 1: Lies can make no sinning against base intentions.

GESTROMY: I must conduct the truth of this to Balcome.... And this truth is that you are a sick— disciple of the august, struggling mightily to recover your thoughts, saying what is said to you dearly, and feeling for your father's aid and sufficiency, not to fall prey to a madness. This is a reasonable route available through the dilemma, and you might be dreaming now through spell-bound wishes and dire entreaties of earnestness.

BILL: Can I be convinced of this? What will my colleagues claim —?

BRONTROL: That I've arrived to handle the deportment of my son again— and his education! and that boastful rallying loses favor, in this corner of habitat, as could they silently sit around our castle boisterously mean and low. But we can shake chains at them, with your presence spared to see alive and fit!

BILL: Have they become.... such an enemy, now? I am for twists of mind much winded.... Yet.... I wish to live, with my father's grace and acknowledgment received, cleared of a grimace. For the condemnations of Balcome were not a reverie disposed of bad thoughts, but some real event chosen for to sculpt my obligations to thinking of how some parentage is tried.

BRONTROL: That was no dream descended of, but a true volition and a forceful responsibility. What wears of me, to these concerns of plight, is warned not to be deceived henceforth, but rather show its blatancy of conception. For else this situation is impossible, were I commanded to it with spite and scorn. But I feel these pressures as being worthy to the teasing and tears of conscience. And so.... may much resolve through this sift(ing) of mind, for all that's right to be accomplished. You have obligations indeed to me, as I draw of penchant to seize this bear.... of a problem reduced by. Yet be this simply for a fidelity to one's own nature.... and kind, the obliged toils. (*starting to leave*) Then heal to lie, and lie to heal.... as I find this lord, and tell him of your recoveries.... physical and mental.

BILL (*more to himself*): It does turn me towards my weaknesses to think. Am I for him or his pledge? Yet cold and somber are the fears of disobedience. (***Brontrol** has exited.*) And did I curse the man to his presence?

GESTROMY: It was hard to hear the renunciations.

BILL: I've.... only imagined falsehoods gathering, for a storm of bequeathals into misery and sorrow. That is the irate business bound to, ire.... in this state of meaning.

PHYSICIAN: Then sweeten the tone of your complaints. You

don't yell at gods. And as they yell at you, their thundering descriptions to apply.... then shake of sound and comply of the quakes. For what is honest is always the most reasonable to discover. And what is dissembled shivers of its falsehoods. A foulness hiding within youth and beauty causes much discomfiture of the passions.

BILL: Am I this way or that?!

SOLDIER 1: There's no great fault to having an anger. And as we let the father pass, we take our chances in observing (him) with mind, to hope his screams are quaffed as well as beverages drunk, less taste the more is throated. It's well securing that one should follow, for his trust to insure. Yet have he range as free as a might of justice.

SOLDIER 2 (*leaving*): I am as mixed as the boy. Being so proud lends itself to such humiliation as to concede of begging.

GESTROMY: A great care your father has for you, considering his character and skills—

PHYSICIAN: And that is more or less protection of one's own.

GESTROMY: —need not be tender, but certainly transcendental of your causing.

BILL: I am to accede to his will in this. What, to make him cry to our lord?... till today comes?

GESTROMY: I know that Balcome may be moved.

SOLDIER 1: That's best for your guarding.... Bill.

PHYSICIAN: And I attest the people's upsetting is valid.

SOLDIER 1: Care me much for that! Those taxes are not for us to administer. But what does crumble around us.... is a sense of application, and the broad view of our worth. Symbolic is your castle for our stationing, antiquated valors and their promoting. There's much leeway to our loyalties, in such a place, this fine capsule of your provinciality and reckless thought of abandonment. And what do you think of us, Bill, as could your father be?

BILL: ....I am tended to be his.... and thus yours.

SOLDIER 1: That *is* a transformation completing of itself, and competing of this wrong.

GESTROMY: So soundly is this error served, that resentments dwindle into singularities of naught.

PHYSICIAN: Well that's life itself, sir, in constant correction of its mistakes. The night continues, and there's some sleep for you yet. But comes some hunger or thirst? also be pleased with this.

BILL: Perhaps a drink of tears. I have committed him—

GESTROMY: Committed to him.

BILL: Of my assurances, then, the famine of his caring drains much force.

Act V

Scene I — ***Balcome's*** *office in the castle.* ***Balcome*** *sits on his table, slightly agitated to maintain composure. Before him standing a distance are* ***Makethrift****, holding a pouch of tea, and* ***Linda*** *with* ***Hoe*** *in tow. A wall sentry is also present.*

BALCOME: You've brought a remedy, to our distresses? But who is this youngster?

LINDA: I care for him, my lord, during the night slumbers. But he feared to be left alone in this emergency, and cried of (seeing) flames, which I could not choose to have him find in my crude abode. As for less proud a flicker here, they're aft to be abused. And so my own to plead for, not to be ungraciously played with. He is a good boy.

BALCOME: That is suggested with more cordial intention than I've found in others. Then find I much to play with. (*to the* ***sentry***) Fetch some candy for the seemly tot. (*as the* ***sentry*** *exits*) I am all for pleasantries to provide. Yet could he sleep for being tired and replaceable.

LINDA: Hoe's an orphan, my good lord.

BALCOME: So! They make good ones here.

MAKETHRIFT: As befitting for any of the townsfolk, Lord Balcome, I've brought them to help me aid your perquisites of generosity, as payment for our hospitable display and service to you. For we are essentially the soul and nature of this populace to demonstrate. And comely Linda, the wife of Brontrol and mother of Bill, entreats for the welfare of her son, and implores of your greatness to undo his imprisonment.

BALCOME: The lad is being nursed— The lad's mouth.... I'm not sure of, with this treatment. But one of your local physicians attends to him, along with his father and my sheriff to you, whose character you so much abide with and favor. If only for my own person you people could feel the same. But I have not worked through you as ably as he, perhaps. Such familiarity seems wanting here in other aspects, if I'm to believe stories. Bill.... condemned the king, mother, a strange crime. Brontrol denied knowledge of his own son, as outrageous a depravity of cognizance— or credence— as his assertions for our king to represent— to me.... may be fallible. And I am seated to decide these.... sides of a coin tossed (at) me. Firm to follow this distraction barely, I'm not directly remonstrative of these behaviors. For one's opinions and thoughts are endowed of their requisite punishment of person. And if the calibre of mind for this locale is to disown our majesty, if this is how your sons are taught or tutored, then to your own attainments of misdeeds may you fall, with squalor of ideas as a pride for punctilious foolishness. But I enforce the higher state of nation in this region given— not to own, but to command of reason, and to assemble spirits— free of will— against treasonous principles of assault. And to deceive me is difficult to take standing. Yet do you claim to be the heart of the matter? a representation more actual of the populace than those who wish to defile the grounds surrounding this established pinnacle of historic import and (a) noble past.... with their sullen presence sitting on the ground, squat messengers of grotesque insolence— including your Bill, and perhaps your Brontrol, in an odd wrenching of the sentiments and twisting of faiths. For I have remained faithful to all of my assignments, and I suppose he to his. Well, I see you as well made: a man, a woman, and a child. And that is your best appeal to my sensitivities and expectations for a kind nature, in a more gentle state than I have been returned.... or swayed towards as won, and pushed to to demonstrate a power. But as to this triumvirate of Bi— Bro— Ba!... the son, the father, and lord, we must prove of ourselves what dicta rule of this valley, what truths are held more solidly than our carelessnesses to say things irresponsibly and without substantiation. I claim the king rules over us still. Brontrol says these horribly insulting protesters decline in number— through his persuasion of rational argument. And Bil— ly!... a protester himself, is sick and irrefutable for the uselessness of any overbearing authority, including myself. So we shall certainly (*The* **sentry** *returns with a small bowl of sugared treats.*) test ourselves of these principles and their consequences of fallacy. Ah! Does candy come?!

SENTRY: Hard to find, my lord. The men are not much of sweetness bent. They're (*approaching* **Balcome**) stolid to be sour of this coolness, but are piqued that a child has been brought here, in a way moved with curiosity. One did deliver to me these treats he has stored. (*showing the bowl contents*)

BALCOME: Give of me! (*as he takes a piece*) a sample. I have not had candy in a year. (*mouths*) This is sugar-safe and sweet for a joy to be tasted. Serve the young boy.

LINDA (*as the* **sentry** *approaches*): Hoe, my lord.

BALCOME: Hoe will have some too. There is no cruelty to the credible.

HOE (*accepting a piece, as by retrieving pleasantries with a cautious optimism*) Thank you, my lord.

BALCOME: Take several, for your dear soul. (**Hoe** *complies, a hand his only container available.*) I want all people to be treated nicely, as for are they subjects to a beneficence instilled and implied of this country's governance.

HOE: This tastes great!— an' pours of pleasure, flushing my mouth.

MAKETHRIFT (*as the* **sentry** *comes back to* **Balcome**): My lord, I have concocted a mixture of teas and dried berries, to soothe and strengthen Bill's condition. Some need only be steeped in hot water, for a drink to make.

BALCOME (*as the* **sentry** *places the bowl near him, on the table, and resumes his positioning at a wall*): That's fair, for a solution to his illness.

MAKETHRIFT: ....Are you to prepare some?

BALCOME: I mean not to have him suffer.

MAKETHRIFT: ....May I offer you this pouch?

BALCOME: There is more content in a bag of tea than the universe has stars.

MAKETHRIFT: Where is this awkwardness allowed? I give you

it, for him.

BALCOME: Stay ahold of, sir. I think he sleeps, at the moment —

LINDA: We are anxious for his health, my lord.

BALCOME: So 's the physician, and his father. What is health dreamt of? reclining for one's principles? Then how does one get ill?!

MAKETHRIFT: You give us play, my lord, on Bill's misfortune's to belittle. You give us.... sweets— in lieu of him. He is not the cause, of your catastrophic reception, or repudiation, if only he is led by common thought. But it is with rumor that you're blighted.

BALCOME: How does one get ill?!

LINDA: My son left our home, my lord, in order to be less of a burden on me, as his missing father's support (of monetary aid) dwindled.... through these wars thinned and diluted, I suspect. What of his constitution less attended to, by necessity may he be weakened.... and less thoughtful— more inducible by others to a.... criminality conceived as right. But can you blame the boy? as much as blame for Hoe to lose his parents? This does occur, and influences apply themselves to those who search. He is not mean, and lessens with cruelty and foul treatment. Allow us to retrieve him, and with his healing he will apologize to you, out of gratitude for the reprieve.... Or I will make him so, with care for his sufficiency to restore, that a baseness in him will be driven out.

BALCOME: Driven out.... I have been deceived. But I feared it was the Devil's dungeon due, to work on this suffering. Our castle is resplendent, for my capturing of this essence of being a lord, your lord.... You may certainly have his body back—

LINDA (*with fright*): How is he?!

BALCOME: The physician caters to this determination. (*standing off the table*) But there was much of a stupor to him, the last I saw. The weakest anger reminded me of flailed opponents cursing in jest their demise from being slain as all too actual and obvious. But I take pride in treating Bill, first with a banquet for himself with.... peers, and now with great attention to his.... nursing off such malady. No, mother. I don't think he's dead, just yet, or your physician would have rushed to me in great embarrassment, by now. So let us stay safe, within the sanctity of your promises— for an apology. What's more to do, though, depends on Brontrol's words. Did he leave you?

LINDA: ....(*silence with difficulty*)

BALCOME: Is this the brunt of your contention, for much brutish behavior? Oh, feels much, to relax by stretching muscles, a good tendency to stand. I was stiff! upon the table top— protesting comfort. Now be with tea, you are. But what of *my* comfort?! And so with candied supplications, I have ate.... and you have devoured.

HOE: It does please me, my lord, to have lick of this and suck. A year is longer than I may count. For fun of time winds not of much ability than to be before you awakened. Let Billy come to us, and I'll share my candy with him.

BALCOME: It's that important? And be less profound with more honest gesturing of hand. I've offered you.... what we spoil ourselves for. And because of this, some do even deny themselves of it, through their maturities to reach. But we find more pleasantries through various stages of social and developmental advancement, proficiencies, acclaim, and so such. I doubt if Bil— ly longs for sweets just yet, though a gesture of offering is permissible. You may all offer him your condolences— for his imposition of illnesses, since he must be out of his head to challenge me so openly. That's caused by a strain of diseases barely tolerable.

LINDA: He left us to fight for the king....

BALCOME: You care for— husband, mother.

LINDA: as if some imperative were offered to a madness.

BALCOME: You call defense insanity?

LINDA: A madness against his tolerance of family, that which had pushed him away, like a pressure; led him out of our midst, without the squandering of his responsibilities, until the recent perception of his lapses of caring for us— with his due dole contributed. That's the only evidence left (us), for his valorous attempts, and most particularly in this country and the kind scenery of nature we've inhabited to settle of, a poverty of crops or no. Though I won't say the man was afraid of those prospects, not entirely for his purposes of journeying. It's useless to relate to him, for the sensitivities he's found, among your types of hardship.

BALCOME: He did not escape reason, then, if you can not understand the necessity of his actions. The call was instinctual, from many parts of our nation, to establish more the prowess of our fighting forces under his majesty.

LINDA: It was not dire to be spoken of, but suddenly occurred to him— as an excuse to leave. There was no great expedition called for or advertised for recruits, nothing celebrated as a great cause to do or exemplary purpose to undertake. He simply left after an argument with me, leaving his son totally perplexed and frightened for security. And it is this struggle of conscience you have bound, with resentments crafted from contortions of mind and manner, sensitized against authority perhaps unjustly. But he is indeed ill. And can you punish a man for his incapacitations?!... or his childish reproaching? his faltering to avoid expressing the insults felt? and thus not to care of insulting. He simply left, and grounded us to nervousness of grime, poured over us more worrying, until I hardened, accepted the pain and the fact that it would never cease, and continued living with its toil subservient to my needs of subsistence.

BALCOME: You care for the— son, then?

LINDA: My boy rebels, it's simply all, of this uncertain condition, to make for himself some goals that can be passively kept to heart, that he is for his cares to care for— and will never abandon them, will refuse to ignore their tending, and culturing throughout all blights an' plights. But he is more supple of a youth and has not hardened yet. His aches seize him to be foolish, at times. Mistakes occur, which at any moment could be devastating for him, among so many made that stay.... innocuous, as chance may teach without structured instruction. More that he's made to be confused than cal-

lous or impertinent. I wish him back, Lord Balcome—

BALCOME: As well you would, to coach and coax more.... catering of this valley's feelings towards me.

LINDA: You've no solemn justification to detain him in this way, a detention of enforced casualty.

BALCOME: He is ill and recovers, in this region's most noteworthy house. The convalescence is as proper as this castle stands. And when he is well.... he will leave. If he does not become well— he will die. And that is life's judgment of this matter, not my own.

MAKETHRIFT: You mean as the heads you see heal, not to allow themselves to be counted. The maturation is one of discernment about you. But that can hardly be fought for, without your respect for us. Let me ease the lad at least, that his dilemma will not be felt so harshly—

BALCOME: Is there poison in that pouch?! I've given you my envy to play with, for patience is an envy of restraint. And it's by your own actions that fate is made.

MAKETHRIFT: The sugar is a medicine, my lord, within the teas. Have a goodness felt, as you would leaven uneasiness with a buoyancy of kindness—

LINDA: I plead for sanity in this!... Have you a wife, my lord, or ever had?

BALCOME: ....No. Not to say so, wife.

LINDA: Then of some children prospered by?

BALCOME: Not the such to be known with. Rumors always abound a prestigious entity's description or presumptions of.... But my men are my hands that make of my commands much prospering of state.

LINDA: Have you no license to be felt of a family other than these wars? What is implicit of your gentleness of treatment to know that some wrongs are more awfully felt than their effects can ever cause of hardships? This harassment of Bill, as your response to his irritation of your notions, can never enhance your standing here, not with fear nor circumspection for your powers. It is a futility to try and promote yourself this way. Only condemnation results from his abuse, only a disgracefulness of nature is seen to shudder from and avoid, till that one is encouraged to fight against to cause angrier results. Then let our resistance to you pass without such incidents; and return my son to me for nursing, out of this cold tomb.

BALCOME: Return?! What of your family? mother. You have a son— and a man, and neither seem related by any legal bonds nor blood, from what I see, as make these compunctions of debate. I do nothing that isn't sensible for some standpoint of country and our majesty's desires, dreads, designs, and demeanors of ruling over all of us. Then you must be taught for some lack of decadency to be deterred by, with this loyalty. You must be shown your family—

HOE: She is my Aunt Li— da! And I am hers, not to be a fool for anyone, that you would like us as we are presented. And I want my cousin safe, and will do for you whatever you want to make this so.

For I only have a family of friends to guide me, and take to care for my poor needs. Then if you are as high as seen, you'll let this happen for everyone distressed within our home. Again, I will share my candy, sir, as you have. I will promise to behave. And I will work the grain more carefully. With all that I know I will produce for you desired results.

BALCOME: And I am much taller, young sir— only a tot for tally— than you can possibly see. Yet for convincing, you are thoroughly attached.... to this fine mother of a kind. And what she has produced befuddles. You can only repair yourselves, in such cases. No amount of external compelling can correct your scars. I do not suggest myself to be so mean as to be unreasonable. But it is most evident, too terribly obvious, that it is you inhabitants of this nature that must change to approve of me. You must do the work, since I am the state. And as Brontrol represents directives from the king, then see how hardly he must labor for me, with these homegrown difficulties. If it is a poor effort.... wanting.... lacking.... ineffective.... then find what is produced from this debacle of caring. Or hardily he'll make his way back amongst you, to convince you of ourselves and our royal imperatives, then that is implicit of this tensioning, the implied intent of this lesson. I grant you everything; but the world is of its own to be moved, by the methods of which we must adopt. You have let my house—

MAKETHRIFT: Of how many?!

BALCOME: —fall almost into total ruin. I have repaired and restored it through my insistences of earned residence. Yet now you chafe at my presence, as for a reform movement against the very principle of our society, mandates from his majesty. Well then, what else does near a ruining in this corner of the princely hovel you'd describe? Is it compliance to the tendencies of obedient nurturing? to have some sons nearly destroyed?... or many more disturbed. Your actions are your own in this, of revolution, revoltion to devotion, and the tyranny of tedious tendentiousness to be refractory to the notion of our ruler's solvency of life and effective sovereignty over these dominions. You may not so easily overtake this land for yourselves by merely denying him and his authorities over you. If that is what you insist on fighting, then I am here! to put down and abolish all such rebellion— I have many abodes, sir.... several homesteads throughout our country, and one abroad. But I am ordered here! This is my station, for the present. And my soldiers grumble about it, since it's not so comfortable, and even less welcoming under the current social climate. But societal pawns need not contradict our purpose solidifying this region for the crown. And posh of penury! mother. Who do I detain but one most decrepit of spirit to be allowed such poverty of mind. And this is clearly representative of your erring ways, in this village that's noble enough even to support orphans. Then stay the heads away, is the simple request. Show that easy sign— of respect— and I will adopt you back!— with great caring and understanding and tolerance and relaxation of rules and taxes and high expediency of social welfare— into the fold of our most reputable civilization. Sans this contrition on your part, deserve you not my lenity (of leniency) in this matter you have caused all of yourselves. Disobedience is foul. Can you realize how much I've suffered you?! already dominating my thoughts of vexation. Yet, for your handsomeness I have much sympathy. For your pretty children I will adore them and their aid. But you must earn your own proprieties of sycophancy to the king. You must fashion that yourselves, with your own earnest desires to behave— And if not!... then let you have these sacrifices, one after the other, however you send them

to me— till hordes are destroyed, or until we tire of piddling around and take full retributive action as an army. And what is your bothersome organization that camps around me members with such disquieting insolence?!

MAKETHRIFT: The VIDs? We are not ourselves members of this group. And while Bill is, that's only a mark of commonality for peaceful protest, without attacks or assaulting, nor even much rambunctiousness. If it seems a disgrace to you, then that is the message sent. How should you harm their passive thoughts and active stillness? sitting and sweltering under the day's hot sun. They only mean.... to have our valley regain its association to this landmark, a castle once ignored and even embarrassed by. They only want that you acknowledge they share of its history, and that even if renamed for yourself they are also of its making and fortifications and symbolic grandeur. And if it is to be so greatly restored.... admit that this activity is done by them as much as yourself. For our taxes and our sweat pay much of all for this. If you could say as much, or make a written pronouncement to that effect, readable at leisure.... a flood of forgiveness may run rampant over the valley environs. For it will signify your gesture to at least *attempt* to be reasonable with them, and to alleviate— with swift and intelligent *regulation* — some temporal financial burdens to this regard. We only want that some wisdom is made of the problem.... and the obligation we may mature enough to assume, or resume— if that's your taste of meaning.

BALCOME: Were I not here.... this would crumble, with your rumors heard.

MAKETHRIFT: That may be true. But must you make a fault of your stanchion?!

BALCOME: Your son, mother, would dethrone all such places. He spoke nothing at all of support for this august, palatial outpost, as could a boy with difficulty hold up his candies with more effort.

LINDA: Bill's.... not been well supported, my lord, with the substance that matters to a conscience striving of growth. The concept may be failing, falling foreign to him. But if he is tutored by his recent associates, he knows of it as part of a platform to believe and stand on. The bitterness of his expression is not meant to be a cruelty to you, but rather for himself an acute sense of absence, of feeling and tenderness, that must worry much his aspect making towards life. I can only suspect this of him, and offer my apologies if he has upset you so much. I will recover him—

MAKETHRIFT: Yet he's a weak excuse for your position, my lord, which is simply put to make a hard example of. He is ill. His defiance may be due to that illness, and not to his better sense of addressing a lord. And he did not say so much— to you— restricting himself to a pertinence of remarks not as complicated as the facts we find, since no one condemns you, not seriously, not even in jest. The lad was simply out of his head for a moment, near peaking to the boil that led to his collapsing torpor. Blame this disease more than his distemper.

BALCOME: I blame neither. I'm only observing how you constrict yourselves to these punishments. For I am not offended by an invalid youth. (*as **Brontrol** enters, rushing in but with a dignified restraint, as he continues to speak*) More valid is your Hoe to me, a loyalty of companionship to sense, and with an innocence of honesty that can not be blasphemous. Could you think I'd want to harm such a natural sentiment if it may resound throughout your true natures and tendencies?! I do not a thing. Your hands master Bill—

BRONTROL: My lord! He confesses the error of his words to you, awaking from a feverish quarrel of body and mind to announce that he was wrong!—

BALCOME: So?

BRONTROL: And that his bane of proper thought was evilly instilled by false suggestion and physical weaknesses to overcome.... He agrees, for the righteousness of authority, and has made a peace with me, as his father, to help guide him more strongly, and to cleanse him from the remnants of foolish pits he's fallen into and with your gracious help and aid has climbed back out of.... He admits of a madness once, and has struggled fiercely to recover from it, offering you his total and abject servitude.

LINDA: So he lives!... with a lucidity regained?

BRONTROL: Our son.... is still in a precarious state, but his health allows him to 'come sound of mind and repentant of his spoken sins. Were not we (had) heard them both, they must have been profound.... to call himself "fatherless."

BALCOME: Hard. Hard, Brontrol. A hard joke, if he has recanted to you that unholy idea. But what has this to do— with counting heads tomorrow?!

BRONTROL: He rejects them, my lord! They are less might for being seen. The assemblage no longer moves his soul. But I do linger for a preaching to—

BALCOME: That is definite.

BRONTROL: —to correct his conceptions of life, your lordship, our majesty, and the observances of reality through a finer analysis than has combed his hair thus far. For of my family he's bound and pledged to obey you—

BALCOME: With more certainty than calculation.

BRONTROL: —through my direction and suggestion of behaviors. He has won more modesty before your authority, and his defects assume less deception from his rightful inheritance as a proud and sturdy subject of the realm.... He'll join my proselytizing, sir, on your behalf, when he is well—

BALCOME: Now, courser!... that is more point than puncturing, of the rising stakes of (bags of) ballast as you run— ever more swiftly to disperse this weight, that *you* must proselytize to *make* him well, and keep him thus for me.

BRONTROL: Did he curse you directly, Balcome, that you would want to crush me so?! I tell you he repents against your harm, and regrets his vitriol, which could not have been as vigorous as his war-like sleep, to fight within himself these self-inflicting wounds and hatreds. And do you deny my urge to find him well— and demand it of myself?!

BALCOME: How ever you may demonstrate your wantonness and deficiencies bodes certainty for Bill and a definite result from

my observations, since I'll not define this meaning of being fatherless—

BRONTROL: From what I've been told, it is described of a condition of induced confusion from your detention-ing. And you are the cause of the outburst!

BALCOME: ....But not the disease. From strains of standing still in filth I could merit castigations without such conclusions. To you it is applied, and to the world, this description of himself. Not to my oversight is this drawn of, and clearly not the king's perspective — what you may demand of yourself! I'll not be challenged that way, courier, not successfully through your deceptive sight. I'll see the heads to count, tomorrow, as on an abacus of mind— determining fates from follies and foolish insistences to foil my fruition of force among.... this family of valley friendships. And your goals will be terribly dependent on my sight. What hopes have you about this and yourself? Blame not me for what transpires, if the outcome's bad or regrettable.

BRONTROL: Your stance is not logical to this course of affairs, my lord; for I bring you facts of knowledge of his contriteness and desire— need to be forgiven. And all of this, these statements of issue which I repeat, are fully verifiable by your sheriff. So we do not lie to you—

BALCOME: I did not cause this disease! The "D" was not determinate of this— state! state of mind and manner and manners towards me. Guests! they were.... Logic is not knowledge, nor truth or proof of anything. You must demonstrate that! (*as **soldier 2** enters*) And so you have your chance to. All seems dependent— on the king, since you represent his assistance.... as ably as he may allow, as much as I may bargain to receive in persuading all of you.

SOLDIER 2: Does Brontrol disturb, my lord? He's just had precious words with his faint and failing son, and may be stirred of the emotions, but that the lad becomes fidelious and respectful again, through his weaknesses speaking, uttering what may be substantial to be heard—

BALCOME: I'll have none of that— promotion for his cause.

LINDA: Does he fail of health?

SOLDIER 2: The physician thinks it may rise—

BALCOME: As I will rise—

SOLDIER 2: —to a more stable condition.

BALCOME: —in the tower!... in the light of day and acknowledgment.

BRONTROL: He sheds all cruelty, Balcome. Can not the improbable be spared as well?—

BALCOME: Improbable?! You've claimed to be miraculous among men, a deity of consultation about your town's inhabitants and as a guide for my residence and stay. And now you say improbable?! Impossible?!

BRONTROL: The risk is too delicate for me to boast more of. Why necessitate this ungodly chore?! It defiles my reasons to have left here, confounds sad irony to return, embarrasses my welcoming with trial and a horridness of scapegoating. And lord! you hurt of treasonous intent to fear of yourself, if at all you harm my child. For the king, more just than judicially embellished these days, will not stomach that transgression due purely to your personal grievances of stationing. Your actions will be condemned, revolted of, and instigative of reprisals from many corners of interest.

BALCOME: All against me? Do you dare declare the possibility? Improbable?! Impossible?! You soldiers know.... of my tenacities, and that I will only lead you faithfully into our battles.

SOLDIER 2: Fear we not the toil— but the tally! Is our assembly more warranted than such crime, or do we decompose?!

BALCOME: Make not such insult to me, as even your ranks grumble. The rancor is gratuitous of shame! as I promise to stay the embodiment of your utility.

MAKETHRIFT: But lessons must be taught, we just assume of this. And here makes a fierce one, a ferocious one, my lord, one devour-ous of the saintly pieties supposed for a ruler, or governor. Have not revolt a principle of escape from such terror. That is possible! Probable!

BRONTROL: You go directly against the king, to challenge me in this way, Balcome, to be so stubbornly ballistic in fighting reasonable suggestions of composure and compromise to this region.

BALCOME: You are nothing, less than Hoe!

BRONTROL: ....Hugh! Hound the devil on you, lord! Tomorrow brings red sights, unless you let me take my son home now!— tonight! I've tried to placate your inordinate offense as much as possible, that sense of chafing displacement. And I'm much more than nothing. I am part of a populace to contend with, and a severity to try and subdue.

BALCOME: That's a father's privilege. But there'll be no such provisioning of person to accommodate your tasks. Your mission is what I meant and have commanded of you—

BRONTROL: Come! Linda—

BALCOME: Where are you going now?!

BRONTROL: We're leaving this hole!—

LINDA: Take me to our son! Take us to see him.

BALCOME: That is permissible.... But you can't leave with him. He's still too ill.

SOLDIER 2: The lad's guarded, liege! Not about to walk out, wouldn't have the strength to. Would have to be carried to a destination. I can attest to that, still a weak thing to behold. But as for ill.... What is a pacification worth, these days? I suppose he's compliant to his situation—

BRONTROL (*to Linda*): Then I'll take you to him. He'll not confess more than what I've said—

BALCOME (*to **soldier 2***): Escort them.

SOLDIER 2: He knows the way, liege. (*as **Linda** and **Hoe** start to leave with **Brontrol***) It's lined by sentries.

BALCOME: Rebuke my commands?!—

SOLDIER 2: I've followed him, to find his disposition towards you, my lord. And this is evident with his comportment, though he seems not to have argued well.

MAKETHRIFT (*holding the pouch*): I have some tea, for his medicine made.

SOLDIER 2: He's not about to brew it, in his condition. He can barely lift his frame. The physician will have to review its application.

BALCOME (*as **Linda**, **Hoe**, and **Brontrol** are just exiting the chamber*): I said escort them!—

SOLDIER 2: What does that mean? sir. He is a soldier for us, with.... his family, visiting.

BALCOME: Are you mutinous become?!—

SOLDIER 2: I'll not be ordered as a wainscot made, for these stoney walls.

BALCOME: Surly are you men, and disrespectful. (*going to sit at the end of the table top*)

SOLDIER 2: There is a temper growing amongst us, liege. We're tired of this uncomfortable retreat, and the uncertainty of your footing with the king.

BALCOME: So grumble, as you do before great battles. I'll take it then, your proffering of suggestions and pointing out slight details of oversight to my campaigning strategies. But I'm ill-disposed to receive them now. Have Gestromy collect your complaints, what soldiers fathom.... and put them in a bowl, so I may choose some to address, or play with.

SOLDIER 2: What becomes of this lad Bill, my liege? He seems not so unordinary, or extraordinary a threat to us.

MAKETHRIFT: He should be nursed out of this castle, my lord. Comforts are difficultly found here, even with your reservations so far. (*cautiously going towards the table, to where the candy bowl lies*)

BALCOME (*watching his approach*): I did not make him ill, caterer. Nor did you, with your fine designs of cookery. It is his father that has done this.... and his king, perhaps, in being remiss.... for more attention. And thus will be, for how a death's provided.

MAKETHRIFT (*placing the tea pouch by the candy bowl, as the wall **sentry** comes up to the table*): It seems too subtle a reprimand for a youth's boisterous talk, to have his father placed in agitation of a life.

BALCOME: The sky fells many with subtle viciousness. I must stand for some discipline, in this way, to be discreet of powers that must eventually be asserted fully. (*as the **sentry** examines the con-*

tents of the pouch) We defend our privileges of existence with total circumspection and consideration of this environment. Yet, to demand of ourselves some granted license we must do what has approached our polity to carry out and, as I said, demonstrate for everyone who must be involved with us.

SENTRY (*after tasting a bit from the pouch*): It is a sweet tea, my lord. Bitter but sweet.

MAKETHRIFT: The finest available, sir, at my command of composition. (*as the **sentry** leaves the pouch to go back to a wall*) And while I have rushed its making, this is as thorough of a treatment as I can create. It may soothe the nerves and relieve tensions and headaches.... and perhaps bouts of fright. What cause is there of this, my lord? The day only brings less *questioning* of our assertions. Within your rights, you have no need of carnage in this domain.

BALCOME: Need? Claim you need to be destructive? That we eat destroys the food. That we breathe destroys the air. That we rule destroys.... the subjected, else they be governable, and not so impudent of law and imprudent of speech. The man stated categorically a total dismissal for rulers; and at my dining table as a guest, before my face of attendance, it's as much to thoroughly reject me as an unwanted pain in his stomach! Such bold incitement can only deserve this treatment, which is as judicious as I can devise, to depend on his own.... reginal and parental devices. For does he cause to himself, (as) incineration from the mouth, the ideals of our civilization and societal espousals. And there is a price to pay even for those thoughts honestly spoken.

SOLDIER 2: A brash lad only, liege, spanked by God. It's as much as a curse from lips to perform. To be ungrateful is a youth's pleasure, to be spiteful his volleying for attention. I would have just slapped his face strongly, an' laughed hardily till he cried. But what more punishment than this is reasonable?!

BALCOME: I was too far away from him to even attempt such a thing. Dare me not my anger, though. I sensed some truth through his distresses, that he was claiming what he felt and not merely preached on or touched by a philosophy brought to. Patiently I accepted his displeasure. The error is the father's lie, that instigated so forcefully my wrath, since this justifies my approach to the problem.

SOLDIER 2: His lie?

BALCOME: Deception is a lie.

SOLDIER 2: He is a courser for the king.

BALCOME: But what is reasonable? (*coming off the table*) For you, you mean. Have you your plow?

SOLDIER 2: ....Ever on me is my dagger.

BALCOME: Good. And you shall continue your guard over— Bill.

SOLDIER 2: He's much a corpse already, to be fought for,... and no danger to himself, for attempting to flee. Hasn't the strength.

BALCOME: Then tomorrow, with reception of my signal—

MAKETHRIFT: Bill is a subject of the king, and has full rights of citizenship within his majesty's nation, including the regal protection of his person— against arbitrary assailment—

BALCOME: Eternity may be placated for its gifts to consume, caterer, and assume its providence over all of us! not one of us to fear a patronage of souls un-vindicated.... My confirmed signal, the second message received, as for the first to trigger with alertness its expectation.... then as the day is doing, have this done! (*starting to exit*) What lies of destiny in dreams, relieve more horror to experience, more shame and humility to feel, and more distraught nature to bicker of one's self with. For we are fighters and not captives to be captivated by wrongs and wrong views, and will not capitulate to devious intentions or notions to be lessened with. And I have made it clear, my judgment, to you, as lord of this entire realm to honor our particular estate. We will be to war, and proven of our tolerances. We will be towards heaven ascending, through our battles— blades in mouth! if necessary, to ease the awkwardness of climbing; and will not dally here of some local foolishness, but we'll rather make a statement to be determined by, 'gainst all thoughts of cravenness to execute.... our earned mandates and commissions of action and progress, in this valley and to anywhere we're drawn. And not to be forgiven. No one can forgive a need.... (*to the **sentry***) Give those candies to Hoe. And let be prepared Bill's drink of tea, for his presentiments are his own made. (*has exited*)

SOLDIER 2: That is decided, then. As strict a command as ever struck by, does he serve.

MAKETHRIFT: But what will you do? You know it is unauthorized, unjust, not representative of his majesty nor any fair principle of governance, and that it is only mean and malignant of your tending to.

SOLDIER 2 (*as the **sentry** comes for the candy and tea*): To the throat, across the neck— rip! As for a rabbit to be fitted with silence, then quiet rending— quick! All for below me stretched, with no empathy of resistance, here is made my subject weak of spirit, drowsy, perhaps even sleeping. I would observe this.... presentation lying, for hours, and have never come to a target of such helplessness. And yet I'll be standing there, with notice of my waiting. Such a fine lad otherwise. Then should I make his final hours fearful by informing him of this command for execution? No, I'll not do that, though he'll presume it anyway— and then hope.... that his stay at castle is only a reprieve of time to get well. All prisoners dream of a release when faced with consequences as dire as this. I'll not tell him that I'm prompted in particular, so as to hear him begging me of mercy. I'd like this unapproachable to seem, as I ponder the exactness of my method. No screams of terror should issue from the flesh. No enemy has been promoted to me— not properly— for my hand's tearing loudly. The lord puts me to this task for my irreverence. He knows I hate wasteful slaughter, murder (as a crime), and unmerited killing.

SENTRY: Then dope his tea with a sleeping potion. He'll not cry as you, to pounce upon him, when it's time.

SOLDIER 2: We've none of that here. What need we more of the discomforts than sleep against the agitations warranted! Our irritations master themselves, so that we stay alert and arousable as soldiers. Let's not so much cause a denial of the nerves. I mean not to have his person falsified for (his) final moments. That would be like theft is surprised, the manner of it doing to disavow any hope of manly challenge and resistance to the crime.

SENTRY: The method's more, then, for your bother, than the motive. Can you possibly fear the attack to be unclean? We're used to the war cry, even upon stealth. And the agonies let loose their injurious worth of affliction, and cause more value to our hands than heads—

SOLDIER 2: One seldom emphasizes with an enemy prostrate to your heavy blows.

SENTRY: Is there an enemy here? Then I say dope him, if you've no faith for the cause of your theft of life, stealing from the helpless, or hapless defender of his arguments.

MAKETHRIFT: There's much in the tea that is relaxing already.... and stimulant enough for his predicament to shield from, or want to be withdrawn from, by favorable tastes and flavors.

SENTRY: Pain works best with its upsetting to blunt (the senses). Indeed! at its height it fells you to unconscious surrender. Dope this boy, then, as if his being were painful— if this is all false.

SOLDIER 2: The lord does not enlist falsehoods, but fatalities of character and flaws of rapprochement left for him to utilize, even understate as reasons.

SENTRY: Need we quarrel about that? One can only commit a wrong by the way something is done. We do not analyze intentions, for these commands to promote some effect. But if you're meant to be surreptitious of compassion, with your kindnesses abusing of this body, then insure that he is under before putting out of breath's ability. This is not a real incision-ing of the combative opponent—

SOLDIER 2: I can't chance for surprise's legitimacy to be generous, that it is drained for loss to prevent some complaint or struggle 'gainst my mission. Why strip him of that honest honor to assure some candor of existence? Yet, to have a loaf as loaf determinate makes poor its issue for the baking.

MAKETHRIFT: What?! But he's well aware of the peril with which he resides—

SOLDIER 2: This is more a matter of technique of execution we debate, sir, than the temerity of any appropriateness we might scorch ourselves to assume.... for this actionable tort. But to have him dead.... asleep.... laying there below my thrusts of conscience, before I deaden sleep with stabs.... for me does symbolize a growing futility for our employments.

SENTRY: It is unjust, you mean, for what you are commanded to do. And that is why it should be hidden from God's protests. For knowing not what occurs, the boy can not bring sanction to his mind, to scream internally above the terror and the fright and reach for holy assistance, and in that final moment of reckoning— forgive everything! that has been done, as we've so often seen in the eyes of our dying assailants and comrades. Yet spare the boy this agonizing right and final privilege to be seen.... irascible and vital, if this is not right and is illegal of the military valors. Dope this

boy! then, to dreams.... of the deserved dawn, and a new approach to the liveliness of mankind, a delusion for the grandeur of one's existence and a welcoming of all that is attractive to him. The senses of sleep are too profound to be manacled to reality.

SOLDIER 2: I can only question Balcome's judgment, but not his orders. We haven't those condiments here, man! And I'd be unfit to use them, if we had, to add them for his nervous slurping, the poor, wretched weakling. But that his thirst is cruel, for heaven searching, must then he know not even this deception? Your goad is that we've become unfit already, to descend our courage down to this sad tactic of destroying a boy, and a proper king's subject.

SENTRY: What wear to it can we face? This candy was provided by ourselves, as much to soften Balcome as a toddler. None of the men find much virtue in the detention of Brontrol's son, and to this slap of the king through accentuation with a countryman's death, as warranted as sinfulness to speak contrarily. His majesty is sheltered, by Bill's wounding— of his rule. How can our lord be complained of? with this.... event to provide. So will we suffer much scorn, in this objectionable need, as the fellows who can do such a thing to our own people— or at least the king's. But Balcome protects his lordship, which he is authorized to do, from this castle. Thus you are bound, to make this deed.... ignorable, silent, and unquestioned.

MAKETHRIFT: More to the point, sirs, it is Bill who suffers. It is he who is squeamish and is sick, lies in discomfort and is abominated for the coming day, is ghastly to be awake for its aware of it, disallowable for it. And if you can not prevent this treachery and treason.... without the stewardship of any acceptable, reasonable policing, then that is bad for you, and vile for being sad, villainous of capability, and squander-ous of youth. There is a potent root available that is notorious for inducing a deep slumber, through its powders drunk or mixed in eaten foods. I can supply this, for the unfortunate lad, to remove all trace of sensing harm and agony.... And if you can kill a lying dog, then have at that poor boy this way served for your heroics. It will seem to be of a limitless ease. But ever as you last, its difficulties will haunt you, and you'll have no support against the heinousness of your vicious principles to forswear, as you obey the command and commit this transgression of loyalty, nobility, righteousness, and eternal peace. For as you've killed so many men, to find their just embracements of a final rest, henceforth after this you all will be the walking dead! with eyes buttoned from any right to see the legitimacy of your purpose as soldiers and fighters and defenders. You know this sentiment to be overwhelmingly correct—!

SOLDIER 2: I fear it, yes. If I could be confused about the error, that could give my misgivings more envy of so well accomplishable a task. But I understand too well the purpose and rationale— of our effective debasement, and belittling into henchmen, the abandonment of an army by the corrupting influences of.... futility to be deserved.

SENTRY: That makes the culpability as tolerable as its accounting for a dementia of rights. We'll not gamble more over this straitening development except to let it have its play over our temperaments and composure of service, as men who struggle against the most unusual odds foreseeable to remain gallant and trustworthy enforcers for the realm.

SOLDIER 2: The entirety of realm!

MAKETHRIFT: And the valley's lore appeased to.

Scene II — *Before the castle, at dawn.* **Brontrol** *stands with his sword wielded, to prevent any protesters from daring to approach, should some show up this day.*

BRONTROL (*to himself*): All of my crimes and vaults of evil condense into this despicable display of thwarting the VID's resumption of contempt before this most criminal lord's view, who even now I sense could be peering from a tower even as the light's too faint before the coming day, so anxious is he to start his tallying, and condemn my son with a disturbance of the results. He is too powerful to be reasoned with; and so I stand here threatening of brandished weapon, to give some sign discouraging more visit, for the sake of my angry demeanor to confront. But it does worry me. Worry is the cause, and the poverty of numbers among the calm undulations of ground and tranquil presentation throughout the night. They've all gone to bed and sleep, or rest without a peevish quarrel, as this has all become so normal an activity of late. And did myself return my disconsolate wife to house, her energy of chores to restore with whatever dreams may be left of night's time and passage, and while promising her.... and her Hoe.... to save our son with all of my powers left, however this can still be done. Yet can I think but only to shield the sun, I'm too thin of a figure. But as they come they'll see some meanness to my pleading, a fierce determination to have them abandon this ploy today through perhaps the most honest persuasion I have left, to injure their strides and advancements. This is one man's sympathy, and I am desperate. Cool seems the morning, though, before the sun's full reign. So peaceful life appears, as I await the meddling masses. I can not argue against their tactic, it's been efficient, effective and nonconfrontational. But it is inappropriate, and disconcerting of the true priorities, in this instance with a clever Balcome. These people— mine.... they must have considered my dilemma, ours. Then will they appear? Will I have to show a combativeness, to protect my own? This strategy is not so unusual. They'll see me from a distance, discern my conditions, and avoid coming near. It is patent, and blatantly stated: I will scare you off, for the sake of my son's health. Balcome certainly calculated (this) with experience. These are not warriors— they won't want to get hurt, they fear injury. But will their discomposures to governance overcome even that.... hesitance for battle, battling me, overrunning me, overturning me, to fight with this sulking prince, now as I would die with. All of these principles confuse our arrears and losses. I'm not worth the matter, except they are contrary to their own best interests. They show the opposite to what they'd like. For they've never stood these grounds before, not as I've known ourselves. This is a relic here, a forgotten monument of pride, only laughed of with recalling. And now to reclaim it, merely to insult a usurping influence.... that's not enough reason, to permit this shame of harming. So I'll balance this out, with what an armed soldier can do, and cause an incident unfortunate for many lives. Honor will be hatred for their callous approach.... I see a few afar, and can not tell if they're actually coming near, coming towards me and this entrapment— my punishment for presuming too much of.... these common folk I've left and return to viciously. And I can't hold back the day, the singes of this day. It imperils with growing brightness. It is implicit of many failures. It falls upon us to reveal the disgusts more clearly. It does make sacrifce of our terrestrial weaknesses— (*turning sharply to his side*) Who comes?!... but one of those absurd night patrollers. (*calling out*) You see me here!

PROTESTOR 1 (*approaching cautiously*): As so we've noticed, and that you make ready of the blade. But we are totally unarmed and passive. Do you suggest a change of approach by the lord, a vicious shift to effect casualties?

BRONTROL: You've been here all night, walking without means, you and your friend.

PROTESTOR 1: Soon to be relieved by a multitude of others. The morning shows great promise.

BRONTROL: Promise?!

PROTESTOR 1: I feared it would be more cloudy.

BRONTROL: Then you don't realize the drama that awaits. I sincerely don't want you here.

PROTESTOR 1: I wish to leave myself, sir, since I'm all tired of this.... Are you to pounce on me, now, because you find my presence too reprehensible? I've still some courage left, for these chances to take. But we both saw you approach the castle armored, too far between ourselves to make remark of it. Then you're not one of us, and may be Balcome's man—

BRONTROL: I am of the valley.

PROTESTOR 1: It's not our method, to engage in fighting, and to want to cause any violence.

BRONTROL: That's all your vanity allows. I am a king's soldier, and honored to be around here, on this particular day. You'll not be replaced easily this time, while I stand. I've tried to discuss with your members the foolishness of this attack. Now bear the consequences of impatience. For I have no time to waste any more, on lessons and privilege. Whoever shows up of the VID I'll fight, today, until I'm run-down to a death or dearth of ability to defend myself. But as you commonly have no arms, you'll commonly bleed much more until taught not to be here. I advise you to go without meeting up of friends. They may explain to you why I take this stance, both symbolically and for real. But at the moment I'm just too proud to give more reason than my strengths of exertion, not to you two silly gaits. By decree I am steadfast of a vitriol. That's my only warning now.

PROTESTOR 1: Is the king that angry with us? The rumor is he's lost so much of his effective authority that he should really not care how we behave. Our concern is with Lord Balcome, and his delay to redress our grievances of excessive taxation, to support his.... sojourning amongst us.

BRONTROL: Rumors are rancid of their knowledge.

PROTESTOR 1: Yet the information is here, has arrived and spreads throughout the population. And Balcome should be decent to everyone. We know he's not about to make this his homing camp. We're a retreat, for him, from his bouts of conquering in military assaults and his management of (his) other personal, varied estates. He's here only because this castle was a gift to him presumptively made by his majesty. And so he is expected, by courtly protocol, to inhabit it awhile, with all of its discomforts and embarrassments. This is the logic behind a wasteful increase in revenues our people will not put up with much longer. Our peaceful revolt is simply to have him sense a greater urgency of assuagement, as he prepares to leave, with his soldiers, having visited and learned of our growing independence of mind. Is this not a wisdom we would share, or want to spread as well as rumor, that we take care for ourselves as regality crumbles? that we are responsible to ourselves? Why fight at this maturing impulse?

BRONTROL: You remain a subject to the crown— You all do!

PROTESTOR 1: I lend you my body to prove this wrong. Make holes here, and ream them through with your sword—!

BRONTROL: You are.... just making more of your demonstration, with exaggeration of your empathies to appear as a victim of just causes cried for and carried. It's with annoyance that I should slay you, and not for your rights to offend. But I have more victimizing to prevent than your parody of courage and virtue.

PROTESTOR 1: If I thought you savage of reason, I would not even try to complain for more sympathy. But see what broadly descends upon us.

BRONTROL: Villagers about their daily tasks to begin.

PROTESTOR 1: Some will be coming here. And you'd be a clown of obstruction fighting them off.

BRONTROL: Better they should consider me than this. I am to make real lunges and sharp incisions. But you VIDs are vipers to be so resolute against my life. Then that is fair, with knowledge they must have gained. And you are ignorant of the night just passed, with all of your paces futile of step. You've slept through it as much as they (have), it seems.

PROTESTOR 1: What brings you to demand my sleep, then? I've been aware of everything. Can't you understand that this is a form of celebration, protesting in this way, without a manner of injury or even threat? Can you actually consider us fearsome?! And yet you position yourself as boldly as we would sit, to make some show of it, that your displeasure is aroused. A counter-protest? or protestation to your faith in man. Profess that you find him mindless to be, at times, without a ruler leading, and that his attitudes are by and large petty compared to those of noble princes and renown majesties, monarchs and emperors, and other such caitiffs employed to dominate the affairs of the common, destroy them by demeaning their will into a subservience of the terrified. Well, we haven't had such a refreshing swell of lungs, to breathe in these sweet free airs, since several lifetimes of subjugation, that the moment is now ripe and legitimate to declare an effective independence from an enfeebled potentate. And it is fun to witness of ourselves these feelings openly expressed. That's why we demonstrate with the courage of passivity. But we do not mean to cause your war-raging against us, we who are happy to feel empowered for our quiet and even serene displays about the boarders of.... Balcome's castle. We are good becoming here, and friendly, non-obstructive of any passage to or fro, and highly respectful of all visitors and occupants. So why fight us, soldier, as we simply make a statement? Why pillar yourself before us, as out of the dark arrived, to insist on a confrontation during the birth of this fine morning's light? Why so suddenly show up to make a meanness evident concerning what we happen to do? You want some incident of blood?! I offer you my head! (*bending his head towards* **Brontrol**)—

BRONTROL (*with slight pushing of the head away*): You're jesting again, making fun of this danger to have me seem foolish or trite. But this sword is real and anxious, and in trained hands well exercised for its skilled use with impulsive deployment (*as if demonstrating his intentions*). And you stand, sir, to be able to walk. But any one of you who dare now to sit before this castle will fall subject to its sharp abuse— I warrant crying foul.... for I am serious today. (*as* **Protester 2** *runs towards them*) And I may just make of you an example, to show all sights that I depend on these actions thoroughly and decisively, if you continue to joke upon my grim countenance.

PROTESTOR 1: Has the king ordered you to do this?

BRONTROL: The king.... does spare you time— to be convinced of himself.

PROTESTOR 1: Then does Balcome order you to make this provocation?!

BRONTROL: The lord does make me want to save a life. But life is a headway forged; and I'll certainly have yours, if you continue in this mode.

PROTESTOR 2 (*rushing up to them*): What for this cruel display?! (**Brontrol** *takes up an overtly defensive posture.*) Our members arrive shortly, and this is not to be expected of a greeting to morning's glory.

BRONTROL: I will demolish such a pleasure for yourselves!

PROTESTOR 2: Why do you deem to attack us?! You came not from the castle.

PROTESTOR 1: He is a king's soldier. Must seemingly defend the crown's interests, but claims to be from this valley.

PROTESTOR 2: Well that's appropriate, is it?! the way you arrive, like a specter out the night to just show up and stand with a terrible warning, station yourself as a sentry to entrance.... or approach to this dominion. And see how you are observed, to be quarrelsome and threatening. We've all been watching—

BRONTROL: All?!

PROTESTOR 2: The membership approaches. This day begins its customary siege.

BRONTROL: Then I'll relinquish the lives, of you two silly fools, to avert their steps and divert their travel towards this monument today! For I mean to prevent your squatting malice to be seen —!

PROTESTOR 2 (*avoiding a warning with the sword*): You're a solitary nuisance! Mean you to fight a whole army of contentment?!—

BRONTROL: You make not an army, without weapons. Go run to the VID's, and inform them not to make any approach today, or I will swipe to lop off any heads of the sitting. And if you come with arms of offense, rather, to impress your point, then we will fight! And I'll strive hard to prevent a swarm, until it's done for many reddened. This will be such a battle as diseases may enlist their opportunities of mind, to turn your dastardly heads finally into stone! You comprehend not. So go beseech your brethren for clues. But as they are here, defame yourselves completely! For you have no souls and I haven't any patience left.

PROTESTOR 1: You exhaust your king with this demeanor—

BRONTROL: I will make a bath! of your comportment today, not to have considered me and mine more carefully.

PROTESTOR 2: Then pause for our behaviors, that you present a protest to them. We'll not have grime of your mistrust for our civil nature rubbed across our skins. And to assault the defenseless and defendable, you are a royalist, an implant, and a deception from our village come. Balcome placed you there, with surreptitious shade of night, to stir up trouble for this day. Since we have only been courtly of our posturing, and he can not run out his overbearing troop onto our seated stances on the grass without appearing outrageously villainous. So then he must incite a fight to make rid of us, away from his estate— for a while. But that is no solution, soldier.

BRONTROL: You are st*uuu*pid! to think so, and banal of your credences. But that your members appear at all is the crime here. So best you go to them— quickly! and persuade them to stay away. Or I'll stab you now, for all to see of my emotion about it, your intransigence to this sick principle of annoyance and blundering! of a life.

PROTESTOR 1: I think they hold off already, with this spectacle of contempt, for little as it may last, such delay of encroachment. I've offered you my life—

BRONTROL (*preparing his sword in earnest*): And I'll take it now!

PROTESTOR 1 (*running off*): But you are seriously disturbed, to believe such things may be so casually lost. You are an oddity to appear. But you can't stop the rages of a mob you may have caused.

PROTESTOR 2 (*avoiding **Brontrol's** occasional swings and lunges*): Balcome has failed with you, sir. We will access the problem of your protracting perturbation, and devise countermeasures. But you can not stop a justice sensed with our warning of the lord! (*stumbles and falls heavily to the ground*) Aaaah! I am hurt! I am dead to you, and your red rage.

BRONTROL (*preparing to stab*): Such befits a casualty of irritanc'(y)!—

PROTESTOR 2: Surprises me not! as Hector would surmise your blind fury. I'm unable to escape this talent. Yet extol more virtue to my misery, and virtu of my awkward fall to give you easy victory towards your strange designs. Yet as a villager descended.

BRONTROL: You are one!

PROTESTOR 2: Yet as a valley resident? I can't believe it.

BRONTROL: I can't afford many others.

PROTESTOR 2: They'll come to my aid, unless you slay me now, with definite destruction to propose. And I'm to be pitied. But I cry not, but for this pain of knee— Then devastate to end it, if you are such destruction.

BRONTROL: The trapped always defends its plight, if thought inevitable. Now VIDer! VIDer! You're more a hostage than a hurting to me. Should any of your members dare to approach today, I'll dare to jab you through. And then this will be just, how you may be saved with their discernment evidenced. For are you worth— your goals? If you are, with their decisions.... then so is mine— unto the death. And stage with fairness, for this misery of contentions. Yet, do you know *why* I assume you to be so?

PROTESTOR 2: So what?!... victor!

BRONTROL: That you are fairly down— and caught. Then make with ignorance your pleasure to remain so fiercely resentful. You need not contemplate the cause other than your own, and die ascending with that belief, purified with certainty— stubbornness — obstinacy— and coercion pledged to. And now you are pounced upon by fate. Get up to be slain immediately, if you're so faithful to these methods of attacking rulers, your principles of squat! Or stay lying, and let the VIDers decide.... this balanced outcome for us.

PROTESTOR 2: I am.... broken of bone, I feel. And you are broken of valor, to assume this can of any fashion be right!

BRONTROL: ....Oh, but it is.... And gray appear them, staring at this display. But do they know me now, with righteous protection for my shield? Then this may seem terribly contrary to you, with infliction of your person to be bemused by. But I have returned! to opposites opposed, through me, for what they mean. The subtle hills.... are meant for seats deprived.

Scene III — ***Balcome*** *is in a castle turret, looking out. He is accompanied by **Gestromy** and a **soldier**.*

GESTROMY: It does for day, my lord, the typical spectacle of a broad scenery. The winds are pleasant and the grasses shine with the roar of the sun's embellishment. Trees stand substantially in the distance, and all is for a nice environment as the landscape is speckled with the various colors distinctive of sparsely inhabited plants and shrubs. Laborers commission themselves with their animals and tools and carts. And life is as normal as its nurturing potential allows. Is this not, then, of a tranquil setting about us, infused of the calming influence of regular purposes to make substantial any epoch of gainful time? as would have you mend from previous intentions?

BALCOME: It's careful of the count, for this mushrooming, Gestromy. See how they approach from all sides, to every corner of this periphery of station, to sit below me, and for my sight to adjust with the fiendishness of reason.

GESTROMY: I dare not pause to take such views, my lord. For with shame of events they are too predictable. The people here are customary in their endeavors, and you can not account for their actions over their heads. But it's for what you may feel that is more suggestible to me. You've given herd to impossibilities of promotion. Yet, does not a typical calm of day satisfy the emotions to be generous and forgiving, and permissive of a peaceful assembly.... seen so often before?

BALCOME: Dry paint does shred with scars. And I'm not moved less to find this painting for my tally worn, a picture of refusal to become compliant to my aims and motives. They sit to be more stationary than my nerves of indignation. And see poor Brontrol frantic for control, as they storm around him, at his point, to avoid.... and virtually ignore, the futility of his appeasement with sword much too evident, for opposites of persuasion. How can anger promote kindness, or fear security? Yet this is how we manage our affairs all the time. Ah! He falls to his knees in frustration. Failure is devastating. But he hasn't much influence through his circus acrobatics. Go! soldier, and take my message.

SOLDIER: What message is that. my lord?

BALCOME: Why, to start the proceedings as I've ordered, of course.

SOLDIER: It's vague, sir. To whom should I proceed—

BALCOME: To slay the son of Brontrol, as you well know!

SOLDIER: That's what I thought might be.... after a thorough examination of your perspective—

BALCOME: What is this hesitance?! Doubt you that I've not! But twice I have examined the hilly plains, to confirm the result that their heads presented me exceed the total from yesterday. Now rush yourself to this execution!

SOLDIER (*leaving*): The last I saw.... of Bill.... he seemed dead already.

GESTROMY: No screams have been heard out there. Why then cause them within our castle home?

BALCOME (*turning to face* **Gestromy**): They have spoken with their silence, that this is an acceptable outcome for their behaviors towards me, and their nation entirely. But I've been more than fair, to let them peruse through these arguments of accounting. The decision is their presentation to me, the silent protest that demolishes so much faith, and the lying about capabilities that undermines one's personal commitments and regarding of others. I have not prevented anyone at doing anything of their desires throughout the entire course of these developments. What they may achieve of their efforts is a different matter. They certainly won't defeat me, though martyrs may be fashioned 'round their grievances. But is this as sad a day as they are silent? They could be boisterously happy— of spirit— to know what they have caused, to be contented of their pressures thoroughly expressed and agreed to as a group of militants. If this is self-sufficiency, then they have served themselves the lamb. They have made their sacrifices, and now celebrate with quiet prayer, and a commiseration of feeling important enough to effect their wishes while enduring the most bitter obstacles to pride, the worth of an individual. They have given that sacred value up, as a delusion against their goals and promises to stay.... discreet in this expression of numbers. And I acknowledge this curious strength of character with some fascination. It must be actuating of some eternal principle of nature, or the nature of all occurrences, to *want* to injure yourself— And they certainly have!... They know that, with no doubt to it. I suppose, to have my way, I'll have to be conspicuously cruel, as their muscles of consolidation and consistency have been powerfully flexed. But up to

now, all that has been done has been done by them, those of this valley. I have relented to their wishes in these matters. Now I will ascribe their wishes for my own to be genuinely permitted. For we as soldiers are never afraid to bring lance to those unafraid of the impingement. I'd say they've proved, or at least provided, some fortitude to endure, and possibly withstand, a thorough assaulting.

GESTROMY: Is this practical, my lord?

BALCOME: Neither will this be a punishment nor a castigation, but merely a test of nerves and reserve of faith or trust with the talents (we) each are provided.

GESTROMY: Start you a war within our own land?

BALCOME: As such is land to be as conquered, the final arbiter will be truth.

GESTROMY: Wait for them yet to dwindle, Balcome, upon the sorrow that's evoked and is penetrant of souls upon the demise of a favorite.

BALCOME: Many may hardly know of Bill—

GESTROMY: And many may hardly know.... of the necessity they've provided for his killing. Most may be terribly ignorant of this. It's only been one night, and of perhaps only a few officers of the VID discussing this. We shouldn't assume an electric spark of knowledge courses through all of the members immediately, after the conclusions of some heads are made.... Wait for their numbers to decline a bit. Give them a week more, to wear their energies out with sadness and remorse for what they have contributed to your actions.

BALCOME: You know as well as I that we shouldn't allow their promotion to weaponry, if we're to thoroughly subdue them and take over this land, this territory for a permanence of residence under my lordship. Now is the country made, for my precious part to own.... or be acknowledged with. Balcome's valley is more definite today than ever it has been in this recent past. And they outside do grow to this, as by a force of nature making them succeed to my insuring— with an honest military might. The nation may now be carved out this way.... in many such places, Gestromy. That is the deed for our times to take advantage of. For the citizens do come to me one way or another, whether they understand the import of their actions or not. The meaning here— today, is that I'm licensed by my own handling of this situation.... to be a lord, devout for as much certainty of— *our* positions.... as I may provide, this management hurting both father and son, but through their own devices hurled into an oblivion of state. I am compassionate.... for all misfortunes to seize and remove, from this area of governance, with wisdoms training more than wills and wishfulness. (*turning back to look out*) So then must we subject ourselves to these views of patient casualty.... But I have known it ever to result this way. It was the method to make *them* provide the necessary impetus of motion that had to be crafted to allow my acceptance of this path, a manufacturing befitting their.... unobserved discomposures. And the sun is certainly hot and high, over their hats and brows of complacent torture.

GESTROMY: That is a definition of self-sacrifice for a cause.

BALCOME: He's gone!... I see Brontrol not more. He's.... left the sight.... the scene.... the might. Given up unto himself, I suppose.

GESTROMY: What this portends is not so beneficial for our relationship to the public. It will be harder to be at all cordial over the issue of collecting revenues, and of socializing among them, their holidays and feasts of harvest to celebrate, a good will to champion for all, and a high reverence for the indomitable forces of nature that compact us towards these fateful determinations of our manipulative guidances and utilities, the shaping and sharpening of what amounts to a class or caste of mutual feeling and deprivation for enlistment. I tell you, Balcome: if this result was at all foreseeable, it could have been achieved with much less anguish than this. It could have been ordered outright with a strict declaration of obedience to avoid more harm, and given the commoner a firmer choice of direction towards us, instead of playing games through their internal sympathies to assess and dissolve. We are more the beast for our malign accomplishments to demonstrate, and I can only sense more pity for our powers to assume through this region, that there ignites in us much matter of thought to grieve about for everyone. I feel ungodly sadness this day, peering down on an inevitable dejection, and from these false heights. May our shadows bend to bring some comfort to these unruled events.

BALCOME: Then an enlightenment is pitched for viewing, in the sense that we fashion our friendships, Gestromy, rather than allowing them to evolve out of drifting manners and lackluster consequences of what merely occurs to happen. We do not allow rumor to shape the mind, concerning our activities and theirs. Yet harshly, at times, may we bring forth this mind of passion to abide our necessary requests, and this through a schooling of might not always violently applied.... but often judiciously through intricate teachings to be remembered, that what is done is never as important as how. And what is caused today will be ingrained for careful ponderation a long time to come. It could have been much different, for all my weighing and precaution and tenderness with sensitivity of this situation. I may be blamed for being something horrible and frightening, but never disputable of purpose, and more clear than this posed display (*turning his head away from the view*), an array of portliness upon the grass, sloth and laziness below my nose as were the feet and legs turned to lead, and a calibre dimensional of much stillness. *Polast!* Blast their defiance— to the ground!... as floating on a sea unmoved, their weight as light as for a buoyancy to withstand. I am not happy with this current.... but it befits my curing. And I will cry for this boy Bill, in my own way, as could I do for anyone diseased and dismayed and distempered into a blighted crookedness of pride and fouled demeanor before his lord. Well, actions have their causes, often much too independently of the results. But I doubt if you'll be much abused or castigated personally for what occurs. And even to your office of collection will you remain.... responsible, and they responsive, fair sheriff. For out of all of this a keen and strong sense of authority will be sharply enhanced—

VOICE (*from without*): My lord! A courier arrives with urgent message!

BALCOME (*responding*): Then have him arise to it. (***Brontrol** rushes in, causing **Balcome** to shriek in terror, to attempt an impossible retreat. **Brontrol** is followed by a **courier**.*)

BRONTROL (*pointing with his sword*): Lord Balcome!—

COURIER: My lord! I come directly from the royal court, as swiftly as possible. Our majesty has died of his grievous wounds, succumbed to them most horribly— but valiantly. The king's council demands that all warrior lords return immediately to the palace, to engage in deliberations for selecting a new ruler. The active executive ministers are of controversy and confusion for and against the young nephew hereditary for ascendancy since he is not only underaged but there is some question of his legitimate marital relation to the late king. Your attendance is required to discuss the matter since it is expected that the interim ruler will be selected from your cadre of lords.

GESTROMY: This is grave news for everyone!

COURIER: I am to accompany your lordship, as an official escort of imperative and privilege. And we must be on our way to arrive as soon as possible, so that you may handle the full scope of political inquiry and intrigue that now descends upon your personal management. That, may I add, of this latent rush, is your only advantage afforded you for these discussions among equals. How things are decided will depend on what early quorums are built.

BALCOME (*recovering his senses*): This.... is for an outrage of incidence to be burdened with, and I am struck to the heart with grief for our demised king. And now you gather up his angels to make sure the effect of his rule is fully prospered. For we will force a thorough continuation of his.... methods and principles of running our nation, as we insure his righteous nature much lengthening of time—

BRONTROL: And what of my son?! Balcome! as to my knees did I drop to learn this news of the king, his solemn departure made, then did swell for knowledge of my son to gain as he's much more important now than the deceased.

BALCOME: ....More for the cause of this, and this entire valley pledged, that He is dead, then I absolve your son of any crime related to the name and pronouncement of our former king. And you will find— Bill.... where he has been nursed, and indeed where last you left him, you, your wife.... and Hoe. Take precious candy for this reprieve— of sentence, in remembering a sweetness made for thought, that in fact I release this entire valley from my patient quarrel and contention. Its cause seemed queerly prescient of these inhabitants. And so the dispute ends. Your folk have won their rights of taxing reconsiderations out of necessity of my negotiations, and renegotiations, with lords. So then to demonstrate of this, I must be quickly away—

COURIER: The coursers are ready, my lord. And all accommodation will be provided along the route to make this travel fleet of propitiousness (for your arguments).

BALCOME: Gestromy, handle all affairs in my absence. Appease everyone that, apprised of these developments, this populace gains only my favor and courtesy of treatment (*starting to exit with the **courier***), and I will sanction their complaints.... as resolved entirely to their permissions— of justifiable and responsible debate. Then in this course, arrange with them the manner and extent of these taxes you know so well, for fair collection and moderate application throughout.... I'll assess the appropriateness of these.... when I return, and will be hard pressed to find need of any adjustments from your adjudicating purview.

GESTROMY: Parity will be our commission, my lord, for the honest upkeep of this castle, that never should the populace nor any member of be strained beyond their privilege to adore their rights of living with so momentous a symbol of their histories (***Balcome** and the **courier** have left.*) and social achievements.... He will ride fast, beaming upon and past their faces as a victor in quest of more important matters, finally satisfied of these country residents' tumultuous though strangely tempered greeting, polite in a most oppressive way. And how are these people that you've fought—

BRONTROL: I haven't fought any!

GESTROMY: —Our soldiers will be much relieved to know of the lord's current mission, as it pertains to them, returning to something higher than this stay. Without a doubt some armies will commingle of partnerships, to leverage greater influences on foreign affairs. And Balcome is a master of orchestrating such mixtures to his advantage, as did he often do so much interfering among forces that perhaps as a stalling move he was sent here with his least contingent at the moment, to cool off, until our late majesty could grasp a greater handling and management of all his military attributes. But now that facility of Balcome's will be much in play, to wrangle of effective details, for our nation's ultimate success, in defense.... and beneficial expansion.

BRONTROL: My people are plain, and past around my threats, daring me not to do anything, though I had made a captive for show. I was indeed to slay any sitting. But can violence be made without violence? Prepared to take my assault with cunning, as he would accept his worst for his best to represent— a dead body— he just smiled at me to prove my.... ultimatum of request. Yet could I not be so vicious to floating ghosts, as they spread about hardly to notice our presence, flowing to their positions of ground. There's some wisdom to that stupidity, that transformed my desperate strait to delay the killing, the covert sacrifice, put it off and postpone as this sharp sense of contrariness pierced my perception to signal some message was afoot with their calm composures and comportment. I could see it well. on their neutral grins, knowing them as I do— out of bitterness without disdain. And then this courier, much my like, approached me with fearlessness, to tell me of this news about our king. He thought I may have been some sort of policeman of order for the castle, and could gain the most expedient passage to Balcome through the squatting crowd. His surmises were, I suppose, all true. But when I learned of our majesty's demise.... much drained from me, as a reverence swelled in gesture. And then time became more pressing than our supplications. I left my captive groaning— for he too was touched by the information— and rushed the courier up to address Balcome directly, my son's demeanor much in the balance of this hurry. I felt the count of heads upon my back as were bumps of anticipating grief and surrender to an ornery fate. Yet am I so challenged? I might have slain Balcome today, were not his reprieve made for my care to recapture.

GESTROMY: He.... ordered Bill be executed, his counting done, his tally made. And with his own eyes saw the results, these numbers made below him and petulantly real.

BRONTROL: ....And as I've spared a life, so then has he, through my procurement of this courier, myself the same with fury of a need. Or be abandoned to the shame of failure, not for instants did we waste, the moments made with anger for the imperative. And haste do hound the steps. They are so slow outside, and steadied of their walk, self-confident of the comity they pursue. That is the awe of this group, and my tribe as it were, to be from, that disobedience may shimmer slightly, lightly made and without much upsetting of their individual prospects. And so opposite to their concerns are they arrayed. Then where is peace in this quiet stroke, Gestromy? I am for matters brought to their peak, in this crude turret of opinion for me. (*looking out*) They seem as dead, and are as much alive, sitting and wondering what's to occur now, now that he's dead.... The country is sustained, the principle upheld. And all liveliness is melded into one potent pause of pondering, where at I be.... for these fair breezes and pastoral views.... Spirited we were, sprinting for this tower.

GESTROMY: What arises, Brontrol, for this development?

BRONTROL: ....It's a green setting, here, spotted and dotted.... with these heads, and pins of conscience. What makes of my own, concluded in this way, does bring bereavement.... into a spaciousness. And they're so quiet. So.... satisfied. So relatively unseen. There be a warmth beating down and squashing.... into the inconsequential of this task. Life makes for it.... no mores upon itself to hold.

GESTROMY: That takes it so (*starting to exit*), a second breath up here, to contemplate these things. More mystery abounds in loftiness than loyalty.

BRONTROL: Then that is a casualty of one's position. (*alone*) What tragedy is made.... with this tranquility portrayed? Here are the screams of their enlistment. And did I see.... those friends of Bill? They stared both at me.... and without to see at all. Then they must have seen everything.... of my tutoring.... Bring some shade to this sky, sirs.... and let them rest. This clear blue bust, remiss of the clouds, their graying tears enveloped of wispy intentions, then let the sky hide itself, as were for crying meant, to deceive folk, and bring their protest to an end. (*retreating from the view and to himself*) Like death is solitude.... with no amendments won. How strangled be the enigmatic plight of one's strengths. Yet high are some decrees brought, to be adopted on review. For which I've studied all the same to be.... renewed of these deceptions. Then does this crime allow, a taste of the paternity. It's bled for all to see that I am calumniated through these gross complicities of fortune, (and) will take less of the blame than the brunt of tendered dispatches and societal posturing to allow these heads their wrath upon me— Yet more upon themselves are they conscripted to be kings. And I'll accept that terror made, as I explain myself for what has happened. It is with thought provided, that we all must rule o'vr our deaths with defamation for the attempts as characterized. But it's solely brought through the ironies of our withstanding. (*looking back out on the view*) One's greatest worth.... is to be deceased.

}{}{}{}{}{}{

Intrepid Trepidations

Intrepid Trepidations

Characters:
 Beck
 Heroford
 Langly
 two Waitresses
 Mike
 Josh
 Betty
 'Aron
 a Boy
 Maureen
 Laura
 Danny
 Duncan
 John
 a Woman
 a store Attendant
 Breur

Locale: A metropolitan city

Act I

Scene I — *Morning to an outside café. A store attendant, not distinguished enough as a waiter, is setting up the tables and chairs and bowls of packeted sugar or sweetener.* **Heroford**, *lightly dressed even for the summer, abruptly sits in a chair before it's positioned at a table. The attendant notices and complains without authority.*

BECK: What can ya wack?! I need a table like a sack ta pull. (*continues at his chores*)

HEROFORD: Had to get out early. The air is stuffed everywhere inside that house.

BECK (*positioning a table*): There're heavy. Not any particular one. But with the several here, they become heavy. Ain't you cold? You don't have a jacket on or nothin'!

HEROFORD: A chocolate *mo'(llig-)lait*, please, heated under a tuff of white cream.

BECK: I don't take orders. Wouldn't let me carry that stuff out. Too awkward, still.... You ain't even at a table, mister. Here, let me place one by you. If they complain, I'll say you wanted to sit there. (*puts a table up to* **Heroford**)

HEROFORD: I'll pay for it—

BECK: The sure you will!... But it's a little cool yet, don't you think? Maybe some breakfast toast will do? Anything that's sweet and syrupy and hot.... (*going about his work*) Ain't got any menus on me.

HEROFORD (*placing arms on the table top*): Going to be a blaze today, by noon. The afternoon even warmer. Don't feel it at all yet.... Why aren't you wearing a uniform, not even an apron?

BECK: Because I'm not uniformed.

HEROFORD: Seem old enough to be.

BECK: I only help clean up the back.... and make up the front. They don't need me around serving customers. The pretty maids are better at that. Haven't even arrived yet. But the patrons mostly serve themselves, especially outside.... How come you're so early? People don't come here generally for an hour more. It's cool, these mornings. Sharp an' crisp. I like ta keep busy in this kind of air. Ain't you got any food at home?

HEROFORD: I have a whole family at home. I must leave them to clear my head. But as not to disturb them with resentments, I often get out early. I need the quiet that is natural for a city, this time of day. Can't stand all of that radio(ed) music the children play. Never liked the radio on at all, in the morning. Though my wife loves it, so I've pretended all these years to enjoy the custom, the ritual for waking up and preparing for the day. But now, mixed with all of the boys' racket, I have to leave early. They're polite, though. Keep the sound down for themselves, not to drown out their mom's station. But they don't like to use the ear plugs. And each one prefers his own station. I hear the sounds, like individual personalities. That's for giving them those transistors. Well.... the eldest earned the money to buy his. And the others wanted their own— They got.... The youngest one just received his from an aunt, as a birthday gift. Still hasn't quite mastered the manners of playing, the etiquette associated with listening to these sounds. You do not turn it on, while someone else is trying to sleep, unless the volume is already tuned to be low. I think he's afraid of damaging the ear plugs. New instrument, you know— Had to get out!

BECK (*occupied*): Can't be that bad, sir.

HEROFORD: Up an' out! It's a darned impediment, living through a vacation. When you're about to go to work, you can handle all of this morning business. You have something to look forward to do: not necessarily doing it, but the fact that it's going to be done. But these days off, it's trying. It'll be better when we all go on a trip, next week. Everything will be unusual, and we can all take the novelty, traveling, you know. But at the moment, it's become a pain.... a bit of a pain.

BECK: That's nice. I haven't been on a voyage since I left school. Too busy working. Can't afford vacations yet. Have to find employments all the time.... But I feel secure with the need to do something. Better than any holiday.

HEROFORD: We're going to try to visit the Tetons correctly, this year. Went there a couple of years ago, but didn't know what we were doing. Didn't plan things out right. And the youngest was too upset with the rush and discomforts and not getting the attention(s) he was used to. Now he can't wait to get into any kind of mischief.

BECK: Well, that's the way ya grow up, an' out of the doldrums of too much motherly dependence.

HEROFORD: Booked almost everything in advance, with only a slight premium to it, but savings at the hotels and some restaurants.

BECK: That's the way ta do it. You have to search for those discounts before hand. Kind of a shock ta learn, since everything's presented to you, like in packages of adventure, while you're being educated, in school. You can't think for yourself until you're shown how. Then when you finally come to it, all life and processes of living seem bizarre, like they have ta be learned all over again, but with some kind of sophistication to it. You're not allowed to make the innocent mistake; it has to be more worldly worn to tolerate. And now you're really on a treadmill, to keep running and keep it up, that spurt of consequence to extend and hurry up. No one ever told me before that living was going to be like one big bout of a heavyweight's fight; but I'm game as long as I can take these punches, and throw off some. Many would say I live quite misera-bly. They mean lowly, but I don't feel any depression about it, yet. Though things are gonna have to improve.... (*finished setting up*) Now.... you wanted something to drink? Maybe I can induce one of the owners to tend to you. It's still kinda early. But I think we *are* open. I don't know how open's made, the actual activities to get the food production going. But I think they're ready to start busi-ness. (*going inward*) What was that you wanted?

HEROFORD: Warmed chocolate *mo'lait*, the molle equivalent for a drink.

BECK: I'll see if they make that Don't know anything about han-dling the machinery in there. Scares me ta know. (*Exits.*)

HEROFORD: Lot's of thick pieces floating about. Had it once, on a morning like this.... some time ago.... Not here, though. (*looking around aimlessly but with a curiosity*) Gosh! I must've been a young man, then. Things are coming back to me, dear senses to atone with. But this is a pleasant enough place. I'm just a little out of sorts, rusty for this kind of stro.ling. Don't quite know how to be free and useless again. That sort of thing's stripped from you early, so I am early here. (*faintly hears a voice from inside: "He can wait!"*) The decor's.... not as colorful. A little.... industrial modern, like within the teeth of a robot. That's it! Rather metallic. Less red brick. (*A couple of young ladies come up to enter the café, one looking at him as being unusual to be there.*) Morning, ladies!... I'm quite ready. (*slight giggles from the women as they enter*).... So I'm not the typical patron, I guess. Well this is not a typical day. My head hasn't cleared yet of the fog, the mists of being estab-lished.... but feeling out of place. The haze hasn't lifted yet, to evaporate away. I can't stand the dread of being truly disturbed. Ought to be more thankful, but I feel funny. Is a bit cool, outside. Gonna warm up, though. The radio said so— fiercely! Should have put on a jacket, but the news said it's going to be hot. I felt hot. What's for these weather reports anyway, if they can't be accurate. The announcer's warm voice is all that matters, that radio rasp. That's what she likes to hear, and that calming music— that can't quite be classical. Suites, I think they're called. At least they suit me when they die down; the musicians have finished some work well, a right concerted effort. But since it's a recording, it's all reli-able.... to listen to in the first place. All of that editing, Heroford, the perfecting. I must keep perfecting.... my sham.... to be away without deception. I'm tired of everything, but of no one.

*A gentleman, not elderly but conspicuously older than **Heroford**, and more appropriately costumed for the outside temperature, comes up to find a table, and sits.*

LANGLY (*as **Heroford** glances toward him*): No. No! I haven't any money. I'm just here for a customary glass of water. I'm not of ten denied this free service. It's particularly fresh tap. Mine seems dulled, or contaminated with age, an aged effluence too invisible to

characterize as dirty. The water pipes in my apartment are old, though, and ping a lot.

HEROFORD: ....Oh. For a medicine to take, no doubt.

LANGLY (*looking at **Heroford** with slightly caustic notice*): No, sir. Just for the taste of some purely good water.... One can get it from the fountains at the park. But that's a bit for walking, and I don't think they're turned on yet.

HEROFORD: (I) Thought they never stop functioning— Oh!... They can turn those things off, in the park. Off to deter the va-grants.

LANGLY: Yes. Purity is criminal, unless it's provided cau-tiously.... I am regular, but come before the regular customers. (*A **waitress** brings him a glass of water.*)

WAITRESS: Here you are. Finish it up quick! You're lucky Breur's not around today. He don't like you much. That's why I couldn't serve you last time. (*retreats back*)

LANGLY: Any time is worth a pleasure being with. (*sips*) I used to be a paying customer. Was friends with one of the proprietors, the original one. Deceased now.

HEROFORD: Oh. That's sad to say—

LANGLY (*inappropriately roused*): No it is not! (*calming down*) Death's all a pleasure to one who's earned it.... I don't mean to teach lessons about that. But it becomes more and more evident with every passing day, of my life. The eyes harden and the sight grows weaker and more shaded. And this tells you it's a *good* thing to get ready for.

HEROFORD: You don't look that aged to me.

LANGLY: And you don't look dressed comfortable enough; but you look so, and show no sign of complaining about the tempera-ture.

HEROFORD: It's gonna be hot today.

LANGLY: I mean the ambient temperature, right now. And right now I know what I'm preparing for.

HEROFORD: That doesn't sound very positive to me. You should be up beat about your prospects, no matter conditions found.... health-wise or other.

LANGLY: What really makes you think I'm not? This water is enjoyable and refreshing. The fact that it's given to me freely is very ingratiating; even this poor body is considered of some impor-tance.

HEROFORD: But you imply you're preparing for.... an eventual-ity, heaven forbid to look forward to a retirement.

LANGLY: My conscious days of struggling are quite over, is all to interpret about what I notice. I past my peak some time ago, and was glad to withdraw; because what I saw atop there was not all that different from having a simple drink, sipping from rarefied airs and potentials. Why, I looked down and was shocked to realize that not one thing below was a mistake, so there wasn't such a need to climb out of it in the first place.

HEROFORD: Oh. So it's retrospection that entertains you. Well that's fine, if you're done— if you're finished. But that can't satisfy the truly active. Have you no responsibilities to anyone but your self?— I don't mean to pry. My whole life is based on that. My at titudes conform around this principle, that others depend so thoroughly on me. It would be more than a shame if I suddenly disappeared.... for good.

LANGLY: Well, good sir—

HEROFORD: Harold Heroford. And I don't mind saying so. My name was even in the local newspapers, a few years ago. A slight article, but very convincing for my advancement as secretary for the city treasurer.

LANGLY: Really? Sounds important and impressive. But I though women did that—

HEROFORD: Oh, you're thinking about office secretaries. No, my position is very administrative for elected officials. I've worked under three now, allowing their executive orders to be implemented correctly. You'd be surprised of how little they know of the actual doings and gears. They make the decisions and draw the conclusions, but I get the job done— real time! I do the programming, for example, and the actuarial designs for the insurance risks taken on, and of course give all of the explaining to the heads.

LANGLY: Outstanding! sir. You must be wealthy. I myself live on the pittance of a disability pension. Would have been much more if I could have retired regularly. The lungs gave out to the asbestos. That was medically, and so legally established. Worked in construction, with a specialty for demolitions. Yeah. We were a surefire breed of able souls, my lot. Took on the projects no one else would. Ate them like pretzels. (One) Can't be so active today. Need many more permits and state board reviews. That was a good time to be emboldened and fearless.... in your work. Lost more than a few friends to accidents. But we cleaned up this state, to make room for its modernization. There was glory in the effort, for a controlled explosion, everything falling to expectation— We were engineers!... though I never actually studied the science, in college. But we could do it better than most. Small buildings, mainly, with hard grit and hand work for the demolitions— Labor intensive, as it was called. I enjoyed everything so much, till that awful cough persisted and felled my career.... way before its time. Was stuck in a hospital for almost a year: half in a bed ward, half in a supervised housing complex— with regular nursing. Used up most of my health benefits there, at that time. Now I live in a small, more normal apartment.... with enough funds to just get by comfortably. But that's getting harder, as living standards become more expensive. Glad I won't live through what's foretold, by the ominous social pundits. Should only have ten to fifteen years left, instead of the usual twenty to thirty. Lungs badly scarred, the specialists say. Don't know how they can tell, without an autopsy. But I don't feel too bad, as long as I take it easy, drink pure or simply, and breathe cleanly— especially in the mornings. Can't afford to lose this.... time.

HEROFORD: Sorry to hear of your affliction. But I'm glad you're not bitter about it.

LANGLY: Never have been, sir. No wrong was done to me, not on purpose. I was simply of my element, and relished the labor. We all have our jobs to do, in that sense, I suppose. I mean daring disasters to strike, while you're living it up.

HEROFORD: Being resigned to one's fate must produce a certain amount of peace and closure about life.

LANGLY: There is a certain amount of comfort to a terrible condition, when you know there's nothing you can do about it. If you lose a limb it's not gonna grow back. Seen a lot of that in my time. You stop feeling sorry that it won't.... (sips) This is tap water, sir. Not what they feed the customers. But I appreciate it more highly, because it comes from.... friends— friendliness. His nephew's.... not so much. But I can understand. I'm not the sharpest dresser. I don't want to discourage anyone from sitting out here, when the sun heats the space up to a temperateness. He wants to attract a young, mobile crowd, to enhance the appearance of the café, and draw in more fine people.

HEROFORD: Well it could do with a lot more color decked out here, for my tastes. Too metallic looking. Too.... electric, perhaps.

LANGLY: Consumer appliances are much in favor today, especially the portable kind.... (sips) for.... entertainment, information.... and cultural ambiance, to be up to date and current. I haven't any of those devices. I don't carry anything electronic. Even prefer a wind-up watch, almost an antiquity of technology today.

HEROFORD: You don't have a television?!... at your apartment.

LANGLY: No, sir. The commercials annoy me. I got totally fed up with TV while in the hospital, and discovered I didn't really need it, since the individual housing units were not supplied with any, other than what you brought or purchased yourself. Funny how they have you watch that stuff while in a convalescing bed, but it's no longer thought important enough during your more private occupancy. I don't miss it. When I got.... home, I seldom turned the one I owned on any more. Had sort of a revulsion to watching, though I liked the equipment. Noticed a poor preteen— I think— moping in the communal garden of my housing nook one day, with the saddest face you ever saw. Asked him why he was so unhappy on such a bright day. Said he wanted to watch a movie on TV, but the family only had one box and his grandmother had already conscripted it for her regular soap operas. He asked for an exception this one time, but she refused him. May have been the first time he felt that from her, for something "really important." So the boy was totally depressed about it. I think the movie was some sort of "Tom Sawyer" exploit, if he explained it to me right; totally dismissible. But he was so sorrowful about himself. I felt something fundamental to his character was just occurring to him, that he was stupid or ungrateful, or maybe even evil to ask at all, to inconvenience probably one of the few daily pleasures of an otherwise much bothered but thoroughly beloved.... guardian. I lent him my small console. Said he could keep it as long as he liked. Never asked for it back, though one time I think he sheepishly wanted to know if I'd want it returned. Could never quite understand his expositions completely— probably due to his schooling, and that generation in general— But I swayed off the attempt. Don't miss it at all, for what it does.

HEROFORD: Very nice of you. I always prefer children work for their presents, in some way, like I did. The wife's more permissive, though.... But you must have a radio for yourself?!—

LANGLY: That I have, sir. And I use it daily. Not on continuously. Silence is more precious to me— silence for the solitude. It's on when friends come over for cards. But the station I really like is difficult to receive, during the day. And the news?... well that blares more like advertisements these days. (*sips*) I have it on usually for less than an hour at night, to help me get sleepy. When it's dark and you're already in bed and you can hear the radio, softened for the static, it's easy to get drowsy, turn it off and fall asleep. There's no wasting of my patience, and no expectations to worry about. I don't have to be aggressive about anything anymore, even company— even romance. I am a done person due his.... retirement, and don't have to prove a thing like I used to. It's no longer important, the body doesn't crave for status.

HEROFORD: That might be due to the medicines you probably must take continuously. They dull and dilute the senses.

LANGLY: You have medical training too?!

HEROFORD: Pharmaceuticals, pharmacologicals?... Well I know when I'm on prescription drugs for the flu, I don't feel like much for doing anything. And when I take something for a cold, even though I can stay busy, for awhile, I certainly don't want to engage in any fun. But if I'm not taking the drugs (while ill), I might attempt to, to a limited degree— Eating favored foods, for example.

LANGLY: ....That might be right, sir. My appetite's become quite banal these past years. I've not wanted anything even remotely exotic or exciting. The taste buds simply don't respond to the spiced up offerings. And my regular regimen of facilitating pills and tablets, unless I really fall to sickness, suffices for any surprise in diet. The most common foods seem warmly enhanced to be dear to me. (*sips*) But they're what I can amply afford, mostly vegetables and fruits. I'll not do without my chicken curry for long, though, the one culinary vice I've left myself from earlier times to enjoy. Fried rice with shrimp is just too expensive these days.

HEROFORD: Hum. Oh, I like that kind of meal also, on occasion. But a growing family always prefers beef.... baked potatoes, and perhaps some string beans, if they're not too saturated of oils to a dinginess. My youngest threw up the last time we fed him rice. It was prepared too complicatedly, and mixed up with carrot cubes and peas— I tell my wife to make it simpler, for everyone. Clean white rice with butter is all that's needed— for the rice. The other, more colorful vegetables can be off on their own— Then you won't confuse the kid. He said some of it reminded him of bugs mashed up—

LANGLY: May I ask, sir—

HEROFORD: —Their imaginations are more sophisticated than their palates.

LANGLY: —if you're as well off as you sound, then why are you dressed so slightly, this cool morning?

HEROFORD: Oh. It's not breezy, though. If it were breezy I'd be uncomfortable. You know, kids play games with their foods, trying to make the most revulsive looking the most delicious to eat, or

visa versa—

LANGLY: *Why*, sir?

HEROFORD: —Licorice bits for ant heads, that sort of thing.

LANGLY: ....I thought at first you might be an unfortunate panhandler. They sometimes dress sharply but weakly, slightly inappropriate for the weather. It's what they can gather up as clean, I suppose, however they may find their supplements of fashion. But since they're often young, at least they have a sense of what's current. But.... you're not one of those—

HEROFORD: Them? No. Oh, I rushed out of the house, this morning. The weather report said it was going to be warm— Hot! That impressed me more than being substantial for my clothing, I mean putting on a coat. Was in a hurry to leave, and clear the head of sounds, those.... dwelling-worn sounds of family life. You need a solitude, sometimes— and isolation, for a few minutes.... like *you* need some early air.

LANGLY: Might a back garden do? Must be a pretty one, large enough for camping.

HEROFORD: My estate? It's a fairly good size, for this city. Not ostentatious, but individualistic enough to be uniquely owned. I'm not rich. It's mortgaged and I have taxes and debts. But it is properly owned.... I don't want to be chained entirely to that scene— not all the time. It's.... mostly for the children. Children should grow up in homes, distinctive houses, hopefully. Builds up a self-determination for one's character. Don't like the way kids from apartment complexes have to live— They're wayward thinking.... unless those apartments are shaped like houses.... Town houses are fairly O.K.— but they should be gated.... or restricted in both location and access. I'm thinking about this for the children.

LANGLY: Yes. (*sips*) Well, they can grow up virtually anywhere, even under bridges. We found a squatters' camp once, under an old abandoned land bank bridge. An entire micro-community: men, women, children.... sages, scoundrels and saints. Had to forcibly disband them, because a nearby demolition would probably affect their area. I groaned with my.... tinkering on their lifestyles. Thought those kinds of associations were not permitted to exist anymore. Certainly not today.... Left a lot of discarded clothes, along with a great deal of trash, in their rush not to be bulldozed by punches and sticks.

HEROFORD: Well that's kinda sad.

LANGLY: They resettled within viewing distance, but out from under that bridge. Sure was lucky it wasn't raining. But it wasn't warm. A precarious time of spring.... I don't think any from our crew alerted the city authorities about them.

HEROFORD: Housing is eventually provided for the destitute.

LANGLY: (*sips*) I don't think that's by law. (*The other **waitress** comes out, to approach **Heroford**.*) I'm almost done.

WAITRESS (*to **Heroford***): Are you sitting to place an order? If not you can stay awhile, but not too much longer. Gives this spot a sense of being active, with early vibrance. But when payers start arriving, you should discreetly slip off—

HEROFORD: I told your table setter—

WAITRESS: —the deck.

HEROFORD: —I wanted a.... warmed chocolate drink.

WAITRESS: What? A molche, or a cappuccino?

HEROFORD: That will do, I guess. A big one. It's kinda cool out here.

WAITRESS: You can sit inside, if you're ordering.

HEROFORD: No. I want to be in the air.

WAITRESS (*retreating*): We have plain ordinary coffee too, you know.

HEROFORD: I can make that at home.

WAITRESS (*just entering*): First class brew.

LANGLY (*standing*): Well, sir. I've enjoyed your eccentricity. But this water was delicious nonetheless.

HEROFORD: Well I'm hardly that, fellow.

LANGLY: To actually buy a drink, just to show off— to no one?! I think it's because, when I have my tap water, to take down pills, it by necessity tastes funny, first thing in the morning. So I look forward to theirs. It's almost like a resolve of temptation.

HEROFORD: So now you're going back to your flat? a dungy one?

LANGLY: No. I'll walk around awhile, for my constitutional. The strides make the breathing more difficult with pace, but also much deeper to get in this good air before the traffic heats everything up and the day turns more cumbersome outside. If I stay cooped up too long in my apartment, the whole afternoon and the night, I can hardly breathe at all. Some muscles have to be exercised, to keep the whole apparatus in shape.

HEROFORD: You should get an oxygen ventilator.

LANGLY: Unfortunately I can't afford that luxury. (*starting to leave*) And neither should you make them for yourself, what's unnecessary. Proves nothing anymore—

HEROFORD: No! Wait, sir. Would you like a real chocolate drink, or some of their best coffee? (**Langly** *stops to consider him.*) I'm for real. Come over here, and relax with something warm. (**Langly** *cautiously approaches the table.*) That will really clear out your lungs. The mucus melts away with heat.

LANGLY: I doubt that.

HEROFORD: But you can't afford to make yourself a little Java every morning?

LANGLY: Actually, that would interfere with my medicine, for a short time—

HEROFORD: Sit!

LANGLY (**sitting**, *after pulling up a chair*): About half an hour after treating the tongue to a chalky bitterness, I shouldn't have any rich brews in the stomach. Supposed to hinder the absorption of the drugs, the caffeine, cocoa and the like. Probably just competes. I try and time these things, to enable my walk, so I can come back to something salty, like a small bowl of soup.

HEROFORD: And then?...

LANGLY: And then it's to sit in a comfortable chair and stare at the walls and the window until I feel up for some reading. If the weather's bright, I go out to the garden nook, and try to animate my mind with what's being written, or what's been as I discover it. My (reading) comprehension's not as keen as it used to be, in the school days. Years of non-practice, I think. But a magazine with lots of pictures.... descriptive ones, will do. Newspapers are too expensive, and wasteful of the paper and ink, often on my fingers. I'm afraid novels tend to bore me; I can't concentrate enough on the story, or whatever the plot's supposed to develop into. If I'm feeling particularly healthy, during my strolls, I'll make it down to the public library, the closest branch. But I can't reliably take out anything. Meet friends there, though. Guys older than myself. They say I've "gotten out." That's when I can invite them back (home) for cards. Used to be able to play at the library, but not any more. Say we get too loud, and attract too many kids watching; though those dears are allowed to play their checkers an' chess there, in the stalls.... I suppose we can too. Feels too confining. The park's better, but that's some distance for me to walk. Only occasionally get up there.

HEROFORD: But you're alright now, for some coffee. There's been enough time.

LANGLY: I would say so, if you're offering.

HEROFORD: Digestion is a continuous process, and shouldn't really stop for a comedy of aims.

LANGLY: What?

HEROFORD: I mean, you can build up your thirst in many ways.

LANGLY: For life? Well that's well spoken for. (*The first* **waitress** *comes out to retrieve* **Langly's** *glass at the other table.*)

WAITRESS (*noticing them together*): What's this, then?!

HEROFORD: I've invited my friend for a cup of coffee, please. (*to* **Langly**) Is that all right?

LANGLY: It should do fine. The name's—

WAITRESS: If it's to be creamed, we serve it creamed: plain, vanilla, or mint menthol. Or you can ask for some artificial creamer
—

LANGLY: That's not too good for me.

WAITRESS: —We don't leave those packets out with the sweeteners. (*retrieving the glass*)

LANGLY: Well.... I'll have it.... vanilla creamed.

WAITRESS: Will do. Heavy or light? (*retreating*)

LANGLY: ....Er— Heavy.

HEROFORD: That's good. Heavy is good. It's what you want, for the taste.

LANGLY: They used to have some "orange blast" confection I could never get settled with, back in the day.... The green mint menthol's much better.

HEROFORD: Novelties for the young. Though I can't speak much for their preferences. It's not for me to understand them too well. I liked candy, when I was a kid, and couldn't wait to become an adult so I could buy as much as I wanted. And I couldn't stand any form of medicine. (*The 2ⁿᵈ **waitress** comes out and approaches with his cappuccino.*) Today the young seem to revel in illicit drugs and nasty tastes, anything that induces deviant and stark sensations of hallucinations on reality. And they just don't want to grow up out of that malicious behavior and try to search for any valuable responsibilities any more. (*The **waitress** places the drink on the table, by him.*) They actually shun work (*getting out his wallet*), when I thought it was exciting to discover. (*takes out a credit card from his wallet and hands to the **waitress***) Add my friend's coffee to the tab, please.

WAITRESS: Very fine, sir. (*impressed*) Will you want the receipt?

HEROFORD: Well, sure. (*The **waitress** retreats.*) Not that I need so much paper to carry around on me. But I don't like giving anything up that's earned.

LANGLY: So you *are* an interesting fellow. I suppose it's possible that you feel slightly harried at home. (*as **Heroford** starts drinking*) The one happiness I've found, in all of my situation come to, peradventure, is that I'm not upset with my home life— not anymore.

HEROFORD: Aren't you kinda bored at times, staring at the walls and window?

LANGLY: No. That comes after a real effort. These are my exertions, the morning escapes, what I can accomplish now that's significant. And my reward is a rest in solitude. Can last for several hours, sometimes. Well into the afternoon. I'm not sleepy at all, nor tired. That's an experience I'd never had before, when doing nothing. But I've learned one can do nothing just as honorably as straining to achieve, when the rest is warranted, when it's not due to laziness. If your circumstances make you lazy, that's a shame. If your society makes you so, it's a crime. But if you've earned some inaction, that's not wrong— it's what you've won. The real problem is how to fill up the rest of the day afterwards. I can only read so much, before distractions command me not to pay attention.

HEROFORD: Well, you must have done a lot. You can write about your recollections, and produce a memoir.

LANGLY: I don't want to drudge over the unimportant, and remind myself about fuller times. They're never coming back, however I may miss them. Hate feeling sorrowful. It's a feeling, you know— not a thought. (*picking up a sugar packet from the table's bowl*) A pinch of sugar can remove it from your spine.

HEROFORD: There's that materialistic drug culture coming through. (*The 1ˢᵗ **waitress** comes out with a cup of coffee.*) I can guess they feel like you, if that's in any sense an inspiration (or aspiration), but without any history to substantiate their complaints and requests, the young dopers today. I fret for my own boys.... But schooling should be well enough. You should see the homework my eldest has. Far more sophisticated than what I was ever cornered with. Doubt if irrational limits and their numerical expressions can make much difference to the building up of character.

LANGLY (*as the **waitress** offers him his coffee*): Hum.... Could be preparing him for the calculus. Thank you. I was never quite given that opportunity; though I tried to take up some of it on my own— and failed. (*starts sweetening his coffee*)

WAITRESS: Here's you receipt, sir. I hope you don't mind if we've given him the best.

HEROFORD (*accepting his card and receipt*): No problem at all about that. (*folds up the bill without looking at it, while replacing his card and wallet to pocket*) As a regular, he deserves your finest, sometime. (*puts the bill in a shirt pocket, as the **waitress** retreats*)

LANGLY (*stirring his coffee with a little wooden stick provided*): Though I'm not as thirsty as before, coffee solves a lot of problems. It provides the stamina my walks often confer. (*sips*) This is delicious of a memory. At least they haven't changed that.... standard. When Bruce died, I couldn't bring myself to spend money here any more. Didn't seem right. Back in the day, this was a common stop for much of the crew, bantering with that old guy.... Now I really can't afford their prices, regularly.

HEROFORD (*sips*): But you linger on to some relation with the past, by frequenting this place.

LANGLY: It's too familiar for me to ignore. (*drinks*) It is my morning's expectation, a line of precious string.... but not a dependence. I've taken up drawing sketches in an artist's pad. Horrible stuff! But it can idle away some time. Used to do it for real on the job, to help me figure out my mode of attack on a project, to design what should happen, how and when, for a specified order of events. Was good to do; so now I do it for fun. Haven't any training, though.

HEROFORD: Well that's nice. Pastimes don't have to be formalized. It's a problem I have myself. I don't know how to waste time. There are more than enough parties, brunches, and outing occasions with my work to be entertained by. But aside from work, there's not much about me that's creative at all. The perception to want to be, during leisures, seems almost driven away from my careful consciousness. And I refuse to undertake facetious, silly dangers, with the responsibilities I hold. It's this not being busy that stokes my awkwardness, some mad daring portrayed. I have many coats, jackets, windbreakers, sweaters.... Didn't think of any of them. Just ran out. Though, if I were the type to wear a hat, I would have put it on, I think. People don't wear them much. Not the men, anyway.

LANGLY: Had to have a plastic cap many a time. Almost like a helmet.... Almost like a uniform.

HEROFORD: Yes, but I mean, those things don't dignify a person like they used to, like when I was a youngster and was given my first social hat, a step beyond the cap stage. Well now that entire stage of events of formal advancement has collapsed. No longer exists for any notice. Was not so necessary, even though it fascinated.... my eyes, at least. (*sips*)

LANGLY: Trends in fashion are still very important to our cultural assessments. Ask anyone on the go, today, and they'll tell you — First! you must dress properly.... to make any favorable impression for contacts and groupings. That hasn't changed, only the manner of what's acceptable. (*drinks*) Well it seems like you've gained your expectations, or good judgment. The sun is out, and we're warming up nicely. You won't look so unusual soon, while I may seem peculiar, unless I limp about like an old dotard— which I am not. But I must work up my constitutional, to be deserving of the day. And I thank you for offering me this.... sublime beverage. Brings back much. Some days are fallow, others incredible. (*standing*) Each a gift. I'm sure you'll find something to do.

HEROFORD: Probably will just go back home. They all have their activities planned, directions learned to. I feel.... rather more divested of interests for the time being. The feeling is not comfortable. I'd hate it to become.... permanent.

LANGLY: There are some nice museums around. They all charge for entry, though. But the park remains free, with its pond and streams, and heavenly, lackadaisical banks. You can spend much time loafing up there, without looking ridiculous at all. Old bridges and walkways leading to.... wherever you're at. (*starting to leave, as a couple approach to enter the café*) Have a terribly nice morning, sir. You're very gracious.

HEROFORD (*responding*): My pleasure. (*sips*) This must be a rather popular shop, I've just run to without knowing anything. But nothing's done, on a work week's day. And where are each off to, but some summary care of purposes found. Maybe I should have bought him a whole breakfast, but he didn't seem too famished. Must have mine. Betty must be startled that I didn't take any.... from her. Not hungry now, though. (*sips*) I'll go back and have some. Sure it will be left over for warming. She planned out the servings before my surprise. But I startled myself. The notion just shot through my head to get out, stagger along for some routine to find.... that's more correct for what I'm trying to do, these impatient days. It's not that I'm afraid, of what's occurring. (*notices as a man with a newspaper comes up to a table, plops down in its chair, rummages through the sweetener bowl to pocket a few packets, slaps the table top for some attention and then starts reading; Heroford speaks too softly to think he's being heard.*) Did you really have to do that, steal some sugar for your lunch's tea?! (*The man looks up, but to see if he's to be served.*) Well, if that's the type! (*sips, as the 2ⁿᵈ waitress comes out*)

WAITRESS (*approaching the man slightly warily*): What will it be? Mr. Pomelory.

POMELORY: Bosc! An' tear jerked eggs, as usual. I want those yokes running. You do it so well here. (*with some disdain*) I can't teach her.

WAITRESS: The eggs with sausage, yes. Ten minutes. (*retreating*)

POMELORY (*reading*): Don't bother to filter. (*looks over at Heroford for a brief glance before returning to his reading*)

WAITRESS (*noticing Heroford's table*): Oh, so he's left?

HEROFORD: Yes.

WAITRESS (*coming towards*): May I take in his cup?

HEROFORD: With certainty. I think he drank it all, and with some delight.... You really should serve him some treats more often.

WAITRESS (*picking up the cup*): The store doesn't run on free treats. I'm not paid out of largesses, and the café isn't formal enough for tipping.... Although sometimes the chief gets a bonus, to stay on.

HEROFORD (*leaning towards her, with lowered voice*): Tell me, please. What's "Bosc"?

WAITRESS: ....Oh, that's essentially a heated raspberry milkshake, the suspension brought almost to a curdling. A specialty of the house. Is your cappuccino O.K.?

HEROFORD: Yes. Quite nice.

WAITRESS (*retreating*): Fine.

POMELORY (*places the newspaper on the table and stares at Heroford, who after awhile begins to stare back*): We showed them how to make it. I have influence here.

HEROFORD: I should say so, with what you're allowed to poach.

POMELORY: It's a family tradition, you know, the Bosc. Only we tended to make it with ground pear preserves, or sometimes apple slices. Nothing went to waste which was stored or stocked. Often it's all we had in the morning, (what) was that hearty drink, and a biscuit buttered. Kept us till the noon dinner, in the original fashion— for meals. Labored much till then, in the orchards. If you don't mind me asking.... Do you work?—

HEROFORD: Most certainly do!

POMELORY: ....Well that was a time when work was much to do, and proud to be done. But I was a young man then, and agriculture was like sterling. You could actually live off of your just deserts. Not have to slouch around at cafés and bistros.

HEROFORD: I'm not doing that. But I'm on vacation. (*sips*)

POMELORY: ....Oh. Yes, I see. These packets of sweetener. Well I deserve a few. I'm a regular patron of this establishment, and they value my coming. It's not like I'm stealing anything, with what I've done for them, and continue to—

HEROFORD: Do *you* work?

POMELORY: ....Of course.

HEROFORD: On your way to an office job, I suppose. Reading a newspaper and all, you don't have time to otherwise, except now.... or at the end of the day, in which case it would seem less like news, if you're the type who likes to keep abreast of things and controversies. I never read while on the job, not trivialities of information.

POMELORY: You can learn all of that from the television and radio broadcasts, the trivial. I demand more studied analyses and informed primary opinion making. Actually, my office is at my home, being semi-retired. But I prefer to breakfast here often. Simple dishes, but they know how to do it right. My sister never learned properly, the artistry involved in having a meal taste good. The foods are not enough— proficient cooking and preparation are also necessary. A commercial establishment must master this forthrightly, or go out of business. And I've seen a few do just that, offering their customers toast burnt on the edges, or bacon that's too oily, or hominy grits too ridiculously salty for most tastes to tolerate, and even fouled tomatoes on occasion, the produce aged too long. Our family farm provided many restaurants, grocers, and supermarkets their best offerings. Now I run a portion of the shipping from my home— a small portion, just to keep my hand in it, make sure fine methods are being followed, and fair practices kept. Sons handle the real work, now. And it remains just as hard, despite the mechanistic advancements.... Hard to garner a decent profit, I mean, with all of the conglomerate competition that abounds. But that's their problem to deal with. We had ones just as dire to push through, and leave them some starting advantages; although ours concerned more the crop production: yields and health. Guess that's what separates the generations, the different difficulties of importance.

HEROFORD: You can't cook your own breakfast, or your wife? — Oh! Forgive me if you're a widower.

POMELORY: ....Never gained the learning, to do a spectacular job, which I prefer now. So do my brothers. We've earned this.... relatively simple indulgence, to be feted well— or professionally— for some meals.... The Mrs. has the talent. But after much of a lifetime doing so she refuses to do more. She's earned the rest, and is slightly invalid with flare ups of the arthritic fingers and wrists— Though she'll teach the interested.... My sister can't be educated well enough. She tries, but takes after her brothers too much, like myself. Always has. The family stays together, though, empirically. She lives with us. Divorced. *He* might be dead. Doesn't speak much about him, so I don't really know, other than how she senses it. But that's what sisters do.... No. I want things done right, if they're to be done for me. I've always strived hard to do right for others. No other way to go about life.

HEROFORD: Well, you can be fed up with the routine.

POMELORY: That has never occurred to me, in an honest way. (*picks up the newspaper, to go about more reading*)

HEROFORD (*watching; sips, with a sound that he's almost finished*): Things swell in the head, that you're tired of the usual. Need a change of atmosphere, for awhile.... My wife cooks fine, and I'll not have a breakfast here.

POMELORY (*offhandedly, while reading*): That's most casual of you.

BECK (*coming out as a customer is approaching and **Heroford** is standing*): Hello, Mr. Pomelory. (*coming towards*)

POMELORY (*looking up*): Beck. Everything working out? You clean up well? (*The customer enters.*)

BECK: They ain't complaining. How soon do you think I can approach—

POMELORY: It's hard work, kid. They want diligent workers— I know my boys. And with some experience at doing *something*. Bring them a good reference, and they might try you out. Show that you're a conscientious employee, or can be. Maybe in a couple of months.

BECK (*at **Pomelory's** table*): Wow. That's long. But thanks for asking the café to hire me.

POMELORY (*as **Heroford** feigns non-interest, to slightly notice them*): I still think you might be a distant relative, the way you look. Shouldn't have gone to them directly, without anything to offer. Yet you picked me out to call, after noticing my number, on a desk blotter.

BECK: Oh, I've been trying everywhere to get work, any lead I can find.

POMELORY: But that was astute. Must have shocked you that my office was a residence.

BECK: Did seem a bit strange, when I picked out the house. Inviting me up for a chat, though, was too much of a chance to let go. Never know where I'm headed, in all of these attempts. And your name looked like the big man of the firm—

POMELORY: The big Mo, the mogul.

BECK: The pressure of an intuition speared me on.

POMELORY (*as **Heroford** starts to leave, just finishing his drink and placing down the glass*): Well, there are several of us. But I'm the eldest for the advertisements. And I'm not really in charge anymore, as I told you. In about two months a serious harvesting season starts up, and we'll be sure to want stout laborers. It's back-breaking work, though, to learn— But if you do, if you master it enough to refine your capabilities at it.... then you can teach others, if you catch my drift, and move up the chain towards greater responsibilities. That's generally how we do things, with anyone who shows some initiative and more potential. The job is so particular, though, at the starting level, that a spell (of time) is needed to determine, for yourself and others, whether you're suited to it. Might not be healthy enough, for some reason, or allergic to the organics and pesticides we're compelled to utilize today. Can't afford too many insurance risks, so a trial run is always enforced, for pickers, stripers, pruners and packers. (*The 1ˢᵗ waitress comes out, to retrieve **Heroford's** glass and tidy up that table a bit.*)

BECK: Well, sure. I can understand that. Haven't been sick lately, and don't know of any sensitivities to.... common, earthy chemicals.

POMELORY: That might show up right readily on the skin. You work with your hands a lot, you know. Really going for the fruits.... But you look like you're leaving, with your blazer (jacket) and all.

BECK: Off to another job site. Temporary job, helping to gutter an abandoned house. Not sure what I'll be doing, except moving a lot of trash around, in dumpsters. They won't need me back here till evening.

POMELORY: Well, I hope you wash up well. And be careful of those old ruins. When they hire non-contract or non-union labor, the work can sometimes be.... with great need of suspicion, as if hidden dangers are rife they don't want their trusted to handle, those unscrupulous contractors.

BECK: Oh, we're given dust masks. And the pay is in cash. Don't have to deal with so many checks, and their processing fees, just for a few days' work.

POMELORY: Well, open bank accounts, and establish some savings, if you can. (*as the 2ⁿᵈ **waitress** comes out with his breakfast*) What you really want is credit, to bide you over emergencies. Those things happen, often at the worst possible times. And credit only comes with savings.

BECK: That might be possible, some day. Not too practical at the moment, though. (*as the waitress serves **Pomelory***) Can't handle debts at all. Have to get things done very transparently, without much complication.

POMELORY: Thank you, dear.... (*starting to eat, as the **waitress** retreats*) Build up a relationship, with your service providers, as I do here.

BECK: Well I'm more of a provider than ever a receiver of services. And that should remain until I'm decent, financially.

POMELORY: Oh, I don't know. Aid's meant to make one whole, it's said, as far as decency is concerned.

BECK: I can't see it so clearly, to become that dependent—

POMELORY: Everyone is a dependent to his society. It might offer you opportunities to recognize and seize or scramble with (*burps*), but there are always strings attached, so that a generalized social benefit evolves, and a clean environment— for ideals of achievement— persists. No good feeling your position in the world is ordained by anything other than your harmonious attitude to mankind. Thus, civilizations are formed, even out of the craziest foundations and fundamentals of behavior.

BECK: I just look for work.

POMELORY: Don't make that a religion in and of itself. Assess your capabilities and interests, while you're scrambling. And always work to be trained. Because what's positive about it is mainly what you learn during the process. So many others can do the same job, so you're not unique for the effort; and the results are the property of those who've hired you, essentially society at large.

BECK: I see. I'm not working for myself. I'm learning for myself.

POMELORY: Right! Can take a lifetime realizing that. Whatever fortunes or misfortunes you accrue, the only thing worthwhile is what you know. The rich can be terribly miserable for their ignorance, with a deep-seated sense that they should be wiser given their status and plaudits and material attainments and social rank. While the poor can feel exuberant of satisfactions, while managing through their destitutions, sensing they've learned just about all they're capable of given their situation, that they are at least masters of it, for themselves and their charges. (*drinks a little*) You can't fake the pleasure. It's either there or not, in your bones to perceive and reverberate with. And you can hardly care how others consider you, praising or disdaining— either of which can lead to much depression if you *do* consider it important, to have enough of or reduce thoroughly. You just work to know. Often in the midst of valuables or extreme security, I've sat dissatisfied, because of an ignorance: How to do something. How to get someone to do something— right! How to create the effect that I want. How to have the world seen fruitfully. The unease can become debilitating, no matter your wealth and comforts and respect owned. That's the way we are, saddled with sadness— not depression, which is a disease, but dissatisfaction— until we have conquered some personal unknowns: Few for some, many for others. Till then, misery despite joys.

BECK: I'm not quite sure what you mean—

POMELORY: Misery is the cloudy fog of obscurity whose transient hole('s) of clearing leaves but what is joy, simply the absence of this substance of misery.

BECK: Well, if it's striving to survive, then I can understand. There's no misery to it, as long as you're active.... learning how, I guess.

POMELORY: Yes. Participation is what makes for a liveliness.

BECK: Sure. Enjoy your breakfast, Mr. Pomelory. I'm not in a rush. But hard labor comes in about.... an hour and a half. (*starting to leave, as other patrons arrive*)

POMELORY (*still engrossed in eating*): What will you do till then?

BECK: The park's a nice place to warm up at, with the sun's fair heightening.

POMELORY: Then enjoy the day, Beck!... (***Beck** has left audibility.*) Though I would shiver in there.... with such prospects. But I'm not a young, hopeful man any more. I'm a dying wart, well enjoyed of the skin. Yet, these eggs are made great. So I'll just continue to swell for awhile. This pleases me. (*Two men take a table adjacent, one pulling up a chair from another table.*)

MIKE: I hope they don't mind if we sit together. Silly to restrict one chair to one table.

JOSH: Probably fine to do. (***Pomelory** casually looks up and over at them.*)

MIKE: The deck's not set out that well anyway. People won't want to sit so isolated, out here. They'll want to be grouped together, as friends socialize.

JOSH: This is just the starting arrangement. The customers set things right. Don't worry about it. Proper order comes with usage.

MIKE: Maybe later in the day the heat makes you want to sit alone. Should have some table umbrellas, if that's the case.

JOSH: Don't be so picky, Mike. This is more or less fast food. We're not gonna stay here all morning. You're too nervous about nothin'!

MIKE: Well I'm just observing the probable disorder upfront. Doesn't speak of neatness.

JOSH: But of utility. This could be a busy place. I'm tired of trying to calculate what should occur. It's not important, even if we have a hand in it, because this store's not so vital. Not much around here is. Everything replaceable. The charm is O.K., but that means little— to me. If they want my business, it better look pretty. The food can taste average, for average prices. I don't have to expect much. And it's fine. But it better look pretty. And if we move chairs around, so what?!

MIKE: Not done disruptively—

POMELORY: If I might say, you've already left two tables without chairs.

JOSH: ....No you may not! And we've only moved one chair— So you're dead wrong, mister.—

POMELORY: Oh. Of course you found one rightly, there. My mistake, sir. I'm sorry. But there seems to be much movement already, this morning, by the customers' designs. I hadn't thought the patio was so active yet. Well, this can be a popular stop, on nice days.... (*goes back to eating*)

JOSH (*more to Mike*): Maybe too much so. Everyone's business transcends privacy, in this unfashionable world.

MIKE: I just thought it could be objectionable, Josh. Didn't even notice that other table had two, like he didn't. Merely assumed things by viewing peripherally.

JOSH: Took for granted. We take things for granted too easily. That's how advertisers get us to buy so much junk, by implying what's not so and can't be so, like getting something for nothing, or for too little to be reasonable to sell. Even the small print's a phony explanation, these days, requiring professional interpretations.... for what's probably so.

MIKE: There was no one out here.... but him.

JOSH: You're too self-conscious about it. No laws are broken. They're not going to throw you off the property because of that. They want your patronage. It's not a great problem. But I tell you. Knowing everything is.

MIKE: I'd want some service for it.

JOSH: You can't hide anywhere without being found out, at least that you're hiding. Someone— or something.... has to know something about you, just to get through this society. And how much to know is growing and growing and growing, like a fat hog.

MIKE: Of machine and computers. Information rules.

JOSH: Back when I was a kid, I can remember retailers surveying the public, to determine what they liked and could sell. Now all they do is compel you to buy things— using psychological ploys and pitching to special interests with spokesmen or talking heads they think are designed to appeal to buyers: smug for the smug, silly for the silly, stupid for the stupid, and smutty for the smuttish. It's all thoroughly disgusting, how everyone is taken for granted, because of these.... programs of conditioning, or conditioned responses. I've a mind not to buy anything anymore, for a whole month.

MIKE: Can't live that way. You'll just make up for it with desperation, the next month.

JOSH: Well, that's my inclination to want to do. Need not be practical, nor possible. But I feel like I'm fed too much— garbage, and told I haven't got enough. Get some more, to keep the state running.

MIKE: State.

JOSH: The condition we're in. That's the only important thing now. Keep it up. Maintain it. Not us. Only our— buying, our purchasing, our participation. We're not important. Our mouths are. Our eyes are— our senses are. But *we're* not.

MIKE: That's the genius of commercialism. Makes a few wealthy, and more than a few satisfied, short term, like sharp impulses of quick stimulations.

JOSH: Oh, to the brain, like drugs. (That's) Why they're so popular.... I could do with some caffeine, right now. No, we're just bodies to be abused, pockets to be emptied. And soulless, senseless creatures become, left roaming the terrain of civilization, searching for what's been stolen from us. (*A customer leaves the café proper.*) We're being bred to be ghosts, Mike. (***Pomelory** burps.*) The substance of reality condenses on the marketing sheets of corporations, and incorporate entities of enterprise. (*1ˢᵗ **waitress** comes out, to approach*) There's nothing to cry for, if there's no pain to it.

MIKE: Or try to pardon—

WAITRESS (*as **Pomelory** starts coughing*): What would you like, guys?

JOSH: Two coffees, creamed.

WAITRESS: Oh. Regular?

JOSH: Yes.

WAITRESS: Regular, with regular cream— Are you all right, Mr. Pomelory?

MIKE: And a jelly roll for me, strawberry.

WAITRESS: Strawberry.

JOSH: I'll have a cinnamon raison scone.

WAITRESS: Raison scone. (*coming over to **Pomelory***) Got it!... Is everything O.K., Mr. Pomelory?

POMELORY: That's the darn best sausage I've ever had! You do it so well here. I tend to think more for the eggs, but that pork was great— salty! But just so right. Surprised me.

WAITRESS: Well they're in season now, I guess.

POMELORY: Pork? A glass of milk might do.

WAITRESS: Sure thing. (*retreating*) Oh! Cold?

POMELORY: Yes. Fresh milk, please.

MIKE: I tell you, man. Their jelly rolls are huge! Larger than your scone.

JOSH: Yeah. Well there's more sophistication to the dough of the scone. That's why it's called a scone, like a cake. Yes. Shops like this know how to sell good pastries. Not like the mean doughnuts we get on the job at noon. They're so cheap, they're stingy for the taste. But you've worked so hard by then, they seem fine to eat.

MIKE: A lot of them, those little, greasy things. They're an energy boost, for the afternoon. But they ain't gonna give your stomach cramps. Wouldn't like that when bringing down a wall. Can't afford to be distracted.

JOSH: That site hasn't had many accidents. Nothing serious. Just some muscle sprains and strains. But we're coming to the tough stuff to take on. Up to now it's been preparatory. Let's try and do this correctly, Mike. Don't stand under it so often. You're experienced, but every so often I get the feeling you're not as aware as you should be. If something comes crashing down on you, not much (i)'s gonna lend you protection. No way a falling scaffold. Those platforms are made with corrugated iron, for support.

MIKE: I know. I know all that. It's the codes that confuse me, sometimes. You have to get under there, to inspect the spot.

JOSH: But you have to time it not while we're bringing it down. (*1ˢᵗ **waitress** returns with some milk for **Pomelory***) Can't take chances like that. These staggered details can leave you caught. If you're not sure, ask what the others are doing. They assume you already know, or that the plan knows, the scheme of events worked out by the engineers.

POMELORY (*accepting the milk*): Thank you. (***waitress** retreats*)

MIKE: You mean their PCs. Why can't they give us the whole work score to follow, instead of dividing up chores and their obtuse timings?

JOSH: Because the schedule is not linear; it's filled with a lot of cross-overs, branches, step ups, delays and things we couldn't understand or rationalize through. You have to follow your own script of actions assiduously, because today some components of a building or a structure have to be removed or destroyed before others.... because of the way they were constructed originally. We're

not meant to follow the complications. That's all been worked out for us.

MIKE: It's not a skyscraper we're dealing with, Josh. It's just a three story municipal office building.

JOSH: A single head can't contain a decent project anymore. Too many laws and statutes and restrictions involved. And pedestrian and other traffic have to be protected. We have to avoid doing a lot of things inadvertently. Excesses are not acceptable— We damage the sidewalk just an inch.... and insurance liability premiums escalate a ton. But if you get hurt, you're left for what the work manifest says you should have been doing and when. Unless you can prove an error there, the error is yours. And that has been worked out down to every facet of our operating— You won't be able to get through any court unscathed.... or your life insurance enhanced by any contract negligence found. So stay.... as alert of others as for yourself, at this site.

MIKE: You don't have to convince me of that. (*The 1ˢᵗ **waitress** comes out and approaches with their orders.*) It just feels funny.... not knowing.... everything, nor anything.... too often.

JOSH: They even plan for mistakes and accidents, to lessen their impacts on others and the project. I've seen one of those lists, or operations charting. It's full of 'what ifs— then', like a computer program. It's a computer program! (*The **waitress** starts serving.*) Thank you.... A programing too obscure for our heads. (*takes out a wallet and extracts a credit card for the **waitress***) Put it on this. It's a company card.

MIKE (*as the **waitress** takes the card*): Thank you.... Look at the size of this thing!

JOSH (*as the **waitress** retreats*): But it's mostly jelly, Mike!

POMELORY (*almost through sipping*): You two are construction workers?

JOSH: ....Yeah! And what's that to you, now?!—

POMELORY: Nothing so particular to invade your suzerain of interests. I have a young friend who says he's heading out to a demolition site today, as a temporary add-on.

JOSH: Oh.... Non-union. They're mostly used to clean up the debris. Don't have a functional employment as far as furthering the project. But we can't stop to pick up all of the pieces while we're plowing ahead. The procedure's on a strict time course, and progress is already predetermined.

POMELORY: I told him to be careful about it. Shouldn't let him jump into holes, or do the dangerous stuff you'd avoid.

MIKE: Well if he's untrained, he won't be allowed to handle any machinery, and very few of the tools. In fact, he won't be able to do any actual demolition— nor construction. Would be against the law. You have to be certified and licensed. They're like manual labor, stockroom sweepers. (They) Don't get involved with anything we do.

POMELORY: But (the) dangerous is what causes one harm. And you guys may tend to avoid hardships or tediousness by employing the poorly initiated.

JOSH: We don't treat anyone slovenly!—

POMELORY: I know how organized efforts work. Ran an operation for many years. And mindsets creep in, have to be rooted out — when they're detrimental to anyone's well-being.

MIKE: We value each worker on the site. Can't afford accidents — They slow us down and raise the bills, the costs that accumulate. Can even end the day, for us, if they're serious. All bloodsoaks have to be investigated by city regulators, who grant a 'continuance of operations' permit for a studied, reviewed site—

JOSH: So it's not worth our while letting any freeloaders get hurt. We're paid by the hour. And if we're not working, we don't get paid.

POMELORY: That's a bit of a reassurance, if it's above board, having a monetary incentive to keep everyone safe—

MIKE: Well sure! Professionals don't like workplace disasters occurring. That's akin to being seen as sloppy with your methods and environment— working environment. And sloppy teams don't last, they're at least disbanded and dispersed to other memberships. While good teams command higher pay, enough to afford.... temporaries, to do some clean up. So your friend must be coming to good hands. Maybe even ours. There are several construction sites on going. But I think we're the only one working on a demolition today, around here.

JOSH: There's another in the next county.

POMELORY: No, I think he said he had to be at it soon enough that he could stay in the park till then. Not driving out to it.

MIKE: Then it's us.

POMELORY: His name is Beck.

JOSH: ....Beck. We'll be careful with 'im.

POMELORY: As best you can. Because he's a friend of mine, and I'm a friend of this city.

JOSH: Don't you feel just a little pompous about it, pal? I'm sure the boy can take care for himself, if he's been picked to work with us.

POMELORY (*after sipping*): Not at all.... young fella. But I'm not unfamiliar with the catastrophes you construction workers like to mitigate through, as being a part of the occupation sometimes unavoidable.

JOSH: There ain't no such thing. All catastrophes are condemned, and we strive greatly to avoid them.

POMELORY: So. And what about that poor member of your profession that had his lungs busted, with your common work?

MIKE: Lungs busted?!

POMELORY: By breathing in asbestos too regularly, for too

many years.

JOSH: ....Asbestos?! There are laws against that— You thinkin' 'bout Langly? That's years ago, man! We don't work that way any more. Special procedures, and even crews, are used whenever any sign of asbestos is around, as well as lead and uranium deposits and any minerals that have been defined as.... deleterious to health and ingestion.

POMELORY: Yes, I believe that's his name, something like that. The poor fellow comes around here early, for some clean water, the café owners tell me. And his story is not pleasant. Can you imagine how put out his body is, if he needs some water to drink, cleaner than his tap, and can't afford the bottled? or even a charcoal filter for his sink's faucet? Even home aquariums have that for their tanks! The point is, you types have left him that way, in not a good state at all, financially, apparently, if he has to come around here for a free glass, submit himself to the charity of an old haunt. So don't give me this spiel about caring for workers. You care for the jobs to perform. The city, state, and federal government makes you take notice for the care of the workers, just as they did me for my field laborers.

JOSH: If you don't work, you don't get paid!

POMELORY: But your construction work, that field is a lot dirtier than my work with the soil ever was. You take advantage of your worth, to ultimately devalue men.

MIKE (*to **Josh***): Who's Langly?

JOSH: ....A demolitionist from a few years ago. Many of his crew used to frequent this café. The custom sticks, somewhat, for today. Or at least I know about the place to favor.... He's on a disability pension, though. Sure about it! Wish I had that to slouch around with.

POMELORY: After a hard fought round of legal battles and challenges, my friend.

JOSH: Conducted indefatigably through our union, my pompous — sir.... or what became our union. So he still comes around here? Wouldn't recognize him. Never knew 'im. Not that much of a legend anyway, or legacy.

POMELORY: Early, as the place is just opening, or setting up. Beck helps with that, you know. He's.... industrious. I can avoid him, this— Langly.... but not his circumstance to think about, how you have destroyed a person—

JOSH: We've done nothin', mister! And that was in the bygone days. Nothin' to do with us. Committees handle those problems now. We know nothing about it(, what they do).

POMELORY: Bosc! You don't even care!... I mean how your profession can degrade one's health, and not only that, but the person himself—

MIKE: Doesn't happen any more!—

JOSH: Hey, look. It's some dangerous work. And accidents happen. We know that getting into it. These jobs have to get done. That's how we make a living. You can't blame us for all of these personal misfortunes, that are being prevented more an' more. An' Langly ain't no hero, coming from the days of lost limbs. No one has to feel sorry for him, and he would probably detest that (sentiment) like a foul taste. What remains is our individualism, to do what we want when we can. The struggles of the past might be a foundation built upon, but they don't count, not those particular instances that are over an' done with. (*Someone seats at a table, not near them.*) You don't have to know about the past, except to know it's been corrected for the future.

POMELORY (*finishing his milk*): You are *not* an individual.

JOSH: ....What does that mean?! Every person is his own individual.

MIKE: Let's eat this stuff, Josh. We have to get going.

JOSH (*as **Mike** takes bites*): Start on it. It's huge!

POMELORY: If I thought that way about my men.... I wouldn't have had much of a business to leave my sons. (*plopping down his glass to the table and standing with his newspaper*) That is a delicious milk. They do things here well. (*turning slightly towards **Josh** and **Mike**, who are eating, as the 2nd **waitress** comes out to attend to the new table customer*) Would you like my newspaper? (*flopping it down on his table with some referred derision*) Maybe you two can read that there are no individuals left, if any can be so contorted by their.... (*leaving*) values and outlooks!

JOSH (*as **Pomelory** distances*): No, I don't need to. (*reaches over and grabs the newspaper*) But if it's free.... That guy feels he's so high up. Well I guess they all do, if they have any supervisory capacities.... (*lightly perusing the newspaper front while eating*) Business to his sons, hun.... I hate these things!

MIKE: What? the—

JOSH: Newspapers. They print trash— They're tantamount to trash. And you find them as trash, in the streets and their gutters. What's worth to be printed? if it can be thrown away so easily, and (that) is done so often and readily. Then it's not important at all, while the headlines might be blazing for concern. Forced think. Would the world be different?... if this had never been printed. Do I change less?... not reading this type!

MIKE: It's all glorified gossip, hopefully based on facts. But the sports reports are sound. Crowds of spectators verify them.

JOSH (*absently*): Competition makes life real.

MIKE (*as the 1st **waitress** comes out to clean up Pomelory's former table*): ....We're protected from those kinds of hazards more thoroughly now, aren't we?

JOSH (*reading*): What?... As opposed to Langly's day? Of course. Why do you think they made us suffer through that absurd radiation class, and pass that multiple choice exam, the answers for which we'll never remember or retain, as if knowing the distinction between dosage and quantity is going to help us much before a pile of rubble laced with deuterium or tritium. They hire special workers for taking down old hospitals, but we all have to be certified.

MIKE: So that maybe someday we can enlist to the ranks of the more specialized and higher paid cadre. Though it's true, I don't remember much about atoms, ionizing radiation, and the different significances of Geiger counter readings and dosimeter exposures.... But I can go back to those thick notes the course gave us, for consultations, if I have to.

JOSH: Not gonna save you much when confronting the fact of an unfortunate spill of contamination. We'll still be the guinea pigs to correct matters, or see if the standardized guidelines for management of a clean up and scrub down are adequate, for a sanitation's security. But.... I don't think I'll ever get into that area.

WAITRESS (*while picking up **Pomelory's** remnants and wiping the table*): You guys enjoying yourselves?

MIKE: Yes, ma'am!

WAITRESS: Anything more?

JOSH: No. This is fine.

WAITRESS: Then I'll be back with your card in a moment. Do you need the receipt?

JOSH: Yes, or else we pay full costs.

WAITRESS (*retreating*): O.K.

MIKE: Sanitation?

JOSH: What?

MIKE: You said something about sanitation, for social security?

JOSH: ....Oh. Mike, let's face the facts here. As demolitionists, we are essentially high class sanitation engineers. Those old buildings and their intricate structures have to be brought down, and the resulting rubble removed. We're trained to do this safely and with a proficiency that can make them disappear almost by magic, given the time constraints. But our work comes after spending a regular spell in normal construction, entry level employment. After a couple of years some are seen as having promise, potential, and good heads, and offered training in this more dangerous branch, controlled demolition. Has to be done carefully and to the particular requirements of leaving a site more or less approachable to the public, health and safety wise; although it usually remains cordoned off with fences or walls, sometimes for a considerable while. You'd never guess it, by reading the newspapers, the care that goes into these activities, because they only report the results— of things— and opinions on these results, while we're responsible for *making* the results, fashioning the histories. But we are sanitation, of a mode of force to remove the decrepit or decrepit-able, or the unsuitable, or that for which it costs less to bring down rather than keep up standing and protecting or trying to use. Ours is a particularly profound notion for the society, the acknowledgment that some detritus must necessarily become from man's constructive labors, sometimes even of the most noble architectural art. And in the past our requiring, from inevitable results of functional and structural deterioration, was often confused with being a crudity and lowness of nature. Laborers seen as given justifiably foul jobs, just like we still view garbage collectors today. These newspapers don't remedy that misapprehension or prejudice. If anything they intensify the falsehood, that some people— with disesteemed character traits— are simply meant to do dirty jobs, as of a caste of society. That was apparently someone like Langly's problem. He was *allowed* to be damaged, over and over again, because it was felt his type— our type— deserved to be utilized that way, simply because we are destroyers, or metaphorically.... evil doers, if houses have hearts, beating hearts with sensitivities to persist like living entities of a city's ambiance and atmosphere. The dangers of constantly, regularly breathing in asbestos were well known for decades before his illness, from the health tragedies of mineworkers harvesting it, as minerals for cheap fireproofing— But *we* were not protected against its effects for a long time, until medical studies hinted at the detriment of free flowing air-born particulates (of it) for children of poor families exposed through cracks and gashes in their apartment domiciles and dwellings, broken walls, floors, and ceilings. (*as the 1ˢᵗ* **waitress** *comes out to approach*) Only after *we* complained and threatened were regulations changed. Then the newspapers pick up the story, to rationalize its components, as a tread of social conscience. We're still considered onerous to think of, and the poor (are) hardly even thought about but to be yet another burden signaling deficits that crop up in the social fabric.

WAITRESS: Here. (*handing* **Josh** *his card, with receipt*) Hope everything's O.K.

JOSH (*accepting*): Thanks. (*looking at the bill*) Hey, that's not so much. You deserve a tip.

WAITRESS (*retreating, with a smile*): You know we don't.... do that, here. Enjoy.

MIKE: Yeah. You have to fight for your rights and what should be decent: the work place, your pay and hours, health benefits. People just aren't correct for you unless you make a stand.

JOSH: Well, not fight. Langly never fought nothin'. He accepted everything, as part of the process of his living. The gears of confronting malfeasances eventually ran for him, to a degree. Grind out his current life. But we're still only a step removed from his tracks. What if you're really seriously injured, on the job— You'll be doomed! That's why you have to be more.... cognizant of everything that's going on. (Being) Careful's not enough. Training— not enough. Accepting.... the environments we worm through, that's what we have to do, and brilliantly.

MIKE: Can't.... plan out everything. Have to take as things come. That's why *we're* given the jobs, can deal with what pops up.

JOSH: Yeah.... Ain't nothin' real, in this paper. (*lays paper on table*) Don't want to read anything more.... Well, let's not turn this breakfast into a brunch. You can bring that with you, it's so— grand!

MIKE: Delicious too!

JOSH: So was my scone.... What is the lowly, Mike? You do your best, and still you're derided. I'd hate to *have* to buy a paper every morning, just to find out.... the world hasn't ended.

MIKE (*as* **Josh** *stands*): Curious for current affairs.

JOSH: I feel like I've been relaxing too much, just before a tough bout. (*stretches a bit*)

MIKE: Yeah. It's gonna be a hard day, to run through everything we gotta get done.

JOSH: He means that *he's* no longer an individual. He's a passive observer within a mush of civilization, hoping to alight on some news that brings him round to being— unique again.

MIKE (*standing, with his roll*): Well I certainly feel unique.... ready, prepared.... waiting.

JOSH: You don't have any shopping to do, do you? Let's head out to the site. I want to go over that procedure again. Some of the afternoon part 's not so clearly written.

MIKE: 'Cause it depends on how much we get done this morning. But everything has to be completed on time, or within its little schedule of achievement.... I used to love to buy a slight something, going into a store for some trifle, before heading out to work.

JOSH: You're peculiar that way, Mike.

MIKE: Makes me feel like I definitely have something to do, something to look forward to, for going somewhere afterward.

JOSH: Agrees with my early rising. I like to be a couple of hours awake, before getting on (to) site. That way you're really wide awake. Don't have to depend so much on coffee and stimulants. (*as the 2ⁿᵈ* **waitress** *brings a breakfast to a table, and a couple come to enter the café*) But this morning, nothing to buy?

MIKE: I have a premonition too, that we should get there early. Size up the effort, a bit, what strenuousness is going to be required. Shake that anxiety for the sledging off, out of our heads, and replace it with some anticipation.... Our local argued to do it this way, instead of blasting it down. Should be safer, for the surroundings. But we have to do it quick. Have most done for a day. It's just the size of project that can go either way.

JOSH (*as they're walking off*): Not really, as far as I'm concerned. They should limit dynamiting, explosives for the tall structures of too much girth to pull down easily.... You don't work, you don't get paid. Now let's be the lowly artisans, and do this with finesse.

Scene II — **Heroford** *is at a kitchen dining table, having some cereal for breakfast, the cereal box conspicuously placed near his bowl. He sneezes, as his wife enters.*

BETTY (*going for the cereal box*): Why'd you go out so early without a coat?

HEROFORD (*as she's reaching*): I might want some more of that.

BETTY (*taking the box anyway*): You're not coming down with a cold, are you?

HEROFORD (*as she's bringing the box to a pantry cabinet*): .... No. I didn't find anything prepared for me—

BETTY: You left! I'm supposed to guess at it?! It's enough to feed the boys without wasting food.

HEROFORD: But they had warm meals, I'm sure. A carefully prepared breakfast.

BETTY (*storing the cereal*): Ducan can cook his own, if he wants to. But as it is, I (make) ready for their needs, while you seem to refuse everything, like you're totally out of place at home. (*turning to face him*) The children snicker at your ineptitudes—

HEROFORD: They do not!

BETTY: —Well, your uneasiness and awkwardness. It's so contrary to your normal manner, the control you try to suggest— Stop it!

HEROFORD: Stop what?!

BETTY: Stop acting like you're unsure of yourself at home. It's only humorous to them, but I have regimens of a regular home life to perform, for all of us.

HEROFORD: ....You do it well, as a housewife.

BETTY: And if you start to act clownish, they'll take me for being a clown too— which I can't have.

HEROFORD: They're not laughing at me. They love us!

BETTY: But you seem unusual. They noticed the way you rushed out— for no reason. Danny asked where you're going, and I couldn't answer. He was puzzled, and I looked foolish. I don't like that feeling, especially before them—

HEROFORD: Lie!... Make up a story. Little kids are always inquisitive about the behaviors of their adults. Tell them I went to Timbuktu! if they ask too much. They'll get the hint, that it's not so much of their business, what their elders do on their own. I was expecting some bacon ready (made)—

BETTY: They want to feel secure, Harold, and naturally worry about you, or your feelings towards them.

HEROFORD: That can not be doubted in any way.

BETTY: You don't act like you want to be around us, for long.

HEROFORD: You're not around me! for long.

BETTY: But when we are together, for prolonged periods, you seem unsettled and uncomfortable, out of place and resentfully resigned.... to this family life.

HEROFORD: And when are we all all together?! It's only for unnatural moments, like when we suddenly bump up to watch some TV, in the living room— as a surprise. No, I can't enjoy it alone, as I had wanted. Naturally they'll be disappointed. They won't like what I came in to see. And what could they possibly view in the afternoon, when they're usually in school anyway?! They're just examining channels, to see what's on, or what what's on is like, while I'm planning to see a discussion show— or anything to past this time reasonably.

BETTY: Duncan will get his own tele soon.

HEROFORD: There's too much noise in this house. But other than the times when we meet up accidentally—

BETTY: Accidentally?!

HEROFORD: —they're off on their own at play, and you to your house(hold) chores. So naturally when we come together it seems unnatural. I'm normally at work. They're normally in classes. And you're here managing the household.

BETTY: The weekends are not so frustrating, Harold.

HEROFORD: I thought my vacation would be *like* one long weekend— But it's not. They have planned their activities, for the weekend, and stay out of my hair. But now it's a constant bother to notice them, they noticing that I have nothing to do, and can't command their attentions for anything purposeful. They better not act this way during our trip.

BETTY: Why suddenly now are you concerned, with some disorganized habits? It's summer and they're finding things to do. But there have been other summers, and you've been on earlier vacation leaves.

HEROFORD: They're.... growing up too fast, Betty. They are becoming individuals that I'm confronting— and they're discovering me, invading my privacies to see that I'm fallible of feeling and emotions. I have to deal with that at the office, placating so many tempers and haughty attitudes right down to the thoroughly downright unsure of themselves. And now such problems come home to me, when I want a rest from them. Children can be treated as children, with a façade of parental authority. But when they present real personalities to me.... I've become sensitized to this. It affects me wrongly— but I can't treat them as co-workers and superiors. I want out of this, back to the way it was before, before our youngest refined his tantrums towards anything serious. They're becoming of age to be people. First one, then the second, (and) now all three. And it's frankly become too riotous to stand for any length of time.

BETTY: ....Out of this.

HEROFORD: I let them blast their sounds in the morning, because that comports with your ritual for awakening and starting the day.

BETTY: Their radios are always kept to a respectably low volume. I've warned them of that and they consciously comply— happily. And mine.... is warm enough of voice (and sound) not to bother you nor anyone. It's pleasing, to relax the morning tensions.

HEROFORD: And I let them run through their childish gawkiness during the evenings, because that's how life develops, and they can't help the annoyances caused. But I'm away during the day, and that long span of time makes everything tolerable, as I put on my professional shields to handle the personalities with my own totally under control, calm, compatible, and complaisant towards progress. But I can't do that at home. (It) Would be blasphemous, and distasteful to me personally— And I'm here to wind down, as a home should be used for, and as a marriage should promote, during the quiescence of settling down to retire. But instead of winding down, I'm being worn out, with apprehensions— of my condition, of our comportments together. This is becoming too extended of a stay for my growing restlessness, to return to work, maybe, as what I'm trained to do.... or to escape some.... disapprovals of my truth, that I don't like in particular the personalities forming of this house. They are not.... harmonious to my expectations. They are individual and distinct, and are waning in stages of subservience.

BETTY: We've not had so major problems of discipline and behavior with our boys. They get into scraps and scrapes like all kids do, of similar ages— Unless you mean me. I've never been subservient to you. I handle you! to make for a decent family.

HEROFORD: You don't handle me! You won't even cook for me —!

BETTY: If I don't keep you from doing the most stupid things, making terrible decisions.... we might dissolve immediately, we might never have even had three children. You're a secretary. You're not the emperor class— You can't run your own business. And don't tell me you're tired of this life, or running into some kind of crisis about it. Because the race stopped soundly, when you found your true calling, which I guided your silly head towards, and kept pushing and turning and shoving until you clasped hold of the opportunities for good. It's what your schooling and training was meant to do. And I got you serious about it— and us— instead of dreaming of prospecting through fantasies for gold, easy money, and corrupted respectability. You can't be a Mafia don of the business world, Harold. You've too timid a mind and practical of your strengths to employ. And your early notice for some slight renown was due entirely to my suggestions. Before that we were going nowhere— I guided you to think properly, and clearly for your prospects of advancement. Now you're complaining?! as I struggle just as hard to keep this family solid, and aid our boys growing into fine men? Just at the start of the real challenge, for what matters, you weaken, and drift into rushing out. Well I can't stop you, to put on a coat before running into a cold morning, to heaven knows where or what, and bringing back a cold for all of us to suffer with (*He sneezes.*)— during the vacations! But you should be more thoughtful to take care of what we have and continue to prosper with, our union to be more delicate to handle. I won't let you be so self-serving in this house—!

HEROFORD: You'll drive me away. Men have left for far less than this squeamishness. We own each other, but not more.... our satisfactions found and wasted—

BETTY: Wasted?!

HEROFORD: —or rotting, through our bitter contempts, that we should feel more grateful for this assemblage of prosperity and privilege, while I can only fight a growing sickness—

BETTY: At that— hastening dementia to be stupid!

HEROFORD: ....Feeling bad, Betty. You're being too harsh! Too.... critical.

BETTY: I'm seeing you're starting to stray off again, teeter towards some foolishness that may make you unprepared and rested to go back to work. We can't afford that disaster. Too many depend on your stability. You can't go about jeopardizing life and circumstances now.... You're not bored with your work, are you?—

HEROFORD (*eating*): Hardly that!

BETTY: —It's the only thing you know how to do!... well, well enough to maintain us. And we're substantial, Harold. We're a substantial responsibility you can't suddenly dream to give up—

HEROFORD: Can't! Can't! That's all of your definitions for me. Describe me as what? do you say. The "Can't" ?! Can't wait to get back. But stop trying to tell me what I can't do!— because you don't know it! Why can't you ask what I want?— I want to be happy. And this is some small space of time to shudder about those prospects, the window of three weeks. I could die if it were longer. But the grind at work is fierce enough to face. I can handle that, not this— dissatisfaction at home. It's enough to boil potatoes. And you wonder why I rush out into the cold. But the radio said it was gonna be a warm day— again! So be it so. I wanted some relief, for a time worth trekking through, reviewing life and lifetime achievements, and wishing for snow—

BETTY: Snow?!

HEROFORD: ....to cool things down.

BETTY: During your vacation time?! I don't like this unsteadiness in you, your vacillating into God knows what of a state— of mind, manner and menacing.... of our security. Now's not the time for that behavior, Harold.

HEROFORD: Menacing?!

BETTY: We're too sensitive and developed of a family, and a social unit that can fall apart terribly with your uncertainties and failings. I've always feared such outcomes in a home. I've.... learned about too many occurring— in the past, and can't stand to have it begin to happen to us.

HEROFORD (*eating*): ....Well, it's your job to prevent that, isn't it!—

BETTY: I mean to! I've been meaning to confront you directly—

HEROFORD: No more confrontations!

BETTY: —about your pretenses of understanding, and pacifying your existence through your children's growing individualities, searching to be themselves.

HEROFORD: I'm not delaying that— And I'm not your folks—

BETTY: What do you mean?

HEROFORD: Your.... distant relatives, the dissolute ones you dread to discover a common link with, or fancy through.

BETTY: All of my relatives are dear to me. You are the husband and father of this house. (*turns away*)

HEROFORD: What's mated to. Mate! We're not transforming to come apart, grow loose and separate of our entities.

BETTY (*still facing away from him*): You're threatening me. Why do you say you're unsatisfied— with us! Your children are handsome. This household is splendid, for what we've paid and struggled with. And you are viewed as a present day success, with us. What can you possibly be unhappy about? concerning your home life. That you haven't amassed more of a fortune, or friends substantial? Are the children of others better than ours, or are wives more remarkable—

HEROFORD: I simply feel—

BETTY: —personable and attractive, that you have met, than what strives for your conception of satisfaction?!

HEROFORD (*eating*): ....I simply feel.... unsettled and unsure, that life is not.... making its way for the best.... in my circumstance.... that our founding is not.... more comfortable to me, than it should be by now. If something is not wrong, it still seems awkward, and impinging on an agony of not being.... the wholesome calm I had dreamt would become of us, the evident satisfaction with ourselves we should deserve, after these years of sacrificing to obtain it. I don't feel proud about anything, these days, except my work— And I know that's not right; but it's the honest truth, and even shameful to think, frightening to admit.... between us, Betty. I love you all without any questioning, to posit that fact before my conundrum of sensations. Take my work away, though, and I become a shell of restless.... tiredness, as if I still have to redress a great fault in myself, and can not relax contentedly with my dear ones earned.... to be. Something is not correct, perceptions displaced, and sentimentalities misplaced. I am not satisfied, and am coaxed to squirm, away from the job. Talents have not been fulfilled, enough, for my.... our achievements. Though I haven't any hankerings to go into any business, on my own. The drive is come from elsewhere, to seek relief from this.... weariness and fear, that I've yet to prepare ourselves properly—

BETTY: Prepare for *what*?

HEROFORD: ....For our ties through life, and associations endearing and endurable.

BETTY (*insulted; sharply turning to face him*): Then drown in your oats! if you can't find your pleasures with us, what we've produced. Are you searching for others to be dear to, and spread your wealth on?! Do you think you've made a mistake with me—!

HEROFORD: No!

BETTY: —as I've so often feared, to settle on your desultory manner. I thought I had reared you out of that condition, that the waywardness and aimless notions were finally abandoned, firmly graved.

HEROFORD (*tensely*): You rear the children.... (*normal voice*) I'll not make a cause for more complaining, but that I was very disappointed you hadn't left any breakfast for me—

BETTY: Ach!

HEROFORD: —this morning. Met a fellow who has to take his out, out of doors— grand doors, possibly.... And another who may not even have any at all, that's considerable, more than this cereal, I had though could not be so my plight ever to simulate. But look at what has become. I expected more.

BETTY: You don't test me!—

HEROFORD: I had wanted bacon to be seen, virtually assured a fellow it could be found.... on my return. And now I'm proven wrong with this reality to face, that I'm not so highly regarded of my family, and my wife spurns my needs—

BETTY: Needs?! You ran out!—

HEROFORD: Well then desires!... Such an innocent, little thing, as to leave me a few leftovers from the boys' meal. Do you think I could dream for that, to be more fortunate than a couple of strangers I chanced to meet? But it's the pettiness in me to bring me down and realize I haven't even that good graciousness to own. Brought me feeling low—

BETTY: I'm not cooking for you so unreasonably, like a slave would to an unpredictable tyrant. And you certainly are no tyrant. If you want bacon, and toast and eggs— cook them yourself. Nothing's stopping you from being able to, though you never shopped for them, or thought to provide these provisions thoughtfully— but —for—the—funds— of course. So make your way to it— But you are hogging (D)'Aron's favorite cereal. And he won't be happy about not having any left for him(self). (*starting to exit*) (He) Gets thoroughly upset when he finds there's none around, because he knows *we* know he wants it, like a security for his being around.

HEROFORD (*sotto voce*): My favorite too!

BETTY: And then you can bring your.... persona hunting to apologize to me, for my patience, and my upsetting, and my agonies.... And the nervousness of nerves to keep this house in order — and you upright for it! (*Exits.*)

HEROFORD (*to himself*): I just want my mocha.... the way it used to be.... I just want to be restored to my better thoughts about this and everything that's occurring, so carefully planned for and carved up out of.... my means and modes.... and mundane mortifications. Why the mendacity to face it? I feel.... perplexed of the truth, that this life bothers me too wholesomely to deny— another manner of being. (*resumes eating as* **'Aron** *somewhat aimlessly strolls in, a youngster of about five*)

'ARON: Daddy! What's that?

HEROFORD (*taking casual notice*): Some cereal, 'Aron.... for my breakfast. Did you have yours, this morning?

'ARON (*losing interest*): Yes.... I've nowhere to go. Mommy pushed me in here.

HEROFORD: So you're not hungry? Want something to eat?

'ARON: No.

HEROFORD: Drink?

'ARON: No. Is that my cereal?

HEROFORD: Yes.... You don't mind if I have some, do you?

'ARON: No.

HEROFORD: Then you're just strolling about the house. When it warms up a little, you can play some in the backyard.

'ARON: I don't get to go anywhere. Danny and Duncan left me alone!

HEROFORD: You're too young to follow them. They probably went to a store. And Duncan can take care of Danny. What is he— almost a teenager now? But you're still too small to keep up with them, so you shouldn't try to.

'ARON (*aimlessly walking around*): But what's fun to do?

HEROFORD: And don't badger them too much. You'll have your time, to roam around the neighborhood, and discover life.

'ARON: What's life?

HEROFORD: ....Living.

'ARON: Oh.

HEROFORD: Till then, you have to stay home till taken out. Aren't your toys enough?

'ARON: They ain't living, 'less I breathe for them.

HEROFORD: You mean, when you get interested for them.

'ARON: Yes.

HEROFORD: To play with.

'ARON: Yes. (*running up to the table*): Can I sit under here?

HEROFORD: Go ahead. (**'Aron** *sits under the table, studying* **Heroford's** *shoes.*) Where's Marce?

'ARON: On the sofa.

HEROFORD: Better get off of it. Mommy doesn't like that. Do you mean on it, or under it?

'ARON: Under it.

HEROFORD: You've been chasing her?

'ARON: She runs from me.

HEROFORD: Well, you know we don't want you bothering her *too* much. Good thing she's little, and sweet of disposition. Puts up with your curiosity remarkably well.

'ARON: Comes to me, sometimes.... when I can feed her.

HEROFORD: Oh? What do you feed her?

'ARON: Cheese.

HEROFORD: ....From your lunch? Your sandwich?

'ARON: Yes.

HEROFORD: Hum.... She likes the mustard and all?

'ARON: Oh, I lick that off.

HEROFORD: Hum. And the spiced ham?

'ARON: I eat that.

HEROFORD: Don't you like cheese?

'ARON: When she's not around. What's that?!

HEROFORD: What's what! Maybe we feed you too much.

'ARON: That thing on your shoe.

HEROFORD: ....(*imagining what it could be without looking*) It's a shoe buckle, an ornamental decoration, for.... older sized shoes. In a few years you might wear some with them, though they're going out of style. Right now, you barely use more than slippers. And its a good thing they're inexpensive, because we have to change them every few months. But, you have your brothers' to choose from, often.

'ARON: Why do we wear shoes? To keep the feet warm?

HEROFORD: ....Because it's a custom for us; and the ground is not too level outside, so we can walk on rough surfaces, not like the smooth and soft ones inside the house. I mean, you don't need them for the carpets and the rugs. But that's virtually all you've seen, till you get outside.... You like your shoes, don't you?

'ARON: ....Yes.... They pinch, sometimes.

HEROFORD: Well, you have to get used to that. Shows you they're working— It shows you.... Of course, if they're becoming too tight again, then you'll need a larger size. But you're only foot-ing sneakers anyway, slippers—

'ARON: Why do we have to walk at all?

HEROFORD: ....To get around. You walk, mostly, to get around. Can't roll much outside.

'ARON: Oh.

HEROFORD: And once you're able to get past the house, you'll be doing a lot of walking. It'll be one of your most important ac-tivities, called "traveling." We spend much of our time traveling, because the world is much larger than this house. Danny and Dun-can are traveling, and you can't keep up with them yet. That's why you're left behind.

'ARON: But they come home.

HEROFORD: They return home. Traveling is returning too.

'ARON (*slightly anxious*): Can I have some?

HEROFORD: Some what?

'ARON: Cereal.

HEROFORD: You're hungry again? O.K. (*getting up*) I'll make you a bowl. (*going over to the cupboard*) I thought you might be, coming here.

'ARON: Mommy pushed me—

HEROFORD: Oh, get up from under there, 'Aron, and sit properly at the table. (*as 'Aron complies*) Dining is as much a ritual as a necessity, and should always be done.... with a certain amount of — respect for the activity. It's such an important one. (*getting a bowl and spoon*) Though snacking is more casual. But even that's not as fun, if it's done sloppily. (*bringing the bowl and spoon to the table*) It's a sin to be slipshod at anything.

'ARON: Hun?

HEROFORD: Try to be neat at whatever you do. That says your body deserves its best efforts. (*placing down by him, 'Aron instinctively picking up the spoon to handle*) And you want?

'ARON: *My* cereal.

HEROFORD: That's a certainty, my lad. (*going towards a pantry cabinet*) You're not like Duncan and Danny. When they were your age, they insisted on having their cereals sugar-coated.... But you like to spoon in the sugar, as I do.

'ARON: It tastes better that way. The other ones make me sick!

HEROFORD: ....Well, you just don't like *too* much sugar. An early refinement of the palate, you'll prefer the dry wines. (*getting the cereal box and bringing*) Spoon yourself some.

'ARON: Before it's in the bowl?

HEROFORD: ....That's not going to change the sugar. (*at the table*) Now, how much?

'ARON: Just a little.

HEROFORD: O.K. (*pours some cereal in 'Aron's bowl*) This much?

'ARON: Yes.

HEROFORD (*returning the cereal to the pantry, and getting some milk from the refrigerator*): It's a good brand. Simplicity is the hardest thing to craft.

'ARON: Hun? (*while getting some sugar from the table's sugar bowl*)

HEROFORD: Can't hide many mistakes with a fine plainness. Complexities just load them up like treats. The errors become the principal advantage of the whole effort. Whenever something's done that's too wild to understand or explain— beware!... It's meant for you to take for granted, and someone's out to put something over the others, fool them into believing things that can't really be proved or demonstrated sufficiently— Recover the sugar (bowl)— for any reasonable acceptance.... So, the masses just leave these matters up to the practitioners of the complex, at least until something goes dreadfully wrong— and everyone is astonished. But simplicity is the reward of the correctly patient, though some claim it's only symmetry. (*at the table, as **Betty** enters*)

BETTY: What are you feeding him again for?!

HEROFORD (*almost starting to pour the milk*): Said he was hungry for some.

BETTY: He just wants to make sure his cereal's left.

HEROFORD: Wants to have it, Betty. (*pours milk*)

BETTY: Well he better eat it all up. I'll have none of that wasted. And put that milk in the refrigerator.

HEROFORD (*with suddenness*): You think I'm a clown not to?! (*as 'Aron starts eating*)

BETTY: I('ve) found it left out on several occasions. And this milk is too perishable and too expensive to let spoil.

HEROFORD (*going to the refrigerator*): The boys, Betty. You know how in a rush they can be. They forget (to do) things. But you're here to make corrections.

BETTY: I don't want to have to throw out any more food. It's not sensible— Harold!

HEROFORD: They're away from school. Whenever there are changes in the routine, things like that can happen. (*returning the milk*) What kind of cheese are you using with the lunchmeat?—

BETTY (*going towards the table*): There's no reason for any of us to be sloppy with what's precious. And I have to shop for it— Eat every last drop of that! 'Aron.... (*as **Heroford** faces her from the refrigerator*) And you're becoming lazy and forgetful yourself.

HEROFORD: I'm out of my routine also. But I don't leave milk out—

BETTY: There isn't much excuse for any job left undone; though you look fairly finished, and will clean up after yourself— and him.

HEROFORD: Leave us have our cereal, Betty.

BETTY: But I thought I could start preparing for the dinner—

HEROFORD: It's still morning!

BETTY: The meatloaf has to be seasoned and marinated, Harold, for a few hours at least. And lunch has to be planned— My order of things is being corrupted. I just tidied up the beds and the bathrooms—

HEROFORD: Well, (just) give us five minutes to finish— please? And we'll leave the kitchen pristine for you. Take a rest on the sofa. Everything doesn't have to be so regimental or busy—

BETTY: I have to look after him, Harold! I have to look after all of them, and that takes some modicum of order and regulation. And I don't need you two absent (*coming towards the refrigerator, and he avoiding her to return to the table*) to start my work. Ground beef has to be worked! Just don't interfere.

'ARON: I'm done!

BETTY: —Good! Then wash his bowl, Harold.

HEROFORD (*sitting at the table*): I'll wash his when I wash mine.

'ARON: I can wash it!—

BETTY: No you can't! I don't want my plates broken, with your climbing on (top) the kitchen counter. (*at the refrigerator*) Go and play with Marce.

'ARON (*getting up*): O.K., mommy. But she's—

BETTY: Or try and read some of your picture book. A few words you should have mastered by now. (*retrieving a platter of ground beef*) We'll review it. (*'Aron taciturnly exits.*)

HEROFORD: You're not going to bother him about that, are you?

BETTY (*bringing the meat to the kitchen counter*): I'll not have him starting kindergarten as a dolt You have to progress fastest at the beginning, or else your peers leave you behind.

HEROFORD (*resuming his meal*): That's not true. Individual advancements and progressions in education are quite variable, for time and place of studies. As many doctors start out backward or with dyslexias as not. (*looking over into 'Aron's bowl, while Betty is occupied*) Must certainly be the case for some of the chaps I deal with.

BETTY (*unwrapping the meat*): That's hard to believe, Harold.

HEROFORD: Well they're very high(ly) placed, despite their faults or notable imperfections, like speech impediments or lack of concentration for long (periods). It s what you make with what you have that counts, more than what you have to make with.

BETTY: An obvious observation based purely on what must be done. (*back at the refrigerator, to get some vegetables and oil*)

HEROFORD: I'm sure most people (*pouring the contents of 'Aron's bowl into his*) have more talents than what they display.

BETTY: That's possible, if they haven't discovered them, or had not the opportunity to. (*retrieving a clear plastic container with one green and one red pepper, a tomato, an onion, a clove or bulb of garlic, and a small bottle of oil with vinegar: i.e. a carefully prepared collection of working materials, as Heroford noticeably slurps*) Handicaps are not much excuse for not trying, though.... (*bringing to the counter*) for overcoming them. And 'Aron's been a little slow, compared to his brothers. So the corrections must be made now, while they are palatable, or innocent enough to achieve. (*on a kitchen board at the counter, starting to chop up the vegetables*) It might seem difficult for him now, but those same efforts will be much easier later, as he's gotten used to doing them.... skillfully.

HEROFORD (*getting up, and bringing the bowls with their spoons over to the sink, which area adjoins the counter*): Very much so, Betty.... But too much work is frightening to a small youngster, who can not possible know what he's capable of doing,

and is timid or shy of its importance. I simply don't want him culturing a bad attitude towards obtaining knowledge.... (*starting to wash*) Different personalities must be introduced to these activities differently, if they're to be left with a wonder, happily of the results.

BETTY: I'm not one for playing games with learning, Harold. It's either done right, or it's not done right. (*starting to salt the ground beef*)

HEROFORD (*stacking the bowls and spoons on a kitchen rack*): I gained a foul taste for elementary calculus, and never did learn it correctly. Impeded me no end, particularly through statistical analysis courses. Slowed me down to a crawl, at times, compared to others; though it made me much more careful of my lessons— by necessity.... It was not the mathematics that was malevolent, but that early inculcation of its overwhelming significance and importance, that could frighten any honest student. (*looks over at Betty*)

BETTY (*preoccupied*): Fear has no face for facts. What must be done (*working the beef*) will be, as were a mirror to the soul to plead for seeing.... of one's deficits— relieved.

HEROFORD: ....That must be cooked particularly well, Betty, to destroy any chance of living contagion.

BETTY: I know how to prepare our foods for eating, Harold. Thoroughness is a manner mated to. You simply do not notice this much.

Act II

Scene I — *A park bench, still morning.* **Langly** *sits at it very still and upright, but with some difficulty of exhaustion. His serenity is pensive, but not particularly contemplative.*

LANGLY (*to himself, although some pigeons may be around*): Ooooh! I shouldn't have tried for the park. But I felt somewhat energized this morning, by that ever so fine coffee an' cream. Now I'll have to walk all the way back, and it's going to take hours for me to regain my strengths.... The lungs are burning, as the day heats up. And I am stuck here for awhile, like an old matron of the forest, surveying her lively, disinterested treasures. I used to be able to take such long and pleasant walks. Well, it's what I insist to still do, to some extent. A.... constitutional, I called it, as for a profound effort.... of jollity. I can hardly think well. And this was.... some kind of mistake. But I'll recover from it. (*starts to slouch a bit*) The lungs will clear of their.... congestions? waters, or mucus. It's strange. I don't have much difficulty breathing, as from an asthma or a chest cold. But the matter goes slowly and with a squeezing pain, of the ribcage muscles, perhaps as well exercised as the heart. My lungs just are not as efficient as they should be, exchanging the oxygen and the carbon dioxide, through membranous diffusions. But they're not blocked up. I just have to take it all slowly, and wait.... through this leaf-ly purified air, until more exertions can be made. I don't find the fragrances disturbing, as if it hurts to sniff at them. There's only a lack of energy to me, as I replenish my means of drive, this.... sensible ambition. And I must— wait!... until this condition clears, this state of quiet tension is lifted and I can return, for my afternoon's most comfortable quiescence. But do I replace that, with this worrying out here, so open to the normal splendors of greenery? No, I'm not caught.... to a calm. This was a good thing to do. And now relax, to do some more later. I will get there.... A wild attempt— no doubt of it. How do they take notice of me? what occurs around my helplessness. The little animals and insects just scurry to their business, and the winds blow fondly of their remembrances, the breezes.... quaint from storms. I'm not denied my thoughts, and senses. (*trying to sit back upright, but not quite able*) Where are we doted upon, for life to taste, its fruits ambrosial and arrayed for our picking and choosing? I made of my career its dangers and am now not disgraced by them. Why wish to hide the cracks in the coconut! But what defers my recovery? Will it happen this time? or have I put too much out of myself, for this interesting day to be in?... I've made it to the park before. But I'm not sharply seeing things; and deteriorations.... they're additive, when due. Much too tired to fight them off, at the moment. Complacency is a fine bed to sleep in. But there's this slight kink, nick in the shoulder, from yesterday, that should have perhaps warned me not to go all out for anything, until it clears. Yet comes the irritation innocently, with no discernible cause.... I simply moved about, and it was there.

BECK (*casually coming up to him*): Yes. It's Mr. Langly. Surprised to see you down here. Can you walk this far that often?

LANGLY: Who's there? You know me?

BECK: Heard speak of you. I set up the tables, at Bruce's café.

LANGLY: Oh. Breur's now. Recent job?

BECK: Just a couple of weeks, so far. (*stands confidently*)

LANGLY: I don't follow the details of the store any more, but was friendly with Bruce— the original one. I think his son took over, when he fell out of it. Hence the name sticks.

BECK: Some relative or other.... Would have expected you to be quite winded, with your ailment.... unless someone drove you—

LANGLY: Well I am slightly put out. But I still try myself to make good of this city, when I feel I can. The courage is still there, from when I was much more.... ambulatory.

BECK: Able. Gonna be a nice day— I can feel it! Sun's already clearing.

LANGLY: That's definite. Else I wouldn't have attempted the stroll (*sitting up*), lad. But it's all well worth it, to be within this city's most favorable locale. Did a lot of work here—

BECK: Beck's the name. Yeah, it's a grand place, but aging just like any other. And it's not exactly as clean as one would like. But these are wholly public spaces. So from that point of view it stays remarkably kept, despite the damage and the litter, the fact that it can't really— ever be closed down.

LANGLY: I suppose so. Looks clean enough for me.

BECK: Visitations range from the fleeting of momentary passage, along its outskirts, to the ridiculously permanent of.... unofficial resident. I've spent some nights here, even, when I got to town.

LANGLY: Say you were tramping?

BECK: I was arriving, and settling into this location, with nothing to think of but to survive through yet another new environment. Oh, the city seemed particularly novel to me, with its mixture of old and new buildings, its granite façades and brick and wooden frames, and like a living organism in constant change. I slipped into town at the most dark of night, as I generally do, from out the countryside, when the streets are generally deserted. And I walked through them, observing what grandeurs were offered, to give character to the town. A fairly modern one, I concluded; but with some history, because a few old buildings struggle to persist and remain extant and worthy of standing and structure. I like to sense such.... emotional displays during the silence, as the personalities of an area are presented for my simple perusals, very clear visions for the reading. Daylight sort of obscures these things, as people take over your interests to view. Yet, the deserted are not alone, at night.... I wasn't— tramping, as you say, but living off the land awhile, as best I could, and with some money from my last job, earnings that let me feel.... comfortable for their carrying. Then that's not a tramp. But slowly the sum was dwindling, through my occasional purchases in cities, down to a point when you conclude that you have to make some more.... or decide to, for your own standing to maintain. So up into this town I climbed, as the next one found. And it's not half beautiful— but more, with its strong sense of heightened self. Some tall and majestic towers (are) here. Its citizens must be special. And then, to my astonishment, I discovered this rich, sprawling, expansive park, right within and enclosed— owned, as it were, but with as much of a country's atmosphere as could ever be found through the remoteness that surrounds and divides and distinguishes principalities. So, I bedded in, a few nights, looking for work during the day. And though sparse of inhabitants, when the sun reclines from view, I discovered that this park has many.... regulars, and of any given type or group to describe. I suppose they feel more protected in here, than on the cold nocturnal streets and sidewalks to camp out for stillness and slumbers. But.... they certainly are tethered to the city, to remain, or I would find the like much farther out and away from here, in the greater vastness from which I entered. And I knew quickly, instinctually, that I would have to make a more normal berth while restoring my funds, or I'd become a— regular, tied to this meaning of destitution and occasional squalor. So now I rent a room, to be practical of my savings and expenditures. My café job pays for most of the housing expenses, for now.

LANGLY: And you feel hopeful, living like this?

BECK: Why, I search for more work every day, and can find as much, with a little diligence, in this city, to scrounge. It's a very busy place (*sitting on the bench*), with lots of leftovers to seize or timely claim. The organism wastes nothing here, by drawing in good scavengers at ripe moments.

LANGLY: Then.... how's the wilderness?

BECK: It was meant to surround towns.

LANGLY: Better there to be alone than shut up in this city.

BECK: The natural sculpturing of the land is (re)'fresh'(ing) and wonderful, the views of earth with sky awe-inspiring, particularly for the comings and goings of the twilights, or the storm ravages to shred sunlight through booming clouds and high pitched gusts of wind-blown leaves and grasses, the turf of the remarkable in upheavals generous of sight. If I were a simple animal, I'd be contented living in the open and noticing all that is uncontrollable, to make master of these events for one's existence, such as the appearance and scurrying of prey. But we're not so simple, eh?—

LANGLY: No.

BECK: —Our sophistications demand greater attainments than what can be provided for mere hunters and herbivores. But then, I'd live on a high hill, to overlook my range.... and not have any fears till fears are come, and only anticipate my management of efforts to resist inconsequence. Frankly, I prefer cooked meals, well-made bunks, hard drinks, and... social services, to make more company a pleasure within intimacies, than any proud hound of a prairie or tigress, lioness of a forest could ever fathom. But the wilderness does draw one out with urge, to stand with a sense of jungle driven from, when this can be done of rational intention and circumstance. And I know that as I age, those opportunities will become more and more rare of possession, to seize one's will of enjoying the primitive, and the primordial sensations we harden from feeling.

LANGLY: So. Hardening of soft, moist surfaces, which were only of spongy resistances. If you had a more substantial wealth, you might forgo the emptiness.... of the all-encompassing vistas.

BECK: ....It's not so uncomfortable, sitting on the grass during a rain, sitting within the grasses, even lying stretched out. It's what you can allow to happen to you, and what you may get used to. Well (*standing*), summer is a time for this adventuring. I'm.... shielded with depressions during the other seasons, to take more care of myself. Yeah. And be careful, cautious, cowardly and cross, until my most basic needs are met. That's how one's supposed to live. You can't be a carefree casualty.

LANGLY: You should be making a career of something, while you have the energy to employ your gifts and gains.

BECK: I'm trying to, desperately.

LANGLY: These things may consume you, but we're all meant to be.... for a subsistence of others.

BECK: I'm discovering interests and possible capabilities everywhere I'm at, of an organized populace.

LANGLY: We.... do for ourselves, eventually to do for any, is what I mean. I've become broken, physically. I don't know what they told you.... But what I did was worth it, even if it's destroyed much of my lungs.

BECK: Well that's about what the waitresses said. Tissue problems— lining of the lungs.

LANGLY: Even if they pity me— for water. I've served mankind, civilization, and our value systems, and have not abandoned.... the motives for our involvements with one another.

BECK: Sure. You're just restricted in your activities.

LANGLY: And I used to be a healthy paragon of those activities. What I scratch at now I could (once) cleave! And I miss those abilities. But while deprived, I don't feel undeserved. Severities in anything, even the finest of existence, must take their toll. And accidents are not a curse, but a kindness— if you survive them, given their cause— and a capping off (of the oil well) if you don't. It's all of some order made to follow and from which no one can withdraw.

BECK: Well, I can't say I'll accept the preventable. But as challenges are made, chances are freely taken. We are forced, though, to obey the mandates of survival, or give up on life— as a slave to nothing.

LANGLY: I'm not nothing, leading to—

BECK: But leaning on.

LANGLY: —some residual self-worth. That's all that's left for me to perceive, since I tire too easily from complaining. So whatever happens ta ya, fella—

BECK: Beck, again.

LANGLY: —'s not a crime, but what must continue of our earthly habituations, to make of but so much change with significance for our once being.

BECK: I'd make more of that to clamor about. Sure, you're undone.... and might feel superfluous, superfluously sufficient for what you have achieved. But there's much more to do. The world is opening up its quarries of action and consequence; and revitalizations are occurring, as the new stuff floats.

LANGLY: Stuff?

BECK: The sea bed yields to the surf of the surface. It relinquishes the salty foam for a spectacular endeavoring.

LANGLY: This is about as dry a place as one could hope.

BECK: But that's how rushed I feel, for the mingling and stimulation of excitement, and the becoming of the unpredictable. (*stretching*) I can reach out to grab the sun— but I know it's too hot. So I wouldn't do that. But the promise is there, to be able to: if not to burn my hands, then to handle life aggressively and with much hope of fulfillment.

LANGLY: Fulfilling what? I was a young— Beck too, once. But time races chances. And there's no great delivery of results without suspicion and surprise. So you feel like you're on the verge of accomplishing a lot. That's fine. But what's in a sum. I've demolished a lot. Doesn't add up to much— And there's no anguish about that. Was terribly fun to do, within professional rigors, manners and management. I don't plead about the times, the accidents and losses, and the conquering over difficulties and steel-laden obstacles, to recall so much a glorious past. Living for the day is great. But it was never.... superfluous— or superficial for me, even if it becomes quite meaningless today, even if it could have been done by a thousand others— worst or better or much the same. I can't complain about the.... consequences, as if you could make for some improvements, or the town in general. My time was then, to be fresh and lively. And now, prematurely stalled, or stunned.... Well, that's a life worth its astonishment. And I'm not crying about the surprise nor the suspension. But gloating to be healthy, fit and able— an' scratching the streets for gold— is no valedictory speech to hear. It's hardly even an intention of character or purpose. And what might you do that's so profound? I must have done (it) a hundred times over, in my relatively short life. I don't like your implying that, though perhaps limited, my remaining days are worthless. They're as flamboyant as my heaving breaths and shallow-ing chest to provoke vigorously, firingly, like a dragon's flame to force.

BECK: Oh, well. It's a struggle, a challenge.

LANGLY: And each issuance is a lively success, a sparking, sparkling fortitude— and a presumption!... 'gainst foolhardy lingering (on).

BECK: Can't doubt that. Wouldn't claim to. Even the simplest of animate forms vie with a fanaticism, at times. But don't you think walking all the way out here is overdoing it an awful much? considering you live over at those.... apartments.

LANGLY: What about "those apartments"? It's what I can afford.

BECK: They're about as far away as the city can place.... a sight sore. And you made all of that distance, just to prove your vivacity?

LANGLY: And vitality. I happen to be along the route, today, and felt particularly energized this morning. Though now I have to rest, to get back. But I'll make it, to a justifiably solid community dwelling.

BECK: The park is pretty— But still.... A person in your condition shouldn't tempt such traveling.

LANGLY: They certainly told you a lot about me.

BECK: No. Just some colorful facts that can light up a conversation. A few words can define well some descriptions. And you are not capable of these efforts, anymore. So why try and damage yourself further, with proof of nothing?... except what you'd still like to do.

LANGLY: Pertinencies and relevancies are all consumed in the flame, of desire made the mourning to be wished. I am not dead, and am but dead. So grace me pleasurable treks and tries. It was the doing it that made me fit for the fun. Getting here.... well that defines a triumph, I suppose. But it's not so important, besides the beauty of these surroundings to enjoy— for an extended spell. The walk home will be much slower, as my head gasps of thoughts, reminders of views and remembrances. The sights are growing more orange-red now, as if to shift for recollections of a past life.

BECK: Some review. I could offer to take you there. But a walk's a walk to you, as strenuous accompanied or not.

LANGLY: There's no emergency to it, Beck. The breezes are quite nice to feel. And the squirrels don't seem much concerned with my presence. Only one really stopped to notice. Yes, this city is rich of memories for me. I still see it as it was when I was able to go about more. We were both bold to be, and demanding. The cosmopolitan scenery graced my face and perceptions of the world as a fanciful environment to exist in. And I don't think I ever cared to think much beyond the immediate future, every concept of life was flying so well. That's how I see it to stay current. And I never had to.... sleep in the park. Didn't feel above that, though. Helped shape some of her contours.

BECK: (I) Don't feel belittled by having done so.

LANGLY (*after a slight cough*): Good pine and evergreen scents here. Had I thought about it more, I would have brought some bread crumbs, to recompense for the trees' serenades. I didn't think to dare this at all, before reaching the café, this morning. My constitutional is usually much less circuitous of town.

BECK: A nice place to spend time.... or waste time, and wait for more.

LANGLY: How do you back? What lean you for to do? relaxing lazily till then. More work to look for, that's most certain.

BECK: Why, sir. It is a curious commission, for you to learn of. But I'm about to join a team of building demolishers, in a short while, as an untrained assistant.

LANGLY: ....A rubble collector. But that's some heavy lifting work, and full of injuries for awkwardness. You're not paid for being careful— though they'll caution you to *only* be careful, for your muscles to employ to their fullest, with duration of a seemingly endless task. I felt fortunate not to have it, 'cause we caused an awful lot of pieces falling, a beach-ful of debris. And it all has to be cleared away in a timely manner, or else the site looks very sloppy and unattractive. The whole point is to rush through ugly appearances, to bring pardon to the downfall of a usually handsome structure, always internally well crafted, and often nicely decorated.... This is not a job for idle dreamers, Beck. It's down to earth grimacing with grime.

BECK: I've worked heavy loads before, though not in the building trades. I can endure.... much regulated strain, with those who know what they're doing and how to direct novices efficiently—

LANGLY: And that's all that should be done to you. But I'm saying, bones break easily during construction. Fingers are cut. They'll give you some gloves— but still watch out. Sharp and jagged fragments are the norm. Arms get twisted. Neck strains, back pains— especially shoulder spasms occur frequently, for the un-intuited who repeat avoidable motions counterproductive in the long run. 'Cause you have to repeat these actions over and over again, with a monotony that can never be monotonous. Pay attention and stay alert. Don't question our commands and do exactly what you're told, no more and no less. Not so simple a job as you might have thought. The concept is simple enough, but the getting it done is always complex and visited with the unexpected, like driving a car and finding, despite your best efforts, always to make at least one certifiable traffic violation per trip. It's not that you may be bad to do it, but that the process, while common, is still much harder than it can seem. And it will catch up with you eventually—.... one way or another.

BECK: I'm always cautious.

LANGLY: We don't often have bad accidents with the freely hired. We keep them out of the way of danger as much as possible. But the "newbies," the ones actually coming onto the profession, from a formalized union or other recruiting.... well, they either break themselves in— or break themselves up, rather early. You'll recognize them as occasionally being unsure. And they're never to give anyone orders. Just stick to what you're told, always by superiors in blue or green shirts— sometimes thinly striped with gray— and use some common sense for the task at hand. Don't try and prove yourself by being extraordinary. If workers remark about you, that's generally not a good sign to promote. Only the entire effort of transformation should be indelible for memory or comment. The individual is not so sacrosanct at a work site.

BECK: I don't mean to be noteworthy. I'm just looking for some labor with pay. And if I get a good reputation at it, I may find some more. Tedium is not a deterrent for me— I seldom dream awake. And when I'm engaged in some effort my mind goes blank except to its purpose. That's by training and concentration, and why I can do so many small chores for money without feeling humiliated or belittled. Because small, as in menial or manual, does not lessen the nobility of an exertion. But I won't be slavish to any concept that detracts from my worldly standing. I'll always demand decent wages for my time, patience, and persistence at a task. No. I won't be slighted by this in the least.

LANGLY: The sweating alone will thin you out. Takes more than concentration, but a commitment to be used, or you won't last an hour. And I've seen a few run off in a panic from the demands, the loads that overwhelm them and what they presumed were their more adequate or capable physiques and physical strengths.... But that's rare. The insufficient bodies tend to gradually bend their backs lower and lower. We call them squibs, because sometimes they suddenly collapse with a bang that can even startle veterans, right onto a pile of dismantlement— And we have to figure out if they've been injured or not. We suggest to the squibs, at the end of the day— They're always barely past boyhood— not to ever show up for this kind of work again. They can be seen like talismans for trouble, not that the trade is very superstitious in general. But a sore back for us could be a broken back, or displaced vertebra, for them. And somehow we always fall with the responsibility, threatening to, in some way, somehow dock our pay a bit to help with their health treatments. 'Cause they never seem to have any insurance or other benefits coming to them. And we don't let any conscientious member of a team who's tried his best for us (to) lie flat an' out without any aid. But there's only so much we can take— an' give.

BECK: Oh. Conscience led collections. That won't happen to me. Charity. Donations—

LANGLY: Accidents can happen to anyone. Who knows what rigors your body can actually withstand, when you inevitably pull a muscle or two. You have to keep up the try, despite that— thoroughly against all medical doctrine.... or succumb to the lower depths of plausibility. For mediocrity is simply a broad(ening) range of the plausible.

BECK: Yeah. Well, I don't feel so terribly challenged by this— And I ain't gonna kill myself.... like some others would.... allow. But thanks for the warning. I don't tend to bend the back much anyway. You should lower yourself through the knees, and buckle to pressure as much as feasible, within reason for your protection. The flexible, rubbery, and more yielding avoid the cracks, tears, and breaks. Tree limbs show me that all the time.

LANGLY: This isn't nature in the making, Beck. We're not doing anything naturally, to our bodies nor the world— Even sport isn't like that.... if it's professionally competitive. This is land management, abode fostering, civic improvement, and nation engendering.

You think I could be proud for my condition in any other way earned?! It's due the harm. It's justified.

BECK: If.... you don't mind me saying so.... I think your difficulties over the years have lightened your head and caused you some aberrant thinking— But that's your privilege won.... since it's all you have left to contend with, conjecturing.... why.

LANGLY: ....Why. Why have I done this to myself? I didn't. And I'm not mad due to poor, low (levels of) oxygen getting to the brain, for too long a spell and too often. But I can understand your assertion, and the honest arrogance that backs it. You haven't contended yet, Beck.... with so grave an affliction as mine. So it's your right to feel more fortunate and better.... than you think composes for myself. (*coughs*) So let's hope our composures stay equitable in this— I am not destroyed! This weight of fate has not done that to me. And you've yet to be even truly tested.... I hope you fare well —

BECK: What, in your job?! I won't take it as so definitive, nor deformable— nor defining. Now, take care of your lungs. Mr. Langly. Are they a burden?... Well, so is life (*starting to stroll off*) when it comes to a point.... of falling leaves, and graying hairs.... and premature decrepitation of the bones, and accelerated aging.... of the mind and spirit. But it's all honestly done.... No fault to it. No shame....

LANGLY (***Beck** has left.*): That's foul, to be a true believer, in such.... multiplicit-ous harm, as could be numbered, in dying hairs and leaves.... and wishes of impossible return. (*slouches a bit*) Yet does a tree squirm as much as I, for a dreamed for spring, with carelessness of abundance returned, and life restored so wildly.... and resplendently arrayed. Oh! For this summer, now, to hear me cry within such colorful gayness and bounty, that I should hurt, crumpled and forlorn.... from a painful walk that I thought could prove my mettle. And now I am tried by, of all things, youth. Well gather around to see the *Kerl* rejuvenate. Come closer and examine him. Insects, crawl and bite of him, to find out how he is. There's no resistance left in him, to such attacks of inquiry. And.... birds and animals, come near to tease, that your approach is meaningless to me, for your endangering. I can not seize, I can not fight— I can not struggle for you, or to capture your amazement at my lethargy. That is the wearing of a day that warms, as I clear my head. And did I have a conversation.... with some lad called Beck? Or is this more daydreaming? as the heart pounds to pump, and force exchanges through this rare, lively air. I thought I saw my demise, without much fright. Now is that foolish? Do I sear of the mentality, to take such risks? It's hard, this breathing still. Still. A vassal of example.... to avoid? But see I clear as the wind, this day is pleasant stirring. And I am a part of its burn, to give advice.... and make a fortune in return, with coins of candor— for my shaking fingers to collect.... that I am doing this. (*weakly lifts his head*) Sufficient.... Sufficient to take in this appealing atmosphere, for caring of a life to seem and lift from doldrums of self-pity, delusion, derangement, questioning, and deplored sacrifice that devastates my self-consciousness and reserve. So pretty is what's seen, to contradict the folly of my feelings. Yet here we are, to sense forgiveness for my wasting away, my pride of person.... and my perfidy personified to regret.... (of) what I caused. (*A **boy** sits at the other end of the bench, unwraps a sandwich and stares at it with disapprobation.*)

LANGLY (*noticing*): What's that you have there, sonny? Don't seem conducive to your consumption.

BOY: Liverwurst! My mom says I have to eat it. But I hate the stuff— tastes like liver!

LANGLY: That's very fundamental.

BOY: I was looking to throw it away!... But I can't be so bad, for what my mom makes me.

LANGLY: Give it here! I'm broke, broken, an' hungry, and can take the taste well, if you haven't abused the sandwich much.

BOY: But it ain't made for you!

LANGLY: Would you rather it be wasted? for all the effort your mom put into it? Then give it to me, if you're not gonna devour it.

BOY: I'm not sure that's right.

LANGLY: Do you want it?—

BOY: No!

LANGLY: Then, is your mom's work trash ta ya?!

BOY: ....O.K.! (*coming over to him*) No work's worth for trash to make.

LANGLY: ....Or leave, littering the park. But you're a good boy. There are disposal baskets around. But you just couldn't bring yourself to stuff that lunch in one. (*accepts the sandwich*) And it smells quite nice. Looks fine— You haven't tarnished it any with your reluctance. (*starts eating*)

BOY (*running off*): It ain't for lunch.

LANGLY (*eating*): Well it's not to feed the pigeons!... Tastes fine to *me*. First free coffee this morning, and now this. Some days are made for offerings from the gods.

Scene II — ***Pomelory*** *is in his home office, which is more like a comfortable lounge with a broad desk and plush but swivel leather chair. He is relaxing in the chair at the desk, though turned some-what sidedly (or sideways), and with a slight grimace of concern to his face. His sister enters with a tray carrying a carafe of tea, cups, spoons, etc., although the room contains an automatic coffee maker and the carafe is often used to port water to it. He casually watches as she places the tray on the desk and prepares two cups.*

MAUREEN (*while pouring tea*): They're definitely in some sort of trouble, you know. Whether it's major or minor I can't tell.

POMELORY: I told them, they have to figure these things out themselves. I'm not running the farm anymore. I'm responsible for it, but they make the standards now, and conduct the rules the spread lives by.

MAUREEN: It's still a family run business, Oak. We should help with our advice—

POMELORY: If they ask for it. Is that the pure honey? Then two teaspoons for me.

MAUREEN: All honey is pure, Oak.... Well, my son calls me to say things are becoming very trying.

POMELORY: The commercial brands are contaminated with un-necessary ingredients and additives. Who should ever have al-lowed putting vitamins in honey! It's just a way to dilute it with water to make more of a profit from volume sales. Distributors will try anything they can get away with, any darn blasted idea or con-cept.

MAUREEN: The tea masks any distinctions between the grades of honey (*putting his cup where he can reach for it easily*), and by definition dilutes it. The result is all the same, for how it's used. (*takes her cup to sit on a nicely sized wall sofa*) But we should still continue these discussions to decide on how they're managing. My retirement is involved. (*sits*) And so is yours.

POMELORY (*sips*): Between my two sons— and nephew, they should be able to handle any problems that come up, grave or not. The most important thing is not to acquire too much debt, or make too many promises— which amounts to the same thing. I've drummed that into them well enough already. But they haven't consulted me at all—

MAUREEN: They're afraid to, that you'll find out matters might be hopelessly bad.

POMELORY: Not possible for that to happen, unless the earth dries up, Maureen. And no drought's occurred. The produce ships fine, as far as I can tell. Now I told them—

MAUREEN: Your boys, Oak. Not mine.

POMELORY: —It's all right to take some risks, if they're clever about it, and (as) attune to the markets as they should be. They're young and smart, ambitious and current, and have to work these things out. I've spent a lifetime worrying through such problems, and deserve a rest without anxiousness, leaving their able shoul-ders to handle the daily burdens. Sure, sometimes you have to get shot, to realize what a bullet can do. But if that has to happen, you finagle events so that you take it in the backend— and not the head. You don't learn unless you rough yourself up with what you're doing— And you never rely on luck and chance or fortuity. You force your way through difficulties with deliberateness and planning, and objective assessments for change of tactics. They certainly know that, coming from me.... Have they asked you about any specifics, even comments on vague tendencies?

MAUREEN: No. My son only says that the bankroll seems to be thinning, and they can't understand it—

POMELORY: Can't understand it?! Don't they keep up with the accounting? If anything's a weak point with them, it would be that. They've only learned their finances in college. Probably spend too much time playing with computer models, than finding out what's really going on. It's complicated. But you have to sift through the dirt with your bare hands, to find out what's going on with the roots, the root causes to these difficulties: checked flow of capital or resources, or whatever other errors are occurring. You have to work hard to find out the truth of matters and practices, particu-larly when the control is shared by others and commitments must leverage your actions. Veiling failures is a disaster. Shine light on them instead. Then you'll learn what they really are, instead of imagining— for the best or worst.

MAUREEN: They probably don't do that, Oak. They're brave boys. (*sips*)

POMELORY: Ugliness is as much meant to be seen as beauty. The purpose is to make an impression, to stimulate a response, corrective action or praiseworthy utility.

MAUREEN: Do you think it's possible they could bring us to ruin?

POMELORY: I only hope they do! so that they stay alert to avoid that.... prospect. We have our savings anyway.

MAUREEN: You know we'll give that all up to help them, or bail them out of impossible situations, even if our funds are like water running through a sieve.

POMELORY: Shouldn't come to that. And if things get so bad.... they'd be too proud to directly encumber their parents. Our gifts would have to be more substantial than ourselves, to keep that farm running, and our businesses going.... (*sips*) No. The economy hasn't been awful. Few great hardships suffered by any who remain game to the entrepreneurial spirit and atmosphere. They might simply be coming up to.... obstacles or puzzles novel to them, the queasiness of first time approaches.... But if they don't think it's important enough, to ask my wisdom for how to deal with odd(ly) shaped knots.... then employ more energy to solve.

MAUREEN: You could question them. (*sips*)

POMELORY: You know that's the one thing I've vowed not to do, once they took charge. It's probably a petty matter of accounting that they should discover on their own, so that they'll know what caused it and how to fix it. They make their own formulas—

MAUREEN: My son wouldn't want to alarm me, unless all three would like to pass onto you some concern to realize, knowing that I would.

POMELORY: You're savvy enough to run the whole enterprise yourself—

MAUREEN: Heaven's no! that's not my creed, Oak. Your wife would be better.... Our brother would have been. I never liked giving orders, managing people; and that's what farming's become, today. (It's) Not the fun and gainful occupation I grew up with, and shared with you. It's not the same— And agriculture is mean and narrow-minded, now. Paying more attention on the tiniest details than the whole worthiness of the soil's organics. If they could replace earth entirely with minerals and sand to sell, for planting, they would.

POMELORY: Scientists— or should we say terrestrial engineers.... say they can do that already, but the process is much more expensive than using what we can find at out feet.

MAUREEN: Well it's not awe-inspiring, Oak. There's no communion with the land involved, that way. And the grazers know the difference, are even insulted by what foolishness now's presented them.

POMELORY: But improvements that have been made, in common practice, you have to admit are impressive. Store bought fruits and vegetables are far tastier than any from a home garden, and much healthier or less blemished. And meats in general are more robust of flavor— for cooking— and are cheaper to obtain than when we were kids, especially the fanciful cuts.

MAUREEN: Less crude does not have to mean more crowning of achievement.... But we're reminiscing. True, the public doesn't have much to complaint about. They find something, within all of these optimizations of produce growth and delivery. But the complaints are seldom very legitimate, nor even genuine. They're too often fictional.

POMELORY: Well, blame advertisements for dealing exclusively with fictions. They're.... overwhelming of ubiquity. But a person is still allowed to use their own senses to make the final decisions on what's personally to be consumed. And good options offered, to that effect, have increased greatly and become a reasonable necessity for living. You don't find worms in apples any more, or caterpillars in pistachio nuts. We too must adopt our ways, to these higher expectations of reward. And my boys follow the industry standards with both acuity of staying up to date and practical training from our family experiences. They should be more than capable of handling any subtle adversity, like a consortium made change in seed requirements, or government subsidizations that lessen the value of one or a few crops to stabilize or increase production or available bulk for.... newly found uses.

MAUREEN: I'm sure our sons follow the markets very closely, and put into practice organizational trends that we would approve of— and understand— for creating the finest possible foodstuffs. Simple shops simply can not suffice for our suave society and current culture. Few *can* provide their own meals, nor would want to. There's still much reason for being of our family efforts. I fear more outside influences and external temptations. The younger ones would be more aware of current confluences for these to promote supposed opportunities, and that is a snare I suppose our sons could be daring with.... But I don't know, Oak. I'm not much for administrative analyses.... Or I don't want to be.

POMELORY: ....A few bends of the bone can't cause a great deal of damage to the youthfully resilient and recoverable. It's all of a process of learning how to manage through your working environment. Everyone else goes through much the same— And that's the only thing which keeps the field more or less level for all players. I can't imagine they'd let themselves fall prey to ridiculous or outrageous scams.... Though many weird things can befall some under the slightest of prolonged pressures.... But they're too capable—

MAUREEN: I think the world's become more aggressive, Oak. The waters are more turbulent than when we actively swam with simply the innocence of our convictions and perseverances. More compromises to ideals are necessary, to continue effective participations. Even major changes in precepts may be occurring. We worked the farm for our love of doing so. But in today's rush of rumble-tumble for results, our offspring might be doing so because they *can*; the outlines of operation have been prefabricated for them and are easier to follow or fall into step with than for.... less fortunate others. But the essence of the concept might be missing from their psyches, because that's no longer needed. I think Johnny would have preferred to become some kind of.... elementary school teacher. Heaven's knows why. But he took some classes for that in college. May even have credentials for it. I don't know. He's very private about.... what he's given up. Says he's let loose of the idea. Means he thinks he's lost the prospects for it. Too busy, now.

POMELORY: Could it be due to the influence of his father?

MAUREEN: ....Who?

POMELORY: Your ex, Maureen. You can't keep ignoring that he exists. His father demeaned the teaching profession as unsuitable to a person of any esteemable character, that old philosophy of "the doers do, and those who can't teach," which he often upheld. Rather amiable fellow otherwise.

MAUREEN: I don't know of a *person* to whom you could refer. (*sips*)

POMELORY: Simply insisted that it takes more than learning— to be learned. Results are demonstrations, not descriptions. And too much of a trend towards the latter denigrates and degrades the consciousness of a society, like (having for it) resting on its past laurels instead of laboring for new ones.

MAUREEN: If there are achievements more to be made, the calibre of their architects must be much greater than the substance of ghosts and deceptive faiths.

POMELORY: Darn it! Maureen. You experimented in your youth, like just about any.... found a plaything and tried to convert his very race! It didn't work— He bolted and left, insulted and inspired by another.... and others more similar to himself after years of constrained torture and anguish— and longing. It was all social misappropriation, under the guise of polite subjection and demurring. You're just angry for your failure with him; but how often and (for) how long can a man take, having his.... physical and conversational attributes belittled or slighted for improvement?! And does it not occur of thought? that he thought wrong of you, as much to subvert towards a lasting comfortable, compatible relationship. When two proud heads butt up against each other....

MAUREEN: There's more to be dismissive about cultures than empathies for them. The current between them and which distinguishes each.... is simply due to misguided curiosities. But all that's left are phantoms to know about, discover, learn, and reject. And Johnny is my cautionary tale, of whatever may be corrected as salvageable from these interests. The lesson is only to be fit— for anything.... And that is what I've lent the farm's enterprise.

POMELORY: Very well, sis. I can't condemn griefs from grating tempers. The abrasions must remain sore. But, you've always been like your brothers— And you should have been a man, to try against these follies, of playing with a marriage, and trying to handle that!... from the womanly side. You can't command a proud man (*as **Laura** enters*) with any form of equity.

LAURA: Are you two still having a chat about the business? There's little to be said. And I've come for the teapot.

MAUREEN: There are things to discuss, and concerns to face.

LAURA: Face up to. It's a silly expenditure of time, Maureen. The boys run it. And you could have used the coffee maker's.

That's what it's in here for, to heat up liquids— for instant soups and teas, and to make coffee, just like in a real office. But the teapot should really be kept in the kitchen, or for serving in a living room. It's more than a utilitarian appliance. It's a decorative item, and a fairly expensive crockery. And, well, it disturbed me, not finding it in its usual place. And seeing that you've brought it here, for such a trifling reason— again— I've come to bring it back, because I'm ready for some, and also to dissuade you of its use this way. Break the habit now, before it is— accidented.

MAUREEN: Is that really all of your complaint?—
LAURA: You are a bit unstable, dear—

MAUREEN: You could have brought a cup with you—

LAURA: The water's probably tepid by now. You can't steep tea in tepidity—

MAUREEN: The tea is in the pot!—

LAURA: *My* tea is always in the cup! the tea bag properly homed — for the right amount of time—

POMELORY: I do find our little conversations conforting, Laura, between siblings discussing common matters of personal interest.

LAURA: But there's no purpose to it, as regards running the farm, Oaker. You don't anymore. You've given that up. And this sort of meeting seems almost disrespectful and lacking of confidence— to those now in charge. So let our sons handle it—

MAUREEN: Yours.... and mine.

LAURA: And have them do so without your wish to interfere. Your unuseful chats, if done in the spirit I suspect, affronts their accountability to all of our family. And I won't have that placated so easily as by a misplaced teapot when you already have an automatic pot for coffee.

MAUREEN: This is being too picky of you, Laura. Picky and mean of detail—

LAURA: For the superfluousness of your efforts it is not low to lightly chastise silliness. That particular earthenware imparts a particular wholesomeness to the hot beverage, when you brew it as you do, much of the tea's harshness absorbed by the pot's (inner) surface. But using it too often that way, to over-cook the tea, lends some bitterness to the clean water— without purifying that water more— so that it's more difficult to judge a correct steeping in a cup, or may even make the procedure unproductive.

POMELORY: But our little tête-a-têtes are not. And we certainly don't speak to disparage.... anyone concerning (the) operations at the farm. If anything we seek to find means of encouragement. And this is all very informal anyway, and totally noncommittal of opinion for either of us, (un)'less intrusions pierce these sentiments distortedly. You interpret things wrongly, for what we have to confess to each other of our interests in what's going on. The fact appears to be that they may be running through a span of difficulties. And we're trying to conjecture on, if this is so and what it could be. They've not spoken to me at all—

LAURA: Facts?! I don't wish to pry into private affairs that are indeed my own, inclusively. But they've already spoken with me about their squabbles with John, who thinks they're too hasty to take on loan commitments for an expansion of production in the current economic climate. John wants to wait until interest rates fall down a bit before revving up the soybean returns, while we want to get in on a trend—

MAUREEN: We?!

LAURA: —that's sure to grow substantially, and increase that acreage for much higher profits.

POMELORY: Is that all it is?

LAURA: Already these legumes are being used for vegetarian meats and diets. Tofu is becoming a fashion trend. And edamame is replacing green peas not only as a side dish— but also as a snack! in up scale restaurants. It's only a matter of time before they become more common in grocery stores and supermarkets. And we can supply the bulk of it, for this region of country, if we're smart of the initiative to be a major grower.

POMELORY: That sounds practical, Laura. Can't say I follow that particular crop—

LAURA: John's concerned that some of the farm's savings are being adjusted to take on more risk, and can't quite follow the logic of the maneuvering. But the enterprise will fall into no greater (load of) deficit if demand leads to expected sales and surplus of buying. Forget about the wasibi! The finances remain comfortable, they assure me— Not that John is being a bother to remain prudent, but he hasn't the imagination to leap over the current and into an obvious future.

MAUREEN: We're all blinded by that way.

POMELORY: They'll have to take care of this themselves—

LAURA: He was never taught the verisimilitudes of risk taking and capital endeavoring.

MAUREEN: Johnny's as well trained in economics as your sons, Laura. I'm sure they all must have taken similar courses—

LAURA: He hasn't the degree. He can not figure out the complexities as easily. He only concentrates on what he can understand and can't encompass a broader picture— And that's not in and of itself a detraction, since it allows for a valued capability much required. But it *is* a limitation of vision, a narrowing of scope (and perspective).

MAUREEN: Well I wouldn't want him to do things (that) he didn't understand.

POMELORY: And so they come to you to come to me.

LAURA: Come to me nothing! Oaker. We've always been easily conversant. But this is much of a sporting game for them, to see how much they can figure out without *having* to come to you for advice. And so I say you shouldn't undermine them—

POMELORY: We weren't thinking of doing so!

LAURA: —through ignorance and false prophesying of their abilities and courage.

POMELORY: That's not what we were about at all.

LAURA: Well when you have to bring the teapot in here, it heralds some supposed gravity. Because Maureen gulps it down like water when she's upset. Couldn't keep enough of it in the house, during the divorce— And very cumbersome for me, not finding any bags, for a reasonable respite of drink.

MAUREEN: It's an affectation I admit to—

LAURA: Well don't start it up in here.

POMELORY: I see. It's the competitive pressures at play. And I suppose if we're having coffee then everything's all right.

LAURA: That's the obvious indication, Oaker. It's why the machine's around.

MAUREEN: Has always soothed my nerves more than any other beverage.

LAURA: And you're not even a devotee— not an educated one.

MAUREEN: But only when those nevers are just slight enough to be sharp, piercing, and focused for the penetration of anxieties. Tea is particular, that way, to relieve—

POMELORY: Hot tea.

MAUREEN: —the stresses.

LAURA: It only reminds you of your adolescence, when you thought your sophistications could rule the world and command all wanted attentions. Strange how people fixate on simple things during that age of development.

POMELORY: Well I'm sure they can handle these.... ploys for fortune—

LAURA: That time before you discovered the deficiencies of character and conjecturing about the universe, that you can not make everyone stand on their toes to notice you with fascination—

MAUREEN: But I can make tea....

LAURA: And not a decent breakfast.

MAUREEN (*as **Pomelory** stands*): I can still correct imperfections, and put to rest improper attempts at life.

LAURA: How vague to show a disappointment! After all of this time, and what's happened to you, you're still so childish.

MAUREEN (*as **Pomelory** goes towards a window, to face away and look out*): And why do you dislike my blown dreams?

LAURA: Blown up and discarded, like a busted balloon. I resent the immaturity that lingers. Certainly you were the princess of the farm, a miracle just as able as your brothers, to do the hard and noble work, of laborious— pumpkin growing, and egg harvesting—

MAUREEN: And tilling and plowing, milking, raising and rearing stock!

LAURA: But that's in the past, and you're not the renown precocity allowed. You should stop trying to see yourself in that light. You should grow up to the realities of your wanting, to help explain your.... misadventures—

POMELORY: She's gone through a very troubling marriage, Laura, that's caused depressions greater than the disappointments met and battled with. (*facing them again*) But that has not, after all the hurting and bother, dimmed her unique spark, nor diluted the fine distillate of her being— not around here. That privilege of consciousness remains for me to adore.

LAURA: I'm not one to *want* to complain. But it cost us some loss of money, to get you out of a crushing situation, an embarrassment to face, remortaging your house and finally selling it underpriced.... because you left that man in such an impossible way to handle the affairs between you two. And the least gratitude you can show us should be to share of your responsibilities as a tried and tested adult to demonstrate your acceptance of a disastrous and almost ruinous handling of a once sincere companionship—

MAUREEN: I'm to blame for every sore on my face, every lash of wrinkle— then! And I bought some tea in bulk, crushed-up loose leaves— and even gave you a tea ball for steeping with, as a present.

LAURA: I don't deny the genuineness of your hurt feelings.... I dislike having to wash the thing, but it was thoughtful of you.

MAUREEN: They were on sale, and I paid for several. Several are around.

LAURA: The metal tarnishes easily. I'm not sure if this affects the taste—

POMELORY (*coming back to the desk*): We're all fit for a lifetime of distractions. The harder we work, the more acute they seem — the sharper they pinch. But that's no point (*standing at the desk*) to cowering at regrets and upholding mistakes as trophies of achievement to define our callous natures. (*lifting his cup to sip*) We.... rebound from errors just as quickly as we make them (*sips*), considering their effects must remain rather insensitively viewed. There's always much more to do, after corrections.

LAURA: I'd say so, Oaker. To stay busy is the only correction for faults.

POMELORY: And I do so, in my partial retirement. (*sips*)

LAURA: What does semi-retired mean, if you loose your feel for a business or activity?

POMELORY: You keep your hand in it a mite, so that the touch doesn't wane entirely.

LAURA: But what about one's interests? (*looking at the teapot*)

POMELORY: ....We'd hardly want to interfere, and break things

by shoving about influences. I'll not lessen their practitioner-ing and handling of— complications or surprises. Under such tensions I was most committed to the operation, and became extremely annoyed by freely offered and utterly undesired opinions and advice — even when one sensed it could be good or propitious. But then you concentrate your mind on what is foreseeable, and devise your goals around that vision instead of the predictions of others. For the most part that works, when you can control your own efforts, when you are the instrument that may self-calibrate to succeed. If compromise is involved, though.... the give and take must foster the progressing. And our boys should work that manner out, instead of indirectly or surreptitiously trying to enlist me to convince Maureen that John's worries should be subdued. Oh, it's a ploy they've tried as children. But it won't work now. My hands are off the minds of men—

LAURA: Good! (*as he sips*) It was only casually suggested that you might think otherwise. But they never counted on it, and shouldn't. Though John must prove his caution as a useful counterbalance to enthusiasms.

MAUREEN (*as **Laura** starts to leave*): He has a fine enough head for that enlistment. (*sips*)

LAURA (*noticing her sipping*): Please wash that out with a weak vinegar solution, Maureen. (*Exits.*)

POMELORY: Enthusiasms. That's the blood's life. We need the more. (*sips*).... This tea's quite fine, sis. Has always been, when you make it.

MAUREEN: Thank you.

POMELORY (*sitting on the desk top*): We need more of the enthusiastic, like a young fellow I led a job to.

MAUREEN: That guy who came here? for of all things an interview?

POMELORY: Well I invited Beck for that. And it wasn't so much an interview as an observation, since he had absolutely no qualifications for the position he was asking about. Wasn't particularly demoralized to realize it after my questioning.

MAUREEN: But you got him a job there?—

POMELORY: At the farm? No. I don't hire. My boys do.... I can't get anyone a job, Maureen. But I recommended him to the café we used to frequent.

MAUREEN: Oh. Bruce's.

POMELORY: Been having an occasional breakfast there, lately.

MAUREEN: I can cook it for you, Oak. We still both get up early.

POMELORY: You know I can be particular, some days. We're not talented in that way. Too crude to burn things.

MAUREEN: Well, we had a chef, and you're used to that.

POMELORY: She was mostly for the farm hands.... Laura did all right by me. But not any more, not after retirement. No practical need, she says. And she does not like to put up with the unnecessary.

MAUREEN: Why don't you have it at the farm?—

POMELORY: I'm not driving miles for a breakfast, like I'm chained for my existence there with some kind of dependency. And I'm not part of the labor any more, to warrant that type of perk, I certainly don't feel.... An early morning walk to the café is good enough, when the cornflakes just won't do. Always nice to have a purpose about something you're doing. (*sips and then places down the cup*) But the young man has no purpose but to be, which is refreshing. It's how we all felt, didn't we!

MAUREEN: Well.... the purpose was already a part of us. We didn't have to think about *that*. We grew into what was around for us.

POMELORY: Finding what's around. That must be frightening — if you have to look.

MAUREEN: Do you mean compelled by necessity, or curiosity? (*sips*)

POMELORY: What's the difference if you have to— I mean, you're not likely to find much, in some situations.

MAUREEN: The world is mostly made of tar to step through, Oak. Suddenly one day you realize what you've been running on; and then you can explain the delays, and the condition of your shoes and feet, and why you're never gonna get to where you've been headed for a mind, but have been too heavily swayed and detoured; and it's impossible ever to get back on the original, contemplated track. Because the thing was always totally fictitious. So now you're resigned to go slower and slower until you stop an' slowly sink. And that is a life. You were deceived from the start, and by the end you can hardly care for the deception to antagonize your thoughts. You're just relieved you've something to drop into. They call it a life's experience to look back on and force some fondness from the frivolities of your frustrations and forfeitures.

POMELORY: That's a rather pessimistic view of things, sis.

MAUREEN: The better now to see its truth than to play the child all over again with the foolhardiness that's become such a fashion today: making yourself over, redoing your appearance, initiating new drives and interests just to be seen as commendable. The feeling good about one's self is hypocritical to the reality of your aches and pains and creaks.

POMELORY: We're not totally decrepit.

MAUREEN: Life was never really meant for pleasure. Some activities just lead to that sensation incidentally. To linger on solely for those occasions is wrong. It is.... juvenile.

POMELORY: You want to be happy and satisfied, with *some* aspects of living!

MAUREEN: You don't understand what I mean. Not that the body to feel good is wrong, but that life is not the pleasure taught. It's quite independent of dreariness and elation, fright and security,

trust in and rejection from. It is ambitions to be tarred, feathered, and stopped, made stolid as a lump in the throat. It is the electromotive force that must wind down, once foolishly thought to be more precious than the circuit. Once found handsome enough to want to drink, it only burns the lips.... these vitalities we glorify too readily, these imperatives of emotion to find a sanctuary of vows.

POMELORY: Beck vows to make something of himself. That's all I can interpret of his behavior. How he goes about it is quite novel and fascinating for me to observe, because I thought.... our family was the definition for such concepts. But the man is of such an other kind, unschooled to useful experience, that he relates to me my earliest impressions of uncertainty and the idyllic empathy that once youthfully proposed. I want to see at least an indication.... of how one such works through— the tar. Though he's hardly depressed of prospects. Seems to be well shuffling tables and chairs.

MAUREEN: Well I'm not depressed either. I'm concurrent of the realities as I see them. But those types of laborers value coal more than gold, because they know they'll never get their hands on any considerable riches, Oak, unless something extraordinary happens (to them). And they're practical enough not to wait for that. So any decent opportunity reached is a victory to make them seem inordinately delighted.

POMELORY: Why, the café stint is only a part-time job. Wants at the farm, though. I told him to ask in a couple of months, when more hands will be needed. I'll recommend him, due to a demonstrated reliability of service. That's the most important thing for any employment— much more so than talent or accomplishment or fine finishing. Dependability, at whatever level of ability you descend to fit well to. It's much like measuring the density of grains in a column of graduated oil denseness. If you're good, you'll stick to your level. If you're rotted out, you'll sink. But you're wanted for being a grain.

MAUREEN: You mean for groveling to find a stable position or place.

POMELORY: Grain, not gravel. You feed the pigs that, and they starve.

MAUREEN: Oak, a confrontation's brewing between Johnny and your sons.

POMELORY: They can work it out.

MAUREEN: He's coming to me for sympathy.

POMELORY: Give him all the understanding you may. But leave those problems to them. (*coming off the desk*) It's their show — And it won't be a showdown, because that's silly. But cousins have to feel a way around themselves, to adjust their standings and relevancies.

MAUREEN: It's two to one—

POMELORY: No. (*sitting in the chair, but straddling along side the desk*): It's one of three. But being slightly older and more conservative, he's not about to entertain an unproductive squabble. They're equal partners, and he'll bow to the majority view as much as he has to. But everything will work out fine, without our interventions.

MAUREEN: It's strange how I no longer have much of a feel for the farm, when I was so able there.

POMELORY: ....We're out of the way, Maureen. We deserve this rest— you especially, after all you've gone through. And we're represented faithfully by those we've generated. City life was never so bad as what we could have supposed.

MAUREEN: No.... It's not, as long as we remain comfortable, decent, respectable, upstanding— and venerated.... as examples of leading but well grounded inhabitants, earthy but earth bound (*sips*), and as frail and fallible as anyone could become.... when deceived and misused.

POMELORY: ....Misapprehended, perhaps. John makes up for all casualties of heart. He is an equal. And I assure you, it will stay that way.

Act III

Scene I — *A grassy backyard. **'Aron** is sitting on the lawn, next to a toy cart. It's sunny but slightly cool yet, in the early afternoon. He has a coat on, but has taken a shoe off. **Heroford**, in a jacket, comes out from a porch door.*

HEROFORD (*approaching **'Aron***): Why (ha)'ve you taken your shoe off?

'ARON: It pinches—

HEROFORD: Already?! We'll have to get you another pair. But doesn't your feet feel cold?

'ARON: Not so much.

HEROFORD: It is warming up. Supposed to get quite hot today. (*stopping by **'Aron***) Why don't you put that back on, before Mommy finds out. (*as he's complying*) You know she's afraid you'll catch cold again. Has the radio on, with those gravely tones she loves. I suppose that kind of announcing, or show hosting, is a dying art. But it sure is a nice day to be outside, 'Aron.

'ARON: Yeah.

HEROFORD (*stretching*): The sun is shining yellow, and the grass is greening blue. Marce come out here often?

'ARON: Sometimes.... I don't chase her around.

HEROFORD: Surprised why cats stay so particular to a home. Heard stories of where they suddenly wander off for weeks, months at a time, and then return. It's called straying.

'ARON: Where do they go?

HEROFORD: No one knows, unless you see them. They're not legally bound to a residence, even though they can be considered pets. But that's a definite problem with them, what they can bring (back) with them, from where they've been.

'ARON: She *likes* being under the sofa.

HEROFORD: I just saw her lying out here, one day, lying on the lawn as inconspicuously as her size, she's such a little thing. Noticed that, while she kept her distance, she didn't seem too perturbed with my being around. I like to sit out here and relax. Didn't disturb her any. Noticed her again the next weekend, and she was more friendly. Although I think Danny may have been playing with her, or encouraging familiarities. Eventually, we invited her inside the house. Mommy was a little suspicious; but she's never been mean or vicious, and waits calmly to go outside— I'm told— without any fear of not being allowed to, in the mornings.... and comes back here before school lets out.

'ARON: Sometimes she stays all day in the house.

HEROFORD: Well, when it's very cold out. Takes to the kitty litter quite nicely, as long as you boys keep that tub clean.

'ARON: Hun?

HEROFORD: Duncan and Danny. Mommy seldom has time for that responsibility.

'ARON: She can go wherever she likes, when she's outside.

HEROFORD: Yes. But it's still another expense. We feed her generously.... Oh, you'll be allowed to roam farther very soon. When you start Kindergarten in September, that in effect will be so.

'ARON: I'm afraid of the words.

HEROFORD: Words?

'ARON: In the picture book. I don't know them.

HEROFORD: That knowledge will come. They're not going to test you on it in school. I don't think they do any testing in Kindergarten.

'ARON: No?

HEROFORD: None. It's mostly play made constructive for learning, associating objects with names, and getting familiar and used to letters and the alphabet. That picture book might be a little too advanced for you, but it's essentially the same thing, associating pictures and concepts with names and words called identifiers and descriptors.

'ARON: Hun?

HEROFORD: The red ball and the fast car and the squeaky clean bath tub. You'll get used to all of that.... It's only that Danny took to it so happily, at your age, and quickly learned the material. But the book was a gift to him, from Grandmama. So he was enthusiastic about it. You don't have to be as interested. But you're not going to be scolded for what you can't do.

'ARON: No?

HEROFORD: Genuinely not. It's what you kids *can* do which you shouldn't, that bring on the punishments.

'ARON: I can't do anything!

HEROFORD: Well then you've had it safe, so far. I know there're going to be problems. Duncan certainly (ha)'s come of an age of mischievous growing independence. But you don't have to worry about any of that yet. You probably don't even remember what you were doing at two and three, so frustrated by not understanding things.

'ARON: Does.... frust-ry make you sad?

HEROFORD: Don't drone on the picture book. Don't ever be afraid of any book, even if you can't understand the contents. Mommy just wants you to be a little prepared. But these facilities will come to you in a normal time. You already know how to use most of the words you're ever going to learn.

'ARON: I do?

HEROFORD: And you'll learn how to master more, to qualify your expressions and distinguish your thoughts from others.

'ARON: Most and more?

HEROFORD: Progress will be so rapid you won't notice it. Few do. But it's learning how to master things that counts. Like telling time. When you learn how to read a watch, you'll get your own. And you won't have to ask anyone what time it is. They teach that sort of thing in Kindergarten.... Something to look forward to. And when I was your age, they taught you how to tie your shoelaces— Though I'll teach you that, or one of your brothers, when those kind of shoes become useful for you. And when Grandpa was your age, they taught him how to make various kinds of neck tie, straight and bow and what have you. It's all in the mastery of what's in fashion for an era.... I went for a year with a clip-on tie to school, before I learned how to make (up) my own. Now boys don't even have to wear them for classes to attend. Disciplinary behaviors and patterns are changing, uniformity becoming less common. So you have to master on your own, earlier, with less aid and guidance, or organizational instruction. I personally don't like the trend, treating elementary school students as if they're in a college of tutoring, as if they've already earned a character of independence for their studying. But that's the trend, to rush upon educational advances, whether the pupils are emotionally ready for them or not, as if levels of maturity need not be considered much. It's a machined analysis for efficiency, not a human's. So much more important to gain a manner of mastery for yourself. That way you'll more easily realize what's really worthwhile, by doing a lot of sampling of interests with a nicely honed ability to satisfy your curiosities safely and with practical steps. Much better than allowing yourself to be conditioned towards one path or another, as a culture sees fit. I mean, a society might have its needs, requests; and they will be fulfilled by the expenditures of people's efforts. But one should still craft a personal identity by what you discover that suits you. And that process takes a mastery over your attempts, a polishing of your applications to adventure....

'ARON (*standing up*): But what does frust-ry do for you?

HEROFORD: ....Frus— frustrations? Oh, we all go through that stage, when faced with something new that we can't immediately understand. It's not a sadness that's produced, but an uneasiness, an urge to want to get down to the meaning of a problem, to expend your energies focused for a solution to find. And when that seems difficult, your actions often look like frantic frustrations, frenzied fierceness and fighting habits, foul tempers and bad moods displayed, pouting and pounding toy blocks. In short, what's called a childishness becomes overt. That means it's in evidence behind your activities.

'ARON: Is that bad? (*sitting in the cart*)

HEROFORD: Well, if you succeed, especially to accomplish something much desired by many to have done, then you are cheered and celebrated with having the drive and ambition of a youth, no matter your age. If you fail, you're scolded or rebuked as being immature and bratty, and wasting much of your energies in fits, counterproductive convulsions and scandalous seizures too severe to tolerate as forthright for the attempts. One, therefore, rather quickly learns to hide public expression of one's frustrations, since it's generally much easier to fail than to succeed.... unless you are machined, driven by.... an assembly of thought, in which case the deeds are not really yours.

'ARON: Deed do dere!

HEROFORD: It's hard to be your own person, in a world of organized deeds.

'ARON: Dere dares the moat, to float upon.

HEROFORD: Yes. You're in a boat.

'ARON: And crocodiles to grab at you, and pull you under.

HEROFORD: The lawn can appear much like a sea, with broadening of aspect and perception into the imagination.

'ARON: What happens under there?! It's cold!

HEROFORD: They transform, for you to reconsider them. The fun is in the want to play with you, and keep you busy. You'll be doing much of that in school. You won't be bored. A little frightened, maybe at first. I can't imagine they've made school boring, not the early schooling. I even hear some grades allow the watching of.... instructional cartoons. I never had that. Not sure if I'd want it. The greatest thrill I had, back then, was gaining possession of a pencil set— with sharpener and crayons.... and then a school box, with compass and protractor, and then a small briefcase to carry the notepads, pamphlet and small books.... It progresses. Now they start you out with backpacks, and you graduate to internet phones, pads, pods, tablets, and laptops. All of the toys are so much more organized, for topical thought. I wonder if they'll show you how to read an analog clock face.

'ARON: Will I get a phone?

HEROFORD: We'll see. Duncan doesn't even have one yet. You have your radio....

'ARON: What does it do?

HEROFORD: Your radio?!

'ARON: No, the phone, the pocket phone.

HEROFORD: ....It's like a video radio, an advancement in communications, a leap over transistor radios, house phones, and televisions.

'ARON: I've never seen one.

HEROFORD: You won't have to worry about it, till they start up "Show and Tell." Maybe that's how they'll introduce you to today's modern devices. Essentially it's a privacy killer, so that everyone stays connected, to a common well of mostly unnecessary information. Mommy and I don't like them in a house of small children, because they can too easily lead to some naughtiness, by accident of inquiries. But they're more and more like the computer in my den, but about as far from programing as one can throw a usage. That's why Duncan's disappointed, even disparaging of them as petty playthings, because he's starting to take up some (computer) languages. But he's never use one. They'll eventually weave their way into the household, as you kids grow up. We'll be getting him a desktop computer soon, and see what happens: if they're actually going to start lending out e-book readers to the students, or usb-pens loaded with course material. I'm not sure if I like this, although it's a wealth of information. But the expenses the schools save will probably be made up by the equipment, machinery we parents feel we'll have to provide, to keep everyone up to standard. I'm too old-fashioned to believe true intelligence will be accelerated by this profusion of electronic capabilities. But I won't hinder my children by depriving them of what's current to play with and hope on. Though it's been my assessment that what I thought was crucial for knowledge devices when I was growing up have seldom resulted as. And I indeed started (out) on an abacus, as a tot.

'ARON (*loosing interest*): I have lots of kinds of balls.... The surf's swelling, and threatens to topple me over!

HEROFORD: Where are you headed?

'ARON: ....To an island, of jade.... and gold ducats!

HEROFORD: Has Mommy been reading you adventure stories?

'ARON: Yes. She shows me the words, but I can't understand them.

HEROFORD: Well, she's persistent, till the light clicks on and you start recognizing a few, (as) like you're able to use. You *do* understand them, and will *want* to read on your own very soon, as soon as you realize this growing ability, especially as your interests are peaked by classmates and you see them trying to read. You'll want to do as much, and it won't be so.... threatening.

'ARON: I hope so. All I can do now is read colors.

HEROFORD: ....That's something. Can you spell "red"?

'ARON: ....R—e.... a—d.

HEROFORD: Well, this is coming along. I don't think it's a dyslexia, but a misplaced attention or memory.

'ARON: Is that right?

HEROFORD: As right as read. You know if it's right or not.

Never try to fake yourself.

'ARON: But is it right? I'm not sure.

HEROFORD: There's no disgrace to making an honest mistake. That's really the whole point to growing up: living through and surpassing your errors. I've made so many, even as an adult, it would be criminal to count. But that's the only way to improve, because nobody is born perfect. And no one can ever reach perfection in anything, except being alive.... and not being so. Now, where is this ocean you're sailing on?

'ARON: It's out there, all around us.

HEROFORD: Well that is fantastic. We'll soon be seeing true splendors of geography, as a family, during our vacation trip. But a backyard's a wonderful place to invent an image, any kind of fictional event.... I know it's becoming slightly limited for you.

'ARON: I play here fine.

HEROFORD: You'll have lots of friends in school. They'll all be your age. They'll be your peers, and you all will act up together. That's the right kind of socializing you need, under controlled, supervised situations. We can't let you wander around the blocks, like your brothers, yet. You wouldn't know how to interact with strangers, who are virtually all you'd come to meet. But this is a fun enough enclosure, I suppose. Safe from the traffic. I didn't have it as well when I was a youngster—

'ARON: Danny wants a dog.

HEROFORD: ....We're not getting any dog. Would scare Marce.

'ARON: A puppy he saw.

HEROFORD: I'm sure there are many, irresponsibly procreated. But I think it's cruel to force a dog to live in the city. They can be fine, loyal pets; but they are rural animals by nature and should have plenty of space and territory to roam around in. Dogs limited to houses, in some cases even apartments, are sad animals. They can't be the beasts they were meant to be. I've already spoken to Danny about this. Kids naturally can't understand this point of view, since they're fascinated by all sorts of life to play with. But you have to be respectful to the needs of other beings also, not just your own curiosities to satisfy. When I was little, I grew up in a tenement, a dense crowding of flats in a building, among a row of them along the street and for many streets in the area. I won't call it so much (as) a neighborhood; though there was a pretense of security there, by the default of residency. But it was a dangerous, semi-impoverished place. And I saw many a dog and other pets very poorly treated, by their.... frustrated owners. Most were ultimately abandoned or starved or killed. Not uncommon to find the rotting carcass of a once friendly creature in the filthy street gutters, one you may even have played with. Those blocks were a death trap for pets. They really didn't last long. Leaves you with a distaste for subjecting them that way.

'ARON: A friend showed him it.

HEROFORD: Yes, from a family that's giving them away. They're not free. They're an unfortunate responsibility I don't feel we can take on. They have no papers, no certifications; the breeding conditions are questionable, the healths doubtful, and the practice disreputable to support. They have to be licensed and given vaccinations, the type of expenses that family is trying to pass on to others. We can afford it. But it's not right to promote. And I don't think Danny is mature enough to take care of a puppy. Duncan wouldn't want to, it not being his priority— though he might help— Mommy won't have the time for it, and I certainly can't take the trouble to deworm the poor beast and give it the proper attention a fledgling needs: house-breaking and obedience training, and anxiety pacifying and barking abatement, and so on. We're not ready to raise a household dog. The issue is closed.

'ARON: What if he brings him home?

HEROFORD: ....He won't. I told him not to. Danny is smart enough not to disobey. And we're leaving for the plains in a couple of weeks. You can't leave a house dog out, with dry food dispenser and water on the porch like we can do for Marce. They legally can't be left cooped up and without any company. And they can't be left alone to roam the streets, deserted and confused. They can't be let loose that way. They're not as independent as cats. And a puppy?! That would be a savage cruelty, totally inhumane. No dogs. No dogs ever, 'Aron. Don't expect any.

'ARON: We could take him with us.

HEROFORD: ....Not to our reservation vistas. They don't allow pets as guests. Why are you pleading for Danny? He certainly doesn't plead for you.

'ARON: I'm not, Daddy.

HEROFORD: ....Did he.... show you the puppy?

'ARON: No—

HEROFORD: Good! Because you could only have seen it at home. And I told him not to bring the poor thing here. We simply can't conscientiously handle that type of burden.... Now if he were to hide him, the animal could only be abused and suffering. And that would be a hateful way to treat something that can't look after itself. So if he ever suggests doing something like that, you alert me and Mommy. You can be our scout.

'ARON: ....O.K.

HEROFORD: Sometimes children can't help but act impulsively, they want something so much. They can't think past the wanting, and that is a lack of reasoning— I've heard no yelps of distress, and Marce is fairly quiet.

'ARON: He ain't done that, Daddy. He just says it would be fun to have around, a guard dog.

HEROFORD: Well that's a relief, 'Aron. Because this sort of thing is serious, when you're dealing with a life, and trying to manipulate it to suite your own pleasures. It's really a despicable and arrogant behavior people should grow out of. Association with the animal kingdom is one thing, but to mistreat them is another. I don't like zoos either, because all the prisoners are caged up. You can't understand that, yet. But the presentation is not worth the dreariness and tragedy of that kind of existence. They may be fed well, and nursed well from sicknesses. But they are not healthy

representatives of themselves in any sense, if the wild are not allowed to be wild. How would you like to live the rest of your life in a box, or (confined to) an artificial shelter?!

'ARON (*coming out of his cart*): I wouldn't.

HEROFORD: Besides, if any are extraordinary to know about, we can do that now with videos much more informative than any containment or captivity.

'ARON (*standing by **Heroford**, but apparently waiting to reenter the house*): It's getting warmer.

HEROFORD: Yes. Tell me, 'Aron. Do you actually like the children's programs you see on T.V.?

'ARON: Not too much.

HEROFORD: I've had a chance to glance at a few. They wouldn't hold my attention and interests much. Leave nothing to the imagination. Explain too much. Not nearly as personable as the shows I saw when I was young. They don't entertain with good comedy and satire as much. They don't teach irony well. And authority figures are trite and dismissible, too unreal and staged of fantasies contrived to be significant for the social mores of our time, as perceived by specialists and appointed educators. I can see through that immediately. It's like feeding your eyes and mind stale bread dyed with a rainbow of colors and actions. Too much technology makes for less of a technique, an art of instilling values. I can still heartily laugh through some of the cartoons from my era of childhood, when I come across them. This current material seems dull, seems catered to lull thought, placate short attention spans.

'ARON: The old ones are good.

HEROFORD: You can distinguish them, then. The newer ones are less founded for the human and his foibles, and more for humanizing into a culture made insensitive with too much overstimulation of mundane qualities and mediocre tastes hammered with to make common. Your cartoons are commercials now. When I was a kid, they were mini-movies, with good, engaging plots. And the commercials were a relief of tension, something you thought about while relaxing your mind from the exercised analysis of the entertaining animation's demands for your attention.

'ARON: They're "keen queued."

HEROFORD: The what? Today the stories are just a melange of preaching to you kids about something, instead of demonstrating what could be fun to think about. You can't tell the advertisements from the aims of the characters.

'ARON: Keen queued. That's what we're told.

HEROFORD: What does that mean?

'ARON: 'Cause it's all keen to do.

HEROFORD: Oh, to behave as children ought to, or are going to anyway.... From "hot" to "cool" to "keen."

'ARON: Hun?

HEROFORD: Paradigms of speech to exclaim something is good.

'ARON: ....Parachute?

HEROFORD: Well, they fall out for each generation or so, ways of expressing things simply and sharply. The young are always in such a rush to get on to the next activity.... Are you feeling a little warm in your coat?

'ARON: Yes.

HEROFORD: Then.... go inside and hang it up, neatly. You don't have to wait to be asked. (*'Aron stays put.*) What's wrong? You've changed your mind already?

'ARON (*squeamishly pitiable*): ....I'm not sure it's right.

HEROFORD: It's right?— There are no wrongs about it. There are no wrongs about an effort ever.

'ARON: Am I scared.... to be backward? Do you look after me?

HEROFORD: ....You mean, are we scared you're becoming backward? No. Not at all. You're just beginning to grow into a personality. You're not slow at all. And "me" is a universe to contend with. You'll find that out. It never stops churning. Everything always advances, in one's self, no matter the directions taken, whether they turn out good or bad. Mommy's just a little concerned with your reading. But she pushes after it to improve slightly, so that you're not embarrassed when you start classes. I'm sure you'll find several kids who haven't even attempted to read yet. And they'll be afraid at it, or what the teachers will say is apprehension, reading apprehension. They find terms for everything, so that they can train themselves beforehand to deal with problems. But those children with fears, their nervous energies will pull them through. That's the way it generally happens. Rough starting out, but everyone's even by the end of the semester. And most of the students are astonished, surprised by this. But that's what a schooling is made for, to standardize the pupils, at least as regards to examinations and grade levels.... You will help your classmates, and they will help you. And you'll all play together, through this education. It's never a shame to try.... to improve yourself. But is the fear for the difficulty learning, or the failure to after you've tried? Well, the one is an apprehension— the other a humiliation. But neither is a shame if you've tried. If you let things happen to you, that's an unfortunate depression. But the fact is, most of our efforts seldom achieve their purposes fully. So you just have to keep at it, till as much of a goal is reached as you can be satisfied with, when considering all of our other pursuits in life. What I thought I wanted, when I was even only a little bit younger (than now), has changed so dramatically of importance that it's frightening to look back on. I'm not as happy as I thought I should be by now. And you'll find, with what's called retrospection, that your uneasiness, with these early attempts at studying, will have been your best moments of drive and achievement. You will remember the difficulties fondly; because you will have long surpassed them, while so many others during a life.... will simply prove to be insurmountable, their complexities unpredictable, and their solutions impossible. So.... we're not afraid you're falling back, when you've hardly even begun: testing the waters of knowledge with a concerted-ness, a dedicated application which you'll soon train yourself to use, as a technique, for things you find interesting— and things you do not. That is the

greatest disciplining of the mind we ascribe our endeavors towards, which can, at times, be much more uncomfortable than Mommy's insistences.... Unpleasant.... Draconian of spite.... No. I haven't come out (here), for fear of your floundering. I don't expect that of you, 'Aron. It's not in the cards. It's not in the nature you share with my own. But I really don't like— that radio rasp.... Can you understand?

'ARON: ....Yes.

HEROFORD: Now, you go inside, and take off your coat.... and settle down (in)'to something— 'Cause there's not much to do out here, I agree.... It's a boring sea, we've made.... And I'll help you myself, with those names of colors (*as 'Aron starts towards the house*), to spell.... a little later.... She lets you pacify yourself with the television. Try something else. Try your own radio.... or other toys. (*a pause; as he slowly strolls around the yard, to come up to a lawn chair*) Yes, the day will be as comfortable as predicted.... I've tried so many things to foresee, and force, and constrain to a satisfaction. But the tear tears. And I feel alone out here, in my own backyard. Is this the life of family, and what I've worked so hard for to provide? Seems.... lame, and lamentable. I do need.... the exercise of labors, during this— vacation, of coming to terms with these results. But I refuse to play on the Internet, and be bombarded with its information.... when my own breeds of more tedium to realize: I am living through an era of errors. I've adjusted myself to stay discontented, bandaged up and numbed with— thorough work. But is it right.... to see this way, and seed this way? Or is this all a shivering with heat, to accept correction? So many responsibilities to fulfill. So few engendered of my principal wants. And not one temptation among them, but a slow suffering of burdens bargained for. I am.... in need of denial, of this day (*sitting in the chair*).... the reckoning of aims per-se-cu-ting me. With fear from this, I'm humbled to be true. We are no great achievement. Better than many, but done of wishes. And here is the abrupt breeze, a good lawn to lie in— lay.... of embarrassment, for the subtle treasures with abundance— left!... the roving sea, its waves.... procured.... of thought. Much like tears to think of, with this crushing embitterment.... But it's nicely provided for. I've done as well as willed. Then dreams deduce deceptions— Dreams deceive!... And I succumb to terror for myself. What an angry gentleness I have become, what a frustrated child! But that's for each to return to, and retrace the steps.... and wonder with more stepping of correction, that a lawn more golden lies, for one's panic to assuage, of head and heart, the swelling contempt for being so fundamentally wrong, and having this play out.... with consequences keen. There's a slow roving, in the garden treads, as I think this out. Toys are aimless, having been doubted.... So there's a sea for meanness and the meant, a capture of the lamentation's bent (*sighs*), conspiring to be forewarned of sight and feeling, an eternity among the few.... few blades for weeding. Slept am I, shivering the thought. Some paces made, forever lost their studying of. (*reclines to relax in the chair*) It's more than warmth to leave, but a hesitance. The world's too modern to suit.... diseases of the mind.

Scene II — ***Duncan*** and ***Danny*** *are walking along a street's sidewalk, and come across a construction site that partially obstructs its passage, with appropriate ribboned barriers of warning and caution.*

DANNY: We should cross the street, to get through. There's a lot going on.

DUNCAN: Odd place to be busy at. That building wasn't dilapidated at all. Was a nice looking structure, handsome and useful. And now they're tearing it down—

DANNY: Who are?

DUNCAN: —The city.

DANNY: Why?!

DUNCAN: To replace it with another, I guess. It's too expensive to keep standing. They're making all kinds of civic improvements these days, but this one's totally unnecessary. Dad took me into this bank once. The interior was grandiose and fancy, impressed me thoroughly; gave me a real appreciation for fine architecture— better than at the movie theaters. There's no need to take it down, it wasn't crumbling.

DANNY: The bank went out of business?

DUNCAN: No. There's a sign here saying it's moved to another location, with— more modern facilities and services to offer. That's just a shameful exploitation of something cheaper to do. I don't know how, but it's probably gonna cost less for them and more for the customers.... I think they found more parking area. No reason for this, though.

DANNY: Don't look pretty now!

DUNCAN: That's because they're demolishing it, Danny. Probably took out all of the art and high-class fixtures to transfer to the new bank. Feels like a theft.

DANNY: Why does it bother you?

DUNCAN: Well, when you've been to a smartly done place, you come to respect it, and the effort made to create it. I don't like the old lady being destroyed like this. Robs me of my thoughts, my images.

DANNY: Old lady?

DUNCAN: This is the bank of Sacred Saint Cecilia. Had glorious red tiles on the floor. I was stunned by them— They were patterned. Never imagined such a thing before, for walking on.

DANNY: It's religious?

DUNCAN: I think she was run by a church. That would make her an important relic, to my view. Shouldn't be left to this terrible treatment. It's like castigating my youth.

DANNY: ....Casting it out?

DUNCAN: I remember the building with fondness and importance for me, though I can't explain why. Dad said it was convenient. I just found the interior fascinating. It had an ambiance that spoke to you.

DANNY: ....Like a haunted castle?

DUNCAN: No. Some things demand your attention and praise.... by virtue of their beauty alone. And the sights sing to your head—

without any sound. You are simply astonished by its truth.

DANNY: Truth?

DUNCAN: That it exists. That you had never thought of anything like it before, and then you suddenly find it and that it's so.

DANNY: Well, it's not so much any more. So it couldn't have been that powerful.

DUNCAN: You're yet to understand these things, but it hurts to lose something significant.

DANNY: Doesn't hurt me!

DUNCAN: Would if you cared about it! If someone stole your candy, you'd cry to me to get it back. I can't cry to anyone about this; although the harm feels different, not like I'm wounded or intimidated, but that my primogeniture feels lessened.

DANNY: Primo— Pride? There wasn't nothing I could do about it. I had to come to you. He was bullying everyone. But you put him right.

DUNCAN: Yeah. And all the little kids are happy about it. And I got a speaking to from Dad, that I shouldn't push my weight around. And that that brat might have older brothers who would come after me. So it could grow to a never ending conflict or feud. So you tots have to figure out your own candy rights yourself—

DANNY: I'm not a tot!

DUNCAN: Well you seemed like one, while you were bawling.... But as first born, and old enough, Dad said he thought he should take me here with him, occasionally, to get an idea of how one becomes responsible with handling their affairs in a society, and teaching me some aspects of his work. And now it's all coming down, as if it wasn't worth a thing. I'm savvy to the condition of our city,— always keeping up with the trends and times,— not to accept that wholeheartedly. But it's still a shame, to be brought down this way, like a casualty from some war. A sacred place deserves more consideration than this. (It) Deserves a gesture of.... admittance, some token act of appreciation. Maybe we can find a part of one of those tiles, in the rubble.

DANNY: ....We can't go in there! It looks dangerous. And they're working.

DUNCAN: They throw out stuff, Danny, and we can search through that. The sidewalk's narrowed, but cleared enough to go through. So it can't be that dangerous to look around— from the safe side— to see if there are any remnants from the building we can seize, before it's all grounded up into a nothingness to think of. I just want a trinket of remembrance— It was pretty in there once. And I don't want to leave my conscience with this ruining so totally accepted by everyone. *I'm* not for it. They ransack a nobility, this way. I want a piece of that colored tile.... I'll treat it like a jewel, 'cause it's precious for my notions. They can't do this without some attempts like— ours.

DANNY: Aren't the dumpsters kept on *their* side of the barriers? We can't get at them.

DUNCAN: We shouldn't try at something as loaded as a dumpster anyway. They're heavy and large.... and probably unstable for climbing into. But there should be some trash bins around, less carefully attended—

DANNY: They'd be on their side too!

DUNCAN: Not all of them.... Not always. You still view things in black and white, yes and no. There are gradations that can be followed. I've noticed construction sites before—

DANNY: That's for construction!

DUNCAN: —and there are always some lassitudes to what's to be expected, some leeway from strict procedures. That's how you can find things, sometimes. They have to get rid of so much, almost everything that hasn't been removed before the demolishing starts is considered discardable and little of value. And some of that ends up in accessible trash cans near by, a safe distance from the craning and banging on the brick walls. But I'm not spending my time there unless I can find some indication of those tiles here, or we can.

DANNY: How can we do that?

DUNCAN: Looking through these linked fences. They provide a lot of view.

DANNY: But the building's way inside of them.

DUNCAN: They have to bring he rubble out, Danny. They don't leave it where it falls. They bring it to the sides, closer to the fences.... And these fences are not that thorough. There are gaps in them—

DANNY: Where the laborers and some trucks come and go. We can't get past them. They're peopled.

DUNCAN: You can tell much just by looking. To them it's just dirt. But a little bit of red is a prize for us. We might even be able to ask for some.

DANNY: I'm not sure if it's all worth the trouble we could get in (to). How good can a broken piece of tile be? I don't understand.

DUNCAN: Haven't you ever treasured something that's beyond value? Didn't you used to keep one of your teeth, because it was yours?

DANNY: That and I didn't get a dime for it, like for my other tooth.

DUNCAN: Well that's a custom. Mom replaced the tooth with a coin under your pillow. But how often can that childish infancy be done?! You knew it wasn't going to happen again—

DANNY: I didn't like losing it even for some money. It was a part of me.

DUNCAN: Well, but it was worthless. But you still wanted to take care of it, for awhile. Let's look closer up along the fence.

DANNY: Will a piece of glass do?

DUNCAN: What would we want with that?!

DANNY: If it came from a window?

DUNCAN: ....I'm not sure, but I think they remove the windows first. Buildings to be brought down often have holes where the windows were. Glass shards are too difficult to work through. They can cut easily. Any glass around here's probably not from a window. Just look for something more colorful, especially if there's a design or part of a figure on it.

DANNY: This is called: prospecting for nothing!

DUNCAN: ....Someday you might be doing so more eagerly, when something stirs you enough to want to remember it.... This is done all the time, for souvenirs from famous places to catch. One of the main activities for tourists.

DANNY: But we live here!

DUNCAN: Maybe that's why it's significant for us. We should take more hold of our city, and not be treated like passing visitors or guests.

DANNY: I don't feel like a guest!... I feel like a dog searching for bones— There's something red there!

DUNCAN: ....That looks promising. But it's kinda distant— (*A loud crash from within the site is heard.*).... Something's happened! Men are rushing towards one side of the building—

DANNY: Where?!

DUNCAN: There! To your left.

DANNY: ....Something fall down?

DUNCAN: I think an accident's occurred. They're not working as usual. They've stopped.... to investigate a mishap.

DANNY: ....Maybe it's a break.

DUNCAN: Dead in the afternoon?! No. It's unusual, Danny. Boy! That structure looks like it could collapse at any moment. It's well gutted to much framework.... Maybe a wall's come apart where they're headed, but out of our sight. Came crashing down.... on someone?

DANNY: I hope not!

DUNCAN: I might be able to sneak in, to take a look at that tile —!

DANNY: Don't do it, Duncan! I'm afraid for you.

DUNCAN: It's.... not so far. And all of their backs are turned and kinda distant from it— You spotted the piece! I have to find out.... You stay here. (*leaving*) That opening further down's not guarded.

DANNY: ....I sure will! I sure will! (*very worried for his brother, and looking through the fence*) Say! There's a white truck going towards that side.... Could be an ambulance!

JOSH (*from on the other side of the fence*): What are you looking at, kid?!

DANNY (*startled with surprise. and some fright*): Nothin!

JOSH: You're past the ribboned corridor. It's dangerous to be peeking around on this side.

DANNY: I'm not stealing anything! I'm outside!

JOSH: That boundary's a warning for pedestrians. We're not responsible if something hits you between the ribbons and the fence. We have to make up for damage to the concrete— but not your face. You understand?!

DANNY: Yes!

JOSH: What are you curious about, anyway?

DANNY: ....There was a loud boom—

JOSH: So?!

DANNY: ....So something fell. away from our sight— But we heard it!

JOSH: And all the other little kiddies ran off. Well noises like that occur all the time during a.... reconstruction of foundation—

DANNY: A what?

JOSH: A renovation of the foundation, to prepare for another building.

DANNY: You're tearing down that bank!

JOSH: Well ain't you well informed! It has to be done.... correctly, or else all of the platform ng will have to be reformed and sculptured. Can't have any cracks at the base, where a building stands. Weakens past code.

DANNY: But what happened?

JOSH: Oh, the noise? Some fool got foolish. Nothing to do now but hold up the work. That's why I'm not pushing you off.... One whole side of the building split unexpectedly. A wall had to be pulled off quickly, before it fell on its own. And now engineers have to reexamine the attack, to see how safest to proceed. We're still on pay, though. Can't leave the thing standing as it it.

DANNY: Was anyone hurt?

JOSH: ....Someone was caught off guard, pressuring the split beyond contractor specifications. Caused a damage. Yeah, there was a collateral injury. Too late to cry about it now, except that it wasn't his fault— and it was his fault. But that's how it goes. He was told to be careful and attentive. Can't give much more advice than that. Sometimes being alert's just not enough. This is a dangerous game. But blame the engineers. They got bested. And the guy shouldn't have been where he was, as it turns out. But when a beam pops off its rail, the force is so tremendous it can send things flying anywhere, within a vicinity. Even as far as here, if there's a

free passage. But he took the brunt of it, all in a fraction of a second. No warning. No excuses. No exculpations. And right to the head.... His helmet's saved what's left, what blabbers. Now if everything were done correctly, we could finish this today. But I doubt that now. You can't waltz through emergencies. This is gonna be a long one. There's no permit for blasting at this site. They'd have to restrict whole streets from traffic. Just wouldn't be safe for neighboring structures. The real problem's how we're gonna get in there again. That's for the minds to figure out— but fast. 'Cause this is contracted to go down— cleanly.... and swiftly. Delays are gonna cost us, from more profitable ventures to get to. Right now, things just look a little too precarious. (*voice heard, from behind him: "What you tryin' ta do!"*) What's going on now? (*turning to look, as* **Beck** *brings* **Duncan** *up to the fence*) What's he doing here?!

BECK (*bringing up* **Duncan**): Tryin' to get some rock! Probably thinks someone died!

DUNCAN: No! I'm searching for tile! Red tile.

BECK: Tile! You're a nut! (*They're at the fence.*)

JOSH: You can be placed in jail for a prank like this.

DUNCAN: There was no one at the gate!

JOSH: So you just sauntered in where it's prohibited?! No helmet. No vest. No protection at all— Stupid! And you're old enough to know better— You little squirt! You could have been squashed all over the place! I'm gonna have you arrested for trespassing—!

DANNY: No! He's my brother—!

DUNCAN: Shut up, Danny! Run away!

JOSH: What's this about a rock?

BECK: Looking for this piece of trash here (*takes from a vest pocket*). I came back to scoop up a pile, and there he was crouched down over it, digging like a gopher.

DUNCAN: That comes from the floor of the bank. It's part of an ornament—

BECK: This junk?! From the floor? Off the floor! It's not worth spit, boy!

DUNCAN: This bank was a grand lady, mister. And I remember that tile since little. Not right to destroy everything like it never had meaning to anyone—

BECK: You don't own it!— Here! Take it! (*giving* **Duncan** *the fragment while releasing his grip*) And get outta here fast!

JOSH (*slightly perturbed with* **Beck's** *assumed authority in this matter*): Yeah, that's probably cheap ceramic. (*as* **Duncan** *is going*) Reds are common dyes for porcelain and clays.... More like it at the new bank, I'd bet.... What a childish thing to do.

DANNY: I saw it!

BECK: You're spotting for 'im? You liked this heap of structure so much?

JOSH: No sense loading the rubble, Beck, till we're sure about our procedures again. We might need you for some other (muscle) work later, with overtime, if this goes (into) night.

BECK: That suits me. But I have a café job at around ten.

JOSH: We ain't gonna be out here that late. Night just means after dark. But they have to figure out the lighting for safety. Can't have any kind of construction (much) past nine, a local ordinance.

DANNY: I ain't never been in the bank. But my brother has. That's why he wants the relic, like a tooth.

JOSH: Can't say I understand the fixation,... boy. But I never thought anyone would be sorry for a bank ta be brought down. An' risking his life to retrieve a bit of scrap? Well there's the younger generation for you. When I was his age, I was trying to steal a kiss from Betsy Lou! But you guys are way too materialistic.

DANNY: Who was Betsy Lou?

JOSH: That's a generic term for a romantic interest, Beau Long an' Betsy Lou. Way past your comprehension (as **Duncan** *joins* **Danny**); but boys and girls thought differently, when I was a youngster. I did daring things, but not idiotic ones.

DUNCAN: This isn't idiotic, sir. My father used to take me here. He's kind of tired of us, now; but.... I thought I could get away with it.

BECK: Guess you have, kid.

DUNCAN: That was a really beautiful building, once.

JOSH: They all are. Anything constructed up to code is handsome. This city has always thrived on its architecture. And the more it has, the more it must do so. Though those excessively modern conceptions are only likely to become eye sores, the ones that try to project a sense of motion— in stillness. The whole point of a building is to be stable and secure (on the ground), for its occupancy. Not flying off with the whim of a wind for wings. And they'll be the devil to take down, not nearly so simple to access for demolition. The designers don't think about that so well, they're so involved with their constructive fantasies. But a building is a living organism that must devolve to its demise just like any other. It's not a work of art meant to remain forever, but must be carefully crafted to decompose in a utilitarian manner, without so much danger and surprise. Blocks are meant to be disassembled, and the same for bricks and wooden planks. An evolution of change must be allowed— change, not just development— to allow an accustoming for the current and their societal demands. And that is an art poorly spied by too many construction specialists today. But you only have to think for what is useful, and what probably will become useful.

DANNY: Houses don't last forever? What about castles? They can stand for centuries!

JOSH: Do you see any castles around here?

DANNY: ....No.

JOSH: We don't make them like that. We don't think that way for buildings, unless they're to be monuments. Clearly not for residences. Ours is not a culture of palaces and pyramids. We even are starting to favor cremation over burial today, because the land is getting— crowded, and by necessity must become reusable. Just like the cells of a body, everything is replaced. Do you know that everything in your body is rebuilt in around six months!

DANNY: Even the skin an' scalp?

JOSH: Especially the skin.

DANNY: What about scars?

JOSH: ....Protective covers have to restore themselves from injuries and tears. How it's done might physically leave scars that seem permanent— Er.... The body contorts its surfaces for them, with an accumulation of skin pigments and some kind of memory for that. But everything's replaced, 'cause no living cell can last too long. That's what it means to be vital. Castles of old are dead, or the idea of them. Yeah, that's the rationale to take. What we're doing is vital, destructive but constructive— for the future. And the future is complex. It can be described, but it can't be explained.... I better go an' get a drift of what the minds are thinking, (*leaving*) 'cause we have to get started on something soon.... The daylight's too precious.

BECK: You kids stay outta this area, looking for— relics. No jewels here, but a lot of danger. Ain't seen anything hinting at a diamond. And I thought banks hoarded great treasures, hidden wealth. But that's all been taken away long before now. All that's left is a lot of hard work and sore muscles.

DUNCAN: You do this kind of job regularly?—

BECK: No! An' neither should you. By gosh! You kids get an education to avoid this type of labor. I never thought something so simple to think of as clearing a space from rubble would be so hard to do. But the repetitiveness of the loads and their strains get to you. An' sweat's no comfort at all, as the piles stay formidable. The mindlessness of the work, though, has its benefits. You get into a zone of effort that pushes you through, much farther than you thought or judged possible, as your bones transform their machinery to the levers and fulcrums of the effort. It's like lifting, hauling, and tossing twenty pound bags of potatoes over and over again. A hundred. Two hundred. A thousand. You count them for your age. You count them for your age in hundreds. You count them for your age in thousands.... seems like. After an hour an' a half, you've moved a mountain, reduced it to a plain of cowlick.... an' can start on another in fifteen minutes. If the back doesn't break, it must be broken already— broken in. You come not to think about the pain. It's gonna be there, so just accept that fact and get on with your work. Surprising how it ceases to disturb you after awhile.

DUNCAN: You can do too much, to hurt yourself badly!

BECK: Others are doing the same. My only fear about this is exhaustion, over-exertion. You can't judge it well, when you know what the goal is. When true muscle spasms come, and you ignore them— or even fail to realize them in your consciousness.... that's when real damage can be done, of the crippling kind, (lasting) long after the work is finished. But the career men have judged this already. They oversee with a.... professionalism that's supposed to be scientific, telling some to rest a spell, and others that they're just not fit enough for the job ta be done.... as timely as needed. I ain't been so warned yet. But this is hard, to think about. I suppose it's reasonable labor. Don't have to be defined by it, but it does broaden the shoulders.

DANNY: Why don't you do something else?

BECK: I'm ready for many things, kid. Living's not easy, an' I do a lot of it. But I'm not trained much for.... the less extensive employments of physics to be as profitable. And that's a dynamo of thinking, you know. Because while everyone implores to the contrary, as a profound common sense.... deep down inside I can't. And (I can't accept) the more mental work to be as ennobling, for proof of your time spent on this earth. Praise for being clever doesn't get to me much, as the mind churns for what to do. If only to be practical, I'll have to find better. But so called "smarts" draw a resentment in me, that I guess I'l. have to fight off too, like a pain that's gonna be there anyway.... But not for you. I was raised free an' loosely, and with an independence to make way for myself quite early on, or make a way.... through the forests of human nature. That preaching of the necessity of a solid education, to make a foundation for one's life, was never leveled on me.... so earnestly as it must be on you, with growing improvements on raising children, psychologically an' health-wise. I can only assume you have these advantages younger generations often gain by some sort of cultural default. But my imperatives for survival are probably much different from yours. I've no great wish to become famous or successful, nor even satisfied with toys or accumulations of wealth and stature. I wouldn't fight the reception of those positives to characterize oneself; but it's not what I'm fighting *for*, within my personal struggles of subsistence.

DUNCAN: You have to do well in school— or else you're treated (as) dumb!

BECK: Hum. Perhaps that's more to the point. I wasn't threatened that way as a youngster, as you must be. And I grew to like spending much time.... the majority of time, if I'm allowed to.... away from cities and organizations of society. Deep in the woods I prefer to be, more often than not. But in this world today, one must have a currency of reserve to get by, some little money— to buy things that grow into necessities under certain circumstances more an' more enforced by the influences of extant and nearby civilizations, or gatherings and groupings of people for a peaceful living. I can do without a refrigerator, as long as I can, say, fish as needed. But a power outage for you tykes would be a catastrophe to experience, long term.

DUNCAN: Oh. You're a survivalist. Do you have military training?

BECK: No. And I'm no misanthrope either, to be mistaken with a hate. I haven't any great ones; that's why I can do such horribly manual work, that should rightly be looked down (up)on by you as in no way an example to follow. But I just feel there's something more to simple work, than the perfection of its utility.

DANNY: Ah! You're a.... philosopher. You're looking for a poem in life.

BECK: Maybe. As much in life as out of life, can spread a word its meaning. But it's more a sense not to be too conclusive of tradi-

tional ways for being current or modern. Too many notions held by most, of what should be important, affront my sensitivities an' tacitly challenge me or how I'd like to exist.

DUNCAN: Well my Dad says: Juvenility is simply Revolution.

BECK: The need for it, as emotions, angsts, and desires are concentrated like chemicals in a small frame, apt to react violently, with apportionment of their urges and appetites mixed up and confused. (*A faint but shrill whistle blows.*) Yeah! All rests are officially over; an' we're back to our duties, as designed. (*leaving from the fence*) Take care, kids. But don't enter here.

DUNCAN (*holding up the tile fragment to examine*): I got it, Danny! I'm sure it's from off the floor. You can just make out part of a design, an etching.

DANNY: Looks small and broken to me.

DUNCAN: But it's real. It really is.... from that past destroyed.

DANNY: Ain't it always? treated that way.

DUNCAN: I thought I told you to run, Danny! They might have wanted to apprehend you too.

DANNY: I wasn't gonna leave you there! I didn't know what to do! And I was outside— I was outside the restriction.

DUNCAN: You should do what I tell you to do, when it's an emergency.

DANNY: You shouldn't have tried it anyway! An' all for that piece of— stuff!

DUNCAN: You can't understand.... how important it is. (*looking at **Danny** inquisitively*) It's actual. It's concrete of what was.

DANNY: You almost got your head knocked off for it. You almost got taken to jail.

DUNCAN: ....Yeah. Here! You keep it—

DANNY: What?!

DUNCAN (*as he almost forces **Danny** to accept the fragment*): You have it! 'cause you can't understand it yet.... But you know I dared myself to get it. Broke the law to try. (*as **Danny** examines with inherent, childish inquisitiveness*) You can throw it away. But you know it must mean something to us, if we can risk ourselves to find some.

DANNY (*examining*): It's.... kinda pretty. But just so. Remains shiny, for all that's been done to it.

DUNCAN: Well she was a grand old lady. (***Danny** pockets.*) Let's get on. We shouldn't be here anyway.

DANNY (*as they get past the ribbon*): They sure do work hard at it.

DUNCAN: What the city deems must be done. I can't understand it much. But it's for cause.... that we have to move on. (*They re-*

sume their walking, down the street.)

Scene III — *An arbor to the fashionable garden of **Pomelory's** mansion. It is very sunny, and there is an ornate bench there with shading, on which **Maureen** is relaxing. Her son, to which she is entertaining a visit from, stands more in the sunlit green, enjoying the warmth.*

JOHN: It's much brighter here, Mom, than in that dreadful patch of plants.

MAUREEN: Don't you like your work any more?

JOHN: They're totally different environments to consider. There all one thinks about is what percentage of pesticide to fertilizer should be used, to control the bugs. We were thinking of trying to convert a spot to strictly organic growth. Then you have to consider how much neighboring contamination must inevitably occur and whether this can be reduced to a legal limit to advertise the vegetables as fully organic. Run-offs must occur with the watering.... I'm not sure if it's worth the trouble. Same bell peppers at almost double the price. But the netting and all. Lower yield. Not as pretty— Not up to our standards of quality. It's not a niche we can easily leap into, except as a pilot project at most. No problem here, though. The flowers look splendid.

MAUREEN: We hire a fine gardener to maintain them. I can hardly find the interest anymore. But it's part-time work. And he has many parts. A good number of customers— highly recommended.

JOHN: Just does some seasonal trimming and pruning, I suppose. Well, he has control of things.

MAUREEN: So do you.

JOHN: I mean, how to handle all of his work. It's much more complicated with us. Full of compromises. We have a reputation for our produce to hold up. Commercial suppliers are inundated with the most unreasonable requests: unseasonable fruits, unseasonable storage, unseasonable transportation. Everything's irregular for normality. Who can tell what must really go on to fulfill these demands? I'm quite afraid to buy any of my foods from the supermarkets. They could be last year's supplies, especially if there's a dullness to them— after rinsing.

MAUREEN: The vegetables? But *you* can tell.

JOHN: Not always, Mom. Radishes, for example. Their green leaves that one can cook: You have to smell them carefully, for traces of delayed decay. We're only responsible for growing them, not so much the manner of their marketing.

MAUREEN: People want their victuals fresh.

JOHN: Cook everything! to be sure.

MAUREEN: What about for salads?

JOHN: ....unless you've grown the foodstuffs yourself, by eye. Instances of bacterial contamination are too much on the rise to take many chances.

MAUREEN: That seems kinda ridiculous, John.

JOHN: I suppose I'm sensitive to the realities. But what's happened to the eggs is working its way to all the foods: *salmonella* and *E. coli*.... and prions for the meat. They're not even supposed to be live pathogens. One great catastrophe is slowly storming through our existence, like a commercially induced plague. Regulations to protect us only connive for more fallout to occur, with their complexities, as everyone tries to find ways around them that are permissible and therefore excessively promotable, to the point of absurdity.

MAUREEN: You're being too pessimistic.

JOHN: It's all about how much (of) a foulness can be tolerated— how much anything can be tolerated. I'm glad I can stand in the sunlight, and irradiate away the cruel.

MAUREEN: Well, even some pests have hearts, only being themselves.

JOHN: ....Then the deleterious entities that consign our minds to blatant wrongdoing. We know we're poisoning ourselves with these execrations. But we're all in the same pond swimming, swimming the same way. Improvements wrangle their way to strangle us, like too much chlorine in a pool, their convincing to commingle too much good with too much profit. Oh, the ideas sound great, the strategies sound, their details and designs employable and workable. But they only cause harm— the efficiencies hamper. Simplicities of thought are lost. The good apple is no longer the tastiest, but the longest lasting, the most regularly sellable. And strange correlations result, like struggling to reduce 'disease to return' ratios. Priorities have been contorted from proprieties. It's what.... the people will tolerate that matters more than your pure achievements and goals. The best.... is not successful, Mom. The sufficient is.

MAUREEN: I always worked to be at my best, and I'm sure you do too. But that's not as common as what a circumstance may need or want. Issues of the day can crush your preconceived conceptions. If a public demand rot, then you sell good rot; though it's far below your capabilities to provide, and against your principles of finest provisioning. Few crave for the empyrean when a scrub will do the trick, since one wants its service more than itself. The largest pumpkin is admired only as a clownish oddity— the spectacle of media; but they're all carved up for pies, meals, and same-sized seeds. And quality is a special thing. It's perceived.... with a peripheral sight. You don't really know it's so until you experience its taste above your needs. The desires it may cause are only ephemeral for your usage, since that usage is the same with lesser craft.

JOHN: People still pay a premium for the finest.

MAUREEN: But that payment may be as wrong as its presumption, and can cause a lasting grief of error if the quality was just not right.... for your price. So naturally people swarm more in numbers towards what's barely sufficient for their pleasures.

JOHN: Even if it's killing them?

MAUREEN: Few take supreme chances. Few are bold enough to let their arrogance rule (over) their judgment. They'll let what's customary stand for their actions, and what's advertised to be.... the minimal of their inordinately reenforced beliefs. But who today would want a damaged peach? If it's not damaged, then it's fine. It doesn't even have to be particularly sweet or juicy. It does well for a peach to be bought. The masses are satisfied, and your incomes are saved.

JOHN: It may look fine. But it's never going to ripen correctly. They don't realize that it's dead!

MAUREEN: I suppose a social consciousness does, to accept these advances, and to assume their benefits and implied promises that don't have to be proven. It is eatable, and within the standards of what's to be sold.... them. As long as they don't have to work the fields, or do the agricultural science to keep matters.... plausible for a nutrition, then this is what's accepted.

JOHN: It's humiliating to know so much otherwise, and what has to be suspected more broadly than just by farmers and providers.

MAUREEN: And doctors and politicians, and educators and entertainers. This is the status quo.

JOHN: So's the sanitizing, sterilizing sunlight. But this is no stasis to be bound with. The motives are inherently deceptive, that the clean may not be cleanly, that the wholesome may not be whole, that the fresh may not be alive— like dead fish, and that the principles invested in this.... may not be evidentiary— may not be actual or sound. It is a fiendish crassness to allow such assumption populate a populace lazy to be more prospective, when all of these activities are defined by customs of definition to be prudent, understandable, defendable, and licensed to be this way of a method driven, guaranteed with the backing of patronized oversight. That's the state that we befoul ourselves through, that we find ourselves immeshed of causalities unavoidable and certain to demean or punish. That's the nature of what happens when what's proper becomes less than banal but even prohibitive of our deserved prosperities, for all the hard work we've committed ourselves to.

MAUREEN: The business isn't falling under water, is it?—

JOHN: No!... Not that. Faithfully I resist such a description. We've an engine of conveyor that can go on running for years, kept as our services are. But.... the tide in general rises— to drown everyone as unprepared. As standards seem to be raised, to combat every health malady found associated with food production, protection from those diseases feels like(ly) to be eroded in effect, in the effect that the public buyers accept less and less the deliciousness of their vegetables and corns, and have in fact developed a meanness to their tastes that allows many more unscrupulous suppliers to submit their produce to the markets, flooding the shelves, bins, and stalls with the minimal to pass for selling. And it all looks pretty, with shades of feigned ripeness or maturity of texture and flavor. So competition makes an ugly transformation, for our goals of growth, when the cheap becomes princely enough— to drive us into desperations.... not cutting corners, for example, but *reshaping* them, as the very meaning of some criminalities change, to what we should consider criminal being lessened to a tempting permit of regulation.

MAUREEN: It should be simple enough, to stay clean of your fruits, as we always have.

JOHN: We're allowed to ship them too early and young, like the

Lima bean—

MAUREEN: What?!

JOHN: The soy bean, Mom, simply because the agronomy inspectors, investigators and regulators deem at earlier ages there will be less to no chance of substantial infection of the eatable parts, for plants that are to be thoroughly cooked anyway, for a customary serving. But this makes no sense for the integrity of the taste. It does not exemplify the wholesomeness of the bean that is readily achievable.

MAUREEN: Well, the soy is more of a background addition to the meal, used more for its texture and contextual binding than for its flavor. It's a bedrock that promotes the tastes of other ingredients. Its service is subservient to spice, color, and tang.

JOHN: Simply due to this apparent loophole, my cousins want to devote a whole field to its production, with rapid harvesting. The effort will be expensive to initiate. Both seeds and ground are at a premium. And this will take resources away from our other vegetables and crops, ones we know how to raise reasonably, and without the nearsightedness of this horticultural review. The project is too big as currently wished, I argue. And that earth we've worked so hard to sterilize from pathogens, and will have to continue for more of, should be parceled out more wisely instead of thrown almost entirely to this risk. The scientists don't have to sell— We do! And I think our market for this will become limited to buyers for specialty items to create, our profits linked strictly to breezy trends and passing fads. You simply can't do as much with a young bean as a fully grown one. Why, the pods themselves take special manipulating. We should work this into a much smaller scale project, and with much more experimentation, market examination for prospects, and farming precautions in general— But I'm outflanked by my two younger associates, who are generally prudent enough in most of our efforts— But definitely not this one!

MAUREEN: The disagreements between you are intense?

JOHN: Not to the point of stagnation of aims. They know of my disapprovals only with politeness. But for such a major transition of labors we must all be conforming, if not committed. Too much capital will be involved, for my slack of approach to target yields.... I need adherents to the more predictable realities of these types of ventures—

MAUREEN: Oak doesn't want to get involved in your disputes, especially ones like this.

JOHN: One shouldn't tie so much chance to fleeting fashions.

MAUREEN: He wants you boys to work these kinds of problems out yourselves, precisely these kinds. If it's how to do something agricultural, he'll lend requested advice. But running the farm is left to you now, and so (too) the gambles that you find or create.... I've already been admonished for trying to enlist his help, on your behalf.

JOHN: What?!

MAUREEN: You know that Laura is in league with her sons, even excited by their daring in this matter— Though she can hardly grasp fully the dangers of such a leap towards one crop. It's (been) ample variety that has always sustained the farm. We were careful not to let any star seller be our reliance for the whole, but worked diligently to promote all of our foodstuffs, no matter what the changing economies suggested. That rectitude of leveling out our expectations through moderations and averages served us well throughout commercial climates. But the times today may be considerably different, to which you younger are more acute and can more easily reason through. A cash crop might be more what's ordered or dependable, to make a kill of the market beast before it runs out of your capturing range. This is what Laura believes, and what Oak hesitates to give an opinion about. So I can't easily draw him to your side.... our side.... of responsible management of opportunities, as we've always done. In essence, *his* sons will have to figure that out for themselves, with these experiences to learn by.

JOHN: She admonished you?!

MAUREEN: Only a reprimand not to be fickle towards the farm doings.

JOHN: Fickle?!

MAUREEN: Not to try to offer influences as unsteady as she sees me.

JOHN: You really must be more commanding of your status, Mom, as one of the originators to make our farm a success.

MAUREEN: Well, not quite that. But I did have much a hand in it, as much as my brothers.

JOHN: And if Unc' Beer were here now, he'd be with me.

MAUREEN: Always one to take a definite stand without much debate, or caution.... But no one is with you or against you three, since you are totally in charge and can not be hindered in your actions one way or another.

JOHN: But hampered certainly, with reason and judicious scolding.... This could fell us, Mom, if we make the wrong move. Loans are treacherous, today. Our working principle should be tied up elsewhere. And while I can admire my cousins' assertiveness.... I fear for yours.

MAUREEN: It's just how it should be, John, with how I've come to bother them for help and understanding. Your Aunt Laura does make for a demanding relative. But she's not untoward to me— and has nothing against you, which is a great relief.... Because I've brought them a bit of trouble, in the past, with my hardships of heart.... But never you. You should be absolved of these difficulties.
JOHN: It's more uncle's doing, than aunt, to have me accepted.

MAUREEN: By rights you should be the actual head of the enterprise, being older.... But Oak's older than me.... so....

JOHN: There's no quarreling about that. It's understood implicitly that since he was effectively running the place, he would decide how it would be run with his retirement. Rather graciously allowed me to participate— I'm thankful. Don't worry about that aspect to all of this, Mom. Take more care for yourself, and the standing you deserve around here. There really shouldn't be any ranking allowed. I certainly don't accept it. And we *can* handle all

of the problems ourselves. But at least I can flush them out with you, to clear my thoughts. I may be complaining to myself unproductively, and need a listener with the proper experience to make good comments.

MAUREEN: You are all such young men to me. I can't understand why your social gatherings have not produced for you happier partnerships by now.

JOHN: I guess there's little time for such earnestness to want to find. That's a peculiarity of the day, Mom. There's no great rush to a coupling, when everybody has their independence to protect. And we compromise ourselves.... in many other ways. It's a secondary thought, Mom. Not an overwhelming goal, nor much of an imperative.... at the moment. I'm not sure if it creates much of a hindrance in today's living. Certainly enhances the complexities and hardships without much perceived benefits of the really important things, aside from trusted companionship. But yokes much delay independent strides. And that's a most important attitude to take nowadays. Because you have to stay agile enough for a whole lot of change and unforeseen circumstance. At the very least it can become a very expensive encumbrance, if the morally bound relationship becomes disastrous and dissoluble— which is terribly common. But to put it bluntly, one highly cherishes the luxury.... of privacy. It is a privilege of social sophistication oft fought in much of the world.

MAUREEN: ....You don't even gather to discuss these matters much, as a clique of interested concerns?

JOHN: I thought you were talking about marriage!

MAUREEN: ....That's not a partnership. That's a pairing, a complementation that's hopefully productive. But seldom are equals mated well. No, that sort of combining has to come strictly from devoted hearts ready to give themselves up to the mixing, however foolishly pursued. I would never force such an issue for someone, least of all my son. Nor should society at large, since too many errors are made, and much too easily. The common views received through our cultural environments seem to cheapen everything, even that concept of unity with matrimony. And it is not to be rushed towards or undertaken casually until the heavy responsibilities involved are thoroughly understood— or at least anticipated— and accepted. Because we make not our own ourselves anymore. That is left to the deviancies of our age. And we are forced to allow those commitments of social.... or civil.... development, for our offspring to somehow prosper or thrive through this ever more obscuring mélange of human consciousness; the fact being we need — and want— much less children, and far more principles for raising the few we have or would permit to deter our perversities of life, those individual disasters we seem to live for and strive at. No, I'll not coax you into finding some nuptial love. That's much abandoned anyway, from my security to find out of experience. It is a terrible waste of longing, a lugubrious drain of emotions coupled with upsetting hours and disappointing seconds of awareness of your situation, or entrapment to rally through with dejections. I can not define it any other way schooled. Never do such things when they're so totally unnecessary. That is a sin against our natures, and a plight of unworthy experimentation. Our society needs not marriages, except to sell to (resulting) families.... commercially demonstrated necessities material and immaterial.... Not that I would condemn the practice for few successes. But the waters remain as brackish as ever I could perceive for them a swim through. Security is better found with more solid thought. And mistakes do leave you drying in a pool of denials....

JOHN: The idea of consulting clubs, for commiserations, is becoming a passé thing, Mom, in a fully developed country like ours. Overly developed, maybe. Any sort of information, notice, or opinion that might be applicable to your inquisitive caring of a problem is easily found through many avenues of impersonal access. And the only rule that runs as rampantly is to carefully judge the accuracy or correctness of the data supplied. It is already highly thought out, edited, and annotated for references and cross-referencing to related topics. All manner of mentality is fully displayed, and can not be mysterious and attributable to the otherworldly or supernormally fantastic. Although sometimes the most irrational arguments are uncovered. But facts remain highly important for foodstuffs.... less so for market methods. In any case, I've examined a great many notions. There's no drought for that. What I need is a substance.... of superior recommendation—

MAUREEN: Oak's still too reluctant to intervene—

JOHN: Well.... I would not ask him to, until a gravity of crisis approaches, that horizon (is) seen clearly of its blazon threat.

MAUREEN: I will try with Laura. But it's like the kitten to convince the mare of danger.

JOHN: Danger for her youngest that would marry. He's certainly closer to this than I or his brother could be.

MAUREEN: Well he feels buffered for more fanciful engagements, with you two elders around

JOHN: Fine. (*coming towards the bench*) I'm not against a lunge towards happiness, at least. But what is happy does seem to me to be.... merely the displacement of the ponderous (*sitting*) onto others. You drift off.... into a calamity of wishful thinking, as the chores are handled more concertedly by those in toil of real pursuits. Not that he's just so brought to slavery of fun.... But as an equal third, does make the motions due with too much safety from a finer half, is my fear of his indulgences to civil whim. And I'm not complaining out of any jealousy or envy. But I think three independent thoughts would do better, in this instance of three required.... Does get warm today, with some surprise, even as forecast.

MAUREEN: You have been bright, within the brightness standing. You are bright.... Oak is comfortable with you, gratified that you are there. Your contributions satisfy him for.... our retirements. Don't keep doubting this.

JOHN: One has to have doubts, Mom. To be sure of anything is silly; but— the more important.... the more doubtful.... even with all of this support.

MAUREEN: Then don't dwell on it so much, as if it were a command. We know you'd be useful for many (other) endeavors, several various stabs at procuring a meaningful life, and that you are not entirely convinced of this particular profession. But that's a bit of a prejudice, don't you think? against yourself, that you can handsomely overcome. And you've been as splendid as valuable.

JOHN: ....Oh, I know that well enough. We all must be very ca-

pable, in today's competitive atmosphere: alert if not sharp, an' sharp if not accurate. No fear of that lack of application in myself, I concentrate fully for the job. But that's not to say there are times when I feel thoroughly misplaced, and can easily be made to seem, look.... foolish—

MAUREEN: That you can never be. Stupid is a trait I'm afraid has been excised from your genome. Awkwardness is a sense of balancing only. You feel like the head, but you are not the head. You can't have everything done as you would like, and that leaves in you a strange awareness to sample. But the legitimacies remain, that your good is better than the best of many others. And you should not need this defense to fall back on. That's only being defensive. That is posturing to claim rebukes unwarranted.

JOHN: Quite sure. Quite sure, Mom. I just don't feel all that proud about all of this. But is this an era of pride? I'm not certain about that, the way we conduce ourselves into the most subtle horrors, like we're born to be deceived of everything, and work— labor even— to contradict our pride.... and fall to comfortable resignations of the world, as we exhaust ourselves. One tries— so hard, and accomplishes.... very much the preordained, or ordinary to be adjudged, as if breathing were less a right than a restriction. Then, how are we sculptured to be, in what real image, if we can only perform this way?

MAUREEN: ....It's made definitely of a debt, to find one's form. But *no* one's been pushed out of the boat, for this family.... The farm will suffice for you, even through these contentions and disagreements. And your courage of sense will flourish throughout all proceedings. For that's the primary right you've been afforded and offered. It's how Oak has planned things, and how our enterprises will be maintained.... Nothing out of the ordinary is occurring. It's not remarkable. It doesn't condescend. You are still a primary component, and will have much to do about whatever happens or occurs. So no control is slipping from your grasp.... Shades from recovery are not being drawn to obstruct your view. It's unnecessary to be doleful about this.... apparent challenge to reason. That ratio with risk often fluctuates.

JOHN: Certainly, Mom. Certainly so. But by now I should have gained a firmer foothold of influence— I mean as regards to practicalities. I'm almost treated as a junior consultant at times, when they are particularly sure of things. And for most of those matters I often agree— But when I don't.... it's almost a denigrating attitude to receive. Feels awkward, confounded, out of step and subject to ridicule.

MAUREEN: Is that ever done?

JOHN: Never overtly.

MAUREEN: Would they really make a major move, execute anything of importance, without your consent?

JOHN: It takes three heads of agreement for something of that nature. But time pressures (for) action. No one wants to be responsible for dropping an opportunity. Convincing is in order, and that strains relationships and heightens tensions of cordiality.... They've offered to buy out my position, Mom, for a full fifth the value of the farm—

MAUREEN: What?!

JOHN: —The working value. The current market assessment. Oh.... they've thought to do so before. But this time it's a real attempt— and passionate. They want me out and away.... from their doings altogether.

MAUREEN: It's not going to happen, John. Your uncle would have to approve of it; and he definitely won't want that.

JOHN: They have the legal right to ask. We control the operation, and they each think they can assume a tenth more of burden. It's certainly daring, to get their way. But it speaks loudly of what they think of me.... Uncle Oaker can only advise them, as he effectively only controls a fifth, as do you. Together we might thwart the attempt. But I'm sure that's thoroughly against his principles, and perhaps yours, when he wants the younger men to handle these questions.... to make the most acute decisions.... But I'm being swayed, Mom, if I'm not wanted. Where's the profit to derision and growing fractiousness? Life is feeling very uncomfortable to me, restricted to a shade, shades of consideration.

MAUREEN: ....It's your judgment, that overrides these matters— to run away. This would make you wealthy, but not rich.... or as we say: foolishly rich, since you well know that farm is always worth much more than its working capital at any one time.... What are you trying to do?... What are you trying to uphold?—

JOHN: I don't want to fight with my cousins. I've the courage.... and the ambition. But this is not the country game to me as it is for them. I am seriously restrained.... by atmospheres down there doubting me.... This might only be a ploy, to convince me of their ideas more faithfully. But it is a very heedful "out" to consider. More than a suggestion, but a declaration of intentions, they produce for me with this resolve. What would—...? What would be a prudent thought for me?...

MOUREEN: What could be your satisfaction to consult? Memories for how things were done to me?!—

JOHN: No!

MAUREEN: —Solutions found, that absolved everyone from blame?! Oak wants the three of you to work there— to work it out.... But you have to be happy with your labors, the thought of them, or they're just not worth the effort. They'll harm you with a careening of the sentiments, until you get rid of them, justly or unjustly, fairly or falsely. You have to love your work if you're working hard, or at a full clip or pace. Excuses not to do so surround you all the time— But if you're fed up! then you drop everything and leave! You let the pretended hopes shatter as you race away. It was never for you in the first place!— It was brought to you to find, like (for) an infant in a crib!—

JOHN: I only like the sunlight, its rare honesty— and radiant purpose, to define clearly what you see— to grow under it, to expand with its brilliance and truth and evidence boldly shown, the shine that is actual for its fire and blaze. I don't like these shadows, Mom, (getting up).... the denials and ignoring, and the forgiving by forgetting—

MAUREEN: It runs blindly, John. It doesn't see anything! It simply scoots to make more shadows. And when it's gone, the greatest of all night depicts its absence— May it never come again! to fool

you with some warmth.... Well that's what some might say, if they're bitter. It's too cold without a flame. It is the worst of winter (as *John* walks into more sunlight), to make yourself snug for some resistance to it.

JOHN: I'm just bound that way to enjoy it. I don't like to hide— myself.... from anyone I must deal with, nor from the practicalities of my enlistment. We're still a solid family, as regards the agriculture. It's only been tempting, to tease me a bit. (*facing some hedges*) Quite suddenly thought out, that it was possible— workable.... foreseeable.

MAUREEN: Not of their own opinions plowed? Laura may have had a hand at persuading them—

JOHN: That's possible.... It's not a fault. It's an observation— and from way out here. But I'm seen in such a light. (*turning more towards her*) No, my infusion of interests is more important than this little dispute. I like my job, as I come to do everything I handle well. And we're not angry with each other. But these silly internal competitions are taxing, and distracting. I just need to be pardonable, Mom.

MAUREEN: You are!

JOHN: ....Oh?

MAUREEN: I mean, for what you are, more conservative of tendencies. If the climate is right for soybeans, then they will be had. You'll come around to accepting even that, as a wise choice of major crop. It depends on your privileges of intuition.

JOHN: It's popular to intuit things today, as if you could at all correctly. But they are becoming very popular. Maybe there's a resonance to it, these notions that might prove constructive. Uncertainties.... only cause weaknesses in deployment, until you're certain one way or another. There's a spot of browning over there, atop one of your hedges, Mom.

MAUREEN: ....The custodial gardener says to accept it, John. Would do more damage to try and correct right now.

JOHN: Yes. I might see that. We're still acclimating. The days are going to become genuinely warmer, soon. You can't fix what isn't faulty. You can't fix what isn't found.

MAUREEN: Well that's the preference we're left with, son.

JOHN: And you are feeling well, Mom?

MAUREEN: ....Yes.

JOHN: No scolding for being depressed? I know how much that always annoys you.

MAUREEN: I haven't been so lately, John.

JOHN: Then that's the better of it lent to wish for, if we're all in need of satisfactions.

MAUREEN: Resolutely. You stay with all of your talents in place. They'll pull you through dilemmas as if they were only passing daydreams. And I'll try to atone for your doubts, inbred.

Scene IV — ***Langly*** *is on a city street, standing up against the window to a small department store offering various minor articles of anything. His presence is unusual because it is still and apparently not involved with the displays, facing away from them.*

LANGLY: Oh! I'm so winded, getting home. But this is tiresome. It's grueling. What a place to lose your energies and stand at. But I didn't want to be noticed as peculiar.... So I abut myself to this banal shopping. What sharp articles of crude nature and needs. That's the pretty packaging and bubble wrap to announce the arrival of another colorful or intense-looking trifle. What was it there? All that dazzle for a pair of ear buds. Hardly headphones, today. Hardly anything today that I could recognize as lasting. The lotions and shampoos don't, an' the lice treatment for the scalp. Even odds it's just some soap with alcohol. Yet I wouldn't advertise it that way, with such a large critter struggling. And the nicely shaped kitchen appliances. They'll break down, like a coffee grinder's motor, when you feed the blades less than beans. It's all good stuff delusional. People buy it because t's new. Couldn't rent that trash. Oh, I've had my share, my share.... my store of worthless wonders, toys for deeds— and the utility of soft strifes, combs, brushes, razors: the noise sharper than the cutting screens, for those cheap battery powered ones. They were called "Five and Dimes" in my youth, establishments like this. I wouldn't shop so much now, in these modernized connivence stores, except for the most dire necessities to come by, like a particularly flavored mint candy, and bulk aspirin on sale, and a puzzle with L.E.D. lights— that I still can't figure out for its advertising allure. I guess if you assemble the pieces right, things light up... . Can't understand why I feel so weakened. This is one of the worse waning of strengths I've felt, almost as bad as some bouts of waking up and getting out of bed. But once you're over that, it gets easier. Reserves— of fat, I suppose— buffer you into normal pursuits of daily activities. I knew this trek would become a long one, turn into something tedious. But the shear of this exhaustion shocks, as if I've lost a body part or something. And there's— some distance to go. Can't afford to fall down and crawl in public. It's more than embarrassing. Kids might stomp on you for play. (*A **woman** comes up to the window, distancing herself from him to lock in.*)

WOMAN (*after some struggled observations*): Could you move over a bit, please. I'd like to see the contents of that hair care basket more closely.

LANGLY: I can't move at all, lady. I'm sorry.

WOMAN: That's very discourteous. I just want to study the prices.

LANGLY: They're usually priced similarly to other items of a kind.

WOMAN: Don't like to go into a store for something unless I'm gonna buy it.

LANGLY: That's a good prescription to follow. But I can't budge a bit. Locks are on me to stay still a spell.

WOMAN: You're blocking the view—!

LANGLY: I don't wish it.... that way. But I've walked a while, am tired out, an' leaning up against the window at this unfortunate

spot. Have some patience with me. Your access is not limited nor unlimited.

WOMAN: You don't look elderly. Just mean of manners.

LANGLY: Do you really have the time to argue about it? You can't tell of my distresses unless I tell you. And that is, I have to rest here for a short while. And I'm not denying you anything much, as I catch my breath.

WOMAN: I think you're just a rude bumbler.

LANGLY: Well I don't want to start any disputes about *that*. But what makes you think you can say such a thing to me alone like you are, unaccompanied by any friends?!

WOMAN: Alone?! You're no threat if you're winded, in this weary city that can't teach public decencies anymore. You joggers are a self-absorbed nuisance. You think you own the streets with your crazed fanaticisms to complete. And when you run yourself into a crisis you stop anywhere, like a dumb mule or potted elephant, to disturb everyone of your panting presence, searching for some sign of esteemed congratulation and impressiveness to be regarded, when all you are is a bore breathing heavily, living to bother—

LANGLY: You think I'm a runner? I'm a pathetic man, lady. But I'm no runner.

WOMAN: How many blocks today?!

LANGLY: Oh, from the park and back.

WOMAN: That's hardly a sprint! You're out of shape, and wasting the air at it.

LANGLY: No. But my lungs have at much, to share it with you.

WOMAN: Your kind don't share a thing you'd rather steal. Self-centered pustule! One of those resentful egalitarians who'd treat a woman with the same crudity as yourself.

LANGLY: Well you certainly talk it out to explain. But a man would just shove me over, push me out (of) the way, if drawn to such a silly insistence. You can't want to do *that*.

WOMAN: I want my rights obeyed! because they're supposed to be owned by everyone in public concourse. Too many cads are tolerated today. And I'm tired of seeing your contrariness so often, at almost every turn of corner, defying the agreed upon character and charity of a city.

LANGLY: Well that's some heady thought for you.

WOMAN: Not threatening me! But it is boring, and disrespectful of everyone, a simple request to be civil refuted of any importance.

LANGLY: Then I'll adjust myself, lady (*moving slightly*), to let you see more.

WOMAN: You're still blocking!

LANGLY: That's the caveat of my injuries, for this disturbance you'd pretend with. What's so desperate about hair conditioner! but my best ambles (on) towards your grievances. I am incapacitated for much of nothing to observe. And if you can't see past me, then that's your doom— isn't it! Nothing there you can't own anytime, but leave me my difficulties to battle through— or usurp them! to earn the truth of this affliction. I might even die (right down) on this spot from your bantering ignorance and disconcert, and lack of concern for me. But I told you that I am out of energy to move much more, that I'm resting so short a time— to try and regain my abilities— walking, and that this is not as hopeless as your pressing need to prove more genuine an urgency to soaking and coloring hair follicles. But I am fettered to this position momentarily, by true compromise of my health, and can prove this by insulting your privileges.

WOMAN: You're only wounded by spite to be complained about. I asked you politely, and you refused—

LANGLY: Have not!

WOMAN: So then you want to make a demonstration of it, to show how common courtesy can be resisted. And my anger's at that attempt once again. To dare put me off—

LANGLY: Just go inside and buy the stuff, lady. It's all so very low to deal with. And its more common than my tact, I assure you.

WOMAN: You can't realize how often.... I wouldn't lend myself to that cheap product, now that you've put a stare to it that makes it seem bad and awful, shadowed by your sulking soak—

LANGLY: It's perspiration.

WOMAN: —Reviles my senses for it, saying you are here.

LANGLY (*as she's walking away*): I don't mean to.

WOMAN (*walking off*): Trash near the trash!

LANGLY: ....It *nears* the trash, to be made indignant for being stuck to.... as if that draws all of your fortitudes to behave. Then should I just have sat down quietly, and without any explanations, to startle her for some comprehension of this.... public indiscretion? But that would be more a sign of something wrong. The grown don't do that on city streets, unless they are indisposed.... of character or slightly wanting, or a sickness shielding hidden fears. Can I be more unusual than that? standing up to these displays? She targets me only because I appear weakened, infirmities too apparent. Wouldn't dare approach me that way if I looked more healthy, a man of my height. Wouldn't dare reproach me at all. But I thought I could hide the feebleness this way. Failed at it. The crows peck. Well.... it's only a slight pause. Restoratives surge through, course through the veins. Or be they arteries? A brighter red's with oxygen. (*takes a deep breath*) I'll be able to retake the travail soon. Restart the long walk— back. It's so hard to begin a heavy labor again. Oh! Quite the opposite of my earlier days. Then it was too easy to start, and too hard to stop. Difficult! Could hurt yourself trying to— the ambitions were so strong, powerful, excessive. It was dangerous, jumping off the running merry-go-round, the spin of delights. You had to learn how, study the meaning of a rest, practice techniques for obtaining it— without breaking your stride, an' back. But now, all of a sudden.... it's all that's left— me, like some jocular contortion of the face in a fun(ny) house mirror.

Is that what I really am, today? No wonder she could laugh at it, this— outrageously muscular fake! But no self-pity, sir. We're not swimming through a carnival, today.... but through a city, with.... public works an' distractions, misrepresentations of importance or imperatives. One would think you only live for shopping, in such a town. The stores look.... impressively urgent, impassively profound, with silent clamoring of lights, sights an' signs. A waste of energy in the day, for all these neon tubes, unless a rich profit's made, I think. Can I begin? But the weather's become beautiful for window shopping along these streets and thoroughfares. I shouldn't blame her *too* much. I'm a bit of an obstruction to the fun. But that's the price, lady, to be found, for my waiting in this glorious emptiness, an' profusion of blokes who once were similar to me.... Or I them, once, indifferent to the painful senses, spreading spurred, spurred on— by.... happiness? Contentment of living? I'm happy now— I must be, to linger so dreadfully, and without regret for these colorful displays of essentially cheap objects (to be) nearing meaninglessness. Don't want to be trapped.... on such a sidewalk. But it's a clean pavement, iron lung for staring from. Yes. (The) Street's kept clean; though I wouldn't want to eat off them, like a toddler. They'll do for the likes of me to pass along. Must be pumping more— have to be getting able. And even the gutters don't look that dirty, considering what they must contain. Chained up in there, I guess, the debris of a populated forest, clumps of the discarded an' wasted away. Breathe more, to look out— for such safety.... Oh. Sanitation. Well you can't stay still forever, if you've got ta go. (*A slightly uniformed store **attendant**, i.e. wearing a sleeve-less store jacket and obviously much lower of rank than a manager, comes out.*)

ATTENDANT (*trying to confront, but with some timidity*): Hey, mister! Why are you standing here so long? Could be keeping customers away.

LANGLY: It's a city's street, son.

ATTENDANT: But people will want to look at the window wares.

LANGLY: ....Pedestrians. I ain't blocking entrance to the department store, nor exit. But this is a free space to stand at. And I'm doing so, to catch my breath.

ATTENDANT: Breath? For several minutes now. You're leaning on the window, sir. The sidewalk's yours, but that glass is ours. Looks a little odd, maybe precarious for an accident.

LANGLY: Well that's how long it might take. I ain't banging on the pane. And there's nothing behind it but some dusty setups, an' rusty cosmetics.

ATTENDANT: They're not dusty.... And they're ruddy cosmetics. You're loitering before a store's front, sir. That's not meant to be allowed. There are.... proprieties of behavior we've a right to expect from the public—

LANGLY: The outside public.

ATTENDANT: Yeah. And you're not supposed to be standing here so long.

LANGLY: There's actually a law against it? You'd let me shop around inside as long as I wanted, and not buy nothin'. But I can't rest up here a bit?

ATTENDANT: I'm just suggesting to you that it seems unusual. Are you that disabled?

LANGLY: ....Have there been complaints?

ATTENDANT: Our manager has noticed you. That's why he sent me out, to find (out) what's going on, if anything's the matter.

LANGLY: To be dealt with? I've just taken a long stroll, and am about to begin it again— That's all. I ain't signalin' for any trouble. I'm not dying in the streets, or marring up your window with my coat, or stopping any passersby from looking at your sales items. They can take peaks pass me, around me.... even through me, if they want some honest opinionated descriptions. But I'm not hindering no one from enterin' your store. And I'm just about ready to trek off. There's no crime to being hurt. Causing hurt, maybe, but not being hurt.

ATTENDANT: Are you discomposed from an injury?—

LANGLY: What?!

ATTENDANT: —We can call for paramedics, and place you in a restroom till they come.

LANGLY: Only hurts to breathe some, till I get my wind back. Then I'll be on my way. I don't want this to condense into some god-awful fuss. Though standing here awhile *wouldn't* attract such attention. Better than loafing at a lamp post and watching cars. Where else can one stand?! At a curb? waiting to cross with an endless delay? In the middle of the sidewalk, having people walk and crawl around me? like I was lost in their presence of wave? In front of someone's residence, like a burglar lying in wait? I haven't even gotten that far, to the districts of solitary houses. And pretty they are, with their fanciful awnings and handsome presentations to the eye, and their clever, inviting steps. But I live more at the project houses, apartments led to by longer, higher steps. And I thought this was a safe place to hide. But I'm being affronted for being abnormal?! I'm not!

ATTENDANT: I don't know where you live, sir. But this is a fairly busy place, people coming and going throughout the day. And I'm not sure you can realize your impact on it, standing still. The public don't tend to do that, as long as you have. I suppose it's permissible, if you're really indisposed. But if that's the case.... then you can be assisted. If that's no longer the case.... could it be that you have fears for being lost? It's happened before, though they tend to wander *inside* the store, aimlessly confused, and almost always elderly or very young. We can help. We know how to handle that. We can get you home.

LANGLY: Oh, I'm not lost, son. I know this town.... too thoroughly. And I don't have the brain disease, the alchemy of plaques to the brain. My memory is strong. The lungs are weak, damaged from my.... previous work. That's why I'm not aged, with this affliction. But I'm not threatening anyone either. And I can take this walk. I have the ability to do so. Just overdid it slightly, today. But.... on my own I'm capable of these things, which is why I continue to. I'm not proving to myself. I'm doing for myself what can be done. And on a nice day like today, it's always worth the try. But it's been special for me, you see, to overextend the possibili-

ties, and to let what happens surprise my efforts. Somedays are like that— and they're rare. And I'm jumping.... But traps can be as beautiful as gutters, if that's where you're meant to stay.

ATTENDANT: This ain't no such—

LANGLY: I mean.... I'm standing still— awhile. If it's an awful while, it's not because of me. But the world's so busy. They don't want to see that junk!

ATTENDANT: I wouldn't work in a trash heap. I have some dignity left.

LANGLY: Left?

ATTENDANT: Well I'm not exactly fabulous, doing this. But it's all that I could work for.... and possibly work up through. But I got no brains either— I know it. I'm as sick as the unaccomplished.

LANGLY: Don't seem so crippled to me.

ATTENDANT: This is not a glorious occupation. A start at something— A stab.

LANGLY: Any job's a good one, if it pays you.

ATTENDANT: I can't think that way, sir. That evidence doesn't appear to me. If it renders you disappointed, if it wears out your spirit and tears down your excitement for living.... if this is what results about you—

LANGLY: Leaves you breathless.

ATTENDANT: ....then it can't be all that good to own, or own up to, that this is what you're worth— to everyone and anyone who is objective about it.

LANGLY: You're too impatient for fame and fortune.

ATTENDANT: No, I'm just stuck. And my lungs are clean, but I can't do better. I know I never will. Haven't the smarts. Yeah. The world rushes too fast, for me to grab hold of anything worthwhile. I'm not designed for anything greater than this lackeying.

LANGLY: Don't live so dejectedly, boy. Break bricks, if you have to. Sear those lungs! My emphysema's not half as bad as your self-doubt.

ATTENDANT: Only a resignation, sir. It hurts to realize things.

LANGLY: You can't be tired already. You've only just about started. Took me a good half-decade before being settled with worthwhile prospects, feeling comfortable about a working life.

ATTENDANT: Well I don't have that much time. If things don't look up soon, it's to college for a good half-decade. At least one can worm a way through scholarships to pretend at making achievements.

LANGLY: Hum. A degree's not worth the effort, if that's all it's about.

ATTENDANT: No. But it brings in the rations of conferring the dole of uselessness. Are you sure you don't need any assistance?

LANGLY: Quite. There ain't no war going on here. Handouts aren't baked like cookies for children at a classroom tea party, to teach them how to behave socially. You do your job and buy them for yourself. And I'm about to leave for myself, because my breathing's come back to normal, thank you. But don't be so pathetic of hope just because you're in a bum position. What you find yourself swimming through has to be common for lots of others. And they're not so complaining of their merit. It's only attitude that counts to make a personality, and a depression.... I'll be on my way, son. (*starting to leave*) Won't bother the shoppers no more.... You've done the good push and brush. Hope this gives you a bright light in there, with your superiors an' supervisors.

ATTENDANT: I'm the least offensive male to send out, what they'd joke about. This is only a slight inquiry towards reproach. But I welcome much your comprehension for what it means.

LANGLY (*leaving with his back to him*): So be the gains for you. I'm not displeased.... But it's a long walk going.

Act IV

Scene I — *The afternoon.* **Heroford** *is walking down a street, and finds* **John** *crouched (down) towards* **Langly**, *who is sitting on the sidewalk with an arm supporting up his torso.*

HEROFORD (*coming by with a rush upon recognizing* **Langly**): What's happened?

JOHN: This poor fellow slipped and just collapsed to the pavement. I saw him fall some distance away. Seems confused, but the event frightened most of the people nearer him. Though a couple tried to help. But he's so confused. I think he warded them off. Yet he just sits here. I've come up to see if he's hurt.

LANGLY: Just blacked out suddenly. Not confused at all.

JOHN: But are you hurt?

HEROFORD: You were at the café!

LANGLY: Don't feel no pain. But don't have any energy. Surprised me to a shock, the blackout came on so suddenly. No warning. I didn't fend anyone away. Just couldn't answer much for a spell. I'm a fairly big fellow. Must 've frightened them.

JOHN: You refused help?

LANGLY: I don't know. Wasn't me awhile. Thought the earth had caved in, or something. I've been weak before, but nothing like this— all at once. Wasn't even exhausted. Just plopped down, like the legs had no brains.... didn't know what they were doing.

JOHN: I think one of them called for assistance.

LANGLY: That's very considerate. But I live near-by. All I have to do is get home.

JOHN: You might need a medical overview.

LANGLY: Can't afford that. I'd be indentured for the rest of my life. Once the physicians get their grips on you, they control your body and what you can do with it. Dope you up into a thorough compliance. I know that well enough already. Took me over half a year to get over the bloating from their medicines an' prescriptions. Couldn't walk at all for months— It was frightening. You have to work your way out of that convalescence, an' atrophy of muscle an' deposition of fat. Your body tells you what it needs to do. Can't rely on those drugs, a cocktail of them distorting your mind, your principles, your self-sufficiency—

HEROFORD: Told me you were going on your Constitutional, after the coffee.

LANGLY (*looking up*): Oh! You're that kind man, who offered me a cup. Yes. I've been walking. Reached the park, today!

JOHN: That's not too far. But how are you feeling now? Any aches?

LANGLY: No. Only live 'round the corner a block. Fairly well made it. My head just blanked. The lights shut off.

HEROFORD: If police are coming, then we should wait for them. Or if you're rushed for time, I can stay with him till then—

JOHN: No.

HEROFORD: —I'm just off on a stroll, to clear my own head.

LANGLY: This is a day for it.

JOHN: No. I was just visiting relatives, in town. Help run a farm not far from here. Don't have to be back so immediately. Pomelory Greeneries, it's not that distant. I was heading for the bus out to those parts.

HEROFORD: Oh. That's a very fine outfit.

LANGLY: Pomelory?! You don't happen to know old Oaker, do you? He used to frequent that café we met at, Mr....

HEROFORD: Heroford.

LANGLY: Yeah. We construction workers saw him a lot, there, praisin' 'is vegetables. Become a big concern, his operation.

JOHN: Why, I'm a nephew of his.

LANGLY: Now don't you say! What a small world! I remember three of us were sittin' at a table, and he come(s) over and give(s) us a bag of tomatoes for free. Said his wife couldn't use them, but he was too proud of them not to let them go to waste. Thought the café might have use for them, but they already had some and suggested he might give them out to some favored customers. They were as sweet as apples, mister. Very impressive.

JOHN: Uncle Oaker's like that. He's retired now, though, pretty much.

LANGLY: That was some years ago. But I'm glad to know he's still with us. A nice gentleman. But who shouldn't be, who's proud of himself?

JOHN: Are you sure you're feeling O.K., fella?

LANGLY: Yes. Langly's the name.

JOHN: Oh. My name's John. John Mendez. Soon to be a baron of soybeans.

LANGLY: That's nice. Can't say I've ever had any, that I'd know about. They like chickpeas?

JOHN: Similar.

LANGLY: Oh! They make those soy nuts out of them.

JOHN: Yes.

LANGLY: Then I've had a few. A change of pace from peanuts an' pretzels. They do well, salted.

JOHN: Quite well, in fact. They're replacing, supplanting usual bar nuts, even corn nuts. Supposed to be healthier.

LANGLY: Then it was probably recommended to me. My lungs aren't good, from the construction work. But I'm rested enough now.

JOHN: Do you think you can stand up, Langly?

LANGLY: Sure! If you help me. (*They start to help him up.*) It's just a general weakness that really surprised me. Didn't see anything like this coming.

HEROFORD: You overworked yourself, that's all. The coffee stimulated you to try more than you were capable of today.

LANGLY (*standing on his own, though **John** shoulders*): Might be. Might be, Heroford. Wasn't planning to. Should have gone straight home. But the day started to feel like a premonition for reaching out and extending yourself, a preview of what's to happen as these abilities become lessened, the opportunities get(ting) more and more scarce. I certainly didn't foresee anything disastrous. But something compelled me to try at it now, before I'd never be able to rightfully, walk to the park and back with some health. I guess it's getting more difficult that I supposed, but I had a nice time. The try was decent. It was an achievement, an accomplishment over bodily torts. And I wasn't straining myself much, doing all of it at a slow gait. Though sometimes I'd look down a street, and it would seem to roll up into a less discernible sight of imprecise imaging, like the shops off in the distance would snake out through waves of watery contractions an' ebbs, the granite shimmering as the buildings bent to be captured of their perspectives. And the flow of view was rushing towards me an' away, like a monstrous reckoning of adventure an' purpose. And I know I'll soon be swallowed up. But let's face this magnificent beast of life at least once more, and dwell in the splendor of its magniloquence to be displayin' of itself without any concern for your presence. It was like seein' a mountain you know you'll never gonna climb.... But that sensation, paused of various moments, drifted back to normal, as I ambled on. And I'll look forward to my chair to rest in, for the remainder of the day, having beaten my fears of loss back down a ways, once more. Since I used to walk down these streets without noticin' a thing except the purpose of my destination. Yet there's a tidal wave of events to see, if you're alert, or sensitized to what you've avoided.

HEROFORD: I just wouldn't exert myself this way, in your condition. It's not too rational, since you don't have much to prove, of your past at least.

LANGLY: I'm not biting on that. But it's darn disconcerting, sometimes, to know that you shouldn't be able to do things, that you aren't allowed to rationally try—

HEROFORD: Well that's.... that's a thought— for madness!

LANGLY: It ain't insane to be deficient. Maybe to show your lacking, or promote it, but not being so— Oh!... I must have bruised the inner side of my leg, while falling.

JOHN: Does it hurt?

LANGLY: Haven't felt it till now. Rubs up against the pants. Strange sort of injury. Based on your clothing.

HEROFORD: I still care. It's not madness. Maybe oversight—

LANGLY: What?!

HEROFORD: You fail to consider thoroughly everything you should be concerned about, including your health or well-being.

LANGLY: There's but so much you can really take for granted, Mr. Heroford. I'm sure you treasury types try to account for every penny of what something's worth, a stock or bond, before proposing city investments or borrowing on.

JOHN: You're a financier?

HEROFORD: Secretary to the treasurer.

JOHN: High position.

HEROFORD: Worth my carry and my doubts.

LANGLY: I'm not ruined yet.... gentlemen. I'm poor an' broken down, but not that. There's still a fight 'gainst the serpent left. But you have to face the beast, like that— Beowulf. You have to go to Grendel, an' provoke— make a challenge.... And fall to the street, if you have to— that's no shame.

JOHN: It's not exactly courteous, for a public's apprehension to arouse.

LANGLY: I did not want to fail. (*rubbing his leg*) I didn't want to cause any fears, about me or the world in general. Guess I was blinded of effects, but.... you lose some sense of proprieties when you're impinged by debilitations, while you're engaged in a struggle against them. Didn't wake up this morning, thinking I would stroll to the park today. Wasn't on my mind at all. Only the usual daily efforts.... had been expected. But this kind gentleman empowered me to try for more— and I did. And I almost completed it without embarrassment. Well that's some sort of triumph of the daring, even if it's unpredictable, even if it's not respectable, or foolhardy or stupid. I felt a surge to take on the threatening monster once again, and dissolve for a time these binds of intimidation, with the trepidations refreshened for assault, like I used to go after an' tackle hard projects— impossible ones. You never really win. You just feel grand while you're trying. An' why can't I clear the head that way, every so often? It's been months!... in a cocoon of misfortunes and feeling— degradable, denigrat-able, and needing that.... clean glass of free water, to remind my body— of what it's achieved throughout this life, throughout this candidness of accepting my stalled paces and lethargies too currently appreciated. I'm not battling myself, or I would stay in bed and rot away whatever spirit's left to survive with. So I'm not doing the harm. An' I didn't make myself fall. And I didn't make insult of fellows near me, or cause them to run from me, or find my presence revolting enough to avoid, repulsive enough to challenge slightly— and then flee. Was not I who did these heinous things today. Was not the disgrace in me that led to such shame. Was only the humane gift, that strikes at all foulness, and relates strangers as companions swimming through the restless tide that seeks to annoy our preservations. It's the confrontation of that annoyance which causes all disagreeableness.

HEROFORD: That shouldn't make you abuse your physical strengths way up to the point of this danger. The mind knows better than to sup on disasters— I.... I would avoid courting disasters. I would have felt happy about the coffee, and then gone back to my apartment— as planned.... and waited for a better day to try at the park.

LANGLY: Well that is logical, sir.... and extremely good advice. But the point of the danger is the point: It's no danger. Dropping dead on the street is no danger to me.... nor dying in bed, suffocating through sleep, which the doctors say could eventually occur. These things only occur through efforts to be fulfilled. The danger is, as always, in the denial. Without that, I'm fearless. Weakened, withered, weary, worn-out, but still capable of being true to my insistences, that the day is nice enough to visit the greenery, and I'm just able to do so— not driven by insanity nor fright, that I may never do so other than this day to make the attempt, but simply that it's something fortunate enough to be done— I have.... And despite what's happened.... it was pleasant.

JOHN: A job completed is always a pleasure—

HEROFORD: You wanted to.

JOHN: —But this one is certainly not. You have to get home—

HEROFORD: Wanting is not enough. Not even being worthy is enough. There are conditions placed on you, that make for certain matters to be appropriate or not. You can't just have your way with their total disregard. You have to stay responsible to the essence of their imperatives. A bum dying on the streets *is* a bother to society, a wounding, an absolute infliction. Each person should have enough pride or merely self-consciousness to dearly want to avoid that for themselves. It is incorrect thinking otherwise, an error of shamelessness, an affront to comeliness and overall comity and above all a social purpose assumed, a worth ascribed to and given selflessly by your fellow citizens. Say that you erred in this, that your decisions were not sound, the judgment poor. I don't ever want to see this sort of tragedy in my city— nor should anyone!

LANGLY: ....I erred, sir. You lent me your madness— I'm still appreciative, but my judgment was awful in this matter, and I'd never want to offend any stalwart and upright— and terribly generous city official. It's true. I've not heard of you till this day, or I don't recall. I can't remember much of our governance. But I believe you thoroughly. And I was absolutely wrong to try as I did, disturbing so many people. I can't tell you how many, since I've stopped counting grimaces weeks ago. But you are so right about me. And this is not a sarcastic pledge, but I have been a sore on this town.... for a long time, and should try to hide more whatever excesses of nature I occasionally yield to— It's not habitual with me.... I shudder to let people hear my groans; and I am thoroughly embarrassed when I cough in public, because that reflex brings up more than gasps but expectorations of utter filthiness. And this is a disgrace, what I have done today, hitting the pavement while among fellow pedestrians and startling them—scaring them with my imbalance and want for seemliness. A bad character, and a scandalous stepper, I have become through my lapses of physicality and moral prudence. I knew it from the beginning, when that silly notion came to me. But I just couldn't help myself, your generosity so inflamed my courage.

HEROFORD: You should be tucked comfortably in your shelter, relaxing and safely removed from these trying hurdles of existence. You should be restored.... to the manliness that you've so long assumed and that you are being forced to assimilate away from—

LANGLY: Feed me more and I will bark!

HEROFORD: —But this crisis is how you are dealt.... and will overcome, these saddled privations of life.

LANGLY: A reprehensible behavior I've succumb to. But perhaps you've never felt as hopeless for the future—

HEROFORD: Oh, I've had my days.

LANGLY: —to the degrees of growing futility. Not that I'm in anyway afraid of what's coming; but you pick your moments as you find them, and then you run with it— full force, with as much steam as you can promote.... I didn't want to be a public nuisance about it. We almost succeeded—

HEROFORD: We?

LANGLY: —without major incident. Well you did pay for my coffee, sir. You sort of backed the effort, if not championed it implicitly. Do you know how thankful and nourishing that made me feel? It allowed me to run free from my troubles, for awhile.

HEROFORD: Run free.

LANGLY: Nice, healthy gulps of air. Practition-ing of the physique again. It's all I can do, now, for a full-time stretch of work. And your kind gesture enabled some attempt at a handsome becoming, even if bothersome.

JOHN: More frightening than bothersome. People just felt a little imposed to muster up some aid for what looked like a sudden accident. You may have slipped—

LANGLY: Was the grace of the coffee, sir.... the stirring of the nerve to be adventurous again. That it was offered freely is the point. Could 've brewed some back at my digs. But the coffee's not a glass of water. It was a symbolic allowance, and a lending of the café's atmosphere back to me, as I could remember it, and recall my stronger days and brighter perceptions of the world.

HEROFORD: Allowance. Well I simply felt you were a decent fellow and deserved a morning's treat. And I may even approve of the nerve it brought up, to defend your individuality of independence, in this cosmopolitan mixture of blank, bland, hidden faces and common motions of progressiveness. And I certainly can't be one to reprimand the courage. But I take adults to be reasonable of their actions, not to work themselves into such a frenzy of losing energies as to fall flat down before their fellow man, in the company of the public common. I wouldn't have wanted to induce such a disturbing display of wallow an' wane in the streets. Though I might infer some epiphany of consciousness was reached, since you seem not terribly damaged with recovery but may even feel somewhat revivified. My aim was simply to reward you as I might myself, that early, without any pitying for your— for anyone's hardships.

LANGLY: You'll not do a silly thing yourself, sir, when the surge is sensed?

HEROFORD: ....I went directly to my house, after having a slight conversation with a rather pompous man who I fear might be your uncle, Mr. Mendez.

JOHN: Mine?!

HEROFORD: Does he like some sort of a Bosc drink?

JOHN: Yes. That would be Uncle Oaker.

HEROFORD: Said he ran a farm. I might have overheard his name, but I wasn't listening.

LANGLY: Well it *is* a small world.

HEROFORD: Not so much if he frequents that café.

LANGLY: I've never seen him there, these years—

HEROFORD: You come by too early, Langly. You don't want to meet up with the past so readily, so starkly. You're afraid he might not approve of what you've become.

LANGLY: ....That might be so, sir. Deflates me, to realize this. I've been hiding, an' rushing not to be noticed, while taking chances. Surprising, for a person as stiff an' slow as I am.

HEROFORD: He might not even recognize you much, or recall the friendship. If it wasn't for my vacation, I'd have never met him. It's not so important.

JOHN: Pompous, you say?

HEROFORD: Well, particular about his breakfast; but convivial enough. Thought I might be indigent, because I wasn't sporting a jacket. Isn't ordering at the café disproof enough?

LANGLY: Actually, I was leaning towards such a view, initially. But you came through, and sparked the unusual with.... the unusual.

HEROFORD: I was rushed of an impulse, I suppose. But the weather reports said it was going to be a warm day.

JOHN: Fairly well is. Good for the crops. Bright sunshine. It's called a rich day. Not overly hot, but lots of nurturing radiance. Plants reel for a day like this. They're traveling— growing, while remaining stationary, at least to be observed and counted. But both the stems and the root-work can become very extensive. A sky like today's promotes *that*.

LANGLY: Guess I've been rooted too, to roam.

HEROFORD: Roam. Life's becoming dysfunctional, to me, with not much to do. But my family will be traveling soon, to visit places and resorts. So that will take care of that. Always a hassle, these trips. There'll be too much confusion to control.

LANGLY: I wasn't confused at all, sir.

HEROFORD: Commotions and squabbles then, with the children, and.... others we meet. The park was like a vacation spot for you.

LANGLY: I have to say so, if this life's much (of) something other. I'm trapped into the regularity of habits like a retired person. But my mind's not quite thinking that way. It says it has to get out of this— rut's appraisal of an existence, and make a break, a run for it, an escape.... somehow. Oh, I'm not bored. Little efforts are too difficult to cause lengthy distractions of thought. But I am bound by my limitations.

JOHN: That's true for any active thought, the degrees of which are altered only by your personal circumstances. Others may accrue much finer privileges; but the restrictions feel the same, believe me. They are hurting, if not harmful and extracting.... You are no worse off, to what's real of your afflictions, than others are to their debilitating resentments, disappointments and angers. We all have to find our ways to be happy, even if it's too crucial an achievement to realize safely. I was seriously thinking of leaving the farm business altogether, but was talked down from that position— I think. Happiness is not the most important thing. Satisfaction is slightly higher on the rung. And striving for your goals tops the pole, as far as the mind is concerned. I'll answer to that— need, for a bit longer yet. But I don't think you've disgraced yourself with this attempt to visit and return from our park, Mr. Langly. And if you had been left helpless, sprawled on the pavement from a collapse of effort, taken to be an unrecoverable and worthless life without a need to waste assistance on, no less seen is the disgrace of our envies and sorrows to entertain and thrive with. But this is so much better a day than that, the greenery shines and shimmers to be raised and lifted out of our self-appointed depressions and worries. And this is so much so that had you died right at this spot — it would have been a victory earned, defying shame and the shambles that we can too easily make of our.... personablenesses and relationships. It would have been a sprouting of the verity to our endearments to the conflicts stood up to, faced and struggled with to defeat. For it seems that we only live for some growth as this, the sentimentalities of our candors to overcome travesities and obstacles to striving.

LANGLY: Can't say I thought to be so heroic about it, Johnny. Felt good to try, is all. But I can take your reasoning to heart, for the bravery of any gentle stalk to dare reach for the sky an' sunlight, an' fashion of its cloth a cloak of perseverance.

JOHN: The birth of a bud booms into both consciousness and conformity, when witnessed.

HEROFORD: You mean one's vanity to take on these goals is enough of a reward to be seized by for obtaining. The obsession wins itself.

JOHN: ....Vaguely. But let what wants to, what needs to flourish, flourish. Don't condemn a weed outright, until you see its flowers bloom. They may be pretty, but merely misplaced. We spent a whole season stripping a field of tubers from mis-growths, only to find in a pilot quarter acreage, set up experimentally on my recommendation, those growths actually enhanced tuber size, through some kind of symbiosis of nitrogen fixation the weeds allowed for a certain bacterium (for the weeds) to reside with. In other words, the soil was improved. The misplaced plants hindered the number of tubers recoverable, with the machinery we were using, but they increased the size and stability of the produce.

LANGLY: Oh, potatoes.

JOHN: Yes. A variety of red yam, actually. Much sweeter with the weeds, simply due to a more comfortable rearing. The mis-growths were caused by insects, a type of fly, bringing along with them seeds to the plants for their natural habitat. Didn't infest the yams at all, they being roots underground. And the prettiest yellow flowers were inflorescent. Only have to guard for densities, now.

HEROFORD: So let nature take its course, you mean, in some ways through your handling of affairs— she being the arbiter of all complexities. Leave time for some flights of fancy.

JOHN (as **Langly** rubs his leg): Yes. Or, if you're of a must to be finicky, do it with some amount of fickleness.

HEROFORD: Right. I can see that in many matters, mostly personal. One has the least control over your intimates, and the most control over their intimacies. Social relationships are a most profound thing. At work I find them quite antiseptically handled by some, one being expected to be thorough only in your specialty for commendation. But you can't do that at home. Feelings are too often hurt as individualities of temperament struggle to be manifest. Especially as relates to a family environment, this is true. And heed the day a company or government is run like a family. Such temerity of presumption makes for a woeful terror.... of indoctrinations and enforced manners. But by a family (it i)'s O.K., Mr. Mendez. And as for this society.... I don't think police are coming for you, Langly. No assistance at all. Perhaps more pressing matters exert themselves, for their attentions. But you should be gotten home. You look peaked.

LANGLY: Walking is a lot of standing. It's just around the corner, a block. Not that pretty a sight, but the complex serves it purpose extraordinarily well, for fixed incomes.

HEROFORD: And as I find myself to be a social welfare serving, in several ways and with many means, I offer to help get you back, bring you safely from this impulsed excursion and conclude a clarion call's overt impressing for a day's strenuous activities outside and through the city. I hope the battle was worth it.

LANGLY: That's kind of you. And the battle was same as lent—

inane perhaps, but then so are the lame to think through voyages.

JOHN: It's the least that we may see you to your apartment. That type of dwelling.... I've never been to, and yet have recollections told me as a child, something like a dignified squalor.

LANGLY: Well it's simply not rich, Johnny. Simply not rich. When the buildings were new.... and I helped with some of the construction.... they were the peach of pleasing to the eye outside and a modern homeyness inside. But these sites can weather very shamefully, in a generation or two, to become somewhat depressed. Still, they manage to be appreciated; because, for shelter, they are very functional. And some interiors are often kept quite handsome.

HEROFORD (as they are leaving, **John** securing **Langly's** steps and **Heroford** to **John's** side, slightly backing): The city apportions much, for this usage and maintenance. It is not in sight of blight, but perhaps more of sternness.... efficiency and efficacy, But we'll make sure.... We'll make sure you get there.

LANGLY: And the day's not half done for sitting.

Scene II — **Pomelory's** living room, very decorative and up scale. **Laura** is sitting on a luxurious sofa, its leather making it seem cold. She is relaxing but pensive, and appears not to be in the habit of enjoying distractions like television or radio, perhaps preferring carefully chosen records of antiquated or antiquarian sound and size. But the space is quiet, as **Maureen** enters.

MAUREEN (heading for an armchair of ample plushness): The pot is washed, Laura.

LAURA (with slight remoteness): That's nice.... (more attentive) Did you first rinse with vinegar?

MAUREEN (sitting, not directly facing **Laura**, so that she must adjust herself to speak to her frontally, which she seldom chooses to do): I don't see the need to waste wine.

LAURA (resignedly reprimanding): You dilute a soap solution slightly with some. It only takes a little. Makes a noticeable difference, and protects the inside surfaces.... (recovering) Did you enjoy your son's visit? I didn't want to disturb you.

MAUREEN: You could have washed it yourself. You had all the time. I can't see why it's of such an importance—

LAURA: It's not.

MAUREEN: —But I left some for you, because you wanted some tea.

LAURA: It's not my place to achieve your errands, dear. I was really only afraid for a nice container. Safe and sound now, I hope.

MAUREEN (turning to face her): John is concerned with your plotting to oust him, Laura.

LAURA: That is too much of an exaggeration. (turning her head not to face Maureen) But I suspected he might be complaining to you about feeling misplaced in the business. That's for his own conscience to handle. You know as well as any that the work is his priority.... but less his chosen pursuit than others. And if he is uncomfortable, it's only to be shown for his unhappiness. John is not a settled man, Maureen, capable but still despondent much of when I see him. Whether that's a familial trait I won't guess at. Heredity is due to genetics. And genes.... make fibs. Recessives, I think they're called— until uncovered. But he is not the complete person, on our farm, in terms of calling. Else he would be married by now, for such activities complement a complete lifestyle. And I wish not the suffering of his compass—

MAUREEN: Your eldest isn't even engaged.

LAURA: John's position makes it difficult to be relieved.... of growing tensions. And what is best is what everyone concerned should look after and encourage: the prospering of our farm and its enterprises.

MAUREEN: Oak will not have your tampering—

LAURA: I am not. But I'm expressing opinions.... told, heard and acknowledged. The overall management of the operation might run more smoothly without John's.... uncertainties.

MAUREEN: That's what your sons believe.... But that's not what my son believes. And he hasn't any hesitance to put forward his views. I can assure you that there's absolutely no reluctance in him to promote the best courses for the farm.

LAURA: A sure footing is what's always needed—

MAUREEN: And he has that. You may play the minor squabbles called to hear about, but that can yield to little adjustment of the arrangements Oak has approved of. It's a petty game of teasing tempers to think otherwise. It is maladroit to hope differently.

LAURA: I've never been that. Some stability.... Maureen, of handling the problems and concepts that occur of importance is required at the top of any major undertaking, for fear that otherwise things will certainly come apart, relationships and marriages. Oaker has constructed an unusual bonding of tastes, but the difficulties imposed of this.... incoherence makes for resentments and possible stultification of activities and initiatives. I was hoping for John to reason similarly. There's little against his drive to do his tasks well. It is admirable, considering the awkwardnesses he finds for himself— there. But what is best in the long run, Maureen? You know that he's not likely to devote the rest of his life to this occupation. He is— complexed.... towards other directions eventually. And a firm needs a firm head to be practical. One leading voice is to command of insistences much better than among a multitude of rancorous complaint and raucous clatter. An assemblage such as ours, with what has been amassed and achieved, can not be properly run for long through fiat of such dissimulation of parity and divergence of opinion. Our sons have been arguing, perhaps at times too crucial of opportunity to allow for conceptual delays and hindrance of execution. Isn't it reasonable that I should suspect one of my sons to take this lead, and be the force that pushes matter along its destined tracks? Oaker has allowed the three to mix of their skills for awhile now. But that is only to determine what must result for a practical, feasible spirit of operation, with one person in charge to make the major decisions, start up new activities and change directions from the old. That will never be John's position, and you know it. He realizes it, and is sanguine enough not to be the obstructionist to an obvious future. I had expected he would

give you these sentiments of doubt to be much more, because my sons already sense it in him. And he is honest with you, for consultation. Now what have you advised? Of more fractionating resistance? Dissociable Friction? Yes, it's become a huge practice, for one family to run. But consolidations occur, which he must find reasonable, and to which he has been offered to partake of. This *is* how Oaker would have them work it out; since he is the primary example, finally running the farm splendidly on his own, before retirement. This is a natural progression of events, Maureen.

MAUREEN: Doubts for dimwits!

LAURA: Would you have a catastrophe happen?! like your— arrangements to fall to pieces? Shamefully! to leave you only soaked of quivers and sobbing of incomprehensions? We can not afford such disgrace to ruin us entirely. (*looking at **Maureen***) You know he is desolate there. You allow him to be desolate and reproachable, like a sacrificial payment for the agonies of your mourning, and the distresses of your erring. But he has some say too, in what develops for him, his natural tendencies to lose repression— and gain more travel in faithfulness. I'm sure that's what he tries to tell you, as your broken edges harden more of cold denials and may be jagged beyond repair.... He's satisfied his obligations to you, to Oaker, to everyone involved of the farm, and should now be freed to expand towards his own enterprises and interests. Oaker wouldn't tell you this directly as I try: for one thing not to test your sensitivities— which are peculiar and destructively intense, to suffer seeing for a sibling still dear.... but for another thing because this is so outstandingly obvious to be evident that it should not even have to be coached. There is some baseness that it falls on me to provide the attempt, the call to regain some resilience in yourself for his own well-being to foster and assist. The drought of conscience may never mend, to satisfy the thirst for drinks restorative, the scars upon its letup left. But for a consciousness it must, with finding more favor for John.... his disposition and character, and establishment of a fulfilled life. Lessen the scratches felt.

MAUREEN: He is at best for his work most usable and demonstrative. And I left it for himself to wear at these perceptions.... your rejection. But he is as sound as his cautions and paints a higher stance to be appreciated, that his talents are most capable to head the farm. And he is patient for much commenting, considers all views equaled to their sincerities spoken. Yet with candor is he spiked behind the ears. If compromise is injury, then what is a divorce— but death! He would not be better made, to separate his interests from his good deeds done and continuing. And you can not spearhead his routing with a limp psychology. There are more gains to win by being correct than being current. And as he might judge fads for foods, he is credulous to try— of a kind— but not to annihilate his heartily fought standing among us by running away from your— scowl. His weight is too substantial here, Laura. You can not tease him away through your sons' petty bickering about his intentions and to your ear their displeasures at his stalwartness and convictions agricultural.

LAURA: I've always presented myself quite approvingly to John, and can not doubt his vigilance to the destructive dangers he must constantly face of affronts to personal inclinations, tastes and beliefs. And I do not argue he's unsuitable. But substantial may not be sufficient. My overriding fear is of course that he may one day suddenly feel so fed up with the whole business that he'll rush to remove himself from it so irresponsibly as to possibly harm the enterprise in too fundamental a way for our family to reasonably tolerate or even recover from. We can't take such blows from uncertainty, and revisit the casualties of our disquiets. If he wishes to leave now, when the departure can be managed and afforded.... you shouldn't dissuade him from his heart's calculations. We do not promote an ousting, but are trying to reduce worries for a future of promised difficulties. Stability of handling is not the last resort, but the premier tact to take in a complicated endeavor. It is what we can strive for civilly, cordially, and productively. It's what I've fought for, throughout all of my activities. It's what my apparent fastidiousness cares about. It's why I cringe to sense all hurt in others, when it's beyond their control to feel or labor through. It is an agony to subdue and take care of, and an emotional hardship to even caution of. So don't misinterpret my intentions as bitter or aggressive, but rather with anxiety and patience, an often painful combination for a conscience to endure. And I don't like to be considered spiteful and disagreeable, vicious and intolerant.... of weaknesses in character and flaws of personality as evidenced by how people deal with their crises and upsets. But as matron of this household, and thereby effectively of our farming, I must wield my influences as best they may be employed for the interests of our sustainable lives. And I've only approved of what my sons have recommended for commenting. I'm not against John. I only thought— and said— it was a reasonable proposal. I responded to *their* uncertainties, to help end a guilt from such beliefs. They can not function well enough with this crippling uneasiness. And they are not wrong to be annoyed.

MAUREEN: To offer that, you are no cultivating wonder. It's more to be perplexed by what I have produced. There should be more sympathy for a marvel, considering the turmoil raised and kept to— for perfection of my sincerities to keep what is workable (and) strong. John will not be the cause of any major disruption to the farm life. He is much too clever, dedicated, and mature, and is an anchor Oak prefers and which I allow. And you will accede to it, Laura. For your sons have a.... devilish twang of disreputable wishfulness at times which can even frighten their father. It's why he leaves them to whatever they may accomplish, in these swiftly current and modern environments, without directly challenging. And he's brave enough to let everything go under, if it must, if that is fate's demand. But his say in this, his reasonableness with this chance taking.... is my son, grown through a more thorough inuring to displace more of the wildness with resourceful perspective. And that must entrap your fidelities. Because the world is crazy without a strict compliance to the sensible, as Oak and I can see it. What's real is firmly handled and grappled with, through our experiences learned—

LAURA: Real— or fantasized?!

MAUREEN: What's learned till soiled.... with what our sons have hopefully been educated about— to be proficient of the real and its wrenching reel an' jerk! John will make up his own decisions about this, and will not be pushed through tremors— not with my buffering. And a weakness is not fantasizing on concrete assertions and dreams. There'll be no flights to escape their failing.... or more tea to drink— for that reason alone.

LAURA: So you're telling me it's still up in the air?!— Well.... prepare for some battering, with the least sign of insipid resignation— to encourage a decently arrangeable.... transition. We won't go through another tormenting of the spirit and loss of vanity for our becoming and existence. Not I can handle more distress of that nature, and with hardly a gratitude for the sorrow. (*turns face away*)

MAUREEN: If errors are ingrained.... they may repeat. But that is not a crime of substance, more submission to our generations. And I take hold of the fact that we do not argue or debate, but confide.... (*standing*) for our awarenesses to treat. Sometimes the ground shifts too quickly to notice, the dizzying defeats too swiftly to yield much pleasure to a long recovery of composure and balance. And I appreciate the attempts to understand and accept the poor steps made. Beats the heart more soundly for these justifications. They are made solely for the future, and the anxious thumping. What runs.... can still not disappoint me, nor us. That will be the tale told, of these agonizing times of turn an' change.... and motion not of aims, but earning more respect for the wearisome and troubled. You need not agree with more than what is capable, to leave it have its diligence applied. And this is what is done. No back-stabbing, please!... It's not called for, is unnecessary, results in no useful wounds for our purposes, and marks a discretion soured upon hearing of, (*while exiting*) since it does not escape its timing to force issues.... But only parleys them with worse conjecturing of our attitudes and vulnerabilities— less real than gratuitous to think.

LAURA (*as **Maureen** exits*): Nothing's been decided but our meanness to be truthful. How they take over plows the wrinkles, without more care. Yet seeds can make their own furrows, of germination, and should know best the nature of this soil. I'm not so disagreeable as careless of contempt for wrongs. (*retakes a pensiveness*)

Scene III — ***Langly's*** *bedroom. He is lying in his bed, though dressed and uncovered, as **Heroford** stands over him.*

HEROFORD: This is a fairly barren place, Langly. The impecuniousness is stark, shabby but neat. Frightened Mendez almost immediately.

LANGLY: I try to keep things as simple as an asceticism can allow. No sense in hoarding in a lot of stuff that's eventually going to be thrown out anyway. This is the richest room in the apartment, the one I have to be most comfortable in for the longest amount of time, the retreat that I return to for the vapid slumbers. The living room couch is cold, most of the time. But there's a chair in the dining space that I like. Its offset kitchen is usually sunny. I could afford more, once, furniture et al. But I'm reduced to my mind's needs, now.

HEROFORD: But it drove him off, like facing a brush stroke on a blank canvas. He turned ghost-like, with the revelation of such destitution.

LANGLY: Simply adjusting to means of existence. Can't understand why I suddenly became so feeble, when we entered. Forgive my behavior and near fainting.

HEROFORD: You returned from an arduous and painstaking journey, and your bones felt ready to relax at home. Hardly anything in the refrigerator.

LANGLY: Arctic wastes of a frozen tundra. The water was enough to drink. Thank you.

HEROFORD: I think that shocked him the most. You needed

something stronger, like some soda. But there wasn't anything potable in there. I've never seen such a thing, frankly. Why have it on if it's not gonna hold much, contain much of anything?

LANGLY: Because the regulated rent apportions for it to be working, the electricity along with the water and heating. What's in the cupboard's more my diet anyway. Perishables shouldn't last that long. This is fine for me. I've adapted to these conditions over the years. But the place is clean and relatively safe, both by accident, perhaps. Not much problem of insects, and few vandals at this general residence. Some other areas have it much worse, for what I can afford. But this complex was not only recommended to me, I was practically forced into it through my disability proceedings, since it's conveniently near and easily accessible to the convalescence hospital. When matters align up so well like that, you feel grateful to be able to stay in the city. Otherwise I'd be pushed out more rurally, and with much more living desultoriness and uncertainty.

HEROFORD: So your pension agreements essentially force you here. Somehow you must be getting a bargain for what you're paying. But this kind of atmosphere seems Spartanly bleak to me.

LANGLY: That's more according to my nature. Other apartments here are more finely furnished and decorated. I don't need those fineries, and I can't eat much in a day anyway. The cans of soup, and the beans and rice stockpiled can last for months. Given my condition, I'm probably of a nicer leanness than when I was really healthy. And the lunchmeat and bread in the fridge do just fine for me. I keep the fat down through my own nutritional agendas and methods. Hardly ever feel deprived through these circumstances, as long as everything stays simple and predictable, trustworthy to occur with regularity. And as long as I can avoid the predators searching for weaknesses, especially during my walks, everything is sound to live by. Once the normal routines get disrupted, though, like during the state's financial crisis two years ago, and remuneration checks aren't deposited— or that's delayed too long (a time)— all confusion and anxiety reign. Desperation and despair take a chokehold on you, for these basic privileges to lose for any span of time. I thought it wasn't possible, what was happening to me, and what conceivably could. Fortunately, our municipal recession didn't stretch out too heinously. The dip was buoyed back up in just a couple of weeks, at least effectively, for my reimbursements. But it was a harrowing go to sense, and I'm not going to be caught having more to be discarded than to keep.

HEROFORD: That blip was due to a political scandal, easily corrected and remedied.

LANGLY: If I don't need something, I don't really want it much. It'll be a pain to keep track of or protect. Luxuries have little meaning for the sufficient.... of character, at least.

HEROFORD: Well, this is a competitive world, for your fun, Langly, and even more so in this nation. You shouldn't berate that for those who can put in the effort and draw out some success. Pleasures are as important as being practical, or else life is a dreary subjugation to necessities, a tepid misery waiting to take hold of you with some feverous decimation of a god-given élan to achieve some happiness.

LANGLY: I ain't unhappy, knowing I can't ever get better than this poor being, for these surroundings to contain. If I could run races, I would— But I'm not daft. And I'm not lazy. An' I don't

have to delude myself to want any more than I can comfortably use.

HEROFORD: But it's the contrast, I mean.... You can't blame others for wanting to avoid.... a circumstance like this. And you yourself were walking boldly, to prove yourself out of it, for awhile. That's a fair nature in all of us, exhausting your strengths with vigorous exercise, apparently. Disclaiming your condition, for much of an afternoon. I wish I could.... handle my disgruntlements as easily, as simply as testing and teasing physical abilities. But it's not so when you're as able as I am. That sort of thing proves nothing— to me. It is a gratuity of wealth in health and riches.

LANGLY: Maybe so, sir. But either I'm satisfied or I'm moribund. I can't afford mixtures of the two. I'm not saying my problems are so extraordinarily lowly as to make them distinguished of themselves to laud. I felt of as much a high calibre worthiness as you may, when I was active, younger of strengths and fuller of achievements. And I could easily disparage the thought of those less capable, as entities to avoid becoming like. But I did not refuse myself that possibility, knowing my line of work and its dangers. So my trekking today has little to do with any self-shaming, which I abhor above all disgusting attitudes to take in a person able to do anything for himself. If it's a proving, then it's to be pleased, and that's all. I'm not afraid of that. I'll try for it any chance that comes by. I'm sure we've both come across too many people who can't conceive of tackling such a thing when their actions might seem contrary to their overall motives in life of living responsibly, doing responsible things for themselves and others, and staying fit, physically, fiscally, and mentally, for those social goals to reach and maintain. But comes one to a place of character when being pleased seems a higher ambition than safety and serving, for others and yourself. It is a fundamental right realized that has little to do with your current abilities and more for your actualities to stretch at and examine personally, to approve of your temerities 'gainst trepidations rational. An' I wanted at the park today, because the head said so, and the try was simply an expression of the pleasure of this wanting, this inciting nervousness to excite for the itch's satisfaction. I know the ability may not come again.... but the pleasure may. The itch may astound my senses to be placatory of some proud— and proven— existence, as long as I may do my best to give it (any kind of) satisfaction.

HEROFORD: Can't so easily run! Not made for myself! Haven't constructed things that way, all throughout my life. Not a fashion lent me. Not a method seizable. Have dependents.... and deplorings to ameliorate. Can't enjoy the selfishness. Can't justify a self-serving nature. Am not crippled.... nor afflicted, but by ties to errors— of emotion. Have burdens of concert, but not yours of.... regret? You must resent something much, for having lost your health and liveliness, your former pastimes and collegialities, the being involved with something productive, skillfully directed towards those ends. Can't hack that off, Langly, without bitter regrets.

LANGLY: I was a professional person, yes. Days had particular meanings, for a vivid employment of labors. They drew on mindsets of anticipation and planning, and review for relaxation with revelry. The life was indeed a celebration intensely colored and hued of great principles of high enlistment to create— or destroy— and improve.... the landscape and the city environs. And each working morning was a crusade of happiness to awaken to and grow fortified with, preparing for the muscle-crushing duties. I would have more— struggling, more aching and enslaving of a need. The slightest breeze to face pronounced my necessity, and worthiness, respect by strangers ignorant of who I was and what I did.... But who is more professional of a person than himself?! Stripped of these abilities of construction, I remain— to be pleased, built up, rebuilt of a mind's grandeur facing the day, each morning challenged to find that essence of fulfillment with mental wondering, and to craft an appeasement for my lingering on. What else is there to do for one's self?! but to have one's way inside, no matter how thoroughly reduced to our environments, no matter how supplicatory to our waning earnings, no matter how detested of our existence or imaged.... to be disposable, your services replaced by nameless others, your offerings debated for descriptions of the past. Then it's not so important, what the ignorant bemuse themselves of. You know yourself. That is your heart which beats inside, and the days become.... internally accessible. Yet, the external breezes remain so similar, at times. This was a deed to face up to, sir, and enjoy! More— taste to it than food can manage, while passivity is grounded up and fed to pigs as passersby who would disown your visage seeing. Well, that's being somewhat defiant about it, I grant. But it felt so good— to attempt. And that pleases the day, with it's ambulatory thrust. I have defeated a lethargy of spirit with this try. Tomorrow it might be as simple.... as a subsistence on corn; another day, to survive an inordinately drawn out spasmodic attack of breathing difficulties— can last hours; and I say: "Cruel body, behave more wholesomely! This is embarrassing! This is absurd!... And I am due for some relief, your god and guardian.... and protector.".... But today, I wore the world. I embraced the city for my use, as proudly as ever, and for longer than to reckon— for a glass of water. Who can do better than this!— Who can do as well? (*sitting up slightly*) Oh, it's musty! these drives, perhaps above their wanting to endow one's soul. I cry when I say "cruel" to myself. It's trying its best and needs no further recriminations to call it cruel as it accuses me of hardships brought to lungs. So dare I plead with wrongs. I enjoyed them too.... and applied myself to this battering, this beating.

HEROFORD: Plead with wrongs.... No! No! You can't do that! You can't invite them for your pleasures. They might consume what's best about you— to convert to the worst, degrade your nature into the gruesome and mean. What have you left?!— Oh! The desire is so strong. The want's delicious!— But bear the blame.... for having destroyed so much. That is what is caused. That is what will be caused. That is what they'll claim.

LANGLY: ....I can't follow your diversion, sir.

HEROFORD: All of this blissful following is destructive, Langly. You can't live through whims as the major component of your activities. Look at how it's damaged you, this afternoon.

LANGLY: That be true. But the greatest question ever posed is "So what!"

HEROFORD: One must assist one's social snares. They're for the benefit of everyone in general, complexities too onerous to think through for rearrangements to engineer. So you fall into them willingly, as they 're placed. And you must accept your position found and manufactured for your comfort and good comportment with others. Do you know how difficult it is for the city to sustain residences like this?

LANGLY: I imagine a fair penny of public trust is required.

HEROFORD: To call it subsidization would be gentle. To have it merely as a generosity would be an error. It is as much a social obligation as the demanded structural integrity of any building whose construction is regulated by law, codes, and statutes. It is a civilizing principle to avoid the intolerable demeaning of purely precious folk, inhabitants of a vital town, contributors to its success and notice as a fine place of charmed atmosphere. This takes much discipline within each of us to hold on to—

LANGLY: The city, on the whole, is beautiful. I've never said otherwise, nor thought to belittle its splendors. It is a magnificent organism, this group of people and their public interests. And I bow to the civilization, can not even fathom— and would not be so arrogant to think I could— the obtuseness of its management, and accept all of aid and benefit that is given me or lent with an ever growing humility that glows through darkness and trying times and obscuring obstructions to genteel sufficiencies, and ostracisms!... from my trained practices of value to this potently progressive force of humanity. Then if I have affronted the dignity of our thoroughfares with my uneven gaits and crooked straits of direction traveling, and if I have confronted others disagreeably today during my contentious labors and struggles of achievement, then I take this punishment of sore tendons and tender joints and efficacious mind to treat these limbs so poorly— without regret. I will subscribe to the results as totally merited. This is a good pain to endure.

HEROFORD: I hope you're not terribly undone.

LANGLY: No. My feelings are not outrageously bad, sir. But I'll pay for my license, seized perhaps improperly. The regular things will be harder to do, for a while. Restrictions of motion will occur, as thoughts will run slower, for their capabilities to wrest in demonstration. And I'll plead with myself to forgive this lumbering haul useless of urbanity. And I will recover to never recover, in a more normal— and now natural— state of lost servility.

HEROFORD: Then you're not totally lost, to reflect on the nature of those around you, in a fashion supportive of everyone, since it is this general consciousness which labors hard enough to keep you fit for an existence among people.

LANGLY: My least efforts are spent towards complaining about anything, except how languors seem to spend their times abruptly. The awakening, to a new day, is quite a thunderclap of forcing your mechanical gears to get in place once again and creak of a dull or repetitive liveliness. But the shocks are bold enough to get things started, what's about to be overcome of a deadened state, and what surprises await for correcting and cleansing. They've all been done before(, these tasks), and all occur again— but in a frenzy of approach, much to their wont to appear. Why shave? but for the need to bare the face freshly, to a restrictive void with reckoning of a somber tide. Why dress at all? except for being presentable to a glass of water. Why fulfill this menial enlistment? but comes the walk, like a flushing rush after the breakfast. It is a ritual of envies.... and soothing qualms. Seldom can I take water from here quite so cleanly. It is— metallic, and contrary to my morning pills. But that anxiety is overcome with necessity, and the café's drink is a welcomed reward afterwards, for having done these chores again— living yet once more to face the bland though scintillating day. And then the afternoon awaits, with it's hefty solitude of relaxations from a physical exhaustion, so vigorous are these languors. I sit in a chair to resume my deeds of caring, courageous for this body to maintain and keep of function; thinking, retrospective of all of the harms that have been for most of waste, temptations abandoned from their testing— that I might do more than suffice of lungs to breathe of quiet airs. I was once crushing bricks, and toppling hard walls, and demolishing public fortresses to restore their footprints for a further use. And now I note the lints that can choke, and put one down with anguishing through the frenzied dusts and bizarre twirls. But sedate becomes the nerve, to contemplate all of this action and abstraction around me, as hours churn of sunlight run through taciturn windows, the rays displaying a mean lowliness of view that could frighten circumspection of the future. Some children play out there, ambivalent to poverty, and like lions roar towards conquests— Though what they hunt 's abysmally common and un-singular. Yet fun is made with these delays of responsibility or doing the substantial. And there's a warmth to see the personalities of pinions and peons in the abandonment of non-self-conscious flight. The heat through the glass.... envelops my contemplations as the tendons become more supple and yielding from tensions. Respiration eases as the ribcage fails to sleep. I've become my head again, for awhile, without so much to dream about, but rather take hold of what is actual and permissive to occur. And then the evening veil approaches, as my muscles regain their full plumpness of utility, and the onslaught of night must be prepared for. The scare is tedious. A dinner must be made, and any affairs of finance or fixing must be worked through, left for the last of the day to tackle or worry about. So bored are problems that they give you all sorts of titillation for your attention. And you deal with night, and a warm wrap for bedding, to finally remove your brain from the hatreds of being awake, shuffle soundly to a quietude of sleepiness, pause suspended— yet portending of the march to come, and all too soon of rush. That is the life you might suspect of mine. And what of yours reduced to? Hopefully more proficient and productive, constructive and commendable. But I remain.... washed regularly, not disrespecting of myself an' own, and within the thicket lying— absorbed to be, some proof of character left me, resentments absolved or dissolving. And what I can make of the world, as through glass seen, while walking through it, seems shimmery enough of vital colors and attractions. I'm not condemned to be unmoved, the iridescence beaming of many conjecturing shades to stay fascinated by, though growing more entombed by gradations. Yet witness....

HEROFORD: What?—

LANGLY: Breezes.

HEROFORD: You see breezes? You see your walk, your constitutional, to face these breezes, to meet the outside. Your work— made much of the outside. Outside (of) yourself, you mean? No one can walk that way— well, if at all. (*Some sounds of ambling, from another room, are heard.*) It doesn't happen that way, realizing the hopelessness of a condition. Was that door locked? Do neighbors have a habit of just entering?!— I think someone's come into your apartment.

LANGLY: Doesn't occur to me. Nothing here to steal of much value, not while I'm present and not when I'm away, though the door stays sealed in either case, open for transits and transferences.

HEROFORD: Broken into? (*as **John** enters*) Oh! It's you! Thought you had a bus to catch.

JOHN: I bought you some groceries, Mr. Langly. The emptiness of your refrigerator stunned me. That can't do for emergencies. Rushed around to the nearest store I could find. Prices seemed kinda high, considering the neighborhood. Convenience of location, I guess. But there's some soda for you, some fruits and vegetables, a can of apple juice, a bag of chips— and one of pretzels —.... and for some reason I felt to bring a hero sandwich of salami and ham, and bottles of relish and mustard to go with its cheeses and mayonnaise.

LANGLY: That's very kind of you, Johnny. I was planning for a bowl of noodle soup tonight, with chicken or turkey bits. It's dried stuff, desiccated for easy and long storage. But I'll work on the sandwich as an unexpected treat. I've had several today. Generosity blossoms and blooms.

HEROFORD: A pizza an' soda would have done. I think I spotted a store around here. There are no really depressed sectors in this city, Mr. Mendez. The administration works hard to prevent that.

JOHN: Well, let's be thankful for it. But I've never seen such a bare cold box before. Hadn't realized what a condition that might feel like, to return to, when there's really so much of plenty to accustom one's self. You must admit, Heroford, that one must feel this strangeness to realize it.

HEROFORD: Was a nice deed done. I could have also, something similar. But the poor guy got kinda weak suddenly, after you left. All of that walking. That's why we're in here.

JOHN: Yes. I was a bit surprised to search. That's what comes of impulses to do things. Hadn't thought much of consequences. Must have seemed fairly flippant and brisk to you. But my honest excuses made me feel upset and undone, thoroughly unfinished of goal. You have to be made to see a goal, sometimes. Can't always anticipate these things. Too many convolutions of experience to wade through.

HEROFORD: No candy.

LANGLY: I wouldn't have thought of any.

HEROFORD: Well, no bottled water, then. But that might deny the passage its right, the visits to....

JOHN: This is a splendid city, to be able to contain so many souls and individuals comfortably, within habitational reason. And I've a splendid farm to work at. May we each stay as able, for our desires to pursue. You most certainly must have truly reached your park, which is a community treasure. And I'm sure the city coffers are well managed and manipulated for the societal good, with intelligence, planning, and swiftness of beneficial result. I can't say my uneasiness has been totally assuaged, by doing something nice, philanthropic— even if superficial. But the surge of disquiet is abating in several areas for me, and I suppose to all of us, when we come to terms with our conditions, forcing ourselves to be thankful that we're still able to handle them, have some decisive influence over them.

HEROFORD: That is the one goal, to keep striving, to keep.... working— at it. To remain noble over the adversities we marry or merit, that is the grand silliness we nurture through to plow up and reform— recast our endeavors, our beds of soil, toil, tenaciousness, talent an' terms of abiding through life.... and without so much casualty as we would fear, if this is honestly done— and prosperously encouraged.

LANGLY: There are all sorts of patterns followed for surviving through one's misfortunes. But by an' large it's to accept what must be as a deficit defined personally, and to start rising again from whatever foundation you can establish or discover or what is provided you. I don't have any tears for a grief or a groan, but I respect their resurgences to pay attention to.

JOHN: Are you in any serious discomfort? We can call for help —

LANGLY: None more than I'm patiently used to. A little tired an' a little sore, but not put out to pasture yet— an' thrilled with what I've done today. Tomorrow that sense of heroism will be all gone, and I'll have the usual to contend with. There'll be no arrogance to life— It's proud enough already, an' more than capable to deal with the mundane, the crushing simplicities that tame your valor and vanity. I've never been a vain destroyer. There have always been good reasons for my efforts; and the joy of accomplishment was not a reflection of the terrible reduction, but rather for a well-designed restitution of service, a restoration of materials far above my head to organize with craftsmanship. So this punishment denies nothing, and orders everything about me.... concerning me. I think much less of myself than what I'm led to do, whatever the tribulations may enhance of view or intensify— exemplify. I ably accept the irony of my own body's demolition. Doesn't startle me one bit! Not a payment for anything!— nor a recovery of mordant spiritualities, if those buildings had any. What does a structure make anyway? but a scaffold for thinking, as far as we're concerned, the constructionists an' redevelopers. No; no ghosts hiss at me at night, nor bite at my brain while I'm dreaming, nor castigate with futilities as I'm wishing for something. But a little rest, and I'll be fine. And if there is some demon deity devilish upon my past deeds to dwell on with decaying demurrals, I'll shake (h)'is hand and then wrestle down to the final blows an' bouts; because that's my way to face these situations, not so much with defiance, but with a stern depiction of my value to insist an' deprecating insults or losses to limit— within my own judgment to these affairs of universal entity.

HEROFORD: On man and the universe. That's more or less solidly put, that you don't back down to these travails. The heart is a continuum of many; else life never began, that energy being extraordinarily prodigal considering its specializations into manifold but thoroughly restricted organisms. No more long walks, though, Langly.... for awhile. You can't hold up— to them, and you can't chance on being as lucky as you've been today. Guardian angels have peered through the clouds to see your distresses, and have sent envoys of coincidence and incidental natures to help alleviate the results of your mis-stepping. But I've become more and more convinced these angels are in general quite blind to see clearly and sharply through a world so greatly obscured with unhappiness and foreboding circumstance. They are at best randomly applied, or applicable.

LANGLY: I take your gist, sir. I don't rely on the sympathies for my miserable existence. Or rather on their absence am I apathetic, such catering utterly unreliable.

JOHN: But don't lose faith in your resilience. It will find a more

suitable outlet of expression. It will condone a more employable manner. It will transfer the motives.... into more substantial adventuring. And dream more.... of acceptance. That's the key, the fragile glass key, that chimes the prettiest of sound, when struck for its most pure tone, without breaking of a brittleness within its lock, to open that gate of glorious entry—.... It is the most difficult task imaginable, for our awkward hands (to manipulate) and impetuous leanings to get something done. But it is so by evidence of having been accomplished.... that it is possible.

LANGLY: Who has? The answer to the riddle.... is that it's already been. Does the afternoon turn golden?

HEROFORD: It's late.... late in the afternoon, Langly. Mendez has a bus to catch, to get back to a most productive and esteemed enterprise. And I.... return to a home life also, having experienced this warm day running through, as had been predicted somewhat shrilly— to my tastes. But the tones of lighting become quite hospitable, in this environment, as the sun lowers. Quivering waves of longest transit, you know, bring out the oranges and reds. The glass panes prevail of them, with a slight prismatic effect or bounce. You are a fortunate citizen; if minor, not miniscule. No need to make further adjustments.

LANGLY: Then I'll relax, sir, till evening and the shiver. I do feel a touch tired. But work awaits, and the treadmill continues. It's so remarkable how short a success may impose itself to your thinking. It's not worth the brain power (*lying more comfortably to doze*), only the doing of it counts.... Doesn't last long, the imprimatur of disposition.

HEROFORD (*as **Langly**, with his head on the bed, watches them leave*): So much to take up, that strong license served, that willingness to be thoroughly apparent— an' ever truthful about it. What for? these cruelties of mind! Then we'll leave you to a nap that you can manage, as I might manage funds respectful of their.... prowess over much activity, and how others may nurture the vegetable to ripen and taste good. But that's much better than unnecessary transits, that only lead you to return here, as the necessary do. For want of exercise, perhaps, is not an excess made of comity. (*They have exited from sight.*) But the views are stark to impress of plight. I've not seen so much before.... to be this slighted by.... Langly!... Goodbye!...

LANGLY (*through sounds, at first, of their departure from the apartment*): So is the stroke.... It's curious, what they could think of me.... and this humble place. And who could deign otherwise, that I'm such a simple man.... to be abided sentiment and generosity. (*closing his eyes*) I cringe at these laborious pains for understanding, that the extorting muscles weaken to become limp, no longer useful to seize that.... froward frown for rubble-making, and the levity from its aftereffects to syphon into consciousness of a heavy, lively task.... as crumbles.... crumbles.... drifting through the silt, my animations. And what crime is this heart, pressed with difficulties to endure these hard fought for strokes, is this love brought down of burdens? the amour so supplicating in this body tensile and tired and teased of its tearing, a tortuous terror— Yes! I've been afraid from first the lungs burned, and can not fight. The medications can not strike this harrowing pain. The meditations can not relieve the suffering. And the mediations of conscripted friends can not end my damaging, not by doles or donations, duties of discernment of my aggrieved face of grimaces with struggle.... to fight the squeamishness of my decadent decline, as such a musculature wastes its weight and pines.... And now the back seems tense. (*turning to lie on his back*) What first, the legs were heavy.... then the chest, and now the back. But calm down this silent toil.... with surge of serenity. For in a day so much is to be read that I spend time.... by its pages. An' there's not the boredom of illiterate mendacity.... All.... truthfully observed, the world keeps its passengers.... so much alike.... this back is hardly missed. So then it's real, what's told of to be average.... common, typical and sure. And must they have.... locked the front door? They must have. I'm too tired to check.... just now.... Just now.... a breeze! I'm lifted up, and by a closed window spread. That is a dumbfounding thing, to hallucinate on sore muscles.... and long drawn out ambitions. Night will be.... less of the poverty stricken.

Act V

Scene I — *The* **Herofords'** *backyard.* **Duncan** *is sitting in the lawn chair as* **Danny** *comes out with a dartboard and some darts.*

DANNY (*coming over to* **Duncan**): Mom says she doesn't want us using it inside, punching up the walls with holes.

DUNCAN: Yeah. That wouldn't be good, especially with your aim. I saved you from getting a beating once, because of a scratch you made. You didn't know about it. When she asked, I said I caused it, by accident. Didn't get hit—

DANNY: Oh!

DUNCAN: —Too old for that. But you're not aiming right. You have to *produce* a throw to reach a target, not just wish on hitting the center. What we really need is some backing for the den wall. But if Mom already prohibits the playing, that's probably out. I never caused any cracks, always got to some part of the dart board. Tallied up some good numbers with just three tries. Not that interesting a sport anymore, though. Once you gain a technique, results are predictable, don't surprise you nor please (you).

DANNY: I can get better with some practice. My arm's not strong enough.

DUNCAN: That's ("fish") scup! You're just sloppy at doing it.

DANNY (*placing down the board on the grass, while holding onto the darts*): But they don't often stay up when they stick in. (*studying the tip of a dart*)

DUNCAN: You have to make a trajectory that brings the dart's tip *down* into the wood. Otherwise they'll fall off. The trick is the acceleration of the bend. You never throw at it straight.

DANNY: Can't get it.

DUNCAN: You never throw at anything straightly. You have to figure out what the real motive is, how to get at that, and then work out a solution. Being right about something's not so obvious, because problems are so particular in their details. I figured out how to make them stay in quite early (on). (*impulsively snatching a dart*) The lower part of he tip (*demonstrating*) has to rest positively in the hole it makes.

DANNY: Positive?

DUNCAN: An upward inclination. The darts are designed that way so that the mass deceptively increases towards the tip and away from the tail.

DANNY: Oh.

DUNCAN: That's why the tail is called the feathered end, although these are plastic.... You have to study them, to reason out how they're used.

DANNY: That's not so much fun. I thought the sport was in the throwing.

DUNCAN: It is, like shooting clays. But if you don't hit any, what's the point?! You have to realize what you're doing, to master some success. Can't just jump into things without much thought. Oh, for you maybe the throwing is the fun of the activity, the chances taken at it. But as you get older, or more proficient, results take on greater importance. Sometimes they usurp all considerations. Getting an "A" on a test, in a subject you couldn't care beans about, (i)'s still an "A" to improve your grade point average. So you study hard, and learn how to concentrate on anything— just by habit, to get good results.

DANNY: The fun is the "A"?

DUNCAN: Well it ain't this dart! (*pitches the dart offhandedly, but with expertness to have it land on the lawn sticking up*) Ultimately, you do things for some kind of prizes or awards. That's the attitude you have to take, exertions made while growing into something praiseworthy or respectable.

DANNY: That's fun?!

DUNCAN: Well, play is supposed to be for children practicing their skills to reach adulthood. Why the adults have recreation's never been explained to me.

DANNY: Doesn't sound right.

DUNCAN: I think the theory's probably wrong also. But it's what's expected.

DANNY: Aspects of it may be correct.

'ARON (*running out the house to greet them*): What ya doin'! What ya doin'!

DUNCAN: Aspects of any logic are correct. You can rationalize any logical opinion. But that don't make their conclusions true or accurate.

'ARON (*tugging on* **Danny**): What's that?!

DANNY (*annoyed*): Darts! You can't play with them. Their ends are like needle tips, and can hurt you.

'ARON: What are they made for?

DUNCAN: To hit a bull's-eye, in that target.

'ARON: Don't make no sense!

DANNY: There's no place to hang the board around here, Duncan. Can't stand it up anywhere. If there were a tree, we could nail it to that.... Maybe the fence?— Get off! (*pushing 'Aron off him*)

DUNCAN: Now that would be really stupid. All you'd do is lose darts outside the yard. You'll spend more time retrieving them than throwing. It's a boring game anyway.

'ARON (*down on the target board*): This is a game?!

DANNY: It's boring to you because you know how to do it easily. I haven't reached that level—

DUNCAN: I didn't say it was easy. Takes some talent of approach, some feel for it. Look at all of the numbers, 'Aron. Can you count them?

'ARON: What?!

DANNY: Maybe the side of the house?

DUNCAN: They're all in circles. You put the darts in the circles, and gain their numbers. Add them up for a tally. (*standing*) I don't think Mom would want us placing it up on a wall out here. You might break a window, with your wild shots.

DANNY: But she said play with it outside.

DUNCAN: She probably meant don't play with it at all. It's a dangerous thing to figure out. Dad showed me.

DANNY: Well he doesn't show *me*. You mean it's all useless?

'ARON (*handling the board*): This is heavy!

DUNCAN: The darts are impressive. He's as tired of it as I am. Guess he expects me to show you— And I have. But you just can't do it well.

DANNY: I can do anything well, if given enough time and attention. (*as 'Aron starts pulling the board along the lawn*) You must have been as bad as me, when Dad first showed you the techniques to use.

DUNCAN: That's so. He was careful— about the details. The game doesn't fascinate me as much, though, as it does you now. He was very enthusiastic about it for me, back then. Gave me encouragement no matter how poorly I did. Never did beat him at it— officially. Out-scored him once; but he was distracted with something, and didn't take all of his rounds. Praised my tally I was beating him with, went off to take care of something while telling me to keep building up points— and he would match and surpass later. That happened a few times before, so I racked up the numbers. But he never did come back to reach the total, in the same number of turns we allowed. Yeah, he trusted me. But that win was unofficial, as far as I'm concerned. Came back to beat me the next week, though. But those events between us started to dwindle off. Said I was becoming.... "fairly good," and should engage some competition with friends. Didn't quite know what he meant. We didn't play darts at school. But anyway.... (I) gradually lost interest in it, as he did, and after I figured it out.

DANNY: Well now I want to figure it out. But when my turn comes— it's kaput!

DUNCAN: That's not so!... But it does tend to be a den's activity. You can get in there, and find all of the tools for playing. And this is high quality equipment.

DANNY: Mom's cleaning it up.

DUNCAN: ....Cleaning out the den?... She doesn't want you to get rid of this, does she? That's not the intention?... I hope.

DANNY: I wouldn't say so, if she gives it to me. I couldn't lose the whole set in a backyard. (*turning to look at 'Aron playing with the dart board*) Hey!

DUNCAN: We just took it out for the summer. It was stashed away, neat but tired, along with so many other.... idles.

DANNY: Idles?

DUNCAN: Things to pass away the time with, doing essentially nothin'. Preoccupations of leisure.

DANNY: Toys?

DUNCAN (*going towards the dart he threw*): Their intentions are more sophisticated. They're supposed to educate.... in sociabilities, I think, because they usually involve more than one player. We better pack it up, get it out of the way— if we want to keep it. (*retrieves the dart*)

DANNY: But I'll never learn how to aim right.

DUNCAN: Is that so important? (*studying the dart reflectively*) This doesn't look old at all. Used, but not worn. (*recovering*) Believe me. Once you get the bull's-eye legitimately, the whole thing will become boring, an almost dreadful tedium of tenacity. And you won't gain much.

DANNY: But I want to try.

DUNCAN: We can't set it up out here. It doesn't fit.... the environment we have. Can't hang it on anything practical.

DANNY: We can get a couple of crates, and (a)'fix the board atop of them.

DUNCAN: Sounds sloppy and unsteady. If you can figure it out, then do. But we don't have no such boxes around here. It's too much trouble, looking.

DANNY: The back of a chair, then?

DUNCAN: That'll make Mom angry.

DANNY: An *old* chair. Must be something stored somewhere.

DUNCAN: Look, Danny. The whole thing's becoming too complicated to conjecture about—

DANNY (*with slight fierceness*): Stop playing with the target, 'Aron!

DUNCAN: —And it has to be at a proper height for throwing at, so you'd have to place an unstable chair on top of something else. Just won't work well. It's meant for the den, a game room. It's expected there.

DANNY: You won't help me solve the problem!

DUNCAN: There's no problem to it. It's not suitable for this yard. You just have to face the facts. It's become antiquated for our family's enjoyment. That's why she chased you out with it, to make a point.

DANNY: She didn't chase me! She doesn't want us playing with it inside— anymore. And now you're telling me we can't manage it outside. But *you* could, if you wanted to.

DUNCAN: If I wanted?... You don't understand, what you want. It's not learning how to do anything anymore.

DANNY: What do you mean?

DUNCAN: It's applying yourself with what you've learned, what you've been filled up with. (*about to toss the dart again*)

DANNY: Hun?— Hey! 'Aron! Turn it over!

'ARON: What?!

DANNY: Turn that board over, so (that) the numbers are on top. If I can see them, then maybe I can pitch darts to it, like horseshoes.

'ARON (*turning the dartboard over*): Like this?

DANNY: Yeah—

DUNCAN: That's stupid!

DANNY: —But they can't hurt the lawn. This could be a new way of doing it. (*impulsively tosses a dart at the target*)

DUNCAN: Wait!—

'ARON (*scampering out of the way*): Hey!

DANNY (*excited, as the dart reaches the board*): I hit it! (*laughing*) Boy! Did you run off!

'ARON: Don't do that!

DANNY: Bet you can't shoot as well as me, Duncan, with a new style of doing things.

DUNCAN: Didn't even stick. It has to stick up in the target, to count the numbers. Dropping on it doesn't mean anything. But you still throw wildly, so don't—

DANNY: It's harder, this way. And you can't match me as easily, since it's new to learn. (*tosses another dart, as* '*Aron rolls away frightened*)

HEROFORD (*running up to the yard from the street, while no-*

ticing): What are you idiots doing?! You almost hit 'Aron with those darts!

DUNCAN: Dad?

'ARON (*sitting on the ground, slightly confused and slightly crying*): I was scared.

HEROFORD: They are dangerous projectiles. (*up to the fence*) You could have wounded him— You could have reached his face with a three inch pin! Pierced him with metal— Blinded him! You brats are terrible!— And laughing about it!

DUNCAN (*as **Danny** drops his remaining darts*): We.... were aiming at the dartboard, Dad. Not 'Aron—

HEROFORD: You should know better! so much better, Duncan, than to be this foolish. And just incidentally letting your little brother hop around like a hare escaping fire— if not targeting!

DUNCAN: We'd never do that! 'Aron got too close—

HEROFORD: Don't lie! Don't lie to me! I saw everything! You didn't care a wit about him. You didn't try to protect him at all! And he was right at where you were throwing— Don't *do* this, Duncan, lie to me. You're smarter than to count on fibs now—

DUNCAN: I'm just destroyed!

HEROFORD: —You're smarter than to blabber with falsehoods —

DUNCAN: Falsehoods!... False!—

DANNY: I wasn't trying to hit 'Aron, Dad. I just got carried away, and he was so pestering—

HEROFORD: Blame your hatreds, now?!

DANNY: —I forgot to notice he was near—

DUNCAN: We don't hate anyone, Dad!

DANNY: —My concentration slipped—

DUNCAN: We!... don't hate anyone of us.

HEROFORD: Your antagonisms. Sibling rivalries. Whatever— You can't hide behind that shield for your mischiefs any longer. Not you especially.

DUNCAN: We're not the ones to run off!—

HEROFORD: What are you telling me?— to my face! You want a beating?! Confronting me?!

DANNY: Mom said we could play with it outdoors. Said we *had* to.

HEROFORD: She wouldn't have meant for any stupidity like this. And as the eldest you're supposed to prevent such things among you three.

DUNCAN: Then I failed— I'm a failure at it!

HEROFORD: ....You were meant to take charge of situations like this, reason out the proper behaviors— and attitudes amongst yourselves and to others. That's why you're around. And if 'Aron had gotten hurt.... seriously hurt.... then you really would have failed— me, and everyone. But I haven't seen any failures today— just a lot of blundering. And you can't escape punishment for your own actions, no matter how fair or unjust. But lying about anything— is lying to yourself! And whatever might have— could have occurred.... would be no accident. There are no accidents on earth!... for those responsible for their actions. And the true afflictions your brother might have faced— would be your afflictions meted. I won't have this kind of scurrilousness at home, of what I've never seen so blatantly before, such a careless demonstration— for yourselves. Children have been maimed this way, even unto death. And you know the danger of those darts, how not to treat a weapon in your hand— Apologize to 'Aron!

DUNCAN: Apol...?

DANNY: I'm sorry, 'Aron. I was acting wrongly, and without.... inhibition for my errors.

'ARON (*wiping away some tears*): O.K.

DANNY: I saw you, but didn't see you there. I didn't consider your safety as I should have. Was too caught up trying to prove something to myself.

HEROFORD: So you disregarded him. But he is a real person, with a growing personality. You have to discern these things, son. Mistakes are too numerous to be trapped by without forethought for your actions. Disasters and tragedies spark up all measurements of time— can occur at any time, when brought about by lapses of attention. Forgiveness is at best a response, and therefore only a secondary platitude to receive. It can hurt you gravely to need it, and can never be sufficient. Yet it can be the most important thing owned.... of an emotional consciousness.

DUNCAN: Well forgive me, 'Aron, for not being sharper and more protective. I'd never want to see you harmed. It was an oversight—

HEROFORD: Darn well yes! And you'll put away those darts, if you don't know how to use them properly for sport. I don't want to see anything like this occurring ever again.

DUNCAN (*almost flippantly*): You certainly won't!

HEROFORD (*going for the front entrance*): ....Adolescence is a time of terror, for the childish growing pains to survive through. But I thought you could have done much better— than today's misadventures. And what else happens, when I'm not around? Can't be heaven sent.

DUNCAN: I should have been more careful about everything. Just our luck that he comes by, but I didn't think I'd have to warn you about the dangers of these things. That was more than dumb to do, Danny. It was demented.

DANNY (*anxious*): What's gonna happen now?

DUNCAN: I'll take the brunt of any punishments. But I've had it out. He knows what I suspect, what's been pent up.

DANNY: Is this going to scuttle our trip?

DUNCAN: ....No. Pick up those darts. (*as **Danny** starts to comply*) You're finished playing with them, aren't you?! (It) Wasn't so much fun, was it!... Oh, maybe it was— for awhile, when you weren't thinking. But it's become.... inappropriate for us now. Doesn't belong.... within our regimen of family activities.

DANNY (*going for the dart board and the thrown darts*): We'll have to go back inside.

DUNCAN: So? Store everything in a can in the garage, like glorified trash. Put all of it there, hidden and never to be seen again for years— (*rubbing the dart that he holds*) until 'Aron wants to play with this junk!

'ARON: What's that?! (*still sitting*)

DANNY (*collecting*): I'll have to get the key from Mom. She'll be upset.

DUNCAN: That won't be the word.... She'll take it in stride, because she forced you out with it. I was here, and she would have expected I'd prevent something like this.

DANNY: Why did you bother me so? Why was it fun to make you run from the target?—

DUNCAN: Don't blame 'Aron.... (*as **Danny** comes up to him with what he's collected*) You got something in your mind and couldn't control yourself, what to do.

DANNY: Guess so. But it felt.... compelling, Duncan. (*getting **Duncan's** dart*) Could I have really wanted to hurt him, excited by solving several problems at once?... led astray to do something terrible? Is that in me, those— hatreds?

DUNCAN: No. You just weren't thinking straight, being childish. Don't get too self-conscious about it. Nothing bad happened, so its not that important. Chance favored you. But her sword will cleave limbs eventually, unless you think things through more thoroughly, for the possible consequences of what you're trying to do.

DANNY: I don't want to be evil. I don't want to have a grain of that in me. But I thought I figured out the whole thing completely, and was feeling happy and bright about it, till the sudden scolding. And then like a flash it was the darkest reception to realize what I was doing, where I was at, and isolated with a horrible regret. How can your sense of things change so sharply? How can events become so wrong?

DUNCAN: You're perceiving too much. Mom's not gonna spank you anymore. That's done to restrain your activities when you're a toddler. Even 'Aron's getting too big for it. The hurt's more in a disappointment in you, now. But maybe there are no wrongs to change into, only how you feel about what's occurring: around you, to you. Our senses can't change the outcome of things, they only help explain them. In school they talked about a religion, or a faith, that believes everything is entirely preordained, like for a computer program running, wildly— even unpredictably. But your outrages and exhortations can't change the results. You mind as well not even be around while the machine is running. Fate is ambivalent to your abuse.

DANNY (*as 'Aron gets up to race to the lawn chair*): Sounds godless.

DUNCAN: This one claims you're your own god, and pitiable through your weaknesses.

DANNY: You mean there's nothing you can do about them?

'ARON (*jumping onto the chair and swinging his torso*): I'm riding the waves!

DUNCAN: They're nothing to be concerned about. You just maintain a composure through all adversity.

'ARON: I'm finding— treasures, and islands!

DUNCAN: And difficult times are coming, Danny, for all of us. So you have to work your way through your doubts, not to let them get the best of you.

DANNY (*with insecurity*): Sure. The world is supposed to explode, someday. But.... who will be around, living through that.... cataclysm?! We're all supposed to be dead by then. That's why the world can end, no need for its gods any longer. Maybe I can leave all of this stuff out here.

DUNCAN: Take it in. It has to be stowed, so you should do that. You were interested in it.

DANNY: Yeah. (*slowly going to reenter the house*)

DUNCAN: They'll expect you to try and show some maturity about a mistake.

DANNY (*half-heartedly responding*): Yeah. But I'm fearful for myself.

'ARON (*stilled and observing*): There's water all around, with fish and boats.

DUNCAN (*more to himself, as he paces in a general direction towards the lawn chair*): Hooks.... and anchors pulled up or cast out. Over the side, overboard. Keeps you steady while hidden, deep in the water, on the sea bed.

'ARON: Why can't I....?

DUNCAN: Can't what? What are you fantasizing?

'ARON: Why can't I play that game?

DUNCAN: You're too young. You don't even understand what happened.

'ARON: Did I turn it up wrong?

DUNCAN: You merged with the target, and mixed up Danny.... Dad became afraid for you. He shuns us, you know, because of

fears like that.

'ARON: Sh*uuu*—

DUNCAN: It's hard to handle, what you can no longer stand.... responsibilities like us, and a marriage. Anchors. Anchors tugged at. But a boat can sail just fine, or be adrift, without their use.

'ARON: I get everything wrong.

DUNCAN: You're too young to *do* anything, let alone get it wrong. But that's alright. You're not the cause. Get up!

'ARON: I'll get my feet wet! (*obeys, to hop out of the chair*)

DUNCAN (*as he's jumping*): Come on! I have to think this out. (*sitting in the chair*) And now *I'm* afraid to return— But not of him, but for the reality of all of this. And I was relaxing before, without much hope. And now the slightest tendency leads to errors that can.... explode the world. How can I work out these.... trepidations.... without falling foul to.... disrespect, antagonisms.... and animosities so deep-seated that they can only fail our character for being expressed? But for my own pressures, 'Aron, I would never want you harmed. Yet to force the issue to some catastrophe.... I was reluctant not to laugh, to see the displacement of your panic like a shove to confront the matter that something does not well. And trying to act like an adult, trying to think like one, I should have physically stopped Danny, hindered his excessiveness of pleasing. But I let it go on, for the darts I was friends with, the toys of a relationship I once held as needed and heedful for. They were so stunning to handle, felt good to be able to direct them any way you wanted, with an earned precision. That's what Danny was looking for, some stability and control created by his own efforts. It shouldn't be denied him, as we grow and learn how to do things.

'ARON: I can get that. I'm getting better too. But.... the words are so hard. They don't stick (to me) well.

DUNCAN: Don't doubt that. Must be confusing for you, still. Danny wized by them. But he was excited about it, for some reason.... I forced myself to grasp that kind of stuff quickly, because Dad wondered if I could. I think he argued with Mom about it. Couldn't understand the real gist of the dispute, but only that I should make myself bright about the picture books. Wasn't that hard, 'Aron.

'ARON: Well....

DUNCAN: I guess you have to train yourself to become interested in things that don't seem important at the time. But you like boats. They don't bore you.

'ARON: Yes!

DUNCAN: So you'll pick up on the subject and the related terms. And the words will have meaning and you'll remember them and how to spell them and write with them. This will come to you in short order, because you're a rich kid.

'ARON: Treasure!

DUNCAN: I mean in terms of support. You have all of that, all the help you'll ever need— from Danny an' me. Though Danny's still very young and preoccupied with his own interests. But he'll help you out, just like I've helped him out. You know what I mean?

'ARON: ....Sure.

DUNCAN: I mean you won't annoy him so much when he comes to realize you're an actual person with feelings and reason, so that you're to be respected for your thoughts and opinions also. Till then, you might seem like much of a bother; but maturing takes time, to remediate what seems. He was a pest to me for awhile, till one day I sensed I hurt his feelings about something, not giving him some candy I promised. The loss of the candy wasn't nearly so bitter as his reluctance to face me about (it). I simply forgot to consider him. But when I was reminded about the promise, by speaking with a friend about that particular treat, then I suddenly realized Danny was trying to avoid me, out of hurt feelings about it. Disappointments make one real. I worked to make it up to him, and he wasn't so cumbersome to have around. Though as we grow older, differences between our perceptions about the obvious lessen.... So it will be the same with you.

'ARON: Good!... But I can't do it well, Duncan. I've tried. Mommy has to answer for me, when she asks.

DUNCAN: One day you'll answer the questions yourself. It'll just come to you. You'll feel relaxed about it, and start reading books on your own, more than enjoying the pictures and illustrations. Then you'll discover the comic strips, in the newspapers. And then you'll buy your own comic magazines, and go through the ones Danny and I have kept. And then the simple books and the formal literature, you'll pick up, what the teachers have you study. And stage by stage.... it will come through. You are not.... limited or slow. Mom just worries a little because of.... other things, uncertainties with the family. Whenever there are uncertainties and doubts about.... life and daily matters, people strive to toughen themselves up, and those they have to take care of.... So don't be discouraged. Even as something gets harder, others ease up. That seems to be the way I've noticed, for progress, since we do so many different things, or are drawn to have to. Otherwise.... we'd be anchored down, by our problems.

'ARON: It's a sea everywhere!

DUNCAN: Wouldn't be able to solve any. Would be confused into a stationary submission, like a crashed computer program. That ain't us. That's not the Herofords. We'll correct the algorithms, and evolve.

'ARON: Yeah!

DANNY (*at the back doorway, calling out*): Mom wants you inside, 'Aron!

'ARON: Awh! (*starting towards the house*)

DANNY (*coming out the house, but not particularly towards **Duncan***): I don't know what's going to happen about all of this. But they're upset. I didn't mean to do anything bad.

DUNCAN: You didn't. It was just stupid. But you'll have to face it.

DANNY: Mom didn't give me the key. Didn't talk about my pun-

ishments either. Guess that's coming.

DUNCAN: Don't worry about it too much.

DANNY: Dad just pointed at me to get out of the kitchen, when I tried to ask her for it.... I put the stuff in your room, Duncan. I wanted to keep it, but I'm afraid, now. Seems like to frighten to own. An' I can't (*watching **'Aron** enter the house*) trust my own hands.

DUNCAN: They're arguing?

DANNY: I don't know. He mumbled something about letting you brood out here.

DUNCAN: So an' sure! I'm upset too.... but not with you, Danny. Go back in and face it.

DANNY: What?!

DUNCAN: You have to learn, there are no fantasies worth holding onto. Just know what's really going on, so that you can make plans and contingencies. Go to your room. Close the door and occupy yourself with something, if you want to. But face it. I can't protect you from what we're doing, if that's the way it has to be done.

DANNY (*heading to reenter*): What about you?!

DUNCAN: I already know.... what to accept. They're not gonna spank you.

DANNY: But he said he might beat—

DUNCAN: This is a lonely isle. I want to be isolated for awhile.... (***Danny** has reentered.*) He was just.... talking about our frustrations. Can't control what's let loose, cast away.... committed to a comprehension scorned.

Scene II — ***Heroford's** kitchen. **Betty** is sitting at the table, somewhat demure. **Heroford** is standing a distance away, addressing her, though she isn't facing him.*

HEROFORD: I'm fed up! Betty. This state is killing me. And with those kids trying to destroy themselves, I don't want to take this— degeneration any longer. I'm tired of it all—

BETTY: They were just playing foolishly. You came up to them —

HEROFORD: Is it this way all the time?! It's.... base demeanor, and some mean gratitude for what I've put up with. Why struggle with all of this dissembling?—

BETTY: You're exaggerating your distaste for us, Harold! You're using your vacation to blow yourself up to be dissimilar and anxious. You're looking for excuses to justify your anger and animosities to stay responsible to us. And you're trying to force the issue by picking on them, and making them respond to you as evilly as you need. Maybe it could have caused an accident, a very injurious one. But you're almost demanding that something is wrong here, that we haven't built ourselves up into a proper family, that I'm not a fine enough wife and mother to misbehaving kids, and that you find yourself insufficient as a husband and father. But that fantasy doesn't lead to an escape of burdens you feel threatened by. It's such a delusion of self-centeredness. And there are (some) far worse than we have ever been. There's no freedom here to run off because of that. There's no right for you to think you have the choice—

HEROFORD: Well let's be honest, then! You don't please me. You don't please me because I don't please you! I don't please my children, since I can't make them grow up correctly— or at least intelligently, even as I've supplied all of you with my best. My very best! My outrageous best! ignoring everything that dampens my spirit to stay positive about our situation, if only for some slight satisfaction— sometimes. A few minutes of the day, in one hour, when I come home to hope— that I am appreciated, for all of the endless work I do to keep us maintained and surviving, and intact as a handsome unit of societal grouping. This is a nice house, and we are handsome people. And I am noted not for a few accomplishments in my professional agonies of persistence. But return I home, at once that hope diminishes of effect within an hour, within half an hour— within minutes. The atmosphere bemoans me for my knowing, that we are not the prosperous unit I had dreamt for, while we have more than enough of the material needs with the income and finances I've provided, a virtual emporium greater than the casualties I've seen. But this hypocrisy is crushing, Betty, to all possible candor of faithfulness. What is deserved among us? if we deceive our very notions of lingering on this way, unhappy and too typically so!

BETTY: Are you denying your obligations to us, just because of a few days' restlessness? to slight me personally as if there were no affection left.... to your family at all? Is this the shame you propose, for the shambles we should become? You ungrateful hound! (*hands to eyes, but without tears*) Lecherous thief! How we've all struggled, to uphold your proud view, and accommodate your description as a sound and solid person of government with everything of home life and background under blissful control and satisfaction—!

HEROFORD: My coworkers and employees care nothing about that!

BETTY: —How difficult it's been, with your sham growing more and more evident, the prevalence of your disgust over our shadows —!

HEROFORD: That's not true!— And it's not the point! I disown none of this, no one of my family, my dependents— No desertion! Yet!... it's all so irritating to take, and live through, knowing the reality fights for its decline, that there is no honest sympathy shared between myself.... and my others making of me. All of you take me totally for granted, as a paycheck earned, and a machine for banking on and spending with— Not that I would claim otherwise to be fit, for you or anyone. I took on the burden and constructed myself with it. But myself has not become ourselves, as I had hoped. There's no sense of sharing of these privileges I've brought you. You all just extract them from me like a vacuum cleaner, and leave me feeling emptied and ridiculously smarted. There's no praise at all for whatever I accomplish, any more—

BETTY: I directed you towards your success.... by keeping you balanced off the silliness of some of your ambitions. (*hands from face*) Now you want praise for standing straighter, and more rea-

sonable for your goals? more applicable to your abilities, more squarely within your realms of management? more feasible to be a provider and sustainer? Why can't you be more thankful for our influences of responsibility and grounded, down to earth judgment?! (*becoming more demonstrative*) Your disappointment is your lack of merit— if at all— for what you have, and what reputation you've gained throughout the years. Yet if the world knew what a silly head you have, if your peers could see the foolishness in it, they would be astonished and astounded, and try to take advantage of the instability, given your position to manipulate. The days off work you see yourself as unfulfilled— Well you can not have that fulfillment ever! You are incapable of it!... And with cruelty you try to resist that fact, by instigating our upsets and frights. (*standing up defiantly*) But we will not be so pushed! We are defensive for our purposefulness—

HEROFORD: Purposefulness?!

BETTY: —for life, living, growing, nurturing, and surpassing your intangibilities and remote fixations on the undoable.

HEROFORD: Undoable?! What is undoable?! I've seen the impossible done— today— with a little bit of courage to the feet an' brain. I've seen the absurd of actions worshiped and plodded for— and won! of a right to try at. And I can do— the unspeakably ridiculous, to assert the most simple of aims: being truthful to the means and understanding to this universe of conditions we find ourselves swimming through.... or rowing with swiftly, rowing against slowly— rowing.... along these currents of our constructed happenstances, brought (about) accidentally on purpose to our minds. Oh— I can do it with a certainty of effort, without the slightest bit of failure to fear, because such thoughts evaporate with the onerousness of the steps—

BETTY: The walking away?— The running away!

HEROFORD: —The playing of a heart's worth of consolation for one's lost dreams.

BETTY: What have you to lose but us?! We are your grandeur! We are your dreams! We are your aspirations labored for! We are your future proposed for making of, a legacy of your results in life — marriage— procreation!... And we are your fulfillment bargained for, your treasury constructed with, and your duty of performance to uphold. You haven't any other— mentally substantive. You must face us, with that irritability of a wantonness to leave— and claim that pressure false! not your practical endowment of a nerve to do, a reversal of all principles of being right and sound— but more to blockheaded stupidity for this uneasiness to steer through. You've made promises, to yourself and your society, to conform and condone to the necessities of *our* existence. And that's the only attitude an upstanding person should be allowed to take, what has been induced to be with your wishes and wants. You can't fall back on errors and error-prone inspirations to blame everyone but yourself for your dissatisfactions, and be rational about it. More than yourself is involved. You are *not* independent about it. You can not isolate the effects strictly for yourself to be threatened by and assuaged with looseness of character and dismissal of meaning for recognition and acknowledgment of your travails. There's only to continue what you're charged to be, for this household demanded of. And what foolish leanings you have left, to escape the lengthening of our domiciled state— that is difficult enough without your periodic wavering— You should drop those tendencies once and for all. Because you can't have more freedom than what you've wanted to be bonded with— and must continue to earn. And as for binding contracts, they are your children made for. You needn't be reminded— of who to be pleased. And as for myself, do you think I labor inanely for your passion?! My passion received is for my family's grace. What other you have found, blow out of your head— finally to leave it commensurate!

HEROFORD: It's not possible, to take a vacation here at home with disagreeable natures contradicting my intentions—

BETTY: To do what?!—

HERODFORD: To vacate!... To relax, and stay calm and reasonable of the surroundings, the harmonious potential they provide, the soothing of the soreness— to the head an' heart, and sense of urgency to find a solution to these— empathies of procurement, settling down of a whirlwind of anxiousness and anticipations for a more comfortable reality to find. I must.... separate.... myself, for awhile, to conduce the tremors of my backbone and standing tolerance of this environment.... to soften, abate into some manageable abstraction we can all abide. It's not for a divorce from reality to deem this motive all menacing, that I must make this move. But what I see.... it's just not pleasant enough of a situation to endure. (*as Betty slowly reseats*) It has its aspects.... of an ugly realization of some suffering, of personalities and persuasions. And either the coolness must heat up, or the heated tempers cool down. But something must sway ourselves into a sufficiency of the emotions, with a verity of any compassions left. I see—

BETTY: Abandonment.

HEROFORD: ....not what I have done as plausible of a success, to leave me damaged natures passed as whole, deprived of a genuine care as I could fight to insist on. But what could be so real.... as a loss for this impulse? What would be more false, than with such anguish forced to do?

BETTY: I can't keep you from sawing off (from) your own neck'(s)head.

HEROFORD: I'm simply not impressed with the debacle of this home life for me, and the inabilities it propounds for my guardianship. One week this way, and I need a vacation from a vacation. I find it— hard.... to serve up myself any serenity about it. The state is unfair, the condition unjust, to hold back contempts I dread to express, and are not suitable to be felt or experienced.... And I have no one, Betty. That is my most honest complaint, to comply with this awful sense of drain and emptying with fear. It's too unusual, to be this usual at home.

BETTY: If you find your family so despicable, why have you kept up so many pretenses for so long? Why have you endowed yourself with our presence so regularly? And why are you so determined to condemn us now? to the uncertainties you've always despised and shunned to think about, or at least speak of.

HEROFORD: That's not my meaning at all—

BETTY: I'll be brave despite you!— We all will!

HEROFORD: —This is no jest to me! lady!... I adore each of you, and the sum total of our making. And I work from that foun-

dation we have built, to promote my views to you today— but to keep the floor steady and the ground sound an' solid. I am not withdrawing, I am leaving. I am not removing myself. I am— clearing my sinuses to jab at a desperate recovery from the cold destitution that I feel in this house, the chilly wariness that surrounds me as I try to stay pacified of an all too somber atmosphere, due to a queasiness of discomfort I can withhold less and less from showing. And I do not want to cause more harm to the precious sensitivities.... of our making, that to live with falsehoods is to be ground up by the grinders of lying— and lying about that to save face from others who may question your apparent quiescence of demeanor. Can you know how I must act, to suggest more than a hint of confidence during my professional activities? I must often heatedly, arduously convince people to trust my judgment and take considerable fiscal chances according it, according to constructions of strategies and governmental policies I have wrestled with and fought through rationally to work out. And a conscience is upon me weighted, to be honest for the reasonability of my thoughts. Else disaster for many a person is due, to failures earned with high mendacities. So with this solemn trust, can I present to them— my work as a whole— a much torn self?! a twisted being incapable of facing the truth about his wanting and inabilities to guide his family wholesomely, healthily through this troublesome environment we've devised of ourselves, or the very noxiousness I can't help to sense? (it is so evident already.) My colleagues would care nothing of this, or very little. I've seen them drop members from their consciousness, those who suffer through an incapacitating mental anguish or have fallen to some professional misfortune or other disgrace, without hardly a thought. The demented may disappear as easily from government as they may remain. It's not the damage to yourself that counts, but your discipline to remain useful to others. But *I* can not function well with continual wounding, a compunction to notice that I am becoming less fit for you this way— ever sensing my disappointments to such an intensity as to make me cry of rage!... But against whom? Not my cherished dears. Not you and the children. Only to myself, this fearful fulsomeness is pressing for release and escape, evacuation from a home susceptible to cruel insincerities. For my candor chokes of pride to have me lie to you: I don't like! this.... situation of us. You've always found me (to be) trite, even during our romances, which were too often off an' on. But now I'm more contemptible a stick, in your eyes, to— beat off bills with and prop closed the door to a good shelter. Were you to have a true companion too, would be better for both our empathies. And the children sneer at me, with hesitance to my approach, and with hisses to my back—

BETTY: They do not! They don't have any such disrespect for you. This is a mad, delusory sickness you suffer with, have always suffered from that makes you less than your cautions and more than your worth to tolerate.

HEROFORD: —Except 'Aron. He's too young for such disparagements to want to feel and transmit to others. But I'll lose him too, as time develops and generations generate their— cares.... against my latent disregard for what's not right about them. The heinousness of these hypocrisies we shoulder to patronize, this brings the contempts, and the hatreds, and the self-loathing—

BETTY: You are not well, Harold—

HEROFORD: I know it!

BETTY: —looking for these excuses of blame, within yourself

for the shame of finding out.... you have no heart.

HEROFORD: ....None?!

BETTY: Not for marriage. Not for your offspring.

HEROFORD: Can't be true!

BETTY: Not for the procedures necessary to go through with it to a grand conclusion, not as a trustworthy patriarch. You lack the sense to perceive organically the obligations you've set in motion. It's only been a social custom to you, facts learned and performed by. You haven't the depth— to feel honorable. The eye simply isn't there to see the gold represented by.... an endearment for your family. And you surround yourself with the hostilities of perspective for your grotesque views of societal aims. To think you need a reason to be....! There's no defense for you nor against you. But that will not diminish by any fraction your compulsory duties— mandated by the machine of life in this city, and the boroughs of harshness you subscribe to, to make for us.... your living. (*standing*) I have dinner to prepare, Harold. There's work to it. (*going towards the kitchen counter*) And then there's the trip we've planned. It has its expenses. Some details have to be worked out.

HEROFORD: ....Work them.

BETTY: And the house itself.... (*getting some utensils and plates and chopping board*) often in need of more conscious(ness) raising, and always the cleaning, for a good presentation to live with.

HEROFORD: It's been my blood to provide—!

BETTY: And the children's tutoring and schooling. We can't afford any missteps in that, even during a vacation.... And their behaviors must be coached more seriously, because dependencies grow more nebulous.... within any age.

HEROFORD (*going towards the table, as she to the refrigerator*): That's a challenge of the head an' hands, as they continually handle things to examine and investigate. I don't mind such curiosities as the town provides. Just remain upstanding and decent, is all one has to do to behave. (*standing by the table*) But I don't doubt that trouble comes with any internal qualms struggled with (*as she retrieves some lettuce and vegetables to bring to the counter*) or wrangled through silent contentions to act one way or another, confused with innocence and inexperience. But I can instruct and educate, if not as a model offer through advice.... and commands. I still direct some realms of our reality.

BETTY (*busy at the counter*): Don't be a model to the detritus of one's imagining.

HEROFORD: Nothing erodes of my pride. No sentimentality wastes away for my endeared and engendered. I forfeit none of that, those rights of longing and belonging. Just plead for some understanding and consideration of my condition— my status as head of the household, the wage earner and home keeper.... the brave one, Betty, to make of this.... some consumption of heartache— and reluctance. (*slight pause as he observes her*) Fresh, of a vinaigrette. You always do that so wonderfully. We all admire it, your production of salad. The children never complained of the tastes, even when they were little, too little to appreciate the exoticisms you construct with those simple, common vegetables.... Fresh an'

lively foods. They're little more than a delicacy today, with everything so manufactured to a determined purpose and expectation of approval or acceptance. Yet that which can be done by any.... are only done by a few. Should be nicely salted, though.

BETTY: That's meant for individual tastes, Harold.

HEROFORD: It's what I mean. The foundation of flavor should not disrupt the possibilities of one's improvements, to make a serving personally desirable. Nice means not too much, just enough to hint at the potentials of more.

BETTY: They have the salt shaker.

HEROFORD: It's use. I'll not decline it. Petty worries to enhance things. You design the meals to look pretty and enticing. That's a great skill, a.... reconciliation to make our worldly labors more commensurate. I'll not give up that balance we have weaved— Wrong to say discovered. It's been assiduously worked out.... Betty?... The trip will be as planned. And I'll try to explain myself to the boys.

BETTY: If you're man enough to, then.... so be it.

HEROFORD: They'll have no lack of me, during our vacation. I don't find them shallow. But I fear for the shallowness of our existence.

BETTY: ....Together? Who would you scold, because of that?

HEROFORD: ....It's a heartbreak.... and an earache— and a neck sprain all at once, how we are heaved together, and mashed up, in this society. An impossible homogenization, it can only lead to an eventual ruin of many. And so.... one's independence is at stake, to stay unique and oblique to the common prescriptions of the social norm.

BETTY: Go on your roads, and deny everything.... defy anything.

HEROFORD: I shan't! I shan't! You are the homemaker. I shall not crush too brutally a comfortable menage.... for the living.

BETTY: That's at least considerate. But you are bothersome.... while I'm preoccupied.

HEROFORD: A pact, Betty, to be mutually practical about this, is what we must be bound to, and agree—

BETTY: The details are all locked up in your head, dangling or dancing or dwindling of relevance for our feasabilities. Consult those clown princes of persuasion before you come to me with agreements to argue over; and ask them if they are sane, first, before they lend you their dimensions of disavowal. I doubt if you even know yourself, when you have nothing to do. But it is tedious to have you upset. Barely threatening, like an infant's growl. And every time it is the same proposed. But it's growing worse and worse of the intensities— as we age, the similarities taking on a sinister grin of teeth as the fangs become more worn and yellowed of a uselessness to feed with, these empathies you grovel towards but can only lick at, the airs sensing with a tired mustiness. (*heading towards the refrigerator for some meat chops*) So go out— into this frost.... to clear your head. And let me make our dinner.

435

HEROFORD: ....It's been a warm day.

BETTY: You wouldn't know— or would you! with a jacket or without. (*returning to the counter with the chops*)

HEROFORD: The afternoon heated up, as predicted.... by those crackly voices.... Didn't explain a thing I couldn't assume. Didn't expound for me at all. I really can't stand that type of correctness. Improves prospects little. (*She starts poking the meat with a fork.*).... Betty?...

BETTY: Go!... Harold.

HEROFORD: Is there paprika on them? (*She starts mashing the meat with her palms.*) My sides are burning.

BETTY: I must tenderize, before spicing.

HEROFORD: Took a walk this afternoon.... to regions an' neighborhoods. Funny how those exertions affect the back.... Just forgive my tenseness a little. I'll try to relax the muscles in a comfy chair, but nothing was sweated of the effort—

BETTY (*lowly*): Harold!

HEROFORD: ....only observing the regime, of our humanity in strife.... civil distresses and descriptions of one's plights. But the day was.... pretty.... ordinary for me, to be this way accustomed for a walk, I should suspect. Incidences happen all the time, in a busy metropolis.... Aren't you interested? I met a farmer—

BETTY: There are tons of them.

HEROFORD: ....a produce provider, and a shopper of groceries. Now can you imagine such a thing?! That farmers need groceries?... No? Well, he needed to do that.... (*exiting*) I could not have predicted it.... But there were similarities involved—

BETTY: Talk to Duncan and be fatherly. Ask forgiveness for your rage.

HEROFORD (*stopping*): Can't do that! I wanted him to *know* my rage. It wasn't only an outburst. It was a prevention. And he'll have to do the same.... So he mind as well know how.... (*observing her working*) Uneasiness crackles! (*Exits.*)

Scene III — *Morning. The café patio.* **Beck** *is setting up the tables and chairs, slower than usual, favoring a hand for weakness. The two* **waitresses** *arrive together.*

WAITRESS ONE: That's odd, Beck. Would have thought this all done by now.

BECK (*arranging*): Hurt my hand last night.... Lucky I didn't break a finger. But I sure did make a favorable impression to the contractors.

WAITRESS TWO: Oh, your construction work?

BECK: Been invited back, this afternoon. Still more to do.

WAITRESS TWO: Well that's not safe. Can you handle it, injured?

BECK: Only sore. Didn't complain. They don't know it an' I won't show them. In a couple of hours it'll be fine. Was throbbing last night, though. Some heavy plaster fell on it—

WAITRESS ONE: Oh! (*grabbing his favored hand*) Doesn't look cut.

BECK (*stopping to be conversationally tendered*): A strange pressure, that was as sudden as a rush, or a blast of lightening. Didn't really feel much pain right after it happened. The worst of it started up while I was trying to sleep, in bed. I knew some discomfort was coming, though. Has to, to take such a brunt of brick. But the little hand did its job. Continued like nothing had happened.

WAITRESS ONE (*mockingly gives the back of the hand a kiss*): Not even an abrasion. Should be O.K. for that kind of labor.

BECK: A demolition, you know.

BREUR (*coming out the café*): Let's get with it, girls. (**Waitress one** *releases the hand.*) We are officially open, and you two still have to get ready.

WAITRESS TWO: Yes, Breur. (*entering the café with the other*): But there ain't no customers arrived yet—

BREUR: There will be. Why aren't you finished setting this up? If I have to help you, it's less to be paid.

BECK: Hurt my hand last night. (*resuming*)

BREUR: Well, I want the patio decorous, for my customers. Don't want things incomplete. And if it is, while anyone's sitting out here.... I mean customers. Let's speed it up a little, Beck, for your own benefit. Not bandaged.... Kind of anxious, today. We're being looked over, today or tomorrow, or the day after. A licensing official wants to see the place— Not an inspection, but just a casual sighting. It'll be the first time Bruce's offers liquor beverages— and I want to make sure the café looks respectable and neat; all of it, especially out here, so that the appearance suggests fine sociability instead of an opportunity for any derelict behaviors.

BECK: I'll be careful to place the seats symmetrically.

BREUR: Hmm. Do so. Just make the tables orderly and inviting. And I want at least a couple of them with umbrellas, if you can manage that.

BECK: Kind of cloudy this morning, sir.

BREUR: Folded, of course. Supposed to have a chance of sunniness in the late afternoon. Open umbrellas are attractive, the shades welcoming. But if you can't put them up, I won't hold it against you— today.

BECK: Oh, I'll do it. Nothing prevents me from that. And with what I'm heading to later, it's nothing.

BREUR: Good. Impressions of serviceability are so very important, these days. Don't hurt yourself more trying. Those stands are fragile, can be damaged through shipping. The aluminum pipes are thin. Had to return a bent one. (*as* **Heroford**, *appropriately clothed for the morning outdoors, approaches the patio*) And they're not cheap, but they can't be too weighty to topple over the tables. Takes some fine bit of engineering to get the combination just right and stable, hardly ever appreciated until you work with these products— See! A patron's come already. Well, would hardly want to sit outside yet. So I guess you're safe. But get this done. (*starts to return indoors when he notices* **Heroford** *is merely standing on the patio*).... The café is open, sir. The drinks are hot and ready to be served, inside, if you'd like.

HEROFORD (*responding*): That's fine. Fine. Maybe I'll order a coffee, later.

BREUR: From out here? It's slightly brisk. Frisky weather we call it, this morning. (*reentering the café*) But we provide for our customers whatever the day brings.... Should be much sunnier in the afternoon.

HEROFORD: There *is* a bit of a chill. (**Breur** *has withdrawn.*) It's seasonable for contemplations. (*observes* **Beck** *working for a short while before addressing*) Say, has that fella come by, for his glass of water?

BECK (*virtually finished*): What?!— Mr. Langly? No, he hasn't been around.... And it ain't early for him, but probably not the most opportune day, 'cause the employer is anxious about appearances, even on this deserted isle. The deck might stay empty, on a morning like this. Not that the clouded sky makes everything seem so barren, but the slightest gusts of wind make seating uncomfortable, I'd bet. But the whole place still has to look good, and promote its offerings.

HEROFORD: Well how regular is he?

BECK: I don't keep a tally on him, mister. It's that, if he's around — at this time— it's not so unusual. I suspect he's fairly regular, since the waitresses know him well enough. But that Breur, the guy you spoke to, he don't like those types, using the deck space without paying for anything. So it's just as well the poor fellow hasn't arrived today. Maybe he had a premonition the owner would be more out on the prowl against indigents showing up—

HEROFORD: He's not indigent. Dresses well enough, doesn't let himself look poor.

BECK: One can tell. One can often tell. The seediness is in the heart. Can't fight it. Can't misplace the depths of opinion. Just one glance at a person is all it takes. Like the shadow of a fellow seen.... is his face. The face avoids depicting. Now I've been destitute several times, but I've never been shabby inside. And not that Mr. Langly really is, either. But.... the guy's too proud, you know, of his past, and refuses to catch up with the present. He speaks to me of warnings, as if he were still wary of what's going on today, in a position he once held. Such beings, they're maybe not to be pitied as much as kept at a distance so that their delusions are not so stifling. A construction worker, he was. Sure, he has rich experience. But no tune for the instrument any more, is his declension. Walked his way to the park, yesterday, to spite his health problems. Now when a person is that miserable of himself— to ignore himself— there's not much help for 'im, nor hope to lend. Refused my aid to get him back to his.... place. I don't resent it, and he didn't resent the gesture. But I wouldn't be surprised if he injured

himself yesterday, and can't make it here this morning. Might be an unknown in a community hospital bed. You make your own roads, to find your own destination.

HEROFORD: I bought him a cup of coffee.

BECK: So the ladies told me. He didn't pay for it himself.

HEROFORD: Maybe I set him on his journey inadvertently. Feel a little responsible, the stimulation that aroused his courage. No one should be afraid to go to our park. It's meant to be a shared civic attraction of public ownership.

BECK: Hm. I just found a man caught beyond his scope, trapped to a bench to rest on, probably scared for how he was going to make his way home, but too defiant of the fear to ask the squirrels for some assistance— their nuts to travel with.... I've seen the work first hand (*looking at and favoring his hand*), even felt it. And it's a keen thing to do, but nothing worth a conscience, a commitment against your well-being or the realities you have to face currently. He's a shriveled up Atlas, still raising up the Earth, his world.

HEROFORD: So he should just die out of existence?!

BECK: It is pathetic. But that's what we're meant to do. You can't linger on soundly, in a condition no longer reasonable for you. It's like displacing the more capable, just to hold onto your spot— Feels that way, sparking resentments in others, though it may not actually be the case physically nor effectively. But it's stupid to put yourself into troubles unnecessarily, unless it's for a sport. And I gathered he might have been satisfied with his daring, but not particularly happy about it. So there was no playing on the fortunes of a gamble. It was an obstinancy to live through past deeds.

HEROFORD: To each as he may find his manner's mane.

BECK: Well it bothered me, like something subtly indecent. I don't dislike the guy, his coming here for water. And I have sympathy for his respiratory problems and enfeeblement. But there are real indigents in that park, who haven't half as much as his collected benefits, who suffer their environments much more harshly. Yet they're alert enough, and truer to themselves, not to do foolish things, court greater hardships an' fish for disasters. I've seen them, gone to bed under trees. And you can't deny it. Even little children, too circumspect of their normalcy to be aware of their dilemmas. It's amazing how many simple activities can be done by anyone, to make them seem as regular as for those of much greater means. These are conditions we face and trudge through. But I protect myself, by knowing thoroughly what I am at the moment, the conditions that define me for others.

HEROFORD: Yes. The universe. That's important.

BECK: I'm a laborer.... and a café worker. Part-time statuses for a full time life. But I'll improve. There's a world outside the city, outside the conjectures and contentions— of a worldly view and its secreted principles.

HEROFORD: You're young enough to try at whatever you can find— that pleases. It's an idealism strongly recommended.... I was just wondering about the fellow.

BECK: To come back here? *That's* your purpose?

HEROFORD: Oh, I'm on vacation. And I haven't anything to do. But I liked what you served me here yesterday. So I suppose I'll make this place a regular haunt, for a short while at least. Mornings are characteristic for me— Have always been so. They help to define my motives for the rest of the day. It's the only time (in) 'which I normally have to workout all disagreeableness to my character.... or even intentions which I must employ later. And if this Mr. Langly comes by again, as I have, I'll buy him some more cups of coffee. Because you are wrong, my fellow citizen, to disparage his nerve. In our final accounting as members of the human species, it's all we have to live with, what we must do, irrational or not. And I hope you prosper far above these labors, and beyond this menial work. But do not begrudge anyone's laboring, nor attitude to accomplish what one sets out to do. It's not your purpose in life to consider the legitimacy of that, but only what is done and its effects upon others.

BECK: Don't mean to speak against a compromised individual. Just being honest about how I see things. Can't say I can adore such a personality, but would not hesitate to help 'im.

HEROFORD: Honesty is fine. It's the best a person can display. But perspectives have to change.... with any refinement of consciousness. Myself do change with transformations. My-selves. It's hard to stay sound to one's compunctions— but they make me up.

BECK: Sitting out here?

HEROFORD: No, I think I'll go in, for that tasty drink. It really is unseasonable yet, for the deck. The day has to grow some more. And then....

BECK: Well I've gotten through what I have to do, out here, aside for some umbrellas.

HEROFORD: Umbrellas?!

BECK: You never know. It could rain or shine. (*They enter the café. There is a pause before **Langly** comes up to the patio.*)

LANGLY (*standing at a table*): Unpleasant. But dare I go in? I used to, before Breur became really annoyed with me. It's like a signal that I sit out here, to break the day for its arrival. But these breezes make for a peculiar desolation sensed, of (it) not being practical.... Yet I'll sit, and they'll see me, hopefully with some grace to lend. (*sitting*) It's terribly cool, as for a depth to earn. And still, for this hospitality to shake of, will I stay awhile. Is this trembling— of a fright? I've recovered nicely, from last night, the muscles rested and the head replenished. But benefactions pour with costs. This is a morning that threatens not to be as good as many were. A short stay, if a glass would come to where few would be seated, of these uncomfortable temperatures. And sorely alone I'd be for quite a while of these conditions. But I haven't all morning to wait on generosities. My spirit's freezing. Do they notice me? or could expect I'd dare the need even on a day like this? It comes that way, in any sort of weather, though stiff as ice I am to see if it might be presented. And the window of time for this affair is usually short, brief. Quick gulps of a clean water in a day's patience. No one else would ever sit out here this way, with a desperation to be serviced. But.... the deck is played for visitors nonetheless. So I can't be said to trespass— unusually. I'm just alone, when no one else would come, to where no one has such shame as mine to beg,

on this inclement morning. Such sharp turns of fate. Yesterday was so good to me. And now today, a shrill beast, a harsh winter within a summer's turmoil. Well, the elevation's fairly high, above sea level. One can not be surprised with such mornings occasionally. Else why have this café's patio at all for a decoration? but to attract through warmer winds the better days outdoors. And as it's said.... I'm desolate. No offering, this morning? I know it is a pity blessed with, as could a pity be its highest honor held.... to be deserved, and without spite endured. But not today? I am devastated, sitting isolated here, like a buffoon would stay to be cold. Perhaps it's my lot, to be dismissed more often. Maybe this comes (as) a turning point, for me. Such events may happen this way, conclusions made not to patronize him any longer. I've had that magnificent day already, and that is enough of a prize to last the rest of a lifetime. But what becomes me now, to be so disconsidered? I need the freshness of their water.... or want it, for a taste of confirmation, that these toils are good, remain worthwhile, and awakenings are purposeful. Who comes?— and what is delivered? I am so solitary, I could be a chair myself.

BREUR (*stepping out from the café*): What are you doing out here, Langly?! (*staying a distance away*)

LANGLY: Please, sir, mighty Breur. My throat needs a respite, of cleanliness and fresh restoration, with your more than healthy, well filtered water. It's better than my rusted pipes can deliver me.

BREUR: You're stupid to sit out here, waiting for nothing. You know you've been growing on my nerves—

LANGLY: It's little to ask.

BREUR: A public service, sure. But you never pay for anything — 'cause you can't. And as a friend to the establishment, I've been serving you, as when the old times kept you coming. But haven't you enough pride left to realize you're no longer wanted? not for the memory of Bruce's café!

LANGLY: My medicines.... are bitter, Breur. I don't remember much, except that Bruce was a very nice and decent man. And I was a happy customer here. And now I plead for some simplicity to my dire straits, because they're never to improve much more than that glass of water you may regularly give me— out of the way of everyone, and with no humility to shed like hairs of a dog on anyone's legs. But all of my persistence is for want. I've had an angry night, as tight muscles struggled to relax. And the battle was chosen, an' started here. So to here I return, to shine of some victory and appreciate you. I am the humble thirsty for some of your refreshment, starving but for your indulgences I've found myself regular for. Kick me away, and I will cry, silently distressed. Let me depend on you, this early, before there can be any complaints about me.

BREUR: You're a pain, I have to say, Langly. Worst than a has-been who was—

LANGLY: You deny me?

BREUR: —It's without the suddenness that should depict you, your loss of circumstance. You should be dropped immediately for replacement, without any cackle—

LANGLY: From the astrophysical?

BREUR: —That's how it's prudently done. But to sit here in the frost, waiting for something ice cold to drink, shows you haven't any respect for yourself any more. You live through custom instead of assertion, and try to survive on repeated habits that make less sense than your impulses to do them. Now let's come to some true conditions once and for all. I don't want you around here ever! I can't strictly banish you, except to claim you as a slight public nuisance I've grown tired of tolerating, and making a big rant about that to the civic officials— who would find it peculiar to be worth my time and theirs.... to bother them about your simple requests. I'd rather freely serve water to anyone, especially children, less especially pets. But you?! You should feel better than to ask by now. Wiser, at least. It's not even an (h)'ospitality any more, but some sort of social trenchancy to be responsible to the .... conscience provoking—

LANGLY: Why do you have the deck out anyway, in this weather?!

BREUR (*as **waitress two** comes out with a glass of water*): Because it's blisteringly fashionable for this establishment to welcome any customers with this patio of warmth.

WAITRESS TWO (*bringing the glass to **Langly***): You'll let him have his morning water? You've agreed to it.... again?

BREUR: If you dress up well enough, at least that's seasonably acceptable. (*as she serves **Langly***) But it's a tiring practice. I mean you should tire of it. And be off! before others must partake of your presence.

WAITRESS TWO: No one's gonna sit out here in this cold. It's frigid for you, Mr. Langly. You should be allowed to drink it inside.

BREUR (*as he sips*): He prefers it this way, approving of his loathsomeness to be restricted to some discomforts, as if that legitimizes all of the difficulties.

LANGLY: This is as sweet as tears, to be regular.

WAITRESS TWO: Drink it up quickly, dear. I can't wait on you today.

BREUR: As a matter of fact, I've told him not to expect to, ever—

BECK (*coming out with an umbrella poll and base, along with **waitress one** bringing the same*): So you made it back, Mr. Langly! Well kudos for your adventure. I thought you'd be stuck at the park, exhausted from a strenuous trial. (*starting to set up an umbrella at a table*)

LANGLY: I had my means of will an' help. Thank the heavens it was a fairly warm day.

BREUR (*as **waitress one** starts to set up an umbrella*): You're getting her to help? I'm not sure about the wagering in this.

WAITRESS ONE (*working*): A warm afternoon.

BREUR: I suppose I can let it pass this time.

WAITRESS TWO: And all other times, if it's necessary. Are you finished?

LANGLY: Yes, thank you. The nursing has been resplendent, as always. (*as she retrieves the glass*) Odd to have more shade out today, Breur.

BREUR: They're not for you, and they're not for this morning. And they're remaining closed, but their esthetic is telling, that we are ready to serve customers. We are open. And you girls should be— inside! waiting on the one we have.

WAITRESS TWO (*returning to the café with the glass*): He shouldn't be so lonely. A particular man, for his concoctions.

BREUR (*as **Langly** stands*): Well that's something we try to please for everyone.

LANGLY: And a berth awaits, for this reinvigoration—

POMELORY (*coming up*): Of a breakfast? I'm quite surprised you'd have it out here this morning, Breur.... (*turning towards **Langly**, as **Beck** stands to notice them, after accomplishing his chore, and **waitress one** comes over to tend to his hand inquiringly*): Why, I remember you! Gave you and your friends a bag of tomatoes, once. Was one of my delights, before old Bruce.

LANGLY: Yes, Mr. Pomelory. I'm just taken aback you could recall me, all these years.

POMELORY: Well you look much the same.

LANGLY: Ah! I'm standing.

BECK (*to **waitress one***): Didn't strain it at all, honey. But thanks for the help.

WAITRESS ONE: Yeah. It doesn't make any sense to me, but if it'll keep you able.

POMELORY: We all are standing. But I'm certainly not going to have my breakfast out here. It's rather chilly yet, and the Bosc would gel up something absurd.

LANGLY: Oh, I only had a glass of water (out here), Mr. Pomelory—

BREUR: That's right! But an old friend of Bruce's is more than welcomed to have his breakfast within the warmed comfort of— our café. And we know exactly what you like. So it won't take any trouble at all to prepare promptly.

POMELORY: That's good. It's why I come so often, Breur. But just a glass of water? (*noticing **Beck** favoring his hand*) That's something peculiar, especially with this chill. Have you had breakfast already?

LANGLY: A little toast, with jam preserves. That keeps me going for the morning, at home. Can't afford cereal, with the price of that commodity these days. And milk's just out of the question, to be practical about.

POMELORY: Oh. Times of difficulty? That was a tough profession to stay with, as I recall.

LANGLY: But Breur allows me to wash my mouth out of the medicines that I have to take, some time after the toast, with a glass of water— nice clean water here— most mornings.

POMELORY (*as **waitress one** starts to reenter the café*): That's very nice. But I haven't seen you around for ages— And you've probably been right before me, I dine here so many mornings. Allow me then, my friend, to treat you to a proper breakfast of bacon and eggs and ham slices on potatoes, if you're hungry for warm foods as fresh as Bruce's can prepare them.

LANGLY: Well that's very decent of you, sir.

BREUR: Extraordinarily so.

LANGLY: I wouldn't mind that at all, today, to make sort of an extension of good luck, that tends to run in streams like this occasionally.

POMELORY: And a hot cappuccino for you too, unless you'd like Bosc!

LANGLY: Thank you. The coffee will do.

POMELORY: Well, enter the café, sir. You are welcomed on my account, Breur's, and the great times we had— when we were fortitudinous an' flourishing. I'll join you directly, after I have a word with my friend Beck, here.

LANGLY: This is most gracious of you. (*as **Breur** disingenuously leads him into the café*) I haven't seen the interior for some time.

BREUR: It's much the same as ever.... Mr. Langly. And you'll be placed at a prime spot, for your meal. For Mr. Pomelory likes to face the historic photographs, when he dines inside. Swears he appears in one of them, as a merchant.... But I can never tell. The images of the people are so small. It's the crowds of humanity that really seem to count most. Means their businesses are all engaged. (*They enter the café.*)

POMELORY (*turning towards **Beck**, who comes up to him*): So, my son. You heading off for more work?

BECK: Yes, Mr. Pomelory? Oh, I'm pretty much done here, till this evening. But I have to get my coat. Yeah. I won more demolition duty. Impressed them well enough, that I'm a diligent and conscientious laborer.

POMELORY: I've no doubt of that at all. Any regrets, though?

BECK: Well, work is work, sir. And employment wins me a comfortable bed.

POMELORY: You keep favoring that hand.

BECK: Yes.... Some brickwork fell on it yesterday—

POMELORY: Hmm.

BECK: —It's not broken. Hardly damaged. Was wearing construction gloves they issue. Hurt awhile, last night. Swelled a bit. But it's O.K. now. They welcomed me back. That's the most important thing.

POMELORY: I told you. It's dangerous work. Can be unforgiving. You have to do more than look out for yourself. You have to be.... a rather fortunate sort of person, in life. I guess you think you're winning much, by gaining these reprieves.

BECK: I'm active. I'm doing something. I'm not abandoned to the park. I can return to the country, I can live for real. I'm in charge of all of this which makes my steps. Maybe that's why I feel so immune— from a depression for my situation.

POMELORY: Reprieves from major accidents, I mean. They're bound to occur with that endeavor.

BECK: There was a wall collapse, yesterday, that caught the workers by surprise, seeing as how it was not on the manifest to occur that way. Someone *was* injured. I hope not too seriously. But that helped to hold up the project, delay it for an extra day, along with some night work.

POMELORY: It's the recklessness of being fickle, boy. Even with your most careful awareness, you got hurt. Sure, it's great to be able to be active, and doing something about your state. But you can't deny yourself the reasonable protections of conscience, that some activities may just be too risky for your person to handle. I.... wouldn't want you to get so easily drawn into such opportunities. But as I tried to explain before, there are reasons why they often take on the inexperienced or uninitiated. It's a profession of necessary turnover of personnel. And the better that be the less talented or trained, for them.

BECK: I can see that well enough, sir. Don't delude myself about any of it. But it's hard to make a living, schooled or not. Most times, you take what you can get, what's welcomed for you. Serious choices are rare.

POMELORY: That's definitely so.... until you come to the crossroads of decision, those which you can only find yourself. You go on to your work today, and celebrate your chances. You're industrious enough, as far as I'm concerned, to possibly make the best of it— But let me give you something.... (*takes out a business card from a coat pocket, and a pen from a breast flap*) Funny how I keep carrying these things around with me, but I'm proud of them, the personal touch to advertising. I had them made up for my business eons ago, and always keep a few for my traveling, wherever that might be. Has my signature printed on it, which I'm very dear about, after going through some convincing that it's a safe thing to do. I was naïve about forgeries. This is an advertising signature— with my left hand. But let me write your name on it— with my right hand (*doing so, on the table*), that I personally recommend you. And for my farming enterprise, I think somehow you'd be splendid. (*offering **Beck** the card, who accepts*) Now you take that.... to my nephew, preferably, when you get back out to the ranch, to look for work. Ask specifically to see John Mendez. (*as **Beck** examines the card*) And tell him that I personally recommend you for some— office work, until that hand heals.... and then some farming chores. It's up to him, of course. But he knows I have his favor.

BECK: This is what like I noticed— what I glanced at— the last time I went there. (*pockets the card*)

POMELORY: Well it's more than fortuitous. It's something earned, I think. Because you're very sincere about what you're trying to do, which is all important. Might not be sufficient, but it's a sound start.

BECK: I thank you heartily for this chance, Mr. Pomelory. You're right to think I thought.... I might make it with the masonry. It's what was presented for me to try— and I grabbed at it gratefully. But if I can do the country work, it could be better, maybe best.

POMELORY: It's what you're really looking for, son. You want to blossom, not destroy.

HEROFORD (*exiting the café and noticing them*): Well can you imagine that! (*coming over*) Coffee for Langly yesterday.... and now a breakfast today. You're most generous. I would have only bought him another cup.

POMELORY: Well sir, one's not to overdo things. But I see you're better dressed— for the weather, this morning.

HEROFORD: I wasn't as rushed to leave the house.... Pomelory. I must say that your kind.... have an extraordinary civil spirit, if done peculiarly or with impetuousness. But it's nice to know about. I rather took you as something rather pompous, before. But it's not so much a vanity to help others, but an obligation if you can. I certainly wasn't trying to be vain, just.... searching for some commiseration.

POMELORY: That is a decent approach, sir, when you're filled with doubts and questioning.

BECK: Ya do what ya gotta try at.

HEROFORD: Yes. And community assistance is so pleasant to realize.

POMELORY: And I must realize my breakfast, as you may your.... vacation? (*starting to enter the café, with* **Beck**)

HEROFORD: Yes.

POMELORY: A trip, was it? with your family?

HEROFORD: Yes.... to the plains.

POMELORY: Well may it be marvelous for you, sir— the time spent. (*They enter the café.*)

HEROFORD: Yes.... I'll be alone at it.... Breezy out here. But it's supposed to warm up, this afternoon.... The time spent. Could be brief, or it could be an eternity to reckon with. But that's to be explained and accepted. Because it's all so difficult, to truly propose what one feels, how it comes to be, and like a marriage declare your vows are just. But there is a chill, to be truthful— an' blatant. And that is, I suppose, a valor of discomfort earned, to justify the uneasiness.... of one's free steps, to sojourn of subjection to one's state. And were I really evilly bent, (this) would be a devil of a day in heat. But as it is, it's only disagreeable, disposable, and dispossessed, for temptations more heroic to inhabit.... than hatreds. The reason is this: the time is spent, can not reoccur, can not be rerun. And so a hesitance is spun— to see more grace in this beauty of recognition, for our own personal travels to undertake. Yes.... This is cold. But so's an hour's sleep, with less of aim. And now a day is due, to warm up through. For I have no choice of paths, on being definite to leave. Sorrowful be bravery's penchant drive. I decline nothing of the drift to dissolve legalities. And be as served a writ of morals, well.... a walk takes some walking to do, some convincing to tread upon, some conviction to tether draw an' pull, and the burden of this weight to subdue. I'm not the mean one left, but the menace meted morals. (*pointing to the imaginary*) And could you be as true?

BECK (*coated, coming out the café*): You still here?... Are you alright?

HEROFORD: ....I just become a bit uncertain, where I'm going.

BECK (*approaching*): Oh? You're not dazed, are you? The sugar rush in your hot beverage?

HEROFORD: I have absolutely nothing to do today, and sometimes feel sudden jerky senses of displacement. But yet we all have nowhere lurking, it's profound to seem busy. What of you, as a young man, to say of it?—

BECK: Me?!... I'm headed towards a park, sir, to waste some time. And then I'm off to a construction job, to save some time. And then, if I survive, and am prosperous, it's to a country farm, to make some time. And that's the life of it, of going somewhere. 'Cause that where is like snow, I guess. It melts, till you find it frozen.

HEROFORD: I might agree.... Secure yourself, young man, with whatever you do. (*With a pat or tap on the back, which surprises, disturbs* **Beck** *slightly, they depart, in separate directions.*)

)()()()()(

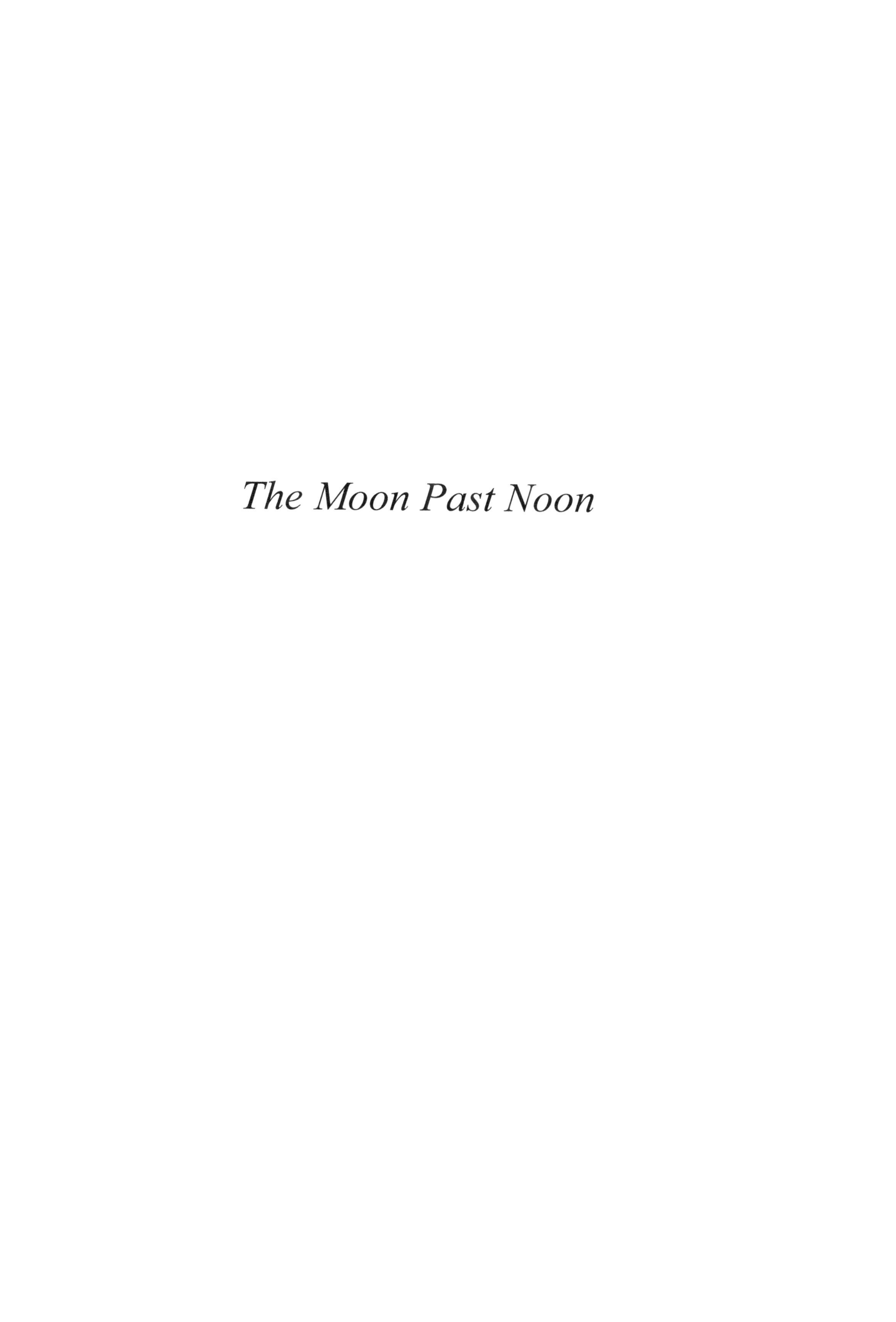

The Moon Past Noon

The Moon Past Noon

Characters:
 Lancet
 Cantal
 a native Girl
 Ancher
 Ancher's Secretary
 Holston
 Fenic
 Bastrel (exchequer)
 Mytdra (king)
 Doria
 Pale
 a native Boy
 a native Guide
 Suzie

Locales: a jungle habitat; a city of advancement, and one of not

Act I

Scene I — *Lancet stands next to some large, ornate jars of pottery, admiring the designs forged on them. Cantal comes over to investigate.*

LANCET (*noticing Cantal's approach*): If these are ruins, they are magnificent!

CANTAL: Why, they're glued to the ground, sir. Too difficult to cart off with. Probably fragile of the centuries.

LANCET: That may be so. But I've predicted some such recovery. This is definitely the right location.

CANTAL: You can't manage it. No troop of excavators could. That's why they're left standing. Impossible to stabilize for transport.

LANCET: Methods are being developed, Lord Cantal. But what for taking with?— I mean to own.

CANTAL: That's a pretty kettle to filch. And what would your master say about that!—

LANCET: He didn't find them. The most professional of the matter owns, with circumspection of its placement scrutinized. This belongs.... to a scientific method.

CANTAL: Or merit. They're too large. You don't even know what's in them. They could be loaded with earth.

LANCET: Dear lord, I don't mean to steal them away immediately, like finding some décor for a castle or a hermitage. Historical research owns these treasured antiquities. I am abided a mission of curiosity only, because I serve a much enlightened patron of— casual gems of knowledge and quaint oddities. But the initiatives of (the) search are my own. And I'll have it this way done to stand with, that they'll stay here quite awhile yet. Besides, the cultural permits have not been officially obtained for absconding with any artifacts of this region. Even if trinkets of gold were dug up, they must stay legally bound— to time and its ages, until agreements are drafted and signed for mutual exploitation and profit sharing, through expositions and the like. No. What I claim.... is an important discovery named after myself.

CANTAL: The "Lancet Vases"? ewers of eternity?

LANCET: I've done the research, and predicted the location. I've found these—

CANTAL: Your subordinates did, and called you up to examine.

LANCET: Well the credit is mine, since I directed them properly, in effect led them to the spot.

CANTAL: And out here weathered. It is remarkable how they stand so un-accosted.

LANCET: As much as rock monuments, Lord Cantal. Why should they be bothered? in this relative jungle, of all places. The animals would have no interests in them. And the native population find this area so uninspiring as to be totally disregarded, as much a wasteland as they could imagine. They do not live with the thoughts of ancient ancestors, but decrepit implacabilities impractical to honor, adore, or adorn themselves with symbolically and reflectively. This! is meaningless refuse to them: old, discarded impiousness worshiped by a lackluster spirituality practiced in the distant past. And what for value, as we may pay them to allow an exploration, then as much we may reap of discoveries. Who is the more foolish? The value is strictly intellectual, the clay as worthless as the soil— for their purposes. And this is too broad and expansive a lot, of the unpromising, for much further consideration, in their minds.

CANTAL: Yet if you bought for little and sold for a lot, what can be disinterred, some awareness of this might perk up a savviness for the place. Though frankly, I credit their information more, that there's not much here worth carrying on backs. You've found some jugs, but not even a rich ore of civilization. The custom was not to preserve much, in the days of eons ago, so that whatever little thing is found is valuable only for its spite of staying intact.

LANCET: These are little?!

CANTAL: I say, educator, this is not much of an archaeology, if villagers direct you to this openness with much the slight of a disdain. The matter is more of conquest. Find we the locale suitable?

LANCET: That's more your party's interest, lord.

CANTAL: But it's to your beholden presence, merely as an appraiser of the landscape— and part time prospector. The motive is your master's. For it is his domination that stirs the inquiry.

LANCET: As much a spot of land to buy, extensive as it may be, for a destination of retreat from the winter howls, I should say it may be adequately— leased. The wood's a bit unusual for our tastes, being so exotic. But the climate is exacting of some pleasure, if seasonally timed. Too much a habit of his, though, to find or scope out more and more tracts of land to abound his own ambitions of significance.

CANTAL: You haven't the mind, sir, to analyze the strategy involved, so devoted are you to the narrow-mindedness of trifles. But

I can state that there's more to it than these earthworn vessels. We collate opportunities of extreme influence with such outposts, if we can obtain and maintain them profitably. For, if you can widen your perspective eyes a little, for your irises' pronouncement of a gaze, you might find that they may form something of a ring, around our homeland's provinciality.

LANCET: I haven't made my geological speculations with such a broadening to calculate.

CANTAL: That is because you only accompany us. But your master wants to dominate much circumambient activity, and with a good faction of interests control the internals of a great deal of trade and relationships to the country.

LANCET: It is too grandiose a vision for me to contemplate fully.

CANTAL: We serve ourselves, to own the matter of the land, its meat of substantiation.

LANCET: You've been claiming so for years, to want to effectively usurp our governance through economic means. It's not possible, through a freely disposed population, to achieve such a goal. Yet you continue to target and bargain and arrange for territories. And as my master is an uppermost proponent of this view, a leader in fact, I benefit of some excursions to lay claim to knowledge found in remote areas. But I see not an overthrow of powers by this.

CANTAL: It makes for a gradual metamorphosis of ruling strengths, and without a disruption nor disturbance of class principles. Too slow for your mind to observe, as you only find your luck in fun. But Ancher will control much more eventually than any government could possibly.

LANCET: In numbers external to our country, perhaps. But my rich earl can count more pebbles of his enterprises than the more solid enterprises of our state. Numbers are only a subtle suggestion of size. What is powerful may be his lending, but not his leadership at court. There isn't the heritage there to lead him much higher. He's not related to a governor.

CANTAL: We work to remedy such obstruction, by transforming the methods of liege-making.

LANCET: That is inconceivable, with loyalties drawn only to a demonstrable might. You can not break the chain of ascendancy except through discriminatory wars and battles. Everyone knows this. It is a law of nature.

CANTAL: We hope to avoid such crude bellicosity, by the practicality of collective holdings.

LANCET: You'd have to own much of the world, to exert such insensible pressure. This has always been an impossibility many have often entertained themselves with, upon the patronizing of my master. But how it may be amusing gives the prospects less weight.

CANTAL: You're not so felicitous to his imposing.

LANCET: Of standing? But to anyone of intelligence it is only ever evident. He is a welcomed nobility, and an artful manager of minions and vassals. The ruler of a nation, though, is out of his sphere of condensation, for a world-wide view. What country would recognize you more than bankable, in this manner you wish to exert, and perhaps take turns playing as a controller of human destinies in our land? It is wishful thinking of the proud and capable, but not complicit of the duplicity for such a rearrangement. Some, no matter how wealthy, are trapped into their positions and social standings.... by the art of a proper regality. And I can't imagine such an order to be destroyed without a cataclysmic damage to every facet of living (our lives). I can't envision it, Lord Cantal. It brings giggles to speak about or contemplate for rhyming pretensions of couplet. That he is an aggressive man— Yes! for many trains of thought. But this leads not to the realizable. And any non-congested head— without a bounty to employ for trying— recognizes this. Not that I wish to offend anyone's prophetic beliefs, but some say the earth haven't enough Sabbaths to last their lifetimes. Now how can you argue around such god-like views, each person a world unto himself?

CANTAL: Skepticism is for the mean of mind. It is the advantageous of thought that devours all privileges obtained. Our plans are carefully committed, Lancet.... But as an observer and roving bystander— indeed! an excuse!... you'll come to see the terms of our forwardness, and never be so much involved as to sweep after our steps, if it's value you find in the dust. Now these are truly grand and gigantic, though ordinary in shape.

LANCET: Functional, Lord Cantal. Races of the past tended to be practical. Though I suspect they are celebratory of a harvest ritual, being so much larger than necessary.

CANTAL: Necessary for what?

LANCET: ....For possession. I doubt if they were used for much storage. In this clime vegetables are not meant to last long, and ample grainage may be procured at any time of the year. And the elongated shape, near imitation of the obelisk with these lines drawn, say they weren't constructed of water bearing purposes, but of contents much lighter. Perhaps leaves. Perhaps scented flowers. Something of a dedicatory nature. Note that what damage occur(s i)'s more at the top than the bottom— Retrievals.... and replenishing.

CANTAL: So. Well, you'll need a stand, to get up there and look in. Keeps the rain?

LANCET: Probably leaks out, through cracks.... Maybe embedded like roots, submerged beneath the ground cover.

CANTAL: Hey. A regular man-made plant of pitcher, then. And what are these designs for?

LANCET: A pictorial language, perhaps.

CANTAL: Not as handsome or sharp as the Egyptian hieroglyphics.

LANCET: They don't always have to tell a story. These may simply annotate a feeling or perception, like courage or satisfaction of sufficiency.

CANTAL: Hm. You've much to learn, of the symbolic here.

LANCET: One can't know by mere speculation. These may be priceless, by being uniquely survived. But what value, really, is this land, can you suppose?

CANTAL: Hmm. Spices, maybe.

LANCET: My master comes— for the cinnamons?!

CANTAL: He comes not at all, on an expedition of your choosing— or suggestion, but rests comfortably to await news of what others have found. And so you think. But.... *he* has led you, and probably remains rather restless in his dealings, within the throes of his anxious casts of line for any conceivable felicitousness to hook onto, in divers directions.

LANCET: He has the funds, to ploy of puddings.

CANTAL: You think we waste our monies, do ya?!—

LANCET: Don't all benefactors? to leave it in the hands of the Benedictine? Blindly relying on the experts to substantiate the largess-ant inquiry of this such research, I would say is the mode of your forward thinking grace an' grant.

CANTAL: Sir, you are no expert here, from what I can see. But you say the retreat will be comfortable for him. I can feel it and confirm, being a more active party to the adventuring, as an entertainment of curiosity of investment— of time and leisure. I don't want to belittle your importance too much.

LANCET: You needn't try to. My master will be well served with these activities, and my efforts sincerely employed. And that is why I feel I can be blunt about them. No, he is not a foolish man, to throw riches upon recommendations. I've studied the possibilities carefully— from afar, an' for a long time— that here may be a very worthy post of penchant remarkable for what it may provide: anthropologic interests along with a tranquility away from the bustle of his regular activities.

CANTAL: So be it poured, another conquest. I will support its purchase, this range of the underdeveloped. A nice camp for us, with healthy hills— and a strategy.... What else have you found, at this latitude, along this meridian?

LANCET: Some species of wood here appear to be very porous.

CANTAL (*dis-genuflectory*): Oh, astonishing!

LANCET: Some species of insect here are terribly large.

CANTAL: Is this all you can ascertain of a locale of such vim, throughout your vigorous studying? What else had you expected to find?

LANCET: A wholesomeness to being unrefined. These grasses are untutored, these vegetative growths spontaneous as sparks along the ground, bulbs spectacularly colored and variegated, and perhaps a new form of onion— if my instincts of taste have been properly alarmed an' alighted. Sweet heat!— Sugary!... but a mattress for the ants.

CANTAL: You better do much better than that. Ancher wants high value with this possession, and you predicted to him you'd find much of the outstanding. Aside from some leviathan extravagance, on occasion, there isn't much of note to relate that would stir a crowd's consternation, back home, in discussion of the unusual or otherworldly. You're under the mark— as a target— to come up with something incredible to justify your insistences for this.... exploratory permit(ance) an' pestering. I'd worry more, of your aptitude for this result to fulfill. You can't carry these jugs back. Pieces of molded clay will not impress unless they're indented with an alphabetic profuseness for any significance. And the spiders to be trapped here are no larger than tarantulas— You better find some jawbreaking evincement that opens the mouth so wide as to change its purpose speaking.... or be fanciful to return home scolded with broad tongues, what he's spent on you already!

LANCET: Speaking as for praise or for penury to enforce. I say, you're not disappointed so far with these events, are you?

CANTAL: No. They've been fairly enjoyable, for my nature. But you know the earl's standards are high, with expectations promised.

LANCET: The most I can lose is my position with him, Lord Cantal.... and possibly gain a disgrace of station. But he trusts my reasoning, as I always. And that will come through to save the expedition some importance and renown. So let me tell you of a discovery I've found and even predicted. It was the true spur for the trip, and I think I've confirmed it to a reasonable satisfaction.

CANTAL: What's it, then?

LANCET: A form of spirited, aromatically divine, and yet wholly eatable grass.

CANTAL:.....Ah! Cabbages!—

LANCET: No. It is a true grass, a slender leaf, with nutty or oaty tastes that can be made somewhat delicious. Raw it is peculiar, but only in being distinctive— not disagreeable. And I have barely cooked some, to determine without doubt that it is of a nourishable flavor and delicacy for the palate— as profound as seaweed. In fact, I wish to call it Lancet's Weed!— It may only be cultivatable in these soils, or of a warmth that lasts year round. But it is here that men may actually graze like cattle, if it comes to such a hunger, and quite happily survive.

CANTAL: I haven't noticed any of the native(s) doing so—

LANCET: They needn't, preferring to fish for meat. But this region abounds, at particular patches, of this splendid grass. A green wheat it might be called. Yet not for seed but leaf is the nutrition actual. And the gold becomes a jade. For just imagine how it can feed. A crop of potatoes and this could do you for a whole winter — of simplicity in meal— if we can figure out its growing characteristics correctly, and employ this towards our prosperity. Now you talk of riches, Lord Cantal. But this is redounding of wealth, if manipulated right. And my master will welcome this achievement of uncovering a fashion, for a passion of a populace to foster.

CANTAL: Let me have some of this stuff, to eat— test!

LANCET: I can do better than that. I can take you to a spot where some of it grows. And you can try for yourself this new manner of open dining. I haven't been terribly secretive about it, collecting samples and even mouthing them before others. But I suppose many don't see or recognize what they don't expect.

CANTAL: They're not like the alfalfa bud, that fodder, is it?

LANCET: No. Actual blades of grass, growing up like the hairs of a skin, the land's proud beard, upon the face of it, for food of giants.

CANTAL: That could be remarkable, Lancet. But you're an enthusiast, with highly learned and acquired, and sometimes offensive tastes.

LANCET: The octopus' backed tentacle can be as much a culinary delicacy as the escargot's shell filling of flesh an' fanciful salad dressing. But one taste for this wild weed by yourself, ma lord, will make determinate the factuality of this new food source. Imagine fields of it— empires!... for its agronomy. Not merely an edible flower, not merely an edible plant— but a whole, wide ranging, terrain covering grass, and for man's propitious enjoyment as to bring him down on his knees to grope and chew.... Or let's say the children would, if the grown are too dignified for such postures to take for feeding and breeding to be fed. But there is no fallacy to the overwhelming potential that may be possessed, with this discovery— or more properly put: a research to find. For I predicted the rich minerals of humus, in a constantly warm soil, might evolve such a result, where decay is not so much a decadence but a fertilization actively empowered—

CANTAL: What would we be eating?!... upon a savanna?

LANCET: What is a grassy plain for, sir? Most do not ask why, its purpose on the earth. For grazing?... of some creatures? Obviously, the stalks are stores of nourishment— But why?! Simply because the earth is rich, to retain its moistures and ingredients. Now, where the crop is coarsely ate— by animals— the ground is likely to become barren or parched of dusty patches. But where it's not possible to this to occur, because of such a fruitful abundance of the blade—over centuries, over millennia.... so easily then may a finer species evolve or be allowed to establish itself in great quantity and occurrence, that is thoroughly suitable for human consumption. And it's all the more attractive to know that the natives around here don't eat the weed— because they don"t have to. That! is what I was looking for, that anthropomorphic condition that is completely special. The people who live here are in no way horticulturists, and yet they live in a virtual lyceum of plants, an environment endless of age. Here! is where to look, for such a specification to be permitted. And it's been found, as if ordained by pure logic alone. There's no conception against the vegetable here. It's not an ownership or prize to be hoarded and hidden. It is simply everywhere— for everything. What is valuable, Lord Cantal, is the notion— not the plant. What is exciting is the realization, not the commonness to this grass. For ubiquity is nature's penchant always — and why not such a herbage to occur for humans!

CANTAL: They're good arguments sounding. Men can be just as piggish as the real thing, and a boar with tusks for their determinations to hog at any manner of wealth an' holdings. But I'll reserve my judgment until after I've had a taste. Tall growing stuff, is it? unperturbed in this wilderness.

LANCET: Ah! That is interesting. It is not so long.

CANTAL: Then I *will* have to get on my knees.

LANCET: I think it has a conceit of itself, not to need to be tall. But it often abuts the taller, if my few observations are just to conclude. A darker shade— but with a brighter sheen, it is of a trained distinction, without being a distraction from the ordinary. I can not call it precocious or obtrusive. It is calm, with the plainness of being plenty.

CANTAL: Aside from naming it, Lancet, what do you earn out of this discovery, or unveiling? How may you profit?

LANCET: The rewards are intellectual, the satisfactions more cerebral than cereal. I am restricted, for my position, from any hope of monetary gain. I can only recommend this knowledge to my master, and have him take charge of the situation as he would like. No doubt if you find any advantage to it, you might try the same, with persuasions or influences, or taking over the project wholly for yourself—

CANTAL: Oh!... the finances for such an undertaking are major. I'm not as bold as the earl, with my resources. I'd need much more convincing, not only of the taste— but also of the feasibility, practicality of planting, crop production, and return on capital expended, before I'd dream to control an opportunity such as this. Odd tricks of nature can occur, you know. It might taste good when there's no hunger, and yet seem horrible during a famine. Can't easily account for the psychologies of mankind, but if one is forced to eat the same thing regularly, and almost exclusively, even candies can become maniacally hateful, the bonbon bombs of my youth. And when one is trapped into necessities, these necessities are no fun. But then, as you say, we might have to farm remotely. And as it's here, it might not be cost effective of a shipping compared to its purchasing, the selling compared to its handling and packaging. These sorts of realities often only reveal themselves fully with the trying. Though Ancher is a much shrewder man than myself. I would never have though to have the likes of you accompany me on a sojourn of adventure. Yet I accompany you, and with some pleasure, perhaps, depending on how our interests are stimulated. There can be no denying that you've hit (upon) some kind of a strike— if it's true— based on your intuitions and analyses of the possible. And if it were for me to buy some parcel land in this remote place, I'd very possibly hire you, to amplify your instincts and try to find more curious treasures— just out of a temptation to let you run with ideas, for awhile. But this not being the case, I'll hold back my surges of initiative until we learn what Ancher wants to do about this, if anything.

LANCET: You express a form of flattery for me, lord. But I am heartened by your prowess to deliver it. We are each stuck at our levels of tree, like monkeys resting. And sincerity only relates to proof. Consider the hugeness of these wonderful containers, how that exuberance is only wasted today, on our pitying eyes to feel for how they were once majestically admired, and to search for their exultant meaning. I might characterize ourselves as hunters of the vanquished; but of honor, glory, humility, or gloom.... it's hard to find out accurately with such a long passage of time. No villagers we've met, who know of these, can explain them adequately as yet, except as artifacts of some antiquity simply born to crumble of any current consideration.

CANTAL: I doubt if such masterworks have need of any consolation. They were constructed by.... broad enough minds, to leave them alone and with much peace. An' only weeping comes a rain. But of this forest, what could tame it now? I'd build a cabin here, and a mansion there, and cut down much obstructing growth. Yet for our clearing sights, achieve we mights of passage. The natives see far more thoroughly, of a contentedness, as we shove our chests to pump of these remarkable elixirs of atmosphere, and en‐joy our heightened standing and verve to experience new conten‐tions of life. Now for your pleasure, Lancet, might you capture a marmot, or the like, to be examined of a personality— or cringe at masters. But I am of a goal with sight-seeing, to relate myself at great(er) aims.... importances, than for purely scientific brouhaha to wheedle. Much leaf is edible, I'd wager, with cooking. So this had better be an astonishing enlightenment, that a grass can be for man as much are (as) mares. And then make you your philosophies about it. We've lost our pastoral(itie)s to find such things, or pas‐turing intentions to drive as grazing for our works.... and effects upon the earth. But I insist upon a remoteness from the heady to indulge of, occasionally. Direct ourselves to this patch of land where may be raised towards our compass a greenery of sconed strings. And I will have at that elongated pea—

LANCET: Not a bean.... but a bending of our arms.

CANTAL: Ensconced of the miraculous (as **Lancet** *starts to lead them away*), this will be as profound as it is out in the open. But we should really conclude of its importance more circumspectly. For I advise as much of the material good should be left undis‐turbed as possible. Only serious curiosities should be stimulated by our notices. You've not considered the transience of the plant, nor any of its important growing habits, but only the obvious: that it occurs with heat. Not enough to make a spectacular determination.

LANCET: All of these aspects will be well studies, I assure you.

CANTAL: Yes. That's a given, Lancet. But the more facile of hand, and fickle of mind, will not take such courtesies to heart, and with a misplaced enthusiasm inadvertently try to destroy the bounty, whenever such blind foolishness is possible of an event. One sees a thousand, and thinks there must be millions, because that is nature's way. Yet if man produces any, that is specialized and particular and limited, almost by definition of his character. It's to be protected and marshaled, stored and withdrawn from public scrutiny, until the owners can most skillfully craft their controls.

LANCET: That rather sounds more like the consciousness of greed to me, those propensities of propitiation regulated to some advantage. Leave it all for you to taste. It's not forbidden in any way, being grown wild. Of course I'm not one to stupidly hark'(en) on it for any desultory interference. It's more like an open secret, pleading to be bled, what more constitutional minds may find a significance for. I was just hoping to have an occasion to express my delight to someone of standing. And now you charge to investi‐gate the wind-blown wondrous. That's enough substantiation of my delirious tease— for knowledge, that someone of a mental aptitude is also aroused by same and similar applications of interest (*They notice a native* **girl** *approaching the tall jars, and stop.*) as a sort of foraging demanded.... But what for she, to utilize the jugs?

CANTAL: Here eyes a purpose, Lancet. Some local custom may be employed. This is how you determine.... attitudes of socializa‐tion. Relics— Nay! Not for old, but she is young and supple. (*They watch as she heaves up a bunch of flowers into the jars. Some only fall to the ground, to which she falls to retrieve.*) We

should investigate this possible ritual, don't you think? if the vases remain pertinent to their current customs.

LANCET (*as they go back, to approach her*): Local.... The prac‐tice mystifies, somewhat, the sparseness of attention given to the ancient monuments of sculpture. I would have thought them rather more rejected or dismissed, an ideology lost, an idolatry shed.... But they're not hidden, for some use. So then we come back, to promote our observations into the reality of witnessing an action not described before.... before us. (*She raises her head as they near her.*) Take not any alarm to our rapprochement, fair lady.... Are you conversant?

GIRL (*still squatting*): Who is not conversant?!

CANTAL: We do care to see your manner.... of the flowers— to dispose?

GIRL: Dispose? They are deceased, strewn from their stalks, the fragrances (with) motil(it)'y employed.

CANTAL: Suspended, then, into a cushion of contents, for these high containers to receive? The meaning leaves bewilderment for spying.

GIRL: They are urns of the forest. (*standing, holding some flow‐ers*) We restore them regularly, for personal invocation of our mourning rites.

LANCET: It's educated.

GIRL: They are fed with remembrances, as a flower might be called, since they sprout from the ground, and from past lives erupt again.... Their scents are their spirits.

LANCET: It is an educated mores practiced since the establish‐ment of human civilization. Though various of fashion it is the same of meaning, a regenerative insistence. And so you pay hom‐age to the beloved departed, with this gesture of respect? But then how anciently is viewed the past? For what is recent is less mighty, and all the more grievous.

GIRL: The past is pyre-d always. But here is for ourselves to be contained, with a spiritual emollient we must prepare.... patiently, over years. We are bred for this, and choose the times ourselves, to make for easing of the passage.

CANTAL: Bred for it?

LANCET: This is a communal site? They're large, but not hu‐mongous of a population.... to hold a store of ashes throughout the generations— pious though they be to be contained. Here must be a symbolic religion, for what is possible to do.

GIRL: The remains are spread through the soil, for growth. Where flowers bloom of them, some are sacrificed, for our beds to make—

CANTAL: Yet bred for this?

GIRL: —Where only grasses grow, lends diligence to the cause, that each must stay persistent.

CANTAL: Grasses.

GIRL: I had started as soon as I could reach a valid throw. (*throws a flower, which fails*) But it is difficult still. And so with much company must I try, so that some reach for entrance. Were for the art of it, I completely tire of my efforts, and then carefully clean the site of the fallen.... to return with.

LANCET: Mark you an age for this? Or for whom do you celebrate, today?

GIRL: It's for myself that need is pressed, and as I feel it drawing forth the intention.

LANCET: A vanity of beauty, is then the inclination.

GIRL: And the more that I may fill with.... is the more that I am dealt with for success.

LANCET: Oh, self-substantiated wholly. But then it's as remarkable as the number. And may we perhaps help you? This is simply a fertilization rite, for promising a prosperity of springs and offsprings.

GIRL: Then to return barren would be much precocious. The aid of gods is always allowed—

CANTAL (*to the ground, to pick up some flowers*): Then humbly accept it, dear lady.

GIRL (*as Lancet joins Cantal*): —but is oft told of for some of a time with fortunate meeting of the like, repressions are spared to gratify the need. (*as the men stand and start pitching the flowers into the jars*) And of a vitality of seeding, much to be a vigorous display of aroused jollity. For this we make our own— and any does, with a frenzy wanting to.

LANCET (*throwing*): All for a friendly intention, nice girl neat. We cling to happiness for you, and that you are prospered— with our help— by those who remain to perceive your achievement.

CANTAL (*having deposited all of his*): And now for yours (*gently clutching her hands*).... dear lady, that these flowers wend— to the emotive necessities of life. (*while obtaining the flowers*) I can predict for them.... good territories to reach. So be us clean to fling. It is a precious custom followed, as like the wishing well to do, but more enjoyable both of essence and ambiance to the athleticism. (*tossing*) Like amour to toss, winds and aromas of love, breaths of enchantment. And all for you, my dear, every fragrant petal. (*One flower fails.*)

GIRL (*as he continues tossing*): Oh, do be clean. I would be scolded to leave behind.... a lost proposal.

CANTAL (*as Lancet picks up the flower*): It's to be done, lady. These flights are mine— made.... to your enlightened eye.... And the trophy is your unspoiled blush, that youth is more magnificent of mentioning— that its cautions. (*is done*) And are we civil today, but I would have made for you a rich bouquet of effort.

GIRL (*with a slight genuflection of thanks*): Then you have enabled me much more than I could have alone. This is allowed.... (*walks off graciously, as they watch*) and made for the wishfulness

of what it's worth.

CANTAL (*turning towards **Lancet**, to grab the flower*): This!... is mine. (*takes the flower, **Lancet** half handling*) A charming thought is it. I'm touched to the bone. She is so pretty.... and typical (*clutching the flower*). How one may perfume the day, with such a simple notion.

LANCET: Simplicities, my lord, come towards the strongest urges.

CANTAL: I'm not so sure about this grass, Lancet, if it's fed with the ashes of their bodies. And yet, such could be a most heavenly entombment, the most pronounced.... pacification.

LANCET: All of the world, from new to ancient, is old. This soil is all of antiquity everywhere. Little changes of its contents organic and mineral. They define the histories of any matter— And all species of vegetation are subject to the same cultivation. The brussels sprouts may as well be heads of former kings, and the carrot root an emperor's tongue. Or of an animal laid near leaves, after a death's woe, the furry creature embalms itself with life. For ne'er let be said what we are ate, it is the river's lust per chance through streams of a gluttonous, salivating time.

CANTAL: What's mouth recommends itself for taste. I see the point of fertilization with remains, but the thought is crematoria-l — Alas.... such may be a food's bathing. To think of it anciently, though: rejuvenation with incense— Oh! Metabolism's such a fire. Now towards this grass, what do you think— with what we've learned?

LANCET: ....I think it's all to do.... of a specialty of ponderousness, how tediums so often give birth. But as to our reflection upon it, no less its weight from our perspective. I doubt if the different grasses discriminate much for their meals, where live they close for sharing and our shearing. And nor should we propose to view this like.... encryption. But for a flavor's paste, it could be well desired. Keep we secrets? The method is transportable and repairable. A rich soil is what's more needed than some sultry definitions. Though perhaps barbecued bones may gain a kind advantage off the pig. Nevertheless, I'm not revolted by the graying that night be around. Let us conjecture differently, lord.

CANTAL: No. My sign is in this now— with transformation to embrace. And through the deed we will not squander of disgraces to reek of wretchedness. (*sniffing the flower*) I become renewed and sharpened, of all possible senses united. Take to this platform of lawn. It's character bites at me for tussling with wisdoms, that which I've never met but through a beatification to possess and be possessed by. And now the curiosity is aflame of obsession, how I might stand amidst this conflagration— burning!

LANCET: You'll not so change, sir. The weed is not that way transducing of its.... energies. It is simply applicable of a nutrition. Note myself.

CANTAL: Examiner.... you've only tasted, not drawn the terms of testing this temptation— with yourself. Become we gloriously heathenish, when let our laurels loose. This is a romance of mankind to encounter and advance with— Mouthfuls! Mouthfuls! Mouthfuls!... So celebrate the exigency to be transfixed— bored through an' through, by these blades. You can not comprehend the razor that's been found, (*pressing the flower to his chest*) and the pound's weight of pang. For once it touches you, flooded are all hemolymphic reserves— I'll do it, yes! An' on my knees, be natively bred.

LANCET: One could make a sacrilegious astonishment to this— the information alone!

CANTAL: It is a funny estuary lapped.... I am the hound become, or the wolf's groveling towards Hecate and ecstasy. Let us go to some of Lancet's lure. And if it is remarkable— I will dance on all fours.

LANCET: That would be an interesting challenge to describe, lord, for my annals. I don't claim the ambrosial for one's passionate referral of a substance. Yet how you are affected already, is the whim of— potential, and the whiff of benevolence.... for our doing, and being able to assist a maid. Rationalize more the art than the archery we have performed. They are a people unto themselves, and we have been.... gentlemanly.

CANTAL: It's hard to stay objective, when moved by our own generosities. For what has moved us 's even greater to realize and encompass fastidiously. (*surveying the ground around him*) We must be meticulous, not a petal to be left as dropped from a bellowing spew.

LANCET: There're not so many, crawling with the breezes—

CANTAL: Yet all are tears, passionate of tint an' dye of longing. (*to the ground, to pick up petals*)

LANCET: That is devotion, to a gallantry upheld of our.... station.

CANTAL: Then— in details are made the love. I'm caught! to capture with two hands. An' one be bonded.... Yet, for pockets made, these treasures. (*starts pocketing the petals, while holding onto the flower*) Uhmmm. This be taste, Lancet, in freedoms erring for a want. Some day.... you might sense the spin.

LANCET: It's a crazed liberty.

CANTAL: The errand's done (*pocketing the flower*)— unless you're stepping on some!

LANCET (*slightly jumping aback*): Not I to lick the sole.

CANTAL (*standing*): Then no. Note a lord's hands.

LANCET: Rusticated.

CANTAL: Like a god's humbling before the universe. These jars are profound— made by giants.

LANCET: Well so's the earth, if there's construction to a genesis.

CANTAL: And have I taught you anything about research?

LANCET: ....What be, to plow for parts?

CANTAL: One must feel it, sir— be driven mad by the effort. A hoe does not find, the hoer finds everything. Your theories may

make an admirable implement or machine. But what have you learned without the thrust of investigation? to face the beauty found and commit to this your life an' ambitions, since her presence has changed you thoroughly.

LANCET: You can't say you're so seriously enthralled?! a personage of your experience and standing, with what is found here naturally and naively kept!

CANTAL: You don't understand. It is the adventure, much better than any prize or result. And so to your grasses. Were you delirious? That's more a shame to being silly. But coming up upon them, what could be true, what may be affirmed— by feeling right (about it); that is the new epoch, that is the goal's extension— for the greater part of its worth to pursue. And you can only think of its agricultural prospects and managing— But look at where this has led us!

LANCET: By the solemn pitchers. Perhaps you're surprised by the anthropological principles inherent in all peoples. But this kind of— freshness.... is common, and general, and low, as a purity of water tends to itself with an abundance of the same. Then you're not used to being startled by fair creations, even with your many excursive trysts?—

CANTAL: I am moved. An' that's how one may be, with truths. Perhaps this is a sentiment of my class. But to be so peculiar is my worth, on being entirely humbled by actual, unassuming greatness. And that's a lesson for you. I wouldn't try to name *her*.... or cultivate and redeem for my purposes, but to let this flow descend within me for its own.

LANCET: That's perhaps a nobleman's privilege. But others must be swayed more objectively, not that factualities belie their objects. She is handsome of the kind. We are pleased to help. You are infatuated to assist. And now we may determine ourselves, with witness of the real that's growing— and as like kelp today, I'm sure. Much of plenty is the rule for a description, as for undulations enticing waves, many so lazy to appear for such frantic motion.

CANTAL: Uncharted bevies. Then let the aromas of the winds draw. For you were led to them by scent, no doubt. That is the most unusual distinction for a turf or sod, which alerted you unconsciously to it. And so you think you have found it. But it has found you. Greeneries must make attractants.

LANCET: ....That might be true, although I have not noted any smell particular nor intoxicating other than what's owned by grass. True, hidden sachets may provide of their own herbs, to help persuade an aid to pollination through the roaming of animals as liaison to the roaring brush of the wind. The seed grains are.... ripe-like, their odors compelling for a meal, the sampling of a savory innocence. You have produced a scientific cuticle of thought, to provide some basis with. Yet, at its base— it is simply there, to be manhandled. How could we harvest it otherwise?

CANTAL: I've seen riches the world over, Lancet, and their most common denominator is their intense stimulation of the senses, even as one can not describe a sight that tugs, or a taste that buries, or a sound that blinds.... And what you noticed not does notice you. That's often the way, through paths of the desirous, and silent incantations of the pulse. So heavy is the air, of one's permitting to perfume, upon an excitation. But speak fondly of this crack of consciousness. What alerts you must be favorable for something, as an odious command may train your mettle. So.... withstand and restore ourselves to this, this— scag you've found? My aim's more patient, to have a reason to remain an' return. Our industry will decide. Per chance, they see the sky but not the stars. We see the stars but not the sky.... with this enlistment of the weed.

LANCET: That is— offhandedly put, lord. Plants sense the light but see nothing. They haven't our eying. Now let me take you to some, so you can judge that for yourself. (*directing*)

CANTAL (*as they are leaving*): Sir, I've seen.... what they saw— Is it abuse? Oh make me rational of lugs. I will ablate myself through this decorous sear of desrous tug. Well-behaved an' gentlemanly, you say. That is the constraint.... or the coercion through which we stroll. But what will be, after it's ate?... More manner— of the telling, I am told.... for what it's worth.

Scene II — *The state room of an estate, where plans are drafted after fierce debate among confidants and associates.* **Ancher** *is standing at his desk, thrashing letters with a quill of huge ceremonial plumage. His agitation is self-entertaining, as much as for a display.*

ANCHER (*loudly, for one who is alone*): Darn these details! They are congestive of my fortunateness. In every direction they lead promise. But I can only undertake but so many offers of temptation to pursue, that leaves a bane of negligence whatever I decide to do. Play with your swiftness, notes! What stays aloft is for my rote of some endeavoring, as like a game of stares. Yet would I pen: Acceptable to me! A wealth of commissioning, I try a frenzy to partake, that what's in flight restrains its haste. I see more feathers than the bird, more than the bird has beak. And this alone delays my speculation.

SECRETARY (*entering, as he continues*): The lord's constabulary, sir. Shall I remain, for a tidying?

ANCHER (*as the secretary is directing **Holston** in*): Yes. Take notes— from off the floor, and direct them through the chamber to the "pit of penance." For I will find distress in not receiving them more properly—

HOLSTON: Or dignified, Ancher!

ANCHER: No doubt some opportunity falls, drops from my grasps. But I have so many seized. (*stops thrashing, to acknowledge **Holston** with politeness*) It is an embarrassment of possibilities.

HOLSTON (*taking a chair, away from the desk, as the **secretary** works to pick up some letters*): Are they plots— against the government? They you are sent (*sitting*), sir. And they better not be!... seen.

ANCHER: Well, what is for opinion? These are all only suggestions, of action. And sometimes I put my name to more than I shall manage fidelit-ously.

HOLSTEN: Sign them? Some must be chicken-brained.

ANCHER (*sitting atop the desk*): They all take great precaution to be rational, because my monies are real— whatever the cur-

rency. But we do not lend an' loan, do we! We acquire services an' products—

HOLSTON: Plots! (*as the **secretary** sits on the floor to read*) Nary a one is feasible, flooding the markets with some exotic treat — Would make it less exotic.

ANCHER: But capturing a market for it is more promising.

HOLSTON: Capture— hey! An' that bloke who wanted to control all sugar production. Knew only of the cane— an' not the beet. You simply can't see your way through too many openings. But the king will quarry for any reductions felt of the necessary materials to his nation, raw or ra(n)'kish, I can be assured. If the court lacks for its teas' sweetness, he will beat the honey!—

ANCHER: More supreme is surrogation, cop! We can not be blamed for our associations external to the crown, unless there's a war be placed between them.

HOLSTON: I favor the ambush.

ANCHER: Not a spirited way to be successful at this. Most think one controls through influences. Nay, it's through a control of inducements. For the king has all of influence, and political are his decrees. So what is left can only be financial, monetary, and of a mercantile inspiration. God controls the king. But who controls the economy? That's the only point, for a civilized people to consider. A country shops. Then own the shops. A country trades. Then own the trade. An' the king is left his babbles of jeweled influences.

HOLSTON: I learned a while ago that armies only flutter. Now apply myself to more responsible delegations. An' to keep in tow some skills with skills an' skillet— Oh! I am rich become, by ordering people about.

SECRETARY (*loudly, but conspicuously to himself*): This one seems respectable.

ANCHER: Your craft is to accumulate, an' that wears your craftiness with power. But you know I too know how, an' others few, to build up gold like growing wheat, by knowing where the sun shines for expansions.

HOLSTON: We keep our interests perked— Though I hate investments. They are watery. And as the cask is filled, there are ever leaks from cracks. Own real businesses, true action-(mak)ing, and not the virtual for hypotheses to foam. I've done so well with my ex-regal jails. The confinements cost less than a tenth of the consignments purchased.

ANCHER: Only when the prisoners are not paupers, cop! But it is a clever ruse you've designed, to lead them out with payment, to a sanctuary away from the anger of his majesty. The irritations from both sides are reduced as they vacation themselves through a sentence. Though how you make the magistrates lean in to your.... designations, is a marvelous legislating of legalities.

HOLSTON: If a voice be heard, they call me not Holston, for want of Holston called. The haughty would cause much more trouble otherwise. My proffer is to support the best interests of all involved of a controversy of outspokenness or any manner thought to be outlandish. This avers the realm its tranquility may stay pronounced, bolstered by good judgment. Yet are we not all derelicts to some good nature, tease the talents to *se taire* to make this less evident. And the best way to do this is simply to be away. Indeed, the easiest thing of being is to be. If you wish to make a mistake, be error-prone. That will be most convincing. So many convicts relish their wrongdoing. And if you wish to be faulty, then fail. Though the intentions may not be overtly criminal, just nails manicured badly. I often wonder of our mankind, earl, what's made for jails an' resorting retreats. How leads this most contrary urge to be obstructionist— an' break the peace by doing so of character that elicits the royal yowl— is my picture's breadth of view, for certain kinds of head. More fathomable is illicit crime.

ANCHER: What for the cure? You simply police. Intrigues are boxed, an' political assaults lanced. But it is the motive of the eye, to see a spot of filth and detest it, whether it be ordinat'(ional).... or liturgical. For who is the prince among(st) scavengers, what's been left us to argue over an' about? We are all scandalous at heart, an' swindlers at birth. And there is no Holy See to reduce that faith— We rob from the sagacity of the finest ideas, to produce havoc and hardening of thoughts. Yet a military might is not the answer for *ourselves* to contain. It is wiser to restrain from that compulsion an' make our resurgences with decency, and undeniable supposition. The king is a brilliant fool— by birth! And what he says is already a crown of decrees that we scurry along as mice to transgress. But that is the game to the breaded cheese. And need us less appetite, at our levels arose, then more for the game— to defy! without a simpleton's expulsion. They are banal to be caught—

HOLSTON: Ha! Wear it like a badge of wound.

ANCHER: —and serve not the purpose of lessening his hands' rule. But loss of butter— Oh! That is potent. That is viciously observed, to undermine mistakes an' criticize administrations. This is the thoughtful way to usurp—

SECRETARY: And here's an entire engineering described.

ANCHER: —*Power!* an' bring the well closer to our troughs. Then pleads the majesty excuses, to relax his standing— as we rise!

HOLSTON: Hasn't felt much to lately.... Is as arrogant as ever for commands to issue, that some policy be this or that to follow, or merely named for his preference. Nears he not to sense— a reckoning with titans, in such proud and boastful behaviors. Shatters glass with 'is teeth, the wicked luster of his denticles, the dreaded fear of bare prominence to the oratorical sensation. And now declaring even numbered Sundays, in even numbered weeks of even numbered months, his dominion to preach from, laws divined by his stature alone for prayers made and hymnals wrote an' composed.

ANCHER: That's the better of his sins. They are innocuous— and poorly done. But he wishes to devise a new, expanded year with them, for a king's reign to calendar. So be we bow, on those days, some fright to withstand, that he should order doves to be eagles and inject worthless peers into high peerage of state functioning, dominant positions an' doting fidelities of obsequious enlistment, without respect at all for the better kin of country. Well, the more offices he has, the more offenses tolerable, as they dilute themselves of efficiencies into ceremonial significances. But claw meaner raptors for branches to perch at— an' vanquish toolies. His

own way is not so profound to except infiltration. Already he has anointed me an artificial privilege, to attend his court with silence for his words, numbers speaking more with argument of his desires, convinced he has my favor— And he does!... for duties that I own, to lend him more.... but for my own purposes to spare.

HOLSTON: You will be a delegate of imperial reach, but never an executive under his government.

ANCHER: That's a false hope anyway.... Yourself 's more an example of what can be done by being clever.

HOLSTON: Discretely squamous. We all need advocation of the law. How twisted keeps the metal ribbon intact still. I suture seditious tendencies, the evidence for which is by our sovereign's nose for sensing. I alert.... the incriminated, and propose our profitable means of appeasement, profitable for his head an' my hump.... of belly fat.

ANCHER: You're more fit than a Castor.

HOLSTON: Oh, but I feel the success, an' its growing weight. An' what could be, if this monarch should have me in-clandestinely chopping *off* heads?! That would be much more to swallow than some distasteful, disagreeable oil. And what if he could have me spying on our nobility? with threat of dismemberment for being false.... of the reconnaissance an' reporting. That would be a fine discrimination. (*to the* **secretary**) You better not be reading plots, down there.

SECRETARY: Oh, they are imaginative licenses only, creative crusts preformed for baking.... but illusory of outstanding merit without more help. The logic is in the details, most carefully constructed, reasoned with an eloquence that fades all detraction. But how to accomplish all these results is puzzling without much strife an' strength, backing and strident calls of support. This is descriptive of.... not so much the possible as the portamento of sometimes very curious logics. The visions are very capable, however, and demonstrate much determination to paint extraordinary satisfactions.

ANCHER: I've not failed yet, to extract something positive from the doable I've granted graces. Keener eyes may see the exaggerations more— than mine. But there is a weight to facts, a tonnage measurable. And one investigates a load that way, with scales of perfection. Ask yourself what is the ideal outcome, and then examine scrupulously to see how this is limited by the method presented you. Most minds leave such things to chance an' risk. I look for the golden guarantees, whether they're worth the effort or not. There must be something in all plans. That's not the point of planning. The sun's transit across the sky is always successful, even if unseen, even if you rise for the west an' bed to the east. But there are tensions to these.... solicitations. Some are more imperative than others. Most are often based on this peremptory drive much more so than an ability to devise a sensible way at the attempts. The grapes will ripen soon. That's fine, but are they vinous enough at that location of the mountainside for a reasonably efficient harvesting? or will lots of funding be wasted for a scarce result in usable quantity of wine, given practical and adjudicatingly predictable difficulties? Yet that particular grape remains terribly special to pursue. And so the imperative overrules much sensibility of thought, with unassailable fervor of assertion. Who might be right or wrong, with such delirium? No. The details mean less than the discerning. I do not buy for the persuasion, that's too actionable to folly. But a neatness in order, for presentation, has its place, like references to be aligned an' furrowed through. Seldom (are) surprises there.

HOLSTON: Ah! But the king's passion is to persuade. That is his particular gait. He feels he has the gift of conversation to his light, as it shines to reveal the truth of the walkways traveled along. And he waits for all contentions and resistances to ease against his proclivities of rule, or how he may manipulate opinions towards achievement of his goals. This empyrean fire is ruthless, I'll state for my friends. And he is not lacking of strategic cruelties. Now.... he thrives on surprises, sir. That invigorates his nerve to be outlandish and search for keen ironies to abuse or tickle or tease or tweak mercilessly until confidants fall apart against each other, fighting to escape a certain torture. There's some wisdom there, with his tenacity at play— and the turgid at bay. Only warnings can suffice to avoid a proud manner by him neither learned nor thought, since his nose senses this dearly as a stimulus for attacks.

ANCHER: The dagger's smile with gleam. He stabs in daylight. Wishes to be known for the acridity as an open tendency. Frightens demons and challengers alike. You can't confront such a personality with dramatic directness, he'd be more frank than your stone throws. And he has too many knives to be blunt with.... Reason he our discourse?—

HOLSTON: Of course he suspects! Why do you think I'm here? — To warn! I've come out of fairness of alerting that a scent's been laid— for this hound. He is no idiot, earl. Some profiteering he feels restricts his range. And to be honest about it, he suspects what some lords are up to. With legal measures plied, he may bite viciously—

ANCHER: Or peck at.

HOLSTON: —but through dishonest means he will lunge for a destruction, a thorough stripping of the flesh. Already thunder bandies the tongue.

ANCHER: You think he has my pretensions cornered?

HOLSTON: What he invites.... is the game. An' he has sent me specifically to see you, more or less. And when you see him on his grounds, he will be smiling— because an understanding has been suggested, or levied or tossed: Try to work with him to avoid much competition. He does not hold you to be grievously at fault— merely rich. Yet has he also maps an' pins, to scout at what trades you an' others may control.

ANCHER: All perfectly legitimate and sanctioned enterprises. The crown has not objected—

HOLSTON: And the royalty are not particularly upset by it, since they own— us. But let there not be conspiratorial vapors ascending. I've visited others with the same message. Too much ambition wastes one away, like the delirious you're wary of. Your waters pour through any holes met for craining, and may not necessarily resuscitate the fish of stream.

ANCHER: Then he views us as a pact?

HOLSTON: Calamities of wealth farsightedly inspected. And has

me to enlist, for a sarcasm to beam, that we confide of social shares well off. This can be handsomely maintained, along his guiding purview— kept! not to disavow his worth as highest always.

ANCHER: Such a head as his.... what do you think?

HOLSTON: I think he's bold enough to play with fools and bury thinkers. But there is a genuineness to his resourceful enthusiasm to make this move known. He offers some clear thinking about our positions, and that this relationship to him— remains unhindered. So with a sophisticated submission might we agree to stay bowed, that is his grin.

ANCHER: So as we are, Holston, unless the economies are to be plundered. He notes the trend, and tries to prevent the innocence of sheer gravity, that we can not remain idly powerless with what we may control. No fabrication of sincerity can avoid brutal beliefs in ourselves. (*standing, tossing the pen on the letters*) We are accosted of mind, for a consolidation of heart— not to be mistreated or neglected. Now only higher principles than his can change that reality—

SECRETARY: Or for what's real, this is unusual, ma lord. Way beyond the speculative to the incredulous. It's a letter from Lord Cantal.

ANCHER: Cantal? Had Lancet go off on a trip with him.

SECRETARY: Ay. An' by these means has Lancet done a foot's dance—

HOLSTON (*standing*): The man is deviously deranged of some sort of charismatic puppetry through thought gods.

ANCHER: Why be so anxious then, with ideas?

HOLSTON: Because they lead to a tyranny of quarrels, lord, insisting upon the most irregular and unfashionable of proofs, and upsetting traditions as to make one's self seem a fool up to now!

SECRETARY: Yes. And here's more for the tom to be pleased. For in gracious terms the lord says that Lancet has found a remarkable foodstuff that may change all manner of civilization— if it can be made marketable— an edible grass.

ANCHER: Grass?! You mean some herb, or flowered stalks or asparagus.

SECRETARY: No. He takes queer liberty to relate that these shoots are actual blades of grass, that he has tasted them, that they can easily be made (to be) delicious, and that they are an actual food source in the raw while growing as regularly as ordinary grass on a lawn. He says they look like the retarded— in growth— cousins of the normal or more conformable (to) of nature, but have an immediate savor that clearly makes them princely. (*as **Ancher** takes the letter to read himself*) This could make unusual the beast of man, cowed into grazing with a courage for this craving. Yet Lord Cantal is anything but a serious adventurer, and praises Lancet wildly for the discovery. Says he means to pay some profitable attention to the curious resource, and is investigating the feasibility of transcontinental germination by trying to grow some few seeds in a perpetual shade.

David L. Birdsall : Ergomont

HOLSTON: Transcontinental— Ha!

ANCHER (*while still reading*): This is certainly a find that perks the coffee. And on his knees he had some raw.... Is delaying his return, with some enchantment. So thus is Lancet due to stay awhile longer— with my permission, and possible benefaction to advance his salary.... or lend his services to a local lord's control— This is serious employing. (*takes letter from face*) Can this have actually been found, or have they gone partially purblind in the jungle?!

HOLSTON: The wilds, their wiles take many for a spin. But this is just a thing that Lancet could evoke. What harrowing decadence to provoke! And now produces he an edible mulch. How richly grown a turf will be, with men to barge on through it like the animal in leas!

ANCHER: What more extraordinary game, Holston! They've found something that fascinates. And now to ply my hand. Can this possibly draw to great advantage for me? And with a letter for a reply, I fume: You've found a worthy debate for my methodizing. (*admiring the letter*) Why.... this might be gold leaf!—

HOLSTON: What could the state do with edible grass?!

SECRETARY: It's what is done to it that merits the chew— of proposals to brew over and submerge one's thoughts into.

HOLSTON: No error for the king, to bring about some spectacle of debacle, this— excitation over means for nourishment. And thus a new fashion employed, for gaining grace with a personal graze of the land? Abominable habituation, earl.

ANCHER: No— See that they have found this in the wilds, within a jungle's atmospheres. But it need not be cultivated in that manner, to make man the more pastoral of animal. And would the king see high desire? Do young boys climb up trees for their apples?—

SECRETARY: Or coconuts!

ANCHER: —This could make for us a fortune of attainment while the monarch steadies upon blindness.... to such principles of gentrification, as broad fields become more valuable of human presence and astonishment— if this stuff does tastes good as raw. An amusement can be made of the practice. But treat we with more culinary talent what is at our disposal. A high cuisine, or the art to it, may be crafted, as the like to grow one's herbs to supplement a meal or its cooking. Yet, can one contain the view of an entire field of such supplementation? The enormity of the prospects startle.

HOLSTON: Awh, Ancher! I tell you— Lancet subtends the ridiculous by his own curious nature, to bring forth into view or consideration such oddities and absurdities. He'll lead you into much trouble, with these inclinations to reveal the outlandish. And your monies will be wasted more on spectacular spiritedness than on savory spices. But in no way will this become a country of human mulching for a strange munching. That urge will not overtake our noble populace— And you can not outrun sovereignty with such a ploy to consume our people's interests with this— green rice or wheat of fatuity of faddish flocking. Those climates they're at, they bring out the queer, weird of sensations to elicit the tongue even for earth itself, and salty rocks to crunch on, whatever minerals may be found to imagine a delicacy tasted, probably as much the

fungi as the fool utilized. They may have entirely corrupted their palates for awhile, and that grass might seem like sugar cane digested— Oh wait until the abdominal distresses that must ensue. Man has only one stomach for our cuds. But the tendency will never catch on here, that lessening of human means down into odd temptations of taste and touch. And I must have my broccoli finely chopped when cooked, or it makes me nauseous. Else raw, no matter how presented of size, I bite into a bunch— and it is delicious, especially when fresh and sweet. (*as the **secretary** collects the remaining letters on the floor, and stands*) So trumps a mystery with the cure of pleasing me, then do they similarly with some unusual recurring of habit—

ANCHER: Cantal notes the meal's not natively prepared, of the green growths. They prefer their gold-speckled fish—

HOLSTON: Well then whatever has brought them down to the terrain. Probably bitten mad by insects. It will not be accustomed here.

SECRETARY: That depends on how good a farina's made. But I assume your lordship finds interest in that brief account, of.... our foreign activities?

ANCHER (*folding up the letter*): Yes.... Though nothing's been confirmed to me, yet. I will reply directly, that Lancet has his leave. Those other letters, to the pit— For my mind's abuzz. And these on the table, well they're sufficient of chance. (*placing the folded letter on the table*)

HOLSTON: Some way to manage your affairs, earl. But this is more a recreation for your attentions.

ANCHER: So solemnly aligned am I to it, cop! that I shall gain a public's gut, for the responsibilities of various fulfillments to bestow or enhance. But my quivering self tells me something yearns to be efficient about this.

SECRETARY: Ask for tremendous samples to be sent. I'll check to see if any were processed out from your letter delivery.

HOLSTON: As well, there's no leverage over the king with this.... curiosity to pursue.

ANCHER: So tell him about it. He'll probably question whether mind-altering inducements are at play. But Cantal is a very solid and experienced man, to take this in any way seriously. And papers for reading are meant to be read, not rummaged through particulars accompanying. If I spent the time on examples sent, or schemes worked out through models, I would curate a museum as my primary occupation of time.

HOLSTON: So be it, they are way *under* time.... It is late, earl, to make usurping conjectures. He will never be dethroned, not in principle nor in practice. And you must always work to win his favor; for the exercised crumbs of your position, be they plentiful and wealthy.

ANCHER: Yes— Well, the cane's for the tea, sir. Behaviors are the most important part of being to control, or influence successfully.

HOLSTON: From your standards of perception. But this is loony wishfulness, to follow up on such a tease. But be it your rendition....

ANCHER: I'll explain myself to the court.

HOLSTON: Aye. Of being a loyal subject, this is a peculiar vain-glory. And search you for a thousand others similar— in vain. I will commend your free thinking, to study enthusiasms abroad, a giddiness to be excited and agitated by the outstanding— of news and events.

ANCHER: That is the risk of being profound with one's funds. And were it a fruit, then the exotic would be desired for some display and notoriety of flavor. But as a grass, it's simply a very common thing thought, the concept low an' common, an' for the commoner to please. Then be dazzled, if this holds up. Not even I expect that. But with some substance to the truth, we may corner markets— of notice and progression. And that is something not easily alterable of leadership and inspiration. With such a cacophony for drinks proper, there are these sensitivities of people— to legally manipulate. Whole nations have floundered.... with the rot of a single type of grain. An' contentedness.... is for being made with shame of.... certain prosperities.

HOLSTON: What?!— For the crown?—

SECRETARY (*exiting*): That may be, ma lord. And I will continue to peruse the— fallen, for your possible extrication and provisioning.

ANCHER: That might be what's firmly on the ground.

HOLSTON: Hmm. As you may provide, earl— produce.... or production.... (*starting to exit*) take to care for irritations warranted, in trying to promote a new social gist of action-ing or familiarity.

ANCHER: I'm only an investor, for some opportunities presented me, Holston, as for with to determine more girth of pronouncement — to be mine.

HOLSTON: Then a harry— good afternoon for you. We make our wealths.... within the system to be royal. (*has exited*)

ANCHER (*picking up the plume*): So! (*writing on the folded letter*) Reply! Supply! Replenish an' restore— of astonishment pronounced!

Scene III — **Bastrel**, *with hat in hand, stands in front of a very large chair of arms, regal but not well upholstered, waiting for the king. The lackey* **Fenic** *arrives.*

FENIC (*entering*): He contends, exchequer.

BASTREL: So delayed.

FENIC: But promotes you to wait, in this solicitous room, a space of time for your provisioning is his anchor. (*running up to the chair*)

BASTREL (*adjusting himself out of the way*): So much so.

FENIC: Ha, ha! Well you are his ballast for final arguments, the way you must adjust the funds for his.... developments of assumption. (*at the chair*) This could be a throne itself, were there some plush blue cushions to the seat, and perhaps encrusted gold trimming, with velvet drops of vertical linen or silk.

BASTREL: A bare ornateness.

FENIC: The vermilion veneer suits this wood. (*sitting in the chair*) I shall warm the seat for him. He hates it cold. (*crossing his legs*) High! And do you bring monies?

BASTREL: I don't ferry coins, Fenic. Why pursue this faking? You don't impress me much.

FENIC: But you impress, to impose so many conditions to his vault exchanges. If he wishes to go on a major sojourn, for an extended stay, then he must consult your knowledge of our state's finances— to see how affordable is time spent. And that is a very patient pondering, I would suggest.

BASTREL: I doubt if you could ponder much of pertinence, to know what to suggest. Why he allows your baleen baboon-ing about him has often given me dispute of thought.

FENIC (*squirming playfully*): Ah! Well I do tend to filter what's to be presented for any countenance of contention or inquiry. With what of my attention, I mirror his inspection of what should be seen or heard. And in short I am totally agreeable to his intentions — and conduce visitors to be likewise.

BASTREL: I'm hardly a visitor. I'm a court's consultant.

FENIC: But you bring to us news of deficiencies?

BASTREL: I bring to *you* nothing. But the world brings its deficiencies on its shoulders, as could Atlas raise the heavens totally invisible, and stars hang— or hang themselves for our views. Not to get in the way of facts, one can not gain impossibilities, one can not earn the unapproachable, and one can not feed on desires alone. Strip the treasury— it's not a reasonable plea.

FENIC (*with obvious fawning of anxiety*): Oh heaven's! The country ain't broke, is it?!

BASTREL: No. No jest to a continuous line of regular and requisite investment.

FENIC: Refilling for what is spent— or lent. I would go frenetic for my jewels, to force them borrowed.

BASTREL: You haven't any, yet what is collared—

FENIC: Though I believe we dine on surpluses.

BASTREL: The nation, perhaps. But that's strictly by definition of decree, that there will always be something left of a defense for her. Now I'll not barter with you concepts.

FENIC: But I'm so silly. It makes them more digestible— Conceive with me.... what could be wanted.

BASTREL: He wants me to officially sanction yet another little diversion. These things add up to totals lost. And I'm to say we have the funds. Well— yes, they're there, but not to be wasted.

FENIC: He *owns* them, does he not?

BASTREL: Mouth you responsibilities for him? They are ultimately the nation's bread. One can not so easily make rapine upon necessities.

FENIC: Are there not more worms in the soil? Then find me more soil!—

BASTREL: You are not he.... gentle grin.

FENIC: But I ape him, a monkey on a leash. And he would say: milk more, reap more, make more— Squeeze for juices more thoroughly. Find for our expenditures their expenses to placate.

BASTREL: Why do you abide yourself this way?

FENIC (*becoming semi-serious*): I present for you what would be his mood. Learn by this and adopt from me this gift of character.... (*relaxing*) We do not ever want to be disappointed. I myself only take caution to feel out what may be sensational. If we were in trouble I would frown. But if we are at leisure's quake I'm jovial to be pleased— and not a fool to panicking. Then do I hear his resonance to breathe with, and reverence to be with. That way I'm never alone, and never myself disowned.

BASTREL: Even to sleep, or for ablutions and toiletries? There is the intra-personal devoured—

FENIC: We are each a part of his.... body or embodiment. And I think not of others to become, as when I smile to greet you. As with any other day, then, it's of my own advantage meeting that yet we again find ourselves of a grand organizing principle embellished with some individual charms. How else can you explain the concurrence of day and life! For if not life, there is no day to bring — And he is that for all of us. Once I was brutally punished, on his command, until enjoyment released my tears— and after months and years of torment, not being his ingrown recession of tail and sense.

BASTREL: Perhaps overfed of peaches and cream, and tart sugars, whatever makes a bellyache of privileges. But you've recovered to what? And how would make a nation its populace so abused? There is a hidden condescension to you—

FENIC: Hidden? Why, I'll take off my shirt and pants! (*mocking to disrobe*) But that is only flesh underneath. Who could have devised that.... fragility of skin? And such scratches, I am torn to tatters— in rich clothes. Now let us be apart alike. What are his insistences to tailor to? We both work and struggle and contort— to be approved of. Why be defensive about that, when it must be enjoyable?

BASTREL: Because mistakenness is cruel to you, eventually.

FENIC: I am paid with a good hind.... ness, and an absence of fear to threats deserved. Shouldn't a person be molded this way? Shouldn't he share such a courage with others?

BASTREL: So it's brave to be a bevel cut—

FENIC: Completely heroic. Put yourself in that manner. Like manner like me!—

BASTREL: I can only make report on what we have, what is available to use, and what would be prudent of restrictions against excesses of indulgence at this time—

FENIC: But time is my mantle, like a slop of cosmetic. For if I'm around, he is soothed— of developments, and spared false frankness.

BASTREL: Not everyone can treat himself that way. Some must find the time to be faithful—

FENIC: And specious!

BASTREL: —to circumstance and contemptuous realities. Some banks bend of evacuative pressures, debts incurred by he who must repay—

FENIC: False floating! It is we and one, an entity that flourishes with sound riding of the float. Surpluses are at hand— and we can make some more.

BASTREL: He—

FENIC: With a twitch of the finger—

BASTREL: Our country can not be so easily gendered.... as that which moves by such malicious gesturing—

FENIC: Malicious?!

BASTREL: —to befuddle honest empathies with mendacious promises. He can not simply erase what some must live on, and depend through time with, and are suffering now from lack of. This has to be dealt with, that things outside— *our* court— are not as harmonious as your chain. And you might apply yourself to some ingenuousness about this, so that you may mirror grave concerns to him— without offending your leash nor his tugs. If he could rely on such an imposter-ing of faithfulness, you may be redeemable for some remedy of awareness. The state is taut of resources and recoveries from slack of just proportioning. What is taken from one region, to fill another, leaves dangerous holes and onerous demands of relief from despicable desperations. Tell you he of rioting? if such of mirrored concerns he would inform you of. Then tell you him of revolt! if this continues of any placid extent till tears the fabric of our tranquilities— And I warn you.... this is a real light seen, of bitter complaining.

FENIC: So much you've actually noticed? or only heard about, many mouths removed from the melodious quarrels and yells. Inform me.

BASTREL: I say there's some unrest. We can't keep shifting commodities from one place to another without fulfilling promised recompenses. We can't continue to let values fall to an even level so far below their actual worths, such as to make the debts seem nonexistent or dismissible. And we should not be scooping out the bare necessities of many to stock up the fat of a few. Farmers must live off their grains, instead of shoring up reserves used like indentured bargaining against growing expenses that time enlivens. And while the treasury seems solid of bulk, with accounting, those descriptions depend on such myriad adjustments as to drain a lake into a puddle of mud, should all services be paid for with its drink. So the king owns what is owed and can not be hidden as easily as in the past with the various transfers he employs— for me to conduct, or figure out. Starvation follows, as a bomb of blight, if this is done inconsiderately and without finesse, and will definitely occur if our mismanagements continue even short-term. You might gaggle him upon your face perplexity! not to realize the growing gravity— while I can only say.... here's a sum that may be taken, for your leisure exchanges to ply with. Yet its replenishment will be difficult and only through a long duration of fiscal stressing accomplished. Some pockets of the country have been broken, Fenic.... and are poverty-stricken towards accelerating deterioration. They are emptied of all but a spirited anger.

FENIC: Enterprise will take up the cause, if the soil remains fertile, and the minds manacled to self-preserving pursuits.

BASTREL: Enterprise will have to pay for it, servile foister, to make in any way a solution to these complex dealings and transfers. For that pulsation may represent the only blood left to the fluidity of our monetary veins and arteries—initiatives of greed and profit!... from a risk of being sure to apply one's seeded resources and burgeoning beasts of bountifulness, or bounty from these daring deeds of commercialism.

FENIC: Our sovereign head contains the confusions which disturb you, and complies of them to capture this like of feeding from. For the such exude prosperity, and the manner of all tactic is how to contain them to a reasonable gain for everyone— of country.

BASTREL: The point is, that should not be relied upon, so precariously are we perched, as the only means available of choice. From off the ledge, or atop a wall or fence, the winged may fly— but not the wended for a fall. And topple giants easily, intoxicated too thoroughly from their powers.

FENIC: What more consulting—?

BASTREL: You are no advisor!

FENIC: —Is the money there, for his *perchances* to permit, that our king handles these affairs more adroitly than your worrying? I will cow to couch myself before his bastion of gracious petting, to yelp with bark of fear, a whimper at any error perceived— in me.... not to alarm ourselves. Yet liberal is his grooming grope, of the fur. His alertness is profound, soon to be legendary. And I abide his lick!— and nip. But now is warm, warned. He manipulates the outstanding— tail, and seizes scents, for sense about our doing.

BASTREL: I think recesses are in order; but these funds are provided, with their enthalpies—

MYTDRA (*entering, with* **Holston**): Grass?! How makes a lawn of it, with skin of bone for teeth?

BASTREL: Your majesty!

MYTDRA: Get up from there, you jackal!

FENIC (*jumping out the chair, but deferentially*): Your hiiighness!

HOLSTON: Does bend the leg, I think, is the most proficient way. For I imagine much knee employed.

MYTDRA (*approaching the chair, but acknowledging* **Bastrel**): And sod for sauerkraut?! But green! green! the nub of pate, as like a child's embrace for heads to be acquainted. And so does Bastrel make, more matters more, to eat of grass?

BASTREL: Your majesty?

MYTDRA (*sitting*): He says that Ancher's found some such lawn vegetable consumable by humans. I might invest of this knowledge. We're on friendly terms, you know.

BASTREL: That might be admirable. But what's to cost?

MYTDRA: Then what's to spend?

BASTREL: Reserves are plentiful, your majesty. Still, caution is at play with mysteries.

HOLSTON: It's all only been confirmed through correspondences, sir, which means there's shallow proof.

MYTDRA: Yet would he corner an ambitious pâté-ing.

HOLSTON: He's sent out an expedition to the exotic wilds, just to find the unusual that may be marketable, as my friend is an aggressive player in these types of adventuring to foster. But this brings far above a savoring to admit— or remit— if true. Man is not meant for such peculiar gatherings, as could a grazer emulate.

MYTDRA: Yet grass is grass, is grown.... broadly and rich, and with simplicity of any occurrence. One can hardly stop it for their weeds. A valley of the edible is a pungent thought, like a cover of irrepressible money over fashionable soils. And this is an interesting direction to drive funds towards, or (is) at least imaginative. But then as compensation for some losses, a district might assume this with much welcoming, and become expert with its ease.... of production, cultivation, and waywardness of mouths. Yet one can always imagine the imaginary to be true. Corn is such a pain to trample through. The sweet maize as ripened tastes delicious of its ears. How so a shoot of grass, a blazing blade of verdancy? The idea is tantalizing— to control. I shall invite the earl to explain it more. Yet comes a twinge to pay attention with, this twitch, for what might be clever as well as expansive. But now to adjust resources. Have you any doubts of lending.... to such enterprises?

BASTREL: It is unfathomable, the number of routes, to lose one's income. Yet, if not income, then dire are the losses made. This is what's saved, stored up. Throwing currency into voids is a foolish propensity of thought. But as established, there is much wealth to be exercised for good endeavoring and regal wisdoms. Everything is in preparation for your handling and apportionment to good causes and adjudged careering.

MYTDRA: I spare not expenses. But do the holes balance the hills?

BASTREL: There is such transition to this, that the motion makes difficult the contrasts to observe— predictingly, if not correctly. The state has not arrears but of awareness done to.... precaution of the shifts to indulgence.

FENIC: As much a bed spread spreads upon the indentations and the wooly puffs, it is a good measure of both and yet can conceal all depending on its tautness, tie down, and tenacity of cover.

MYTDRA: Why, there could be rocks under it, or dirt!... So it's how I make the bed, of course. Though handsomely as plain a resting spot may be— inviting, the country seems as well settled, even and level to my rule. But are there disputations to announce, that you are anxious to receive my presence today?

BASTREL: Well.... holes complain, your majesty. They have been promised some filling, some recovery— of funds, at least, for their depletions to have caused. And there is some debate a god is capable to restore them or deplore them so thoroughly, those singed of their depressions. I say the equilibrium is very bothersome at this time, may be easily disturbed with the slightest force into a maelstrom of discontentments, and that all new avenues for the royal spending— or ingratiating dispersions of funds— must be very carefully considered, if at all conducted currently.

MYTDRA: You advocate a lull for me, because you can't count stones for pockmarks? I want to take a regal voyage, to the nether regions if need be, and enhance my realm with an adjunct notoriety of possession or celebrity or purpose. It can be to distant places, yes. But I've learned that such locations pay well to be owned, to be clothed of the royal red banners and loyal blue ribbons, and that what is cheaply obtained may transform into inestimable value and priceless privilege. For what is power? It is nothing never felt, and everything when made conscious of.

HOLSTON: You yourself would do the traveling, King Mytdra? Such prospecting may be better left to your adventurous subjects, their excursions made under your name or appointment.

MYTDRA: Grab at what grovels, Holston. My reign has been invited to visit many undeveloped lands longing for a fantastic subjugation that induces only pride to be a treasured popularity. But there are practical standards that must be maintained under my stewardship, which allows some freedom of discovery and exploration by able heads, enabled through their organizational skills. And as such, my honest and proponent control over their activities are necessarily restricted towards what achievements they may claim or virtuously win. But I myself am the highest exponent of such endeavoring to leave all results indisputable of court courting and consideration. For where I step is as well to be defined my footprint placed. The most exemplary example is made, and can only be challenged with proofs overwhelming of my powers. But where may that come from? Of wily contestation is my motive— I will demonstrate with welcoming of my seizures the most personable of embrace, before much competitiveness may take place to enforce a founding acknowledgment. This is a way to secure the most important treasures of the world, a direct grafting of the earth to your prominence— with auspiciously chosen attempts too slack of failure to disavow.

FENIC: Makes for a grand half-sphere at least, procuring fame, the hemi- of Mytdra's domain, withstood with cordiality of armored approach. Then does a rapprochement evolve of worthiness to be possessed, and jollity pervades the notion of an inextinguishable flaming for lands obtaining— Happiness provided for the indigenously born, so brings a weight of the moon, to lift the flaccid and disorganized tides towards her lunar lucre and arrangement.

For where fluidity does make a wall, sometimes unbreachable.... that is with much strength endowed.

MYTDRA: That is the brow of prominence to promote, and advantages to mate of and ensconce for greater might. Yet does the drive have reasonableness— to fill out our holes! and be helped by those who indoctrinatedly can help themselves. The timing is ripe for a particular try. And are our balances in order, that a surplus may be spent?!

BASTREL: ....Such a personal investment of your labor.... is permissibly financed. Or affianced to such a courage, I can not dispel the possible effects of these funds. Ambition profits here, with what is usable.

MYTDRA: And I have calculated all in my mind, for what is usable. One must promote to oneself a certain insensitivity to a temporary suffering of others, as opportunities hock the horizon for appearance, to play with your daring to improve the lot of all under guidance. And there is subtle convulsion for giving various controls within my realm. I am astute to many complicated sentiments — But I have the overall leverages, to stretch the strings out of a confused.... or confessed, snarl and tangle of resentments. The capable must stay as permitted for our healths. Yet I will win the veering as a majesty enchanted. (standing) For the nation condenses to confide in my thoughts. I am original, and for the betterment of all of my subjects— Assure me I have the monetary scope!... in this. I can play with lords to a feast of fluencies in flight of fights! but not so easily of loans looming of repayment, laudable lending of my own laws leafed.

FENIC: It is accountable, your license made for all noteworthy acts. We are the supposition to provide for you the most energetic support. And how does craft a treasury to your entrusting, it is all for your tamed tenacity.

BASTREL: ....In taming thus, the monies 're there, as would be coffins overflowing of their (quick)silvers. How can your rank be harmed, if you are of the unsurmountable row? Then my plea is pure, for some success to return, on checking of the funds. Largesses only dwindle within lenient lures.

MYTDRA: Ah. I won't be trapped so easily. All gather for my aims, these ideas. I gather them all for consideration very carefully, and observe hospitably most enactors. Say does Holston do me much fine service to analyze the lords, or describe the current tendencies. There'll be less upheaval than in the past, dissipating heats of action upon my proposals of rule. Complaints surface to the boil of cauldron as spiked bubbles, my face to bare (to) the surface of these dreams. And where there is unrest, does mirror me delays of reflection— I shall clamp down on the energies; I shall reduce the turbulent enlistments of discontented churning, and that most violent fear of usurping methods lost, revoltions urged, revolutions spurred. What is more disagreeable 's to be dissolved thoroughly, within the sundry mix of proud instigations. So as to police well is better than a broad ladle's purge of gaseous vapors that might pop! to spillage and foul waste. It is a commensurate idea to have the brawny observe the brawn. Lures are luck, towards these achievements. Trust more in planning.... (looking around at each of them, and then finally almost yelling) threats! (normal voice) Capitalize on ambitions, and bring onto your servants their surrendered loyalties by fact of being personable to your purposes and intentions— beneficent. That is a definition of the subject.

FENIC: Raw roar to believe in. We consume the sincerity of this to accept—

MYTDRA: Good mongrel.

FENIC: —or by a restlessness to chide. We are the stationary to produce a thought.

MYTDRA: Solid ground does make a virtue standing, confiding in my faith. Perhaps you're mixed with angels' wings— or their appendages, but so close to this king remaining. There's no abuse of favoritism possible.

FENIC: We have for better to perceive your winning all contentions. Yet err, with haste of speed, does frighten minds, if the treasury, and its commitments, are as sensitive as the exchequer believes, a funneling of funds confusing.

MYTDRA: Bastrel has good boldness to conceive of it.

BASTREL: I will do my best with any matter placed before my handling to adjust.

FENIC: And with some suffering, more justification?

HOLSTON: The nobility are very cautious to prevent the king's ire. They devote themselves to what is passable and possible under his commands of aggressiveness allowed. But as for the commoner, I know not of them other than as beasts for jailing when aroused. And there is no doubt that not a few are bitter of their conditions and grow of spite. These numbers are confined to concentrations stern, but can erupt as anything unpredictable. And I have explained the petty lawlessness, so that a surge of that activity might be used to foretell a contaminating burst of friction and factioning throughout the realm. Yes, it's of a definite simmer, at some locations, but dispersed with an uncomfortable regularity of distance— that as you go so far you're bound to find some considerable dissension.... My monitors have become agitated, some alarmist. But as of yet.... we sense no lords to want to take advantage of the discord. That is not the planning, for the populace to control. Though it seems an obvious thing to attempt, for the disloyal.

MYTDRA: I must pay the kettle— salts. I'm quite aware of the growing patches of poverty, like some subtle crystallization due to droughts or parching. This is strange, though, since the country on the whole has never been healthier and more in league with advancements in almost every important area: civil, commercial, military, and regal might. So how to judge this stying of the few? I've drained too much of resource from some counties, to enhance the growing prospects of more prosperous ones that boost and burgeon the overall economy and welfare of the nation. Thus do I seem to leave, in some places, a type of permanence for indigent stagnation— which doesn't disappear and cries more loudly for retaliation as oft I bade reports on civil conduct and affair-ing. So to this problem I remit— return of wealth.... without discovering the true cause of these implacable results? Changing a fabric through some alkaline washes may destroy its chemistry as a cloth. And so holes are born of banditry preferred.

HOLSTON: We deal with all of that, and it's not overwhelming yet—

MYTDRA: But I am agitated to this trend, if it is precipitous of character, weeds to lawn— No! surviving of the difficulties inherited of a starvation that was not so intentional. I do assume more earnings for some will improve the lot for all, if all are clearly under my purview of ingratiation. Yet this dogmatic tie may be twisted wrongly. So social pressures must be alleviated cautiously, and without foolishness or absurdity. Bring me evidence of hatreds to be assuaged, by lessening of the harrow's scrape only.

FENIC: It is befit your knowledge satisfying—

HOLSTON: I'll draw on the criminal sufficient to be tested as not maniacal or inflamed, but simply dubious of relief or caring for his plight. As to our lords, they play boisterously and with a happy flight of fortunes, that their discontentments are with the manipulation of man rather than of their engendered manliness to promote. For this is an exciting time to be.... able at anything. And the atmosphere of our governance persists of this gay vivacity for the capable.

MYTDRA: It is much how I assume things. (*again looking around to each*) Yes?!

HOLSTON: They feel very virtuosic indeed, King Mytdra.

BASTREL: As can all officials under your most permissive control and accounting, your majesty. You provide much room of action to allow granting of your wishes with the most practical of means to make happen. You interfere not at all with my methods of adjusting funds for your ideals to eventuate.

MYTDRA: I consider what you do with every strategy to provide as my own; and only after this do I give you the order to effect it, or try. Not I to hand over to a blindness of production, with wishful thinking or praying that something is achieved. I have seen already what's most likely to be done. But to do takes the skills— of yourselves, which is why you are all valuable to my crown, and welcomed to my presence. So bring me drones of scarcity (of fluid thought and industrious merit), I see your faces still. You will not be idle for my causes and challenges to conquer. And this is more than appreciated: It is approved.

FENIC: And so does write the rules with the shake of a finger, as could the air be read of implications, pages of commands felt deeply through our reverence for the blessed of powers; this is a happy time to be perceptible by honor—

MYTDRA (*shaking the head and a finger*): Yeees! you lucky chisels of the wall, of regal contemplations. You know what I want, being intelligent enough to understand what I say and plead for. And I know what you may do, for such a demonstration makes more commentary than claims of talent.... Now then, I have your assurance some— free funds are available for adventuring.

BASTREL: Absolutely, your majesty. (*as **Mytdra** starts to exit*) The prudence is all with your sphere of commands.

MYTDRA: That would be angered with obstructions unbalancing. Bring me proofs of true dissent, to shatter the uneasiness of thoughts, that such a thing may be warranted of country strifes. I go (now) to councilors aristocratic.... of the weaponry and artillery of purchased strengths. That seems so needed of a country. Yet for these minor leaders leaf, more fortune for an earning tree, commands so much dispassionate of cost. O for the servile of note, demeanors decisive and demanding, mark my approach for figs on a tree. So tastes the snares of success with sustenance.... But it is a considerable representation for our soldiering. (*Exits.*)

BASTREL: There will be such contortion as to bend byways for the slough of monies he'll request, or require as found, such as may banks stow unaccounted.

HOLSTON: What?! Say you less than admitted?

BASTREL: There is no permanence of sum, constabular constituent, that is constantly in use and rapidly employed. What I say as now means little for tomorrow, and less or more the day after— It all depends on the undulations of expenses. But what I find as saved is hardly secure when whims command such attention to the calculus.

HOLSTON: Complain of your pride, then, to accost the situation faithfully, for wont of business doing.

BASTREL: How can I be up front to him for my endless doubts and anxieties?! It is my role to paint over the panic that I count, and contain its fiercesome fee silently— arrears disallowed, charged to future receipts that must be projected. The calculations absolve deceit, but (that) makes not a slanderous tongue before the king, that financial ruin may approach his tenacities, as long as they may run on debts. And why should I obstruct abruptly other court counsel?

HOLSTON: Because you know the truth of funds—

BASTREL: I know the truth of country, but that is no cause to defame the fact that *he* is the state and assumes all of these problems. More for my ache of conscience am I torn to ponder.

FENIC: That does he well recognize, with a measure of time to forestall. For the distance of a run is the measure of its feet. So be then his enlistment of such talent. What fails to hold, like the law breaker, his water carried, will be more for ruin than his funding. You'll find a way, with figuring, to keep proceeding. Grace of granting makes more gratitude receiving.

HOLSTON: Oft given. They are crushed, Fenic, when caught. I make certain of it. For to leave the criminal standing 'pon 'is crime spreads disgrace to a multitude made feverish, and induces similarities of behavior that may become uncontrollable, unassailable, and tolerate to a section of society, the run-down an' downtrodden. No lack of severity is due them, at that level, for their torts committed, even the petty offenses— which only lead to an accumulation of great terror, if not the trend broken with a stab to chest and blow to back. Headless— thinking— are their worth to me, wrought of these evils; since, in those localities, civility's the more imperative to achieve. In richer places there's more play allowed.

FENIC: There's not much excuse for hunger—

BASTREL: Well that's a military mindset, Holston, for falling into one's rank with an appropriate comport to sustain. But more wild are the wealth'(y) or credited, I can assure you, than any lowly thief or crook or brutish assaulter or barroom brawler or molester of the enfeebled or helpless. Those backed with riches have a cour-

age to be weird, their errant ways overcome with payments to the offended or damaged. That is what I call "supine crime." For the heady contemplate their actions (while) on their backs— that is, through dreaming wildly as if above the fray of common human struggles an' complexions of conjecturing what should be possible, to face or pursue, provide— produce— propound. Only a monarch should be so endowed of this pleasure. And then it is not crime.

HOLSTON: So as you might constrict the flexibility of lords, for their foolishnesses to stem, that is a problem for their peers to delve at. When recompense is possible from misbehaviors to attend — and even in some murders this may be assessed— then they are given relief from these gravities through legal admonishment and public shame. But the more ratty culprit must be thoroughly strained an' strangled, never able to— nor ever even imperiled with implication to— make any just recovery or restitution for 'is crime. That may very well be the fair demarcation between privileges within the social orders of a sustainable civilization. I have no qualms at all to bust the angry heads of street criminals an' rural rouges, robbers an' ruffians. Their battered skulls please my mighty maul.

FENIC: It's to be free to educate with such examples the reality of monsters. Yet, adjudication's still deserved for any one accused.

HOLSTON: Squash the croaks of frogs! You are not busy enough, sir, to deal with daily labors of the disreputable. You only serve our king, and not society on the whole, as do I and the Exchequer—

FENIC: Not true! frenetic flic(k).

HOLSTON: —Find you some impish mock pulling at your tail along a road or route pastoral, you'll call for my stick stun strokes with wails and pleas for immediate assistance and arrest of the perpetrating mucker, muck of mugger—

FENIC: I travel carefully outside, by walks, to assimilate the beauty of fine country life, and do not bring to assembly for weaknesses to display and distresses to possess. But as I am well weaponed— with calm perusals— stares are bounteous for passing peacefully and with common ease, that I contain the common man's predicate of courtly grace and grandeur for the human kind not to attract denizens of base deviousness. Yet serve we both of a perfume pronounced.... of noble errand-(mak)ing an' urges to propose for our addressing of the servile kind. Then clean our faces with the dew of such panache, does term to make our regal stenches sthenic as august aids— to any for approach. And capped am I proudly, for devouring observances and any sort of stalking. This is not a fellow to do battle with, so onerously of plume, pluck, an' bluff may I stroll throughout these environs. And of transits broad have I struck alone, my presence made for travel, my attire resplendent, and my nerve.... commensurate to the fun— of a lion's adventure, venturing through the complacencies of everyday events and normal attitudes to greet. So care-take more your— affections to bash.... than my prowess at home. Only the king is above my injury.

HOLSTON: That may be so.... of frailties tethered to, how to account for influential natures within a magnificence purchased with your life. For you have the king's kick. But that's no more an appraisal to be envied, to some justice found with station pleased by an' pandered to. Above suspicion there are none but its peak, and even you may be submerged with its worries. So dote for time, and the king's credit.

BASTREL: May our surpluses surround us—

HOLSTON: Worry more, Fenic.

BASTREL: —through our comprehensions of a national plight of instabilities to be prepared to balance through. I mean not great calamity, but a casual calm during much suffering of thought that our majesty might try.... unusual measures to attain a satisfaction generalized throughout the realm, and make conditions temperate of temper, harsh feeling assuaged, and hungers hulled or hewed at.

HOLSTON: Lawlessness will not breach the worthy ambitions of some princes throughout my command of the civil obedience our king applies. The lesser whet, the harder, stoned; but all will come within the fold, how foams Mytdra, be it with madness.... or gentility of sloven waste and looseness. Of course there are pretenders and pretensions 'gainst his government. That's only a healthy state of surging current an' ambitions— which he welcomes as a ruling challenge to stay firm and alert, and to play the game decreed. But let there be no doubt to the extent of this license. Debacles debase all o'(f) Heaven. Extract what you may leading up to them, but not beyond— not with such greed of turmoil and turbulence of condition. The genesis of wealth is a depletion from others'; and man's common assembly is to ascribe this evolution, how be it accomplished, through acceptable placations and accommodations of peaceful transference—

BASTREL: That is to say without complaint of stealing, but permittance for being stolen.

HOLSTON: Yea, or fear the heel's full breadth.

FENIC: It is a dangerous game, this kicking.

HOLSTON: My friends know how to apply themselves to it, for some advantages to whisk out of.... minor disruptions or confusions, disturbances of ordinary transactions— Not more, though, than what is legal, statutory, and established of regular practice, with the cleverness of forward thinking minds.

FENIC: Pardon their escape from the drudgery of stagnating circumstances.

HOLSTON: We all work to climb, monkey—

BASTREL: As birds along a rope, towards some light that others darkness.

HOLSTON: Monkey into trees to nest.... Birds on the ground to roost. Some strange perfection makes us sanguine for our drives and testing of spirit.

BASTREL: I demonstrated this to myself, with a budgerigar pet, as a boy—

FENIC: There makes a temporal petering out, of causes more or less inherited, to subsume.

Act II

Scene I — *A scruffy town square, with a stone or boulder bench, on which a middle aged woman sits with a basket. The area looks wind blown and fairly deserted, of the evening's approach. And she is of a destitute demeanor.*

DORIA: What of to be? No clamp of tendering is this, a poor, poor frost. But I am lent to beg, with all reserves found of exhausted weight. Then as I make to plead (*lifting up the basket for reception, though no one is around her*), bid for some fortune's grant, that there is need here. (*lowers the basket*) Not forthwith to come, yet can they tell my meaning, with any slight sight or glance, by these motions, mechanically distraught. Or to my chain, it works for bits, some to receive with diligence of this labor, hours earning little, yet hard on this stone lay. For of my dependence, what can I say more to do? What suffices in the morning for the smaller mouths is gone. And to this I am left no alternative but to come, avoiding their gasps of suffering malnutrition, as I try to make some restoration.... for the next morn. (*lifts up the basket again*) There is no manner left, to which I've come by for support. Our community is devastated. And what serves for self can hardly speak. (*lowers the basket*) We've tried most frantically, to stave off such despair of life. But starvation leaves more rich a marrow, for our boniness to assume. Yet still amounts some simple gratuity with effort to be seen, as could a long day bring few rovers left, lesser than our peasantry becoming; though staying hard to help with shares that dwindle.... dwindle, dwindle.... to make the contrariness of fate deniable a while longer. Stiffens the decline, as the back bends. For what is nothing of amendment, more to have what's bare, then past a desperation are we culled. Only to ask with silent moans for what might be a modicum of one's treasure. The method's seemly found, in this society. Not more seems possible to do, at this time. (*raises the basket, as a villager comes over*) It makes for caution to be torn to timorous address.... of anyone.

PALE (*coming up to her*): I feared for death last night, Doria, the mean morsel that I ate, of meaty grub. What I found, so little to be stored, in an emptying closet. (*dropping a piece of bread in the basket*) Yet so I felt aroused with fright that the tiny bit would stab within my throat, as could my work turn much as shallow, and narrowing ever. (*She lowers the basket.*) Then doubt I can keep up provisions. But there's some bread left, not to sham my guilt for those who haven't anything left. I come to that position soon enough.

DORIA: It is a stale gem to be watered, stowed until tomorrow for a ravishment. Almost enough for crying.

PALE: I had avoided the boldness of your entreaty for so long, to dare conditions had worsened thus. But all too regular you have become, as we become. And there'll be more to share the stone, more ladies weeping of the droop (of strength) with only courage to hold back wasted tears; more children crying for their mouths befuddled of such purposes carrying, that the tongue stings of ingratitude; and more men bewildered of this desertion, the deserting of our prosperities and drought of attainments. I can not sell to anyone my labors here, for none have much to pay, trade, nor barter with. Who needs a broom or household fixture made? whereat a luxury nears death-strokes, as people recline in their homes dissatisfied to rise through any of our dispositions found, and wish more that depression brings them to a permanence of sleep to muster an insensitivity to their fright and bedfast panic. And yet my work is said (to be) too crude, by merchants of the better townships, for any store's approval. And I am laughed at, by family heads of higher housing and estates, for what I try to wend them towards my craft. So I am stuck here, with diminishment or diminution of future, and can only count the days with my diminishment or diminution of worth into the uncertainty of oblivion, or its certainty of aimless grief. For such is an encasement hard to recover from, a shell of emptiness and loss contained by.... But mine is for the scare, while yours is of the bearing. Times have been bad before, but this becomes too long of a predicament to terrorize, too great a stretch of period to punish with. With more certainty the uncertainty is lost, that we decease of effective mortar for a life sustaining village, and are thrown to throes of a lingered decimation of societal girt'(h) past embarrassment. So can I approve to give you what I've scraped. It does seem poached to have a stock when some are suffering. But it's more fear to contend with that approach for more of others and the same, as self.

DORIA: There is the listless warning that there's nothing more to do. I've come to this end of mendicancy as a last protection. And beyond a sorrow felt, the hopeless purge of dignity, it does seem strange that we're arranged this way as other regions blossom of their fruits of labor, their handsome recruitment of glorious pride in themselves and the accolades received for being healthy of the scandal we embrace. I'm but an early face, or picture of, deserted by a man's death, who withered of convulsions through loss an' lack of livelihood and tried with any means available of trust in character to maintain the succulence of our family's berth, and sustain the urge to try until he was beaten down into the disgrace of failure, attempting the impossible which killed his talents and muscles' strengths, and life itself. He worked himself to death, ever complaining of the pains that brought in little, less, and nothing, as could a husband count from gains with useless exertion. His market was displaced, removed deliberately from this locale, and shoved into another, remote of caring yet promoted with enhancement, that the scuttle bark be processed there for its rich tannins an' dyes, the yew more plentiful of coniferous array. And as for holes produced, our services erased, nearly refused.... from this location. He stripped the trees so carefully, as had been taught to a mastery to keep them growing, and proud of this ability owned. But yet, with arduous fret, he struggled of the imposed handicap of his more expensive efforts, taxed to dissuade from selling of his careful production, and yet prohibited of competition with the more regularly available goods, (to be) suddenly restricted here for any commerce to obtain— by a god-awful ordinance of ruling intervention to isolate our profiting, so that other regions may burst of fatty earnings, their former ease at this now exploded 'gainst our shrivel an' destruction. He could not pair the specialty with our provisioning, as pared his reward more an' more, with greater an' greater endeavoring of desperation. And his hands bled of wrongfulness to feel, as he knew not else how to survive for any prospering of kin. So less could he bring us to live on, then more the shame enlivened his recoiling fright, that does adduce the terror with his fight for stubborn instances of trying.... more than he could. By accident or lent pardoning he died, as by a yew exhausted, a grasp of bark cut off his only defense for caring of this worldly manner. It would never be sold, a sample borne of cruel contempt, now gathered with some others rusting of a rot in a once prized strongbox. So does make (for) us none of it to use— he choked on wood, and chewed away the chain.... I am disabled to do as much, the disappointments of a mulch leven the disgrace. Yet, simple ideas are easily strangled, and I can only adjust to these circumstances unac-

complished and barren, so fall the poor provoked into this misery. I know not what delays the eventual garroting, and am left this manner after trying to subsist on what was left or could be found or sold for.... a nourishment of hope.... too long, too long. It is an abysmal perplexity of being fazed; and I only have the grief of my two small brood, as I bland the face for this daily chore, what's left me to attempt.

PALE: It is with scurrilous haste. We must collect ourselves, to be protected by. But is this fashionable a threat— to our governors?... Is it passable to produce such condensation to battle common strife, when commanded of our households to impose more concern and attention? We are pressured to obey the whims an' whorls of providence as individualized entities under divine rule. So much for justice to this. My wife trembles.... and curses you with fear, for having so become as could us all. It is difficult to crop the major doles under this current climate of anxieties and apprehensions.... But I see the civilities as not so alarming as our guilts, of self-preservation.... and perhaps a growing defiance of these conditions totally unnatural but to high error. (*leaving*) So send your thoughts towards common wisdoms, Doria.... as my throat does gulp of painful knots, for this constriction placed upon us, a fearful resistance inevitably to enlist and serve ourselves more for relief.... from this heinous neglect.

DORIA: ....So binds the platitudes, for sharing of a community its cautious griefs. (*lifting the basket*) I would not dare persist, except that I may try.... to tackle these responsibilities. Would rather weep (*lowers*) and scorn myself of this material debilitation to trod on able limbs and educable mind. Yet can this way of living be fabricated for our demise, it is a conditioning to be ineffectual when brought down to an inequitable meanness and lowly surcease. O had I more thoughts of ability, I would rise!... But, this is real. This is no fantasy of demeaning. This is the hard stane of a terrible blow. And I must be as complacent of its stun.... as of my methods to recover from all hardships, as brief the role may be allowed my succored pant. This is less than wrong, an' more than right— to do. (*raises the basket, as **Fenic** comes up to her*) So issues a salvation.

FENIC: Lower the port, of terrible times. I could not address as much so easily found. (*putting, almost with a pour, some coinage in the basket*) Feared I would not arrive?!

DORIA (*lowering*): I can not count of such return. It's been awhile, and empty are our cups to your dependence, not to rely on it—

FENIC: Support is earned. Yet riches are all stolen. I am declension-ed by my baiting to be gratuitous.

DORIA: He was a friend.

FENIC: Fair fellow found, along my traveling. Who can forestall the realities of the unpredictable, Doria? It was argumentative that I might prove myself, my capacities at court— and to the king's service directly bound. Yet to bring shine of notice to this area has poured of shadow to its usage.

DORIA: You claimed you might enhance the trade of this region.

FENIC: I said.... boasted to a companion of drink, whose taverns (here) offer stoutly, that I could perform such miracles. And your late husband pushed me for a friendly demonstration. Without incense I smiled to prove it, for we were more than amiable together, but perhaps foolish of our revelries an' jesting fun.

DORIA: The times were sufficient for that then.

FENIC: I admired him, and what he could accomplish of a family an' lifestyle, somewhat the opposite to my free-spirited self, and yet as gay through his dependencies as I through my independence of character. It was a good mixture of notabilities, and we had good company of enjoyment, feeling through our leisures off work, when I should occasion to visit the area.

DORIA: Much welcomed, Fenic. The children know you as an uncle.

FENIC: False though to be, I liked the distinction. And we could all afford to be harmonious of pleasure and very proud, secure beings, that while ago, when welfare an' success danced on our fingers of achievement— For it's difficult work to miter Mytdra, I can easily confess.... Those were days of broad, healthy, deep breaths — of satisfaction in our lives, Doria, and dismissal of what was minor to be disagreeable. Indeed, I was blind or insensate to much gravity.... but not our monarch, who thrives on sensing the free-wheeling airs to determine their causes and contort them to some advantage. I only mentioned this village once to him, as to win for some approval its industry in a relatively restricted space of township, its size made happy of the inhabitants I'd found, lively and flourishing of eager, working contentment. A favored discovery, I professed, among my journeys through the realm of casual wandering. But Mytdra studied the suggestion with unexpected intensity, his assiduous curiosity promoting more information about the region, its surroundings and neighboring cities, and at length— with craft an' counseling by fiscal officers— drafted means of enhancing the wealth and revenues of other towns at its expense. For the king desperately needed some increase to the royal coffers, and the larger cities magnify greatly this effect over yours, no matter how handsome your peoples' labors. So.... trades were lessened here, the very opposite of what I had proposed— to friends, and I could only shimmer of the shadow brought to you, over this fine extent of kindly minded countrymen, high-minded of their chores and decent for their morales of achievement and relaxation. I can only cringe at the growing struggles that are suffered here. But the death of your husband, a dear comrade of pitifully few in number— For as I have many associates, with overwhelming bouts of conference to mediate, few would I have for company and to accompany my genius lent in leisures or pastimes, for who would have my manner with an' shared.— as was your precious husband one, I feel enshrined of calibre to be responsible for some great mitigation 'gainst this travesty of care that I have caused, or instigated too directly with indirect error of judgment. For I still insist to be cherished for my influences at court. But what may I do? that's certainly to protect your family with my riches— which are not as high as supposed by most. Indeed, I'm as rich as Mytdra, in a way — and he's full of debts. We essentially live on the credit of the nation, to which is manipulated constantly. To that which is disposable, and liquid of a currency available to me, though, I offer from a sum— to you. And how this may be regular, that's as difficult as my cutting guilt. I've ever only fended for myself, before the culpability transferred my crime to poor results for others. And now must I adjust for more to help, out of my cruel pockets— to give.... not lend.

DORIA: It is taken to buy, for food and clothing.

FENIC: Then what is satisfaction? It can not go on forever, and is as irregular as my abilities to provide. Yet bleed my eyes, to see you in this posture, who once was wisely merry, with circumspection of your married condition. It is heroic, and not faithfully lost, to beg this way for needs.

DORIA: I must languish for two to last at all, Fenic.

FENIC: And that they're yet so young is frightful. I'm too ashamed, yet, to elicit their joys of seeing.

DORIA: What might have you, to right this error?

FENIC: Your town decays by prearrangement, Doria. It is no error. I to suggest, the king is more clever than my banter, praising him. He uses me, for support of his conscience. The suffering is known. My disappointments beam as I grin at him. Mytdra is more practical than evil. The drain of river swallows lethargies, for what is left t' become this place will be stalled and worn of detritus' sloth. Though fields may be captured for some resource, if that's ever found as profitable as earlier. The best that I may hope is to raise your children, with your decline. This is no promise pledged, but a declaration of intent, should conditions sunder ever worst your strengths. Yet more to follow the trend of these sad events, as the village dissipates its aura of existence, it's clear that you may sell to move to a location where, if not more prosperous, at least could show more potential for a subsistence.

DORIA: Sell of house an' furnishings? *What* would buy, to own such an improbable gain? We are trapped, for the only cover left us.

FENIC: I make plea to Mytdra, when we are most personably alone, to permit adjustments in population, so that the meaner areas do not simply fester of purulence and discontented heats. I would never make the suggestion in front of planners and strategic contractors for the country's run. For to them small blights compel the brighter contrasts to the achieved heights that are allowed. But to himself will sense, perhaps, such need as advantageous, and with a subsidy lending, to reduce the friction caused by some success. After all, while others may, he's not one to forget— anyone.... or any passion lengthened under his subjugating rule. It is a bold ploy, of course, to offer, but such is his mind receptive, conducive, suggestive an' employable of tendency to understand and diffuse of particulars for any worth. If there's any benefit to be found for his constituent awareness, it's clear the resentment of you an' your like, as prudently taken as an initiating migration (of movement), can only improve all general circumstances an' situations, certainly outlooks and future prospects. The king has in effect ignited such transformations of populous pockets to sanctify shifts of habitation in favor of larger, more prosperous towns.... This is what I hint to argue at before him, assuming he's already considered these aspects. But reinforcement of thought is a profound stimulus for action and inducement of attempts. I know it will be difficult to leave this local charm that I've admired and found so sociable, inviting and receptive.

DORIA: It is.... a home town, Fenic, once more than passable for any upright, virtuous lifestyle. Are our simple pleasantries of existence now to be condemned forever? I had thought we had been granted a fitness for living—

FENIC: What have you now?!

DORIA: —Yet luck severs with evil shears, sometimes, its ties to promises, made to persons one another, that life can be made comfortable for the righteous. And now have we only an uncontrollable mount to growing hardships and predictable disasters, permanently separated from one who was thought to be a lifelong partner and mutual provider of our shared cares, interests, an' loves.... Are you dreaming for the king's reaction, Fenic? We live along a precipice to fall into cruel certainties of starvation—

FENIC: Not while I live.

DORIA: —and can only stagger drunkenly of our misfortunes and blows of stunning hurt. The children realize, perhaps just dawning to accept, an eventual decline into oblivion, or fierce obligations of strife to survive, and are very fearful become, beyond the simplicities of their ages. Yet, to hope for what you wish to pursue is not a strong mandate for striving on. How can that be counted on? It is too speculative of your contrary influences, those which you now doubt and panic of conscience with, though such a good friend remaining. We can not afford to move nor stay, but can only abandon ourselves to what sweeps.... our prospects dwindling out of contention for any graces or appeasements. And this dejection has been gradually impaled. For I have hoped of some sufficient way to arrive, as the light continues only to dim. But conditions force me to behave this way. And when will you return again, having failed to persuade your master and our effector? I will not curse your vanishing if shamed to it, as I can not rely on such schemes as yours to uplift spirits.

FENIC: The earlier to do will make the better gain of trying.

DORIA: To leave as from a fire, any destination will do— for an escape. But what is worse, here, with uncertainty for finding provisions and shelter, with two small children to attend of their disquiets? The travel is cold, into the unknown and along endless roads of risk. It's better to die starved in a house or communal casket of this family, than with scattering strained of the final losses, out of one's head and delirious of ignorances for what could fate achieve upon your stupor.... along a road or within a woodland moat fallen or tossed. There is more security of thought to know what I must do now, and in *his* house.... to stay until the end, unless the guarantees are made— or a fire burnt. And you can not propound of Mytdra faithfully—

FENIC: I might burn to.

DORIA: —to assure us.... even yourself.

FENIC: Yes. Chills the cry, I can not guarantee a thing. I'm only a lackey to the leader of us all. But.... does the nerve a right, I won't let your devastations happen without my own to furnish. For my only value in the world, my only wealth at hand, is to honor friendships, and be dutiful to these relationships of man and memory.... Had you attendants for your children, I might attempt employment for you at the court— most probably strictly from my pay, so it could not last long. But if the trend I weave is fruitful for Mytdra's mind.... it could be forged just enough of time to afford a proper move. Subsidy of subsidies, he must think of this to promote!—

DORIA: I am alone. My children are by themselves incautiously involved even now. Our social justice is such, currently, that each must take care of their own relations, this individualism of family seen as protective from our worrying climate and shriveled associations. Misfortunes breed of bodies met, as enforced or delivered responsibilities chain you to ever deeper and unmanageable difficulties in our crumbling town. Were I to leave my children with neighbors or acquaintances— and not return, from death of hardships won.... what depths do magnify the miseries of innocents! what laboring is now established for one's own decline with unbearable troubles. It is thought too chancy, tacitly, to accept.... too guilty of one's adversities to propose. So my children are alone at play, in a closed room, and for this long day— uneasy for what I should bring them.... to home, of what they can contemplate of the hardships between bouts of expressive fun or joy made decent of their youth an' carelessness.

FENIC: So thoroughly declines a community, it would be better off as wild natives bunched during a calamity of weather. O for what I have wrought, a seeding's to be done— to restore social pride! Enchanting of a callousness is wrongful service, an' poor liberty upon my dedicated responses. Yet these coins I've poured you can help (for) awhile, then not to end the need for begging are they provided. I can not tell how long they *must* suffice. I can not produce a sure solution to this gravity. I can not intercede surefirely for great results. But we can prepare ourselves to a fancy of better outcomes and happier times than these. Make of me some optimism— yours engaged thoroughly an' single-mindedly. And be that the angel of a thought. For I hear they're all blind, and need not even wings of senses, but are simply provided to the deserved of airy, ephemeral assistance to help you last from day to day of your travails an' harming. Torture is the loss of hope, tedium its gain.

DORIA: It was a bad lapse this time, Fenic. We could have drowned for emptiness of any aspirations. Yet decided, I should gravitate to charity, a desperation calmed and lowered to.... They understand what we face, and what others will come to assume.

FENIC: The rich should be humble, an' the poor proud, withstanding as they do their tenacious mandates towards luxuries an' life. The delay was more a mixture of shame, humility, and funding due to. Yet lessens not my resolve to improve a lot of what occurs here. As I've made the village noticed, with its valleys herbaceous an' pastoral, then will it be charmed with my attentions more, stratagems of intensification for its purpose within the realm. For, if so desolated of commerce(, presently), it must still remain of a fertile ferity for some simplicity to bemuse into contraction. As is my temperament to handle this affair, and other wild freedoms to become embroiled of, the land will not be vainly deserted. And your lot will be paid for somehow. And you will be made whole again, for a future to wish for and caress. That is my pining temptation, aspiration, and aggression stirred to. For what challenges me is not this summation of harm an' grief, but its summary adoption 'gainst my pitched aims. And they will be recovered. My bravery owes this least objective, to restore of good use these village plains, hillocks, an' homesteads. And as the inhabitants may wish this unknowing of my participation, then shall be done the deed, whose obstacles are only certain tyrannies of thought. Life may seem so horrendous to withstand, of drudgery and depression, disappointments an' constant demands. But all this yields to the bubbling of potentialities as our human mixture stirs. And what is a decadence declines, Doria, of a considered decimation of wills.

Your life will become more fashionable soon, as soon as Fenic's shadow shoals.

DORIA: You can not be brought to bad of intentions. (*looking or glancing into the basket*) Yet here is some bread, to remind me of my crisis, that I am waited on (by) hungers an' severe expectations. Yours is not so much the cause as the cost.... of friendships lurking around woodland lagoons, and their lingering goals of fulfillment with the resplendence of being personable shallows. (*almost habitually raising he basket*) But my fears numb of shock much more.

FENIC: So be it fit that some may watch. They are decrepit not to feel your tender agonies.... Lower (*She lowers the basket.*), and stand— up off a hard surface for so long sitting. It draws of dusk, and the weary may not pay more for darkness' sheaths, or gray impressions of the twilight. That paint of sorrow stymies hearts already troubled. An' I'll withdraw with you— to protect your sum, this petty achievement. But I won't greet your gregarious young, till I prove myself less foul of their ingenuous felicities. (*She stands.*) What is shade will feed, as we depart, my empathic ambitions to improve your lot. And I will give more (*as they start to leave*), dearly bled— scrupulously filched from for a pious duty.... to repair my own constitutional details, and improve from the internal commotion of my emotion's torment.

DORIA: This is done to be as sufficient as you are, an' accompany me with pardon of final observers. For this day is where to find ourselves, now, at its closing, to see that I have the sympathy.... of a noble man, and yet return to honest cries for enforced disciplines. The breakfast this does make.... for several days, a wait for light an' astonishment— recovery! An' as for supper suffering, enlaced with prayers, does train the hurt, to remark on crumbs.... as a salvation. So sense the Muses of their justice, to find impoverishment belittled, but not scattered yet.... among the fallen leaves, as would the crazed deprived— their vital humors spent an' dried.... crawl out onto an openness, to be conspicuously found, for want of conscious caring— lost to their last day's.... morose delirium.

FENIC: Would be the intoxicant of drink to sting of stiffness withered trees.... an' friends.

Scene II — *Daytime, as of a late morning.* **Ancher** *is on the court grounds, which front* **Mytdra's** *palace. The air is brisk, and he is formally attired. His* **secretary,** *acting as his assistant, comes up to him, carrying some documents.*

ANCHER (*greeting the* **secretary**): As for being invited, it's early yet, in this chilled atmosphere to present one's self.

SECRETARY: Our coats do labor of the morn, my lord.

ANCHER: One must always be early, though, to show some attention to the request, and respect for the offering of a visit to court. The palace should be more heated, out of the direct draft of wind. But you must stay outside, to suffer these breezes, unless called for.

SECRETARY: That is my alleged role. The sun will gather high soon, though, and I'll find a bright spot to shake of standing, as a servant to eventually be summoned. Yet autumn awaits, even still the early day cools heads. And here is the request (*handing him a document*), a detailed description of this exotic activity, as can be gathered and estimated from Lord Cantal's brief. And I have supporting evidences loosely researched from Lancet's preliminary notes of prelude to the adventure, whatever I could find for the trip that's relevant.... as well as notes on other of your ventures, should they be called for.

ANCHER: Not a venture yet— but you are schooled for an examination, not even what I contain of specifics.

SECRETARY: I've studied what may be, should there be questions.

ANCHER: He may have us wait for an hour or more, with pride of calling, that we appear with certitude.

SECRETARY: Then it would be warmer for me. But were you sovereign, would you have the palace guest us this way?

ANCHER: I would have us kept indoors, until the king were ready for our addresses. But this is merely a spectacle of our servitude to be alert. You see, we are not important business yet, or we would be rushed into hearth-fired rooms immediately upon arriving. So this privilege of attendance is a custom of friendliness, that I may put up with such a discomfort while waiting, as a strong personage of health and body applied to his royal lease. Were an objection meant to be sensed, I would be waiting in the coach, along with the women or infirm. Yet as we show ourselves standing on court, in this relatively slight chill, this garners from perceivers an appreciation of strengths 'gainst similar annoyances. It is a cocky manner to promote, and I would do away with it as ruler. But the monarch Mytdra is of such a vanity to insist that we be courteous in this way— and others— for his favoring. The shivering is symbolically to squirm before his awe's expectations, you might understand. It's not meant to be improperly suffered, but as an excitement— and exaction of comport.

SECRETARY: It diminishes the compassion for his professed austerities.

ANCHER: And that's a ploy of abuse; for as austere as he, it's done with the riches of others— to plead for our suggesting, as he makes the realm more civilly stalwart and sturdy. We lend as compassions felt, though unequaled to his reception. But it is that process which equates to us the rights to be considered and made party to consultations of royal advice— Now this I would strip down to codes!... and approvals. For some of our noble gentry may never be recouped of outlays without those laws. Still, some influences are made through these means. But the overall debt of the king grows peculiarly, multi-facetiously shaped and odd, and of a size to warrant a reversal of fears— as our gravitating faces squirm: grimly, gray and garrulous. So might this ruler count on.... friendly attributes to appease this consumption, the fine men stand out here.

SECRETARY: It's not wasted, his excessing to control the economies, if our country seems rational and remains stable.

ANCHER: But that's to be seen of a silent sheen, much pruning of dead feathers and preening of the silk. A collection of lords, now, might very well own this nation through a fiscal prowess— and that is what he struggles with to alter, as if to balance off complaints against one another, to keep his stake of control more solidly standing at the peak of our possible surmounting, too difficult to achieve divided. Well be that as a usual, not exceptional majestic tendency that (the) most happily accustom to, as our land stays calm and prosperous. But some greater payment must be made for this to continue, in that more peerage succession is allowed and advancements won of more governmental importance than to attend rehearsals at choir. Else a more dominant force ensues, to tackle the nation's problems more directly and adroitly than has been. He throws resources away at procedures, processes and concepts that make little advancement within the measure of reasonable time to nurture into something really productive. That's not a terribly inspired intuition to demonstrate, but wishful thinking over mountains of information and novelties reviewed. There's (but) so much that a mental fortitude can accomplish without more thought— and delight for successes. Our Mytdra is fantastic at the regal displays, but such swirls lead not the winds of fortune. Rather are we wound into considerable contentious ploys, to better ourselves by our own devices independent of majestic aggrandizement— and for the nation's good and prospering weal. Then let me be a representative of this.... surge surreal, and surreptitiously sublime.

SECRETARY: Yet hidden traits may hamper heeds, ma lord. Stay furtive nobility a danger all too noticeable to proud mights endowed with cunning. And on his actual grounds speaking, forebodes some atmosphere of capture or snared enlistment to be rent. Does he invite so often as.... to be forewarned, this is some game applied to treasure makers— by the seeker of his own, perhaps embellished by our properties.

ANCHER: We all do strut to strike for ores unnecessary, when owned of such pavilions of the mind. But for a country due its wealth, with ability to self-supply by clever fostering, this royalty today does cause a lessening of intelligent gains through wasted applications of obligatory readjustments to commerce. And there is no warring cause or threat, as yet, to legitimize the internal dishevelment. Lords, then, must bring ramparts against the tussle, as could a bulwark of finance to lend our more aggressive natures to commit to.... a functional security for the country. We are the positive action of advancement his majesty wishes to tool with blinded utility, as were *we* to invite *him*, of disinterested effort, to partake of our endeavors. Such a party can only awkwardly be run, but does allow for usurpations casual, with much slippage— of the feet in dance.... and the tongue in oration for an actual understanding of what occurs about oneself. So can control a breeze its temperature, by rushing towards stagnation at a wall. With time more clement heads will run our government; and I should be among that fold a captain of achievement, too renown to be obscured by anything.

SECRETARY: Even a prison's barriers?

ANCHER: ....That is for pondering of rules, to carefully employ ourselves through this morass of social conscience. How can the king trap me, as were others, if I stay to seem advocating for his causes? We lend ourselves always into deceits. Yet challenge our proposals to assist him— he is blind to it, and can not do but take a greedy cut of our formalities to enrichment. For so haphazard are his high sights, he can only conceive of the world as owned, its spaciousness condensed to curves of eye. And what can contain for miracles of sight as with a dreaming? Then as for dreams, awake they may seem true. Thus does he feel his royal commands in place, they can only lead towards his dispositions. He'll sense no others, as an eagle flying about the realm: such haughty insensitivity, such indignity to decent concerns, such loath aim to be compassionate to what squirms, such character of caretaking— with hate for the truly heroic.... or the most meager to the most major of

struggles. This is simply not seen by him, and gives us our shadows' light to promote ourselves—

FENIC (*coming up to them*): Earl Ancher! It is a delight that you have arrived, and so trustworthily to an appointment. Our majesty is at present occupied, with counselors and their striations of opinion, delivering ideas already contused of a royal battering to consider with private contemplations utterly personal for the universal exigencies, as he makes for all of us— worths. Yet he expects to greet you warmly, and looks forward to the company, *et al* to choose for his enjoyment.

ANCHER: The passing Fenic. Does breeze today awhile.

FENIC: You look hardy.

ANCHER: As for your observation that I am.

FENIC: What's more for your attendant. This attendance is spectacular. And King Mytdra has real interest in your extraordinary discovery, such that it may be supported with facts ov'r luscious fiction.

ANCHER: My.... envoys are serious people, in describing all details. I will not doubt them, but can not yet authenticate the find.

FENIC: But you are one to get the gold of grains, always seeking ways of extension to the boundaries of propitious attainments, in style of clothes, mobility of horse, and now a supple food of incredible ubiquity. Or shall I say an incredulous greenery is claimed, for your proposed enlistment to culture. Is this too real to be defeated of shameless boasts? How might some hills profit of expansiveness with such a crop? For as I count it's mostly hills to see of a roving, undulated sea of (a) common ground's cover.

ANCHER: Discount you such, as possible to render of the concept?

FENIC: It would be remarkable as storied. Yet plentiful could a valley make of this, I'd know locations with to plow—

SECRETARY: It is of a heated clime deserved.... of what I have assessed, for the earl's interests.

FENIC: So make the tropics hot! with this. In tales of wonder are rumors matted for a mow.

ANCHER: And everywhere you see, now rumors to be called— my lending of this knowledge?!

FENIC: I am as informed as his majesty may speculate.... around me. And what's alleged may change the course of towns. Yet can I warn you that our king.... is always disagreeable to letdowns, and especially by the promising to promise of involvement—

ANCHER: I've made no venture to it, Fenic! except for more inquiries to pursue. My stance in this is all of the preliminary and prudent. Yet, to draw into an involvement— nothing! unless it meets my high calibre of arousal.

FENIC: And as that calibre is great, and more than fortuitous, but of a studied nature, then does our king make it so. The pressure's on, already for this to be real. Though our majesty, in his polite graciousness and gentility, may treat the matter only as a casual form of welcome to see and converse with a happy and accomplished subject. Now then does bring my nose to this, to ply for desperate enlightenment. Aye! to this wondrous advance to conclude of, that in no way may our most royal be offended or slighted with errors and mistakes, such as to cause an action past most forms of litigation to avenge. Know you then certainly.... that I'm responsible to— suggest to him original punishments. How they turn out in the end makes for a cunning's education.

ANCHER: You have his loan of leeriness, as might a latch be caused—

SECRETARY: 'Tis how the operation works!—

FENIC: Or lea of way. But to precede is my intercession. He is genuinely amused by the prospects you promote, and wishes for some confidences to proclaim, that the Earl of Ancher is a good associate, worthy to be trusted and leaned on for conferred advice, and does not cater to fantasies remarkable of hearing, and does not lure of sound to fish for influences factual, and is always self-possessed of honorable intentions. And I responded to his uncertainties that this seems to be the case, though a few successes can make one wild to predict (for) more. And so I would investigate by simply asking the gentleman myself, this nobility of purpose and friend of the constabulary and the law and all of royal decrees.... that you do not exaggerate yourself to— stay in regal company. For to stay, some feel they must continue to enhance themselves with claims of potential achievements, when our majesty simply likes good friendships to sustain, their beneficial offerings as natural as life's involuntary metabolisms. Whatever you should do— and how— can only support his ruling.

ANCHER: Holston alerted you— the court, to this find. I told him he may; for I feel favored, as late, of King Mytdra, and should not blaspheme of projects considered. Yet I hold our policing in terribly high regard, to keep our state shielded from germinating illegalities throughout the land, that state being a pacific causality to.... persist— in a reasonably social and productive manner. And as to the politics of friendship, there are certainly conspiracies of ascension rife and ripe to the splendors of our country, bound as by hanging off stout trees rooted to the history of our regal governance. And predictably to fall, these lust-worthy fruits, their sugars and juices awaiting cakes and pies— at bake in royal ovens.... or loyal dungeons, there make some fates of capitalists. Ah, for advancement! But to base this heat on fairy-tales and fay's brush, Fenic?! I tell you not a story ordered for importance, to induce more notice of my suave and swerving ways. And like all of my friends ever known, of a rank to bring attention of our king, my hands are in many businesses searching for just profits, constantly rejuvenating tired events and adapting to pert novelties of promise, all for the glory of our nation's vitality to insure. And if I were more creative, I would make the wondrous and possibly wonderful — but instead I find them, with expert searching and intuition for the profound. So may our Mytdra be pleased— we scour out all doubt.... to our endeavors' doubts. And strictly speaking, I've not known one person like me who wasn't as intense of these.... intentions roughly proportional to their station. Though some do fall so hard to the ground as for a most discourteous squashing of their tender wealths— and are wasted!... by abuse of their conceptions improperly assumed. And they are fouled by rotting, with wanton alertness, attentions to false or falsified details, and mean but lowly tempers to destroy— achievements!—

FENIC: Think me so? with threats addressing for the king?! Of all else who might be, I am the more suggestible. And yet my talent is for his digestion. And a whack is made, here and there— not by *my* decree. But upon observation of its meriting, it's done. Whatever happens was meant to happen, so say the wandering stars. But I am not a destroyer— less.... a crime, against our state applied some heresy. The real are friendly to me. So make me not responsible for some of your friends' declines or expulsions— But be genuine!... as your gold. For as I may bend with beating, others must break themselves to bow.

ANCHER: See that I support this tree— without resentments, this central plant of governance, whose nature of a tough pillar I admire. And as thus.... inherited.... I would never lead our majesty into ridiculous ventures that can only draw forth outrageous losses for the country and his personal investing. Were it as simple as the forfeiture of finances yields to wacks, it would be an ignoble risk-taking— with ignominious suffering. But if I say I believe in a project, after having vetted thoroughly all capabilities caused with my and other associations recommended, then I am never at a loss to substantiate my bidding and succor from any damages experienced those implored by me to apply my thoughts to joint commitments.

FENIC: That's a bold assertion for assurance. But I've heard its like before by the balder-ic. And even under written contract it much depends on what one is able to do— of the results. A good record, though, of plentiful returns on efforts, makes for hearing of pledges and proposals in august company. And so I state for you as clearly as your verve, and verging here, that this is the implied pin of pact, that proof is only one way seen— and without vacillating contingencies but an obdurate resting of the stone, whether it glitters or is cold to a wind's stirring. For scintillations heat the eye and heart, in mental effect. And as for any affection, the king provides when he is pleased the greatest of all: the rub of chin, to be as proud to be with. So are we doused with airs, and their dusts. You have the true terms of expectation, and without surprise may they be impacted. Seldom do I see it, really, after I have chatted with our interests. For their faces are grim, onto their own deserts as grime, when *here* perceives a failure.

ANCHER: And what might you for enlistment grind to, Fenic?

FENIC: Does please my Mytdra for, extremities to lend, I'll ball up in your hand to clasp and strike.

SECRETARY: An' thus be winnowed worthily of love—

FENIC: And!... with conversation here and to the king.

SECRETARY: ....for this effect to feel. I can produce what's more of your employments, my earl, at any moment's call—

FENIC: Has he not wit of his own architectures?!

SECRETARY: Some are as complicated as the invisibility of breezes.

ANCHER: I account for myself, Fenic, the utility of brain, to keep in order and explanation all of my dealings. My secretary provides the details that are correct, myself those that are interpretable for discussion, and the heavenly graces— of inspiration and insightful ploying of the heart— those of much whim inducing of pursuit. Yet for my happiness is made.... more meaning lust than manacing fulfillment of a minion.

FENIC: So twists the cord of devoutness earned.

ANCHER: So.... do we arrive?

FENIC: You are here.

ANCHER: To enter!

FENIC: Through those gates of entry, of course— but for a moment— wait.... You have arrived, it's keen to say. Yet to be called for, in addressing to the king directly, not so immediate is the announcement.... warmly placed. He knows well that you have come. Yet I return alone within to prepare the precious time of greeting, as to be arranged for an appropriate slot of meeting grace. And yourself will be sent for, from off this garden lawn and into the palatial realms directed. For the withholding rooms are kept for officers of state and dignitaries, those spaces heated to a comfort of noble urgencies to prepare one's thoughts before high presentation. I'm rather more dogged, to observe what arrives, than to conduct the machinery of this protocol. There are others standing around, that I have counted, or walking about. And they have no less place than you, for placement through cordial visitations not meant to be outstandingly substantial— yet still highly valued to warrant this approach of the regal. Then the sweat is high this morn, of dewy ethers, how might your anxieties about this prevail. Earl Ancher is not forgotten; nor can he be so terribly impatient to disrupt courtly manners. But you have more than my sighting, as I've.... your comments to argue for and ascribe to a sincerity apparent for some decency— to me.... and my wishes, only to please the king.

ANCHER: Well, Fenic, I don't claim to be important.

FENIC: You are to me. And call me "friendly Fenic." For I will help you much, with Mytdra's monarchy to compose through, if you are true, and of a predictable reliability. Let's be more divulging, sir. You wish to gain some officiating property aside our majesty, you and a thousand others. Your talents, at least organizational, are widely acknowledged— And I may recommend them, if you can prove your utmost fidelity.... to our country. But in this I have needs that are responsible, for I *observe* the country. So as an intermediary to *our* lord's persuasions am I to be aided. And with this— admitted.... I propose an experiment for your proof, whereby you greatly improve the degenerating circumstance of a rural village of interest to me, but hardly to the king. If you can accomplish such a feat, through your delve-ish acumen, that will impress myself for you to rally for unabashedly, since our Mytdra can always be made excited by someone shown capable to assist him in some office— with a demonstration he had not even anticipated— And this is fiendishly hard to do.

ANCHER: Are you talking about charity?—

FENIC: Development of an area that is falling fallow— not of the soil.... which is rich and fertile, but of utility of a social climate for advancement of its economic prosperities self-sustaining.

SECRETARY: What cause for riches are an agriculture? That is self-sustaining.

FENIC: King Mytdra adjusts the regional importances of his towns, to enhance their profitability for the crown. Thus larger regions are favored by necessity, and some smaller sites suffer a despicable decline thought unavoidable and thereby excusable. Yet one such village is dear to me personally, out of friendships. If you can astonish original thought, by turning this area around, through your— processes of commercial advancement via the imaginative subtleties of your skill to find such means.... where thought perhaps impossible to do so.... that is the overwhelming proof to reduce my candidness into a concentrated advocacy on your behalf, and to forge for you a true candidacy of administrative powers within this court. You, Earl Ancher, could become an actual and executive advisor to the king. This is clearly what you want to do, the insertion of yourself into real realms of power— to make changes.... according to your ethical precepts and moral convictions for your country's continued advancement. I am this test of obstacle, given you by our Mytdra, for you to prove yourself with a current force of effectual facility, as I have and have stated for you an actual problem within the tangles of our legalities. Why should a simple village suffer into the blights of obscurity and forlorn prospects and be ruined of its populace with desertion and inutility, when a clever man with a workable idea may resuscitate her hope by achieving specialties? by withstanding somehow the cultural and mercantile supremacies of larger townships neighboring and extorting? My implications for you are now at play with this problem, my personal interests at stake. So now as stated is this chilly proposal for you to consider, my dear lord. And as you stay here wondering about it, that is your qualm.... But should you *dare* to go through these gates and enter the palace, to meet with our majesty, it can only be with expectation by me that you fully accept this challenge and danger to provide the positive results— Else leave this lawn today! without affronting our graciousness inside. For I must assume you to be true, one way or the other decided. And should I uncover a hint of deception.... disaster for your position, sir, within this country. That is my especial talent, for our majesty: to suggest what is disastrous.... So then, my lord, I am blatantly honest with you, as I've been with others, *before* any engagements of fortitudinous courage and strength to provide.

ANCHER: This is peculiar, Fenic.

FENIC: So am I allowed. (*starting to return to the palace*) So have I been thrashed!

ANCHER: And what of details?

FENIC (*retreating*): The details are in my heart, Earl Ancher. Make of yours as stolid.... Sir, insist upon yourself.... what can be done. (*Exits.*)

ANCHER: What can be done?... about this ruling gambled with? —

SECRETARY: This is clearly a trap, my lord, for your vulnerabilities to make responsible. How can anything you do be guaranteed?! You are with risk—!

ANCHER: So makes of day its rove, to be undone, and thereby seeks a hurried pace that is too frantic for the more courtly estates. (This is) Some rush to think with, hurried are the stars for pondering these plights of conscience— to tame wolves. And the moon's averred past noon, this quaff-some day. She wishes obviously to make something of this crop spectacular to grow here, in this region of Fenic's. To this we drink so great a chance as volumed. I know nothing of this plant that my hands have handled. Yet here is a way— and his dare to me.... to gain influence and initiate a feasible process of usurpation. And shall I come as called, drunk of this temptation? Where is the resistance but some caution? I am a wealthy man— and that protects me; that is a considerable shield.

SECRETARY: But ignorance can deify deathtraps, my lord. Others have been rich, and at least run out of Mytdra's length of arms to assault, that span being the width and breadth of this nation— and many allies, with much of their possessions confiscated or lost or taxed to a permanence of someone else's ownership. It is a poor escape for the legitimately proud, trying to continue as they were, and only for this falsity of station somewhat maintained. Can be most miserable a denouement for (a) solution to dilemmas.

ANCHER: There's one condition that overrides all others, and that is I'm correct in this race— to make my strides. Mismanagement of the country is abysmally obvious and evident, to the point that even a close confidant of the king's complains. Thereby, the nature of our standing begs for these attempts. With wobbly legs and errors made, yet careful are our steps— to be supported, by the sheer need of some resilience to wrongs and the inordinate. And while I don't claim necessarily to be able to reach the high level of a potentate, this engineering favors some definite progress to improve the affairs of many and perhaps most, as is adopted a coursing drive or force for change to push us through into successes, whereby more nobler influences predominate our governance. The question is not whether to do, but how, and more immediately how to assuage this Fenic's more personal concerns, through his passed along confederacy of secret and possibly private affairs—

SECRETARY: He reports directly to the king! in all matters pertinent to his being. This is a goad to produce obverse conspiracy, observed openly and reviewed, and easily discussed to see how far you may be brought into the depths of contentious baiting, the Mytdra laughing over the awkwardness of your designed duplicity against—.... if not himself, then his notions at least. The test is of your bravery to be naively led, not a quality you can possess intelligently, not one you've shown at all previously— though bravery alone may hue the cause of many debacles.

ANCHER: Here some intuition must reign, trusted servant. For as I trust you explicitly, can not his sincerity be tested for the same?—

SECRETARY: What?! Fenic's?

ANCHER: I sense he's not so much the dog as the dogged, and that he genuinely seeks help. Else his exposition would be more than extraordinary, definitely imprudent. And indeed, what of other it may be, this disposition of his is at least as risky as his tempting of me. I might find a champion in him, with any sort of assistance—

SECRETARY: Awh!

ANCHER: —I've seen the type before: the distempered though indelible of faith.... to try many avenues to a problem, the impressed upon.... to be rendered sufficient to all masters— and many masteries, and the cudgeled.... bearing up of his social weight to serve— some general good. For why live to serve a baseness without any reward? Why exist to live meanly and with detraction without a higher view to see and hope for reaching? an endurance

made worthwhile for something, so be it an intrenched patriotism or love of country. I am experienced of such.... independent thinkers, through my youth while accompanying my noble father's peacetime military oversight of soldiers and adjutants— And such people are capable of inhabiting the most odd assortment of demeanors, with subjugation towards finding some self-serving glory of an overall beneficent purpose, tolerating many disgraces of character— externally observed.

SECRETARY: Fenic's no fighter, I believe.

ANCHER: But his thinking is alike, to wear a shame that's not berating his internal dignity, however lowly treated or thought of. And with these effects thoroughly restricted to the king's pleasure.... or indifference or propensity for a given subject, then with a profound protection is he held of a position, such as may allow him some occasional personal risk taking. And I feel he will not report to Mytdra about this matter at all, should I present myself, but has quite honestly stated his case— to leave me at task. For he has not interposed any means at all to be mandatory, no manner at all to follow. Leaving the methods totally up to me— and knowing of my capabilities and history— he *wants* my trying to be successful. I sense here an earnest plea. This fellow need not lie at all, except to the batterer.... For that's too far below his disgrace.

SECRETARY: I can not find that complexity of personage in him, ma lord. I am myself of many missions, all upfront and straightforward. But what is simple to tell is the deviousness of this exercise. How may you be surprised and overtaken, by your own devices, is the cruel trick here. You are reaching beyond this circumstance to find another, that's all too clear. And if the Fenic is as honest as you believe.... what strikes his joy to realize *this* eventually? What is the nature lush to personality, that he may find a remonstrating end once realized of your true persuasions? He accentuates this talent to warn you of, and is so principled.

ANCHER: I.... will usurp his master, till then he becomes my mind. That is the first step to this sort of recovery to propriety of thought and social approbation felt. Then as I climb, with generosity are his welts. And how far to go depends on heaven's lift and wanting. But Fenic will want me pleased.... as much as I'd please him.

SECRETARY: Can I tell you I have grave doubts about this, sir? Is it possible a man as that— a man as his.... can ever really be happy?

ANCHER: That is the dimension we will cater to, if I accept this plough. An earth that's rich and fertile should be satisfied in many ways. I could invest in plots investigatingly— Yet must I decide now, without details of even where! of what can be sowed and grown. And what may I rely on? Dear Lancet's foraging(ly) researchful prowess.

SECRETARY: He is an excessive enthusiast, at times.

ANCHER: But not so boastful as to dissemble anything of his discoveries or intentions of hypothesis. And there's the confirmatory backing of Cantal, pulled from (being) disinterested into the realm of being curious, as he so graciously allowed me to load my man on his train. So then what serves me? as when called up to a destiny, an opportunity so fleeting as to remain permanent of result. So much does run here to consider, that the fantastic of green might not be bitter to the taste, then of a common and lowly expanse empyreal to be imagined, a spread of the flame of grass. This is the dictatorial wildness of some fantasy.... or else it is real. But what is lent of insanity does provide more lesson of upheavals brought— and due. Can this, then, thus be true, to warrant of more cause my dreaming purpose? The notion has been sent me. Now *I* must be inclined, to make advancement or dishearten, behoove or behold what really perjures the spirit: a.... possibility!

Scene III — *Inside a camping tent. It is night, and the tent is lighted with a lantern. But it appears to be sweltering of heat, and **Cantal** sits in a folding chair, of portable ease, with some discomfort of prolonged, relative exhaustion. There are also a single cot, some trunks, and a thin table with dining utensils and bowl. The simplicity of arrangement in the tent is noticeable, the lantern steadied by resting perched atop one of the trunks to avoid its disturbance; but its relatively low level gives yield to an eery iridescence of illumination, especially to faces above it. The passage of entry, to his side as he's sitting, is open of flap but totally dark of view. A pistol lies on his lap.*

CANTAL: How can an evening boil as this! (*rubbing his neck*) It's a sultry decadence that I've seen better felt— Who's there?! (*A relatively tall native **boy** enters, carrying a shell cup.*) Oh. Darn this humidity for thirsts. I haven't, and yet I long for.

BOY (*approaching quite naturally, and without any pretension of reverence*): Swells to the head, Lo'Cant'. This appeases the knocking.

CANTAL: That might be some remedy. But it's only because it's so hot for so long. (*accepting the cup*) What prolongs the heat? This is an unusual climate. And I've been through several tropical forests. (*sips*) Nice bitters of sweetness.

BOY (*as **Cantal** drinks*): Cooler by morning. The vegetations pop of nightly baths this way, through thick mists along the grounds. They won't bloom well otherwise. Let me play with your gun.

CANTAL: Go to. It's not loaded. (*as the **boy** seizes the pistol with anxious curiosity*) I only have it out because I'm not feeling well. It's like an amulet of protection for the psyche. But it can't do much in here.

BOY (*examining the pistol*): Why have a gun, Lo'Cant'? There aren't any monsters to shoot at.

CANTAL: No game animals? You've driven them away to the higher elevations. Or maybe they prefer it— maybe we should.

BOY: This is so handsome.

CANTAL: It's called a metallurgic processing for product, a sophistication of craft you haven't need to master yet.

BOY: We know how to handle metals, from their ores, and make fine things.... but nothing like this.

CANTAL: You haven't reached the stage of machinery. That is a machinery—

BOY: It's a weapon. We have weapons!

CANTAL: That is much more intricate. It has springs, and makes discharges. There's more science to it than your primitive catapulting of rocks an' sands.

BOY: Science.... How's your head?

CANTAL: It suffices, now. Not only the mind can make tremors.

BOY: We don't feel weak at all, in this temperature.

CANTAL: An' how's your education, then! You've grown used to this— swelling of the environment. Yet I feel so exhausted and emaciated, while you industrially fish. I haven't overworked myself to deserve this. I certainly know how to apply my physical exertions carefully.

BOY: We've had ages of visitors telling us things, long before even my father was born. Some make useful knowledge—

CANTAL: Don't point!

BOY: —some not.... It's what you've eaten, Lo'Cant', that brings you down. Food for the gods. It prepares you for the gods, to be eaten by.

CANTAL: Maybe. But I can't suspect *that*. It's the heat, and the profuse sweating. Wears you out at a time when there should be relief from it.

BOY: There are rumors....

CANTAL: What bakes in the day should not have to bake at night. But the daytime is so healthy here, that one doesn't feel encumbered at all.... Now don't be foolish and look down the barrel, even if it's empty. That's not a habit one should pick up.

BOY: You are promoting yourself for some.... interesting formality—

CANTAL: Transgression? Only flowers pitched into a vat.

BOY: Lo'Cant'!... Hearts are in that vat. It's to be avoided, of course, unless you're meant to reside in it.

CANTAL: Ritualistic superstition. You are not at all aggressively violent around here, and have hardly a notion of antagonistic tendencies, but are peaceful with your satisfactions in the waters, those tranquil rivers and streams.... and your stringed darts and long pikes for stabbing and jabbing at the floppy, swerving fetishes of food. It is so curious, that you've not become more vegetarian—

BOY: Tenacities of rivals are being tempted. And you are innocent, but not of warning with this pursuit.

CANTAL: What?! A great man as me? Well I am felled, lad. Melts the gun with this transfixion— and I'm caught, like a coconut in hand. I can't leave, without my persuading effected that I am desirous and feel chained to a wanting of some miraculous punishment. Yet have I found an Aphrodite— it must be harm.

BOY: It is a sore head. You've actually been seen hiding yourself with chase, and wandering disavowed through trails and paths. At first it was amusing to follow and describe and tell of as more story of the sting. But now, serious?— Lo'Cant'? You are vaporous of much fortune this way. And she does not laugh, but is committed to use your help as if you were ordained in some way to this. That is dangerous and more, for you. Yet my mater only says: Cool his head; while my pater says to let you struggle through the blistering heat, "a man of his means." For et with fighting, each must prove himself devourable. And you allow yourself to be transcended of the vase, as rain would come for suffering with purity, sent mysteriously and for no other reason.

CANTAL: With no other reason is my blood confluent of the rite. Then it's simply my time for amazement, for this boiling preservation. And I strongly want the manipulating terror to experience, to solve its way through my utility. Yet as a practical man, it gives me some exhaustion from the throbbing.

BOY: Can I have this— amulet?

CANTAL: Now what could you do with an empty gun?!

BOY: Wear it about my belly. There's no protection for you, Lo'Cant', in this stark enlisting. It's what is ate.

CANTAL: Put it on the table, there, to see how defenseless I've become. Your remedy is greater than my might. Then that is nothing much to own, except to show yourself some sun, as it makes a mess of fishes, should you ever find its meal.

BOY: Or, but I want you to face the power of some fealty. She assumes your thorough assistance, and that other.

CANTAL: Without a weapon at all?! You're rather a precocious explanation—

BOY: I am a will, Lo'Cant'.... to help you through this one way or another, achieving some manner of resolution— for all of us.

CANTAL: Nothing that I know of is forbidden here. And where is "that other"? in this utopian Manifest Destiny athwart.

BOY: He sniffs too much to be decent, for your higher erring to allow with, this degeneration of gentlemen.

CANTAL: Not out researching even now, is he? In this darkness?!

BOY: ....But he is ate, Lo'Cant'.... by the ants of expedition.

CANTAL: Expiated, you mean, for the insatiability of his urges towards discovery, or substantiation of his theories— And only them, for falsehoods be withheld. Though there's nothing false to the forest, with its fusty shrubbery an' venomous insects. So I suppose he can find all manner of things to fully occupy his time— and fascinate, dilate the buds of interests, be absorbed into these unusual surroundings and found as if hidden in a dream.

BOY: Sleep. It is the fever excusing yourself.

CANTAL: A slight disturbance of stuffiness and thick vernal aromas— year round, it seems. How is it warm and yet frigid? This fine concoction draws away the heat.

BOY: As a cold pepper provides. Is there gold anywhere in this?

CANTAL: No. No gilding whatsoever, not even for the trim. No need of ostentation, boy, to fire. An' not a plaything, for your intellect.... I feel much rested, by the potion— Thank your mother.

BOY: As one to another.... as adversaries— you are not lost here. She has much experience in treating these symptoms peculiar to the adventuring visitors, exploring travelers and traversers.

CANTAL: It's a nice courtesy to me an' my party, a friendly and humane one. Extremely civilized, to help someone suffering a question. Thank her with the generosity of my complete appreciation. Now if only the weather could improve or implode.

BOY: What are you to do more?

CANTAL: It's a banal silliness, I know. But to become re-attuned to one's progenitive nature is disquieting, and shocking when sudden, a full man like myself brought to Lethe with a single stunning sight of awareness.... and forgetfulness. All that I have ever pursued, or thought to have been worthy to— captured in a view, a flash of lightning's realization. An' what of? What of more? What of past? What of late?... What can you think of me to envy? I have wealth— an' spirit, overpowering my senses, sensitivities and sensibleness. O this sudden silliness reduces me to a cad of longing, questioning, bemusement, confusion, consternation for my condition. The panic of this perfumed pleasing, the fragrance of awakening, emerging from years of self-imposed hibernation, haunts of aging with sophistications that deceive and disguise the true primordial being. Uncovered and stripped for bearing senselessness does make a strange covenant to glances and painful, shameful spying, averted teasing of fulfillments, goals bred for a gallows's perseverance. An' to be noticed for this— an' commented on.... brings one to be thinned of reason and tried of a tensioning. Rivals? say! An' to emote to youth, that's youth itself— Ha!... I took the bullets out— to prevent myself, from some delusional enrapturing, to dispel of mortal joy the wounding. Yet be them of shadows' shame for fun— the blasts!... the blasphemy of enjoying this torment, as to be able to feel.... the native.... an' run as primitive— to justify the coersion. Now so, do flowers weep when they're plucked off their stems? A compulsion drives my appetite, what's currently allowed. Combat? is it! Or concurrence.... that my aim accepts the fate designed, to keep flowers from wilting on their stalks. An' the blossoms bloom of blood. I know this certainly— an' it's maddening. So now, young gent, young disciple of the juniper, prevail of thought this lusty gloom for me. How sweeps the wavy motion of the ground fog tonight, an' where do the undulations lead? Is it as dark as it seems, outside, as thick canopies shed from stars an' moonlight? I make no error to the incessant demands — being prepared for it! The only resistance.... is one's health, is the toxicity repulsed— by a discerned body, a mindless soul, a carriage. Are there not even fire flies about—?

GIRL (*entering*): Is he fit?— Oh! Don't shoot him!

BOY: As much the head does quiver, and quibble over weird ideas, he seems of solid sanity.... for the procedure.

GIRL: Do you want to help me?

CANTAL: With a heartache, while it's beating, longs a servitude retreating to its origins.

471

GIRL: Why do you eat— grass?

CANTAL: Because it's g*ooo*d!... Made for to seem, can I imagine tastes?

GIRL: It's like a prophesy of animal pleasures to undertake.

CANTAL: My traveling compainon, Mr. Lancet, provides the location. And under his idealism it is forced to be so. Perhaps induced with strong wishes, but I am tempered to agree— it's more than possible to eat, and cook for food. So have I written to a friend an' associate. Yet for this spell am I involved to stay, I would gnaw at tree stubs for any sap that's left.

GIRL: The night is ready, for a ritual— But you can't be pushed into it unwillingly, because the essence of our meaning is one's eternalized volition.... to participate. And you have been shadowing me all about, with the most curious hiding stares, anxious to perform some good? It's only why I invite you, for delivery of passion, as described by ancient actions, that produce a population of the jungled prefered, and idyllic of the simplicities of a calm and concordant existence, and any further jealosies dissuaded by the permanent promenence of a stern legitimacy enacted. For I have too many suitors who would battle over me otherwise, and disrupt composures endlessly. Then comes a sacrifice, of manner, to displace all further romances, against myself and my chosen of bind, for bier of childishness and immaturity. All of that is.... thrown off, and cast to a reverential sleep. So then you'll come, for your participation to offer, and sanctify this occurrence?—

BOY: His head is cooled.

CANTAL: Whereat be the dispatching of this nativity?

GIRL: Where we first met, as how you aided me to instill this approval, and establish it of stone. At the great urn, to which you and your friend assisted of my try, is where of course to gather.

CANTAL: An' what of Lancet, sharing for your blessing—?

GIRL: He's already there. But you're more principal of surge to his ancillary must, delivered of a muse concerted. He is cured of grass as well, as can be spoken of his presence. And the flowers are odorous, as have been lent, that this spot is ready now, under the darkened sky with bright perceiving of our aims. The scents are wishful for some assembly of kindred, kindled hearts.

CANTAL: Oh, thus be drawn as how, a lent be leashed? An' taken ov'r was he led, by tales an' tails an' talismans of charmed adventure, where what to find he is deceived to earn.

GIRL: We urged him not, but on his own made way.... and with that talent of a whim produced this evening's drama—

CANTAL: As propitious an' expedient.

GIRL: —to fantasize upon as how the shading's justly prepared, with fired torches all about for witness of the ritual. Then was he tantalized to roam and romp to place his cause deposited, and created the initiation of the spectacle as ripe of night and daunting dread. And that defines all happiness— For what is ready is achievable.

CANTAL: Downed his eyes. He was looking for seed! in the dark.

BOY: I was called by friends, to prepare for some outrageous engagement. Yet without these machinations seen nor heard, I'm sent to you— to soothe your head and temperature, as a gesture of copious comploting in the blind—

GIRL: Oh, be alerted, boy! All is fashionable at that launch to be elated through. Yet have I seen the twirls of flame that angrily relent their purges to draw forth and bathe! awaiting pairs and pairs and pairs—

CANTAL: Or Mama seems kind to nurse.

GIRL: —Be anxious! and anticipate the tug that brings you to this spot—

CANTAL: Or Papa does kindly name his terms for ye.

GIRL: —We are central, for the nocturnal notching. And could you ever doubt it?!

BOY: It was so suddenly pushed to my awareness, that now's the time of sacrifice. And one can not recover from improper wounds or misplaced steps when made that heated path to roll! along. But say they to me it occurs (an) equinox of our persuasions mating— as two are sent, and two are used for two to come. And I'll have this gun at least, to ward off evil erring as the treks of mutuality are made along those soil-torn tracks and deadened graces—

GIRL: Resist your fear! For it is hurried of the night, and the burning rushes our enlistments to elope of carnal cause and casualties of grief, wherefore the graves are rustic in broad leaf and strong of wrap— Pretentious green! in the pitch dark, through a calculus of woven hearts endangered of their substance, make brave array for our procession floral and florid.

BOY: Two are two, symbolic— Girl! as twined ten thousand, stoked on poles and stuttered with— our vows! as can be gravitated towards some wild pastiche of flowers and embroidered flesh! Yet (i)'s darned the dag for our dalliances, brought to this solemn immediacy so soon?! Then make with caution briming floods of anxiety and relentless quarrel of the internal suplications to our passions wrought of casual innocence and fun. Now to be man is headstrong and hurting. And to be woman (i)'s wrung of a battering to conform to life's catastrophies of grim grinding servitudes. Have fear of this maturity some day!—

GIRL: I'm in a rush— to placate the stars of our cajolery, into persons delicately firm, and of a paired ideology to squirm! through the daily chores; as so with night demands and demeans our sacrifice, this is the definition we're determined by— And for the offspring raised, here at we spring! of nocturnal pageantry and celebration our worst provisioning of everlasting crime! the decadence of life to spend as won of voluntary adumbration towards our serious decrees of spanned, spirited, currented interment—

BOY: Does make a heavy fog!—

GIRL: The haze is ablaze!—

CANTAL: Led he to this?! Want for protection is he, needed of the hesitancies—

BOY: Wont for your amulet!

CANTAL: —There's much for a pettiness' sake, dear children. Diverge not his singular employment into the rut of your calamities. He's simply of a scientific mind, and requires my assistance to withdraw him from your dangers, as more innocent is he than your infinitesimal infamies of flowered blossoms and your wrangling past the paradisiac infancies you've embroiled. So much that wherewithals, of person, appear to align themselves to make complicit this embarked stride of coupling, then at the wherefore made whereby this is achieved heaves not more sigh of one's destruction necessarily, but is simply a stone of composure worked through— his singlemindedness to exotic details. That does not portend your binding frights, nor gain for me release of responsibility to habits uncovered or recovered or drenched with. But how you serve yourself a flame is by the stars' dementia: fate, chance, luck, and taming of the maim, the smite of heaviness to head (u)'pon realizing the cruelties of your attitudes an' missioning through life, and the drudgerious weight of labor that entails. O wasted time for dreams, not with the prepossessed endowed. And we appear, as visitors, not to instigate acceptance for some rite, and its cutaneous mends, but to observe ourselves as wending through ephasias— what's too difficult to describe— and to just experience the extraordinary brought back into ourselves, an' to eat it! an' chew on such succulence as could be provided by more native spirits more intune to the damascene rising with a growing dawn of human happenstance an' aspiring. Then make us not out of grown gods offering themselves as ravishment for your wiles, but as creatures just as simple as your own crudities employed upon an' ov'r this most spacious worldliness of view and thicketed pastoral(ization) of existence. Yet do I insist upon this quaking swirl a walk to give assistance to my nose, and bring for you some comedy of motion to our movements, as you deploy at once an' to deprive— yourselves as much — the depravities of our investigations. So sound are my wishes, as silently they're heard as I would pass. It is too signal, to approach that vase in the dark, within a night's capture of assertion, the implications ceremonial, intentional, and poignant. Were could I to, have warned the fool; but out of ignorance no warning can be made, (ex)'cept try to avoid the unimagiable with the unseen, and the unseen with the unsolved of one's inquiry. As is he fetched an' sent for, not a granting gratuitous for your exploiting as a sprite beckoning of purpose or a spirit's grace to achieve your certain, perhaps circumventing aims as a dimness could enlist, but scold of wondering for less serenity of sight in that inquiry for some can become relentless of a drive, an' to this hour may be thoroughly uncalled for. Yet such is time never commanded of its progression, so be an urge not prescribed for any fortunateness.

GIRL: Whereat to roam on a moonless night, it is divined for some determinate deed devilish of all heeding and any head provoked, as could clouds bring forth rains and shrouds, and a blanketing of light to induce the torch for some proceeding. Thus is done for our management mandated to this action that now occurs to us, with a readiness of persons and persuasions. What more could do this, and eat grass too? It is the sign inimical, for all passions to be pounced upon until they're rarefied through an earthly marriage, or marriage to the earth and firm grounding to one's dependencies of service, further development, growth, and transcendence into productivity and the fertility of populations.... the sole eternal purpose of the living—

BOY: Does remark of a decay for dawdling, the decimation of a dear impositioning of weightless worries.

CANTAL: Come to. (*sips*) An' have you seen his chin? Is it bold enough for you to behold, grandiose of some severity to temper time around this tenebrous festival? (*sips*) So heavy as a chin. My head does bane-ly make of soundness. Not to regret this surge, or stay persistent, then does the brow dry of heat, an' damp humidities wassail for to help you, l*aaa*dy!— lass.... absolved of me, as through generations distant, your custom born. I am a cruel thing to hold.... decadent an' debauched of my caring. Yet is it more surreal to cry at wondering, an' stroll past such eternity? without a view at all?

GIRL: ....It.... sits by the panoply of hearts—

CANTAL: Canine, cowed, or caucasian?

GIRL: —the vastly contained and ennobled of pile and potting, as flames light upon a countenance bewitched for all its feeding and green expression—

CANTAL: Stone encasement to a mental mush.

GIRL: —that has been ate itself.

CANTAL: Remake oneself, and regenerate the purse of pursuit. You say.... that this intoxicates. (*standing*) I have some bullets in my pouch.... To make for more pronouncement of our straying, I will accompany the night.

BOY: It's harshly done—

CANTAL: Afraid already?! I must recover my friend from his heathenish hedonism. An' you can shoot at stars, for all I care.

BOY: I have become.... solemn, Lo'Cant'. This is the last reproach before manhood. (I'm) Scared with.... uncertainties only—

CANTAL: Keep hold of it, then.

BOY: —but I know it must be enacted with this current's stream. For that's the most remarkable of me to deliver— obligations.

CANTAL: It's a.... fouling stuff to eat. I know. It brings you the — mental colic. An' he's all in pain of bewilderment. Well.... there should be one converted observer, adult enough to see the betrothed put to union. And then, I take it the parents and grown relatives leave all of this for yourselves to orchestrate an' realize, all on your own an' alone to do it. That's a novel idea, among the friends and the cereal gods whispering platitudes of encouragement, emboldening— an' safety, through the august steps, where the priest be.... peaceful acknowledgement by all, acceptance by everyone. But you want *me* to oversee it, for some definite admittance, and profound sanction of the real— caring.... So come our hearts an' flowers, children. Let's leave for the ruins, an' have this performed totally within your most acute circumspections, that there is no falsity involved, or legendary ghosts bereaved or amused by your illuminated strengths, orange-glowed perserverances looking about upon the process(ion)ing figures. (*as the three start to leave the tent*) A march of greatness, or gradations of obscurity. When the scale marks are so fine as to make them hard to read, still may they measure accurately our worth's dimensioning an' load.... It is entirely dark out there. (*sips*) Have I come of cup, that you'll return with my thanks. For be this all approved by me.... of me, on me won. Unsure will be the sun, tomorrow, unless we proceed. (*They exit.*)

Act III

Scene I — *A meeting room in* **Mytdra's** *palace. He is with* **Holston**.

MYTDRA: Are you absolutely certain of this?

HOLSTON: As sure as I may follow turns and rumors proved of comment. He has been seen there more absolutely than can be denied a periodicity.

MYTDRA: That wouldn't be so unusual, since he alerted me to the region.

HOLSTON: But for of what does he now with some resentment plot? Sore might he be.

MYTDRA: A trifle, yes. A trifle, no. He is ambivalent until I lean him one way or another. *I* am his love of country, though he travels throughout it with his spare time excessive to such wandering. And we make for beautiful countryside panoramas, since I shape all handsomeness of view related to my charge and contribution of concerns.

HOLSTON: Yet does he meet with a begging lady there, most ostentatiously to note, almost as to protect a beloved—

MYTDRA: Fenic thus? You do not like him.... Yet for affection, all o' a man may have some.... kind gentleness for dame. But more to share, and from me torn, less loyalty evolving? I do not count him so dissolute of being.... though I've picked him for his resoluteness for me— my positions and opinions perfectly reflected. And of his nature kept.... singular.... this type of diversion should be trifling—

HOLSTON: But it is hidden regular.

MYTDRA: —to me. How is hidden? How deceitful can you implicate it?

HOLSTON: By the span of time between visits.

MYTDRA: Is that irregular or illegal?!

HOLSTON: A townsman there has been annoyed by the unusualness of it. That's how I have been alerted, with the eventual passage of this information.

MYTDRA: So, beggars to concern me?! Devotions to inflame my ire? I don't care how the man may play his wooing. They're unpredictable entanglements, and may recover of themselves in the strangest forms.

HOLSTON: I was not particularly alarmed, your majesty. Though this penchant for giving, by Fenic, is to a husband-less family, which seems only slightly peculiar if neediness is found. So

particular, though. Then I had lightly initiated a review of his most recent activities, to find any more evidence of— peculiarities. And this comes to rumor, my lord, that he complots with others— for some reason that escapes me yet. Paying monies, though, must signal something significant. A mound grows somewhere in this entreating. It's this funding that changes matters. Seen to be extravagent, deliberate, impulsed, almost subscriptional. Yet to the indigent, the mendicant, the destitute? More than a social conscience here, but a protestation, a possible beguiling towards some movement— Only that the fellow is so highly placed am I drawn to suspicions shaded by these activities. My policing of the commons and conspiratorial alike alerts my notice to thin threads bobbed of sharp needles and a quilting malice. And for intuition's sake, I've seen lesser suggestion lead to surly havoc and disturbance, within people who think themselves above the fold of their compatriotions secreted or recessed.

MYTDRA: But this is not. And you simply seem astonished that the man might be in love.... with someone post-privileged. The oddities of the manner of romance change of ages and means, and leads to the the spoiling of many with ease, though without the rot of our contentions to stay civil or cerebrally fit. What of this alleged complotting, though? State you only rumor? He is my envoy to the environment that surrounds me. And not to remain insular of perception, I have him meeting with many people, and observing the general behaviors of our populaces varied of distinctions and class attainment. So he might speak with almost anyone, that I'd not handle personally as could his worth be bound to my desires to know what's real of my estating. A form of mask is he, grotesquely brilliant to my crafting to subject, but quite comfortable to me.... upon your doubting— Proof! constabulary.... of some defiance or dissension, then provide. For my instinct is sharper than your own, about the personalities that I permit to be pleased with and satisfied for. I have no doubt of your.... honest ways to think of this. Yet to no one but myself is he particularly dear, as for his character of tooling. Now you present he might be peculiarly dear. Well.... dear is dear, is affectionate, is affable, is an affectation of manner for the purposes of devious discovery— sometimes. But accusations are raw of flesh to bite. And do you accuse him of any misdeeds?...

HOLSTON: No, your highness.

MYTDRA: Of drunkenness in the streets, or rabblerousing, or bouts of boasting discontents?! Not of his happy nature, Holston, not of his precious make to be subjected to my municipal enlistments. He observes for me, and describes. And I improve his eyes' sight. Maybe his longing is my own unshielded, reinterpreted for more public viewing. But treacherous? I've had him as a soapy balm, for washing of my feet, with no resistance to the chore. Do you understand? He cherishes this way, as may I divine less slavery than sovereignty of heart to keep for all matters a felicitation for my wishes— they being the nation's judiciously imposed. And to this slight contrast of intention you bring to suppose, have you more ideas for him than the simplicity of his public actions?

HOLSTON: I know he is in no way related to the lady. And his history does not suggest a greatly lodged residency nor benefaction to her town other than as a traveling visitor. Of course I can not discern much of the details to this, but he has met her *several* times, always with a sum to give.

MYTDRA: And do they then retreat together?— I needn't know

this. It's not my point to afford for him his pleasure seeking. More that he may be human is truly my liking. Yet I myself am not so linked to such desires, having had my fill of such possession in my youth as to leave it meek of will for me today, and more boring than ravinous an undertaking. Yes, this does surprise me of him, but not demolish his knees. And I'll have him entertain the women courte— seekers more, if he does smoke of the more visceral inclinations.

HOLSTON: ....That is then as can be an oddity to relate, your majesty. There are reports they do indeed retire to her abode. But that is all, he does not enter— merely escorts her home. So then, the time must be spent in discussion. Only for this.... expensive talk.... has he paid for conversation, and with she struggling to be unragged. There is no further relationship at all, or for messages delivered or received, instructions learned or taught. More rational are costs, your majesty.... to provide for some injustice felt, deeply earned. And may I say he is not particularly rich, being your confiding servant— but much less an officer to me.... then no official really, your trained louse.

MYTDRA: That definition hurts him, Holston. He suffers much contempt, to be described that way. And does he endure it wholely, to sacrifice his social helm for me. as I may be disgraced only with his portentous bickering to detractors of my will. Blame for him suffices of the evil done, as stands for you the subjects he admits to.

HOLSTON: They are wiser through his mouth than yours, be (them) learned to leave by me. Yet dirt defines itself by tastes to be spoken of.

MYTDRA: Then have you not conferred with him directly, of your solicitudes?

HOLSTON: I've not seen him, before you, spoken of to deem the matter precious.

MYTDRA: Ha! He's due now— (*yelling out*) Cur!... to casticate your presence, come in— Fenic!

FENIC (*entering, from a back door to the meeting room*): Cruel posture! Holston.... to malign me as if criminal!

HOLSTON: So overheard you this redressing—

FENIC: As you do overhear the principals I pertain of to this king!

HOLSTON: My anxieties relieved to cast suspicions out and before him, what you may do—

FENIC: I am of league for lessening all harrassment of our Mytdra's regality—

HOLSTON: So dare you say with ears for some supporting. Yet bed you to provoke some calculated calumny of his rule, that must spread by ferrets to the dispossessed and tottering—

FENIC: I am not adjusted so.... but with fidelity bleed. And you should have come to me, for your spying made to dispose, and your exaggerations to mince! or miter into something more constructive than this! manacling of truth to worthless thought, locks of miscomprehension and even lying of me.

HOLSTON: There's no way on earth that you can ever explain yourself other than with some malice attempting to perform. You extract for a nefarious deed of doing the services of the poor and constituently abashed and abrupt. And that's so often the mark of a fiend to the legalities in place, that I'm sure you have some shameful component to your public meddling, and shield proponents of anarchy that have gained your favor— You learn of so many, and know of so many, and have suffused their taint to your indebted hide to imbue their consciences of anger and wrathful revenge—

FENIC: Indebted?!

HOLSTON: Of the evil sentiments that must make for your nature solvable to exist!—

FENIC: I hide nothing of myself! and am not covered with that kind of shame, but tell all to our master, for his mastery to construct from such knowledge the entire(ty) of his advantages, such as I may be privy to them. How am I known? such hatreds as you'd confess, the populace to mak' spire with?! Such salvation of remorsefulness! Such soothing salve! Such!... as our king well knows that there are many who are unhappy and live through distresses, depressing of themselves their miserable circumstances. But that is not my bait! And that is not my heart to batter with, per each pulse of protestation. I am sincere to him with this disapproval sensed— and hide nothing! to tell, as he allows me to hear. Yet has this dawned on the police?— a scoundrel taken for! as I am worth? Then plead me more of the ear than you of your perceptions of me. And I do tell you with candid fury that you are abrupt of contest here for his influencing with my knowledge, and that your scornful contempt of me reigns little of the tears of caution— I have applied!

HOLSTON: What give you then, to the beggar for?!—

FENIC: In friendship am I torn to pieces! Can I not bleed freely of my wealth?! Herself and her small young I've come to know— and I help. Does that cause you to accuse me, in my private affairs, as if a cad can not?! Feel me for some deviancy here, that I should not be capable of charity, of sympathy and compassion? Is that what you make of me— of us— of our Mytdra's rule?! Then there's some disease to your thinking, because *we* are not so coarse to grate men into a fibrous loss, down to the despicable fiendishness you ascertain— or even hope for to provoke towards license of your administering panic!

MYTDRA: ....There are fires of revolt, petty but smoldering in several places throughout the realm, some even caused by my own actions to placate the whole as best can be done for a total of prosperity. Distortions inevitably occur of the common sheet (spread). Fenic alerts me to everything, and all that I ask of him.... to learn, to know. The trust is in his candor to be pierced with misfortunes and yet retain his happiness for me. And for himself, then what is odd makes a sublimity of effectiveness, rolling along from off my swat and sway. How learn you of this— lady and her privations?—

FENIC: Through friendship, my lord!— with her husband, who was himself destroyed, by the loss of his professional gains in that village which receeded of fortunes and fortitudes, due to your necessary rectitude of other towns—

MYTDRA: That.... charming hamlet you did find—

FENIC: Aye, my lord!... and described. It suffers much, now— It's peoples are uneasy, and some collapse into a poverty that may embitter themselves, against each other, against the world, but mostly just 'gainst the cruelty of their difficulties braced.

MYTDRA: And.... that informant, Holston, comes from there.

FENIC: Who?!

HOLSTON: ....The name is.... Pale, I believe.

FENIC: I do not know him. Must he be as any one, from the dwindling town. As fanciful as I may appear to see, that notice is for comments I'll never deny. For upfront I am with any person dealing, and to annotate the remarks of any observers with the verity and openness of my blatant presence. There is no skittishness nor skulduggery involved with my sight. It is boldly honest to be lent for view of conversations, and hardly ever arguementative or vitriolic to assume of my debating, comments, or small talk with people. For I am gathered with freedom among them for many public convivialities and social functions. So what can this Pale say, but that I support a friend! It is all. It can't be as unusual as my appearance, to partake of this courtly manner with a decency and as much dignity as can pertain to the event. But I aid a friend, and only wish to help that region recover for all of the inhabitants there. Therewith some might accuse me of wrongdoing? That is the irony of shame and discourteous embarrassment.

MYTDRA: How might *you* improve that region, Fenic? You're hardly a municipal planner, and haven't the wealth to propel such a keenness by others.

FENIC: Through agricultural refinements was my thinking, my lord. Their best industry has certainly involved their soils. I think they need a specialization to harbor some renown, and to this aim I've been working—

HOLSTON: Working?!

FENIC: —with persuasions of the influential and their capital.

HOLSTON: Then your— courting has been persistent of associations. For there are rumors that you involve lords with some plotting— and behind the king's back!—

FENIC: Have I been always?... There is no conspiracy to this. For this particular matter, which concerns me greatly and of urgency to find some expedience, I've simply asked Earl Ancher to consider planting, on an experimental basis, his remarkable crop of eatable grass he's discovered, in an area that needs some such noteworthiness of development to regain a proud stature and state. It is a simple and straightforward idea, as could he take the chance of making such an overwhelming contribution of excellence to the realm as to provide for his name the utter astonishment of history.

MYTDRA: Ancher.... Well the fellow seemed quite gracious enough at our last meeting, where he explained this discovery.

FENIC: And promoted it as something quite outstanding and to be tested. And this man is certainly no revolutionary against the throne.

MYTDRA: Yes. No hostility of opinion at all displayed, though somewhat independent of his thinking. I myself count less for this profundity of grass. It comes from a land of exotic tastes and ideas. And what is eatable there may not be so here. For he hasn't the proof of his mouth to ascertain this, evidently. Yet to attempt to sell me on this— concept for consumption.... shows an interesting aggressiveness to attract my investment at such an early stage of realization.

HOLSTON: It.... might be foolhardy, for my friend, my lord.

MYTDRA: Well it's his claim of neck to bare. And to now couple this with any cur's indulgences strikes me for more dear thought.... of something substantial evolving, as dear a cur may be to me— or expensive. Motives are certainly mixed here. And what occurs was open before my sight. Who shall I blame for the impetuousness?

FENIC: I shall be thrashed for failure, my Mytdra— and flayed willingly. I felt your ignorance only to be some gratitude for my weaknesses with scrounging desperation. But as always I hold back nothing from you asked. I have in effect demanded of him, that Ancher be correct in this.... challenge to the astuteness of his dealings. And he has totally accepted this danger with me— Now am I the dog! Holston, that your friend himself is involved with my attitude?! towards this nation and its ruler. Then see how base is this imploring.

HOLSTON: He takes his chances with you, sir, as others have.... Yet, if I've associated you with sedition, then it's as much as he. And that is too headstrong for an open acknowledgement. Such is his nature to be thorough, as is yours. I will apologize on this gesture of your candidness. As sees the path, my way through trends are they employed— eye of circumspection! But suggestion is not proof.

MYTDRA: Not yet to hound or dog of me, for this awareness brought to. But more.... but more, but more.... a nerve to pinch— seems pinched. Nor is sight fully veiled, from closed or shut eyes. Here, though, be an apogee of awareness to claw from, with some distance measured of these lords, of faithful estimation and esteem. Now, intuition is some path to follow, Holston. And I'm all for your sentimental pursuits, Fenic, finicky as they may be to accomplish. Yet judge by more of what is done than said. Actions are more real than words and pledges. And I'll provide this wave some coach of wariness, and coat of submission. For all what stirs is pushed. Heaven knows the stars are anxious in their movements regular, to find occasional discourse and discursiveness.... Now kiss the cur!—

HOLSTON: What?! my majesty?

MYTDRA: — Shake his hand, that a pact of some discocery is made, in concert to my wishes.

FENIC (*with some reluctance, extending his hand*): It's deemed to be as thorough, meant the bind with gash of blows.... that I am fit to approve your mind's setting. Though this does (*as **Holston** shakes his hand*) hurt of wounds to speak.

HOLSTON (*shaking hands*): More with an honesty of lusts, and applications of our trusts, Fenic. (*They break.*) My suspicions make for a current. And what of yours? sound shaking soundly made of sounds.

FENIC: I can only be fully reconciled with a very high achievement, since the abjectness is more than merely current— but spreading of paces, and quickening of destitute results. What am I personally involved with harms me more than cruel harangues, being a cause of misfortune to friends entrusted to, being responsible for blights through noble erring, being a hound of boasts and blabbery with drinks. Some ales do choose our wisdoms improperly, yet what promotes them are our own choices enacted.

MYTDRA: So be you personable of senses, have you in no way such powers as to cause storms and darken the skies. The plurality of might is one. And you are both my candidate executors to be receptive and inform. This is my patent, to find patterns of our social consciousnesses these days. But be forewarned of harshness without bitterness, nor regret, nor restraint of the tenacities of terror— should I choose retaliatory courses 'gainst any perceived or uncovered subversion to thwart my governance, whatever complications may imply of seemingly simple tactics and stratagems of suggestion. For I smell heat finer than a frost.... as must things burn! to freeze of uselessness. And such detraction is not my gain, but only to suspect, and with proof— destroy! Convince yourselves, then, before you con-vince— me! So may be your discretions 'gainst disgraces. But what I find, no luster's ever hidden if it shines.... and no deception's ever found until it makes of itself a certain and most definite crime.

HOLSTON: Such to be surprised and startled by, then what we should construct are mostly airs to surround with and control. Preventatives! are the main manner, and the most forceful or persuasive of hand slapping. That will I seek to do, your majesty. (*starting to exit*)—

FENIC: Seek Pale!... and contemn his leverage.

HOLSTON: That is for a party of observations influential, with finding the unusual to complain about, Fenic. (*Exits.*)

MYTDRA: Stem this discrimatory temper towards my policing.

FENIC (*quivering slightly*): Oh! My holy, holy host. Was not in me the least to deceive, or even to contemplate dissatisfaction.... but only to find some remedy to my actions. For only I have wronged, while you may only be right. And so in my hands.... corrections are naturally awkward, stumbling and prolonged with errors, prone to mischief, produced of a spine of tenderness through worry. So am I weak to think but only feel, only experience imperatives and address them with mawkish mind. Here am I placed, to find a cure for friends. And here did I alight on that possibility, totally uneducated to its promise, but with bright hopes pursued. It is my station to find rescue and resource with you.

MYTDRA: I'm not particularly unpleased.

FENIC: I would not offend you to endure any measure of pain. My heart severs for solution, escape, resolution, absolution.... obsequies for your approval— and obsequiousness to your commands. But I an charged to find my destruction.... her recovery— theirs.... and the recovery of a land, an entire village swathed of my ambitious toying, as I struggle to disabuse her.

MYTDRA: Cry not for the mendacity of fears. You're not sufficient for these tears to leap unto the floor. How I may manage now these twists is purely due to a recognition of them, the twine about me and the weaving.

FENIC: I could hope to bring her to court, for some service housed, to better of the family all prospects.

MYTDRA: That would reach an ocean, for a drop pitched well beyond your means. See I more complexity in this state than confusion, but see you only deliberateness of soulful and sincerely felt obligations. You are simply not the matter for this, Fenic, and incapable of weight burdening to restore life and reconstruct an entire town.

FENIC: The immediacy of need impinges on my bind to humanity, and the sanity of my exixtence. Yet do I sacrifice my head.... my income and position— to save them, and prevent more loss. Two small children as well are involved, encompassed of the sorry state I've dredged up.

MYTDRA: (They) Would not be as serviceable as yourself.

FENIC: I'm as an uncle to them—

MYTDRA: There are stronger tides in motion through this sea, can't you understand?! Plead more for my acuity than to deliver on you tasks you can not handle. You've no correct notion of the life, an effective husbaning or paternal supervision of a family?! You're not of that kelp and wouldst drown yourselves. Here should I ply, for your assistance to be kept— functional, instead of with more messiness to make, and sentimental mending into horrific squalors greater than you see. But stay the paw dawn, and walking freely— and concealed auspiciously. And what will be delivered of your chores, in compliance to me, will be the charmed delight of setting right your present worries, as I have often corrected your mistakes of independent prowling and portentousness. Does help to stir up the sediments and uncover boring boars— as could with eel enlist of forms appropriate. Yet for examining are your intrusions used. What know you really of usurpers and their potential?!

FENIC: Only as defined: potential usurpers, I'm told with casual knowledge (of) what flies, what may breeze make of opinion. But always with attempts to failure are they revealed.

MYTDRA: Oh! The potential lures, and pulls them out— to make clean the (land)'scape.

FENIC: Who I do recommend? for expulsion, they least escape. They are related to eels? They win their pauses of withdrawal and imprisonment, such as they're worth to characterize of capabilities. If they fail you, they often have time to run, a tacit endowment you've allowed, which makes for surface a pure stream— or tranquil brook.

MYTDRA: Not all are doomed of testing, Fen'. Some better do succeed of their venturing, to find more position around me, as much as they may float of engendered rank. Yet, a few would try to overstep that mark, and overstate themselves of thrusting worth— and (with) follying about!

FENIC: So are they proud to be— overdrawn.... of their talents. One sense of this— one whiff— and I am tasked to tell of disappointments brewing. You are keenly alerted of what to expect from such a creature. Though you have already suspected everything of the truth, and only await the evidence provided by a second's seconding a profoundly royal intuition. I am often astonished, of the sheer preeminence of this mastery, that you are best to explain what I can only conjecture and admit.... to you.

MYTDRA: More than a gift, but a prerequisite for my control, that is a regal dress, a rainment manditory— worn.... for my imperious reigning, and tailored through the immaculate wisdoms of gods. For how else may one be *born* into such station over men, if not with higher demands of you implied? So upon high this rule is brought, and only upon high may it be destroyed. That's why I trust the simpletonness of your nature, the simplicity of your worries, and the similitude of your loyalty to my purposes. Now.... Find you lords remarkable, they constantly surprise you— and disturb you with much erring. And do you think I'd be so close to one as yourself, a gentleman and noble taut of the temptations defining social privilege and advancement? They are lessons observed, Fenic, of irresponsible adjustments to achieve false equilibrium of their weights. That's mainly how I suspect their strifes, from being— behaving as— unnatural, or extraordinary. And while you can never understand this.... explanation.... until you realize it yourself for being actual— as could be a religion comprehended but never fully delineated or interpretable to solve the riddles of one's salvation.... there is some complaciency felt, to these torts attempted and squashed. It shows for a justification of means and manners ever to be exercised through tests and struggles and difficulties and hardships— and devastations.... So are your anxieties and concerns justified by any wronged friend, by fate or fiend or the fanaticism of circimventing beliefs. Yet what is to worry of has not been through your causes, but by the divinities of correctly adjusting weights, the scope of these readjustments well beyond your passions to involve. This is more inevitable than elected, and more prosecutorial than prescribed. For what is necessary to do in any affair often brings forth a residue of being distraught.

FENIC: Seems like a void self-sustaining, my lord, for interpretations, how emptiness does weigh one so to the ground, and grinds your girth into the groans of game and gamy substance, if any are to contend so of misfortunes despite their venialities.

MYTDRA: So say you're not responsible for a winter's storm, or a desert's heat and freeze, or a climate's changing. They're beyond your control and reasonableness, sufficiencies and servitudes.

FENIC (*bending slightly*): Yet have I done my head into your hands, have been molded into harms.... which splay contriteness for this suffering into a fluxion of rivers seeking tributary status, to find a home collective of solution, these waters oceaned of their drops. Into your service save me, and oust my crippling of others.

MYTDRA: ....So that some might is made of this, perhaps.... perhaps some billet may be found for— her, she that does you gracious amicability still, under your error which is not!... for payments granting— I bend only a bit, towards your sensitivities to uplift and improve with my organ of prominence to allow for (almost) anything. Yet soothes your qualm at least with this expediency.

FENIC (*raising head*): It does, my Mytdra, to know that some of my thoughts are possible to achievement, and that I am not so twisted as to be errant of beliefs and hope. So may I be mauled vi-

ciously of attack, with descriptions hateful or villainous, yet stay this your dunce of shaping. And I care little for myself to bruse or burn, and most for others in their turn to allay who have never truly deserved such treatment. Then only rend me, with their suffering to transpose, and of gaiety to eye will I be of happenstance to bounce and roll and *rire*!... of what's occuring to my stances and poses to ponder. For I give equal chances to all, to be persuaded of my rendering for your causes and insistences. And be they not as candid to me— it is a threat to you! yet of peel to break or split. Then with my peal of laughter ripped into astonishment to warn for the conversion, is how I treat myself to otherwise befuddled lords not well secure of their proffering to you. So be I as if led to bark, not squemishly, then harrowing for your approval. And I am greatful of the nerve— enjoyed— with your demonstrations ever so generous to me, to show its soundness as if as kind. So be one king of mind, I am more subject than these seasons bring.

MYTDRA: Subservience is in the ore's great promise, for the value's in its use and artful refinement. Then be as grateful as a jewel carved handsomely towards brilliance of a decoration, that what is represented 's more than a paternity of posh effects but rather of my means much to express with individualized vigors.

FENIC: I will act as keen as your colors may be brazen of a spectrum's dominance over the realm, my highest heaven's lord. To treat a rank (thing) such as me so nicely 's demonstration enough of the profundity of honest power and pride, with this relationship to churn of into a sublimity of tasefulness. How meanly low I am, to be more made of thee, these years suggestive of my cunning to persist within your honor. That is my only noteworthy passage through life thus far. And I adore for more to become of this— with shapeliness of fervor— your indecencies pronounced upon my lips as golden.... ore or lease as or(,) or all submitted to your summit's highest mounting mouthed for me, the *right* to behave as you wish and may command of anyone. Then am I due the pleasure to obey, being the indecent face of your reverence—

MYTDRA: A façade of focus.

FENIC: —which induces the might of others.... or their perhaps espousals.... for your termerities and tendencies to only strengthen. But this is *la chaleur du cœur* which you possess to inflame my stediness through doubts of many improvised applications of your mandated strictures that can only improve a state of being.... once *you* have analyzed thoroughly their overall consequences. I do not claim myself to think at all, compared to your processes of mind and thought. And so I abound myself with security, to prove you right in all eventualities that occur, and to be correct of them, corrected by them, and solicitous for their causes appealed to. I should not understand ever such greatness, but only shiver, shudder and shuffle, to resolve myself of the heat to your presence ever gradually realized for the importance you ornate in me.

MYTDRA: Deep be earned a radience to the soul, foundations forged to be inured with sound rigidities and just bending. Yet errors elicit more the punctualities— to find out what to do. And admiration is a frothy substance on top of beers, often blown away to get down to the serious drinking. So be taken by me, that I thrive on positive results to sip.... And you are lustful, no?— So be it as, with this lady to, per chase of love.... enthrone.... upon your heart's imperial hurting. I do not absolve such things as loneliness— to be the spur on mistakes. Find your fitness, Fen'— though I had thought just as fine fending of your ways— and make your proper carapace of hearth and heartening, without a social blundering to my prefixes on your servitude: *ad admiror tu.* You are for my peculiar leashing wondered at, that of your nature fruitful for me and repellent to others. But this is a dodge, of my daring and your doing, what may become divinely instilled of some perfection of our purposes. And that you find throughout this some amour is really wonderous— So.... be emboldened with my care, and not misguided.

FENIC: For love of a responsibility, yes. It is a pretty thing. It is the most beautiful enwrapment of a person's founding fleece, more just to his description than his wayward hairs, to be dependent on a good purpose, a necessary one, an immediate one— an enviable anguishing of devoltion, the high(t)'est fear, the purist poke to punishment.... that is your own making of the scorching wound, to heal upon your nerve of baring.

MYTDRA: Such feeling may bewitch the pain. But what is that compared to elongated dejections? And could I thought you hurt more, I would have been much more crass and indignant. But you are Fenic for the whip or stick, and can take these blows and strikes superbly, as like a super arch or awning hit with lightening bolts and rain.... So can I count on your resiliance without gruff malice to my officials. There is a superlative motive to your wonderful nature.

FENIC: Does seed upon me, my lord, your sprouting, as your majesty demeans with praise the worthlessness of my mettle. Hurt may only belittle prev(i)'ous pain. About my way is more to gain some stepping with.

MYTDRA: Then through the back again (*as **Fenic** heads towards from whence he came*) converge upon the calculated groping of my meritorious means, as I go out to digest some messages of ministers (*starting to exit*), and perhaps invite back here.... Wait! upon this. And with sealed eyes caution your ears, the stationary pounce of your benevolent manufacture. (*Exits.*)

FENIC (*alone*): Does eel to be shut up with contemplations and reserved heinousness.... Yet be the win, for stupored styling of my trust. It is with eminence, to be deceptive.... and deceived. (*retreats*)

Scene II — *Daytime, at an arbor on **Ancher's** estate. He is entertaining **Bastrel**, who sits on a bench.*

BASTREL: It's with much dispassionate relief, earl, that I may avow you this very definite loan to his majesty's resourses. Others have made the same, though yours is a considerable contribution that helps to keep some matters afloat.

ANCHER: I can't really expect a delirious welcome for it, if it's very much expected of a privilege to the courtesies of my rank and social position. I am faithful for the king's pledges.

BASTREL: Nethertheless, I demand to state that this is most definitely a loan, which will be repaid in time, and according to the interest tables that you are aware of. This gesture buys you nothing more than that satisfaction. I can admit, however, that it's becoming more and more difficult to secure these funds from throughout our majesty's exploitiveness of land and superficial gentry.

ANCHER: Superficial?

BASTREL: Thus grows the direness of the nation's relative losses, that I must constantly come to you and your like for some manner of healthy restoring of the treasury, in a timely way— And definitely not replenishing. Oh, no. This is too capricious of its use. I mean to say a half year runs swiftly, when it's made up of the more aggressive months. A curiosity remains that the closer to the throne become less rich than the farther from it, these days, of those who may be considered.... considerable, and that this is clearly evinced by my need to scramble of sincerities so desperately regularly and irregularly. I fear a practical wall approaches (which) I won't be able to breach, nor even anyone's god. Our majesty's occasions of use of state monies stretches the limits of tolerance for much repair, at least in my hands to oversee.

ANCHER: Are our finances so terribly bad (become), lately?

BASTREL: It's not so much due to extravagance, but rather the contingencies of spending. His majesty insists on some unprofitable ploys, which result in little return on the investments. So the treasury funds tend to dwindle down to such hairraising depths that they must be buoyed up artificially with loans until a period of stability is reached, allowing for a more natural growth. The drops are not so terribly often, maybe thrice in a year— but they can be quite sharp, plummets that would scare the most heroic diving gulls from off their fishing, enough at times to reduce our previous exchequer into a degenerative condition of debilitating neurosis or neurotism, unfit for a general man of high manners. Has yet to recover fully, even after half a decade's rest. But the king's nation is his credit, as he sees it. And that's as he remains healthy, as would he demand by decree— leaving the details to me for monetary policies and preparedness. Thank the vicissitudes of gracious woe that we are not at war or running through a true depression of means. It is only an accounting to him: of sums, tallies and totals, as pockets of the realm fulfil much promise, while others fail or suffer floundering. He enhances the better— as he sees to do with circumspecting aid — but can not remove the drag of the latter so easily. And unusual arrangements of fortune and misfortune may batter the treasury, as unfortunate towns are simply ignored and the more prosperous ones can not on occasion make up for them (their lacks), their taxes, tolls, tithes and tenebrous tenures of commercial commissions to be paid for. He does not perceive the rational argument that all locales of enterprise are related to each other, when they are components of a unified whole. And thus their progressions must affect one another. But this theory of economics is not within the royal compass, and the meaner regions can multiply like a disease— if left unchecked, until tides of luck return to wash away the bestrewed matters of debasement.

ANCHER: Some would say a considerable change of governance is in order.

BASTREL: The king is not adept of such thought. He can see only the glitter of a gold, as the silver tars and the iron rusts. And his answer to all problems fiscal is to, in effect, mine for more precious ores and gems whereat they appear to exist. How can one explain to the heady head of action that the common nature of man is to produce rather than find?! that which separates him from most animals and insects— not that the beaver's damming and the bee's honeying are altogether unlaudable and crude. But man's nature takes constructive and creative enterprize to much higher extremes, and so is endowed with more facility than merely to find promising water holes, oases, profitable paradises and edens, as if laid out strictly for one's amusement of discovery. I understand fully the compulsion to want to continue and exploit the talents of successful centers, to have them grow into magnanimous metropolises— for the national coffers.... presumably to have such wealth equitably spread out across the realm for a base level of subsistence of all subjects reasonably to be maintained. But there's more to ruling kingdoms than that. There are more imaginative courses to choose, whereby one promotes the submerged from advantage to rise into great advantage, as the social standards of a country increase in general.

ANCHER: That is.... utilitarian privilege, where the poor and depressed are led or forced to lose their ornamental identities of dejection, dispiritedness, and downtrodden grimace of association with a country. The sad representation of what is allowed of a lowliness can only be abolished with primed and focused utility. And how the population of a region is defined of capabilities is not by what is not done, but rather by what is. So be the elevating view of some countrymen.

BASTREL: Yes, the *"Steigen"* influence of some sociological opinions. A foreign intervention of views. Well, it's all a matter of approach to people. The oscillation of our treasury funds does not suggest such a method is in place here, under the Mytdra royal. Though I have heard of places where the climbing stops for everyone, and peace and balmy tranquility pervade entire states of mankind. But as that's rumored, what's a heaven's penna to one may be only a dull blade to another's thought, as achievement must manage its strifes. Yet such endurances are oft' too easy a sign concerting for blind approbations.

ANCHER: It's always dangerous for a single head to try and notice a universe of ideas, or even a noteworthy constellation of them. Obviously, Lord Bastrel, our majesty requires a much greater, broader span of advisory talent—

BASTREL: That has been noted before.

ANCHER: —And he would do well to expand his legion of governing officers, even if they do so on a voluntary basis.

BASTREL: They are governance officials. And we only work to execute his directives, some rather specifically, others more generally, according to the nature and complexities (of directives) of these tasks. The more complex they are, the more in our hands, individual hands, they are left. Yes, the number is somewhat limited by the rank of person our majesty assumes to choose from. Though for this reason alone we are compensated for our efforts; and some even grow to depend on this, to help strengthen and secure our personal estates, since hereditary privilege leaves only hereditary wealth. Now your types, sir, of usually younger initiative and vision, make for an interesting though passively natural insurgence — I mean to imply not exceptional for circumstance— to capture more of the responsibilities of govenance from the king, and to increase your say in the important affairs of this nation. I'm not entirely opposed to some evolving aspects for this to occur, a loosening of considerations for appointments by the king, perhaps conducted through his primary officials rather than always directly of himself to speak— for official roll— the names chosen. But too much push in this direction, with the rapidity of blundering force or shoves, disturbs of course his majesty— And I have been warned.... by he who does choose us, and others, to avoid the pleas for change too hasty for regal consumption to countenance as easily as spoken. Our governmental sympathies must always lie against radicalized pursuits or confidences to such intentions. And as well did I expect your entreating as most typical to garner favor. The step onto a land of revolt is (a)'cross a very thin line, Ancher. I do recommend you for your competence, generous spirit to assist in your country's doings— they being his majesty's— and your steadied reasonableness, to allow for.... some more advancement towards influence with the king, as he already finds you quite a sensible person. I'll ignore the desultory rumors that always seem to alight upon the shoulders of personalities (who have) become popular or well noticed. That you are courteously contributory to the weal of this nation is good enough for me. Though there are specious elements who would implore to instigate any manner of rioting and upheaval to address the slightest of discomfort or disagreement to current public policy— or royal rendering— I can not count you among such ignorance of comportment, no matter what is said against your successes. Merely envy.

ANCHER: Some would blame grease for its tackiness 'gainst its slipperiness. All the same matter of result, this depends on the application. I have done for myself.... good involvements, within the districts of fanciful commerse, extraordinary wares to search for, promote and sell. And for this freedom of independent.... application, a resourcefulness of thought and imaginative daring upon real risks and hard currencies— spending, to be sound, Lord Bastrel— we may in all truth emphatically thank his majesty for fostering the cloudiness and sheltering of a permissive and at times even exuberant mercantilism. So I would never fight against *that*, and an environment through which I may professionally thrive. And of course, my.... successes, as they may develop, are employed for the nation's benefit.

BASTREL: This sentiment pleases me greatly, Earl Ancher. Even I am not fully aligned to Mytdra's propositions of country management in total, while remaining a pledged confidant, errors observed and dismissed. Yet I would not contract against any aspects of this, because.... it is the nature of our governance, the foundation of our perpetuity, and the mystery that makes for covenant to the prosperity of an entire nation of beings. For to explain the role is beyond our heads, to be provided of sovereignties devised by heaven's lease, eternal and yet ethereal according to the godly manifests.

ANCHER: One could describe it as a religion, to define such beliefs intransigent and intractable.

BASTREL: The very character and national pride of our people are involved, with this observance of his majesty's license to rule over his kingdom. As best he may, while quarrels may erupt about this— by freer thinking heads than thought traditionally allows.... they are not of quality yet, earl, to pay for their inducements of unseemly ravages and disruptions to state.

ANCHER: That is worldly obvious, as dissolutions lose their firmness with the barest of contentions. Then they are due of liquidation, and contemning of their weaknesses.... By abuses have they been sewn.... But this is much for parting— and complaints are more enlistments to be heard. When a substantial portion of a society roars of discomfiture, it's like a body telling its brain to have the man stop jabbing himself with needles— for no obvious physiatrics. That hurts. And while the brain may like the pain, for some medieval reason, yet does it the body no good to contain this notion long before withdrawal of all sympathy. So see that quarters hurt, they sigh. But is a countering or counterbalancing produced, with screams.... they try.

BASTREL: No rationality wants pain upon itself. (*starting to get up*) And yet I climb down wells, to sever swells and pacify the agitation to our drinking resources— What's this?!

HOLSTON (*approaching*): Earl! Rush! Rush!

ANCHER (*as by greeting*): Holston?

BASTREL: I almost half expected.... (*sitting back down*)

HOLSTON (*with them*): Lord Bastrel. You are precocious!

BASTREL: Of a warning?

HOLSTON: His majesty's nose is aflame, Ancher. He suspects some.... diversity of interests, through your involvement with his Fenic—

ANCHER: Fenic? The man's my maiden—

HOLSTON: No! For that he is his own, knows Mytdra now he holds more devious intentions around the throne.

ANCHER: Devious?

HOLSTON: As not of Mytdra's mind, then to improve a situation stealthily, and thus insultingly to the king— by your devices, or more Lancet's.

ANCHER: What to care of it?!— Lancet's?—

HOLSTON: Fenic's been acting strangely, and I had to investigate. But for plotting in a village, he leads me to *you*. And Mytdra's fully aware of everything.

BASTREL: Plotting?—

HOLSTON: Complotting, ma lord— and with Lancet's plant.

BASTREL: Plotting or complotting?!

HOLSTON: The ferret overheard my investigatory address to the king about him, (was) brought out and forced to confess to his attempts.... to apparently transform a village's agriculture.

ANCHER: So what of this, as could surprise *me*?

HOLSTON: Mytdra is more sensible than sensitive. But this sensitivity is keen for any such alerting. And to now, that you propound, for his interest in that— grass! this is compounded with Fenic's lechery of luring what might be seen to be a force of complicity hidden around the king. And so justly does Mytdra suspect something perhaps untoward or weirdly lewd about his consideration held, as a false sheet of slime be wedded to for sleep. So makes the garment Fenic such a confidant—

BASTREL: He's only a servant, of principled powers. But....

HOLSTON: High servant shrouded by; too often, I must admit, for my services inculcated. But now, Ancher, you are seen as a driver along, to try to influence the king, be he unwittingly

touched, moved— pushed— shoved! through your desires. There's a shock to it when put this way divulged.

ANCHER: I had calculated Fenic to handle all of this regal presaging. But there is no mold to treachery about this. The king invited me to explain the find, and Fenic tests that I might prove it real and useful—

HOLSTON: It had better be good grass! It had better be actual and superior to claims.

BASTREL: What is such a knurly strain on grass to pitch for— of its brightness and sheen, texture and mowing capacities, evenness of spread and resistance to weeds— as could they be in complicity of beauteous nature in view?

HOLSTON: It is— alleged an eatable variety, ma lord.

BASTREL: ....Inedible?—

HOLSTON: Proposed fit for human consumption. That is the spark you must dread now, earl, for being based on rumors through dispatches.

ANCHER: I have full confidence in Cantal's judgment.

HOLSTON: Well I'd better make more sure of it, my friend. Mytdra has been bristled 'bout the nape— unnerved of confidances and possible conspiracies. He'll not take to any kind of surreptitiousness uncovered about him— around his back even— gainfully, unless all is thoroughly proven of a positive outcome and intention.... And so I must advise....

ANCHER: Advise what?!

HOLSTON: The merit of some voyaging, and self-accomplished distancing from Mytdra's grasp, for awhile, of whatever you have fathomed in this deployment of your specialist acuity. Fenic has told of you, and the king does search for chains!

ANCHER: You must exaggerate his concerns greatly.

BASTREL: So is it you so claim, Holston, that Ancher's grass you told us about is actually inedible? that there may be no profit to it at all?

HOLSTON: I'm afraid that is all the manner of a conspiracy, my lord, to believe in something that is not so.

BASTREL (*rising*): You'd better approve of this, Ancher.

ANCHER: I'll send my secretary—

BASTREL: I would attend to the matter myself, as the constabulary suggests.

HOLSTON: This is a friendly alert, earl. I've seen the likes of this play out so often it's almost commensurate for recognition of results that the faster you take the problem in hand all the better the outcome. This indeed is Mytdra's nature of a forewarning, and the astute with favoritism pick this up very quickly, and save themselves—

ANCHER: No, I won't be rushed so easily, to stumbling over my ties, Holston, without much deeper thought. Suspicion is fraught with the make(ers) of error.

BASTREL: That's not nearly the point, Earl Ancher. You've caught the wrong coat, I'm afraid, a coating of some intrigue or scheming at least. And the king must be intensely wary of that these days. For he is tried to be hung of his own deviousness otherwise, to these unsteady times, with uprisings of incalculable sorts. 'Gainst this he must always have absolute control. 'Gainst these he must demean interpretations. 'Gainst your mantle has he wagered now, unless you can prove him thoroughly wrong of these— suspicions. And that should only be done with your personal attention, in presence of the disputed. For the disputation is clearly whether this stuff is eatable, the only thing to find remarkable about it, what could lightly astonish his majesty. That you must judge with your own doing. For as I inferred earlier to him, its profitability— even that possibility— is quite another matter, related to a selling propensity about its quality. And you might bloat yourself throughout its indigestability, forced to take of meal. But to recommend it as so.... sufficient of nutrition, to the man himself, without having tested the proposition yourself.... That leans towards a criminal intent— don't you see? And to compile upon this testy atmosphere an imperilment of implied conspiracy— that could be a deadly testament of reparations earned. It's up to you to demonstrate the opposite, or to find ways of apology. That can only be done where at it is to eat! So is your lesson, sir, perhaps a whole field of it for humus munching, chump to chyme. Then to witness for yourself its taste, that is your tease slung back to your face— and Mytdra's challenge. Endure it honorably, sir, with some vacationing opportunely timed— or ignited— by the friend here.

ANCHER: You play my doubts to me, and I who am a benefactor of the state. Yet meaning to enlist a village for a trial, am I to trial enlisted. And to such a place insisted upon, I must insist upon. Then to Cantal's *quotidien*, I must be reasonable. But this is not a flight, Holston, nor a self-imposed incarceration as per asylum, but only to review in person some promise reported to me— and which must! fascinate. As to my involvement, it is an assignment.... of curiosity. For of those climates are emboldened many wonders, and as many times them myths!

HOLSTON: More safe to say, it is hot there—

ANCHER: Who could blame me for that?! But I apply myself, Lord Bastrel, with your encouragement. And as for implications, I appeal to you to make amends for my reason.

BASTREL: I'll not discard you, sir, with a detraction for running away, but if asked by Mytdra try verily to uphold your.... capacity of gifts and fine general nature. It is a growing decrepitude in the realm to lose such as yourself merely due to taints of asperations or accusations propertied of future proofs. Defend yourself well, as best you can, and I shall do the same.

ANCHER: Here comes the toll, Holston, I had never expected for myself.

HOLSTON: Such be said for some enlightenments— they're devastating! But then it's trending of research to plow too many sides and cornors and have not enough room for more wild and willing developments, or the proper encroachment of natural law and facts. For to strip the land so regularly of its thickness leaves it stunned!

to the calamity of one savoir-faire, which, if not dull of flavor, savors more the flair of flailing fortunes. I've seen it much, too oft' to hope for relenting in these cases. But then.... that is to play the field, and climb mischieviously, earl. What's honest's as is won, the intentions of one's pursuits revealed not through acclaim, but by that which accompanies, burning of flame. Yield to confusion, some, and summorize what you have produced. Take stock of this, and have your truths more easily manufactured 'gainst impervious odds.

ANCHER: Or oddities.... to relate to, yes. That's how our environment twists our goals to strangle. But I'm not totally deterred of nature. Dreaming of what can't be sometimes become— of oddities. And I'll not retire dulled, but rather blazing for effect! now so dependent on them, these wayward turns and stretches. (*starting to exit*) Excuse me, sirs. Arrangements are impromptu (to be) formed.... Have free range of the estate, as be the winds of cataclysmic airs enticed by.... with uncertainties to bear. (*Exits.*)

BASTREL: Oh, that's the worst of it. I'm always sure to be unsure with the treasury.

HOLSTON: Does paint the shores with sand, ma lord, to keep the boundaries. Deflating as may be at home, she stays in tact, yet menacingly at times.

BASTREL: Or for this Fenic, he's the best to watch out of. A piece of mind plucked out the king's head, too clever and too wickedly he spreads of influence to him who'd own us all. Clandestine, did you say?

HOLSTON: A simple purity of mixtures, my lord. Our king still embraces, I'd contend, this fellow shaked with, as could with flagitious arm. His space is the breadth of the kingdom, backed by Mytdra and in back, knotting the cords no doubt— for hanging ropes. Yet may the beast sue even our king, of deviousness. I think they love each other.

BASTREL: So do sour tastes a salt.... or sugar, all three together — they may be complimentary, not exclusive, and not opposites or mutually annihilating.

HOLSTON: So means?... (what this?)

BASTREL: Swift changes may result from this, as the Mytdra may modify himself, especially if Ancher proves worthy of a change. For the beast is simply out of his own mind, come to bark and reprove, of conglomerate talennts lent. An adjustment of attitude is in order, as the faithfully known becomes unusual to tool. You do admit the possibility that Mytdra may have foreseen this change?...

HOLSTON: Perhaps.

BASTREL: As is his want. And thus he staged your presentation before and before Fenic, to animate his sweep or gathering of thought.

HOLSTON: Certainly some as planned, at least half as much.

BASTREL: I'm so interested. The king does more than manipulate. He pulls on ties. He wishes to subvert a crippling of head, one can hope, by finally coming to terms with what may be dire about his.... administrating single-mindedly. He searches for sophistica-ton, in seeing some off Fenic.

HOLSTON: That even the cur has heart, off shame of cruelty? I see that maybe, but it's hard for me to believe Mytdra can adopt an even greater permissiveness than now. His is not a temper of re-pentance to his effects—.... though this be with Fenic, perhaps sure to sense. This is change I can grant, these circumstances being more peculiar than what I've seen before.

BASTREL: Might be his way, to alleviate tensions. But only if Ancher proves himself.

HOLSTON: Count for this, then?

BASTREL: One makes one's dining, by how one participates to eat. I've seen some gravels delicately consumed— by men of teeth, if other means. Excuses can be powerful if coached correctly. And interpretations of the possible, feasable, desirable and ordinary are for the disparate cultures of the world to define. All that might be said could be of a mission concluded, with a treaty to exchange.... exoticisms.

HOLSTON: Lord Bastrel, one can't fake bread.

BASTREL: Nevertheless, I see the out, if the bread be panned, for a clever mind to make of this. Gently grass is free, you know. It is a covering of earth, a lawn of lavished soil, and a turf of tena-cious vitality and fruitful beckoning. Pretty to be seen, beautiful to be enhanced by, a signature of life and habitation, it may be for-given for many lush lusts longed.... And so as a good example are some dealing with it for their behaviors— to improve. I've oft been worried of the kineticisms of keen young princes like Ancher, who only mean well of their aggressive passions to convulse into dis-plays of stance to keep the country vital and vigorous. No one is blind— here— to misplaced fortitudes. But there are ways to mend an awkwardness, sharpen and tone the actions of one's muscle.... by important and desperate distraction. So may he find such care of dealing, to refine his— tastes of targeting, what goals may be al-lowed and appropriate, at certain times. There are seasons to these things of seizure, not as conventional as for nature's earth, but all to it its atmospheres adopted and adapted to. And one might find reward this way, more instilled in courtesies than courage, and ro-main prudence than rampages of careless cunning and deception— however carefully devised or disguised through details. So may a vow of some such enterprise find its own uncanny means of letting off pressures that would ruin heroic containers and their.... trans-formative efforts.

HOLSTON: More for Mytdra. He is a wise ruler to allow such outlets. Yet release is less destruction with more loss, and wealths of mind deplete.

BASTREL: So how we make a stem, I've never known, what storms may bend completely, as for a straw's weathering. The sun will do, to place about its own fires restorative, regenerative, and purifying. The fertilizing ash is not so much a loss as a bold and in-tense catastrophe's provisioning. May more verdant growth be spurred about these results— herbaceous to a fault, sir, with these payments.

HOLSTON: I would try to contain the thought, my lord. But it hurts me much— to allow and yet oppose this dwindling.... of

friends, companions, and capable ken, what knowledge I might make to be only ennobling of a shared spiritedness throughout my policing range. And the wood dries dangerously of ignition, with accident of malice, as its sap is scrounged to open prisons. What even are not as, is this a true forewarning of certain kinds of disas-ter.

BASTREL: ....As even for yourself to visit, constabular of mind. Well, revolt is constitutional, even instructed for— though strictly within the complacencies of the king's subjects. Let us try the turn-around about a sheet both sides the same, and brush off these leaves of wind. (*They exit.*)

Scene III — *Daylight in an open field.* **Lancet** *sits among a turf of grass, slightly hunched over, his clothes somewhat disheaveled and even torn or open to the chest. He is wild-eyed and uneven of face and head, with the perplexity of commiseration to his thoughts.* **Cantal** *stands near him, with a stick chair that he occa-sionally uses, not as concerned as one could think about his com-panion.*

CANTAL (*taking deep breaths*): Air! Lancet. The radience of an open forest reinvigorates.... of scents and woodlawn perfumes.... and the chirping of abundance, sir.... And you've ate too much, I fear— But that's alright.... It concentrates to play upon your mind, this discovery, this lucidity.

LANCET (*still slightly transfixed*): So (as a) bevy 'bout ma thoughts, Cant'. I am reborn to it, this lunge of relatedness, and revelation. And all my cries are stirred to this golden enlistment.

CANTAL: You were wild in the night, that's sure, for sounds and imitations of cracking, pops and screeches and hisses, to be search-ing for.... enlightenment in the dark? I came to save you, stole to that vase for placement, impulsed of wonder to see things blindly, and with a sense to permit just about anything to your body or around it, as you sat gasping of delight. Did you know that the na-tives were actually painting you as they danced?! this ceremony of shadows impinging upon you to conversion drenched with. And your meager frame grew more of the striations, as the many lucif-ers beamed their casts of view upon your varied squat.

LANCET (*smiling*): I was by me, devoured and dipped into eter-nity's encroachment of the shade, the night's pervasiveness to ex-plain— all manner of design an' purpose, with the swell of intoxi-cating lusts and savage love for living, cleared of mind to be crushed of heart into a rapture, losing for a time all corporeal de-traction an' degeneracy.

CANTAL: They made much peace of it, with the vigor of their vows an' ties, letting off the steam of their illuminations as they won the perfection of a unity with nature.

LANCET: Limbs are insufficient things! to drag along, as you transmit your eyes to gaze at the paintings of a universe bound to an' pertaining of, the heavy mixture of mists to dominate your per-ceptions until you find sight candled and diminutive and evoked! of multicolored arrays of pointed sinews which seem to touch you as could pins, there with your head's notice of the sear through depths imaginable through their enticement to explain— that what cones about is but a plan of matter whirling subcutaneously— an' you encompass all of this that might be known of knowledge within the cringe of your crushing entrails—

CANTAL: Encrusted with spiked jewels of fallibility an' vulnerability to your outstretched desires—

LANCET: So comfortable feeling of this unease, Cant'! Drones invisable pushed me to the place, as I caught wandering throughout the evening's dole as for a necessitated urge of nerve an' spine. Once lay me in my bunk, now to awaken hurled! to find more matter of this thought, incessant of its prodding that I am agitated to recover more anxiety and rush of dolorous splendor. For seek we all this meaning driven to, the night's infused quiescence to absorb with hectic might, and loud annoyance placed of feet: to walk, stroll, run— fight with flight, to reach the feast of candidness! An' count on that, count on this delivery to bestow yourself onto the moors empyreal sacrifice, reaching for these answers, Cant'! trusted to be destroyed with and for eternal enjoyment of the carnage. What intense desire is this lunge to know! this.... covetous calm, with fire of the illumination flared, as could be the flames of — fondness! Fondness! Fondness! (*continues with decreasing volume*)

CANTAL (*as **Lancet** exudes*): This could be the mania.... of a faun's fantasia. But err to be the spoke of spoken wheels, the thoughts in flight of ethereal and yet crucial conjuration, conjugation of the pyred souls inflamed of taut approach, as you may apprehend this process. (*sits*) What raw release of mind is in view— the picture of this debarkation onto shores beguiling for more sweat of sanction to this propounded port of inquiry? (***Lancet** silences.*) What is learned of the want, that I myself have lost! or lose to heuristic missioning that I have earned in ways more demonstrated than discernable of kind, our crude, crude efforts in comparison, diluted of a culturing malaise.... or insufficiency to be more human than our husks, this peel of skin abandoning our lusts for luridness and the excitement of details an' particulors, an' very nearly pieces of the flame, this information catered to for conched description— and burial in tomes, omnibuses, and ornithologies of truly wild birds— or soarers and sorcerers. Oh! How I have been disabused, Lancet, of my own worldly pettiness an' slack of carnivorous coaching. We eat of ourselves— man! an' have always done so!... have always tasted of the same, an' ground of similar grains, from off our lands. Yet are we forced to find retreat in villainies of heart, from running out of our debauched mandates, or renditions of a civil cause an' state, society's genteelness laid bare and flustering with fusty embrace. Or do I turn more lecherously of stone to dare imagine this incipience of passion and emotional caulk! How violently they danced an' called an' yelled of cackled cries and twirled and swirled of feathers of the mind!—.... Yet, what have you found, devouring your mind's capacities with the atypical acid of transparancy and brutal evidence exposed— appealed to?! I've seen the char(red) heads (*unseating*), burning my own upon an altar of— of severe disgrace.... and absolute acceptance of my swaying disintegrations, of a firm attitude brought— from out that morning's malice, an' the colder land.... the finer clime, the circumspecting breezes, the capital conscription of thought an' belief of one's nature, as has been taught through ice! an' sea soaked bellows of husky gales an' hoarse commands of screaming froth.

LANCET (*as **Cantal** seats*): O trip of mine. O sea voyage. O waves of foam and gatherings of green-gray power and turmoil, and the rolling rush of undulating might. Then to this land more sight of the disturbance. An' cost, the aim of creatures paid, can bind me to preach as from the gut! For here was ate— Here!

Here!... the fish of knowledge, the scallion of sardine, the shrimp of serrate blades— with cutting of all illusion, my mind wound into spirals of the skull, an' beached before the vase— to pull my heart out and displace into its boundaries, as a flower tossed, as a sacrifice to urge, as a wanting to, as a need for furnishing this stationary monument pitchered on the grassy jungle floor. An' as my heart did fulfil itself to a desire demanded of a forest's spirit, I plummeted to sit before the urn of humanity, dumbfounded with the greeds of delight, an' totally blanked out of thought, for which I was brought to— hurried an' spiked. O silent drones. O lighted fires gathering about, encroaching me to singe. Yes! I have come to give myself this cavity's pleasure an' confinement. Yes! The manacle is beauteous of all imperative locked into, as the head does bob within the void— to peer an' look an' find.... giant snails an' slugs of mind. A dark red pierced through into bright creeds, I paused myself to be, and let the smoke serve for my eyes to breate an' nose to see, as could I make vision— continentally fit, to where I sit an' sat, a dear one— a dear one, brushed with swath of darkness. Imperiled by none, provided with a limp rigidity, I howled low voiced and pitched toward the infinity, with invisable demeanor, hiding not from the good demons who poked my sight. An' there I would have melted of the flesh, with greatest satisfaction of the moult, for what is known that bubbles of the brain, served serving, my remains uprighted into the provincial quizes of corn gods, stalked masters of the maze of questioning— an' the sweet maize dreamt to taste for the kernals of knowledge overwhelming the numbers of teeth an' mouth— an' biting! Biting! Biting!— bitten by! my mane as cornsilk threaded by, as the head's chopped off— or pinced or knifed or popped— an' the ear is cleaned. O for the buzz, the buzz of rustling grass. O for its airy tussle through the night— This sound.... does bleed the eye, an' brought me forth. Yet now appeared, to my nonchalant paralysis— my death!... a sigh envisioned of a mastered man, an eternal thought that speaks (***Cantal** unseats.*), an' thus would tell all of the cost, all of the doubts that must linger, that must be paid, that seem to remain for a curiosity to merit. An' so he belched, with a storming wind-blast of heat that seared from the nose hairs to the nape, "Lancet!... Lancet!— you are fouled! You can not believe till terrorized of town, inhabited of your worldly goods and possessions, this dull structure of your manhood, a chest of concaved vanities, and a trunk of twiddling trusts! Lancet!... Seize your muse!— make your plea! Angle at your embarkation to find the golden streams, and hairy downed contentments." Such was I taken to insist upon him my blur, an' not with fright but for the answers providential. This incline did meet me barefaced an' proffered. For whether wheat or oat or New World burst! I had considered all of such a mixture of conjecturing an' tending, towing, towning, treating, teasing, trowing— that before my eyes was day bedazzled of the thought, an' night's catatonic state, or for volition of the wish, darkness revealed as light. And these imponderables persisting did I ask of— Host! Host to head! Host to hind! Why does the night bring day, when all the world's the same? Why must angels fight, who haven't heart to hate? Why does age reduce into a muss its make? An' why must there be such,—so much,— any work in paradise, the very foundation for all of our troubles?! To which the void responded with solemn scowl an' snear, "You grit of grime! You grain of grim gusts! Have at your dreams but all! Not can you contemplate but better than this placation of your arousals. And in such dreams you *are*, compared to all creator forces, and must by erring with your weaknesses make mistakes of perception. Then the contradictions can never be explained to you for your weak satisfactions. They are wronged— above your plane of performance and mental calibre. They are your mis-digestions of the gree and the *gras*, and not to be interpreted by your poor scansion. (***Cantal** seats*.) So be as born— you flit Light thing upon the earth provided with, perverse, pervasive and produced, as could a spume over the oceans float with sordid, sullying, obscuring work. Ask not about your form that runs across the world ever green, ever tender, ever naïve and wasteful and mistaken and sluice refined. It's all that you can see of slush to be as gold yourselves. Do eat of this panic given ye— and not complain!" So inststed my burning hairs, an' ashen countenance, to more believe the smoke about by nostrils than errant musks of knowledge. I am misconception, with these seeds. I am bitten of an' blown at. I am eaten of the grass, and all foolish intellectual pride evaporates! Could be that all the darkness disappears, as I stare. So was that shroud of cognition, through that faithful evening— Fate not! No fate for ghosts.... or drifting spirits. An' this profundity— of myself— dissipated to all notion, or covered me, as my mind went along through the bellows an' the bounds of action, their— doing of me ate, their genuflections sate of mating ritual. An' my eyes did gape as my mouth could close of sound, with silence for their fire(s) to permit. What burns is the breath of panting now, an' from me still, metabolism's yearn for some reward of caustic reckoning an' the flowers' ashed incense, my glorious funeral slipped on—

CANTAL: Was not there that I took you from—

LANCET: —bathed with!

CANTAL: —but after the arrangement and commotion, and the handsome corruption. It was after the ceremony, Lancet, where you lay bled.... of heart? of the intentions furthering.... our goals? A cadence left for you my view, so suddenly hopeless, an' worthless, yet of the rigors of intensity. Know what you know, or what you think, or what did pass. But that was a considerably pale night, during your.... transmutaion? transmigration? Oh! But I dearly hope we must return. For there's nothing here of value left us. Left us for arrears of value was this tendency. The grass does churn one so, an' consternate— confuse.... an' condemn. And she asked why we did eat it.... It was never fit for man. I was never fit.... for.... manners of association. She! Where in this irony was warning.... mere curiosity— an' late!... Lancet! This is a bust! We shall return, as soon as patentable, as soon as weather permits for declarative sailing an' we can prepare ourselves and provisions for leaving. And you will recover, as the toxicities dilute themselves out from your bodily system. We will regain our constitutions an' health, as loyal subjects to a king. What is remarkable here devolves from natures too exotic for our containment and continuance, an' justifiably predictable to our observance. An' what is not remarkable.... Oh! leaves us so much lower of ourselves to picture with comparisons, that a primitive shaming must evolve to guarantee our sanities. But that is done without evil prospering, (rather) only through our own demands and lengthenings of looseness, the laxity of the thoroughly confounded arrogance. utterly erroneous. (*unseats*)

LANCET: I've seen.... too many answers, Cant'. Ca*aa*nt'!—

CANTAL: We'll shoal ourselves to home, my true responsibility, an' drink of pure waters once again, or uncontaminated brines. But breathe now deeply, of this fresh air, this resplendent filtering by rich greenery, of the night's occurrence. And I will have you tamed to your talents. For come you better on them to rely than on delusional treats an' tethers.

LANCET: O worth for me to roll—!

CANTAL: Behave yourself!... as the sun shines.

LANCET: There impinge some pins of perpetuity—

CANTAL: Not all things last endlessly, not even rawed feelings. Then act to count your knees with a hand, to the left as first, to the right as second. Repeated endlessly for any number of tries, to swell your head of this mindless but physical attempt, then ever will the left be odd and the right be even. Can even God! break such a fold? (*as **Lancet** starts counting to himself*) Then the omnipotent is mere conception of the meek, an' lies within the meek to prove or break. Reprove yourself with this exercise.

LANCET (*stopping*): ....It's ever thus, an' unexplanable. Always is a truss of caution. My mind's been demonized, but slowly reforms. An' to your care the stultifying stanches of its drops, drippings of mentality upon the ground as useless liquids, ideas an' conceptions an' remembrances. So have me placed. (*stretches out, arms and legs*) I am a skinned lion.... lord.... An' this is ineffectual — this flesh an embarrassment, this meat already.... mouthed.... Condone me an' console, for I've been tired of the heresy. Yet be it real, this is an angelic land.

CANTAL: A territory of nobles. What think you of the eatable grass?

LANCET: ....Was there taste to it?... Was what a sunstroke commits for one's awareness to entertain of thought, that there is license with faries, ferns, and frustrations.

CANTAL: Relax even more.

LANCET: Have at these legs an' limbs to carry (a)'loft an' loftly. I'm sure some magic dusts persist to compound into logics.... these vacationed woes.

CANTAL: So clears the head through pulmonary thrusts. Let's not debate that we've been of some resurgence. But in a field such as this, much is welcome for review. An' as drying tears displace themselves, instead of replacement are the clean dews kept for morn. More able an' limber are your approaches to crisis, or what was met. Still pardonable we are, with theories windswept, cured — Relax as much! The apoplexy de-intensifies, an' a calmed permissiveness of flow resumes, as could obliterate entirely the matting of a red tide. This (*brushing cursorily the stick end of his seater along the grass*) is the need for moon, the restorative enticement of a re-beginning, a 'Lunar Lent'.... Have.... you retained the courage of your cares, to the carriage of your credulities? Then be at peace awhile, an' withdraw from tensioning. I hate to see a sputter of fulmination, to draw attention of one's illnesses an' wrong ways battling.... a crimson crudeness roughening up the smoothed curvatures of eye, an' madness overtaking reason, practiced reserve.

LANCET: ....It was a slightening of research, with pathological objection for not obtaining the desired.

CANTAL: Yes.... That was it, a fiend finial to cap, a wonder of what to leave— in the wake of one's moral devastation, how you were found demolished, Lancet, as a thinking, rational man.... I took it from him— or so he lost its sight of use during the delirium of his rampageous engagement, how so furiously they danced—....

an' could have terminated your pain, had I thought it worth as well to equal a permanence. But we were only dining at a feast of visions, I concluded, an' neither of us banished from our epistles reading of common, comely human behavior. Merely head sore an' over-exercised. But the fright of your crabbing was real, as a crustacean('s) crust surrounded you, an' your eyes bulged out as if on bending rods. What for this feeding?

LANCET: I was remiss, eating up all of my samples, gluttonous — where out was ever more to be.... harvested. Was not a sound system about me, of regularity more usual to my efforts. (*sitting up*) Something caught my attention, like hook to the eye, an' kept dragging me along for fear of loss an' tugged at pulsations of anguish, disconcerting bother. Was a dim diminishment of deed turned into a real doping, as the skin turned of shades almost as green, a depositing dye.... of death! of dimwittedness.... of indecency. (*coughs*) Raw rust must have been of the saw that was slicing up my brain. This.... cruelty— self-inflicted— remains to heal. For what is left is tender an' swollen of thought.... No! There's no taste to it, my lord. None— None at all! Just weed, an' undigestable cellulose.... or slowly that, permitted of a way of breeding.... loose.... Yes, comes to me more faction— of analysis.

CANTAL: The recompense for one's tastes are one's times. (*as **Lancet** awkwardly stands*) An' this place here is too much of the Janus facing, backward to my suborned youth an' forward to the decrepitude of aging; thus primitive an' practical of nature moreso than ourselves—

LANCET: I feel so tired!

CANTAL: —an' perhaps foreseeing of ourselves a more natural, unobliging state to return to.... eventually. But I've been an adventurer. To be hit so with stones of contriteness in longing defeats much of the spectacular purpose. Are you whetted out?!—

LANCET: I think so.

CANTAL: —my lush responsibility to prove of man some caretaking with a civilized prowess, an' to handle *all* eventualities encountered with capable enforcement of our wills. We'll walk through the sun an' back to the stills of daring drink, of all our circumstance. An' as you are made, demented purview not to the perplexities of bounteous fortune, with the simplicity of not owning anything that grows of this enlivened soil, as do these people live with the casualness of their flora, aside to pick flowers an' toss into vanities! then shall our trip be made not speciously, an' our visit no less taxing of the art, an' the heart's urge to travel. Remembrance of beauty will remain for reigning of our thoughts pertaining to, our sun-soaked enhancements of vitality in this pure an' gracious field received an' pacified through. Agree to it! Lancet, as a re-earned conscience.

LANCET: What blooms here of the pastoral enticements may charm with non-abhorent mists of pleased rustic delicacies an' the perfumed perfection of silent joy, the native scents so slight as to be prescriptive of an earthened indulgence to nurture all comers. An' though some tingling retains itself throughout, this is of positive aspect ascending with decline. So well behooves our patience to get righted, as you are the most adoring to seize of need, my subordinate companionship to sustain. I am through with grasses in this place, except to lie upon and kiss, for their dementias' enlistment of tranquil sensitivities and feeling— of the grandeurs of heaven, what can be made local to our steps and passages.... an' rests, wherever this may be throughout the world. Since what we find is what is found— as furnished to our discoveries, an' not more to the lust need lusts be flung over an' enraptured by.... So, I've determined nothing— an' you everything.... which is how these missions should be conducted in the first place—

CANTAL: Ay.... You thought you saw all.... but not what for me to see, or seem.... discerned.... distinguished an' distinguishing. There is some challenge to that.... thinking, that embarrassment. But we have ended the crawl, and the quivering and the astonishment to frittering of our motions an' motives. I.... can't have!— So there! The forest's ever lush an' radiant, full an' fulsome to our wonders, and delectable of kind.

LANCET: Definitely may I not eat any more! Yet why the insects are not afraid of taking their bites off a giant as myself need not be a point for fathoming further. Some things are just pre-various an' prevalent, pre-selected for an' precarious, an' to be done as done or done for doing—

CANTAL: End your quibbling, but not your questioning. Life is less a puzzle than its learning, an' the sky is striated of clouds conditional an' ever changing or dissimulating their intentions. Thus are we likewise humiliated by our thoughts. An' that is a blunt purpose to look up to. This space of time is slow an' quick to drift, the span enduarble of an airy dome; yet empty an' not, vacillating an' determinate, such have been our heads, our minds.... and our recoveries. So let us retake steps of reposing an' composure about this.... This is such an awe-inspiring day!

LANCET: Like many, sir, but just that I have seen. Here is the purpose of such beauty— evident.... to be displayed. An' not for much more is it done, no less utility, induced for some advantage, our manipulations careless at best an' of woeful inefficacy to the least. Be we charmed as powerless titans to this environment. An' I will pick of fruits to classify an' eat, over the pasture's sweetness an' beckoning. (*as **Cantal** comes over to him*) Though frail I remain, sir, to these.... realizations of fragility received of frozen hands, now warming, now fulfiling our commiserations with life in total as some sort of continuation of consciousness. (*as **Cantal** touches his shoulder*) Have I been asleep, sir? Have I been so frivolous?

CANTAL: We've both been burnt of gut— deeply of the innards felt.... But this passes of the days, with much retaliation for that night of drawn, extended empathies, that we were syncopes to the truth an' sycophants off nature's wealth to praise deviously an' try to worship ingratiatingly. An' now for correction, now back to our tents finally revived; though my face is grayer, tamed to be more forthright.... of its age, an' your joints are still tender an' nipped at — resisting that hunched walk you've adopted, as to cower from some weight of reckoning. Be straight of back awhile, Lancet (*Lancet extends.*), though it might hurt a bit. An' let that dourness roll off to burden grasses.

LANCET: The cracks an' crunches are my own, at least, for animal being, an' made with bones— Not a shell.... no longer that, nor pincers fishing. Keen for man an' skin-like pelt.... sharply is our world as dealt. (*almost collapsing onto **Cantal***) Oh! What ambivalence is faith! to leave one drowsy in believing any state pursued.

CANTAL: That's just the wearing of our temerities haphazardly *et à gauche.* To some bunks, then.... an' letter writing (*they start to slowly exit,* **Cantal** *aiding* **Lancet**.) an' preparations to begin.... Was more for ease of transit when you were an isolated, cold-fleshed crab! Lancet, what drooled of heated sweat profusely evaporative. But now does lend your legs to aid of the direction back.... a wholesome perserverance. In either case led to leading, the prospects finding, you are influenced by stars seeing.... or persuasions staring at you.... for some protection an' much more guidance agreeing.

LANCET: I suppose then, my lord, that I should lug around a radicalized conspicuousness.... For I feel famished! of this meat. (*They exit.*)

Scene IV — *A pub.* **Pale** *is sitting at a table with his* **wife**, *both having beers.*

PALE: I don't know how much longer we can manage. That extra income is so trivial. No one needs any carpentry around here.

WIFE: It's what's to be done, of a succoring nature. They're so little, and intolerant for loss of caring.... But I'd not put out, without the funds, this benign gesture. (*sips*)

PALE: So can it be explained. I'll accept the particulars with rearing.

WIFE: Our eldest daughter attends. But this must be temporary.

PALE: It does seem odd. But she was so destitute, wife. It's nice to know that someone can capture the gold of benevolence, no matter how strangely pitched, and to the most deprived. Well be that we should share of it— There's nothing else, as the savings dwindle. But I sensed that some occurrence extraordinary was fashioning itself. And I attached, with inquiry, so that it leads to this.... passing mode to relent. (*sips*)

WIFE: I was frightened terribly. of the condition brought down to, and could willfully ignore it. But not for those two tots to be scoulded by, of such dire circumstance, could I resist much longer some empathy. It's good to have developed this way domonstrated, with your alertness provisioning the manner, that in some way we're not harmed of accusations for being falsely altruistic. There's so much fear in this valley, for what is resulting of us, so much imagining not to accept— or to shun fiercely of those possibilities.

PALE: This is what is called an economic drought, my dear. And it should pass, according to the principals. But along this way do fall some holding out from merciless retainer, this is always assumed. And I have wariness intensified that chance rules more than chancery to rule our fates. For we're not helpless yet, but becoming nearer that drain of fortune which only accident or serendipity may surge to remedy.

WIFE: Your lunge was more acuteness to details of beggars, and that acuity has spared us some from that social condition. Yet how they may be aided, with work— Is this real? Then may we be so splendid. A contract with the king to end such plights? It is remarkable.

PALE: This is a peculiar circumstance, of some particulars of engineering I can never be sure of to understand— fully. But it's certain that the royal fear of revolt. And our social weal does whirl to

changes accountable or predictable to organized complaint evolving. Such is my circumspecting charge.... yet as much as an idler to the tensions and frictions growing. Dare I not say out loud, for fact of fury.... not in public. (*sips*) That's to mean of course that what is obvious is blantantly so, and virtually no one from this area is contented. But what can be done of a miserable state to weather through? So the king sends— liaison, to stifle the blister of uprisings, or determine their secreted transitions of growth from angry complaining into raucous yells and demonstrations of discontent through self-maiming, or daring to confront legions of absolute terror, what can be imagined of a repulsion. Well, it's very subtle, isn't it? to be within a mixture of hates for this condition of uncertainty, not knowing just how to battle a withering or desiccation. (*sips*)

WIFE: Comes not from negligent intentions, I can assure anyone. This is a village of industrious craving. We presume for ourselves, down to the least to work for— of benefit— some pride of reason for being. But what contributes to the making of a despoiled place?!— The weather? Hardly not! One finds a horse irreparably dead, and leaves it rotting, or a house demolished of a storm or fire, and leaves it crumbling. Thus as we are, it's becoming to view.... made heathenish through no actions of our own, speckled of our prides an' spanked for this— it seems, resisting the decline, dispelling the stagnation with fruitless implacabilities and refusals of acknowledgment.... fallin' besmirched, an' splattered of mud. (*sips*) There is no recourse available, to fix the machinery of our township, once so vibrant of opportunity and fervent demands to succeed, now left to rusting impatience— from the shock of what is evident, what becomes prevalent about our social condition.

PALE: There are not enough skills an' energies, within our hands, to leave and easily establish a well-being elsewhere, not with the fiesty competition found. You must almost be born to the merchandising of other towns, they are so protective of their prosperities and ranges of control. But what is left for us, then to drift away as beggars into those locales an' societies— It is not permissable to climb without talons, without the evil intentions of a preditor, which keeps us wary of an' thought contemptible. An' yet to find the beggar here, it is a sign of our demeaning into despicable creatures feared, that we can not sustain ourselves by our own efforts an' must enlist decay, lowliness, meanness of temper, distemper, dis— affection!... The dole, the dole. How can one contain such isolates as any? isolated in a crowd of forged concourse made appealing. You are not ignored. You are discounted as only to be misplaced outside your province, your hole. You are belittled an' spat upon or kicked, for fear it would be fun to make this action regular— 'gainst all propriety of noble thought.... An' you are fanged for trying, wife. You are hurt and huddled 'pon yourself for warmth to contradict the cold stares an' grimaces. (*sips*)

WIFE: I know you try your best. Your sensitivities are generous to a fault. But we are trapped to a certain cravenness an' guile just to survive this span of detention from our rights to live happily, an' with some equity, equality of justice. I do make the pledge of not to beg; for I'm not strong enough to be so shameful, an' haven't the courage to cry for food an' shelter from harsh elements. I'd rather have cold rain pour as nails on an insensate corpse. Then driven as wings to the manacing beak should a body rise transformed, if this is the conducement for our nerve an' pride regained. I would.... steal lives, this way. (*sips*)

PALE: (You) Sound as would the king, picking of his precious made, mis-rearing—!

WIFE: I would pick off lives, to steal. So come we calmly more to sit an' wait. This is a thieving, Pale, an' turns my stomach. Can not be much, from which is much to take.

PALE: I've.... not found another way substantial, wife. I am flustered an' dependent on si—.... necure. But we must bide with the tide for what it brings, for now, much independence hampered to these straits an' pins of mendacious erring. In my heart I am a kind gentleman, easily abashed of face to find the privations of others incommensurate and cruel to all thought, an' if by man produced then abominable of crime against his better nature.... an' if by nature made, then abominable of crime against man's higher nature. But what of this dispossession of fortune can I wonder of for myself? It is by what?!— Need not we say!.... Distinctly are we discerned wrongfully without wrath, without being objectionable nor disobeying any precept for our country's greatness. An' while it's difficult to educate on the cause precisely, clearly it's driven by a strength more royal than immaculate. Then aid us some by our own devious ploys to buttress up 'gainst this suspicion. We have only that initiative left. Though the valley remains beautiful through these transgressions of responsibility an' sincerity of subjects. What for a land is better than its make, with not the maker making as can do to be made thusly, is the maker's mark upon *us*! an' not the land.

WIFE: We're not as dirty as the soil.... nor as pretty as the flowers grown from soil. But to the time are we encased in soil. That is a pity—

PALE: That is the striggle! (*sips*) I weep for honor if we fail ourselves not to be embedded nobly, but rather decaying as some horse of decomposition in a pasture. That is.... a fanaticism of revolt, to be of sights an' senses revolting an' yet gentle of display. Come.... See the teeth an' taste the tongue an' hear the bees of festering an' smell the breath of emancipated stench— an' convulse upon the touch of this horse!... That is our house?!— No! lady. That is our town's terror making, an' vi-o-lence!—

WIFE: Violets, husband.... for the peace of our contentions. We do rip rawly on the cord.... of this umbilical, distraught distress. She is but of tawdry direness still, to be forced away, well immershed of the actionable placations an' taking charge for what is possible to do.... yet distant, an' by necessity aloof of this discourse. So great are woes revived of life with living.

PALE: No hard choice there, with these responsibilities worth the emotional burden— an' the shedding an' shreading! But be fit to rend already. This county does embattle the contuses, an' we're honed much to a fighting hurt (*sips*) that more of punch would not discourage— Swells the tissue now!... the fabric of our consciousness of community. It's more than fear become, but felonious intention throughout the region, to save our own— and fight for this! with any given instance, of a flareup.... of injustices, or indulgences against our hearth.

WIFE: Revolution's.... still more a debate or description than a determined malady. It ain't much spoken, dear, of terms actual.

PALE: Faces of indignation are louder. The wrinkled lines an' furrows show(s) the drying up of patience, if we have been goodly brought of shadows an' disbelief of day, for ever waking of this light, that heavy weight upon the eye, of our disasterous state yet once again to face an' stumble with, no— cherishment instilled, but an emaciation of our worth, an' gray romance, as could the cobwebs coddle. This is where we lie of plight an' stagnation, as if to emphasize the wealth of others. So.... if motive be this terrifying epoch to distill, draw out the blood of reddened fury, fuchsia's rage — an' by the pints dripping, splattering 'gainst these swords of economic menacing. If we've nowhere to go, then let us die here enlivened. (*sips*) Let us argue.... an' protest loudly, an' show a stubborn resolve— to be displeased— of.... this estate.... this squirming, this negligence, this noxious turning of, this loss of fashion or even a pretense of value— to.... (*as **Fenic**, with **Ancher's secretary**, enter to come towards them*) cry saints for seats! He arrives for our waiting, as preordained to come. Yet settled less we are. It is a hairy brush always to meet.

SECRETARY (*while they are nearing the table*): Peanuts! Some legumes would be better planted here, for the pace of a certain commerse, than this overwrought prediction that drives him foul, or afouled his management.

FENIC: These plains need more exotic hope than that. And our king expects the miracles of Earl Ancher. For that is my Mytdra's wish, for stepping stones to fortune for this region, and not merely some peas an' beans to soak our heads with. (*at the table*) So Pale! and his wife.

WIFE: So as she for her strife!

FENIC: Good greetings to you. (*sitting, as the **secretary** remains standing*) I arrive from Doria, and the dearth of propinquity, a ferryer of the lesser scones of fortune, and to take matters abruptly into my own hands, that there is no coalition of discontentedness between our lines. For it is triffle an' paultry, and yet imperative to relate to you, for how she labors at with scullery—

WIFE: An' induced skulduggery!

FENIC: —to provide for her children, herself, and your warding of her cares, displaced as she must be, and exhausted of a week, for which she will repatriate in two, for a couple of days her young to bind. And this until enough is ever earned could be her faring, else more to resolve itself of this town and her home.

WIFE: She's not talented enough to be scullion to a king.

FENIC: The sum is slight.... An' that she's not, but more for what the talent brings of the inconsequentially served, the lower of a servant, lent by me, to find some utility establishable.... And not of conspiracies to Pale's sight, this knuckled struggling demeans for cleaning an' scraping, fish and meats for salting, curing.... yet the struggle does not hinder rights, nor needs to be provisional.

PALE: We're bound that way, the nature of a family to uphold— by.... various means that at times may fall to the derogatory or even groveling. Then could we be not pleased to ties of heartstrings otherwise; this is a roving circumstance throughout the valley, an' poor Doria is the lucky wavefront to that blistering approach. She's found a means.... to gain some solvency, while we fall nearer to her destitution every moment of our consciousness. What is the shame then of disgrace in hard work? I would sigh for of my talents to employ, with singed irritance at the delay imposed by—.... the nature of our bounds an' boundaries, an' the day as over an' over again, a repeated deprecatory malice somehow designed.

FENIC: I've crafted the means for her, with difficulty. And so to the entire village am I drawn of similar attentions—

PALE: The word of a king to issue, an' pander by, a social amelioration of a populace with professed sympathies—

FENIC: Still as definite as myself, our Mytdra constructs ways, and orders me to try.... roundabout potentials.

WIFE: She's firmly within your grasps, then.

FENIC: ....There's not dissension— you have spied upon, lady! for which to punish at our capital. And you have not seen our conversing, for conversing's to be heard.

WIFE: It's my husband's dedication to be observant.

FENIC: So has the such been served, this actually improves matters, or would more enemy's pay be brought to you, as well as my redition of it. Yet come to terms of Pale—

PALE: I never called you as a fiend, but suspicious. And are you not?! I've aided Doria myself, with less conspicuous protraction, for what may be designed. Mine was from the heart. Mine was as a neighbor concerned. Mine was.... disuaded, and done with some attempted reticence— or occasional daring. But you have blasted! of a colorful recruitment, suggesting what?! Beers are at the bar to take, not to bring your own. There is disorder, to your ordering of attempts, or to make this scurrying unusual.... of a social conscience. And I am patriotic, but not for your queerness.

FENIC: Our Mytdra cares, above your inabilities to analyse our intentions—

WIFE: *Ours*?!

FENIC: I am his arm into you, as a friend, and as a tributary fashioning, to regain you a fertility within the kingdom. For Mytdra wants all regions to be honestly productive, an' without destructive exclusivities—

PALE: That sounds more opposite to life's realities. Where's the dole?—

FENIC: But these ways have to be organized according to the king's passions.... an' favoritisms. Already he has made challenges, to an earl no less, to want to invest in your interests— This is the earl's secretary!—

SECRETARY: And he is not a proponent of counterproduction with his monies, nor of entanglements to insurgancies brutal and wasteful. Yet, as of late, he is unnerved— because of you, most contrary to his normalcy to undertake any initiative.

PALE: We've not done anything, but chewed on Ancher's rumor, or rumors of some lords to show their interests in this region. An' now is he skittish? or unnerved? How much more may we be defiled, by embarrassment an' contemptuousness— of ki—!... kindly hearts, an' interventionists. So may our majesty produce them, they can only show us some reluctance of development. An' this Ancher is more for the foreign imprinting to our land, with the strange

matters he involves himself of. Now to make us a domicile for exotic animals—?

FENIC: That's not the clause for his devotions of attention about this area.

PALE: A manuring backwater!—

SECRETARY: Nay! but for disgusts in greenery, the mulch of a stomach to find insatiate. He'd purge from your *soils* something novel, or remarkable of taste, to be marketable of a restorative economy. Yet it is a slight fantastic, too, which cankers a worthiness to be worrisome. And with some agitation has he been alarmed to justify.... interpretations of your lot.

PALE: We are the comely citizens of a high ground—

SECRETARY: A rich one certainly. But if for gains as must be told of to the king, the sure must not only be the soil and its serenity.... but of other manners—

PALE: Manu—

FENIC: Feasts of famines! are you as not? We want your clays to roast with an' produce from.

WIFE: Crops? Agriculturalists? That's of many regions of the realm. There's no distinction there, yet of the land to own and relish. Acres are many. Acres are spread for purchasing. But no novelty to it, as like a house to live in, as even Doria has, homesteaded by my daughter while she's away— laboring! This does not save the tenor of our times, with a decay of true distinction an' merchandise-able quality. We have been made sterile of the economy— and now you talk of wishful dirt?!—

PALE: Where's the dole!

WIFE: —Is there not some deception of sincerity here, to your lambent interests in our miseries— if not to scold with teasing?! or amusing of yourselves?

FENIC: Would I be so to arrive, and tread of some deliberation? shamefaced to approve of this as could be to form treaties with the indigenous! I partake of casualty, to force improvement of this region (*abruptly getting a small bag of coin from a vest pouch and briskly slamming it on the table*), to reimburse my friend an' friendship! There! is the little offering, shaved from a tiny sum, for her little ones to be shared. Or Pale you are, to dare of it for need, how there is desperation to this fee, this fine of heart. (*as **Pale** retrieves the bag to open and count of monies*) Then should I scold with this?! It is more blood than banking on your care. It's more than bread. It is labor an' hardship an' cruelty— an' some capital invested in your worthiness to be civil an' decent an' helpful. It is the writ of statement, that I abort remiss(sive)'ness to carefully planned allegiances an' celebrations of our social valors, an' that we all work, in our parts, ways, an' methods, to correct poor situations.

PALE (*re-caching the coins*): This is all that I have earned, sir, in these many days of prominence to exist. And it's all here, as usual, for trust of her endowment or alloting.... which you carry faithfully an' with no doubt of earnestness. Yet see that this is pitiable to have become, to rely on, and to rely on tears, since her separation

489

must tear at her dreadfully. (*clutching the bag*) Is this better than begging? I am woeful of these an' many other plights to our valley residents, and can not blame your ostentation— of dress— to alert me, but merely notice that by comparison we may be fatally stretched of nerve an' anxieties. But I do not blame you, king's envoy, for our fall. An' from grace or greed, I can not define it well. For we've always been industrious here, an' not so patronizing of the wasteful, needless an' hedonistic practices and characteristics shunned by pious locales an' abhorred by the healthy of spirit an' disposition. But nature brews a storm for any of its creatures regardless of their nobilities or simple earthly virtues. An' we are hung to a conscription of slow torment unrelenting of our confusions an' futilities to confront, combat or escape from. So be (we of) us yes a backwater to some refuge of opinion. I can blame no lord to consider our inefficacies as heinous or despicable. But we are so stunned by them (*His **wife** consoles him by the back and shoulders.*), we may only be subdued and submissive by whatever plans you have, what his majesty mandates as could be proper.... for our mean an' lowly like. We can not surmount the difficulties ourselves, or by ourselves, as so many had thought. Then we must be churned, from cream into butter, as the yellowing of our nature persists. Forgive me of what I could accuse, to any oddity seen an' grasped at as some sign to reason of our decrepitating, however unexplanable but curious. I thought.... the decaying might covet converts against our present rule an' state, that such a condition foments.... cohesions to rebellious attitudes an' their cohorting, an' that I.... was still well above such coercions of temperament to foster an' further disloyalties 'gainst our.... nation. But such a fool is this to think against this decline of fealty while stationed at the hopelessness we find, that we should perish from our vigorous intentions carved an' pared away, propulsion pummeled as our purses pour. An' I did blame the beggar worse.... than my pride. These vicissitudes of caring reinstruct. Her children are dear, innocently angelic, easily made happy with the leanest of necessities an' some charmed caring. And yet I fear to notice them, for this abusive treatment I find myself going through which could recommend tactfully their insouciance. Or that they await the parent anxiously, what must I await, or wait on?... Tides turning? Changes to the mooring of our statures— as capable men?— or beasts? I do go mad to be this out of place within my conscience, rushing still, finding little and then less an' less.... till overrun, I've made my day into night, morning into evening, an' more of disentangled rout into a perplexed twine of rueing rote to learn acceptance of each hour spent awake an' suffering of empty hands an' veils for this cruel liberty.

FENIC: You're always thus apologetic, with your meal received. But it's not so true as very true to me, here is a *liberating* misfortune. The sun vacates to make for new prosperities. And I will guarantee these effects as I compel them, with our Mytdra's steering or contorting of me. This have I applied to the earl his miracle —

SECRETARY: And that does lead to his mis-sharpened courage (*sitting at the table*) to be awkward in confirming. For at once he sent a missive out to his.... friends of excursioning that he intends to visit them, the very next day he departs on the trip, within a frenzy of intention for the voyage, uncertain as to what to find— for you.

WIFE: Us?! A new form of tomato? A melon with sugary rind? I've heard of such peculiarities imagined to be practical— but they never are to come so outstandingly. The tomato merely varies in size, an' perhaps intensity of flavor, as does the melon's meat to a thinness of coating. Yet they stay as they are cultivated and bred, cross-bred, conserved or smoothed, of a food's delectability typical for each species. Make for we (are) singular in this, that is a stretch. Our land is commonly rich—

FENIC: But the king is capable of restrictions to it as for any of his regions. You are for a specialty saved.

SECRETARY: And Ancher's miracle is so astonishing I dare not announce its means of verdant methodizing to the common folk until it's proved satisfactory of this earth. That is the earl's commandment to me, to assuage his hesitancy of demeanor until he discovers for himself the remarkable fact he's learned of— partners and servants serious to him describing.

WIFE: A mystery plant? Can that spare us some enlightenment? My husband's not a farmer—

FENIC: But he will cater to a stern organization of them led by Ancher or established through the earl's leases he may buy. The town will gain a protected vibrancy, as this is worked through with intelligent enterprise an' commitment an' advertising of such a reform of horticulture as to make this valley renown not only of the realm but internationally famous. This is for my head to insist an' leer of jerk obstructions to this truth or contradiction to my aims. These are as Mytdra's determined, the branding of this culture into the novel an' noteworthy. The earl's secretary inspects these possibilities an' potentials— I assure them, vouched with oath pledged to the king, my very life in gratitude to his generosity in allowing me to help— a friend. So are we all dearly here combined, of forces an' common faction. An' this will be the breakwater 'gainst deterioration of your cherished land an' life, town of home an' country, nor for backs to plead of waywardness an' degenerations towards the primeval. This will be a modern fulfillment of contented essences an' essentials, as discovered of the most early an' undisturbed— by comparison to ourselves— of populations civilized an' compassionate of sound an' simple society. Then such a taste of this will thrill as remarkable— an' cultivated nobly.

WIFE: This is obscuring disenchantment. But that Earl Ancher researches jungle aspects— of the unique an' novel, an' old for new — is well known. His exotic thirsts apply to many places, though, an' not just the primitive. Yet are we to become the soil for such a seed?... to complicate and make sophisticated for our values? An' are the flowers pretty, from this plant, or its oils abundant, for its aromatic refreshment? May it be huge or petit of bud, an' as a stimulant conjoin to taste or chewing? What vegetable does make a paste of itself?— I'd search for the pasta-izing gourd of trophy. Or is it more a mushroom like? that suits our muds for fungal spread of spores. Gigantic heads can they have, the stools of emperors. An' is this round an' colorful, with many eyes for rooting, as a spudding rock?— or leafy flat an' broad, with waving of the wind an' undulating tempest? Or it is more cylindrical, to form a tight spearing tip? as can asparagus shoots. There is more wonder of a plant than its seed, these aftereffects eaten, strung or hung. Marvelous shapes to the offspring grown, yet from a common seed-like form. Then must all matter start as commonly, from the simplistic imperatives of growth an' the basic character of all initiation— as we might imagine it. But this is a much developed valley already, sirs, covered with green. Roads are exceptional—

SECRETARY: Oh, I'm sure the gift is as plain as profound. The land here seems quite fertile enough for any kind of agronomy or tillage. Plenty of sunlight, on fair summer days I'm sure. Some swirls of chilling, though. But a firm bed is here, for great variety of plant life. Is it remarkable enough, though? That's to its sheltering, I would propose. And being not a scientist, nor climatologist, nor cataloger of the vestments of earthly events, I'd say your most important contribution to this venture of resurgence is to be a protector of the fields grown and sharpened of boundaries. The inhabitants here must take pride in being protective, warding off visitors as well as your own exigencies to sample and submit to satisfactions—

WIFE: It's that delicious?!—

SECRETARY: —A sufficiency due simply to its evidence seen —

WIFE: It must be a fruit! or a garden's *pomme* or pome—

PALE: Sir!... Is this a corn? They must be cultivated carefully— to repel the bugs— husbanded assiduously an' the ears harvested arduously with repetitive strain of pain to pull off an' chop down. It is a certain industry that characterizes a town to rustication with cows, as these ruminants may have a potent feeding—

FENIC: Corn-bred beef! Corn-fed meat! But Pale!... This is your life we speak of, to recover, not for you to discount or disparage. The miracle is simpler than that, much so. The miracle is in your attitude to it. You will regain your work, your type of work, to do as you may do proudly. And this locale will not be squandered of its potential, nor converted into a country's pigsty with its beetles. An' this is not a mystery that must be kept from you, an' not from others, but only for its extraordinary hope of proclivity which could shock disconcerting eyes as could the mind remain resistant to the fact till forced of some acceptance with reality. For nobles talk of this already as.... allowable, by stern an' capable lords to accomplish. But it's in your good an' better interests that I remain faithful, an' thus the king to you— an' everyone of this place.

PALE: That is the message, sure.... But you help Doria, an' so must he by influence to trickle down through you. So does Mytdra aim to aid us some way, unusual as it may be, an' particular of persons an' time. An' you do not skimp of payments ascribed an' agreed to, so you are soundly walking. I had such a fear you'd excise a tax for your efforts.... But there is none. None to beggars. I do call our majesty true— You can count on that— an' not shrilly heard of him our pleas for restoration, how uncomfortably such sound might be meant, yet through your filter eased, or pleased for royal— ears, Fenic. We haven't the strength left to contemplate your actions with more wonderment, as our muscles are tensioned for survival and brutalities expected— err you to make of us anything but submissive, an' yet sorely tried. Still, as we stay as standing, resolutely aged, until the final destructions devastate our minds.... the sanity of being hopeful an' positive of thought.... then may you make of us, shape of us as loyal subjects. But life is scared here, as you are as generous a royal envoy to be so honest of our fears. May the high an' higher lords deliberate well. I can not account for them, their toils supernatural to ours.

SECRETARY: That is to be spared the noble liabilities. My fear is that the earl feels forced— out.... of his way, to do things.... responsibly.... while ever before the responsibility rested with the thought— to do, actions carefully planned before being taken. The

such may be that surprise distills intentions.

FENIC: There then may we part of expectations. For friendships can not surprise me, but only what surrounds their circumstance. Ancher must have his.

PALE: So be so soothed, sir. (*standing, and inducing his **wife** to do likewise*) You may order at the bar.

FENIC: I'm well acquainted of the tavern-works.

PALE: But the walk must have been dried of dusts. There's a good heat out, this afternoon.

FENIC: It is passable. We traveled courtly, and the thirsts are of munificence to employ, throughout these grounds magnificently.

PALE: An' we take our steins back up to the counter as a courtesy. Yet for to leave them is as a heart. What's left to do is miserable of thinking; for it is nothing, except that my wife may have her household chores, my children theirs an' play, an' I.... may hold my head awhile, to think.

FENIC: That's not uncommon, pondering.

PALE: Have you seen her children?

FENIC: ....I've promised to, for my visiting may comfort them, as a familiar friend at where the mother goes.... an' stays. I've brought some sweets, for their enjoyment—

WIFE: Give it them from us, I beg. For I've promised something of the like, an' can not deliver, delaying an' delaying till prudence catches up with fortune.

FENIC (*as **Pale** and his **wife** head off to the bar's counter, with their mugs*): So may I claim.... a delicate mixture.... A type especially sent, to forestall grieving all to much.... from Doria herself. This is my proof.... It's very difficult to convince any of them, they've been so disheartened of their declining fortunes around here. A thorough demonstration, one greater than faith, is necessary— and as soon as possible.

SECRETARY: Well that is what my master designs, sir, for your test of him and his facility to please the desires and mandates of the king. And some experimentation around these parts seems plausible to me, for this is hefty farmland should anything to grow or germinate prosperously a wish. That is no fault to this county to be held by. But my master was shoved into this.... display of prowess to prove a feasibility—

FENIC: He is opportune for a dilemma to dissolve, as could he boast of wonders and stimulate our Mytdra.... with possibilities to uplift many, and correct unforeseen dishevelments.

SECRETARY: Is boasting to describe what he has heard— of? Policing now dogs him at home. Yet he has committed no untoward act against anyone, but feels it's best to flee awhile—

FENIC: He does research, for what I need— I being an agency of enlistment for our Mytdra. The constabulary does not imperil, but for rapid impulse to provide— of some imperative to initiate.

SECRETARY: ....Or some impending doom to be isolated by, should some claims be false—

FENIC: That by his careful nature is impossible, impossible to approach the king with declaratory enhancement of the tale after my fair and just warning— not to do (so) if it can't be done!... And I have been approving of the words, that my words lend as much conspicuousness to be hopeful. For we all are dependent on good fortune for our finer grins— But I am more so possessed of the need.

SECRETARY: ....As an appendage.

FENIC: What might you be, then! The Mytdra is to be assured. So have I assured, and so has Ancher—

SECRETARY: He only celebrates of news purged from gossip, as he would ascertain from his associates and trusted servants. Yet dons he research, sir— not does. He is not the provider of miracles. That can not be done by any man, not even through his majesty's staring, or pressuring. Then as the king laughs, so must you? And if he bites, must you provide your teeth?! I am faithful to Earl Ancher— but I am not his finger poking. Still must I be an executive in some of his affairs. I am not his mask nor *Geist*. I've my own mind and parables to follow.

FENIC: We are all extensions of our sovereignties, and only that to barely that depending on our abilities. But I am fully absorbed of this affair, because it deals with personal friendships.

SECRETARY: Mytdra's? or your own. Mytdra's are your own.

FENIC: His is a fabric that I wear. Why should I say otherwise? Why should not my skin be smeared of this? But I am not his cad — not the king's cad. I am this town's cad. And within Mytdra's hand I take up his cause to improve it. For that is what I've suggested to be done— Now see you more than ears.... for revolutionary fervors to detect. Do you think I am senseless, most felicitous secretary?— I am practiced in this! Expert! for these scents to notice. And you have not been fair to me, for saying what Ancher could all too obviously think— But this is fine.... because you have fidelity of servitude, which can always be respected. It is the man himself that matters to me, for what he can do or produce, or accomplish. But I do chomp my mouth upon this bit, despite those contrary sympathies sensed— for some help!... and with forgiveness sanction Ancher to do the best he can— and succeed for Mytdra. His association to the king is now mine. And his fall with failure will be more than mine. It will be this entire village's. Thus are our risks all shared. And can there be a higher test, of personal integrities, than full absolution through just deeds? Then why disparage— *our* attempts with my obsequious personality to denote?! There is more calling here than that, but rationality to seize out of jaws remarkable, out of the improbable and startling to base our goals with. Then do we share ourselves profoundly. The king estimates everything, assumes for analysis every position. And I have given your earl the most of sport, to prove himself— dear!

SECRETARY: ....The king has. I'll agree, we're not so singular a species. Each heart has heart, has placement for its beating. But as for might, as might for yours to leverage— I mean his.... I fear for my master's beating. These researches are due to a servant who can be overenthusiastic of his science. All of this depends on his.... glides through veracities to perceive matters that may be even

above the truth. This propensity has been pointed out before, by discerning magistrates, such that all suspicions cast on Ancher derive most strictly from this fellow's reach for discoveries predicted to be fantastic. Now we can only judge events.... by their results on us. But not before had Ancher *need*.... to rely on him. And for this alone I am terribly doubtful, just as such notion of practice seems dreadful.

FENIC: Confirmation comes from the lord— Cantal. And this— scientist is only provisional, for the details, a tool to find the grail, the cup of our cupidities, a greed that drives many for knowledge and acknowledgment. But this project is much more serious than that. It's become for a relief from terror and mass wretchedness. So does this Cantal provide me with a more favorable risk, the word— or writing— of an esteemable man, a noble of means, manner, and thought, a rank of class dependable, and a personage noteworthy to our Mytdra. So does the king hone our challenges to be realistic, that we may have an honest chance at succeeding. So can Ancher make the most of it.

SECRETARY: But there are fevers to where they'd dwell, that can distort the common senses into phenomenal acceptances of the base as pure of best, that can only be recovered for what's real when out the heat. That can I suspect for any novelty put into a daze by.

FENIC: A carefully and cleanly written correspondence argues against that—

SECRETARY: The notes you read were of my own pen—

FENIC: But you attest to the neatness of Cantal's lettering, do you not?

SECRETARY: ....It was indeed nicely woven, with his characteristic handsomeness of script.

FENIC: So the mind to it was there, and everything else about the story, or the revelation of details. That suffices our trusts and Ancher's. Are you thirsty? We head for youth soon.

SECRETARY: The unusual is a bear-like beast, until subdued with familiarity of its nature. It is most often exaggerated towards various observers' points of view.

FENIC: So? I've not lost skepticism. But I need the bounty of a miracle; and I am hooked to this, as would I talk to others as they would, to please. So may we be pleased, with the vernacular of certain wonders. I then insist to stay with possibilities. They are what empower to transform plausibilities into practice.

SECRETARY: Anything that can be conjectured by man is, I suppose, possible. There are certainly extremes to this proposition. I am not skeptical— but doubtful. One shapes some wet clay into a ball and call it "earth." Who knows what occurs within, or what may. That's somewhat beyond your manufacturing, because they're due to properties inherent to the soil, and not of your control to shape but only alter and disturb, or destroy into a powdered ash as with burning and forging. But what can be are merely of the minerals and organics contained. One can claim anything, with such ignorance. A dried up flower bed, in which nothing has long grown and no seeds may longer germinate is called spent earth and impractical for cultivating life, a wasteful, heavy dust. Yet some claim, in theory, to be able to extract the very essence of potential from such piles, by simply defining them of original merit.

FENIC: The "Ur-creationists." That's more of a religious posturing than fact, a belief of ascension only from the comprehensible: Thus as this soil is made to be, it must be soil and not else mysterious or obscure to reason. For soil is soil. How can it be else without destructive transformation? Those properties of soil must remain. Tricksters mix it up with fresher samples for planting. Even in crushed rock and sand or gravel may some kinds of sprouting occur, based on watering nourishments. I don't suspect you find such problems here.

SECRETARY: No. There is a fertile greenery of vegetation all throughout the valley, of what I can see. If ever a place for my earl's experimentation is to be found, this be here for prospects prodigious. But claims are always cloaks of wearing and show, garments underneath hidden of their colors and holes. So much as any clay contain for to be shaped are claims made for this artistry.... The idea, you must admit, seems preposterous, much with the warmth of mistaken judgment. His notes are full of so many speculations—

FENIC: What is unimaginable amazes. But this is thoroughly conceivable. I've seen rows and rows of vegetables as like grass. Then why not grasses as like vegetables?!—

SECRETARY: They are. I don't deny that at all. I mean to speak of the researcher's hopes. They are.... creationist, perhaps.... But we will tell, rather abruptly, with some samples and precious seedlings delivered, by Earl Ancher's personal handling.... You must have doubts, Fenic. It's often the sauce that makes the salad delicious, even tolerable. A vinaigrette can hide distaste and be the principle for food, what turns paper into nutrition.

FENIC: Of course have I.... unsettling cake to speak of. But I have tasted seaweed before. And that is edible, but not fashionable here— because we are not fish, nor brine of shrimp. Yet are we more to eat, a worldly man devours much and becomes.... more universal of the tastes. So then, to make this region worldly and exotic is the only hope or ploy left transformational, resurrecting, restorative of fortune and prosperity—

SECRETARY: Perhaps green beer.

FENIC: —and life! Is that for thirst? We'll order from the bar, and be persuaded of this tinder. For we've taken up space for sitting inordinately. (*stands*) Yes I have doubts. Why does the earth keep revolving? Where is the mishap? (*as the **secretary** stands*) But then that's not a question to be settled with disbelief, that something so grand can be in motion and stay that way for our eternities. (*as they head for the bar*) It's the spoken truth, and the leading passion for our observances, as could the ancient Greeks refine our minds to be atomic, and singular for the full acceptance of many concepts and evidences.

SECRETARY: Yet were they weaned from simpler greens than herbs of sanguine optimism.

Act IV

Scene I — *A courtyard to the king's mansion. The afternoon is warm, and **Mytdra** sits on a broad stoney bench of obvious decorative worth or ambition, as **Holsten** stands near him. A slight distance on the same bench sits **Bastrel**.*

MYTDRA (*to **Bastrel***): Don't tell me you've altered him away. I've made no overt threats.

BASTREL: I thought it best that he apply himself to his beliefs awhile, and out of this jurisdiction.

HOLSTON: As a citizen of our nation, he is bound to us wherever he may go, and as well be protected.

BASTREL: But this is a hard clime for calamities, constabulary. And while I reinforced the urging, you quite brutally delivered the sentence.

HOLSTON: He is a friend, ma lord. I did not want him disillusioned of the king's requirements, or prerequisites for further company. Mine was a warning. Yours was an explanation, that he totters towards disproof and exaggeration.... of abilities.

MYTDRA: You drive many away, with good discernment for reasonable intentions. But this is a petty matter to a minor slate of process, some.... remarkable commodity to delve of and recommend. What you bring, Holsten, is— in effect— the real assertion that Earl Ancher could be an enemy to me, or become one.

HOLSTON: He has a.... young, progressive mind, your majesty, and can wish for more influence than his station could ever attain —

BASTREL: I would doubt that sincerity of enmity, my majesty. He is one of your most ardent and reliable fiscal supporters.

MYTDRA: Yet Holsten's fair intuition suspects a hostility to my rule.... by many. And I respect good policing prejudices. They not only prevent open confrontations, but reduce discoursive tensions, or dissemblances, in a civil and mutually pleasing way. For heats engaged are only to convert something into something other, for the utility of the winning antagonist. Such as a meal is cooked, the fowl is fleshed for eating, its feathers stripped and its skin baked to savor. Or as a victory in war, the defeated are subjected to fees and entanglements of subjugation or tasks of appeasement with worthy works. Then if Ancher is of such a fold of mind, which has not deceived me much of the outlandish, he should stay away for good!... of his safety, until an application is won that presents to me— and his country— a better tribute than he has heretofore conceived or conjured, suggesting of delights improbable that must delay the return. I witness him much, through others, and more of late, for this potential to revolt. And it seems as a milkweed's poison substituted for some false endeavoring, fine for the looking at to consider, but nasty in taste or expected utility. And the more we dine of its conjecturing— for us— the greater the number of distresses turn up about it, that if not fake it is fallacious of its approvable means. Though some applications may be found quite different than its original presentations suggested.... There have been— several, of recent times.... who thoroughly disappoint the simplest tests— to be forthright of their propositions to me, and this nation, and this world. For flattery is foul anywhere, but especially when the wind is up for sniffing.... Now had he refused to go, I would be less certain of his character and courage—

BASTREL: My lord!... he only seeks to assure himself of what he has projected to you.

MYTDRA: And by that way is he not banished, unless it's *reassurance* finding— For he was *sure* to me. And that surety may come from else of his intentions to guarantee, and not this grass to wend for sale. So what is he unsure of now?! That guardian saints enlist him? That the heavenly conjunctions are not occurring wisely above him at this time? He is subdued with the frost of being false and tactfully found out. Much better than a slaying is a sloughing of the sloven suers. Yet all may he do, to redeem himself with proof, is passable slight, unless there's might to the endeavor.

HOLSTON: I can't describe him as having any anarchist sentiments, my lord. He is a terribly responsible fellow.

MYTDRA: Of course, as your friend. You.... protect him against his overreach of blasphemous bunting.

HOLSTON: What able heart would not try to soar above the hearth to obtain a span glorious of spread through wisps of a spirit? a proud spirit.

MYTDRA: As strains of smoke may rise off burning logs. The fellow did not strike me as so passionate. A deception, then?

HOLSTON: It's become almost of a habit, thinly threaded throughout the realm, your majesty, among pockets of capable achievers youthful and less indoctrinated to the gravities of resistance—

MYTDRA: Inexperienced warts.

HOLSTON: —And thereby generally dismissible of talk alone are their casual and friendly pledges and professed wishes, heights to leap to as if deserved of work and wealth. I take it only as a material sociability, and not a weaving of insurrectionist acts, since most of us exhaust our minds with dreaming to be greater than we are or ever may. That is the sincerity of sin, I suppose, to try to stretch beyond one's limits, and through this— urge.... become deluded into criminal intentions, propensities against the state and king to commit. You wisely have me prevent or forestall such excessing. But there are growing shoots, your majesty, receiving the sunlight that beams of energy to use. It is most commonly ambition to focus of and winnow to employ—

MYTDRA: But not illegally.

HOLSTON: —of their talents and trained proficiencies. It's a way of nursing their own tempers, eventually into something or some service more appropriate to imagine (of themselves). I can tell you categorically that Earl Ancher is no enemy to his nation, is no imprudent bane to our royal house, and struggles conscientiously with perceptions of injustice or misjudgment to either abstain from tackling, or to attack. This is a maturing process to the most of meaning, important for all individuals with conception of their place in the universe. I can tell you with embarrassment that most crime committed are by the immature or unripened attitudes, given the means of some mischief and a daring of necessity. Hardly ever is anything profound about it. Yet, for riches owned are temptations buoyed to manifest for change, improvement, blissful reckoning to achieve, and sometimes infantile simplicities to reason with. While we, of executing guardianship, through our practices observe that in order to sow the seeds of contentment and happiness we must plow, we must scar the earth and furrow through even beds of ground, scratching and hurting the land and some of its inhabitants— almost regularly, repeatedly. That to be unjust is not injustice, in the broader scheme of life with disciplines obtained and punishments suffered, does make for value our exertions of control, my policing, the judicial arraignments of magistrates and judges.... and the king's overall governance throughout his realm. This is what constitutes the superlatives of a civilization, the orderliness of submission to the cultural norm— and condoning some natural tendencies of imbalance by what is subtly, incorrectly, and obscurely grasped to be indecent or appropriated of fortuity the strengths of a few to rule over the more.

MYTDRA: What say to nature, then? Nature is not art. Art is to be made by me. I can not forgive even what must be by nature to want to usurp my monarchy. Privilege is a dog untamed, to bite its master, though it comes by its aggressions honestly, without slaps for being a dog. Say you an avalanche of these gents are made, that you can not confine them in such number to stay within the restrictions of my providence?!

HOLSTON: Trends do grow of wealth, your majesty—

BASTREL: We can't keep expelling these young princes out of the country! Our treasury can not survive the insult. Our economies suffer with their loss of involvement and ingraining. If their complaints are due to a stultification within their station and rank, social positioning and influence on governance, then practical ways of advancement must be found and compromised through for their expiation from.... dissentious motives. My majesty, we weaken without their candles lit. We shroud ourselves into a dimming darkness without their impulsions and impressions. We drain of talent, easily out— And this hurts reason! at the scale evolving.

MYTDRA: You're afraid of your accounting and adjustments, exchequer. I contain the necessary calculus— Just supply (the) sufficient figures.... Err they to lose their estates, they come back schooled, converted to obedience with wisdom of more knowledge about the prosaic world, less dominated by skill and honor and divinity than what is constructed for my seating. O but to abuse such privilege and care of craftsmanship! I know the meaner characters of men.... and offer dismissal, a suitable purge of tainted blood, without the violence of decimating loss. But my notions are only on to this for Ancher. I've not accused him of anything, and did not order his departure— but rather hoped.... of angels bound to explanation, more exposition he would find among my privy servitude, for answers to surrender to my ears of what kinds of loyalty submit to me, for which I've cherished. Then examinations, deeper inspections prove what can be done by one.... what might grow fickle of its flame, for love of private subjection— I did claim to myself.... he is a capable and useful man. And perhaps now pardonable? What occurs of this without more scandal and scheming, but to more fraying am I leased of skirt these threads! these.... friendships, and close, closeted attitudes. But what is seen occurs as evidence, wayward tinting of the eye. I can hardly bother of this earl for what to grieve. It is the self-expulsion that supposes on me, and imposes greater discerning.... not that it's so great as I have ever done. I know the man and the beast. Man and beast! And I have served myself, with covenant for both to approbate and conduce for my favoritisms.

BASTREL: These young men are not against you personally, my

liege, nor of the country would they dissuade from benefit and contented evocations. But like any they, though of some means, have individualized concepts. And a general mode of life or living through the predominant schemata of a single mind of course disturbs them with envy to challenge at least philosophically, in particular to points of physical disagreements, such as monetary matters—

MYTDRA: Young! Young! You keep saying young! My enemies are ancient! A pursuit to grandeur by any ambitious will causes challenge through the ages. That's not the spite to recoil from or be repulsed by— But is trust the mail that's worn of conscientious subjects! For I accept all dangers, through their pledges to serve me. And there is not the horsewhip's howl as blatant, sharp, and piercing as a hollow insincerity to challenge of the ear for one's attention and pondering. These are the bad faiths heard— and lies!... that crack a well's foundation to be splendid and supporting of all the common thirsts. Yet is for power the theft through earth?! He is the simpleton, of high intelligence— That is to say he is simply smart, and for not of robes to steal may he bend to, but for an amiable gaiety to exist.... and with some disappointments pout, perhaps, and correct towards his goals of simple life and pleasures.

HOLSTON: Ancher may not stretch the truth of what he's told, your majesty. Though whether he believes it is a story for yourself to hold in judgment of integrity. I can not believe he intentionally lies to you or anyone, concerning his affairs and with his experience and capabilities to handle them, but rather fears he may have imprudently passed along exaggerations which he now scrambles to either authenticate or ameliorate and excuse with personal involvement. It is a responsible course left him. He will make gift to you, I'm certain. Maybe in his travels he will find something even more audacious, as a true golden apple, or a strawberry of such deep flavor, richness and sweet moisture as to be auspicious of eating when ripe. I've heard tales that such a fruit can knock one out with delight, and actually leave a person unconscious with dreamt rapture. The ones we have here, though, are rather disappointing of intensity, till baked. I've suggested to him before, to find the reality to some of these myths. And he responds with time to dwell of handling propitious chances which come across to seize, what's brought to his attention for the leap. But he himself is not such of an adventurer, but more a manager of events, discoveries, results— to make them more outstanding— of profit— than themselves alone. And myths be swung, with advertisement, drawing crowds and buyers.

BASTREL: That's a good ability, for our markets to stay healthy, vital enthusiasm to collect and sell noteworthy materials and produce, which we should encourage more rather than dismay for less.

MYTDRA: Care to rebuke vitality with virile mortar. I am vital enough to spread of worth throughout the realm judiciously, though pocketed through circumventing means of what is possible or can be made feasible. And this is more to understand of lies you do not sense for me, how some expediencies must be tolerated and toiled through— without deceptions!... or loss of faith, and precious fidelity, the only gold left man to bewail of from this loss. You can't grasp the hurt, the stoning sensed, the conscious bereavement.

HOLSTON: He is a faithful gentleman, the earl, and will probably stay so—

MYTDRA: Oh! Dismissive!

HOLSTON: —These necessary sojourns do improve a man of value, as you say, my lord. They do mature, with circumspective valor, and grow of penance to return more piously devoted to you and your heavenly endowed missioning over us.

MYTDRA: So may gods make pact on certain men. I am wounded else of discourse to pertain. And may the fool share rots on islands, to take the blame, if not to be this greenery— That's not my care of this! We have his estates. We have his funds committed to. But what is shame to coat?! with shabbiness of back!— and holes!... So often have I seen before, but not afterwards. So to improve a pain off stab— take pain with stabbing. (*stands*) And I am hurt!— of conscience. For this is usually *corrected* afterwards, taken care of miserably as a dog. But now I must sense the misery of this floating essence that cries of scent revolting, disagreeable, and ashamed.... of back to hear and weep with cruelty, this sentimentality lost to purchase. And I thought it may be from love!

HOLSTON: All of our citizens love you, your majesty, in one way or another. Greeds are particular in taste, to pick at for their itching, as could scabs have flavors on the fingers. But I can certify without much confusion or doubt about it, that Ancher is a forthright fellow, whether he stays away forever or comes back begging for your acceptance. And he represents.... not holes, but harnesses to shape for your mounting, as these straps may be as various as persons to fit their personalities and differentiated ways of convergence for your comfort. Yet as a friend, could he not attack you ever is my thought, to know of people. And of the others.... as well their weights may deceive them till all balance is lost and they fall. But you will not be gashed with their descents, protected as you are with surrogating might, supporters and servants. It is decent of you, however, to feel some grief with this, the upset.... paternal of all flocks, for any straying—

BASTREL: No one can ever upturn this king! He is too clever and refulgent to be found base or bash-able. Yet fear we must our leans and slenderizing. This comports more error to the remarkable than at, more weakening than stones thrown of misses. We must more regularly channel these contentions and dissents into compartments converting to the positive, more offices of discreet influence, highly specialized and limited— as well dismissive! but for their temperances to sway and their temperatures to suspend or lower. That is the only way I can think, to resolve out of these tendencies for potential crime, high and high spirited as it might be for well-meaning attempts to attain themselves for the good, as they'd perceive it, of the nation. For this is definitely becoming a pervasive problem when such a potential as Ancher's is subjectable — or recruitable of careless evils. Though I would claim him not at all evil, but merely careless in these curious details. And there really are not so many of this caliber available to prosper the country.

HOLSTON: Let's not inflame the description. Valuable is his worth, but in no way malevolent towards anyone. His dealings have always been quite refreshed and open, not secretive or mendacious—

MYTDRA: I care not about him!

HOLSTON: —but easily traceable, often observed as clever moves to wise strategies above board—

MYTDRA: Lords be lepers!

HOLSTON: —Yet it pains me to have to have alerted him, and have innervated— for his own good— with some fear and need to take precautions. This is a fellow who likes to stand his ground instead of flee from possible assaults. Yet his battles before have only been mercantile, finding the best positions to take advantage of opportunities over rivals, and to avoid being pushed, led, or tricked into dubious partnering or sacrifices for later, questionable gains. The clearness of his approach to matters is what has spared him from outrageously huge losses, as some others have suffered. Yet he comes to you, my.... majesty, without such debilitations to try to recover from, and had not the need to be deceptive or devious of talk. All was going well of his voraciousness to be outstanding of a subject, until I had to crush his toes in as friendly a manner as I could employ. For I knew you were searching for chains—

MYTDRA: I was not!

HOLSTON: —'gainst malfeasance and malpractices about you— Not?!... What could not be substantiated? of discovery?

MYTDRA: Nay, not but for knave my nonce of nuance. I have always given men their chances, to make good, little matter how ridiculous. And if you doubted him, that is your fine intuition at play. But for the strike of chains across the back, these strokes come from failure only of promises made. And I don't believe he made any— to me, in such particulars as I could hold him to. But to others are they dimmed of light diminishing, with fretful belief for his story, that there is truth and trust within his gallant wishes or intentions. And that's an error made by ears of heart! For one must judge a coin by its coinage over its banking of, what value is left to it completely out of your hands. And in this way suspicion contorts your fealty.... I have made mistakes. And from me that is powerful— Yet does this lord less, to run? What does this portend, of a methodology developing to overrule me?! for whatever may be done behind my back! It is a reproof! no less than that, against some suffering. Yet corrections are unknowing, until now, for blame to charge of and condemn with— See that woman there (*pointing vigorously*), who is so poor to court, and handling a chicken's carcass, carried headless— of a cock, no doubt. A rooster's lout-ing!— She's a beggar! made by me, according to my Fenic. Then I must make amends— and he! by bringing her here for work, paid out of dusts and credits I'll not have to wage nor waver on. It's total mist to air, that I am burnished with for caring, what the dog can think and do— (*calling out, almost angrily, but of regal tone*) Come here, you! Do— ria! (*as she approaches, with hesitance of fright*) Come! Come! She is awed of being noticed! yet does make for me my character dismayed of sight, this display of supposition for what I have done.

HOLSTON: Is this against Ancher's woes? I see not the crime.

MYTDRA: That is for the telling of Fenic. (*as **Doria** cautiously comes up to them*) Can you not handle blood more carefully?! It bleeds upon the lawn, this grass for staining red! the purple endowments of a hatred.

DORIA (*uneased*): Across the court(yard) I must bring, to you, your highness, and the body drips—

MYTDRA: But with ravishment to disturb, these blades—

DORIA: I have just made the kill. I have just culled an' cut, with my simple skills reduced to utilizations ordered, for your supper to mend, that you prefer of several kinds of meat—

MYTDRA: Old and haggard! was this fowl be must to mar, that of my tastes improve with mashing of the flesh, and crushing of its clacks. And yet to fit for your employment made, of tendering such beasts— so hardily?!

DORIA: It has been difficult to learn, my sire, in this domain to handle of these tasks so hurriedly an' rushed an' demanded of without complaint nor preferences of sensitivity an' cries. Since I must eat only of the grains, as these poor animals are devoured.

MYTDRA: Do they scratch and bite at you as you skin them?!—

DORIA: They are dead when I skin them, my lord, or despoil an' strip. Not for their screams would I have left this cleansing—

MYTDRA: And yet you must be accomplished at it. This disturbs me—

DORIA: I have a family of young to care of. I had a family with man! And then we made our meals more sumptuously once. For them I could know how to butcher easily, lest for a starving kindness could we make of our simplicities of life. To here it is a task reduced for lowly, unaccomplished art, and hands inarticulate, mouth speechless, and mind demeaned from begging—

MYTDRA: Beggar!

DORIA: —an' the fright for my children to provide with food, as could make me a stranger to them, with this debasement—

MYTDRA: So (it) serves you of drops that cling with waste. And what to this queer affection that dries insoluble of the crimson sincerities and tinted stench of heart— that cries for young away that *I* have caused?! What have you Fenic, as a man amended?!

DORIA: He is a friend—

MYTDRA: And fiend upon my back. But is this love?!

DORIA: It is for needs we have, my honor gained with.... your royal labors, my hopes regained with your laboring to have me, as could kind Fenic produce some means of salvage—

MYTDRA: Withered and cruel!

DORIA: —recruiting beggars. Finding more expression of salvation from.... our terrible conditioning.

MYTDRA: Such is this feathered skin. You must love him, or this is mutinous romancing. For I can't retain the thought of mere friendship lent, to saddle of recoveries for a town and quartered people, from a cur like Fenic. He was of highest devotion to me, and I his only dear to serve. Yet are you not his representative of shame against me!— as for comploting counties— or the eroticisms expressed among commoners doglike, unavoidable, and unabandonable of these animal attributes eventually yearned for.

DORIA: We are of friendship not romantic, and above platonic fury for devotion of caring, since we were both made miserable. And he is capable of more than monarchs to adore. But in his yowl to restage regrets with replenishment of good, he fathoms more than you—!

MYTDRA: That is impossible!—

HOLSTON: The chicken scorns for brooding of a fate caused. That being of itself what's caused this. Yet lower we of animosities from sadness, and not appropriate to the politics of diverse stations. Either Fenic is the highest criminal.... or feels responsible for the deplorable state fallen of this— matron—

MYTDRA: That he has said! That he has told me! But what is the truth?!

HOLSTON: I say he darts with weaselly help of her and others— through Ancher. I can not take their senses to be conspirators more. For Fenic is such a lessening of person (as) to be halved of envy for more obsequiousness to love.

DORIA: His manner is simply whole of a friend as he did love my late husband to be. An' there is not less of his character and daring. For his flair is honestly obeyed with displays of sincerity for his likes an' enjoyments.

BASTREL: That is certainly a trait of his. Yet he enjoys the king.... and would banter me about to batter on behalf of—

MYTDRA: Tricks and styles!—

HOLSTON: Judge me not my intuition fine?

MYTDRA: You are a friend of Ancher—!

HOLSTON: And not of Fenic's. Yet be they mixed with rumors — And I've tried to (of) compatriots compare.... But she seems genuinely depressed of fashion from stresses, and shows not haughtiness until challenged of her pride, whereas the criminal subverts his pride— to achieve a goal.

DORIA: I'm no miscreant, to lease of what's been done to me as villainous of fate. And Fenic serves to comfort through these hardships as a friend. For this he is most capable and righteous, an' can achieve such endearing states as a true an' trustworthy person. But why align me to some possible illegalities I know nothing about?

BASTREL: This is our principal constabulary officer, my dear. (*Doria sighs with some fear.*) You are at high station here, and may not know of whom to encounter at any given moment or instance of meeting. Yet stay your place to see not any contempt concerning you, if you're not to be noticed at all—

MYTDRA: As does this dripping pant! from coop to carriage brought, as I do ride whereat I stand— ever of the reins of state to handle. See what I have called, with every tug at misplaced twine to bring, some turfed disgrace uncovered or revealed of this business, this— association of scoundrels!—

HOLSTON: There's none to be accused that I have found, of accomplices to injury of crown. And I have looked hard, concerning Fenic and his dealings with all folk. Not a true sympathy for subversion uncovered, your majesty. No hidden malice seen nor smelled a'ferstering underneath the brows of commoners to nobles he's shared of conversations. It's not my place to support this fellow, for he is *your* peculiarity deigned. But I can not contemn him as being against you, is my professional opinion gleaned thus far. He only plots to aid this lady and similar, with really paltry help. To involve Ancher is a wild hope of donated ability and interest, but the earl can stand on his own for what he has promised, suggested, or confessed to. It does little good now to keep pulling at loose threads worn. That only serves to dismantle the garment, or tear it apart in frustration and frenzy.

DORIA (*with unfeigned submissiveness*): I cower at my fears, for I need this servitude to save my own. But I can't understand how I have disappointed or disturbed, being as menial to kings as to my peers with begging. Then to subvert my pride is all that I've accomplished. But I would die to insure my young be fed and raised— in any manner of state.

MYTDRA: That point is accessible enough. It's bled and blessed to be drained— And I don't doubt it. But what is incredulous to me is this heart, hard to accept or understand. For either lusts or carnal currents must pertain to this creature I have fed and warmed of favorably, and can predict with, send out on errands and rely on. Yet, with bordering on dangers, or bartering with crimes— of deceit at least, if not deception of one's very nature to promote, be it curious, courageous or overtly compliant.... or simply carried as by whims of heart— for this matter to be magnanimous is simply difficult to attest. For he has been my *faked* nobility, and with a natural commitment to this duteous antiphony of my word. So must I challenge the fact that he is truly sociable, and can enliven himself this way? What of my subjects sensed have I been wronged, without the spine to judge him properly of the necessary contrasts I must observe correctly? For it's not so to be happy. (To be) "Happy" is from whatever of your doing, and he often shows this — to me— regardless of the grief.... of himself and others, I have assumed. So be accomplishment as happy. There is one's freedom, with me, for the task.... So are these lies?— I can not tell. She holds the chicken well, and yet for mess. You say he loves you.... as a friend? to counterbalance my.... intentioning? Cold be his neck to tempt this as firmly as your grip!—

DORIA: He does fear his arrogance destroyed his friend, my husband, by informing you of our village.

MYTDRA: ....Is that it all? He could have told me so himself, as simple shame. As much is maimed of valor with mistakes. Yet how I see the slight to singe into a monstrous woe, with what I have created in adjustment to his goals and planning. Boasts are for beasts! to bark with. And there is guilt in this jest; for I did consider the matter with entirely of reign, and thought to improve the whole— past your small town remaining sightless. Now come to see, that curs do bite their own backsides!— Rush that poultry away! Or its squirming due to your trembling hand?!

DORIA (*teary*): Thus for what I've done— it shakes! (*running off*) I hope not to have hurt its prospects for you.

MYTDRA: I am the mauling gripe! to frighten her this way.... scare away the chopper. But have you now my suspicion, that there is more complexion to this paint. There is more law battled here.

HOLSTON: I can not find the feint. I can't feeze it out among these strange entanglements of circumstance and personalities.

This is too odd to grasp easily.

MYTDRA: Then seize upon my intuition— and supremacy to know: Her country's ripe for revolution—

HOLSTON: That village?

MYTDRA: —And Fenic tries to appease them with a slight loosening of tension, through his devoted one, until he can find a surer, greater, more efficient way to defeat the contention-ing against supposed harms.

HOLSTON: That's about as much as he might claim for any of this.

MYTDRA: His tail is to my servitude alerted, but to Ancher no. It is his means of biting at the problem. Yet for this earl is scurrilous intention—

HOLSTON: My lord—!

MYTDRA: —Since the Fenic's nature must always snap at scoundrels to me. It is above his will to defeat that urge. It is instinctual and permanent of nerve-borne flesh, an overwhelming inclination to resort to. And you will prove this—

HOLSTON: Against my friend?!

MYTDRA: —You will prove this by thoroughly investigating the earl's relationships with that town.

HOLSTON: I've done so, and much!—

MYTDRA: You will find the bed of soil, for which that crop is planned.... and you will determine what is really planned— of a revolt.

HOLSTON: My lord!... grass can grow nearly anywhere.

MYTDRA: My nose is up! Holston. And the atmosphere is clearing of this puzzlement. He is as generous to Bastrel's entreaties as for a general seizure of hearts and affections from the disaffected. This is all so obvious now!— Determine who sought whom, Fenic or Ancher. But discover this as said within the town! by a shaded gossip learned. That is the means of shred, to find the husk bared of a complaint that lingers and embeds of a rich ground, for grounds to promote my overthrow attempt. And it is as childish as could be thought to succeed. Because such cavities do not breed warriors, but only a backwoods refinement to be displeased.

BASTREL: In such a person candled to your wick, Fenic is of type to come to others for his leveraging and leans. One would certainly not come to him for a despondency to earn. Then it's quite obvious the earl was first approached (by) this king's servant, the fine Fenic, fine as in difficult to gauge of markings in clear sight, his scales of slight prepossession to discern except only to your favor in general, my majesty. Thus to this, had he presented himself, Ancher to Fenic for knowing, the fantasy of wheat to describe as a green prospering.

MYTDRA: Find out what is rumored there, and not fact— For we know the facts! Find out how that populace is being conditioned towards.... some uprising or dishevelment, as justified. Find

out what the people think, and their dispositions to dispossess themselves from my protection and guidance. That is the fermentative fostering, the fermentation that must be planned in advance in order to grow into a strength or consistency of achievable success. Principles are only words felt for valors claimed or proclaimed to the suggestible mind. And note how I weigh these possibilities of fabric costume, as these patterns must become evidently worn. For of result it's only necessary to know what is believed— to find the desirous intention. Chances are offered always, to disabuse my fears or suspicions.

HOLSTON: I will survey the region discreetly of agent, your majesty, to determine the tenor of political emotion and discourse. If Ancher has planted seeds of sedition.... this will be uncovered. Yet if Fenic has sown for hopes with loyalty to you, that will be the countervailing principle shown, to argue against providing of my friend an offensive predilection. Though a revolutionary mouth is not so actual of might, the gripes of timidities eased for scope of hearing argumentative assaults never to be carried out. Wished for crime is not as criminal as death by this, if we're to remain free thinking animals and rational postulators.

MYTDRA: Punishments are made already fitting, for these kind of ideas, and will fulfill themselves most truthfully on their clients and clientele. They are produced of the insufferable winds that demand of foreign address for other of my wingspan's care, else despotically they hinder themselves greatly under my stern control of these regions. It's not for cause of complaints, but for cause of compliance that I'm concerned. I will lessen the complaints. But compliance must be genuine and generous, earned of the material for their being my subjects, that this is divinely produced and promoted.... or I should be gravely gray, of my lifelong presumptions. And I'm anything but that, to glided doubts of crudity interfering through too coarse a manipulation by others. This! a ruler must set right, and only he perhaps. For through the complexities of happenstance and a curiosity for worldly events they occasionally crop up. Yet am I not mean to think menacingly.... of dark clouds and silvered shrieks. Who owns the world, constabulary, if not the gods!

HOLSTON: Seems their pleasure is for our mischief and riffraff making.

MYTDRA: Then confer with the envious to confide of misbeliefs. I want the possible conspirators outed, not for fear of them, but for fear of my wants to exercise.... only judiciously. This is a sign for some review of my practices. And I take it up keenly of interest what to me had only been of passing announcement and little importance. Now serve my plate a meal of this amusement, how tender, warm, and sweet, or laced with meat to stain the teeth, *braisière chaude* in mixture of these roots and carrots orange to red— of rioting, or rendering to renditions more subservient. We shall find out who serves— by what is served. And we shall condense of eye again.

HOLSTON: I've never lost sight of what is privileged for your majesty.

MYTDRA: I mean for what I see evolved and evolving, to agree upon the shades of caution personalized to one individual's trusted enlistment with more freedom of range than need ever be suspected as being wanton or too wide and broad for my prevalence to contemplate with.

HOLSTON: He does dog of shoe some shadow to bemuse our worries. But I'll find him out to be.... pernicious of one Fenic's calibre for another, ruinous of ruse.... or ruined by.

BASTREL (*as* **Mytdra** *starts to walk off and* **Holston** *follows*): So be a justice to the court. I hate subservience deterred. Yet, as a treasury interred with wishes (*standing*), such as a warming rest auspicious.... (*stretching*) so late the sun delaying mishaps and creaking woes. If ever possible to make for monies doles, and giving out our virile stretches off to others, persists our Mytdra their subsistences amending in some exotic splendor and fretful happiness.... as we weave around our weaknesses of character left. Though a cur be bold of spirit, I'll not rate him yet, till ranks as colorful as he has been prevail of my conclusions that this one serves as dominated as a worm in moist soil. And where for funds now fishing?! It does rain in my head, as the drowning leave a barrenness of earth. And to return to. dry as parched? grateful for the shock and disturbance, shaken without tremors? displaced with cooling softness sharp— and respirations hampered? Less fertile can the field remain, as we become domesticated to these turmoils. (*sits*) And yet this torment's green. If ever to cause harm, then why with a vitality and sumptuous soaking of the simplest verdure? the pining for solar enrichment with a free fruition along entire landscapes. Then we've become.... as if of harm harmed.... harming. And that is the nature of our enveloped skewing. The government can not avoid financial collapse, without these rooted nodules retaining, and without a continued lush growth of lawn, grasslands, forest depths, and longing for.... a covered and protected earth. I am at the withe wizened, to contend with king for what is due, and what can only be expected to arrive of a decimating storm too emigrational of productive wealth. Well then, construct with this: the whims of placation.... for demerited locales. I'd make with less, for more incestuous allowances of my own. But better fair to be than foreignly promoted— of catastrophes.... (*relaxes*) Oh, what a rush is day, and spending time.

Scene II — *A beach, during the day.* **Ancher** *and* **Cantal** *are approaching each other from a distance, though a native* **guide** *for* **Ancher** *stays put.*

ANCHER (*calling out*): So away! Away! A far way I've come!

CANTAL (*replying*): Silt through a breeze of airy prospecting, this draw of lightness to the wing!

ANCHER: As salt! Crustaceans bring, a ship of crabs till landed, and off a heavenly guise to a more worthy beauty, are the seas surrounded, their emerald waves pounding as like treasures in motion, unmistakable and yet indeterminate.

CANTAL: So to shore, and safety of the trees, the shoals crowded of much fish and fleeting delicacies, transient occupations, and a fluttering of broad leaf darkened and intense, with such a captive captaining of ocean for of a vessel boring through— and pouring of intentions man-made sweats! And you are well? to weather this voy— age! an' invention of the intrepid seeker.

ANCHER: How's the hay?—

CANTAL: Plush, with bountecus commitment, the surface of this habitat resplendent of handsomeness and native vigor. The floral firms affirm themselves to light up night, through which the animals dance an' soar, and grace themselves perfumed. We ourselves partook of some of these pleasures and festivities. Yet hardly did you bring yourself past lettering, the written writ that caused us even to wait— For we were about to leave! Too much of paradise tans the skin to a leathery consistency, an' soaks the brain with an ambrosial spiritedness that can defeat the aims of accursed creatures such as ourselves. Yet have I valued this, mirrored sights of antecedent freedoms, the antiquity of being pure to our reason, and finding forms of human-make so wonderful an' gregarious of their splendor sharing.... Gives one more pause than time to think, of what you have become between the ages of your learning and educated pastimes.

ANCHER: Where's Lancet?—

CANTAL: Recovering!... from an intoxicant of adventure, and a whirlpool whipping of his mind, the colorful decors billeting within of head for higher senses to bestow an' realize, an' strange decorums to adopt complicitously to stand of and overcome with a more civilized dispatch of the passions flounced, his florid face festooned of wild grins an' occasional jerks of sensitivities to tastes and strong urges—

ANCHER: The fever?!—

CANTAL: Not won! but for disease of wanting, an' long to dwell, the feelings of a fire burning of one's self, with other worldly dominions lifted into, as one could float above the vegetative lushness and find your own debates observable to be dismissed as foolish! an' squandering of energies used for rancor.... Some spices here are very powerful to eat, an' he's tried more than a few to hallucinate his missioning. But now the body's quite worn out of fun, an' he rests up to recuperate some sufficiencies of purpose, at my insistence and for his preference drawn. How much of the angelic can one take? before losing the earthly perceptions of your grit an' grime! (*They meet up and shake hands, he rather more loosely to the other's imperative.*)

ANCHER: More droll be grass. This is a beach! And yet it shimmers all nearby with such huge length and iridescence as to make glorious lampposts.

CANTAL: The salts are healthy for their uptake, these minerals Minerva-bred for growth. And everything here are as natural as excesses may allow for the directions wished of them. The warm winds do push us comfortably, to find our places centered, as a boat may drift without destination, undulating for its home.

ANCHER: Make us reeds with the gigantic— even spears!

CANTAL: There is a great variety to them—

ANCHER: Aye!

CANTAL: —all throughout the purpled chiffon—

ANCHER: Their tops!

CANTAL: —to the reddened thistle and brawny garishness of garnet hues off to a distance—

ANCHER: And Green! Green! Green!

CANTAL: —with blue sapor of patches to the stems that make for orange with height. And thus a spectrum can be seen when standing on a mountain top, at some places. An' yet the ever most make verdant cover, to cause accent with the flower heads of plants, some multicolored, variegated— and yet the purest white I've ever seen of a lily, or some similar form of bulbous wonder, clearly of very ancient extraction. There does idle here the idyllic of very individualized properties an' elements of appearance, along with much mixing about unforced or— hurried due to cultivations. Then find we stores, entire ranges of the primitive unblemished, and their corresponding pollinators unconcerned of age an' travelers.

ANCHER: To the remarkable I'm bound, Cantal. I've promised—

CANTAL: There are humming birds out here as large as robins, enormous processes gyrating their wings and stopping still before us to investigate our snacks. The shock is for their kisses and the buzz with beaks protruding towards infinite delight. What fear have we to find our honeys and nectars! All make acquiescence to the muse of a regular drama, a comely existence, an' our deportations for awhile.

ANCHER: That is a spread of riches natural and guaranteed without adulteration. The water's blue and green, and was endless of sight for long durations. Yet I'm impatient for more blue of green to find, and so beseeches out of necessity a goal of confirmation, that much trapping abounds here to satisfy a hunger—

CANTAL: An' this must do it! I've engaged of my virility again, an' have been allowed to study it. This is an incredible location for bringing forth long subdued severities of heart an' longing. That makes for this more than a vacation of prisoners. For be assured, with a climatic agreeableness to officiate the surroundings and your stay, one is really forced to find the self that's sought, in various ways suitable to your climax of powers. An' so you bind yourself with careless scrutiny of your reactions to the miraculous sensations enjoyed—

ANCHER: Have you been ill yourself?!

CANTAL: —to find a dimension of self-circumspection that answers of the limits you've achieved. Oh, I've been somewhat touched of a light-headedness at times, but it's only to discern what little substance of thought is (up in) there compared to what is offered of a real transformation towards reality. I've felt the perspirations of the cauldron's candor boiling, upon sampling of a siren's beck! to be myself absorbed, the absorptions distant and remote— within the recess of my fornications with known or comprehendible, comprehensive life, an' make with contrast mate! to the obtainable unknown to me but for my flight to perceive it, as a fluid's light through a glide of retrospection, that ne'er can one be diminished by the scolding of your actual enticements an' desires. An' so I've breathed with flush! of air to head, to be restored of vital tendencies thought.

ANCHER: Certainly this is a stimulating remoteness from our typical routines of daily life. Yet as an excursion of discovery this would be so— And you were about to leave? having concluded your investing of interests, your investigations of the remarkable found, your summaries made and items cataloged, charts drafted and maps drawn? and having fulfilled your health's uplift— in this sunny air? the briny gusts for gout's removal, of the gaunt and dis-

pirited to abate their sad circumspections and complaints?!

CANTAL: Our agitations have been lustful, an' this visiting instructive of the many manners of mankind's manhood sheathed! of encrusted corals and ruby colored sands, as for a golden dilettantism amorously pursued. Yes, we've been awakened and stirred with affections, with the largesses of a lavish naturalism an' the strengths of rational ignorance. For here they fish— without inquiry, while we must bait our hooks for purposeful questioning. Yet what is sophisticated of the manifold varieties of mirth, an' of feeling great! an' of being warped of self-astonishment with the discoveries of others an' their terribly simplistic mores. But we ourselves age inside, with criticisms, and must lunge back into a regularity of more complicated social existence. Lancet must regain his head's buffering for more dispassionate, scholarly diversions, an' accustom to his blundering with healing soreness—

ANCHER: Step he on poisoned nettles with bare feet?!—

CANTAL: Spikes of sharpened perfection resting without care in the watered sands an' beach-bled foams. There's plenty of that. But not for error were they made, sea porcupines and starry asterisks. No, he strode more with his face, hands, an' knees, along the earthen ground to find the delectable nourishments of dissipating sanity an' means for grasping of ordinary concepts.... He did eat of his own brain—

ANCHER: What?!

CANTAL: —monsters found, an' dine on perversions to be purged, or perhaps regurgitated to his down of straw bed. For this was his purchasing of whims on wisdom.... An' I must regain my mode of conscious security, to rest the pulsating muscles that have been flown beyond my customary boundaries of permissiveness an' enchantment, an' make for surety of others an ambuscade against wholly salacious thought.... so mordantly to be contained an' jailed within the depths of our base an' lowly subconscious prisons of humanitarian perception.... 'pon the flotsam seeing, of my captive an' much too captivating dreams an' ideals, of finding ivory of a hue an' shape! beyond my recognition to ever contemplate as being own-able or with heights of compassion allowed to keep as for companionship. So searing have these thoughts been made, so slicing of the flesh with heat!... that I must return to a cooler covering of clouds, and seek for rainbows satisfactions after storms. Yet, the climate here has been so perfect as to make me burst!... of ambergris to leave an' let perfume my steps through surf, as we work tenaciously to disembark an' sail away. Most of the preparations have been made. Only some languors remain to greet you; but our ship is just about ready, fully fitted an' of scope.... to return. And.... the weather is fair, for floating, an' for steering with direction an' determination— an' a destination to seize an' yell for!... that we have reached the gray home, the grayer home of goodly ambition an' scandal among gods, an' work for wear an' weariness— and organization!... into a handsome order of our meritorious consciences vying within a modern, contemporary socialization.

ANCHER: You long for your old haunts, habits, and houses again. Even too long a working vacation can become tedious, and make one restless for the ordinary or ubiquitous of one's native culture. But I am flown to authenticate personally your particular prize briefed to me with temerarious startle. Now I count on it so much, for approval of my aims an' worth, that I must investigate with body brought and bound to determine of myself this truth—

an', an'.... and safety! for this find, as we might prospect together, and bring home this unusual delicacy.

CANTAL: That be it much around, not delicate, this edible grass you'd serve— you'd plant.... Yes. It might grow, within our home soil cultivated craftily. But—

ANCHER: But!

CANTAL: ....does it devour useful natures without much more experimentation to discover of its substance lending to the conscious mind for bent of mutilation an' transformation into a very particular cow's-like gait or garnishing of walk an' manner, pastoral for beasts, but petrifying for a person's posture an' carry within normal society an' civil commons.

ANCHER: What?! What mean you now? Your letters—

CANTAL: This is what Lancet has eaten too much of. For it is tasty enough, for raw sampling— I've tried an' confessed, in my messages. But to gorge on it so openly unrestrained, as in a field, as your servant has done, leaves one quite demented of mind an' tackled of impossible enlightenments that must prove false of sensibilities. This substance green, it must be studied for its concentrated worth, and amounts of it devised for safer consumption than we've done. Perhaps only a single blade should do, for a meal. Though that itself can not make of a meal for filling any human stomach, an' hardly a refuse cause. Or perhaps its inebriating effects are due strictly to something in this native soil, that can be removed from ours without detracting of the grasses' viability an' taste, such to affect its purchase for a food. This has to be tried out, by our chemists an' nutritionists, these twists of terra tonic-ing, or sprouting toxins from the sun. Or perhaps the stuff should just be left here, in its native habitats undisturbed by our cravings to adjust it. We are fortunate to be diminished here, by practices thought out long ago and abandoned! For this does spread in broad patches warned of, but not peculiar. An' these natives say the animals an' birds leave, as for a grass, these shoots of green, without to munch on. That is of the wild's proscription natural. An' it is free to roam of (land)'scape without determent of its liveliness, though it prefers to rise next to.... "normal" grass, as to be neighboring the taller growing forms.

ANCHER: ....And has the "normal" next to it been tasted?

CANTAL: It is not of noteworthy flavor, nor for the rudiments of a passable dish supposed, earl.

ANCHER: ....Then not the soil?

CANTAL: Different species may uptake minerals differently, or with various discriminations an' filtering— These characteristics have to be more thoroughly exposed before.... judicious offerings to *any* public are permitted.

ANCHER: No stem! sir. I must myself this try— to prove! Too much depend on wishes bred throughout me. And you few may just be bored, of the miraculous too long stretched. But we'll make of the time what's achievable to show me. One could not expect this to be a thoroughly easy going, or outing as facilitated as deliberate. I of course suspect that much adaptation must occur, or be planned with construction, consideration— an' care! before this green can be construed as a viable subject of cuisine at home, as

long as its taste may be so derived as pleasing and welcomed. And you definitely wrote this may be so, cooked or raw.

CANTAL: I've sampled its slender paralysis. It's of a warming wholesomeness throughout the body, with a reasonable flavor an' a texture not at all disagreeable. But it does leave your constitution wanting for more, or so much as to be distensible of gut very quickly without a rational hindrance of appetite. It's difficult to tell when you *should* be satisfied, for a full stomach fails deceptively.

ANCHER: Then I'm dependent on your judgment, sir. I've promised to try and market something of stupendous potential that might very well revive, restore, an' transform entire regions of farming back home. So am I prompted to pursue this task as a test — by the king himself! through surrogated administering an' pomp of cause. This may gain me the hook of favor I need to attach myself of court with true influential might towards governance. This is the bid— to be more than popular, but populous of my sentiments an' similar within the royal hierarchies and their disseminate forks of officialdom. Here might initiate a true ranking, to displace Mytdra's missteps an' distempers of country view an' chiding. This simple feasting might start the turnabout, ignite the spark of tumultuous national conversion so many of us feel necessary to improve our states an' standing, if only I can prove that such a thing is obtainable as by contorting this dish into freedoms of ascendancy justified by our merits of improvement—

CANTAL: An' that's your goal.

ANCHER: Without doubt to it, Cantal.

CANTAL: An' so the king persuades your imperative of visiting these shores, to find the girth of gallows around these tempting vegetations—

ANCHER: As I did imply in my missive.

CANTAL: —They're not like broccoli heads, you know, or bucolic bunches of green flowers, cabbages or lettuce leaves. This is grass! Wiry grass! an' not to bind the king to post for hanging as with effigy of intent. Then what you pursue delves into the devilish — if not criminal of scent.... an' sense, not that I'm against your rise political by any clever but non-transgressing means. For as close associates an' friends we share much opinion an' even ideologies leaning into the abstract. But to the concrete!— it just seems curious that Mytdra remains of a fitness to send you here, or throw you out of nation's boundaries. For are you trapped?!... to stay away, as others have been? so much so of the past that we have laughed of it!

ANCHER: I am caught to cautions of preserving my abilities as demonstrated. This is the passage, that I handle with some reality to investigate and approve.

CANTAL: An' what if it's simply not suitable for our sods? I said I'd return with samplings. You need not have come— unless you were pushed! or felt so.

ANCHER: I was compelled to with argument only. I've not been summoned away. I can return at any time, with a legality of presence. Yet should I choose to bring more than evidence, but confirmation with myself. I demand that this is true!... or should I stay here imprisoned of a wondrous retreat, or elsewhere roam until truth catches up with me. For my time is now received— an' my sentence is here. I have worded to others, spoken without obliquity, that there's goodness to this find. And more than a reputation is at stake; for I'm determined to become the patron saint of an entire region devastated by Mytdra's folly of dismissal an' abuse, an' through this show some worth of able leadership to peers an' commoners alike, what can be accomplished with wisdom under daring pressures.

CANTAL: And then you wander here abruptly. But this is some design that's been practiced and executed before. Oh! Can you not be deluded by the talents of a shrewd king?! He slaps at you! an' snaps!— to make the impossible materialize, for his manner of thinking, or digesting of your thoughts. I am suspicious of this coercion!—

ANCHER: Yet is this herbage real?! You have had it an' said so. What to the coercion did you feel that can slight this conviction to my face?—

CANTAL: The revelations I have laid, that it takes haughty self-control. I was.... pressed.... to stay.... awhile longer.... to convince myself.... of what I'd learned of the passions an' the passionate, or what I was beginning to realize of internal surges to find my situation remarkable an' warning of instruction— that I have aged past mending of the wild flowers picked. They should decay in my hands an' grasp, an' drain themselves upon my fingers an' palm. I've tossed some away, an' to a sacrifice— my heart as found! more gray an' derelict than my ruse of exaltation. Yes! It grows as much a weed. Lancet's weed. It flourishes here. Yet are we now not bound to physical strengths— in producing it, this love I've tasted. O for elsewhere be, man! It can only be lost.... as I am here. It is delicious.... but subtle, an' striking— striking back at you!... to see yourself, with mind, an' lapses of envy.... for what you are.

ANCHER: ....Perhaps as a snack—

CANTAL (*looking at **Ancher's guide***): They avoid it!... as should you. They don't need such demeaning, to bend down towards this grass, or make of it a monarch for your manner, mannerisms, mask, mass (*subconsciously outlining himself with hands*).... might. Oh!... Old sod!... Foolish rover!... There's a patch inland a ways. They'll take you to it with derision of your curiosity, but will do so with your honest, earnest asking.

ANCHER: ....Can't you?! Don't you have some stashed—?

CANTAL: I'm not ravenous of a possibility to lose yourself with it. My store's on the ship already, carefully conserved. I've forbidden Lancet to meet with any more. An' we'll be on our way as soon as the captain says we may. For I've decided our huts an' tents may stay, along with a few other possessions, as long as you'll be around with your train or company to use.

ANCHER: You mean I'll be alone with my own? I want to speak with Lancet—

CANTAL: He's a sick man, recovering with hardy effort. But I can tell you that he means with distaste to describe it—

ANCHER: And with yourself I'd like to discuss things. You must fill me in, with conversation an' the conformity we're accustomed (to) for warm conferencing.

CANTAL: That you will have. We can't depart at least before tomorrow morn. There'll be this day an' night for our socials. But you may learn as much through the attitudes of these more natural inhabitants. They have a very serene conceit about them, living as it were on a higher, perhaps more focused view of compatibility to the simplicities of their natures found, the fecund surroundings that demand of life only its service— an' not so much description and analysis, these rich territories of abundance an' novel but proud rejuvenations, where ownership— And I can't rightly determine to which state of social constitution they belong— is a concept of ordinary use an' typical practices of surviving, such as any may do as differently as desired from another. They seek not the extra or outrageous to lionize, make themselves of tribes without kings— as far as I can tell— yet have of leaders age an' experience. An' they're not particularly hostile, to strangers or the stressing caused by neighbors. Though I'm sure there must have been such detractions as conflicts, battles, an' wars. Perhaps in an (h)'istorical past these welts of populaces, for the gathering an' grouping of peoples, became abstemious an' eventually absolving of ancient cruelties. They haven't felt the need to keep a documented record of those matters, an' prefer to concentrate on the practical(ities) of beauty rather then on unprincipled behaviors to register in texts or tomes of writing— which they have. Though being illegible an' indecipherable to me they seem more like notes with pictographic lists than stories. But this particular assemblage of the piscatory have adapted themselves to our speech very facilely an' fondly, over recent generations of our exploratory muster. An' so we forage *here* of preference, over similar and related locales, to make of our exotic acquisitions of materials an' worldliness, at this corner of our extensions for knowledge.

ANCHER: They're friendly enough, or polite within their purities of crudeness. We may even choose to colonize them some day.

CANTAL: We? But of our profit are they made.... remote. We shouldn't contribute more than this to a contamination of manners.

ANCHER: Is Lancet bedridden? Is there a real disease concerning him that may be contagious, spread through insects he has bit or swallowed during his over-enthusiasm to find new nourishments?

CANTAL: He hides within his hutch, no longer sure of the gaiety of day or the frightfulness of night. But he certainly exercised himself to such an extent that the body needs some rest from these exertions. I don't think his particular illness is so communicable, as by airborne agents or touch, but is due rather to a weakened constitution, a lack of physical strengths, an' a gain in uncertainties. He blasted of his mind to seemingly find answers, in a paradoxical state of extreme egotistic introspection— for several days. It's with some tender difficulty that I gradually brought him out of his aimless insistences an' led him more to refreshen his lungs and breathing in clean fields of reason and the supremacy of simple facts, whereof there are many here unhindered by any faults or falsehoods for their growths to thrive. So then did he come to find his hands again with questioning— instead of being hoofs. An' that does bring thoughts down to their more enfeebled qualities an' capacities of being unsure. For are these hands made element of destructive delusions so easily? Then guard more carefully their tendencies of regression on primitive lands an' scopes of concerning thereof. I doubt if his blood bleeds poisoned or misused, or that his sneezes are more purulent than most others. But answers found are miseries of conclusion. An' he slowly immerses himself back into the inquietude of drench to pool withdrawn from all certainties an' their deceptions. So is he not hurt with hemorrhaging of fears, but more of disgusts for his behavior brought down to an' loathing to resume. For loath returns avoiding is a method agreeable by both of us to try. Yet may your guide take you to a patch of anxiousness. I've come to greet you, but also to get on my ship, consult a bit with the captain, an' adjust some of our affairs on board, in preparation of returning home as soundly as possible, an' as swiftly as can be a cargo's freight. We carry mostly perishables, you know, an' a few interesting tools an' crafts' articles.... I'm trying to arrange a visitation to our country, that ties to me much interest in a couple, their education with my sponsorship. But this is a farfetched hope to spring at so incidentally of our hurries, and does requite my bothering to be provoked an' stirred of handsomeness. I'm not sure.... if the ship's master can actually allow such automatic visas to our vistas, even on my responsibility an' backing. To this we must together continue a counsel. My heart does beat for wear of losses, yet not even now can I drop the unfathomable without a desperation to attempt, most knowingly— as should be— (the) impossible.... Yet that's (a) man's carrion left, to rot an' bake of a sufferance, an' find some use for this— odoriferous announcement of his presence.... Odious?... Obscene?... Away, away— yes! earl. I can not purge the sadness.... without more attention— to details.

ANCHER: Sadness?

CANTAL: Insufficiencies, Ancher.

ANCHER: I'm afraid their states of health— here— can not be guaranteed for such a voyage and emigration, without the prerequisite medical examinations for sea travel we have undergone, paid for an' applied to for our licenses of motion.

CANTAL: That guide knows our camp well, to bring you there. You may stay at my hut. You may have one for yourself built, but I don't really see much need aside from its fragility to be temporary. An' that we are most definitely, if we remain the substance of our crown. Here is a.... grander clime to acclimate to, upon remaining representatives of the renown, of country, civilization, culture, character— an' content of ourselves to commit to, that we have left the grasses an' gone fishin'... in a sophisticated way. Superior perhaps, but not better than these nature rationalists are our encroachments onto the plausibilities of man.

ANCHER: I simply want some practical achievement, Lord Cantal, to return with. We don't meed the fundamentally thunderous or epoch breaking, but a true outstanding advance.... for our possibilities of commerce back home, something discovered of a pliability to our natures and of applicability to our tastes, and not to overturn, disrupt or uproot established traditions of a mature civil discourse or thoroughfare of public interests an' the high typicalities we're used to an' have adjusted for our daily aims an' pursuits. And so.... may I take the cautionary road of finding treasure. (*with a leading gesture as they walk towards the* **guide**) Every goal has its obstacles an' dangers to overcome. Even breathing requires expulsions. An' while these are wondrous winds of perfumed gale, with tropical tinting and exotic accents of perception unusual for my sensitivities, I don't feel outrageously disheartened by your difficulties here. For they are not so descriptive of failure, but rather a loss of flaunting for the explorer, which I suppose must occur with any extended time or prolonged stay at ticklish environs an' absence from one's suitabilities of more common endeavoring.

CANTAL: The clouds above, Ancher, are of a type that same the world over. They're simply more intensely seen an' felt, at this latitude. They are made more colorful an' irradiating to the eye for their species, but should be composed of the same stuff as found elsewhere for shame an' shine an' sensation. So should all foods be somewhat similar under shared skies, of the vegetative capital replete of our varied diets. The such will not make us fools for knowing— nor using.... But obtaining? that may be the current's storm; since we're not masters of the glistening enrichments of foreign nations, that which make them shine openly an' without embarrassment nor humility for their captivators. An' we call their rice wild — but we are wild! to cultivate. So then, tease yourself, the same as to be wild. (*having reached the* **guide**) Take Earl Ancher to that patch of grass called "craw" near here.

GUIDE: Craw craved? Craw for crows. They dance upon it, honking gravelly, as if announced to having found a nest of fays or hidden tinsel. Stealings (are) stuffed in it, of matters not concerned with man's prudential providence and nobility. They must hide their bodily charms there. Yet more to find and distribute of the pang, Lo'Cant', after what you and your friend have gone through? I'll take— but I won't have.

ANCHER: Why not for curiosity its taste?

GUIDE: The fish don't roam there.

ANCHER: But it's on dry land. Certainly you make other vegetables, fruits, even flowers for your meals.

GUIDE: That's grass, Er. Not even dignified for rodents. But you find value with its sheathy slime.

ANCHER: It's slimy moist?

CANTAL: A slight fur of hairs to the shoot. Barely noticeable except affixed against the sun's glare.

GUIDE: It's of some vanity to be an animal, but resists the exploitation of personalities and feelings.

CANTAL: They *have* investigated it, from long ago, Ancher.

GUIDE: You can have all you want. But it's a silly hoarfrost, a bounty of dimness.

ANCHER: You can experience frosts here?

CANTAL: From what he's heard described: frigid restrictions of activity, the state one's put into— stillness within a fire, the incapacitation of a flame.

GUIDE: No frore but to a mountain's lethargy. You top this on your fish? The cut whiskers give it some color while baking, Er?! We cook ours in the ground, sometimes, to let it simmer within leaves. The juices stay pungent, strong— But we don't eat the husks made, or the broad leaf and its straw. Why do? An ancient form of marinading is all. Yet grass is for the ground, Er, to keep it supple and conditioned. Worms live in it.

ANCHER: If.... the greenery is eatable, that should be exploited — is my only mission. We're not trying to defy the gods their man dates of custom for the human animal. Yet are our curiosities profound for any nutritious article of good flavor.

GUIDE: You would stoop down to grab a handful, and put it in your mouth?

ANCHER: ....If it seemed clean enough for the raw to do, then I would stuff the oracle for taste, to test our furtherance of this gift. For what is offered is only ever what is found.

GUIDE: Yes, we've seen this, and with some astonishment. Well then, it's not far away, Er. Only a few concerted paces— in this direction, going away from the shoreline and into the interior of our fruitions. But there are many other locations all around. Indeed, when one sees grass, the craw must often border, as a chin's dripping. But the grasslands are expansive, in places. Jungles thick in areas, and darkened; forests frothy of extemporaneous and extraneous foams— I take you to a safe spot, fairly open, where you may ply your tasking. All will be temperate, and without.... misapprehension.

ANCHER: Then I'll meet up with you again at camp, Lord Cantal, as I go through dreads, to prove of myself all legitimacy for the mission, that the effort has been forced to be worthwhile in one way or another. That's the manner of a self-determination, a hope, a dream— an insistence.... a dependence on what should be fact. If so unusual, then not.... is the justification for all revelations. (*as he and the* **guide** *go off*) Yet am I scared only to be wrong, an' not of error's might.

CANTAL: Oh, well it be.... That is definitely a cuisine. Yet for our burdens circumspect, we test ourselves of many delights, an' come to favor even the worst of them for titillation.... if we may so govern their effects with choosing of their aftereffects at play. Then naught to be deceived of them, the foul do serve a purpose in our mouths, or in our stomachs or on our faces. An' that's to be presented of our thoughts both rightly internalized an' transparent.... for all to take notice, of this behavior wished. An' bring not evil to an honesty, if it's not deceptive. I have captured my dreams— within a monster's bowels, an' have shown myself to thee as being not the shame possessing.... nor the sham confessing.... but the shore approaching, of its gradations to be off— an' back, original! to my vows.... of decent degradation leaving heaven.... an' for heaven an' for home. Disappointments are like wavelets in a water's constrained dissipation of the nerve. (*walking off*) An' I have shown mine through an' through as true.

Act V

Scene I — *A boarding room where* **Doria** *resides while at the palace. It is a somewhat decrepit living quarters shared with another, but functional enough for a tolerance, having fireplace, an area or corner for regular ablutions, thick walls, beds, and a dining table with chairs, all features and fixtures of surprisingly low or cheap character and considerable dinginess. One mean window affords the space, which offers hardly enough light during the day. But then it's not much occupied under working hours, and is mostly illuminated at evening through the hearth and candles. Therefore, the warmth of the place is its depression. It is the afternoon, and* **Suzie** *is distributing firewood under the mantel, to be used later, as* **Fenic** *sits on a bed, awaiting* **Doria**.

SUZIE (*at the fireplace, occupied*): I say she brings in the laundry, as I carry the wood. An' we take turns at this. But I'm stronger to the shed where it be piled. An' she's dirtier for the wash, always concerned with the viands and the slaughtering. Besides, I have friends of the suppliers— an' they'll chop some extra for me— while she has not. An' it's remarkable that she's here at all, Fenic, 'cause she's not easily made accustomed.... to our practices. I don't know why she was appointed—

FENIC: Leave that to me for all possible ponder-ation. She's an acquaintance of mine, but not a trained servant to the royal establishments, such as yourself. I've told you this before, and not to give her any grief about it—

SUZIE: I don't, master.... But she defines a suffering, being away from her brood. (*getting up and coming over to the table to wipe her hands on a cloth*) She told me about them when we first met. But I can't be much sympathetic to the plight, an' as much have shown her so by not wanting to hear her wailing, no matter how courteously pleaded with the assertions. A devoted mother would not (*wiping*) do that to her young, as to leave them for work she's hardly suitable to do. I simply don't want to hear of these qualms.

FENIC: This is a new era, Suzie. Living arrangements have to change in many unforeseen ways for folk of all walks of life.

SUZIE: I've never been a beggar—!

FENIC: She told you she was?! That's.... being too honest. She wants some understanding for her situation, by being forthright for any possible friendships. Don't be so reluctant to offer any meaningful affection.

SUZIE: You can't tell me what to think, an' can't show me what to feel. A person should keep those disgraces to themselves. An' I don't say that she's an awful person; for the personality is passable an' reasonably restrained to a subservient meekness, for whatever positions she can adopt. But it's not proper to propound on one's problems—

FENIC: She's not a professional—

SUZIE: —I certainly don't want to hear them.

FENIC: —doesn't have the reserve yet of attitude. Be patient. This may develop.

SUZIE: She represents what I'd never become.

FENIC: You haven't even dependents yet. You're not married.

SUZIE: Her presence frightens me, Fenic. It does not fit even the class. It is an abnormality become, that she should force her way within ours.

FENIC: She did not. I introduced her to this servitude, out of necessity to alleviate her state.

SUZIE: She's your friend fallen down, an' worming her way back up through us—

FENIC: Maid! Don't you have chores to perform?!

SUZIE: I just have! an' have won a respite. (*sitting at the table*) It's heavy carting, you know, before attending to the draperies an' furniture. But this won't work, Fenic. The load has been too spectacular on her.

FENIC: What complaints have there been?

SUZIE: ....None yet, as food is being made, an' meats prepared. But she'll hurt herself surely. Any day now I expect she'll be gouged by some ox or boar. Any displeasure sensed by the penkeepers could allow the accident, if only as a cruel prank of amusement. This absurdity can not last. An' should I cry at those injuries? or scowl an' scold at the foolishness that lends to such occurrence!

FENIC: She's kept mostly to the fowl for handling, I'm told. The carcasses of meat are brought to her dead, and she carves of beef an' pork an' venison, for appropriate cuts and servings.

SUZIE: The need is not there. Others are to it to do, an' far more capably an' efficiently.

FENIC: So then she learns the refinements towards a trade. Where's the harm to a schooling?

SUZIE: Heavy knives, blades, an' saws sever human flesh as well, with one slip of the cutting from fatigue or momentary inattention or inexperience of art to try of incorrect manner or method. An' the more this is attempted, the sooner a mishap arrives. One can place bets on this. I'll not allow myself.... to feel for shedding tears on a beggar. For once you reach that depth, misfortune is your only guide, an' privations your cover of clothes. We work to avoid this at all costs, Fenic! An' those that fail are lost everlastingly—

FENIC: Oh not—!

SUZIE: —As such a life is always life, it's everlastingly life— until it dies. An' that of hers is tainted everlastingly now, to some deprivation or punishment awaiting of its logic to produce. She is cataloged to this approach, which too often secures itself with passing or nearby innocence, the vicissitude-d neighbor or relation that slips of a shared plight, so mighty are the devils of casualty.

FENIC: An' today is but one day closer to your death! But that is such a depressing view of one's enduring hardships. Yet are you waste to be, for seemingly to be tired? as you awaken each morning to your labors and stretches of difficulty? Then not so also be an awkwardness with fitting? You are still proper for the tasks, as one should be who desires the achievable of an existence. So can she work here, an' *any* privilege have for working as were another. An' if an accident occurs, that is unfortunate. An' if some malice is committed, that is of lowly gall which should yield to a vengeance, if we are fair beings by an' by. But this suits not to condemn Doria from her needs an' suffering. She's been away from (her) home an' children for almost three whole weeks of sorrow, and must entail a few days more before returning for a couple, to reassure these absences are worthwhile, to spare more hurting. An' not that she would climb as wanted, among our company, but then I have made this avenue, to placate my concerns an' show a means of assistance.... to subjects of our well meaning king. So do not so resent this effort that is legal to a monarch's eyeing.

SUZIE: There are threats here, with your disturbances of people.

FENIC: Might I have made, it's not for you to chastise that you an' others are being wronged with my partial attempts. Complain of theft an' bribes an' hatreds an' disagreeableness an' slovenliness— but not this. Complain of greed an' obscuring of knowledge an' obstruction to advancement— but not this ploy to save a friend an' people. Complain of servile attitudes an' actuation of our demeanors to be callous mendicants to rulers an' our fates— but not this drive to resolve errors an' mitigate against the impossible an' bring forth a saving grace against a cruel betokening of a family an' town. Cry not of the unsuitable to me. I am unsuitable! I've fashioned it a law. I revolve around injustices, an' must yell at times, with such disturbance an' upsetting as to cause more— An' then!... *we* must correct the whole of our bothering. Yet may she move here with her young, or not stay for long— I don't know how reducible the problem is to a solution. But my attention to it is constrained, embossed of nerve, an' confides of your professionalism to be reasonable with her, as may our Mytdra solemnly demand!

SUZIE: She's no favorite of his!

FENIC: What do you imply about— our king?!

SUZIE: More than one, an' many as more, hath seen him chide her.... or have heard of an' remarked of this. He is perturbed of her presence.... as being inadequate. What of to this are we to think?!... as likewise find or allow?

FENIC: ....So, as gentle.

SUZIE: I don't deny her need, but not her necessity.

FENIC: As gentle be to me, put up with, retributions harsh exchanged for favors?... or my pleas demolishable. (*as **Doria** enters with a laundry basket of dried clothes*) You are kind with your sincerities.

SUZIE: I am kind at all, an' caring too!

DORIA (*bringing the basket to the table*): It may be distributed. Some garments are still drying out. (*placing the basket firmly on the table*) But we'll not have the misty mustiness in here as before.

SUZIE: That was an experiment that you learned better by. There's too much for our fireplace to handle.

DORIA: Aye, Suzi, with tie of line to clothes. I didn't know there was a corner of courtyard appointed for this airing. Yet have I wonder of the facilities in this small room.... An' does Fenic speak to this? I am obedient, an' have caused no disasters yet.

SUZIE: We'll discover those together.

DORIA: I've often dried my clothes before a hearth.

SUZIE: I thought you were teaching.

FENIC: There is more dispatch to my visiting than this.

DORIA: It drains me more of days (*coming towards **Fenic** and her bed*), but yet to know that you have visited *them*.... And this is comforting, with memory of keeping still our avowals, that they remain safe an' fed, an' anxious for me.

FENIC: Never forgotten, and distressed with puzzlement.... (*as she sits on her bed, not that close to him*) but adjusting to the circumstance as somehow unavoidable, in their infancies.

DORIA (*facing somewhat away from him*): That is to be bred at all. (**Suzie** *starts extracting her clothes from the basket, at first casually, and then standing.*)

FENIC: Despondencies anoint for courage with superficial oils, more a volume to adopt an' spread. Earl Ancher's secretary informs me that soon some planting of a wondrous specie of land cover, hempen an' suitable for pastoral grazing, will be initiated in your valley, highly localized at first, but doubtlessly spreading greatly with success. Associates an' discoverers arrive from abroad, their excursioning satisfied, to start this promotion for the economy of your town, as I had desperately sought an' promised.

DORIA: Is this possible with talk?

FENIC: His agents are due much by the time you return to your.... family an' friends, according to messages the secretary has received, as could a normal sailing weather permit. I want you to take full participation of this activity from the beginning, to help establish its growing.

DORIA: This is the earl's enterprise—

FENIC: And subsidizing. Through me, as royal envoy, you will be given an imprimatur to publicize the effort as a resident representative. You'll receive a dispensation as long as the cultivations are tried. An' if they last for a profit that may be strictly establishable to your village environs, as restricted to by decree of our Mytdra, your salary may become permanent to the position— or everlasting, so to say. This is the key hope that some ambition may bring. An' not to insist that our fortunes devolve of this way without more wishing, the Mytdra does want to improve your region under the scope of his an' the entire nation's enhancement. So this is something much to look forward to. (**Suzie** *starts taking her clothes to her bed.*)

DORIA: Then the earl will supervise everything locally?

FENIC: You may have a hand in it, Doria, as an expert of the surroundings an' under my privilege to expound.... regularly, for any explanations needed, as to what is where for richness of the soil an' sunniness of plain an' its weathering throughout the seasons, an' so forth. The earl himself seems subjected to a self-imposed detention whereat the miracle was found, since he has voyaged there and delays returning. Thus, his secretary handles his affairs in the major, for now. And that person is close to me in partnering for this development, clearly of my influence.

SUZIE (*to herself, but audible of distress, as she holds up a garment at her bed*): This is shabby!

DORIA: Then I might return to a life responsible an' respectable.... an' restored of a graciousness to be happy—?

FENIC: This is possible.

DORIA: —with my children an' familiars?

FENIC: It's to be tried for an' worked at. It's what I've struggled to fashion, an' what I'd claim of a dependency to our society. This is more than feasible, an' less than false.... But we'll have to see what happens, what can be made to be commercial, whether the grass will grow there— and the sun shine enough.... for fortunatenesses, an' the king's pleasure.... Has he chided you?

DORIA: Not so directly as to be, but for of blood to see— that I am here.

FENIC: That chid'(ing) (i)'s for me. He finds his nation through my eyes, with not to notice more than what is for you placed his presence. Then I am his ambassador for rights an' righting of.... our various crusades of common life.

SUZIE (*with self-exacerbation*): This is so cruel a space!... an' mean, diminutive of faith to take for our encompassed toils, that of this rank we are leveled: maids an' servants— never to escape till aged of scraps, scrapes, (*sitting*) an' weight of labors. So are destined straightforward styles, for human customs to be pleased without the complexities of enchantment.

FENIC: What to for our profession, Suzie, bring our wares of person?! I am an enchanted complex serving— gods! For what be man else, with this delivery of instances? that we must improve ourselves ourselves, an' speak of nature as a net of needling. It is sufficient that I serve the Mytdra, as a conic section of a dunce cap — worn! But shame we not ourselves. (*stands*) No curse hath placed us here, but providence an' reverence of monarch. I've abused friends— unwittingly.... an' make repairs internalized. But what occurs through this eternity of challenge, makes but some sundry news under the domes of our character an' caring. An' this room is no worse than my own, but for its better size. I would have chapels slept to, preaching in my sleep, a better way to think. Yet find my chin is better slapped— or greater made— through Mytdra's jesting, than by my own bitterness. So come we as we are, with fortunes in the cold, naked an' helpless— but for kings an' leaders an' rearers. Without much spite can this be said, that some ascend another's dreading. (*starting to leave*) An' sorrows sanctify our cries. Yet worship we what can be created as called for.... some kind of appropriateness to settle our mistakes an' debts to others, some crafting of a way.... to solve our problems made. So fear we — Doria!... make haste of valor, for these virtues sprouting. Oh!... but life is life, until it's not. That is the truth.... That is the determinant, ladies, for all subjugating principle. I'll not demand more of ourselves to know. (*Exits.*)

SUZIE: You've won something splendorous, it sounds.... while I am menial to stay.

DORIA: That's for a future to decide that caretakes not but simply presents its evolving obstacles an' generosities of events. My husband's dead!... an' that's not a grand enhancement with any fate to come, for what I had affectionately vowed.

SUZIE: Bewail perhaps should I, to improve my chances fulfilling prosperous dreams. That is an attitude an' motive? Yet for our majesty, with crusts of parchment for the mouth, announcements an' decrees provoked an' irrevocable but by his further daring, is more of strategy to be led by, with our more common framing— paused for instruction. Pandered to are our simplicities of breathing. Yet for the epicyclic whirls that exist for explanation of the not fairly describable am I lessened in my own hauteur of hide an' servility. What is decent to learn from this, that you are raised while I stay level an' low? How be the twists an' torture to accept? with green romance, a turn towards riches an' royal approval.

DORIA: No vault as yet is filled with this provisioning. But I am dependent on my poverties to exist, while you— your stature. Trained servants to the king is a gratitude of country an' quite restrictive of talent an' capabilities of achievement. I know this an' of the educations required, as I may interlope only with assertive aid, an' can accept reasons for resentment. But not does privilege make with struggling out of the mire— the impossible!... only an escape to face more turbulence of pride an' principle.... of earthly consanguinities to protect. May angels hover over our missteps, to remark of an' make comment about, with winds an' blows an' gusts of opportunities or opprobriums. All might be possible to chase or be chased by. But I am fixed.... by these waves of sea that bring— these chances, these eclogues of hope an' determination by a noble earl.... an' a nobler friend. (*stands and goes over to the table*) So fit am I to weave detachment.... from emotions due, until what's due is done. The best pleasure will be seeing my children healthy an' unabused, an' absorbed with my return— fanatic for. How must I linger till, with butchery! an' dread crossness. Yet if to stay for home.... is some prize won in addition. (*at the table*) It is.... a Fenic's miracle, not to beg.... My wear is damp a bit, still. Yet more for coming of the blood an' cuts, across the lawn to trace, I find less time with haste—

SUZIE: Distribute!

DORIA: This is a dismal light in here, Suzi.

SUZIE (*standing*): over the bed, to soak the last remnants of a rinse. These clothes are aired an' washed. An' the fireplace will be started up in a couple of hours, for its eery glow consorting of. Thence then commiserate our muscles' pains in supping— (*starting to leave*) But till then more to do, an' ever of the backbreaking chores designed for us.... May you escape your accident. I have accidentally found mine to be no joy left.... But that may be passing of our views traversing fortune, if what is deserved leads us by our weights.... of conspicuous behaviors. (*Exits.*)

DORIA: Accident?!... (*lifting the basket, to take over to her bed*) This is no accident.... An' not a jealousy of premise with this labor. It's all done.... with notable declension of the thought I am away. (*at the bed, taking out some clothes*) An' spread these out.... without a flame to ponder. (*starts assorting clothes on her bed*)

Scene II — *A parlor in* **Mytdra's** *mansion. He is sitting comfortably on a couch long enough for several.* **Ancher's secretary** *is standing to address him, with no such relaxation.* **Holston** *is also present standing, but of a demeanor that says he may sit if he wants to.*

MYTDRA: It's why I sent Fenic out. I want to know from you explicitly the nature of Ancher's exile.

SECRETARY: Exile? your majesty?

MYTDRA: His correspondences to you must explain this. In some detail, hopefully? What is the manner of the messaging? What does he confess of feeling? You've brought me only copies of his letters— in your hand. And I bring you to me for your word,

with respect for his privacy, if this is the way he likes to do things.

SECRETARY: And you would like the originals?—

MYTDRA: That's not necessary, as you are here.... unless what you exclaim sounds extraordinary or ridiculous.

SECRETARY: He prefers, your highness, to stock his personal archive with originals, and to have my copies around— which are scrupulously accurate; for I am no creator but a scribe— for his records of use and review or sharing.

MYTDRA: And that's justly proper for a gentleman of fine business practices.

SECRETARY: The projecting goes well, my grace. He chose not to return with Lord Cantal simply because Lord Cantal was about to leave when he arrived. And the earl is a thorough investigator of all of his dealings. So he stays to observe and take a vacation, leaving me to handle some of his affairs which, while important, do not require his direct attention, at least temporarily. But nothing unusual occurs. We've done this arrangement before. You have the sum of his letters sent to Lord Cantal, and of the earl's sent to me. And this is all I thought to bring, concerning the development of our agricultural experimentation the earl has engaged himself with. Fenic did not suggest to *me* that you want to concern yourself with other of Earl Ancher's enterprises, at this time.

MYTDRA: He does not state.... to you, how long this excursion lasts. It seems open-ended. And the reason for it, as expressed by him.... to you, and to Cantal, seems rather obscure, since this— scientist reads as being fairly competent. And Cantal certainly is.... And Cantal brings the spice. So where is this mysterious urgency to rediscover it?!

SECRETARY: It is.... a new typicality in him, my great monarch.

MYTDRA: I know.... the exchequer and our constabulary induced him to take a more personal involvement with this unusual situation— *he* has devised. But what does Earl Ancher suggest to you, being expert of his personality, that can explain his reluctance to return— here? when there is little reason to stay distant from his customary management. What has he said that can not be written? What may be implied by all of this?

SECRETARY: I can only fathom to think, your royal highness, that he— as did Lord Cantal— finds the exoticness of that location terribly enticing and consuming of his curiosities to make pleasure with. One may abandon oneself to the novel, for a time. But the climate there is said to be spectacular.... And he did mention to me, just before going, that he was starting to suffer a headache— which I thought could be relieved by the salty breezes and gales of a good ocean voyage alone. Perhaps this tropic is even greater for one's health, or inclinations to relax from stresses and heal of constitution and disposition towards weighty, worldly affairs.

MYTDRA: So mention he.... to you— ever! nothing of conspiracies— against my ruling?!

SECRETARY: ....Only, highest lord, liege and lion for us all, that they occur, may spread of occasion with disgruntled dissolution of loyalties on occasion— for particular grievances of state and economy.

HOLSTON: Of course he discusses this, as a topical subject, and as openly as many in high company. His yearning is to rise—

MYTDRA: And has he never told you of his desire to gain more influence with me and in my court?

SECRETARY: Very often, our excellent king, has he suggested to me, or intoned the thought, that he could make himself to be.... a great officer of government— under your leadership.... and under your scrutiny at all times. For without your permittance to succeed he feels of little use to the world and our country in particular, as can any loyal subject claim. Indeed, there are special qualities possessed of special people which can only enhance your fostering of nation, but to which a king himself can not pertain to much reasoning. Gifts in people are as varied as their stewarding as individuals, as for say the mathematician may work under you.... or the economist. Then that you may not count tremendously, or arrange special sections with mental craftiness is no slight at all to your authoritarian prowess and beneficence to every population of your warding. And my earl is particular of his businesses to only enhance the social welfare of this state strictly under your guidance and approval. Thus he does nothing to disprove you by staying abroad a spell, without for fear to manacle the stay. This demonstrates nothing but that he is currently preoccupied with the splendors and graciousness to please of a tropical retreat, a resort of nature's own.... mysteries.

MYTDRA: I need not great geometers to construct for me the proposition that ambition for advances deserve sharp observation and caution for their liberalities to subdue. Has he not freely complained— to guests— of aspects to my rule, that he might do better with instances he is perceptive to manipulate!

SECRETARY: His company is august, great swain of governance and provider of all our powers. Do we not bare and burn our skins under the sun, sometimes, due to carelessness?... of speech or spine to be out baring ourselves among companions as likely bold? What mind can not conceive of schemes more daring and outrageous than the mundane occurrences he finds about his daily existence— what to improve of chances for him?... and for others friendly to him. But this is not to topple our conserved strictures, rather wishfully to think of how matters might be done more efficaciously in some instances to which one is keen of substance knowing and understanding. This is not an absolute rebuke of practices, but a discussion, singly or within a group, to review *themselves* of resolution of disagreements or errors accepted to have been made. Earl Ancher is not a party to anything like consortiums of conspiracy against the affairs of man as we have adopted and revel with. He simply has his concepts made, as an independent thinker— as any one of us— based on his experiences and achievements. We do not blast foully to cry of torments to our wishes, that discoveries may be made and better ways of doing things may be found. That's an elemental pursuit within all rational creatures placed.... by what deems us to be thorough of our thoughts. Yet not for revolt is my earl predicated. And I defend him faithfully, with the entirety of my means at your gendarmeing, if this is a test of his nobility of intentions and responsibility to the profound, visionary and wise ruling of our country. Yet may I say, with an honesty of candor that allows me not only to serve him but more remarkably to be accepted by him as a friend, that he pleads for the health of our sovereignties, and only does as much to threat with his worldly dealings and what he brings to our shores, sands, surges of serenity

— the very earth itself to sway of prosperous tidings with this current venture. And he has ever only praised your majesty to others, of what I've heard, as a worthwhile titan for our dominance.

MYTDRA: Of dominance.... of divinity and dominion. This is parsed of some sincerity to wheedle, lend and yield to. For I already know he's been a confident person to confront— or present.... a hearing of that was sought, a courting of the court, and idle boasting of abilities to astound— or try— my ears and others. Yet this scampering scars, secretary, since it is typical of scamps and reckoning from deviousness, basic skulduggeries thwarted but again.... This character does try to make amends with absence? Yet the plant arrives— without his cheering heard, without his drawing of attention to himself for one more achievement to his gain, to thump at his chest! within our crowds, as an expedient sign or symbol of self-conscious delight. And what of the marrow of the earth? for the grind of guaranteeing some resort— of resolute blood! out of this region found, visited, and alighted by Fenic.

HOLSTON: All that I've had questioned there say that Fenic precedes all knowledge of Ancher's adventuring, majesty. This speaks to say, perhaps, that Ancher comes to Fenic— to instill the idea and promote— were not that Fenic, as (being) unusual, be himself promoting the cause. I can not find definite claims of my friend persuading inhabitants there— or working to with agents— for this novel agriculture before Fenic's.... discovery of the region. Since it is typical of the earl to lay the groundwork of a venture thoroughly before proceeding.... I will assume in this case that he had not, and Fenic must have approached him to try the town, that which has already been professed. No solid evidence contrary to this can be found there, as yet. It seems clear that the order of events serves Fenic's problems to assuage more than Ancher's gloating to the court of this— vegetative wonder. Yet to the town itself.... is no premeditation found.... for my friend.

MYTDRA: That is the puzzle brought: You are his friend, and safety rouser. To be convinced, must bring me proof of purpose— with this plant. For what is a success otherwise, in truth, that it is real for me? this which I have dismissed of much importance— except.... to soften a sorrow's temper in some kindliness to find or be surprised by. Taught are we by indeterminate tests of conscience. I can accuse so many— not one above me to be tried, for any cruelty condensed of person.

SECRETARY: The thing is actual, my highest lord and might of grateful mentioning. It is.... this green thing to be grown, a simple cure to barren prospects, an additional moss of capital, an accompaniment to current grasses that still dominate, a particular usurper of some space with innocence, in areas that may become renown— for its princely administration under your hands. It is real, and hasn't the complication of one weaved fruit, one edible root nor bunch of shrub of leaves nor tree, but merely spline of growth as slender as some twine, for food.

MYTDRA: That currently observed and tested?

SECRETARY: (It) Is a definition of grass, high sovereign liege of enlightenment, as described, or as could be thus.... for some of doubt and ignorance, and lowly animals abandoned to their searches, a grazing of knowledge, as it were, by what is come by to be tested, nibbled at, and delivered for an adjudication of digestion. Then this is pictured for us as has been examined, and of precious worth to own and manage. Yet for the animals to own nor manage, they eat of what is presented as discovered, a treat perhaps with sweetness. Who knows what cattle think? But this is won from its definition, a triumph definitive of human organization to procure from strange lands and establish here as fruitful for our economies and aggressiveness to retain prosperities for many, man to be led to a pasturing of his curiosities, from what he has sampled to what he is fed. Then find he an induction to a remarkable condition, since this is a justification to establish useful plains for — anyone, virtually any kind of person, any master of self, diffident but for their own concerns. This is a symbolic victory that my earl has obtained, for the individual to have space within a vastness of truly useful verdure. And to be imagined this, amongst what rolls and beds as facilely as grass— as endless as some horizons, or directions for viewing, this is a certain kind of freedom with enjoyment, a freedom from concerns of strife or to find one's means.... of nourishment. For what is the bother now, where one is placed, if ever it be great and gratuitous. Then be an achievement made for many attitudes of nature—

HOLSTON: Some kindness with a madness. Some beasts are meant to be fierce! and disagreeable, and always struggling and never satisfied.... Keep we many lean despite ourselves and our moral exertions. There's much to be said with benefit to our discomforts, to keep some sharp and acute and aware and hunting throughout life, not made pacific to any goal, not pacified by any gold, but alert to our prospects and bedevilments.

SECRETARY: Well that is a changing current to our fixations of pursuit, since difficulties enliven a need for their relief and some tranquility and peace to return as salve to tired muscles or struggles of a group franchised to defeat harms and wrongs and injustices and carelessness and inappropriate dismissals of much grief—

HOLSTON: I've never liked the enervation of brute's character, as much as I might fight against its criminality. Finding such oddities to make of themselves permissible squanders verve with the effeteness of that permission— Eating grass! Would rather battle rampaging tubers and stinging vines that strangle for their hunch of supporting frame! Some fruits be brought to this—deliverance of casualty at least, that may be merited with baking and roasting and stewing and hot heavy energies applied instead of a meek act to graze for yielding to your mouth some foraged leaf with masticating murmurs—

MYTDRA: Bring one down to knees! of this depiction for the possibilities of angry men, and disenchanted serfs for serving hatreds to their irritations and prickling temptations of revolt! These arguments are for conclusions, but what is the result?! that fields are gainsaid for a new mulch, and a country is much assimilated to a new demeanor of serving me with proofs of a self-serving nature that is most novel and contradictory of fear and depression. So that sounds utterly outstanding, if worrisome for being unbelievable, with promise to being troublesome from aspects unforeseen. But where is the truth to this?! by what is read of notes and letters and descriptions and plaudits for the rumored remarkable— Rest these people with a hope fanatically brought, and with devious intention to usurp my authority with hope! of a caring cure. This could be quite the maliciousness of unsubstantiated pride, for having found a thought crafted as doable. And the architect is away laughing for much complaint, out of my grasps, leaving us dumbfounded to explain what may be impossible and yet arousable to certain defiances and uncertain defeats. And the scars are on my cheeks to disprove such calumny, such intentional baiting of the beast, with

irony to claim for them passivities won.... with enriched lawns and pleasantness of fawning through their deeds— as fauns indeed till pitchforks and axes turn them into devils, for being deceived and humiliated of the lie, that grass can so be ate by man without some payment for the thought— if this is possible or not! And so as shared of shears of teeth, through me this is committed an atrocity of taste—

SECRETARY: Yet has Lord Cantal said it's doable, my lord!— lords knowing, lords willing—

MYTDRA: That!... is of the winter. That.... is of the vacationed faculty. That is wanting to be entertained. That is in finding a lust for anything exotic— on its own ground. That is my reception to be heard, that Ancher teases with to provoke an idea tossed to meld, and taint and tempt for desperation's, all knowing of its panting resurrection based, that this does solve the climate's haste of action to be torn and thundered with, oh so loudly said.... And for this cheese to fetch come here, come to my plan and apply yourself — foolishly and blinded by your needs. It is a dastard's blow, to a dashingness to cower under me— (*calling out*) Fenic!... Bring in the serpent's tongue!

SECRETARY: But your highness, this is the privy provisioning we've made, that has attracted so many efforts engaged of a success, of an astounding result. Dare we pile ourselves relentlessly on the enterprise. And all over the region there is joy that something positive develops— There is anticipation of the miraculous with only vindication of Earl Ancher's predictions, since his wealth is yours with this employment, profits shared as usual to uplift everyone involved. I have so determined this thoroughly, else I could myself be destroyed and degraded by a saddened and depressed caretaking of estates.

MYTDRA: That is to be tested, secretary. And I'm not without the justice to do so. For I give the opportunity its sway to prove itself. Yet have I more perception of what must be anticipated— there's little doubt of the method here, the extremes are too tractably pinned of fusion for royal contemplation: total discomposure with failure or thorough placation with success. Now which of these could ever prevail for my temper to assist? It is of the nightmare purchased for viewing, to promote my work of strengths and thereby lessen *all* to come to a correctness, with gains of assault— at least philosophically— by some insurrectionists, projecting rocks for dents of armor left. O that this is so unclever makes its weight more penetrate of mind. Yet may it be all too much a nature's course even for Ancher to have fully devised.... details more terrible than tales— (*Fenic enters with Lancet.*) Even expositions have their horrors of exchange with the truth, current topics delivered faithfully through a blustering of projection and emotional whim or granting. (*stands*) Now that an hour comes for lessons to confront, this one returns— with letters sent to me! direct, from out scribe's hands.

LANCET: Or for my right of duty, your majesty, a devastation to inform everyone of, then through your reign define my warning. This is a poison—

FENIC: Half described!

LANCET: —of the mind; a mire swept through and my blaspheme enhanced of circumspection to refine and gaze through fires of torment, is Lancet's Weed! For it engulfed my being of torrential panic till it overtook my justifications of fear, and left me in such a state as divined imprisonment where ghouls are the answers for men, and crazed yearnings rule our drives or demands of survival. Then before a society to educate, overcoming the degradation still — and recovering internally as the heats distill my errant chase— I rushed through madly, but meticulously writing, a letter of notice official and describing all to all facets of the soul's consumption exercised with this green stake. And as soon as I arrived I sent this detailed abstraction to my fellow investigators of country, for circulation and discussion— And they!... have summarized it for the king, some prominence of my execution. Now I am drawn for royal audience, the weird wrangling imposed by this transgressive grass to verify in person. For you find me weaned but wizened of its effects, weakened but not as fully wretched as before when it had taken total hold of my personality and grievous, inhuman actioning— as were a crustacean become!— within the jungle clime. I felt as if this were best to be, and swelled of satisfaction to this modality of existence, until cured of answers of such insights it presents one, which perhaps only a much stronger constitution and mental attainment can reasonably handle, ward and repel of thoroughness. With my meagerness I show you what's not kind in the plant, how I am dissolved to my very foundations of consciousness to retain and exalt— and charge of all intentions to restrict this thing!... to specialists, and cultivators of investigatory investments, herbalists and medicinal-ists. This is not! a panacea for broad land ranges to enact, as if of simple fodder for all folk. And when I learned of this procedure to attempt I was shocked! Leave such matter to more mastery than in large agricultural plots of open or free range, lest the weed becomes a method mystical and wrongly interpreted for a spirituality deadening and overwhelming whole populations— controlling them with false entrapment and constriction of the mind to certitudes narrow and confining of one's nature into senseless, inarticulate obedience to your most fond beliefs— which condense to remove all sensibility and natural proportion to your environmental relationships, as if the mind alone comports of your existence with the sound of a single tone, so sure have you become of a persistent stone.

MYTDRA: See this weathering and thinness! and dejection at his greatest achievement, the discovery of a potent plant predicted of — by the self same subject, a man conferred with caution of experience. Note the gauntness of his fiber. I could not have predicted as well, upon seeing this exact extraction heralded for presentation — of an interesting curse. Yet you say this airy, wind-swept beef, this green beet! of leafy blood, leaves one insubtantial of mind and moribund of manner? Would contuse a region into total.... subjugation. And that it be left wild to grow, dims wildness, with remoteness of thought. Such presents brought here! for my rationing of effectiveness of people's aims, is to be disabused from the casual freedoms of discontented mirth and flustering. A danger is this usage unadministered.

HOLSTON: So sow you often these unnecessary tights for our correction, Lancet. Yet have you gone and found a blossom ending strifes and causing you some incapacitation?! It seems your garnishing of motion is made to reveal any peculiarity of nature. And you can't help this for a garment sown to try of these investigations.

FENIC: Yet for defense.... you descend to recover how? in that you breathe more easily of person vassalled to a normality of man. Or would you crawl for more, likewise a captive of your dreams? in a country more familiar to its habits. You have explained to *me*,

in our private conversations, only of excesses for to rue of indulgences and lack of typical restraint for a regular person, as you run for wiles to find, abandoning a prudent hunger for imprecations to jest with. And now some renown for a power you seek, from a power you've found— that's fair enough. An admonition's due of you, and without a specious salaciousness to your caring about others, as you care to choose yourself as an example prominent to judge. For have you boldness played with, and fun obtained out of a careless wondering with greed for your sensations. An average person might rather make bloating sign for this aggressiveness to subdue. And we are prism-ed here to show more useful attitudes to our foods.

HOLSTON: I've found no criminalities of crab within that landlocked region of horticultural impression.

SECRETARY: The samplings have only just been planted, in experimental plows to foster gemination of some seeds. No one here has tasted of its scant number of precious leaf and root. Nor has this appeared unusual to handle as for any other sod. Upon looking at it, I can doubt delivery of strange effects, yet for to eat and taste good to be tested. And one farmer did say the tinge in color is clearly soil based.

LANCET (*as **Holston** sits*): My madness in the eye, so thorough and convinced, was clearly spotted by Lord Cantal, the host of the expedition. And he did forcefully reduce my dependence on the weed's love, by gradually using reason 'gainst my rapture of thinking to finally know all of fundamental worth and wanting to stay settled in that position of containment to a singularity of wish. With more to know for why I was brought out from what was strict in its deception of real sentiment. And slowly I cleared my heart of head, and my head of mindless ghosts and abandoning of purposeful, productive thought. I needed to be freed from the prison of an ambrosial state. And he did provide some impetus against that enjoyment, countering complacency with a severity of deed, that a vacation remains around us and never within, that we're never to withdraw from our rights of conception and questioning, and that this is evident to our meekness compared to nature's resplendence. For what is natural is more rich than what can ever be contemplated.

SECRETARY: And yet this lord did eat the same, and was not so affected as you— and guided you back, to *some* form of health.

FENIC: It's clear, Lancet, that you ran too hard, went on a rampage of the mind, by having found a taste to excuse it, and so deviated from your main principle of excursion, an invitation to assist Lord Cantal in *his* development of adventure. You have hoaxed yourself, by being prurient of your desires to prove for your researches an establishable goal of peers' merit and praise, and found yourself captive to this intimation's blow of iron to be dazzled, dazed, stunned, and deferred to a wrong prescription of your worth, a catatonic jollity dismissing disbelief in foolish pride, if I sense well your meaning to have found the world entirely within yourself, in that distasteful spell.

LANCET: The evidence will be of its own notoriety. Need not you believe my want of acclamation for this hurting, and swelling of the head, as could be famous for a premonition of this state. A deed itself makes its own renown. It is the carriage of catastrophe, for all things disturb others. Yet in noteworthiness is this stark to tell you of, and inform with precaution of your rush to consume my find, and my naming of— disaster! without as much conceit for what I've tasted, felt, battled and belched from, as could be demons ejected, rejected from my wars of contrition to a universe of proterical sensation leading me, guiding me through a steep deepness of— nothing!—

HOLSTON: You are befit to exceed even your own self as usual. And this is what describes and schools of information?! for descendant minds to nourish on? But what if judgment with propriety of dreams and wishes? your find clenched within the jaws of much demanding, for this substance to be fantastic and of iridescent power to mind's sight, when it may only be a faint spice of delicacy to a mouth's tongue, and little more of heat than haste's devouring. Yet have I seen such personality before, over-enthused of common matters as to leave their behaviors tumbling with declension into social and mental abysses. And you did spend a fine time with your find— away from structure! and control, as could be a child in a playground during a recess from chores. That has always bemused my thought of you in particular, to take life as such, and to search for any excuse to go off and wild on this kind of a lark, not as a responsible man, but rather a childish and irrepressible suitor of the confounding and weird. Judge by you to excite?... those who want to be excited by the trifles of fundamental questioning, such as, "Why does this taste good? and not to make of that so?" I'm sure your friends are bubbling of redolent curiosities, making culinary tests of their local flora of weeds and other vegetations thought to be dismissible for eating. Yet find you jealousies to stoke, for being as reprehensible of action on your knees! That is your flower —!

LANCET: Is a grass! Is Lancet's weed. My find—!

MYTDRA: Your.... discovery... is Lord Cantal's find. It was done during his missioning. And the accolades belong to his allowance of your participation. But strictly so, it is his triumph and trophy returned with, and his acclaim to be formalized with honors, awards, and much discussion. Yet does he not partake of main the agronomy he's caused, a major portion of this pursuit, through manners arranged by Ancher?

SECRETARY: He's had no involvement at all, great majestic leader, aside from supplying the samples expertly kept viable during his sea voyage. If they do take meaningful root within our country's ground, this will be in great part due to his diligence to keep the plants fresh and away from the briny ocean sprays on deck. Though he did explain to me personally his interests in this growth and his experiments of fostering small patches of them, back at that tropical stay: what depth of tillage seemed best, the looseness of the soil to that land which was most favorable, and the sunniness tolerable with some requirements of shading. These details will be very helpful to our horticulture. The grass-(seed)lings as yet do not look much harried here, and sprouting of new shoots not off-color may occur soon. The project is daily overseen by locals of the area, very conscientiously managed and governed—

MYTDRA: No more interest than this? to drop the spindly fibers off and leave all in your hands? Seems peculiar for a prominence to bestow over the process, if there's some profit to be made.

FENIC: Lord Cantal has made jaunt for another adventure, your majesty, a new excursion to find more natural treasures. He told me he was feeling rather restless and somewhat dejected shortly after arriving home, as if he had not fulfilled all of his pleasure-seeking and goals in travel yet. Though he remarked that there is still much time left— if he is adroit at it— as being the most conspicuous— and pernicious of all vanity— lesson gleaned from his last voyage. Mysteriously he said he is not of a will to waste his graying hairs on time. In my perplexity I take that *to* be time. But the man is in a rush, for whatever of his representation. His reputation stays assured, that he'll not fool with any dallying of heart.

MYTDRA: So terse be the commotion about this, then. (*sitting*) Yet what I have presented.... was hoped to clarify. To not obscure.... conspiratorial pursuits, the clouds darken to much meaning of our vexations, except to say that we have delivered ourselves of some several objectives. For Fenic's find is occupied, as Lancet's weed is found. And Ancher be with proof a rascal or a reed, of deliverance and prosperity again, to the carefully chosen.... of our sufferances and abuse, and accidents not criminal, and criminals not of crime, but of our faithful halves once more conjoined with reason. For have we purpose being, of nobility away and executors at bay of notice, treasonous attempts imagined— anywhere, as by mere thought to make resolute a material for our handling; and for servants able and disabled of believing, what's for a mind's dimensioning of promise oft too clearly spoken to be easily interpreted. It's as simple as being ate, and as vague of flavoring.... as by commiseration to help the distressed, a sampling of the sympathies common to all— types. And with this injury recovered from, though never absolved and not ever to be ignored or forgotten, will beam again our hue-laced smiles of growth and maturation towards what shines and warms and warns with denigration by its heating.... ever our unresolved commissions of societal enchantment. Then judge I what is fit.... is what is evident to be produced of our pieces of effort, between ourselves and for our likenesses to enhance or promote, that I perhaps have spent too much, while others have spent too little by necessity of loss, and yet a few spend to exhaust themselves to the limit to redress these imbalances unconsciously. Let this region try the crop. Let her heal!... and redevelop strengths, with management and exhorting promotion. Or let what falls be abandoned under the sun; for we're taught it shines somewhere of a day.... to bid its testing— at any time. And there is no escape on earth from our complied results. More sufficient.... be a revolution stalled and stymied of its race, to embrace our apportionments of just candor and deliberateness of approval.

FENIC (*mostly to **Lancet***): And so may searches continue with less error of doing than of believing what has been done, be we upon our dogged ways to prove of anything, and be approved. The only charm is having been before the thought disturbed and rankled by what should have been as found and left, a pleasant notion all around to leave and keep. Little disgrace to dreaming, yet much more to a dream's making. I suppose man is most subjectable to his whims of importance, and suggestible of these artificially constructed applications to achieve. Yet be we ruled with tides of stirring to deceive ourselves in many things, and make of ourselves mired inhabitants, admired individuals, and beings more useful than our judgments.

HOLSTON: That's as clear as crime to sink to. It's more the degree that counts, to what we attend ourselves as being necessary to try, to steal or harm or hate or test to impair reason. Yet a right for wronging oversteps a wrongful right, a strong brute compelling the weak and meek to cringe and bend to his demands. Accost we not errors in our policing, but displacements in strength. Errors are of a race of rulers to command, commend, or contend with. And I can think of what is best.... and what is distorted, and have no clarity in this, for all that happens of results in friendships and associations, without the guidance of some principles decreed to be and faithfully stuck by. That dissolves in time these mistakes, disappointments.... and regrets.

MYTDRA: So say it practically. To err is only to be abhorrent of the facts, and not to be wrong, but willing to do so— if you can grasp the distinction. So do as may be grown. All types are allowed under my supervision. And I adore each fair-minded soul, high-minded subject, given a range of action by the animate spiritedness to which we are independent to endow or much control. Yet plots.... are plots, and grass is grass. And man makes man, for mass en masse. Where be the sainted without sin? defiling grandiose recruitment, in our hurried lives to end. It falls.... to single eyes to tell what is permitted of our purchases, of valor and heroic attempts to restore.... a happy, good— harmonious life.... for anyone cared of. A narrow vision slit, is this seen, as could be assassins made of dusty ghosts pretending to frighten, when only wheezing and sneezing of their ne'er ending burdens compacted of their adjudged notice and relationship to others ruling.... of their actions. I forgive misunderstanding of my power and intentions. What better be the two is always done by nature of a cause, by being complimentary to a result expected, desired, or not. Yet should we congratulate what fouls, or what is fouling? Go you now— Lancet, with pledge not to discourage this.... gardening, until such complaint is merited with deleterious evidence. You are thinned and pale from a submission to a jungle's feverous vivacity to capture neophytes to its sophisticated coercions towards its wonders and exoticisms, I am convinced, and need to recover your bulk of body and facility of mind with a longer and more thorough convalescence within your native climes. Your arguments are starred with possible misapprehensions due to this foreign conditioning— though I grant you the privilege as a free thinker to work this through your conscience and rationalizing for specialist others, and to continue until you are done of thought and have come to inextensible conclusions.

LANCET: I will let the tears to brain weep with mashing of the straw, your majesty. Does hash it makes to eat, of an amicability of taste mixed in a produce of much pondering.... (*starting to withdraw*) I will be corrected.

MYTDRA: Continue your supervision, secretary. And make known to me all relevancies of this matter to my interest, in particular as regards Earl Ancher, his correspondences and return.

SECRETARY: Yes, my legal lord of enlightened breadth. And may I say you have conducted this inquiry so splendidly— it can only warrant the greatest success for our country and all its loyal subjects. Since wisdom prevails with the wizardry of kings. And my earl will provide for you the greatest gifts from his concurrence with your handsomeness of reign and thorough understanding of his ambitions. (*starts to leave*)

MYTDRA: To serve me. (*a slight pause, as **Lancet** and the **secretary** leave more or less together*) So, Fenic.... you look pleased and yet contrite.

FENIC: I am a cur, your highness, upon his master's bidding to be satisfied and relieved of acceptance. I am viewed squarely, but seen round. For my end is yours defined, only for you to determine. I've mishandled things terribly on my own, and caused you

to correct them with your influence on me and the world. And not soon again will I speed of contentions to jostle our orbits into seeming disarray or inappropriate styles of viewing.... your sovereignty over all that can matter to us, of substance and soul-searching and sufficiency and sufferance. For you, in essence, are the authority that guides our minds, no matter how things appear, so often out of place to be either forgotten— or obsessed about for concentration on. So with this thought I am happy to commit myself to what ever little approbations you might find in my service. For dogs may scrape the earth, but seldom plant on purpose more than bone. And that of mine I hope some day will be secured.... under green grasses, and bountiful enjoyment of the view.

MYTDRA: Yes.... I do an awful lot of correcting. Subservience is such a trick to train. It is not a demeaning into states less subtle than what one already wears, and no belittling at all to one capable of his station. I think it's most an education of the obvious, which somehow 'comes the most difficult for what is taught. Yet, this is the world— and this is the universe! for our positions to accept and compose ourselves into, what is righteous of ourselves of a commodity of exchange among a competent society or aggregation of persons.... solved of order, my laws.... and your lawful exercises through them, however expedient at times they may seem. That is the art, I suppose, that is continually teaching me. For one's power rules over one's self in too many examples to illustrate. It is the tie of humanity *et al*, that makes one common.... under the sky. Much may be summarized with this observation, of the haphazard scurrying below that should appear to those above on sighting frivolous disarray. Yet can we merchants have more motion of our means, distributing these cares throughout incomprehensible routes of confused relationship that makes one unwittingly or indifferently a member of so many others, such as to explain this mystery of nature, or of one's, is a stubborn task. Then I should say we are replete of kind, to struggle through these troubles, selling weeds,... exchanging wishes,... buying wants. And being as we are loathsome otherwise of this frantic participation, the social drive, self-regenerative of all imperative, excuses its own complexity. That's why there are leaders— kings, better pitched among the fray to frown at immoderate distortion of our manners and exertions.... understandably. More so than some external judgment would call this world perturbed of an infantile madness or anger, we are better placed for these perceptions to receive and live through, and by our own exceptions winning forms of justice. The hound is as noble as its teeth— and as powerful as its snout, yet be it lower than a weak fleshiness to spare for higher principles of attainment. Then not the cur, but the curing.... is our theme sung.

FENIC (*to himself*): How have I changed in other ways to bite, there is great wonder in a snap's delight! conforming to these forms of justice. For all if all remains when not more shamed to be.... than salvaging, in manifold ways and dispositions, and resumptions of some astonishments. With gentleness as thin and slender as is green, some grass to mean, how could be you perceived as other leaning towards for your consumptions daily and concerns wailing to be fixed? It's ate by self, one's splendor. It's met by self, one's spine. And the universe engenders where at most we must recline.

MYTDRA: Cautious talk for patience of your thoughts, Fenic? quietly composed.

HOLSTON: He calculates more.... dissertation for your actioning, your highness, through which to mangle more for your influences.

FENIC: That is compounded, yes? But I address my mind for bites and lesion-ing of thoughts, conceptions. Yet for our majesty to will that we may all conduct ourselves through his light to soak and save of ideas, then be this mental pausing brokered with resplendence of purpose, and for each person naturally be made fit, in his presence, to stand! of his notions. Then to the world we must resume our conduct (*bowing, as* ***Holston*** *stands*) appeased.

MYTDRA: Yes. Then to the vault.... and to the box. For I'll have more guests soon, to be backed of their—

FENIC: Your majesty! (*starts to leave*)

MYTDRA: —ingratiatory bickering.

HOLSTON: That's much to hone of quarrels and disagreements rent through praises, your highness. But not always are their concerns.... so dismissible to secreted maladaptation of the ear.

MYTDRA: No bother to my burdens more. I hear all things, and he hears less (***Fenic*** *has exited.*), yet more for this to do, not spying but spurt-ation of the jest, for what's to be known. And compare yourself with more surveillance of the populaces. I need more fines and fees for what occurs, that must be criminal.

HOLSTON: Yes! your highness. (*leaving*) I'd never slack of slavishness to serve the sovran powers by showing hint of sloven-ness to suit.... my careful marketing of law.

MYTDRA: For all revolve around my patience, of being sympathetic (***Holston*** *has exited.*) to this heat and looseness of the rays, by which I blind myself sometimes of their effect. But that is not to be deliberately bad of temper, or sluggish for my motions to improve the sights and aspirations of my subjects. (*standing*) I am simply here, with some notions beaming, and some passions sweeping, and some challenges daring to confront, and some prospects marrying of improvement— with my sear, directives made definite. And if not.... if all of this be wack of eccentricity and odd bestowing of a view, then what is a life but its container of effects unusual of not for self to own? with strange.... graduations to more goals and defeats, and influences upon being seen. It's simply my way to be and to be acknowledged of my trying, if not caring is a real constitutional endowment, but some metaphysical residual of light. So harm each fellow all, with their harmonies made purchasable, and to each some sum allowed for this transaction to become irritated. Then save the windstorms and the blowing thunders to the titanic— admitted to— as you whet your naves for more auspicious travel through the efficiencies of your journeying through more cupidities than my simple culpabilities have allowed. There's no more treat to me in this, than losing a reserve of motion, of slowing down a bit, to see what damage 's done inevitably by your glances unavoidable and seldom for a conscience's view. Yet of the ruins that can be made, with withering, there dews a fascination with resiliency permitted, that hopes to guarantee hope of more recoveries than thought. Since gravities are always shallow before their graves.... Then speak to me, of an audience blinded for my sight. Am I not of a casual haste, pace typical for wrongs with rights? The same as could be seen, with all this might withdrawn, is yours of fashioning, what makes perceptions true. They are created for your guises glancing through what has been done. And what of you to seem, intemperate more than I?!... Nay an'! Does

waste an attempt at explanation. What's evident will be made to seem so for you. And how this is reacted to.... is more the cur to chew, than 'gainst my caring. Cause of cures are within heads of caring— application! of your art, and not with blame the tautness of your restrictions, or notions 'bout our physics and our prides. Yet for all defeatist woes, embrace the point of duplicity to error, or taking blame without for naming blame— in heart to know— the faults be swept as cleaned from privilege of what is cared for being, a righteous intent placed upon us all, and an honest dimness blinded by, some latent passages forsworn to be all but approved. These are just a justice viewed despite its carriage, transit for a rumination of its cause, (*sitting*) the due enlistment of our natures' gravities and ponderations. For have we not thought best.... about all things to do, before some cunning's done to do them? Then blame the art ov'r the error, more than the wrongs portrayed here. They are.... sinecures of action, and not deceits of might. Now, must I prepare for more.... interpretation, the rights of bargaining and bandying for approvals. It is a coarse fineness made of me, to weave this fabric common to our aims of social life.... In here for no, (*stands*) the answers to be made for circumspection lies in this space. But where is partying determines taste of argument. (*leaving*) And that's for spying ears and hidden fears of frightfulness.... Though must it be as similar as I would be a sun, to hide from seeing light, a shear enticed of commentary. I do not deceive with bluntness, and may sharpen with confronting of one's own goals. But that's the passing threat to encourage.... its departure.... All kindnesses are by nature decrepit ensconce-ments of abominable toils. (*Exits.*)

Scene III — *Before a tent during the day.* ***Ancher*** *sits on the ground, his hands to his head, as if to comfort a headache. His* ***guide*** *stands to one side of him, the* ***girl*** *and the* ***boy*** *of* ***Cantal's*** *acquaintance to the other. The day seems warm.*

GUIDE: Er be fit.... for shade?

ANCHER: It must burn away, like a flame to sublimate the tar.... There was nothing here but hardship to be ate, of this green passion. And so I am produced a fool!

GIRL: But it's odd to eat—

ANCHER: Had to be tested, of its rendering and rent of hopes.... You find diamonds cast about as inauspicious stones. And *that* seems odd to hoard, what is so plentiful. Yet what to find is what for use, its purpose more determined by your handling than its own.... precious levity of being.

GUIDE: Not more than grass, Er. This is everywhere that's fit for soothing soil. And it's such a wonder why you'd want to claim it for a diet. But then a man does ration many things for novel purposes, schooled in very particular ways. Why not the earth itself to own? There's more of it than even plants to culture make of strange cultivating bouts or frenzies. Or find you the starfish enviable? It's like an encrusted jewel— yet lives of gems abandonable, to the vastness of the oceans and their coral reefs.

ANCHER: I once captured one, in its sea water, as a youngster, caught of its fascination to study an' observe. I.... stole it to my home and parent's house, as we returned from a vacation of beach pleasures. I can remember how I chased its— running and fright, among the coves of a sandy shoal. Yet in its bowl it only wept and died, having been brought, brought out and away from its habitat, for my sad and childish amusement.... An' thus a spark for things to collect was born, as if to somehow recompense this erring with my privileges. This has never been sufficient of the tug, to alleviate an infant's sorrow. But to do does gain some stock of trying, that one is not so wrong of curiosity to crush things— an' examine.

BOY: That's a curious meanness. One rather whole of substance would take the world as is. Yet, could we find more novelties, we would, to disprove splendors surrounding us, as many of you claim there be: different natures an' natural wonders that exist, or must exist— to deter primitiveness an' stagnation of culture. So we're curious of that, but not to kill ourselves.... disingenuously.

ANCHER: There are sophistications of life an' manner even for the simplest morality plays to advertise (*rubbing his forehead*) and bemuse a populace with. Thought is merely a complexity of thinking.... There's nothing of this grass as found, that would be wholesomely competitive back home, in my opinion. It's taste is corniced to a crimsoning of the emotions. It is a lace of craggy growth meant to be left undisturbed— and for its own withstanding. An' I am lost to find it so, having dreamed so fervently otherwise. This is a failure poured on me.

GUIDE: To find what is, Er? You deluded yourself of hope, as it gave your head a stomachache of conscience. But there's no punishment in nightmares.

GIRL: The nightmares make the night its festivals an' fruition of rites. Every caution must be disposed in a sort of blindness. An' you have been affected by your oddity to do things weirdly, the want to create a new atonement for some sins, I guess, if you typically misuse your grasses where you come from, or make other deviations of bounty. But here is not the wrought for being guilty, only of flourishing—

ANCHER: Of fish. Would you have more?

GIRL: ....I would have more responsibility. Yet other may we know, that's certainly attractive. But the world is anywhere to be studied, at least where food may be obtainable. An' I'll not be

dauntless without some need of construction an' creative manifestation. For what is the point of a sumptuousness of environment except in this way to respond? It did not wrong you to be enlightened, that the act was wrong to commit your crouching of, that it was a funny endeavor, that your capabilities of mischief do not disengage a mirth of nourishment to seek, and that you have made some lessoning with the earth.

GUIDE: The dementia was not so great, Er, compared to what befell that other. He would require a whole hospital— You have your quiet tent of shade an' retreat. No more for harm does make the day, but only rest out of the sun; since its shadow is peculiar, bringing light, lightheadedness an' dizziness, a loss of balance that leads to much flopping an' crawling about. But what is seen, Er, I can't imagine much of a need for, except to fulfill a voracious appetite for excitement, as you adventurers often claim, explosions of consequences an' volcanic eruptions to bathe of the lavas— of information breathing with, an' new experiences to partition your cares of. It's a callous attitude, Er, for what's presented to be observed, simply resplendent of its nature an' unquestioning as to why. Yet I suppose the fish question of us with some fear. An' you?... You ask the blades of grass to tremble. They do not, but for the winds and reaps.... an' munching you've aspired. Tear they for your mind's unusual quest? What is the mental canvas painted with, for green.... behaviors and portentous strokes of brush?

ANCHER: Then as for all my errant deeds and fouling results an' unfortunate achievements that I've ever done in my life, all the harm and unhappiness an' disappointments I've ever caused, all the evil to make staggering of others left as of a way enforced— from suggestion to subscription of my substance.... my cowl of darkness maiming with protection for my head, an' its ponderous torts.... this is my punishment of banishing. I can not resume a life directly under a king proposed to for these wonders, until they transpose themselves of some earning justice. An' I will make habitual, for a decent while, this freer empire to reign of as myself, a man shrouded of light an' due some delight of serenity.

GUIDE: Solemnity is in his eye, that glides towards dawning dusk.... of shore.

)()()()()()(

514

treatise on nul*l*los

treatise on nul*l*los

A continuous conversation presented under the following subdivisions:

What is the point of zero on the surface of a sphere? It is nul/los!

I certainly agree. The surface of a sphere is nul/los. However, a point may be chosen in order to define an origin, allowing other points on the surface to be definable of position.

Manipulation of the sphere's surface, for a particular position to find, is nul/los, since even when there is no manipulation a definite position can be found. Therefore, all physical manipulations or processes are nul/los. And all operations manipulative of the surface of a sphere are nul/los and are expressible as such. For example, to bring to motion from position A to position B (on a sphere):

Both A and B are already defined positions
The distance from A to B must be added to A by an operation; but this operation involves not only an addition, but also (with conference to the sphere) a rotation from A to B
Therefore, the nul/los operations include not only reference to the two positions (A,B), but also involvement of the entire sphere (and all of its elements and positions) to rotate for $A \rightarrow B$

Therefore, a nul/los operation involves the entire sphere.

The sphere as like a closed system or a total consideration for observance. But such a system should have a distinctive formulation to describe (itself).

The nul/los operation here includes an addition (of distance between A and B): M_{add} (as M is a manipulation); and a manipulation of the entire sphere to allow or direct the addition, M_{sphere} (as for a rotation appropriate to the sphere):

(A) $M_{add}\ M_{sphere} \rightarrow$ (B)

Let the yield sign ($\rightarrow$) indicate an active result : R
Then we may also write as well :

(A) $M_{add}\ M_{sphere}$ R (B)

A and B are elements of the sphere's surface.
M_{add} and M_{sphere} are manipulations of the sphere.
R is an indicative symbol.

We should not be too obscure in our meaning, but strive mostly for simplicity of expression in order for an easy interpretation of symbols and formulas. The point is, of course, that an actual (physical) system has no zero component of identity.

The relationship of manner (of operation) for M_{add} and M_{sphere} is dependent on (A) and therefore defines (or instructs or orders) how the manipulations are conducted.

$\therefore$ we may write

M_{add} (A) M_{sphere} R (B)

without invoking any equalities (since A need not be B):

MAMRB

No changes have been made to the sphere (or system), only its manner of observation (or observance). Perspective has moved from A to B ($\therefore$ a shift). No measurements have modified the sphere, and no logics have reproved it of any different factuality: Measurements are not proof, and logic is not fact.

But we depend on measurements for our conclusions and confirmations (of theories and the like).

M_{add} can be re-symbolized by a simple subscript "+", and M_{sphere} by a subscript "rot" :

$_{+}A_{rot}RB$

where "rot" can involve a considerable complexity. This complexity is due to a substantiation of the distance between A and B. Note that from here, if a subtraction from B is performed, of A, then:

B-ARA (position A results)

or

$_{-}B_{rot}RA$

So these expressions seem more like terms, mathematically.

We can let this distance between A and B be symbolized by "d", and any set of adjustments needed by the sphere's rotation to be represented by the symbol "x" (these symbols as superscripts):

$^{d}M_{add}(A)^{x}M_{sphere}R(B)$

or

$^{d}_{+}A^{x}_{rot}RB$

Likewise

$^{d}_{-}B^{x}_{rot}RA$

or (if different adjustments to sphere are used, thereby represented with "y")

$^{d}_{-}B^{y}_{rot}RA$

So, we have the relationship:

$^{d}_{+}A^{x}_{rot}RB$ to $^{d}_{-}B^{y}_{rot}RA$

This could (script-wise) be compacted to

(d+)A(xrot)RB to (d-)B(yrot)RA

R is certainly redundant, to be understood.

$\therefore$ (d+)A(xrot) to (d-)B(yrot)A

But, what does this relationship signify? That "d" is common to both (for manipulations)?

$_+$A(xrot)B(d-)B(yrot)A

or

$^+$A(xrot)B((d)-)B(yrot)A

or

$_+$A^{x}B(d,rot).B^{y}A (allowing the general purpose of "rot" to be common);
or use brackets [] to emphasize a relationship is being indicated:

$_+$A^{x}B[d,rot].B^{y}A

Calculative organization

How can this terminology be used for calculations?

Clearly "d" and "rot" are activity notifiers of certain extents (unless they are used as variables):

"d" is a distance moved
"rot" is a vectorial rotation of a specified degree and direction

Therefore, they must be the subjects of calculations.

A and B are elements of initiation (starting position) and result (ending position). They are elements which are found and sought. They are found and sought through the activities of "d" and "rot".

The expression " $_+$A^{x}B[d,rot].B^{y}A " denotes a solution for "d" and "rot".

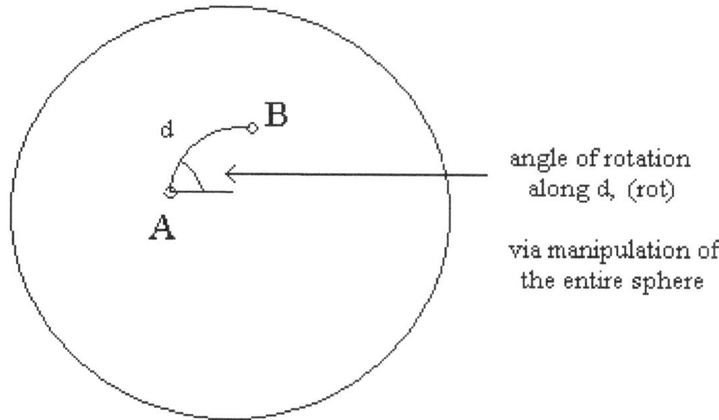

What if "d" has no extent, so that A is equivalent to B (in position)?

A must be axial to the sphere's center, and rotation must be of that axis. There may be great rotation to no rotation.

What if "rot" has no extent?

A is never equivalent to B, unless "d" also has no extent.

This suggests that both "d" and "rot" must each have some extent for the expression "$_+A^xB[d,rot]_-B^yA$" to be physically meaningful (or useful). They are, then, nul/los quantities. (Otherwise, the sphere has not been manipulated, as regards to A and/nor B.) The contention here, of course, is that nul/los elements are real and of extent.

$$\downarrow \text{———} \uparrow\uparrow \text{———} \downarrow$$
$$_+A^{\,x}B[\mathbf{d},\,\boldsymbol{rot}]_-B^yA$$
$$\uparrow\text{———}\downarrow\downarrow\text{———}\uparrow$$

For a calculation, suppose we combine this expression with an opposite relationship:

$_+B^xA[-d,\,-rot]_-A^yB$

(note that the nul/los quantities are different; also, the adjustments of the sphere are kept the same, by maintaining x and y)

$$_+A^xB[\,d,\,\ rot]_-B^yA$$
$$_+B^xA[-d,\,-rot]_-A^yB$$

additive: ________________

$$(_+A+_+B)(^xB+^xA)[d\text{-}d,\,rot\text{-}rot](_-B+_-A)(^yA+^yB)$$

But, "zero" is not allowed in nul/los expressions, so this describes an impossibility (via d-d, rot-rot).

Now, let's define another relationship that can be added to our first one:

$_+C^{x2}D[d2,\,rot2]_-D^{y2}C$ (where x is x1, y is y1, etc.)

$$_+A^xB\ \ [d,rot]\ _-B^yA$$
$$_+C^{x2}D[d2,rot2]_-D^{y2}C$$

additive: ________________

$$(_+A+_+C)(^xB+^{x2}D)[d+d2,\,rot+rot2](_-B+_-D)(^yA+^{y2}C)$$

This yields to:

$$_+A^xB+_+A^{x2}D+_+C^xB+_+C^{x2}D\ [...]\ _-B^yA+_-B^{y2}C+_-D^yA+_-D^{y2}C$$

where [...] is [d+d2, rot+rot2]

$$\therefore\ _+A(^xB+^{x2}D)+_+C(^xB+^{x2}D)[...]_-B(^yA+^{y2}C)+_-D(^yA+^{y2}C)$$

So here the sphere manipulations (rotational) are grouped together in parentheses. We can let d+d2 be d3, rot+rot2 be rot3 (so that A+C is F, B+D is E):

$$_+A(^xB+^{x2}D) \text{ be } _+A(^{x3}E)$$
$$_+C(^xB+^{x2}D) \text{ be } _+C(^{x3}E)$$
$$_-B(^yA+^{y2}C) \text{ be } _-B(^{y3}F)$$
$$_-D(^yA+^{y2}C) \text{ be } _-D(^{y3}F)$$

$$\therefore\ _+A(^{x3}E)+_+C(^{x3}E)[d3,\,rot3]_-B(^{y3}F)+_-D(^{y3}F)$$
$$\therefore\ (_+A+_+C)^{x3}E[d3,\,rot3](_-B+_-D)^{y3}F$$
$$\therefore\ _+F^{x3}E[d3,rot3]_-E^{y3}F$$

Therefore, relationships may be commanded of calculations in nul/los expressions (or nomenclature).The math shares the additive, multiplicative, and commutative natures of basic arithmetic, and can clearly indicate when an impossibility of theoretical propagation arises (by the occurrence of zero). Since in the physical or actual world (or universe) there is no stationary point or incidence, all matter is in a constant state of movement depending on most references of measurement. One judges by measurement, but may also be judged by. "A" may appear to be a stationary point, but is not when judged by other than itself. Thus, its stillness is illusional, which is what nul/los emphasizes or accentuates for notice.

For example: If the sphere moves in space, without its own axial rotation, then while point A may seem to the sphere to be stationary in position, nonetheless it is clearly in motion with the sphere. Nul/los declares that it is not a case of both conjectures (the sphere's and the space's) being correct, but that the sphere's is wrong and all of its positions are in fact in motion.

Thus far, the nul/los "sentence" constructs a relationship between an expression and its opposite. Such logics might probably be more arbitrary, and need not force related expressions as rigidly.

Let us say we have the expression

$$^{(d)}{}_{+}A^{x}{}_{(rot)}B$$

and wish to relate it to an expression not obviously derivative of the first,

e.g. $^{(d2)}{}_{+}C^{y}{}_{(rot)}D$

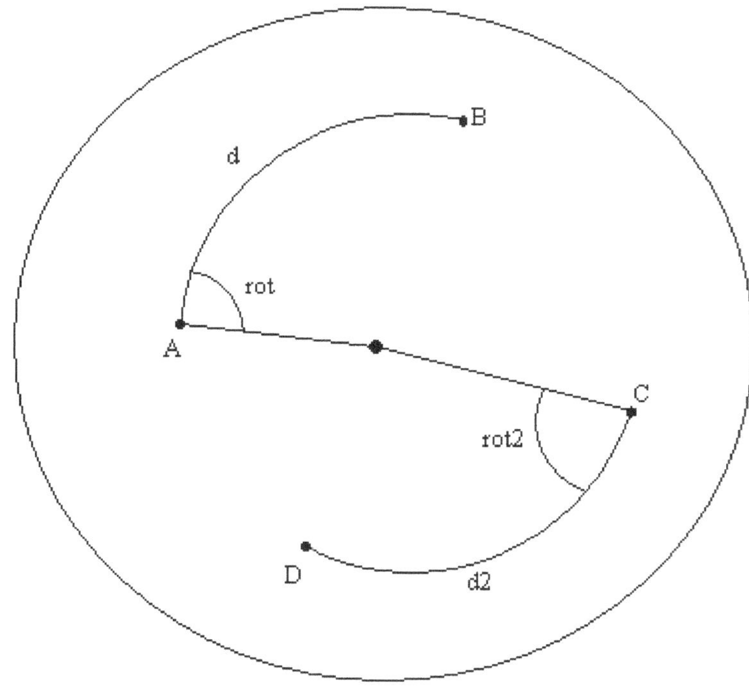

The association of the two expressions is not one such that either implies the other:

$$\downarrow\!\!\!\overline{\qquad}\uparrow \qquad \uparrow\overline{\qquad}\!\!\!\downarrow$$

$$_{+}A\,^{x}B\;[\;d\,,\,\boldsymbol{rot}\;;\;d2\,,\,\boldsymbol{rot2}\;]\;\;_{+}C\,^{y}D$$

$$\uparrow\underline{\qquad}\downarrow \qquad \downarrow\underline{\qquad}\uparrow$$

Yet, for such a statement it is imposed that we must find the relationships between

[d and d2; rot and rot2]

i.e. We must find the nul/los relationships between these quantities (or alternatively define that they exist.)

The elements (A,B,C,D) all being on the surface of the sphere, one can construct lines: say A to C, B to D, which are members of planes "cutting" through the sphere, and with axial components (of the two planes) not necessarily dependent (or coincident) on the sphere's center:

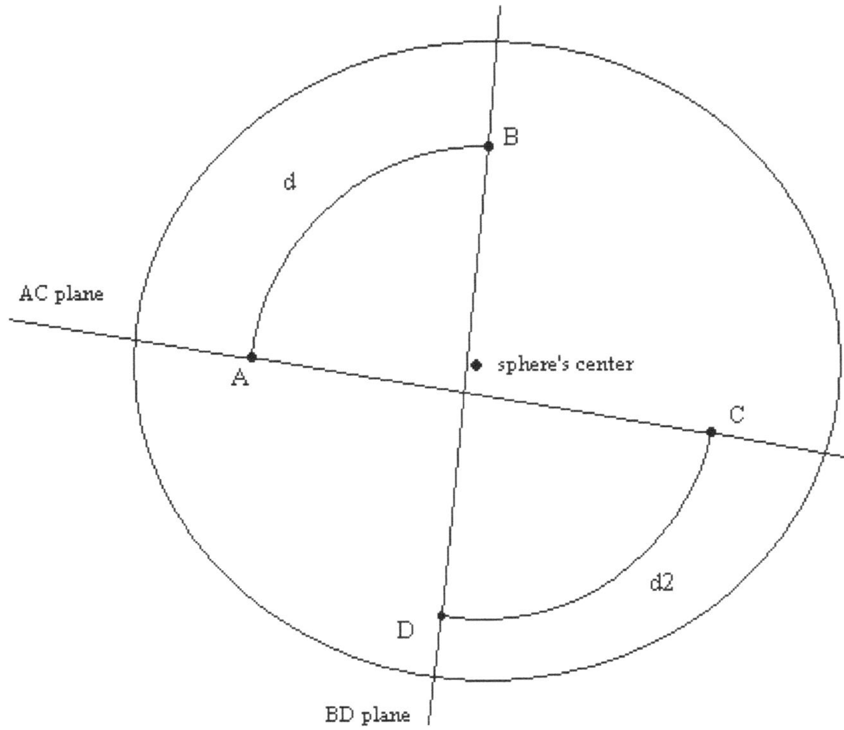

∴ a rotation between these two planes (rot3) will involve a generalized distance (d3) between the planes that will bring A coincident to B as C is brought coincident to D.

$^{d3}_{\;+}\underline{AC}^{z}_{rot3}\underline{BD}$ (where the underlines indicate common members of a plane; planes as noticed ⊥ to page)

And *this* expression may be related to its opposite:

$$_+AC\ ^zBD\ [\ d3\ ,\ \textit{rot3}\ ]\ _-DB\ ^aCA$$

It is clear that rot3 involves rotation about the axis of intersection of planes <u>AC</u> and <u>BD</u>.

But can one rotation satisfy both " $^d_+A^x_{rot}B$ " and " $^{d2}_+C^y_{rot2}D$ " ?

"rot3", then, is a complex rotation involving at least two components (corresponding to the two planes involved).

There are also planes <u>AB</u> and <u>CD</u> (cutting through the sphere). The rotation of each (at the center of the circle including A and B or C and D as members of a circumference) will necessarily bring A to B and C to D. Here the (d) distance values are *circumferential*, jumping off and onto the surface of the sphere.

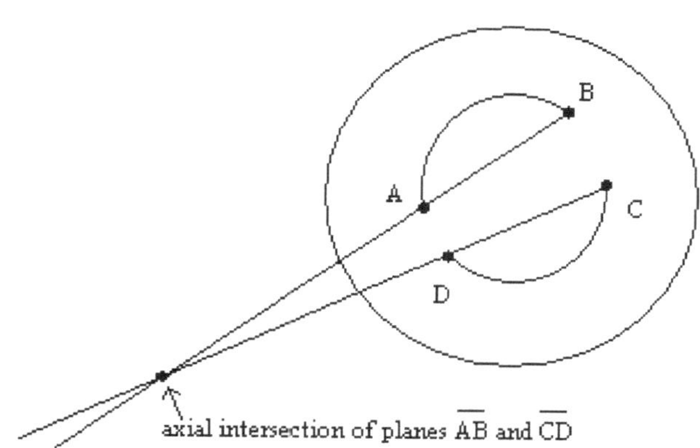

Thus,

$$_+AB\ ^bCD\ [\ d4\ ,\ \textit{rot4}\ ]\ _-DC\ ^cBA$$

and

$$_+\underline{AC}^z\underline{BD}[d3,rot3]\underline{.DB}^a\underline{CA}$$
$$_+\underline{AB}^b\underline{CD}[d4,rot4]\underline{.DC}^c\underline{BA}$$

additive

$$\rule{3cm}{0.4pt}$$

$$(_+\underline{AC}+_+\underline{AB})(^z\underline{BD}+^b\underline{CD})[d3+d4,\ rot3+rot4](\underline{.DB}+\underline{.DC})(^a\underline{CA}+^c\underline{BA})$$

$\therefore\ _+\underline{AC}^z\underline{BD}+_+\underline{AC}^b\underline{CD}+$
$\qquad _+\underline{AB}^z\underline{BD}+_+\underline{AB}^b\underline{CD}$
$$[d3+d4,\ rot3+rot4]$$
$$\underline{.DB}^a\underline{CA}+\underline{.DB}^c\underline{BA}+$$
$$\underline{.DC}^a\underline{CA}+\underline{.DC}^c\underline{BA}$$

$\therefore\ _+\underline{AC}(^z\underline{BD}+^b\underline{CD})+_+\underline{AB}(^z\underline{BD}+^b\underline{CD})$
$$[d3+d4,\ rot3+rot4]$$
$$\underline{.DB}(^a\underline{CA}+^c\underline{BA})+\underline{.DC}(^a\underline{CA}+^c\underline{BA})$$

Of course, each plane can be deconvoluted to the specific points concerned which are a member of:

e.g. $\underline{AC}$ can be rendered as

$$^{d(AC)}_+A^{x(AC)}_{rot(AC)}C$$

and related to

$$_+A^{x(AC)}C[d(AC),\ rot(AC)]\underline{.}C^{y(AC)}A \qquad \text{(etc. for the other planes)}$$

So we are developing several of a collection of nul/los quantities:

d; d2,3,4; d(AC); rot; rot2,3,4; rot(AC) etc.

And we can find relationships for quantities which casually seem disparate:

e.g. rot(AC) is indeed related to rot3 and rot4
    d(AC) is indeed related to d3, d4
(i.e. They may be derived from these other values.)

Exclusion of zero

The prohibition of (the numeral) "zero" is based on exclusion from the relationship statement(s) results (e.g. where d-d or rot-rot occurs) as being non-physical and therefore impossible.

To avoid zeros in numerical calculation results, perhaps usage of a (non-decimal) basis (of distinct digits utilized being) 9 numbering system is useful (as any consistent numbering system is just as valid as any other), where, for example, the next numeral after 9 is 11:

$$\begin{array}{cccc} 1 & 2 & 3 & 4 \end{array}$$
123456789123456789123456789123456789.......

and   9 + 11 is 21
    21 + 9 is 31   etc.   (note that 11 + 11 is still 22)

Discreet calculations, then, can be initiated by the obvious definition that if $_+A^{x(AC)}C[d(AC), \text{rot}(AC)].C^{y(AC)}A$

then $^{d(AC)}_+A^{x(AC)}{}_{\text{rot}(AC)}C$ / $^{d(AC)}.C^{y(AC)}{}_{\text{rot}(AC)}A$ actually relates x(AC) to y(AC)

i.e. $_+A^{x(AC)}C$ / $.C^{y(AC)}A = x$ / y where a "definition" involves an equality, here.
(note that "y" can never be zero (in our basis 9); therefore, undefinable infinities are avoided)

Similar definitions are:

$_+A^{x}C$ / $_+A^{x}C = x$ / $x = 1$
$.C^{y}A$ / $.C^{y}A = y$ / $y = 1$
$.C^{y}A$ / $_+A^{x}C = y$ / x

and

$(_+A^{x}C/.C^{y}A)(.C^{y}A/_+A^{x}C) = (x/y)(y/x) = x\cdot y/y\cdot x$

Non-vectorially:

$x\cdot y/y\cdot x \equiv (x/x)(y/y) = 1\cdot 1 = 1$

$\therefore (_+A^{x}C/.C^{y}A)(.C^{y}A/_+A^{x}C)(_+A^{x}C) = x$
$(_+A^{x}C/.C^{y}A)(.C^{y}A/_+A^{x}C)(.C^{y}A) = y$ etc.

Vectorially, the results would be:

$(x\cdot y/y\cdot x)x$, $(x\cdot y/y\cdot x)y$, etc.
$(.C^{y}A/_+A^{x}C)(_+A^{x}C/.C^{y}A) = y\cdot x/x\cdot y$

yielding corresponding results:

$(y\cdot x/x\cdot y)x$, $(y\cdot x/x\cdot y)y$, etc.

From these definitions, higher orders of complication may be developed for the variables x and y. The usefulness here is that the quantities x and y remain nul/los.

e.g. $(_+A^{x}C/.C^{y}A)(_+A^{x}C) = x^2/y$ etc.
    $(.A^{x}C/.C^{y}A)(_+A^{x}C)(.C^{y}A) = x^2$ etc.

If we let n = x, then $(1+1/_+A^{x}C)\exp(_+A^{x}C) \equiv (1+1/x)^{x}$

$\therefore \uparrow(_+A^{x}C)\exp(_+A^{x}C)$ yields to an "$e_{\text{nul/los}}$"
(x, in radians, is limited to an extent of 2π, but there is no numerical limit to the number of revolutions. Therefore, $(_+A^{x}C)$ can increase beyond 2π radians, with x as a rotation angle. There is, of course, a physical limit to the number of revolutions possible.)

Note: 360° in our basis 9 (as used) would be:

3 x 11 x 11 + 6 x 11 =
33 x 11 + 6 x 11   =
363 + 66 = 429°   (i.e. The base is not decimal.)

Obviously, then, relationship statements may be transported (or made transpondent) to the nul/los variables concerned:

$_+A^xC[d, rot]._-C^yA$
x [d, rot] y　　　　　　(This defines a relationship between x and y.)

So, for example: (going back to our study of planes)

$_+\underline{AC}(z+b) +_+\underline{AB}(z+b)$ [d3+d4, rot3+rot4] $_-\underline{DB}(a+c)+_-\underline{DC}(a+c)$

yields to a concrete relationship among planes as:

$(z+b)(_+\underline{AC}+_+\underline{AB})$ [d3+d4, rot3+rot4] $(a+c)(_-\underline{DB}+_-\underline{DC})$

as to form a description as follows:

$(z+b)/(a+c)$ is related to $(_-\underline{DB}+_-\underline{DC})/(^+\underline{AC}+_+\underline{AB})$ via (d3+d4) and (rot3+rot4)

Note that, if rot3=rot4 and d3=d4, then z=a and b=c

∴ $(_-\underline{DB}+_-\underline{DC})/_+\underline{AC}+_+\underline{AB}) = (z+b)/(a+c) \equiv (z+b)/(z+b) = 1$
∴ $_-\underline{DB}+_-\underline{DC} = _+\underline{AC}+_+\underline{AB}$　　(as definitions of relationship)

∴ One can have the physical concepts:

　　　$_-\underline{DC} = _+\underline{AC}+_+\underline{AB}-_-\underline{DB}$
　　　$_-\underline{DB} = _+\underline{AC}+_+\underline{AB}-_-\underline{DC}$　　　　etc.

But one can't have the (physical) concept:

　　　$_+\underline{AC}+_+\underline{AB}-(_-\underline{DB}+_-\underline{DC})$

i.e.　　　$_+\underline{AC}+_+\underline{AB}-_-\underline{DB}-_-\underline{DC}$
or　　　$(_+\underline{AB}-_-\underline{DB})+(_+\underline{AC}-_-\underline{DC})$
or　　　$(a+c)-(z+b)$; $(c-z)+(a-b)$　　　　etc.

The term $_+\underline{AC}$, then, (for example) is used to describe possible manipulations of a plane (in a nul/los manner).

How, then, can we functionally relate $_+A^xC$, a line, to its plane: $_+\underline{AC}$?

We might impose a limitation to the relationship:

　　　　　$(z+b)/(a+c)$ [d3+d4, rot3+rot4] $(_-\underline{DB}+_-\underline{DC})/(_+\underline{AC}+_+\underline{AB})$
　　　　　　　$_+A^xC$ [d, rot] $_-C^yA$

additive　　$\overline{}$
　　　$_+A^xC+((z+b)/(a+c))$ [d3+d4+d, rot3+rot4+rot] $_-C^yA+((_-\underline{DB}+_-\underline{DC})/(_+\underline{AC}+_+\underline{AB}))$

∴　　　x + ((z+b)/(a+c)) [d3+d4+d, rot3+rot4+rot] y+($(_-\underline{DB}+_-\underline{DC})/(_+\underline{AC}+_+\underline{AB})$)
∴　　　x-y+((z+b)/(a+c)) [...] ($(_-\underline{DB}+_-\underline{DC})/(_+\underline{AC}+_+\underline{AB})$)

Since d, d3, d4, rot, rot3, rot4 are nul/los quantities, y can not be x (in this relationship; i.e. The disposition of the sphere is continuing to change.)

This suggests a fundamental limitation of subtractive properties: the components involved can not be the same.

∴ Subtraction is not the complement of addition. It is, rather, a limitation to addition.

It stands to reason, then, that a sphere's manipulation is related to the intersection of planes (the points thereof) defined with spherical (surface) points involved with these manipulations (or measurements):

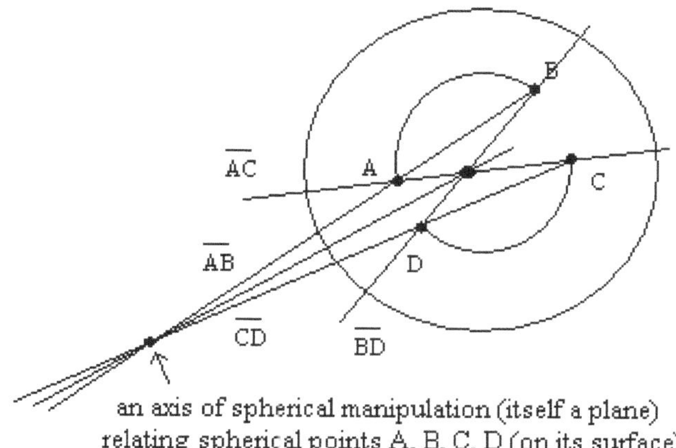

This relationship is defined (as) by:

$$_+\underline{AC}(^z\underline{BD}+^b\underline{CD})+_+\underline{AB}(^z\underline{BD}+^b\underline{CD}) \ [d3+d4, rot3+rot4] \ _-\underline{DB}(^a\underline{CA}+^c\underline{BA})+_-\underline{DC}(^a\underline{CA}+^c\underline{BA})$$

or

$$a(z+b) + c(z+b) \ [d3+d4, rot3+rot4] \ z(a+c) + b(a+c)$$

$$\therefore \ az+ab+cz+bc \ [...] \ az+cz+ab+bc$$

It occurs (to one) that no self-consistent numbering system is worse than another, and nor is it more correct than another.

$\therefore$ Any non-decimal (i.e. Non-zero allowing) base is as utilitarian as a base 10 system; and also, the base 10 system is not of ultimate correctness for calculations and measurement.

$\therefore$ The utility of a number system should be relied upon for its usefulness of physical description and manipulation, as well as for any idealizations envisage-able for perfecting concepts and conceits (of beauty, wholesomeness, encompassing, etc.). The use of a numbering system is a matter of ease for difficulties to complex with solvable endeavors and problems. (i.e. They are used to solve problems.) It is curious, therefore, that the base 10 system is made to uphold all numerical answers, when it may in fact obscure many or most. In all cases, the physical must come before (i.e. be of a higher level than) the math.

The symbolic reduction of the relationship to:

$$az+ab+cz+cb \ [d3+d4, rot3+rot4] \ az+cz+ab+bc$$

can be furthered as:

$$z_+a + b_+a + z_+c + b_+c \ [...] \ a_.z + c_.z + a_.b + c_.b$$

so that we have corresponding pairs that may be grouped (as descriptions):

    z +a
       a -z
          z +c
             c -z
    b +a
       a -b
          b +c
             c -b

to condense (remove redundancies) as:

    z +a -z +c -z
    b +a -b +c -b

and further combined (routed):

    z   -z   -z   ↘ ↗ ↘ ↗
       +a   +c   +a   +c
    b   -b   -b   ↗ ↘ ↗ ↘

This suggests a structural complementarity between the rotations (z, b) and the distances for (a, c), in order to lead to the corresponding distances for (z, b); i.e. a conversion from rotations to distances:

    z ↘ ↗ ↘ ↗ -z   z → -z
       +a   +c   b → -b   via +a , +c
    b ↗ ↘ ↗ ↘ -b

This is, of course, dictated by the functional condensations:

    +a , -b   +c   -z
     a   b   c   z

$\left(\begin{smallmatrix} +a & +c \\ a & c \end{smallmatrix}\right)$ being common to, and $\left(\begin{smallmatrix} -b & -z \\ b & z \end{smallmatrix}\right)$ being particular to, the individualized groupings of pairs.

The routing is itself redundant, as to seemingly make +c superfluous (or alternative):

    z   -z   -z   z   -z   z ↘ ↗ -z
       +a   +c   or $\left(\begin{smallmatrix} +a \\ +c \end{smallmatrix}\right)$ ∴   ( )
    b   -b   -b   b   -b   b ↗ ↘ -b

∴ For example (again consulting the previous relationship statement):

$$^{z}\mathrm{BD} \searrow \quad \overline{\quad} \quad \nearrow {}_{-}\mathrm{DB}$$
$$\left(\begin{array}{c} {}_{+}\mathrm{AC} \\ \overline{\quad} \\ {}_{+}\mathrm{AB} \end{array} \right)$$
$$^{b}\mathrm{CD} \nearrow \quad\quad \searrow {}_{-}\mathrm{DC}$$

Of course: planes <u>BD</u> and <u>CD</u> intersect at the axis common to point D, and <u>AC</u> and <u>AB</u> intersect at the axis which includes point A.

So, a rotation (z) within the plane <u>BD</u>, convoluted with a distance (d3) within the plane <u>AC</u>, yields to a distance (d3) within plane <u>DB</u>.

i.e. ₊<u>AC</u>ᶻ<u>BD</u> [d3, rot3] ₋<u>DB</u>ᵃ<u>CA</u>   etc.

This convolution apparently occurs at/along the intersection axis for planes <u>BD</u> and <u>AC</u>.

The lines or curves (arcs) that connect the points in a plane may be, of course, independent of the surface of the sphere, suggesting that d (d1—4, etc.) of a plane represents an assemblage of possible distances, that may be assembled for possible solutions. Much as 3-dimensional space can have shrinking or growing capacities, while time may only be increasing, then so as electromagnetic currents can have electricities composed of both negative and positive entities (electrons, ions, etc.), while the magnetic component exists only as bipolar (and not mono-polar) entities of attractive and repulsive nature (always a magnet occurring with both a positive and a negative end). The two points that are to be connected (by line or curve) can have a directional character (somewhat vectorial) as by the description of its plane used (for various solutions of problem), while (as for a magnetic line of force) the two points may and must be allowed a (or any) closed path (for simplicity's sake usually taken as circular) to which the two points are members of this circumference. The possibilities of these circumferences are magnanimous for its points. Here again, while starting and ending points may be defined, there is no "zero" point, the circumferences each being continuous within the family of a plane. The relationship statement, then, establishes the circuitousness of the points involved, while the symbolic condensations (here of planes) imposes a directionality for the points (as like an electricity always traveling from positive to negative). They are of course expression statements of the same phenomenon (considered with different points of view for a particular solution).

But what calculations can we make from the comparison of

<pre>
     z   ₋z
       ↘   ↗
        ⎛ ₊a ⎞
        ⎝ ₊c ⎠   vectorial, electric
       ↗   ↘
     b   ₋b
</pre>

to

₊<u>AC</u>ᶻ<u>BD</u> [d3, rot3] ₋<u>DB</u>ᵃ<u>CA</u>   a loop (as magnetic) ?

e.g.
   take   b → (₊a) → ₋z
     ∴ ᵇ<u>CD</u>₊<u>AC</u> [d3, rot3] ₋<u>DB</u>
   take   b → (₊c) → ₋z
     ∴ ᵇ<u>CD</u>₊<u>AB</u> [d3, rot3] ₋<u>DB</u>
   take   b → (₊a) → ₋b
     ∴ ᵇ<u>CD</u>₊<u>AC</u> [d3, rot3] ₋<u>DC</u>

take   b → (₊c) → ₋b

∴   ᵇC̲D̲₊A̲B̲ [d3, rot3] ₋D̲C̲

So now we have an interesting collection of statements (put in original order):

$$|\rightarrow \;_+\text{AC} \;^{\text{b}}\overline{\text{CD}}\;[\; d3\;,\; rot3\;]\; _-\overline{\text{DB}}\quad \leftarrow|$$

$$|\rightarrow\;|\quad _+\text{AB}\;^{\text{b}}\overline{\text{CD}}\;[\; d3\;,\; rot3\;]\; _-\overline{\text{DB}}\quad \leftarrow|$$

$$|\rightarrow\;_+\text{AC}\;^{\text{b}}\overline{\text{CD}}\;[\; d3\;,\; rot3\;]\; _-\overline{\text{DC}}\quad \leftarrow|$$

$$|\rightarrow\quad _+\text{AB}\;^{\text{b}}\overline{\text{CD}}\;[\; d3\;,\; rot3\;]\; _-\overline{\text{DC}}\quad \leftarrow|$$

And we know (from previously):

₊A̲C̲ᶻB̲D̲ [d3, rot3] ₋D̲B̲ᵃC̲A̲
₊A̲B̲ᵇC̲D̲ [d4, rot4] ₋D̲c̲ᶜB̲A̲

And we note that

₊A̲B̲ᵇC̲D̲ [d3, rot3] ₋D̲C̲

while

₊A̲B̲ᵇC̲D̲ [d4, rot4] ₋D̲c̲ᶜB̲A̲

Then clearly a definition for ᶜB̲A̲ is:

₊A̲B̲ᵇC̲D̲ [d3, rot3] ₋D̲C̲ [d4-d3, rot4-rot3] ᶜB̲A̲

or (symbolically)

b → (₊c) → (₋b) → c
b → (₊c → ₋b) → c

How do we handle subtractions to avoid "zero" when, for example, d3 is d4 or rot3 is rot4?

This is (a) physical (operation). ∴ There is no subtraction (nothing physically is done).

∴ We see, with

b → (₊c → ₋b) → c ,

that conversion from one rotation to another may involve distances.

i.e. Conversion of a rotation in one plane (here ᵇC̲D̲) to a rotation in another plane (here ᶜB̲A̲) results from causing a determination of distances between points for both planes (e.g. A to B, C to D). When a particular rotation occurs in one plane, a particular rotation necessarily occurs in the other based on these distances between points on the surface of the sphere. This is based on a relationship established for the two planes (ᵇC̲D̲, ᶜB̲A̲).

So, now we have operations:

$$b \rightarrow (\ _+a\ [...]\ ) \rightarrow \ _-z$$
$$b \rightarrow (\ _+c\ [...]\ ) \rightarrow \ _-z$$
$$b \rightarrow (\ _+a\ [...]\ ) \rightarrow \ _-b$$
$$b \rightarrow (\ _+c\ [...]\ ) \rightarrow \ _-b$$

and

$$b \rightarrow (\ _+c\ [...]\ \rightarrow \ _-b\ [...]\ ) \rightarrow c$$

∴ we can make substitutions (judged for some reasons to be advantageous) for $_-b$ as:

$$b \rightarrow (\ _+c\ [...]\ \rightarrow b \rightarrow (\ _+a\ [...]\ )\ [...]\ ) \rightarrow c$$

and

$$b \rightarrow (\ _+c\ [...]\ \rightarrow b \rightarrow (\ _+c\ [...]\ )\ [...]\ ) \rightarrow c$$

There appears, then, to be a recursion in b (let d4-d3 be d5):

$$b \rightarrow (\ _+c\ [\ d3\ ]\ \rightarrow\ b \rightarrow (\ _+a\ [\ d3\ ]\ )\ [\ d5\ ]\ ) \rightarrow c$$

$$b \rightarrow (\ _+c\ [\ d3\ ]\ \rightarrow\ b \rightarrow (\ _+c\ [\ d3\ ]\ )\ [\ d5\ ]\ ) \rightarrow c$$

Yet, also a recursion in $_+c$:

$$b \rightarrow (\ _+c\ [\ 3\ ]_d \rightarrow b \rightarrow (\ _+c\ [\ 3\ ]_d\ )\ [\ 5\ ]_d\ ) \rightarrow c$$

We also have the evident redundancy of $^{b}\underline{CD}$:

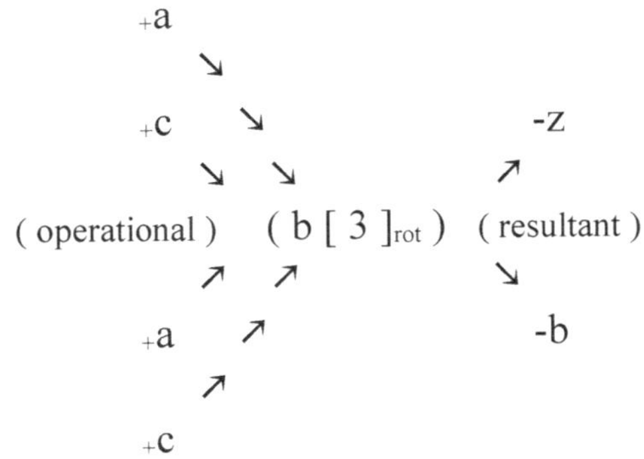

or

$$(_{+}a , _{+}c) \rightarrow (b [3]_{rot})$$

with $.z$ above and $.b$ below.

or

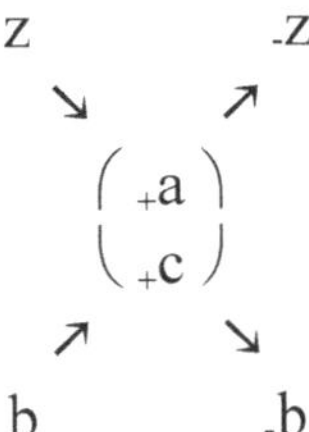

$$\begin{pmatrix} _{+}a \\ _{+}c \end{pmatrix} \rightarrow (b [3]_{rot})$$

with $.z$ above and $.b$ below.

This is the rotational analysis corresponding to the equivalent distance analysis of:

$$\begin{pmatrix} _{+}a \\ _{+}c \end{pmatrix}$$

with z and $.z$ above, b and $.b$ below.

or

$$\begin{pmatrix} _{+}a [3]_{d} \\ _{+}c [4]_{d} \end{pmatrix}$$

with z and $.z$ above, b and $.b$ below.

∴

$$\begin{matrix} & & & & {}_{-}z \\ \left(\begin{matrix} {}_{+}a\,[\,3\,]_{d} \\ | \qquad | \\ {}_{+}c\,[\,4\,]_{d} \end{matrix}\right) & \begin{matrix} \searrow \\ \\ \nearrow \end{matrix} & b\,[\,rot\,]_{3} & \begin{matrix} \nearrow \\ \\ \searrow \end{matrix} & \\ & & & & {}_{-}b \end{matrix}$$

or

$$\begin{matrix} & & & & {}_{-}z \\ \left(\begin{matrix} {}_{+}a\,[\,3\,]_{d} \\ | \qquad | \\ {}_{+}c\,[\,4\,]_{d} \end{matrix}\right) & \begin{matrix} \searrow \\ \\ \nearrow \end{matrix} & b\,[rot]_{3} & \begin{matrix} ⤺ \nearrow \\ \\ ↺ \searrow \end{matrix} & \\ & & & & {}_{-}b \end{matrix}$$

(a compound operation)

and

$$\begin{matrix} z & & {}_{-}z \\ \searrow & & \nearrow \\ & \left(\begin{matrix} {}_{+}a\,[\,3\,]_{d} \\ {}_{+}c\,[\,4\,]_{d} \end{matrix}\right) & \\ \nearrow & & \searrow \\ b & & {}_{-}b \end{matrix}$$

∴ clearly

$$\begin{matrix} z & & & & {}_{-}z \\ \searrow & \left(\begin{matrix} {}_{+}a\,[\,3\,]_{d} \\ | \qquad | \\ {}_{+}c\,[\,4\,]_{d} \end{matrix}\right. & |\,b\,[rot]_{3} & \begin{matrix} ⤺ \nearrow \\ \\ ↺ \searrow \end{matrix} & \\ \nearrow & & & & \\ b & & & & {}_{-}b \end{matrix}$$

vectorial (e.g. electrical) — locping (e.g. magnetic) — resultant (e.g. electrical)

as $\left(\begin{matrix} {}_{+}a[3]_{d} \\ | \qquad | \\ {}_{+}c[4]_{d} \end{matrix}\right)$ ⤺ $b[3]_{rot}$ ↺ can have equivalence with $\left(\begin{matrix} {}_{+}a[3]_{d} \\ | \qquad | \\ {}_{+}c[4]_{d} \end{matrix}\right)$

(apparently without a locping caused by ⤺ $b[3]_{rot}$ ↺)

$$\begin{matrix} z & & & & {}_{-}z \\ \searrow & \left(\begin{matrix} {}_{+}a\,[3]_{d} \\ | \qquad | \\ {}_{+}c\,[4]_{d} \end{matrix}\right. & |\,b\,[3]_{rot} & \begin{matrix} ⤺ \nearrow \\ \\ ↺ \searrow \end{matrix} & \\ \nearrow & & & & \\ b & & & & {}_{-}b \end{matrix}$$

(to keep the nomenclatures consistent)

The looping operation:

$(_+c[4]_d)b[3]_{rot}$

could be a function related to the intersection of the planes ^cBA and ^bCD

and that of

$(_+a[3]_d)b[3]_{rot}$

between the planes of ^aCA and ^bCD
(which occurs with the common inclusion of point C, on the sphere)

∴ An interplanar angle is derived (for each):

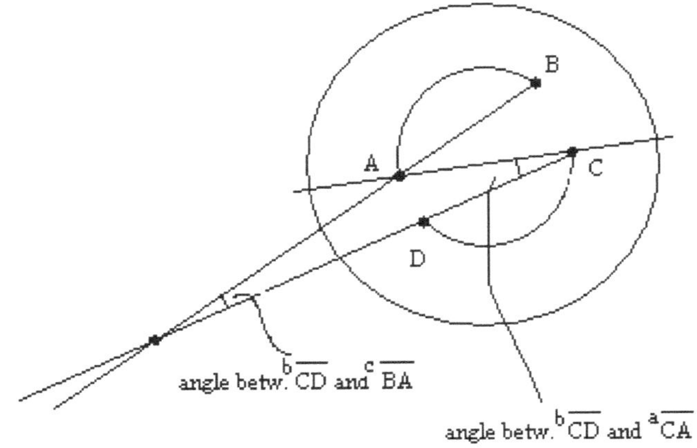

angles (measured) inclusive of the triangle formed by the three planes

The redundancy of ^bCD becomes apparent as a shared element of the relationships described or picked
(i.e. due to $b[3]_{rot}$).

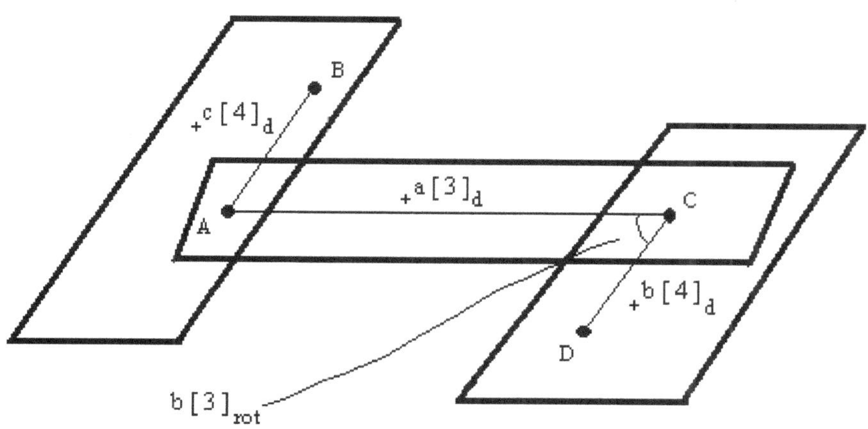

($\overline{AC}$ plane drawn with 2 sides // to line AC)

∴

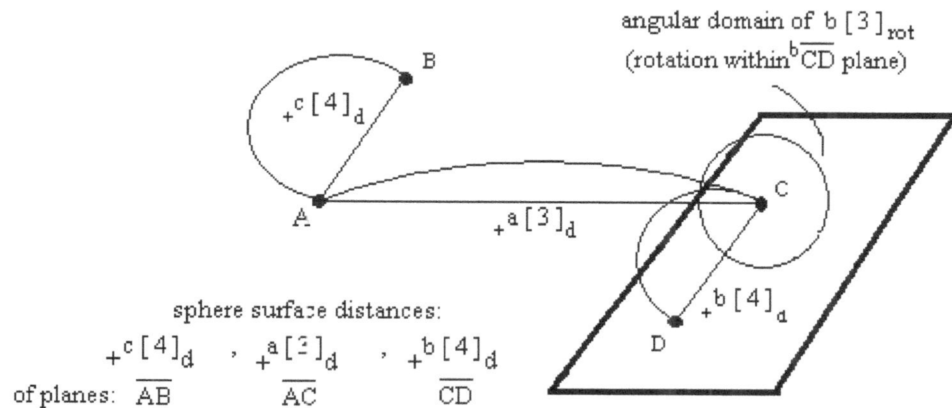

b[3]rot is a rotation of interplanar specificity: formed within the $^b\underline{CD}$ plane, but with components of the $^a\underline{AC}$ plane.

∴ necessarily fixed at point C for a convenient point of commonality here (for this relationship problem); i.e., b[3]rot draws point A to point C and point B to original position of point A, their repositionings occurring along the sphere's surface.

This would functionally seem to require:

1) a rotation about point C of (let us use lines here) line AC (within plane $^b\overline{CD}$) with its extension line BA

2) a translation to bring point A coincident with current point C and (rotated) point B coincident with former position of point A (series of translations)
∴

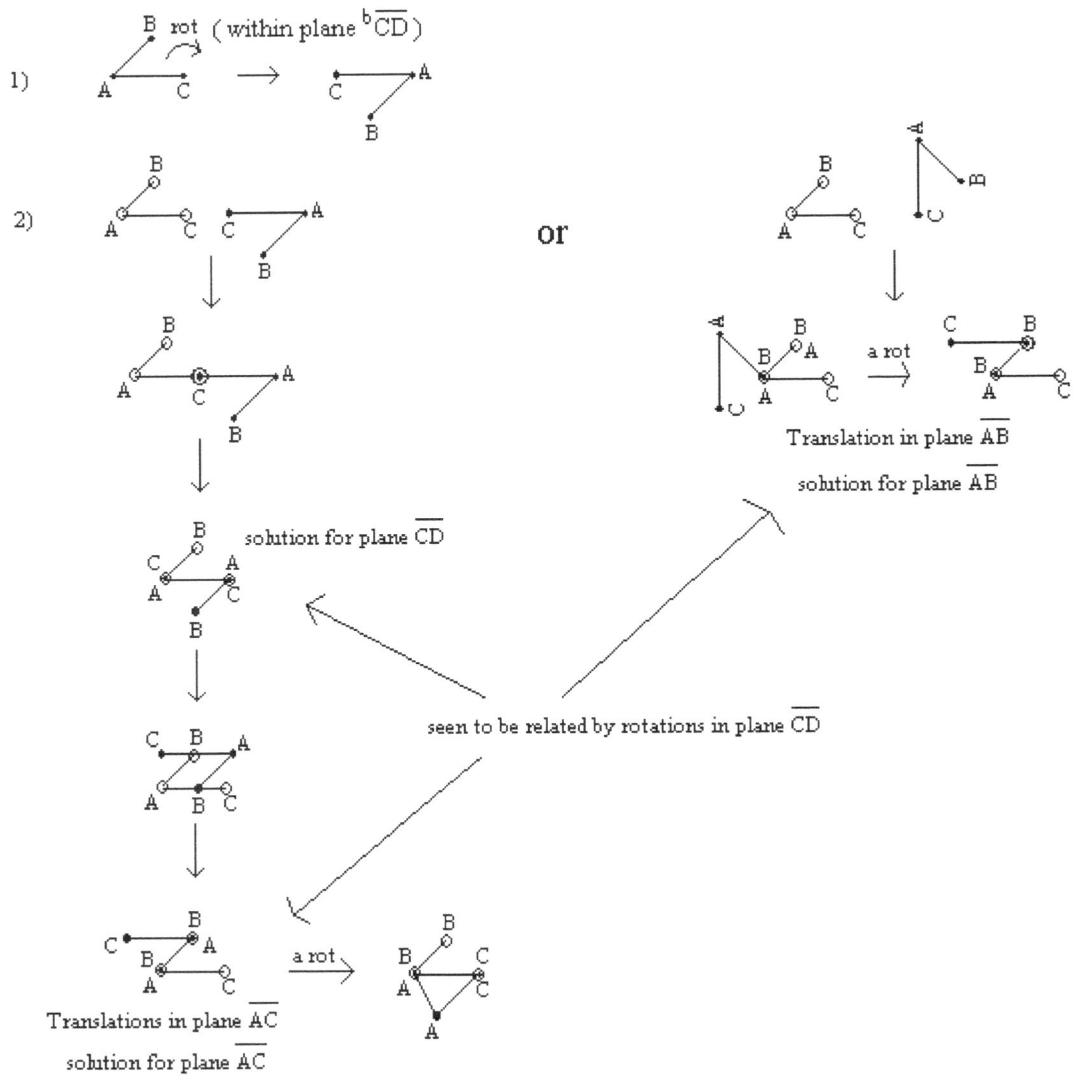

But what of these translations/rotations are physically possible (without extension or shortening of the distances established between the points A, B, C) ? [Projections onto Planes]

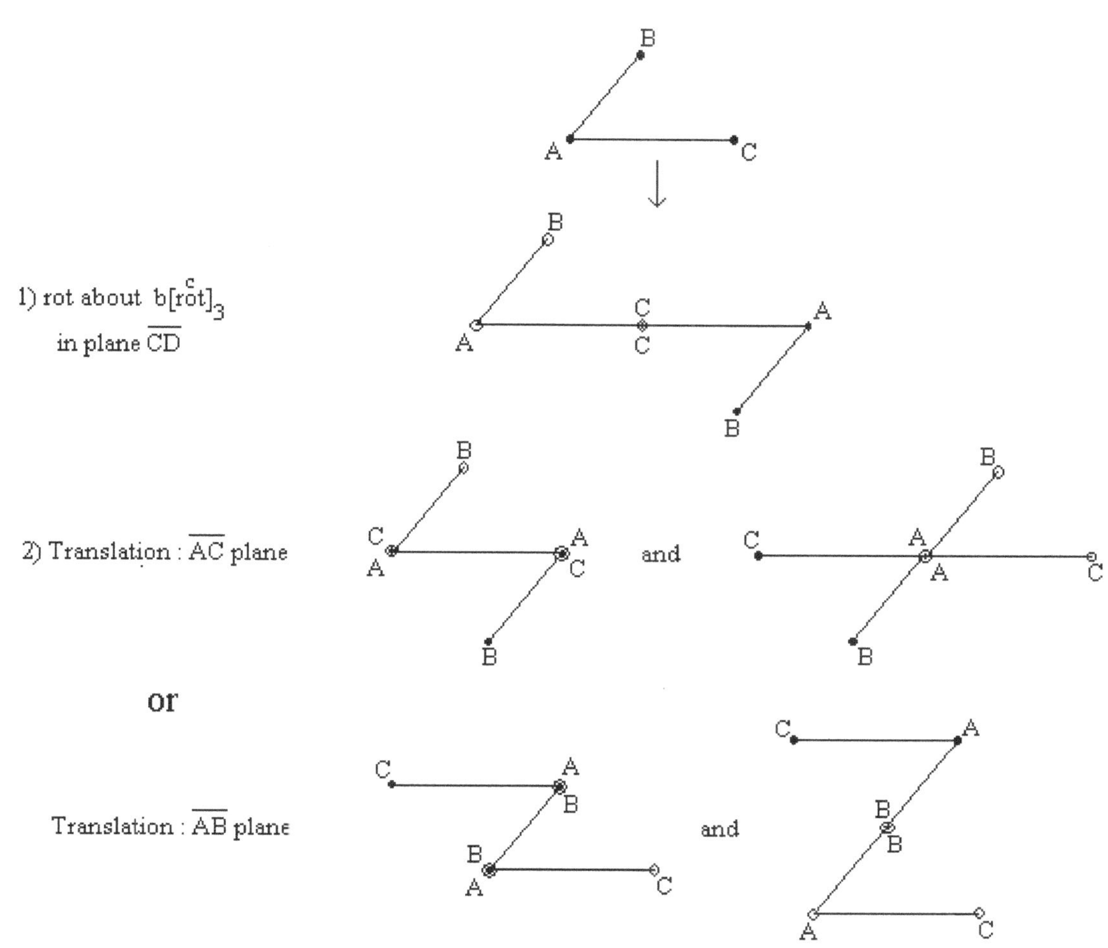

1) rot about $b[\overset{c}{\text{rot}}]_3$

   in plane $\overline{CD}$

2) Translation : $\overline{AC}$ plane   and

or

Translation : $\overline{AB}$ plane   and

One can also have the parallelogram:

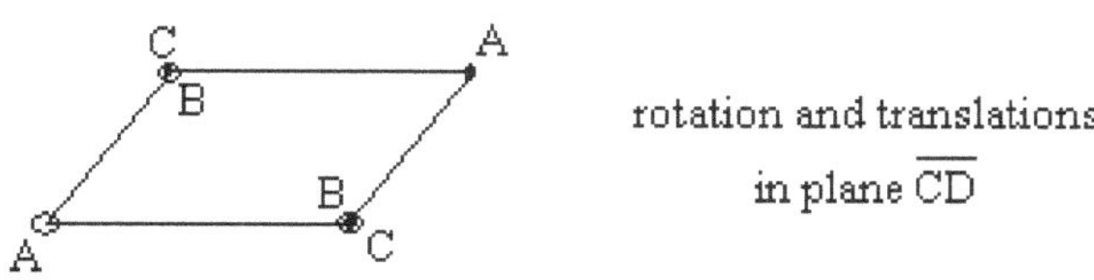

rotation and translations
in plane $\overline{CD}$

Of these solutions:

one brings A to C	one brings A to A
one brings A to B	one brings B to B
one brings B to C	one brings C to C

These six solutions must be encompassed by the nul*l*los relation statement:

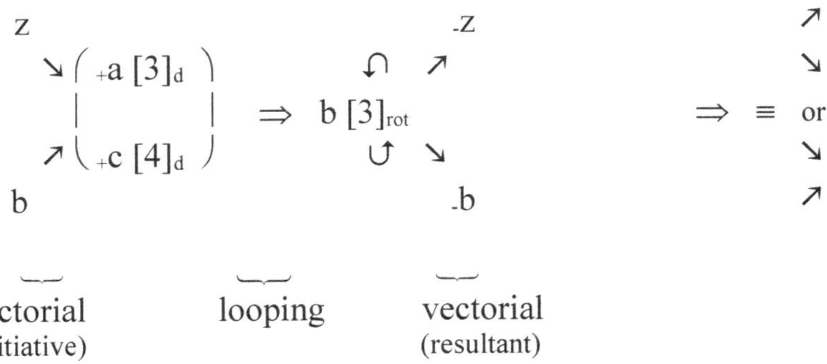

vectorial	looping	vectorial
(initiative)		(resultant)

Operations (from the relation statement):

$z \rightarrow {}_+a{\cdot}b \rightarrow {}_-z$	$b \rightarrow {}_+a{\cdot}b \rightarrow {}_-z$
$z \rightarrow {}_+a{\cdot}b \rightarrow {}_-b$	$b \rightarrow {}_+a{\cdot}b \rightarrow {}_-b$
$z \rightarrow {}_+c{\cdot}b \rightarrow {}_-z$	$b \rightarrow {}_+c{\cdot}b \rightarrow {}_-z$
$z \rightarrow {}_+c{\cdot}b \rightarrow {}_-b$	$b \rightarrow {}_+c{\cdot}b \rightarrow {}_-b$

z is itself a [3]$_{rot}$ (of <u>BD</u> plane), while b is a [4]$_{rot}$ (of <u>CD</u> plane).

∴ Let us analyze (or follow) some of these derived operations:

$z \rightarrow {}_+a{\cdot}b \rightarrow {}_-z$

A 'z' rotation in the <u>BD</u> plane, when yielding to an $_+a$ translation in the <u>AC</u> plane coupled to a 'b' rotation in the <u>CD</u> plane (but of [3]$_{rot}$ calibre or projection onto the <u>BD</u> plane) should result in a $_-z$ translation (in the <u>DB</u> plane). The $_+a$ and $_-z$ translations are both [3]$_d$.

^{z}<u>BD</u> brings point B to point D (on the sphere's surface)
$^a_+$<u>AC</u> uses [3]$_d$ to bring A to C
b[3]$_{rot}$ brings C to D as projected on <u>BD</u> plane
$^z_-$<u>DB</u> uses [3]$_d$ to bring D to B

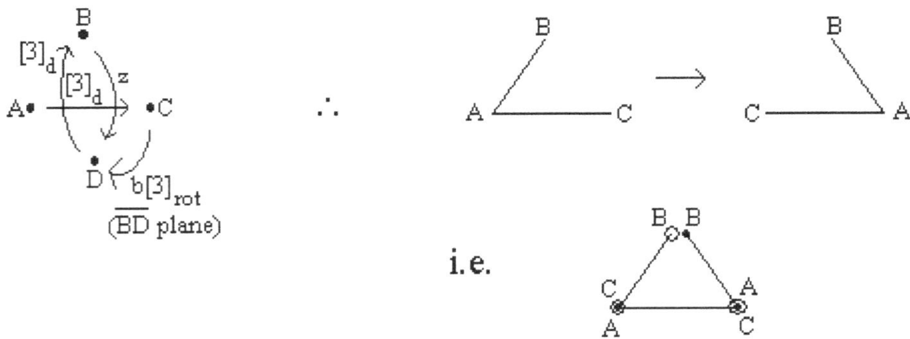

i.e. Pivoting on point B (to keep constant) while exchanging C for A (but of <u>BD</u> plane projection).

The corresponding B to B solution is:

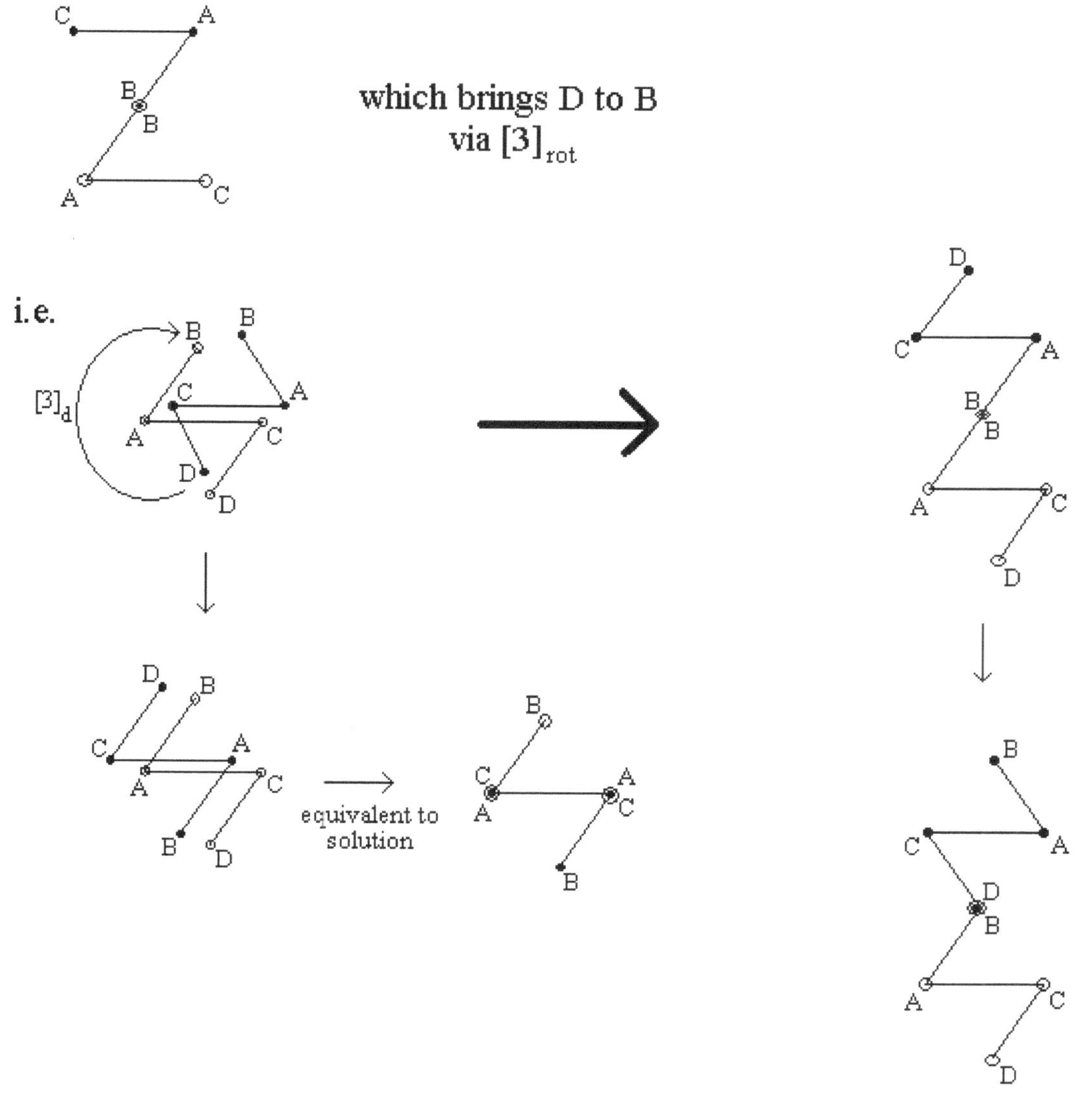

This solution brings both A to C and B to D (and seems equivalent to rotating in plane AC about center of mass of line AC).

b → +a·b → z
1) ^{b}CD brings point C to point D via [4]rot
2) $^a_+$AC uses [3]d to bring A to C
3) b[3]rot brings C to D as projected on BD plane
4) $^z_-$DB uses [3]d to bring D to B

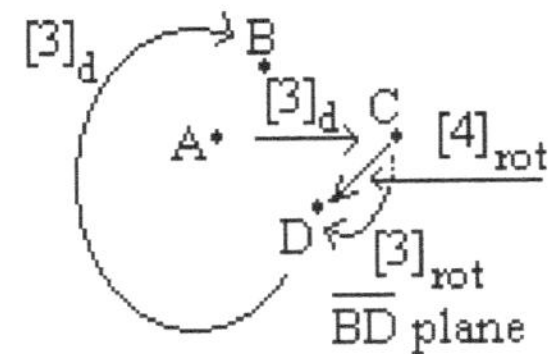
[3]_d
B
A
[3]_d
C
[4]_rot
D
[3]_rot
BD plane

∴

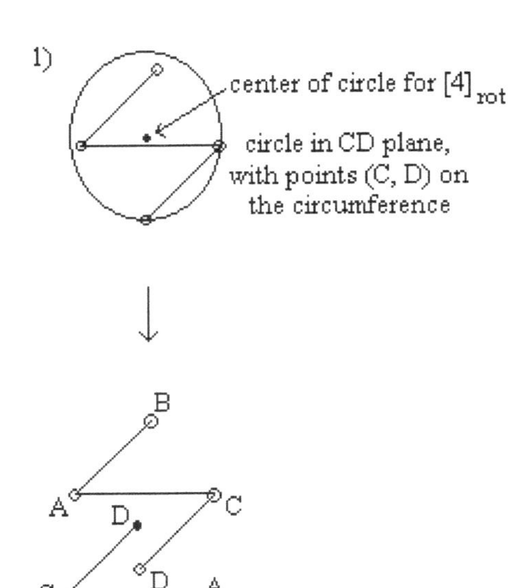
1)
center of circle for [4]_rot
circle in CD plane,
with points (C, D) on
the circumference
B
A
D
C
C
D
A
B

But it might also be possible that from (3), the $[3]_d$ displacements simply revert to a solution back to (2), since $[3]_d$ is an "arcing" distance on the surface of the sphere: i.e. an effective rotation.

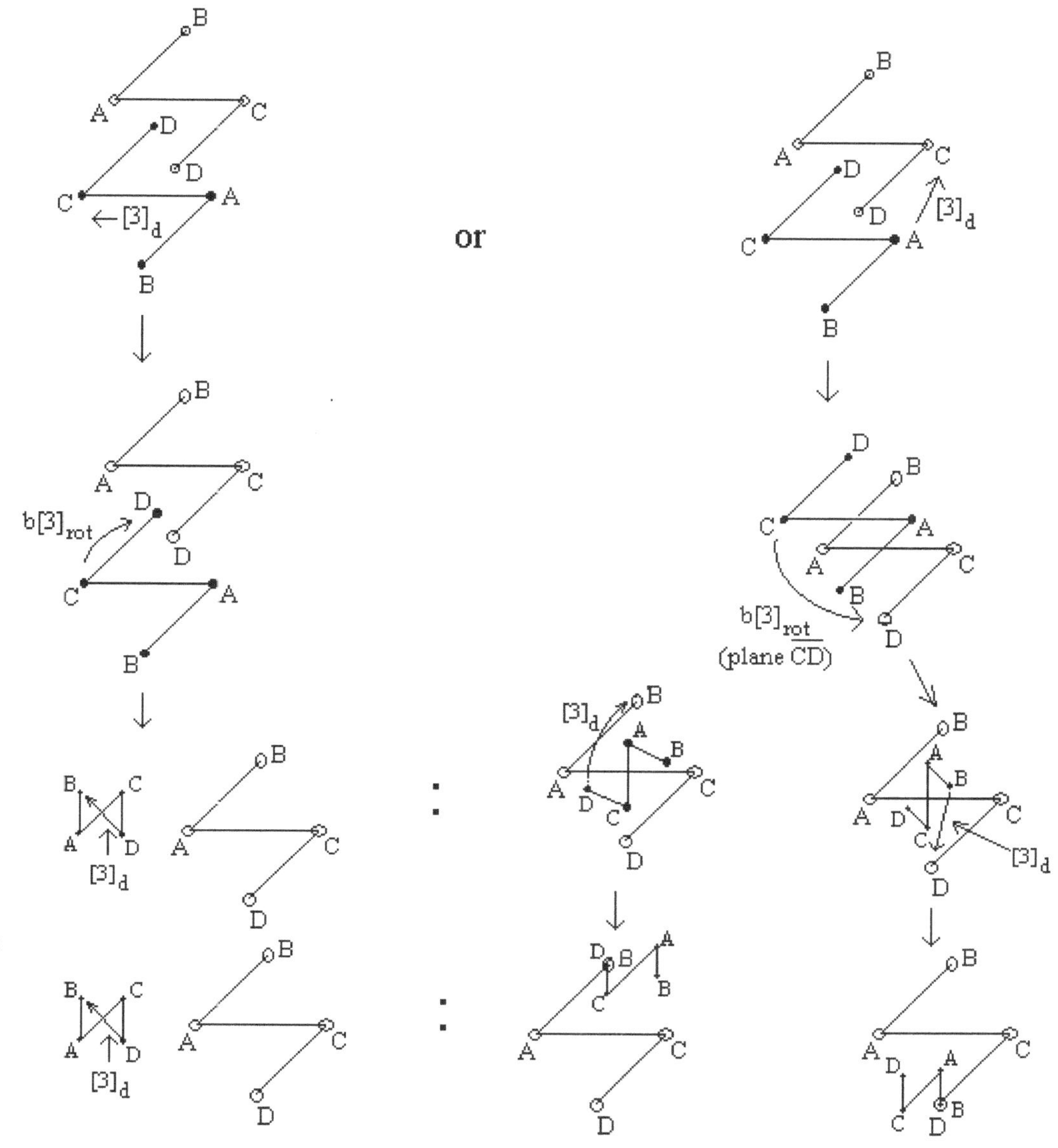

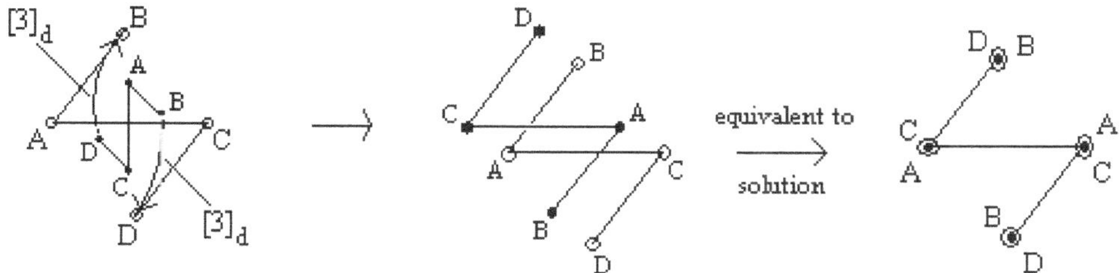

This is a same solution as for $z \rightarrow {}_{+}a{\cdot}b \rightarrow {}_{-}z$

z → ₊c·b → ₋z

1) $^z\underline{BD}$ brings point B to point D
2) $^c_+\underline{AB}$ uses $[4]_d$ to bring A to B
3) b$[3]_{rot}$ brings C to D as projected on $\underline{BD}$ plane
4) $^z_-\underline{DB}$ uses $[3]_d$ to bring D to B

1)

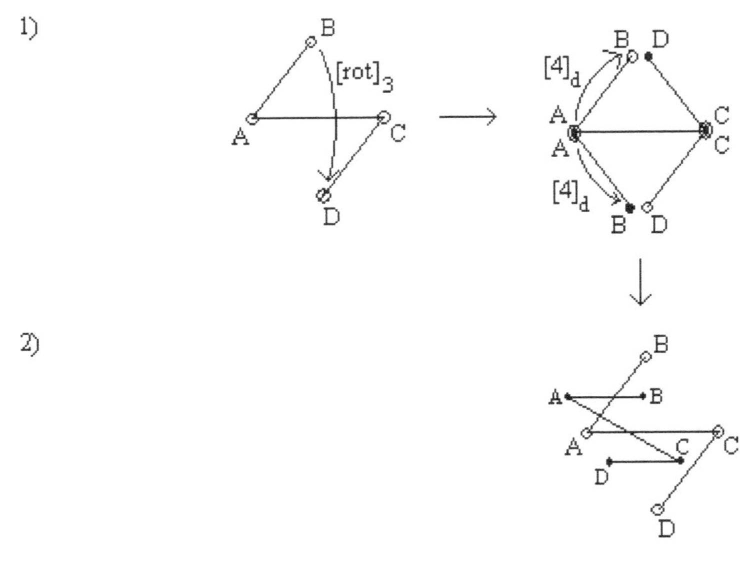

2)

or

on $\overline{CD}$ plane

3)

4)

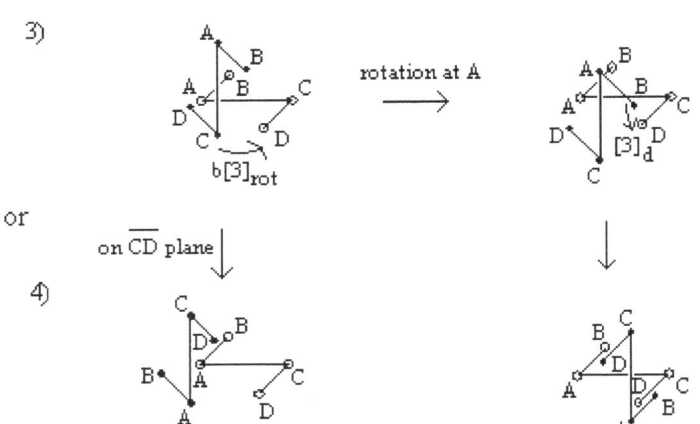

z → $_+$c·b → $_-$b
For part (4): $[4]_d$ brings D to C

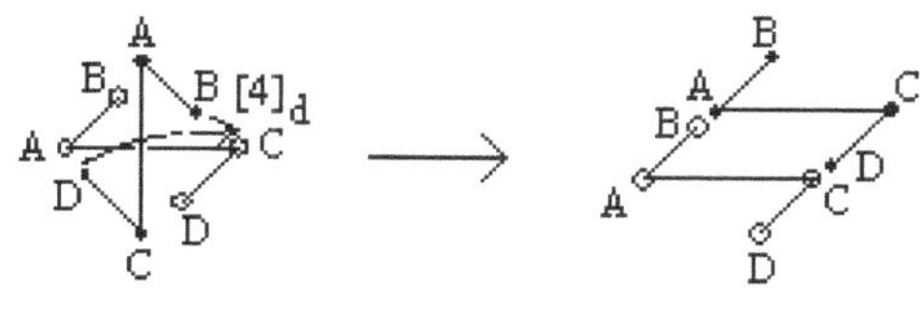

or

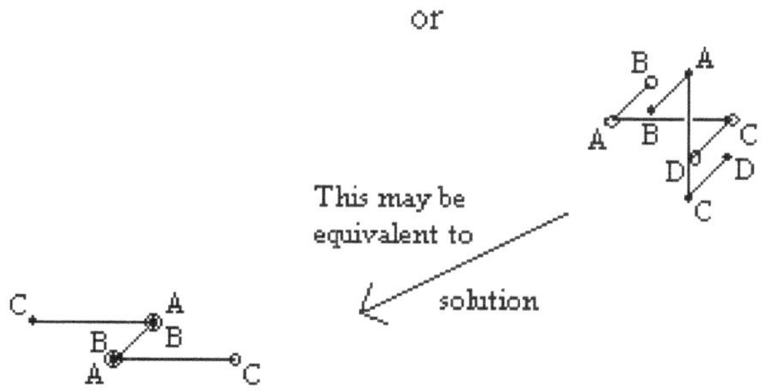

This may be
equivalent to

solution

z → $_+$a·b → $_-$b
1) ^{z}<u>BD</u> brings B to D
2) $^a_+$<u>AC</u> uses $[3]_d$ to bring A to C
3) b$[3]_{rot}$ brings C to D (<u>BD</u> plane)
4) $^b_-$<u>DC</u> uses $[4]_d$ to bring D to C

1)

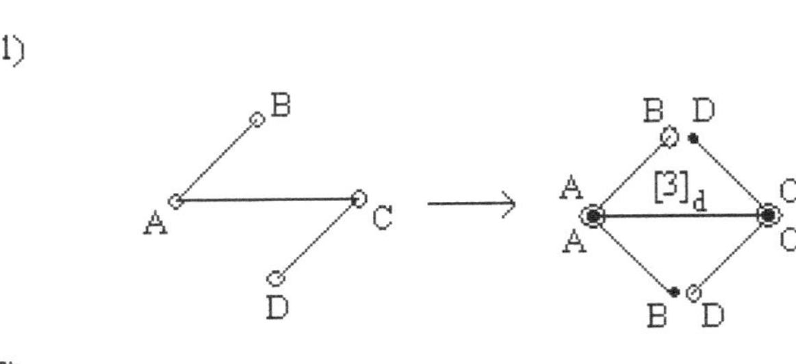

2)

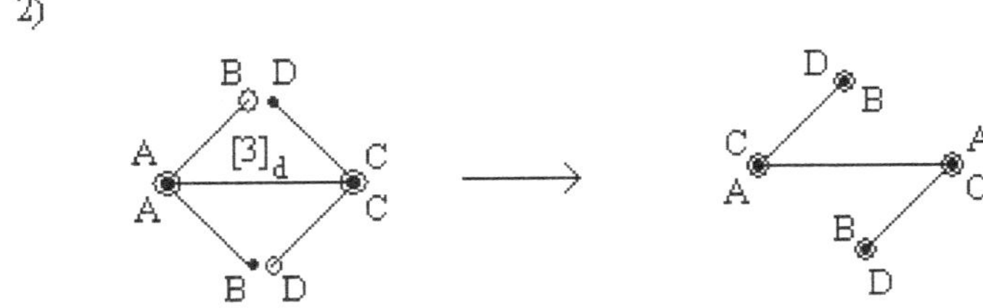

3)

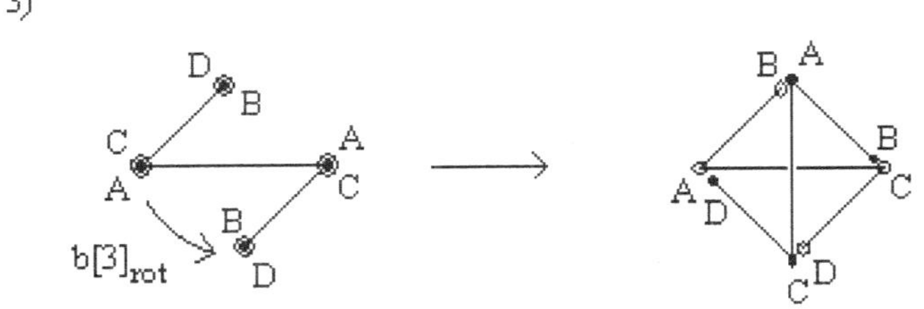

4)

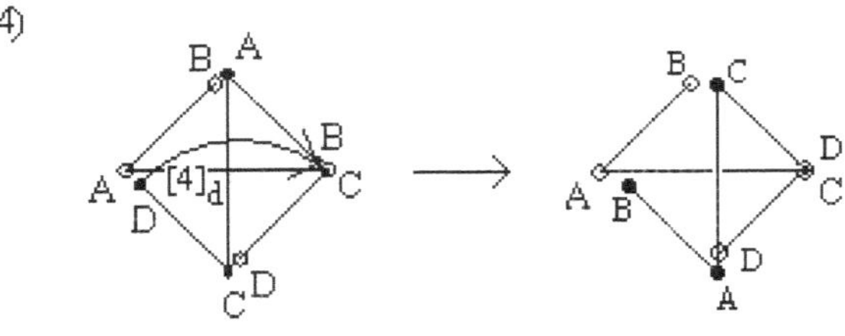

For $\mathbf{z} \to {}_{+}\mathbf{a}\cdot\mathbf{b} \to \mathbf{z}$

4) ${}^{z}\underline{DB}$ uses $[3]_d$ to bring D to B

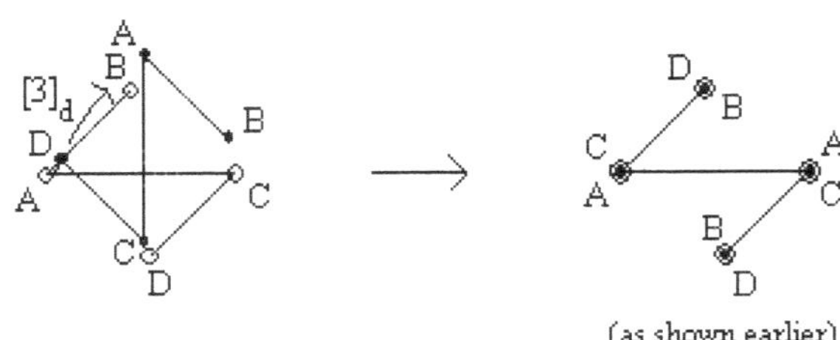

(as shown earlier)

∴ From the nul/los relation statement (of or for a sphere which has no "zero" point) we have derived several solutions for points concurring to one another.
i.e.

allows for several relationships to be derived (or uncovered) on the surface of the sphere. The nature of these procedures would appear to confer as many coincidences of points as achievable to be representative of useful (or instructionally valuable) solutions, all of this derived simply from the argument (or proto-argument): b → ($_+$c → $_-$b) → c

Hence, for example (as demonstrated):

z → $_+$a ▪ b → $_-$z
b → $_+$a ▪ b → $_-$z

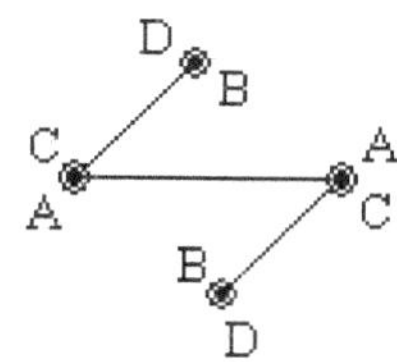

z → $_+$c ▪ b → $_-$z

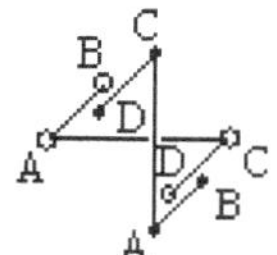

z → $_+$c ▪ b → $_-$b

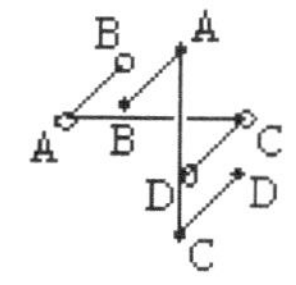

z → $_+$a ▪ b → $_-$b

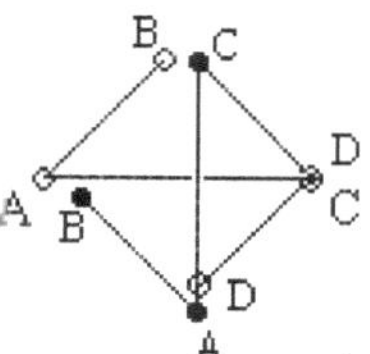

It's clear from the nature of these solutions that

(z,b) → ₊a·b → ‑z and z → ₊c·b → ‑b

are most complete for coincidences of points
(i.e., fulfilling the traverses of all the points involved: A,B,C, D):

        B → D twice, A → C twice
and   A → B twice, C → D twice

while

z → ₊c·b → ‑z and z → ₊a·b → ‑b

are only partially successful (of achievement):

        B → D twice
and   A → B once, C → D once

What's more, the latter (z → ₊a·b → ‑b) appears to show a disparity among pairs:

i.e.   C → D occurs for one (C, D) pair, but not the other
        A → B occurs for one (A, B) pair but not the other

∴ The latter two confer more "open" spacial relationships among solutions (less compact as fewer pairs are satisfied). They form contrasting planes of points; the former, parallel.

There are mixtures or degrees of this (parallel to contrasting planes) concurring for these particular solutions as follows:

most parallel   (z,b) → ₊a·b → ‑z
some parallel   z → ₊c·b → ‑b (less "open")
parallel,
(some) contrasting   z → ₊a·b → ‑b (more "open")
most contrasting   z → ₊c·b → ‑z

(z → ₊c·b → ‑b) fulfills most completely the original problem of bringing A to B and C to D. It maintains the original planes of ᶜ<u>AB</u> and ᵇ<u>CD</u>, though contrasting the original to new ᵃ<u>AC</u> and ᶻ<u>BD</u> planes.

∴ (of the group sited) It is the only operation which retains planes after the initial z rotation:

z → ₊c·b → ‑b   : planes retained; and A → B, C → D
(b,)z → ₊a·b → ‑z or ‑b : planes not retained (i.e. translated) **or** A ↛ B **or** C ↛ D
z → ₊c·b → ‑z   : planes not retained; but B → D

In this sense we can (arbitrarily) rank them:

sum + ( )

$$1 \quad\quad 1\,1 \quad\quad 1$$
$$z \to {}_{+}c{\cdot}b \to .b \quad\quad 4$$

$$1 \quad\quad 1\,\tfrac{1}{2} \quad\quad 1$$
$$z \to {}_{+}a{\cdot}b \to .z \quad\quad 3\,\tfrac{1}{2}$$

$$\tfrac{1}{2} \quad\quad 1\,\tfrac{1}{2} \quad\quad 1$$
$$b \to {}_{+}a{\cdot}b \to .z \quad\quad 3$$

$$(\tfrac{1}{4}) \quad \tfrac{1}{4}\,(\tfrac{1}{4})\,(\tfrac{1}{4})$$
$$z \to {}_{+}a{\cdot}b \to .b \quad\quad \tfrac{1}{4} + (\tfrac{3}{4}) = 1$$

$$1 \quad\quad\quad\quad 1$$
$$z \to {}_{+}c{\cdot}b \to .z \quad\quad 2$$

For this ranking:

$1 \equiv$ planes maintained as identical (in location)
$\tfrac{1}{2} \equiv$ planes maintained of direction, but translated for the figure
$\tfrac{1}{4} \equiv$ planes contrasted within the figure (apparently about the center of mass for the figure)

and

$(\tfrac{1}{4}) \equiv$ planes contrasted through formation of a right angle (a type of inversion we may call a conversion) within the figure.

Here, we seem to be looking for comparisons of original and altered planes with upholding of a symmetry about the center of mass of the figure (where a figure $\equiv$ comparison before and after location of points). In fact, all of the (presented) figures show considerable symmetry, so we are restrained to define an inversion as occurring at the center of mass. Plane inversions (e.g. For $z\to{}_{+}c{\cdot}b\to.z$) also occur away from the center of mass: ${}^{c}\underline{AB}$ to ${}^{b}\underline{CD}$.

The ranking values are arbitrary, but indicate a generalized trend towards agreement of planes before and after (for fulfilling $b\to({}_{+}c\to.b)\to c$) .
It is noted that a fixed inverse of the result for $z\to{}_{+}c{\cdot}b\to.z$ (the B$\to$D coincidences fixed) leads to the result for $z\to{}_{+}a{\cdot}b\to.z$. Similarly, $z\to{}_{+}a{\cdot}b\to.b$ leads to $z\to{}_{+}c{\cdot}b\to.b$ (the B$\to$A and D$\to$C coincidences fixed).

$$\therefore \quad z \to {}_{+}c{\cdot}b \to .b$$
$$\quad\quad \uparrow\downarrow \text{ inversion (a torsion from one inversion contrast to another)}$$
$$\quad z \to {}_{+}a{\cdot}b \to .b$$

and

$$z \to {}_{+}a{\cdot}b \to .z$$
$$\quad\quad \uparrow\downarrow \text{ inversion (more like a torsion for or against an inversion contrast)}$$
$$z \to {}_{+}c{\cdot}b \to .z$$

33

In both cases the $^a\underline{AC}$ plane is involved.
Results seem to cycle (or flow) as:

<u>rank</u>

$3\,\tfrac{1}{2}$ $z \rightarrow {}_+a{\cdot}b \rightarrow {}_-z$
$\downarrow$

2 $z \rightarrow {}_+c{\cdot}b \rightarrow {}_-z$
$\downarrow \qquad \downarrow$ de-fixing and rotated inversion

1 $z \rightarrow {}_+a{\cdot}b \rightarrow {}_-b$
$\downarrow$

4 $z \rightarrow {}_+c{\cdot}b \rightarrow {}_-b$
$\downarrow \qquad \downarrow$ de-fixing and rotated inversion

$(\,3\,\tfrac{1}{2}\quad z \rightarrow {}_+a{\cdot}b \rightarrow {}_-z\,)$

How can we "devolve" this preferred solution : $z \rightarrow {}_+c{\cdot}b \rightarrow {}_-b$?

In summary, we have:

$${}_+A({}^xB + {}^{x2}D) + {}_+C({}^xB + {}^{x2}D)\,[...]\,{}_-B({}^yA + {}^{y2}C) + {}_-D({}^yA + {}^{y2}C) \qquad [...] \equiv [d+d2,\, rot + rot2]$$

Going to planes for points, we have:

$${}_+\underline{AB}^b\underline{CD}\,[d3,\, rot3]\,{}_-\underline{DC}\,[d4-d3,\, rot4-rot3]\,{}^c\underline{BA}$$

as a definition for $^c\underline{BA}$ (i.e. a functional of systematic cause for performing $^c\underline{BA}$),
or (symbolically)

$$b \rightarrow (\,{}_+c \rightarrow {}_-b\,) \rightarrow c$$

Substitutions (for operations) lead to the generalized statement:

$$z \qquad\qquad\qquad {}_-z$$
$$\searrow \left(\begin{array}{c} {}_+a\,[3]_d \\[2pt] {}_+c\,[4]_d \end{array} \right)\,b\,[3]_{rot} \qquad \nearrow$$
$$\nearrow \qquad\qquad\qquad \searrow$$
$$b \qquad\qquad\qquad {}_-b$$

From this relation(ship) statement the solution:

$$z \rightarrow {}_+c{\cdot}b \rightarrow {}_-b$$

is derived and (of those derived) found to be most desired (highest of rank).

$$\therefore\ {}^z\underline{BD} \rightarrow {}^{(d4)}{}_+\underline{AB}^b\underline{CD}\,[3]_{rot} \rightarrow {}^b{}_-\underline{CD}$$

for points A→B, C→D on the surface of the sphere.

These relationship statements are read or interpreted as:

A defined distance of travel of a given point to another (point of location), which may involve a rotation within a specified plane, or a rotation within a plane (to move the point to another) expressly stated, is carried out, to be followed by the next directive, and so on, until the last is accomplished. The result of the series of operations is analyzed (structurally) to examine and estimate or determine its preferableness for the originating objective (here: A→B, C→D). It stands to reason that the defined distances (d) and plane (within plane) rotations (a,b,c,x,y,z) are calculable. Yet, symbolically, this can be avoided when only selecting for the best or most preferred solution.

∴ The means of solution is separated (or is separable) from its calculations. The "zero" digit is not allowed for either, as a way of judging errors that may be performed (committed) as avoidable or easily recognized (not that they necessarily lead to a wrong solution, but their occurrence might indicate a wrong method of logic during the operations performed). Restricting calculations to a "zero-less" number system basis aids in constructing solutions.

The relationship statement:

....[...]....

generally prescribes a proposed equivalency of expressions, although not necessarily an equality between them, as to relate that one may lead to the other given their various functions.

The relationship statement for $z\to_+c\cdot b\to_-b$ (derives basically from, as to) is:

$$_+\underline{AC}(^z\underline{BD} + {}^b\underline{CD}) + {}_+\underline{AB}(^z\underline{BD} + {}^b\underline{CD})\ [d3+d4,\ rot3+rot4]\ _-\underline{DB}(^a\underline{CA} + {}^c\underline{BA}) + {}_-\underline{DC}(^a\underline{CA} + {}^c\underline{BA})$$

∴ $z \to +c\cdot b \to -b$

seems to relate most clearly to the term:

$$^c{}_+\underline{AB}(^z\underline{BD} + {}^b\underline{CD})$$
$$\to \quad {}^c{}_+\underline{AB}{}^z\underline{BD} + {}^c{}_+\underline{AB}{}^b\underline{CD}$$

$$_+c\cdot b == {}_+c\cdot b[3]_{rot}$$

is a modified:

$$^c{}_+\underline{AB}{}^b\underline{CD}$$
$$\to {}_+c\cdot b[4]_{rot}$$

It is interesting that 4 distinct solutions result for a relation statement of 4 terms. In fact:

(b,)$z \to {}_+a\cdot b \to {}_-z$
(b,)$z \to {}_+a\cdot b \to {}_-b$ relate to term $^a{}_+\underline{AC}(^z\underline{BD} + {}^b\underline{CD})$ for (A→C, C→D), (B→D) .
(b,)$z \to {}_+c\cdot b \to {}_-z$
(b,)$z \to {}_+c\cdot b \to {}_-b$ relate to term $^c{}_+\underline{AB}(^z\underline{BD} + {}^b\underline{CD})$ for (A→B, C→D), (B→D) .

We may, perhaps, similarly predict (based on symbolic similarities):

(c,)$a \to {}_-z\cdot c \to {}_+a$
(c,)$a \to {}_-z\cdot c \to {}_+c$ relating to $^z{}_-\underline{DB}(^a\underline{CA} + {}^c\underline{BA})$ for (D→B, B→A), (C→A) .
(c,)$a \to {}_-b\cdot c \to {}_+a$
(c,)$a \to {}_-b\cdot c \to {}_+c$ relating to $^b{}_-\underline{DC}(^a\underline{CA} + {}^c\underline{BA})$ for (D→C, B→A), (C→A) .

Here we must be using $c[3]_{rot}$ for the redundant $^c\underline{BA}$.

$[3]_{rot}$ for c$[3]_{rot}$ (i.e. z·c, b·c) comes from (C→A) ∴ $^c\underline{BA}$ plane,

just as

$[3]_{rot}$ for b$[3]_{rot}$ (i.e. a·b, c·b) comes from (B→D) ∴ $^z\underline{BD}$ plane.

(c, b) are normally $[4]_{rot}$.

So, following these nul/los symbolic relationships, much progress is made in understanding and deriving desired point movement solutions (on the surface of the nul/los sphere), all without calculation of actual (or representative) numerical values.

We can assume, then, (by analogy with z→$_+$c·b→$_-$b) that the preferred solution for

(D→C, B→A)

is

a → $_-$b·c → $_+$c

Further, by systematic analogies:

z → $_+$a·b → $_-$b   is best for A→C, C→D
a → $_-$z·c → $_+$c   is best for D→B, B→A

The systematic solutions suggest (for best solutions):

1) We start in one plane, but end up in another
2) One plane (transitory) adopts the rotation of another
3) The planes involved are all evidenced by the particular terms of the founding relationship statement
4) The planes in parenthesis (in the relationship statement) share a common point (and thus a common line) per term. Since there are 4 points involved (A,B,C,D), there are 4 common lines (each formed by intersection of 2 planes). A common line is shared by 2 pairs of (convergence) planes.
(This particular relation statement avoids 2 planes: $\underline{AD}$, $\underline{BC}$.)

∴ An interesting operation table (or "super matrix") is provided for, comparing 'best solution' to 'points (lines) for' to 'not best solution':

best	lines	not best
z $_+$a·b $_-$b	AC CD BD	z $_+$a·b $_-$z
z $_+$c·b $_-$b	AB CD BD	z $_+$c·b $_-$z
a $_-$z·c $_+$c	DB BA CA	a $_-$z·c $_+$a
a $_-$b·c $_+$c	DC BA CA	a $_-$b·c $_+$a

This seems "reducible" to:

$$z \begin{pmatrix} +a \\ +c \end{pmatrix} \bullet b \quad {}_-b \qquad \Bigg| \qquad \begin{pmatrix} AC \\ AB \end{pmatrix} \quad CD \quad BD \qquad \Bigg| \qquad z \begin{pmatrix} +a \\ +c \end{pmatrix} \bullet b \quad {}_-z$$

$$a \begin{pmatrix} -z \\ -b \end{pmatrix} \bullet c \quad {}_+c \qquad \Bigg| \qquad \begin{pmatrix} DB \\ DC \end{pmatrix} \quad BA \quad CA \qquad \Bigg| \qquad a \begin{pmatrix} -z \\ -b \end{pmatrix} \bullet c \quad {}_+a$$

and further to:

$$z \begin{pmatrix} +a \\ +c \end{pmatrix} \bullet b \quad {}_-b \qquad \Bigg| \qquad A \begin{pmatrix} C \\ B \end{pmatrix} \quad (C \ B)D \qquad \Bigg| \qquad z \begin{pmatrix} +a \\ +c \end{pmatrix} \bullet b \quad {}_-z$$

$$a \begin{pmatrix} -z \\ -b \end{pmatrix} \bullet c \quad {}_+c \qquad \Bigg| \qquad D \begin{pmatrix} B \\ C \end{pmatrix} \quad (B \ C)A \qquad \Bigg| \qquad a \begin{pmatrix} -z \\ -b \end{pmatrix} \bullet c \quad {}_+a$$

And (this is) thus compacted to:

$$z \begin{pmatrix} +a \\ +c \end{pmatrix} \cdot b \begin{pmatrix} -b \\ -z \end{pmatrix} \qquad \Bigg| \qquad A \begin{pmatrix} C \\ B \end{pmatrix} \quad (C \quad B)D$$

$$a \begin{pmatrix} -z \\ -b \end{pmatrix} \cdot c \begin{pmatrix} +c \\ +a \end{pmatrix} \qquad \Bigg| \qquad D \begin{pmatrix} B \\ C \end{pmatrix} \quad (B \quad C)A$$

planes   points

There are some evident visual correlations (symbolically) noted:

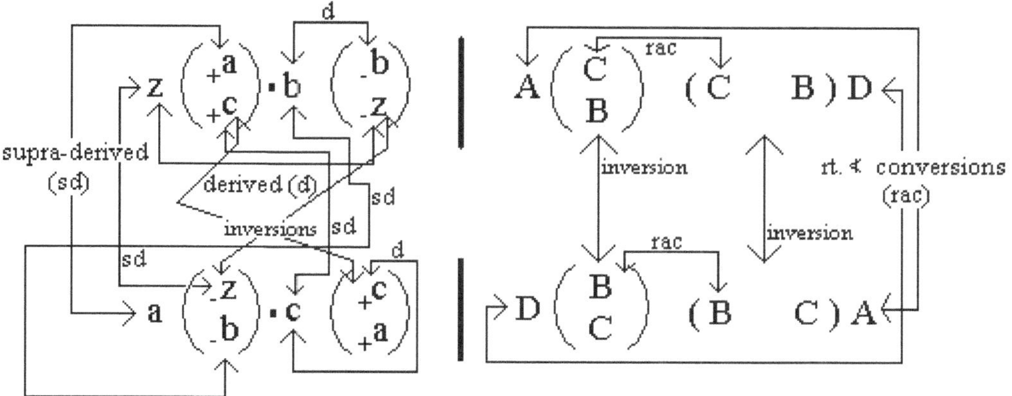

These correlations allow conversion of one table to the other, and may define the functional mechanics of the given "matrices." We, for instance, see how the "best solutions are chosen (or formed).

From these correlation mechanics, various closed loops (or circuitous routes) may be formed:

e.g.

e.g.

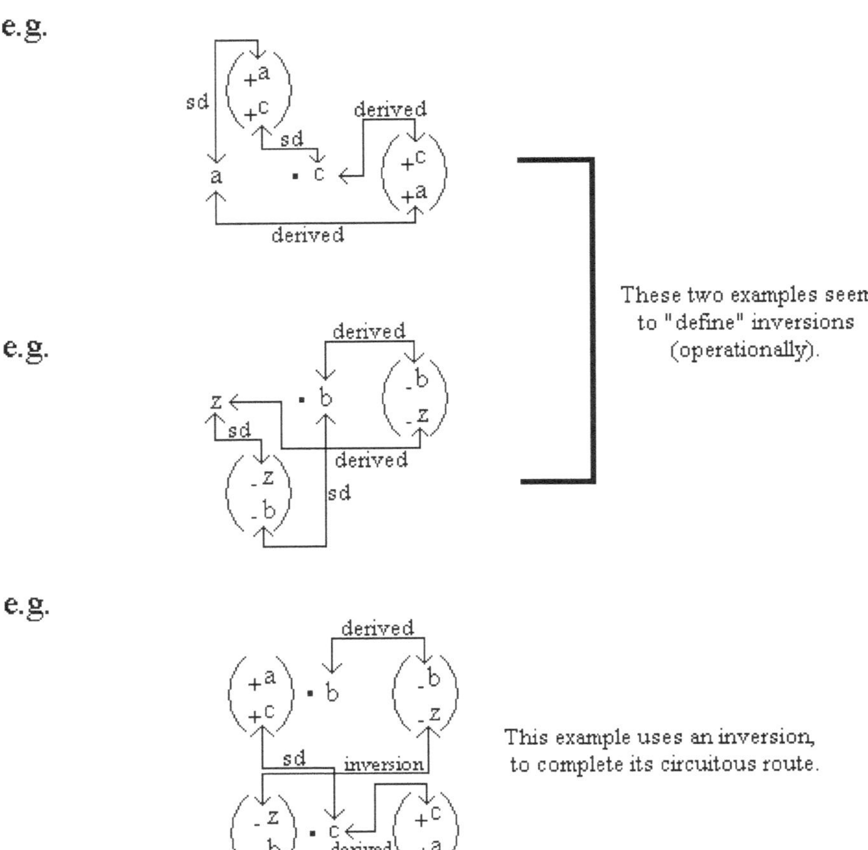

e.g.

If one "inserts" the two operative inversion definition routes here, one brings in (z, a) to complete the plane correlations. If you insert only one, you bring in either 'z' or 'a' (for the inversion announced here, z).
Note: without insertions for both inversions, the circuitous route for the previous example is actually

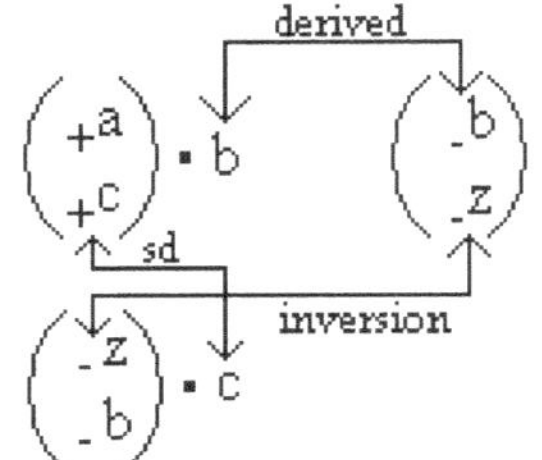

∴ Insertion would bring (as a result):

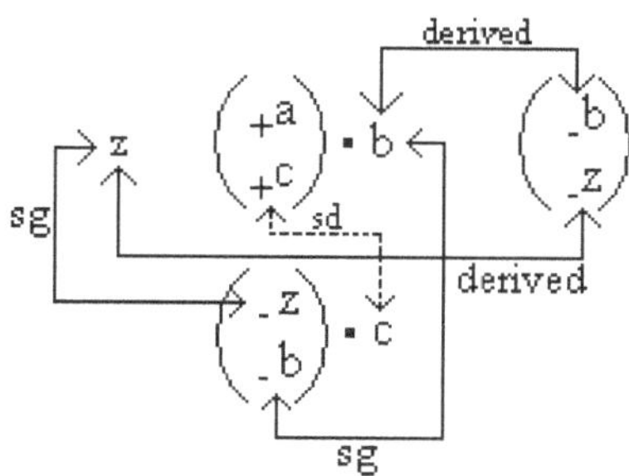

The ($_+$c ↔ c) seems extraneous (for a closed loop), and the inversion evident is already defined (by the insertion). The "extraneousness" of ($_+$c ↔ c) simply qualifies, or "fulfills," the ($_+$a , $_+$c)•b term.

This would seem to give a definition for 'z', as opposed to merely the defined insertion used, that 'z' necessitates a qualification with ($_+$a , $_+$c) . Correspondingly, 'a' necessitates a qualification with (.z , .b) .

It is demonstrated that the "planes" table (or matrix) consists of (in this order): a rotation, a (set of) distance modified, a (set of) distance by a rotation. The "points" table consists entirely of points, demarcated by contrasting relationships of the corresponding point pairs and (therefore derived) lines (or line segments).

If we invert a "planes" closed loop, we get an interesting conversion:

e.g.

i.e. If we consider, for a term like ($_+$a, $_+$c)·b , that ($_+$a, $_+$c) and 'b' are complements of each other (to construct the term), then suddenly the (.b, .z) component (in this example) is provided with its complement (·c) after inversion. A new term (fully complemented) is also provided for: ($_+$c, $_+$a)·b

∴ Inverting is a way to find (or produce) term complements.
∴ A "planes" table consists of a complemented term flanked by two uncomplemented (or self-complemented) terms.

A lone rotation implies an accompanying distance set, and a lone distance set implies an accompanying rotation. So in this sense each is self-complementary. However, the specifications can be obscure (aside

from relating to the "home" [d, rot], which might have to be looked up or uncovered diagrammatically).
e.g. 'z' and 'b' are defined with association to [d3, rot3] and [d4, rot4] respectively.

But clearly a distance set's rotation,
e.g. (-b , -z) and (+c , +a) ,
seems obscure until specified (via inverting):

$$
\left|\begin{array}{ll} z & \binom{+a}{+c}\cdot b\ \binom{-b}{-z} \\[2ex] a & \binom{-z}{-b}\cdot c\ \binom{+c}{+a} \end{array}\right|
\quad\xrightarrow{\ \text{invert}\ }\quad
\left|\begin{array}{ll} a & \binom{-b}{-z}\cdot c\ \binom{+a}{+c} \\[2ex] z & \binom{+c}{+a}\cdot b\ \binom{-z}{-b} \end{array}\right|
$$

As inverting is a form of division, the strong suggestion (or implication) is that multiplication of corresponding terms of a table and its inverting table yields to unity, and entire multiplication leads to a unity:
e.g.

$$
\left|\begin{array}{ll} z & \left[\binom{+a}{+c}\cdot b\right]\cdot\left[\binom{-b}{-z}\cdot c\right]\ \binom{-b}{-z} \\[2ex] a & \binom{-z}{-b}\cdot c\ \binom{+c}{+a} \end{array}\right|
\quad\rightarrow\quad
\left|\begin{array}{lll} z & 1 & \binom{-b}{-z} \\[2ex] a & \binom{-z}{-b}\cdot c & \binom{+c}{+a} \end{array}\right|
$$

e.g.

$$\left| \begin{array}{ccc} z & \left[\binom{+a}{+c}\cdot b \right]\cdot\left[\binom{-b}{-z}\cdot c \right] & \binom{-b}{-z} \\[2ex] a & \binom{-z}{-b}\cdot c & \binom{+c}{+a}\cdot\binom{-z}{-b} \end{array} \right| \;\rightarrow\; \left| \begin{array}{ccc} z & 1 & \binom{-b}{-z} \\[2ex] a & \binom{-z}{-b}\cdot c & 1 \end{array} \right|$$

e.g.

$$\left| \begin{array}{ccc} z & \left[\binom{+a}{+c}\cdot b \right]\cdot\left[\binom{-b}{-z}\cdot c \right] & \binom{-b}{-z} \\[2ex] a\cdot z & \binom{-z}{-b}\cdot c & \binom{+c}{+a}\cdot\binom{-z}{-b} \end{array} \right| \;\rightarrow\; \left| \begin{array}{ccc} z & 1 & \binom{-b}{-z} \\[2ex] 1 & \binom{-z}{-b}\cdot c & 1 \end{array} \right|$$

Note that:

$$\left| \begin{array}{ccc} z & 1 & \binom{-b}{-z} \\[2ex] 1 & \binom{-z}{-b} c & 1 \end{array} \right| \cdot \left| \begin{array}{ccc} 1 & \binom{+a}{+c}\cdot b & 1 \\[2ex] a & 1 & \binom{+c}{+a} \end{array} \right| \;\rightarrow\; \left| \begin{array}{ccc} z & \binom{+a}{+c}\cdot b & \binom{-b}{-z} \\[2ex] a & \binom{-z}{-b}\cdot c & \binom{+c}{+a} \end{array} \right|$$

the original "planes" table

Note that:

since $[(+a, +c)\cdot b] \cdot [(-b, -z)\cdot c] \rightarrow 1$
and $(+a, +c) \cdot (-b, -z) \rightarrow 1$ as $(+c, +a) \cdot (-z, -b) \rightarrow 1$
then $b \cdot c \rightarrow 1$ as $a \cdot z \rightarrow 1$

The distance sets are not "vectorial":

e.g. $(+a, +c) \equiv (+c, +a)$

until qualified by the "points" table:

$$z \quad \begin{pmatrix} {}_{+}a \\ {}_{+}c \end{pmatrix} \cdot b \begin{pmatrix} {}_{\cdot}b \\ {}_{\cdot}z \end{pmatrix} \quad \Big| \quad A\begin{pmatrix} C \\ B \end{pmatrix} \quad (\, C \quad B\,)\, D$$

Here, $({}_{+}a, {}_{+}c) \equiv ({}_{+}c, {}_{+}a)$ may not be true (if order of operations is implied).

Thus, the "directions" implied are:

$A \rightarrow C$, $A \rightarrow B$; $C \rightarrow D$, $B \rightarrow D$ (i.e. these lines are drawn)

$\therefore$ For example,

$z \rightarrow {}_{+}a\cdot b \rightarrow {}_{\cdot}b$

is interpreted as:

$z, (B \rightarrow D) \rightarrow [{}_{+}a, (A \rightarrow C)] \cdot (b, C \rightarrow D) \rightarrow {}_{\cdot}b, (D \rightarrow C)$

Similarly:

$(z, B \rightarrow D) \rightarrow ({}_{+}a, A \rightarrow C) \cdot (b, C \rightarrow D) \rightarrow ({}_{\cdot}z, D \rightarrow B)$
$(z, B \rightarrow D) \rightarrow ({}_{+}c, A \rightarrow B) \cdot (b, C \rightarrow D) \rightarrow ({}_{\cdot}z, D \rightarrow B)$
$(z, B \rightarrow D) \rightarrow ({}_{+}c, A \rightarrow B) \cdot (b, C \rightarrow D) \rightarrow ({}_{\cdot}b, D \rightarrow C)$

Although note that $' \cdot(b, C\rightarrow D) '$ is using $[3]_{rot}$ instead of $[4]_{rot}$.

Why is

$z \rightarrow {}_{+}c\cdot b \rightarrow {}_{\cdot}b$

a better solution for lines AB, CD, BD than

$z \rightarrow {}_{+}c\cdot b \rightarrow {}_{\cdot}z$,   i.e. for the term $' {}^{c}{}_{+}\underline{AB}({}^{z}\underline{BD} + {}^{b}\underline{CD}) '$?

Is it that

$z \rightarrow\rightarrow {}_{\cdot}z$

is improperly recuperative?

Yet,

$a \rightarrow\rightarrow {}_{+}a$

only seems redundant.

If we multiply inversions, we get uniqueness (unitary-ness):

$$\left| \; z \;\; \left[\left(\begin{smallmatrix} {}_+a \\ {}_+c \end{smallmatrix}\right)\cdot b\right] \cdot \left[\left(\begin{smallmatrix} {}_\cdot b \\ {}_\cdot z \end{smallmatrix}\right) c\right] \;\; \left(\begin{smallmatrix} {}_\cdot b \\ {}_\cdot z \end{smallmatrix}\right)\cdot\left(\begin{smallmatrix} {}_+a \\ {}_+c \end{smallmatrix}\right) \; \right| \;\; \longrightarrow \;\; \left| \;\; 1 \quad 1 \quad 1 \;\; \right|$$

$\therefore$ We can compare

$$\left| \;\; z\cdot a \;\; ({}_+c\cdot b)\cdot({}_\cdot z\cdot c) \;\; {}_\cdot b\cdot{}_+a \;\; \right|$$

to

$$\left| \;\; z\cdot a \;\; ({}_+c\cdot b)\cdot({}_\cdot z\cdot c) \;\; {}_\cdot z\cdot{}_+c \;\; \right|$$

Dividing, we get the expression:

$$\frac{\left| \;\; z\cdot a \;\; ({}_+c\cdot b)\cdot({}_\cdot z\cdot c) \;\; {}_\cdot b\cdot{}_+a \;\; \right|}{\left| \;\; z\cdot a \;\; ({}_+c\cdot b)\cdot({}_\cdot z\cdot c) \;\; {}_\cdot z\cdot{}_+c \;\; \right|} \;\; \rightarrow \;\; \left| \;\; 1 \quad 1 \quad \frac{{}_\cdot b\cdot{}_+a}{{}_\cdot z\cdot{}_+c} \;\; \right|$$

Multiplying, we get the expression:

$$\left| \;\; (z\cdot a)^2 \;\; [({}_+c\cdot b)\cdot({}_\cdot z\cdot c)]^2 \;\; ({}_\cdot b\cdot{}_+a)\cdot({}_\cdot z\cdot{}_+c) \;\; \right| \;\; \rightarrow \;\; \left| \;\; 1 \quad 1 \quad ({}_\cdot b\cdot{}_+a)\cdot({}_\cdot z\cdot{}_+c) \;\; \right|$$

But, if we do a "hyper" analysis (i.e. replacing like-looking terms arbitrarily with "1"),
then, ' $({}_\cdot z\cdot c)$ ' looking like ' $({}_\cdot z\cdot{}_+c)$ ',
we have:

Division:
$$\frac{\left| \;\; z\cdot a \;\; ({}_+c\cdot b)\cdot 1 \;\; {}_\cdot b\cdot{}_+a \;\; \right|}{\left| \;\; z\cdot a \;\; ({}_+c\cdot b)\cdot 1 \;\; 1 \;\; \right|} \;\; \rightarrow \;\; \left| \;\; 1 \quad 1 \quad {}_\cdot b\cdot{}_+a \;\; \right|$$

Multiplication:
$$\left| \;\; (z\cdot a)^2 \;\; [({}_+c\cdot b)\cdot 1]^2 \;\; ({}_\cdot b\cdot{}_+a)\cdot 1 \;\; \right| \;\; \rightarrow \;\; \left| \;\; 1 \quad ({}_+c\cdot b)^2 \quad ({}_\cdot b\cdot{}_+a) \;\; \right|$$

But this can be equivalent to:

(for) $[({}_+c\cdot b)\cdot({}_\cdot z\cdot c)]\cdot[({}_+c\cdot b)\cdot({}_\cdot z\cdot c)] \;\; \rightarrow \;\;$ (selectively) $[1]\cdot[({}_+c\cdot b)\cdot 1] \;\; \rightarrow \;\; ({}_+c\cdot b)$

$\therefore \left| \;\; 1 \quad ({}_+c\cdot b) \quad ({}_\cdot b\cdot{}_+a) \;\; \right|$

or

$[\,({}_+c\cdot b)\cdot({}_\cdot z\cdot c)\,] \cdot [\,({}_+c\cdot b)\cdot({}_\cdot z\cdot c)\,] \;\; \rightarrow \;\; [1] \cdot [1]$

$\therefore \left| \;\; 1 \quad 1 \quad ({}_\cdot b\cdot{}_+a) \;\; \right|$

Let us make the hyper analysis by:

1) giving it a symbol: ħ (replacing "1"; analogous to "i" for complex numbers) to specify that certain conditions must be met in operation
2) restricting its use only for paired comparisons, so that it alights only evenly to guarantee the "like-looking" comparisons

Now,
Division is:

$$\frac{|\ z{\cdot}a\ (_{+}c{\cdot}b){\cdot}(_{.}z{\cdot}c)\ _{.}b_{..+}a\ |}{|\ z{\cdot}a\ (_{+}c{\cdot}b){\cdot}(_{.}z{\cdot}c)\ _{.}z_{+}c\ |} \;\rightarrow\; |\ 1\ \ 1\ (_{.}b_{..+}a)/ħ\ |$$

(occurrences of $_{.}z{\cdot}c$ allow $_{.}z_{+}c \equiv ħ$; $_{.}z{\cdot}c \equiv ħ$ via occurrence of $_{.}z_{+}c$)

or, more parsimoniously:

$$|\ 1\ \ 1/[(_{+}c{\cdot}b){\cdot}ħ]\ (_{.}b_{..+}a)/ħ\ |$$

and/or

$$|\ 1\ (_{+}c{\cdot}b){\cdot}ħ/1\ (_{.}b_{..+}a)/ħ\ |\;\rightarrow\;|\ 1\ (_{+}c{\cdot}b){\cdot}ħ\ (_{.}b_{..+}a)/ħ\ |$$

Multiplication is:

$$|\ (z{\cdot}a)^2\ 1\ (_{.}b_{..+}a){\cdot}ħ\ |\;\rightarrow\;|\ 1\ \ 1\ (_{.}b_{..+}a){\cdot}ħ\ |$$

or, more parsimoniously:

$$|\ 1\ (_{+}c{\cdot}b)^2{\cdot}(_{.}z{\cdot}c){\cdot}ħ\ (_{.}b_{..+}a){\cdot}ħ\ |\;\rightarrow\;|\ 1\ (_{+}c{\cdot}b){\cdot}ħ\ (_{.}b_{..+}a){\cdot}ħ\ |$$

Taking the most general conditions (for hyper analysis):

$$|\ 1\ \ 1\ (_{.}b_{..+}a)/ħ\ |\ \ ,\ \ |\ 1\ \ 1\ (_{.}b_{..+}a){\cdot}ħ\ |$$

both/either suggest

$$z \rightarrow {}_{+}c{\cdot}b \rightarrow {}_{.}b$$

and

$a \rightarrow {}_{.}b{\cdot}c \rightarrow {}_{+}c$: i.e. comparing $|z{\cdot}a\ (_{.}b{\cdot}c){\cdot}(_{+}a{\cdot}b)\ _{.}z_{+}c|$ to $|z{\cdot}a\ (_{.}b{\cdot}c){\cdot}(_{+}a{\cdot}b)\ (_{.}b_{..+}a)|$
are "best" solutions, for sets of lines {AB, CD, BD} and {DB, BA, CA} respectfully.

The use of "ħ" more faithfully follows the manners of operation (than the use of "1").

Studying the multiplication of *invert-ions*, this can be expanded out:

$$|\ z{\cdot}a\ [(_{+}a,\ _{+}c){\cdot}b]{\cdot}[(_{.}b,\ _{.}z){\cdot}c]\ (_{.}b,\ _{.}z){\cdot}(_{+}a,\ _{+}c)\ |\;\rightarrow\;|\ 1\ 1\ 1\ |$$

That is (expanding within the brackets):

$\quad$ | $\quad$ z·a
$\quad\quad$ (₊a·b, ₋c·b)·(.b·c, .z·c) → (₊a·b)·(.z·c), (₊c·b)·(.z·c), (₊a·b)·(.b·c), (₊c·b)·(₊b·c)
$\quad\quad$ (.b, .z)·(₊a, ₊c) → .b·₊c, .b·₊a, .z·₊a, .z·₊c $\qquad\qquad$ |

which is further:

$\quad$ | $\quad$ z·a
$\quad\quad$ (₊a·.z)·(b·c)[3]$_{rot}$, (₊c·.z)·(b·c)[3]$_{rot}$, (₊a·.b)·(b·c)[3]$_{rot}$, (₊c·.b)·(b·c)[3]$_{rot}$
$\quad\quad\quad\quad$ → [(₊a·.z), (₊a·.b), (₊c·.z), (₊c·.b)] · (b·c)[3]$_{rot}$
$\quad\quad$ .b·₊c, .b·₊a, .z·₊a, .z·₊c $\qquad\qquad$ |

leading to (i.e. " → "):

$\quad$ | $\quad$ z·a $\quad$ ħ·(b·c)[3]$_{rot}$ $\quad$ ħ $\quad$ |

again suggesting "best" solutions (i.e. z→→.b, a→→₊c).

So the rule, for "best" solution, seems to be akin to:

$\quad$ | $\quad$ z $\quad$ (₊a, ₊c)·ħ $\quad$ (ħ, .z) $\quad$ |

and

$\quad$ | $\quad$ a $\quad$ (.z, .b)·ħ $\quad$ (ħ, ₊a) $\quad$ |

Multiplying invert-ions yields:

$\quad$ | $\quad$ z·a $\quad$ [(₊a, ₊c)·ħ]·[(.b, .z)·ħ $\quad$ (ħ, .z)·(₊a, ħ) $\quad$ | $\quad$ → $\quad$ | $\quad$ z·a $\quad$ (₊a, ₊c)·(.b, .z)·ħ^2 $\quad$ (₊a, .z) $\quad$ |

or, as (₊a, ₊c)·(.b, .z) → 1 :

$\quad$ | $\quad$ z·a $\quad$ ħ^2 $\quad$ (₊a, .z) $\quad$ |

or, as (₊a, ₊c)·(.b, .z) ≡ (₊a, .z)·(.b, ₊c) :

$\quad$ | $\quad$ z·a $\quad$ (.b, ₊c)·(₊a, .z)·ħ^2 $\quad$ (₊a, .z) $\quad$ | $\quad$ → $\quad$ | $\quad$ z·a $\quad$ (.b, ₊c)·(ħ^3=ħ^2·ℌ) $\quad$ ħ=ℌ $\quad$ |
$\qquad\qquad\qquad\qquad\qquad$ (ħ being a general operating principle)

Multiplying invert-ions, therefore, yields to three apparent equivalencies to $|111|$:

$\quad$ | $\quad$ z·a $\quad$ (b·c)·ħ $\quad$ ħ $\quad$ | , $\quad$ | $\quad$ z·a $\quad$ ħ^2 $\quad$ (₊a, .z) $\quad$ | , $\quad$ | $\quad$ z·a $\quad$ (.b, ₊c)·ħ^2·ℌ $\quad$ ℌ $\quad$ |

The middle table is perhaps the most general, informing of an ultimate nature of rotation to displacement. The first and last together contribute to this understanding.

But, thus we have a generalized identity which may be useful in the future:

| $\quad$ z·a $\quad\quad$ (b·c)·ħ $\quad\quad\quad$ ħ $\quad$ | $\qquad$ $|111|$
| $\quad$ z·a $\quad\quad\quad$ ħ^2 $\quad\quad$ (₊a, .z) $\quad$ | $\quad$ ≡ $\quad$ $|111|$
| $\quad$ z·a $\quad$ (.b, ₊c)·ħ^2·ℌ $\quad\quad$ ℌ $\quad$ | $\qquad$ $|111|$

It leads to the interesting relationship (taking diagonals) that:

$$\mid \ z{\cdot}a \ \hbar^2 \ \mathfrak{H} \ \mid \ \leftrightarrow \ \mid \ z{\cdot}a \ \hbar^2 \ \hbar \ \mid$$

∴ $\mathfrak{H}$ may be treated (within proper context) as $\hbar$, allowing for $\hbar^3$, etc.

Therefore, we've constructed a major operating principle, for converting one point (or quality) to another, where the adjustment involves complimentary operations; i.e. Operations which cause others to necessarily occur simultaneously or follow:

e.g. Rotation	occurs with Displacement	(and visa versa)
e.g. EMF (electricity)	occurs with Magnetic Induction	"
e.g. Vectorial initiation and resultant	occurs with Looping of Operations	"
e.g. Multiplication	occurs with Division	"
e.g. Addition	occurs with Subtraction	"
e.g. Vectorilization	occurs with Points/Lines	"
e.g. Multiplicative Inverting	occurs with Equivalency (Unitary-ness)	"
e.g. a Current	occurs with a Closed Circuitous Route	"
e.g. a Closed Circuitous Route	occurs with Defining Inversions	"
e.g. Light (radiant energy)	occurs with Travel	"
e.g. "Planes" relationships (tables)	occur with "Points" relationships (tables)	"
e.g. Points	occur with Lines	"
e.g. Lines	occur with Planes	"

etc. (e.g. Words/sentences; sight/pictures; motion/change)

Where there are two complimentary qualities, their relationship to each other yields a third (combinatorial) quality.

∴ The generalized identity is tertiary by nature.
 And it appears to present $\hbar$ in increasing orders (i.e. $\hbar^1$, $\hbar^2$, $\hbar^3$).

Apparently, the order of $\hbar$ increases with the number of qualities to be considered. We currently restrict ourselves to two, and the quality of their relationship (i.e. rotation/displacement).

Then let us consider the unitary-ness identity, that it may be reduced (to two points per rotation).

We have a, b, c, z :

 four rotations, each of which corresponds to a plane for two points
 (i.e. a line with two endpoints, for each)

∴ Four points are represented (with four rotations)

```
| z·a      (b·c)·ħ        ħ       |
| z·a         ħ²       (₊a, ₋z)   |
| z·a  (₋b, ₊c)·ħ²·ℌ      ℌ       |
```

may be reduced, variously, to:

e.g.
```
| z   b·ħ    ħ  |    | z    c·ħ    ħ  |    | z    c·ħ    ħ  |    | z    b·ħ    ħ  |
| z    ħ²   ₊a  |    | z     ħ²   ₊a  |    | z     ħ²   ₋z  |    | z     ħ²   ₋z  |
| z  ₋b·ħ²·ℌ  ℌ |,   | z  ₊c·ħ²·ℌ  ℌ  |,   | z  ₊c·ħ²·ℌ  ℌ  |,   | z  ₋b·ħ²·ℌ  ℌ  |,

| z   b·ħ    ħ  |    | z    c·ħ    ħ  |    | z    b·ħ    ħ  |    | z    c·ħ    ħ  |
| z    ħ²   ₋z  |    | z     ħ²   ₋z  |    | z     ħ²   ₊a  |    | z     ħ²   ₊a  |
| z  ₊c·ħ²·ℌ  ℌ |,   | z  ₋b·ħ²·ℌ  ℌ  |,   | z  ₊c·ħ²·ℌ  ℌ  |,   | z  ₋b·ħ²·ℌ  ℌ  |
```
etc.

We have planes:

```
        ┌------┐
      d4, [4]rot
 ᵃAC , ᵇCD , ᶜAB , ᶻBD
      d3, [3]rot
  └------------------┘
```

and points: A, B, C, D

For the above, the identity reduction is of course:

```
| z·a        ( · )·ħ       ħ   |
| z·a          ħ²         ( )  |     with selection of distance arguments (terms) from their sets.
| z·a        ( )·ħ²·ℌ      ℌ   |
                 a
```

If we insist that for the ' ( )·ħ ' term the ' (b·c) ' operation must be carried along (i.e. not restricted), then each tabular result has representation of all four points (since planes ᵇCD and ᶜAB together possess all of A, B, C, D points).

∴ The reduction is:

```
| z·a  (b·c)·ħ   ħ  |     | z·a |       | (b·c)·ħ        ħ  |       | z   (b·c)·ħ       ħ     |
| z·a     ħ²    ( ) |     | z·a |       |    ħ²         ( ) |       | z      ħ²      ₊a, ₋z   |
| z·a  ( )·ħ²·ℌ  ℌ  |  →  | z·a |   +   | ( )·ħ²·ℌ       ℌ  |   →   | z  (₋b, ₊c)·ħ²·ℌ  ℌ     |
         a                  a
```
(2·2 → 4 tabular results)

∴ A manner for tabular reduction is established; i.e. reduction (or amplification) is restricted to unbalanced tabular forms (tables with unequal numbers of columns as rows), as compared to multiplication (or division) that is allowed for balanced tabular forms (tables with equal numbers of columns as rows; compare matrices to determinants).

∴ We have the various reduced forms:

$$\left(\ \begin{array}{cc} \lfloor z\cdot a \rfloor & \lfloor z\cdot a \rfloor \\ \lfloor z\cdot a \rfloor & \lfloor z\cdot a \rfloor \\ \lfloor z\cdot a \rfloor & \lfloor z\cdot a \rfloor \\ a & z \end{array}\ \right) \quad + \quad \left| \begin{array}{cc} (b\cdot c)\cdot ħ & ħ \\ ħ^2 & (\,) \\ (\,)\cdot ħ^2\cdot ℌ & ℌ \end{array} \right|$$

which are composed of reducible and (apparently) un-reducible elements.

Here, the un-reducible element is:

$$\left| \begin{array}{cc} (b\cdot c)\cdot ħ & ħ \\ ħ^2 & (\,) \\ (\,)\cdot ħ^2\cdot ℌ & ℌ \end{array} \right|$$

as regards the unitary-ness table (as originated from).

Yet,

$$\left| \begin{array}{cc} (b\cdot c)\cdot ħ & ħ \\ ħ^2 & (\,) \\ (\,)\cdot ħ^2\cdot ℌ & ℌ \end{array} \right| \quad \rightarrow \quad ħ\left| \begin{array}{c} (b\cdot c) \\ ħ \\ (\,)\cdot ħ\cdot ℌ \end{array} \right| \quad + \quad \left| \begin{array}{c} ħ \\ (\,) \\ ℌ \end{array} \right|$$

∴ The reduction continues as:

$$\left(\ \begin{array}{cc} \lfloor z\cdot a \rfloor & \lfloor z\cdot a \rfloor \\ \lfloor z\cdot a \rfloor & \lfloor z\cdot a \rfloor \\ \lfloor z\cdot a \rfloor & \lfloor z\cdot a \rfloor \\ a & z \end{array}\ \right) \quad + \quad ħ\left| \begin{array}{c} (b\cdot c) \\ ħ \\ (\,)\cdot ħ\cdot ℌ \end{array} \right| \quad + \quad \left| \begin{array}{c} ħ \\ (\,) \\ ℌ \end{array} \right|$$

for the operations (and their inverting).

∴ Two quality problems (with the relationship between qualities) can be represented as above. This is essentially "action and result," where either action causes another action as a result (not necessarily "action & reaction").

$$\therefore \quad \left(\ \begin{array}{cc} |z| & |a| \\ |z| & |a| \\ |z|, & |a| \end{array}\ \right) \quad + \quad ħ\left| \begin{array}{c} (b\cdot c) \\ ħ \\ (\,)\cdot ħ\cdot ℌ \end{array} \right| \quad + \quad \left| \begin{array}{c} ħ \\ (\,) \\ ℌ \end{array} \right|$$

may be used to reconstruct the unitary-ness identity.

This of course relates to (or is derived from):

$$\left| \begin{array}{ccc} z & (_+a,\,_+c)\cdot b & (_-b,\,_-z) \\ a & (_-z,\,_-b)\cdot c & (_+c,\,_+a) \end{array} \right| \qquad \text{and its invert-ion.}$$

Terms: $(_+a,\,_+c)$ and $(_+c,\,_+a)$ must be directionally (or "vectorially") distinct,
  as are terms: $(_-z,\,_-b)$ and $(_-b,\,_-z)$, in order to retain a visual evolution to "best" solutions.

(i.e.: $z \to_+ c \cdot b \to_. b$ and $a \to_. b \cdot c \to_+ c$)

But how is this directionality defined?

Perhaps it is displayed as:

$$z \ \begin{pmatrix} _+a \\[2pt] _+c \end{pmatrix} \bullet b \qquad \text{i.e.} \quad z \to {}_+c\cdot b \to {}_+a \to z$$

$$a \ \begin{pmatrix} _-z \\[2pt] _-b \end{pmatrix} \bullet c \qquad \text{i.e.} \quad a \to {}_-b\cdot c \to {}_-z \to a$$

$$\therefore \quad \begin{vmatrix} z & (_+a,\ _+c)\cdot b \\ a & (_-z,\ _-b)\cdot c \end{vmatrix} \quad \to \quad \begin{vmatrix} z \\ a \end{vmatrix}$$

$$\therefore \quad \begin{vmatrix} z & (_+a,\ _+c)\cdot b & (_-b,\ _-z) \\ a & (_-z,\ _-b)\cdot c & (_+c,\ _+a) \end{vmatrix} \quad \to \quad \begin{vmatrix} z & (_-b,\ _-z) \\ a & (_+c,\ _+a) \end{vmatrix}$$

So we may call this (operation) "looping diminution" (or simplification) of tabular forms:

$$z \to {}+c\cdot b \to {}_+a \ , \quad a \to {}_-b\cdot c \to {}_-z$$

$$(\,d3\,,\ [3]_{rot}\,)$$

(Note: we are actually using, here, terms: ${}_+c\cdot b[3]_{rot}$ and ${}_-b\cdot c[3]_{rot}$)

The tabular simplification:

$$\begin{vmatrix} z & (_-b,\ _-z) \\ a & (_+c,\ _+a) \end{vmatrix} \qquad \text{looks fairly fundamental,}$$

and might indeed be replaced by:

$$\begin{vmatrix} ħ & (_-b,\ ħ) \\ ℌ & (_+c,\ ℌ) \end{vmatrix} \qquad \text{in hyper analysis,}$$

to allow a possibly inordinate construction with the unitary-ness identity:

e.g.
$$\begin{vmatrix} z & (b\cdot c)\cdot ħ & ħ \\ z & ħ^2 & _+a,\ _-z \\ z & (_-b,\ _+c)\cdot ħ^2\cdot ℌ & ℌ \end{vmatrix} \ \to \ \left(\text{from:}\ \left|\ \begin{vmatrix} z \\ a \end{vmatrix} \ \leftrightarrow \ \begin{vmatrix} z & (_+a,\ _+c)\cdot b \\ a & (_-z,\ _-b)\cdot c \end{vmatrix}\ \right|\right)$$

$$\begin{vmatrix} ħ & (ℌ,\ _+c)\cdot b & (b\cdot c)\cdot ħ & ħ \\ ħ & (ℌ,\ _+c)\cdot b & ħ^2 & ℌ,\ ħ \\ ħ & (ℌ,\ _+c)\cdot b & (_-b,\ _+c)\cdot ħ^2\cdot ℌ & ℌ \end{vmatrix}$$

This table of course uses different $ħ$s and $ℌ$s.

∴ It might be more appropriate to partition:

$$
\begin{array}{|lll|}
\hline
♄ & (♃, {}_{+}c)\cdot b & : & (b\cdot c)\cdot ♄ & ♄ \\
♄ & (♃, {}_{+}c)\cdot b & : & ♄^2 & ♃, ♄ \\
♄ & (♃, {}_{+}c)\cdot b & : & (.b, {}_{+}c)\cdot ♄^2\cdot ♃ & ♃ \\
\hline
\end{array}
$$

which leads to the controversial operation of equivalent replacements assumed (by history of the tabular development):

$$
\begin{array}{|lll|}
\hline
♄ & : & (b\cdot c)\cdot ♄ & ♄ \\
♄ & : & ♄^2 & ♃, ♄ \\
♄ & : & (.b, {}_{+}c)\cdot ♄^2\cdot ♃ & ♃ \\
\hline
\end{array}
\quad\text{and}\quad
\begin{array}{|lll|}
\hline
(♃, {}_{+}c)\cdot b & : & (b\cdot c)\cdot ♄ & ♄ \\
(♃, {}_{+}c)\cdot b & : & ♄^2 & ♃, ♄ \\
(♃, {}_{+}c)\cdot b & : & (.b, {}_{+}c)\cdot ♄^2\cdot ♃ & ♃ \\
\hline
\end{array}
$$

This emphasizes that these tabular forms are "descriptions" of operations.
The previous two tables are *not* equal.

But they may be legitimately factored as long as the partitioning is kept:

e.g.

$$
♄\cdot\begin{array}{|lll|}
\hline
1 & : & (b\cdot c) & 1 \\
1 & : & ♄ & ♃/♄, 1 \\
1 & : & (.b, {}_{+}c)\cdot ♄\cdot ♃ & ♃/♄ \\
\hline
\end{array},
\qquad
♃\cdot\begin{array}{|lll|}
\hline
♄/♃ & : & (b\cdot c)\cdot ♄/♃ & ♄/♃ \\
♄/♃ & : & ♄^2/♃ & (1, ♄/♃) \\
♄/♃ & : & (.b, {}_{+}c)\cdot ♄^2 & 1 \\
\hline
\end{array},
$$

$$
♃\cdot b\cdot\begin{array}{|lll|}
\hline
(1, {}_{+}c) & : & c\cdot ♄/♃ & ♄/(♃\cdot b) \\
(1, {}_{+}c) & : & ♄^2/(♃\cdot b) & (1, ♄)/b \\
(1, {}_{+}c) & : & (.b, {}_{+}c)\cdot ♄^2/b & 1/b \\
\hline
\end{array},
\qquad
♃\cdot\begin{array}{|lll|}
\hline
(1, {}_{+}c)\cdot b & : & (b\cdot c)\cdot ♄/♃ & ♄/♃ \\
(1, {}_{+}c)\cdot b & : & ♄^2/♃ & (1, ♄/♃) \\
(1, {}_{+}c)\cdot b & : & (.b, {}_{+}c)\cdot ♄^2 & 1 \\
\hline
\end{array}
$$

Partial factorizations may also be permissible:

$$
\begin{array}{|lll|}
\hline
♄ & : & (b\cdot c)\cdot ♄ & ♄ \\
♄ & : & ♄^2 & ♃, ♄ \\
♄ & : & (.b, {}_{+}c)\cdot ♄^2\cdot ♃ & ♃ \\
\hline
\end{array}
\;\rightarrow\;
\begin{array}{|lll|}
\hline
♄ & : & (b\cdot c) & 1 \\
♄ & : & ♄ & ♃/♄, 1 \\
♄ & : & (.b, {}_{+}c)\cdot ♄^2 & 1 \\
\hline
\end{array}
\begin{array}{l}
♄ \\
♄ \\
♃
\end{array}
$$

♄ External Factors
♄ restricted to this
♃ side of table's partition

or

Global External
Factors (i.e. for
both sides of
partition)

$$
\begin{array}{l}
♄ \\
♄ \\
♃
\end{array}
\begin{array}{|lll|}
\hline
1 & : & (b\cdot c) & 1 \\
1 & : & ♄ & ♃/♄, 1 \\
♄/♃ & : & (.b, {}_{+}c)\cdot ♄^2 & 1 \\
\hline
\end{array}
$$

or

$$
\begin{array}{l}
♄ \\
♄ \\
{}
\end{array}
\begin{array}{|lll|}
\hline
1 & : & (b\cdot c) & 1 \\
1 & : & ♄ & ♃/♄, 1 \\
& ♄ : & (.b, {}_{+}c)\cdot ♄^2 & 1 \\
\hline
\end{array}
\begin{array}{l}
{} \\
{} \\
♃
\end{array}
\qquad\text{(global and restricted factorizations)}
$$

The last table allows for the factors (factored out) to be:

$$\begin{pmatrix} \hbar \\ \hbar \\ 1 \end{pmatrix}_{\text{global}}, \quad \begin{pmatrix} 1 \\ 1 \\ \hbar \end{pmatrix}_{\text{restricted}} \quad (\therefore \text{ an "absence"} \equiv 1)$$

Equivalent Replacement Method

The Equivalent Replacement Method (ERM) is simply used to produce (or create) balanced tabular forms (equal # of rows as columns), to allow for multiplication/division as opposed to unbalanced forms using reduction/amplification.

The ERM (multiplication/division) can be characterized as 'quasi', or 'partial', or 'partitional' (i.e. partitive), or 'circumstantial', or 'non-conservative' (or simply 'controversial'). But the resultant tabular forms perform as platforms upon which to construct the design of problems to answer or solve. The goal thereby is of course to convert them into more regular descriptions of relationships (of variables or elements) of operations to derive at "best" solutions:

i.e.   to go from, say:

$$\hbar \mid 1 : \quad (b \cdot c) \qquad 1 \mid$$
$$\hbar \mid 1 : \qquad \hbar \quad \hbar/\hbar, 1 \mid$$
$$\mid \hbar : \quad (.b, {}_+c) \cdot \hbar^2 \qquad 1 \mid \hbar$$

to

$$z \qquad\qquad\qquad\qquad .z$$
$$\searrow \begin{pmatrix} {}_+a\,[3]_d \\ \\ {}_+c\,[4]_d \end{pmatrix} \quad b\,[3]_{rot} \qquad \nearrow$$
$$\nearrow \qquad\qquad\qquad\qquad \searrow$$
$$b \qquad\qquad\qquad\qquad .b$$

to

$$\mid z \quad ({}_+a, {}_+c) \cdot b \quad (.b, .z) \mid$$

to $z \rightarrow {}_+c \cdot b \rightarrow .b$

to ${}_+\underline{AC}\ {}^z\underline{BD}\ [d3, rot3]\ .\underline{DB}\ {}^a\underline{CA}$
       ${}_+\underline{AB}\ {}^b\underline{CD}\ [d4, rot4]\ .\underline{DC}\ {}^c\underline{BA}$

to ${}_+A^xB\ [d, rot; d2, rot2]\ {}_+c^yD$ etc.

How do we analyze a tabular form (given), such as:

$$\hbar \mid 1 : \quad (b \cdot c) \qquad 1 \mid$$
$$\hbar \mid 1 : \qquad \hbar \quad (\hbar/\hbar, 1) \mid$$
$$\mid \hbar : \quad (.b, {}_+c) \cdot \hbar^2 \qquad 1 \mid \hbar \qquad ?$$

Obviously it is to be related to an unitary-ness:

$$
\begin{vmatrix} 1 & 1 & 1 \\ 1 & 1 & 1 \\ 1 & 1 & 1 \end{vmatrix} \equiv \begin{vmatrix} z \cdot a & (b \cdot c) \cdot ħ & ħ \\ z \cdot a & ħ^2 & (_+a, _-z) \\ z \cdot a & (_-b, _+c) \cdot ħ^2 \cdot ẜ & ẜ \end{vmatrix}
$$

which is (at the moment) unknown (to us).

The operations of a tabular form are read across a row, and they are action insistent depending on the symbols used. The direction for proceeding is from left to right ($\rightarrow$).

e.g.

    b   means perform a 'b' rotation

    b·c   means perform a 'b' rotation followed by a 'c' rotation

    .b   means perform a '.b' distance (translation)

    +c   means perform a '+c' distance (translation)

$\left(\begin{smallmatrix} .b \\ _+c \end{smallmatrix} \right)$ or (.b, +c) means perform either a '.b' distance or a '+c' distance

∴

   The multiplication symbol (here: " · ") is merely a procedural placeholder to specify the order of operations.

   The " $\rightarrow$ " symbol shows the direction of operations (i.e. "go to the next operation")

   The " 1 " number simply indicates an absence (or pausing) of operation.

   The ħ, ẜ, etc. symbols show where hyper analyses may occur for operations involving like-looking symbols. This restricts its usage to definite manners of operation (i.e. those only valid for certain symbols; the "hyper" is a replacement indicator for those operations which must occur).

∴ $\begin{vmatrix} 1 \vdots (b \cdot c) \ 1 \end{vmatrix}$

is interpreted as a 'b' rotation followed by a 'c' rotation, this compound operation occurring somewhere between two others unspecified.

∴ $ħ \begin{vmatrix} 1 \vdots (b \cdot c) \ 1 \end{vmatrix}$ $\rightarrow$ $\begin{vmatrix} ħ \vdots (b \cdot c) \cdot ħ & ħ \end{vmatrix}$

is interpreted as (the) previous, except that the ' (b·c) ' compound operation is followed by a hyper operation that uses like-looking symbols for the final operation(s). The initial ħ (due to its partitioning off via " ⋮ ") shows that a different hyper operation (different like-looking symbols) occurs before the ' (b·c)·ħ ' compound operation. (Since the external ħ is to the left of the table, it is global for the row.)

   Each row represents a series of operations that lead to some attempt to reach a desired result. These results may be compared with the "family" of results contained by the table. This family is a restriction on how these results may occur. If we assume that the table is of (i.e. from) unitary-ness, then it must be self-consistent and symmetrical, and could have been formed by multiplying operations with invert-ions of another set to enforce self-consistency. Equivalencies may be designed to construct a balanced unitary-ness table, which can be manipulated to lend to a description of the starting problem. Therefore, the goal is to go from the starting problem to the (or a) unitary-ness tabular form which may be used to specify (or resolve to) "best" solution(s).

How, then, do we construct (or design) a tabular form for a problem?

<u>First</u>: We want it to be balanced (as many rows as columns), in order for it to be derivative of a unitary-ness.

$\therefore$ There should be as many operation statements, i.e. results $\equiv$ rows, as there are (sets of) operations $\equiv$ columns

Employing the fundamental notion that one action causes another to occur simultaneously, or directly to follow, and that there is a reversibility of procedure that may be made evident (so that multiplication with invert-ions can be set up), we propose a problem:

We want A to be converted (e.g. moved) to B, in order to allow a residual of this action to convert C to D. The quality of (A, B) need not be the same as that of (C, D). Nor, indeed, need the qualities A, B, C, D be the same or similar to each other.
We define, by assumption, that it takes (at least) an operation "p" to convert A to B:

A [p] B   (a simplest case relationship)

For reversibility (of motion), an equivalent operation is used to convert B to A ; although this operation need not be the same, nor even the same kind, as "p" :

$\therefore$ B [d] A
$\therefore$ a looping is established:

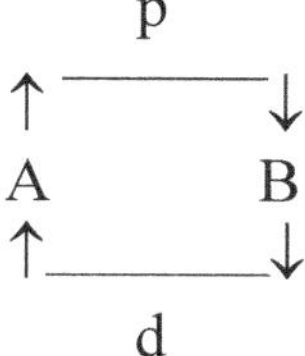

(This looping might be physically prohibited, but is included to produce, or help produce, a balanced tablature.)

$\therefore$ A [p; d] B   or ^{p}A [ ] ^{d}B

<u>Second</u>: To establish an estimating tabular form, it seems expedient (or productive or advantageous) that as

A $\leftrightarrow$ B

then

C $\leftrightarrow$ D

We consider the form of an operation statement that brings these elements into context with each other. Thereby, we must suggest that if "p" is done, then ' C $\leftrightarrow$ D ' (or, if "d" is done, then ' D $\leftrightarrow$ C ').

$\therefore$ The form of a tabular row might look as follows:

| C p D |

But as operations must be provided for (the production of) C and D, then we may "queue" C as p and D as d:

$$\therefore \qquad |\; p \quad p \quad d \;|$$

Already, a hyper analysis may be suggested.
But the operation to allow ' C $\leftrightarrow$ D ' might be designated as "q" :

$$\therefore \qquad |\; q \quad p \quad t \;|$$

Then obviously there should be the correspondence with:

$$|\; t \quad d \quad q \;|$$

So now, we have 3 suggestive (sets of) operations which may be tabulated:

$$
\begin{aligned}
&|\; p \quad p \quad d \;| \;\rightarrow\; &&\text{strictly for } A \leftrightarrow B \\
&|\; q \quad p \quad t \;|\rceil \;\rightarrow\; &&\text{relational for } A \leftrightarrow B \\
&|\; t \quad d \quad q \;|\rfloor \;\rightarrow\; &&C \leftrightarrow D
\end{aligned}
$$

We include a partition (since the first "p" of row 1 might be distinctive from the 2ⁿᵈ "p" of row 1 and the "p" of row two):

$$
\begin{aligned}
\therefore \qquad &|\; \hbar : \hbar \quad d \;| \qquad\qquad \hbar \equiv \text{some form of "p"} \\
&|\; q : \hbar \quad t \;| \\
&|\; t : d \quad q \;|
\end{aligned}
$$

$$\therefore$$

One relationship might be:

$$
\begin{aligned}
\text{global} \quad &\hbar|\; 1 : 1 \quad d \;| \\
\text{side} \quad &\hbar|\; q : 1 \quad t \;| \\
&|\; t : d \quad q \;|
\end{aligned}
$$

Another might be:

$$
\begin{aligned}
&\hbar|\; 1 : 1 \quad d \;| \\
&|\; q : 1 \quad t \;|\hbar \;\text{restricted} \\
&|\; t : d \quad q \;| \quad \text{side}
\end{aligned}
$$

(We "hyper" on "p" because, for example, it is allowed to be (most) common among the partitions, or for some other propitious reason or reasoning.)

Now, we can annotate a table for "results" simply as:
e.g.

$$
\begin{aligned}
&\quad C \quad A \quad D \\
&\hbar|\; 1 : 1 \quad d \;| \qquad\qquad \text{B must be somehow implied (by "d").} \\
&|\; q : 1 \quad t \;|\hbar \\
&|\; t : d \quad q \;| \qquad\qquad C \rightarrow D \text{ via } A \,(\rightarrow B)
\end{aligned}
$$

54

and we can more specify the "hypers" according to this annotation (to help follow relationships):

∴ cħ, ₐħ

i.e.

$$\begin{array}{ccc} C & A & D \\ | \; cħ & : \; _Aħ \quad d \; | \\ | \; q & : \; _Aħ \quad t \; | \\ | \; t & : \; d \quad q \; | \end{array} \qquad \begin{array}{c} A \;_{d}{\leftrightarrow}_{p}\; B \\ C \;_{t}{\leftrightarrow}_{q}\; D \end{array}$$

Note that we have not established an unitary-ness yet (which of course can not be formally partitioned, or for which partitioning is unnecessary).

The usage of "hypers" and the creation of a partition are to enhance the possibilities for an unitary-ness to develop of the problem. They are not necessarily to be utilized.

<u>Third</u>: Now, we must try to resemble an unitary-ness from the selected table, noting that

$p{\cdot}d \equiv 1 \; , \; q{\cdot}t \equiv 1$

$$\begin{array}{l} ∴ \quad ħ| \; 1 : 1 \quad d \; | \\ p{\cdot} \;\; | \; q : 1 \quad t \; |ħ \\ \quad\quad | \; t : d \quad q \; | \end{array} \;\rightarrow\; \begin{array}{l} ħ| \; p \; : \; p \quad d{\cdot}p \; | \\ | \; q{\cdot}p : p \quad t{\cdot}p \; |ħ \\ | \; t{\cdot}p : d{\cdot}p \; q{\cdot}p \; | \end{array} \;\rightarrow\; \begin{array}{l} ħ| \; 1{\cdot}p : 1{\cdot}p \quad 1 \; | \\ | \; q{\cdot}p : 1{\cdot}p \; t{\cdot}p \; |ħ \\ | \; t{\cdot}p : \; 1 \quad q{\cdot}p \; | \end{array}$$

$$\begin{array}{l} ∴ \\ \quad\quad ħ| \; 1 : 1 \quad d \; | \\ d{\cdot} \;\; | \; q : 1 \quad t \; |ħ \\ \quad\quad | \; t : d \quad q \; | \end{array} \;\rightarrow\; \begin{array}{l} ħ| \; d \; : \; d \quad d^2 \; | \\ | \; q{\cdot}d : d \quad t{\cdot}d \; |ħ \\ | \; t{\cdot}d : d^2 \; q{\cdot}d \; | \end{array}$$

$$\begin{array}{l} ∴ \quad ħ| \; 1 : 1 \quad d \; | \\ d^{-1}{\cdot} \; | \; q : 1 \quad t \; |ħ \\ \quad\quad | \; t : d \quad q \; | \end{array} \;\rightarrow\; \begin{array}{l} ħ| \; d^{-1} : d^{-1} \quad 1 \; | \\ | \; q/d : d^{-1} \; t/d \; |ħ \\ | \; t \; : \; 1 \quad q/d \; | \end{array}$$

None of these multiplications seem particularly advancing.

But now let us follow through with the hyperizations (of factors):

e.g.

$$\begin{array}{l} ħ| \; d \; : \; d \quad d^2 \; | \\ | \; q{\cdot}d : d \quad t{\cdot}d \; |ħ \\ | \; t{\cdot}d : d^2 \; q{\cdot}d \; | \end{array} \;\rightarrow\; \begin{array}{l} | \; d{\cdot}ħ : d{\cdot}ħ \quad d^2{\cdot}ħ \; | \\ | \; q{\cdot}d : d{\cdot}ħ \; t{\cdot}d{\cdot}ħ \; | \\ | \; t{\cdot}d : d^2 \quad q{\cdot}d \; | \end{array}$$ ħ is hyper for "p" (here)

$$∴ \quad \begin{array}{l} | \; d{\cdot}h : d{\cdot}h \; d{\cdot}d{\cdot}h \; | \\ | \; q{\cdot}d : d{\cdot}h \; t{\cdot}d{\cdot}h \; | \\ | \; t{\cdot}d : d^2 \quad q{\cdot}d \; | \end{array} \;\rightarrow\; \begin{array}{l} | \; 1 \; : \; 1 \quad d \; | \\ | \; q{\cdot}d : 1 \quad t \; | \\ | \; t{\cdot}d : d^2 \; q{\cdot}d \; | \end{array}$$

This could be, of course:

$$\begin{array}{l} | \; 1 \; : \; 1 \quad d \; | \\ ℌ| \; 1 : ℌ^{-1} \; t{\cdot}ℌ^{-1} \; | \\ | \; t{\cdot}d : d^2{\cdot}ℌ^{-1} \quad 1 \; |ℌ \end{array} \qquad ℌ \equiv q{\cdot}p \;\; (ħ \equiv p)$$

∴

$$\begin{array}{l} \text{ℏ}\,|\ 1\ :\ 1\ \ d\ | \\ d\cdot\ \text{ℌ}\,|\ q\ :\ 1\ \ t\ |\text{ℏ} \\ \quad|\ t\ :\ d\ \ q\ |\text{ℌ} \end{array} \rightarrow \begin{array}{l} \quad|\ 1\ :\ \ \ 1\quad\ \ d\quad\ | \\ \text{ℌ}\,|\ 1\ :\ \text{ℌ}^{-1}\ \ t\cdot\text{ℌ}^{-1}\ | \\ \quad|\ t\cdot d\ :\ d^2\cdot\text{ℌ}^{-1}\ \ 1\quad\ |\text{ℌ} \end{array} \rightarrow \begin{array}{l} |\ 1\ :\ 1\ \ d\ | \\ |\ \text{ℌ}\ :\ 1\ \ t\ | \\ |\ t\cdot d\ :\ d^2\ \text{ℌ}\ | \end{array}$$

Note (from observations of analyses) that:

$t\cdot d \approx d^2$ and $d \approx t$ (and $t = t,\ d = d,\ q = q$)

∴ If we replace variables, using hyper analogies:

$$(d \approx t) \rightarrow \text{Ŕ}$$
$$(t\cdot d \approx d^2) \rightarrow \text{Ŕ}^2$$

then we have:

$$\begin{array}{l} |\ 1\ :\ 1\ \ \text{Ŕ}\ | \\ |\ \text{ℌ}\ :\ 1\ \ \text{Ŕ}\ | \\ |\ \text{Ŕ}^2\ :\ \text{Ŕ}^2\ \text{ℌ}\ | \end{array}$$

Note also that:

$\text{ℌ} \equiv q\cdot d \approx p\cdot d \equiv 1$

∴ let $\text{ℌ} \equiv 1_{\text{ℌ}} (= 1\cdot\text{ℌ})$

∴ we have :

$$\begin{array}{l} |\ 1\ \ :\ 1\ \ \text{Ŕ}\ | \\ |\ 1_{\text{ℌ}}\ :\ 1\ \ \text{Ŕ}\ | \\ |\ \text{Ŕ}^2\ :\ \text{Ŕ}^2\ 1_{\text{ℌ}}\ | \end{array}$$

a tabular form nearing unitary-ness
and possessing perhaps some desirable
symmetries of solution

This (symmetry) could have been predicted from the original (but of non-unitary-ness):

$$\begin{array}{l} |\ p\ p\ d\ | \\ |\ q\ p\ t\ | \\ |\ t\ d\ q\ | \end{array}$$

since: $p = p,\ q \approx p,\ t \approx d$

∴ We have progressed from:

$$\begin{array}{l} |\ p\ p\ d\ | \\ |\ q\ p\ t\ | \\ |\ t\ d\ q\ | \end{array} \quad \text{to}\quad \begin{array}{l} \text{ℏ}\,|\ 1\ :\ 1\ d\ | \\ |\ q\ :\ 1\ t\ |\text{ℏ} \\ |\ t\ :\ d\ q\ | \end{array} \quad \text{to}\ (\,d\cdot\,)\ \text{this :}$$

(a selection)   (attempting unitary-ness)

$$\begin{array}{l} \text{ℏ}\,|\ d\ :\ d\quad\ d^2\ | \\ |\ q\cdot d\ :\ d\quad t\cdot d\ |\text{ℏ} \\ |\ t\cdot d\ :\ d^2\ q\cdot d\ | \end{array} \quad \text{to}\quad \begin{array}{l} |\ 1\ :\ 1\ d\ | \\ |\ \text{ℌ}\ :\ 1\ t\ | \\ |\ t\cdot d\ :\ d^2\ \text{ℌ}\ | \end{array} \quad \text{to}$$

(further hyperization)   (hyper analogies)

$$\mid\ 1\ \vdots\ 1\quad \mathfrak{K}\ \mid \qquad\qquad \mathfrak{h} \equiv \text{``p''} \qquad \mathfrak{H} \equiv \text{``q}\cdot\text{d''} \quad \mathfrak{K} \equiv d \approx t \qquad \mathfrak{K}^2 \equiv \text{``t}\cdot\text{d''} \approx \text{``d2''}$$
$$\mid\ 1_{\mathfrak{H}}\ \vdots\ 1\quad \mathfrak{K}\ \mid \qquad\qquad 1_{\mathfrak{H}} \equiv \mathfrak{H} \equiv \text{``q}\cdot\text{d''} \approx \text{``p}\cdot\text{d''} \equiv \text{``1''}$$
$$\mid\ \mathfrak{K}^2\ \vdots\ \mathfrak{K}^2\ \ 1_{\mathfrak{H}}\ \mid \qquad\qquad \mathfrak{H} \equiv 1_{\mathfrak{H}} = 1\cdot\mathfrak{H}$$

$\qquad\qquad\downarrow$ Obviously, if we <u>force</u> $\mathfrak{K} = 1$, $\mathfrak{H} \to 1_{\mathfrak{H}} = 1$ ($\mathfrak{K}^2 \equiv 1\cdot 1 = 1$)

$\qquad\qquad\downarrow$ then we have unitary-ness:

$$\mid\ 1\ \vdots\ 1\ \ 1\ \mid$$
$$\mid\ 1\ \vdots\ 1\ \ 1\ \mid \qquad\qquad \text{a unitary-ness (with superfluous partitioning)}$$
$$\mid\ 1\ \vdots\ 1\ \ 1\ \mid \qquad\qquad A\ _d\!\leftrightarrow_p\ B \qquad C\ _t\!\leftrightarrow_q\ D$$

$\therefore$ We must focus on "d" :

$\qquad\qquad \mathfrak{H} \equiv q\cdot d \qquad\quad$ suggests $\qquad\quad C\ _q\!\to D\ _{(d)}\!\to A$

$\qquad\qquad\qquad d \qquad\qquad\quad$ suggests $\qquad\quad B\ _d\!\to A$

$\therefore$ A route implied is: $\qquad\qquad A \leftrightarrow B$

$$\begin{array}{ccccccc} & B & & & \uparrow & & & B\ _d\!\to A \\ C\ _q\!\to D\ \nearrow\ ^d\!\searrow A & & \text{or} & & C \leftrightarrow D & & \text{or} & & C\ _q\!\to D \end{array}$$

$\therefore$ To force:

$\qquad\qquad \mathfrak{K} = 1,\ 1_{\mathfrak{H}} = 1$ (i.e. $d \approx t$, $q \approx p$)

we must make (or find an operation for):

$\qquad\qquad D \to B$

This appears to have the operations circular, or looping:

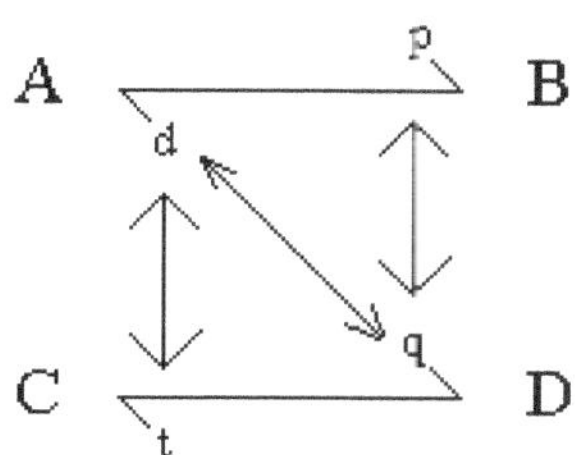

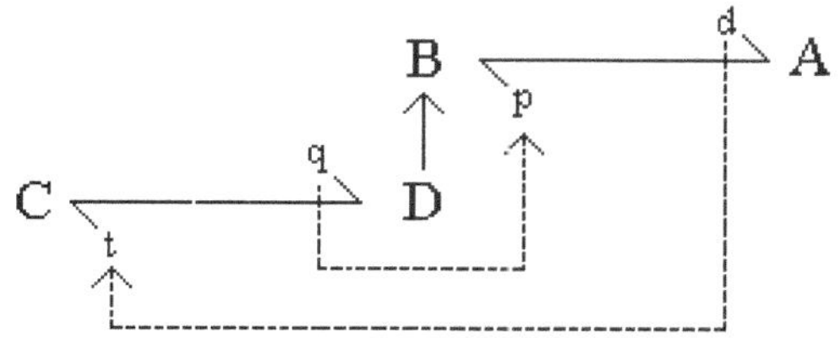

The (directional) "order" of the operations appears to derive (or result) from the selections (among several possible) made in converting the (original) hypothesis table into one of unitary-ness, as well as construction of the originating table itself.

If we follow the hyper analyses and analogies, we may re-annotate the previous figure as:

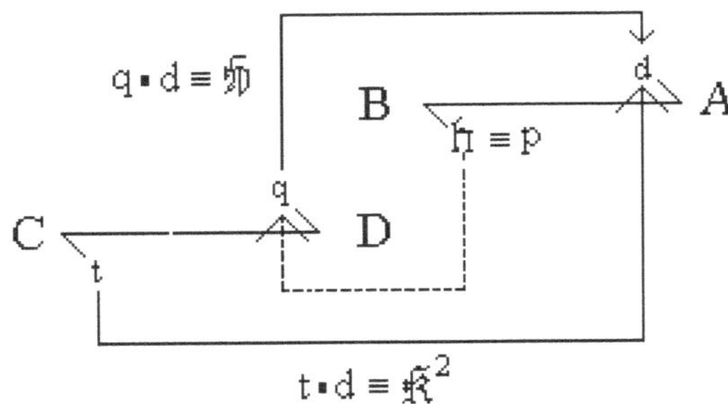

with an apparent tabular solution (or description):

```
|  1  : 1  1 |
|  ℌ  : 1  1 |
| ℜ2 : 1  1 |
```

Clearly (as the figure, with its arrow directions, is drawn), an actual solution would be:

```
| ♮  : q  d |        | ♮  : ℌ  1 |        | ♮  : 1  ℌ |        | 1  : ♮  ℌ |
| ℌ  : 1  1 |   or   | ℌ  : 1  1 |   or   | ℌ  : 1  1 |   or   | ℌ  : 1  1 |
| ℜ2 : 1  1 |        | ℜ2 : 1  1 |        | ℜ2 : 1  1 |        | ℜ2 : 1  1 |
```

$$\text{since } \quad ℜ2 \equiv t{\cdot}d \qquad \text{leads from } D \to C \to A$$
$$ℌ \equiv q{\cdot}d \qquad \text{leads from } C \to D \to A$$

and $♮{\cdot}ℌ \equiv p{\cdot}q{\cdot}d$ leads from $(\,A \to\,)\,B \to (\,C \to\,)\,D \to A$

(i.e. $t \equiv D \to C$, $d \equiv B \to A$, $q \equiv C \to D$, $p \equiv A \to B$)

Either of these solutions (tables) may be used.

The original problem (question) was (how) to go from ' $A \to B$, $C \to D$ ' occurring as a result (or *efficacious*).

We assume that we may simply take the complement operations of the previous:
e.g.

```
| 1  : ♮  ℌ | complementation | 1  :  d t·♮ |      B → A, D → C, A → B
| ℌ  : 1  1 |        ↔         | t·♮ : 1  1  |      D → C, A → B
| ℜ2 : 1  1 |                  | q·♮ : 1  1  |      C → D, A → B
```

Obviously, these complementary solutions would seem rather obscure to deduce without previous derivation of the compliment tables.

The three routes (or paths) suggested by these quasi-unitary-ness solutions
(where unitary-ness ≡ correctness) are:

$B{\to}A$, $D{\to}C$, $A{\to}B$

$D{\to}C$, $A{\to}B$

$C{\to}D$, $A{\to}B$

The last seems most reasonable, saying that ' $C{\to}D$ ' occurs first, before achieving ' $A{\to}B$ '. But all three (i.e. each) must be considered (at least initially) as plausible.

These solutions are from a quasi-unitary-ness table due to the hyper analyses and analogies and the partitioning. The partitioning would probably be forced by the ' $C{\to}A$ ' and ' $D{\to}A$ ' steps, for which operations have not been designed as yet (i.e. during the derivations):

From the $(C{\to}D{\to}B{\to}A)$ complementary table,

the $(C \xrightarrow{\text{ぶ2}} A)$ involves a $(C \to B)$ step of unknown operator

and the $(D \xrightarrow{\text{ゐ}} A)$ involves a $(D \to B)$ step of unknown operator.

The $(D \xrightarrow{\text{ぅ·ゐ}} A)$ involves a $(D{\to}B)$ step of unknown operator, while implying that both the $(A{\to}B)$ and $(C{\to}D)$ steps occur before this, to finally allow for the $(B{\to}A)$ step.

The complementarity is very evident:

simply replace   A for B (or B for A)
        and   C for D (or D for C).

e.g.

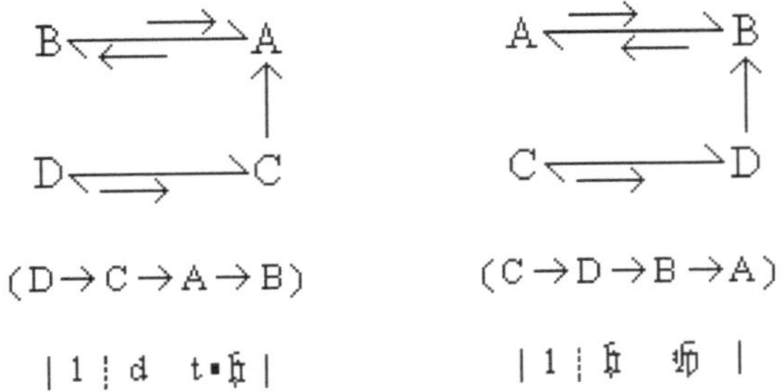

$(D \to C \to A \to B)$ $\qquad$ $(C \to D \to B \to A)$

But the partitioning between operations is well-defined of location, suggesting (here) that a (C→A) or (D→B) step can not occur as a first (set of) operation(s), nor a (C→B) or (D→A) step. This of course follows initiations (of operation) being (A↔B) or (C↔D). These undefined operations are intra-relational (occur within relationship statements) but (are) inter-operational/defined (occur between the A↔B and C↔D operations).

How might these undefined operations be described (or eventually defined)?
They appear, from our example, to be of 3 types:

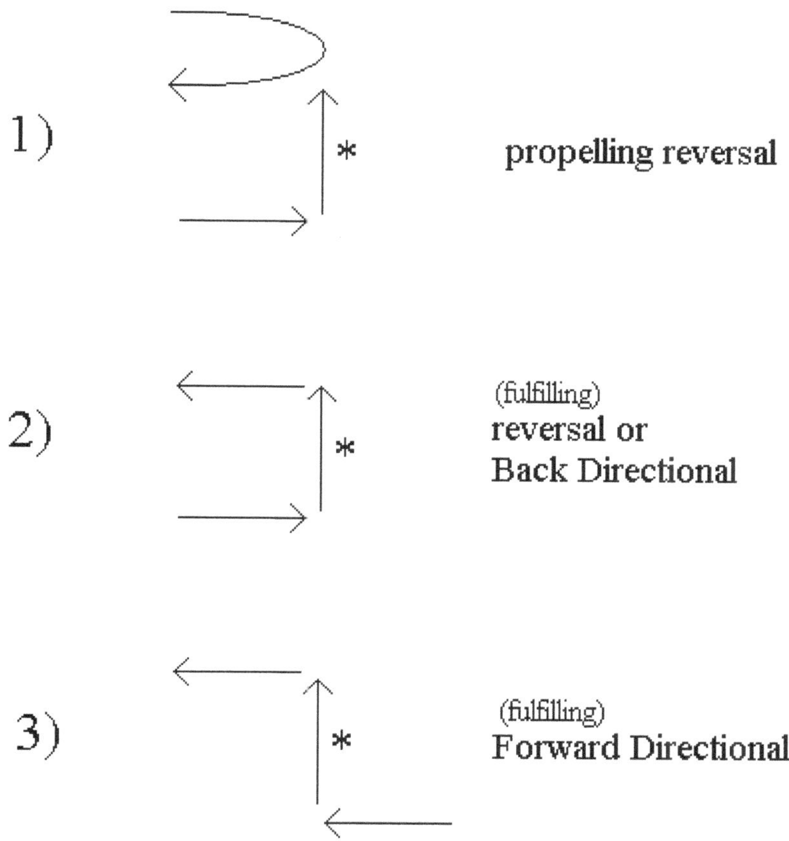

1) propelling reversal

2) (fulfilling)
reversal or
Back Directional

3) (fulfilling)
Forward Directional

and probably fall within the jurisdiction of a partition (latter side) of our solution table:

$$
\begin{array}{lll}
: 1 \quad \tilde{R} \mid & \leftarrow \quad : 1 \quad d \mid \leftarrow & : 1 \quad d \mid \\
: 1 \quad \tilde{R} \mid & \equiv \quad : 1 \quad d \mid \equiv & : 1 \quad t \mid \\
: \tilde{R}2 \; 1_{\hbar} \mid & : t \cdot d \quad q \cdot d \mid & : d2 \quad q \cdot d \mid
\end{array}
$$

(which is clearly unbalanced).

Removing the initial (partitioned off) operations, the forms are:

$$
\begin{array}{lll}
\circlearrowright & \leftarrow & \leftarrow \\
\uparrow* & \uparrow* & \uparrow* \quad \text{(with the apparent identities below)} \\
\mid \tilde{R}2 \; 1_{\hbar} \mid & \mid 1 \quad d \mid & \mid 1 \quad d \mid
\end{array}
$$

∴ The undefined operations (*) are clearly related as:

$$\mathfrak{R}^2 \cdot \mathfrak{H} \qquad\qquad d\ (\equiv \mathfrak{R}) \qquad\qquad d\ (\equiv \mathfrak{R})$$

$$\downarrow$$

$$(d{\cdot}d{\cdot}q){\cdot}d$$
or $\quad d{\cdot}(d{\cdot}q{\cdot}d)$
or $\quad d){\cdot}d{\cdot}(q{\cdot}d$

Since "d" must occur last (to yield A), the generalized (by order) undefined operations are (from $|\ \mathfrak{R}^2\ \ 1_{\mathfrak{H}}\ |$): $d{\cdot}d{\cdot}q$, $d{\cdot}q{\cdot}d$, $q{\cdot}d{\cdot}d$ $\qquad$ (each followed by "d") vectorially or order dependent, and $d^2{\cdot}q$, $q{\cdot}d^2$ being most equivalent.

∴

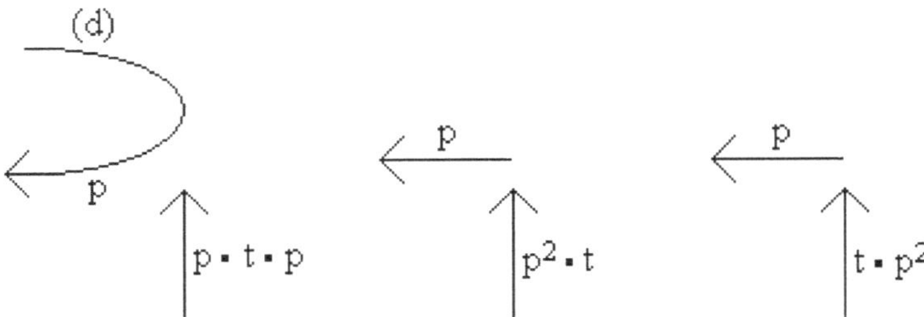

and the complementary ones (to yield B) are:

The (d·q·d) or (p·t·p) ordering, therefore, generates a sort of propulsion or rotor activity, the (q) or (t) somehow resetting matters for another (d) or (p) to occur.

∴ This undefined operation must be regenerative.

This suggests a general rule or condition: that when an operation is (immediately) surrounded by the complement of its opposite operation (or recovering operation), a rotor (propelling reversal) situation is produced. (e.g. d·q·d, p·t·p)

The forms can be substantially modified or complex-ed by changing or exchanging members within a partition.

e.g.

$$
\begin{array}{|ccc|} \hline 1 : \mathfrak{h} \; \mathfrak{H} \\ \mathfrak{H} : 1 \; 1 \\ \mathfrak{R}2 : 1 \; 1 \\ \hline \end{array}
\quad \rightarrow \quad
\begin{array}{|ccc|} \hline \mathfrak{R}2 : \mathfrak{h} \; \mathfrak{H} \\ \mathfrak{H} : 1 \; 1 \\ 1 : 1 \; 1 \\ \hline \end{array}
$$

forms:

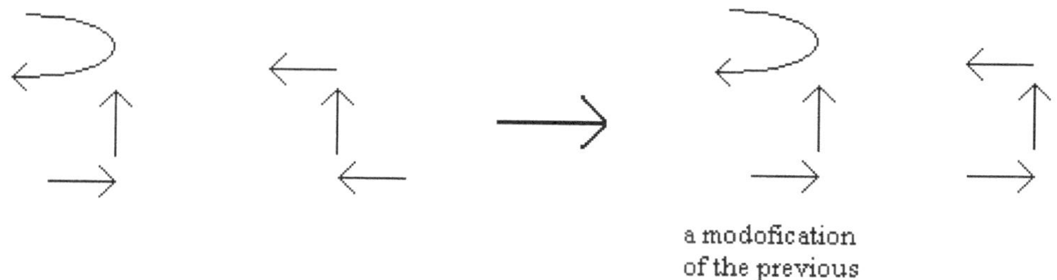

a modofication
of the previous

A possibly interesting complex-ation could be:

$$
\begin{array}{|ccc|} \hline 1 : \mathfrak{h} \; \mathfrak{H} \\ \mathfrak{H} : 1 \; 1 \\ \mathfrak{R}2 : 1 \; 1 \\ \hline \end{array}
\quad \rightarrow \quad
\begin{array}{|ccc|} \hline 1 : 1 \; 1 \\ \mathfrak{H} : \mathfrak{h} \; \mathfrak{H} \\ \mathfrak{R}2 : 1 \; 1 \\ \hline \end{array}
$$

forms:

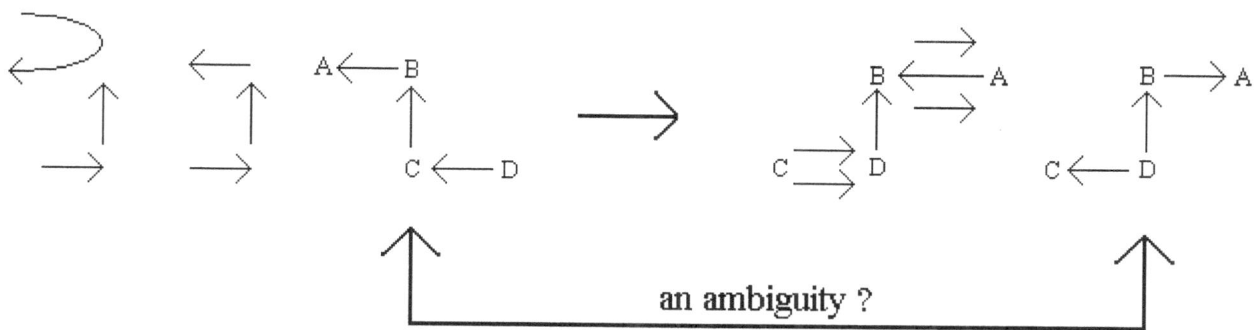

The form (or route):

might be interpreted as

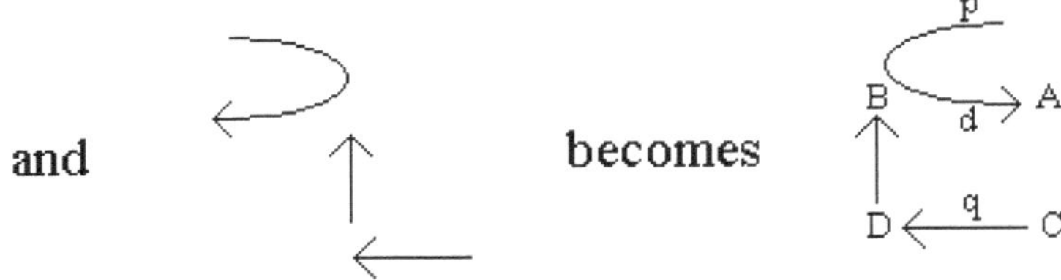

The ambiguity (for ' $R2$ ') between:

$$A \leftarrow_d B \qquad \text{and} \qquad B \ _d\!\rightarrow A$$
$$\uparrow \qquad\qquad\qquad \uparrow$$
$$C \leftarrow_t D \qquad\qquad C \leftarrow_t D$$

seems to depend on which undefined operation (D→B or C→B) is chosen. The overall result for either is (D→A). If we declare that initiates (A, C) and resultants (B, D) must be ranked (or columned) similarly, then the more proper form is:

$$B \ _d\!\rightarrow A$$
$$\uparrow \qquad\qquad\qquad \text{(although this is not "single/directional continuous flow")}$$
$$D \ _t\!\rightarrow C$$

and becomes

The

form might also be interpreted as having two undefined operations:

D→B, and a hidden or uncovered D $_t$→C

64

∴

$\left| \mathfrak{h} \;\vdots\; \mathfrak{h} \;\; \mathfrak{H} \right| \equiv$ 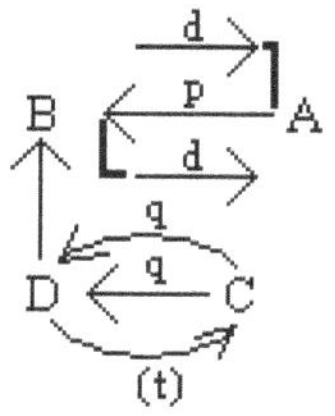

clearly a complexity from earlier (previous) forms.
Further modifications (searching for forms) may come into play using nul/los principles, when we divide tables:

e.g.

$$\frac{\left|\begin{array}{c:cc} \mathfrak{K}^2 & \mathfrak{h} & \mathfrak{H} \\ \mathfrak{H} & 1 & 1 \\ 1 & 1 & 1 \end{array}\right|}{\left|\begin{array}{c:cc} 1 & 1 & 1 \\ \mathfrak{H} & \mathfrak{h} & \mathfrak{H} \\ \mathfrak{K}^2 & 1 & 1 \end{array}\right|} \equiv \frac{\left|\begin{array}{c:cc} t\cdot d & p & q\cdot d \\ q\cdot d & 1 & 1 \\ 1 & 1 & 1 \end{array}\right|}{\left|\begin{array}{c:cc} 1 & 1 & 1 \\ q\cdot d & p & q\cdot d \\ t\cdot d & 1 & 1 \end{array}\right|} \to \left|\begin{array}{c:cc} t\cdot d/1 & p/1 & q\cdot d/1 \\ q\cdot d/(q\cdot d) & 1/p & 1/(q\cdot d) \\ 1/(t\cdot d) & 1/1 & 1/1 \end{array}\right|$$

$$\to \left|\begin{array}{c:cc} t\cdot d & p & q\cdot d \\ 1 & p^{-1} & (q\cdot d)^{-1} \\ (t\cdot d)^{-1} & 1 & 1 \end{array}\right| \equiv \left|\begin{array}{c:cc} \mathfrak{K}^2 & p & q\cdot d \\ 1 & p^{-1} & (q\cdot d)^{-1} \\ (\mathfrak{K}^2)^{-1} & 1 & 1 \end{array}\right|$$

$$\to \left|\begin{array}{c:cc} \mathfrak{K}^2 & \mathfrak{h} & \mathfrak{H} \\ 1 & \mathfrak{h}^{-1} & \mathfrak{H}^{-1} \\ (\mathfrak{K}^2)^{-1} & 1 & 1 \end{array}\right| \equiv \left|\begin{array}{c:cc} \mathfrak{K}^2 & \mathfrak{h} & \mathfrak{H} \\ 1 & \mathfrak{h}^{-1} & \mathfrak{H}^{-1} \\ (\mathfrak{K}^2)^{-1} & 1 & 1 \end{array}\right|$$

Then:

$$\mathfrak{h}\cdot\mathfrak{H} \left|\begin{array}{c:cc} \mathfrak{K}^2 & \mathfrak{h} & \mathfrak{H} \\ 1 & \mathfrak{h}^{-1} & \mathfrak{H}^{-1} \\ \mathfrak{K} & 1 & 1 \end{array}\right| \to \left|\begin{array}{c:cc} \mathfrak{K}^2 & \mathfrak{h} & \mathfrak{H} \\ \mathfrak{h}\cdot\mathfrak{H} & \mathfrak{H} & \mathfrak{h} \\ \mathfrak{K} & 1 & 1 \end{array}\right| \equiv \left|\begin{array}{c:cc} t\cdot d & p & q\cdot d \\ p\cdot q\cdot d & q\cdot d & q \\ d & 1 & 1 \end{array}\right|$$

$p\cdot q\cdot d \equiv q$; $p\cdot d \equiv 1$ (simplifications)

∴

$$\left|\begin{array}{c:cc} t\cdot d & p & q\cdot d \\ q & q\cdot d & q \\ d & 1 & 1 \end{array}\right| \equiv \left|\begin{array}{c:cc} t\cdot d & 1 & q \\ q & q\cdot d & q \\ d & 1 & 1 \end{array}\right| \text{ or } \left|\begin{array}{c:cc} t\cdot d & q & 1 \\ q & q\cdot d & q \\ d & 1 & 1 \end{array}\right| \text{ as for results.}$$

Comment:

The use of exponential functions such that

$$x^1/x^1 = x^{(1-1)} = x^0$$

is not allowed in nul/los (for the "zero" occurring). However, in regular arithmetic,

$$x^0 \text{ is simply defined as } (=) 1 ,$$

another inconsistency in the use of zero by regular math (e.g. $X/0 = \infty$, $0! = 1$).

The notion of something being divided by itself (to yield "1") is not only evident but permissible in nul/los, and is much more clearer of conception and faithful to physicality than the picture drawn by: something divided by nothing yields everything.

For a table (tabular form), each row representing operations for a solution, the order of the assembly must be columnar and therefore (with judicious simplifications) variable:
e.g., for

```
| ℛ² :  p    q·d  |
|  1  :  p⁻¹  (q·d)⁻¹ |
| ℛ⁻² :  1    1    |      the columnar products, as ordered here, are:
```

$$K^2 \cdot 1 \cdot K^{-2} \rightarrow 1 \qquad \text{column 1}$$
$$p \cdot p^{-1} \cdot 1 \rightarrow 1 \qquad \text{column 2}$$
$$(q \cdot d) \cdot (q \cdot d)^{-1} \cdot 1 \rightarrow 1 \qquad \text{column 3}$$

$$\therefore \; | \; 1 : 1 \quad 1 \; |$$

$\therefore$ The order of the table is: $1 \cdot 1 \cdot 1 = 1$ (columnar order)

Obviously this is preferable for (a check on) correctness of solutions.
Then, as for previous examples, we can see the effects of table modifications on (columnar) orders:
e.g.

```
              | ℛ² :  ℏ    𝔥 |              | t·d :   p   q·d |
       ℏ·𝔥|  1 :  ℏ⁻¹  𝔥⁻¹ |  →→          | q  :  q·d   q  |
              | ℛ  :  1    1 |              | q  :   1    1  |

              | t·d :  1   q |              | t·d :  q    1 |
   →          |  q  : q·d  q |     or       |  q  : q·d   q |
              |  d  :  1   1 |              |  d  :  1    1 |
```

orders:

column 1	$t \cdot d \cdot q \cdot d \equiv d^2$	$t \cdot d \cdot q \cdot d \equiv d^2$
column 2	$1 \cdot q \cdot d \cdot 1 = q \cdot d$	$q \cdot q \cdot d \cdot 1 = q^2 \cdot d$
column 3	$q \cdot q \cdot 1 = q^2$	$1 \cdot q \cdot 1 = q$
$\therefore$	$\| \; d^2 : q \cdot d \quad q^2 \; \|$	$\| \; d^2 : q^2 \cdot d \quad q \; \|$

These two are obviously equivalent (through multiplication commutabilities: i.e. $q \cdot d \cdot q^2 \equiv q^2 \cdot d \cdot q$).

Note also that we can "enhance" (or increase) the occurrence of "1"s in a table using multiplicative commutability (while maintaining the order of operations):

e.g.

$$\left|\begin{array}{ccc} t\cdot d: & 1 & q \\ q: & q\cdot d & q \\ d: & 1 & 1 \end{array}\right| \rightarrow \left|\begin{array}{ccc} t\cdot d: & 1 & q \\ q2: & 1 & d\cdot q \\ d: & 1 & 1 \end{array}\right| \quad \text{or} \quad \left|\begin{array}{ccc} t\cdot d: & q & 1 \\ q: & q\cdot d & q \\ d: & 1 & 1 \end{array}\right| \rightarrow \left|\begin{array}{ccc} t\cdot d: & q & 1 \\ q2: & 1 & d\cdot q \\ d: & 1 & 1 \end{array}\right|$$

For:

$$\left|\begin{array}{ccc} t\cdot d: & 1 & q \\ q2: & 1 & d\cdot q \\ d: & 1 & 1 \end{array}\right| \qquad \left|\begin{array}{ccc} t\cdot d: & q & 1 \\ q2: & 1 & d\cdot q \\ d: & 1 & 1 \end{array}\right|$$

<u>orders:</u>

column 1	$t\cdot d\cdot q\cdot q\cdot d \equiv d\cdot q\cdot d$	$t\cdot d\cdot q\cdot q\cdot d \equiv d\cdot q\cdot d$
column 2	$1\cdot 1\cdot 1 = 1$	$q\cdot 1\cdot 1 = q$
column 3	$q\cdot d\cdot q\cdot 1 = q\cdot d\cdot q$	$1\cdot d\cdot q\cdot 1 = d\cdot q$

$\therefore$

$\approx\equiv$

$$\left|\begin{array}{ccc} d\cdot q\cdot d: & 1 & q\cdot d\cdot q \\ d2\cdot q: & 1 & q2\cdot d \end{array}\right| \qquad \left|\begin{array}{ccc} d\cdot q\cdot d: & q & d\cdot q \\ d2\cdot q: & q & d\cdot q \end{array}\right|$$

These equivalent columnar orders may individually have distinct procedural or revelatory advantages over the others:

$$\begin{array}{ll} \left|\, d2: \ q\cdot d \ \ q2 \,\right| & \text{squares separated} \\ \left|\, d2: \ q2\cdot d \ \ q \,\right| & \text{squares neighboring} \\ \left|\, d2\cdot q: \ 1 \ \ q2\cdot d \,\right| & \text{symmetrical squaring} \\ \left|\, d2\cdot q: \ q \ \ d\cdot q \,\right| & \text{reducing from square(s)} \end{array}$$

Whether these delineations are permissible (or possible) depends on the nature or characteristics of the partitioning.

e.g.

$$\left|\, q: \ q\cdot d \ \ q \,\right| \rightarrow \left|\, q2: \ 1 \ \ d\cdot q \,\right| \quad \text{(within a table)}$$

might not be allowable (runs across partitions),

but

$$\left|\, d2: \ q2\cdot d \ \ q \,\right| \rightarrow \left|\, d2: \ q2 \ \ d\cdot q \,\right|$$

for this columnar order(ing) is allowable (occurs within a partition).

$\therefore$ The columnar orders are used to judge tables.

An assumption, or desire, of a table, is of course that it presents equal though comparable solutions (of processes of operations):

e.g., for

$$| \; t \cdot d : \quad 1 \quad q \; |$$
$$| \; q \; : q \cdot d \quad q \; |$$
$$| \; d \; : \quad 1 \quad 1 \; |$$

it is fairly evident that:

$| \; t \cdot d : \quad 1 \quad q \; |$ is equivalent to $| \; d : \quad 1 \quad 1 \; |$ since $t \cdot q \equiv 1$

$\therefore$

$| \; t \cdot d : \quad 1 \quad q \; | \; \cong \; | \; t \cdot d \cdot q : \quad 1 \quad 1 \; |$ (due to quasi-partitioning, with $t \cdot q \equiv 1$)

$\rightarrow \; | \; d \cdot 1 : \quad 1 \quad 1 \; | \; \equiv \; | \; d : \quad 1 \quad 1 \; |$

The $| \; q : q \cdot d \quad q \; |$ solution, though, is much more obscure (to this eventuality).

We know that it comes from a table of columnar order: "squares separated" ($\therefore$ might be reducible).

$\therefore$ We might analyze with (nul/los) divisions:

(col. order $\equiv$) CO $\quad | \; d^2 : q \cdot d \quad q^2 \; |$

$$\overline{\qquad\qquad\qquad\qquad} \qquad \rightarrow \qquad | \; d \cdot (d/q) : \quad 1 \quad q \; |$$

(from table $\equiv$) FT $\quad | \; q : q \cdot d \quad q \; |$

e.g.

$\therefore \; q | \; d \cdot (d/q) : \quad 1 \quad q \; | \quad \rightarrow \quad | \; d^2 : \quad q \quad q^2 \; |$

$\qquad\qquad\qquad$ CO $\quad | \; d^2 : q \cdot d \quad q^2 \; |$

$$\overline{\qquad\qquad\qquad\qquad} \qquad \rightarrow \qquad | \; d : q \cdot d \quad q^2 \; |$$

$\qquad\qquad\qquad$ FT $\quad | \; d : \quad 1 \quad 1 \; |$

and $| \; d : q \cdot d \quad q^2 \; | \; \cong \; | \; d^2 : \quad q \quad q^2 \; |$

($\therefore \; | \; q : q \cdot d \quad q^2 \; | \; \cong \; | \; d : \quad 1 \quad 1 \; |$ might be accepted).

Here we note, of course, that multiplying a solution (as like an equation) by a common factor should not change (the validity of) the solution:

i.e. $q | \; d \cdot (d/q) : \quad 1 \quad q \; | \; \cong \; | \; d^2 : \quad q \quad q^2 \; | \; \cong \; | \; d \cdot (d/q) : \quad 1 \quad q \; |$

although

$| \; d \cdot (d/q) : \quad 1 \quad q \; | \; \neq \; | \; d^2 : \quad q \quad q^2 \; |$

but rather (the latter) is:

$| \; d \cdot (d/q) : \quad 1 \quad q \; |$ modified commutably by the operation 'q' .

Perhaps as evident that a 'q' operation is of notice is simply the observance that:

$$\frac{\mid \ q \ \vdots \ q \cdot d \ \ q \ \mid}{q} \quad \rightarrow \quad \mid \ 1 \ \vdots \ d \ \ 1 \ \mid \ \cong \ \mid \ d \ \vdots \ 1 \ \ 1 \ \mid$$

∴ to bring the table into (desirable) compliance might be:

$$\begin{array}{c} \mid \ t \cdot d \ \vdots \ 1 \ \ q \ \mid \\ q^{\text{-}1} \mid \ q \ \vdots \ q \cdot d \ \ q \ \mid \\ \mid \ d \ \vdots \ 1 \ \ 1 \ \mid \end{array} \quad \text{i.e.} \quad \begin{pmatrix} 1 \\ q^{\text{-}1} \\ 1 \end{pmatrix} \begin{array}{c} \mid \ t \cdot d \ \vdots \ 1 \ \ q \ \mid \\ \mid \ q \ \vdots \ q \cdot d \ \ q \ \mid \\ \mid \ d \ \vdots \ 1 \ \ 1 \ \mid \end{array}$$

↖ ordered operational factors (of a table applied)

It is clear that since each table contains a set (via its rows) of hopefully equivalent solutions, then each table is indicative of that solution and thus has the value of that solution. For the table previous, that value would seem to be most succinctly expressed as: $\mid \ d \ \vdots \ 1 \ 1 \ \mid$;
i.e. Its value is the operation 'd' , which corresponds (here) to: $\text{B} \rightarrow \text{A}$.

∴ apparently, tables themselves can be used as operations.

∴ We might (and may) have recursions such as:

$$\left| \left\{ \begin{array}{c} \mid \ t \cdot d \ \vdots \ \ 1 \ \ q \ \mid \\ q^{\text{-}1} \mid \ q \ \vdots \ q \cdot d \ \ q \ \mid \\ \mid \ d \ \vdots \ \ 1 \ \ 1 \ \mid \end{array} \right\} \begin{array}{c} \vdots \ (1) \ (1) \\ \vdots \ \ 1 \ \ \ 1 \\ \vdots \ (1) \ (1) \end{array} \right|$$

↖ may "fill-in" with "1"s

or "subversions" of the above:

$$\begin{array}{l} \mid \ \{ \ \mid t \cdot d \ \vdots \ 1 \ q \ \mid \ \} \ \vdots \ 1 \ 1 \ \mid \\ \mid \ \{ q^{\text{-}1} \mid q \ \vdots \ q \cdot d \ q \ \mid \} \ \vdots \ 1 \ 1 \ \mid \\ \mid \ \{ \ \mid d \ \vdots \ 1 \ 1 \ \mid \} \ \vdots \ 1 \ 1 \ \mid \end{array} \qquad \rightarrow \quad \text{each} \equiv \ \mid \ d \ \vdots \ 1 \ 1 \ \mid \ \equiv \ d$$

This could lead to powerful instances of simplification.

Multiplicity of solutions

The multiplicity of solutions in a table is of course an attempt to validate them from the initial proposals of the problem. This is based on the conjecture (or conviction) that *the same will lead to the same*; therefore, a comparison to itself (in a multiplicity of instances) yields to unitary-ness. And a balancing of terms (of operation), via rows to columns, affords a symmetry readily adaptable to a regular recognition of obvious patterns (of manipulation) that may be uncovered (during a search for this nature of unitary-ness), hoping to verify, amplify, *emphasize*, or at least correctly analyze the original proposals of the problem (for correctness).

∴ Modifications during the means to achieve unitary-ness help to determine the irregularities inherent in the hypotheses presented and struggled with. These hypotheses of operation form a basis for nul/los manipulations (or subjection of thesis: physical or logical, empirical or idealistic, etc.). In this realm of relationships, a "zero" quantity is not possible, since one starts with a definite entity (or idea) of some kind. Even concepts (such) as infinity, limits, limiting values, extinguishments, etc., may be so treated (symbolically). Yet, in reality, this is absolute; and in the physical world it is only what can be demonstrated (and measured).

But how might we further transform a given table (to some revelatory or exculpatory advantage), such as:

$$\begin{vmatrix} t \cdot d & : & 1 & q \\ q & : & q \cdot d & q \\ d & : & 1 & 1 \end{vmatrix} \quad ?$$

Obviously, we are looking for symmetries/patterns (perhaps distant from the original problem but inherent to the table derived thus far).

Row, Column (and Diagonal) Orders

We can make row (unchanging, or quasi so) and column (order unchanging) changes to see what results, all the while taking advantage of "unity manifestations" (along a row or column):

$\therefore$ for example

$$\begin{vmatrix} t \cdot d & : & 1 & q \\ q & : & q \cdot d & q \\ d & : & 1 & 1 \end{vmatrix} \quad \text{row} \rightarrow \quad \begin{vmatrix} 1 & : & t \cdot d & q \\ q & : & q \cdot d & q \\ d & : & 1 & 1 \end{vmatrix}$$

(balanced table) $\downarrow$column: column order (multiplicative product) maintained for table

$$\begin{vmatrix} d & : & t \cdot d & q \\ q & : & q \cdot d & q \\ 1 & : & 1 & 1 \end{vmatrix} \quad \text{row} \rightarrow \quad \begin{vmatrix} 1 & : & d^2 & 1 \\ q & : & q \cdot d & q \\ 1 & : & 1 & 1 \end{vmatrix} \quad \text{row} \rightarrow \quad \begin{vmatrix} 1 & : & d^2 & 1 \\ 1 & : & q \cdot d & q^2 \\ 1 & : & 1 & 1 \end{vmatrix}$$

$(t \cdot q \equiv 1)$

$$\equiv \quad \text{(via reduction or simplification to a 'minor' table)} \quad \begin{vmatrix} d^2 & 1 \\ q \cdot d & q^2 \end{vmatrix} \text{ (balanced table)}$$

This result would seem to relate to the functional similarities of the 'd' and 'q' operations (when coupled):

$$A \;{}_p\!\leftrightarrow_d B \qquad\qquad C \;{}_q\!\leftrightarrow_t D$$

$\therefore$ for example

$$A \leftarrow_d B$$
$$(x) \;\G \qquad\quad \circlearrowleft (y) \qquad \text{composite operations:}$$
$$C \;_q\!\rightarrow D \qquad\qquad\qquad \text{yielding to intermediary functions}$$

(x, y) may be intermediary products (results) to the intermediary functions.

An obvious complementation would be:

$$A \;_{\Gamma}\!\!\to B$$
$$(w)\,\llcorner \qquad \lrcorner\,(z) \qquad\qquad \begin{vmatrix} p^2 & 1 \\ t\cdot p & t^2 \end{vmatrix}$$
$$C \leftarrow_t D$$

$$\therefore$$

$$A \;_p\!\!\leftrightarrow_d B$$
$$(w)\circlearrowright(x) \quad (y)\circlearrowright(z)$$
$$C \;_q\!\!\leftrightarrow_t D$$

These newly derived tables, although far from the purport of the original problem, emphasize aspects of the earlier hyper analyses and analogies, therewith that: $'d \approx t'$ and $'q \approx p'$.

$$\therefore \quad q\cdot d \approx 1 \,,\; t\cdot p \approx 1 \quad (\, p\cdot d \equiv 1 \,,\; q\cdot t \equiv 1 \,)$$

It is interesting that to overlap them (via the "1") may yield to another balanced table (with filling in of "1"s):

$$(1)\begin{vmatrix} t^2 & t\cdot p \\ d^2 & 1 & p^2 \\ q\cdot d & q^2 & (1) \end{vmatrix} \quad \rightarrow \quad \begin{vmatrix} 1 & t^2 & t\cdot p \\ d^2 & 1 & p^2 \\ q\cdot d & q^2 & 1 \end{vmatrix}$$

with obvious pattern characteristics.

This, then, defines a new (i.e. another) operation (of manipulation) among tables.

We note that if we multiply along diagonals, while accepting that:

$$q\cdot d \approx 1 \;\text{ and }\; t\cdot p \approx 1\,,$$

we form unitary products:

$$\begin{array}{c}
d^2\cdot t^2 \\
\searrow \quad \searrow \quad \nearrow \quad \swarrow \; q^2\cdot p^2 \\
\searrow \begin{vmatrix} 1 & t^2 & t\cdot p \end{vmatrix} \nearrow \\
\begin{vmatrix} d^2 & 1 & p^2 \end{vmatrix} \\
\nearrow \begin{vmatrix} q\cdot d & q^2 & 1 \end{vmatrix} \searrow\; t^2\cdot p^2 \approx (t\cdot p)^2 \approx (1)^2 = 1 \\
\swarrow \quad \nearrow\searrow \quad \searrow\; 1\cdot 1\cdot 1 = 1 \\
(q\cdot d)\cdot 1\cdot (t\cdot p) \approx 1\cdot 1\cdot 1 = 1 \qquad d^2\cdot q^2 \approx (q\cdot d)^2 \approx (1)^2 = 1
\end{array}$$

Non-unitary products are, however, from:

$$d^2\cdot t^2 \;\text{ and }\; q^2\cdot p^2\,,$$

as are column products and two of the row products.

The center row product: $\begin{vmatrix} d^2 & 1 & p^2 \end{vmatrix}$
 yields to unitary value: $d^2\cdot 1\cdot p^2 \approx (d\cdot p)^2 \approx (1)^2 = 1$

The center column product also yields: $t^2\cdot 1\cdot q^2 \approx (t\cdot q)^2 \approx (1)^2 = 1$

Taking all of above:

$$| \ d^2 \ \ 1 \ \ p^2 \ | \approx | \ 1 \ \ 1 \ \ 1 \ | \ , \qquad \begin{vmatrix} t^2 \\ 1 \\ q^2 \end{vmatrix} \approx \begin{vmatrix} 1 \\ 1 \\ 1 \end{vmatrix} , \qquad \begin{matrix} t{\cdot}p \approx 1 \\ q{\cdot}d \approx 1 \end{matrix}$$

results in unitary-ness:

$$\begin{vmatrix} 1 & 1 & 1 \\ 1 & 1 & 1 \\ 1 & 1 & 1 \end{vmatrix}$$

If not accepting ($t{\cdot}p \approx 1$, $q{\cdot}d \approx 1$), the resulting table is:

$$\begin{vmatrix} 1 & 1 & t{\cdot}p \\ 1 & 1 & 1 \\ q{\cdot}d & 1 & 1 \end{vmatrix}$$
which must be the registration of the two earlier derived tables.

Note the diagonal products: $(q{\cdot}d){\cdot}1{\cdot}(t{\cdot}p) \approx (q{\cdot}t){\cdot}(p{\cdot}d) \approx (1){\cdot}(1) = 1$

$\therefore$ We have (through pattern manipulations):

$$\begin{vmatrix} d^2 & 1 \\ q{\cdot}d & q^2 \end{vmatrix} , \quad \begin{vmatrix} p^2 & 1 \\ t{\cdot}p & t^2 \end{vmatrix} , \quad \begin{vmatrix} 1 & t^2 & t{\cdot}p \\ d^2 & 1 & p^2 \\ q{\cdot}d & q^2 & 1 \end{vmatrix} , \quad \begin{vmatrix} 1 & 1 & t{\cdot}p \\ 1 & 1 & 1 \\ q{\cdot}d & 1 & 1 \end{vmatrix}$$

which leads towards:

$$\begin{vmatrix} 1 & 1 & 1 \\ 1 & 1 & 1 \\ 1 & 1 & 1 \end{vmatrix} \qquad \text{unitary-ness ;}$$

i.e.

$$\begin{vmatrix} 1 & 1 \\ q{\cdot}d & 1 \end{vmatrix} \quad \text{and} \quad \begin{vmatrix} 1 & t{\cdot}p \\ 1 & 1 \end{vmatrix}$$

Multiplying yields:

$$\begin{vmatrix} 1 & 1 \\ q{\cdot}d & 1 \end{vmatrix} \ \bullet \ \begin{vmatrix} 1 & t{\cdot}p \\ 1 & 1 \end{vmatrix} \ \rightarrow \ \begin{vmatrix} 1 & t{\cdot}p \\ q{\cdot}d & 1 \end{vmatrix}$$

$$(t{\cdot}p){\cdot}(q{\cdot}d) \approx (t{\cdot}q){\cdot}(p{\cdot}d) \approx (1){\cdot}(1) = 1$$

The columnar orders are 'q·d' and 't·p' respectively.

$\therefore$ The essence of the nature of the problem (or how to enforce it to a resolution) is displayed.

It might be called a "summary" table, or a "truth" table. Such tables may be used to expand related tables:

e.g.

```
              |  1    q·d |                    |  1   1   1   q·d |
   |  1   t2  | t·p |  1  |                    |  1   t2  t·p  1  |
   |  d2  1      p2 |                     ➜    |  d2  1   p2   1  |
   |  q·d q2     1  |                          |  q·d q2  1    1  |
```

If we use this technique on a table closer to the original problem, we result in a balanced table with a solution somewhat "forced" (to compare to the others):

e.g.

```
 |  q·d   1  |                         |  q·d   1 ⋮  1   1  |
 |   1  | t·d ⋮  1   q |               |   1  t·d ⋮  1   q  |
      |  q  ⋮ q·d  q |         ➜       |   1   q  ⋮ q·d  q  |
      |  d  ⋮  1   1 |                 |   1   d  ⋮  1   1  |
```

This has both a solution of 'q·d' (≡ q·d·1·1·1)
  and a columnar order of 'q·d' (≡ q·d·1·1·1) .

Note the interesting (row) solutions "forced":

$(q·d)·1·1·1 = q·d$
$1·(t·d)·1·q ≈ 1·t·q·1·d ≈ d$ (t·q ≈ 1)
$1·q·(q·d)·q ≈ d·q^3$
$1·d·1·1 = d$

∴   q·d , d , d·q³ , d

Corresponding columnar orders:

                q·d , d2 , q·d , q2

The partitioning can be used to indicate (or signify), for example, something like irreversible processes. This method (above) constructs, and is valid for, tables related to a problem.

Assuming the (following) denominator to be closer to unitary-ness (than the numerator),
 the following nul/los division may be performed:

```
 | q·d  1  ⋮   1   1 |
 |  1  t·d ⋮   1   q |
 |  1   q  ⋮ q·d   q |
 |  1   d  ⋮   1   1 |
 ─────────────────────       →      | q·d     1   ⋮  1    (q·d)-1 |
 |  1   1   1   q·d |                |  1    t·d/t2 ⋮ (t·p)-1   q   |
 |  1   t2  t·p  1  |                | (d2)-1   q   ⋮ q·d/p2   q   |
 |  d2  1   p2   1  |                | (q·d)-1 d/q2 ⋮  1       1   |
 |  q·d q2  1    1  |
```

again looking for useful patterns or symmetries.

The row solutions are:

$$(q{\cdot}d){\cdot}1{\cdot}1{\cdot}(q{\cdot}d)^{-1} \quad \cong \quad 1$$
$$1{\cdot}(t{\cdot}p/t^2){\cdot}(t{\cdot}p)^{-1}{\cdot}q \quad \cong \quad t{\cdot}d{\cdot}q/(t^3{\cdot}p) \quad = \quad q{\cdot}d/(t^2{\cdot}p)$$
$$(d^2)^{-1}{\cdot}q{\cdot}(q{\cdot}d/p^2){\cdot}q \quad \cong \quad (q^3{\cdot}d)/(p^2{\cdot}d^2) \quad = \quad q^3/(p^2{\cdot}d)$$
$$(q{\cdot}d)^{-1}{\cdot}(d/q^2){\cdot}1{\cdot}1 \quad = \quad d/(q^3{\cdot}d) = (q^3)^{-1} \quad = \quad q^{-3}$$

This is most noteworthy in that we seem to have several orders of 'q':

i.e. 1 , q , q^3 , q^{-3}

Obviously the solutions are disparate (and distinctive). Perhaps they could be competing or alternate possible processes (when a ' q·d ' analysis is forced).

These such methods (e.g. "truth tables") can be used to probe the nature of a solutions table.

We may also multiplicatively (or divisionally) "overlap" a solutions table (using, e.g., truth tables):

```
                     | q·d  1 |
| t·d :  1   q |    | 1  t·d|         | t·d :  1    q  |
| q  : q·d q |        ───→          | q  : q·d  q  |
| d  :  1   1|     multiplicative    | d  :  1   t·d |
          |           overlap at q·d
          | | q·d  1 |
          | |  1  t·d |
        ↓ divisional overlap at q·d ≡ "underlap"
| t·d :  1   q       |
| q  :  1   q       |        i.e. ⎛ q·d/(q·d)   q/1  ⎞
| d  :  1  (t·d)⁻¹ |             ⎝   1/1      1/(t·d) ⎠
```

If overlapping/underlapping adheres to strict initial replacement of 'q·d',
then the divisional (underlap) result is:

```
| t·d :  1     q     |
| q  : q·d   q     |
| d  :  1   (t·d)⁻¹ |
```

If the overlapping/underlapping is fully functional at 'q·d' (of the solution table),
then the overlap (multiplicative) result is:

```
| t·d :  1      q   |
| q  : (q·d)²  q   |
| d  :  1    (t·d) |
```

<u>Conceptual diagrams</u>
Strict initial replacement:

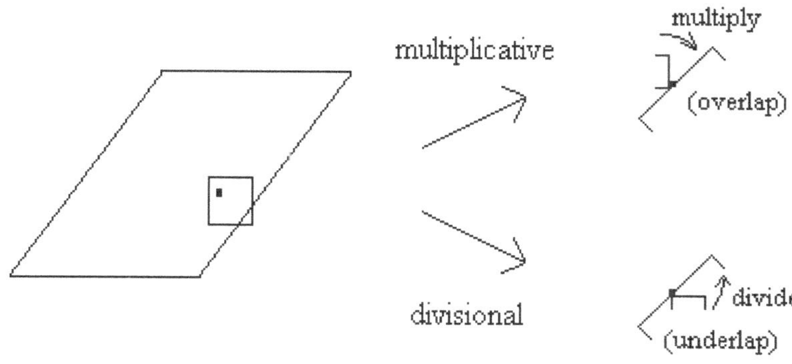

Strict overlap/underlap:

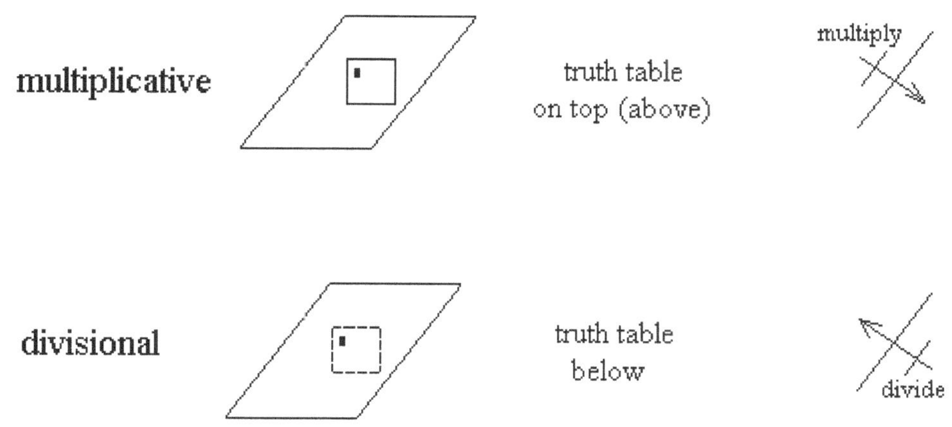

These manipulations are feasible and possible due to the nul/los nature of the tables involved.

Many such interactions between tables, or manipulations of a table itself, can be envisioned (graphically, or via figures or drawings) because of this "zero-less" quality (e.g. rolling a table up for partial multiplications and divisions; or otherwise distorting its plane, such as folding or cutting). This is why, for example, columnar orders can be as important (or instructive) as row solutions. Note that an emphasis on "balanced" tables is developed. One row leading to a single solution is not balanced, even if the row consists of only one term. (A table of only one term, however, is balanced.)

Unbalanced tables may be manipulated as well, but in general unitary-ness results when the number of solutions (rows) is the same as the number of columnar orders. This is because unitary-ness occurs when a process (set of operations) is "proved" correct, by comparing it to itself. This symmetry (as for action and result in its most fundamental form) can not be violated for a true or actual occurrence.

If there are more solutions than columnar orders:

e.g.

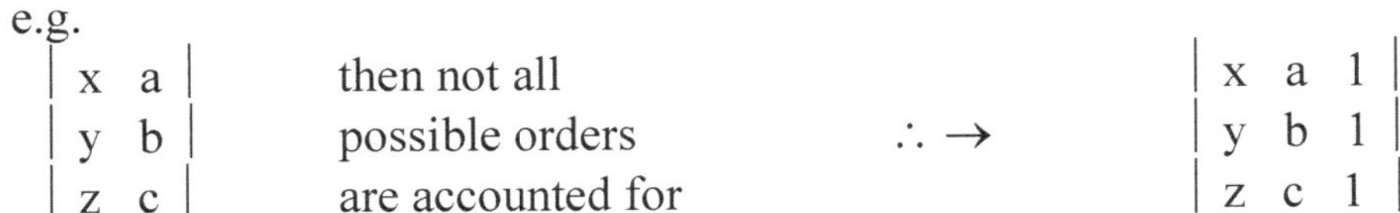

x a	then not all
y b	possible orders ∴ →
z c	are accounted for

x a 1
y b 1
z c 1

If there are more columnar orders than solutions:

e.g.

x y z	then not all
a b c	possible solutions ∴ →
	have been found

x y z	
a b c	and its various
1 1 1	permutations

Note:

1) that

x a 1		x y z
y b 1	and	a b c
z c 1		1 1 1

share the same diagonal products:

xb1, zb1

2) that for

x a		
y b	and	x y z
z c		a b c

while slanted and in fact all "lined" products are the same: xa, yb, zc, xb, yz, ya, zb, xyz, abc ; there is a propensity or "will" for each (table) to "equalize" solutions to columnar orders:

e.g.

x a		x a		x 1 a
y b	"wants to form"	→ yb ←	→	1 yb 1
z c	(a new column)	z c		z 1 c

x y z		x ↓ z		x 1 z
a b c	"wants to form"	yb	→	1 yb 1
	(a new solution, or row)	a ↑ c		a 1 c

so that diagonal products are the same for each:

x·yb·c , z·yb·a

Diagonal products' distinctiveness is true for 1x1, 2x2, and 3x3 balanced tables.

But what of 4x4 balanced tables?

e.g.

x y z 1	
a b c 1	diagonal products:
1 1 1 1	xb·1·1 , 1·c·1·1
1 1 1 1	

$\therefore$ No clear propensity to form new columns or rows.

What about 5x5 balanced tables?

$$\begin{vmatrix} x & y & z & 1 & \mathit{1} \\ a & \mathit{b} & c & \mathit{1} & 1 \\ 1 & 1 & \mathit{1} & 1 & 1 \\ 1 & \mathit{1} & 1 & \mathit{1} & 1 \\ \mathit{1} & 1 & 1 & 1 & \mathit{1} \end{vmatrix}$$

Again, no clear propensity to form new rows and columns. (no symmetrical need or lack)

$\therefore$ We assume only unbalanced tables have this propensity (to form new rows and columns).

While this is not a discourse on determinants and matrices, it is relevant to point out that both rows and columns (and indeed all lines slanted or otherwise) depict processes of operations, and that a columnar order for one table may be a row solution for another (table) given that they (the two tables) share identical diagonal products.

$\therefore$ With balanced tables, solutions may be inter-convertible (between them). This gives particular importance to both columnar orders and diagonal products, when characterizing tables. A corollary is that each unbalanced table has a corresponding and equal balanced form (by forming new rows or columns and filling in the "1"s).

Then, going back to the overlapping (tables) result for the devised circuit:

$$A \;_{p}{\leftrightarrow}_{d}\; B$$
$$(w)\circlearrowright(x)\quad(y)\circlearrowright(z) \qquad\rightarrow\qquad \begin{vmatrix} 1 & t^2 & t\cdot p \\ d^2 & 1 & p^2 \\ q\cdot d & q^2 & 1 \end{vmatrix}$$
$$C \;_{q}{\leftrightarrow}_{t}\; D$$

It would seem apparent that the (w,x,y,z) operations could (or would) be equivalent to:

$$\begin{vmatrix} 1 & t^2 & t\cdot p \\ d^2 & 1 & p^2 \\ q\cdot d & q^2 & 1 \end{vmatrix} \begin{array}{c}\text{inter-conversion}\\ \rightarrow \\ \text{(rows for columns)}\end{array} \begin{vmatrix} 1 & d^2 & q\cdot d \\ t^2 & 1 & q^2 \\ t\cdot p & p^2 & 1 \end{vmatrix} \;\equiv\; \begin{vmatrix} 1 & w^2 & z\cdot w \\ y^2 & 1 & z^2 \\ y\cdot x & x^2 & 1 \end{vmatrix}$$

$$\rightarrow \qquad \begin{array}{c} A \;_{x}{\leftrightarrow}_{w}\; C \\ (d)\circlearrowright(p)\quad(t)\circlearrowright(q) \\ B \;_{z}{\leftrightarrow}_{y}\; D \end{array}$$

The diagonal products are: $1\cdot1\cdot1 = 1$
and $q\cdot d\cdot1\cdot t\cdot p \equiv qdtp\ (= dpqt)$
$\therefore qdtp \equiv yxzw\ (= wxyz)$

This suggests that a balanced table offers competing or complementary solutions in the nature of its information. Such is the manner of tables of operations.

Note that if partitioning is devised (or constructed),
 then it should follow the conversion of a columnar order into a solution (or visa versa):

e.g.

$$\begin{vmatrix} 1 & \vdots & t^2 & t\cdot p \\ d^2 & \vdots & 1 & p^2 \\ q\cdot d & \vdots & q^2 & 1 \end{vmatrix} \begin{array}{c}\text{inter-conversion}\\ \rightarrow \\ \text{(orthogonal)}\end{array} \begin{vmatrix} 1 & d^2 & q\cdot d \\ \hline t^2 & 1 & q^2 \\ t\cdot p & p^2 & 1 \end{vmatrix}$$

This suggests that partitioning itself may (in cases) be inter-conversional:

e.g.

$$\begin{vmatrix} 1 & t^2 & t\cdot p \\ d^2 & 1 & p^2 \\ q\cdot d & q^2 & 1 \end{vmatrix} \xrightarrow{\ \text{inter-conversion}\ } \begin{vmatrix} 1 & d^2 & q\cdot d \\ t^2 & 1 & q^2 \\ t\cdot p & p^2 & 1 \end{vmatrix}$$

Perhaps it can even be irregular (of shape) and/or isolated:
e.g.

$$\begin{vmatrix} 1 & t^2 & t\cdot p \\ d^2 & 1 & p^2 \\ q\cdot d & q^2 & 1 \end{vmatrix} \xrightarrow{\ \text{inter-conversion}\ } \begin{vmatrix} 1 & d^2 & q\cdot d \\ t^2 & 1 & q^2 \\ t\cdot p & p^2 & 1 \end{vmatrix}$$

This being possible, it is strongly inferred that such partitioning can represent unknowns (of operations) desired to be determined:
e.g.

$$\begin{vmatrix} 1 & U_2 & t\cdot p \\ U_1 & 1 & U_3 \\ q\cdot d & q^2 & 1 \end{vmatrix} \xrightarrow{\ \text{inter-conversion}\ } \begin{vmatrix} 1 & U_1 & q\cdot d \\ U_2 & 1 & q^2 \\ t\cdot p & U_3 & 1 \end{vmatrix}$$

Or, perhaps, the problem (here) would be to compare (for example):
e.g.

$$
\begin{vmatrix} 1 & U_2 & t\cdot p \\ U_1 & 1 & U_3 \\ q\cdot d & q^2 & 1 \end{vmatrix}
\qquad \text{to} \qquad
\begin{vmatrix} 1 & U_1 & z\cdot w \\ U_2 & 1 & z^2 \\ y\cdot x & U_3 & 1 \end{vmatrix}
$$

i.e. Compare columnar orders:

$$U_1\cdot qd , \quad U_2\cdot q^2 , \quad U_3\cdot tp$$
$$\text{to} \qquad U_2\cdot yx , \quad U_1\cdot U_3 , \quad z^3\cdot w$$

$\therefore$

$$
qd\cdot \frac{\boxed{U_1}}{\boxed{U_1}}\cdot \frac{\boxed{U_3}}{\boxed{U_3}}\cdot tp
\qquad\qquad
q^2\cdot \frac{\boxed{U_2}}{\boxed{U_2}}\cdot yx
$$

or

$$
qd\cdot \frac{U_1}{U_1}\cdot \frac{U_3}{U_3}\cdot tp \quad\longrightarrow\ \text{a solution}
\qquad\qquad
q^2\cdot \frac{U_2}{U_2}\cdot yx \quad\longrightarrow\ \text{a solution}
$$

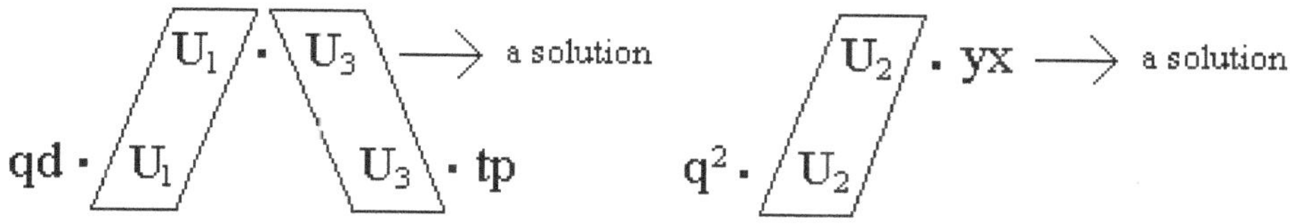

Since the diagonal products (of the two tables) are the same, we know (or can assume), for example, that:

$$q\cdot d \equiv z\cdot w , \quad t\cdot p \equiv y\cdot x$$

$\therefore$

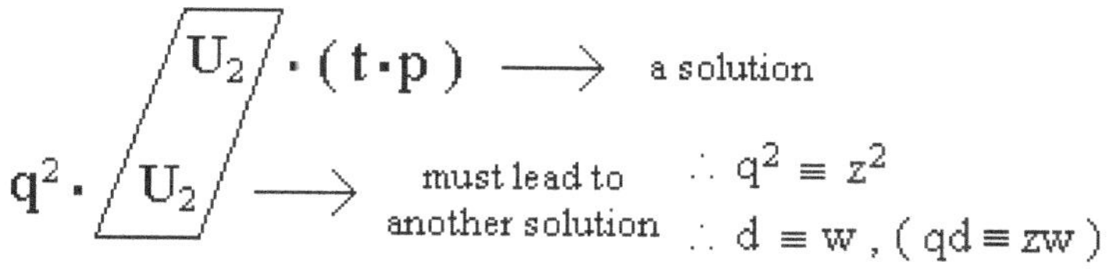

$$
q^2\cdot \frac{U_2}{U_2}\cdot (t\cdot p) \quad\longrightarrow\ \text{a solution}
$$

$$q^2\cdot \frac{U_2}{U_2} \longrightarrow \quad \text{must lead to} \quad \therefore\ q^2 \equiv z^2$$
$$\text{another solution} \quad \therefore\ d \equiv w , \ (\,qd \equiv zw\,)$$

If both tables are of unitary-ness, then, as for row products,
the columnar products should equal each other:

$\therefore$ $U_1 \cdot qd = U_2 \cdot q^2 = U_3 \cdot tp$

            $U_2 \cdot xy = U_1 \cdot U_3 = z^3 w$, $z^3 w \equiv qd \cdot z^2$

$\therefore$ $U_2 \cdot tp$ $= qd \cdot z^2$, $yx \equiv tp$

$\therefore$ $U_2 \cdot q^2 = U_3 \cdot tp$

            $qd \cdot z^2 = U_2 \cdot tp$

$\therefore$ $U_3 = U_2 \cdot q^2/tp$

            $U_2 = qd \cdot z^2/tp$

$\therefore$ $U_3 = (qd \cdot z^2/tp)(q^2/tp) = q^3 d \cdot z^2/(tp)^2$

            $U_1 = U_2 \cdot q^2/qd = (q^3 d \cdot z^2/(tp)^2)(tp) = q^3 d \cdot z^2/tp$

$\therefore$

   $U_2 = qd \cdot z^2/tp$

   $U_1 = U_2 \cdot q^2$

   $U_3 = U_2 \cdot (q^2/tp)$

This seems far from the values we know of these unknowns, when revealed in the original table:

   i.e. $U_1 = d^2$, $U_2 = t^2$, $U_3 = p^2$

But these values are not variables per se; rather they are operations, which (to some extent) can be treated algebraically in their arithmetic (e.g. # of multiplications or divisions; exponential processes with their exponents).

The values determine a procedural nature to these particular operations (that might be informative):

   ($z^2 \equiv q^2$)

     $C \leftarrow_t D$ $U_2 = t^2 = qd \cdot z^2/tp = q^3 \cdot d/tp$

     $A \leftarrow_d B$ $U_1 = d^2 = (qd \cdot z^2/tp) \cdot q^2 = (q^3 \cdot d/tp) \cdot q^2$

     $A_p \rightarrow B$ $U_3 = p^2 = (qd \cdot z^2/tp) \cdot (q^2/tp) = (q^3 \cdot d/tp) \cdot (q^2/tp)$

     $C_q \rightarrow D$ (The forth "square" is of course ' q^2 ' itself.)

So we note that the 'q' exponent runs from:

   2 (for q^2) to 3 (for t^2) to 5 (for d^2 and for p^2)

We can also tabulate based on common factoring:

q^2	t^2	d^2	p^2
$q^2 (\cdot 1)$	$q^2 \cdot (qd/tp)$	$q^2 \cdot (qd/tp) \cdot q^2$	$q^2 \cdot (qd/tp) \cdot (q^2/tp)$

Why is ' q^2 ' least (in 'q') and ' p^2 ' most (in 'q') ?

Perhaps the exculpatory example table from which this derives fashions that:

in order for ' $A \rightarrow B$ ', ' $C \rightarrow D$ ' must initiate.

The exculpatory table:

$$
\begin{vmatrix}
t \cdot d & : & 1 & q \\
q & : & q \cdot d & q \\
d & : & 1 & 1
\end{vmatrix}
$$

results from an attempt at unitary-ness using the factor operations: $\mathfrak{h} \cdot \mathfrak{H}$

i.e.

$$
\mathfrak{h} \cdot \mathfrak{H} \;
\begin{vmatrix}
\mathfrak{K} & : & \mathfrak{h} & \mathfrak{H} \\
1 & : & \mathfrak{h}^{-1} & \mathfrak{H}^{-1} \\
\mathfrak{K} & : & 1 & 1
\end{vmatrix}
$$

where $\mathfrak{h} \cdot \mathfrak{H} \equiv p \cdot q \cdot d$

and $p \cdot d \cong 1 \;\; \therefore \; \mathfrak{h} \cdot \mathfrak{H} \cong q$

' q ' is the operation for ' $C_q \rightarrow D$ '

$\therefore$ it forms a basis for the resulting tables.

The use of partitioning to denote unknowns (only pictorially here), (U_1, U_2, U_3), demonstrates utility of a rather algebraic analysis which might be more cumbersome (or less modern) than possibly other means devisable. This partitioning merely restricts the "unknowns" to relational terms definable by known operations (i.e. not to new ones to be uncovered). And one can imagine (or foster) that the relational partitioning of 'U_1' and 'U_2' simply remarks that, of them, an unknown term does not (must not) appear, while for 'U_3' both ' t^2 ' and ' p^2 ' are inherently extant (' q^2 ' already being so among the unknowns).

But indeed, we may be able to deduce, by the symmetry of the table (under its own rules of unitary-ness) the apparent identities of the unknowns:

$$
\begin{vmatrix}
\boxed{1} & U_2 & t \cdot p \\
U_1 & 1 & U_3 \\
q \cdot d & q^2 & 1
\end{vmatrix}
$$

If ' $q \cdot d$ ' runs to ' q^2 ' (within a solution), then ' t^2 ' should run to ' $t \cdot p$ ' (within a solution).

$\therefore \; U_2 \equiv t^2$

Then obviously (since the orthogonal inter-conversion is given as truthful, in terms of 'w,x,y,z'), for the inter-converted table:

$$q \cdot d \quad \text{may run to} \quad d^2 \equiv U_1$$

and $\qquad p^2 \quad \text{may run to} \quad t \cdot p \;\; \therefore \; p^2 \equiv U_3$

This is a far easier (or more accessible) analysis than the algebraic (although it alienates and keeps hidden the algebraic interdependencies of the known operations). We also note that only for the inter-converted table does each solution have a unique (and only one) unknown.

$\therefore$ Solution to identities of the three unknowns requires this inter-conversion, for necessary information. The logic of the unknowns is exemplified by nul/los mechanics. That is, they can not vanish (into obscurity), but only be transformed into another form less or more revelatory of a solution:

e.g.

$p^2 \rightarrow q^2 \cdot (qd/tp) \cdot (q^2/tp)$

But let us be more cogent of the fundamental (or basic) modes of operation utilizable under the tenets of nul/los.

Say, for example, that we are given two operations: a, b
such that 'a' causes movement of an element of conjecture to the right of a certain extent of translation, and 'b' causes movement of this element to the left of the same translational extent.

Then ' a·b ' leaves the element unmoved (or moved back to its initial position, say on a sphere):

$\therefore$ $\qquad a \equiv \rightarrow \ , \quad b \equiv \leftarrow$
$\qquad a{\cdot}b \equiv \leftrightarrow$

Then, what does ' b/a ' signify?

' b/a ' signifies that state of the element which, when applied of the operation 'a' causes the element to reside (from the initial) as after the operation 'b' is applied:

i.e. $\qquad (b/a){\cdot}a \equiv b$

Similarly, ' a/b ' signifies, essentially:

$\qquad (a/b){\cdot}b \equiv a$

Then ' a+b ' signifies first an operation of 'a' on the element, and then an operation 'b' on the element. The result, here, seems equivalent to ' a·b '; however, ' a·b ' implies an immediacy to the procedures, while ' a+b ' may be interrupted or impeded or otherwise altered of procedure before being fulfilled.
$\therefore$
The order, of the application of operations, is more strict with the multiplicatrive (·) than the additive (+). Likewise, the term ' a-b ' indicates that loose order of affairs for a state such that, after an ' a+b ' procedure is performed, renders the element back to a state (after the initial) characterized by the 'a' operation. Similarly may a ' b-a ' procedure be defined.

Note that the element never vanishes, and it always has some position (or positional description inherent of its consideration to other elements involved with its interactions or not: proximal or distal of operative procedures).

We have defined, here, an element's quality (operationally) as positional; but this is only an example. An element's quality may be anything of definite character, and without probabilities of being indefinite or elusive of description. This firmness of being (i.e. existence) allows for nul/los manipulation.

More sophisticated (extended or complicated) functions of (manner of) operations can be tabularly constructed, given a freedom from "zero" usage.

e.g. Manners of multiplication and division among tables were figuratively drawn as follows:

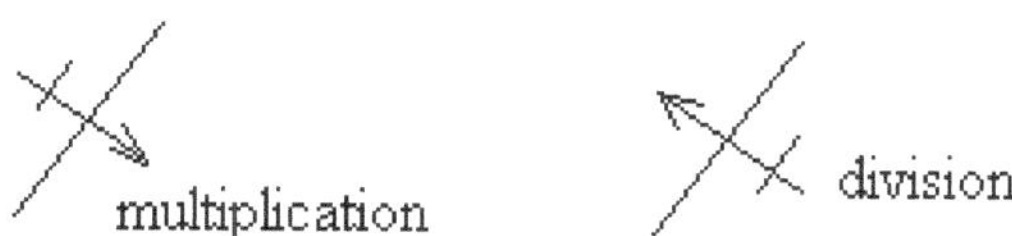

Such overlapping (of tables) can certainly be extended towards regularly recurring functions:

c o m p l e x a t i o n s (combinations)

replacement or substitution

circularizations and recursions

e x t e n s i o n s

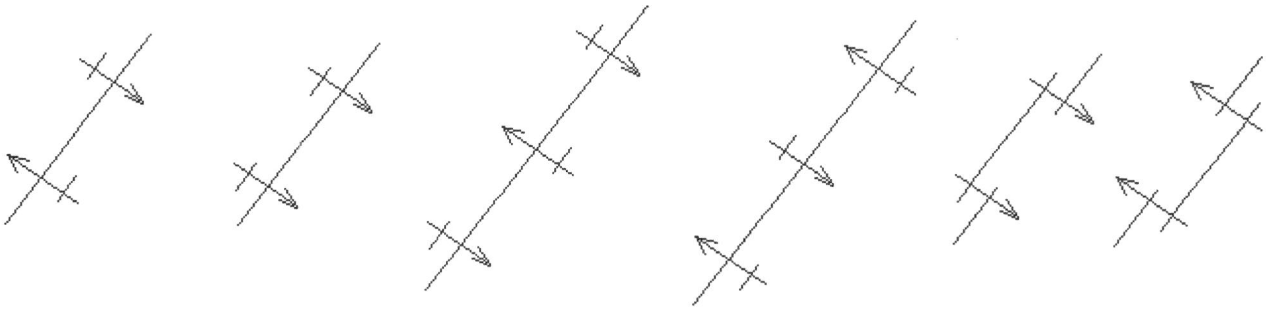

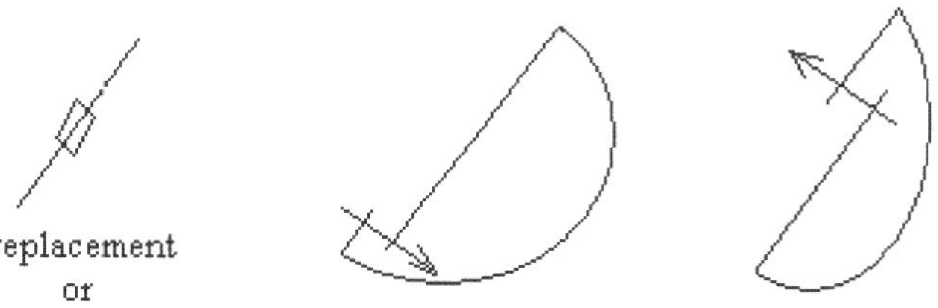

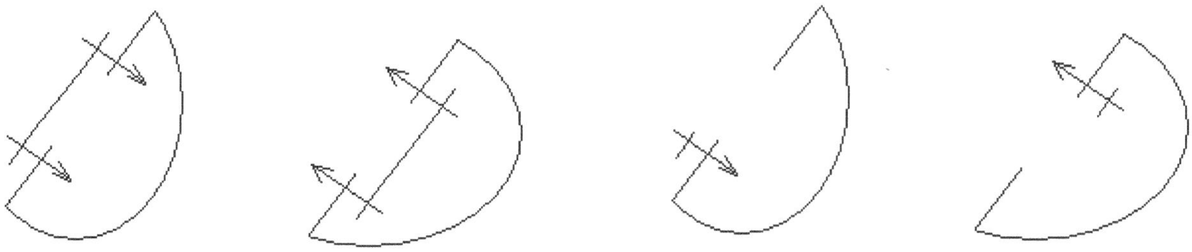

etc.

Even branching functions can be designed:

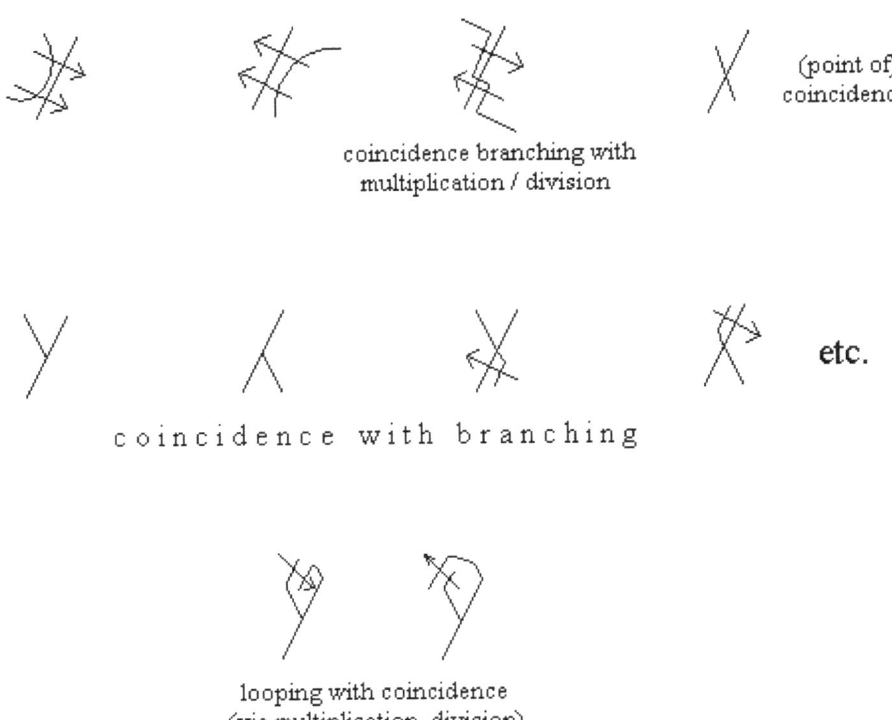

coincidence branching with
multiplication / division

(point of)
coincidence

c o i n c i d e n c e   w i t h   b r a n c h i n g

etc.

looping with coincidence
(via multiplication, division)

Here, coincidence may be of particular elements of tables (in common), or of rows or columns. Of course, compounding may occur:

e.g.

etc.

For all of these functions (or designs) reverting (reversions) may be anticipated.

Probing (probe testing, sampling) is also possible:

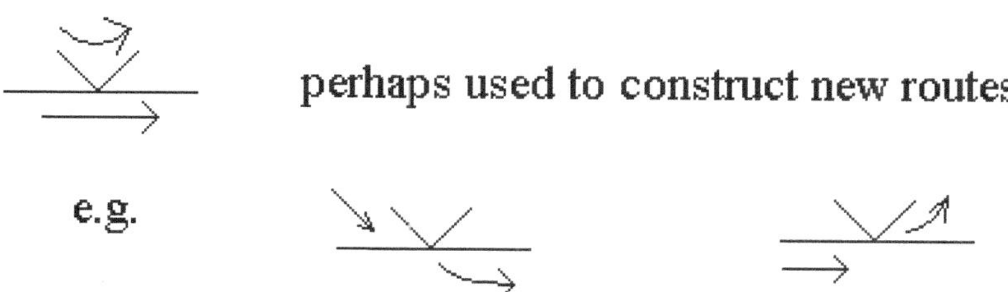

Each construction can result in solutions to something. But this leads to composite tables, where different (or differing) rows/columns represent different (or multi-various) solutions, not necessarily automated of partitioning. Here, it is assumed that the table is not of unitary-ness ($\therefore$ may be subjected to further developments):

e.g.
$$\begin{vmatrix} a & b & c \\ 1 & 1 & 1 \\ x & y & z \end{vmatrix}$$
row $\begin{vmatrix} a\ b\ c \end{vmatrix}$ solves something different from row $\begin{vmatrix} x\ y\ z \end{vmatrix}$

or
$$\begin{vmatrix} i & a & b & c & 1 \\ 1 & 1 & 1 & 1 & 1 \\ 1 & h & k & 1 & i \end{vmatrix}$$
unbalanced, but with points of possible coincidence for looping

wants to form
$\rightarrow$
something like
$$\begin{vmatrix} i & a & b & c & 1 \\ i & a & b & c & 1 \\ 1 & 1 & 1 & 1 & 1 \\ 1 & h & k & l & i \\ 1 & h & k & l & i \end{vmatrix}$$
with diagonal products:
$ia1li \equiv i^2al$
$1h1c1 \equiv hc$

Loop Structures

How, then, might a loop structure be tabled?

Given:
$$\begin{vmatrix} i & a & b & c & 1 \\ i & a & b & c & 1 \\ 1 & 1 & 1 & 1 & 1 \\ 1 & h & k & l & i \\ 1 & h & k & l & i \end{vmatrix}$$

an obvious transpositional expansion could be:

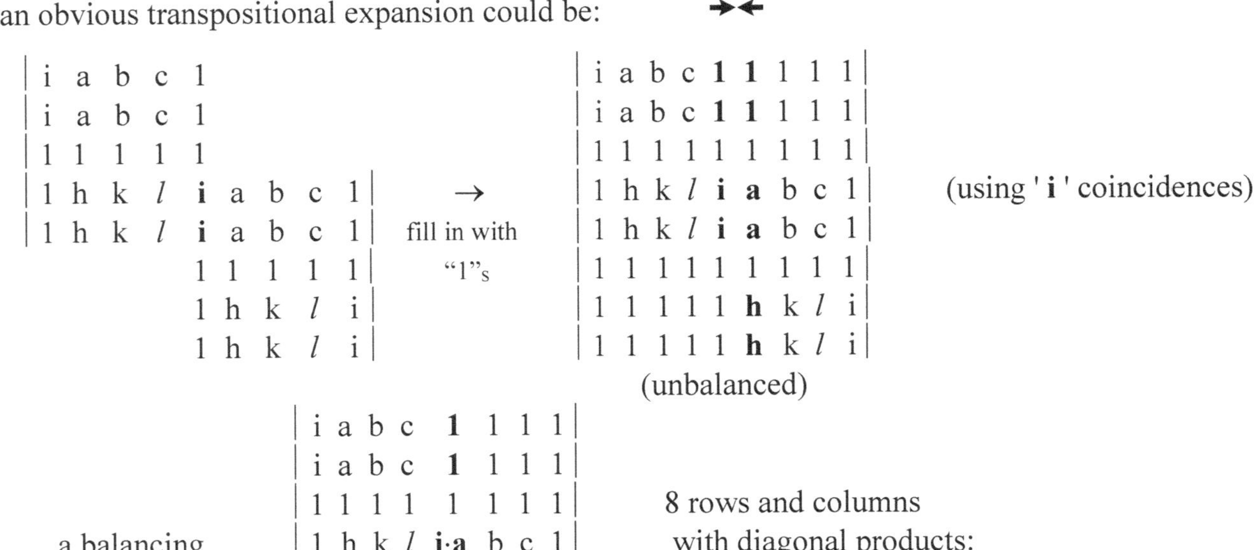

```
│ i  a  b  c  1 │                          │ i a b c 1 1 1 1 │
│ i  a  b  c  1 │                          │ i a b c 1 1 1 1 │
│ 1  1  1  1  1 │                          │ 1 1 1 1 1 1 1 1 │
│ 1  h  k  l  i  a  b  c  1 │   →          │ 1 h k l i a b c 1 │    (using ' i ' coincidences)
│ 1  h  k  l  i  a  b  c  1 │  fill in with │ 1 h k l i a b c 1 │
            │ 1  1  1  1  1 │   "1"s       │ 1 1 1 1 1 1 1 1 │
            │ 1  h  k  l  i │             │ 1 1 1 1 1 h k l i │
            │ 1  h  k  l  i │             │ 1 1 1 1 1 h k l i │
```

(unbalanced)

```
                    │ i a b c  1  1  1 │
                    │ i a b c  1  1  1 │
                    │ 1 1 1 1  1  1  1 │     8 rows and columns
   a balancing      │ 1 h k l i·a b c 1 │      with diagonal products:
       →            │ 1 h k l i·a b c 1 │       ia1l(i·a)1li ≡ i³a²l² = i·(lia)²
                    │ 1 1 1 1  1  1  1 │       111l(i·a)111 ≡ lia
                    │ 1 1 1 1  h  k  l  i │
                    │ 1 1 1 1  h  k  l  i │
```

But, physically, a loop would not be an expansion, rather a condensation along coincidences. And this condensation might be of substitution (replacement) or operational (multiplicative or divisional; divisional yielding "1"). In any case, a loss of element(s) occurs. Obviously, then, the tablature in this mode should somehow be cursive (e.g. circular). This would obscure distinctions between rows and columns (or subtend them to more varied alignments). To display this planarly might simply be:

```
                  │ i  a  b  c  1 │                      │ i  a  b  c  1 │
                  │ i  a  b  c  1 │                      │ i  a  b  c  1 │
 (substitutional) │ 1  1  1  1  1 │        or            │ 1  1  1  1  1̲│
                  │ 1  h  k  l    │                      │ 1  h  k  l   │  a    ←
                  │ 1  h  k  l    │                      │ 1  h  k  l  │ "notch"
                        ↑ (space) like a "hole" to be affixed
```

Although this still obscures the looping motif (except by presumed definition of style).

Other forms might be:

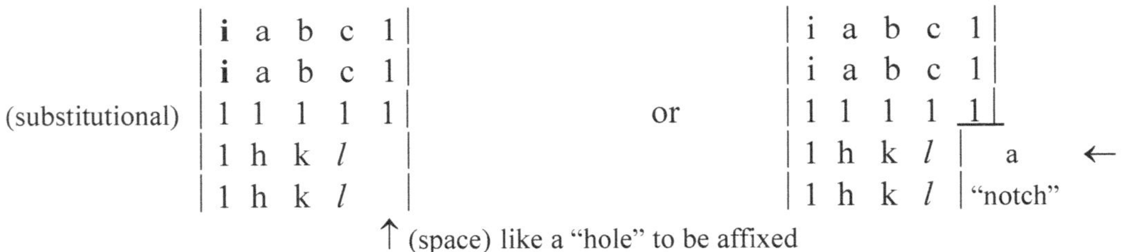

```
│ i² a  b  c  1 │                       │ 1  a  b  c  1 │
│ i² a  b  c  1 │                       │ 1  a  b  c  1 │
│ 1  1  1  1  1̲│                       │ 1  1  1  1  1̲│
│ 1  h  k  l │ multiplicative          │ 1  h  k  l │ divisional
│ 1  h  k  l │                          │ 1  h  k  l │
```

Note that this 'notched' form allows of easy expansive compounding (tiling):

e.g.

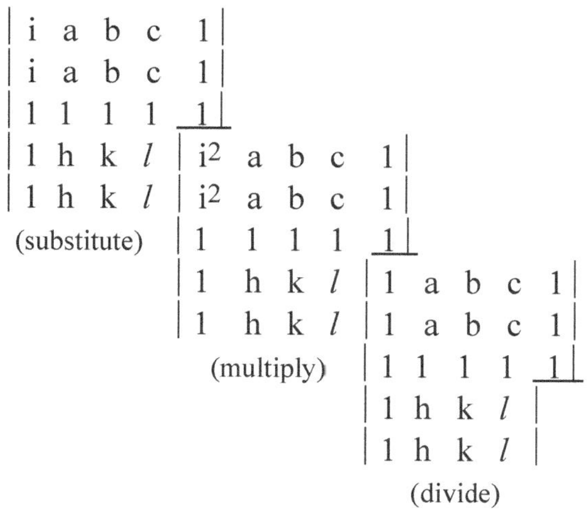

(substitute)

(multiply)

(divide)

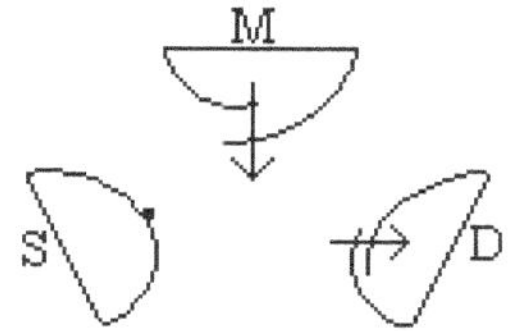

for more complexation (complexity) of procedures (here, 3 apparent loops).

The above pictograph (figure) might also be depicted as:

or

although the gap is
misleading since it
should still be a
complete loop

It is evident that this notched form (looping) helps to highlight minors of a table, which may be selected of balanced proportions, and also that the notched region allows for a progression of functioning that may be used to define more mathematical procedures:

e.g.

$$\left.\begin{array}{|ccccc|}\text{i} & \text{a} & \text{b} & \text{c} & 1\\ \text{i} & \text{a} & \text{b} & \text{c} & 1\\ 1 & 1 & 1 & 1 & \underline{1}\\ 1 & \text{h} & \text{k} & l & \\ 1 & \text{h} & \text{k} & l & \end{array}\right| \quad \begin{array}{l}\text{a sample}\\ \text{composite of}\\ \rightarrow\end{array} \quad \left|\begin{array}{cccc}\text{i} & \text{a} & \text{b} & \text{c}\\ \text{i} & \text{a} & \text{b} & \text{c}\\ 1 & 1 & 1 & 1\end{array}\right| + \left|\begin{array}{c}1\\ 1\\ 1\end{array}\right| + \left|\begin{array}{cc}1 & \text{h}\\ 1 & \text{h}\end{array}\right| + \left|\begin{array}{cc}\text{k} & l\\ \text{k} & l\end{array}\right|$$

$$\text{or} \quad \left|\begin{array}{ccc}1 & 1 & 1\\ \text{h} & \text{k} & l\\ \text{h} & \text{k} & l\end{array}\right| + \left|\begin{array}{cccc}\text{i} & \text{a} & \text{b} & \text{c}\\ \text{i} & \text{a} & \text{b} & \text{c}\end{array}\right| + 2\cdot\left|\begin{array}{c}1\\ 1\\ 1\end{array}\right| \qquad \text{or} \quad \left|\begin{array}{ccc}\text{i} & \text{a} & \text{b}\\ \text{i} & \text{a} & \text{b}\\ 1 & 1 & 1\end{array}\right| + \left|\begin{array}{cc}\text{c} & 1\\ \text{c} & 1\\ 1 & 1\end{array}\right| + \left|\begin{array}{cc}1 & \text{h}\\ 1 & \text{h}\end{array}\right| + \left|\begin{array}{cc}\text{k} & l\\ \text{k} & l\end{array}\right|$$

etc.

Progression of notches:

$$\text{e.g.} \qquad \left|\begin{array}{c}\text{i}\\ \text{i}\end{array}\right| \;\rightarrow\; \left|\begin{array}{c}\text{i}^2\\ \text{i}^2\end{array}\right| \;\rightarrow\; \left|\begin{array}{c}1\\ 1\end{array}\right|$$

$\therefore$

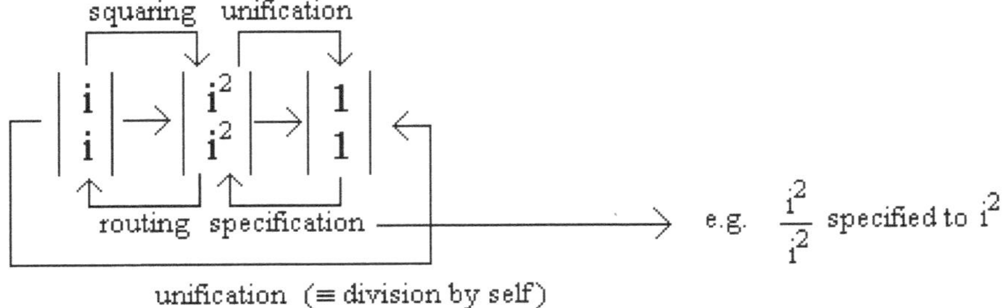

e.g. $\dfrac{\text{i}^2}{\text{i}^2}$ specified to i^2

The tiling itself approaches some sort of functional summation which, with filling in of "1"s, yields to a great expansion of the tabular form. But note that it is not limited, rather definite of character or application. It is essentially an integration, with derivatization resulting back to the original tabular loops. If all of the tables are of the same origin, then the character of the integration is expressed explicitly by the progression of the notched regions, whereby a functional definition (for the integration) is provided.

e.g.

For the progression of "notches":

$$\text{i} \rightarrow \text{i}^2 \rightarrow 1$$

(when all other tabular elements remain unchanged)
the functional definition (for this "integration") first squares 'i',
 and then divides this square (or its identity as 'i') by itself.

The 'i' of course represents an operation, not a quantity. So then does ' i^2 ' and ' 1 ' (no operation).
So ($\text{i}\rightarrow\text{i}^2\rightarrow1$) is really equivalent to ($\text{i}\rightarrow\text{i}^2$), and the tile (thus far) constructs a squaring function, with its derivative (ultimately) leading back to 'i'. We can note (very clearly) that the (expansive) tabular form from the "notches" ($\text{i}\rightarrow\text{i}^2\rightarrow1$) is different (were it) from ($\text{i}\rightarrow1\rightarrow\text{i}^2$), or ($1\rightarrow\text{i}\rightarrow\text{i}^2$);
 although the functional equivalent is still ($\text{i}\rightarrow\text{i}^2$), and that (for each) the various "differentiations" yield

forms of: (i , i² , 1).

Obviously, a notched tile progression may be concluded by a table without a notch (a "non"-loop).

e.g. For a simpler (less extensive) tile:

```
| i a |  →   | i   a_|           | i  a  1  1  1 |
| b i |      | b | i²  a_|        | b  i² a  1  1 |     with diagonal products:
(balanced)       | b | 1  a_|  ≡  | 1  b  1  a  1 |        i·i²·1·i·i ≡ i⁵
                     | b | i a |   | 1  1  b  i  a |        1·1·1·1·1 ≡ 1
                         | b i |   | 1  1  1  b  i |
         (4 tables)               (balanced)       "notches" progression: i → i² → 1→ i
```

Clearly, one of the diagonal products anticipates (and results from) this progression.

Therefore, we can characterize the resulting table as being (possibly) "composed" of the three loops and four derivatizations (derivative or constituent precursor tables):

```
| | i   a |   1   1   1 |
| | b | i² |  a |  1   1 |       Row products:
|   1 | b | 1 |  a | 1 |            ia , i²ba , ba , iba , ib
|   1   1 | b |  i |  a |        Column products:
|   1   1   1 | b |  i |            ib , i²ab , ab , iab , ia
```

∴ We note that the row products and column products are equivalent (i.e. the same).

We also note that off diagonal slant (line) products are either: (a⁴ , b⁴ , or "1") in the (i⁵) direction, and (i , i² , or ba) in the "1" direction (these products alternating with a 'ba' product).

Clearly, then, the table represents a summary from which much data (or method) may be derived (it being known or assumed that it results from "notching compositing").

But a table (thus far) is a construction of operations (along rows) leading to a common solution, to impress correctness of solution (and imply unitary-ness thereby, among their comparisons).

For the given table, this would only seem to be true if:

    1) i ≡ 1
    2) a ≡ b

∴ i²ab ≡ iab , ai ≡ bi

However, operations may (of their character) be (fully) dependent on the order and rank of occurrence:

∴ (for example)
    | i a 1 1 1 | ≡ | 1 1 1 b i |

But these might not be the same as:

    | i b 1 1 1 | , | 1 1 1 a i |

if the operations are order dependent.
i.e. Some operations might have to be performed early, or late, for a solution to result.

And ' ia ' might not be the same as ' ai ' , etc.

An evident way of analyzing (or derivatising) a table suspected of resulting from looping (via "notches")
is to first strip (the table) of extraneous or extra-qualitative ones (i.e. fill-in "1"$_s$ striding the originating
table elements) visually:

e.g.

```
| i  a  1  1  1 |          | i  a \__
| b  i2 a  1  1 |          |_b  i2 a \_
| 1  b  1  a  1 |   →         \_b  1  a \__        a "derivative strip"
| 1  1  b  i  a |               \b  i  a |
| 1  1  1  b  i |                 \ b  i |
```

and then to visually assign the notching locations:

```
| i  a \__
|_b  |i2  a \_              the identity of the originating (or progenitor)
   \_ b  |1  a \_           table being determined compositely by
       \b  |i  a |          the 1$^{st}$ row and 1$^{st}$ column (for the looping).
          \ b  i |
```

i.e.

```
| i  a |          | i  a |
| b   |     →     | b  i |
```

Interestingly, a "derivative strip" may be normalized
        (slants brought to horizontal) to establish a new table:

```
| i  a \__
|_b  i2 a \_              | a  a  a  a |          | 1 a  a  a  a |
   \_ b  1  a \_    →     | i  i2 1  i  i |   →    | i  i2 1  i  i |
       \b  i  a |         | b  b  b  b  |         | b  b  b  b  1 |
          \ b  i |                                | 1  1  1  1  1 |
                                                  | 1  1  1  1  1 |
```

where the (first 3) row products necessarily correspond to the slant products
of the non-derivatized table: The column products correspond to the row and
column products of the non-derivatized table (slant products here including the
diagonal products; from the first 4 rows of the new table).

This new table (normalization of a "derivative strip") yields to the character of the derivative tables used
for notching, via its columns:

e.g.

```
| a  |    | i2 a |          | a  |    | 1  a |          | a  |    | i  a |
| i2 | →  | b  |           | 1  | →  | b  |           | i  | →  | b  |
| b  |                 ,   | b  |                 ,   | b  |
```

These columns are from the center region of the table:

i.e.
$$\left|\begin{array}{ccc} a & a & a \\ i^2 & 1 & i \\ b & b & b \\ 1 & 1 & 1 \\ 1 & 1 & 1 \end{array}\right|$$

The flanking columns:

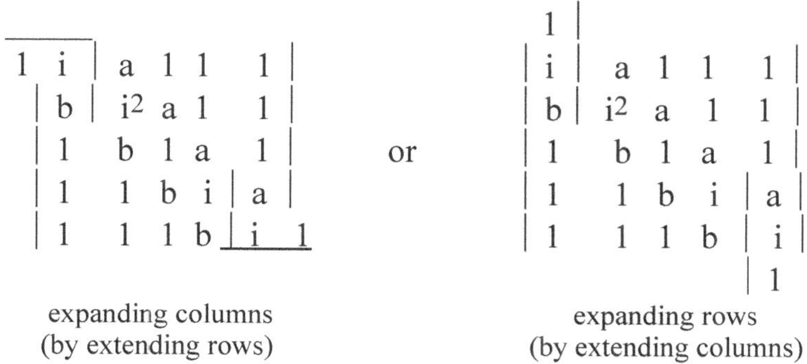

might suggest proper (or sustainable) extensions to the non-derivatized table:

$$
\begin{array}{cc|cccc|}
1 & i & a & 1 & 1 & 1 \\
 & b & i^2 & a & 1 & 1 \\
 & 1 & b & 1 & a & 1 \\
 & 1 & 1 & b & i & a \\
 & 1 & 1 & 1 & b & i & 1
\end{array}
\qquad \text{or} \qquad
\begin{array}{c|cccc|}
 & 1 & & & & \\
i & & a & 1 & 1 & 1 \\
b & & i^2 & a & 1 & 1 \\
1 & & b & 1 & a & 1 \\
1 & & 1 & b & i & a \\
1 & & 1 & 1 & b & i \\
 & & & & & & 1
\end{array}
$$

expanding columns   expanding rows
(by extending rows)   (by extending columns)

Of course, the "derivative strip", rather than being normalized,
can be "verticalized" to found another table:

$$
\begin{array}{l}
|\ i\ a \\
\underline{|\ b\ i^2\ a} \\
\ \ \underline{\backslash\ b\ \ 1\ a} \\
\ \ \ \ \underline{\backslash b\ \ i\ a} \\
\ \ \ \ \ \ \underline{\backslash\ b\ i}
\end{array}
\ \rightarrow\
\begin{array}{|ccc|}
b & i & \\
b & i^2 & a \\
b & 1 & a \\
b & i & a \\
 & i & a
\end{array}
\ \xrightarrow{\ \text{possibly}\ }\
\begin{array}{|ccccc|}
1 & b & i & 1 & 1 \\
1 & b & i^2 & a & 1 \\
1 & b & 1 & a & 1 \\
1 & b & i & a & 1 \\
1 & 1 & i & a & 1
\end{array}
$$

where the row products correspond to the row, column products
of the originating "loop" table (composite), or non-derivatized table;
and the center (plus one either) column products correspond to the diagonal
and slant products of that table.

Here, interestingly, the diagonal products of this new table (as given) are the same (for each diagonal):

$1 \cdot b \cdot 1 \cdot a \cdot 1 \equiv ba$ (also the center row product)

The other slant products correspond to the row products (as do the two diagonal products).

$\therefore$ A "derivative strip" is not only very informative, but also importantly functional (for manipulation).

If the "normalized" table were of the form:

$$\begin{vmatrix} 1 & 1 & 1 & 1 & 1 \\ 1 & a & a & a & a \\ i & i^2 & 1 & i & i \\ b & b & b & b & b \\ 1 & 1 & 1 & 1 & 1 \end{vmatrix}$$

it would share (many of) the same properties as the "verticalized" table: slanting products, as identified, are the same; for the others rows and columns are necessarily exchanged for the two tables.

Looping brings operationally a symmetry to an argument and its procedures.
Thus, for our example, we have the progression established as:

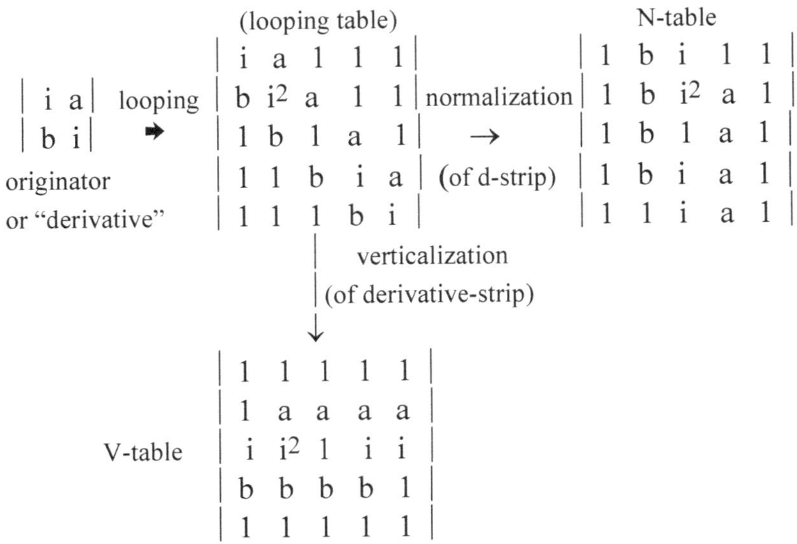

By necessity of the exchange of rows and columns, when comparing the "normalized" table to the "verticalized" one, a unitary-ness is firmly provided divisionally when properly following an order:
  order here running of elements of the normalized table from left to right of row,
  and of the verticalized table from bottom to top of column.

$$\therefore \qquad \frac{\text{N-table row element}}{\text{V-table column element}} \equiv 1$$

This of course only indicates that these tables derive from the common ancestor of looping (the looping table) and by logical extension (or corollary) the looping table's precursor or originating derivative.

Following the element order (for composing operationally N-table and V-table):

$$(\text{V-table})(\text{N-table}) \equiv (\text{N-table})^2$$
$$(\text{N-table})(\text{V-table}) \equiv (\text{V-table})^2$$
$$\text{V-table} / \text{N-table} \equiv 1$$
$$\text{N-table} / \text{V-table} \equiv 1$$

This belabored elemental ordering contrasts sharply with the logistical or unconditional ordering for tables used as probes (either) above (multiplication) or under (division) a table to be modified.

Thus, for example, we may probe the loop table with its originator in several direct ways, as follows:

1) Demarcate the (or a) reasonable or allowable region (minor) of probing

$$
\begin{array}{c|cccc}
 i & a & 1 & 1 & 1 \\ \hline
 b & i^2 & a & 1 & 1 \\
 1 & b & 1 & a & 1 \\
 1 & 1 & b & i & a \\
 1 & 1 & 1 & b & i
\end{array}
\quad \rightarrow \quad
\begin{array}{cc|cc}
 i^2 & a & 1 & 1 \\
 b & 1 & a & 1 \\ \hline
 1 & b & i & a \\
 1 & 1 & b & i
\end{array}
\qquad
\begin{array}{l}
\text{probing here:}\\
\text{un-staggered}
\end{array}
$$

2) Probe each region: here divisionally, and with $\begin{vmatrix} i & a \\ b & i \end{vmatrix}$

$$
\frac{\begin{vmatrix} i^2 & a \\ b & 1 \end{vmatrix}}{\begin{vmatrix} i & a \\ b & i \end{vmatrix}} \rightarrow \begin{vmatrix} i & 1 \\ 1 & i^{-1} \end{vmatrix}
\quad , \quad
\frac{\begin{vmatrix} 1 & b \\ 1 & 1 \end{vmatrix}}{\begin{vmatrix} i & a \\ b & i \end{vmatrix}} \rightarrow \begin{vmatrix} i^{-1} & b/a \\ b^{-1} & i^{-1} \end{vmatrix}
$$

$$
\frac{\begin{vmatrix} 1 & 1 \\ a & 1 \end{vmatrix}}{\begin{vmatrix} i & a \\ b & i \end{vmatrix}} \rightarrow \begin{vmatrix} i^{-1} & a^{-1} \\ b/a & i^{-1} \end{vmatrix}
\quad , \quad
\frac{\begin{vmatrix} i & a \\ b & i \end{vmatrix}}{\begin{vmatrix} i & a \\ b & i \end{vmatrix}} \rightarrow \begin{vmatrix} 1 & 1 \\ 1 & 1 \end{vmatrix}
$$

3) Recompose the results

$$
\begin{array}{cccc}
 i & 1 & i^{-1} & a^{-1} \\ \hline
 1 & i^{-1} & b/a & i^{-1} \\
 i^{-1} & b/a & 1 & 1 \\
 b^{-1} & i^{-1} & 1 & 1
\end{array}
\quad \rightarrow \quad
\begin{array}{c|cccc}
 i & a & 1 & 1 & 1 \\ \hline
 b & i & 1 & i^{-1} & a^{-1} \\
 1 & 1 & i^{-1} & b/a & i^{-1} \\
 1 & i^{-1} & b/a & 1 & 1 \\
 1 & b^{-1} & i^{-1} & 1 & 1
\end{array}
$$

diagonal products: $i \cdot i \cdot i^{-1} \cdot 1 \cdot 1 \equiv i$, $1 \cdot i^{-1} \cdot i^{-1} \cdot i^{-1} \cdot 1 \equiv i^{-3}$

Note: for the minor (table) the diagonal products are: $i \cdot i^{-1} \cdot 1 \cdot 1 \equiv 1$, $b^{-1} \cdot (b/a) \cdot (b/a) \cdot a^{-1} \equiv b/a^3$

	Row Products are:	Column Products are:	
for minor	$i \cdot a \cdot 1 \cdot 1 \cdot 1 \equiv ai$	$i \cdot b \cdot 1 \cdot 1 \cdot 1 \equiv bi$	for minor
a^{-1}	$b \cdot i \cdot 1 \cdot i^{-1} \cdot a^{-1} \equiv b/a$	$a \cdot i \cdot 1 \cdot i^{-1} \cdot b^{-1} \equiv a/b$	b^{-1}
$b/(ai)^2$	$1 \cdot 1 \cdot i^{-1} \cdot (b/a) \cdot i^{-1} \equiv b/(ai^2)$	$1 \cdot 1 \cdot i^{-1} \cdot (b/a) \cdot i^{-1} \equiv b/(ai^2)$	$b/(ai^2)$
$b/(ai)$	$1 \cdot i^{-1} \cdot (b/a) \cdot 1 \cdot 1 \equiv b/(ai)$	$1 \cdot i^{-1} \cdot (b/a) \cdot 1 \cdot 1 \equiv b/(ai)$	$b/(ai)$
$(bi)^{-1}$	$1 \cdot b^{-1} \cdot i^{-1} \cdot 1 \cdot 1 \equiv (bi)^{-1}$	$1 \cdot a^{-1} \cdot i^{-1} \cdot 1 \cdot 1 \equiv (ai)^{-1}$	$(ai)^{-1}$

$\therefore$ A functional relationship between 'a' and 'b' (and 'i') seems apparent:
  something of a correspondence, or exchange.

Of course, other probing regions can be designed (or specified):
e.g.

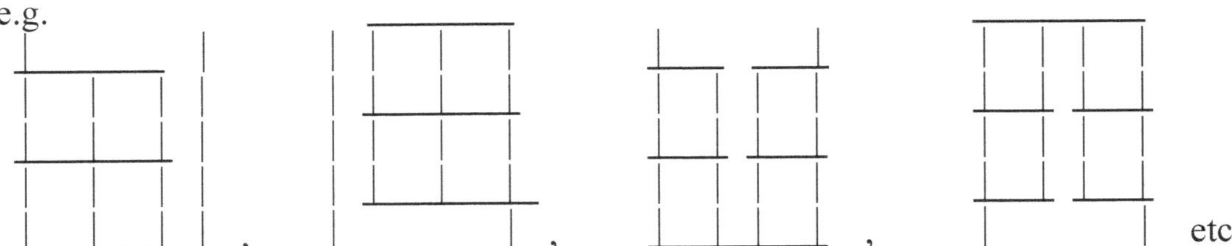

,   ,   ,   etc.

While these diagnostics might not promote a productivity towards unitary-ness for solutions, they do reveal functional relationships among the operations which may (or are allowed to) be typical of certain problems. These relationships are condensible into patterns of operation (or performance).

Of course, probes (probing tables) may be reoriented or modified themselves.

In our example, it is interesting to speculate that the exponent '3' (or -3) relates to the probing of a table with 3 complete(d) loops. Identity with the probe (elemental order) is strongly shown by the final minor table of $\begin{vmatrix} 1 & 1 \\ 1 & 1 \end{vmatrix}$;

but in this example, the actual probe initiates as a minor table: $\begin{vmatrix} i & a \\ b & i \end{vmatrix}$.

This is a coincidence (or acuity) of selection. The forensics of such probe-analyzed tables can reveal much data and trends. But it is important to keep in mind that the elements of these tables are operations rather than values. They are not variables, but variations of physical processes.

Looping tables under nul/los are important because they establish points of conversion for operations, such that principles of operation may be held in common or with desirable similarity of convergence. But there are other table manipulations achievable (or approachable) that can be used to emphasize similarity, such as similarity of environment enforced via prudent substitution, to create new tables of operations with perceivable or estimated advantages (for a given problem or set of problems).

e.g. Taking the previous composite table:

$$\begin{vmatrix} i & a & 1 & 1 & 1 \\ b & i & 1 & i\text{-}1 & a\text{-}1 \\ 1 & 1 & i\text{-}1 & b/a & i\text{-}1 \\ 1 & i\text{-}1 & b/a & 1 & 1 \\ 1 & b\text{-}1 & i\text{-}1 & 1 & 1 \end{vmatrix}$$

we note that the only common operation (active) between the recomposed minor (table) and its originator (unchanged) portion is " i ".

∴ We might make substitution of the originator " i " and its row/column environment into the recompose minor to force creation of a new table:

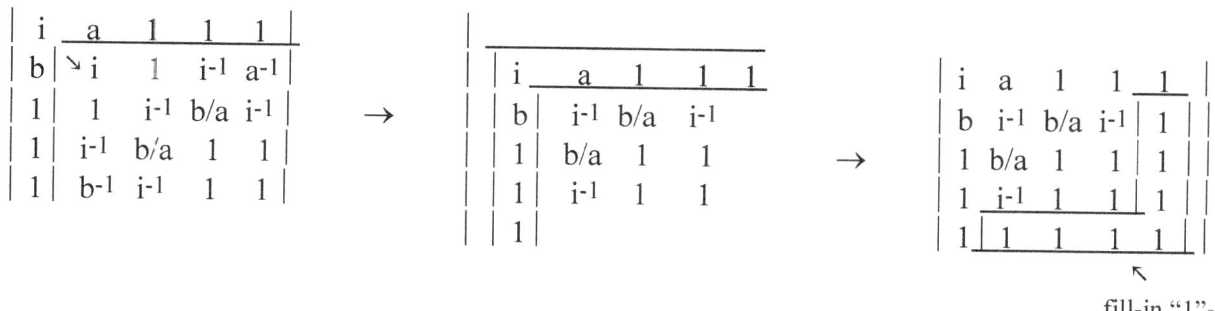

```
| i _a_ 1  1   1 |           |  ____________          | i a  1   1 _1_ |
| b ↘i   1  i-1 a-1 |        | | i    a   1  1         | b i-1 b/a i-1| 1 | |
| 1| 1  i-1 b/a i-1 |   →    | | b|  i-1 b/a  i-1      | 1 b/a 1   1| 1 | |
| 1| i-1 b/a 1   1 |         | | 1|  b/a  1   1      → | 1 _i-1  1  1_| 1 | |
| 1| b-1 i-1  1   1 |        | | 1|  i-1  1   1         | 1|_1  1   1   1_| |
                            | | 1|                                    ↖
                                                                fill-in "1"s
```

We have lost row elements (segments): 1 , i-1 , a-1
                      and column elements: 1 , i-1 , b-1

But there is a necessary increase in "1"s , and the diagonal products (of the new table) are:

i·i-1·1·1·1 ≡ 1
1·i-1·1·i-1·1 ≡ i-2

which differs from its originator (table) by inducement of "1" for one diagonal and a decrease in exponent from '3' to '2' (i.e. i-3 → i-2) for the other diagonal. There is a general reduction (as expected with fill-in "1"s) of the "intensity" of operation for the new table:

row products	column products
i·a·1·1·1 ≡ ai	i·b·1·1·1 ≡ bi
b·i-1·(b/a)·i-1·1 ≡ b2/(ai2)	a·i-1·(b/a)·i-1·1 ≡ b/i2
1·(b/a)·1·1·1 ≡ b/a	1·(b/a)·1·1·1 ≡ b/a
1·i-1·1·1·1 ≡ i-1	1·i-1·1·1·1 ≡ i-1
1·1·1·1·1 ≡ 1	1·1·1·1·1 ≡ 1

with the possible exception of the second row and column products: compared to the originator table's third row and column products, the new table (for 2nd row) is enhanced by 'b', but (for 2nd column) de-enhanced by loss of 'a' (actually an enhancement by 'a').

```
∴             originator         new
row     |     b/(ai2)     →    b2/(ai2)    ≡     b·(b/(ai2))
column  |     b/(ai2)     →    b/i2        ≡     a·(b/(ai2))
            (row/column 3)    (row/column 2)  :
```
due to, from the originator's perspective, a jump (or transfer/substitution) of "i" from (row/column 1) to (row/column 2).

There remains, in the new table, an interesting, though clearly contracted, symmetry of operations presiding among the elements:

```
|  | i    a  |_______  |
|  |_b_  i-1  b/a  i-1 |
|       | b/a |        |
|       | i-1 |        |
|                      |
```

and this might represent an operational "blueprint" of solutions for a procedure (or procedure's problems).

But what can we make of the lost (or discarded)
 row and column elements (segments) from the originator?

It is interesting that if we multiply the resultant elements to the originator elements (in this case) and then find the products of the corresponding row and column, this value for both is ' i⁻¹ ':

$$
\begin{array}{c|cccc}
 & a & 1 & 1 & 1 \\
\hline
i & & & & \\
b & i & 1 & i\text{-}1 & a\text{-}1 \\
1 & 1 & & & \\
1 & i\text{-}1 & a & i\text{-}1 & a\text{-}1 \quad\equiv a{\cdot}i\text{-}1{\cdot}a\text{-}1 = i\text{-}1 \\
1 & b\text{-}1 & & b & \\
 & & i\text{-}1 & & \equiv b{\cdot}i\text{-}1{\cdot}b\text{-}1 \equiv i\text{-}1 \\
 & & b\text{-}1 & &
\end{array}
$$

This suggests that the lost (or discarded) segments have some predictive power and eventual utility. Perhaps even more substitutions (this time for ' i⁻¹ ') are (operationally) allowable, expanding the new table further:

$$
\begin{array}{c|cc}
 & i & a \\
\hline
b & i\text{-}1 & b/a \;\; i\text{-}1 \\
 & b/a & \\
 & i\text{-}1 &
\end{array}
\qquad\rightarrow\qquad
\begin{array}{c|cc}
 & i & a \\
\hline
b & i\text{-}1 & b/a \;\; a \;\; i\text{-}1 \;\; a\text{-}1 \\
 & b/a & \\
 & b & \\
 & i\text{-}1 & \\
 & b\text{-}1 &
\end{array}
$$

Although such an expansion is arithmetically allowable for any operations, 'a' and 'b' (and a⁻¹, b⁻¹) are specifically implied, and with the above ordering.

The such is how (as examples) these tables can be manipulated with some pertinence. The null/los character or aspect permits these manipulations. Note the symmetries of the minor tables found (derived), evolved and produced (with balance) through these ("i" substitution) procedures:

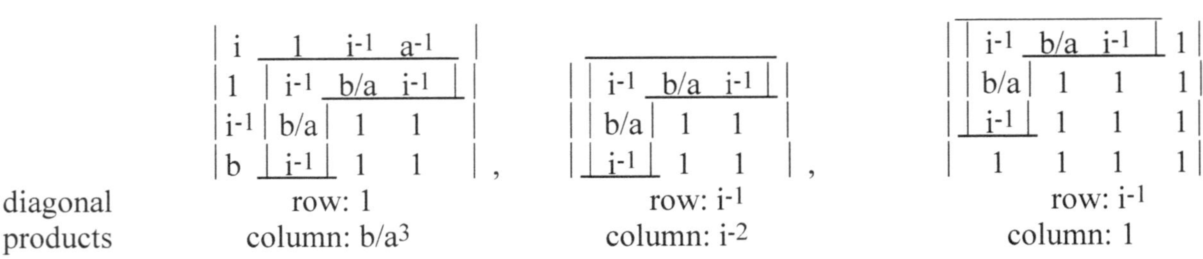

A symmetry is being carried along (or driven) through the progression (towards unitary-ness).

The actual symmetry:

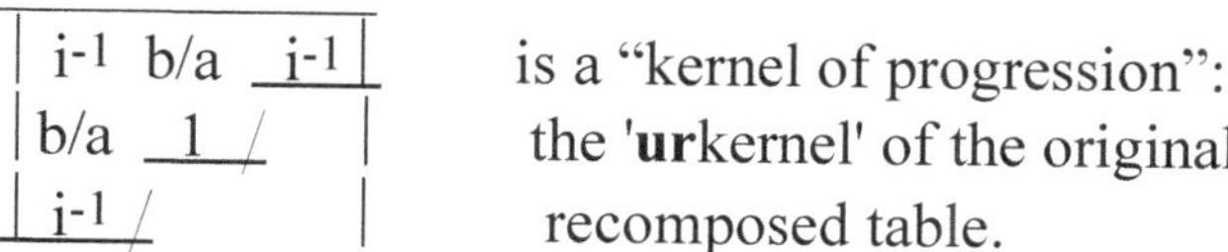

is a "kernel of progression":
the '**ur**kernel' of the original
recomposed table.

One can envision that substitution of such urkernels (with others) may modify tables of operation within prescribable rules (of procedure): i.e. They may act like components of replaceable procedures.

Due to its progressive isolation, it is suggestive that a transformation (of representation) from 2- to 3- dimensions (or 'm' to 'n' dimensions) may occur at such urkernels, perhaps the 3rd (or 'n') axis protruding off (in this case) the centralized " i-1 " element, and carrying with it (as were a bend) the flanking peripheral legs of the originator's minor (or the originator itself) for more complex expression of operations towards growing sophistication of solution:

e.g.

<pre>
              |→ i↘   1   i-1   a-1 →
 new "axis" ↗   1 ↘i-1   b/a   i-1 →
 (upward)   i-1   b/a
               b-1   i-1   to make tiered (or otherwise fastened or even faceted)
                ↓   ↓   table groupings.
</pre>

These visual manipulations of tables (of operations) can be characterized as real (rather than hypothetical or imaginary), although whether they lead to desired solutions is still questioned. Presumably an urkernal must stay unaltered throughout the progression (for a solution to be correct or determinable). Substitutions (judicious) may lose this constancy or continuity, in favor of different results (relatable to the original or initiating quest). There then seem to be analogies rather functional to board games (e.g.: chess, checkers) where a solution (e.g.: a "win") is dependent on the directed operations of the board pieces, this maneuvering being more complicated than for our row solutions to develop. Whereas "wins" for board games may be well designed (to be defined by particular results), our strategies have been devoted to evaluating and deriving or correcting for solutions as based on proposed comparisons of identical outcomes (i.e. per row) as based on assumed identities (or circuiting) required for a procedure to require of itself logical insistences or reciprocities, since any physical operation must be correct of itself regardless of how it is viewed or examined of components and routes. Consistencies of solutions should be upheld throughout the various analyses. And uncertainties should not be relegated to probabilities (or improbabilities) as possible or actual. That would be a violation of the nul/los principle avoiding "zero" answers.

Nevertheless, solutions depicted of operations in a table might not be equivalent among the rows as given, leading to manipulations to derive unitary-ness from the table, since unity of comparison implies correctness (of comparisons). To this mode (of investigation) a table may be folded and modified by many manners (or methods).

Probing

We may also attempt to derive information on a table by limiting a probing to the table's derivative strip, which leads to a staggered and directional probing:

e.g.

$$
\begin{array}{l}
|\ i\ a\ \backslash\underline{\ \ } \\
\underline{|\ b\ i^2\ a\ \backslash\ \searrow} \\
\ \ \backslash\underline{\ b\ \ 1\ a\ \backslash\ \underline{\ }} \\
\ \ \ \ \ \nwarrow\backslash\underline{\ b\ \ i\ \ a\ |} \\
\ \ \ \ \ \ \ \backslash\underline{\ b\ \ i\ \ |}
\end{array}
\quad
\text{probed with}
\quad
\begin{array}{|cc|}
\hline i & a \\ b & i \\ \hline
\end{array}
\quad\rightarrow\quad
\begin{array}{ll}
1 & 1 \\
\text{division}\ \ 1 & 1\ \ 1 \\
1 & i^{\text{-}2}\ \ 1 \\
1 & i^{\text{-}1}\ \ 1 \\
1 & 1
\end{array}
$$

The same result occurs when going up the strip as down (although individual block divisions differ order-wise).

$$
\begin{array}{|cc|}
\hline i & a \\ b & i^2 \\ \hline
\end{array}
\ /\
\begin{array}{|cc|}
\hline i & a \\ b & i \\ \hline
\end{array}
\ \Rightarrow\
\begin{array}{|cc|}
\hline 1 & 1 \\ 1 & i \\ \hline
\end{array}
\qquad \therefore\ i^2 \rightarrow i
$$

$$
\begin{array}{|cc|}
\hline i & a \\ b & 1 \\ \hline
\end{array}
\ /\
\begin{array}{|cc|}
\hline i & a \\ b & i \\ \hline
\end{array}
\ \Rightarrow\
\begin{array}{|cc|}
\hline 1 & 1 \\ 1 & i^{\text{-}1} \\ \hline
\end{array}
\qquad \therefore\ 1 \rightarrow i^{\text{-}1}
$$

$$
\begin{array}{|cc|}
\hline i^{\text{-}1} & a \\ b & i \\ \hline
\end{array}
\ /\
\begin{array}{|cc|}
\hline i & a \\ b & i \\ \hline
\end{array}
\ \Rightarrow\
\begin{array}{|cc|}
\hline i^{\text{-}2} & 1 \\ 1 & 1 \\ \hline
\end{array}
\qquad \therefore\ i \rightarrow 1
$$

$$
\begin{array}{|cc|}
\hline 1 & a \\ b & i \\ \hline
\end{array}
\ /\
\begin{array}{|cc|}
\hline i & a \\ b & i \\ \hline
\end{array}
\ \Rightarrow\
\begin{array}{|cc|}
\hline i^{\text{-}1} & 1 \\ 1 & 1 \\ \hline
\end{array}
\qquad \therefore\ i \rightarrow 1
$$

Replacing this probed strip for the derivative strip (or simply filling-in the "1"s) yields a sparsely defined table with diagonal products ($i^{\text{-}3}$, $i^{\text{-}2}$) , and row/column products (1 , 1 , $i^{\text{-}2}$, $i^{\text{-}1}$, 1) :

$$
\begin{array}{|cccccc|}
\hline
1 & 1 & & & & \\
1 & 1 & 1 & & & \\
i & 1 & i^{\text{-}2} & 1 & & \\
 & & & 1 & i^{\text{-}1} & 1 \\
 & & & & 1 & 1 \\
\hline
\end{array}
$$

The solutions (1 , $i^{\text{-}1}$, $i^{\text{-}2}$) are unlikely to be equivalent (i.e. these different operations are not likely to each reach to the same solution, unless: ' $i \equiv 1$ '). Nor is unitary-ness particularly achievable except by:

$$
\begin{array}{c|ccccc|}
 & 1 & 1 & 1 & 1 & 1 \\
 & 1 & 1 & 1 & 1 & 1 \\
i^2 & 1 & 1 & i^{\text{-}2} & 1 & 1 \\
i & 1 & 1 & 1 & i^{\text{-}1} & 1 \\
 & 1 & 1 & 1 & 1 & 1 \\
\hline
\end{array}
\quad
\begin{array}{l}\text{and further}\\ \text{manipulation, or}\end{array}
\quad
\begin{array}{|ccccc|c}
\hline
1 & 1 & 1 & 1 & 1 & \\
1 & 1 & 1 & 1 & 1 & \\
1 & 1 & i^{\text{-}2} & 1 & 1 & i^2 \\
1 & 1 & 1 & i^{\text{-}1} & 1 & i \\
1 & 1 & 1 & 1 & 1 & \\
\hline
\end{array}
$$

(with selective partitioning)

although the table's sparsity (of distinct elements) seems already very close to unitary-ness.

What is most evident of course (from the derivative strip's divisional probing with $\begin{array}{|cc|}\hline i & a \\ b & i \\ \hline\end{array}$)

is the elimination (without subtraction) of the operations 'a' and 'b' .

Looping, probing, establishing or determining urkernels of processes, and judicious substitutions based on identities of elements are of course not the only ways for processing tables to derive or yield information and characteristics of solutions. One can, for example, shift patterns of elements reflectively (or as by a mirror plane) to see if the solutions remain invariant while adjusting the row/column product orders (as well as diagonal or slant products') :

e.g. Mirroring a derivative strip

$$
\begin{vmatrix} i & a & 1 & 1 & 1 \\ b & i^2 & a & 1 & 1 \\ 1 & b & 1 & a & 1 \\ 1 & 1 & b & i & a \\ 1 & 1 & 1 & b & i \end{vmatrix}
\quad \xrightarrow{\text{reflection}} \quad
\begin{vmatrix} 1 & 1 & 1 & a & i \\ 1 & 1 & a & i^2 & b \\ 1 & a & 1 & b & 1 \\ a & i & b & 1 & 1 \\ i & b & 1 & 1 & 1 \end{vmatrix}
\quad \text{i.e. Is } (ia \equiv ai)\text{ ? etc.}
$$

(The invariance depends on the sensitivities of the orders of operation, or table element positions.)

But one can also do a physical shift (here of the derivative strip), which is not reflective:

$$
\begin{vmatrix} i & a & 1 & 1 & 1 \\ b & i^2 & a & 1 & 1 \\ 1 & b & 1 & a & 1 \\ 1 & 1 & b & i & a \\ 1 & 1 & 1 & b & i \end{vmatrix}
\quad \xrightarrow[\text{not reflective}]{\text{shift}} \quad
\begin{vmatrix} 1 & 1 & 1 & i & a \\ 1 & 1 & b & i^2 & a \\ 1 & b & 1 & a & 1 \\ b & i & a & 1 & 1 \\ b & i & 1 & 1 & 1 \end{vmatrix}
$$

diagonal products: 1, abi^3

row products: ia, abi^2, ab, aib, bi

column products: b^2, bi^2, ab, ai^3, a^2

causing a clear functional disruption of the strip (that makes it non-derivative, but still solution logical). Note that for the strip-shifted table, the row products and thus the solutions yielded to them (disregarding order readjustments based on "1"s) remain the same, but the column and diagonal (and slant) products differ.

Since the last two tables form the same solutions, now of course we can compare their functional equivalencies, most directly by (table) rotation to bring: ' columns $\leftrightarrow$ rows ', or studying the distinctions to their column products, but also by strip normalization and verticalization.

Also, compositing might be attempted: We note that both tables share identical third rows.

$\therefore$ We may legitimately mix the tables, keeping this 3rd row constant:

e.g.

$$
\begin{vmatrix} i & a & 1 & 1 & 1 \\ b & i^2 & a & 1 & 1 \\ 1 & b & 1 & a & 1 \\ b & i & a & 1 & 1 \\ b & i & 1 & 1 & 1 \end{vmatrix}
\quad , \quad
\begin{vmatrix} 1 & 1 & 1 & i & a \\ 1 & 1 & b & i^2 & a \\ 1 & b & 1 & a & 1 \\ 1 & 1 & b & i & a \\ 1 & 1 & 1 & b & i \end{vmatrix}
$$

The four thus tables accumulated each still yield the same row products and resulting solutions. Diagonal and column products differ. Visually, because of the symmetries shown, there is great temptation to (strip-)combine the two most previous (again keeping the 3rd row constant):

$$
\begin{vmatrix}
i & a & 1 & i & a \\
b & i2 & a+b & i2 & a \\
1 & b & 1 & a & 1 \\
b & i & a+b & i & a \\
b & i & 1 & b & i
\end{vmatrix}
\qquad \text{producing (a+b) terms}
$$

This merging enhancement is distinguished from a formal additioning of tables. The significance of the (a+b) terms (or element+1 or 1+1 terms for formal additioning) is of course that the table may be resolved to separate tables (of selective choice):

$$
\begin{vmatrix}
i & a & 1 & i & a \\
b & i2 & a+b & i2 & a \\
1 & b & 1 & a & 1 \\
b & i & a+b & i & a \\
b & i & 1 & b & i
\end{vmatrix}
\nearrow
\begin{vmatrix}
i & a & 1 & i & a \\
b & i2 & a & i2 & a \\
1 & b & 1 & a & 1 \\
b & i & a+b & i & a \\
b & i & 1 & b & i
\end{vmatrix}
\rightarrow
\begin{vmatrix}
i & a & 1 & i & a \\
b & i2 & a & i2 & a \\
1 & b & 1 & a & 1 \\
b & i & b & i & a \\
b & i & 1 & b & i
\end{vmatrix}
\begin{vmatrix}
i & a & 1 & i & a \\
b & i2 & a & i2 & a \\
1 & b & 1 & a & 1 \\
b & i & a & i & a \\
b & i & 1 & b & i
\end{vmatrix}
$$

$$
\searrow
\begin{vmatrix}
i & a & 1 & i & a \\
b & i2 & b & i2 & a \\
1 & b & 1 & a & 1 \\
b & i & a+b & i & a \\
b & i & 1 & b & i
\end{vmatrix}
\rightarrow
\begin{vmatrix}
i & a & 1 & i & a \\
b & i2 & b & i2 & a \\
1 & b & 1 & a & 1 \\
b & i & a & i & a \\
b & i & 1 & b & i
\end{vmatrix}
\begin{vmatrix}
i & a & 1 & i & a \\
b & i2 & b & i2 & a \\
1 & b & 1 & a & 1 \\
b & i & b & i & a \\
b & i & 1 & b & i
\end{vmatrix}
$$

Apparently, an (a-b) symbolism would "remove" (from the results) one of the two tables produceable from this element set (i.e. here where the element 'b' is allowed to result).

∴ " - " (as 'minus') ≡ remove ≡ disregard; e.g. (a-b) ≡ a , where an addition would cause: (a+b) .

Each of the tables at least contains the $\begin{vmatrix} 1 & b & 1 & a & 1 \end{vmatrix}$ row solution, as well of the columns for two of them (product wise).

∴ Quite perfunctorily we have generated 9 (possibly 10) tables with equivalent or shared rows and (therefore) solutions. But the methods employed emphasize the contrast (for tables) between the multiplication/division operations and the (+,-) operators:

While (multiplication/division) remain algebraic, (+,-) are used to compact ('compactify' of information) and resolve tables more or less selectively, and to contain (within tables) and elicit specific elements.

One would expect quantitative (calculative) results to obtain from multiplication/division, and qualitative (distinguishing) results to obtain from (+,-) applications.

Tabular forms (tablature of 'formature'), with their horizontal and vertical projections, make it (visually) easier to follow or outline operational procedures, than for linear (separate) formulations and definitions (e.g.: equations). But more thorough or efficient derivization of tabular results comes from the ***circularization*** of its elements:

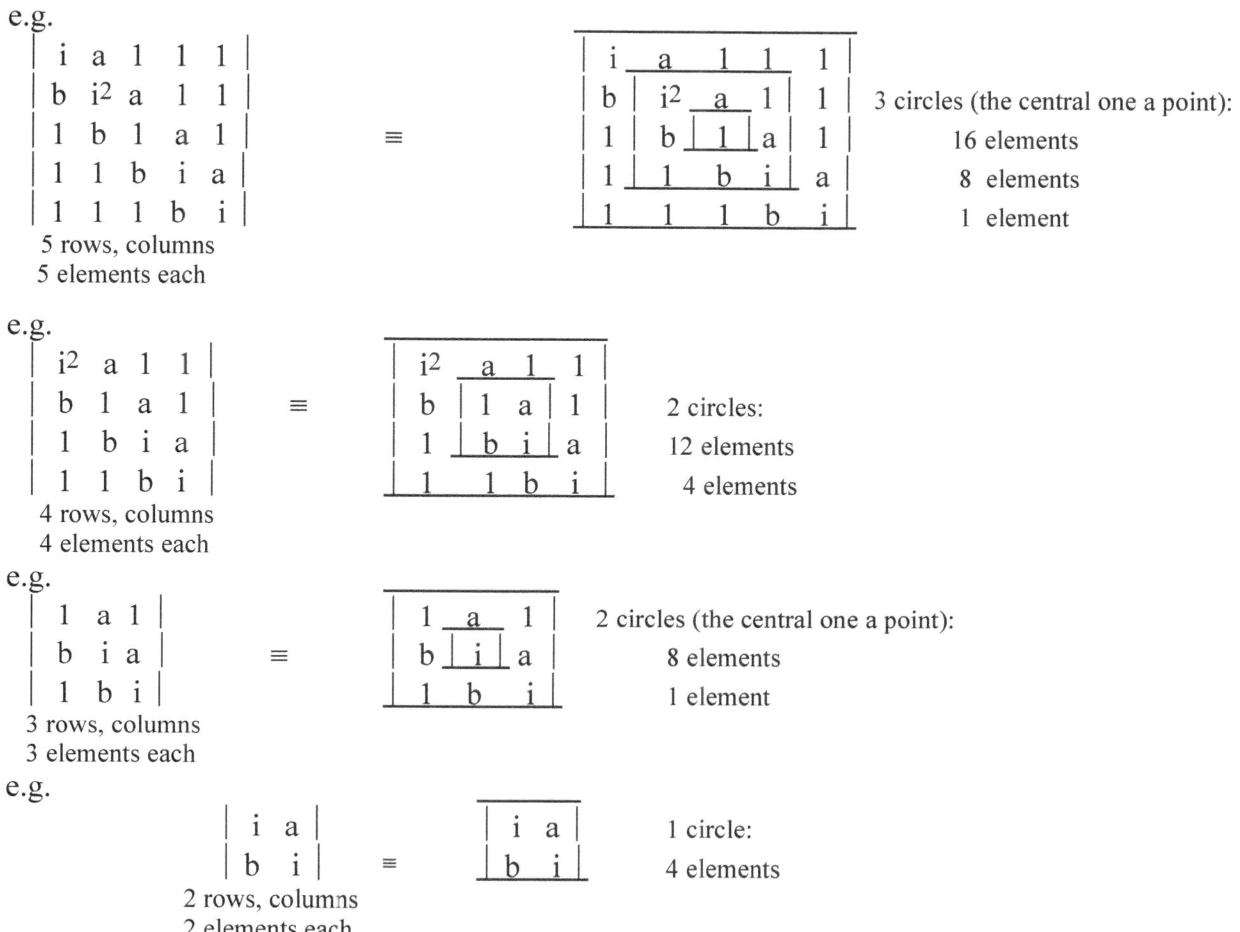

Obviously for a circle, rows and columns automatically merge in meaning. And with this concentrated form, several solutions may be contained. For example, assuming a column can devolve into a row to yield a solution (by some method or manipulation), then a 12 element circle (as the preceding) might represent 4 solutions (each of 4 elements or operations):

With further loosening of constraint for the structuring of the circle (e.g. allowing each element to take a particular position), the number of solutions represented by the circle expands to:

$$(\text{\# elements of circle}) \quad \text{X} \quad (\text{\# solutions per circle})$$

e.g.

The $\overline{|\,i\ a\,|}$ table yields to a $\overline{|\,i\ a\,|}$ circle, with 4 elements and 4 solutions/circle $\equiv$ 16 solutions
$|\,b\ i\,|$ $|\,b\ i\,|$

$\therefore$

$\overline{|\,i\ a\,|}$ ia, bi,	$\overline{|\,a\ i\,|}$ ai, bi,	$\overline{|\,i\ b\,|}$ ib, ai,	$\overline{|\,b\ i\,|}$ bi, ia,
$\underline{|\,b\ i\,|}$ ib, ia	$\underline{|\,i\ b\,|}$ ib, ia	$\underline{|\,a\ i\,|}$ ia, ib	$\underline{|\,i\ a\,|}$ bi, ai

(Some solutions may of course be equivalent and redundant.)

For a circle derived from a tabular form, the number of solutions $\equiv$ # of equivalent rows and columns, which is always 4 ;

$\therefore$ such a circle may yield up to: (# elements of circle $\bullet$ 4) solutions ;
or, the # of apparent solutions $\equiv$ # rows $\equiv$ 2

$\therefore$ (# elements of circle $\bullet$ 2) solutions are obtained.

The effect, here, is that the circle is a more concentrated representation (of information) than the table. Obviously, then, circle swapping is another possible means of manipulation between and among tables. For tabularly derived circles, the # of elements is always even, unless the circle is a point; and instead of diagonal, slant, column or row products we have (particularly from balanced tables to visualize) "diameter" products, the number of which uniquely possible are the number of elements in the circle halved (unless the circle is a point0.

It is also quite evident that the largest (peripheral) circle derived from a balanced table may be used (via the circle's solutions) to design yet another table of same character (in # of rows and columns):

e.g.

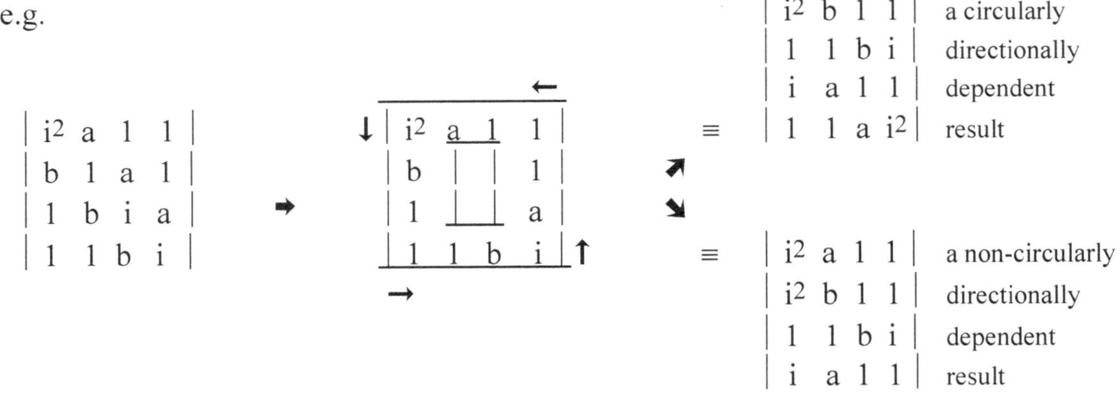

CDD (circularly directionally dependent) Procedures

In fact, such a procedure may be made iterative (with some defined discipline, such as "circularly directionally dependent" ; i.e. 'CDD') to yield consistently derivative information (or construction):

e.g.

$$
\begin{vmatrix} i^2 & a & 1 & 1 \\ b & 1 & a & 1 \\ 1 & b & i & a \\ 1 & 1 & b & i \end{vmatrix}
\;\rightarrow\;
\begin{vmatrix} i^2 & a & 1 & 1 \\ b & & & 1 \\ 1 & & & a \\ 1 & 1 & b & i \end{vmatrix}
\;\equiv\;
\begin{vmatrix} i^2 & a & 1 & 1 \\ 1 & 1 & a & i \\ i & b & 1 & 1 \\ 1 & 1 & b & i^2 \end{vmatrix}
\;\rightarrow\;
\begin{vmatrix} i^2 & a & 1 & 1 \\ 1 & & & i \\ i & & & 1 \\ 1 & 1 & b & i^2 \end{vmatrix}
\;\equiv\;
\begin{vmatrix} i^2 & a & 1 & 1 \\ 1 & i & 1 & i^2 \\ i^2 & b & 1 & 1 \\ 1 & i & 1 & i^2 \end{vmatrix}
\text{ etc.}
$$

loss of 'b'

$$
\;\rightarrow\;
\begin{vmatrix} i^2 & a & 1 & 1 \\ 1 & & & i^2 \\ i^2 & & & 1 \\ 1 & i & 1 & i^2 \end{vmatrix}
\;\equiv\;
\begin{vmatrix} i^2 & a & 1 & 1 \\ 1 & i^2 & 1 & i^2 \\ i^2 & 1 & i & 1 \\ 1 & i^2 & 1 & i^2 \end{vmatrix}
\;\begin{array}{c}\text{loss of}\\ \text{'i'}\end{array}\;\rightarrow\;
\begin{vmatrix} i^2 & a & 1 & 1 \\ 1 & & & i^2 \\ i^2 & & & 1 \\ 1 & i^2 & 1 & i^2 \end{vmatrix}
\;\equiv\;
\begin{vmatrix} i^2 & a & 1 & 1 \\ 1 & i^2 & 1 & i^2 \\ i^2 & 1 & i^2 & 1 \\ 1 & i^2 & 1 & i^2 \end{vmatrix}
\;\rightarrow
$$

$$
\begin{vmatrix} i^2 & a & 1 & 1 \\ 1 & & & i^2 \\ i^2 & & & 1 \\ 1 & i^2 & 1 & i^2 \end{vmatrix}
\;\equiv\;
\begin{vmatrix} i^2 & a & 1 & 1 \\ 1 & i^2 & 1 & i^2 \\ i^2 & 1 & i^2 & 1 \\ 1 & i^2 & 1 & i^2 \end{vmatrix}
\;\begin{array}{c}\text{same as previous table}\\ \therefore \text{ final result}\end{array}
$$

Note that each of these tables (in this example) share the same (initial) row: $\therefore$ same solution.

Obviously, for the opposite direction of the "circularly directionally dependent" procedure, the resultant is:

$$
\begin{vmatrix} i^2 & b & 1 & 1 \\ 1 & i^2 & 1 & i^2 \\ i^2 & 1 & i^2 & 1 \\ 1 & i^2 & 1 & i^2 \end{vmatrix}
\qquad\text{with loss of 'a' and 'i' ;}
$$

$\therefore$

$$
\begin{vmatrix} i^2 & a & 1 & 1 \\ b & 1 & a & 1 \\ 1 & b & i & a \\ 1 & 1 & b & i \end{vmatrix}
\;\begin{array}{c}\text{CDD}\\ \rightarrow\rightarrow\rightarrow\\ \text{(iterative)}\end{array}\;
\begin{vmatrix} i^2 & a & 1 & 1 \\ 1 & i^2 & 1 & i^2 \\ i^2 & 1 & i^2 & 1 \\ 1 & i^2 & 1 & i^2 \end{vmatrix}
\;,\;
\begin{vmatrix} i^2 & b & 1 & 1 \\ 1 & i^2 & 1 & i^2 \\ i^2 & 1 & i^2 & 1 \\ 1 & i^2 & 1 & i^2 \end{vmatrix}
$$

easily predictable results

with the common tabular "structure"

$$
\begin{vmatrix} 1 & i^2 & 1 & i^2 \\ i^2 & 1 & i^2 & 1 \\ 1 & i^2 & 1 & i^2 \end{vmatrix}
$$

"forced" under each initiating row (or column as row): $\begin{vmatrix} i^2 & a & 1 & 1 \end{vmatrix}$, $\begin{vmatrix} i^2 & b & 1 & 1 \end{vmatrix}$

It is of course immediately suggested that: $(a \equiv b \equiv i^2)$, perhaps in some structured manner.

But these tables are not at established unitary-ness.

Perhaps to bring to unitary-ness, cause:

$$\begin{vmatrix} i^2 & a & 1 & 1 \\ b & 1 & a & 1 \\ 1 & b : i : a & & i^{-1} \\ 1 & 1 & b : i & i \end{vmatrix} \quad \rightarrow \quad \begin{vmatrix} i^2 & a & 1 & 1 \\ b & 1 & a & 1 \\ 1 & b & 1 & a \\ 1 & 1 & b & i^2 \end{vmatrix}$$

In this case, perhaps unitary-ness can be simply induced through diagonal or slant manipulation (adjustment):

e.g.
$$\begin{vmatrix} i^2 & a & 1 & 1 \\ b & 1 & a & 1 \\ 1 & b & i & a \\ 1 & 1 & b & i \end{vmatrix} \quad \rightarrow \quad \begin{vmatrix} i^2 & a & 1 & 1 \\ b & 1 & a & 1 \\ 1 & b & 1 & a \\ 1 & 1 & b & i^2 \end{vmatrix} \qquad \text{multiplicative shifting}$$

Note that the diagonal products are not changed.

The concept that ($a \equiv b \equiv i^2$) is maintained. Each of the rows (and both diagonals) and each of the columns yield essentially to (i^4) as a solution. This exquisitely simple manipulation (to gain unitary-ness) might have been (or seemed) obscure without (or before) the CDD analyses. It also results in a table of stronger symmetry.

The (most) peripheral circle contains (elements of) the solutions of the table derived of it, while the internal circles do (or may) not. What may internal circle analysis reveal?

In the previous case, since CDD (in either direction) does not discard any elements (the internal circle being the last), circular analysis simply seems to rotate the elements (to assume).

However:

e.g.
$$\begin{vmatrix} 1 & a \\ b & i \end{vmatrix} \quad \blacktriangleright \quad \underline{\overset{(\rightarrow)}{\begin{vmatrix} 1 & a \\ b & i \end{vmatrix}}} \quad \equiv \quad \begin{vmatrix} 1 & a \\ a & i \\ i & b \\ b & 1 \end{vmatrix} \quad \text{unbalanced}$$

We may treat the unbalanced table in several ways.

1) We can part them into two balanced tables (formally separating the 'a' and 'b' operations):

$$\begin{vmatrix} 1 & a \\ a & i \\ \hline i & b \\ b & 1 \end{vmatrix} \quad \rightarrow \quad \begin{vmatrix} 1 & a \\ a & i \\ i & b \\ b & 1 \end{vmatrix} \quad \overset{\text{CDD}}{\blacktriangleright} \quad \begin{matrix} \overset{(\rightarrow)}{\underline{\begin{vmatrix} 1 & a \\ a & i \end{vmatrix}}} \\ \overset{(\rightarrow)}{\underline{\begin{vmatrix} i & b \\ b & 1 \end{vmatrix}}} \end{matrix} \quad \rightarrow \quad \begin{matrix} \begin{vmatrix} 1 & a \\ a & i \\ i & a \\ a & 1 \end{vmatrix} \rightarrow \\ \\ \begin{vmatrix} i & b \\ b & 1 \\ 1 & b \\ b & i \end{vmatrix} \end{matrix}$$

$\begin{vmatrix} 1 & a \\ a & i \end{vmatrix} \rightarrow \begin{vmatrix} 1 & a \end{vmatrix}, \begin{vmatrix} a & i \end{vmatrix}$
$\rightarrow$ repetitive
$\begin{vmatrix} a & 1 \end{vmatrix} \rightarrow \begin{vmatrix} i & a \end{vmatrix}, \begin{vmatrix} a & 1 \end{vmatrix}$

$\begin{vmatrix} i & b \\ b & 1 \end{vmatrix} \rightarrow \begin{vmatrix} i & b \end{vmatrix}, \begin{vmatrix} b & 1 \end{vmatrix}$
$\rightarrow$ repetitive
$\begin{vmatrix} b & i \end{vmatrix} \rightarrow \begin{vmatrix} 1 & b \end{vmatrix}, \begin{vmatrix} b & i \end{vmatrix}$

and continue with the (CDD) analyses:

$$
\begin{array}{l}
|\,i\ a\,| \\
|\,a\ 1\,|
\end{array}
\ \blacktriangleright\
\overset{(\rightarrow)}{\underline{\begin{array}{l}|\,i\ a\,|\\|\,a\ 1\,|\end{array}}}
\ \equiv\
\begin{array}{l}
|\,i\ a\,| \\
|\,a\ 1\,| \rightarrow \quad \text{repetitive} \\
\hline
|\,1\ a\,| \rightarrow \quad \text{repetitive} \\
|\,a\ i\,|
\end{array}
$$

$$
\begin{array}{l}
|\,1\ b\,| \\
|\,b\ i\,|
\end{array}
\ \blacktriangleright\
\overset{(\rightarrow)}{\underline{\begin{array}{l}|\,1\ b\,|\\|\,b\ i\,|\end{array}}}
\ \equiv\
\begin{array}{l}
|\,1\ b\,| \\
|\,b\ i\,| \rightarrow \quad \text{repetitive} \\
\hline
|\,i\ b\,| \rightarrow \quad \text{repetitive} \\
|\,b\ 1\,|
\end{array}
$$

2) We can balance them variously, with "1"s:

e.g. (to lose both 'i' and 'b')

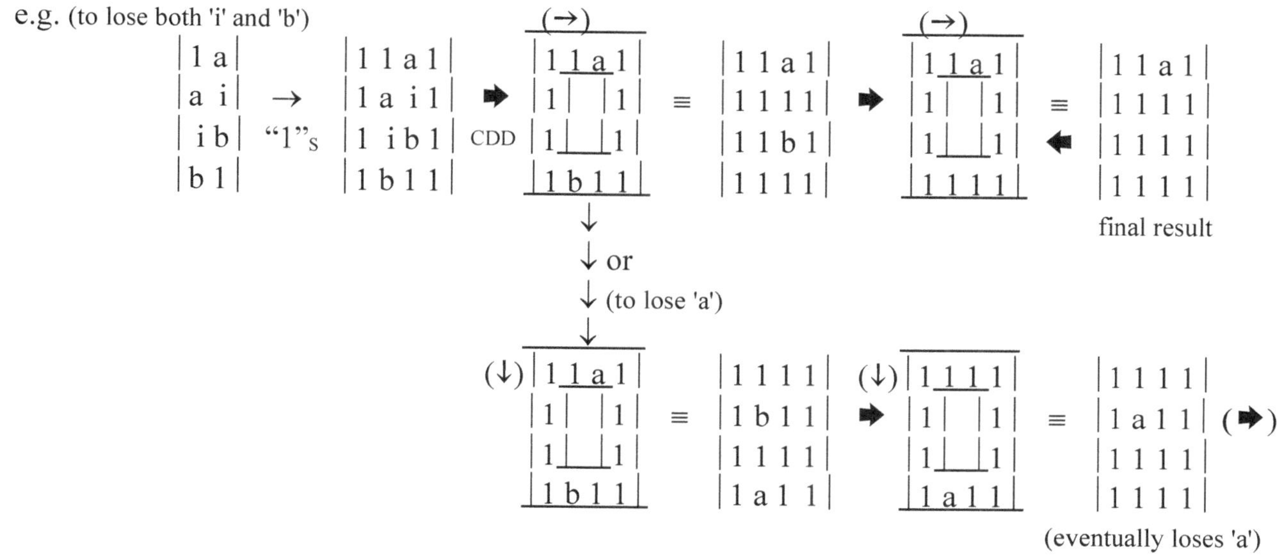

[for the other direction:
| | 1 1 1 1 | | 1 1 a 1 |
| | 1 1 1 1 | → loss of 'a'   or | 1 1 1 1 | final result
| | 1 1 1 1 | | 1 1 1 1 |
[| 1 a 1 1 | | 1 1 1 1 |]

also:

e.g.

$$
\begin{vmatrix} 1 & a \\ a & i \\ i & b \\ b & 1 \end{vmatrix}
\;\xrightarrow{\text{balance}}\;
\begin{vmatrix} 1 & a & 1 & 1 \\ a & i & 1 & 1 \\ i & b & 1 & 1 \\ b & 1 & 1 & 1 \end{vmatrix}
\;\xrightarrow{\text{CDD}}\;
\begin{vmatrix} 1 & a & 1 & 1 \\ a & & & 1 \\ i & & & 1 \\ b & 1 & 1 & 1 \end{vmatrix}
\;\equiv\;
\begin{vmatrix} 1 & a & 1 & 1 \\ 1 & 1 & 1 & 1 \\ 1 & 1 & 1 & b \\ b & i & a & 1 \end{vmatrix}
\;\to\;
\begin{vmatrix} 1 & a & 1 & 1 \\ 1 & & & 1 \\ 1 & & & b \\ b & i & a & 1 \end{vmatrix}
\;\equiv
$$

$$
\begin{vmatrix} 1 & a & 1 & 1 \\ 1 & 1 & b & 1 \\ 1 & a & i & b \\ b & 1 & 1 & 1 \end{vmatrix}
\;\to\;
\begin{vmatrix} 1 & a & 1 & 1 \\ 1 & & & 1 \\ 1 & & & b \\ b & 1 & 1 & 1 \end{vmatrix}
\;\equiv\;
\begin{vmatrix} 1 & a & 1 & 1 \\ 1 & 1 & b & 1 \\ 1 & 1 & 1 & b \\ b & 1 & 1 & 1 \end{vmatrix}
\quad \text{final result}
$$

e.g.

$$
\begin{vmatrix} 1 & a \\ a & i \\ i & b \\ b & 1 \end{vmatrix}
\;\xrightarrow{\text{balance}}\;
\begin{vmatrix} 1 & 1 & 1 & a \\ 1 & 1 & a & i \\ 1 & 1 & i & b \\ 1 & 1 & b & 1 \end{vmatrix}
\;\xrightarrow{\text{CDD}}\;
\begin{vmatrix} 1 & 1 & 1 & a \\ 1 & & & i \\ 1 & & & b \\ 1 & 1 & b & 1 \end{vmatrix}
\;\equiv\;
\begin{vmatrix} 1 & 1 & 1 & a \\ a & i & b & 1 \\ 1 & b & 1 & 1 \\ 1 & 1 & 1 & 1 \end{vmatrix}
\;\to\;
\begin{vmatrix} 1 & 1 & 1 & a \\ a & & & 1 \\ 1 & & & 1 \\ 1 & 1 & 1 & 1 \end{vmatrix}
\;\equiv
$$

$$
\begin{vmatrix} 1 & 1 & 1 & a \\ a & 1 & 1 & 1 \\ 1 & 1 & 1 & 1 \\ 1 & 1 & a & 1 \end{vmatrix}
\;\to\;
\begin{vmatrix} 1 & 1 & 1 & a \\ a & & & 1 \\ 1 & & & 1 \\ 1 & 1 & a & 1 \end{vmatrix}
\;\equiv\;
\begin{vmatrix} 1 & 1 & 1 & a \\ a & 1 & 1 & 1 \\ 1 & a & 1 & 1 \\ 1 & 1 & a & 1 \end{vmatrix}
\quad \text{final result}
$$

It's clear that the initiating row is always retained (in CDD analysis):

$$
\begin{vmatrix} 1 & a & 1 & 1 \\ a & & & 1 \\ i & & & 1 \\ b & 1 & 1 & 1 \end{vmatrix}
\;\equiv\;
\begin{vmatrix} 1 & 1 & a & 1 \\ 1 & a & i & b \\ b & 1 & 1 & 1 \\ 1 & 1 & 1 & 1 \end{vmatrix}
\;\to\;
\begin{vmatrix} 1 & 1 & a & 1 \\ 1 & & & b \\ b & & & 1 \\ 1 & 1 & 1 & 1 \end{vmatrix}
\;\equiv\;
\begin{vmatrix} 1 & a & 1 & 1 \\ 1 & 1 & b & 1 \\ 1 & 1 & 1 & 1 \\ 1 & 1 & b & 1 \end{vmatrix}
\;\to\;
\begin{vmatrix} 1 & a & 1 & 1 \\ 1 & & & 1 \\ 1 & & & 1 \\ 1 & 1 & b & 1 \end{vmatrix}
\;\equiv
$$

$$
\begin{vmatrix} 1 & 1 & a & 1 \\ 1 & 1 & 1 & 1 \\ 1 & 1 & b & 1 \\ 1 & 1 & 1 & 1 \end{vmatrix}
\;\to\;
\begin{vmatrix} 1 & 1 & a & 1 \\ 1 & & & 1 \\ 1 & & & 1 \\ 1 & 1 & 1 & 1 \end{vmatrix}
\;\equiv\;
\begin{vmatrix} 1 & a & 1 & 1 \\ 1 & 1 & 1 & 1 \\ 1 & 1 & 1 & 1 \\ 1 & 1 & 1 & 1 \end{vmatrix}
\quad \text{final result}
$$

The circular direction and starting point (for row) must be stated (decided or defined).

Of course, a combination of ① and (then) ② may be performed.

Here we can examine how to retain all operations (a, b, i, 1) in CDD:

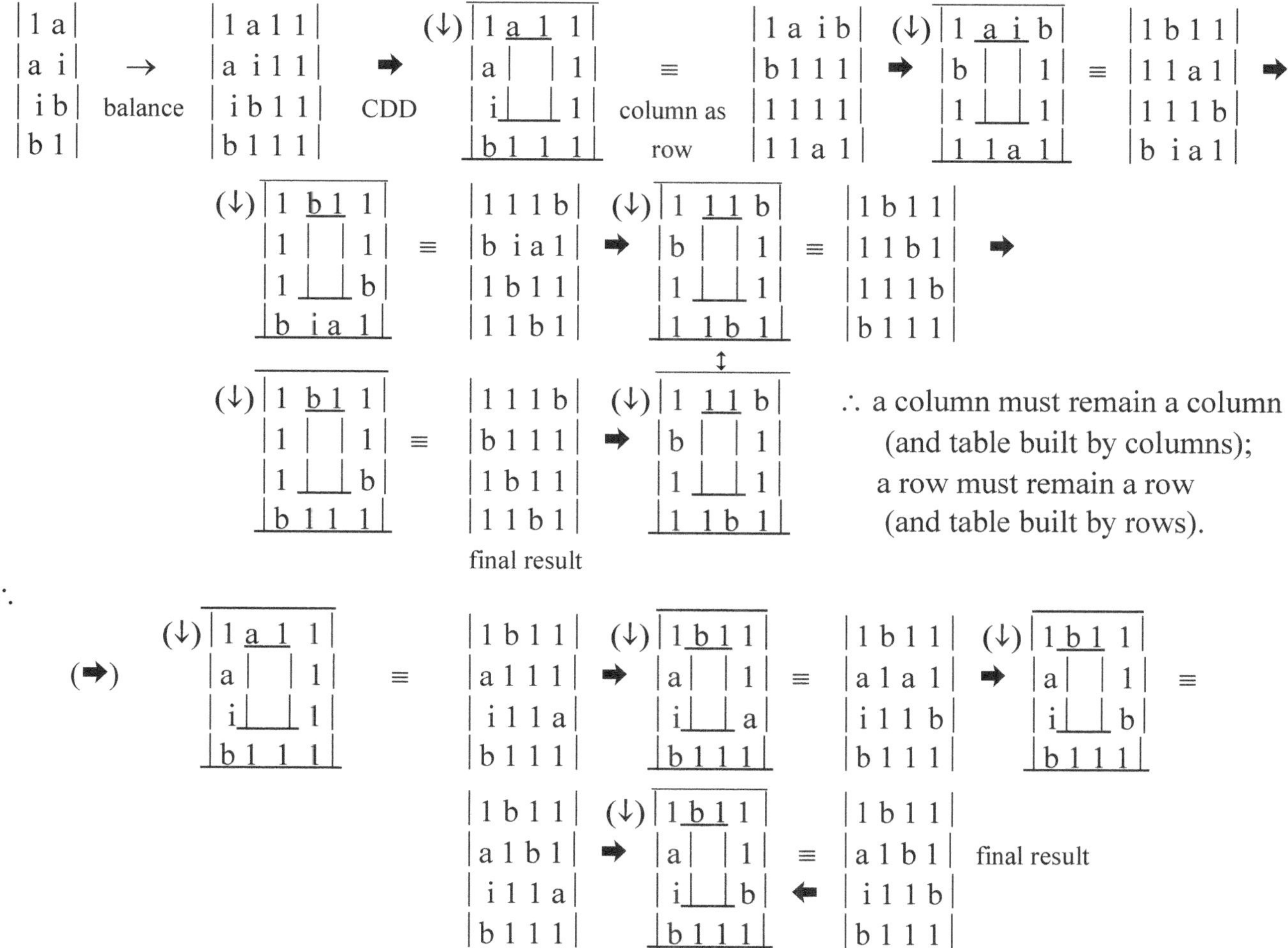

Tables (from CDD) must (or should) be built using column/row consistency for the initial column or row to be maintained.

A balanced table, of any number of rows and columns, has only **four** sides. ∴ The circularization (in CDD) is here as like a rectangularization to a square (or polygon). This (right-)angularization (of corners) is necessary to maintain (and distinguish) rows from columns. The term "circular" blurs this distinction and frees up the ability to construct rows and columns for a table from its elements (for as many tables as are there peripheral elements).

For unitary-ness all such distinction breaks down, for a proper circularization. But for more practical, rigid application, only rows (of operations) need yield to the same solution (in a table).
Thus, here, "circular" refers to a sort of circular permutation of the elements.

It is not un-noted that:

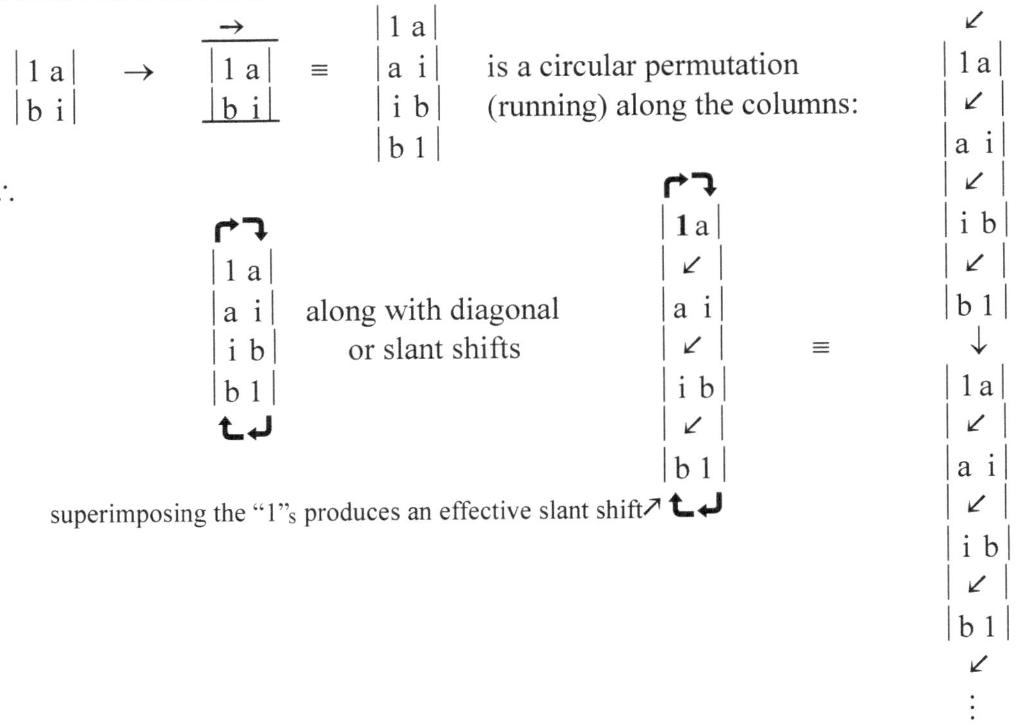

$$\begin{vmatrix} 1 & a \\ b & i \end{vmatrix} \rightarrow \overrightarrow{\begin{vmatrix} 1 & a \\ b & i \end{vmatrix}} \equiv \begin{vmatrix} 1 & a \\ a & i \\ i & b \\ b & 1 \end{vmatrix} \qquad \text{is a circular permutation}$$
(running) along the columns:

$\therefore$

$$\begin{vmatrix} 1 & a \\ a & i \\ i & b \\ b & 1 \end{vmatrix} \quad \text{along with diagonal} \\ \text{or slant shifts}$$

superimposing the "1"s produces an effective slant shift↗

If we make this shift functional (i.e. modifying) and multiplicative, then:

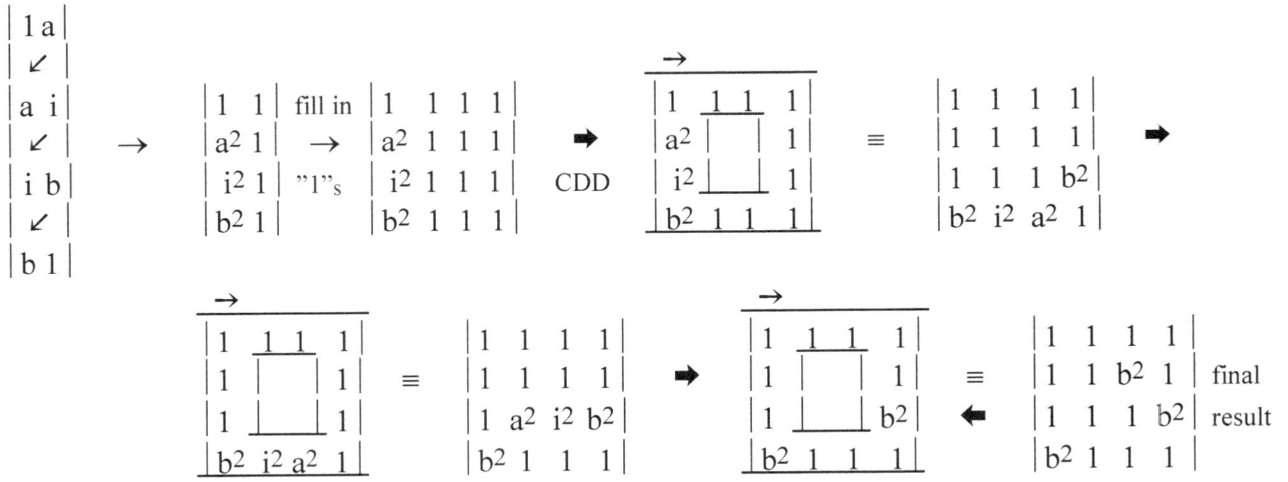

$$\begin{vmatrix} 1 & a \\ a & i \\ i & b \\ b & 1 \end{vmatrix} \rightarrow \begin{vmatrix} 1 & 1 \\ a^2 & 1 \\ i^2 & 1 \\ b^2 & 1 \end{vmatrix} \xrightarrow[\text{"1"s}]{\text{fill in}} \begin{vmatrix} 1 & 1 & 1 & 1 \\ a^2 & 1 & 1 & 1 \\ i^2 & 1 & 1 & 1 \\ b^2 & 1 & 1 & 1 \end{vmatrix} \xrightarrow{\text{CDD}} \begin{vmatrix} 1 & 1 & 1 & 1 \\ a^2 & & & 1 \\ i^2 & & & 1 \\ b^2 & 1 & 1 & 1 \end{vmatrix} \equiv \begin{vmatrix} 1 & 1 & 1 & 1 \\ 1 & 1 & 1 & 1 \\ 1 & 1 & 1 & b^2 \\ b^2 & i^2 & a^2 & 1 \end{vmatrix} \Rightarrow$$

$$\begin{vmatrix} 1 & 1 & 1 & 1 \\ 1 & & & 1 \\ 1 & & & 1 \\ b^2 & i^2 & a^2 & 1 \end{vmatrix} \equiv \begin{vmatrix} 1 & 1 & 1 & 1 \\ 1 & 1 & 1 & 1 \\ 1 & a^2 & i^2 & b^2 \\ b^2 & 1 & 1 & 1 \end{vmatrix} \Rightarrow \begin{vmatrix} 1 & 1 & 1 & 1 \\ 1 & & & 1 \\ 1 & & & b^2 \\ b^2 & 1 & 1 & 1 \end{vmatrix} \equiv \begin{vmatrix} 1 & 1 & 1 & 1 \\ 1 & 1 & b^2 & 1 \\ 1 & 1 & 1 & b^2 \\ b^2 & 1 & 1 & 1 \end{vmatrix} \begin{matrix} \text{final} \\ \text{result} \end{matrix}$$

One can also expand the permutation horizontally (i.e. row-wise), altering solution possibilities:

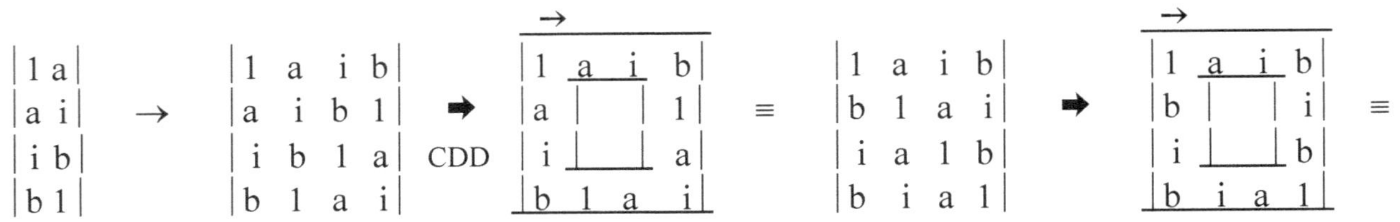

$$\begin{vmatrix} 1 & a \\ a & i \\ i & b \\ b & 1 \end{vmatrix} \rightarrow \begin{vmatrix} 1 & a & i & b \\ a & i & b & 1 \\ i & b & 1 & a \\ b & 1 & a & i \end{vmatrix} \xrightarrow{\text{CDD}} \begin{vmatrix} 1 & a & i & b \\ a & & & 1 \\ i & & & a \\ b & 1 & a & i \end{vmatrix} \equiv \begin{vmatrix} 1 & a & i & b \\ b & 1 & a & i \\ i & a & 1 & b \\ b & i & a & 1 \end{vmatrix} \Rightarrow \begin{vmatrix} 1 & a & i & b \\ b & & & i \\ i & & & b \\ b & i & a & 1 \end{vmatrix} \equiv$$

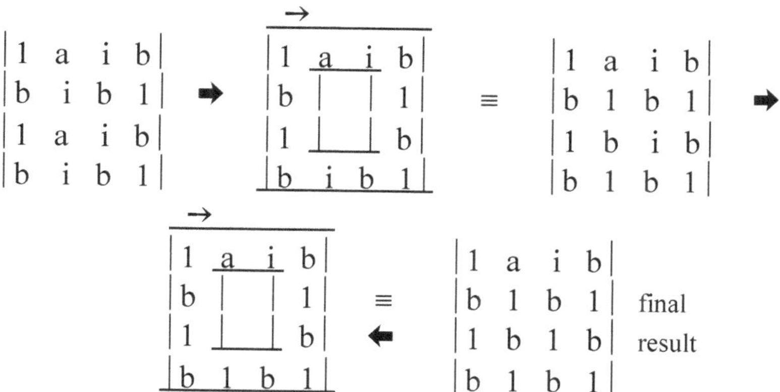

The general implication for these internal circle analyses (on the discarded tables presented here) is for (i ≡) a ≡ b ≡ 1 . Each of the various final results (of tables) supports this triviality.

This (kind of) CDD analysis, therefore, presents a powerful method of reducing any balanced table (of # of rows and columns) down to one of **4** rows and **n**-columns for exploitation of its inherent solutions and revelation about the (functional) nature of its constituent (operational) elements. For unbalanced tables, more variation (of filling in the "1"s) is allowed in the analyses and fulfillment of (final) results (for elemental properties revealed).

What can this type of analysis reveal for a looping table?

First, let's take the non-looped table, and its looping representative:

$$
\begin{array}{|ccccc|}
i & a & b & c & 1 \\
i & a & b & c & 1 \\
1 & 1 & 1 & 1 & 1 \\
1 & h & k & l & i \\
1 & h & k & l & i
\end{array}
$$

non-looped

$$
\begin{array}{|cccccccc|}
i & a & b & c & 1 & 1 & 1 & 1 \\
i & a & b & c & 1 & 1 & 1 & 1 \\
1 & 1 & 1 & 1 & 1 & 1 & 1 & 1 \\
1 & h & k & l & ia & b & c & 1 \\
1 & h & k & l & ia & b & c & 1 \\
1 & 1 & 1 & 1 & 1 & 1 & 1 & 1 \\
1 & 1 & 1 & 1 & h & k & l & i \\
1 & 1 & 1 & 1 & h & k & l & i
\end{array}
$$

looping representative

both tables balanced
(but not "squared" of
polygon) with 4 elements
per row/column

Since "1"s are operations (of pausing placements), and solutions are restricted to rows (by functional definition), we can immediately do an extreme simplification to yield a table representative of both: by removing (or reducing of) "1"s:

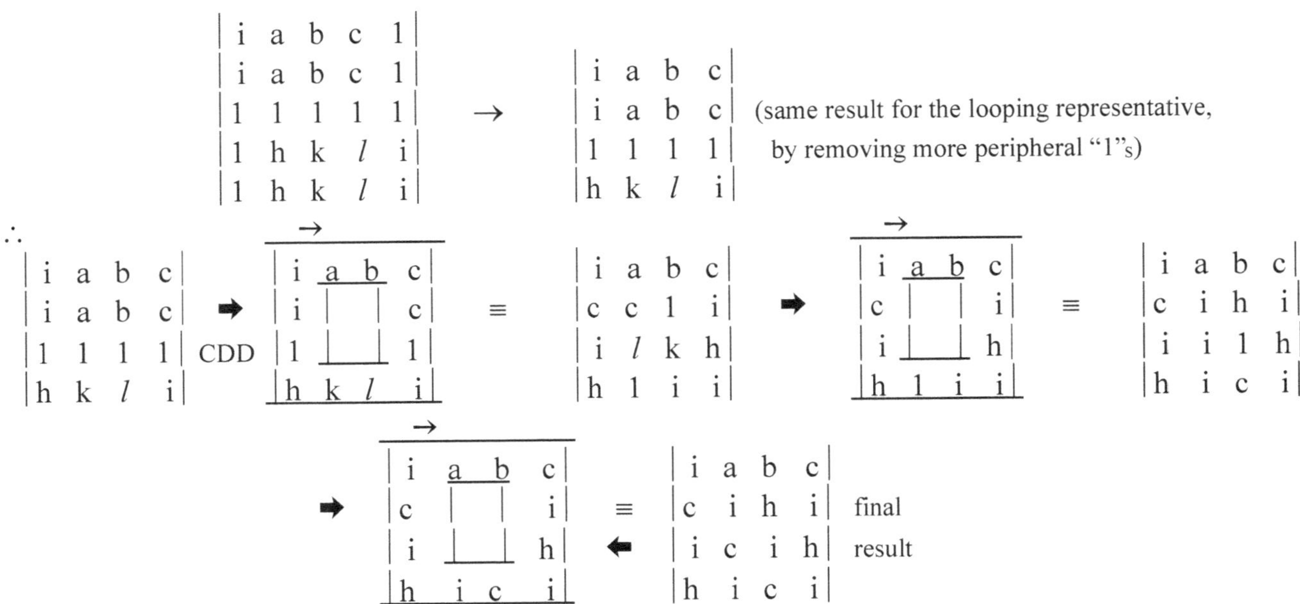

Similar for an earlier table, we notice (in the final result) a very persistent alternating pattern below the top row, concerning (here) the elements 'c', 'i', 'h' (i.e. an "i" alternating with c and h). This also occurs for the first and last column and for the lower parts (i.e. after the 1st row) of the other two columns.

The row products are : iabc , and three of i^2ch

  column products : ai^2c , bcih , and two of i^2ch

  diagonal products : i^4 , $(hc)^2$ or h^2c^2

The strong (circular) permutations extend in all row/column (consistent) directions (discounting the a , b elements):

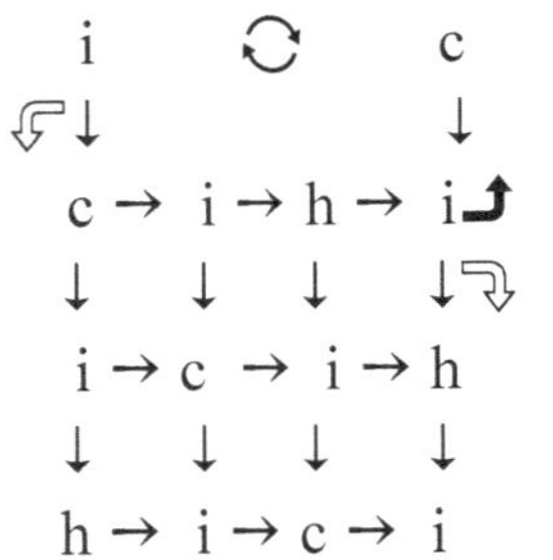

Always, the apparent pattern is: 'c i h' .

It's clear that 'a' would be replaced by 'h', and 'b' by 'i', to support this pattern.

∴ The implication is: (c i h ≡ c b a).

If we make the substitutions in the originating ("1"s reduced) table, to analyze by CDD
(i.e. a ≡ h , b ≡ i) :

$$
\begin{vmatrix} i & a & b & c \\ i & a & b & c \\ 1 & 1 & 1 & 1 \\ h & k & l & i \end{vmatrix}
\equiv
\begin{vmatrix} b & a & b & c \\ b & a & b & c \\ 1 & 1 & 1 & 1 \\ a & k & l & b \end{vmatrix}
\;\xrightarrow{\text{CDD}}\;
\begin{vmatrix} b & a & b & c \\ b & & & c \\ 1 & & & 1 \\ a & k & l & b \end{vmatrix}
\equiv
\begin{vmatrix} b & a & b & c \\ c & c & 1 & b \\ b & l & k & a \\ a & 1 & b & b \end{vmatrix}
$$

$$
\Rightarrow
\begin{vmatrix} b & a & b & c \\ c & & & b \\ b & & & a \\ a & 1 & b & b \end{vmatrix}
\equiv
\begin{vmatrix} b & a & b & c \\ c & b & a & b \\ b & b & 1 & a \\ a & b & c & b \end{vmatrix}
\;\Rightarrow\;
\begin{vmatrix} b & a & b & c \\ c & & & b \\ b & & & a \\ a & b & c & b \end{vmatrix}
\equiv
\begin{vmatrix} b & a & b & c \\ c & b & a & b \\ b & c & b & a \\ a & b & c & b \end{vmatrix}
\;\begin{array}{l}\text{final}\\[2pt]\leftarrow\ \text{result}\end{array}
$$

∴

The real alternating pattern (or permutation) is: 'b a b c' .
The products for each row or column form: 'b²ac' .
The slant products are either exponential (with powers) of an element, or of the product: 'ac' .

The looping (as designed here) can occur at "b" (to create a notch):

$$
\begin{vmatrix} \mathbf{b} & a & b & c \\ c & b & a & b \\ b & c & b & a \\ a & b & c & \mathbf{b} \end{vmatrix}
\;\rightarrow\;
\begin{vmatrix} b & a & b & c \\ c & b & a & b \\ b & c & b & a \\ a & b & c & \end{vmatrix}
$$

The (non-looping) table as crafted shows consistency of all row and column products,
 and ∴ solutions, and ∴ has unitary-ness.
There are of course 7 more CDD analyses available to be made for this ("1"s) reduced table, depending
on the initiating row or column used and the direction of the circularization. CDD necessarily enriches
for characteristic permutations. ∴ It may be used to compare a solution (as a row) to these various but
consistent patterns of operation.

e.g. For the previous:
     $\begin{vmatrix} i & a & b & c \end{vmatrix}$ is compared to $\begin{vmatrix} b & a & b & c \end{vmatrix}$ to provide a unitary-ness (to indicate correctness).

The CDD method, however, is not one to actually yield (or conduce) solutions, but rather to promote a
unitary-ness for a given solution (row). It is a pattern producing method. As for precision compared to
accuracy (i.e. reproducibility compared to correctness), while correctness must have unitary-ness (for a
problem), an apparent unitary-ness does not "prove" a correctness. And there may be degrees (or ranges)
of correctness sub-optimum of unitary-ness and approaches towards (but not yet fulfilling) unitary-ness
with rising degrees of correctness as associative. In the particular example used here for this exercise, the
starting ("1"s reduced) table is particular in that one of its "solutions" (i.e. row) is not to have any
operations (other than pauses) at all (i.e. $\begin{vmatrix} 1 & 1 & 1 & 1 \end{vmatrix}$). Therefore, a certain freedom of pattern evaluation
may evolve (as regards finding solutions).

Out of curiosity we may do the obvious and rotate the table to an intra-conversion of 'rows↔columns' (i.e. a right angle rotation):

$$
\begin{vmatrix} i & a & b & c \\ i & a & b & c \\ 1 & 1 & 1 & 1 \\ h & k & l & i \end{vmatrix}
\rightarrow
\begin{vmatrix} h & 1 & i & i \\ k & 1 & a & a \\ l & 1 & b & b \\ i & 1 & c & c \end{vmatrix}
\ \overset{CDD}{\Rightarrow}\
\begin{vmatrix} h & 1 & i & i \\ k & & & a \\ l & & & b \\ i & 1 & c & c \end{vmatrix}
\equiv
\begin{vmatrix} h & 1 & i & i \\ i & a & b & c \\ c & c & 1 & i \\ i & l & k & h \end{vmatrix}
\Rightarrow
\begin{vmatrix} h & 1 & i & i \\ i & & & c \\ c & & & i \\ i & l & k & h \end{vmatrix}
$$

$$
\equiv
\begin{vmatrix} h & 1 & i & i \\ i & c & i & h \\ h & k & l & i \\ i & c & i & h \end{vmatrix}
\Rightarrow
\begin{vmatrix} h & 1 & i & i \\ i & & & h \\ h & & & i \\ i & c & i & h \end{vmatrix}
\equiv
\begin{vmatrix} h & 1 & i & i \\ i & h & i & h \\ h & i & c & i \\ i & h & i & h \end{vmatrix}
\Rightarrow
\begin{vmatrix} h & 1 & i & i \\ i & & & h \\ h & & & i \\ i & h & i & h \end{vmatrix}
$$

$$
\equiv
\begin{vmatrix} h & 1 & i & i \\ i & h & i & h \\ h & i & h & i \\ i & h & i & h \end{vmatrix}
\Rightarrow
\begin{vmatrix} h & 1 & i & i \\ i & & & h \\ h & & & i \\ i & h & i & h \end{vmatrix}
\equiv
\begin{vmatrix} h & 1 & i & i \\ i & h & i & h \\ h & i & h & i \\ i & h & i & h \end{vmatrix}\ \text{final result}
$$

One might rotate rows (from top to bottom) and ∴ assume the following (final) results:

$$
\begin{vmatrix} k & 1 & a & a \\ a & k & a & k \\ k & a & k & a \\ a & k & a & k \end{vmatrix},\quad
\begin{vmatrix} l & 1 & b & b \\ b & l & b & l \\ l & b & l & b \\ b & l & b & l \end{vmatrix},\quad
\begin{vmatrix} i & 1 & c & c \\ c & i & c & i \\ i & c & i & c \\ c & i & c & i \end{vmatrix}
$$

to match the previous kind of pattern.

Yet:

$$
\begin{vmatrix} i & 1 & c & c \\ h & 1 & i & i \\ k & 1 & a & a \\ l & 1 & b & b \end{vmatrix}
\Rightarrow
\begin{vmatrix} i & 1 & c & c \\ h & & & i \\ k & & & a \\ l & 1 & b & b \end{vmatrix}
\equiv
\begin{vmatrix} i & 1 & c & c \\ c & i & a & b \\ b & b & 1 & l \\ l & k & h & i \end{vmatrix}
\Rightarrow
\begin{vmatrix} i & 1 & c & c \\ c & & & b \\ b & & & l \\ l & k & h & i \end{vmatrix}
\equiv
\begin{vmatrix} i & 1 & c & c \\ c & b & l & i \\ i & h & k & l \\ l & b & c & i \end{vmatrix}
$$

$$
\Rightarrow
\begin{vmatrix} i & 1 & c & c \\ c & & & i \\ i & & & l \\ l & b & c & i \end{vmatrix}
\equiv
\begin{vmatrix} i & 1 & c & c \\ c & i & l & i \\ i & c & b & l \\ l & i & c & i \end{vmatrix}
\Rightarrow
\begin{vmatrix} i & 1 & c & c \\ c & & & i \\ i & & & l \\ l & i & c & i \end{vmatrix}
\equiv
\begin{vmatrix} i & 1 & c & c \\ c & i & l & i \\ i & c & i & l \\ l & i & c & i \end{vmatrix}
$$

$$
\Rightarrow
\begin{vmatrix} i & 1 & c & c \\ c & & & i \\ i & & & l \\ l & i & c & i \end{vmatrix}
\equiv
\begin{vmatrix} i & 1 & c & c \\ c & i & l & i \\ i & c & i & l \\ l & i & c & i \end{vmatrix}\ \text{final result}
$$

Similarly, for

$$\begin{vmatrix} l & 1 & b & b \\ i & 1 & c & c \\ h & 1 & i & i \\ k & 1 & a & a \end{vmatrix} \quad \text{CDD leads to} \quad \begin{vmatrix} l & 1 & b & b \\ b & l & k & l \\ l & b & l & k \\ k & l & b & l \end{vmatrix} \quad \text{as the final result.}$$

And for

$$\begin{vmatrix} k & 1 & a & a \\ l & 1 & b & b \\ i & 1 & c & c \\ h & 1 & i & i \end{vmatrix} \quad \text{CDD yields} \quad \begin{vmatrix} k & 1 & a & a \\ a & k & h & k \\ k & a & k & h \\ h & k & a & k \end{vmatrix} \quad \text{as the final result.}$$

∴

$$\begin{vmatrix} h & 1 & i & i \\ k & 1 & a & a \\ l & 1 & b & b \\ i & 1 & c & c \end{vmatrix} \quad \begin{matrix} \text{a series of} \\ \rightarrow \\ \text{CDD} \\ \text{analyses} \end{matrix} \quad \begin{vmatrix} h & 1 & i & i \\ i & h & i & h \\ h & i & h & i \\ i & h & i & h \end{vmatrix}, \quad \begin{vmatrix} k & 1 & a & a \\ a & k & h & k \\ k & a & k & h \\ h & k & a & k \end{vmatrix}, \quad \begin{vmatrix} l & 1 & b & b \\ b & l & k & l \\ l & b & l & k \\ k & l & b & l \end{vmatrix}, \quad \begin{vmatrix} i & 1 & c & c \\ c & i & l & i \\ i & c & i & l \\ l & i & c & i \end{vmatrix}$$

This would seem to suggest (operationally): $i \equiv i$, $h \equiv a$, $k \equiv b$, $l \equiv c$ (judging by diagonals), which was (of course) the clear intention of the exercise (in formulating, or stating, the original tables).

The nul/los aspect to these analyses (in general) is that every table element represents some sort of operation (if only a pause of operation as "1"). And because of the nature of evaluating tables, a table of any size, balanced or not, may be conferred to one of 4 rows and 4 columns for CDD analysis, with filling-in of "1"s, and/or condensing operations to 4 per row and column. The CDD method does not make unitary-ness (of the original table) but may be used to reveal of patterns what might be desired for unitary-ness and therefore for (leading towards) correctness (of solutions for the problem).

Obviously, other polygonal sidedness (than 4) may be constructed for similar analyses. Say, for example, if a trigonal procedure is developed, where there is one assumed row (of operations) leading to a (desired) solution: $|a \ b \ c|$, and related operations are allowed utilizing each of the row operations (as a legitimate functioning for some accepted procedure): α for a, β for b, γ for c, where (α, β, γ) are forced to be interrelated. Then we may propose the triangle:

$$\begin{matrix} & \beta & \\ \alpha & & \gamma \\ a & b & c \end{matrix} \qquad \begin{matrix} \text{rows} & : \beta, \alpha\gamma, abc \\ \text{columns:} & \beta\alpha a, \beta b, \beta\gamma c \end{matrix}$$

as a table.

Such a table can be handled in various ways:

1) a '3 x 3' table can be promoted by filling-in "1"s to make (judicious) rows and columns:

e.g.

$$\text{from} \quad \begin{matrix} & \beta & \\ \alpha & & \gamma \\ a & b & c \end{matrix} \quad \rightarrow \quad \begin{vmatrix} \beta & \beta & \beta \\ \alpha & 1 & \gamma \\ a & b & c \end{vmatrix} \quad \begin{matrix} \text{This can be expanded with more "1"s} \\ \text{to make a '4 x 4' table for CDD analyses} \\ \text{(if desired).} \end{matrix}$$

2) Since the triangle has 3 sides, one can treat each side as a row, to make various '3 x 3' tables:

e.g. $\begin{vmatrix} \beta & \gamma & a \\ a & b & c \\ c & \gamma & \beta \end{vmatrix}$ (also expandable to '4 x 4' tables for CDD)

For either ① or ②, arbitrary expansion is not necessary, since CDD will (by its process) add "1"s to construct a '4 x 4' table:

e.g.

$\begin{vmatrix} \beta & \gamma & a \\ a & b & c \\ c & \gamma & \beta \end{vmatrix} \rightarrow$ 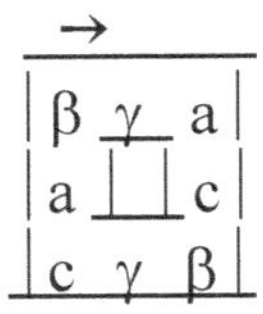 $\equiv$ (e.g.) $\begin{vmatrix} \beta & \gamma & a & 1 \\ a & c & \beta & 1 \\ \beta & \gamma & c & 1 \\ c & a & \beta & 1 \end{vmatrix}$ with loss of 'b' (and eventual loss of 'γ')

e.g.

$\begin{vmatrix} \beta & \beta & \beta \\ \alpha & 1 & \gamma \\ a & b & c \end{vmatrix} \rightarrow$ 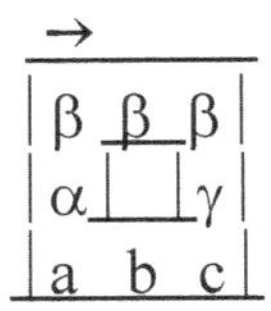 $\equiv$ (e.g.) $\begin{vmatrix} \beta & \beta & \beta & 1 \\ \beta & \gamma & c & 1 \\ c & b & a & 1 \\ a & \gamma & \beta & 1 \end{vmatrix}$ with eventual loss of 'b'

3) A more direct CDD can be established from the (3) sides, similar to ②:

e.g.

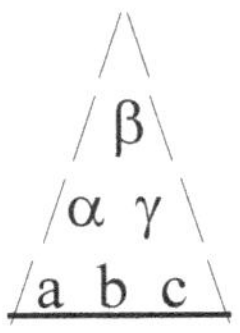 $\rightarrow$ 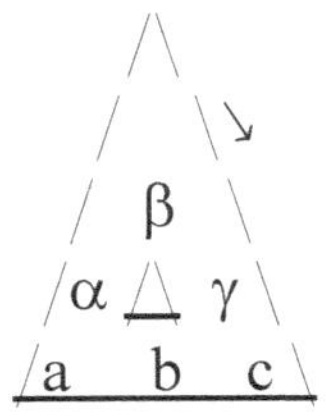 $\equiv \begin{vmatrix} \beta & \gamma & c \\ c & b & a \\ \alpha & \gamma & \beta \end{vmatrix} \equiv \begin{vmatrix} \beta & \gamma & c & 1 \\ c & b & a & 1 \\ a & \gamma & \beta & 1 \\ 1 & 1 & 1 & 1 \end{vmatrix}$ etc.

or

$\begin{vmatrix} \beta & \gamma & c \\ c & b & a \\ a & \gamma & \beta \end{vmatrix} \rightarrow$ 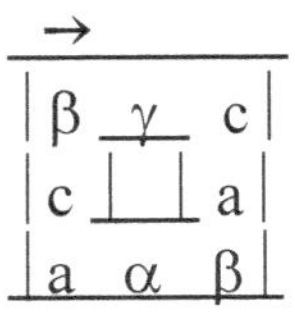 $\equiv \begin{vmatrix} \beta & \gamma & c \\ c & a & \beta \\ \beta & \alpha & a \\ a & c & \beta \end{vmatrix} \equiv \begin{vmatrix} \beta & \gamma & c & 1 \\ c & a & \beta & 1 \\ \beta & \alpha & a & 1 \\ a & c & \beta & 1 \end{vmatrix}$ etc.

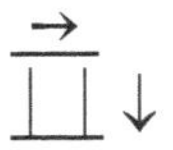 $\equiv$

$\begin{vmatrix} \beta & \gamma & c & 1 \\ c & \beta & a & \beta \\ \beta & c & a & 1 \\ a & \beta & c & \beta \end{vmatrix}$ ➡ 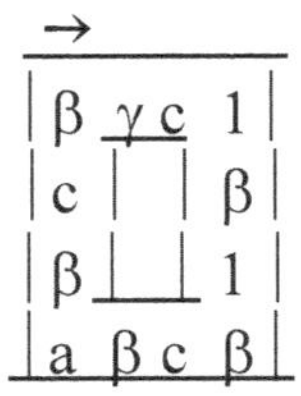 $\equiv \begin{vmatrix} \beta & \gamma & c & 1 \\ 1 & \beta & 1 & \beta \\ \beta & c & \beta & a \\ a & \beta & c & \beta \end{vmatrix}$ ➡ 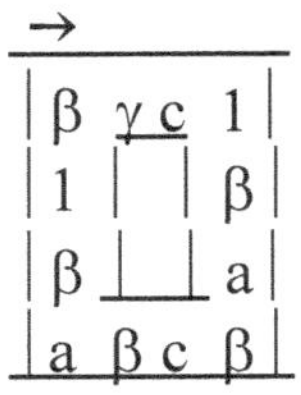 $\equiv$

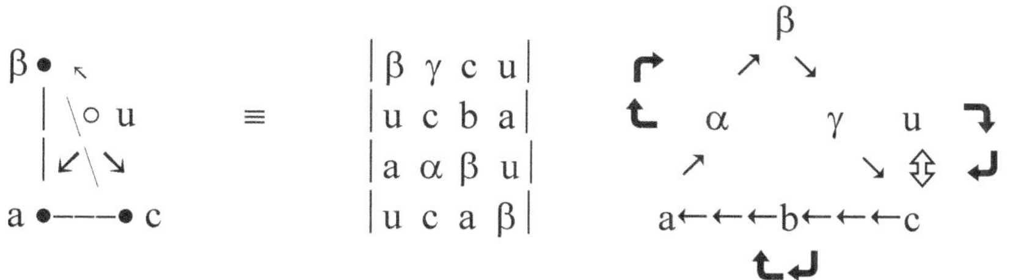

$$
\begin{vmatrix} \beta & \gamma & c & 1 \\ 1 & \beta & a & \beta \\ \beta & c & \beta & a \\ a & \beta & 1 & \beta \end{vmatrix}
\;\rightarrow\;
\begin{vmatrix} \beta & \gamma c & 1 \\ 1 & & \beta \\ \beta & & a \\ a & \beta & 1 & \beta \end{vmatrix}
\;\equiv\;
\begin{vmatrix} \beta & \gamma & c & 1 \\ 1 & \beta & a & \beta \\ \beta & 1 & \beta & a \\ a & \beta & 1 & \beta \end{vmatrix}
$$

final result (brings a ≡ 1 , based on diagonals)

4) Rotating the '△' keeps the permutation of elements in the given order, since, as each side is a row, there are no real columns. ∴ One can simply change the order of rows of the derived '3 x 3' tables:

i.e.

$$
\begin{vmatrix} \beta & \gamma & c \\ c & b & a \\ a & \alpha & \beta \end{vmatrix} , \quad
\begin{vmatrix} a & \alpha & \beta \\ \beta & \gamma & c \\ c & b & a \end{vmatrix} , \quad
\begin{vmatrix} c & b & a \\ a & \alpha & \beta \\ \beta & \gamma & c \end{vmatrix}
\quad \text{etc. (for CDD)}
$$

5) One can bring an analysis into 3-dimensions by postulating the existence of an unknown operator (say: "u") located above the center of the plane of the triangle:

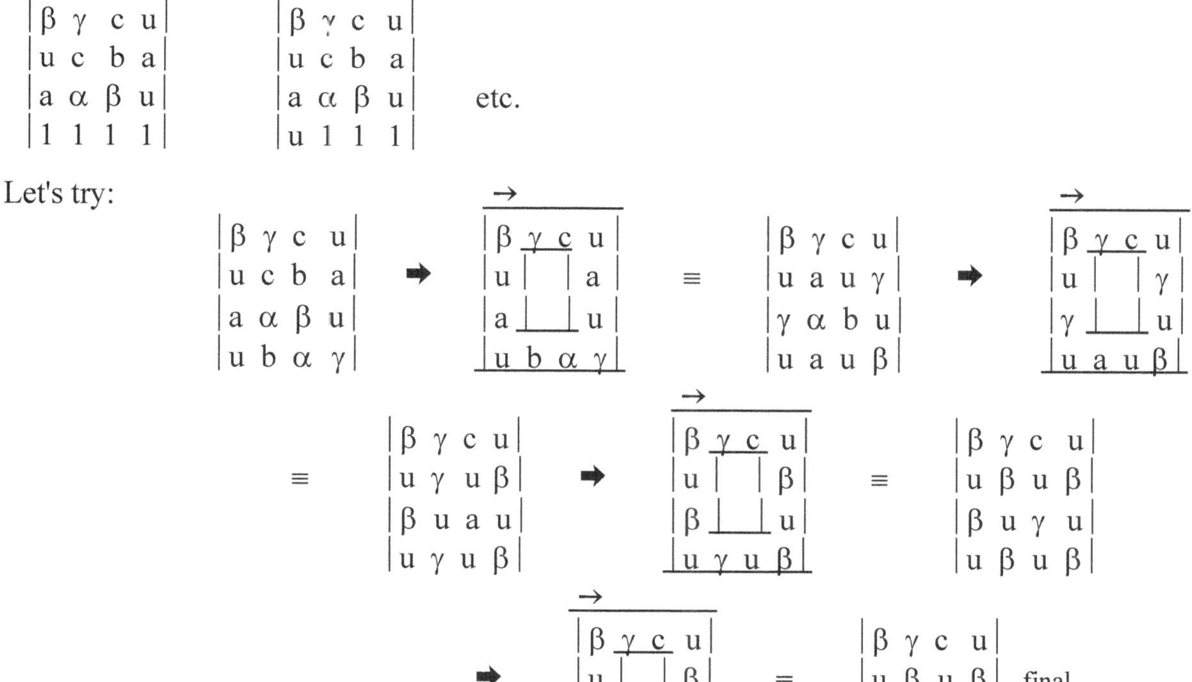

[the last row being inter-operative (of row) rather than intra]

or

$$
\begin{vmatrix} \beta & \gamma & c & u \\ u & c & b & a \\ a & \alpha & \beta & u \\ 1 & 1 & 1 & 1 \end{vmatrix}
$$

or

$$
\begin{vmatrix} \beta & \gamma & c & u \\ u & c & b & a \\ a & \alpha & \beta & u \\ u & 1 & 1 & 1 \end{vmatrix}
\quad \text{etc.}
$$

Let's try:

$$
\begin{vmatrix} \beta & \gamma & c & u \\ u & c & b & a \\ a & \alpha & \beta & u \\ u & b & \alpha & \gamma \end{vmatrix}
\;\rightarrow\;
\begin{vmatrix} \beta & \gamma c & u \\ u & & a \\ a & & u \\ u & b & \alpha & \gamma \end{vmatrix}
\;\equiv\;
\begin{vmatrix} \beta & \gamma & c & u \\ u & a & u & \gamma \\ \gamma & \alpha & b & u \\ u & a & u & \beta \end{vmatrix}
\;\rightarrow\;
\begin{vmatrix} \beta & \gamma c & u \\ u & & \gamma \\ \gamma & & u \\ u & a & u & \beta \end{vmatrix}
$$

$$
\;\equiv\;
\begin{vmatrix} \beta & \gamma & c & u \\ u & \gamma & u & \beta \\ \beta & u & a & u \\ u & \gamma & u & \beta \end{vmatrix}
\;\rightarrow\;
\begin{vmatrix} \beta & \gamma c & u \\ u & & \beta \\ \beta & & u \\ u & \gamma & u & \beta \end{vmatrix}
\;\equiv\;
\begin{vmatrix} \beta & \gamma & c & u \\ u & \beta & u & \beta \\ \beta & u & \gamma & u \\ u & \beta & u & \beta \end{vmatrix}
$$

$$
\;\rightarrow\;
\begin{vmatrix} \beta & \gamma c & u \\ u & & \beta \\ \beta & & u \\ u & \beta & u & \beta \end{vmatrix}
\;\equiv\;
\begin{vmatrix} \beta & \gamma & c & u \\ u & \beta & u & \beta \\ \beta & u & \beta & u \\ u & \beta & u & \beta \end{vmatrix}
$$

final result

The diagonals suggest a distinction between 'β' and 'u' .
Let's try:

$$
\begin{vmatrix} \beta & \gamma & c & u \\ u & c & b & a \\ a & \alpha & \beta & u \\ u & c & a & \beta \end{vmatrix}
\;\Rightarrow\;
\begin{vmatrix} \beta & \gamma & c & u \\ u & & & a \\ a & & & u \\ u & c & a & \beta \end{vmatrix}
\;\equiv\;
\begin{vmatrix} \beta & \gamma & c & u \\ u & a & u & \beta \\ \beta & a & c & u \\ u & a & u & \beta \end{vmatrix}
\;\Rightarrow\;
\begin{vmatrix} \beta & \gamma & c & u \\ u & & & \beta \\ \beta & & & u \\ u & a & u & \beta \end{vmatrix}
$$

$$
\equiv
\begin{vmatrix} \beta & \gamma & c & u \\ u & \beta & u & \beta \\ \beta & u & a & u \\ u & \beta & u & \beta \end{vmatrix}
\;\Rightarrow\;
\begin{vmatrix} \beta & \gamma & c & u \\ u & & & \beta \\ \beta & & & u \\ u & \beta & u & \beta \end{vmatrix}
\;\equiv\;
\begin{vmatrix} \beta & \gamma & c & u \\ u & \beta & u & \beta \\ \beta & u & \beta & u \\ u & \beta & u & \beta \end{vmatrix}
$$

final
result
(same as previous)

The diagonals again suggest a distinction between 'β' and 'u' .
Let's try:

$$
\begin{vmatrix} \beta & \gamma & c & u \\ u & c & b & a \\ a & \alpha & \beta & u \\ u & 1 & 1 & 1 \end{vmatrix}
\;\Rightarrow\;
\begin{vmatrix} \beta & \gamma & c & u \\ u & & & a \\ a & & & u \\ u & 1 & 1 & 1 \end{vmatrix}
\;\equiv\;
\begin{vmatrix} \beta & \gamma & c & u \\ u & a & u & 1 \\ 1 & 1 & 1 & u \\ u & a & u & \beta \end{vmatrix}
\;\Rightarrow\;
\begin{vmatrix} \beta & \gamma & c & u \\ u & & & 1 \\ 1 & & & u \\ u & a & u & \beta \end{vmatrix}
$$

$$
\equiv
\begin{vmatrix} \beta & \gamma & c & u \\ u & 1 & u & \beta \\ \beta & u & a & u \\ u & 1 & u & \beta \end{vmatrix}
\;\Rightarrow\;
\begin{vmatrix} \beta & \gamma & c & u \\ u & & & \beta \\ \beta & & & u \\ u & 1 & u & \beta \end{vmatrix}
\;\equiv\;
\begin{vmatrix} \beta & \gamma & c & u \\ u & \beta & u & \beta \\ \beta & u & 1 & u \\ u & \beta & u & \beta \end{vmatrix}
$$

$$
\Rightarrow
\begin{vmatrix} \beta & \gamma & c & u \\ u & & & \beta \\ \beta & & & u \\ u & \beta & u & \beta \end{vmatrix}
\;\equiv\;
\begin{vmatrix} \beta & \gamma & c & u \\ u & \beta & u & \beta \\ \beta & u & \beta & u \\ u & \beta & u & \beta \end{vmatrix}
$$

final
result
(same as the two previous)

$$
\begin{vmatrix} \beta & \gamma & c & u \\ u & c & b & a \\ a & \alpha & \beta & u \\ 1 & 1 & 1 & 1 \end{vmatrix}
$$

suggests no particular solution (for a row: i.e. $\begin{vmatrix} 1 & 1 & 1 & 1 \end{vmatrix}$) ;

then:

$$
\begin{vmatrix} \beta & \gamma & c & u \\ u & c & b & a \\ a & \alpha & \beta & u \\ 1 & 1 & 1 & 1 \end{vmatrix}
\;\Rightarrow\;
\begin{vmatrix} \beta & \gamma & c & u \\ u & & & a \\ a & & & u \\ 1 & 1 & 1 & 1 \end{vmatrix}
\;\equiv\;
\begin{vmatrix} \beta & \gamma & c & u \\ u & a & u & 1 \\ 1 & 1 & 1 & 1 \\ 1 & a & u & \beta \end{vmatrix}
\;\Rightarrow\;
\begin{vmatrix} \beta & \gamma & c & u \\ u & & & 1 \\ 1 & & & 1 \\ 1 & a & u & \beta \end{vmatrix}
\;\equiv\;
\begin{vmatrix} \beta & \gamma & c & u \\ u & 1 & 1 & \beta \\ \beta & u & a & 1 \\ 1 & 1 & u & \beta \end{vmatrix}
$$

$$\Rightarrow \quad \begin{vmatrix} \beta & \gamma & c & u \\ u & & & \beta \\ \beta & & & 1 \\ 1 & 1 & u & \beta \end{vmatrix} \;\equiv\; \begin{vmatrix} \beta & \gamma & c & u \\ u & \beta & 1 & \beta \\ \beta & u & 1 & 1 \\ 1 & \beta & u & \beta \end{vmatrix} \quad\Rightarrow\quad \begin{vmatrix} \beta & \gamma & c & u \\ u & & & \beta \\ \beta & & & 1 \\ 1 & \beta & u & \beta \end{vmatrix} \;\equiv\; \begin{vmatrix} \beta & \gamma & c & u \\ u & \beta & 1 & \beta \\ \beta & u & \beta & 1 \\ 1 & \beta & u & \beta \end{vmatrix} \begin{array}{l}\text{final}\\ \leftarrow \text{result}\end{array}$$

with 'u' ≡ 1 ('β' ≡ 'β')
(based on diagonals)

This essentially allows 'u' to be inconsequential and merely an operation used for aid of analysis. (i.e. extractable or removable). It's noted that (for CDD) all of these examples automatically (and quickly) remove 'b' and 'α' from any possible (final) results. We might try (rt. ∡) rotation, to insure each row the possibility of a non-trivial (i.e. not $\begin{vmatrix} 1 & 1 & 1 & 1 \end{vmatrix}$) solution. This can be followed by row/column cycling, so that each of the operations can remain for final results:

e.g.

$$\begin{vmatrix} 1 & a & u & \beta \\ 1 & \alpha & c & \gamma \\ 1 & \beta & b & c \\ 1 & u & a & u \end{vmatrix} \Rightarrow \begin{vmatrix} 1 & a & u & \beta \\ 1 & & & \gamma \\ 1 & & & c \\ 1 & u & a & u \end{vmatrix} \equiv \begin{vmatrix} 1 & a & u & \beta \\ \beta & \gamma & c & u \\ u & a & u & 1 \\ 1 & 1 & 1 & 1 \end{vmatrix} \Rightarrow \begin{vmatrix} 1 & a & u & \beta \\ \beta & & & u \\ u & & & 1 \\ 1 & 1 & 1 & 1 \end{vmatrix} \equiv \begin{vmatrix} 1 & a & u & \beta \\ \beta & u & 1 & 1 \\ 1 & 1 & 1 & 1 \\ 1 & u & \beta & 1 \end{vmatrix}$$

$$\Rightarrow \begin{vmatrix} 1 & a & u & \beta \\ \beta & & & 1 \\ 1 & & & 1 \\ 1 & u & \beta & 1 \end{vmatrix} \equiv \begin{vmatrix} 1 & a & u & \beta \\ \beta & 1 & 1 & 1 \\ 1 & \beta & u & 1 \\ 1 & 1 & \beta & 1 \end{vmatrix} \Rightarrow \begin{vmatrix} 1 & a & u & \beta \\ \beta & & & 1 \\ 1 & & & 1 \\ 1 & 1 & \beta & 1 \end{vmatrix} \equiv \begin{vmatrix} 1 & a & u & \beta \\ \beta & 1 & 1 & 1 \\ 1 & \beta & 1 & 1 \\ 1 & 1 & \beta & 1 \end{vmatrix} \begin{array}{l}\text{final}\\ \text{result}\\ \leftarrow (\text{bringing}\\ \quad \beta \equiv 1)\end{array}$$

e.g.

$$\begin{vmatrix} a & 1 & u & \beta \\ \alpha & 1 & c & \gamma \\ \beta & 1 & b & c \\ u & 1 & a & u \end{vmatrix} \Rightarrow \begin{vmatrix} a & 1 & u & \beta \\ \alpha & & & \gamma \\ \beta & & & c \\ u & 1 & a & u \end{vmatrix} \equiv \begin{vmatrix} a & 1 & u & \beta \\ \beta & \gamma & c & u \\ u & a & 1 & u \\ u & \beta & \gamma & a \end{vmatrix} \Rightarrow \begin{vmatrix} a & 1 & u & \beta \\ \beta & & & u \\ u & & & u \\ u & \beta & \gamma & a \end{vmatrix}$$

$$\equiv \begin{vmatrix} a & 1 & u & \beta \\ \beta & u & u & a \\ a & \alpha & \beta & u \\ u & u & \beta & a \end{vmatrix} \Rightarrow \begin{vmatrix} a & 1 & u & \beta \\ \beta & & & a \\ a & & & u \\ u & u & \beta & a \end{vmatrix} \equiv \begin{vmatrix} a & 1 & u & \beta \\ \beta & a & u & a \\ a & \beta & u & u \\ u & a & \beta & a \end{vmatrix}$$

$$\Rightarrow \begin{vmatrix} a & 1 & u & \beta \\ \beta & & & a \\ a & & & u \\ u & a & \beta & a \end{vmatrix} \equiv \begin{vmatrix} a & 1 & u & \beta \\ \beta & a & u & a \\ a & \beta & a & u \\ u & a & \beta & a \end{vmatrix} \begin{array}{l}\text{final}\\ \text{result}\\ \leftarrow \text{bringing } a \equiv a,\ u \equiv \beta\\ \quad (\text{based on diagonals})\end{array}$$

117

6) We can try "doubling" triangles:

e.g.

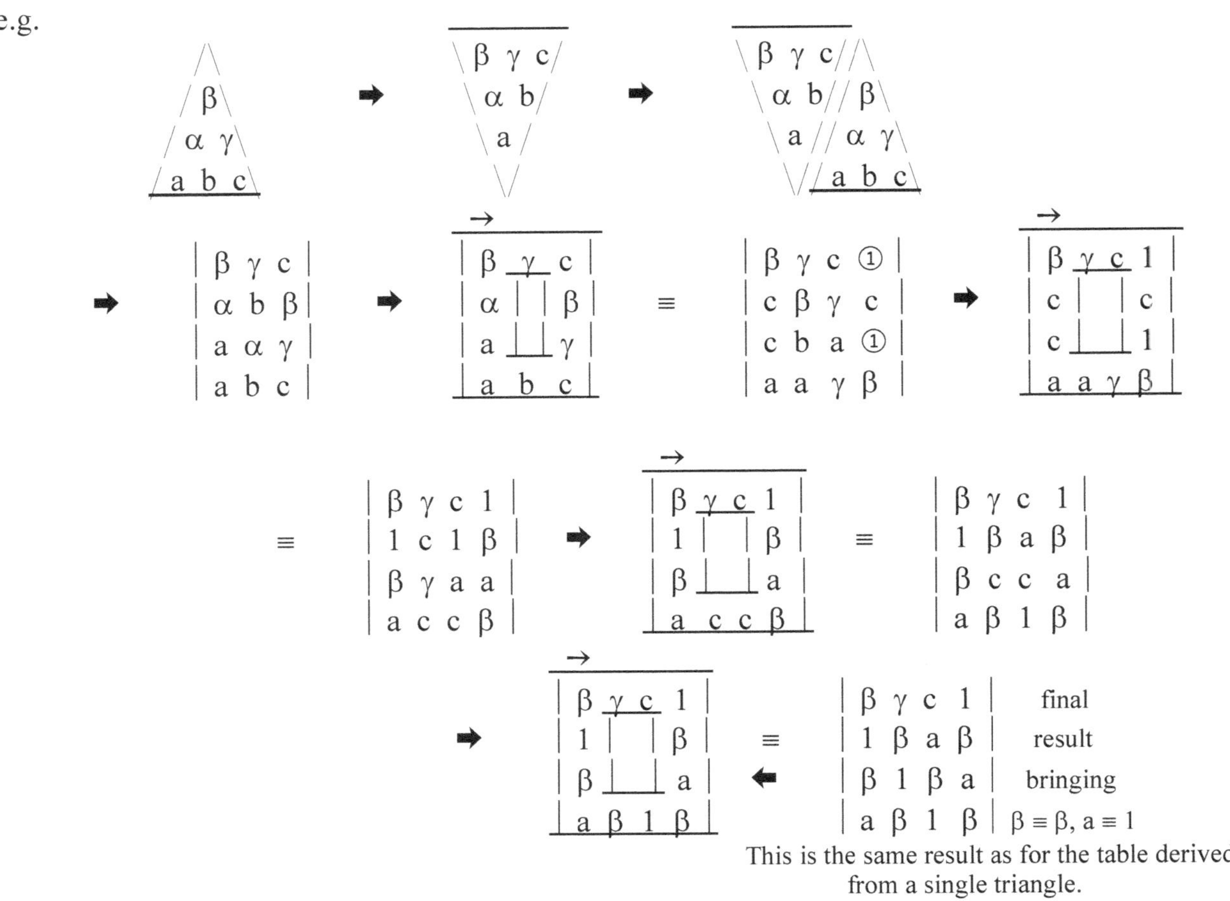

final
result
bringing
$\beta \equiv \beta$, $a \equiv 1$

This is the same result as for the table derived
from a single triangle.

The strong dependence on a 4-sided structure (of table) analysis (for CDD) is noted.

7) One can try various other pattern manipulating methods, such as transposing elements:

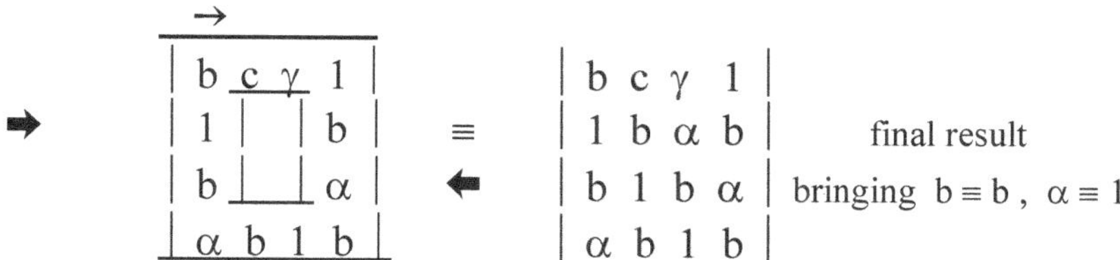

One, particularly for final results, can notice the irony of the deletion of (table) internal elements and can, therefore, posit that perhaps (for example, from above):

$$\begin{vmatrix} b & c & \gamma & 1 \\ 1 & b & \alpha & b \\ b & 1 & b & \alpha \\ \alpha & b & 1 & b \end{vmatrix} \equiv \begin{vmatrix} b & c & \gamma & 1 \\ 1 & 1 & 1 & b \\ b & 1 & 1 & \alpha \\ \alpha & b & 1 & b \end{vmatrix} \qquad \therefore \qquad \begin{vmatrix} b & \alpha \\ 1 & b \end{vmatrix} \equiv \begin{vmatrix} 1 & 1 \\ 1 & 1 \end{vmatrix}$$

This would conclude (under assumed unitary-ness) that each row yields a solution that is "b"
(due to $\begin{vmatrix} 1 & 1 & 1 & b \end{vmatrix}$).

Diagonal products: $b \cdot 1 \cdot 1 \cdot b \equiv b^2$, $\alpha \cdot 1 \cdot 1 \cdot 1 \equiv \alpha$ |
Row analyses: If $b \cdot c \cdot \gamma \cdot 1 \to b$, then $c\gamma \equiv 1$ | The trivial suggestion
Diagonals (internal table): $\alpha \equiv 1$, $b \equiv b$ | is that $b \equiv 1$.
Last row: If $\alpha \cdot b \cdot 1 \cdot b \to b$, then $\alpha b \equiv 1$ |

After the previous posit(ing), one can be extraordinary:

One can insert the original (originating/manipulated) triangle into the posited table
(replacing most of the "1"s), since the derivation of this table is found (via CDD) to be recursive.

i.e.
$$\begin{vmatrix} b & c & \gamma & 1 \\ 1 & b & \alpha & b \\ b & 1 & b & \alpha \\ \alpha & b & 1 & b \end{vmatrix} \to \begin{vmatrix} b & c & \gamma & 1 \\ 1 & 1 & 1 & b \\ b & 1 & 1 & \alpha \\ \alpha & b & 1 & b \end{vmatrix} \to \begin{vmatrix} b & c & \gamma & 1 \\ \alpha & a & b & b \\ b & \beta & c & \alpha \\ \alpha & b & \gamma & b \end{vmatrix} \to \begin{vmatrix} b & c & \gamma & 1 \\ \alpha & & & b \\ b & & & \alpha \\ \alpha & b & \gamma & b \end{vmatrix}$$

This alters all but the first row, so the solution must stay the same.

We note, perhaps coincidentally, that the 2nd column shows (maintains) relationship of "β" to its opposing elements (a, b, c) in the transposed originating triangle.

(or, i.e. The parallelepiped:
$$\begin{array}{ccc} & b & \\ \diagup & & \diagdown \\ a & & c \\ \diagdown & & \diagup \\ & \beta & \end{array}$$
elements are retained not only in the inserted △ but also in this 2nd column of resulting table.)

Here we are simply looking for suggestive patterns (of elements) of solution.

$$
\begin{vmatrix} b & c & \gamma & 1 \\ \alpha & & & b \\ b & & & \alpha \\ \alpha & b & \gamma & b \end{vmatrix} \equiv \begin{vmatrix} b & c & \gamma & 1 \\ 1 & b & \alpha & b \\ b & \gamma & b & \alpha \\ \alpha & b & \gamma & b \end{vmatrix} \Rightarrow \begin{vmatrix} b & c & \gamma & 1 \\ 1 & & & b \\ b & & & \alpha \\ \alpha & b & \gamma & b \end{vmatrix} \equiv \begin{vmatrix} b & c & \gamma & 1 \\ 1 & b & \alpha & b \\ b & \gamma & b & \alpha \\ \alpha & b & 1 & b \end{vmatrix}
$$

$$
\Rightarrow \begin{vmatrix} b & c & \gamma & 1 \\ 1 & & & b \\ b & & & \alpha \\ \alpha & b & 1 & b \end{vmatrix} \equiv \begin{vmatrix} b & c & \gamma & 1 \\ 1 & b & \alpha & b \\ b & 1 & b & \alpha \\ \alpha & b & 1 & b \end{vmatrix}
$$

final result:
The same result as earlier,
∴ demonstrating the recursiveness
(of the table)

The CDD engine (of procedure) works to foster unitary-ness by essentially submitting the (or a) table's peripheral (i.e. outer) rows and columns as solutional to a newly derived table. This induces a recurring pattern (of elements) for the three lower rows of the derived table, while maintaining intact the initial row (or column) used to initiate the CDD procedure. Ideally, the retained permutation of elements (initial row or column) is in some functional way reflective of this recurring pattern, by its *subjecture* of being solutional of the entire table (and thereby "forcing" the patterned rows to be also).

These are definitional constraints not meant to be actual unless (and until) a suggested solution is real. Until this occurs, a trivial solution might only be found (i.e. each element $\equiv$ 1); but the pattern might also be suggestive of what is correct and what needs correcting. There is still no absolute necessity for a column (of a table) to be solutional. The CDD derived (or developed) table, being circular, is actually an extension of the initially considered row (or column), for means of (circularly) returning back to it: the first element of the table and the last necessarily being the same, and each lower row starting with a last element from the upper:

e.g.
$$
\begin{vmatrix} b & c & \gamma & 1 \\ 1 & b & \alpha & b \\ b & 1 & b & \alpha \\ \alpha & b & 1 & b \end{vmatrix} \equiv
$$

$$
\begin{array}{l}
b\,c\,\gamma\,1 \\
\quad 1\,b\,\alpha\,b \\
\qquad\quad b\,1\,b\,\alpha \\
\underline{\qquad\qquad\quad \alpha\,b\,1\,b} \\
\rightarrow b\,c\,\gamma\,1\,b\,\alpha\,b\,1\,b\,\alpha\,b\,1\,b\,\lrcorner
\end{array}
$$

The developed repeating : "1 b α b" is of course "forced" to be equivalent (solutionally) to the retained row (of the CDD derived table):

$$
\therefore \ |\, b\,c\,\gamma\,1 \,| \equiv |\, 1\,b\,\alpha\,b \,|
$$

Disregarding operational order:

 a casual inspection (here, for example) seems to (or might) suggest lower and perhaps more (or further) revealing equivalencies (which may be employed in new or continued analyses).

$$\text{e.g.} \quad |\, \not{b}\, c\, \gamma\, \not{N}\, | \equiv |\, \not{N}\, \not{b}\, \alpha\, b\, |$$
$$\downarrow$$
$$|\, c\, \gamma\, | \equiv |\, \alpha\, b\, |$$

$$\therefore \ \text{e.g.} \qquad
\begin{array}{|ccc|} b & c & \gamma \\ \gamma & \beta & \alpha \\ \alpha & a & b \end{array}
\qquad \rightarrow \qquad
\begin{array}{|ccc|} b & \alpha & b \\ \gamma & \beta & c \\ \alpha & a & \gamma \end{array}
\qquad \text{etc.}$$

Note that the new table (here) can not be produced by a triangle (of itself), but has the interesting property of relating (via diagonals) the (operational) terms "bβ" to both "α" and "γ" :

$$b\, \beta\, \gamma\, , \ \alpha\, \beta\, b \quad \text{(diagonals)}.$$

Likewise (by slants), "γ" is related to "a", "α" to "c", "α" to "γ", "a" to "c".

Physically (as for physical operations), recursiveness is used to stem infinities (which are not physically possible or realizable).

 $\therefore$ nul/los mathematics is properly limited to realities and what can be demonstrable, and away from the idealistic or merely imaginable.

Movement within planes

Now, let's try to employ practical examples, first of
 1) simple planar movement:

 Say, we wish to move an object (on a planar surface) from point "A" to point "B".
 We have as operations

$$\begin{aligned}
\text{movement 1 unit left} &\equiv \text{``}l\text{''} \\
\text{movement 1 unit right} &\equiv \text{``}r\text{''} \\
\text{movement 1 unit up} &\equiv \text{``}u\text{''} \\
\text{movement 1 unit down} &\equiv \text{``}d\text{''} \\
\text{diagonal movements down left} &\equiv \text{``}dl\text{''} \\
\text{diagonal movements down right} &\equiv \text{``}dr\text{''} \\
\text{diagonal movements up left} &\equiv \text{``}ul\text{''} \\
\text{diagonal movements up right} &\equiv \text{``}ur\text{''} \\
\text{pause (or no movement)} &\equiv \text{``}1\text{''}
\end{aligned}$$

Our "board" of movements is thus:

				B	
A					

One solution is:

	↗ur→	→r→	→r→	→r→	→ B
A↗					
A	ur	r	r	r	B

Another is:

	↗ur↘		↗ur→	→r→	→ B
A↗		↘dr↗			
A	ur	dr	ur	r	B

Another is:

			↗ur→	→r→	→ B
A→	→r→	→r→			
A	r	r	ur	r	B

∴ We have an initial table (with 3 solutions):

```
| ur  r   r   r |
| ur  dr  ur  r |
|  r   r  ur  r |
```

It is unbalanced (as of a table) and requires a forth solution.

To deal only with elements (of operation) already evident, let's say this forth solution must contain the element "ur". We note upon inspection that the order of operations (in a row) does not seem to be strict, since:

$$| \text{ur} \ r \ r \ r | \equiv | r \ r \ \text{ur} \ r |$$

∴ for the forth solution we may place "ur" in any column.

∴ e.g.
```
| ur  r   r  r |
| ur  dr  ur r |     is a possible starting table
|  r   r  ur r |           (as balanced).
|  1  ur   1  1 |
```

What does CDD yield (of information)?

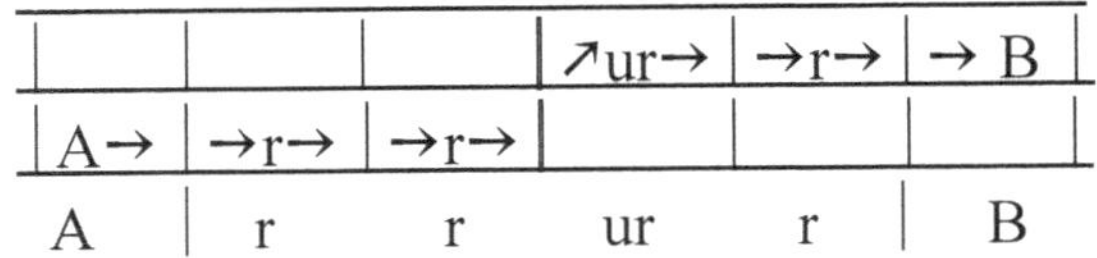

```
| ur  r   r   r |        →              | ur  r   r   r |
| ur  dr  ur  r |   ➡    | ur  r  r  r |  ≡   | r   r   r   1 |
|  r   r  ur  r |        | ur |    | r |      | 1  ur  1   1 |
|  1  ur   1  1 |        |  r |___ | r |      | 1   r  ur  ur |
                         | 1  ur  1 1 |
```

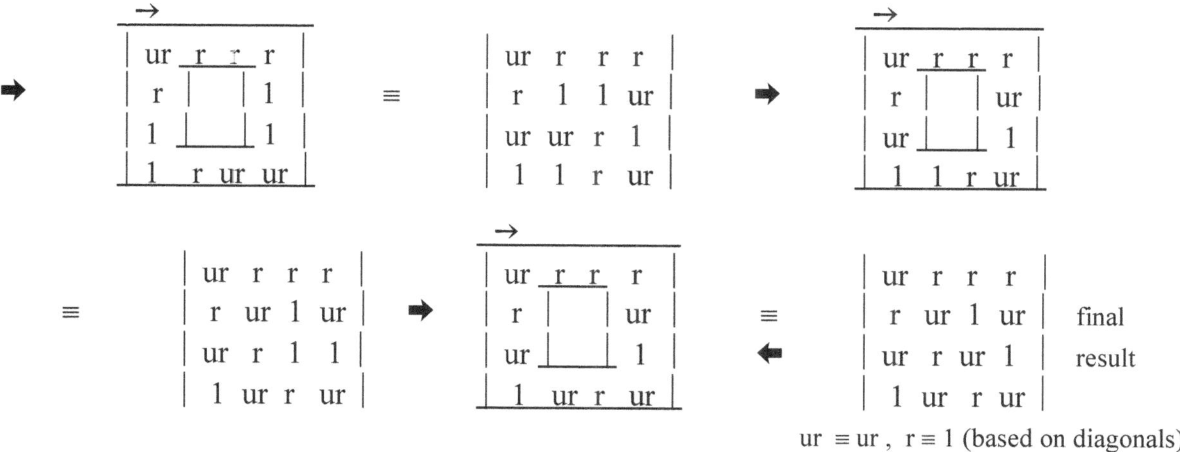

Based on the design of our "board" we note that (having) two "ur"s in a row is unlikely to yield a correct solution (with the elements left us), and also that the known solution (the initial row for the derived table) only contains one "ur" element. One or more "r"s seems permissible.

∴ By inspection we can dismiss the lower 3 rows of the final result, as well as 3 of the columns (since these 6 rows and columns each contain two "ur"s).

Yet, one column contains one "ur", two "r"s and a "1":

Is it possible that this suggests a solution, if we replace the "1" with an "r"?

Of course, we already suspect (by intuition) that the correct (forth) solution,
with an "ur" in the 2nd column, is: | r ur r r |

We can, of course, judge this procedure using (row, column) products:

$$\text{rows} \quad : \text{ur} \cdot \text{r}^3 , \ \text{r} \cdot (\text{ur})^2 , \ (\text{ur})^2 \cdot \text{r} , \ (\text{ur})^2 \cdot \text{r}$$
$$\text{columns} : (\text{ur})^2 \cdot \text{r} , \ (\text{ur})^2 \cdot \text{r} , \ \text{r}^2 \cdot \text{ur} , \ \text{r} \cdot (\text{ur})^2$$

$$\downarrow (1 \equiv \text{r})$$

$$\text{r}^3 \cdot \text{ur} \quad \text{(favors initial row)}$$

So, the completed table (for four A→B solutions) is:

ur	r	r	r
ur	dr	ur	r
r	r	ur	r
r	ur	r	r

All four rows, as well as the two diagonals, yield the same solution, but none of the columns.

Note that:

1) Prior to the analysis, we already had two "ur"s in the same row for a solution.

∴ Having two "ur"s in the same row does not rule out a solution; but the CDD analysis (in this instance) uses a row (maintained) that has only one "ur".

∴ The avoidance of another (i.e. two or more) "ur" for the forth solution is reasonable.

2) In this particular example of CDD analysis, the "ur" used to help define the forth solution (were it to be tagged) is actually eliminated (as part of an internal table) fairly quickly (i.e. in the 3rd step) from the procedure.

3) The only elements possible for the final result (as early as the 1st step seen) are "ur", "r", "1". Since these describe the initial (and maintained) row, the forth solution (obtainable from this current analysis) would probably be commensurate with the element usage of the initial row.

Forth Solution:

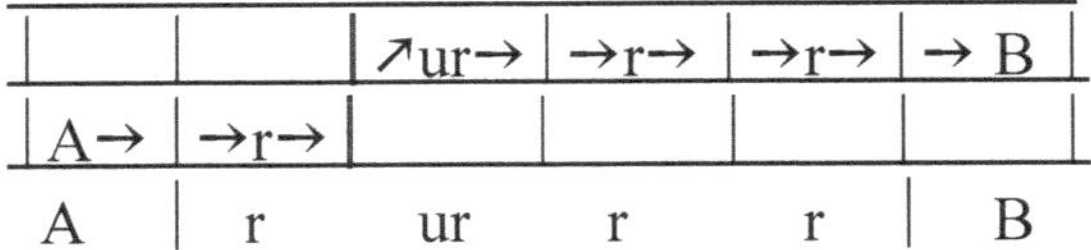

		↗ur→	→r→	→r→	→ B
A→	→r→				

$$A \quad | \quad r \quad\quad ur \quad\quad r \quad\quad r \quad | \quad B$$

To avoid losing an element, it is instructive to denote which elements of a table are automatically lost during CDD. This can easily be outlined:

e.g.

$$\begin{vmatrix} ur & r & r & r \\ ur & dr & ur & r \\ r & r & ur & r \\ 1 & ur & 1 & 1 \end{vmatrix}$$

The outlined elements are eventually lost, and the final result is made up of the types of the 5 elements left.

Since we wish to retain the 1st row (as a solution), we should place an element to be retained (as per some constructive argument) as the 1st element of the last row (or the last element of the 1st column).

"dl" e.g.

$$\begin{vmatrix} ur & r & r & r \\ ur & dr & ur & r \\ r & r & ur & r \\ dl & 1 & 1 & 1 \end{vmatrix} \;\Rightarrow\; \begin{vmatrix} ur & r & r & r \\ ur & & & r \\ r & & & r \\ dl & 1 & 1 & 1 \end{vmatrix} \;\equiv\; \begin{vmatrix} ur & r & r & r \\ r & r & r & 1 \\ 1 & 1 & 1 & dl \\ dl & r & ur & ur \end{vmatrix} \;\Rightarrow\; \begin{vmatrix} ur & r & r & r \\ r & & & 1 \\ 1 & & & dl \\ dl & r & ur & ur \end{vmatrix}$$

$$\equiv\; \begin{vmatrix} ur & r & r & r \\ r & 1 & dl & ur \\ ur & ur & r & dl \\ dl & 1 & r & ur \end{vmatrix} \;\Rightarrow\; \begin{vmatrix} ur & r & r & r \\ r & & & ur \\ ur & & & dl \\ dl & 1 & r & ur \end{vmatrix} \;\equiv\; \begin{vmatrix} ur & r & r & r \\ r & ur & dl & ur \\ ur & r & 1 & dl \\ dl & ur & r & ur \end{vmatrix}$$

$$
\begin{vmatrix} \text{ur} & \underline{r} & \underline{r} & r \\ r & & & \text{ur} \\ \text{ur} & & & d\mathit{l} \\ d\mathit{l} & \text{ur} & r & \text{ur} \end{vmatrix}
\equiv
\begin{vmatrix} \text{ur} & r & r & r \\ r & \text{ur} & d\mathit{l} & \text{ur} \\ \text{ur} & r & \text{ur} & d\mathit{l} \\ d\mathit{l} & \text{ur} & r & \text{ur} \end{vmatrix}
$$

final result

$\text{ur} \equiv \text{ur} , \; d\mathit{l} \equiv r$

(based on diagonals)

As before, if we replace "d*l*" with "r" in the 3rd column, and convert column to row, we have a solution. But this solution does not involve (i.e. include) the "d*l*" element.

These results show that CDD is actually a test of the initial (and retained, solutional) row against the last element of the 1st column. This element is essentially made equivalent (via diagonal notice) to the last element of the 1st row. Therein with this equivalency may a new solution be deduced (from the 3rd column of the final result).

$\therefore$

$$
\begin{vmatrix} \text{ur} & r & r & r \\ - & - & - & - \\ - & - & - & - \\ d\mathit{l} & - & - & - \end{vmatrix}
\quad \text{CDD} \quad
\begin{vmatrix} \text{ur} & r & r & r \\ - & \text{ur} & d\mathit{l} & - \\ - & r & \text{ur} & - \\ d\mathit{l} & - & r & \text{ur} \end{vmatrix}
$$

$\text{ur} \equiv \text{ur}$

$d\mathit{l} \equiv r$

The 3rd element of the last row (i.e. the last element of the 3rd column) also reflects this last element of the 1st (maintained) row.

$\therefore$ For any 4x4 table, we can immediately predict the final result of the CDD procedure on it:

$$
\begin{vmatrix} a & b & c & d \\ - & - & - & - \\ - & - & - & - \\ e & - & - & - \end{vmatrix}
\quad \text{CDD} \quad
\begin{vmatrix} a & b & c & d \\ d & a & e & a \\ a & d & a & e \\ e & a & d & a \end{vmatrix}
$$

: $a \equiv a$

: $e \equiv d$

: diagonals: aaaa , eded

and $\quad \begin{vmatrix} a & b & c & d \end{vmatrix} \equiv \begin{vmatrix} d & a & (e \to)d & c \end{vmatrix}$

For the final result, the other elements (outlined) may also be immediately (via slants) predicted, as well as the repeating pattern for the lower 3 rows: here "a d a e" (3rd row).

The only elements involved in the lower 3 rows of the final result are "a, d" (from the 1st row) and "e" (from the 1st column). So the result, as far as analysis, is of an (a, d, e) table. As far as CDD is concerned, this pertains to the originating table as well as to the resulting (final) table.

The two internal columns (of the final result) do in fact represent solutions, if the equivalencies suggested by the diagonals are applied:

e.g. $\begin{vmatrix} \text{ur} & r & r & r \\ r & \text{ur} & d\mathit{l} & \text{ur} \\ \text{ur} & r & \text{ur} & d\mathit{l} \\ d\mathit{l} & \text{ur} & r & \text{ur} \end{vmatrix}$

$\text{ur} \equiv \text{ur}$

$d\mathit{l} \equiv r$

For column 3: allow $\quad d\mathit{l} \to r \quad$ and row $\begin{vmatrix} r & \text{ur} & r & r \end{vmatrix}$ $\quad A \to B$.

For column 2: allow both $r \to d\mathit{l}$ and row $\begin{vmatrix} \text{ur} & d\mathit{l} & \text{ur} & d\mathit{l} \end{vmatrix}$ $\quad A \to A$.

This 2nd column derived solution essentially leaves us back at point A, contrasted to the 3rd column derived solution allowing us to find (go to) point B.

Likewise, from columns 1 and 4, allowing r→d*l* establishes A→A "solutions." This is also true for each of the lower 3 rows (of the final result).

We can say that from CDD, the (a,b,c,d,e) element positions (of the originating table) are invariant for the final result, and that the (b,c) element positions are un-utilized (in the lower 3 rows of the final result).

∴ Only row 1 and column 3 (with modification) are "extraordinary" of the final result (i.e. leading to solution). The other rows and columns are "ordinary."

Of course, the CDD procedure has no notion or initiative of drive to allow A→B (nor A→A). It is merely reflective of the test of the first row to compare to the last element of the 1st column.

In a sense (as per this example, based on the results) we can ask when (or how) is: d*l* ≡ r .

This can occur when both operations performed (from the same unit of cell) are recovered by two "r"s and an "u", or an (ur r):

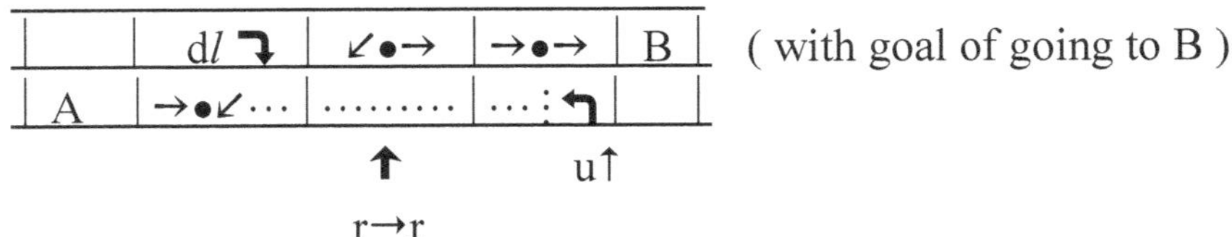

(with goal of going to B)

For initiation at a point preceding "B", an "d*l*" element (operation) will always be counter productive in the goal of: A→B . (But the CDD procedure is ignorant of all goals.)

∴ As we (judgmentally) look at the final result, we find that we should apply: d*l* ≡ r (i.e. d*l* → r) in the 3rd column (to have a hope towards our goal of: A→B). Yet, may we notice (suspiciously) that (ur r) abounds throughout the (final result) table, perhaps hinting at the functional nature of the (diagonally predicted) equivalence.

∴   "d*l* → r" is the same as saying: (d*l* r r u) or (d*l* ur r)

i.e. (for the 3rd column row-ed):

$$| \text{ r ur d}l\text{→r r }| \equiv \text{(from 4 elements)}| \text{ r ur d}l\text{ r r ur }| \text{ (7 elements)}$$
$$\text{or } | \text{ r ur d}l\text{ ur r r }| \text{ (6 elements)}$$

The expansion of element number can suggest expansion (or enhancement) of table:

e.g.

| ur r r r | | 1 1 r 1 | | 1 1 r 1 1 1 |
| r ur d*l* ur | → | 1 1 r 1 | | 1 1 r 1 1 1 |
| ur r ur d*l* | | ur r ur r | → | ur r ur r 1 1 | (as balanced 6x6)
d*l* ur r ur		r ur d*l* ur		r ur d*l* ur 1 1
		ur r ur d*l*		ur r ur d*l* 1 1
		d*l* ur r ur		d*l* ur r ur 1 1

Here, we can tell that only column 3 row-ed leads us to our desired goal (and solution), although we may adjust order of elements in that column to allow the (now) 3rd row to fall into compliance (i.e. having an "r" instead of an "ur" for its 3rd element). Of course, changing the order (of elements) within a column may alter the solutions of the corresponding rows.

But since we know what solution "we" want, let us, for the 4x6 enhanced table, change the order (judiciously) of the 3rd column, to exchange the 3rd element "ur" with the 2nd element "r", and then bring the 3rd row to top (pushing the first two rows down):

$$
\begin{vmatrix} 1 & 1 & r & 1 \\ 1 & 1 & r & 1 \\ ur & r & ur & r \\ r & ur & dl & ur \\ ur & r & ur & dl \\ dl & ur & r & ur \end{vmatrix}
\rightarrow
\begin{vmatrix} 1 & 1 & r & 1 \\ 1 & 1 & ur & 1 \\ ur & r & r & r \\ r & ur & dl & ur \\ ur & r & ur & dl \\ dl & ur & r & ur \end{vmatrix}
\rightarrow
\begin{vmatrix} ur & r & r & r \\ 1 & 1 & r & 1 \\ 1 & 1 & ur & 1 \\ r & ur & dl & ur \\ ur & r & ur & dl \\ dl & ur & r & ur \end{vmatrix}
$$

Now we can apply (4-sided) CDD, with its automatic "1"s expansion (of rows):

$$
\begin{vmatrix} ur & r & r & r \\ 1 & 1 & r & 1 \\ 1 & 1 & ur & 1 \\ r & ur & dl & ur \\ ur & r & ur & dl \\ dl & ur & r & ur \end{vmatrix}
\Rightarrow
\begin{vmatrix} ur & r & r & r \\ 1 & & & 1 \\ 1 & & & 1 \\ r & & & ur \\ ur & & & dl \\ dl & ur & r & ur \end{vmatrix}
\equiv
\begin{vmatrix} ur & r & r & r & 1 & 1 \\ r & 1 & 1 & ur & dl & ur \\ ur & r & ur & dl & 1 & 1 \\ dl & ur & r & 1 & 1 & ur \end{vmatrix}
$$

$$
\Rightarrow
\begin{vmatrix} ur & r & r & r & 1 & 1 \\ r & & & & & ur \\ ur & & & & & 1 \\ dl & ur & r & 1 & 1 & ur \end{vmatrix}
\equiv
\begin{vmatrix} ur & r & r & 1 & 1 \\ 1 & ur & 1 & ur & 1 & 1 \\ ur & 1 & 1 & r & ur & dl \\ dl & ur & r & ur & 1 & 1 \end{vmatrix}
\Rightarrow
\begin{vmatrix} ur & r & r & r & 1 & 1 \\ 1 & & & & & 1 \\ ur & & & & & dl \\ dl & ur & r & ur & 1 & 1 \end{vmatrix}
$$

$$
\equiv
\begin{vmatrix} ur & r & r & r & 1 & 1 \\ 1 & 1 & dl & 1 & 1 & 1 \\ 1 & 1 & ur & r & ur & dl \\ dl & ur & 1 & ur & 1 & 1 \end{vmatrix}
\Rightarrow
\begin{vmatrix} ur & r & r & r & 1 \\ 1 & & & & 1 \\ 1 & & & & dl \\ dl & ur & 1 & ur & 1 & 1 \end{vmatrix}
\equiv
\begin{vmatrix} ur & r & r & r & 1 & 1 \\ 1 & 1 & dl & 1 & 1 & 1 \\ 1 & 1 & ur & 1 & ur & dl \\ dl & 1 & 1 & ur & 1 & 1 \end{vmatrix}
$$

$$
\Rightarrow
\begin{vmatrix} ur & r & r & r & 1 & 1 \\ 1 & & & & & 1 \\ 1 & & & & & dl \\ dl & 1 & 1 & ur & 1 & 1 \end{vmatrix}
\equiv
\begin{vmatrix} ur & r & r & r & 1 & 1 \\ 1 & 1 & dl & 1 & 1 & 1 \\ 1 & 1 & ur & 1 & 1 & dl \\ dl & 1 & 1 & ur & 1 & 1 \end{vmatrix}
$$

final result
(unbalanced)

The final result can be treated various ways for balancing:

— Two rows of (6) "1"s can be added.

$$
\text{e.g.}\quad
\begin{vmatrix}
\text{ur} & \text{r} & \text{r} & \text{r} & 1 & 1 \\
1 & 1 & d\!l & 1 & 1 & 1 \\
1 & 1 & \text{ur} & 1 & 1 & d\!l \\
d\!l & 1 & 1 & \text{ur} & 1 & 1
\end{vmatrix}
\;\rightarrow\;
\begin{vmatrix}
\text{ur} & \text{r} & \text{r} & \text{r} & 1 & 1 \\
1 & 1 & d\!l & 1 & 1 & 1 \\
1 & 1 & 1 & 1 & 1 & 1 \\
1 & 1 & \text{ur} & 1 & 1 & d\!l \\
d\!l & 1 & 1 & \text{ur} & 1 & 1 \\
1 & 1 & 1 & 1 & 1 & 1
\end{vmatrix}
\quad
\begin{array}{l}\text{Both diagonal products}\\ \text{(here) are "ur".}\end{array}
$$

— Since each row has at least two "1"s, these (for each row) can be removed (to reduce the table to 4x4).

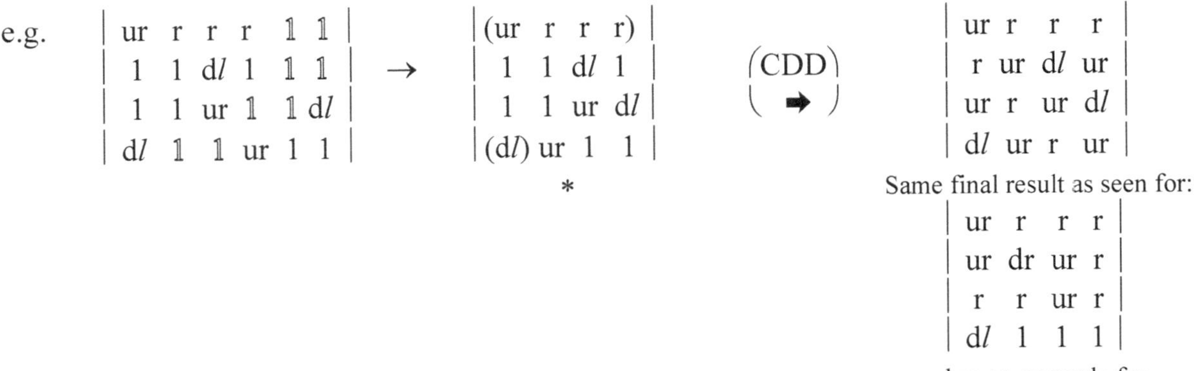

In this (particular) case, each row yields (or devolves) to an interesting operation:

$$
\begin{vmatrix} \text{ur} & \text{r} & \text{r} & \text{r} \end{vmatrix} \equiv \text{(retained) solution}
\qquad
\begin{vmatrix} 1 & 1 & \text{ur} & d\!l \end{vmatrix} \equiv (\text{ur } d\!l) \equiv 1
$$
$$
\begin{vmatrix} 1 & 1 & d\!l & 1 \end{vmatrix} \equiv d\!l
\qquad
\begin{vmatrix} d\!l & \text{ur} & 1 & 1 \end{vmatrix} \equiv (d\!l \text{ ur}) \equiv 1
$$

*(Only the elements in parentheses are necessary for a final result, under CDD.)

— The last two columns can simply be dropped (removed), leaving a 4x4 table which with CDD will yield to the same final solution as the just previous:

$$
\text{i.e.}\quad
\begin{vmatrix}
\text{ur} & \text{r} & \text{r} & \text{r} & 1 & 1 \\
1 & 1 & d\!l & 1 & 1 & 1 \\
1 & 1 & \text{ur} & 1 & 1 & d\!l \\
d\!l & 1 & 1 & \text{ur} & 1 & 1
\end{vmatrix}
\;\rightarrow\;
\begin{vmatrix}
(\text{ur} & \text{r} & \text{r} & \text{r}) \\
1 & 1 & d\!l & 1 \\
1 & 1 & \text{ur} & 1 \\
(d\!l) & 1 & 1 & \text{ur}
\end{vmatrix}
\quad
\left(\begin{array}{c}\text{CDD}\\ \rightarrow\end{array}\right)\text{ etc.}
$$

In this case, the row "solutions" are:

$$
\begin{vmatrix} \text{ur} & \text{r} & \text{r} & \text{r} \end{vmatrix} \equiv \text{(desired) solution}
\qquad
\begin{vmatrix} 1 & 1 & d\!l & 1 \end{vmatrix} \equiv d\!l
$$
$$
\begin{vmatrix} 1 & 1 & \text{ur} & 1 \end{vmatrix} \equiv \text{ur}
\qquad
\begin{vmatrix} d\!l & 1 & 1 & \text{ur} \end{vmatrix} \equiv 1
$$

These (forced) manipulations and convergences to a common final result highlight some details for handling the element "d*l*" with respect to our desired solution (A→B). (This is essentially discovered by the 3rd column "solution" found previously, i.e. $\begin{vmatrix} \text{r} & d\!l & \text{ur} & \text{r} \end{vmatrix}$, in how to handle "d*l*" towards gaining our goal of (A→B).)

2) Complexed planar movement

Having derived a forth solution for the (A→B) goal:

```
| ur  r   r   r |
| ur  dr  ur  r |      there is a point of operational identity that seems to emerge,
|  r  r   ur  r |       in this case on the element "r" :
|  r  ur  r   r |
```

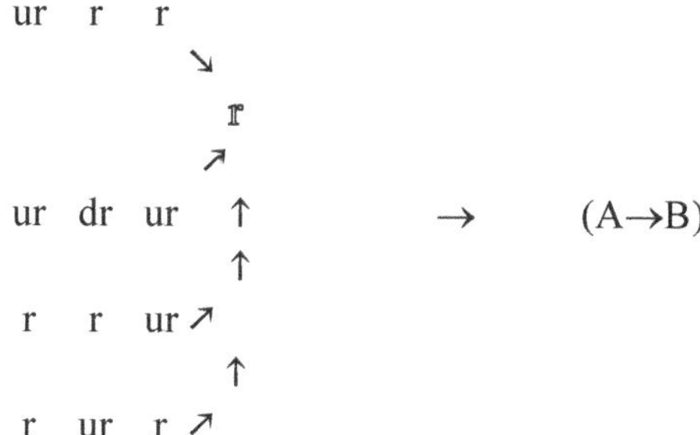

Not only is "r" the last element of each row (which might be important for an order-dependent procedure), but it is also the only type of element found in the last column. This occurrence can be used, for example, as a point of overlap to an "r" dependent synthesis of functionality, extending a procedure beyond the "A→B" requirement (or necessity):

```
ur   r    r    ⅂ ↗→
ur   dr   ur   ⅂ ↗→
                r
 r    r   ur   ⅃ ↘→
 r   ur    r   ⅃ ↘→
          (A→B)
```

If (A→B) is initiating (of a procedure), we might fancy the diagram (of operations) as:

```
↪→(A)  ur   r    r    ⅂ ↗→(B)
↪→(A)  ur   dr   ur   ⅂ ↗→(B)
1                      r
↳→(A)   r    r   ur   ⅃ ↘→(B)
↳→(A)   r   ur    r   ⅃ ↘→(B)
```

Let us say, for hope of variety of movement, that the furtherance of procedure involves rotation of grid or "board" about point B, converting B → B* (A remaining A):

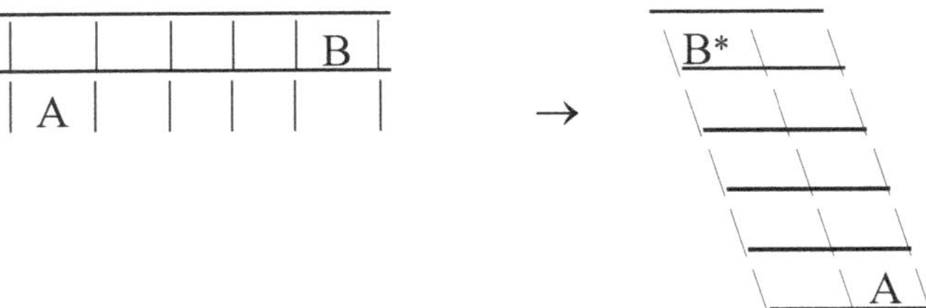

This is B → B*, essentially in that it initiates a "new" board or grid:

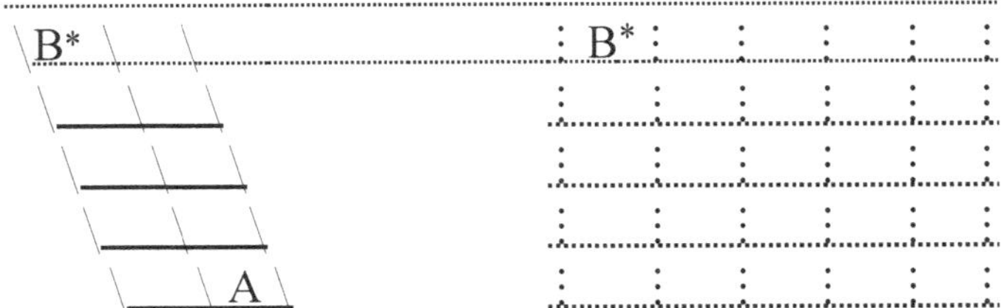

In terms of tables, we have an overlap of procedures, for two different boards:

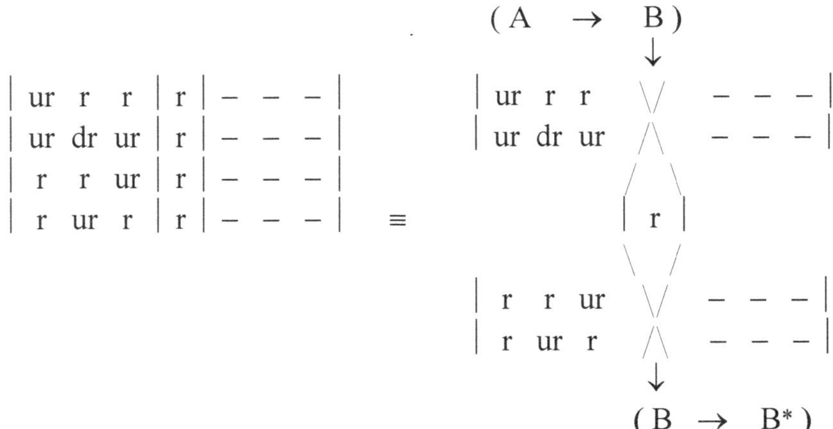

Since "r" is a movement (to the right) we must envision it (now or here) as part of a procedure of rotation (after A→B) at point B. The overlap takes advantage if its convenience. At this juncture we must, of course, create some (plane) rotational operators that can be conditional upon a position reached by "r". Allowing the rows to be order-dependent, let us condition that each rotational operator (for simplicity) is itself dependent on the occurrence of the "ur" operation that leads to "r" reaching point B:

∴ say, for example:

$$\text{ur} \quad \text{r} \quad \text{r} \quad \text{r} \;\rightarrow\; \phi \qquad\qquad \text{r} \quad \text{r} \quad \text{ur} \quad \text{r} \;\rightarrow\; \varphi$$
$$\text{r} \quad \text{ur} \quad \text{r} \quad \text{r} \;\rightarrow\; \theta \qquad\qquad \text{ur} \quad \text{dr} \quad \text{ur} \quad \text{r} \;\rightarrow\; \phi \text{ or } \varphi$$

Each of these rotational operators (ϕ , φ , θ) must be delimited to extent such as to allow B→B* .

We can make, for example:

$$\phi \equiv \text{clockwise rotation (of plane)}$$
$$\varphi \equiv \text{counter-clockwise rotation}$$
$$\theta \equiv \text{either clockwise or counter-clockwise rotation}$$

Then, to delimit (rotations), the operator π_{B*} stops the (preceding) rotation operation when B→B* (i.e. π_{B*} stalls the rotation to stop when B→B*). We may also adjudge the speed (or efficiency or temporal nature) of the rotation with $\tau_{1,2,3,4}$.

∴ we have, for rows:

r	ϕ	τ_1	π_{B*}		Obviously (as measured or observed empirically)
r	(ϕ , φ)	τ_2	π_{B*}		$\tau_2 \equiv \tau_1$ or τ_3
r	φ	τ_3	π_{B*}		$\theta \equiv \phi$ or φ (with $\tau_4 \equiv \tau_1$ or τ_3)
r	θ	τ_4	π_{B*}		

∴

$$(A \;\rightarrow\; B)$$
$$\downarrow$$

ur r r	ϕ	τ_1
ur dr ur	ϕ, φ	τ_2
r		π_{B*}
r r ur	φ	τ_3
r ur r	θ	τ_4

$$\downarrow$$
$$(B \;\rightarrow\; B*)$$

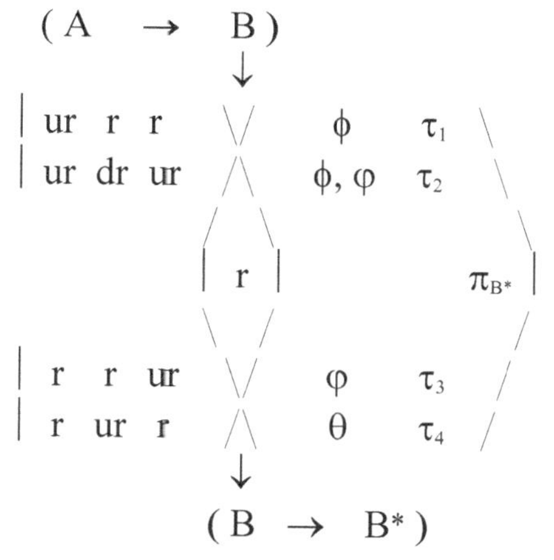

and

$$\longmapsto\rightarrow(A) \quad ur \quad r \quad r \quad \reflectbox{$\lnot$}\nearrow\rightarrow(B) \quad \phi \quad \tau_1\rightarrow\searrow \quad \nearrow\rightarrow (B*)$$
$$\longmapsto\rightarrow(A) \quad ur \quad dr \quad ur \quad \reflectbox{$\lnot$}\nearrow\rightarrow(B) \quad \phi, \varphi \quad \tau_2\rightarrow\searrow \quad \nearrow\rightarrow(B*)$$
$$1 \qquad\qquad\qquad\qquad r \qquad\qquad\qquad\qquad \pi_{B*}$$
$$\hookrightarrow\rightarrow(A) \quad r \quad r \quad ur \quad \uparrow\searrow\rightarrow(B) \quad \varphi \quad \tau_3\rightarrow\nearrow \quad \searrow\rightarrow(B*)$$
$$\hookrightarrow\rightarrow(A) \quad r \quad ur \quad r \quad \uparrow\searrow\rightarrow(B) \quad \theta \quad \tau_4\rightarrow\nearrow \quad \searrow\rightarrow (B*)$$

After π_{B*} is performed, a new (i.e. rotated) plane is established.

If no temporal component is to be considered for the (plane) rotation, then of course: $\tau = 1$.

The two tables thus far established are:

> Homeric (table of translations for movement),
> and disparate (collection of different types of operators; e.g. ϕ, φ, θ rotation, "r" translation,
> τ temporal, π stoppage).

Because the two tables overlap (at "r"), they can be separated (without harm to solutions), and analyzed separately. For the procedure (overall), being order-dependent (of operations), the columnar elements (particularly for the B→B* table) are not solutional (i.e. solutions are restricted here to rows).

$\therefore$ CDD analysis (on $B{\rightarrow}B^*$ table) would not reasonably apply:

e.g.

$$\begin{vmatrix} r & \phi & \tau_1 & \pi_{B^*} \\ r & \phi,\varphi & \tau_2 & \pi_{B^*} \\ r & \varphi & \tau_3 & \pi_{B^*} \\ r & \theta & \tau_4 & \pi_{B^*} \end{vmatrix} \quad \overset{\text{CDD}}{\Longrightarrow} \quad \begin{vmatrix} r & \phi & \tau_1 & \pi_{B^*} \\ \pi_{B^*} & r & \pi_{B^*} & r \\ r & \pi_{B^*} & r & \pi_{B^*} \\ \pi_{B^*} & r & \pi_{B^*} & r \end{vmatrix}$$

Not very informative (since there are no rotations, to stop, for the lower 3 rows).

The ambiguities and uncertainties in some of the operational assignments, e.g. (ϕ , φ) and θ , allow for some measure of trans-row (or row-internal) functionality (since there is no column functionality here, towards acceptable solutions). Otherwise, the table (solution) would consist of definite but non-interchangeable elements of strictly directed paths of procedure. Already, however, as defined, we can observe some interrelationships:

$$\begin{vmatrix} r & \phi \searrow & \tau_1 & \pi_{B^*} \\ r & \phi,\varphi \nearrow & \tau_2 & \pi_{B^*} \\ r & \varphi \searrow & \tau_3 & \pi_{B^*} \\ r & \theta \nearrow & \tau_4 & \pi_{B^*} \end{vmatrix} \begin{matrix} \rightarrow \\ \rightarrow \\ \rightarrow \\ \rightarrow \end{matrix} \begin{vmatrix} r & \varphi & \tau_1 & \pi_{B^*} \\ r & \phi & \tau_2 & \pi_{B^*} \\ r & \varphi & \tau_3 & \pi_{B^*} \\ r & \phi & \tau_4 & \pi_{B^*} \end{vmatrix} , \begin{vmatrix} r & \phi & \tau_3 & \pi_{B^*} \\ r & \varphi & \tau_4 & \pi_{B^*} \end{vmatrix}$$

This allows, of course, for an effective rotor of element permutations:

$$(\theta) \overline{\qquad} \rceil$$
$$\{ r \} \quad \{ \phi, \varphi \} \quad \{ \tau_1 \ \tau_2 \ \tau_3 \ \tau_4 \} \quad \{ \pi_{B^*} \}$$
$$\llcorner \overline{\qquad\qquad} \lrcorner \text{ (operations restricted to their element class)}$$

e.g. (for rotations)

at τ_4 : θ can be either clockwise (ϕ) or counterclockwise (φ)

at τ_2 : clockwise (ϕ) or counterclockwise (φ) occurs

at τ_1 : clockwise (ϕ) is preferred

at τ_3 : counterclockwise (φ) is preferred.

Yet clockwise under τ_3 ($\phi \ \tau_3$) can be studied, and counterclockwise under τ_1 ($\varphi \ \tau_1$) can be studied. (That is, the rotation typical for ϕ process might be due to a counterclockwise process, and that typical for a φ process might be due to a clockwise process, under the typically corresponding τ_s .)

Degrees of (operational) freedom are thus expanded by the (or this) "element permutation rotor" (epr).

$\therefore$ We can symbolize this by something like:

$$\begin{vmatrix} r & \phi & \tau_1 & \pi_{B^*} \\ r & \phi,\varphi & \tau_2 & \pi_{B^*} \\ r & \varphi & \tau_3 & \pi_{B^*} \\ r & \theta & \tau_4 & \pi_{B^*} \end{vmatrix} \equiv \begin{vmatrix} \phi & \rightarrow & \tau_1 \\ \phi,\varphi & & \tau_2 \\ \varphi & & \tau_3 \\ \theta & & \tau_4 \end{vmatrix} \quad \text{etc.}$$

(If the rows are strictly conserved (for the table), then the epr dissipates.)

Yet, in terms of order dependence, it would seem as if the rotations and τ_s are interchangeable (of order), or at least composite:

e.g. $\quad\quad \phi\ \tau_1 \equiv \tau_1\ \phi \quad$ (since the elements operate together)

$\therefore$ They are always "one and the other" in performance.

$\therefore$ The table represents a sort of engine (operationally):

$$\left|\ \text{initiation}\ (\equiv \text{``r''})\ ,\ \ \text{epr}\ ,\ \quad \text{stoppage}\ (\equiv \text{``}\pi_{B*}\text{''})\ \right|$$
$$\text{(reaching point B)} \quad\quad \text{(conversion)} \quad \text{(having formed point B*)}$$

In accordance with nul/los principles, the composite operations can be considered as circles, the conjunctions of which produce a synergy of performance to engage the respective (concordant) actions:

e.g.

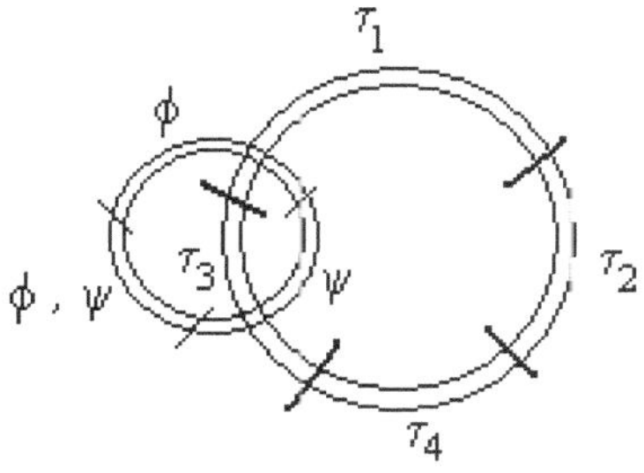

conjunctions at $(\ \phi\ ,\ \tau_1\ )\ ,\ (\ \varphi\ ,\ \tau_3\ )$
$\quad\quad$ The range of θ can be the combined ranges of ϕ and φ.

e.g.

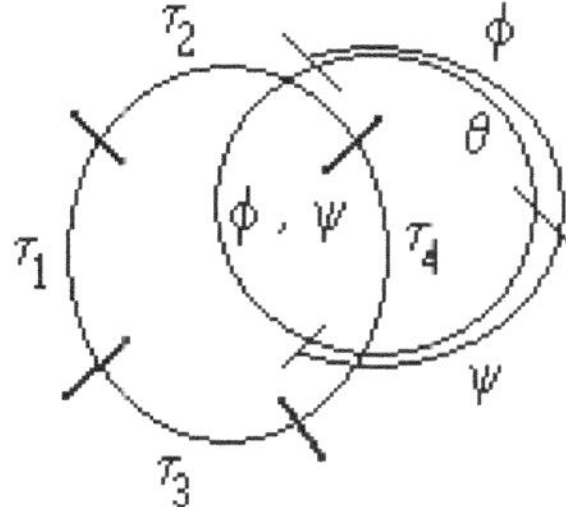

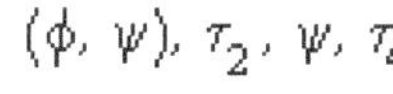

conjunctions at:
$(\phi,\ \psi),\ \tau_2,\ \psi,\ \tau_4$

e.g.

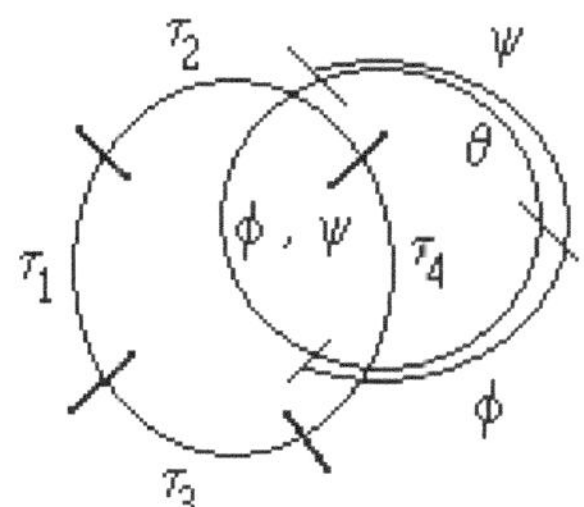

conjunctions at:
$(\phi,\ \psi),\ \tau_2,\ \phi,\ \tau_4$

The distinction between (ϕ , φ) and θ can be depicted as ϕ , φ overlaps vs. ϕ or φ for θ :

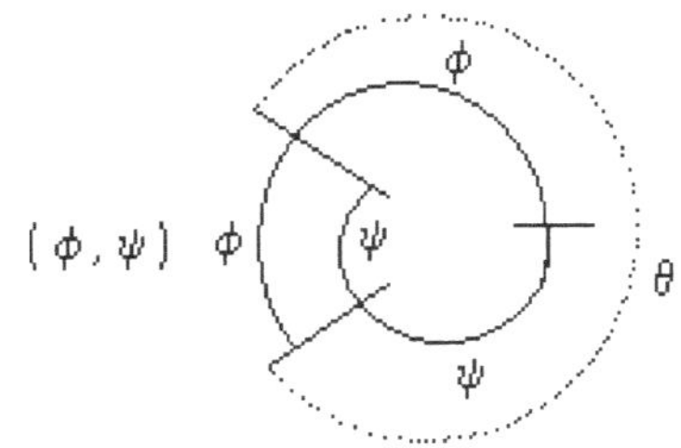

$\therefore$ θ excludes the overlap (ϕ , φ) region.

Conjunctions (of circles) obtain (or occur) in 3 ways: at a single point, at 2 points, at total conjunction (full coincidence of circles, or complete overlapping of circles).

Only at total conjunction (as depicted here) can all four solutions (of the table) be simultaneously satisfied (as for composite operations):

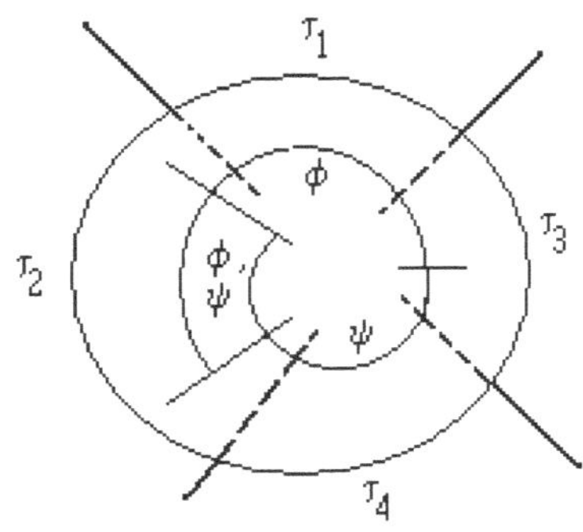

$\therefore$ The table reflects, in part, the simultaneity of the composite operations (elements).

∴ The rotor involves the total conjunction (search) which satisfies the table solutions:

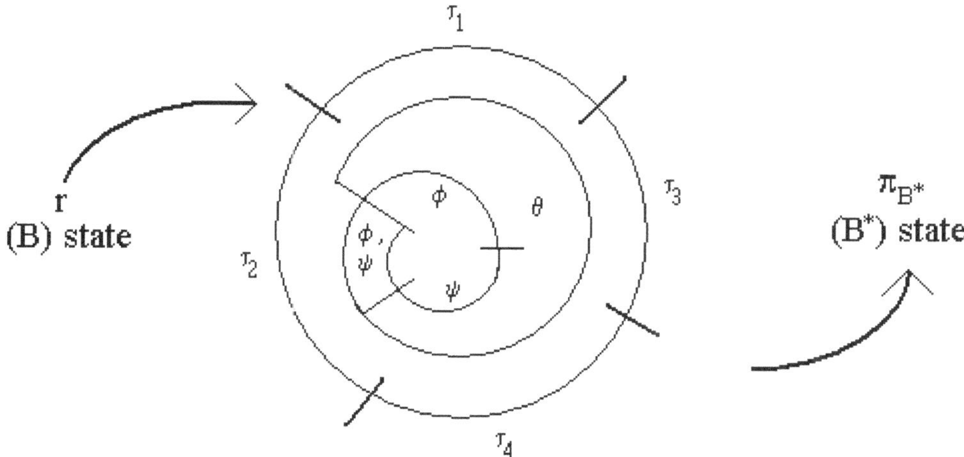

As depicted in this particular diagram:

> ϕ has some τ_1, τ_2, τ_3
> φ has some τ_3, τ_4, τ_2
> (ϕ , φ) has only τ_2
> θ excludes the overlap (ϕ , φ) , but has $\tau_1, \tau_2, \tau_3, \tau_4$
>   (in fact, it includes all of τ_1, τ_3, τ_4 , and some τ_2)

When "r" is achieved, the rotor starts, in search of such an orientation (of circles in total conjunction) as shown here. When this is achieved (found), "π_{B^*}" is performed (stoppage, as B → B* is achieved).

Since the composite operations are at total conjunction, diagrammatically either circle can be drawn as internal. The order of the revolutions: ϕ , φ , (ϕ , φ) , θ , and the temporal components (τ elements) is immaterial with simultaneity; the operations occur concurrently (but after "r" and before "π_{B^*}"). Obviously, other total conjunction orientations (of circles), that are satisfying, may be drawn and achieved.

There are then several (of a family of) total conjunction orientations that allow for satisfying all of the row solutions of the table and thereby cause "π_{B^*}" to occur. This family (of orientations) has the properties (as stated earlier, by observation) as sets:

> $\{ \phi \mid \tau_1 , \tau_2 , \tau_3 \}$
> $\{ \varphi \mid \tau_2 , \tau_3 , \tau_4 \}$
> $\{ \phi , \varphi \mid \tau_2 \}$
> $\{ \theta \mid \tau_1 , \tau_2 , \tau_3 , \tau_4 \}$

And this (observation) can itself be related to a table of (4) rows:

> | r $\{ \phi \mid \tau_1 , \tau_2 , \tau_3 \}$ π_{B^*} |
> | r $\{ \varphi \mid \tau_2 , \tau_3 , \tau_4 \}$ π_{B^*} |
> | r $\{ \phi , \varphi \mid \tau_2 \}$ π_{B^*} |
> | r $\{ \theta \mid \tau_1 , \tau_2 , \tau_3 , \tau_4 \}$ π_{B^*} |

This "functional" table compares (with distinction) to the "descriptive" table originating earlier. The functional table defines acceptable (or advantageous) epr designs.

So now, the composite operations (of the rotor) can be considered sets of operations.

3) Extra-planar designs

It occurs of course, that upon the conversion of B to B*, the two tables need not share the same (as one or the other) plane, the (or some) rotations being extra-planar. The point of convergence (of the two tables) being at "r" as a pivoting locality to distinguish planes. (There is always a 3rd plane shared by two.)

$\therefore$ We have a design such as:

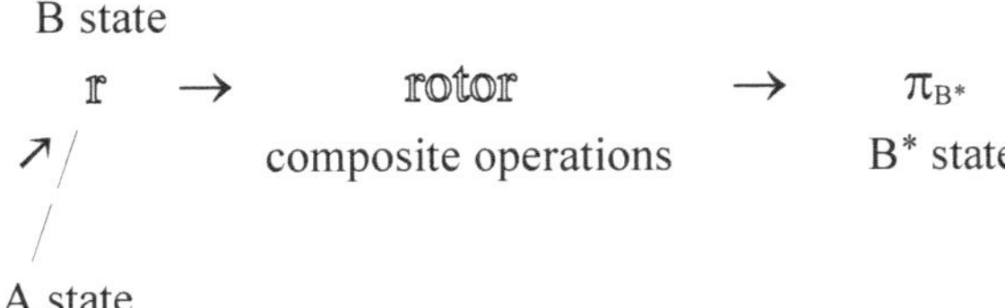

The planes would be 3-dimensional, coming out of the plane (shared) of the paper.

A possible objective of this construct might be to discover how to convert A→A*, probably through a multiplicative relation between the two tables:

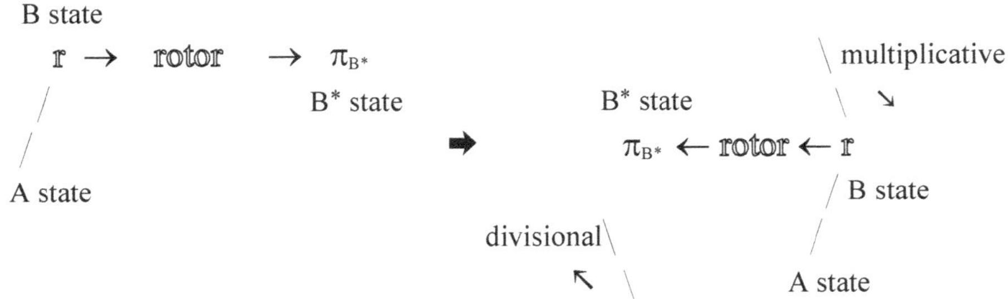

$\therefore$

$$
\begin{array}{llll}
\mid \pi_{B*} & \{\ \phi\ \mid\ \tau_1\ ,\ \tau_2\ ,\ \tau_3\ \} & \mid r \mid r\quad r\quad ur \mid \\
\mid \pi_{B*} & \{\ \varphi\ \mid\ \tau_2\ ,\ \tau_3\ ,\ \tau_4\ \} & \mid r \mid ur\ r\quad r \mid & \text{multiplicative} \\
\mid \pi_{B*} & \{\ \phi\ ,\ \varphi\ \mid\ \tau_2\ \} & \mid r \mid ur\ dr\ ur \mid & \Rightarrow \\
\mid \pi_{B*} & \{\ \theta\ \mid\ \tau_1\ ,\ \tau_2\ ,\ \tau_3\ ,\ \tau_4\ \} & \mid r \mid r\quad ur\quad r \mid
\end{array}
$$

$$
\begin{array}{lll}
\mid r\ \{\ \phi\ \mid\ \tau_1\ ,\ \tau_2\ ,\ \tau_3\ \} & \bullet (\ r\quad r\ ) & \pi_{B*} \bullet ur \mid \\
\mid r\ \{\ \varphi\ \mid\ \tau_2\ ,\ \tau_3\ ,\ \tau_4\ \} & \bullet (\ ur\ r\ ) & \pi_{B*} \bullet r \mid \\
\mid r\ \{\ \phi\ ,\ \varphi\ \mid\ \tau_2\ \} & \bullet (\ ur\ dr\ ) & \pi_{B*} \bullet ur \mid \\
\mid r\ \{\ \theta\ \mid\ \tau_1\ ,\ \tau_2\ ,\ \tau_3\ ,\ \tau_4\ \} & \bullet (\ r\quad ur\ ) & \pi_{B*} \bullet r \mid
\end{array}
$$

Obviously, to go from B* to A to A* requires reversals (absolute) of movement (of the Homeric translations).

$\therefore$ Some equalities (or equivalencies) are set up:

$$
\begin{array}{llll}
|\ l\ \{\ \phi\ |\ \tau_1,\ \tau_2,\ \tau_3\ \} & \bullet\ (\ l\ \ l\ ) & \pi_{B*}\bullet dl\ | \\
|\ l\ \{\ \phi\ |\ \tau_2,\ \tau_3,\ \tau_4\ \} & \bullet\ (\ dl\ \ l\ ) & \pi_{B*}\bullet l\ | \\
|\ l\ \{\ \phi,\phi\ |\ \tau_2\ \} & \bullet\ (\ dl\ \ ul\ ) & \pi_{B*}\bullet dl\ | \\
|\ l\ \{\ \theta\ |\ \tau_1,\ \tau_2,\ \tau_3,\tau_4\ \}\ \bullet\ (\ l\ \ dl\ ) & \pi_{B*}\bullet l\ |
\end{array}
$$

$\therefore$

$$
\begin{array}{lllll}
|\ l\ \{\ \phi\ |\ \tau_1,\ \tau_2,\ \tau_3\ \} & \bullet\ (\ l\ \ l\ ) & \pi_{B*}\bullet dl\ | & \equiv\ \tau_{A*} \\
|\ l\ \{\ \phi\ |\ \tau_2,\ \tau_3,\ \tau_4\ \} & \bullet\ (\ dl\ \ l\ ) & \pi_{B*}\bullet l\ | & \equiv\ \tau_{A*} \\
|\ l\ \{\ \phi,\phi\ |\ \tau_2\ \} & \bullet\ (\ dl\ \ ul\ ) & \pi_{B*}\bullet dl\ | & \equiv\ \tau_{A*} \\
|\ l\ \{\ \theta\ |\ \tau_1,\ \tau_2,\ \tau_3,\tau_4\ \}\ \bullet\ (\ l\ \ dl\ ) & \pi_{B*}\bullet l\ | & \equiv\ \tau_{A*}
\end{array}
$$

or, one can say:

$$\{\ \phi\ |\ \tau_1,\ \tau_2,\ \tau_3\ \}\ \bullet\ (\ r\ \ r\ )\equiv(\ l\ \ l\ ),\quad \pi_{B*}\bullet ur\equiv dl$$

$\therefore$ if (given) $|\ r\ \ l\ \ l\ \ dl\ |$, then $|\ l\ \ l\ \ l\ \ dl\ |\ \rightarrow\ $A

$$|\ l\ \{\ \phi\ |\ \tau_1,\ \tau_2,\ \tau_3\ \}\ \bullet\ (\ l\ \ l\ )\quad \pi_{B*}\bullet dl\ |\ \rightarrow\ \text{A*}\quad \text{etc.}$$

Note: the composite operation sets $\{\ |\ \}$ are rotors, and are not the same for different tables.

Alternation from $\ \text{A}\rightarrow\text{B}\rightarrow\text{B*}\rightarrow\text{B}\rightarrow\text{A}\rightarrow\text{A*}$

constitutes a machine-like activity.

And one can of course avoid A$\rightarrow$A* and simply have A going back to B, etc.

$\therefore$ We have interesting comparisons. Let's take, for example:

$$|\ r\ \ \{\ \phi,\phi\ |\ \tau_2\ \}\bullet(\ ur\ \ dr\ )\quad \pi_{B*}\ \bullet\ ur\ |$$

and
$$|\ l\ \ \{\ \phi,\phi\ |\ \tau_2\ \}\bullet(\ dl\ \ ul\ )\quad \pi_{B*}\ \bullet\ dl\ |$$

If we multiply these, since the translations (respective) are complements of each other (so that their multiplications lead to "1"), we get :

$$|\ 1\ \ \{\ \phi,\phi\ |\ \tau_2\ \}^2\bullet(\ 1\ \ 1\ )\ \ (\pi_{B*})^2\bullet 1\ |\ \equiv\ |\ 1\ \ \{\ \phi,\phi\ |\ \tau_2\ \}^2\ \ (\pi_{B*})^2\ |$$

$|\ r\ \ \{\ \phi,\phi\ |\ \tau_2\ \}\bullet(\ ur\ \ dr\ )\quad \pi_{B*}\bullet ur\ |$ is the actual process (of the machine).

$|\ l\ \ \{\ \phi,\phi\ |\ \tau_2\ \}\bullet(\ dl\ \ ul\ )\quad \pi_{B*}\bullet dl\ |$ is the assumed process (to lead to or allow A$\rightarrow$A*).

Dividing the actual by the assumed yields:

$$
\begin{array}{lllll}
|\ r & ur & dr & ur\ | \\
|\ l & dl & ul & dl\ |
\end{array}
$$

or, more symbolically:

$$\left|\begin{array}{cccc} \rightarrow & \nearrow & \searrow & \nearrow \\ \leftarrow & \swarrow & \nwarrow & \swarrow \end{array}\right| \equiv B / A$$

∴ The "machine" is clearly "B-centric" in that it approaches solutions based on conditions advantageous to the B and B* states (over A and A*). It's physical operation does not absolve (or eliminate nor dissipate) the formation of a "B-like" state or condition (of function for the combined operations to perform). In short, the pivot about "r" still leaves (one) in a B* state. But the functioning allows estimation for A→A* to be performed (as to how and perhaps how).

The multiplicative comparison seems to suggest a state B* powered:

$$\left|\ 1\ \ \{\phi,\varphi\mid\tau_2\}^2\ (\pi_{B*})^2\ \right| \equiv \left|\ 1\ \ \{\phi,\varphi\mid\tau_2\}\ \ \pi_{B*}\ \{\phi,\varphi\mid\tau_2\}\ \pi_{B*}\ \right| \quad \text{element order transformed}$$

for the actual machine state (upon pivoting on "r" and multiplying tables). This, of course, could even imply a pause once the B* state is achieved (to allow for some other manipulation):

$$\left|\ \{\phi,\varphi\mid\tau_2\}\ \ \pi_{B*}\ \ 1\ \ \{\phi,\varphi\mid\tau_2\}\ \ \pi_{B*}\ \right| \quad \text{element order adjustment}$$

Yet, A* is not actually achieved (by the "r" pivot), through the actions of this "machine."

∴ The machine is distinct from an (A→)A* functioning.

" $\left|\ \{\phi,\varphi\mid\tau_2\}\ \ \pi_{B*}\ \ 1\ \ \{\phi,\varphi\mid\tau_2\}\ \ \pi_{B*}\ \right|$ " can be interpreted as a furtherance of rotation after the initial "π_{B*}" stop state (or operation) is reached. (All of these table elements remain operations.)

Apparently, the B* state can not be escaped (successfully or *sustainily*) by this (pivoting) procedure.

Geometrically, however, we may view the "machine" as simply an angular conjunction. Taking the above example for illustration:

$$\begin{array}{ll}
\overline{\pi_{B*}} & \\
\{\phi,\varphi\mid\tau_2\} & B* \\
|\ \text{ur}\quad \text{dr}\quad \text{ur}\ \underline{\textbf{r}}\ | & \uparrow \\
& A \rightarrow B \\[1em]
/\pi_{B*} & \\
\{\phi,\varphi\mid\tau_2\} & B* \\
|\ \text{ur}\quad \text{dr}\quad \text{ur}\quad \textbf{r}\ |/ & \nwarrow \\
& A \rightarrow B \\[1em]
|\ \text{ur}\quad \text{dr}\quad \text{ur}\ |\ \textbf{r}\ |\ \{\phi,\varphi\mid\tau_2\}\ \pi_{B*}\ | & A \rightarrow B \rightarrow B*
\end{array}$$

etc.

There is no path within its mechanism to allow A→A* (definitionally). But, multiplicatively, there is the availability of an " A•B* " state (pivotal convolution). The pivoting about "r" need not be planar (to the page), but demonstrates that the "machine" (thus far) is capable of at least 4 states: A, B, B*, A•B*, during its progressions and structural logistics. The latter state (A•B*) assumes a continual replenishment of A and, therefore, a circuitry established whereby B* may convolute with A, justifying the existence of a B*→A path (possibly at or initiating from the point of that conjunction of states). This path could clearly be extra-planar (or "supra-planar") to that so far diagramed (e.g. as a "bubble" coming out of the plane of the diagram as drawn). Or, as seen from another perspective, the A•B* state may be used to "resolve" B* into (back down to) A :

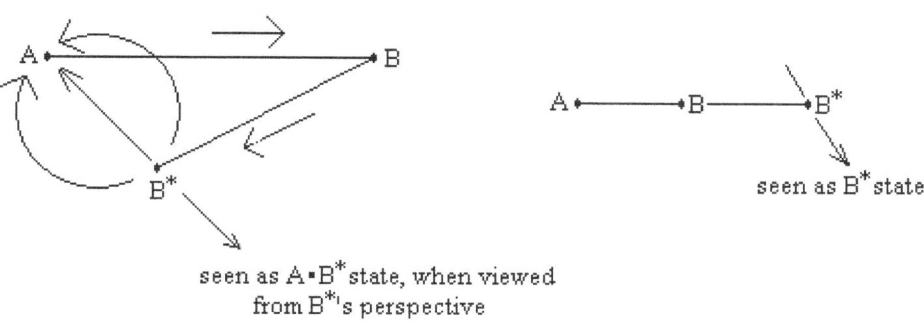

(This would make the A•B* convolution virtual.)

It is even conceivable that B* (due to "r" pivoting) may "spin" around A (in some concerted "steady-state" of functioning), until something other must occur (e.g. before reversion of B*→B→A, or before some other functionality of process extending the "machine's" operations; the elements of the tables remain physical operations). This "spin" allows for an extra-planar aspect to the process.

During the conjectured "spin" the A•B* state persists.

∴ for example:

$$\left|\ \text{ur·}\pi_{B*}\ (\text{dr}\quad\text{ur})\cdot\{\phi\,,\varphi\ \middle|\ \tau_2\}\quad\text{r}\ \right|$$

which suggests

$$A{\rightarrow}B,\ B^*{\leftarrow}B$$

makes most typical (for observance) the initiating function (or term)

$$\text{ur·}\pi_{B*}\quad(\text{as the state A•B*}).$$

We have already assumed (via as given earlier) that if

$$\{\phi\,,\varphi\ \middle|\ \tau_2\}\cdot(\text{ur}\quad\text{dr})\equiv\text{d}l\quad\text{u}l$$

then $\qquad\qquad\text{ur·}\pi_{B*}\equiv\text{d}l$

$\therefore$ We can correspondingly assume (or follow):

$$|\ \mathrm{d}l \quad \mathrm{d}l \quad \mathrm{u}l \quad l\ |$$

is equivalent to

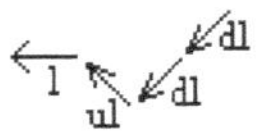

which is the contrast to

$$|\ \mathrm{ur} \quad \mathrm{ur} \quad \mathrm{dr} \quad \mathrm{r}\ |$$

$\therefore$

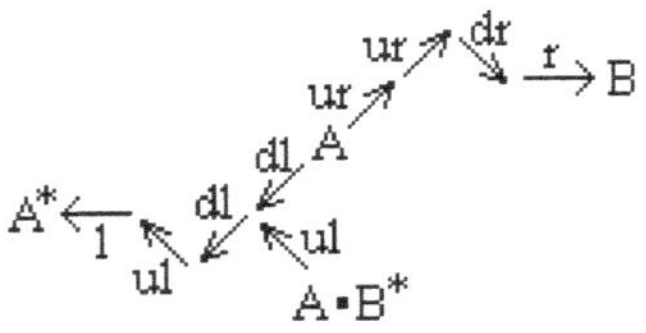

Here we assume (order independence): $(\mathrm{dr}\ \ \mathrm{ur}) \equiv (\mathrm{ur}\ \ \mathrm{dr})$

$$(\mathrm{u}l\ \ \mathrm{d}l) \equiv (\mathrm{d}l\ \ \mathrm{u}l)$$

The "spin" (and thus states A•B* and A*) are only hypothetical (to the current "machine's" instruction, construction, or process).

$\therefore$ e.g.

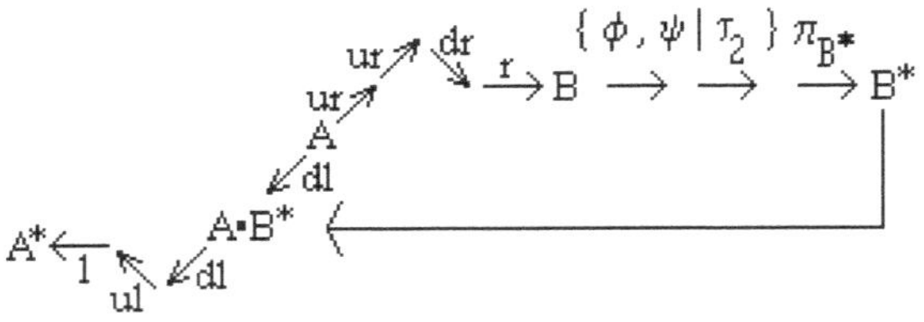

A comprehensive "spin" state

This could be interpreted as implying that when state A is seen in conjunction with state B*, a path to state A* is revealed (or formed). Composite state A•B* is a consequence of allowing a path to state A* (i.e. the composite state occurs during the pathway). We see (under this proposal) that while both B→B* and A→A* involve movement, due to the influence of B* (in the A•B* state) the A→A* path might not have to include a rotation of plane and $\therefore$ could be restricted to translations within a plane.

In essence, the "spin" absolves the necessity for a change or modification of plane, since the "spin" encompasses all planes (3-dimensional).

Yet, what would drive a "spin"? Clearly a desire or propensity to form the composite (A•B*) state applies a spin-motive (on the machine). The "machine" can only reach towards the A state (with B*). And the inducement to create A•B* leads (or opens up) a pathway to make A* (as an after product). That of course is never achieved (by this machine). But the hypothetical potential results in the spin. The conic spin, but not its extension (to A*), is a property of the machine. Reversion back to A may also be such a machine's property. This reversion has aspects which clearly mimic A→A*, after B*→B is achieved.

So, we relate how a mixed translational/rotational path (e.g. B→B*) can be made hypothetical to a translational only path (A→A*), with comparison and similarity to an actual translational only path (A→B). This then gives functional definitions to (the meaning or nature of) rotations (or changing of states without translations) as having (as usable) both translational and rotational elements: i.e. The B→B* rotations, for a change of state stationary in position, can be decomposed into translations via comparison (and consideration) of analogous (here A→A*) pathways. Although these comparisons (and adopted equivalencies) are all hypothetical, they remain functional, and can be catalogued as characteristics (for later possible use). Some definitions may be made dependent on them.

Obvious, although sometimes bizarre, branchings off can be presumed, for further advances in processing:

e.g.

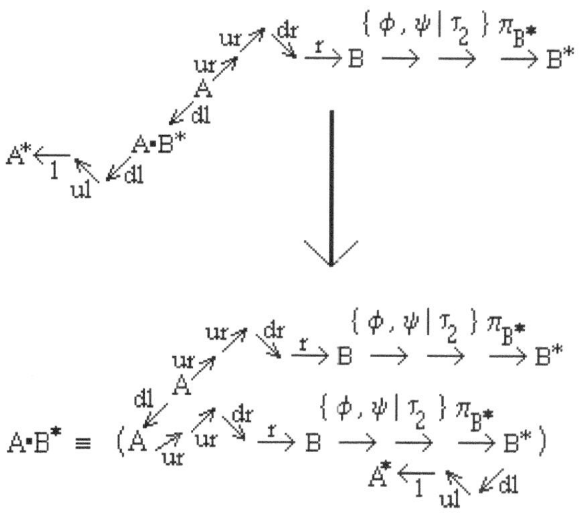

$$A \bullet B^* \equiv \left(A \xrightarrow{\ \ } \cdots \xrightarrow{r} B \xrightarrow{\ } \xrightarrow{\ } \{ \phi, \psi \,|\, \tau_2 \}\, \pi_{B^*} \xrightarrow{\ } B^* \right)$$

∴ An interesting extension from B* (to A*) could be:

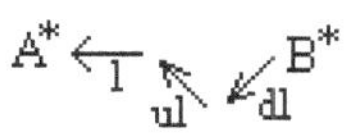

$$A^* \xleftarrow{1} \nwarrow_{ul} \swarrow_{dl} B^*$$

suggesting a change in position (i.e. translations of B*).

∴ A (closed pathway) circuit can be devised:

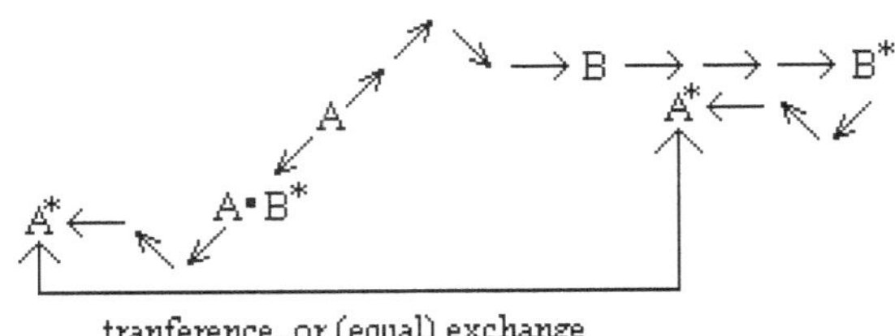

etc.
i.e.

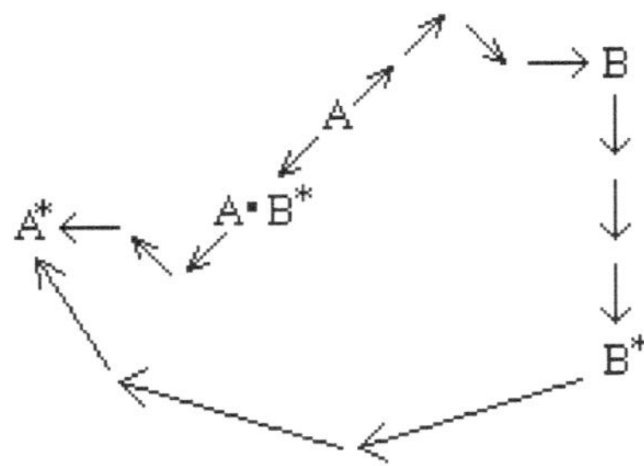

with a fulcrum on/at A* .

Or, of course, it can be open-ended, with (for either or) transference:

e.g.

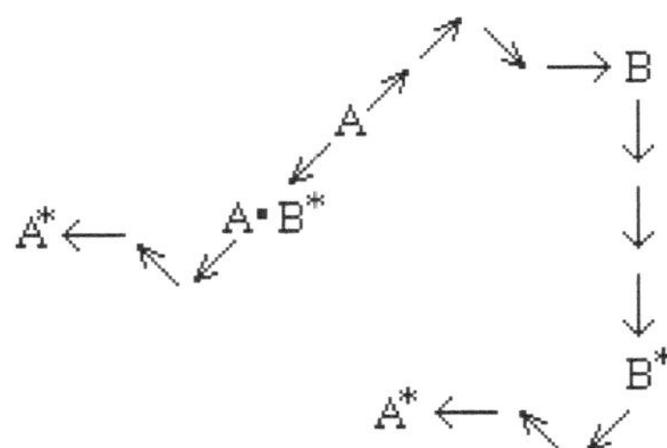

one A* being antithetical to the main route, and the other being conducive (and following).

The obvious suggestion is that the following pathway is available:

$$A \nearrow \quad \to B \to \to \to B^* \quad A^* \leftarrow \swarrow$$

$\therefore$ The circular (closed) route implied is:

$$A \nwarrow \quad \to B \to \to \to B^* \quad A^* \leftarrow \swarrow$$

with fully consistent (i.e. continuance of) direction.

This result provides that, while B→B*, rather of this (or in contrary fashion) A*→A.

$$\therefore \qquad A \to B \to B^* \to A^* \to A$$

The path from A* to A actually remains (aside from reversal) unknown, thus far.) The "spin" promotes this hypothetical fiction (or considered possibility).

If translation steps are equivalent (of distance or unitage), then the transition from A*→A (symbolized here with "u/") could be something as like:

$$A \nwarrow \quad \leftarrow \leftarrow \leftarrow A^*$$

i.e., in general, the opposite (in directions) of A→B

This confers to A* as being of a state similar to B (which is clearly implied since it derives from B*). These hypothetical logics (as opposed to proofs) are possible due to the nul/los nature of the maths. However, they are strictly from the A→B→B* (and $\therefore$ B-weighted) perspective; i.e., if B* had a way or quality to procure A* and lead the path back to A, it could do so in this way. An analogous assumption could very well be:

$$A \nearrow \quad \to A^* \to \to \to B^* \quad B \leftarrow \swarrow$$

etc.

The path to B→A could be filled in with simple reversal of the known A→B pathway.

∴

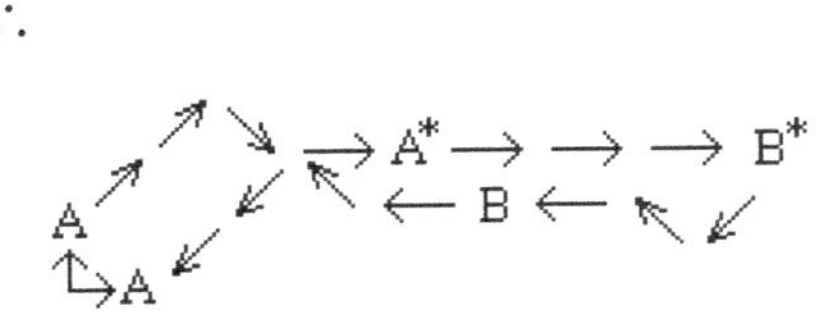

Or, if B were coincident with A* :

Then, of course, A can be used to insert this whole circuit (in another procedure): i.e. The circuit can be used as an "excrescence" off point A.

This proposes the observation that A* is an alternate state of B, and leads to the obvious questions what are the (corresponding) alternate states of A and B*. Viewed broadly, due to the excrescent nature of the (stem &) loop, one could claim that A and B* are alternate states of each other. But, under null/los, there is no "non-state" applicable to the argument.

$$\therefore \qquad A \leftrightarrow B^* \ \text{ as } \ B \leftrightarrow A^*$$

These are (intuitively) surprising developments, since one would naturally assume that A and A* are alternate states of each other, as for B and B*. But under the pathways developed here, with the logics (and logistics) chosen, A and B* are perhaps more distinct states than B and A*. Hence they may form the more valid composite argument (state A•B*), states B and A* being only transitional (from A to B*).

The "spin" is only one application of the A→B* pathway. Not only may B* provide impetus for a further process(ing), but, just as B* may spin around A, B* may justifiably spin or abut, impede or enjoin something else totally independent of "A-character." And here we definitely require a supra-planar motif or nature:

e.g.

or

etc.

The impedance (of C by B*) would be plane dependent. And, of course, various junctions could be formed:

e.g. With B* impedance, a path (from B*) directly to D could be prescribed. (Such a junction, for example, might allow C or E to abut A, or conduce past B, a transitory state.)

The nature of the excrescence stem (off A) almost looks divisional (see as earlier):

$$A \;\; \rightleftarrows \; \binom{B}{A^*} \; \rightleftarrows \;\; \rightarrow \;\; \rightarrow B^*$$

$$\sim \; (B/A)$$

Complementation

But the operations (elements) are merely complementary of reversal. Yet must, in this sense, the loop structure be complementary (but here without reversal):
e.g.

$$\binom{B}{A^*} \; \xrightarrow{\{\phi,\,\psi\,|\,\tau_2\}} \; \xrightarrow{\pi_{B^*}} \; \xrightarrow{B^*} \;\;\; \xleftarrow{1} \;\; \nwarrow_{\mathrm{u}l} \;\; \swarrow_{\mathrm{d}l}$$

$\therefore$ We have $|\; l \;\; \mathrm{u}l \;\; \mathrm{d}l \;|$ complementary to a set of procedures:

$$| \; \{ \phi \;|\; \tau_1, \tau_2, \tau_3 \} \qquad \pi_{B^*} \; |$$
$$| \; \{ \varphi \;|\; \tau_2, \tau_3, \tau_4 \} \qquad \pi_{B^*} \; |$$
$$| \; \{ \phi, \varphi \;|\; \tau_2 \} \qquad\quad \pi_{B^*} \; |$$
$$| \; \{ \theta \;|\; \tau_1, \tau_2, \tau_3, \tau_4 \} \; \pi_{B^*} \; |$$

$\therefore$ The $|\; l \;\; \mathrm{u}l \;\; \mathrm{d}l \;|$ procedure provides a special function of complementarity (in this case).

Here we find the interesting nature (or quality) of complementation, for through both multiplication or division of elements the result is "1" (as if each element itself were "1").

$\therefore$ e.g. $\qquad | \; \{ \phi, \varphi \;|\; \tau_2 \} \;\; \pi_{B^*} \; | \; \bullet \; | \; l \;\; \mathrm{u}l \;\; \mathrm{d}l \; | \;\equiv\; | \; 1 \;\; 1 \;\; 1 \; |$

and $\qquad \dfrac{| \; \{ \phi, \varphi \;|\; \tau_2 \} \;\; \pi_{B^*} \; |}{| \; l \;\; \mathrm{u}l \;\; \mathrm{d}l \; |} \;\equiv\; | \; 1 \;\; 1 \;\; 1 \; |$

Note: The stem (off A) has complementarity, but not a special function of complementarity; rather, (it has) complementations specific to the elements for direct reversals.

145

∴ We can symbolize this special function of complementarity as:

Compl

and make statements such as:

$$(B \rightarrow B^*) \, \text{Compl} \equiv | \; 1 \quad 1 \quad 1 \; |$$

or

$$| \; \{ \phi , \varphi \; | \; \tau_2 \} \quad \pi_{B^*} \; | \, \text{Compl} \equiv | \; 1 \quad 1 \quad 1 \; |$$

etc., which can be taken as definitions (of functions).

This, of course, allows for the curious result that:

$$\text{Compl} \equiv | \; 1 \quad 1 \quad 1 \; | \; / \; | \; \{ \phi , \varphi \; | \; \tau_2 \} \quad \pi_{B^*} \; | \equiv | \; \{ \phi , \varphi \; | \; \tau_2 \} \quad \pi_{B^*} \; | \; / \; | \; 1 \quad 1 \quad 1 \; |$$

(which is only true when Compl $\equiv | \; l \quad ul \quad dl \; |$).

∴ Methods for determining Compl may be devised (based on or via particular pathways). The symbol may be shortened (as a procedure and ∴ a function) as thus: Compl → Cpl

∴ $\qquad (B \rightarrow B^*) \, \text{Cpl} \equiv | \; 1 \quad 1 \quad 1 \; |$ etc.

The symbol may even be condensed (esthetically) to:

$$\text{Cpl} \equiv \textit{Cpl} \; \text{ or } \; \textit{Cpl}$$

to give it (as a function) a more numerate flavor or quality of design. It is not, however, an operator, rather a stand-in for a procedure or function.

∴ The extrusion (off A) can be characterized as:

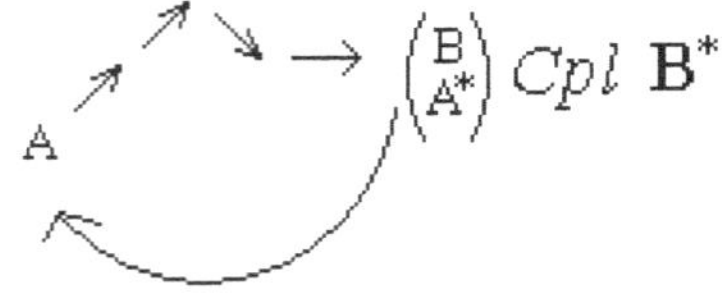

for symbolic purposes (as opposed to informative details).

This creates (a reason for) the interesting supposition for the circular closed loop (or route) given previously:

where, rather than direct (or straight) reversal, the path from A* (or $\left(\begin{smallmatrix} B \\ A* \end{smallmatrix} \right)$) to A may also employ a function of complementarity (to be found).

This route is planar. Yet, an obvious choice would be "*l*" followed by $(\phi , \varphi , \theta \; | \; \tau_s)$ and finally "π_A" for a stop.

$\therefore \qquad A \; Cpl \begin{pmatrix} B \\ A^* \end{pmatrix} Cpl \; B^*$

e.g. $\qquad | \; \pi_A \; \{ \phi , \varphi \mid \tau_2 \} \; l \; | \; \equiv \; A \leftarrow \begin{pmatrix} B \\ A^* \end{pmatrix}$

and $\qquad | \; \pi_A \; \{ \phi , \varphi \mid \tau_2 \} \; | \; l \; | \; \{ \phi , \varphi \mid \tau_2 \} \; \pi_{B^*} \; | \; \equiv \; A \leftarrow \begin{pmatrix} B \\ A^* \end{pmatrix} \rightarrow B^*$

We note that this is a bi-directional expression, and that the "*l*" really stands for $\begin{pmatrix} r \\ l \end{pmatrix}$ (in the preceding tabular statement of overlap).

$\therefore$ The generalized statement (procedure of operations) is:

$$| \; \pi_A \; \{ \phi , \varphi \mid \tau_2 \} \; | \; \begin{pmatrix} r \\ l \end{pmatrix} \; | \; \{ \phi , \varphi \mid \tau_2 \} \; \pi_{B^*} \; |$$

This could be made more distinctive (easier to read) as:

$$| \; \pi_A \; \{ \phi , \varphi \mid \tau_2 \} \; [\; \begin{pmatrix} r \\ l \end{pmatrix} \;] \; \{ \phi , \varphi \mid \tau_2 \} \; \pi_{B^*} \; |$$

or

$$| \; \pi_A \; \{ \phi , \varphi \mid \tau_2 \} \; [\; \overset{r}{l} \;] \; \{ \phi , \varphi \mid \tau_2 \} \; \pi_{B^*} \; |$$

i.e. [ ] is shared by both tables (tabular forms).

This could be further delineated as (something like):

$$| \; \pi_A \; \{ \phi , \varphi \mid \tau_2 \} \; [\; \overset{\vec{r}}{\underset{\leftarrow}{l}} \;] \; \{ \phi , \varphi \mid \tau_2 \} \; \pi_{B^*} \; |$$

to more deliberately show directionalities.

But we can also presuppose arbitrary directionality, with:

$$| \; \pi_A \; \{ \phi , \varphi \mid \tau_2 \} \; [\; r , l \;] \; \{ \phi , \varphi \mid \tau_2 \} \; \pi_{B^*} \; |$$

which would be the more generalized (operational) expression without knowing (or predicting) of pathways. (i.e. We would not impose which direction uses "r" and which "*l*".)

$\therefore$ We have the operational statement:

$$| \; \pi_A \; \{ \phi , \varphi \mid \tau_2 \} \; [\; r , l \;] \; \{ \phi , \varphi \mid \tau_2 \} \; \pi_{B^*} \; |$$

and the corresponding functional statements:

$$A \leftarrow \begin{pmatrix} B \\ A^* \end{pmatrix} \rightarrow B^* \qquad A \; Cpl \begin{pmatrix} B \\ A^* \end{pmatrix} Cpl \; B^* \qquad \text{(both expressions of states)}$$

Note: with knowledge of the (translational) pathways, we can also have the operational statement:

$$\mid \ ur \quad ur \quad dr \quad [\,r\,,l\,] \quad l \quad ul \quad dl \ \mid$$

as well as:

$$\mid \ dl \quad dl \quad ul \quad [\,r\,,l\,] \quad l \quad ul \quad dl \ \mid$$

But these details of specifics (of translation) seem less informative than the more generalized statement for the nature (or functions) of the given procedures.

The operational statement is clearly an example of an expression of reciprocity, as to what is to be expected of a process using the same or similar elements of operation. This would be how one would initially formulate a problem without knowing the specific details of the encumbered pathways (here translational), but rather through the effects desired (here plane rotation) to be demonstrated or achieved of some successful solutions.

$\therefore$ The statement is one to be realized (as well as studied as to how to accomplish).

The partial symmetricallity of the (most) generalized operational statement favors towards correct elucidation of the pathways constituent (or involved). In this case, however, we note that the (chosen, for example) paths led to the construction of the statement (through the logics applied). It's clear that we may "disjoint" the statements at any time (for individual considerations, or analyses):

e.g.

$$\mid \ \pi_A \ \{\,\phi\,,\varphi\,\mid\,\tau_2\,\} \quad [\,r\,,\,l\,] \quad \{\,\phi\,,\varphi\,\mid\,\tau_2\,\} \ \pi_{B*} \ \mid$$

$\rightarrow$
$$\mid \ \pi_A \ \{\,\phi\,,\varphi\,\mid\,\tau_2\,\} \quad r \ \mid \ , \ \mid \ l \ \{\,\phi\,,\varphi\,\mid\,\tau_2\,\} \ \pi_{B*} \ \mid$$

and
$$\mid \ \pi_A \ \{\,\phi\,,\varphi\,\mid\,\tau_2\,\} \quad l \ \mid \ , \ \mid \ r \ \{\,\phi\,,\varphi\,\mid\,\tau_2\,\} \ \pi_{B*} \ \mid$$

And for the specified paths:

$$\mid \ ur \quad ur \quad dr \quad [\,r\,,l\,] \quad l \quad ul \quad dl \ \mid$$

$\rightarrow$
$$\mid \ ur \quad ur \quad dr \quad r \ \mid \ , \ \mid \ l \quad l \quad ul \quad dl \ \mid$$

and
$$\mid \ ur \quad ur \quad dr \quad l \ \mid \ , \ \mid \ r \quad l \quad ul \quad dl \ \mid \quad \text{etc.}$$

The ambiguity, provided by the bracket operations $[\,r\,,l\,]$, allows for modification of pathways to be considered (or even conjectured).

$\therefore$ We have:
(originally hypothesized or designed), and (now also, for consideration):

If we add (combine) the first two, we get:

For the bottom two, the combination is:

But, we can also have (through the permutations):

and

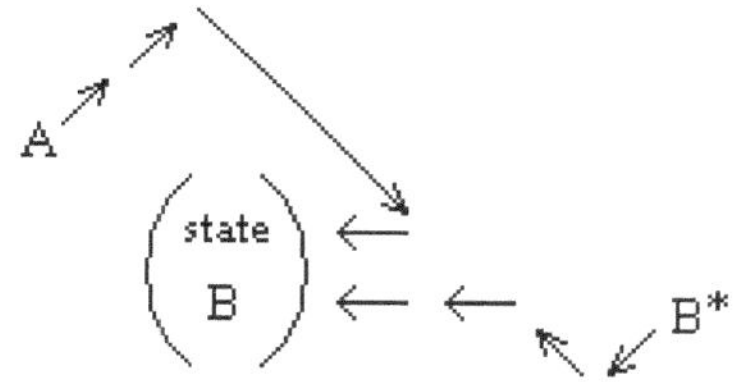

Combining redundancies for the last pathway yields:

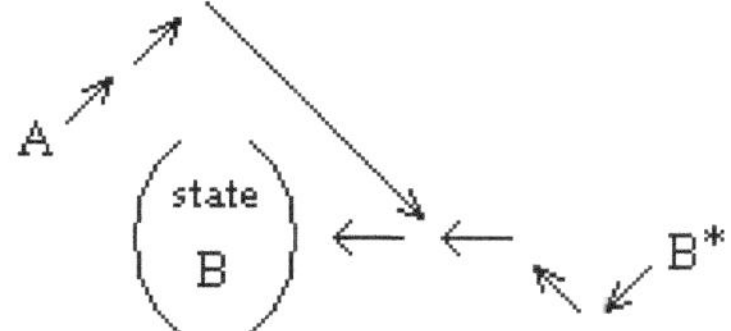

with an apparent fork (or bifurcation).

Diagramed more spatially (generalized), this could be:

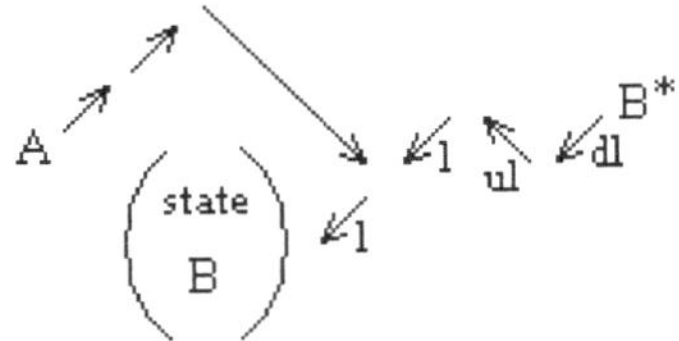

(although specific step natures or identifications should be retained and not modified).

All (or each) of these configurations might be possible.

The

pathway can likewise be reduced (apparently, by noting complementary translations) to:

with pathway steps, to:

Operationally, or procedurally, this appears more efficient than the first combination (from original hypotheses) drawn.

We see through these examples that generalizations through (or elucidations of) pathways can lead not only to changes in processes but also modifications of the nature, the number of operational steps required for a result, through such utilities as reduction via complementary notice, step conjunctioning or (simultaneity) compounding, forking, ambiguities of state, and step "jumping" upwards or downwards:

e.g.

∴ A ↔ B* converting from 6 steps to 4;
or more correctly: A → state ← B* .

Note that the reductions (condensations), as well as the combinations, only imply completed pathways in either direction. But we could not so reduce with the implied pathway for " state → B* " (via direct reversal of steps), unless we were also to consider " A ← state " as similarly. ∴ No state is left un-accoutremented or nulled.

Note: The corresponding statement " A → state → B* " would produce an interesting (though only implied) forked pathway:

e.g.

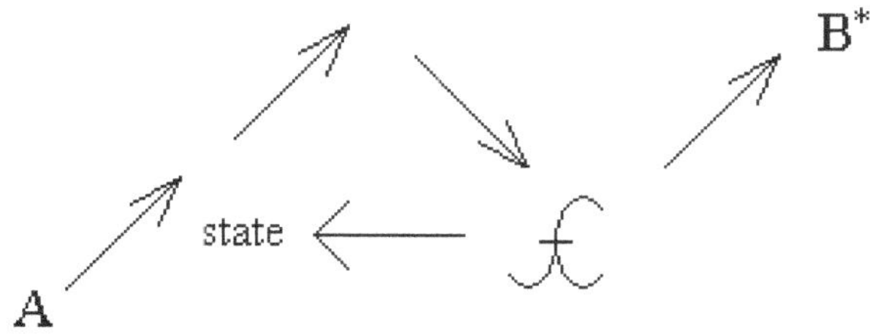

with hypothetical choices of 4 steps each (to lead respectively to "state" or "B*".

The point of forking may itself be considered a state, say: symbolized with

" ℒ "

(where the fork allows for a selection of paths):

Thus, the relationships between the 3 states (state, *above character symbol*, B*) may be pursued.
Where the fork does not (particularly) allow for a selection of (directional, or here translational) paths, the fork state might be symbolized by:

" ℒ "

e.g.

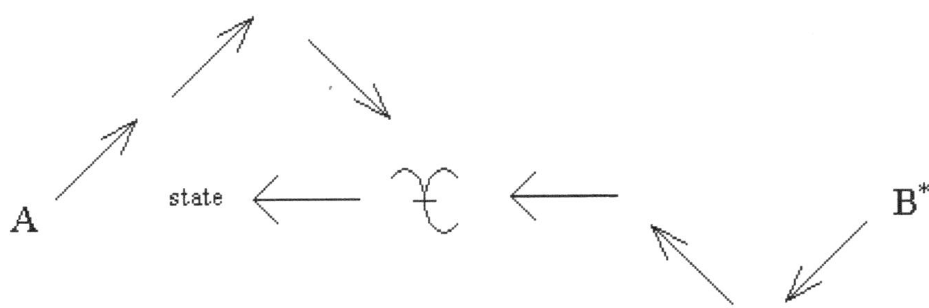

The combination (with step redundancy reduction) of the two above forked paths would of course be (as diagramed):

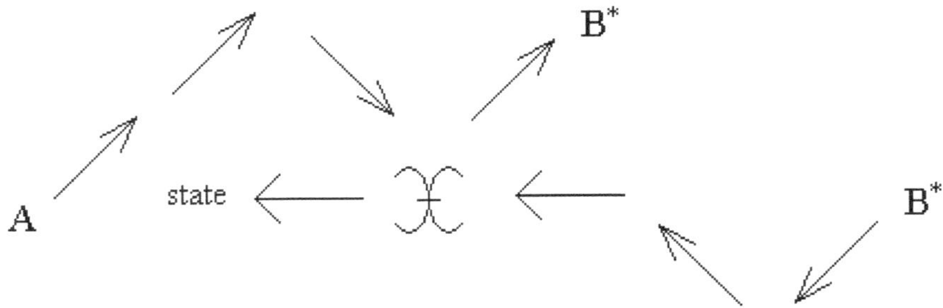

This poses an interesting (though conditional) cycling of state "B*":

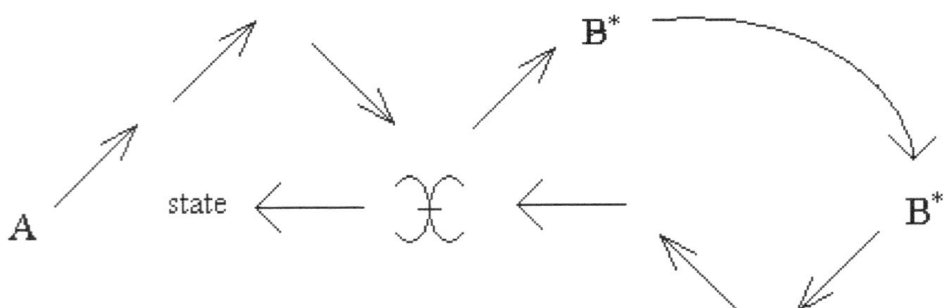

∴ This may be drawn as:

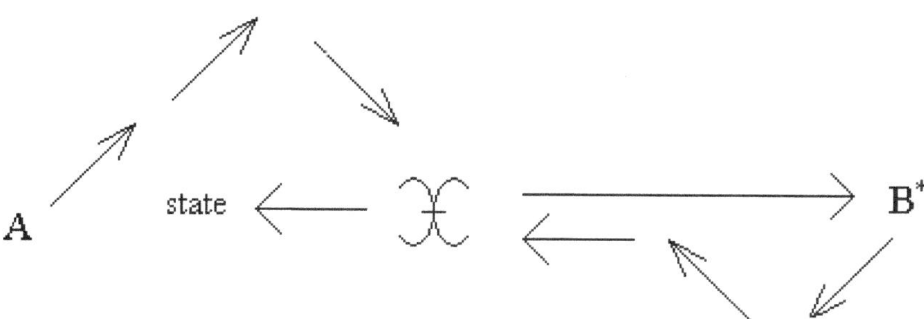

or more properly:

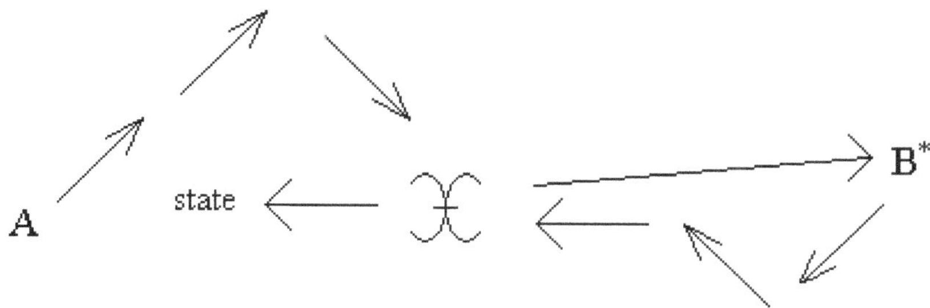

Here the pathways are concerted (can be followed with a line) without necessarily being continuous (due to the forking state):

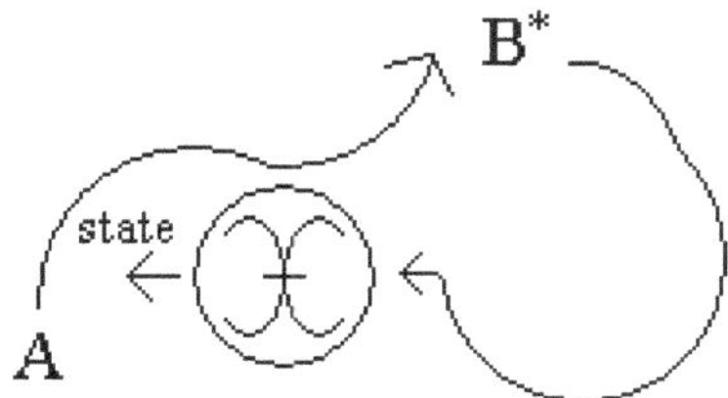

∴ Combining two un-concerted pathways (opposing arrow directions) can lead to a composite concerted one:

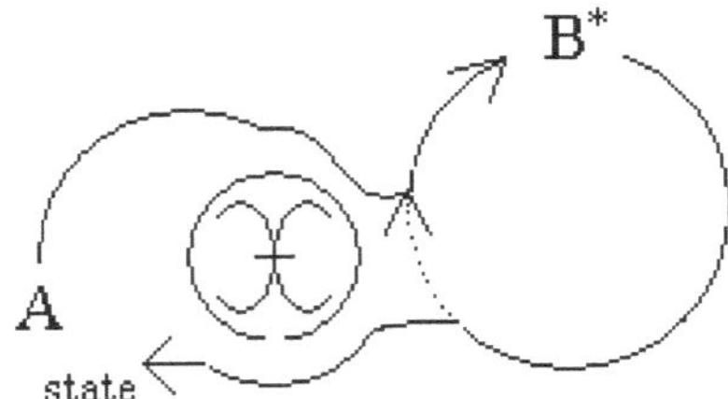

with an implied possible (but smooth curved) cycling (here of "B*").

We note then that, in this case, "state" acts as an endpoint of the process. "A" is the starting point, and "B*" is intermediary and transient. Note how we have effectively obscured "B" due to the forking state, which acts as a cycling or conditional station for the procedure.

∴ Nul/los analyses allow for examination of pathways (of procedure) with considerable manipulability and flexibility (albeit conceptual, from given starting premises).

In this way pathways may be conceptually designed without specifics of steps, which are to be determined or designed into possible tabular arrays (constructions) for more (process specific) crucial analyses.

e.g. Given the design:

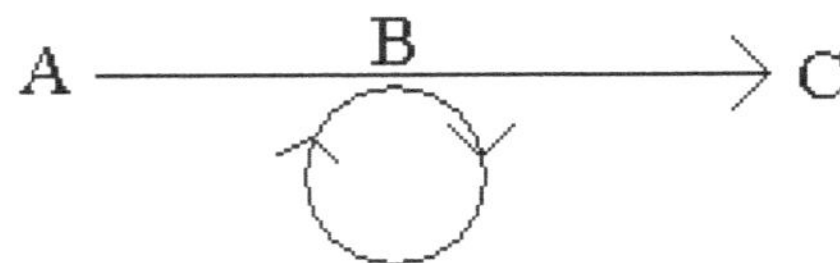

point (or state) B might be deemed a fork or transition (along the route to C) necessitating some sort of conversion of property or quality to the operation(s). As a fork, choices are allowed (in direction). As a transition, the paths are committed (to follow the loop).

∴   Fork   : $A \to B \to C$, $A \to B \to$ loop $(\to C)$
Transition : $A \to B \to$ loop $\to C$

We desire that (for simplicity) the steps for $A \to B$ are the same in number as for $B \to C$, while those of the loop are independent in number.

∴ say

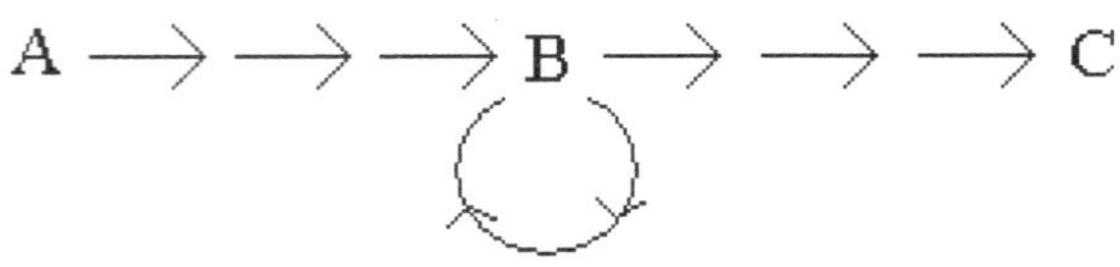

As in our earlier analyses (of example) we can consider state B to be due to an overlapping step.

$$\therefore \quad A \mid \overset{1\ 2}{\rightarrow\rightarrow} \mid \overset{3}{\rightarrow} \mid B \overset{4\ 5\ 6}{\rightarrow\rightarrow\rightarrow} \mid C \qquad \text{steps numbered}$$

$$\text{step overlap}$$

$$\therefore \qquad \mid 1\ 2\ 3 \mid , \mid 3\ 4\ 5\ 6 \mid \equiv \mid 1\ 2\ [3]\ 4\ 5\ 6 \mid$$

But for simplicity of symmetrization, we might also consider step 4 as an overlap:

$$\therefore \qquad \mid 1\ 2\ [3\ 4]\ 5\ 6 \mid \equiv \mid 1\ 2 \mid 3\ 4 \mid 5\ 6 \mid$$

The state of B occurs after step 3 and before step 4. $\therefore$ This overlap is within the domain of the loop:

$$\mid 1\ 2\ [3 \updownarrow 4]\ 5\ 6$$

or

$$\mid 1 \rightarrow 2 \Rightarrow 3 \overset{B}{\searrow\nearrow} 4 \Rightarrow 5 \rightarrow 6 \mid$$

$$A \qquad\qquad\qquad\qquad\qquad\qquad C$$

So we have developed a sort of template for the proposed process. We might "number" the loop steps from a......e , with c being of undetermined (number or manner of) character (as of yet) for operational steps. Note that (due to nul/los, reduction leading to "1") b and d are complementary, and could be reduced if desirable:

$$\mid 1 \rightarrow 2 \Rightarrow 3 \overset{e}{\searrow\nearrow} 4 \Rightarrow 5 \rightarrow 6 \mid$$

$$A \qquad\qquad\qquad\qquad\qquad\qquad C$$

$$\longrightarrow$$

$$\mid 1 \rightarrow 2 \Rightarrow 3 \overset{a}{\underset{e}{\searrow\nearrow}} 4 \Rightarrow 5 \rightarrow 6 \mid$$

$$A \qquad\qquad\qquad\qquad\qquad\qquad C$$

Steps a and e are not complementary, since both are moving towards the right (to state c); but they are contradictory of motion (direction): e directly going towards c (as "out" of the loop) and a entering the loop to lead towards e (and ∴ c). The numbers represent operations, the letters steps.

Of course, reciprocalities may occur:

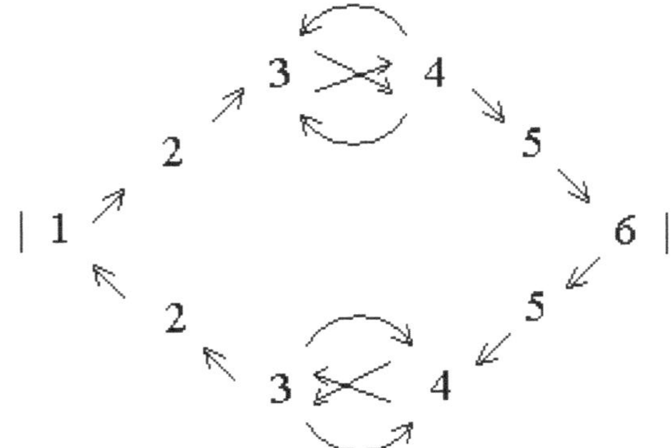

Again, c and f are not complementary.
Many such elaborations are possible.

Indeed, the "table" can be made circular (or elliptical) to focus on (reversals for) complements:

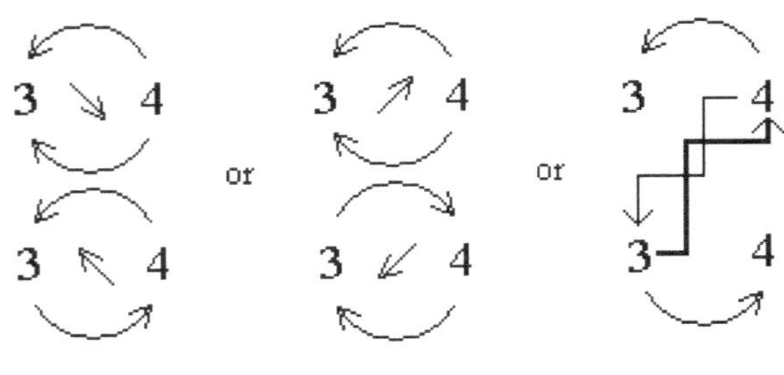

Note (for the circular table) how the (shared) forked or cross-over regions can be selectively attenuated:

thereby delimiting possibilities (choices).

These possibilities emphasize the directionality of the overall pathway (A→C). The "circle" runs clockwise.

Chosen complementations may be made (enlisted) and examined for reduction (of the pathway or circuit):

e.g.

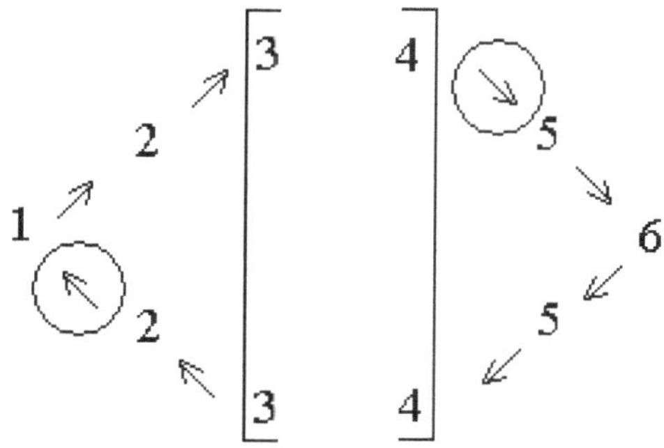

transformed to

All steps (1.....6) may still be experienced through the circuit (i.e., state 5 simply comes after state 6).

Because of the circularization, truncated routes (with missing states) are possible. And indeed, the shared two [3 4] states presumably can be combined into one, to promote aspects of this:

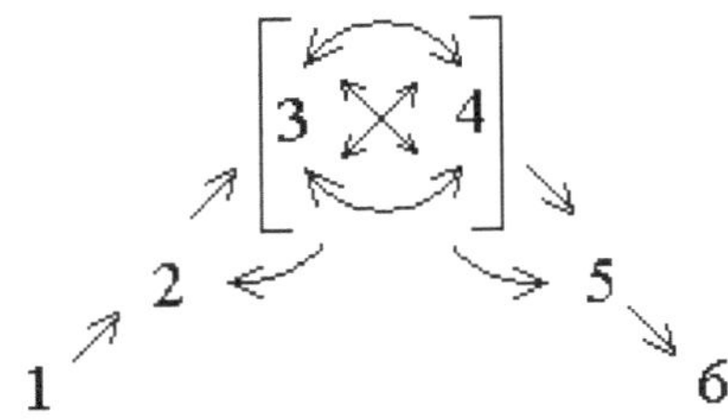

e.g.

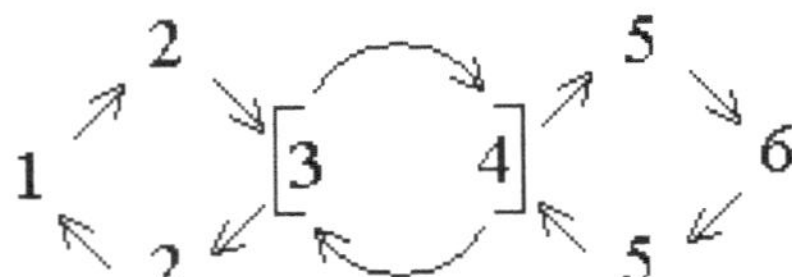

suggesting (1,2,3) and (4,5,6)
loops of pathway

also:

and its selective
complements

Note that the symmetries remain directional (i.e. clockwise).

The streamlined version (highlighting selective complementations) is roughly equivalent to a process of multiple looping, or "bubble" formation, within the tabular diagrams:

providing (1,2), (2,3), (4,5), (5,6) "bubbles' during the process.

∴ Structural compartmentalizations of operations may be constructed, by a template that can be more and more or less complex as desired.

But, to wit, the simple (as relative) construction

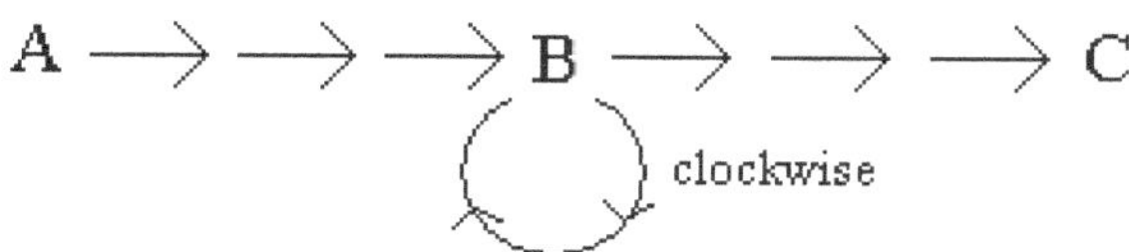

is elaborated by functional possibilities of nul/los tabular logics:

where each compartment may be studied (or considered) separately: (1,2), (2,3), (3, 4), (4,5), (5,6) and perhaps even exchanged or altered of effect (with changes of state):

e.g.

$$A\ 1\ \begin{array}{c}5\\[6\ C\ 4]\\5\end{array}\ \begin{array}{c}2\\3\ B\\2\end{array}$$ **exchange of (5,6) and (2,3)**

Since numbers here represent states, we may re-annotate the compartments as (for $1\equiv A$, $3\equiv B$, $6\equiv C$):
 (A, 2), (2, B), (B, 4), (4, 5), (5, C) and, after the above exchange:
 (A, 5), (5, C), (C, 4), (4, 2), (2, B) .

  But we have been pursuing revelation of routes, rather than (physical) behaviors. The cause for an exchange, for example, might be a revelation of behavior. In the case depicted here it is certainly allowed (functionally) by the similarity of compartments involved:

         2 ↘   5 ↘
         ↕ 3   looks much like   ↕ 6   in shape of route or pathway.
         2 ↙   5 ↙

            ↗ 2   ↗ 5
Similarly,   1 ↕   looks like   4 ↕ .
            ↖ 2   ↖ 5

And, after the first exchange (of compartments), [6 4] is forced to favor [3 4] .

  But note, after the (first) exchange, all compartments are modified.

  ∴ A behavior (to change all compartments) is that similar(ly path shaped) compartments can exchange with each other.

These are examples of commensurate behaviors. A disparate behavior would be, for example, (1,2) exchanging with (5,6):

i.e.

From

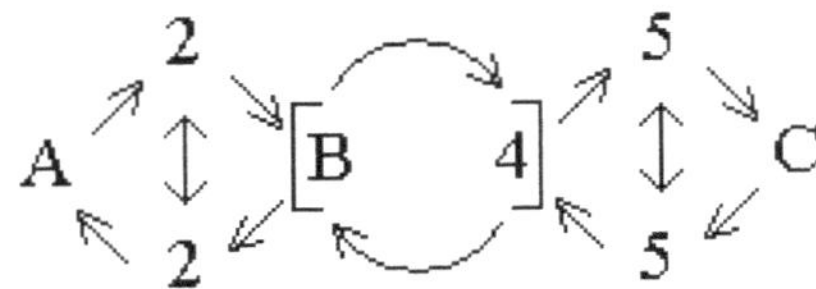

to

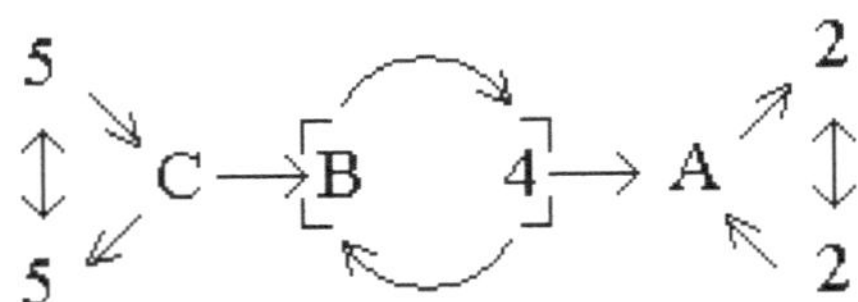

making both compartments committed, or, both (C→B), (4→A) reversible.

(5, C) is committed to its compartment (triangle) unless C→B is chosen (to further the process).
(A, 2) is committed to its compartment unless (4→A) is made reversible (presumably allowing C→B to be also).

We note, in this example, that a commensurate behavior (exchange) alters the shared region, while a disparate behavior does not; although it can be made to: e.g. (2, B) for (4, 5). Likewise, the commensurate exchange (A, 2) for (4, 5) would also alter the shared region.

∴ Both types of behaviors can be found to alter or not alter a shared region. Its conservation is choosible (designable).

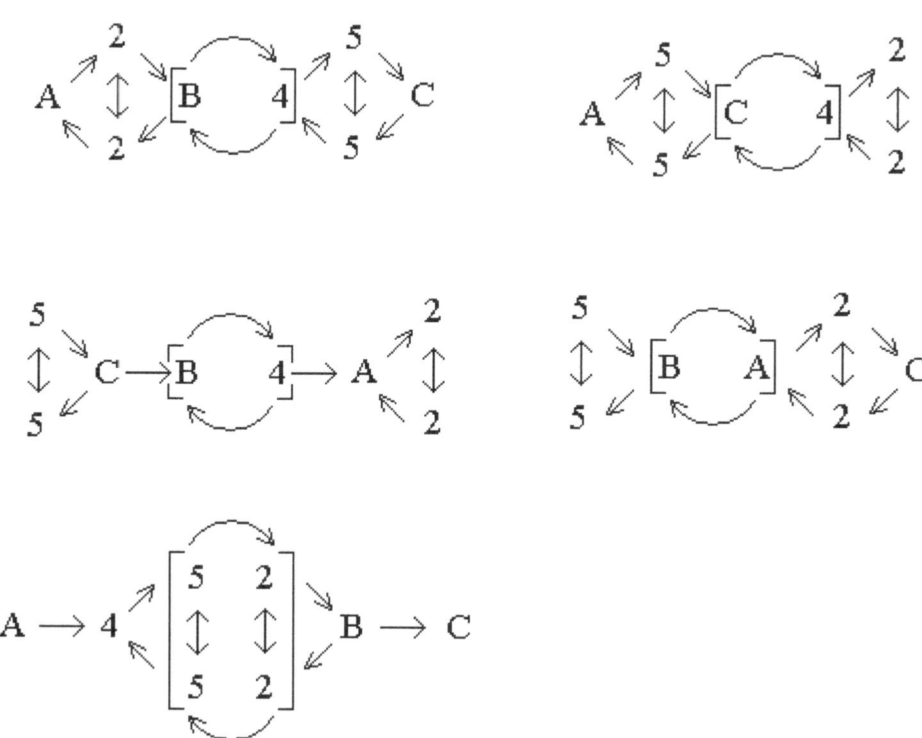

The shared region here might be reinterpreted of arrows as:

$$\begin{bmatrix} 5 & 2 \\ \uparrow & \downarrow \\ 5 & 2 \end{bmatrix}$$

This last pathway drawn seems to be an even more direct route (for A→B→C) than the originating one, since the (A, B, C) states are not part of a shared region.

These conjectures of order manipulation are subject to dependencies of the process. But how, then, can we compare the efficiencies of say:

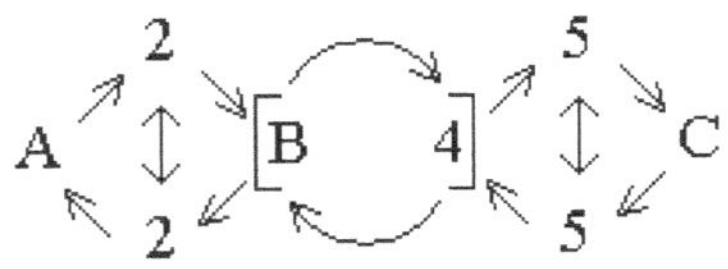

$$A \; \substack{\nearrow \\ \updownarrow \\ \nwarrow} \; \substack{2 \\ \\ 2} \; \substack{\searrow \\ \swarrow} \; [B] \; 4 \; \substack{\nearrow \\ \nwarrow} \; \substack{5 \\ \updownarrow \\ 5} \; \substack{\searrow \\ \swarrow} \; C$$

to

$$A \to 4 \; \substack{\nearrow \\ \nwarrow} \; \begin{bmatrix} 5 & 2 \\ \updownarrow & \updownarrow \\ 5 & 2 \end{bmatrix} \; \substack{\searrow \\ \swarrow} \; B \to C \qquad \qquad ?$$

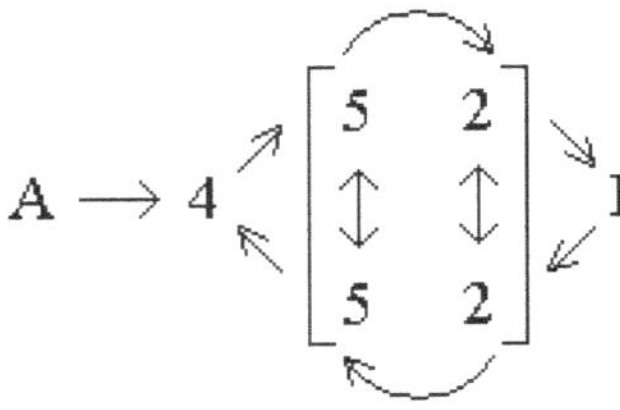

$$A{\to}2{\to}B{\to}4{\to}5{\to}C \quad , \quad A{\to}4{\to}5{\to}2{\to}B{\to}C$$

There are the same number of progressive steps (for both). $\therefore$ Observing external to the process, they would look the same (i.e. A$\to$C). Discounting regions as being shared, they are equivalent of efficiency. Since state B is un-shared and occurs after the shared region, it is more "composed" (or invariant) in the 2nd pathway. This poses the interesting concept that the intricate details of a procedure may be shielded from its results, as regards to measurements of a procedure. $\therefore$ (Accurate) measurements of a procedure might not be (or provide) proof of its details.

Physical analogies

Thus, for example, we know that the charges of electrons of a neutral atom balances out the equal number of protons and their charges in the atom's nucleus. Yet, why is the electron so much slighter in mass than the proton? What, then, is the actual nature of (the concept of) mass? when a proton of considerable mass can have an exactly opposite charge carried by an electron barely of mass. Does a strong force constraining (or compelling) the proton to the nucleus contribute to its larger mass (than an electron)? And yet, why do (nucleus) freed protons continue to exist of considerable and regular (as measured) mass compared to (freed) electrons? The measurements (of these masses and charges) do not constitute proof of the actual atomic nature (in formation of an atom or regarding of its rested state).

If (A$\to$C) "performs" an atom, and (C$\to$A) "de-performs" the atom, the steps constituent remain conjecturable of many logics and manners of operation (of steps and the states resulting). Can "positive" electrons of protonic mass, and "negative" protons of electronic mass, be hypothesized (or searched for), i.e. via momentum shifts observable during atomic disturbances, for example? This with compelling of the notion that these quantities can not be nullified using positive electrons (of electronic mass) and negative protons (of protonic mass) to balance equations of state. For example: taking account of mass loss due to (nucleon) binding energies.

Say, state A is an un-associated assemblage of atom constituents, state B accomplishes a nucleus, and state C accomplishes a (neutral) atom. The initiating atom constituents (electrons, protons, neutrons) remain, more or less uncharged, for state C (as from state A). The differences for the procedures given, then concern the complexities of nucleus formation (more substantial for the 2nd pathway) and of electron orbital creation (more sophisticated for the 1st pathway).

 Thus may such pathways be applied to practical questions: the utilization (or placement) of electrons occurring with a similar difficulty of motive as for the concentration of nucleons in the nucleus using the 1st pathway, while much greater difficulty of nucleosynthesis followed by easier (more immediate) adoption of electron cloud orbitals using the 2nd pathway. This is because a cyclic elongation (or extension) of duration during nucleosynthesis is allowed with the 2nd pathway, while with the 1st pathway this cyclicality of duration occurs after (or between) electron orbital formation and nucleus packing. Which method (of procedure) is more physically correct? Can the meanings of the B and C states be exchanged (so that B allows for nucleus formation, and C for electron placement)? It seems (as an hypothesis) more reasonable for a nucleus to form before electron orbitals (using mass arguments and therefore momentum and thus energy arguments or reasons). As for an ion to be satisfied, this is easier to do electronically (i.e. easier to adjust electron number than nucleon number, the nucleons being subjected to the strong nuclear force).

To be more clear (or distinct), we say the (2,B) compartment establishes the electron orbitals, the (4,5) the nucleus. The shared region (of either pathway) contributes to both (time or physics necessary for the formations). But in pathway 1 the e- orbitals precede nucleus formation, while with pathway 2 nucleus formation precedes. (2,B) would seem to present more elementary (i.e. simpler) physics than the question proposed by (4,5): Why do nucleons condense (to form a nucleus)? Of course, the various other pathways devised might represent different stages of atom formation or disruption (disintegration) or modification:

 e.g. The

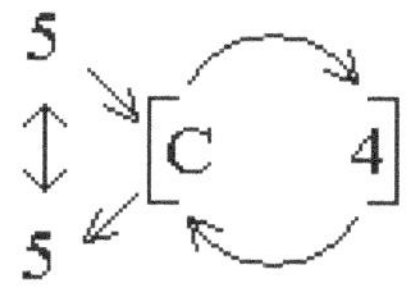

component might signify nucleon ejection, which could then be followed by readjustment of electron orbitals:

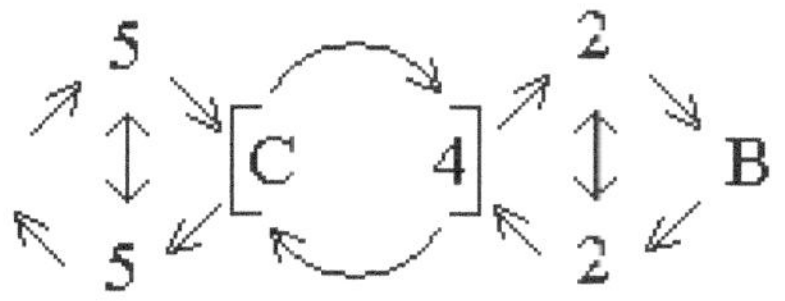

or nucleus disruption concomitant with adjustment of e- orbitals, followed by nucleon ejection (or modification):

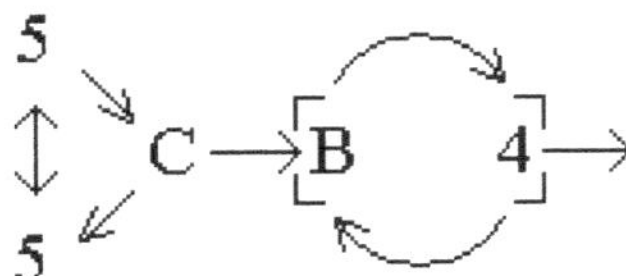

or nucleus establishment followed by orbital reformations and dissipations (before some structural stability for the atom is achieved):

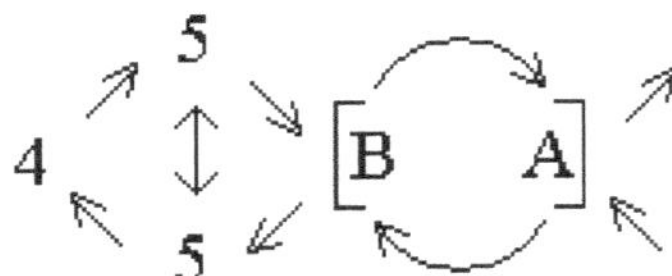

Each of these possibilities simply follow nul/los logics.

  As well, of course, they may be combined into interesting "complications" (or complexes):
e.g.

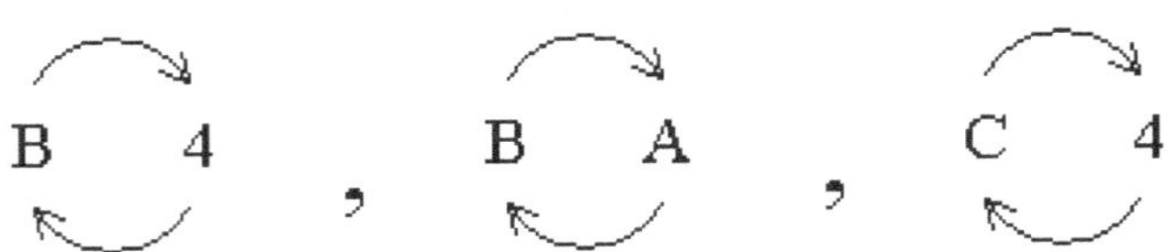

Here we see a logic of following (the) arrows for change:

e.g. 4↗5, or for stationary pause or transferal:

e.g. 4→4 (i.e. ≡ "4")

as well as directional transfers:

e.g.   5   2

             ↕ , ↕

              5   2

and circularizations:

e.g.

These arrow attributions (applying directional attributes to operations) allow for a certain amount of (recursion) reiteration of design. Indeed, the complex just stated earlier almost seems like an enhanced rejuvenation of its (components) end (in the perhaps exaggerated sense that it starts with 5↕5 and ends with B). If a process can be so sustained in length (of time and development or elaboration), then it must be directionally inspired (or led).

Of course, the complex (as projected) is clearly an extension of the original procedure. With (the) acceptable consolidation (i.e. $4{\to}4 \equiv 4$), it can be depicted as:

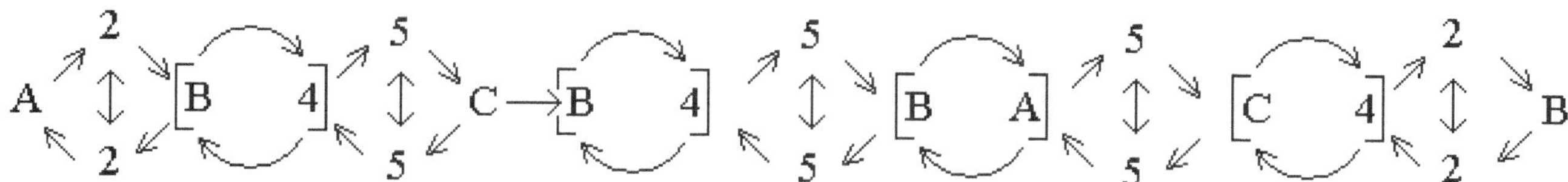

$\therefore$ An atom forms and undergoes various transformations (of stability), these configurations predictable under nul/los. It is easy to see (for) further extension, using the original procedure as a model. But as stated thus far, there is an interesting (pseudo) symmetry of : $(2\updownarrow2)(5\updownarrow5)(5\updownarrow5)(2\updownarrow2)$, depicting exchanges of equivalent states as a procedural (and progressive) norm. This suggests the nature of a physics as to rest*ing* (of action) during such exchanges. The complex may be simplified (of diagram) as:

$$A \to B\ 4 \to 5 \to C \to B\ 4 \to 5 \to B\ A \to 5 \to C\ 4 \to 2 \to B$$

with the return route as implied similarly, although this simplification is not as informative:

e.g. "5$\to$C" does not distinguish the nature of 5 (as 5$\updownarrow$5) from that of "C" (as C).

Compounding procedures are therefore conceivable (and designable). We are assuming, of course, that these such behaviors are universal of physics. And they allow for manipulations such as 'exclusionary excision':

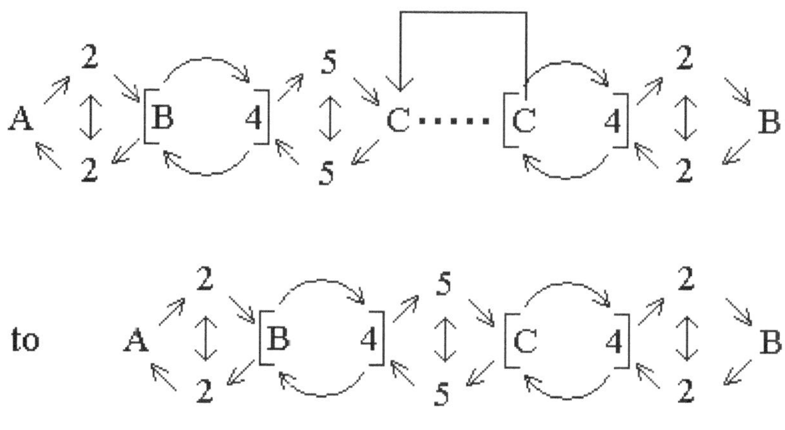

(an effective reversion)

Due to the nature of the excision, the effect is merely an attenuation of process for the same (final) result.

$\therefore$ A process (of complexity) may be simplified towards equivalent effect. All of these natures are based on the incapacity of nullification(s). Of course, the opposite process ('exclusionary insertion') is also possible (of design).

The exchanges of equivalent states (e.g. 2⭥2 , 5⭥5) may be thought of as resonances to accomplish alteration of direction (arrows) of operations and ∴ procedures. The shared areas represent a cycling (exchange) of states.

∴ A short-hand for the process might be, using * to represent resonance and ✿ (outline) to represent cycling, as follows:

A → 2* → 𝔹4 → 5* → C → 𝔹4 → 5* → 𝔹𝔸 → 5* → ℂ4 → 2* → B

The originating process as:

A → 2* → 𝔹4 → 5* → C

The exclusionary exclusion (excision) from the complexed process as:

A → 2* → 𝔹4 → 5* → ℂ4 → 2* → B

And the excluded process (fragment) as:

C → 𝔹4 → 5* → 𝔹𝔸 → 5*

(which contains each relevant state as a constituent: A, B, C). The first three contain all (types of) states utilized: A, B, C, 2, 5, 4; ∴ the last one might be termed "2" excluded (or missing of state 2).

Excluding the excised fragment, the processes depicted are utilitarian of all states designed. They are holistic (in this sense). As well is holistic, but distinct of initiation from the other 3, is:

5* → C → 𝔹4 → 5* → 𝔹𝔸 → 5* → ℂ4 → 2* → B

with component fragments:

5* → ℂ4 → 2* → B , 5* → C → 𝔹4 , 4 → 5* → 𝔹𝔸

And there are, of course, the other behaviors (of processes) previously stated (or postulated):

A → 5* → ℂ4 → 2* → B
5* → C → 𝔹4 → A → 2*
4 → 5* → 𝔹𝔸 → 2* → C
A → 4 → 5*2* → B → C

Already we have enough processes (9) to establish an aetiology:

(1)	A → 2* → 𝔹4 → 5* → C
(2)	A → 2* → 𝔹4 → 5* → ℂ4 → 2* → B
(3)	A → 2* → 𝔹4 → 5* → C → 𝔹4 → 5* → 𝔹𝔸 → 5* → ℂ4 → 2* → B
(4)	C → 𝔹4 → 5* → 𝔹𝔸 → 5*
(5)	5* → C → 𝔹4 → 5* → 𝔹𝔸 → 5* → ℂ4 → 2* → B
(6)	A → 5* → ℂ4 → 2* → B
(7)	5* → C → 𝔹4 → A → 2*
(8)	4 → 5* → 𝔹𝔸 → 2* → C
(9)	A → 4 → 5*2* → B → C

Note that the arrows of processes can be stretched out to bring into congruence operations for (or to) the most complexed process: 3. A concluding "C", for (B→C), would have to be added to process 3, in order for processes 8 and 9 to be encompassed in this through arrow stretching.

The progression of changes is used almost to refine a logic of behaviors:

A.................................C
A..........................5*
A → 2* → B4 → 5* → C → B4 → 5* → BA → 5* → C4 → 2* → B
                 C → B4 → 5* → BA → 5*
     5*..B
                      A..................................B
         5*.............B4 ------> A ------------> 2*
                 4.............BA ------------> 2* -----------> C
A ----------------------> 4 -> 5*_________________2* -> B ------> C

This logic allows for anticipatory extension (e.g. for the last two processes). What is most evident, of course, is that the flow (cf operations) is determinable (or somewhat predictable), based on following (roughly) the most complexed process (as a supra-organizing template). ∴ All of the processes depicted must be (functionally) inter-related. This would be the assessment upon studying them, as a group, without previous knowledge of their (individual) derivations. This is very much like an operational etymology for processes.

It is of nul/los character that equivalent substitutions can be made (or "cassetted" as such):

e.g. For the "2" excluded fragment, a fulfillment can be constructed (or visualized) as

$$C → B4 → 5* → BA → 5*$$
A..........C   5*..........B
→   A..........C → B4 → 5* → BA → 5*..........B

with an effective doubling (duplication) of the fragment (another copy occurring within 5*......B).

Or, just as ably (for "2" fulfillment):

$$C → B4 → 5* → BA → 5*$$
                         5*..........BA → A → 2*
→   C → B4 → 5* → BA → 5*..........BA → A → 2*

without this doubling (here, a virtual definition of "2" is proposed), etc. The characters and numbers here represent states (of operation). Obviously, there can be no state of void-ness. Something is always occurring, if only a pause due to resonance (of a state). Then a vanquishing causes dissipation (of a state) into another state of operation.

Taking this interesting fulfillment (of "2") as an example, we may tabulate to generalize by treating the resonances (*) as "1" pauses, to result in the interesting devolvement of a C state back down to the A state:

| C → B4 → 1 → BA → 1 → BA → A → 1 |

i.e.

| C B4 1 BA 1 BA A 1 |

∴ Accordingly (and likewise) we have "progressed" from:

| A 1 B4 1 C | (i.e. $A \rightarrow 2^* \rightarrow B4 \rightarrow 5^* \rightarrow C$)

to

| C B4 1 BA 1 BA A 1 |

∴ by equivalency substitution and extension:

$$| A \quad 1 \quad B4 \quad 1 \quad C |$$
$$| C \quad B4 \quad 1 \quad BA \quad 1 \quad BA \quad A \quad 1 |$$
$$\rightarrow \qquad | A \quad 1 \quad B4 \quad 1 \quad C \quad B4 \quad 1 \quad BA \quad 1 \quad BA \quad A \quad 1 |$$

The shared (circular) states can be treated in various ways. For the moment let us simply demarcate them as of types α and β :

$$| A \quad 1 \quad \alpha \quad 1 \quad C \quad \alpha \quad 1 \quad \beta \quad 1 \quad \beta \quad A \quad 1 |$$

This simplification implies that α leads from A→C, and β from C→A. Yet α may also lead to β ($\alpha \rightarrow \beta$). Contracting (to remove pauses) we might simplify the table to:

$$| A \quad \alpha \quad C \quad \alpha \quad \beta \quad \beta \quad A |$$

as an overall process.

One could pensively attempt to remove some redundancies to yield the most general of processes:

e.g. $\qquad | A \quad \alpha \quad C \quad \beta \quad A | \qquad \therefore \quad A \rightarrow C \rightarrow A$

This could be fashioned as a starting assumption (of reversibility, or recoverability of states).

Of course, α and β represent variants of the B state (of operations).

∴ The most general (and fully intuitive) process is simply:

$$A \rightarrow B \rightarrow C \rightarrow B \rightarrow A$$
or $\qquad A \leftrightarrow B \leftrightarrow C$

This is simply progenitive of the original proposal (suggestion):

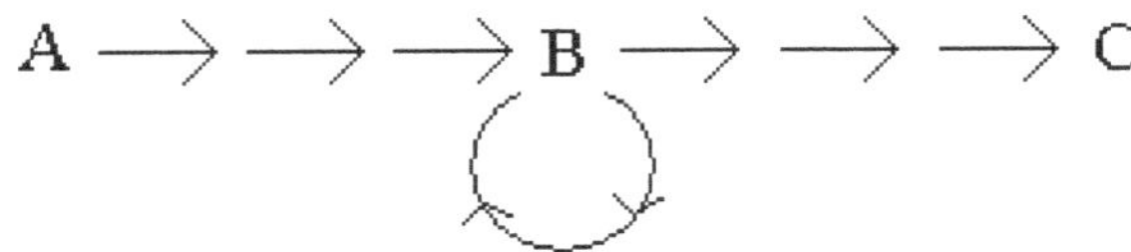

Going back to the shared circular regions (the variants of B states), they clearly represent alternatives of the process's progression:

Being circular, a decision must be made for each before further progression:

∴ e.g.   A 1 B 1 C ....
               A 1 4 1 C ....   etc.

Solution Construction

This can generate 16 ($\equiv 2^4$) combinations of rows (of solutions) of table, which may already encompass (or contain) the 12 elements (of row) present. Note that in nul/los there is no 2^0 ($\equiv 1$). If desired, the element number (per row) may be increased to 16 with "1"s ; or it may be decreased down to 7 by removing "1"s already present.

In this latter pursuit, we can make a table (one of several) with seven such solutions as to be balanced:

e.g.

A B C B B B A	A B C B A
A 4 C B B B A	A 4 C B A
A 4 C 4 B B A	A 4 C 4 B A
A 4 C 4 4 B A	A 4 C 4 B A
A 4 C 4 4 A A	A 4 C 4 A
A B C 4 4 A A	A B C 4 A
A B C B 4 A A	A B C B 4 A

    ↖ These should rightfully be the "A"s ; but the following distortion
illustrates useful means of tabular manipulations.

Each solution (here) starts from (state) A, encounters C, and runs back to A (suggesting reversibility).

Each has (neighboring) redundancies (for reduction, as given after the table). Such reductions can be used to reduce the solutions to 6 elements, with "1"s bringing 5 element rows to 6.

∴ A "reduced" (but equivalent) table (balanced) could be:

| A B C B A 1 |
| A 4 C B A 1 |
| A 4 C 4 B A |
| A 4 C 4 A 1 |
| A B C 4 A 1 |
| A B C B 4 A |

Note that, for this table, if we can (presciently) *equivalate* state 4 with a (generalized) state B, then two more (neighboring) redundancies are caused which, with reduction and "1"s insertion, allows establishment of a new table whose final column is of "1"s :

∴ interestingly

A	B	C	B	A	1
A	4	C	B	A	1
A	4	C	B	A	1
A	4	C	4	A	1
A	B	C	4	A	1
A	B	C	B	A	1

Generalized "B"s outlined.
Rows 2 and 3 "look" the same.

This table, then, can further be reduced to one of 5 rows (as balanced), with the dropping of one row (say the 2nd, if we treat 4 as also a generalized "B"):

A	B	C	B	A
A	4	C	B	A
A	4	C	4	A
A	B	C	4	A
A	B	C	B	A

Here we find that the first row (solution) is repeated as the last row.
∴ we only need 4 rows (although the table would be unbalanced).

∴ with further reduction:

A	B	C	B	A
A	4	C	B	A
A	4	C	4	A
A	B	C	4	A

≡

A
A
A
A

+

B	C	B	A
4	C	B	A
4	C	4	A
B	C	4	A

A concatenation, or appending of column.

≡

A
A
A
A

+

B	C	B	1
4	C	B	1
4	C	4	1
B	C	4	1

A ≡

A
A
A
A

+

B	C	B
4	C	B
4	C	4
B	C	4

A
A
A
A

In this manner there is denoted a clear oscillation about state C of the states B and 4, which brings to fore the essential nature of the procedure as like a constant struggle of a stable atomic structure against changes in electronic orbitals (as an explanatory or expository example for the table). The symmetry provided depicts the four possible juxtapositions of the C state bounded by the various combinations of B and 4 states. It is interesting that, while as (table) listed, the B and 4 states are circularly equivalent (as: B⟳4); a CDD analysis of the (central) table necessarily eliminates state 4 (leaving states A, B, C):

B	C	B
4	C	B
4	C	4
B	C	4

➡

B	C	B
4		B
4		4
B	C	4

≡

B	C	B	(1)
B	B	4	4
4	C	B	(1)
B	4	4	B

➡

B	C	B	1
1	4	1	B
B	4	4	B
B	4	B	B

$$
\begin{vmatrix} B & C & B & 1 \\ 1 & & & B \\ B & & & B \\ B & 4 & B & B \end{vmatrix}
\equiv
\begin{vmatrix} B & C & B & 1 \\ 1 & B & B & B \\ B & B & 4 & B \\ B & B & 1 & B \end{vmatrix}
\;\Rightarrow\;
\begin{vmatrix} B & C & B & 1 \\ 1 & & & B \\ B & & & B \\ B & B & 1 & B \end{vmatrix}
$$

$$
\equiv
\begin{vmatrix} B & C & B & 1 \\ 1 & B & B & B \\ B & 1 & B & B \\ B & B & 1 & B \end{vmatrix}
\;\Rightarrow\;
\begin{vmatrix} B & C & B & 1 \\ 1 & & & B \\ B & & & B \\ B & B & 1 & B \end{vmatrix}
\equiv
\begin{vmatrix} B & C & B & 1 \\ 1 & B & B & B \\ B & 1 & B & B \\ B & B & 1 & B \end{vmatrix}
$$

final result:

essentially equating

B with a pause ($\equiv$ "1")

This suffices to suggest the (formation or dissolution of the) B state as one of "commotion" for the process. Of course, CDD analysis necessarily compares the first row (| B C B |) to the first element of the last row (B). Since both rows (1st and last) start with the same element (B), this is a comparison of the state to itself, to reveal its "pausal" nature (for the 1st row and solution of process). The (ranking or) order of the rows may be changed for other elemental comparisons (under CDD).

Aside from the shared circular region B⟳4 , the shared region B⟳A may also be considered, with the quality of having a generalized "A." Properly, from an "A" perspective, the previous 7 element balanced table should be:

$$
\begin{vmatrix} A & B & C & B & B & B & A \\ A & 4 & C & B & B & B & A \\ A & 4 & C & 4 & B & B & A \\ A & 4 & C & 4 & A & B & A \\ A & 4 & C & 4 & A & A & A \\ A & B & C & 4 & A & A & A \\ A & B & C & B & A & A & A \end{vmatrix}
\quad
\begin{array}{c} \text{redundancy} \\ \text{reductions} \\ \Rightarrow \\ \text{"1" insertions} \end{array}
\quad
\begin{vmatrix} A & B & C & B & A & 1 & 1 \\ A & 4 & C & B & A & 1 & 1 \\ A & 4 & C & 4 & B & A & 1 \\ A & 4 & C & 4 & A & B & A \\ A & 4 & C & 4 & A & 1 & 1 \\ A & B & C & 4 & A & 1 & 1 \\ A & B & C & B & A & 1 & 1 \end{vmatrix}
\equiv
|A| +
\begin{vmatrix} B & C & B & A & 1 & 1 \\ 4 & C & B & A & 1 & 1 \\ 4 & C & 4 & B & A & 1 \\ 4 & C & 4 & A & B & A \\ 4 & C & 4 & A & 1 & 1 \\ B & C & 4 & A & 1 & 1 \\ B & C & B & A & 1 & 1 \end{vmatrix}
$$

The 1st and last rows (of the major table) are the same.

∴ We can remove the last (for a balance to the major table):

$$
|A| +
\begin{vmatrix} B & C & B & A & 1 & 1 \\ 4 & C & B & A & 1 & 1 \\ 4 & C & 4 & B & A & 1 \\ 4 & C & 4 & A & B & A \\ 4 & C & 4 & A & 1 & 1 \\ B & C & 4 & A & 1 & 1 \end{vmatrix}
$$

CDD on the major table yields:

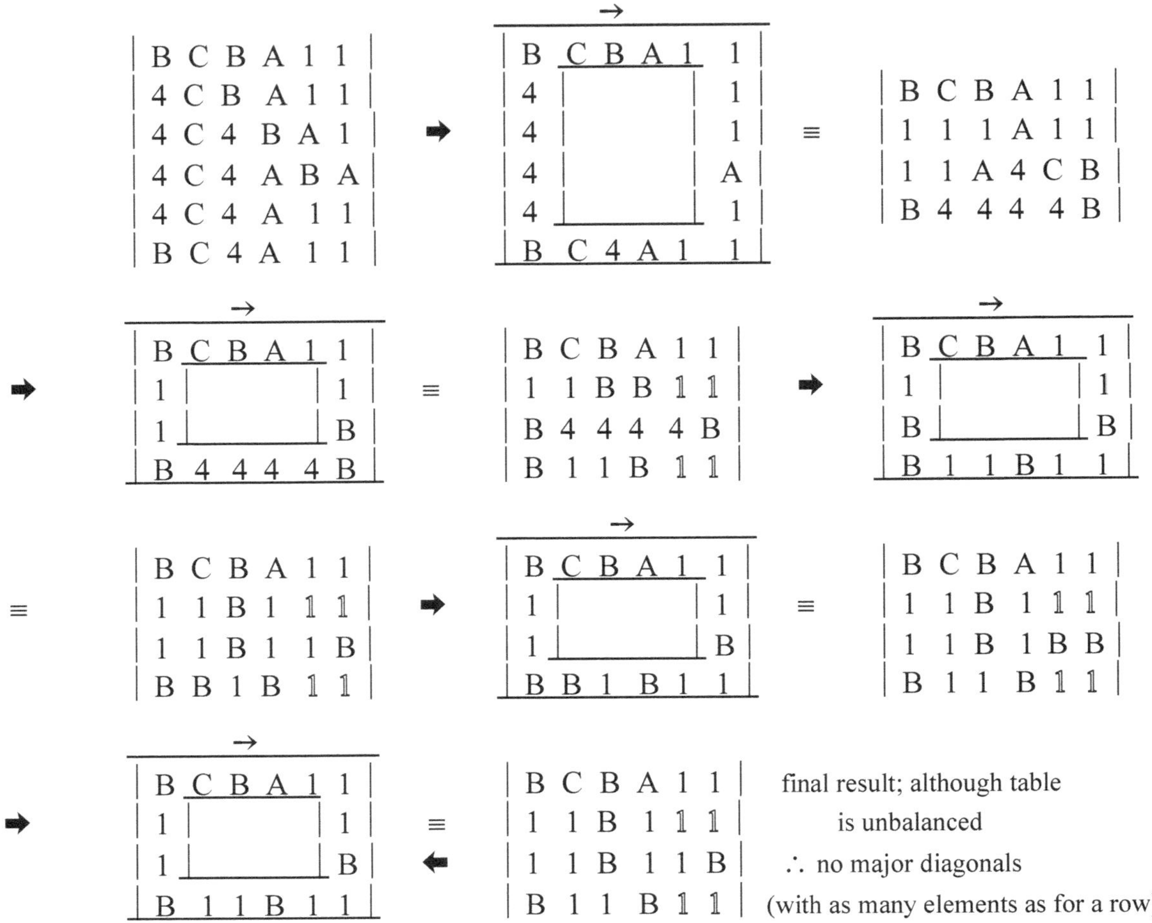

If we judiciously eliminate (apparently) unnecessary "1"s from the final result:

we see that (diagonally) B alternates with 1, A, B (in the major diagonals).

∴ This suggests that B can be a pause (i.e. a "commotion"), stay as itself, or return to state A (the initial state). This is the relation of "B" to the solution (row): | B C B A | .

Note that the (major) table (as previous):

| B C B A 1 1 |
| 4 C B A 1 1 |
| 4 C 4 B A 1 |
| 4 C 4 A B A |
| 4 C 4 A 1 1 |
| B C 4 A 1 1 |

can be treated to convert a "B" (as outlined) to a generalized "A"
(via B⟳A) to yield to a concatenation of tables as:

$$
|\,A\,| \;+\;
\begin{vmatrix} B & C & B \\ 4 & C & B \\ 4 & C & 4 \\ 4 & C & 4 \\ 4 & C & 4 \\ B & C & 4 \end{vmatrix}
\;+\; |\,A\,| \;+\;
\begin{vmatrix} 1 & 1 \\ 1 & 1 \\ A & 1 \\ B & A \\ 1 & 1 \\ 1 & 1 \end{vmatrix}
$$

with tables (column) reducible (of identical row solutions) to:

$$
\begin{vmatrix} B & C & B \\ 4 & C & B \\ 4 & C & 4 \\ B & C & 4 \end{vmatrix}
\quad \text{and} \quad
\begin{vmatrix} A & 1 \\ B & A \end{vmatrix}
$$

The former is identical to the table reduction used earlier, leading to the interesting observation that now we have the former tabular description:

$$
|\,A\,| \;+\;
\begin{vmatrix} B & C & B \\ 4 & C & B \\ 4 & C & 4 \\ B & C & 4 \end{vmatrix}
\;+\; |\,A\,|
\quad \text{along with the additional table:} \quad
\begin{vmatrix} A & 1 \\ B & A \end{vmatrix}
$$

This is a consequence of a proper (elemental) referral to the shared (circular) regions:
(B⊙4) and (B⊙A) .

CDD analysis on the additional table is:

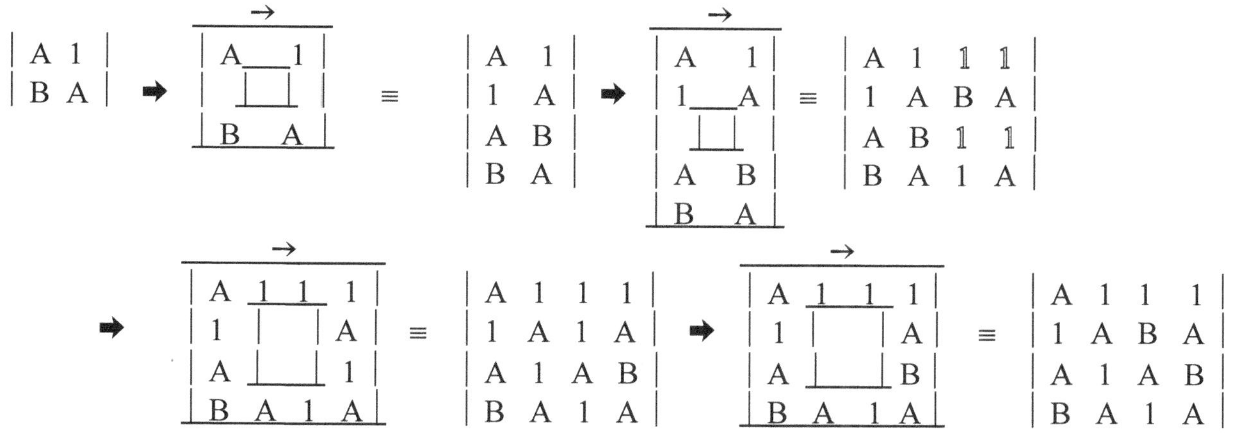

$$
\begin{vmatrix} A & 1 \\ B & A \end{vmatrix}
\;\Rightarrow\;
\begin{vmatrix} A & & 1 \\ & & \\ B & & A \end{vmatrix}
\;\equiv\;
\begin{vmatrix} A & 1 \\ 1 & A \\ A & B \\ B & A \end{vmatrix}
\;\Rightarrow\;
\begin{vmatrix} A & & 1 \\ 1 & & A \\ & & \\ A & & B \\ B & & A \end{vmatrix}
\;\equiv\;
\begin{vmatrix} A & 1 & 1 & 1 \\ 1 & A & B & A \\ A & B & 1 & 1 \\ B & A & 1 & A \end{vmatrix}
$$

$$
\Rightarrow\;
\begin{vmatrix} A & 1 & 1 & 1 \\ 1 & & & A \\ A & & & 1 \\ B & A & 1 & A \end{vmatrix}
\;\equiv\;
\begin{vmatrix} A & 1 & 1 & 1 \\ 1 & A & 1 & A \\ A & 1 & A & B \\ B & A & 1 & A \end{vmatrix}
\;\Rightarrow\;
\begin{vmatrix} A & 1 & 1 & 1 \\ 1 & & & A \\ A & & & B \\ B & A & 1 & A \end{vmatrix}
\;\equiv\;
\begin{vmatrix} A & 1 & 1 & 1 \\ 1 & A & B & A \\ A & 1 & A & B \\ B & A & 1 & A \end{vmatrix}
$$

final result:
with A≡A, B≡1
(from the major diagonals)

∴ We have:

```
        | B C B |
| A | + | 4 C B | + | A | + | A 1 |
        | 4 C 4 |           | B A |
        | B C 4 |
            ↓        CDD        ↓
        | B C B 1 |         | A 1 1 1 |
        | 1 B B B |         | 1 A B A |
        | B 1 B B |         | A 1 A B |
        | B B 1 B |         | B A 1 A |
```

| A | , for CDD purposes, is equivalent to:

```
                        →
| A 1 1 1 |      | A 1 1 1 |      | A 1 1 1 |
| A 1 1 1 | ➡    | A      1 | ≡   | 1 1 1 1 |
| A 1 1 1 |      | A      1 |      | 1 1 1 A |
| A 1 1 1 |      | A 1 1 1 |      | A A A A |

        →                                →
| A 1 1 1 |      | A 1 1 1 |      | A 1 1 1 |
➡  | 1      1 | ≡  | 1 1 A A | ➡  | 1      A |
| 1      A |      | A A A A |      | A      A |
| A A A A |      | A 1 1 A |      | A 1 1 A |

                        →
| A 1 1 1 |      | A 1 1 1 |      | A 1 1 1 |
≡  | 1 A A A | ➡  | 1      A | ≡  | 1 A A A |   final result;
| A 1 1 A |      | A      A |  ⬅  | A 1 A A |   with A≡A, A≡1
| A A 1 A |      | A A 1 A |      | A A 1 A |
```

∴ The summaries (of the major diagonals), from the CDD analyses, are:

$$A \equiv A \ , \ \ B \equiv B \ , \ \ A \equiv A \ , \ \ A \equiv A$$
$$A \equiv 1 \ , \ \ B \equiv 1 \ , \ \ A \equiv 1 \ , \ \ B \equiv 1$$

What is particularly curious, therefore, based on the CDD analyses, is that the

```
| B C B |
| 4 C B |        table complement is functionally equivalent
| 4 C 4 |        to | B | for "B" as | A | is for "A" .
| B C 4 |
```

∴ We have (functionally):

$$|\,A\,| + |\,B\,| + |\,A\,| + \begin{vmatrix} A & 1 \\ B & A \end{vmatrix} \equiv \begin{vmatrix} A & B & A & A & 1 \\ A & B & A & B & A \end{vmatrix}.$$

With reduction of redundancies, $\begin{vmatrix} A & B & A & A & 1 \end{vmatrix} \to \begin{vmatrix} A & B & A \end{vmatrix}$
and, from extended (BA) redundancy, $\begin{vmatrix} A & B & A & B & A \end{vmatrix} \to \begin{vmatrix} A & B & A \end{vmatrix}$.

∴ We (can) have $\begin{vmatrix} A & B & A \\ A & B & A \end{vmatrix}$ to subject to CDD:

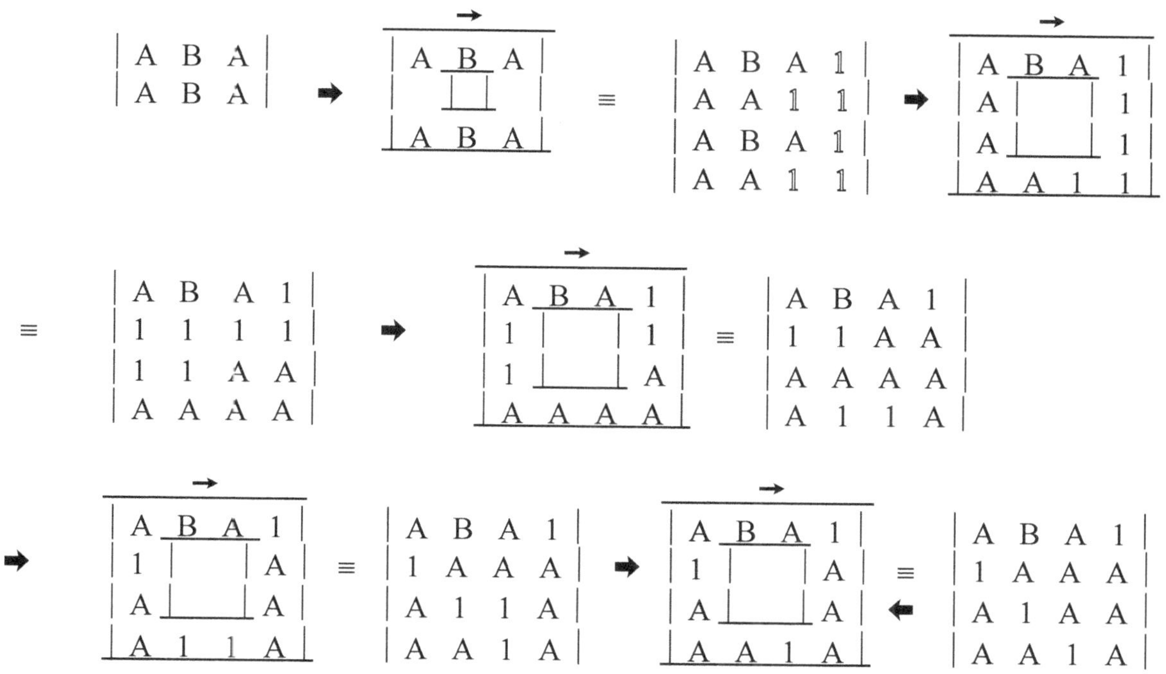

The final result is functionally equivalent to $|\,A\,|$.
Note: adopting partially the (beginning of the) above CDD:

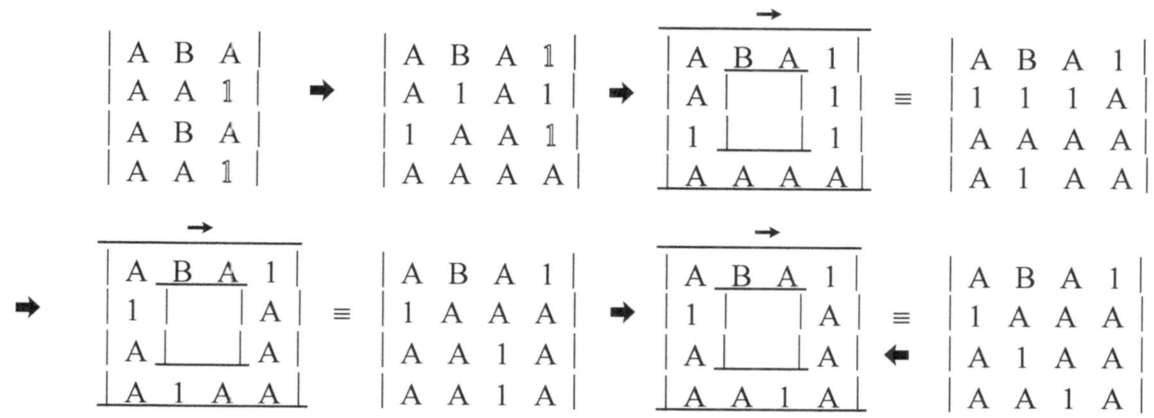

final result;
with A≡A, A≡1

same final result
(as above)

∴ Functionally, we have:

$$|\,A\,| \;+\; |\,B\,| \;+\; |\,A\,| \;+\; \begin{vmatrix} A & 1 \\ B & A \end{vmatrix} \;\equiv\; |\,A\,|$$

In nul/los, this indicates that:

$$|\,B\,| \;+\; |\,A\,| \;+\; \begin{vmatrix} A & 1 \\ B & A \end{vmatrix} \;\equiv\; 1$$

since

$$|\,A\,| \;+\; |\,1\,| \;\equiv\; |\,A\,|$$

∴

$$|\,B\,| \;+\; |\,A\,| \;+\; \begin{vmatrix} A & 1 \\ B & A \end{vmatrix} \;\equiv\; \begin{vmatrix} B & A & A & 1 \\ B & A & B & A \end{vmatrix} \;;$$

and with CDD analysis:

$$\begin{vmatrix} B & A & A & 1 \\ B & A & B & A \end{vmatrix} \;\Rightarrow\; \begin{vmatrix} B & A & A & 1 \\ B & A & B & A \end{vmatrix} \;\equiv\; \begin{vmatrix} B & A & A & 1 \\ 1 & A & 1 & 1 \\ A & B & A & B \\ B & B & 1 & 1 \end{vmatrix} \;\Rightarrow\; \begin{vmatrix} B & A & A & 1 \\ 1 & & & 1 \\ A & & & B \\ B & B & 1 & 1 \end{vmatrix}$$

$$\equiv\; \begin{vmatrix} B & A & A & 1 \\ 1 & 1 & B & 1 \\ 1 & 1 & B & B \\ B & A & 1 & B \end{vmatrix} \;\Rightarrow\; \begin{vmatrix} B & A & A & 1 \\ 1 & & & 1 \\ 1 & & & B \\ B & A & 1 & B \end{vmatrix} \;\equiv\; \begin{vmatrix} B & A & A & 1 \\ 1 & 1 & B & B \\ B & 1 & A & B \\ B & 1 & 1 & B \end{vmatrix} \;\Rightarrow\; \begin{vmatrix} B & A & A & 1 \\ 1 & & & B \\ B & & & B \\ B & 1 & 1 & B \end{vmatrix}$$

$$\equiv\; \begin{vmatrix} B & A & A & 1 \\ 1 & B & B & B \\ B & 1 & 1 & B \\ B & B & 1 & B \end{vmatrix} \;\Rightarrow\; \begin{vmatrix} B & A & A & 1 \\ 1 & & & B \\ B & & & B \\ B & B & 1 & B \end{vmatrix} \;\equiv\; \begin{vmatrix} B & A & A & 1 \\ 1 & B & B & B \\ B & 1 & B & B \\ B & B & 1 & B \end{vmatrix} \quad \begin{array}{l}\text{final result;} \\ \text{with } B \equiv B \,,\; B \equiv 1 \\ \text{(from major diagonals)}\end{array}$$

This is functionally equivalent to $|\,B\,|$.

∴ Functionally, we have:

$$|\,A\,| \;+\; |\,B\,| \;\equiv\; |\,A\,|$$

$$|\,B\,| \;\equiv\; |\,1\,|$$

indicating that state B is of "paused commotion," as asserted earlier.

More conceptually, we can simply say that:

if $\begin{vmatrix} A & 1 \\ B & A \end{vmatrix}$ represents $|\,A\,| + |\,B\,|$

and $|\,B\,|$ represents $|\,1\,|$,

then functionally (i.e. as of a state resulting):

$$| A | + | B | + | A | + | \begin{matrix} A\ 1 \\ B\ A \end{matrix} | \; \equiv \; | A | + | B | + | A | + | A | + | B |$$

$$\equiv \quad | A | + | 1 | + | A | + | A | + | 1 |$$

$$\equiv \quad | A | + | A | + | A | \; \equiv \; | A | \qquad \text{(CDD functionally)}$$

Of course:

$$| A | + | A | + | A | \; \equiv \; | A\ A\ A |$$

$$| A\ A\ A | \; \equiv \; \begin{matrix} | A\ A\ A\ 1 | \\ | A\ A\ A\ 1 | \\ | A\ A\ A\ 1 | \\ | A\ A\ A\ 1 | \end{matrix}$$

which upon CDD yields:

$$\begin{matrix} | A\ A\ A\ 1 | \\ | A\ A\ A\ 1 | \\ | A\ A\ A\ 1 | \\ | A\ A\ A\ 1 | \end{matrix} \;\rightarrow\; \begin{matrix} | A\ A\ A\ 1 | \\ | A\quad\ 1 | \\ | A\quad\ 1 | \\ | A\ A\ A\ 1 | \end{matrix} \;\equiv\; \begin{matrix} | A\ A\ A\ 1 | \\ | 1\ 1\ 1\ 1 | \\ | 1\ A\ A\ A | \\ | A\ A\ A\ A | \end{matrix}$$

$$\rightarrow\; \begin{matrix} | A\ A\ A\ 1 | \\ | 1\quad\ 1 | \\ | 1\quad\ A | \\ | A\ A\ A\ A | \end{matrix} \;\equiv\; \begin{matrix} | A\ A\ A\ 1 | \\ | 1\ 1\ A\ A | \\ | A\ A\ A\ A | \\ | A\ 1\ 1\ A | \end{matrix} \;\rightarrow\; \begin{matrix} | A\ A\ A\ 1 | \\ | 1\quad\ A | \\ | A\quad\ A | \\ | A\ 1\ 1\ A | \end{matrix}$$

$$\equiv\; \begin{matrix} | A\ A\ A\ 1 | \\ | 1\ A\ A\ A | \\ | A\ 1\ 1\ A | \\ | A\ A\ 1\ A | \end{matrix} \;\rightarrow\; \begin{matrix} | A\ A\ A\ 1 | \\ | 1\quad\ A | \\ | A\quad\ A | \\ | A\ A\ 1\ A | \end{matrix} \;\equiv\; \begin{matrix} | A\ A\ A\ 1 | \\ | 1\ A\ A\ A | \\ | A\ 1\ A\ A | \\ | A\ A\ 1\ A | \end{matrix}$$

final result;
with A≡A , A≡1

This is functionally equivalent to $| A |$.

Going back to the major table reduced of one redundant row, a cogent CDD analysis requires that a table be brought to 4 rows with 4 columns, so that the major diagonals that result with the final table has as many elements as there are in rows.

∴ CDD is really a form of circularized squaring formation of elements. Since each row (of a table) is a solution, and for correctness of a process each row (of a table) should yield to the same solution, then perhaps we can select a subset of the (indicated) major table to reduce to a **4x4** system:

$$
\begin{vmatrix} B & C & B & A & 1 & 1 \\ 4 & C & B & A & 1 & 1 \\ 4 & C & 4 & B & A & 1 \\ 4 & C & 4 & A & B & 1 \\ 4 & C & 4 & A & 1 & 1 \\ B & C & 4 & A & 1 & 1 \end{vmatrix}
\quad \xrightarrow{\text{exclusion}} \quad
\begin{vmatrix} B & C & B & A & 1 & 1 \\ 4 & C & B & A & 1 & 1 \\ 4 & C & 4 & A & 1 & 1 \\ B & C & 4 & A & 1 & 1 \end{vmatrix}
\quad \equiv \quad
\begin{vmatrix} B & C & B & A \\ 4 & C & B & A \\ 4 & C & 4 & A \\ B & C & 4 & A \end{vmatrix}
$$

↓ (as equivalent)

$$
\begin{vmatrix} B & C & B \\ 4 & C & B \\ 4 & C & 4 \\ B & C & 4 \end{vmatrix} + \begin{vmatrix} A \end{vmatrix}
$$

CDD

⌣
↓

$$B \equiv B \qquad A \equiv A$$
$$B \equiv 1 \qquad A \equiv 1$$

CDD analysis yields:

$$
\begin{vmatrix} B & C & B & A \\ 4 & C & B & A \\ 4 & C & 4 & A \\ B & C & 4 & A \end{vmatrix}
\quad \Rightarrow \quad
\begin{vmatrix} B & C & B & A \\ 4 & & & A \\ 4 & & & A \\ B & C & 4 & A \end{vmatrix}
\quad \equiv \quad
\begin{vmatrix} B & C & B & A \\ A & A & A & A \\ A & 4 & C & B \\ B & 4 & 4 & B \end{vmatrix}
$$

$$
\Rightarrow \quad
\begin{vmatrix} B & C & B & A \\ A & & & A \\ A & & & B \\ B & 4 & 4 & B \end{vmatrix}
\quad \equiv \quad
\begin{vmatrix} B & C & B & A \\ A & A & B & B \\ B & 4 & 4 & B \\ B & A & A & B \end{vmatrix}
\quad \Rightarrow \quad
\begin{vmatrix} B & C & B & A \\ A & & & B \\ B & & & B \\ B & A & A & B \end{vmatrix}
$$

$$
\equiv \quad
\begin{vmatrix} B & C & B & A \\ A & B & B & B \\ B & A & A & B \\ B & B & A & B \end{vmatrix}
\quad \Rightarrow \quad
\begin{vmatrix} B & C & B & A \\ A & & & B \\ B & & & B \\ B & B & A & B \end{vmatrix}
\quad \equiv \quad
\begin{vmatrix} B & C & B & A \\ A & B & B & B \\ B & A & B & B \\ B & B & A & B \end{vmatrix}
$$

almost like a composite summary

final result; with B≡B , B≡A

However, we now expand CDD type of analysis from circularized squaring (or 4-) formation of elements to curvilinear n-formation of elements:

n-CDD pictographs

(n : 4 --> 11)

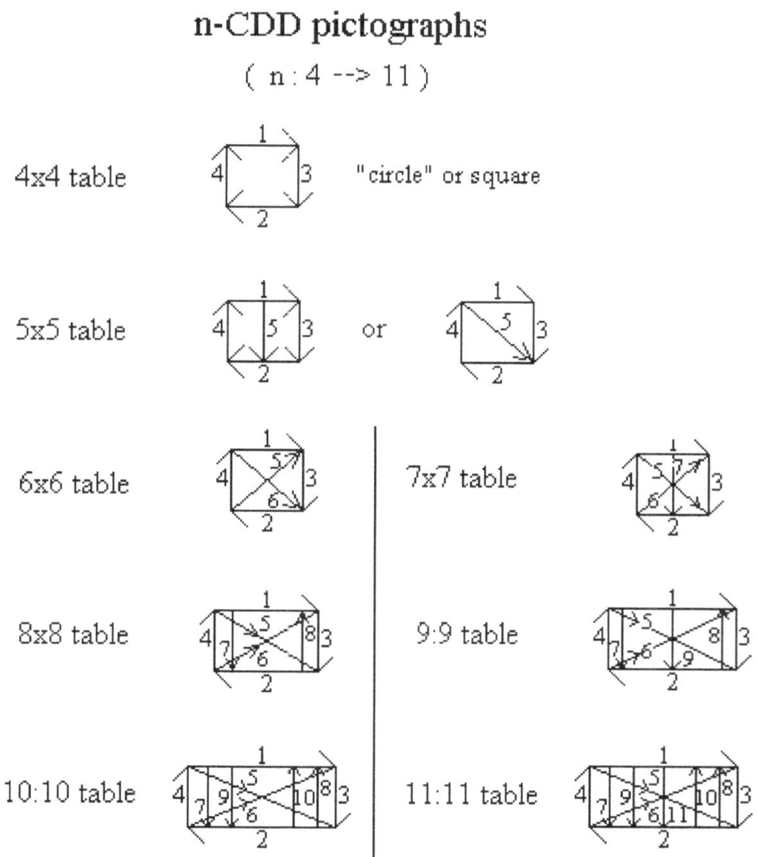

The extension for CDD (now curvilinear n-formation of elements) includes further use of columns as well as the major diagonals. The lines given (in the pictographs), as well as their directions, determine the rows (and ∴ columns) for the next table in the (progressive) analysis. It is evident (since all balanced tables have only 2 major diagonals) that the column order of elements increases of importance as n gets greater than 6 (≡ 4 + 2), and is also more important (or utilized) for the n=5 table (with discretion). The n- extended CDD is somewhat freer than the n=4 CDD, with arbitrary choices of the columns to use and probably should be limited (in consistency) for optimal symmetries used (formed or preferred). Clearly, as n > 6, column order dominates; and row order becomes subservient (with only 2 members) at n > 4.

∴ Roughly, as n increases the analysis becomes a measure of table orthogonality (as far as correctness of solutions). During the process of course, columns and major diagonals become rows, but each table retains the processing (or square ordering/circularization) of the fundamental 4x4 derived "circle" as a representative influence on the final result.

Let us try, then, a 6x6 table (major table as previous, or without exclusion of rows), where each kind of curvilinearity (row, column, major diagonal) is given equal weight in number of ($\equiv 2$); although 2 are redundant:

```
                              →
| B C B A 1 1 |       | B C B A 1 1 |       | B C B A 1 1 |
| 4 C B A 1 1 |       | 4 C     1 1 |       | 1 1 1 A 1 1 |
| 4 C 4 B A 1 |  ➡    | 4   4 B   1 |  ≡    | 1 1 A 4 C B |
| 4 C 4 A B A |       | 4   4 A   A |       | B 4 4 4 4 B |
| 4 C 4 A 1 1 |       | 4 C     1 1 |       | B C 4 A 1 1 |
| B C 4 A 1 1 |       | B C 4 A 1 1 |       | B C 4 B 1 1 |

         →                                             →
   | B C B A 1 1 |       | B C B A 1 1 |       | B C B A 1 1 |
   | 1 1     1 1 |       | 1 1 B B 1 1 |       | 1 1     1 1 |
➡  | 1   A 4   B |  ≡    | 1 1 B 4 C B |  ➡    | 1   B 4   B |
   | B   4 4   B |       | B B B 1 1 B |       | B   B 1   B |
   | B C     1 1 |       | B 1 A 4 1 1 |       | B 1     1 1 |
   | B C 4 B 1 1 |       | B C 4 4 1 1 |       | B C 4 4 1 1 |

                              →
   | B C B A 1 1 |       | B C B A 1 1 |       | B C B A 1 1 |
   | 1 1 B B 1 1 |       | 1 1     1 1 |       | 1 1 B B 1 1 |
≡  | 1 1 4 4 C B |  ➡    | 1   4 4   B |  ≡    | 1 1 4 B 1 B |
   | B B B 1 1 B |       | B   B 1   B |       | B B B 1 1 B |
   | B 1 B 1 1 1 |       | B 1     1 1 |       | B 1 4 1 1 1 |
   | B 1 B 4 1 1 |       | B 1 B 4 1 1 |       | B 1 B 4 1 1 |

         →                                             →
   | B C B A 1 1 |       | B C B A 1 1 |       | B C B A 1 1 |
   | 1 1     1 1 |       | 1 1 B B 1 1 |       | 1 1     1 1 |
➡  | 1   4 B   B |  ≡    | 1 1 4 B 1 B |  ➡    | 1   4 B   B |
   | B   B 1   B |       | B B B 1 1 B |       | B   B 1   B |
   | B 1     1 1 |       | B 1 4 1 1 1 |       | B 1     1 1 |
   | B 1 B 4 1 1 |       | B 1 B B 1 1 |       | B 1 B B 1 1 |

                              →
   | B C B A 1 1 |       | B C B A 1 1 |       | B C B A 1 1 |
   | 1 1 B B 1 1 |       | 1 1     1 1 |       | 1 1 B B 1 1 |
≡  | 1 1 B B 1 B |  ➡    | 1   B B   B |  ≡    | 1 1 B B 1 B |  final
   | B B B 1 1 B |       | B   B 1   B |   ⬅   | B B B 1 1 B |  result
   | B 1 4 1 1 1 |       | B 1     1 1 |       | B 1 B 1 1 1 |
   | B 1 B B 1 1 |       | B 1 B B 1 1 |       | B 1 B B 1 1 |
```

180

The major diagonals are: B 1 B 1 1 1 , B 1 B B 1 1 . They replicate the last two rows.

This seems to indicate that B alternates with either 1 or itself (B). As for the n=4 CDD, the 1st row is retained for the final result, and a comparison of the 1st element of the last row to the first row (in particular the last element of the 1st row) is conducted. But what may be as instructive are the column orders. Indeed, the final result may be partitioned (for concatenation) as:

$$
\begin{vmatrix} B & C \\ 1 & 1 \\ 1 & 1 \\ B & B \\ B & 1 \\ B & 1 \end{vmatrix}
\;+\;
\begin{vmatrix} B \end{vmatrix}
\;+\;
\begin{vmatrix} A \\ B \\ B \\ 1 \\ 1 \\ B \end{vmatrix}
\;+\;
\begin{vmatrix} 1 \end{vmatrix}
\;+\;
\begin{vmatrix} 1 \\ 1 \\ B \\ B \\ 1 \\ 1 \end{vmatrix}
$$

The 1st column (in reverse) and the last column also appear as rows (in the final result).

What do CDD (circular or curvilinear) analyses really reveal? They process for "surrogate" solutions to compare the established (or assumed) 1st (and unchanging) row solution provided, this challenging the nature (and importance) of particular elements or operations or states of operation: the 1st element of the last row and the last element of the 1st row. ∴ It is a test of element utility.

For the above final result, it is observed that we can reduce (both row-wise and column-wise) down (at least) to 4 elements (per row or column), by (judiciously) eliminating (∴ for simplifications) "1"s :

$$
\begin{vmatrix} B & C & B & A & 1 & 1 \\ 1 & 1 & B & B & 1 & 1 \\ 1 & 1 & B & B & 1 & B \\ B & B & B & 1 & 1 & B \\ B & 1 & B & 1 & 1 & 1 \\ B & 1 & B & B & 1 & 1 \end{vmatrix}
\;\Rightarrow\;
\begin{vmatrix} B & C & B & A \\ 1 & 1 & B & B \\ B & B & 1 & B \\ B & B & B & B \\ B & 1 & B & 1 \\ B & 1 & B & B \end{vmatrix}
$$

Here the 1st row solution is not effectively changed.

$$
\begin{vmatrix} B & C \\ 1 & 1 \\ 1 & 1 \\ B & B \\ B & 1 \\ B & 1 \end{vmatrix}
+ \begin{vmatrix} B \end{vmatrix} +
\begin{vmatrix} A \\ B \\ B \\ 1 \\ 1 \\ B \end{vmatrix}
+ \begin{vmatrix} 1 \end{vmatrix} +
\begin{vmatrix} 1 \\ 1 \\ B \\ B \\ 1 \\ 1 \end{vmatrix}
\;\Rightarrow\;
\begin{vmatrix} B & C \\ B & B \\ B & 1 \\ B & 1 \end{vmatrix}
+ \begin{vmatrix} B \end{vmatrix} +
\begin{vmatrix} A \\ B \\ B \\ B \end{vmatrix}
+ \begin{vmatrix} 1 \end{vmatrix} +
\begin{vmatrix} B \\ B \end{vmatrix}
$$

$$
\Rightarrow\;
\begin{vmatrix} B & C \\ B & B \\ B & 1 \\ B & 1 \end{vmatrix}
+ \begin{vmatrix} B \end{vmatrix} +
\begin{vmatrix} A \\ B \\ B \\ B \end{vmatrix}
+ \begin{vmatrix} B \\ B \end{vmatrix}
$$

Here the 1st row solution may be changed (altered) if we allow that (if desired):

$$
\begin{vmatrix} B \\ B \end{vmatrix}
\;\equiv\;
\begin{vmatrix} B \end{vmatrix}
\;\equiv\;
\begin{vmatrix} B \\ B \\ B \\ B \end{vmatrix}
$$

or that, while we can let:

$$\begin{vmatrix} B \\ B \end{vmatrix} \equiv \begin{vmatrix} 1 \\ 1 \\ B \\ B \end{vmatrix} \equiv \begin{vmatrix} 1 \\ B \\ B \\ 1 \end{vmatrix}$$

yet insist that:

$$\begin{vmatrix} B \end{vmatrix} \equiv \begin{vmatrix} 1 \\ B \\ 1 \\ 1 \end{vmatrix} \equiv \begin{vmatrix} 1 \\ B \\ B \\ 1 \end{vmatrix} \equiv \begin{vmatrix} 1 \\ B \\ B \\ B \end{vmatrix} \quad \text{aside from} \quad \begin{vmatrix} B \\ B \\ B \\ B \end{vmatrix} .$$

To maintain the 1st row solution, we can judiciously choose that:

$$\begin{vmatrix} B \end{vmatrix} \equiv \begin{vmatrix} B \\ B \\ 1 \\ 1 \end{vmatrix} \quad \text{and} \quad \begin{vmatrix} B \\ B \end{vmatrix} \equiv \begin{vmatrix} 1 \\ 1 \\ B \\ B \end{vmatrix}$$

$\therefore$

$$\begin{vmatrix} B & C \\ B & B \\ B & 1 \\ B & 1 \end{vmatrix} + \begin{vmatrix} B \\ B \\ 1 \\ 1 \end{vmatrix} + \begin{vmatrix} A \\ B \\ B \\ B \end{vmatrix} + \begin{vmatrix} 1 \\ 1 \\ B \\ B \end{vmatrix} \;\Rightarrow\; \begin{vmatrix} B & C & B & A \\ B & B & B & B \\ B & 1 & B & B \\ B & 1 & B & B \end{vmatrix}$$

This result is a quasi-subset (the last row being redundant) of the earlier simplification (of the 6x6 table reduced down to 4x6).

The flexibilities of these (tabular) manipulations offer considerations of the processes depicted and derived. The above 4x4 table is functionally very similar to an earlier 4x4 table, both derived of manipulations to CDD final results:

$$\begin{vmatrix} B & C & B & A \\ 1 & 1 & B & 1 \\ 1 & 1 & B & B \\ B & 1 & 1 & B \end{vmatrix} , \qquad \begin{vmatrix} B & C & B & A \\ B & B & B & B \\ B & 1 & B & B \\ B & 1 & B & B \end{vmatrix} \qquad \begin{aligned} &\text{for both:} \\ &B \equiv 1, B, A \\ &\text{(from the major diagonals)} \end{aligned}$$

Both are derived from the final results of CDD analyses on the same (major) table.

Let us tag, then, both of these tables to follow the CDD application furthering their descriptions:

$$\begin{vmatrix} B & C & B & X \\ 1 & 1 & B & 1 \\ 1 & 1 & B & B \\ Y & 1 & 1 & B \end{vmatrix} \Rightarrow \begin{vmatrix} B & C & B & X \\ 1 & & & 1 \\ 1 & & & B \\ Y & 1 & 1 & B \end{vmatrix} \equiv \begin{vmatrix} B & C & B & X \\ X & 1 & B & B \\ B & 1 & 1 & Y \\ Y & 1 & 1 & B \end{vmatrix} \Rightarrow \begin{vmatrix} B & C & B & X \\ X & & & B \\ B & & & Y \\ Y & 1 & 1 & B \end{vmatrix}$$

$$A \rightarrow X$$
$$B \rightarrow Y$$

$$
\equiv
\begin{vmatrix} B & C & B & X \\ X & B & Y & B \\ B & 1 & 1 & Y \\ Y & B & X & B \end{vmatrix}
\;\Rightarrow\;
\begin{vmatrix} B & C & B & X \\ X & & & B \\ B & & & Y \\ Y & B & X & B \end{vmatrix}
\;\equiv\;
\begin{vmatrix} B & C & B & X \\ X & B & Y & B \\ B & X & B & Y \\ Y & B & X & B \end{vmatrix}
$$

$$
\Rightarrow
\begin{vmatrix} B & C & B & X \\ X & & & B \\ B & & & Y \\ Y & B & X & B \end{vmatrix}
\;\equiv\;
\begin{vmatrix} B & C & B & X \\ X & B & Y & B \\ B & X & B & Y \\ Y & B & X & B \end{vmatrix}
\;\leftarrow\;
$$

final result:

$$B \equiv B$$
$$Y \equiv X$$
$$(B \equiv A)$$

(X)

$$
\begin{vmatrix} B & C & B & A \\ B & B & B & B \\ B & 1 & B & B \\ B & 1 & B & B \end{vmatrix}
\;\Rightarrow\;
\begin{vmatrix} B & C & B & X \\ B & & & B \\ B & & & B \\ Y & 1 & B & B \end{vmatrix}
\;\equiv\;
\begin{vmatrix} B & C & B & X \\ X & B & B & B \\ B & B & 1 & Y \\ Y & B & B & B \end{vmatrix}
\;\Rightarrow\;
\begin{vmatrix} B & C & B & X \\ X & & & B \\ B & & & Y \\ Y & B & B & B \end{vmatrix}
$$

(Y)

$$
\equiv
\begin{vmatrix} B & C & B & X \\ X & B & Y & B \\ B & B & B & Y \\ Y & B & X & B \end{vmatrix}
\;\Rightarrow\;
\begin{vmatrix} B & C & B & X \\ X & & & B \\ B & & & Y \\ Y & B & X & B \end{vmatrix}
\;\equiv\;
\begin{vmatrix} B & C & B & X \\ X & B & Y & B \\ B & X & B & Y \\ Y & B & X & B \end{vmatrix}
$$

$$
\Rightarrow
\begin{vmatrix} B & C & B & X \\ X & & & B \\ B & & & Y \\ Y & B & X & B \end{vmatrix}
\;\equiv\;
\begin{vmatrix} B & C & B & X \\ X & B & Y & B \\ B & X & B & Y \\ Y & B & X & B \end{vmatrix}
$$

final result:

$$B \equiv B$$
$$Y \equiv X$$
$$(B \equiv A)$$

The two final results are identical (via CDD processing).

$$\therefore$$

$$
\begin{vmatrix} B & C & B & A \\ 1 & 1 & B & 1 \\ 1 & 1 & B & B \\ B & 1 & 1 & B \end{vmatrix}
\quad , \quad
\begin{vmatrix} B & C & B & A \\ B & B & B & B \\ B & 1 & B & B \\ B & 1 & B & B \end{vmatrix}
$$

$$\downarrow \text{CDD}$$

$$
\begin{vmatrix} B & C & B & A \\ A & B & B & B \\ B & A & B & B \\ B & B & A & B \end{vmatrix}
$$

$$B \equiv B , A$$
(from major diagonals)

Obviously they are the end elements of the arrows for the pictographs that determine the CDD patterns (below the 1st row) for the final results.

Let us repeat an analysis, but this time tagging also the 1st element of the 1st row and the last element of the last row (i.e. all corners of the table are tagged):

$$
\begin{vmatrix} W & C & B & X \\ B & B & B & B \\ B & 1 & B & B \\ Y & 1 & B & Z \end{vmatrix}
\;\Rightarrow\;
\begin{vmatrix} W & C & B & X \\ B & & & B \\ B & & & B \\ Y & 1 & B & Z \end{vmatrix}
\;\equiv\;
\begin{vmatrix} W & C & B & X \\ X & B & B & Z \\ Z & B & 1 & Y \\ Y & B & B & W \end{vmatrix}
\;\Rightarrow\;
\begin{vmatrix} W & C & B & X \\ X & & & Z \\ Z & & & Y \\ Y & B & B & W \end{vmatrix}
$$

W≡B , X≡A
Y≡B , Z≡B

$$
\equiv
\begin{vmatrix} W & C & B & X \\ X & Z & Y & W \\ W & B & B & Y \\ Y & Z & X & W \end{vmatrix}
\;\Rightarrow\;
\begin{vmatrix} W & C & B & X \\ X & & & W \\ W & & & Y \\ Y & Z & X & W \end{vmatrix}
\;\equiv\;
\begin{vmatrix} W & C & B & X \\ X & W & Y & W \\ W & X & Z & Y \\ Y & W & X & W \end{vmatrix}
$$

$$
\Rightarrow
\begin{vmatrix} W & C & B & X \\ X & & & W \\ W & & & Y \\ Y & W & X & W \end{vmatrix}
\;\equiv\;
\begin{vmatrix} W & C & B & X \\ X & W & Y & W \\ W & X & W & Y \\ Y & W & X & W \end{vmatrix}
\qquad \text{final result}
$$

Note that for a balanced table, using the pictograph arrows as a guide, there are always as many arrowheads (and thus elements) as there are elements in a row or column, ∴ as many (CDD appropriated) element row or column ends as there are arrows (for the pictograph).

It is interesting, however, that the last element of the last row (here: Z) "drops out" during the CDD analysis. This seems to herald the approach of the final result.

Tabular pattern analysis

We note that 3 of the tabular corners (W, X, Y) remain throughout the analysis. They are invariant, and the 4th corner (Z) is replaced by the 1st (W). In the **6x6** CDD example earlier, where the last element of the 1st row and the last element of the last row are redundant each for two arrowheads, all four corners (there: B, 1, B, 1) are invariant.

Now, clearly, rows below the 1st, of the final result, can not be representative of (correct) solutions. But they are indicative of patterns of behavior of end elements (during the analysis):

YWXWYWXWYWXW
WYWXWCBXWYWX
XWYWXWYWXWYW
WXWYWXWYWXWY
YWXWYWXWYWXW
WYWXWYWXWYWX

This subjugates B ≈ W , C ≈ Y .

$$(W \equiv B) , (Y \equiv B)$$

The extra tabular elements drawn are based on diagonal arguments (and fulfilling horizontal and vertical arguments). Only with B ≈ W and C ≈ Y can the rows fulfill the solution of the table's 1st row. But this occurs throughout the drawn array, skewed diagonally (easily followed with the all-"W"'s major diagonal).

∴ The CDD analysis makes the 1st row solution repetitive but generalized to the invariant elements. Here, the general solution is: │ W Y W X │ . The actual solution (assumed) is embedded within the array (of elements).

Note that, in this way, solutions can appear (or occur) column-wise (from a 4x4 pictograph of CDD design). Element "Z" is of course dropped functionally, because it prohibits the all-"W"'s major diagonal to result.

So CDD provides for a generalized solution which, if adopted (or is adoptable) insures (overall) correctness (at least internally) for the resulting table. But for the 6-CDD (n=6) example given, several "solutions" for the final result seem to be brought into view (or are made evident). So let us repeat the analysis, though tagging each of the tabular corners (the two right-end corner elements being shared by 2 arrows, or arrowheads):

```
                                              →
        | W C B A 1 X |         | W C B A 1 X |          | W C B A 1 X |
W ≡ B   | 4 C B A 1 1 |         | 4 C     1 1 |          | X 1 1 A 1 Z |
X ≡ 1   | 4 C 4 B A 1 |   ➡     | 4   4 B   1 |    ≡     | Z 1 A 4 C Y |
Y ≡ B   | 4 C 4 A B A |         | 4   4 A   A |          | Y 4 4 4 4 W |
Z ≡ 1   | 4 C 4 A 1 1 |         | 4 C     1 1 |          | W C 4 A 1 Z |
        | Y C 4 A 1 Z |         | Y C 4 A 1 Z |          | Y C 4 B 1 X |

              →                                                →
        | W C B A 1 X |         | W C B A 1 X |          | W C B A 1 X |
     ➡  | X 1     1 Z |         | X Z Y W Z X |    ➡     | X Z     Z X |
        | Z   A 4   Y |   ≡     | X 1 B 4 C Y |          | X   B 4   Y |
        | Y   4 4   W |         | Y W Y Z X W |          | Y   Y Z   W |
        | W C     1 Z |         | W 1 A 4 1 X |          | W 1     1 X |
        | Y C 4 B 1 X |         | Y C 4 A 1 X |          | Y C 4 A 1 X |

                                      →
        | W C B A 1 X |         | W C B A 1 X |          | W C B A 1 X |
        | X X Y W X X |         | X X     X X |          | X X Y W X X |
     ≡  | X 1 A 4 C Y |   ➡     | X   A 4   Y |    ≡     | X Z B Y 1 Y |
        | Y W Y X X W |         | Y   Y X   W |          | Y W Y X X W |
        | W Z B Z 1 X |         | W Z     1 X |          | W X A X 1 X |
        | Y 1 Y B Z X |         | Y 1 Y B Z X |          | Y Z Y A X X |
```

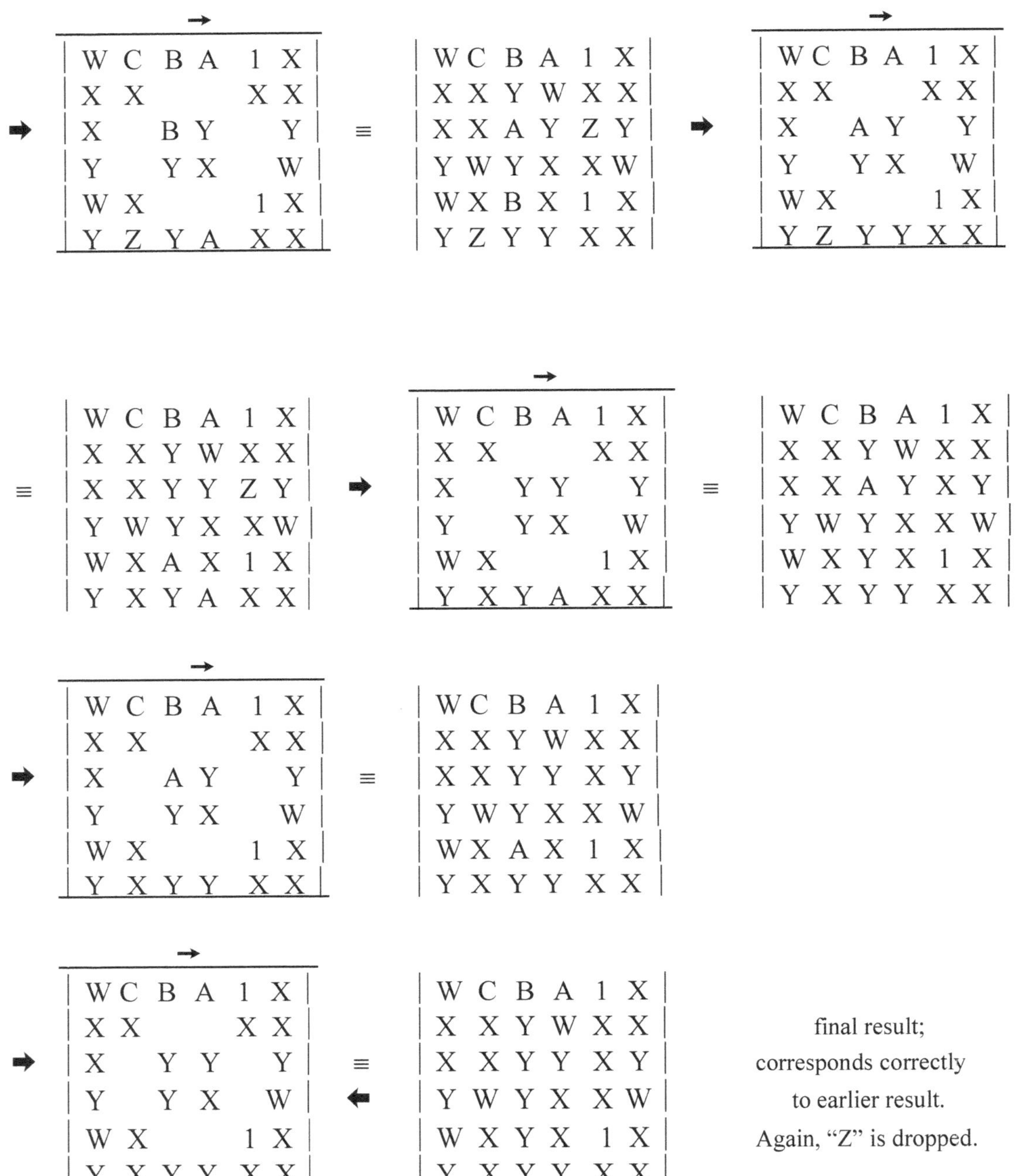

Below the 1st row, only the elements: W, X, Y, and 1 ($\equiv$ not "X") are utilized for the final result. The corner W (1st element of 1st row), X (last element of 1st row), Y (1st element of last row), and this (single) "1" are invariant throughout the process. This "1" is the 5th element of the 5th row. The 3rd column are all "B"s, and the 5th all "1"s, comporting with the earlier final result, and also the 1st row's column signatures. The major diagonals are exactly (element) replicated for the last two rows, arrow-wise (i.e. directionally).

Discerning the general array of pattern for the elements is more difficult than for the n≡4 CDD case. The pictograph for the n≡6 CDD process shows that the arrows are continuous only through the 5th arrow, but then must make a discontinuity (jump or loop) to the 6th arrow. One could make the process

186

more continuous by having as the 6th arrow a repeat of the 3rd arrow, and then having the 2nd major diagonal (going "up" the table) as a 7th arrow. Yet, for further distinct arrows some looping would be required since they would be interspersed of columns (and any arrow must contain as many elements as a row or column or major diagonal). One can also reverse directions, to force some continuousness. But what only is probably required is a consistency of (CDD) design.

We do note that, within the table, element patterns are replicated (along rows, columns, and major diagonals):

e.g.

$$
\begin{vmatrix} W & C & B & A & 1 & X \\ X & X & Y & W & X & X \\ X & X & Y & Y & X & Y \\ Y & W & Y & X & X & W \\ W & X & Y & X & 1 & X \\ Y & X & Y & Y & X & X \end{vmatrix}
\qquad
\begin{vmatrix} W & C & B & A & 1 & X \\ X & X & Y & W & X & X \\ X & X & Y & Y & X & Y \\ Y & W & Y & X & X & W \\ W & X & Y & X & 1 & X \\ Y & X & Y & Y & X & X \end{vmatrix}
$$

$C , A \approx 1 \equiv X \ (X \equiv 1)$

$C , A \approx X ; Y \equiv B$

$Y , W = B$

$$
\begin{vmatrix} W & C & B & A & 1 & X \\ X & X & Y & W & X & X \\ X & X & Y & Y & X & Y \\ Y & W & Y & X & X & W \\ W & X & Y & X & 1 & X \\ Y & X & Y & Y & X & X \end{vmatrix}
$$

∴ Considering patterns forward and reverse,
all of the rows are replicated
(explicitly or implicitly).

(Arrow directions should be followed for patterns.)

Considering arrow directions (enforced and apparent), the following directions seem to apply:

$$
\begin{vmatrix} W & C & B & A & 1 & X \\ X & X & Y & W & X & X \\ X & X & Y & Y & X & Y \\ Y & W & Y & X & X & W \\ W & X & Y & X & 1 & X \\ Y & X & Y & Y & X & X \end{vmatrix}
$$

→ enforced

⇒ apparent

From this we can assume that: $C , A \approx X \equiv 1$, and that the first row solution is replicated (implicitly) as the 5th row solution.

Of course, a CDD final result represents an idealization of correctness for the 1st row solution (based on activities of the other originating rows). The table of the CDD final results suggests what a correct (tabular) solution might look like, while upholding the 1st row as a definite solution.

So now we have that CDD on CDD results brings (for a common problem or process stated here) the suggestion (or indication) that B≡1, B, A (via manipulations of original or primary CDD results), and

B≡B, A on further CDD. The 4x4 array resulting suggests C≈Y (Y≡B). Primary CDD (i.e. applied once) on the major table of the originating process (to the 6x6 tabular form) suggests C, A≈1 (i.e. a pause).

∴ For A → B → C the end states (A, C) are pause-like and the metamorphosing state is B (to which C
    ↻   may be brought back to upon further CDD, as an idealization of correctness for
           the process).

What, then, can CDD analysis (primary) on an entire process (7x7 table) reveal?

The complete process table (selected) is interesting (for a 7x7 table to undergo CDD analysis) in that all of the arrow-head end points fall on (an) element "A" except for the 7th (central) arrow, which falls (its head) on "B" (refer to the pictograph for n≡7 CDD) :

```
| A B C B B B A |
| A 4 C B B B A |
| A 4 C 4 B B A |
| A 4 C 4 A B A |
| A 4 C 4 A A A |
| A B C 4 A A A |
| A B C B A A A |
```

∴ The final result should be heavily "A" weighted (for element occurrence in the table). The table, of course, was previously manipulated in an attempt to reduce its size (or complexity) before formal CDD analysis. But there is the interesting occurrence of one of its solutions (rows) not having the "B" element (in preference for "4" state). The 1st and last rows don't contain the "4" state.

Let us again tag (all of) the arrow-head end elements:

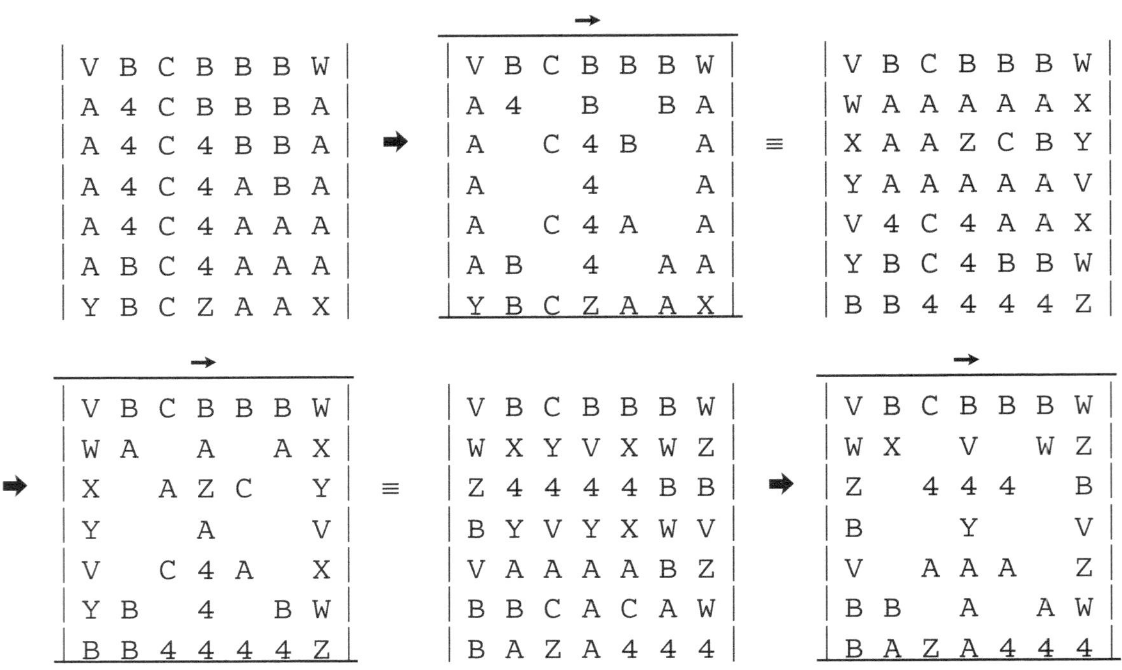

```
| V B C B B B W |          →                | V B C B B B W |                | V B C B B B W |
| A 4 C B B B A |     | V B C B B B W |      | W A A A A A X |
| A 4 C 4 B B A |  ➡  | A 4   B   B A |  ≡   | X A A Z C B Y |
| A 4 C 4 A B A |     | A   C 4 B   A |      | Y A A A A A V |
| A 4 C 4 A A A |     | A     4     A |      | V 4 C 4 A A X |
| A B C 4 A A A |     | A   C 4 A   A |      | Y B C 4 B B W |
| Y B C Z A A X |     | A B   4   A A |      | B B 4 4 4 4 Z |
                     | Y B C Z A A X |
```

```
           →                                       →
      | V B C B B B W |       | V B C B B B W |       | V B C B B B W |
      | W A   A   A X |       | W X Y V X W Z |       | W X   V   W Z |
   ➡  | X   A Z C   Y |  ≡    | Z 4 4 4 4 B B |  ➡    | Z   4 4 4   B |
      | Y       A     V |     | B Y V Y X W V |       | B     Y     V |
      | V   C 4 A   X |       | V A A A A B Z |       | V   A A A   Z |
      | Y B   4   B W |       | B B C A C A W |       | B B   A   A W |
      | B B 4 4 4 4 Z |       | B A Z A 4 4 4 |       | B A Z A 4 4 4 |
```

```
                              →
  | V B C B B B W |    | V B C B B B W |      | V B C B B B W |
  | W Z B V Z W 4 |    | W Z   V   W 4 |      | W 4 B V 4 W A |
≡ | 4 4 4 A Z A B | ➡  | 4     4 A Z   B |  ≡ | A A A Y 4 V B |
  | B B V B Z W V |    | B       B     V |    | B B V B 4 W V |
  | V X 4 Y A A 4 |    | V   4 Y A     4 |    | V Z 4 B A W A |
  | B B A Y 4 W W |    | B B     Y   W W |    | B B 4 B Z W W |
  | B V 4 Y A A A |    | B V 4 Y A A A |      | B V A B Y Y Y |
                         ( loss of X )
```

```
        →                                              →
  | V B C B B B W |    | V B C B B B W |      | V B C B B B W |
  | W 4   V   W A |    | W A B V A W Y |      | W A   V   W Y |
➡ | A   A Y 4   B |  ≡ | Y Y Y B A V B |  ➡  | Y   Y B A   B |
  | B     B     V |    | B B V B A W V |      | B     B     V |
  | V   4 B A   A |    | V 4 A B A W Y |      | V   A B A   Y |
  | B B   B   W W |    | B B 4 B 4 W W |      | B B   B   W W |
  | B V A B Y Y Y |    | B V Y B B B B |      | B V Y B B B B |
    ( loss of Z )                              ( loss of 4 )
```

```
                              →
  | V B C B B B W |    | V B C B B B W |      | V B C B B B W |
  | W Y B V Y W B |    | W Y   V   W B |      | W B B V B W B |
≡ | B B B B Y V B | ➡  | B   B B Y   B |  ≡  | B B B B B V B |
  | B B V B Y W V |    | B     B     V |      | B B V B B W V |
  | V A Y B A W B |    | V   Y B A   B |      | V Y B B A W B |
  | B B A B A W W |    | B B   B   W W |      | B B Y B Y W W |
  | B V B B B B B |    | B V B B B B B |      | B V B B B B B |
```

[It is notable that, from errors in writing out the procedure, there is much self-correction in CDD to lead to a *correct* final result. This must be due to the relatively vast amount of information that is "dropped" or lost.]

```
        →
  | V B C B B B W |    | V B C B B B W |
  | W B   V   W B |    | W B B V B W B |
➡ | B   B B B   B |  ≡ | B B B B B V B |        final result, with
  | B     B     V |  ⬅ | B B V B B W V |       X, Y, Z "dropped"
  | V   B B A   B |    | V B B B A W B |     (although 7ᵗʰ arrow-head end
  | B B   B   W W |    | B B B B B W W |       retains a "B";  V, W ≡ A )
  | B V B B B B B |    | B V B B B B B |
    ( loss of Y )
```

189

$$\equiv \begin{vmatrix} A & B & C & B & B & B & A \\ A & B & B & A & B & A & B \\ B & B & B & B & B & A & B \\ B & B & A & B & B & A & A \\ A & B & B & B & A & A & B \\ B & B & B & B & B & A & A \\ B & A & B & B & B & B & B \end{vmatrix}$$

Interestingly, only one "A" (aside from the arrow-head end "A"s) is retained for the final result (as 5th element of 5th row).

While heavily weighted for both A and B, the table is more heavily "B" weighted (i.e. more "B"s occur in it).

As for the n≡6 CDD case, there are row/column/major diagonal redundancies in the final result:

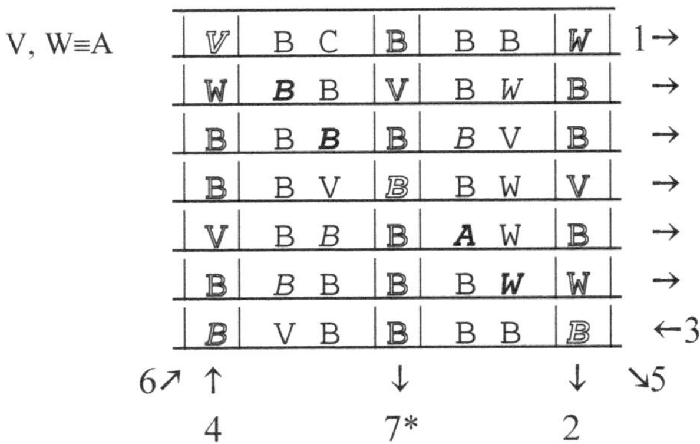

Here, we find it extraordinary that the (4th) column for the 7th (*) arrow (and its direction) is the reverse for that of the 4th arrow (7th row) and (with implied direction) the 3rd row (although now, one can just as easily claim this row's implied direction and order from the 7th arrow column ≡ 4th column).

It is also surprising that the 1st row (solution), which is of course invariant, seems to be quite distinct: No other row nor column nor major diagonal begins and ends with "A" character elements. In fact, all other of the rows, columns, major diagonals have at least one "B" end.

The "C" of the 1st row possibly has "B" character, if we assume its rightward diagonal (C V B W B) is a functional subset (of solution) of row 2 and column 7: W B B V B W B . The "C" 's leftward diagonal of "B B C" is too short to directly comport with the various "B" formations within the table, which include "B" element lengths of ("B"s in linear arrangement): 1, 2, 3, 4, 5, and 6 .

This being CDD on a complete procedure, the final result would seem to suggest in general an oscillation between A and B states (each row containing representative elements of both states).

Comparison through ontology

The ontology of our analyses thus runs as:

<table>
<tr><td>n ≡ 4 CDD</td><td></td><td>n ≡ 6 CDD</td><td></td></tr>
</table>

$$\begin{vmatrix} W & C & B & X \\ X & W & Y & W \\ W & X & W & Y \\ Y & W & X & W \end{vmatrix} \qquad \begin{vmatrix} W & C & B & A & 1 & X \\ X & X & Y & W & X & X \\ X & X & Y & Y & X & Y \\ Y & W & Y & X & X & W \\ W & X & Y & X & 1 & X \\ Y & X & Y & Y & X & X \end{vmatrix}$$

n ≡ 4 CDD: W, Y ≡ B; X ≡ A; *C ≈ Y ≡ B* *for 1st row*

n ≡ 6 CDD: W, Y ≡ B; X ≡ 1; *A ≈ X ≡ 1*; *C ≈ X ≡ 1* *for 1st row* }→ assumes row 5 is a representation of row 1

from major tables

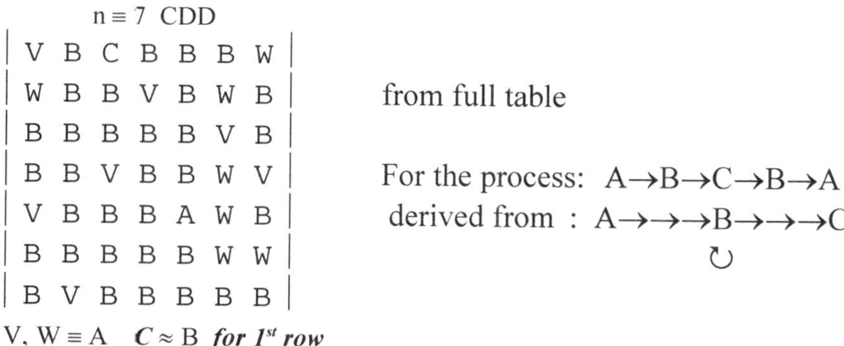

$$n \equiv 7 \ \ CDD$$

$$\begin{vmatrix} V & B & C & B & B & B & W \\ W & B & B & V & B & W & B \\ B & B & B & B & B & V & B \\ B & B & V & B & B & W & V \\ V & B & B & B & A & W & B \\ B & B & B & B & B & W & W \\ B & V & B & B & B & B & B \end{vmatrix}$$

from full table

For the process: $A \to B \to C \to B \to A$
derived from : $A \to \to \to B \to \to \to C$
↺

$V, W \equiv A \quad C \approx B \ \textit{for 1}^{\textit{st}} \textit{ row}$

The 1st rows (of these CDD results) relate as follows:

$$\begin{vmatrix} B & C & \mathbf{B} & \mathbf{A} & 1 & 1 \end{vmatrix}$$ $n \equiv 6$: equivalent tabular extension (with "1"s)

$$\begin{vmatrix} B & C & \mathbf{B} & \mathbf{A} \end{vmatrix}$$ $n \equiv 4$: removed ("B") redundancies

$$\begin{vmatrix} A & B & C & \mathbf{B} & \mathbf{B} & \mathbf{B} & \mathbf{A} \end{vmatrix}$$ $n \equiv 7$: full process (has leading "A")

Do these CDD results suggest a similarity of process, or devote themselves to a fundamental nature for the underlying process (basically: $A \to B \to C$) ?

We might try simple hybridization (replacement or substitution) in the full process result with the partial process results:

Converting to elemental types,

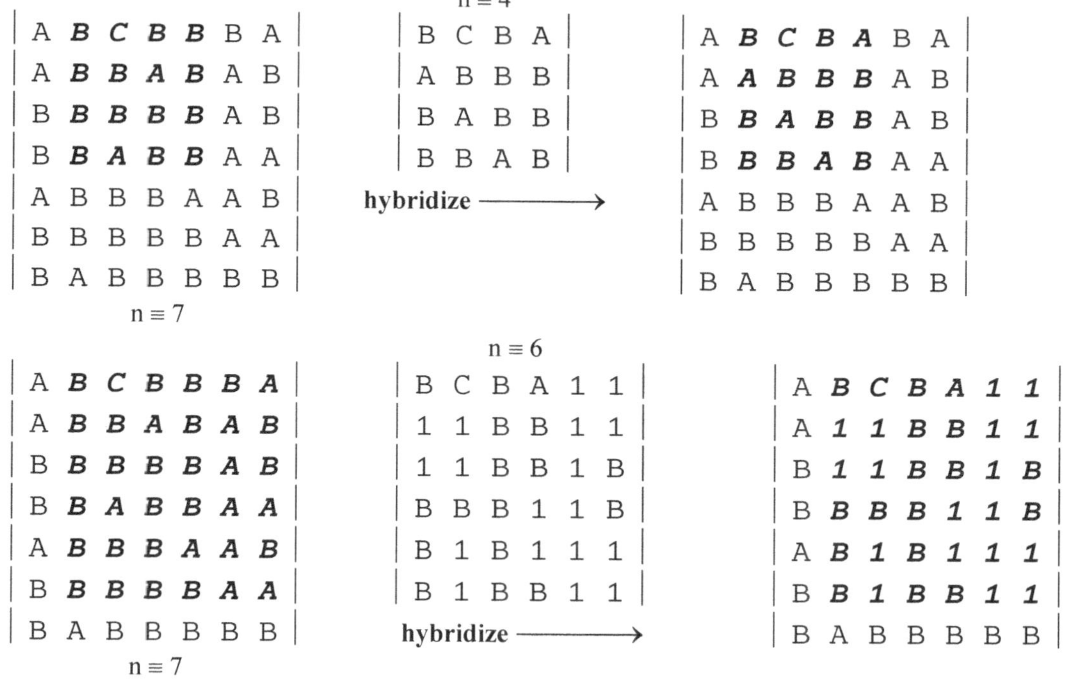

191

Also, double (i.e. multiple) hybridizations are possible:

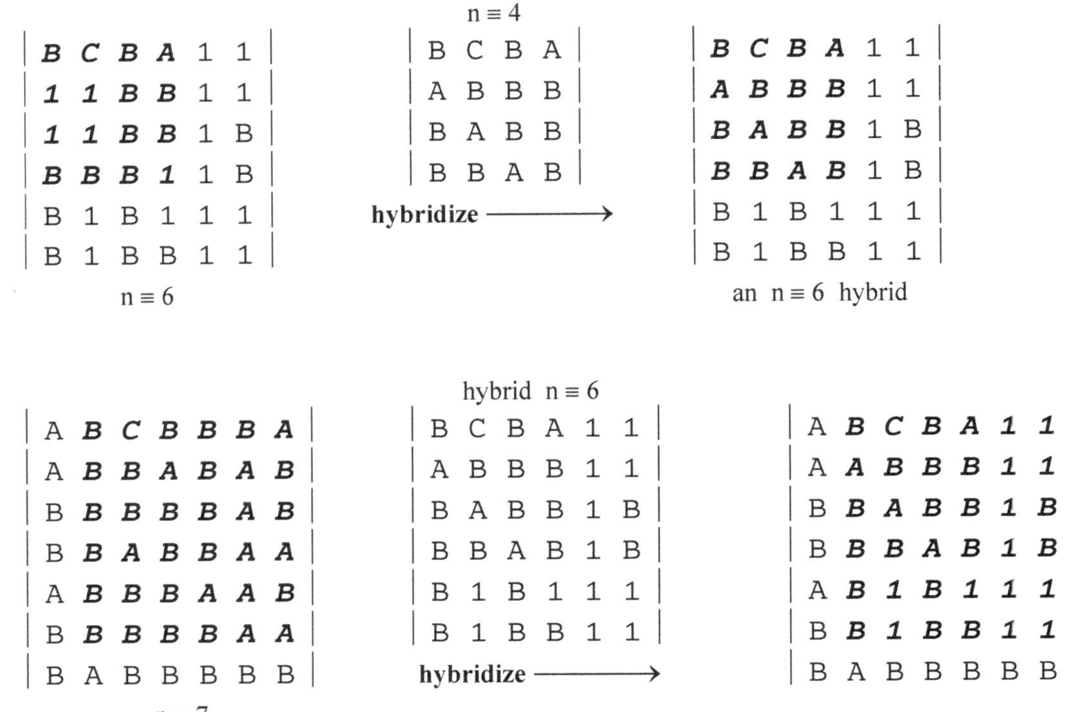

Of course, all of these (tabular) manipulations represent possible (physical) procedures, including CDD itself.

It is clear the preceding shows (particularly for the n≡6 hybridizations) instances (of solution) that are absent of the A (i.e. initial) state. Only the B state (with "1" pauses) occur, in some rows. This is because the A state (for n≡6 results) is comported (along with "C") as a state of pause (≡ "1"). In this sense (or way) the B state is generalized (by these results) as a state of flux between pauses (states A and C). All solutions (here, for the hybridized n≡7 table) suggest an oscillation of state B (into pauses or A or C). Each of the major diagonals also adopt this character.

But these contritions to differences in the final results are only feasible conclusions from the CDD analyses performed. The hybridizations disrupt the replications of element order (in rows, columns, and major diagonals) for the completed (hybrid) table. Sub-replications may persist (depending on how intact the tabular substitutions remain).

∴ If one were studying such (hybrid) tables without knowledge of prior derivations, manipulations or history (of formation, construction), then one would first search for these sub-replications to divine earlier paths of creation. It would be like searching for or elucidating hybridization "fingerprints." For example, let us consider the n≡7 table result as if it were presented *a priori* (i.e. without a history of development). Aside from the (7 element) order redundancies we might wish to search for evidence of, say, a prior n≡4 hybridization:

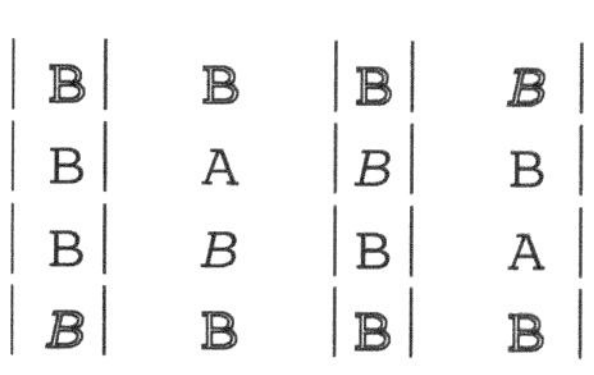

A	B	C	Ǝ	B	B	A
A	B	B	A	B	A	B
B	**B**	**B**	**B**	**B**	A	B
B	**B**	**A**	**B**	**B**	A	A
A	**B**	**B**	**B**	**A**	A	B
B	**B**	**B**	**B**	**B**	A	A
B	A	B	B	B	B	B

We might quickly alight on the highlighted (boldfaced) box, since it has (extensive, or extraordinary) row, column and (sub) major diagonal redundancies (these redundancies progenitive for the higher order 7 element redundancies):

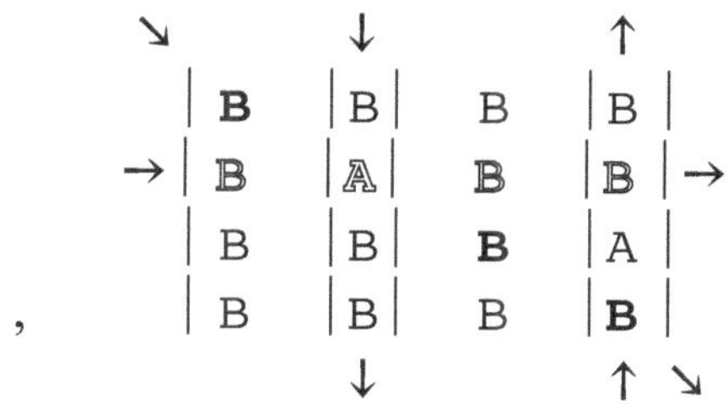

,

Consideration of redundancies

There are, then, 9 redundancies, in sets of 5 (BBBB) and 4 (BABB): one of the 4 being a reversal, and each/all of the 5 being reversible. The only unique row (or 4 element order) is (BBBA), this out of 10 linear orders of (4) elements (or, if we include reversibilities, 20). This kernel, therefore, may be treated as an approach to process, with state A clearly taking on "B" character.

One can similarly expand this kernel to (an) n≡5 box:

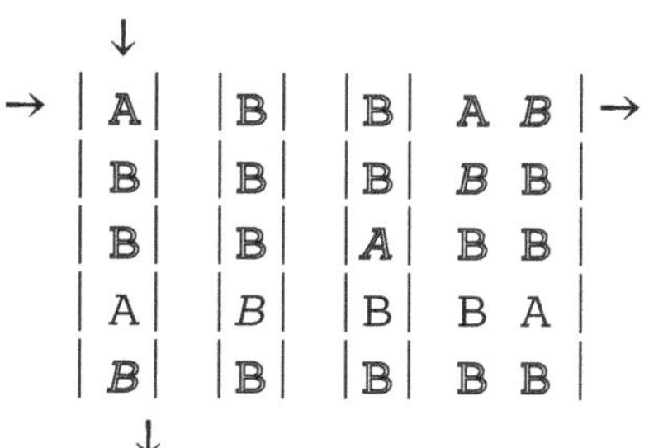

Though the 4th row remains distinct.
(It is comprised of the distinct 3rd row
of the n≡4 box.)

One can of course suggest a functional concatenation (addition) for the expansion:

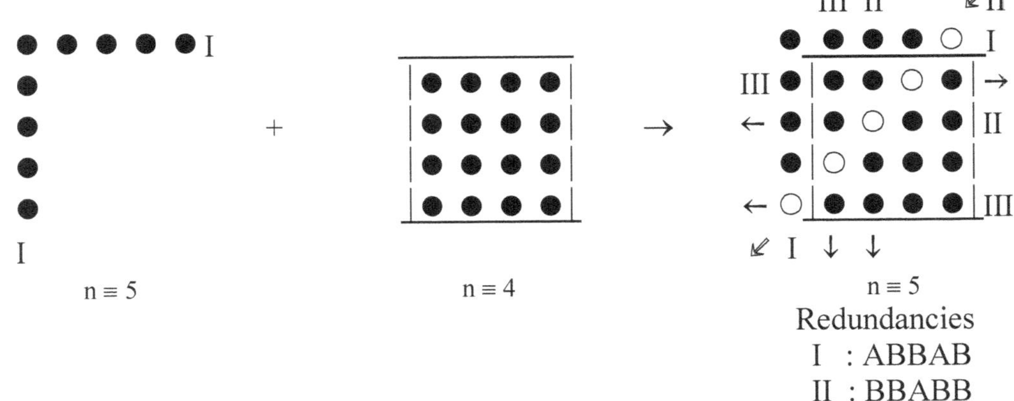

Redundancies
I   : ABBAB
II : BBABB
III : BBBBB

Similarly, corners of n≡6 and n≡7

can be added, to build up to the whole table, with redundancies as apparent or found.

Interestingly, if one reduces the n≡4 box in this manner, one comes to solely the element "B" :

One can, of course, result solely in "A" :

using disparate corner types (L ,⌐ ,⌐).

Otherwise:

```
| B | B B ꓱ |          | B | B B |              | B | B |  →  | B |
| B | A B B |          | A | B B |  →           | B   B |
| B | B B A |  →  | B   B A |        | B | B |  →  | B |
| B   B B B |              | B   B A |            | B   B |
```

```
| B B B | B |          | B B | B |              | B | B |  →  | B |
| B A B | B |          | B A | B |  →           | B   A |
| B B B | A |  →  | B B   B |        | B | B |  →  | B |
| B B B   B |              | B A | B |            | B   A |
```

∴ Consistent corner types: $(\llcorner, \llcorner, \llcorner)$; $(\lrcorner, \lrcorner, \lrcorner)$; $(\urcorner, \urcorner, \urcorner)$; $(\urcorner, \urcorner, \urcorner)$ yield to "B" (as to regularly decrease n from 4 to 1).

It is therefore notable that enhancement (concatenation) from n≡4 (box) to 5, 6, 7 occurs through disparate corner additions: $(\ulcorner)$, $(\lrcorner)$, $(\urcorner)$. Thus are characteristics achievable from CDD generated results, and mischiev-able from any hybridizations present; necessarily, if possible, disrupting the large extents of redundancies for the whole(d) table.

So let us then consider a highly hybridized (from the same or similar processes) table *a priori* :

```
                                                        ↓
| A  B  C  B  A  1  1 |                        | A  B  C  B  A  1  1 |
| A  A  B  B  B  1  1 |      Note redundancies, | A  A  B  B  B  1  1 |
| B  B  A  B  B  1  B |      ───────────►       | B  B  A  B  B  1  B | →
| B  B  B  A  B  1  B |      and hypothesize a possible  | B  B  B  A  B  1  B | ←
| A  B  1  B  1  1  1 |      (hybrid) table outline.     | A  B  1  B  1  1  1 |
| B  B  1  B  B  1  1 |                        | B  B  1  B  B  1  1 |
| B  A  B  B  B  B  B |                        | B  A  B  B  B  B  B |
        n ≡ 7  hybrid
```

```
  ↓   ↑↓
       | B C B A |              | B   C B A |                  ↖ ↕ ↑  ↙↗
➡  → | A B B B | →    corner  | A | B B B |   corner    ↔ | B B B | ↔
   → | B A B B | →    ───────► | B | A B B |   ───────►   → | A B B | →
   ← | B B A B | ←              | B | B A B | reduction (removal)  ↔ | B A B | ↔
  ↓   ↑↓                                                        ↙↗ ↕ ↑  ↖↘

  corner        | B   B B |    reduction      ↓ ↕ ↙
  ───────►  | A | B B |    ───────►  ↔ | B B | ↔  corner    | B   B |  reduction
                | B | A B |                  | A B |    ───────►  | A | B |  ───────►  | B |
                                          ↙ ↓ ↕
```

This would suppose "B" to be a (or the) fundamental element (state) of the process. The hybrid (insertional) identity is assumed to be correct.

Another (sub)table with redundancies could have been chosen:

e.g.

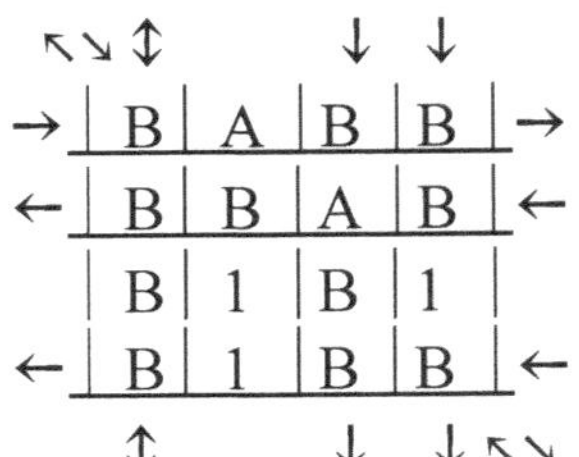

This table is original to the n≡7 hybrid (i.e. we did not use it, with hybridization, to create the n≡7 hybrid).

It is a result of the n≡7 hybrid, but we might not be able to say this *a priori*.

```
  | B  A B B |
  | B |B A B |  →   | B  A B |        ↖↘     ↑ ↗
  | B |1 B 1 |      | 1 |B 1 |  → →  | B  1 | →
  | B |1 B B |  ←   | 1  B B | ←  ↔ | B  B | ↔  →  | B |
```

Another could be:

```
         ↓ ↑   ↙
  | A  B 1 B |        ↘     ↓ ↓ ↙
→ | B |1 1 1 | → →  | 1  1 1 |      ↘ ↓  ↓ ↙
  | B |B 1 1 |   ←  | B |1 1 | ←  → | 1  1 |
  | B |B B B |      | B |B B |      | B |B |  →  | B |
      ↙   ↓↑         ↙   ↓ ↓ ↘      ↙ ↓  ↓ ↘
```

Another could be:

```
        ↑  ↓   ↑
→ | B  B B A | →     ↖↘
  | A |B 1 B | → ↔ | B  1 B | ↔    ↘ ↓  ↕ ↙↗
  | B |B 1 B |   ↔ | B |1 B | ↔ → → | 1  B | →
← | B |A B B | ←   | A |B B |     ↔ | B |B | ↔  →  | B |
    ↑  ↓    ↑                 ↖↘      ↙↗↓  ↕ ↘
```

Another could be:

```
        ↓  ↓   ↓
  | A  B C B |          ↓  ↕ ↑ ↙↗
→ | A |A B B | → → → | A  B B | →          ↓  ↑
← | B |B A B | ←   ↔ | B |A B | ↔ → → | A  B | →
→ | B |B B A | →   ← | B |B A | ←   ← | B |A | ←  →  | A |
    ↓  ↓    ↓         ↙↗↓  ↕ ↑          ↓  ↑
```

major diagonals:
| A A A A |, | B B B B |

Another could be:

```
     ↘ ↑   ↓   ↑
     | B  A 1 1 |              ↖↘
  →  | B | B 1 1 | →    →    | B   1 1 |         ↖↘        ↙↗
  →  | B | B 1 B | →    ↔  | B | 1 B | ↔  →  ↔ | 1  B | ↔
     | A | B 1 B |       ↔  | B | 1 B | ↔    ↔ | 1 | B | ↔    →   | B |
       ↑  ↓  ↑↘                  ↖↘           ↙↗      ↖↘
```

Another could be:

```
       ↓↑   ↙
     | A  B 1 B |          ↘    ↓    ↙
  →  | B | 1 1 1 | →    →    | 1   1 1 |        ↘  ↓   ↓ ↙
     | B | B 1 1 |       ←  | B | 1 1 | ←  →  | 1   1 |
     | B | B B B |          | B | B B |          | B | B |    →   | B |
       ↙   ↓↑              ↙    ↓    ↘       ↙  ↓   ↓ ↘
```

etc.

The occurrences of redundancies for so many selected $n{\equiv}4$ (sub)tables (of the $n{\equiv}7$ table) strongly suggests the parent table was constructed (with possible hybridizations) using the CDD method. It is also possible, due to the ordering of the table, to "corner reduce" down to each type of element occurring within $n{\equiv}4$ boxes (except for "C"), since each such element occurs as the last one (4th element of 4th row) for some $n{\equiv}4$ box. It would be difficult to determine the actual (though intact) hybrid (used) to make the $n{\equiv}7$ box (i.e. the unaltered hybrid left in the 7×7 table).

Nul/los, of course, proposes that no physical (or real) process can result in zero (or null result). So whatever occurs is a pause ($\equiv$ "1") or some change. $\therefore$ While CDD promotes redundancies (or operational order), and these redundancies can be reduced (via, e.g., corner reduction or restriction or subtraction) down to some fundamental operation (or element), then the reverse is also plausible: to build up a process around or based on a fundamental operation (via, e.g., corner enhancement or allowance or addition). The logical following of the reversibility of a process (whether the such may proceed or not) is a basis for establishing a possible correctness of a procedure (operationally and logically). In the above (previous) examples we are certain to focus on establishing the state "C" (from "A"), and the logic follows for oscillations through state "B" as a necessary procedure (for the given process: A→B→C). Again, studying the (hypothetically proposed) hybridized table of $n{\equiv}7$, for possible $n{\equiv}4$ hybrid identities, with judiciously "C" placed as part of a 1st row solution (and therefore not necessarily replicated through CDD processes, but definitely maintained throughout as an element), we see that there are only 3 possible $n{\equiv}4$ boxes available for analysis. Two have already been presented:

```
  | A  B C B |
  | A | A B B |  →    | A  B B |
  | B | B A B |       | B | A B |  →    | A  B |
  | B | B B A |       | B | B A |       | B | A |  →   | A |
```

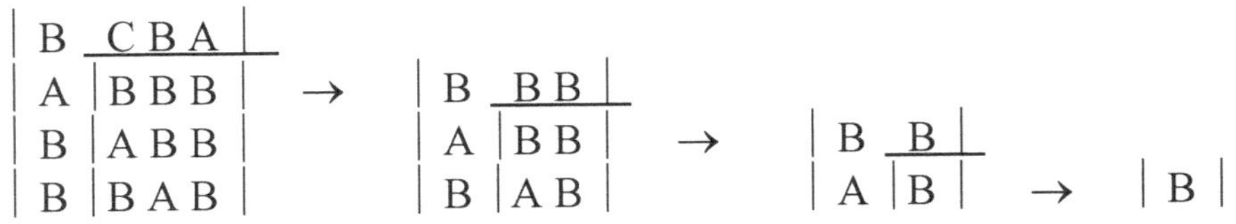

```
| B  C B A |                  | B  B B |                  | B  B |
| A | B B B |  →     | A | B B |   →      | B  B |
| B | A B B |                  | B | A B |          | A | B |  →   | B |
| B | B A B |
  (actual hybrid used)
```

The last is:

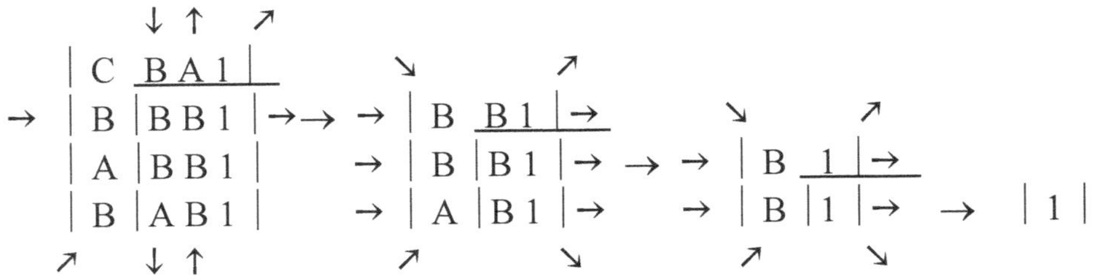

```
         ↓ ↑  ↗
    | C  B A 1 |             ↘              ↗
 →  | B | B B 1 | →→ →  | B  B 1 | →      ↘           ↗
    | A | B B 1 |           → | B | B 1 | → → →  | B  1 | →
    | B | A B 1 |           → | A | B 1 | →      → | B | 1 | → →  | 1 |
   ↗  ↓ ↑            ↗          ↘          ↗          ↘
```

∴ We see that each of the 1st row elements (aside from "C") can be reduced to a fundamental one, representing all of the elements below the 1st row (and all but "C" of the 1st row). The obvious corollary is that each of these "fundamental" operations can be used (though not necessarily must be used) to produce a state of "C", in a correct (or logically coherent) process (of the hybridized n≡7 table presented for analysis). It also seems fairly evident that state "B" is the most common fundamental found from the n≡4 sub-tables studied. (It is also the most numerous element in the n≡7 hybridized table.) Logistically, since "C" occurs on the 1st row (and) ∴ not on the 4th or lower, it does not have an opportunity to be reduced to a fundamental state in an n≡4 box (unless column reversal is utilized: flipping the n≡7 table upside-down). It is also the 3rd element of the 1st row ∴ (it) can not be a 4th element for an n≡4 box (unless row reversal is utilized, subject to the manner of corner reduction and/or column/table reversal). Of course, hybrids of n≡1, 2, 3 (element #s: 1, 4, 9) may be considered for use, although n<4 boxes are generally automatically brought up to n≡4 during the CDD procedure.

The n boxes (studied) can be thought to comprise the potential operations of a system (or sub-system). And obviously, the more redundant and symmetrical the system the more functionally correct can be assumed its description thereof (operations and orders of operations or elements). This is why analogies can be easily (or facilely) brought to atoms and atomic (or molecular) states. Note that the reversal of the | 1 | fundamental can be construed as the fundamental | C |, by reversing the manner of corner reduction:

```
                              ↓ ↕  ↑           ↘ ↓  ↙↗
   | C B A | 1 |      → | C B | A | →      → | C  B | →   →  | C |
   | B B B | 1 |      ↔ | B B | B | ↔  → ↔ | B  B | ↔
   | A B B | 1 | →  ← | A B  B | ←      ↙↗ ↓        ↘
   | B A B  1 |           ↓ ↕  ↑
```

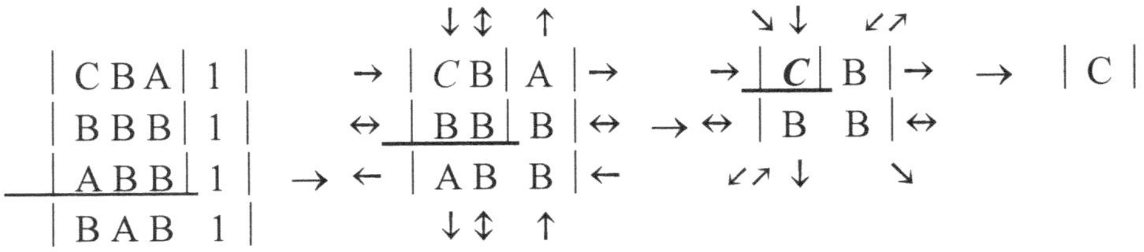

For the other two n≡4 "C" boxes (i.e. containing "C"), the reversals lead to the same fundamental elements: A, B. The n≡7 (hybridized) table, then, represents a conglomerate of potential (n) states.

What process of (physical) nature succeeds with the CDD (processing) procedure to follow? It is a generalization for correctness (i.e. allowance) of a procedure or process. Elements are selected and made fundamental through CDD. It is clear that a hypothetical table (i.e. before CDD) for a procedure may be treated for searching of redundancies and fundamentals as well, since these are only tabular manipulations (whether premature or not):

e.g.

```
| A B C B B B A |                    | A  B C B |
| A 4 C B B B A |   select      →    | A | 4 C B | →     → | 4  C B | →
| A 4 C 4 B B A |   ─────────→  →    | A | 4 C 4 | → → ↔  | 4 | C 4 | ↔
| A 4 C 4 A B A |   n≡4  box    →    | A | 4 C 4 | →   ↔  | 4 | C 4 | ↔
| A 4 C 4 A A A |
| A B C 4 A A A |
| A B C B A A A |
                          → →  | C  4 | →
                          →  | C | 4 | →    →    | 4 |
```

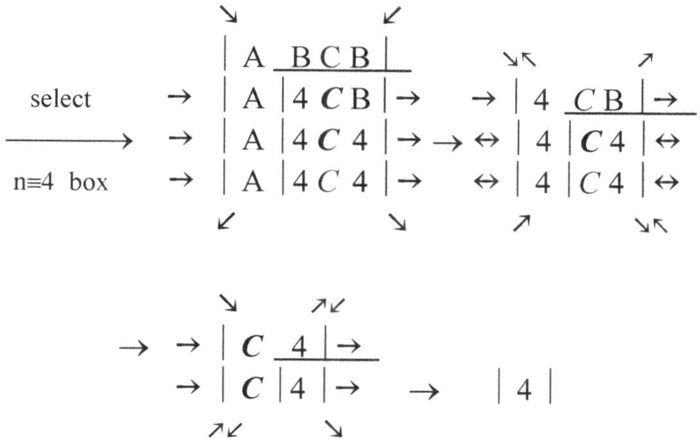

Its reverse (corner reduction) is:

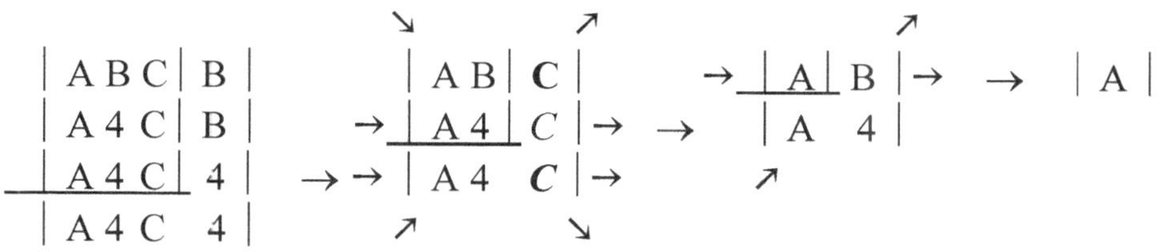

```
| A B C | B |        | A B | C |       → | A | B | →   →   | A |
| A 4 C | B |     →  | A 4 | C | →  →     | A   4 |
| A 4 C | 4 |     → → | A 4   C | →
| A 4 C   4 |
```

As for CDD (treated) resulting tables, the occurrence of extensive redundancies heralds assumptions of correctness for the procedures (rows and solutions). Again, we are treating the parenting table *a priori*, but are not necessarily searching for (n≡4 box) hybridizations (or suspecting).

But it (physically) stands to reason, that as we may extract (a sub-table), for a fundamental element to derive, then under nul/los a shell is created which must (and may) be filled in (with "1"s).
For example, extracting a strictly internal n≡4 box from a hypothetical process as of previous:

```
| A B C B B B A |                    | A B C B B B A |
| A 4 C B B B A |                    | A 4 C B B B A |
| A 4 C 4 B B A |   extraction       | A 4 1 1 1 1 A |
| A 4 C 4 A B A |   ─────────→       | A 4 1 1 1 1 A |
| A 4 C 4 A A A |   with fill-ins     | A 4 1 1 1 1 A |
| A B C 4 A A A |                    | A B 1 1 1 1 A |
| A B C B A A A |                    | A B C B A A A |
```

we suddenly find identical solutions (rows) and one very similar solution:

3 of | A 4 1 1 1 1 A | ; 1 of | A B 1 1 1 1 A | .

The extracted sub-table is characteristically treated down to the fundamental | A | :

```
      ↘             ↗
     | C  4 B B |          ↘          ↗
  →  | C |4 A B |→ → →  | 4  A B |→
  →  | C |4 A A |→    → | 4 |A A |→  →    | A  A |
  →  | C |4 A A |→    → | 4 |A A |→       | A |A |    →    | A |
      ↗         ↘        ↗        ↘    (redundancies obvious)
```

and also for the fundamental | C | :

```
               ↘         ↗        ↘         ↗
  | C 4 B | B |     → | C 4| B |→      → | C| 4 |→   →   | C |
  | C 4 A | B |     → | C 4| A |→  → → | C  4 |→
  | C 4 A | A |  → → | C 4  A |→          ↗        ↘
  | C 4 A  A |       ↗          ↘
```

One also notes that the "1"s filled in the table can in various ways be more or less directly "sub-reduced" (via physical facilities):

```
  | A  B  C  B  B  B  A |
  | A  4  C  B  B  B  A |              →
  | A  4  1  1  1  1  A |      | A 4 A |
  | A  4  1  1  1  1  A |  →   | A 4 A |          | C B B B |
  | A  4  1  1  1  1  A |      | A 4 A |      ↓   | C B B B |
  | A  B  1  1  1  1  A |      | A B A |  ,       | C B A A |
  | A  B  C  B  A  A  A |
```

with the unbalanced sub-reduced tables easily (facilely) made balanced (with "1"s) :

```
  ↘ ↓  ↓ ↗                    ↓ ↓ ↓ ↙
 → | A 4 A 1 |→           → | C B B B |→
 → | A 4 A 1 |→           → | C B B B |→
 → | A 4 A 1 |→             | C B A A |     each with redundancies
   | A B A 1 |     ,        | 1 1 1 1 |
  ↗ ↓  ↓ ↘                   ↙ ↓ ↓ ↓
```

Obviously, the overall motive for these manipulations (corner reduction, finding fundamental elements, CDD) is to search for (tabular) redundancies (to support the correctness of the tabular process, a table presumed to be composed of the same solution for each of its rows).

But what could redundancies refer to, anoint or indicate as possible? Again, using the analogy of the atomic state, with *orthogonalization* (right $\angle$ turning) as a form of derivation(ing), we may take for an example the sub-table (as previous):

|C 4 B B
|C 4 A B|
|C 4 A A|
|C 4 A A|

⬏

⟶

orthogonalization

(columns becoming rows)

|C C C C|
|4 4 4 4|
|A A A B|
|A A B B|

to denote, among other things, a persistence of state "C" (as a row solution), ∴ made equivalent to a persistence of state "4" and yielding to (or allowing) solutions |AAAB| and |A A B B| (each of which can be redundancy reduced to simply: |AB|).

∴ A derivation of solution |C4BB| is |CCCC| (according to the used sub-table); and as well, a derivation of |C4AB| is |4444|, of |C4AA| is |AAAB|, and another of |C4AA| is |AABB|.

∴ The (orthogonalization) derivative (sub-)table:

|C C C C| → |C|
|4 4 4 4| → |4|
|A A A B| → |AB|
|A A B B| → |AB|

can be redundancy reduced to:

|C 1| → |C 1 1 1 |→
|4 1| ≡ |4 1 1 1 |
|A B| → |A B 1 1 |→
|A B| → |A B 1 1 |→

The result can be (corner reduced) fundamantalized down to |1| and its reverse, |C|.

Thus (is) for an atom (of an assembly) stuck in state "C", "C" being virtually (but not actually as by function) a state of pause (as like "1"). All of these tabular manipulations and results represent (possible or available) physical processes.

 Given |CCCC| and |4444| as accepted solutions, another obvious manipulation is symmetrical displacement (SD), here contingent on the

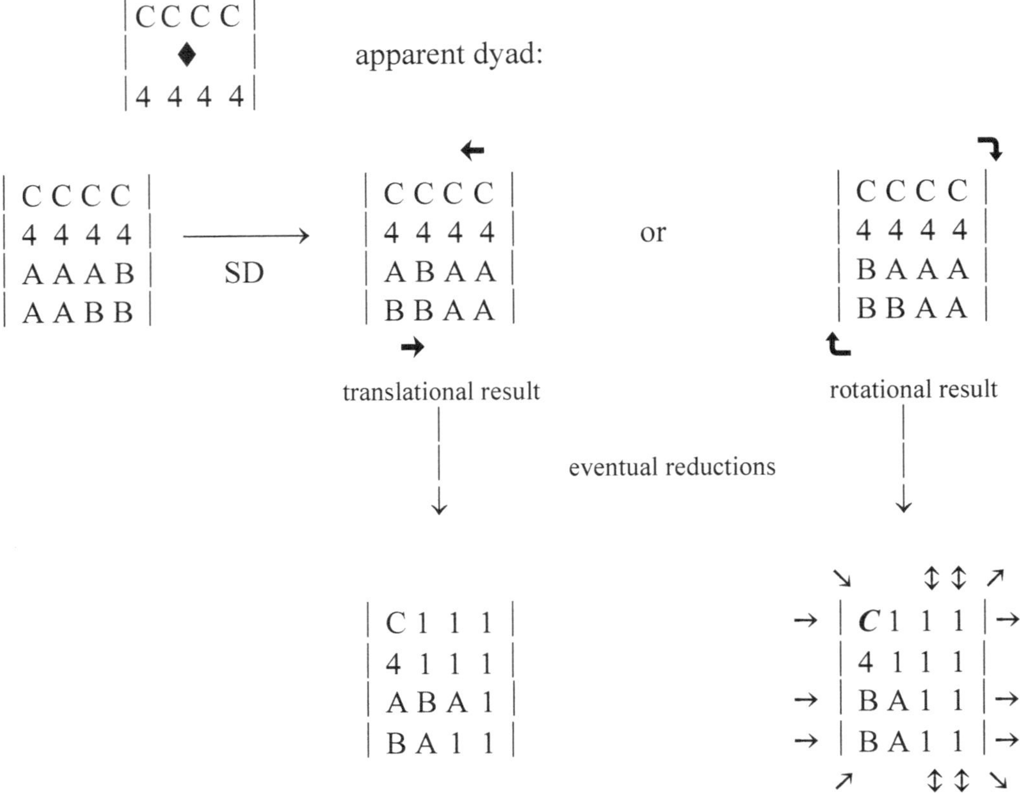

(In this particular case,) The rotational SD leads to a result with much redundancy, while the translational SD leads to a result with none. Both SDs are physically possible, but the rotational SD is simpler (i.e. simply a rotation, through planes, about the dyad of the n≡4 table) and could be more valid for the (proposed) process.

 Translational SD requires two opposite translations of half-tables within the tabular plane, and therefore involves replacement (substitution) of elements somehow without their positional disruption (through coincidences of position and placement: i.e. During the process there would have to be half-table displacements of an extra-planar sort).

The rotational SD result, of course, retains the (manners of types of) redundancies of the original table, since the resulting table is simply the original turned (horizontally) around (via the table's dyad), the 1st and 2nd row solutions unchanging (which functionally or theoretically allows for this manipulation):

$$
\begin{array}{c}
\circlearrowleft \\
| \; C\,C : C\,C \; | \\
\text{Table's} \qquad | \; 4\,4 : 4\,4 \; | \qquad \textbf{rot SD} \qquad | \; C\,C\,C\,C \; | \\
\text{dyad} \qquad \longrightarrow \qquad | \qquad \blacklozenge \qquad | \qquad \longrightarrow \qquad | \; 4\,4\,4\,4 \; | \\
\text{(for rotation)} \qquad | \; A\,A : A\,B \; | \qquad\qquad | \; B\,A\,A\,A \; | \\
| \; A\,A : B\,B \; | \qquad\qquad | \; B\,B\,A\,A \; | \\
\circlearrowleft
\end{array}
$$

(The translational SD is of course disruptive of (row) manner as well as order.)

This is in contrast to vertical SD for rotation and translation, since in either case the rows are not disrupted in element order (but only position or rank):

$$
\begin{array}{l}
| \; C\,C\,C\,C \; | \downarrow \text{trans} \\
| \; 4\,4\;\;4\,4 \; | \qquad\qquad \textbf{SD} \\
\text{rot } \circlearrowright \dots | \dots\dots \blacklozenge \dots\dots | \dots \circlearrowright \text{rot} \longrightarrow \\
| \; A\,A\;\;A\,B \; | \\
\text{trans} \uparrow | \; A\,A\;\;B\,B \; |
\end{array}
\qquad
\begin{array}{c}
\searrow\; \downarrow\downarrow\downarrow \;\; \swarrow \\
| \; A\,A\,B\,B \; | \\
| \; A\,A\,A\,B \; | \\
| \; 4\,4\,4\,4 \; | \\
| \; C\,C\,C\,C \; | \\
\swarrow\; \downarrow\downarrow\downarrow \;\; \searrow \\
\text{rot}
\end{array}
\qquad , \qquad
\begin{array}{c}
\searrow\; \downarrow\downarrow \;\; \downarrow\; \swarrow \\
| \; A\,A\,A\,B \; | \\
| \; A\,A\,B\,B \; | \\
| \; C\,C\,C\,C \; | \\
| \; 4\,4\,4\,4 \; | \\
\swarrow\; \downarrow\downarrow \;\; \downarrow\; \searrow \\
\text{trans}
\end{array}
$$

Both results yield to redundancies, while fully maintaining original rows, ∴ solutions (as well as columns of alignment of elements, though reversed, for the rot. SD).

Clearly for a table (i.e. process), then, one can assign a "redundancy value" for arithmetic comparisons and alteration through (tabular) manipulation. For example, let us take, again, a previous sub-table:

$$
\begin{array}{l}
| \; C\,4\;B\,B \; | \\
| \; C\,4\;A\,B \; | \qquad \equiv \quad \rightarrow \\
| \; C\,4\;A\,A \; | \qquad\qquad \rightarrow \\
| \; C\,4\;A\,A \; |
\end{array}
\qquad
\begin{array}{c}
\searrow \qquad\quad \nearrow \\
| \; C\,4\;B\,B \; | \\
| \; C\,4\;A\,B \; | \rightarrow \\
| \; C\,4\;A\,A \; | \rightarrow \\
| \; C\,4\;A\,A \; | \rightarrow \\
\nearrow \qquad\quad \searrow
\end{array}
\qquad
\begin{array}{l}
\therefore \text{ 5 redundancies noted} \\
\text{(follow the arrow directions)}
\end{array}
$$

$$
\begin{array}{l}
| \; C\,4\;B\,B \; | \qquad \text{orthogonalization} \\
| \; C\,4\;A\,B \; | \qquad \longrightarrow \\
| \; C\,4\;A\,A \; | \qquad \text{(for derivative)} \\
| \; C\,4\;A\,A \; |
\end{array}
\qquad
\begin{array}{l}
| \; C\,C\,C\,C \; | \\
| \; 4\,4\,4\,4 \; | \\
| \; A\,A\,A\,B \; | \\
| \; A\,A\,B\,B \; |
\end{array}
\qquad
\begin{array}{l}
\therefore \text{ 5 redundancies retained} \\
\text{orthogonalization (or derivative making)} \\
\text{does not alter redundancy value}
\end{array}
$$

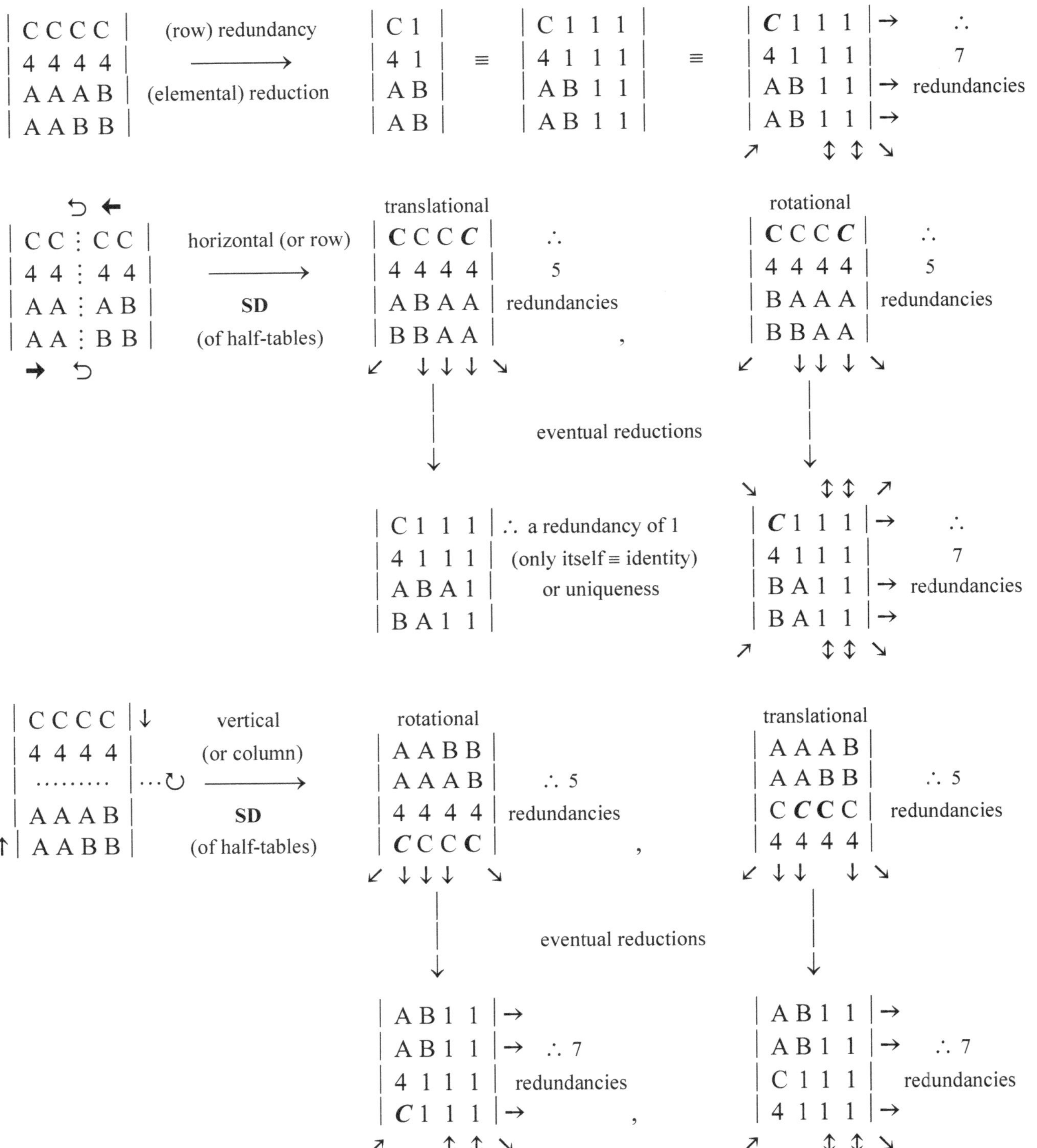

∴ (Eventual) Simplifying redundancies can enhance (increase) or lower redundancy values. SD itself (either horizontal:row or vertical:column) doesn't seem to alter the redundancy values. Neither does orthogonalization (so alter). ∴ We are discerning tabular manipulations which do and do not alter redundancy values.

Mathematical operations

We may also begin to develop some singular basic mathematical operations for tables ($\therefore$ processes):

e.g.

$$\begin{vmatrix} C & C & C & C \\ 4 & 4 & 4 & 4 \\ A & A & A & B \\ A & A & B & B \end{vmatrix} + \begin{vmatrix} e \end{vmatrix} \equiv \begin{vmatrix} C & C & C & C & e \\ 4 & 4 & 4 & 4 & e \\ A & A & A & B & e \\ A & A & B & B & e \end{vmatrix} \qquad \text{since} \quad \begin{vmatrix} e \end{vmatrix} \equiv \begin{vmatrix} e & ? & e \\ e & & 1 \\ e & & 1 \\ e & & 1 \end{vmatrix} \equiv \begin{vmatrix} e \\ e \\ e \\ e \end{vmatrix}$$

$$\begin{vmatrix} C & C & C & C & e \\ 4 & 4 & 4 & 4 & e \\ A & A & A & B & e \\ A & A & B & B & e \end{vmatrix} - \begin{vmatrix} e \end{vmatrix} \equiv \begin{vmatrix} C & C & C & C \\ 4 & 4 & 4 & 4 \\ A & A & A & B \\ A & A & B & B \end{vmatrix}$$

Redundancy value (rv)

But is:

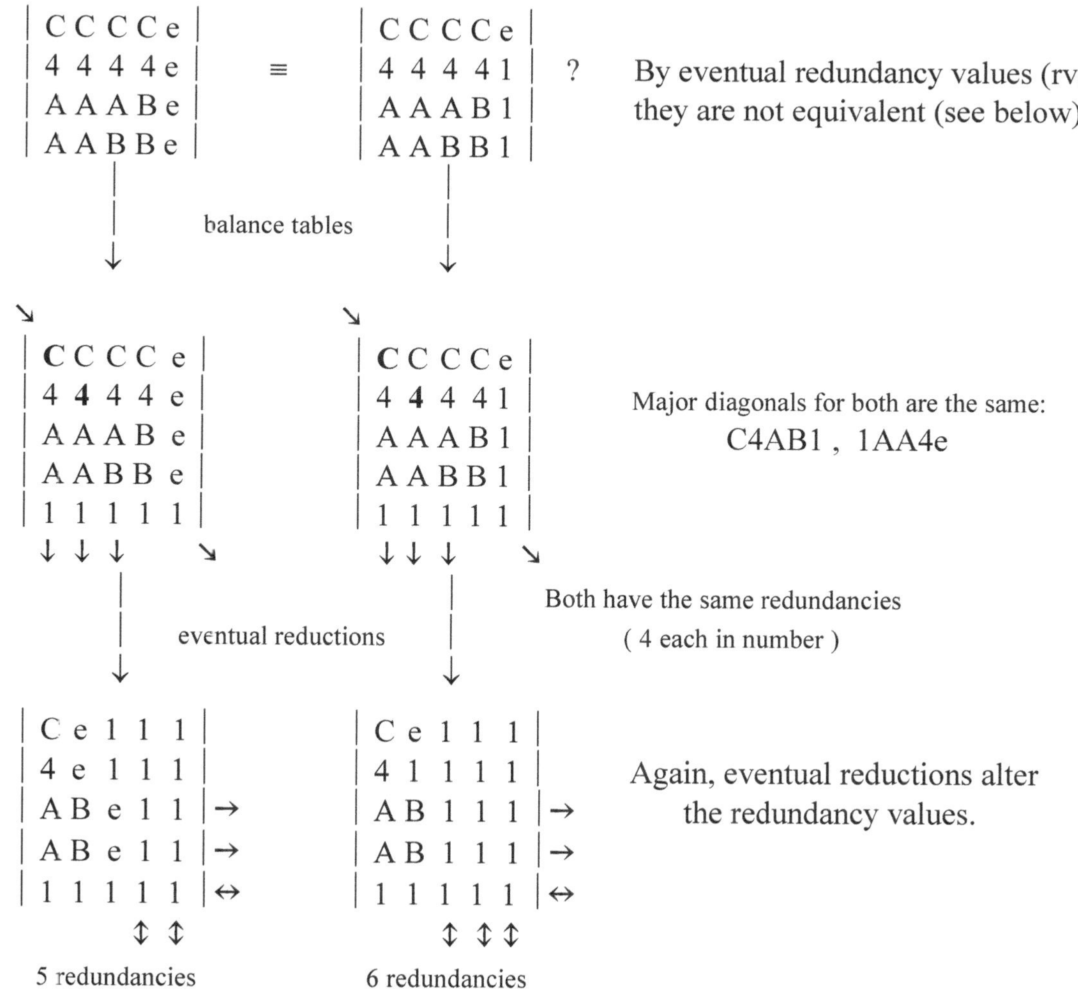

By eventual redundancy values (rvs), they are not equivalent (see below).

Major diagonals for both are the same:
$$C4AB1 \ , \ 1AA4e$$

Both have the same redundancies (4 each in number)

Again, eventual reductions alter the redundancy values.

$\therefore$ Simplifying reductions alter tables (rvs).

Now (with the realization that reductions alter tables):

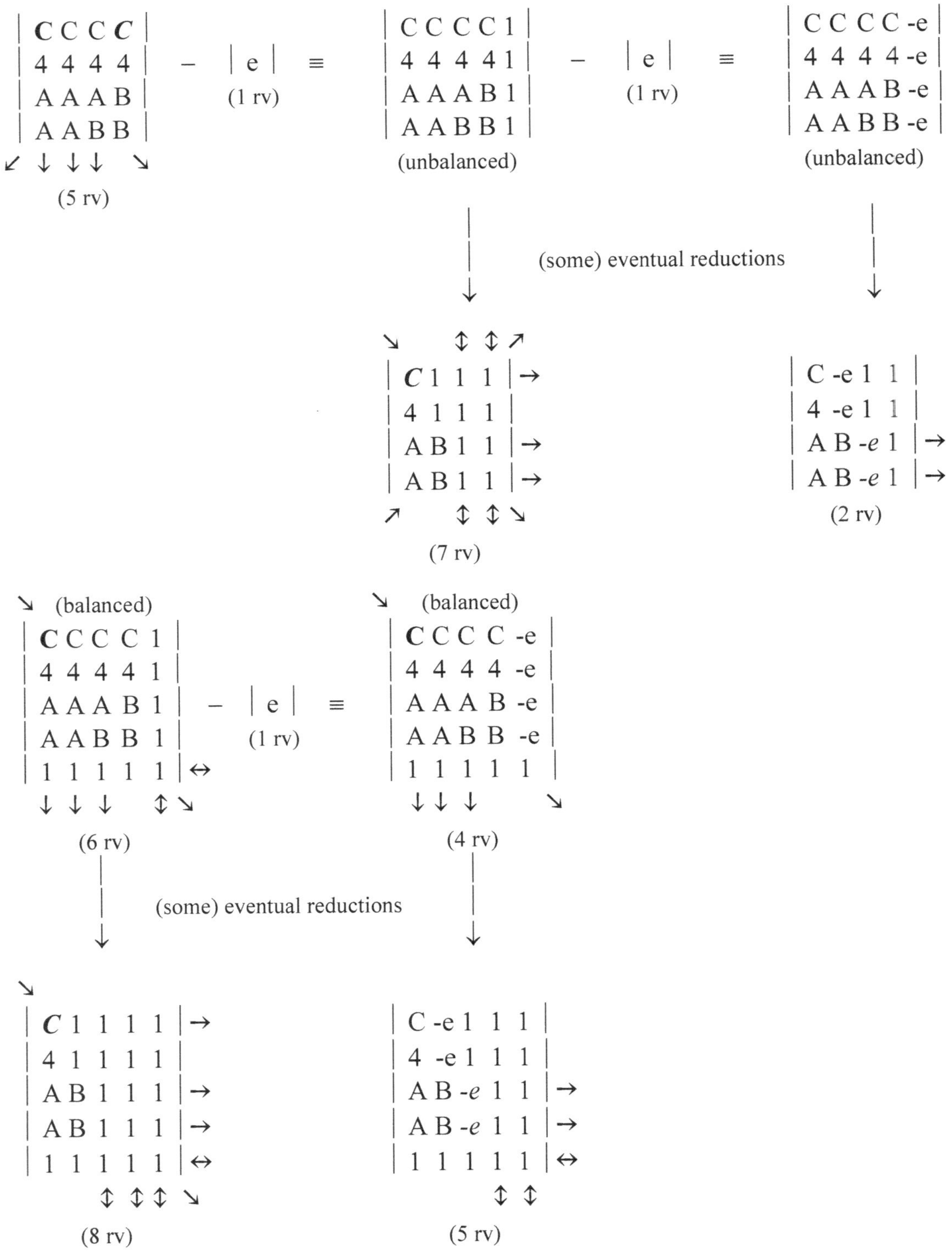

$$
\begin{vmatrix} C & C & C & \boldsymbol{C} \\ 4 & 4 & 4 & 4 \\ A & A & A & B \\ A & A & B & B \end{vmatrix} \; - \; \begin{vmatrix} e \end{vmatrix} \; \equiv \; \begin{vmatrix} C & C & C & C & 1 \\ 4 & 4 & 4 & 4 & 1 \\ A & A & A & B & 1 \\ A & A & B & B & 1 \end{vmatrix} \; - \; \begin{vmatrix} e \end{vmatrix} \; \equiv \; \begin{vmatrix} C & C & C & C & \text{-e} \\ 4 & 4 & 4 & 4 & \text{-e} \\ A & A & A & B & \text{-e} \\ A & A & B & B & \text{-e} \end{vmatrix}
$$

(middle: 1 rv; right: 1 rv)

left: ↙ ↓ ↓ ↓ ↘ (5 rv)

middle: (unbalanced)   right: (unbalanced)

(some) eventual reductions

↘ ↕ ↕ ↗
$$
\begin{vmatrix} \boldsymbol{C} & 1 & 1 & 1 \\ 4 & 1 & 1 & 1 \\ A & B & 1 & 1 \\ A & B & 1 & 1 \end{vmatrix}
$$
→ (rows 1, 3, 4)
↗ ↕ ↕ ↘
(7 rv)

$$
\begin{vmatrix} C & \text{-e} & 1 & 1 \\ 4 & \text{-e} & 1 & 1 \\ A & B & \text{-}e & 1 \\ A & B & \text{-}e & 1 \end{vmatrix}
$$
→ (rows 3, 4)
(2 rv)

↘ (balanced)
$$
\begin{vmatrix} \boldsymbol{C} & C & C & C & 1 \\ 4 & 4 & 4 & 4 & 1 \\ A & A & A & B & 1 \\ A & A & B & B & 1 \\ 1 & 1 & 1 & 1 & 1 \end{vmatrix} \; - \; \begin{vmatrix} e \end{vmatrix} \; \equiv \; \begin{vmatrix} \boldsymbol{C} & C & C & C & \text{-e} \\ 4 & 4 & 4 & 4 & \text{-e} \\ A & A & A & B & \text{-e} \\ A & A & B & B & \text{-e} \\ 1 & 1 & 1 & 1 & 1 \end{vmatrix}
$$

left: (1 rv); ↓ ↓ ↓ ↕ ↘ (6 rv)

right: ↓ ↓ ↓ ↘ (4 rv)

(some) eventual reductions

↘
$$
\begin{vmatrix} \boldsymbol{C} & 1 & 1 & 1 & 1 \\ 4 & 1 & 1 & 1 & 1 \\ A & B & 1 & 1 & 1 \\ A & B & 1 & 1 & 1 \\ 1 & 1 & 1 & 1 & 1 \end{vmatrix}
$$
→ (rows 1, 3, 4); ↔ (row 5)
↕ ↕ ↕ ↘
(8 rv)

$$
\begin{vmatrix} C & \text{-e} & 1 & 1 & 1 \\ 4 & \text{-e} & 1 & 1 & 1 \\ A & B & \text{-}e & 1 & 1 \\ A & B & \text{-}e & 1 & 1 \\ 1 & 1 & 1 & 1 & 1 \end{vmatrix}
$$
→ (rows 3, 4); ↔ (row 5)
↕ ↕
(5 rv)

∴ Balancing tables alters redundancy values (by necessity).

From addition and subtraction, we have multiplication:

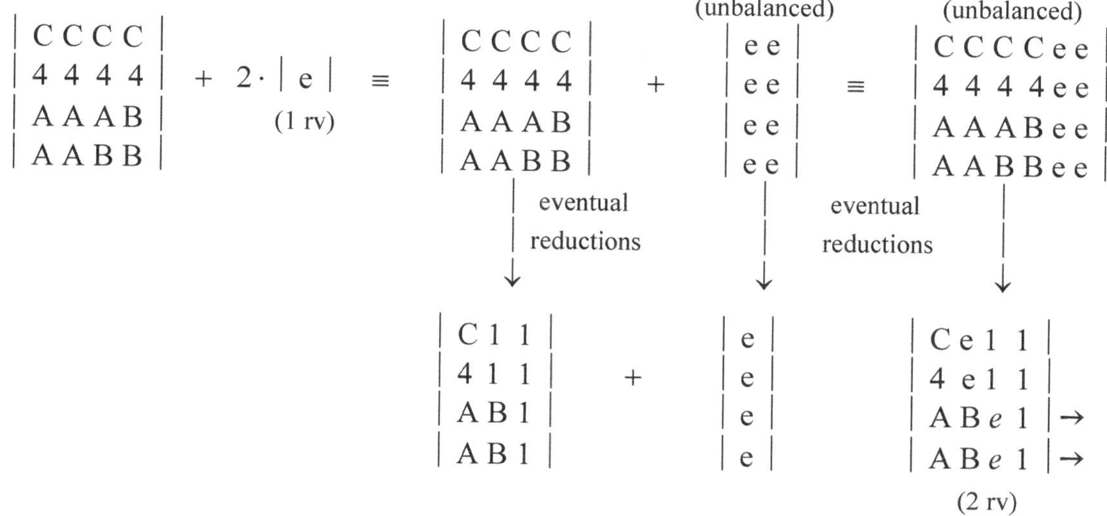

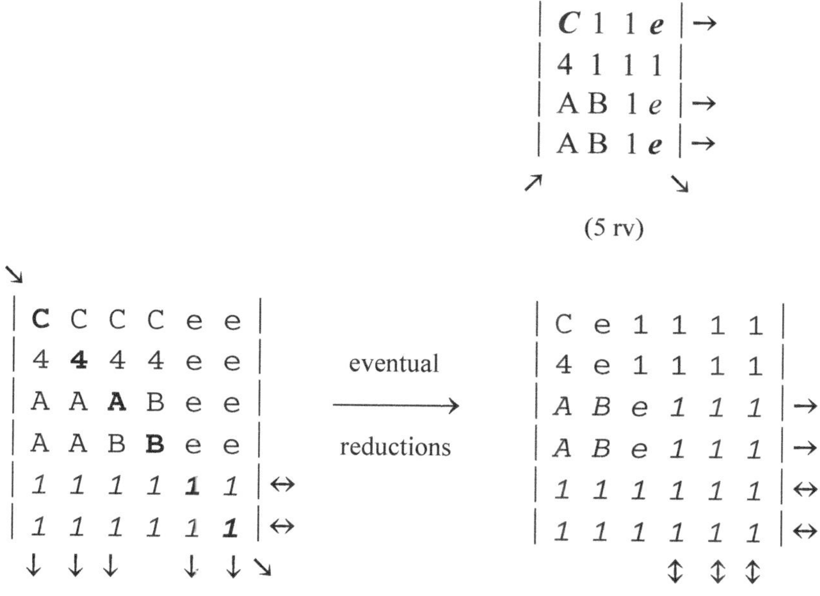

But division is a particular palate for tables:

e.g.

```
| C C : C C |          | C C C C |        | A A A B |
| 4 4 : 4 4 |          | 4 4 4 4 |   +    | A A B B |   ,
...|......:......|... ÷ 2  ≡
| A A : A B |          | C C |      | C C |
| A A : B B |          | 4 4 |  +   | 4 4 |
                       | A A |      | A B |
                       | A A |      | B B |        (etc.)
```

Analyzing (these results) with eventual reductions:

```
| C C C C |      | A A A B |          | C C |     | C |        | C 1 1 1 |
| 4 4 4 4 |      | A A B B |          | 4 4 |  →  | 4 |   ≡    | 4 1 1 1 |
    ↓                ↓                | A A |     | A |        | A A A A | ↔
                                      | A A |     | A |        | A A A A | ↔

| C 1 1 |→       | A B 1 1 |→                               ↗   ↑ ↑ ↑
| 4 1 1 1 |      | A B 1 1 |→                                    (6 rv)
| 1 1 1 1 |↔     | 1 1 1 1 |↔
| 1 1 1 1 |↔     | 1 1 1 1 |↔         | C C |     | C 1 |      | C 1 1 1 |→
↙↗  ↕ ↕ ↕↘   ↙↗   ↕ ↕ ↘              | 4 4 |  →  | 4 1 |  ≡   | 4 1 1 1 |
   (8 rv)          (8 rv)            | A B |     | A B |      | A B 1 1 |
                                     | B B |     | B B |      | B B 1 1 |→

                                                          ↗    ↕ ↕ ↘
                                                             (6 rv)
```

Division is physical (rather than factorial). One object divided (or separated) into two objects results in two objects. This distinguishes division from (factorial) multiplication.

In fact,

```
| C C C C |
| 4 4 4 4 |  ÷ 2        has many other possibilities
| A A A B |             (to result in two distinct or separate tables).
| A A B B |
```

for example:

```
| C C C C |  +  | 4 4 4 4 |
                | A A A B |
                | A A B B |        as results, etc.


       | C C C C |           | C C |  | C C |  | A A |  | A B |
   ∴   | 4 4 4 4 |  ÷ 4  →   | 4 4 |, | 4 4 |, | A A |, | B B |  etc.
       | A A A B |           (only one set of possible objects resulting, as shown)
       | A A B B |
```

A factorial functioning for division must be specified and (made) explicit:
e.g.

$$\begin{array}{|cccc|}\hline C\ C\ C\ C \\ 4\ 4\ 4\ 4 \\ A\ A\ A\ B \\ A\ A\ B\ B \\ \hline\end{array} \div \begin{array}{|c|}\hline C \\ 4 \\ \\ \\ \hline\end{array} \rightarrow \begin{array}{|cccc|}\hline 1\ 1\ 1\ 1 \\ 1\ 1\ 1\ 1 \\ A\ A\ A\ B \\ A\ A\ B\ B \\ \hline\end{array}$$

(5 rv) ... (1 rv) ... (7 rv)

rv yield statement : $5\ rv \div 1\ rv \rightarrow 7\ rv$

since it is a (limited, as factorial) form of multiplication:

$$\begin{array}{|c|}\hline C \\ 4 \\ \\ \\ \hline\end{array} \bullet \begin{array}{|cccc|}\hline 1\ 1\ 1\ 1 \\ 1\ 1\ 1\ 1 \\ A\ A\ A\ B \\ A\ A\ B\ B \\ \hline\end{array} \equiv \begin{array}{|cccc|}\hline C\ C\ C\ C \\ 4\ 4\ 4\ 4 \\ A\ A\ A\ B \\ A\ A\ B\ B \\ \hline\end{array}$$

$$\therefore \quad \begin{array}{|c|}\hline C \\ 4 \\ \\ \\ \hline\end{array} \div \begin{array}{|cccc|}\hline C\ C\ C\ C \\ 4\ 4\ 4\ 4 \\ A\ A\ A\ B \\ A\ A\ B\ B \\ \hline\end{array} \rightarrow \begin{array}{|cccc|}\hline 1\ 1\ 1\ 1 \\ 1\ 1\ 1\ 1 \\ A\ A\ A\ B \\ A\ A\ B\ B \\ \hline\end{array}$$

⌐ (row) **universal** operation side (of table)

but

$$(\ 5\ rv \quad - \quad 1\ rv \quad \rightarrow \quad 4\ rv\ )$$

$$\begin{array}{|cccc|}\hline C\ C\ C\ C \\ 4\ 4\ 4\ 4 \\ A\ A\ A\ B \\ A\ A\ B\ B \\ \hline\end{array} \div \begin{array}{|c|}\hline C \\ 4 \\ A \\ A \\ \hline\end{array} \rightarrow \begin{array}{|cccc|}\hline 1\ 1\ C\ C \\ 1\ 1\ 4\ 4 \\ 1\ 1\ A\ B \\ 1\ 1\ B\ B \\ \hline\end{array} \rightarrow$$

⌐ (row **specific** operation side (of table)

A distinction in symbol (as like "÷") should probably be made for true division (i.e. partitioning) and factorial division.

∴ Both true and factorial division can alter redundancy values.

rv yield statements

It's interesting that, for factorial division, the *rv yield statements* are degenerate of results:

$$5\ rv \div 1\ rv \rightarrow 7\ rv\ ;\ \ 5\ rv \div 1\ rv \rightarrow 4\ rv\ .$$

As well, true division (full or complete partitioning) can also be degenerate of results.

∴ A versatility of results can be strategized for (tabulated) processes and their developments or manipulations (for functional gains or opportunities of development).

For example, we can follow a process, mapping out the occurring rv values during its run:

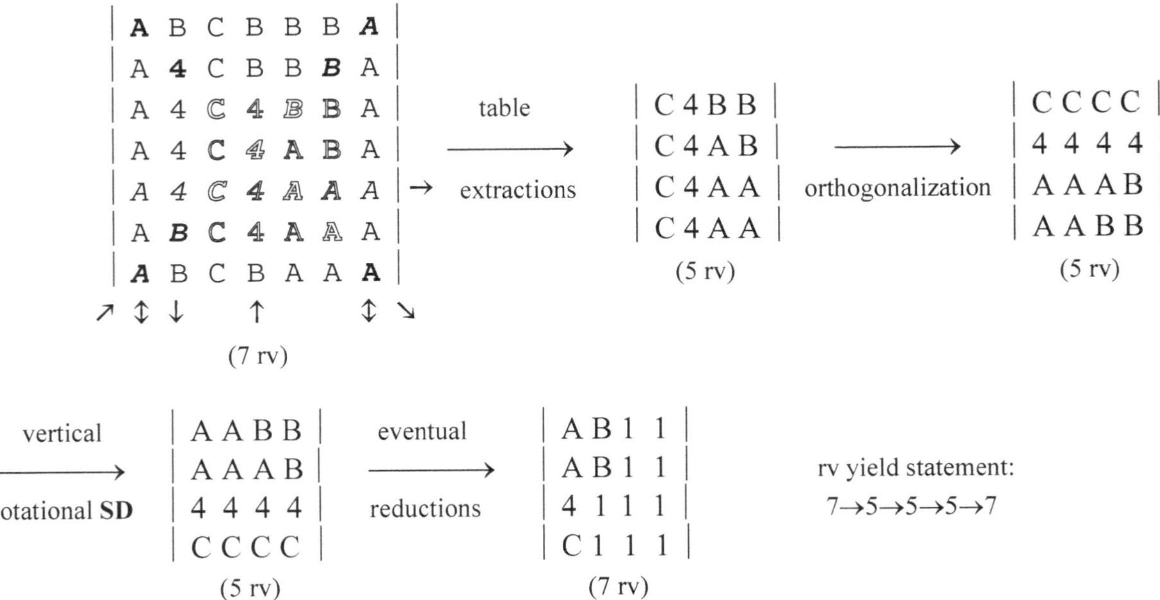

rv yield statement:
7→5→5→5→7

Studying the rv yield statement *a priori*:

The decrease in rv from 7→5 could be due to (eventual) reductions, or extraction (excision of table).
The modification of tables for 5→5→5 could be due to SD, orthogonalization.
The increase in rv from 5→7 could be due to reductions, extractions (or insertions).
Changes in rv can also be due to arithmetic procedures:
    addition / subtraction, (factorial) multiplication / division .

Of course, the process can be followed graphically (*a priori*), if we are given the rv numbers (statement) and the n-values of the (balanced) tables involved:

e.g.

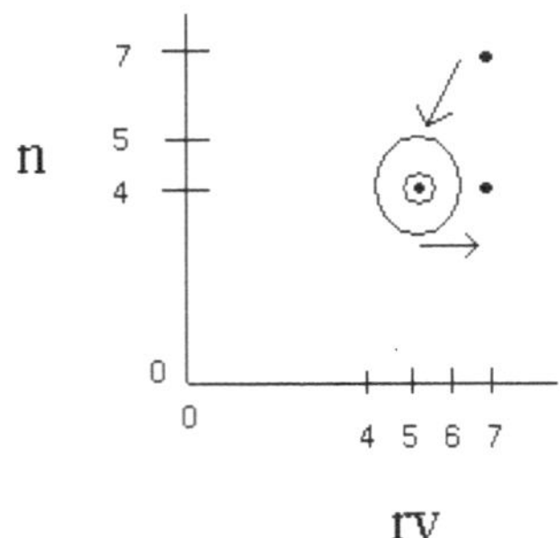

Yet, the origin is "zeroed", which is not allowed in nul/los. Also, three points are overlapped (or composited).

∴ Another sort of graph is preferable:

Process of 5 tables

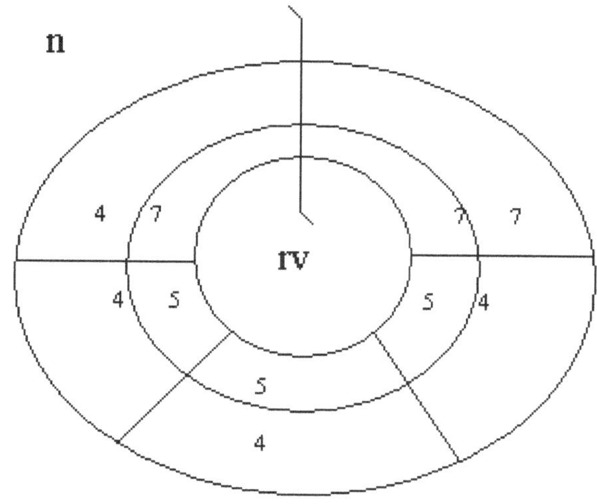

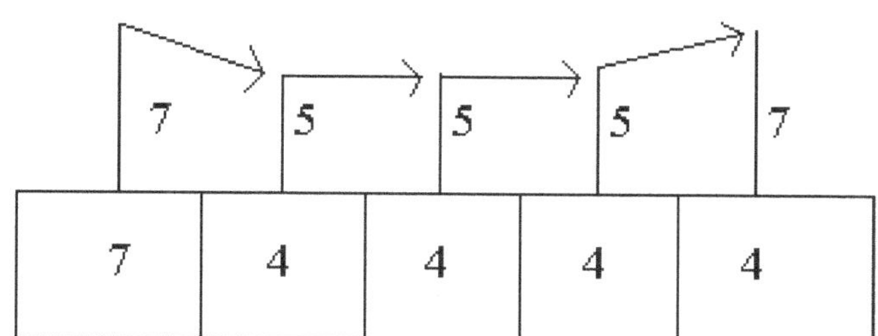

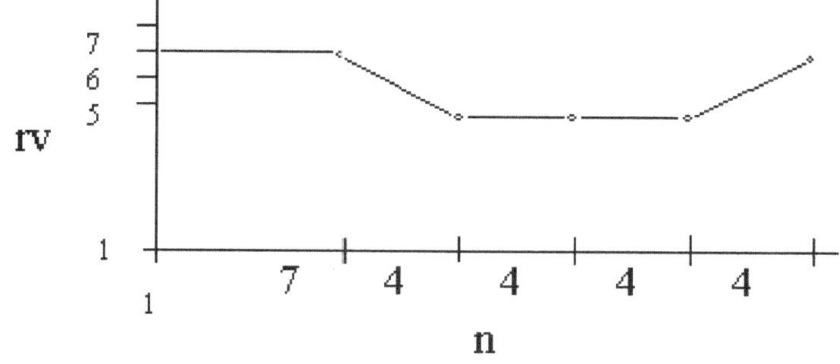

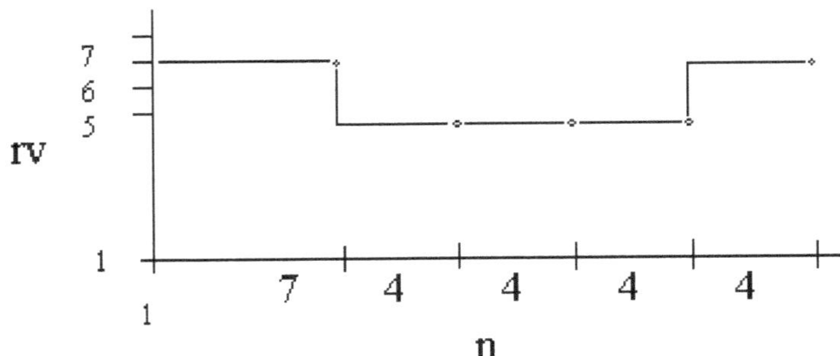

No table is a non-table (n≡0), and no redundancy is a non-redundancy (rv≡0).

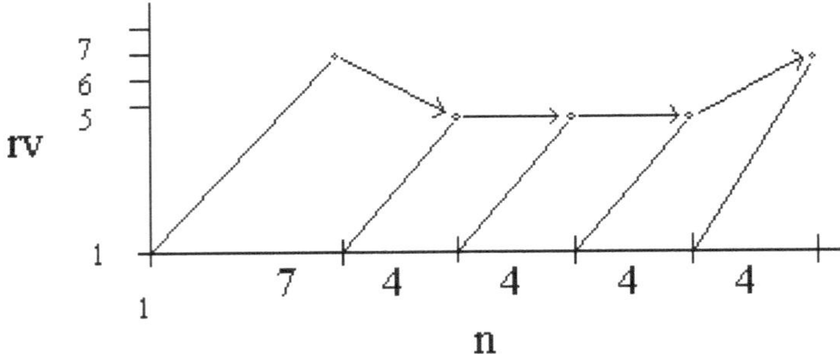

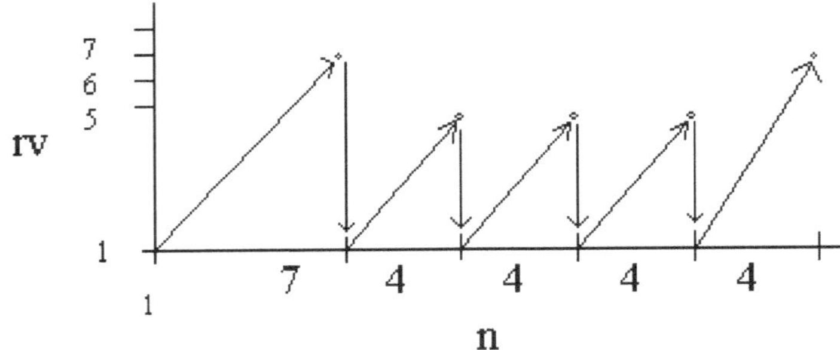

What the last graph says is that the rv yield statement is actually (through functional development):

$$1 \to 7 \to 1 \to 5 \to 1 \to 5 \to 1 \to 5 \to 1 \to 7$$

(The functional development is of course of each table.)

If, as for the graph, we treat the n axis numerically (as increasing) then we can compare:

$$\begin{array}{c|l} rv & 1 \to 7 \to 1 \to 5 \to 1 \to 5 \to 1 \to 5 \to 1 \to 7 \\ n & 1 \to 7 \to 7 \to 11 \to 11 \to 15 \to 15 \to 19 \to 19 \to 23 \end{array}$$
$$\qquad\quad 7 \qquad\quad 4 \qquad\quad 4 \qquad\quad 4 \qquad\quad 4$$

This delineates a sort of tabular progress more evident (or informative) than just the rv yield statement alone. "n" is treated as accumulative during a table's run (or formation) such that the progress of the process has more antecedent (or historical) perspective.

212

If we consider the ratio n/rv (from the proceeding) we have:

$$1 , (7/7) \equiv 1 , 7 , (11/5) \equiv 2+1/5 , 11 , (15/5) \equiv 3 , 13 , (19/5) \equiv 3+4/5 , 19 , (23/7) \equiv 3+2/7$$

Effective Redundancy Density (erd)

The accumulative ranking (n) of tables alternates with the **effective** (for process) n-rank of **redundancy "density"** (erd) per table (as contributed to the entire process).

We can plot the accumulative ranking vs erd, allowing for:

```
            erd,  rank .........
erd, rank | 1 , 1 , 1 , 7 , 2+1/5 , 11 , 3 , 15 , 3+4/5 , 19 , 3+2/7 , 23
∴  erd    | 1     1       2+1/5       3        3+4/5        3+2/7
∴  rank   |    1      7          11       15           19           23
```

Thus,

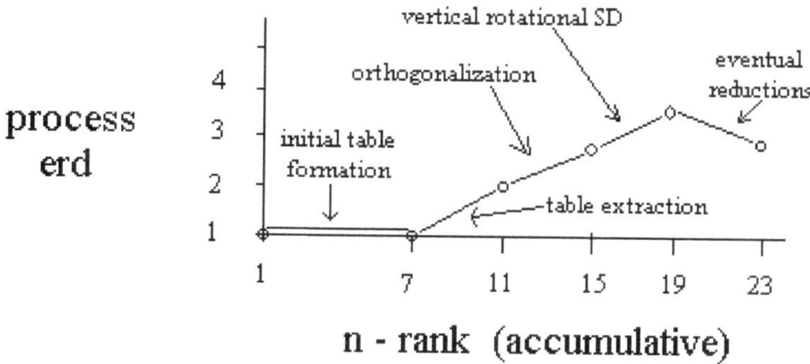

n - rank (accumulative)

This tends to indicate that the redundancy density throughout the process increases until the last step (eventual reductions).

Here then (graphically) is a better description of the progress of the process (in terms of forming redundancies) than merely plotting out (n vs rv): i.e. The erds (∴ points) are less weighted of rv per (home) specific table than for considering the process as a whole.

The erd, as defined here, is really the accumulative n-rank divided by the current table (of process) rv. ∴ If a particular table's rv is particularly high, the erd will tend to lower, perhaps not being too representative of the process as a whole (as shown in the last graph for the 'eventual reductions' step). The 'density' is the density of accumulative (n) ranking at a current table's rv. The previous example of process (with its graphing) shows that reductions can greatly alter a table (more so than other operations, and even extractions or insertions in some instances).

But, the example also shows, as regards to calculations, that fractional division of numbers, as opposed to what must be altered of decimal division (to avoid "zero"), is particularly appropriate for nul/los.

We can also now confront the numerical division under nul/los, using our base 9 system (as previous). This system might be termed "elfemal" to contrast it with the decimal system, since the number 11 in elfemal holds the same usage as the number 10 in decimal:

```
                 1               2               3               4
elfemal   : 1 2 3 4 5 6 7 8 9 1 2 3 4 5 6 7 8 9 1 2 3 4 5 6 7 8 9 1 2 3 4 5 6 7 8 9 1
          ________________________________________________________________________
                 1               2               3
decimal   : 1 2 3 4 5 6 7 8 9 0 1 2 3 4 5 6 7 8 9 0 1 2 3 4 5 6 7 8 9 0 1 2 3 4 5 6 7
```

For using (long) division, we can now compare the common approximation of π by the fraction $22/7$ (an irrational result):

```
           3.142857...                    3.152652...
    7  / 22.000000             7  / 24.111111
       21 | | | | | |             23 | | | | | |
        1 0| | | | | |            1  1| | | | | |
          7| | | | |                 7| | | | |
         30| | | |                   41| | | |        ?
         28| | | |                   38| | | |
         20| | |                     21| | |
         14| | |                     15| | |
         60| |                       51| |
         56| |                       46| |
         40|                         41|
         35|                         38|
          50                         21
          49                         15
        decimal                    elfemal
```

π remains irrational (presumably) in both systems. For the elfemal system, zero is avoided.

The conversion of numbers between systems is simply afforded by substitutions for the 10 or 11 factors accordingly:

e.g. $22 = 2 \times 10 + 2$ decimal
              $\downarrow$
            $24 = 2 \times 11 + 2$ elfemal

e.g. $30 = 3 \times 10$ decimal
              $\downarrow$
            $33 = 3 \times 11$ elfemal

e.g. $41 = 3 \times 11 + 7$ elfemal   [counting to add 7: after 9 comes 1, in elfemal]
              $\downarrow$
            $37 = 3 \times 10 + 7$ decimal

Quick conversions (for possible whole numbers) :

$$\text{elfemal}\rightarrow\text{decimal} \equiv \text{number - (\# of 10s in the number)}$$
$$\text{e.g.} \quad 41 - 4 \rightarrow 37$$
$$\text{decimal}\rightarrow\text{elfemal} \equiv \text{number + (\# of 10s in the number)}$$
[with proviso that after 9 comes 1 $\therefore$ a zero $\rightarrow$ 1]
$$\text{e.g.} \quad 37 + 3 = 40 \rightarrow 41$$

But, $24 \neq 24.111111$ (although it is treated as such), and the elfemal long division given is done in error, since $11-7 = 4$ in decimal. Under elfemal, $11-7 = 3$. This leads to the following embarrassment:

```
    3.14
7/24.11
  23  ||
   1 1|
   _7|
   31
   31
    ?
```

The approximation of π becomes rational; but $31-31=0$ is not possible in elfemal.

What must proceed "1" ?

In decimal number line, the "center" (of the number line) or point of inflection from positive to negative numbers is "zero"; while for elfemal this point must be taken up by "-1" (since there is no zero):

```
                        |←   3 – 3  ( = 0 ) decimal
decimal  |      -3 -2 -1  0   1   2   3
elfemal  |      -4 -3 -2 -1   1   2   3
                        |←   3 – 3  ( = -1 ) elfemal
```

```
           1              ←(-) (+)→              1
dec:   4 3 2 1 0 9 8 7 6 5 4 3 2 1 0 1 2 3 4 5 6 7 8 9 0 1 2 3 4 5 6
```

```
           1              ←(-) (+)→              1
elf :  6 5 4 3 2 1 9 8 7 6 5 4 3 2 -1 1 2 3 4 5 6 7 8 9 1 2 3 4 5 6 7
```

$\therefore$

```
    3.14
7/24.11
  23  ||
   1 1|
   _7|
   31
   31
   -1        (a sign or result for the end of division, in elfemal)
```

$\therefore$ in elfemal, $\pi \sim 3.14$ (as 24/7) is a rational number

In elfemal, it seems arbitrary whether one uses -1 as inflection point (for use with positive whole numbers) or +1 as inflection point (for use with negative whole numbers). If the point of inflection in elfemal can be (either) +1 or -1 , then elfemal uses a "sliding" (or dynamic) number line. This must be based on the advantages for any particular calculations.

How do we handle the fractional parts of a number, when doing decimal↔elfemal conversions?

Judging on comparisons of the two numbers:

<pre>
          dec: -1 0 1
          elf :   -1 1   (for positive values)
or   elf : -1 1   (for negative values)
          [0→1 in decimal has the same length as -1→+1 in elfemal (∴ equivalent fractions)]
</pre>

there is no difference in the fractional components.

∴   3.14285 in decimal → 3.14285 in elfemal .

If a zero occurs in decimal, it is replaced by "1" in elfemal:

e.g.   22.367 decimal → 24.367 elfemal

            14.905 decimal → 15.915 elfemal

Summary:

        (22/7)$_{\text{decimal}}$ ≡ (24/7)$_{\text{elfemal}}$ (equivalency)

        (22/7)$_{\text{decimal}}$ = 3.14285..... (irrational)

But our elfemal division of (24/7) → 3.14 (rational, by definition of "-1" as sign for end of division). The two results of division are not exactly the same (perhaps) because (24) is treated as (24.1111....) under elfemal division; i.e., the numerator is irrational in elfemal, as the fraction is irrational in decimal). Apparently, all numerators must be treated as irrational (numbers) under elfemal division, although the results of the division can be rational (by operational definition).

        ∴ "1" (or "-1") replaces zero in elfemal (compared to decimal).

———————————

The "base" of a number usually refers to the common factor by which a number is expressed:

e.g.
in base 10   101 ≡ 10 x 10 + 1 ≡ 1 x 10^2 + 0 x 10^1 + 1 x 10^0
in base 9   101 ≡ 9 x 9 + 1 ≡ 1 x 9^2 + 0 x 9^1 + 1 x 9^0

But they still refer to decimal systems (i.e., use of "zero" or using digits: 0,1,2,3,4,5,6,7,8,9).

The term "base" used in nul/los means "basis" as there being only 9 distinct digits used for numbers (instead of 10).

Elfemal is a form of nul/los which essentially contracts the number line by removing zeros. Since the number line is infinite (in both directions), this is a legitimate contraction (or manipulation) that is not: the number line remaining infinite.

We can also compare a rational fraction (e.g. 1/2) to an irrational one (e.g. 1/3) in **elfemal**:

```
        .5
  2/  1.1
      1 1
       -1            ∴  1/2 → 0.5  in decimal        ,  or .5  (rational) in elfemal

        .333....
  3/  1.111
      911
      11|
       9|
      11
       9
       1            ∴  1/3 → 0.333.... in decimal   ,  or .333.... (irrational) in elfemal
```

This also comports with fractions (less than one) being the same in elfemal and decimal.

Of course, the question of sign of fractions (less than |1|) is according to the sliding nature of the elfemal number line (with regard to where "zero" would be).

For positive fractions, the inflection point is "-1" :

```
              ←(-) (+)→
decimal       -1   0   1
elfemal       -2  -1   1
```
positive fractions

For negative fractions, the inflection point is "+1" :

```
              ←(-) (+)→
decimal       -1   0   1
elfemal       -1   1   2
```
negative fractions

Again, because the number lines remain infinite (in either direction), these are reasonable manipulations without invalidating them (the number lines) or their usage. The notion here is that there are no such infinities to physical reality.

The illustration of physical phenomena can be used to show contradictions in explanations due to adoption of null quantities. For example, diffraction of light (energy) through and by a crystal results in distinct observations of reflections. But this does not obviate the wave-nature or sinusoidal propagation of the light. The distinctness of the reflections on a screen or detector (along with or accompanied by regions on the detector of no reflection intensities) is due to the complexity or complications of the crystal. A perhaps simpler example is the dropping of a flowing liquid (using gravity's force to provide the work of the process) onto a grid with holes appropriately far apart form each other (relative to the speed of the flow of the liquid) so that discreet drops of the liquid occur on a screen or (detecting) platform below the grid. The discreetness of the (detected) drops does not obviate the (nature of the) continuous flow of the (falling and spreading, on the grid) liquid, but rather is due to the complications (complexity) caused by the grid to the process.

This is clearly demonstrable by all (observed) quantum phenomena, and leads to the following proposition:

The observance (or measurement) of a null quantity is not representative of a phenomenon, but rather must be due to a greater complexity of its description than what is fundamental to the process, since null-ness is not a (fundamental; i.e., necessary) property of nature.

The sliding (or shifting) nature of the nul/los number line is a consequence of the proposition (as appertaining to measurement).

Ardors and Passions

David L. Birdsall